FOUNDATION ENGINEERING HANDBOOK

**Design and Construction with the
2009 International Building Code**

Robert W. Day

*Principal Engineer
American Geotechnical
San Diego, California*

Second Edition

New York Chicago San Francisco Lisbon London Madrid
Mexico City Milan New Delhi San Juan Seoul
Singapore Sydney Toronto

The **McGraw·Hill** Companies

Library of Congress Cataloging-in-Publication Data.

Day, Robert W.
    Foundation engineering handbook : design and construction with the 2009
international building code / Robert W. Day. — 2nd ed.
        p.   cm.
    ISBN 978-0-07-174009-8
    1. Foundations—Handbooks, manuals, etc.   2. Soil mechanics—Handbooks,
manuals, etc.   3. Standards, Engineering—Handbooks, manuals, etc.   I. Title.
    TA775.D39   2010
    624.1′5—dc22   2010024748

McGraw-Hill books are available at special quantity discounts to use as premiums and sales pro-
motions, or for use in corporate training programs. To contact a representative please e-mail us
at bulksales@mcgraw-hill.com.

Foundation Engineering Handbook, Second Edition

2 3 4 5 6 7 8 9 0   DOC/DOC   1 9 8 7 6 5 4 3 2 1

ISBN 978-0-07-174009-8
MHID 0-07-174009-0

The pages within this book were printed on acid-free paper.

Sponsoring Editor
Larry S. Hager

Acquisitions Coordinator
Alexis Richard

Editorial Supervisor
David E. Fogarty

Project Manager
Nancy W. Dimitry, D&P Editorial

Copy Editor
Nancy W. Dimitry, D&P Editorial

Proofreaders
Donald L. Dimitry
Donald Paul Pomeranz

Production Supervisor
Pamela A. Pelton

Composition
D&P Editorial

Art Director, Cover
Jeff Weeks

With love to my wife, Deborah

ABOUT THE AUTHOR

Robert W. Day is a leading geotechnical engineer and the principal engineer at American Geotechnical.

CONTENTS

Part 2 Foundation Design

Chapter 5. Shallow and Deep Foundations   5.3

Chapter 6. Bearing Capacity of Foundations   6.1

Chapter 7. Settlement of Foundations   7.1

Chapter 8. Consolidation   8.1

Chapter 19. International Building Code Regulations for Foundations   19.1

Appendix A. Glossary   A.1

Appendix B. Example of a Foundation Engineering Report   B.1

Appendix C. Solutions to Problems   C.1

Appendix D. Conversion Factors   D.1

Appendix E. Bibliography   E.1

PREFACE

The goal of the book is to present the practical aspects of geotechnical and foundation engineering. While the major emphasis of college education is engineering analyses, this often represents only a portion of the knowledge needed to practice geotechnical engineering. One objective of this book is to discuss the engineering judgment that needs to be acquired through experience. An example is the application of sufficient redundancy in the design and construction of the project.

In California, structural engineers typically perform the actual structural design of the foundation based on the recommendations supplied by the geotechnical engineer. Foundation design, in terms of determining the type and spacing of steel reinforcement in concrete footings, is not covered in this textbook. This book deals only with the geotechnical aspects of foundation engineering. In addition, this book is only applicable for the analyses of clean soil, which does not contain any known or suspected hazardous materials. Such environmental issues are outside the scope of this book.

Because of the assumptions and uncertainties associated with geotechnical engineering, it is often described as an "art," rather than an exact science. Thus simple analyses are prominent in this book, with complex and theoretical evaluations kept to an essential minimum. For most projects, a limited number of borings or test pits are used to investigate the soil and geologic makeup of a site. Hence, except for cases where the site consists of solid rock, there will usually be uncertainty in the final analyses. Because of this, when dealing with foundations bearing on soil, it is always best to take a conservative approach.

Part 1 (Chapters 2 to 4) deals with basic geotechnical field and laboratory studies, such as subsurface exploration and laboratory testing of soil, rock, or groundwater samples. Part 2 (Chapters 5 to 14) presents the geotechnical aspects of foundation engineering, including the conditions commonly encountered by the design engineer, such as settlement, expansive soil, and slope stability. Part 3 (Chapters 15 to 17) provides a discussion of the performance or engineering evaluation of foundation construction, and Part 4 (Chapters 18 and 19) consists of concluding chapters dealing with the application of the building code for foundation engineering.

The book presents the practical aspects of geotechnical and foundation engineering. The topics should be of interest to design engineers, especially Part 4 that deals with the *International Building Code*. In this second edition, Part 4 has been revised to be in conformance with the 2009 *International Building Code*. The remainder of the book is essentially unchanged from the first edition.

Robert W. Day

ACKNOWLEDGMENTS

I am grateful for the contributions of the many people who helped to make this book possible. Special thanks are due to the American Society of Civil Engineers (ASCE) and the International Code Council (ICC), who jointly sponsored this work. The continued support of Mark Johnson at ICC is greatly appreciated as well as the effort of Hamid Naderi at ICC for his review and comments concerning Chapters 18 and 19.

Portions of this book are reproduced from *Geotechnical and Foundation Engineering: Design and Construction* and I would like to thank Professor Charles C. Ladd, at the Massachusetts Institute of Technology, who reviewed that text and offered many helpful suggestions. I would also like to thank Professor Ladd for the opportunity to have worked on his Orinoco clay project. Several figures, especially those in the section on shear strength, are reproduced from my M.I.T. thesis (*Engineering Properties of the Orinoco Clay*). In addition, I would like to thank Dr. Ladd for the opportunity to have worked with him on various projects over the years.

Numerous practicing engineers provided valuable assistance during the development of the text. In particular, I am indebted to Tom Marsh, principal engineer at American Geotechnical, who provided extensive technical support and assistance. The contributions of my fellow engineers, including Robert Brown, Eric Lind, Rick Walsh, and Scott Thoeny, are also appreciated. Thanks also to Dennis Poland, Ralph Jeffery, and Todd Page for their help with the geologic aspects of the book, and Rick Dorrah for drafting various figures for the book.

For contributions to the chapter on laboratory testing, I would like to thank Professor Timothy Stark, at the University of Illinois, who performed the ring shear tests, provided a discussion of the test procedures and the photographs, and prepared the ring shear test plots. Thanks also to Kean Tan, principal of Kean Tan Laboratories, who performed the triaxial compression tests and provided the photographs. I also appreciate the help of Sam Mahdavi, laboratory manager at American Geotechnical, who provided the direct shear test results and photographs.

For contributions to the sections on geotechnical earthquake engineering, I would like to thank Professor Nelson, who provided Figure 13.22 and additional data concerning the Turnagain Heights landslide. The help of Thomas Blake, who provided assistance in the use and understanding of his computer programs, is also appreciated.

I would also like to thank Gregory Axten, president of American Geotechnical, who provided valuable support during the review and preparation of the book. Thanks also to Carl Bonura and John Pizzi for their longtime friendship and technical inputs.

Tables and figures taken from other sources are acknowledged where they occur in the text. Finally, I wish to especially thank Larry Hager at McGraw-Hill who has always supported my work and had the patience to see my books completed. Thanks also to Pamela Pelton, David Fogarty, and others in McGraw-Hill.

Finally, I would also like to thank the production team at D&P Editorial Services, for managing the production of my book, providing editorial services, and refining it to its final form.

CHAPTER 1
INTRODUCTION

1.1 DEFINITIONS

A foundation is defined as that part of the structure that supports the weight of the structure and transmits the load to underlying soil or rock. In general, foundation engineering applies the knowledge of geology, soil mechanics, rock mechanics, and structural engineering to the design and construction of foundations for buildings and other structures. The most basic aspect of foundation engineering deals with the selection of the type of foundation, such as using a shallow or deep foundation system. Another important aspect of foundation engineering involves the development of design parameters, such as the bearing capacity or estimated settlement of the foundation. Foundation engineering could also include the actual foundation design, such as determining the type and spacing of steel reinforcement in concrete footings. Foundation engineering often involves both geotechnical and structural engineers, with the geotechnical engineer providing the foundation design parameters such as the allowable bearing pressure and the structural engineer performing the actual foundation design.

Foundations are commonly divided into two categories: shallow and deep foundations. Table 1.1 presents a list of common types of foundations. In terms of geotechnical aspects, foundation engineering often includes the following (Day, 1999a, 2000a):

- Determining the type of foundation for the structure, including the depth and dimensions
- Calculating the potential settlement of the foundation
- Determining design parameters for the foundation, such as the bearing capacity and allowable soil bearing pressure
- Determining the expansion potential of a site
- Investigating the stability of slopes and their effect on adjacent foundations
- Investigating the possibility of foundation movement due to seismic forces, which would also include the possibility of liquefaction
- Performing studies and tests to determine the potential for deterioration of the foundation
- Evaluating possible soil treatment to increase the foundation bearing capacity
- Determining design parameters for retaining wall foundations
- Providing recommendations for dewatering and drainage of excavations needed for the construction of the foundation
- Investigating groundwater and seepage problems and developing mitigation measures during foundation construction
- Site preparation, including compaction specifications and density testing during grading
- Underpinning and field testing of foundations

TABLE 1.1 Common Types of Foundations

Category	Common types	Comments
Shallow foundations	Spread footings	Spread footings (also called pad footings) are often square in plan view, are of uniform reinforced concrete thickness, and are used to support a single column load located directly in the center of the footing.
	Strip footings	Strip footings (also called wall footings) are often used for load-bearing walls. They are usually long reinforced concrete members of uniform width and shallow depth.
	Combined footings	Reinforced-concrete combined footings are often rectangular or trapezoidal in plan view, and carry more than one column load.
	Conventional slab-on-grade	A continuous reinforced-concrete foundation consisting of bearing wall footings and a slab-on-grade. Concrete reinforcement often consists of steel rebar in the footings and wire mesh in the concrete slab.
	Posttensioned slab-on-grade	A continuous posttensioned concrete foundation. The posttensioning effect is created by tensioning steel tendons or cables embedded within the concrete. Common posttensioned foundations are the ribbed foundation, California slab, and PTI foundation.
	Raised wood floor	Perimeter footings that support wood beams and a floor system. Interior support is provided by pad or strip footings. There is a crawl space below the wood floor.
	Mat foundation	A large and thick reinforced-concrete foundation, often of uniform thickness, that is continuous and supports the entire structure. A mat foundation is considered to be a shallow foundation if it is constructed at or near ground surface.
Deep foundations	Driven piles	Driven piles are slender members, made of wood, steel, or precast concrete, that are driven into place by pile-driving equipment.
	Other types of piles	There are many other types of piles, such as bored piles, cast-in-place piles, and composite piles.
	Piers	Similar to cast-in-place piles, piers are often of large diameter and contain reinforced concrete. Pier and grade beam support are often used for foundation support on expansive soil.
	Caissons	Large piers are sometimes referred to as caissons. A caisson can also be a watertight underground structure within which construction work is carried on.
	Mat or raft foundation	If a mat or raft foundation is constructed below ground surface or if the mat or raft foundation is supported by piles or piers, then it should be considered to be a deep foundation system.
	Floating foundation	A special foundation type where the weight of the structure is balanced by the removal of soil and construction of an underground basement.
	Basement-type foundation	A common foundation for houses and other buildings in frost-prone areas. The foundation consists of perimeter footings and basement walls that support a wood floor system. The basement floor is usually a concrete slab.

Note: The terms *shallow* and *deep* foundations in this table refer to the depth of the soil or rock support of the foundation.

1.2 PROJECT REQUIREMENTS

For some projects, the foundation design requirements will be quite specific and may even be in writing. For example, a public works project may require a geotechnical investigation consisting of a certain number, type, and depth of borings, and may also specify the types of laboratory tests to be performed. The more common situation is where the client is relying on the geotechnical engineer to prepare a proposal, perform an investigation, and provide foundation design parameters that satisfy the needs of the project engineers and requirements of the local building officials or governing authority. The general requirements for foundation engineering projects are as follows (Tomlinson, 1986):

1. Knowledge of the general topography of the site as it affects foundation design and construction, e.g., surface configuration, adjacent property, the presence of watercourses, ponds, hedges, trees, rock outcrops, and the available access for construction vehicles and materials

2. The location of buried utilities such as electric power and telephone cables, water mains, and sewers

3. The general geology of the area with particular reference to the main geologic formations underlying the site and the possibility of subsidence from mineral extraction or other causes

4. The previous history and use of the site including information on any defects or failures of existing or former buildings attributable to foundation conditions

5. Any special features such as the possibility of earthquakes or climate factors such as flooding, seasonal swelling and shrinkage, permafrost, or soil erosion

6. The availability and quality of local construction materials such as concrete aggregates, building and road stone, and water for construction purposes

7. For maritime or river structures, information on tidal ranges and river levels, velocity of tidal and river currents, and other hydrographic and meteorological data

8. A detailed record of the soil and rock strata and groundwater conditions within the zones affected by foundation bearing pressures and construction operations, or of any deeper strata affecting the site conditions in any way

9. Results of laboratory tests on soil and rock samples appropriate to the particular foundation design or construction problems

10. Results of chemical analyses on soil or groundwater to determine possible deleterious effects of foundation structures

Often the client lacks knowledge of the exact requirements of the geotechnical aspects of the project. For example, the client may only have a vague idea that the building needs a foundation, and therefore a geotechnical engineer must be hired. The owner assumes that you will perform an investigation and prepare a report that satisfies all of the foundation requirements of the project.

Knowing the requirements of the local building department or governing authority is essential. For example, the building department may require that specific items be addressed by the geotechnical engineer, such as settlement potential of the structure, grading recommendations, geologic aspects, and for hillside projects, slope stability analyses. Examples of problem conditions requiring special consideration are presented in Table 1.2. Even if these items will not directly impact the project, they may nevertheless need to be investigated and discussed in the geotechnical report.

There may be other important project requirements that the client is unaware of and is relying on the geotechnical engineer to furnish. For example, the foundation could be impacted by geologic hazards, such as faults and deposits of liquefaction prone soil. The geotechnical engineer will need to address these types of geologic hazards that could impact the site.

In summary, it is essential that the geotechnical engineer know the general requirements for the project (such as the 10 items listed earlier) as well as local building department or other regulatory requirements. If all required items are not investigated or addressed in the foundation engineering report, then the building department or regulatory authority may refuse to issue a building permit. This will naturally result in an upset client because of the additional work that is required, delays in construction, and possible unanticipated design and construction expenses.

TABLE 1.2 Problem Conditions Requiring Special Consideration

Problem type	Description	Comments
Soil	Organic soil, highly plastic soil	Low strength and high compressibility
	Sensitive clay	Potentially large strength loss upon large straining
	Micaceous soil	Potentially high compressibility
	Expansive clay, silt, or slag	Potentially large expansion upon wetting
	Liquefiable soil	Complete strength loss and high deformations caused by earthquakes
	Collapsible soil	Potentially large deformations upon wetting
	Pyritic soil	Potentially large expansion upon oxidation
Rock	Laminated rock	Low strength when loaded parallel to bedding
	Expansive shale	Potentially large expansion upon wetting; degrades readily upon exposure to air and water
	Pyritic shale	Expands upon exposure to air and water
	Soluble rock	Rock such as limestone, limerock, and gypsum that is soluble in flowing and standing water
	Cretaceous shale	Indicator of potentially corrosive groundwater
	Weak claystone	Low strength and readily degradable upon exposure to air and water
	Gneiss and schist	Highly distorted with irregular weathering profiles and steep discontinuities
	Subsidence	Typical in areas of underground mining or high groundwater extraction
	Sinkholes	Areas underlain by carbonate rock (karst topography)
Condition	Negative skin friction	Additional compressive load on deep foundations due to settlement of soil
	Expansion loading	Additional uplift load on foundation due to swelling of soil
	Corrosive environment	Acid mine drainage and degradation of soil and rock
	Frost and permafrost	Typical in northern climates
	Capillary water	Rise in water level which leads to strength loss for silts and fine sands

Source: Reproduced with permission from *Standard Specifications for Highway Bridges*, 16th edition, AASHTO, 1996.

1.3 PRELIMINARY INFORMATION AND PLANNING THE WORK

The first step in a foundation investigation is to obtain preliminary information, such as the following:

1. *Project location.* Basic information on the location of the project is required. The location of the project can be compared with known geologic hazards, such as active faults, landslides, or deposits of liquefaction prone sand.

2. *Type of project.* The geotechnical engineer could be involved with all types of foundation engineering construction projects, such as residential, commercial, or public works projects. It is important to obtain as much preliminary information about the project as possible. Such information could include the type of structure and use, size of the structure including the number of stories, type of construction and floor systems, preliminary foundation type (if known), and estimated structural loadings. Preliminary plans may even have been developed that show the proposed construction.

3. *Scope of work.* At the beginning of the foundation investigation, the scope of work must be determined. For example, the scope of work could include subsurface exploration and laboratory

testing to determine the feasibility of the project, the preparation of foundation design parameters, and compaction testing during the grading of the site in order to prepare the building pad for foundation construction.

After the preliminary information is obtained, the next step is to plan the foundation investigation work. For a minor project, the planning effort may be minimal. But for large-scale projects, the plan can be quite extensive and could change as the design and construction progresses. The planning effort could include the following:

- Budget and scheduling considerations
- Selection of the interdisciplinary team (such as geotechnical engineer, engineering geologist, structural engineer, hydrogeologist and the like) that will work on the project
- Preliminary subsurface exploration plan, such as the number, location, and depth of borings
- Document collection
- Laboratory testing requirements
- Types of engineering analyses that will be required for the design of the foundation

1.4 ENGINEERING GEOLOGIST

An engineering geologist is defined as an individual who applies geologic data, principles, and interpretation so that geologic factors affecting planning, design, construction, and maintenance of civil engineering works are properly recognized and utilized (*Geologist and Geophysicist Act*, 1986). In some areas of the United States, there may be minimal involvement of engineering geologists except for projects involving such items as rock slopes or earthquake fault studies. In other areas of the country, such as California, the geotechnical engineer and engineering geologist usually performs the geotechnical investigations jointly. The majority of geotechnical reports include both engineering and geologic aspects of the project and both the geotechnical engineer and engineering geologist both sign the report. For example, a geotechnical engineering report will usually include an opinion by the geotechnical engineer and engineering geologist on the engineering and geologic adequacy of the site for the proposed development.

Table 1.3 (adapted from *Fields of Expertise*, undated) presents a summary of the fields of expertise for the engineering geologist and geotechnical engineer, with the last column indicating the areas of overlapping expertise. Note in Table 1.3 that the engineering geologist should have considerable involvement with foundations on rock, field explorations (such as subsurface exploration and surface mapping), groundwater studies, earthquake analysis, and engineering geophysics. Since geologic processes form natural soil deposits, the input of an engineering geologist can be invaluable for nearly all types of foundation engineering projects.

Because the geotechnical engineer and engineering geologist work as a team on most projects, it is important to have an understanding of each individual's area of responsibility. The area of responsibility is based on education and training. According to the *Fields of Expertise* (undated), the individual responsibilities are as follows:

Responsibilities of the Engineering Geologist

1. Description of the geologic environment pertaining to the engineering project
2. Description of earth materials, such as their distribution and general physical and chemical characteristics
3. Deduction of the history of pertinent events affecting the earth materials
4. Forecast of future events and conditions that may develop
5. Recommendation of materials for representative sampling and testing

TABLE 1.3 Fields of Expertise

Topic	Engineering geologist	Geotechnical engineer	Overlapping areas of expertise
Project planning	Development of geologic parameters Geologic feasibility	Design Material analysis Economics	Planning investigations Urban planning Environmental factors
Mapping	Geologic mapping Aerial photography Air photo interpretation Landforms Subsurface configurations	Topographic survey Surveying	Soil mapping Site selections
Exploration	Geologic aspects (fault studies, etc.)	Engineering aspects	Conducting field exploration Planning, observation, and the like Selecting samples for testing Describing and explaining site conditions
Engineering geophysics	Soil and rock hardness Mechanical properties Depth determinations	Engineering applications	Minimal overlapping of expertise
Classification and physical properties	Rock description Soil description (Modified Wentworth system)	Soil testing Earth materials Soil classification (USCS)	Soil description
Earthquakes	Location of faults Evaluation of active and inactive faults Historic record of earthquakes	Response of soil and rock materials to seismic activity Seismic design of structures	Seismicity Seismic conditions Earthquake probability
Rock mechanics	Rock mechanics Description of rock Rock structure, performance, and configuration	Rock testing Stability analysis Stress distribution	In situ studies Regional or local studies
Slope stability	Interpretative Geologic analyses and geometrics Spatial relationship	Engineering aspects of slope stability analysis and testing	Stability analyses Grading in mountainous terrain
Surface waters	Geologic aspects during design	Design of drainage systems Coastal and river engineering Hydrology	Volume of runoff Stream description Silting and erosion potential Source of material and flow Sedimentary processes
Groundwater	Occurrence Structural controls Direction of movement	Mathematical treatment of well systems Development concepts	Hydrology
Drainage	Underflow studies Storage computation Soil characteristics	Regulation of supply Economic factors Lab permeability	Well design, specific yield Field permeability Transmissibility

Source: Adapted from *Fields of Expertise* (undated).

6. Recommendation of ways of handling and treating various earth materials and processes

7. Recommendation or providing criteria for excavation (particularly angle of cut slopes) in materials where engineering testing is inappropriate or where geologic elements control stability

8. Inspection during construction to confirm conditions

Responsibilities of the Geotechnical Engineer

1. Directing and coordinating the team efforts where engineering is a predominant factor

2. Controlling the project in terms of time and money requirements and degree of safety desired

3. Engineering testing and analysis

4. Reviewing and evaluating data, conclusions, and recommendations of the team members

5. Deciding on optimum procedures

6. Developing designs consistent with data and recommendations of team members

7. Inspection during construction to assure compliance

8. Making final judgments on economy and safety matters

1.5 *OUTLINE OF CHAPTERS*

The purpose of this book is to present the geotechnical aspects of foundation engineering. The actual design of the foundation, such as determining the number and size of steel reinforcement for footings, which is usually performed by the project structural engineer, will not be covered.

The book is divided into four separate parts. Part 1 (Chaps. 2 to 4) deals with the basic geotechnical engineering work as applied to foundation engineering, such as subsurface exploration, laboratory testing, and soil mechanics. Part 2 (Chaps. 5 to 14) presents the analysis of geotechnical data and engineering computations needed for the design of foundations, such as allowable bearing capacity, expected settlement, expansive soil, and seismic analyses. Part 3 (Chaps. 15 to 17) provides information for construction-related topics in foundation engineering, such as grading, excavation, underpinning, and field load tests. The final part of the book (Part 4, Chaps. 18 and 19) deals with the *International Building Code* provisions as applicable to the geotechnical aspects of foundation engineering.

Like most professions, geotechnical engineering has its own terminology with special words and definitions. App. A presents a glossary, which is divided into five separate sections:

1. Subsurface exploration terminology

2. Laboratory testing terminology

3. Terminology for engineering analysis and computations

4. Compaction, grading, and construction terminology

5. Geotechnical earthquake engineering terminology

Also included in the appendices are example of a foundation engineering report (App. B), solutions to the problems provided at the end of each chapter (App. C), and conversion factors (App. D).

A list of symbols is provided at the end of the chapters. An attempt has been made to select those symbols most frequently listed in standard textbooks and used in practice. Dual units are used throughout the book, consisting of:

1. Inch-pound units (I-P units), which are also frequently referred to as the United States Customary System units (USCS)

2. International System of Units (SI)

In some cases, figures have been reproduced that use the old metric system (stress in kg/cm^2). These figures have not been revised to reflect SI units.

GEOTECHNICAL ENGINEERING

CHAPTER 2
SUBSURFACE EXPLORATION

2.1 INTRODUCTION

As discussed in Chap. 1, the first step in the foundation investigation is to obtain preliminary information on the project and to plan the work. The next step is typically to perform the subsurface exploration. The goal of the subsurface investigation is to obtain a detailed understanding of the engineering and geologic properties of the soil and rock strata and groundwater conditions that could impact the foundation.

Specific items that will be discussed in the chapter are as follows:

1. Document review (Sec. 2.2)
2. Purpose of subsurface exploration (Sec. 2.3)
3. Borings (Sec. 2.4), including a discussion of soil samplers, sample disturbance, field tests, boring layout, and depth of subsurface exploration
4. Test pits and trenches (Sec. 2.5)
5. Preparation of logs (Sec. 2.6)
6. Geophysical techniques (Sec. 2.7)
7. Subsurface exploration for geotechnical earthquake engineering (Sec. 2.8)
8. Subsoil profile (Sec. 2.9)

2.2 DOCUMENT REVIEW

Prior to performing the subsurface exploration, it may be necessary to perform a document review. Examples of the types of documents that may need to be reviewed are as follows:

Prior Development. If the site had prior development, it is important to obtain information on the history of the site. The site could contain old deposits of fill, abandoned septic systems and leach fields, buried storage tanks, seepage pits, cisterns, mining shafts, tunnels, and other man-made surface and subsurface works that could impact the new proposed development. There may also be information concerning on-site utilities and underground pipelines, which may need to be capped or rerouted around the project.

Aerial Photographs and Geologic Maps. During the course of the work, it may be necessary for the engineering geologist to check reference materials, such as aerial photographs or geologic maps. Aerial photographs are taken from an aircraft flying at prescribed altitude along preestablished lines. Interpretation of aerial photographs takes considerable judgment and because they have more training and experience, it is usually the engineering geologist who interprets the aerial photographs. By viewing a pair of aerial photographs, with the aid of a stereoscope, a three-dimensional view of the land surface is provided. This view may reveal important geologic information at the site, such as the presence of landslides, fault scarps, types of landforms (e.g., dunes, alluvial fans, glacial deposits such as moraines and eskers), erosional features, general type and approximate thickness of vegetation, and drainage patterns. By comparing older versus newer aerial photographs, the engineering geologist can also observe any man-made or natural changes that have occurred at the site.

Geologic maps can be especially useful to the geotechnical engineer and engineering geologist because they often indicate potential geologic hazards (e.g., faults landslides and the like) as well as the type of near surface soil or rock at the site. For example, Fig. 2.1 presents a portion of a geologic map and Fig. 2.2 shows cross sections through the area shown in Fig. 2.1 (from Kennedy, 1975). Note that the geologic map and cross sections indicate the location of several faults, the width of the faults, and often state whether the faults are active or inactive. For example, in Fig. 2.2, the Rose Canyon Fault zone is shown, which is an active fault having a ground shear zone about 1000 ft (300 m) wide. The cross sections in Fig. 2.2 also show fault related displacement of various rock layers. Symbols are used to identify various deposits and Table 2.1 provides a list of geologic symbols versus type of material and soil or rock description for the geologic symbols shown in Figs. 2.1 and 2.2.

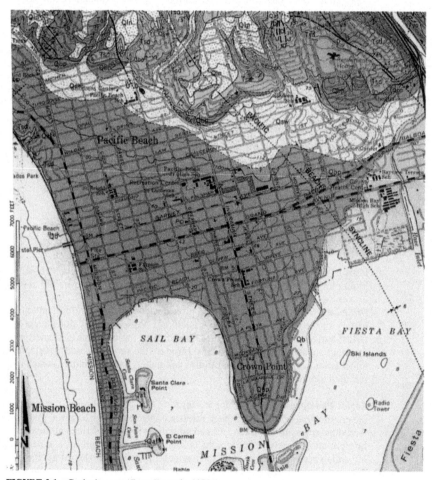

FIGURE 2.1   Geologic map. (*From Kennedy, 1975.*)

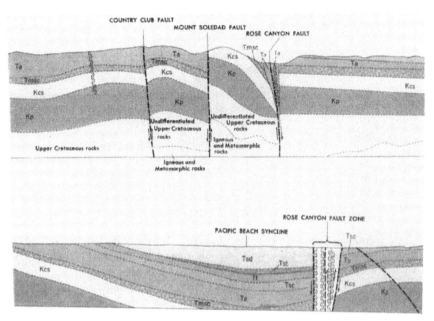

FIGURE 2.2   Geologic cross sections. (*From Kennedy, 1975.*)

TABLE 2.1   Symbols and Descriptions for Geologic Map and Cross Sections Shown in Fig. 2.1 and 2.2

Geologic symbol	Type of material	Description
Q_{af}	Artificial fill	Artificial fill consists of compacted earth materials derived from many sources. Only large areas having artificial fill have been delineated on the geologic map.
Q_b	Beach sand	Sand deposited along the shoreline derived from many sources as a result of longshore drift and alluvial discharge from major stream courses.
Q_{al}	Alluvium	Soil deposited by flowing water, including sediments deposited in river beds, canyons, flood plains, lakes, fans at the foot of slopes, and estuaries.
Q_{sw}	Slope wash	Soil and/or rock material that has been transported down a slope by mass wasting assisted by runoff of water not confined to channels.
Q_{ls}	Landslide	Landslides are mass movement of soil or rock that involves shear displacement along one or several rupture surfaces, which are either visible or may be reasonably inferred.
Q_{bp}, Q_{lb}, Q_{ln}	Formational rock	Various sedimentary rock formations formed during the Pleistocene epoch (part of the Quaternary Period).
$T_a, T_f, T_{sc},$ T_{sd}, T_{st}	Formational rock	Various sedimentary rock formations formed during the Eocene epoch (part of the Tertiary Period).
K_{cs}, K_p	Formational rock	Various rock formations formed during the Cretaceous Period.

Note: For geologic symbols, Q represents soil or rock deposited during the Quaternary Period, T = Tertiary Period, and K = Cretaceous Period.

A major source for geologic maps in the United States is the United States Geological Survey (USGS). The USGS prepares many different geologic maps, books, and charts and these documents can be purchased at the online USGS bookstore. The USGS also provides an "Index to Geologic Mapping in the United States," which shows a map of each state and indicates the areas where a geologic map has been published.

Topographic Maps.   Both old and recent topographic maps can provide valuable site information. Figure 2.3 presents a portion of the topographic map for the Encinitas Quadrangle, California (USGS, 1975). As shown in Fig. 2.3, the topographic map is to scale and shows the locations of buildings, roads, freeways, train tracks, and other civil engineering works as well as natural features such as canyons, rivers, lagoons, sea cliffs, and beaches. The topographic map in Fig. 2.3 even shows the locations of sewage disposal ponds, water tanks, and by using different colors and shading, it indicates older versus newer development. But the main purpose of the topographic map is to indicate ground surface elevations or elevations of the sea floor, such as shown in Fig. 2.3. This information can be used to determine the major topographic features at the site and for the planning of subsurface exploration, such as available access to the site for drilling rigs.

Building Code and Other Specifications.   A copy of the most recently adopted local building code should be reviewed. Usually only a few sections of the building code will be directly applicable to

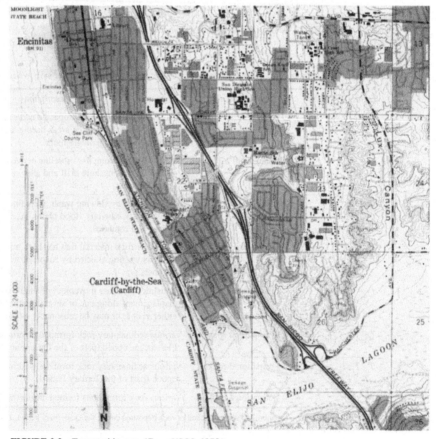

FIGURE 2.3   Topographic map. (*From USGS, 1975.*)

TABLE 2.2 Typical Documents that may Need to be Reviewed for the Project

Project phase	Type of documents
Design	Available design information, such as preliminary data on the type of project to be built at the site and typical foundation design loads
	If applicable, data on the history of the site, such as information on prior fill placement or construction at the site
	Data (if available) on the design and construction of adjacent property
	Local building code
	Special study data developed by the local building department or other governing agency
	Standard drawings issued by the local building department or other governing agency
	Standard specifications that may be applicable to the project, such as *Standard Specifications for Public Works Construction* or *Standard Specifications for Highway Bridges*
	Other reference material, such as seismic activity records, geologic and topographic maps, aerial photographs and the like.
Construction	Reports and plans developed during the design phase
	Construction specifications
	Field change orders
	Information bulletins used during construction
	Project correspondence between different parties
	Building department reports or permits

foundation engineering. For example, the main applicable geotechnical section in the *International Building Code* (2009) is Chap. 18, "Soils and Foundations." Depending on the type of project, there may be other specifications that are applicable for the project and will need to be reviewed. Documents that may be needed for public works projects include the *Standard Specifications for Public Works Construction* (2003) or the *Standard Specifications for Highway Bridges* (AASHTO, 1996).

Documents at the Local Building Department.   Other useful technical documents include geotechnical and foundation engineering reports for adjacent properties, which can provide an idea of possible subsurface conditions. A copy of geotechnical engineering reports on adjacent properties can often be obtained at the archives of public agencies, such as the local building department. Other valuable reference materials are standard drawings or standard specifications, which can also be obtained from the local building department.

Forensic Engineering.   Reports or other documents concerning the investigation of damaged or deteriorated structures may discuss problem conditions that could be present at the site (Day, 1999b, 2000b, 2004).

Table 2.2 presents a summary of typical documents that may need to be reviewed prior to or during the construction of the project.

2.3 PURPOSE OF SUBSURFACE EXPLORATION

The general purpose of subsurface exploration is to determine the following (AASHTO, 1996):

1. Soil strata
   a. Depth, thickness, and variability
   b. Identification and classification
   c. Relevant engineering properties, such as shear strength, compressibility, stiffness, permeability, expansion or collapse potential, and frost susceptibility.

TABLE 2.3 Foundation Investigations, Samples, Samplers, and Subsurface Exploration

		Foundation investigations
Three types of problems	Foundation problems	Such as the stability of subsurface materials, deformation and consolidation, and pressure on supporting structures
	Construction problems	Such as the excavation of subsurface material and use of the excavated material
	Groundwater problems	Such as the flow, action, and use of groundwater
Three phases of investigation	Subsurface investigation	Consisting of exploration, sampling, and identification in order to prepare rough or detailed boring logs and soil profiles
	Physical testing	Consisting of laboratory tests and field tests in order to develop rough or detailed data on the variations of physical soil or rock properties with depth
	Evaluation of data	Consisting of the use of soil mechanics and rock mechanics to prepare the final design recommendations based on the subsurface investigation and physical testing
		Samples and samplers
Type of samples	Altered soil (nonrepresentative samples)	Soil from various strata that is mixed, has some soil constituents removed, or foreign materials have been added to the sample
	Disturbed soil (representative samples)	Soil structure is disturbed and there is a change in the void ratio but there is no change in the soil constituents
	Undisturbed samples	No disturbance in soil structure, with no change in water content, void ratio, or chemical composition
Types of samplers	Exploration samplers	Group name for drilling equipment such as augers used for both advancing the borehole and obtaining samples
	Drive samplers	Sampling tubes driven without rotation or chopping with displaced soil pushed aside. Examples include open drive samplers and piston samplers
	Core boring samplers	Rotation or chopping action of sampler where displaced material is ground up and removed by circulating water or drilling fluid
		Subsurface exploration
Principal types of subsurface exploration	Indirect methods	Such as geophysical methods that may yield limited subsurface data. Also includes borings that are advanced without taking soil samples
	Semidirect methods	Such as borings that obtain disturbed soil samples
	Direct methods	Such as test pits, trenches, or borings that are used to obtain undisturbed soil samples
Three phases of subsurface exploration	Fact finding and geological survey	Gathering of data, document review, and site survey by engineer and geologist
	Reconnaissance explorations	Semidirect methods of subsurface exploration. Rough determination of groundwater levels
	Detailed explorations	Direct methods of subsurface exploration. Accurate measurements of groundwater levels or pore water pressure

Source: Adapted and updated from Hvorslev (1949).

2. Rock strata
   a. Depth to rock
   b. Identification and classification
   c. Quality, such as soundness, hardness, jointing and presence of joint filling, resistance to weathering (if exposed), and soluble nature of the rock.
3. Groundwater elevation
4. Local conditions requiring special consideration

In terms of the general procedures and requirements for subsurface exploration, Hvorslev (1949) states:

> Investigation of the distribution, type, and physical properties of subsurface materials are, in some form or other, required for the final design of most civil engineering structures. These investigations are performed to obtain solutions to the following groups of problems:
> Foundation problems or determination of the stability and deformations of undisturbed subsurface materials under superimposed loads, in slope and cuts, or around foundation pits and tunnels; and determination of the pressure of subsurface materials against supporting structures when such are needed.
> Construction problems or determination of the extent and character of materials to be excavated or location and investigation of soil and rock deposits for use as construction materials in earth dams and fills, for road and airfield bases and surfacing, and for concrete aggregates.
> Groundwater problems or determination of the depth, hydrostatic pressure, flow, and composition of the ground water, and thereby the danger of seepage, underground erosion, and frost action; the influence of the water on the stability and settlement of structures; its action on various construction materials; and its suitability as a water supply.

There are many different types of subsurface exploration, such as borings, test pits, or trenches. Table 2.3 presents general information on foundation investigations, samples and samplers, and subsurface exploration. Table 2.4 (from Sowers and Royster, 1978, based on the work by ASTM; Lambe, 1951; Sanglerat, 1972; Sowers and Sowers, 1970) summarizes the boring, core drilling, sampling and other exploratory techniques that can be used by the geotechnical engineer.

As mentioned earlier, the borings, test pits, or trenches are used to determine the thickness of soil and rock strata, estimate the depth to groundwater, obtain soil or rock specimens, and perform field tests such as Standard Penetration Tests (SPT) or Cone Penetration Tests (CPT). The Unified Soil Classification System (USCS) can be used to classify the soil exposed in the borings or test pits (Casagrande, 1948). The subsurface exploration and field sampling should be performed in accordance with standard procedures, such as those specified by the American Society for Testing and Materials (ASTM, 1970, 1971, and D 420-03, 2004) or other recognized sources (e.g., Hvorslev, 1949; ASCE, 1972, 1976, 1978). App. A (Glossary 1) presents a list of terms and definitions for subsurface exploration.

2.4 BORINGS

A boring is defined as a cylindrical hole drilled into the ground for the purposes of investigating subsurface conditions, performing field tests, and obtaining soil, rock, or groundwater specimens for testing. Borings can be excavated by hand (e.g., hand auger), although the usual procedure is to use mechanical equipment to excavate the borings.

During the excavation and sampling of the borehole, it is important to prevent caving-in of the borehole sidewalls. In those cases where boreholes are made in soil or rock and there is no groundwater, the holes will usually remain stable. Exceptions include clean sand and gravels that may cave-in even when there is no groundwater. The danger of borehole caving-in increases rapidly with depth and the presence of groundwater. For cohesive soils, such as firm to hard clay, the borehole may remain stable for a limited time even though the excavation is below the groundwater table. For other soils below the groundwater table, borehole stabilization techniques will be required, as follows:

Stabilization with Water.   Boreholes can be filled with water up to or above the estimated level of the groundwater table. This will have the effect of reducing the sloughing of soil caused by water rushing into the borehole. However, water alone cannot prevent caving-in of borings in soft or cohesionless

TABLE 2.4 Boring, Core Drilling, Sampling, and Other Exploratory Techniques

Method	Procedure	Type of sample	Applications	Limitations
Auger boring, ASTM D 1452	Dry hole drilled with hand or power auger; samples preferably recovered from auger flutes	Auger cuttings, disturbed, ground up, partially dried from drill heat in hard materials	In soil and soft rock; to identify geologic units and water content above water table	Soil and rock stratification destroyed; sample mixed with water below the water table
Test boring, ASTM D 1586	Hole drilled with auger or rotary drill; at intervals samples taken 36-mm ID and 50-mm OD driven 0.45 m in three 150-mm increments by 64-kg hammer falling 0.76 m; hydrostatic balance of fluid maintained below water level	Intact but partially disturbed (number of hammer blows for second plus third increment of driving is standard penetration resistance or N)	To identify soil or soft rock; to determine water content; in classification tests and crude shear test of sample (N value a crude index to density of cohesionless soil and undrained shear strength of cohesive soil)	Gaps between samples, 30 to 120 cm; sample too distorted for accurate shear and consolidation tests; sample limited by gravel; N value subject to variations depending on free fall of hammer
Test boring of large samples	50- to 75-mm ID and 63- to 89-mm OD samplers driven by hammers up to 160 kg	Intact but partially disturbed (number of hammer blows for second plus third increment of driving is penetration resistance)	In gravelly soils	Sample limited by larger gravel
Test boring through hollow-stem auger	Hole advanced by hollow-stem auger; soil sampled below auger as in test boring above	Intact but partially disturbed (number of hammer blows for second plus third increment of driving is N value)	In gravelly soils (not well adapted to harder soils or soft rock)	Sample limited by larger gravel; maintaining hydrostatic balance in hole below water table is difficult
Rotary coring of soil or soft rock	Outer tube with teeth rotated; soil protected and held by stationary inner tube; cuttings flushed upward by drill fluid (examples: Denison, Pitcher, and Acker samplers)	Relatively undisturbed sample, 50 to 200 mm wide and 0.3 to 1.5 m long in liner tube	In firm to stiff cohesive soils and soft but coherent rock	Sample may twist in soft clays; sampling loose sand below water table is difficult; success in gravel seldom occurs
Rotary coring of swelling clay, soft rock	Similar to rotary coring of rock; swelling core retained by third inner plastic liner	Soil cylinder 28.5 to 53.2 mm wide and 600 to 1500 mm long encased in plastic tube	In soils and soft rocks that swell or disintegrate rapidly in air (protected by plastic tube)	Sample smaller; equipment more complex

Rotary coring of rock, ASTM D 2113	Outer tube with diamond bit on lower end rotated to cut annular hole in rock; core protected by stationary inner tube; cuttings flushed upward by drill fluid	Rock cylinder 22 to 100 mm wide and as long as 6 m, depending on rock soundness	To obtain continuous core in sound rock (percent of core recovered depends on fractures, rock variability, equipment, and driller skill)	Core lost in fracture or variable rock; blockage prevents drilling in badly fractured rock; dip of bedding and joint evident but not strike
Rotary coring of rock, oriented core	Similar to rotary coring of rock above; continuous grooves scribed on rock core with compass direction	Rock cylinder, typically 54 mm wide and 1.5 m long with compass orientation	To determine strike of joints and bedding	Method may not be effective in fractured rock
Rotary coring of rock, wire line	Outer tube with diamond bit on lower end rotated to cut annular hole in rock; core protected by stationary inner tube; cuttings flushed upward by drill fluid; core and stationary inner tube retrieved from outer core barrel by lifting device or "overshot" suspended on thin cable (wire line) through special large-diameter drill rods and outer core barrel	Rock cylinder 36.5 to 85 mm wide and 1.5 to 4.6 m long	To recover core better in fractured rock, which has less tendency for caving during core removal; to obtain much faster cycle of core recovery and resumption of drilling in deep holes	Same as ASTM D 2113 but to lesser degree
Rotary coring of rock, integral sampling method	22-mm hole drilled for length of proposed core; steel rod grouted into hole; core drilled around grouted rod with 100- to 1500-mm rock coring drill (same as for ASTM D 2113)	Continuous core reinforced by grouted steel rod	To obtain continuous core in badly fractured, soft, or weathered rock in which recovery is low by ASTM D 2113	Grout may not adhere in some badly weathered rock; fractures sometimes cause drift of diamond bit and cutting rod
Thin-wall tube, ASTM D 1587	75- to 1250-mm thin-wall tube forced into soil with static force (or driven in soft rock); retention of sample helped by drilling mud	Relatively undisturbed sample, length 10 to 20 diameters	In soft to firm clays, short (5-diameter) samples of stiff cohesive soil, soft rock, and, with aid of drilling mud, firm to dense sands	Cutting edge wrinkled by gravel; samples lost in loose sand or very soft clay below water table; more disturbance occurs if driven with hammer

(Continued)

2.11

TABLE 2.4 Boring, Core Drilling, Sampling, and Other Exploratory Techniques (*Continued*)

Method	Procedure	Type of sample	Applications	Limitations
Thin-wall tube, fixed piston	75- to 1250-mm-thin-wall tube, which has internal piston controlled by rod and keeps loose cuttings from tube, remains stationary while outer thin wall tube forced ahead into soil; sample is held in tube by piston	Relatively undisturbed sample, length 10 to 20 diameters	To minimize disturbance of very soft clays (drilling mud aids in holding samples in loose sand below water table)	Method is slow and cumbersome
Swedish foil	Samples surrounded by thin strips of stainless steel, stored above cutter, to prevent contact of soil with tube as it is forced into soil	Continuous samples 50 mm wide and as long as 12 m	In soft, sensitive clays	Samples sometimes damaged by coarse sand and fine gravel
Dynamic sounding	Enlarged disposable point on end of rod driven by weight falling fixed distance in increments of 100 to 300 mm	None	To identify significant differences in soil strength or density	Misleading in gravel or loose saturated fine cohesionless soils
Static penetration	Enlarged cone, 36 mm diameter and 60° angle forced into soil; force measured at regular intervals	None	To identify significant differences in soil strength or density; to identify soil by resistance of friction sleeve	Stopped by gravel or hard seams
Borehole camera	Inside of core hole viewed by circular photograph or scan	Visual representation	To examine stratification, fractures, and cavities in hole walls	Best above water table or when hole can be stabilized by clear water
Pits and trenches	Pit or trench excavated to expose soils and rocks	Chunks cut from walls of trench; size not limited	To determine structure of complex formations; to obtain samples of thin critical seams such as failure surface	Moving excavation equipment to site, stabilizing excavation walls, and controlling groundwater may be difficult
Rotary or cable tool well drill	Toothed cutter rotated or chisel bit pounded and churned	Ground	To penetrate boulders, coarse gravel; to identify hardness from drilling rates	Identifying soils or rocks difficult
Percussion drilling (jack hammer or air track)	Impact drill used; cuttings removed by compressed air	Rock dust	To locate rock, soft seams, or cavities in sound rock	Drill becomes plugged by wet soil

Source: Adapted from Sowers and Royster, 1978. Reprinted with permission from *Landslides: Analysis and Control, Special Report 176.* Copyright 1978 by the National Academy of Sciences. Courtesy of the National Academy Press, Washington, D.C.

soils or a gradual squeezing-in of a borehole in plastic soils. Uncased boreholes filled with water up to or above the groundwater table can generally be used in rock and for stiff to hard cohesive soils.

Stabilization with Drilling Fluid.   An uncased borehole can often be stabilized by filling it with a properly proportioned drilling fluid, also known as "mud," which when circulated also removes the ground-up material located at the bottom of the borehole. The stabilization effect of the drilling fluid is due to two effects: (1) the drilling fluid has a higher specific gravity than water alone, and (2) the drilling fluid tends to form a relatively impervious sidewall borehole lining, often referred to as mud-cake, which prevents sloughing of cohesionless soils and decreases the rate of swelling of cohesive soils. Drilling fluid is primarily used with rotary drilling and core boring methods.

Stabilization with Casing.   The safest and most effective method of preventing caving-in of the borehole is to use a metal casing. Unfortunately, this type of stabilization is rather expensive. Many different types of standard metal or special pipe can be used as casing. The casing is usually driven in place by repeated blows of a drop hammer. It is often impossible to advance the original string of casing when difficult ground conditions or obstructions are encountered. A smaller casing is then inserted through the one in place, and the diameter of the extension of the borehole must be decreased accordingly.

Other Stabilization Methods.   One possible stabilization method is to literally freeze the ground and then drill the boring and cut or core the frozen soil from the ground. The freezing is accomplished by installing pipes in the ground and then circulating ethanol and crushed dry ice or liquid nitrogen through the pipes. Because water increases in volume upon freezing, it is important to establish a slow freezing front so that the freezing water can slowly expand and migrate out of the soil pores. This process can minimize the sample disturbance associated with the increase in volume of freezing water.

   Another method is to temporarily lower the groundwater table and allow the water to drain from the soil before the excavation of the borehole. The partially saturated soil will then be held together by capillarity, which will enable the soil strata to be bored and sampled. When brought to the ground surface, the partially saturated soil specimen is frozen. Because the soil is only partially saturated, the volume increase of water as it freezes should not significantly disturb the soil structure. The frozen soil specimen is then transported to the laboratory for testing.

   From a practical standpoint, these two methods described earlier are usually uneconomical for most projects.

   There are many different types of equipment used to excavate borings. Typical types of borings are listed in Table 2.4 and include:

Auger boring.   A mechanical auger is the simplest and fastest method of excavating a boring. Because of these advantages, augers are probably the most common type of equipment used to excavate borings. The hole is excavated through the process of rotating the auger while at the same time applying a downward pressure on the auger to help penetrate the soil or rock. There are basically two types of augers: flight augers and bucket augers (see Fig. 2.4). Common available diameters of flight augers are 2 in. to 4 ft (5 cm to 1.2 m) and of bucket augers are 1 to 8 ft (0.3 to 2.4 m). The auger is periodically removed from the hole, and the soil lodged in the blades of the flight auger or contained in the bucket of the bucket auger is removed. A casing is generally not used for auger borings and the hole may cave-in during the excavation of loose or soft soils or when the excavation is below the groundwater table.

Hollow-stem flight auger.   A hollow-stem flight auger has a circular hollow core, which allows for sampling down the center of the auger. The hollow-stem auger acts like a casing and allows for sampling in loose or soft soils or when the excavation is below the groundwater table.

Wash boring.   A wash boring is advanced by the chopping and twisting action of a light bit and partly by the jetting of water, which is pumped through the hollow drill rod and bit (see Fig. 2.5). The cuttings are removed from the borehole by the circulating water. Casing is typically required in soft or cohesionless soil, although it is often omitted for stiff to hard cohesive soil. Loose cuttings tend to accumulate at the bottom of the borehole and careful cleaning of the hole is required before samples are taken.

FIGURE 2.4   A flight auger drill rig (top) and a bucket auger drill rig (bottom).

Rotary drilling.   For rotary drilling, the borehole is advanced by the rapid rotation of the drilling bit that cuts, chips, and grinds the material located at the bottom of the borehole into small particles. In order to remove the small particles, water or drilling fluid is pumped through the drill rods and bit and ultimately up and out of the borehole. Instead of using water or drilling fluid, forced air from a compressor can be used to cool the bit and remove the cuttings (see ASTM D 2113, 2004). A drill machine and rig, such as shown in Fig. 2.6, are required to provide the

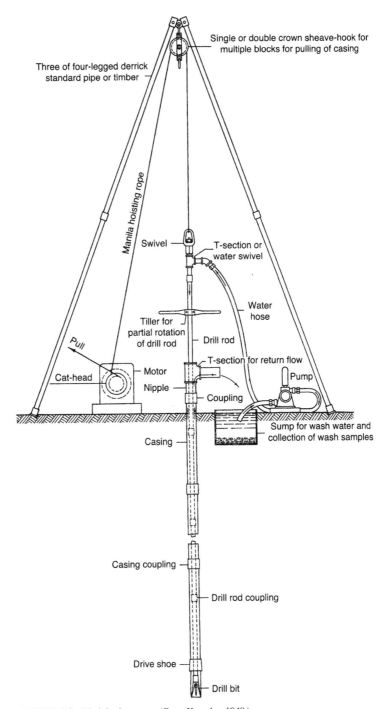

FIGURE 2.5 Wash boring setup. (*From Hvorslev, 1949.*)

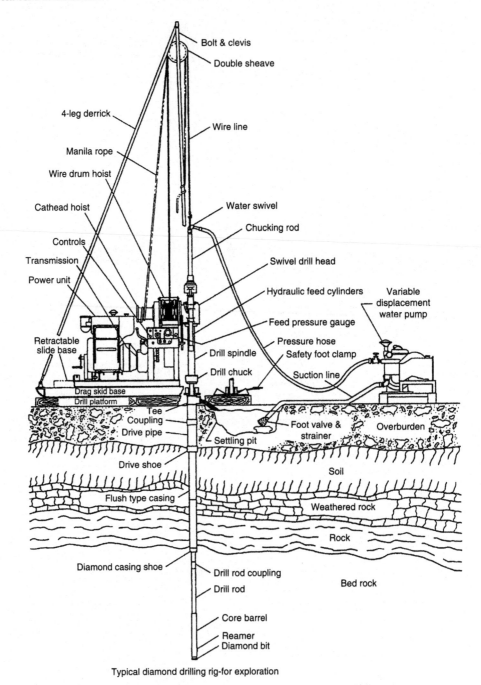

Typical diamond drilling rig-for exploration

FIGURE 2.6   Rotary drilling setup. (*Reprinted with permission of ASTM D 2113-99, 2004*).

rotary power and downward force required to excavate the boring. Other rotary drilling details are provided in Table 2.4.

Percussion drilling. This type of drilling equipment is often used to penetrate hard rock, for subsurface exploration or for the purpose of drilling wells. The drill bit works much like a jackhammer, rising and falling to break-up and crush the rock material. Percussion drilling works best for rock and will be ineffective for such materials as soft clay and loose saturated sand.

It takes considerable experience to anticipate which type of drill rig and sampling equipment would be best suited to the site under investigation. For example, if downhole logging is required, then a large diameter bucket auger boring is needed (Fig. 2.4). A large diameter boring, typically 30 in. (0.76 m) in diameter, is excavated and then the geotechnical engineer or engineering geologist descends into the borehole. Figure 2.7 shows a photograph of the top of the boring with the geologist descending into the hole in a steel cage. Note in Fig. 2.7 that a collar is placed around the top of the hole to prevent loose soil or rocks from being accidentally knocked down the hole. The process of downhole logging is a valuable technique because it allows the geotechnical engineer or engineering geologist to observe the subsurface materials, as they exist inplace. Usually the process of the excavation of the boring smears the side of the hole, and the surface must be chipped away to observe intact soil or rock. Going downhole is dangerous because of the possibility of a cave-in of the hole as well as "bad air" (presence of poisonous gases or lack of oxygen) and should only be attempted by an experienced geotechnical engineer or engineering geologist.

The downhole observation of soil and rock can lead to the discovery of important subsurface conditions. For example, Fig. 2.8 provides an example of the type of conditions observed downhole. Figure 2.8 shows a knife that has been placed in an open fracture in bedrock. Massive landslide movement caused the open fracture in the rock. Figure 2.9 is a side view of the same condition.

In general, the most economical equipment for borings are truck mounted rigs that can quickly and economically drill through hard or dense soil. It some cases, it is a trial and error process of

FIGURE 2.7 Downhole logging (arrow points to top of steel cage used for downhole logging).

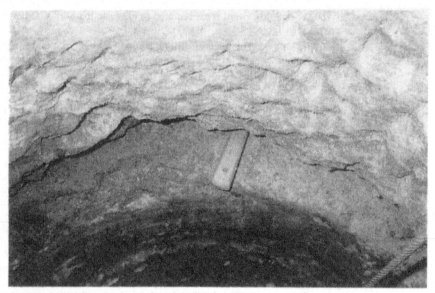

FIGURE 2.8 Knife placed in an open fracture in bedrock caused by landslide movement (photograph taken downhole in a large diameter auger boring).

using different drill rigs to overcome access problems or difficult subsurface conditions. For example, one deposit encountered by the author consisted of hard granite boulders surrounded by soft and highly plastic clay. The initial drill rig selected for the project was an auger drill rig, but the auger could not penetrate through the granite boulders. The next drill rig selected was an air track rig, which uses a percussion drill bit that easily penetrated through the granite boulders, but the soft clay plugged up the drill bit and it became stuck in the ground. Over 50 ft (15 m) of drill stem could not be removed from the ground and it had to be left in place, a very costly experience with difficult drilling conditions.

Some of my other memorable experiences with drilling are as follows:

1. *Drilling accidents.* Most experienced drillers handle their equipment safely, but accidents can happen to anyone. One day, as I observed a drill rig start to excavate the hole, the teeth of the auger bucket caught on a boulder. The torque of the auger bucket was transferred to the drill rig, and it flipped over. Fortunately, no one was injured.

2. *Underground utilities.* Before drilling, the local utility company, upon request, will locate their underground utilities by placing ground surface marks that delineate utility alignments. An incident involving a hidden gas line demonstrates that not even utility locators are perfect. On a particularly memorable day, I drove a Shelby tube sampler into a 4 in. (10 cm) diameter pressurized gas line. The noise of escaping gas was enough to warn of the danger. Fortunately, an experienced driller knew what to do: turn off the drill rig and call 911.

3. *Downhole logging.* As previously mentioned, a common form of subsurface exploration in southern California is to drill a large-diameter boring, usually 30 in. (0.76 m) in diameter. Then the geotechnical engineer or engineering geologist descends into the earth to get a close-up view of soil conditions. On this particular day, several individuals went down the hole and noticed a small trickle of water in the hole about 20 ft (6 m) down. The sudden and total collapse of the hole riveted the attention of the workers, especially the geologist who had moments before been down at the bottom of the hole.

FIGURE 2.9   Side view of condition shown in Fig. 2.8.

Because subsurface exploration has a potential for serious or even fatal injury, it is especially important that young engineers and geologists be trained to evaluate the safety of engineering operations in the field. This must be done before they supervise field operations.

2.4.1 Rock and Soil Samplers

There are many different types of samplers used to retrieve soil and rock specimens from the boring. For example, three types of soil samplers are shown in Fig. 2.10, the California sampler, Shelby tube, and SPT sampler. One of the most important first steps in sampling is to clean-out the bottom of the borehole in order to remove the loose soil or rock debris that may have fallen to the bottom of the borehole.

For hard rock, coring is used to extract specimens (see Table 2.4). The coring process consists of rotating a hollow steel tube, known as a core barrel, which is equipped with a boring bit. The drilled rock core is collected in the core barrel as the drilling progresses. Once the rock core has been cut and the core barrel is full, the drill rods are pulled from the borehole and the rock core is extracted from the core barrel. A rotary drill rig, such as shown in Fig. 2.6, is often used for the rock coring operation. For further details on rock core drill and sampling, see ASTM D 2113-99 (2004), "Standard Practice for Rock Core Drilling and Sampling of Rock for Site Investigation."

For soil, the most common method is to force a sampler into the soil by either hammering, jacking, or pushing the sampler into the soil located at the bottom of the borehole. Soil samplers are typically divided into two types.

Thin-Walled Soil Sampler.   The most common type of soil sampler used in the United States is the Shelby tube, which is a thin-walled sampling tube consisting of stainless steel or brass tubing. In order to slice through the soil, the Shelby tube has a sharp and drawn-in cutting edge. In terms of dimensions, typical diameters are from 2 to 3 in. (5 to 7.6 cm) and lengths vary from 2 to 3 ft (0.6 to 0.9 m).

The typical arrangement of drill rod, sampler head, and thin-wall tube sampler is shown in Fig. 2.11. The sampler head contains a ball check valve and vents for escape of air and water during the sampling process. The drill rig equipment can be used to either hammer, jack, or push the sampler into the soil. The preferred method is to slowly push the sampler into the soil by using hydraulic

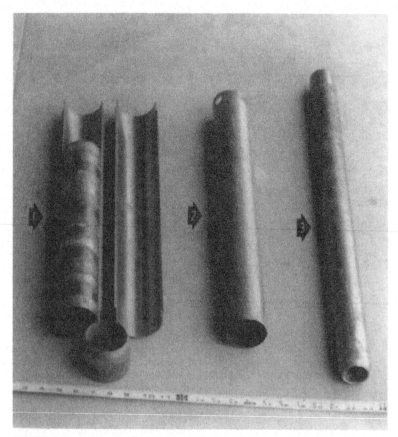

FIGURE 2.10   Soil samplers (no. 1 is the California sampler in an open condition, no. 2 is a Shelby tube, and no. 3 is the standard penetration test sampler).

jacks or the weight of the drilling equipment. Thin-walled soil samplers are used to obtain undisturbed soil samples, which will be discussed in the next section. For further details on thin-walled sampling, see ASTM D 1587-00 (2004), "Standard Practice for Thin-Walled Tube Sampling of Soils for Geotechnical Purposes."

Thick-Walled Soil Sampler.   Thin-walled samplers may not be strong enough to sample gravelly soils, very hard soils, or cemented soils. In such cases, a thick-walled soil sampler will be required. Such samplers are often driven into place by using a drop hammer. The typical arrangement of drill rod, sampler head, and barrel when driving a thick-walled sampler is shown in Fig. 2.11.

Many localities have developed thick-walled samplers that have proven successful for local conditions. For example, in southern California, a common type of sampler is the California sampler, which is a split-spoon type sampler that contains removable internal rings, 1.0 in. (2.54 cm) in height. Figure 2.10 shows the California sampler in an open condition, with the individual rings exposed. The California sampler has a 3.0 in. (7.6 cm) outside diameter and a 2.50 in. (6.35 cm) inside diameter. This sturdy sampler, which is considered to be a thick-walled sampler, has proven successful in sampling hard and desiccated soil and soft sedimentary rock common in southern California. Another type of thick-walled sampler is the SPT sampler, which will be discussed in Sec. 2.4.3.

For further details on thick-walled sampling, see ASTM D 3550-01 (2004), "Standard Practice for Thick Wall, Ring-Lined, Split Barrel, Drive Sampling of Soils."

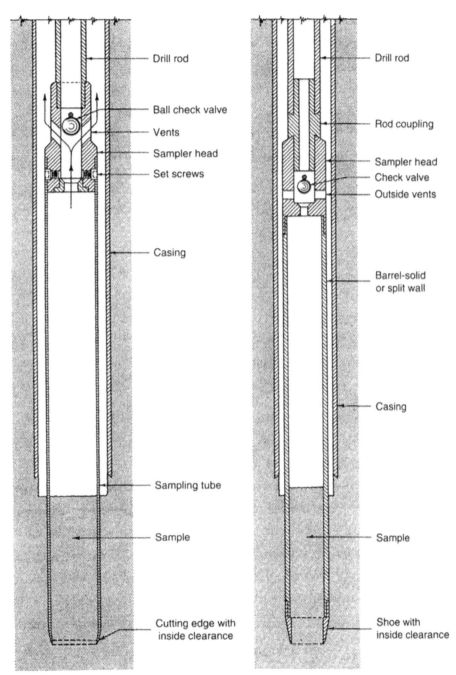

Thin-wall open drive sampler   **Thick-wall open drive sampler**

FIGURE 2.11   Thin-wall and thick-wall samplers. (*From Hvorslev, 1949.*)

2.4.2   Sample Disturbance

This section will discuss the three types of soil samples that can be obtained during the subsurface exploration. In addition, this section will also discuss sampler and sample ratios used to evaluate sample disturbance; factors that affect sample quality; x-ray radiography; and transporting, preserving, and disposal of soil samples.

Soil Samples.   There are three types of soil samples that can be recovered from borings:

Altered Soil (also known as Nonrepresentative Samples).   During the boring operations, soil can be altered due to mixing or contamination. Such materials do not represent the soil found at the bottom of the borehole and hence should not be used for visual classification or laboratory tests. Some examples of altered soil are as follows:

> *Failure to clean the bottom of the boring.*   If the boring is not cleaned out prior to sampling, a soil sample taken from the bottom of the borehole may actually consist of cuttings from the side of the borehole. These borehole cuttings, which have fallen to the bottom of the borehole, will not represent in situ conditions at the depth sampled.
>
> *Soil contamination.*   In other cases, the soil sample may become contaminated with drilling fluid, which is used for wash-type borings. These samples are often called *wash samples* or *wet samples* because they are washed out of the borehole and allowed to settle in a sump at ground surface. These types of soil samples that have been contaminated by the drilling process should not be used for laboratory tests because they will lead to incorrect conclusions regarding subsurface conditions.
>
> *Soil mixing.*   Soil or rock layers can become mixed during the drilling operation, such as by the action of a flight auger. For example, suppose varved clay, which consists of thin alternating layers of sand and clay, becomes mixed during the drilling and sampling process. Obviously laboratory tests would produce different results when performed on the mixed soil as compared to laboratory tests performed on the individual sand and clay layers.
>
> *Change in moisture content.*   Soil that has a change in moisture content due to the drilling fluid or from heat generated during the drilling operations should also be classified as altered soil.
>
> *Densified soil.*   Soil that has been densified by over-pushing or over-driving the soil sampler should also be considered as altered because the process of over-pushing or over-driving could squeeze water from the soil. Figure 2.12 shows a photograph of the rear end of a Shelby tube sampler. The soil in the sampler has been densified by being over-pushed as indicated by the smooth surface of the soil and the mark in the center of the soil (due to the sampler head).

In summary, any soil or rock where the mineral constituents have been removed, exchanged, or mixed should be considered as altered soil.

Disturbed Samples (also known as Representative Samples).   It takes considerable experience and judgment to distinguish between altered soil and disturbed soil. In general, disturbed soil is defined as soil that has not been contaminated by material from other strata or by chemical changes, but the soil structure is disturbed and the void ratio may be altered. In essence, the soil has only been remolded during the sampling process. For example, soil obtained from driven thick-walled samplers, such as the SPT spilt spoon sampler, or chunks of intact soil brought to the surface in an auger bucket (i.e., bulk samples) are considered disturbed soil.

Disturbed soil can be used for visual classification as well as numerous types of laboratory tests. Example of laboratory tests that can be performed on disturbed soil include water content, specific gravity, Atterberg limits, sieve and hydrometer tests, expansion index test, chemical composition (such as soluble sulfate), and laboratory compaction tests such as the Modified Proctor.

Undisturbed Samples.   Undisturbed samples may be broadly defined as soil that has been subjected to no disturbance or distortion and the soil is suitable for laboratory tests that measure the shear strength, consolidation, permeability, and other physical properties of the in situ material. As a practical matter,

FIGURE 2.12   Densified soil due to overpushing a Shelby tube.

it should be recognized that no soil sample can be taken from the ground and be in a perfectly undisturbed state. But this terminology has been applied to those soil samples taken by certain sampling methods. Undisturbed samples are often defined as those samples obtained by slowly pushing thin-walled tubes, having sharp cutting ends and tip relief, into the soil.

Undisturbed soil samples are essential in many types of foundation engineering analyses, such as the determination of allowable bearing pressure and settlement. Many soil samples may appear to be undisturbed but they have actually been subjected to considerable disturbance of the soil structure. It takes considerable experience and judgment to evaluate laboratory test results on undisturbed soil samples as compared to test results that may be inaccurate due to sample disturbance.

Sampler and Sample Ratios Used to Evaluate Sample Disturbance.   Figure 2.13 presents various sampler and sample ratios that are used to evaluate the disturbance potential of different samplers and of the soil samples themselves. For soil samplers, the two most important parameters to evaluate disturbance potential are the inside clearance ratio and area ratio, defined as follows:

$$\text{Inside clearance ratio} = \frac{D_s - D_e}{D_e} \tag{2.1}$$

$$\text{Area ratio} = \frac{D_w^{\,2} - D_e^{\,2}}{D_e^{\,2}} \tag{2.2}$$

where D_e = diameter at the sampler cutting tip (cm or in.)
  D_s = inside diameter of the sampling tube (cm or in.)
  D_w = outside diameter of the sampling tube, see Fig. 2.13 (cm or in.)

So that they can be expressed as a percentage, both the inside clearance ratio and area ratio are typically multiplied by 100. Note in Fig. 2.13 that because common terms cancel out, the area ratio can be defined as the volume of displaced soil divided by the volume of the sample.

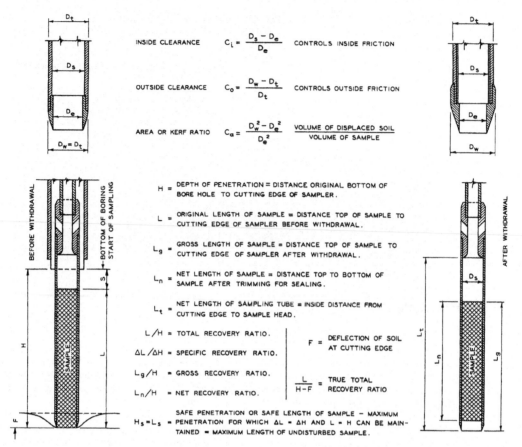

FIGURE 2.13 Sampler and sample ratios used to evaluate sample disturbance. (*From Hvorslev, 1949.*)

In general, a sampling tube for undisturbed soil specimens should have an inside clearance ratio of about 1 percent and an area ratio of about 10 percent or less. Having an inside clearance ratio of about 1 percent provides for tip relief of the soil and reduces the friction between the soil and inside of the sampling tube during the sampling process. A thin film of oil can be applied at the cutting edge to also reduce the friction between the soil and metal tube during sampling operations. The purpose of having a low area ratio and a sharp cutting end is to slice into the soil with as little disruption and displacement of the soil as possible. Shelby tubes are manufactured to meet these specifications and are considered to be undisturbed soil samplers. As a comparison, the California sampler has an area ratio of 44 percent and is considered to be a thick-walled sampler.

Figure 2.13 also presents common ratios that can be used to assess the possibility of sample disturbance of the actual soil specimen. Examples include the total recovery ratio, specific recovery ratio, gross recovery ratio, net recovery ratio, and true recovery ratio. These disturbance parameters are based on the compression of the soil sample due to the sampling operations. Because the length of the soil specimen is often determined after the sampling tube is removed from the borehole, a commonly used parameter is the gross recovery ratio, defined as:

$$\text{Gross recovery ratio} = \frac{L_g}{H} \tag{2.3}$$

where L_g is gross length of sample, which is the distance from the top of the sample to the cutting edge of the sampler after removal of the sampler from the boring (in. or cm). H is depth of penetration of the sampler, which is the distance from the original bottom of the borehole to the cutting edge of the sampler after it has been driven or pushed in place (in. or cm).

The closer the gross recovery ratio is to 1.0 (or 100 percent), the better the quality of the soil specimen.

Factors that Affect Sample Quality.   It is important to understand that using a thin wall tube, such as a Shelby tube, or obtaining a gross recovery ratio of 100 percent would not guarantee an undisturbed soil specimen. Many other factors can cause soil disturbance, such as:

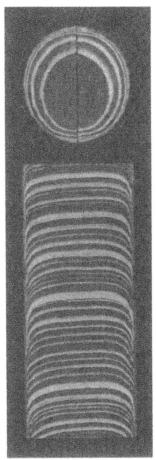

- Pieces of hard gravel or shell fragments in the soil, which can cause voids to develop along the sides of the sampling tube during the sampling process
- Soil adjustment caused by stress relief when making a borehole
- Disruption of the soil structure due to hammering or pushing the sampling tube into the soil stratum
- Tensile and torsional stresses which are produced in separating the sample from the subsoil
- Creation of a partial or full vacuum below the sample as it is extracted from the subsoil
- Expansion of gas during retrieval of the sampling tube as the confining pressure is reduced to zero
- Jarring or banging the sampling tube during transportation to the laboratory
- Roughly removing the soil from the sampling tube
- Crudely cutting the soil specimen to a specific size for a laboratory test

The actions listed earlier cause a decrease in effective stress, a reduction in the interparticle bonds, and a rearrangement of the soil particles. An "undisturbed" soil specimen will have little rearrangement of the soil particles and perhaps no disturbance except that caused by stress relief where there is a change from the in situ k_o (at-rest) condition to an isotropic *perfect sample* stress condition (Ladd and Lambe, 1963). A disturbed soil specimen will have a disrupted soil structure with perhaps a total rearrangement of soil particles. When measuring the shear strength or deformation characteristics of the soil, the results of laboratory tests run on undisturbed specimens obviously better represent in situ properties than laboratory tests run on disturbed specimens.

Some examples of disturbed soil are shown in Figs. 2.14 to 2.16 and described as follows:

Turning of edges.   Turning or bending of edges of various thin layers show as curved down edges on the sides of the specimen. This effect is due to the friction between the soil and sampler. Turning of edges could also occur when the soil specimen is pushed out of the back of the sampler in the laboratory. The turning of

SOFT VARVED CLAY - DEPTH 5 FT.
4¾" MOHR SAMPLER - HAMMERING
SPEC. RECOVERY OF SECTION = 100%

SERIOUS DISTORTIONS
WITH FULL RECOVERY

FIGURE 2.14 A type of sample disturbance known as turning of edges. Note that a Mohr sampler is also known as a Shelby tube. (*From Hvorslev, 1949.*)

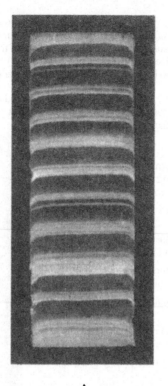 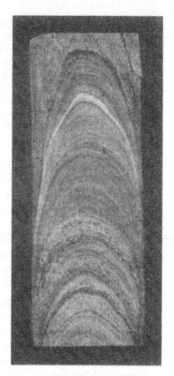

A   B

VARVED CLAY – DEPTH 5'   SANDY AND SILTY CLAY
2" PIGGOT SAMPLER   DEPTH 18' – 2" SHELBY TUBING
SHOOTING   HAMMERING

DRAG AND DISTORTION BY INSIDE FRICTION

FIGURE 2.15   More examples of sample disturbance due to the friction between the sampler and soil. (*From Hvorslev, 1949.*)

edges can also be created when the sampler is hammered into the soil. Examples of turning of edges are shown in Figs 2.14 and 2.15.

Shear failures.   Figure 2.16 shows four examples of shear failure of the soil within the sampler. This sample disturbance occurred during the pushing of Shelby tubes into medium soft silty clay.

X-ray Radiography of Soil Samples.   Although rarely used in practice, one method of assessing the quality of soil samples is to obtain an x-ray radiograph of the soil contained in the sampling tube. A radiograph is a photographic record produced by the passage of x-rays through an object and onto photographic film. Denser objects absorb the x-rays and can appear as dark areas on the radiograph. Worm holes, coral fragments, cracks, gravel inclusions, and sand or silt seams can easily be identified by using radiography (Allen et al., 1978).

Figures 2.17 and 2.18 present two radiographs taken of Orinoco Clay contained within Shelby tubes. These two radiographs illustrate additional types of soil disturbance:

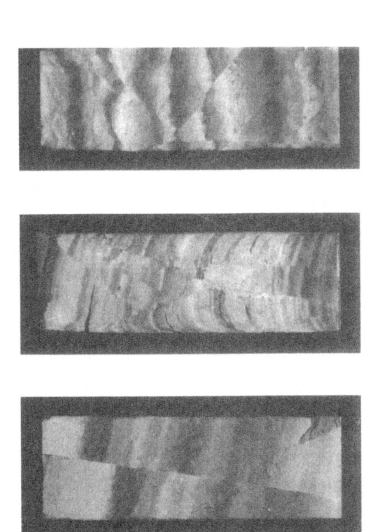

FIGURE 2.16 Four examples of shear failures caused by the sampling operation. (*From Hvorslev, 1949.*)

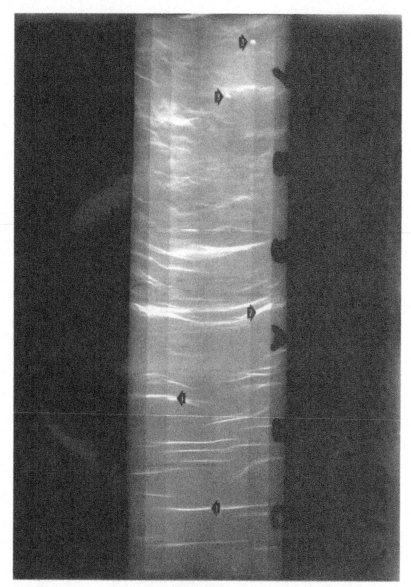

FIGURE 2.17   Radiograph of Orinoco clay within a Shelby tube. (*From Day, 1980; Ladd et al., 1980.*)

Voids.   The top of Fig. 2.17 shows large white areas, which are the locations of soil voids. The causes of such voids are often due to sampling and transporting process. The open voids can be caused by many different factors, such as gravel or shells which impact with the cutting end of the sampling tube and/or scrape along the inside of the sampling tube and create voids. The voids and highly disturbed clay shown at the top of Fig. 2.17 are possibly due to cuttings inadvertently left at the bottom of the borehole. Some of the disturbance could also be caused by tube friction during sampling as the clay near the tube wall becomes remolded as it travels up the tube.

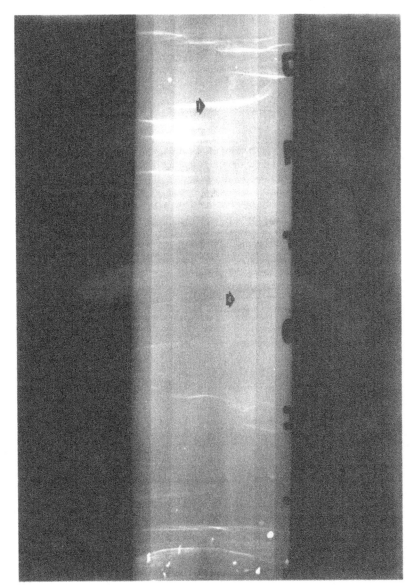

FIGURE 2.18   Radiograph of Orinoco clay within a Shelby tube. (*From Day, 1980; Ladd et al., 1980.*)

Soil cracks.   Figures 2.17 and 2.18 show numerous cracks in the clay. For example, the arrows labeled 1 point to some of the soil cracks in Figs. 2.17 and 2.18. Some of the cracks appear to be continuous across the entire sampling tube (e.g., arrow labeled 2, Fig. 2.17). The soil cracks probably developed during the sampling process. A contributing factor in the development of the soil cracks may have been gas coming out of solution, which fractured the clay.

Gas related voids.   The circular voids (labeled 3) shown in Fig. 2.17 were caused by gas coming out of solution during the sampling process when the confining pressures were essentially reduced to zero.

In contrast to soil disturbance, the arrow labeled 4 in Fig. 2.18 indicates an undisturbed section of the soil sample. Note in Fig. 2.18 that the individual fine layering of the soil sample can even be observed.

For further details on x-ray radiography, see ASTM D 4452-02, 2004, "Standard Test Methods for X-Ray Radiography of Soil Samples."

Transporting Soil Samples.   During transport to the laboratory, soil samples recovered from the borehole should be kept within the sampling tube or sampling rings. In order to preserve soil samples during transportation, the soil sampling tubes can be tightly sealed with end caps and duct tape. For sampling rings, they can be placed in cylindrical packing cases that are then thoroughly sealed. Bulk samples can be placed in plastic bags, pails, or other types of waterproof containers. The goal of the transportation of soil samples to the laboratory is to prevent a loss of moisture. In addition, for undisturbed soil specimens, they must be cushioned against the adverse effects of transportation induced vibration and shock. Protection may also need to be provided against adverse temperature changes, such as overheating or freezing of the soil.

The soil samples should be marked with the file or project number, date of sampling, name of engineer or geologist who performed the sampling, and boring number and depth (e.g., B-1 @ 20-21 ft). Other items that may need to be identified are as follows (ASTM D 4220-00, 2004):

1. Sample orientation (if necessary)
2. Special shipping and laboratory handling instructions
3. Penetration test data (if applicable)
4. Subdivided samples must be identified while maintaining association to the original sample
5. If required, sample traceability record

Preserving Soil Samples.   For ordinary metal sampling tubes, water and oxygen from the soil sample can cause the formation of rust within the sample tube leading to sample disturbance. This formation of rust could adversely affect laboratory test results. Thus if the soil samples are to be stored for any length of time in their sampling tubes, then the tubes should be made of brass, stainless steel, or galvanized metal in order to inhibit corrosion.

Soil samples can also be extruded from the sampling tubes and then sealed in moisture resistant containers. One option is to extrude the soil from the sampling tube and then completely seal the soil specimen in wax. Aluminum foil, cheesecloth, or plastic wrap is first placed around the soil in order to reduce the possibility of the penetration of molten wax into the fissures. Then molten wax is brushed onto the soil specimen in order to completely seal the soil. It is important that the wax is only heated to a temperature that is slightly above its melting point. Using molten wax at too high a temperature could dry-out the soil specimen or it may be so fluid that it penetrates into the pores and cracks in the soil sample.

Some laboratories may come equipped with a humid room. This typically consists of a room that has a low temperature and humidity at or near 100 percent. To reduce the possibility of drying of the soil specimens, the sealed sampling tubes containing soil or soil samples contained in moisture resistant containers can be placed in the humid room. The humid room could be used to store soil samples that have a high water content or those soil samples obtained below the groundwater table. However, certain types of soil samples should never be stored in humid rooms. For example, in the desert southwestern United States, the soil may be in a dry and desiccated state. Storing samples in a humid room could cause the desiccated soil to absorb water and hence reduce the swelling potential of the soil. As a general rule of thumb, it is best to store soil samples in an environment that as closely as possible matches the field conditions.

For further details on preserving and transporting soil specimens see ASTM D 4220-00 (2004), "Standard Practices for Preserving and Transporting Soil Samples." For preserving and transporting rock core samples, see ASTM D 5079-02 (2004), "Standard Practices for Preserving and Transporting Rock Core Samples."

Disposal of Soil Samples.   Although this textbook only deals with the laboratory testing discussion of *clean* soil which does not contain any known or suspected hazardous materials, there may still be regulations concerning the transportation, storage, and disposal of soil. For example, the U.S. Department of Agriculture regulates the transportation, storage, and disposal of soil in the United States. In addition, there may be state or local regulations for quarantine areas. The main purpose of these regulations is to prevent the spread of *pests*, such as fire ants, insect larvae, fungus, spores, and other undesirable plant and animal life.

Especially for near surface sampling, the soil samples taken during the subsurface exploration could contain such pests. The best procedure to prevent the spread of pests is to transport all soil specimens in sealed containers and keep the specimens within the containers during storage. When the laboratory tests are complete, the soil can be returned and sealed within its original container. Then the sealed containers could be discarded in a dumpster for eventual proper disposal in a municipal landfill. Soil samples should not be placed outside (such as for air drying), but rather a drying oven can be used to reduce the water content of the soil. Under no circumstances should soil ever be dumped outside or taken home and used as garden material.

2.4.3   Standard Penetration Test

There are several different types of field tests that can be performed at the time of drilling. For example, the SPT consists of driving a thick-walled sampler in order to determine the driving resistance of the soil (see Fig. 2.10).

Test Procedure.   The SPT can be used for all types of soil, but in general, the SPT is most often used for sand deposits. The SPT can be especially of value for clean sand deposits where the sand falls or flows out from the sampler when retrieved from the ground. Without a soil sample, other types of tests, such as the SPT, must be used to assess the engineering properties of the sand. Often when drilling a borehole, if subsurface conditions indicate a sand strata and sampling tubes come up empty, the sampling gear can be quickly changed to perform SPT.

The system to drive the SPT sampler into the soil, known as the drive-weight assembly, basically consists of the hammer, hammer fall guide, anvil, and a hammer release system.

Hammer.   The metal hammer is successively lifted and dropped in order to provide the energy that drives the SPT sampler into the ground.

Hammer Fall Guide.   This part of the drive-weight assembly is used to guide the fall of the hammer as it strikes the anvil.

Anvil.   This is the portion of the drive-weight assembly which the hammer strikes and through which the hammer energy is passed into the drill rods.

Hammer Release System.   This is the part of the drive-weight assembly by which the operator lifts and drops the hammer. Two types of systems are commonly utilized, as follows:

1. The first hammer release system is the trip, automatic, or semiautomatic system, where the hammer is lifted and allowed to drop unimpeded.

2. The second hammer release system is commonly referred to as the cathead release system. It is a method of raising and dropping the hammer that uses a rope slung through a center crown sleeve or pulley on the drill rig mast and turns on a cathead to lift the hammer. The cathead is defined as a spinning sleeve or rotating drum around which the drill rig operator wraps the rope used to lift and drop the hammer by successively tightening and loosening the rope turns around the drum. The drill rig operator should use two rope turns on the cathead when lifting the hammer because more than two rope turns on the cathead impedes the fall of the hammer.

There are many different types of hammers utilized for the SPT. A commonly used hammer type is the safety hammer, which is defined as a drive-weight assembly consisting of a center guide rod, internal anvil, and hammer that encloses the hammer-anvil contact. Typical internal designs of safety hammers are shown in ASTM D 6066-96 (2004).

Per ASTM D 1586-99 (2004), "Standard Test Method for Penetration Test and Split-Barrel Sampling of Soils," sampler dimensions and test parameters for the SPT must be as follows:

- Sampler inside tube diameter = 1.5 in. (3.81 cm), see Fig. 2.19
- Sampler outside tube diameter = 2.0 in. (5.08 cm), see Fig. 2.19
- Sampler is driven by a metal drop hammer that has a weight of 140 lb. (63.5 kg) and a free-fall distance of 30 in. (0.76 m)
- Sampler is driven a total of 18 in. (45 cm), with the number of blows recorded for each 6 in. (15 cm) interval

The *measured N value* (blows per ft) is defined as the penetration resistance, which equals the sum of the number of blows needed to drive the SPT sampler over the depth interval of 6 to 18 in. (15 to 45 cm). The reason the number of blows required to drive the SPT sampler for the first 6 in. (15 cm) is not included in the N value is because the drilling process often disturbs the soil at the bottom of the borehole and the readings from 6 to 18 in. (15 to 45 cm) are believed to be more representative of the in situ penetration resistance of the sand.

It is desirable to apply hammer blows at a rate of about 20 to 40 blows per min. After performing the SPT, the minimum recommended borehole cleanout is 1 ft (0.3 m). Thus, since the SPT itself requires 1.5 ft (0.46 m) of penetration, the minimum vertical spacing between tests is 2.5 ft (0.76 m). Often a larger vertical spacing of 3 to 5 ft (0.9 to 1.5 m) is used between each SPT.

Factors that Can Affect the SPT.   The measured N value can be influenced by the type of soil, such as the amount of fines and gravel size particles in the soil. Saturated sands that contain appreciable fine soil particles, such as silty or clayey sands, could give abnormally high N values if they have a tendency to dilate or abnormally low N values if they have a tendency to contract during the undrained shear conditions associated with driving the SPT sampler. Gravel size particles increase the driving resistance (hence increased N value) by becoming stuck in the SPT sampler tip or barrel.

A factor that could influence the measured N value is groundwater. It is important to maintain a level of water in the borehole at or above the in situ groundwater level. This is to prevent groundwater from rushing into the bottom of the borehole, which could loosen the sand and result in low measured N values.

Besides soil and groundwater conditions described earlier, there are many different testing factors that can influence the accuracy of the SPT readings (see Table 2.5). For example, the hammer efficiency, borehole diameter, and the rod lengths could influence the measured N value. The following equation is used to compensate for these testing factors by multiplying together four factors as follows (Skempton, 1986):

$$N_{60} = C_b \, C_r \, N \left(\frac{E_m}{60} \right) \qquad (2.4)$$

where N_{60} = standard penetration test N value corrected for field testing procedures.
C_b = borehole diameter correction (C_b = 1.0 for boreholes of 65 to 115 mm diameter, 1.05 for 150 mm diameter, and 1.15 for 200 mm diameter hole).
C_r = rod length correction (C_r = 0.75 for up to 4 m of drill rods, 0.85 for 4 to 6 m of drill rods, 0.95 for 6 to 10 m of drill rods, and 1.00 for drill rods in excess of 10 m).
N = measured standard penetration test N value
E_m = hammer efficiency in percent, as described later

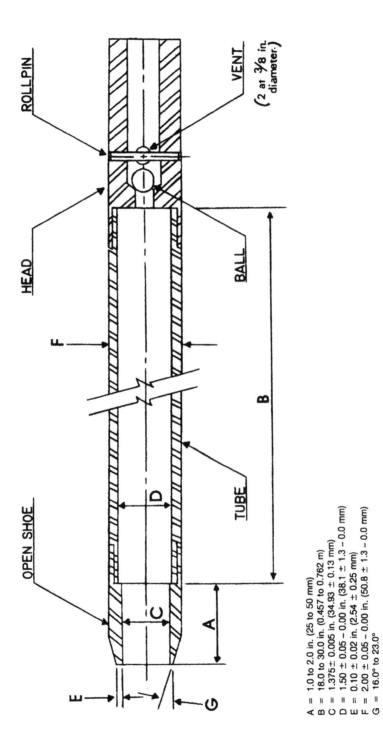

A = 1.0 to 2.0 in. (25 to 50 mm)
B = 18.0 to 30.0 in. (0.457 to 0.762 m)
C = 1.375 ± 0.005 in. (34.93 ± 0.13 mm)
D = 1.50 ± 0.05 – 0.00 in. (38.1 ± 1.3 – 0.0 mm)
E = 0.10 ± 0.02 in. (2.54 ± 0.25 mm)
F = 2.00 ± 0.05 – 0.00 in. (50.8 ± 1.3 – 0.0 mm)
G = 16.0° to 23.0°

FIGURE 2.19 Standard penetration test sampler. *(Reprinted with permission from the American Society for Testing and Materials, 2004.)*

TABLE 2.5   Factors that can Affect the Standard Penetration Test Results

Factors that can affect the standard penetration test results	Comments
Inadequate cleaning of the borehole	SPT is only partially made in original soil. Sludge may be trapped in the sampler and compressed as the sampler is driven, increasing the blow count. This may also prevent sample recovery.
Not seating the sampler spoon on undisturbed material	Incorrect N value is obtained.
Driving of the sampler spoon above the bottom of the casing	The N value is increased in sands and reduced in cohesive soil.
Failure to maintain sufficient hydrostatic head in boring	The water table in the borehole must be at least equal to the piezometric level in the sand; otherwise the sand at the bottom of the borehole may be transformed to a loose state.
Attitude of operators	Blow counts for the same soil using the same rig can vary, depending on who is operating the rig and perhaps the mood of operator and time of drilling.
Overdriven sample	Higher blow counts usually result from overdriven sampler.
Sampler plugged by gravel	Higher blow counts result when gravel plugs the sampler. The resistance of loose sand could be highly overestimated.
Plugged casing	High N values may be recorded for loose sand when sampling below the groundwater table. Hydrostatic pressure causes sand to rise and plug the casing.
Overwashing ahead of casing	Low blow count may result for dense sand since sand is loosened by overwashing.
Drilling method	Drilling technique (e.g., cased holes versus mud-stabilized holes) may result in different N values for the same soil.
Not using the standard hammer drop	Energy delivered per blow is not uniform. European countries have adopted an automatic trip hammer not currently in use in North America.
Free fall of the drive weight is not attained	Using more than 1.5 turns of rope around the drum and/or using wire cable will restrict the fall of the drive weight.
Not using the correct weight	Driller frequently supplies drive hammers with weights varying from the standard by as much as 10 lb.
Weight does not strike the drive cap concentrically	Impact energy is reduced, increasing the N value.
Not using a guide rod	Incorrect N value is obtained.
Not using a good tip on the sampling spoon	If the tip is damaged and reduces the opening or increases the end area, the N value can be increased.
Use of drill rods heavier than standard	With heavier rods, more energy is absorbed by the rods, causing an increase in the blow count.
Not recording blow counts and penetration accurately	Incorrect N values are obtained.
Incorrect drilling	The standard penetration test was originally developed from wash boring techniques. Drilling procedures which seriously disturb the soil will affect the N value, for example, drilling with cable tool equipment.

TABLE 2.5   Factors that can Affect the Standard Penetration Test Results (*Continued*)

Factors that can affect the standard penetration test results	Comments
Using large drill holes	A borehole correction is required for large-diameter boreholes. This is because larger diameters often result in a decrease in the blow count.
Inadequate supervision	Frequently a sampler will be impeded by gravel or cobbles, causing a sudden increase in blow count. This is often not recognized by an inexperienced observer. Accurate recording of drilling sampling and depth is always required.
Improper logging of soils	The sample is not described correctly.
Using too large a pump	Too high a pump capacity will loosen the soil at the base of the hole, causing a decrease in blow count.

Source: NAVFAC DM-7.1 (1982).

The theoretical energy that should be delivered to the top of the anvil is 350 ft-lb of energy (i.e., 140 lb times 30 in. drop). However, the SPT theory has evolved around the concept that about 60 percent of the hammer energy should be delivered to the drill rods, with the rest being dissipated through friction and hammer rebound. Using the cathead release system and a safety hammer will deliver about 60 percent (i.e., $E_m = 60$) of the hammer energy to the drill rods. Note in Eq. 2.4 that if $E_m = 60$, no correction is required to the N value for hammer efficiency.

Studies have shown that the cathead release system and a donut hammer can impart only 45 percent of the theoretical energy to the drill rods (i.e., $E_m = 45$). At the other extreme are automatic systems that lift the hammer and allow it to drop unimpeded and deliver higher energy to the drill rods with values of E_m as high as 95 percent being reported (ASTM D 6066-96, 2004). For other types of release systems and hammers, values of E_m should be based on manufacturer specifications or previously published measurements.

Even with this hammer energy uncertainty, the SPT is still probably the most widely used field test in the United States. This is because it is relatively easy to use, the test is economical as compared to other types of field-testing, and the SPT equipment can be quickly adapted and included as part of almost any type of drilling rig.

Correction of N Value for Field Testing and Overburden Pressure.   For geotechnical earthquake engineering, such as liquefaction analyses, the standard penetration test N_{60} value (Eq. 2.4) is corrected for the overburden soil pressure, also known as the effective overburden pressure or the vertical effective stress (σ'_{vo}). The vertical effective stress will be discussed in Sec. 4.4. When a correction is applied to the N_{60} value to account for the vertical effective stress, these values are referred to as $(N_1)_{60}$ values. The procedure consists of multiplying the N_{60} value by a correction C_N in order to calculate the $(N_1)_{60}$ value. Figure 2.20 presents a chart that is commonly used to obtain the correction factor C_N. Another option is to use the following equation:

$$(N_1)_{60} = C_N N_{60} = \left(\frac{100}{\sigma'_{vo}}\right)^{0.5} N_{60} \tag{2.5}$$

where $(N_1)_{60}$ = standard penetration test N value corrected for both field testing procedures and overburden pressure

C_N = correction factor to account for the overburden pressure. As indicated in Eq. 2.5, C_N is approximately equal to $(100/\sigma'_{vo})^{0.5}$ where σ'_{vo} is the vertical effective stress, in kPa. Suggested maximum values of C_N range from 1.7 to 2.0 (Youd and Idriss, 1997, 2001).

N_{60} = standard penetration test N value corrected for field testing procedures. The N_{60} is calculated by using Eq. 2.4.

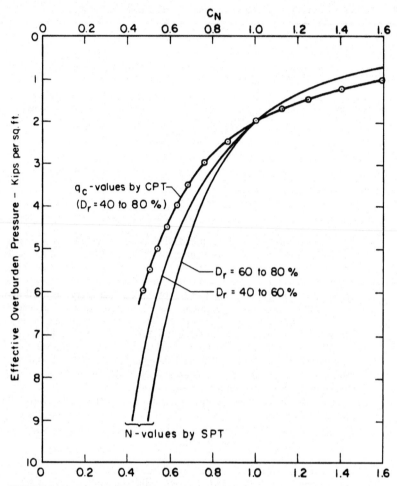

FIGURE 2.20   Correction factor C_N used to adjust the standard penetration test N value and cone penetration test q_c value for the effective overburden pressure. The symbol D_r refers to the relative density of the sand. (*Reproduced from Seed et al., 1983; with permission from the American Society of Civil Engineers.*)

The $(N_1)_{60}$ value (blows per foot) can also be used as a guide in determining the density condition of a clean sand deposit (see Table 2.6). Note that this correlation is very approximate and the boundaries between different density conditions are not as distinct as implied by Table 2.6. If $(N_1)_{60} = 2$ or less, then the sand should be considered to be very loose and could be subjected to significant settlement due to the weight of a structure or due to earthquake shaking. On the other hand, if $(N_1)_{60} = 35$ or more, then the sand is considered to be in a very dense condition and would be able to support high foundation loads and would be resistant to settlement from earthquake shaking.

For further details on determining the $(N_1)_{60}$ value for use in liquefaction studies, see ASTM D 6066-96 (2004), "Standard Practice for Determining the Normalized Penetration Resistance of Sands for Evaluation of Liquefaction Potential."

TABLE 2.6. Correlation between $(N_1)_{60}$ and Density of Sand

$(N_1)_{60}$ (blows per foot)	Sand density	Relative density D_r, percent
0–2	Very loose condition	0–15
2–5	Loose condition	15–35
5–20	Medium condition	35–65
20–35	Dense condition	65–85
Over 35	Very dense condition	85–100

Source: Tokimatsu and Seed (1987).

2.4.4 Other Field Tests

Besides the SPT, many other types of field tests can be performed during the subsurface exploration. Other common types of field tests are as follows:

Mechanical Cone Penetration Test.   The idea for the mechanical cone penetration test is similar to the SPT except that instead of driving a thick-walled sampler into the soil, a steel cone is pushed into the soil. The most common type of mechanical penetrometer is the Dutch mantle cone, which is shown in Fig. 2.21. This test is often referred to as the Dutch cone test or the cone penetration test

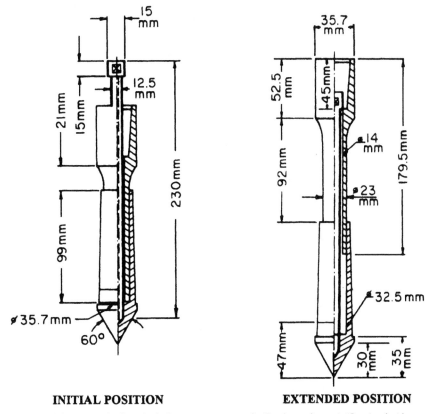

INITIAL POSITION   **EXTENDED POSITION**

FIGURE 2.21   Example of mechanical cone penetrometer tip (Dutch mantle cone). (*Reprinted with permission from the American Society for Testing and Materials, 2004.*)

and is abbreviated CPT. The cone is first pushed into the soil to the desired depth (initial position) and then force is applied to the inner rod, which moves the cone downward into the extended position. The cone is pushed into the soil at a rate of about 2 to 4 ft/min (10 to 20 mm/sec). The required force to move the cone into the extended position (Fig. 2.21) divided by the horizontally projected area (10 cm^2) of the cone is defined as the cone resistance q_c, also known as the cone bearing or the end bearing resistance. By continually repeating the two-step process shown in Fig. 2.21, the cone resistance q_c is obtained at increments that ordinarily do not exceed 8 in. (20 cm). Special features of the cone penetration test are as follows:

1. *Cone resistance versus depth.*   A considerable amount of work has been performed in correlating cone resistance q_c with subsurface conditions. Figure 2.22 presents four examples, where the cone resistance q_c has been plotted versus depth below ground surface. The shape of the cone resistance q_c plots versus depth can be used to identify sands, clays, cavities, or rock.

2. *Friction ratio.*   Figure 2.23 illustrates the two-step process that can be used to obtain the soil friction along a side sleeve f_s. In the first step, the cone resistance is obtained (q_c) and then in the second step, the cone plus sleeve friction is determined ($q_c + f_s$). Subtraction gives the sleeve friction. The friction ratio (FR) can then be calculated, defined as FR = sleeve friction divided by cone resistance = $100 f_s/q_c$. By knowing the friction ratio (FR) and cone resistance q_c, the type of soil can be estimated by using Fig. 2.24.

3. *Liquefaction studies.*   Much like the SPT, the cone penetration test can be corrected for the vertical effective stress. One option is to multiply the cone resistance q_c by the C_N value shown in Fig. 2.20 in order to obtain the cone resistance q_{c1} corrected for vertical effective stress (i.e., $q_{c1} = C_N q_c$). The corrected cone resistance q_{c1} is often used in liquefaction studies (Day, 2002).

A major advantage of the cone penetration test is that a nearly continuous subsurface record of the cone resistance q_c can be obtained. This is in contrast to the SPT, which obtains data at much larger intervals in the soil deposit. Disadvantages of the cone penetration test are that soil samples cannot be recovered and special equipment is required to produce a steady and slow penetration of the cone. Unlike the SPT, the ability to obtain a steady and slow penetration of the cone is not included as part of conventional drilling rigs. Because of these factors, in the United States, the CPT is used less frequently than the SPT.

For further details on the mechanical cone penetration test, see ASTM D 3441-98 (2004), "Standard Test Method for Mechanical Cone Penetration Tests of Soil."

Other Cone Penetrometers.   Besides the mechanical cone, there are other types of cone penetrometers, such as:

Electric cone.   A cone penetrometer that uses electric-force transducers built into the apparatus for measuring cone resistance and friction resistance.

Piezocone.   A cone penetrometer with the additional capability of measuring pore water pressure generated during the penetration of the cone.

Special devices.   The cone can even be equipped with a video camera to enable the type of soil to be viewed during the test (Raschke and Hryciw, 1997).

For more information on these cone penetrometers, see ASTM D 5778-00 (2004), "Standard Test Method for Performing Electronic Friction Cone and Piezocone Penetration Testing of Soils."

Vane Shear Test (VST).   The SPT and CPT are used to correlate the resistance of driving a sampler (N value) or pushing a cone q_c with the engineering properties (such as density condition) of the soil. In contrast, the vane test is a different in situ field test because it directly measures a specific soil property, the undrained shear strength s_u of clay. The undrained shear strength of clay will be discussed in Sec. 3.5.

The vane test consists of inserting a four-bladed vane, such as shown in Fig. 2.25, into the borehole and then pushing the vane into the clay deposit located at the bottom of the borehole. Different

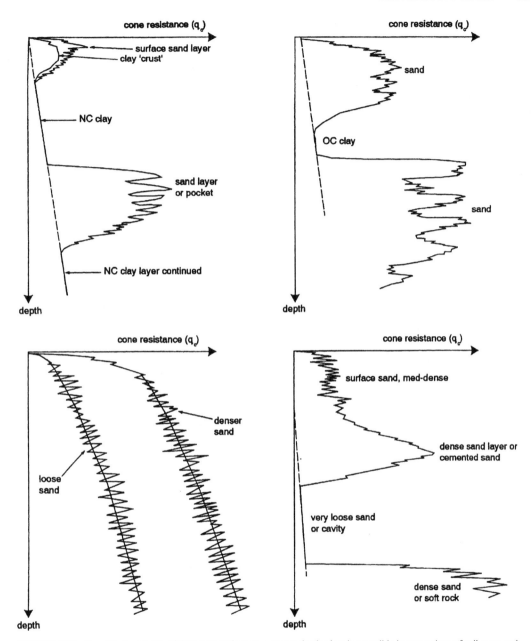

FIGURE 2.22 Simplified examples of CPT cone resistance q_c versus depth, showing possible interpretations of soil types and conditions. (*From Schmertmann, 1977.*)

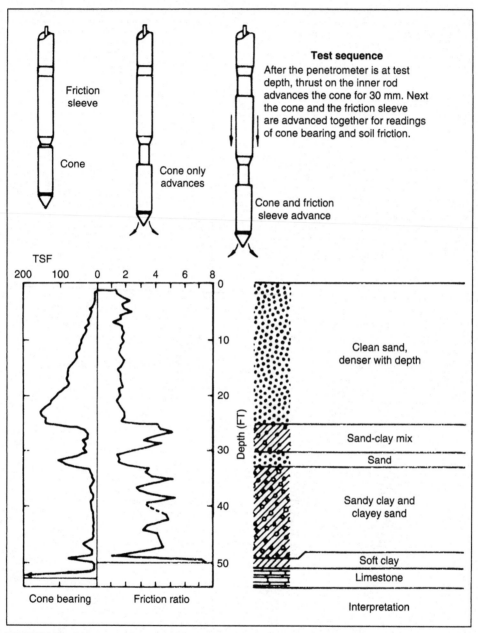

FIGURE 2.23 Test sequence for obtaining the sleeve friction from the Dutch cone penetrometer and an example of the test data plotted versus depth. (*From NAVFAC DM –7.1, 1982.*)

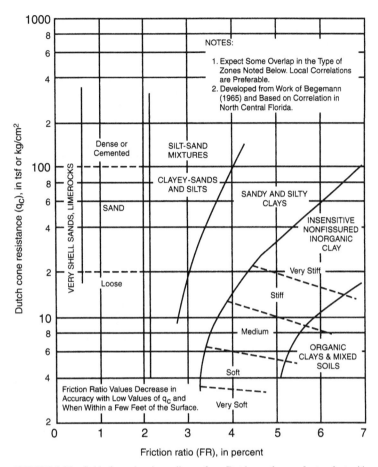

FIGURE 2.24 Guide for estimating soil type from Dutch mantle cone [enter chart with cone resistance q_c and friction ratio (FR = sleeve friction divided by cone resistance = 100 f_s/q_c). (*From Schmertmann, 1977.*)

types of vanes are available, such as a rectangular vane, tapered vane at both ends, which is shown in Fig. 2.25, and a vane that only has a taper at the bottom. Once inserted into the clay, the maximum torque T_{max} required to rotate the vane and shear the clay is measured. The undrained shear strength s_u of the clay can then be calculated by using the following equation, which assumes uniform end shear for a rectangular vane. (*Note:* The following equation is valid only for a rectangular vane with shear failure along the entire perimeter and at both ends of the vane).

$$s_u = \frac{T_{max}}{\pi(0.5D^2H + 0.167D^3)}$$

(2.6)

where s_u = undrained shear strength of the clay (psf or kPa)

T_{max} = maximum torque required to rotate the rod which shears the clay, corrected for apparatus and rod friction (lbf-ft or kN-m)

H = height of the vane (ft or m)

D = diameter of the vane (ft or m)

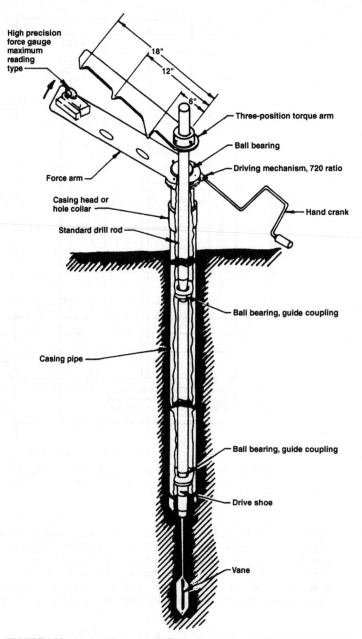

High precision
force gauge
maximum
reading
type

18"

12"

6"

Three-position torque arm

Ball bearing

Driving mechanism, 720 ratio

Force arm

Casing head or
hole collar

Hand crank

Standard drill rod

Ball bearing, guide coupling

Casing pipe

Ball bearing, guide coupling

Drive shoe

Vane

FIGURE 2.25   Diagram illustrating the field vane test. (*From NAVFAC DM-7.1, 1982.*)

In addition to obtaining the undrained shear strength s_u, the undrained shear strength for remolded clay s_{ur} can also be measured. The process consists of first remolding the clay by rotating the vane about 5 to 10 times. Then the torque is measured and Eq. 2.6 is used to obtain the undrained shear strength for remolded clay s_{ur}. The sensitivity S_t of the clay can be calculated as the undrained shear strength divided by the undrained shear strength of remolded clay, or $S_t = s_u/s_{ur}$. Sensitivity will be further discussed in Sec. 4.6.

The undrained shear strength s_u is often needed for many different types of engineering analyses, such as foundation bearing capacity and slope stability. However, it has been stated that for the vane shear test, the values of the undrained shear strength from field vane tests are likely to be higher than can be mobilized in practice (Bjerrum 1972, 1973). This has been attributed to a combination of anisotropy of the soil and the fast rate of shearing involved with the vane shear test. Because of these factors, Bjerrum (1972) has proposed for field vane shear tests performed on saturated normally consolidated clays, that the undrained shear strength s_u be reduced based on the plasticity index of the clay (see Sec. 3.2.6 for the definition of plasticity index). Figure 2.26 shows Bjerrum's (1972) recommendation, where the in situ undrained shear strength s_u is equal to the shear strength determined from the field vane test times the correction factor determined from Fig. 2.26.

Note that the sensitivity S_t is based on the ratio of raw measured peak and remolded undrained shear strengths and is not corrected. For the equations needed to calculate the undrained shear strength s_u for the tapered vane, as well as further details on vane shear test, see ASTM D 2573-01 (2004), "Standard Test Method for Field Vane Shear Test in Cohesive Soil."

Miniature Vane Test. There are miniature vane devices, which can be used in the field or laboratory. The miniature vane device is not inserted down a borehole, but rather the test is performed on clay specimens brought to the surface from undisturbed samplers. An example of a miniature vane is the Torvane device, which is a hand-held vane that is manually inserted into the clay surface and then

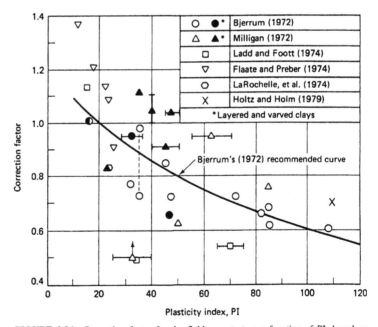

FIGURE 2.26 Correction factor for the field vane test as a function of PI, based on embankment failures. Note: in situ $s_u = s_u$ from the field vane test times the correction factor. (*After Ladd, 1973; Ladd et. al., 1977; reproduced from Holtz and Kovacs 1981.*)

rotated to induce a shear failure of the clay. On top of the Torvane there is a calibrated scale that directly indicates the undrained shear strength s_u of the clay. The Torvane device has a quick failure rate, which could overestimate the undrained shear strength s_u. Because the miniature vane only tests a very small portion of the clay, the strength could be overestimated for fissured clay, varved clay, or clay containing slickensides. Also the miniature vane provides unreliable readings for clays having an undrained shear strength s_u in excess of 1.0 tsf (100 kPa) because the actual failure surface deviates from the assumed cylindrical failure surface, resulting in an overestimation of the undrained shear strength (ASTM D 4648-00, 2004). Because of these factors, the results of the miniature vane test should be used with caution and the results should not be solely relied upon for foundation design.

For further information on the miniature vane test, see ASTM D 4648-00 (2004), "Standard Test Method for Laboratory Miniature Vane Shear Test for Saturated Fine-Grained Clayey Soil."

Pressuremeter Test (PMT).   The pressuremeter test (PMT) is an in situ stress-strain test performed on the wall of a borehole using a cylindrical probe that is expanded radially. The PMT is usually performed by inserting the equipment into a predrilled borehole. In order to obtain accurate results, it is essential that disturbance to the borehole wall is minimized. To offset this limitation, a self-boring pressuremeter has been developed, where a mechanical or jetting tool located inside the hollow core of the probe drills the hole.

Once the pressuremeter is in place, the probe is expanded while measuring the changes in volume and pressure within the probe. The test is terminated when the yielding of the soil becomes disproportionately large. The test provides a stress-strain curve for horizontal loading of the soil and this data could be of use in the design of piles subjected to lateral loads. For further details, see ASTM D 4719-00 (2004), "Standard Test Method for Prebored Pressuremeter Testing in Soils."

Other Field Tests Performed in Boreholes.   There are many other types of field tests that can be performed in boreholes. Examples include the Screw Plate Compressometer (SPC) and the Iowa Borehole Shear Test (BST) (Holtz and Kovacs 1981; Mitchell 1978). These types of tests are used much less frequently than the SPT, CPT, and VST.

The SPC is a field test where a plate is screwed down to the desired depth, and then as pressure is applied, the settlement of the plate is measured. The (BST) is a field test where the device is lowered into an uncased borehole and then expanded against the sidewalls. The force required to pull the device towards ground surface is measured and much like a direct shear device, the shear strength properties of the in situ soil can then be determined.

2.4.5   Boring Layout

The required number and spacing of borings for a particular project must be based on judgment and experience. Obviously the more borings that are performed, the more knowledge obtained about the subsurface conditions. This can result in an economical foundation design and less risk of meeting unforeseen or difficult conditions during construction.

In general, boring layouts should not be random. Instead, if an approximate idea of the location of the proposed structure is known, then the borings should be concentrated in that area. For example, borings could be drilled at the four corners of a proposed building, with an additional (and deepest) boring located at the center of the proposed building. If the building location is unknown, then the borings should be located in lines, such as across the valley floor, in order to develop soil and geologic cross sections. Table 2.7 provides guidelines on the typical boring layout versus type of project.

If geologic features outside the building footprint could affect the structure, then they should also be investigated with borings. For example, if there is an adjacent landslide or fault zone that could impact the site, then they will also need to be investigated with subsurface exploration.

Some of the factors that influence the decisions on the number and spacing of borings include the following:

Relative Costs of the Investigation.   The cost of additional borings must be weighed against the value of additional subsurface information.

TABLE 2.7 Guidelines for Boring Layout

Areas of investigation	Boring layout
New site of wide extent	Space preliminary borings 60 to 150 m (200 to 500 ft) apart so that area between any four borings includes approximately 10 percent of total area. In detailed exploration, add borings to establish geological sections at the most useful orientations.
Development of site on soft compressible soil	Space borings 30 to 60 m (100 to 200 ft) at possible building locations. Add intermediate borings when building site is determined.
Large structure with separate closely spaced footings	Space borings approximately 15 m (50 ft) in both directions, including borings at possible exterior foundation walls, at machinery or elevator pits, and to establish geologic sections at the most useful orientations.
Low-load warehouse building of large area	Minimum of four borings at corners plus intermediate borings at interior foundations sufficient to define subsoil profile.
Isolated rigid foundation	For foundation 230 to 930 m^2 (2500 to 10,000 ft^2) in area, minimum of three borings around perimeter. Add interior borings, depending on initial results.
Isolated rigid foundation	For foundation less than 230 m^2 (2500 ft^2) in area, minimum of two borings at opposite corners. Add more for erratic conditions.
Major waterfront structures, such as dry docks	If definite site is established, space borings generally not farther than 15 m (50 ft), adding intermediate borings at critical locations, such as deep pump well, gate seat, tunnel, or culverts.
Long bulkhead or wharf wall	Preliminary borings on line of wall at 60-m (200-ft) spacing. Add intermediate borings to decrease spacing to 15 m (50 ft). Place certain intermediate borings inboard and outboard of wall line to determine materials in scour zone at toe and in active wedge behind wall.
Cut stability, deep cuts, and high embankments	Provide three to five borings on line in the critical direction to provide geological section for analysis. Number of geologic sections depends on extent of stability problem. For an active slide, place at least one boring upslope of sliding area.
Dams and water-retention structures	Space preliminary borings approximately 60 m (200 ft) over foundation area. Decrease spacing on centerline to 30 m (100 ft) by intermediate borings. Include borings at location of cutoff, critical spots in abutment, spillway, and outlet works.

Source: From NAVFAC DM-7.1, 1982.

Type of Project. A more detailed and extensive subsurface investigation is required for an essential facility as compared to a single-family dwelling.

Topography (Flatland versus Hillside). A hillside project usually requires more subsurface investigation than a flatland project because of the slope stability requirements.

Nature of Soil Deposits (Uniform versus Erratic). Fewer boring may be needed when the soil deposits are uniform as compared to erratic deposits.

Geologic Hazards. The more known or potential geologic hazards at the site, the greater the need for subsurface exploration.

Access. In many cases, the site may be inaccessible and access roads will have to be constructed. In some cases, access may cause considerable disruption to the environment, such as shown in Fig. 2.27. In other cases, such as shown in Fig. 2.28, the access road was relatively easy to construct because the site is an open pit mine. Creating access roads throughout the site can be expensive and disruptive and may influence decisions on the number and spacing of borings.

FIGURE 2.27   Construction of an access road for drilling equipment.

FIGURE 2.28   Access road constructed for subsurface investigation at an open-pit mine (arrow points to bucket auger drill rig).

Governmental or Local Building Department Requirements. For some projects, there may be specifications on the required number and spacing of borings. For example, *Standard Specifications for Highway Bridges* (AASHTO, 1996) states:

> A minimum of one soil boring shall be made for each substructure unit [note: a substructure unit is defined as every pier, abutment, retaining wall, foundation, or similar item]. For substructure units over 100 ft (30 m) in width, a minimum of two borings shall be required.

Oftentimes a preliminary subsurface plan is developed to perform a limited number of exploratory borings. The purpose is just to obtain a rough idea of the soil, rock, and groundwater conditions at the site. Then once the preliminary subsurface data is analyzed, additional borings as part of a detailed exploration are performed. The detailed subsurface exploration can be used to better define the soil profile, explore geologic hazards, and obtain further data on the critical subsurface conditions that will likely have the most impact on the design and construction of the project.

2.4.6 Depth of Subsurface Exploration

Similar to the boring layout, the depth of subsurface exploration for a particular project must be based on judgment and experience. Borings should always be extended through unsuitable foundation bearing material, such as uncompacted fill, peat, soft clays and organic soil, and loose sands, and into dense soil or hard rock of adequate bearing capacity. In a general sense, the depth of subsurface exploration will depend on the size and loading of the proposed foundation, the sensitivity of the proposed structure to settlements, and the stiffness and coefficient of compressibility of the strata that will underlie the foundation.

In terms of additional general rules, Hvorslev (1949) states:

> The borings should be extended to strata of adequate bearing capacity and should penetrate all deposits which are unsuitable for foundation purposes, such as unconsolidated fill, peat, organic silt, and very soft and compressible clay. The soft strata should be penetrated even when they are covered with a surface layer of higher bearing capacity.
>
> When structures are to be founded on clay and other materials with adequate strength to support the structure but subject to considerable consolidation by an increase in the load, the borings should penetrate the compressible strata or be extended to such a depth that the stress increase for still deeper strata is reduced to values so small that the corresponding consolidation of these strata will not materially influence the settlement of the proposed structure.
>
> Except in cases of very heavy loads or when seepage or other considerations are governing, the borings may be stopped when rock is encountered or after a short penetration into strata of exceptional bearing capacity and stiffness, provided it is known from explorations in the vicinity or the general stratigraphy of the area that these strata have adequate thickness or are underlain by still stronger formations. When these conditions are not fulfilled, some of the borings must be extended until it has been established that the stiff strata have adequate thickness irrespective of the character of the underlying material.
>
> When the structure is to be founded on rock, it must be verified that bedrock and not boulders have been encountered, and it is advisable to extend one or more borings from 10 to 20 ft (3 to 6 m) into sound rock in order to determine the extent and character of the weathered zone of the rock.

For localized structures, such as commercial or industrial buildings, it is common practice to carry explorations to a depth beneath the loaded area of 1.5 to 2.0 times the least dimension of the building (Lowe and Zaccheo, 1975). Table 2.8 presents additional guidelines for different types of geotechnical and foundation projects.

Another commonly used rule of thumb is that for isolated square footings, the depth of subsurface exploration should be two times the width of the footing. For isolated strip footings, the depth of subsurface exploration should be four times the width of the footing. These recommendations are based on the knowledge that the pressure of surface loads dissipates with depth. Thus, at a certain depth, the effect of the surface load is very low. For example, a common guideline is to perform subsurface

TABLE 2.8 Guidelines for Boring Depths

Areas of investigation	Boring depth
Large structure with separate closely space footings	Extend to depth where increase in vertical stress for combined foundations is less than 10 percent of effective overburden stress. Generally all boring should extend to no less than 9 m (30 ft) below lowest part of foundation unless rock is encountered at shallower depth.
Isolated rigid foundations	Extend to depth where vertical stress decreases to 10 percent of bearing pressure. Generally all borings should extend no less than 9 m (30 ft) below lowest part of foundation unless rock is encountered at shallower depth.
Long bulkhead or wharf wall	Extend to depth below dredge line between 0.75 and 1.5 times unbalanced height of wall. Where stratification indicates possible deep stability problem, selected borings should reach top of hard stratum.
Slope stability	Extend to an elevation below active or potential failure surface and into hard stratum, or to a depth for which failure is unlikely because of geometry of cross section.
Deep cuts	Extend to depth between 0.75 and 1 times base width of narrow cuts. Where cut is above groundwater in stable materials, depth of 1.2 to 2.4 m (4 to 8 ft) below base may suffice. Where base is below groundwater, determine extent of previous strata below base.
High embankments	Extend to depth between 0.5 to 1.25 times horizontal length of side slope in relatively homogeneous foundation. Where soft strata are encountered, borings should reach hard materials.
Dams and water retention structures	Extend to depth of 0.5 base width of earth dams or 1 to 1.5 times height of small concrete dams in relatively homogeneous foundations. Borings may terminate after penetration of 3 to 6 m (10 to 20 ft) in hard and impervious stratum if continuity of this stratum is known from reconnaissance.

Source: From NAVFAC DM-7.1, 1982.

exploration to a depth where the increase in vertical pressure from the foundation is less than 10 percent of the applied pressure from the foundation. There could be problems with this approach because as will be discussed in Chap. 7, there could be settlement of the structure that is independent of its weight or depth of influence. Settlement due to secondary influences, such as collapsible soil, is often unrelated to the weight of the structure. Especially when geologic conditions are not well established, it is always desirable to extend at least one boring into bedrock to guard against the possibility of a deeply buried soil strata having poor support characteristics.

For some projects, there may be specifications on the required depth of borings. For example, the *Standard Specifications for Highway Bridges* (AASHTO, 1996) states:

> When substructure units will be supported on deep foundations, the depth of subsurface exploration shall extend a minimum of 20 ft (6 m) below the anticipated pile or shaft tip elevation. Where pile or shaft groups will be used, the subsurface exploration shall extend at least two times the maximum pile group dimension below the anticipated tip elevation, unless the foundation will be end bearing on or in rock. For piles bearing on rock, a minimum of 10 ft (3 m) of rock core shall be obtained at each exploration location to insure the exploration has not been terminated on a boulder.

At the completion of each boring, it should immediately be backfilled with on-site soil and compacted by using the drill rig equipment. In certain cases, the holes may need to be filled with a cement slurry or grout. For example, if the borehole is to be converted to an inclinometer (slope monitoring device), then it should be filled with weak cement slurry. Likewise, if the hole is to be converted to a piezometer (pore water pressure monitoring device), then special backfill materials, such as a bentonite seal, will be required. It may also be necessary to seal the hole with grout or bentonite if there is the possibility of water movement from one stratum to another. For example, holes may need to be filled with cement or grout if they are excavated at the proposed locations of dams, levees, or reservoirs.

2.5   TEST PITS AND TRENCHES

In addition to borings, other methods for performing subsurface exploration include test pits, and trenches. Test pits are often square in plan view with a typical dimension of 4 ft by 4 ft (1.2 m by 1.2 m). Trenches are long and narrow excavations usually made by a backhoe or bulldozer. Table 2.9 presents the uses, capabilities, and limitations of test pits and trenches.

Similar to the down-hole logging of large diameter bucket auger borings, test pits and trenches provide for a visual observation of subsurface conditions. They can also be used to obtain undisturbed block samples of soil. The process consists of carving a block of soil from the side or bottom of the test pit or trench. Soil samples can also be obtained from the test pits or trenches by manually driving Shelby tubes, drive cylinders (ASTM D 2937-00, 2004), or other types of sampling tubes into the ground.

Backhoe pits and trenches are an economical means of performing subsurface exploration. The backhoe can quickly excavate the trench that can then be used to observe and test the in situ soil (see Fig. 2.29). In many subsurface explorations, backhoe trenches are used to evaluate near surface and

TABLE 2.9   Use, Capabilities, and Limitations of Test Pits and Trenches

Exploration method	General use	Capabilities	Limitations
Hand-excavated test pits	Bulk sampling, in situ testing, visual inspection.	Provides data in inaccessible areas, less mechanical disturbance of surrounding ground.	Expensive, time-consuming, limited to depths above groundwater level.
Backhoe-excavated test pits and trenches	Bulk sampling, in situ testing, visual inspection, excavation rates, depth of bedrock and groundwater.	Fast, economical, generally less than 4.6 m (15 ft) deep, can be up to 9 m (30 ft) deep.	Equipment access, generally limited to depths above groundwater level, limited undisturbed sampling.
Dozer cuts	Bedrock characteristics, depth of bedrock and groundwater level, rippability, increase depth capability of backhoe, level area for other exploration equipment.	Relatively low cost, exposures for geologic mapping.	Exploration limited to depth above the groundwater table.
Trenches for fault investigations	Evaluation of presence and activity of faulting and sometimes landslide features.	Definitive location of faulting, subsurface observation up to 9 m (30 ft) deep.	Costly, time-consuming, requires shoring, only useful where dateable materials are present, depth limited to zone above the groundwater level.

Source: From NAVFAC DM-7.1, 1982.

FIGURE 2.29 Backhoe in the process of excavating a test pit excavation.

geologic conditions (i.e., up to a 15 ft deep), with borings being used to investigate deeper subsurface conditions. They are also very useful for the investigation of sites where there is a thin veneer of soil overlying hard bedrock.

Backhoe trenches are also especially useful when performing fault studies. For example, Figs. 2.30 and 2.31 show two views of the excavation of a trench that is being used to investigate the possibility of an on-site active fault. Figure 2.31 is a close-up view of the conditions in the trench and shows the

FIGURE 2.30 Backhoe trench for a fault study.

FIGURE 2.31 Close-up view of trench excavation.

fractured and disrupted nature of the rock. Note in Fig. 2.31 that metal shoring has been installed to prevent the trench from caving-in. Usually the engineering geologist performs the fault investigations in order to determine if there are active faults that cross the site. In addition, the width of the shear zone of the fault can often be determined from the trench excavation studies. If there is uncertainty as to whether or not a fault is active, then dateable material must be present in the trench excavation in order to determine the date of the most recent fault movement. Examples of dateable materials are as follows (Krinitzsky et al., 1993):

- Displacements of organic matter or other dateable horizons across faults
- Sudden burials of marsh soils
- Killed trees
- Disruption of archaeological sites
- Liquefaction intrusions cutting older liquefaction

2.6 PREPARATION OF LOGS

A log is defined as a written record prepared during the subsurface excavation of borings, test pits, or trenches that documents the observed conditions. Although logs are often prepared by technicians or even the driller, the most appropriate individuals to log the subsurface conditions are geotechnical engineers or engineering geologists who have considerable experience and judgment acquired by many years of field practice. It is especially important that the subsurface conditions likely to have the most impact on the proposed project be adequately described. Table 2.10 lists other items that should be included on the excavation log. Another source of information is ASTM D 5434-03 (2004), "Standard Guide for Field Logging of Subsurface Explorations of Soil and Rock," which describes the type of data that should be recorded during field subsurface explorations in soil and rock.

TABLE 2.10 Types of Information to be Recorded on Exploratory Logs

Item	Description
Excavation number	Each boring, test pit, or trench excavated at the site should be assigned an excavation number
Project information	Project information should include the project name, file number, client, and site address. The individual preparing the log should also be noted
Type of equipment	Include on the log the type of excavation, such as hand dug pit, backhoe trench, and the like, and the total depth and size of the excavation. For borings, indicate type of drilling equipment, use of drilling fluid, and kelly bar weights. Also indicate if casing was used
Site specific information	The exploratory log should list the surface elevation, date(s) of excavation, and ground surface conditions
Type of field tests	For borings, list all field tests, such as SPT, CPT, or vane test. Also indicate if the boring was converted to a monitoring device, such as a piezometer
Type of sampler	Indicate type of sampler and depth of each soil or rock sample recovered from the excavation. For driven samplers, indicate type and weight of hammer and number of blows per foot to drive the sampler. Indicate sample recovery and RQD for rock strata
Soil and rock descriptions	Classify the soil and rock exposed in the excavation (see Sec. 4.2). Also indicate moisture and density condition of the soil and rock
Excavation problems	List excavation problems, such as instability, sloughing, groundwater induced caving, squeezing of the hole, hard drilling, or boring termination due to refusal
Groundwater	Indicate depth to groundwater or seepage zones. At the end of the subsurface exploration, indicate the depth of freestanding water in the excavation
Geologic features and hazards	Identify geologic features and hazards. Geologic features include type of deposit (see Table 2.11), formation name, and fracture condition of rock. Geologic hazards include landslides, active fault shear zones, liquefaction prone sand, bedding, shear surfaces, slickensides, and underground voids or caverns
Unusual conditions	Any unusual subsurface condition should be noted. Examples include artesian groundwater, boulders or other obstructions, or loss of drilling fluid, which could indicate an underground void or cavity

Note: Additional information may be required for subsurface explorations for mining or agricultural purposes, for the investigation of hazardous waste, or other special types of subsurface exploration.

Figure 2.32 presents a boring log. The boring log lists the observed soil and rock layers versus depth. Basically the boring revealed the presence of 11 ft (3.4 m) of soil overlying rock that has been classified as sandstone. In the upper 11 ft (3.4 m) of the boring, four different soil layers were observed. The soil classification was based on the Uniform Soil Classification System, which will be discussed in Sec. 4.2. The most accurate method to classify soils is to use laboratory tests (Sec. 3.2), although visual classification can also be performed, e.g., see ASTM D 2488-00 (2004), "Standard Practice for Description and Identification of Soils (Visual-Manual Procedure)."

The boring log shown in Fig. 2.32 also lists the location of soil and rock specimens obtained as well as the types of samplers, i.e., split spoon sampler, Shelby tube sampler (undisturbed sample), and rock core (NX type, 2-1/8 in. diameter). Although not shown in Fig. 2.32, laboratory test results such as the water content and dry unit weight of the soil or rock are also frequently listed on the boring log.

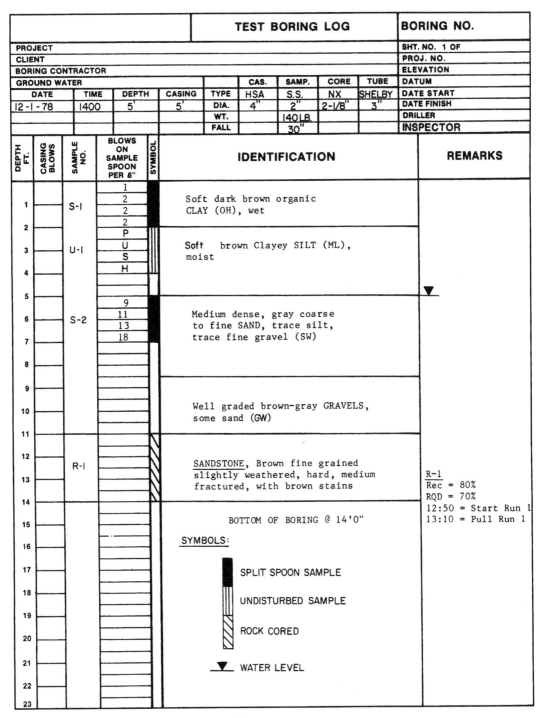

FIGURE 2.32 Boring log. (*From NAVFAC DM-7.1, 1982.*)

An important part of the preparation of logs is to determine the geologic or man-made process that created the soil deposit. Table 2.11 presents a list of common soil deposits encountered during subsurface exploration. Usually the engineering geologist is most qualified to determine the type of soil deposit. As indicated in Table 2.11, the different soil deposits can have unique geotechnical and foundation implications and it is always of value to determine the geologic or man-made process that created the soil deposit.

Another example of a boring log is presented in Fig. 2.33. As indicated on this boring log, frozen soil and rock was observed from a depth of 1.8 ft (0.5 m) to 16 ft (4.9 m). The bold black vertical line indicates the zone of observed frozen soil and rock. The symbols (e.g. N_f, V_s, etc.) refer to group symbols and subgroups used to describe frozen soil per ASTM D 4083-01 (2004), "Standard Practice for Description of Frozen Soils (Visual-Manual Procedure)." For a description of these group symbols, see Table 4.4 of this book.

TABLE 2.11   Common Man-Made and Geologic Soil Deposits

Main category	Common types of soil deposits	Possible engineering problems
Structural fill	Dense or hard fill. Often the individual fill lifts can be identified.	Upper surface of structural fill may have become loose or weathered.
Uncompacted fill	Random soil deposit that can contain chunks of different types and sizes of rock fragments.	Susceptible to compression and collapse.
Debris fill	Contains pieces of debris, such as concrete, brick, and wood fragments.	Susceptible to compression and collapse.
Municipal dump	Contains debris and waste products such as household garbage or yard trimmings.	Significant compression and gas from organic decomposition.
Residual soil deposit	Soil deposits formed by in-place weathering of rock.	Engineering properties are highly . variable.
Organic deposit	Examples include peat and muck which form in bogs, marshes, and swamps.	Very compressible and unsuitable for foundation support.
Alluvial deposit	Soil transported and deposited by flowing water, such as streams and rivers.	All types of grain sizes, loose sandy deposits susceptible to liquefaction.
Aeolian deposit	Soil transported and deposited by wind. Examples include loess and dune sands.	Can have unstable soil structure that may be susceptible to collapse.
Glacial deposit	Soil transported and deposited by glaciers or their melt water. Examples include till.	Erratic till deposits and soft clay deposited by glacial melt water.
Lacustrine deposit	Soil deposited in lakes or other inland bodies of water.	Unusual soil deposits can form, such as varved silts or varved clays.
Marine deposit	Soil deposited in the ocean, often from rivers that empty into the ocean.	Granular shore deposits but offshore areas can contain soft clay deposits.
Colluvial deposit	Soil transported and deposited by gravity, such as talus, hill-wash, or landslide deposits	Can be geologically unstable deposit.
Pyroclastic deposit	Material ejected from volcanoes. Examples include ash, lapilli, and bombs.	Weathering can result in plastic clay. Ash can be susceptible to erosion.

Note: The first four soil deposits are man-made; all others are due to geologic processes.

Depth,		Symbol	Soil Description	Ice Features
m	ft			
0.0	0.0*	OL	Organic, sandy SILT, not frozen	None
0.15	0.5	GW	Brown, well-graded, sandy GRAVEL, medium compact, moist, not frozen	None
0.6	1.8	GW N_f	Brown well-graded, sandy GRAVEL, frozen, poorly bonded	No visible segregation, negligible thin ice film on gravel sizes and within larger voids
1.1	3.7	GW N_{bn}	Brown, well-graded, sandy GRAVEL, frozen, well bonded	No visible segregation
1.6	5.4	ML V_s	Black, micaceous, sandy SILT, frozen	Stratified horizontal ice lenses averaging 4 in. (10 cm) in horizontal extent, hairline to ¼ in. (0.6 cm) in thickness, ½ to ¾-in. (1.2 to 1.9-cm) spacing. Visible excess ice ~20 ± % of total volume. Ice lenses hard, clear, colorless.
2.4	7.7	ICE		Hard, slightly cloudy, colorless, few scattered inclusions of silty SAND
2.8	9.1		Dark brown PEAT, frozen, well bonded, high degree of saturation	~5 % visible ice
3.2	10.5	MH V_r	Light brown SILT, frozen	Irregularly oriented ice lenses and layers ¼ to ¾ in. (0.6 to 1.9 cm) thick on random pattern grid approx. 3 to 4-in. (7.6 to 10-cm) spacing. Visible ice ~10 ± % of total volume. Ice moderately soft, porous, gray-white.
4.4	14.3			
4.9	16.0			
			Bedrock. Laminated SHALE. Top few feet weathered	¹⁄₁₆-in. (0.2-cm) thick ice lenses in fissures to 16.0 ft (4.9 m). None below.
6.3	20.6		Bottom of exploration	

* Surface elevation 963.2 ft

FIGURE 2.33   Boring log for frozen soil. (*Reprinted with permission from the American Society for Testing and Materials, 2004.*)

2.7   GEOPHYSICAL TECHNIQUES

Geophysical techniques can be employed by the engineering geologist to obtain data on the subsurface conditions. Use of geophysical techniques involves considerable experience and judgment in the interpretation of results. Common types of geophysical techniques are summarized in Table 2.12.

Probably the most commonly used geophysical technique is the seismic refraction method. This method is based on the fact that seismic waves travel at different velocities through different types of materials. For example, seismic waves will travel much faster in solid rock than in soft clay. The test method commonly consists of placing a series of geophones in a line on the ground surface. Then a metal plate is placed on the ground surface and in line with the geophones. By striking the metal plate with a sledgehammer, a shock wave (or shot) can be produced. This seismic energy is detected by the geophones and by analyzing the recorded data, the velocity of the seismic wave as it passes through the ground and the depth to bedrock can often be determined. This is accomplished by developing a time-distance plot, where the horizontal axis is the distance from the shock wave source to the geophones and the vertical axis is the time it takes for the shock wave to reach the geophone. Figure 2.34

TABLE 2.12 Geophysical Techniques

Name of method	Procedure or principle utilized	Applicability and limitations
Seismic methods 1. Refraction	Based on time required for seismic waves to travel from source of energy to points on ground surface, as measured by geophones spaced at intervals on a line at the surface. Refraction of seismic waves at the interface between different strata gives a pattern of arrival times at the geophones versus distance to the source of seismic waves. Seismic velocity can be obtained from a single geophone and recorder with the impact of a sledge hammer on a steel plate as a source of seismic waves.	Utilized for preliminary site investigation to determine rippability, faulting, and depth to rock or other lower stratum substantially different in wave velocity than the overlying material. Generally limited to depths up to 30 m (100 ft) of a single stratum. Used only where wave velocity in successive layers becomes greater with depth.
2. High-resolution reflection	Geophones record travel time for the arrival of seismic waves reflected from the interface of adjoining strata.	Suitable for determining depths to deep rock strata. Generally applies to depths of a few thousand feet. Without special signal enhancement techniques, reflected impulses are weak and easily obscured by the direct surface and shallow refraction impulses. Method is useful for locating groundwater.
3. Vibration	The travel time of transverse or shear waves generated by a mechanical vibrator consisting of a pair of eccentrically weighted disks is recorded by seismic detectors placed at specific distances from the vibrator.	Velocity of wave travel and natural period of vibration gives some indication of soil type. Travel time plotted as a function of distance indicates depths or thickness of a surface stratum. Useful in determining dynamic modulus of subgrade reaction and obtaining information on the natural period of vibration for the design of foundations of vibrating structures.
4. Uphole, downhole, and crosshole surveys	(a) Uphole or downhole: Geophones on surface, energy source in borehole at various locations starting from hole bottom. Procedure can be revised with energy source on surface, detectors moved up or down the hole. (b) Downhole: Energy source at the surface (e.g., wooden plank struck by hammer), geophone probe in borehole. (c) Crosshole: Energy source in central hole, detectors in surrounding holes.	Obtain dynamic soil properties at very small strains, rock mass quality, and cavity detection. Unreliable for irregular strata or soft strata with large gravel content. Also unreliable for velocities decreasing with depth. Crosshole measurements best suited for in situ modulus determinations.
Electrical methods 1. Resistivity	Based on the difference in electrical conductivity or resistivity of strata. Resistivity is correlated to material type.	Used to determine horizontal extent and depths up to 30 m (100 ft) of subsurface strata. Principal applications are for investigating foundations of dams and other large structures, particularly in exploring granular river channel deposits or bedrock surfaces. Also used for locating fresh/salt water boundaries.

Method	Procedure	Applications
2. Drop in potential	Based on the determination of the drop in electrical potential.	Similar to resistivity methods but gives sharper indication of vertical or steeply inclined boundaries and more accurate depth determinations. More susceptible than resistivity method to surface interference and minor irregularities in surface soils.
3. E-logs	Based on differences in resistivity and conductivity measured in borings as the probe is lowered or raised.	Useful in correlating units between borings, and has been used to correlate materials having similar seismic velocities. Generally not suited to civil engineering exploration but valuable in geologic investigations.
Magnetic measurements	Highly sensitive proton magnetometer is used to measure the Earth's magnetic field at closely spaced stations along a traverse.	Difficult to interpret in quantitative terms but indicates the outline of faults, bedrock, buried utilities, or metallic trash in fills.
Gravity measurements	Based on differences in density of subsurface materials which affects the gravitational field at the various points being investigated.	Useful in tracing boundaries of steeply inclined subsurface irregularities such as faults, intrusions, or domes. Methods not suitable for shallow depth determination but useful in regional studies. Some application in locating limestone caverns.

Source: Adapted from NAVFAC DM-7.1, 1982.
Note: Also see AGI Data Sheets 59.1 to 60.2 (American Geological Institute, 1982) for a summary of the applications of geophysical methods.

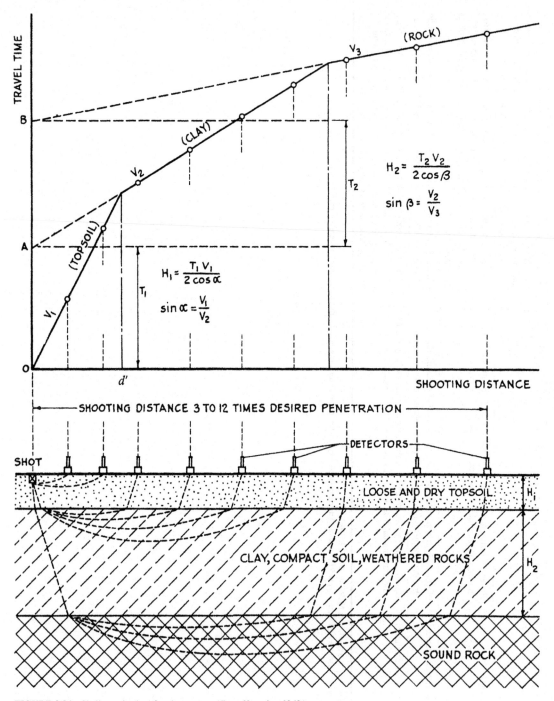

FIGURE 2.34   Shallow seismic refraction survey. (*From Hvorslev, 1949.*)

presents an example of a time-distance plot obtained from a shallow seismic refraction survey and the equations needed to determine the thickness of the layers. The procedure is as follows:

1. Calculate the seismic wave velocities V_1 and V_2. The velocities are easy to calculate, for example V_1 is simply the distance divided by the time.
2. Determine the value of α. Obtain α from the equation: $\sin \alpha = V_1/V_2$
3. Obtain the thickness of the upper layer from: $H_1 = [(T_1 V_1)/(2 \cos \alpha)]$, where T_1 is obtained from the time-distance plot (see Fig. 2.34).
4. Repeat the above steps to obtain the thickness of the deeper layers (i.e., H_2 and H_3).

As an alternate to the above analysis, the following equation can be used to determine the thickness of the upper stratum:

$$H_1 = \frac{1}{2} d' \left(\frac{V_2 - V_1}{V_2 + V_1} \right)^{1/2}$$

(2.7)

where H_1 = thickness of the upper stratum (ft or m)
d' = distance (ft or m) from the shot to the intersection of the straight line segments, such as shown in Fig. 2.34
V_1 = seismic wave velocity of the upper stratum (ft/sec or m/sec)
V_2 = seismic wave velocity of the lower stratum (ft/sec or m/sec)

It should be mentioned that the seismic refraction method could only be used when the wave velocity is greater in each successively deeper layer. In addition, the seismic refraction method works best when there are large contrasts in materials, for example, soil overlying rock or loose dry sand overlying sand that is saturated by a groundwater table. For inclined strata, only the average depths can be determined and it is necessary to reverse the position of the seismic wave source and geophones and shoot up-dip and down-dip in order to determine the actual depths and the dip of the strata.

Very dense and hard rock will have a high seismic wave velocity, while soft or loose soil will have a much lower seismic wave velocity. Typical seismic wave velocities are as follows (Sowers and Sowers, 1970):

Material type	Seismic wave velocity
Loose dry sand	500–1500 ft/sec (150–450 m/sec)
Hard clay, partially saturated	2000–4000 ft/sec (600–1200 m/sec)
Water or loose saturated sand	5200 ft/sec (1600 m/sec)
Saturated soil and weathered rock	4000 to 10,000 ft/sec (1200–3000 m/sec)
Sound rock	7000–20,000 ft/sec (2000–6000 m/sec)

The more dense and hard the rock, the higher its seismic wave velocity. This principle can be used to determine whether the underlying rock can be excavated by commonly available equipment, or is so dense and hard that it must be blasted apart. For example, the *Caterpillar Performance Handbook* (1997) presents charts that relate the seismic wave velocities of various types of rocks to the type of equipment (Caterpillar D8R, D9R, D10R, and D11R tractor/ripper) and their ability to rip or not rip apart rock. An example of these charts is presented in Fig. 2.35. This information can be very important to the client because of the much higher costs and risks associated with blasting rock as compared to using a conventional piece of machinery to rip apart and excavate the rock. As shown in Fig. 2.35, a Caterpillar D11R tractor/ripper can not rip rock that has a seismic wave velocity around 10,000 to 12,000 ft/sec (3000 to 3700 m/sec), with the lower value applicable to massive rock such as granite and the higher value applicable to foliated and jointed rock such as shale.

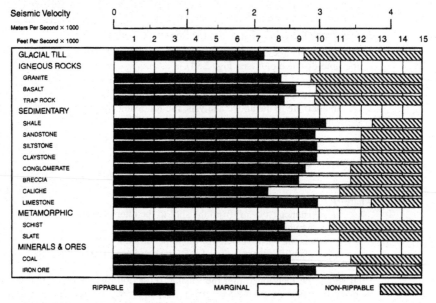

FIGURE 2.35 Rippability of rock versus seismic velocity for a Caterpillar D11R tractor/ripper. (*From Caterpillar Performance Handbook, 1997.*)

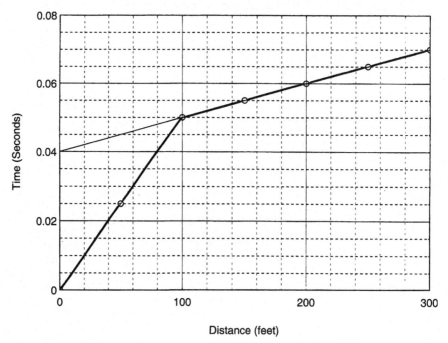

FIGURE 2.36 Example problem.

Example Problem 2.1   Using the data shown in Fig. 2.36, determine the thickness of the upper stratum (i.e., H_1). Can a Caterpillar D11R rip the upper and lower stratum?

Solution

$$V_1 = \frac{100 \text{ ft}}{0.05 \text{ sec}} = 2000 \text{ ft/sec} (600 \text{ m/sec})$$

$$V_2 = \frac{300 - 100 \text{ ft}}{0.07 - 0.05 \text{ sec}} = 10,000 \text{ ft/sec} (3000 \text{ m/sec})$$

Since the upper stratum has a low seismic wave velocity of 2000 ft/sec (600 m/sec), it will be easy for a Caterpillar D11R to remove this material. For the lower stratum, the seismic wave velocity is 10,000 ft/sec (3000 m/sec) and it could be ripped if it is shale, but granite would probably be nonrippable (see Fig. 2.35).

$$\sin \alpha = \frac{V_1}{V_2} = \frac{2000}{10,000} = 0.2, \text{ or } \alpha = 11.5°$$

From Fig. 2.36, $T_1 = 0.04$ sec

$$H_1 = \frac{T_1 V_1}{2 \cos \alpha} = \frac{(0.04)(2000)}{2 \cos 11.5°} = 41 \text{ ft} (12 \text{ m})$$

Checking using Eq. 2.7, where $d' = 100$ ft (30.5 m)

$$H_1 = \frac{1}{2} d' \left(\frac{V_2 - V_1}{V_2 + V_1} \right)^{1/2}$$

$$= \frac{1}{2} 100 \left(\frac{10,000 - 2000}{10,000 + 2000} \right)^{1/2} = 41 \text{ ft} (12 \text{ m})$$

2.8   *GEOTECHNICAL EARTHQUAKE ENGINEERING*

The purpose of this section is to discuss the special subsurface exploration requirements that may be needed for geotechnical earthquake engineering analyses. In terms of the investigation for assessing seismic hazards, the *Guidelines for Evaluating and Mitigating Seismic Hazards in California* (Division of Mines and Geology, 1997) states:

> The working premise for the planning and execution of a site investigation within seismic hazard zones is that the suitability of the site should be demonstrated. This premise will persist until either: (a) the site investigation satisfactorily demonstrates the absence of liquefaction or landslide hazard, or (b) the site investigation satisfactorily defines the liquefaction or landslide hazard and provides a suitable recommendation for its mitigation.

Thus the purpose of the subsurface exploration should be to demonstrate the absence of seismic hazards or to adequately define the seismic hazards so that suitable recommendations for mitigation can be developed.

The scope of the subsurface investigation depends on many different factors such as the type of facility to be constructed, the nature and complexity of the geologic hazards that could impact the site during the earthquake, economic considerations, level of risk, and specific requirements such as local building codes or other regulatory specifications. The most rigorous geotechnical earthquake investigations would be required for essential facilities.

The scope of the investigation for geotechnical earthquake engineering is usually divided into two parts: (1) the screening investigation, and (2) the quantitative evaluation of the seismic hazards (Division of Mines and Geology, 1997). These two items are individually discussed later.

2.8.1 Screening Investigation

The first step in geotechnical earthquake engineering is to perform a screening investigation. The purpose of the screening investigation is to assess the severity of the seismic hazards at the site, or in other words, to screen out those sites that do not have seismic hazards. If it can be clearly demonstrated that a site is free of seismic hazards, then the quantitative evaluation could be omitted. On the other hand, if a site is likely to have seismic hazards, then the screening investigation can be used to define those hazards before proceeding with the quantitative evaluation.

An important consideration for the screening investigation is the effect that the new construction will have on potential seismic hazards. For example, as a result of grading or construction at the site, the groundwater table may be raised or adverse bedding planes may be exposed that result in a landslide hazard. Thus when performing a screening investigation, both the existing condition and the final constructed condition must be evaluated for seismic hazards. Another important consideration is off-site seismic hazards. The first step in the screening investigation is to review available documents, such as those listed in Sec. 2.2, as well as the following:

Seismic History of the Area.   There may be many different types of documents and maps that provide data on the seismic history of the area. For example, there may be seismic history information on the nature of past earthquake-induced ground shaking. This information could include the period of vibration, ground acceleration, magnitude, and intensity (isoseismal maps) of past earthquakes. This data can often be obtained from seismology maps and reports that illustrate the differences in ground shaking intensity based on geologic type; 50, 100 and 250 year acceleration data; and type of facilities and landmarks.

Geographical maps and reports are important because they can identify such items as the pattern, type, and movement of nearby potentially active faults or fault systems, and the distance of the faults to the area under investigation. Historical earthquake records should also be reviewed in order to determine the spatial and temporal distribution of historic earthquake epicenters.

Special Study Maps.   For some areas, special study maps or other documents may have been developed that indicate local seismic hazards. For example, Fig. 2.37 presents a portion of the *Seismic Safety Study* (1995) that shows the location of the Rose Canyon Fault Zone. Special study maps may also indicate other geologic and seismic hazards, such as potentially liquefiable soil, landslides, and abandoned mines.

After the site research has been completed, the next step in the screening investigation is a field reconnaissance. The purpose is to observe the site conditions and document any recent changes to the site that may not be reflected in the available documents. The field reconnaissance should also be used to observe surface features and other details that may not be readily evident from the available documents. Once the site research and field reconnaissance are completed, the engineering geologist and geotechnical engineer can then complete the screening investigation. The results should either clearly demonstrate the lack of seismic hazards or indicate the possibility of seismic hazards, in which case a quantitative evaluation is required.

It should be mentioned that even if the results of the screening investigation indicate no seismic hazards, the governing agency might not accept this result for essential facilities. They may still require that subsurface exploration demonstrate the absence of seismic hazards for essential facilities.

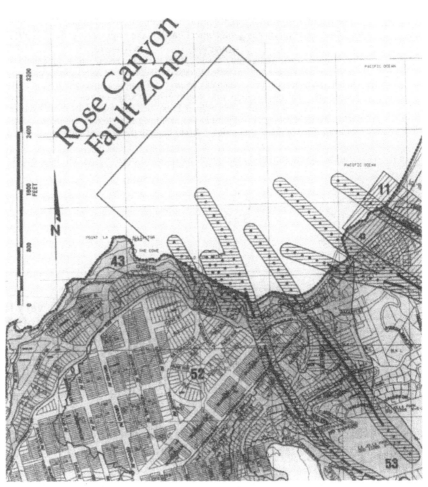

FIGURE 2.37   Portion of *Seismic Safety Study*, 1995. (*Developed by the City of San Diego.*)

2.8.2   Quantitative Evaluation

The purpose of the quantitative evaluation is to obtain sufficient information on the nature and severity of the seismic hazards so that mitigation recommendations can be developed. The quantitative evaluation consists of geologic mapping, subsurface exploration, and laboratory testing. The main objectives of the subsurface exploration are to determine the nature and extent of the seismic hazards. In this regard, the Division of Mines and Geology (1997) states:

> The subsurface exploration should extend to depths sufficient to expose geologic and subsurface water conditions that could affect slope stability or liquefaction potential. A sufficient quantity of subsurface information is needed to permit the engineering geologist and/or civil engineer to extrapolate with confidence the subsurface conditions that might affect the project, so that the seismic hazard can be properly evaluated, and an appropriate mitigation measure can be designed by the civil engineer. The preparation of engineering geologic maps and geologic cross sections is often an important step into developing an understanding of the significance and extent of potential seismic hazards. These maps and/or cross sections should extend far enough beyond the site to identify off-site hazards and features that might affect the site.

The depth of subsurface exploration has been discussed in Sec. 2.4.6. In terms of the depth of the subsurface exploration for geotechnical earthquake engineering, Seed (1991) states:

> Liquefaction investigations should extend to depths below which liquefiable soils cannot reasonably be expected to occur (e.g., to bedrock, or to hard competent soils of sufficient geologic age that possible underlying units could not reasonably be expected to pose a liquefaction hazard). At most sites where soil is present, such investigation will require either borings or trench/test pit excavation. Simple surface inspection will suffice only when bedrock is exposed over essentially the full site, or in very unusual cases when the local geology is sufficiently well-documented as to fully ensure the complete lack of possibility of occurrence of liquefiable soils (at depth) beneath the exposed surface soil unit(s).

Further discussion of geotechnical earthquake engineering will be presented in Chaps. 13 and 14.

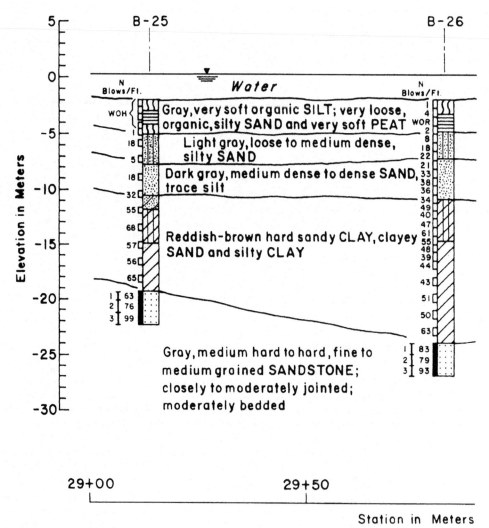

FIGURE 2.38 Subsoil profile. (*From Lowe and Zaccheo, 1975; copyright Van Nostrand Reinhold.*)

2.9   SUBSOIL PROFILE

The final section of this chapter presents examples of subsoil profiles. The results of the subsurface exploration are often summarized on a subsoil profile. Usually the engineering geologist is the person most qualified to develop the subsoil profile based on experience and judgment in extrapolating conditions between the borings, test pits, and trenches.

Figures 2.38 to 2.41 show four examples of subsoil profiles. The results of field and laboratory tests have been included on these subsoil profiles. The development of a subsoil profile is often a required element for foundation engineering analyses. For example, subsoil profiles are used to determine the foundation type (shallow versus deep foundation), calculate the amount of settlement or heave of the structure, evaluate the effect of groundwater on the project, develop recommendations for dewatering of foundation excavations, perform slope stability analyses for projects having sloping topography, and prepare site development recommendations.

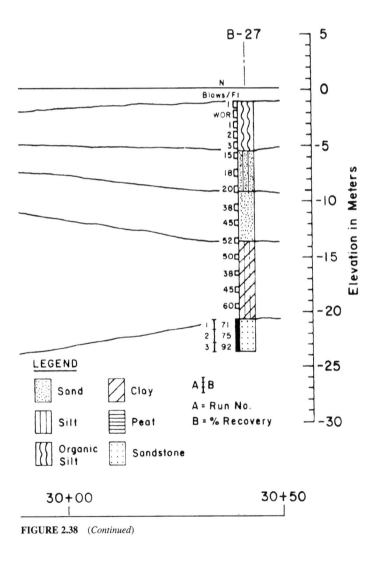

FIGURE 2.38   (*Continued*)

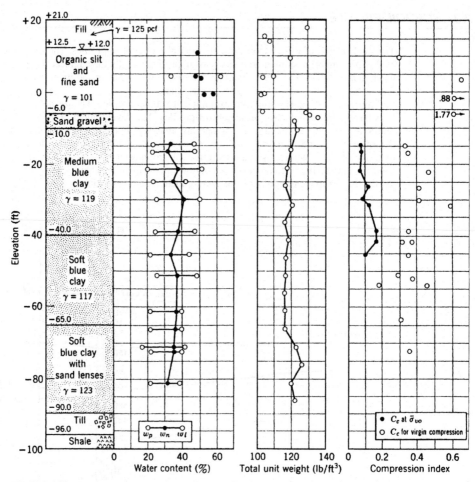

FIGURE 2.39 Subsoil profile, Cambridge, Mass. (*From Lambe and Whitman, 1969; reprinted with permission of John Wiley & Sons.*)

NOTATION

The following notation is introduced in this chapter:

C_b = borehole diameter correction for the SPT

C_N = for SPT and CPT, correction factor to account for the overburden pressure

C_r = rod length correction for the SPT

d' = defined in Fig. 2.34

D = diameter of the vane

D_e = diameter at the sampler cutting tip

D_s = inside diameter of the sampling tube

D_w = outside diameter of the sampling tube

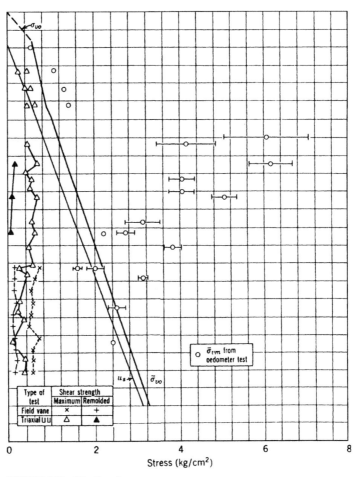

FIGURE 2.39   (*Continued*)

E_m = hammer efficiency for the SPT

f_s = soil friction along a side sleeve (CPT)

FR = friction ratio (CPT)

H = depth of penetration of the sampler (Sec. 2.4.2)

H = height of the vane (Sec. 2.4.4)

H_1, H_2 = thickness of different soil strata

k_o = at-rest earth pressure

L_g = gross length of sample

N = SPT N value

N_f = group symbol for frozen soil

N_{60} = SPT N value corrected for field testing procedures

$(N_1)_{60}$ = SPT N value corrected for field testing procedures and vertical effective stress

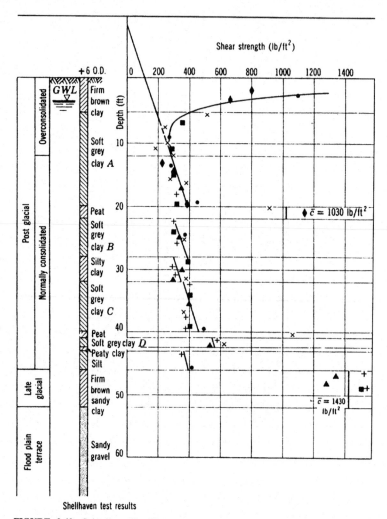

FIGURE 2.40 Subsoil profile, Thames estuary clay, England. (*From Skempton and Henkel, 1953; reprinted from Lambe and Whitman, 1969.*)

q_c = cone resistance (CPT)

q_{c1} = cone resistance (CPT) corrected for vertical effective stress

s_u = undrained shear strength of clay

s_{ur} = undrained shear strength of remolded clay

S_t = sensitivity of the clay

T_{max} = maximum torque required to shear the clay for the field vane test

V_1, V_2 = seismic wave velocities of different soil strata

T_1 = defined in Fig. 2.34

α = defined in Fig. 2.34

σ'_{vo} = vertical effective stress

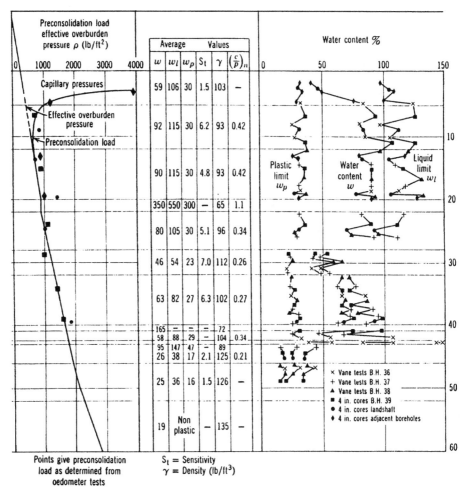

FIGURE 2.40   (*Continued*)

PROBLEMS

Solutions to the problems are presented in App. C of this book

2.1   A sampling tube has an outside diameter D_t of 3.00 in, a tip diameter D_e of 2.84 inches, and a wall thickness of 0.065 inches. If $D_w = D_t$ (see Fig. 2.13), calculate the clearance ratio, area ratio, and indicate if the sampling tube meets the criteria for undisturbed soil sampling.

ANSWERS: Clearance ratio = 1.06 percent, area ratio = 11.6 percent, and it is close to meeting the criteria for undisturbed soil sampling.

2.2   A SPT was performed on a near surface deposit of clean sand where the number of blows to drive the sampler 45 cm was 5 for the first 15 cm, 8 for the second 15 cm, and 9 for the third 15 cm.

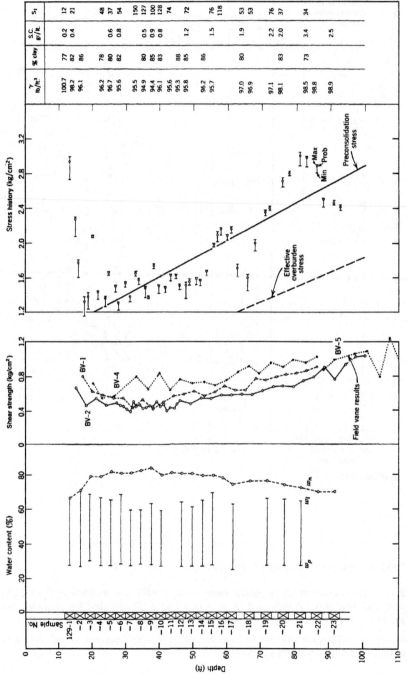

FIGURE 2.41 Subsoil profile, Canadian clay. (*From Lambe and Whitman, 1969; reprinted with permission of John Wiley & Sons.*)

Assume that E_m = 60 percent, the borehole diameter is 100 mm, and the drill rod length is 5 m. Calculate the measured SPT N value (blows per foot), N_{60}, and $(N_1)_{60}$ assuming that the vertical effective stress (σ'_{vo}) = 50 kPa. Also indicate the density condition of the sand.

ANSWER: Measured SPT N value = 17, N_{60} = 14.5, and $(N_1)_{60}$ = 20.4. As per Table 2.6, the sand is in a dense condition.

2.3   A field vane shear test was performed on a clay, where the rectangular vane had a length H of 4.0 in. and a diameter D of 2.0 in. The maximum torque T_{max} required to shear the soil was 8.5 ft-lb. Calculate the undrained shear strength s_u of the soil.

ANSWER: 500 psf.

2.4   Use the data in Fig. 2.34. Assume that $T_1 = T_2 = 0.04$, $V_1 = 800$ ft/sec, $d' = 50$ ft, and the intersection of the clay and rock portions of the graph occur at a distance from the shot = 120 ft. Determine H_1 and H_2.

ANSWERS: H_1 = 17.1 ft and H_2 = 46.0 ft.

2.5   A construction site in New England requires excavation of rock. The geologist has determined that the rock is granite and from geophysical methods (i.e., seismic refraction), the seismic velocity of the in situ granite is 12,000 to 15,000 ft per sec. A Caterpillar D11R tractor/ripper is available. Can the granite be ripped apart?

ANSWER: No, blasting will be required.

CHAPTER 3
LABORATORY TESTING

3.1 INTRODUCTION

In addition to the subsurface exploration, an essential part of the foundation investigation is laboratory testing. The laboratory testing usually begins once the subsurface exploration is complete. The first step in the laboratory testing is to log in all of the materials (soil, rock, or groundwater) recovered from the subsurface exploration. Then the geotechnical engineer and engineering geologist prepares a laboratory testing program, which basically consists of assigning specific laboratory tests for the soil specimens. Experienced technicians, who are under the supervision of the geotechnical engineer, often perform the actual laboratory testing of the soil specimens. Because the soil samples can dry out or there could be changes in the soil structure with time, it is important to perform the laboratory tests as soon as possible.

Usually at the time of the laboratory testing, the geotechnical engineer and engineering geologist will have located the critical soil layers or subsurface conditions that will have the most impact on the design and construction of the project. The testing program should be oriented towards the testing of those critical soil layers or subsurface conditions. For foundation engineering, it is also important to determine the amount of ground surface movement due to construction of the project. In these cases, laboratory testing should model future expected conditions so that the amount of movement or stability of the ground can be analyzed. During the planning stage, specific types of laboratory tests may have been selected, but based on the results of the subsurface exploration, additional tests or a modification of the planned testing program may be required.

Laboratory tests should be performed in accordance with standard procedures, such as those recommended by the American Society for Testing and Materials (ASTM) or those procedures listed in standard textbooks or specification manuals (e.g., Lambe, 1951; Bishop and Henkel, 1962; Department of the Army, 1970; Day, 2001a; *Standard Specifications for Public Works Construction,* 2003).

For laboratory tests, Tomlinson (1986) states:

> It is important to keep in mind that natural soil deposits are variable in composition and state of consolidation; therefore it is necessary to use considerable judgment based on common sense and practical experience in assessing test results and knowing where reliance can be placed on the data and when they should be discarded. It is dangerous to put blind faith in laboratory tests, especially when they are few in number. The test data should be studied in conjunction with the borehole records and the site observations, and any estimations of bearing pressures or other engineering design data obtained from them should be checked as far as possible with known conditions and past experience.
>
> Laboratory testing should be as simple as possible. Tests using elaborate equipment are time-consuming and therefore costly, and are liable to serious error unless carefully and conscientiously carried out by highly experienced technicians. Such methods may be quite unjustified if the samples are few in number, or if the cost is high in relation to the cost of the project. Elaborate and costly tests are justified only if the increased accuracy of the data will give worthwhile savings in design or will eliminate the risk of a costly failure.

Table 3.1 presents a list of common soil laboratory tests used in geotechnical engineering. As indicated in Table 3.1, laboratory tests are often used to determine the index properties, shear strength,

TABLE 3.1 Common Soil Laboratory Tests Used in Geotechnical Engineering

Type of condition	Soil properties	Specification
Index tests (Sec. 3.2)	Water content test (moisture content)	ASTM D 2216–98 and D 4643–00
	Specific gravity test	ASTM D 854–02 and D 5550–00
	Relative density	ASTM D 4253–00 and D 4254–00
Particle size and Atterberg limits (Sec. 3.2)	Sieve analysis	ASTM D 422–02
	Hydrometer test	ASTM D 422–02
	Atterberg limits test	ASTM D 4318–00
	Soil classification (USCS)	ASTM D 2487–00
Settlement (Chaps. 7 and 8)	Collapse test	ASTM D 5333–03
	Consolidation test	ASTM D 2435–03
Expansive soil (Chap. 9)	Expansion index test	ASTM D 4829–03
	HUD swell test	HUD specifications (1971)
	Intact swell test	ASTM D 4546–03
	Oedometer test (method C)	ASTM D 4546–03
Shear strength tests (Secs. 3.4 and 3.5)	Direct shear test	ASTM D 3080–03
	Unconfined compressive strength	ASTM D 2166–00
	Miniature vane test	ASTM D 4648–00
	Unconsolidated undrained triaxial	ASTM D 2850–03
	Consolidated undrained triaxial test	ASTM D 4767–02
	Torsional ring shear test	ASTM D 6467–99[1]
Compaction (Sec. 3.6)	Standard proctor test	ASTM D 698–00
	Modified proctor test	ASTM D 1557–02
	Sand cone test	ASTM D 1556–00
	Drive cylinder test	ASTM D 2937–00
Permeability (Sec. 3.7)	Constant head test	ASTM D 2434–00
	Falling head test	ASTM D 5084–00

Note: [1]This specification is in the ASTM Standards Volume 04.09 (2004). All other ASTM standards are in Volume 04.08 (2004).

compressibility, and hydraulic conductivity of the soil. App. A (Glossary 2) presents a list of laboratory terms and definitions.

3.2 INDEX TESTS

Index tests are the most basic types of laboratory tests performed on soil samples. Index tests include the following:

- Water content (also known as moisture content)
- Wet density determinations (also known as total density)
- Specific gravity tests
- Sieve analysis, hydrometer test, and Atterberg limits tests (used to classify the soil)
- Laboratory tests specifically labeled as index tests, such as the expansion index test that is used to evaluate the potential expansiveness of a soil and will be discussed in Chap. 9

To fully understand index testing, phase relationships will first be introduced. Phase relationships are the basic soil relationships used in geotechnical engineering. To assist in the understanding of phase relationships, soil can be separated into its three basic parts, as shown in Fig. 3.1 and described below:

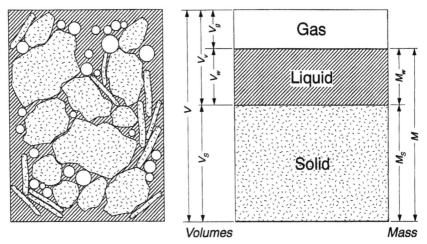

FIGURE 3.1   Soil element and the soil element separated into phases.

1. *Solids.*   Which are the mineral soil particles
2. *Liquids.*   Which is usually water that is contained in the void spaces between the solid mineral particles
3. *Gas.*   Such as air that is also contained in the void spaces between the solid mineral particles. If the soil is below the groundwater table, the soil is usually saturated and there are no open gas voids.

As indicated on the right side of Fig. 3.1, the three basic parts of soil can be rearranged into their relative proportions based on volume and mass. Certain phase relationships can be determined directly from laboratory testing, such as the water content (Sec. 3.2.1), unit weight (Sec. 3.2.2), and specific gravity (Sec. 3.2.3). Other phase relationships cannot be determined in a laboratory, but instead must be calculated, and they will be discussed in Sec. 4.3.

3.2.1   Water Content Test

The water content, also known as moisture content, is probably the most common and simplest type of laboratory test. This test can be performed on disturbed or undisturbed soil specimens. The water content test consists of determining the mass of the wet soil specimen and then drying the soil in an oven overnight (12 to 16 h) at a temperature of 110°C in order to determine the mass of the dry soil solids M_s. By subtracting the initial wet mass from the final dry mass, the mass of water M_w in the soil can be calculated. The water content w of a soil can then be calculated as:

$$w(\%) = \frac{100 M_w}{M_s} \tag{3.1}$$

where w = water content expressed as a percentage
   M_w = mass of the water in the soil (lb or g)
   M_s = mass of the dry soil solids (lb or g)

Table 3.2 presents guidelines on the amount of soil that should be used for a water content test based on the largest particle dimension. Water content values are often reported to the nearest 0.1 percent or 1 percent of measured value. Values of water content w can vary from essentially 0 percent up to

TABLE 3.2   Minimum Soil Sample Mass for Water Content Determination

Maximum particle size	Sieve size corresponding to maximum size	Recommended minimum mass of wet soil specimen for water content to be reported to the nearest 0.1 percent (g)	Recommended minimum mass of wet soil specimen for water content to be reported to the nearest 1 percent (g)
2 mm or less	No. 10	20	—
4.75 mm	No. 4	100	20
9.5 mm	$^3/_8$ in.	500	50
19.0 mm	$^3/_4$ in.	2500	250

Note: Table based on ASTM D 2216-98 (2004). For soil containing particles larger than $^3/_4$ in., first sieve the sample on the $^3/_4$ in. sieve, record the mass of oversize particles (plus $^3/_4$-in. sieve material), and then determine the water content of the soil matrix (minus $^3/_4$-in. sieve soil).

1200 percent. A water content of 0 percent indicates a dry soil, such as a clean gravel or sand located in a hot and dry climate like Death Valley, California. Soil having the highest water content is organic soil, such as fibrous peat, which has been reported to have a water content as high as 1200 percent (NAVFAC DM-7.1, 1982).

The water content data are often plotted with depth on the soil profile. Note in Fig. 2.39 that the water content has been plotted versus depth (darkened circles) and the water content varies from about 30 to 40 percent for the clay. Water content data are also plotted (open circles) in Fig. 2.41, and for this Canadian clay, the water content varies from about 60 to 80 percent.

Water content can provide valuable information on possible foundation problems. For example, if a clay layer located below a proposed shallow foundation has a water content of 100 percent, then it is likely that this clay will be highly compressible. Likewise if the same clay layer below the shallow foundation has a water content of 5 percent, then it is likely that the clay layer is dry and desiccated and could subject the shallow foundation to expansive soil uplift.

Many soils contain dissolved solids. For example, in the case of soil located at the bottom of the ocean, the water between the soil solids may actually have the same salt concentration as seawater. Another example is the presence of cations, which are attracted to clay particle faces (i.e., double layer effect). Once the soil is dried, these dissolved ions and minerals will become part of the mass of soil solids M_s. For most soils, this effect will have a minimal impact on the water content. An exception is ocean bottom sediments having both salt water as the pore fluid between the soil particles and a very high water content (Noorany, 1984). Another exception could be lake bottom sediments where the lake contains a high salt concentration, such as the Salton Sea, California. For further details, see ASTM D 2216-98 (2004), "Standard Test Method for Laboratory Determination of Water (Moisture) Content of Soil and Rock by Mass."

An alternate method for determining the water content is to use a microwave oven, i.e., ASTM D 4643-00, 2004, "Standard Test Method for Determination of Water (Moisture) Content of Soil by the Microwave Oven Heating." In this method, the wet soil is placed in a porcelain or glass dish and then dried in the microwave oven. The advantage of this method is that the water content can be obtained very quickly. The disadvantage of this method is that the soil will be subjected to a temperature that is well in excess of 110°C. This overheating of the soil can result in inaccurate results for soil containing hydrated water, such as microfossils and/or diatoms (e.g., diatomaceous earth). Inaccurate results have also been reported for soil containing significant amounts of halloysite, mica, montmorillonite, gypsum, or other hydrated materials, and highly organic soils.

3.2.2   Total Unit Weight

The total density, also known as the wet density, should only be obtained from undisturbed soil specimens, such as those extruded from Shelby tubes or on undisturbed block samples obtained from test pits and trenches. When extruding the soil from sampler tubes, it is important to push the soil out the back of the sampler. Soil should not be pushed out the front of the tube (i.e., the cutting end) because this causes a reversal in direction of soil movement as well as the possible compression of the soil because the cutting tip diameter is less than the internal tube diameter. In order to determine the total density of a soil specimen, both the mass and corresponding volume of the soil specimen must be known. One method is to extrude the soil from the sampling tube directly into metal confining rings of known volume. Once the volume of the wet soil is known, the total density ρ_t can be calculated as:

$$\rho_t = \frac{M}{V} \qquad (3.2)$$

where ρ_t = total density of the soil (pcf or g/cm³)
   M = total mass (lb or g) of the soil which is the sum of the mass of water (M_w) and mass of solids (M_s)
   V = total volume (ft³ or cm³) of the soil sample as defined in Fig. 3.1

Since most laboratories use balances that record in "grams," and the gram is a unit of mass in the International System of Units (SI), the correct terminology for Eq. 3.2 is *density* (mass per unit volume). The next step is to convert the wet density ρ_t to total unit weight γ_t. In order to convert wet density to total unit weight in the SI, the wet density is multiplied by g (where g = acceleration of gravity = 9.81 m/sec²) to obtain the total unit weight, which has units of kN/m³. For example, in the SI, the density of water ρ_w = 1.0 g/cm³ or 1.0 Mg/m³, while the unit weight of water γ_w = 9.81 kN/m³.

In the United States Customary System, density and unit weight have exactly the same value. Thus the density of water and the unit weight of water is 62.4 pcf. However, for the density of water ρ_w, the units should be thought of as pounds-mass (lbm) per cubic foot, while for unit weight γ_w, the units are pounds-force (lbf) per cubic foot. In the United States Customary System, it is assumed that 1 lbm is equal to 1 lbf.

The total unit weight is defined as the wet soil weight per unit volume. In this and all subsequent unit weight definitions, the use of the term weight means force. In the SI, unit weight has units of kN/m³, while in the United States Customary System, the unit weight has units of pcf, but this implies pounds of force (lbf) per cubic foot.

Similar to water content, it is common to plot the data versus depth on the subsoil profile. For example, in Fig. 2.39, the total unit weight (pcf) has been plotted (open circles) versus depth and the total unit weight is about 120 pcf (19 kN/m³) for the clay.

By using the water content w of the soil and the total unit weight γ_t, the dry unit weight γ_d can be calculated, as follows:

$$\gamma_d = \frac{\gamma_t}{1+w} \qquad (3.3)$$

where γ_d = dry unit weight of the soil (pcf or kN/m³)
   γ_t = total unit weight of the soil (pcf or kN/m³)
   w = water content of the soil, expressed as a decimal (dimensionless)

The buoyant unit weight γ_b is also known as the submerged unit weight. It can be defined as follows:

$$\gamma_b = \gamma_t - \gamma_w \qquad (3.4)$$

where γ_b = buoyant unit weight of the soil (pcf or kN/m³)
   γ_t = total unit weight of the soil for a saturated condition (pcf or kN/m³)
   γ_w = unit weight of water (62.4 pcf or 9.81 kN/m³)

The buoyant unit weight can be used to determine the vertical effective stress for soil located below the groundwater table (Sec. 4.4). Note that the total unit weight γ_t used in Eq. 3.4 must be for the special case where the void spaces are completely filled with water, such as a soil specimen obtained from below the groundwater table.

3.2.3 Specific Gravity Test

The specific gravity is a dimensionless parameter that relates the density of the soil particles to the density of water. By determining the dry mass of the soil M_s and using a pycnometer to obtain the volume of the soil solids V_s, the specific gravity of the soil solids can be determined. A pycnometer can simply be a volumetric flask or stoppered bottle that has a calibration mark and a volume of at least 100 mL (see Fig. 3.2). The specific gravity of solids G_s is defined as the density of solids ρ_s divided by the density of water ρ_w, or:

FIGURE 3.2   Test apparatus for determining the specific gravity of solids. The pycnometer, which is partly filled with soil and distilled water, is shown on the left side of the photograph. A vacuum pump is shown on the right side of the photograph and it is used to remove trapped air bubbles from the soil-water solution.

$$G_s = \frac{\rho_s}{\rho_w} = \frac{M_s / V_s}{\rho_w} \qquad (3.5)$$

where G_s = specific gravity of soil solids (dimensionless)
  ρ_s = density of the soil solids (g/cm^3)
  ρ_w = density of water (1.0 g/cm^3)
  M_s = mass of soil particles used for the test (g)
  V_s = volume of the soil particles determined by using the pycnometer (cm^3)

The specific gravity test takes a considerable amount of skill and time to complete and therefore the test is often only performed on one or two representative soil samples for a given project. Then the specific gravity value is used for the remaining soil specimens that are believed to be representative of the tested soil. For further details concerning the specific gravity test, see ASTM D 854-02 (2004), "Standard Test Methods for Specific Gravity of Soil Solids by Water Pycnometer."

Rather than performing specific gravity tests, for many projects an assumed value of the specific gravity is used. Table 3.3 presents typical values and ranges of specific gravity of solids versus different types of soil minerals. Because quartz is the most abundant type of soil mineral, the specific gravity for inorganic soil is often assumed to be 2.65. For clays, the specific gravity is often assumed to be 2.70 because common clay particles, such as montmorillonite and illite, have slightly higher specific gravity values.

Soils that contain soluble soil minerals, such as halite or gypsum, should not be tested using a pycnometer. This is because the soluble soil minerals will dissolve in the distilled water when tested in the pycnometer (hence the volume of solids will be underestimated). An alternate approach has been developed for testing of soil containing soluble soil particles. Rather than using distilled water to determine the volume of soil solids, a gas pycnometer is used where a gas of known volume is placed into the pycnometer containing the dry soil. The increase in gas pressure is related to the volume of soil within the pycnometer, see ASTM D 5550-00 (2004), "Standard Test Method for Specific Gravity of Soil Solids by Gas Pycnometer." Major disadvantages of this method are the increased test difficulty and the need for specialized test apparatus, which may not be readily available.

The specific gravity laboratory test as outlined earlier (i.e., ASTM D 854-02, 2004) is only applicable for sand, silt, and clay-size particles. Oversize particles are often defined as gravel and

TABLE 3.3 Formula and Specific Gravity of Common Soil Minerals

Type of mineral	Formula	Specific gravity	Comments
Quartz	SiO_2	2.65	Silicate, most common type of soil mineral
K feldspar	$KAlSi_3O_8$	2.54–2.57	Feldspars are also silicates and are the second
Na feldspar	$NaAlSi_3O_8$	2.62–2.76	most common type of soil mineral
Calcite	$CaCO_3$	2.71	Basic constituent of carbonate rocks
Dolomite	$CaMg(CO_3)_2$	2.85	Basic constituent of carbonate rocks
Muscovite	Varies	2.76–3.0	Silicate sheet-type mineral (mica group)
Biotite	Complex	2.8–3.2	Silicate sheet-type mineral (mica group)
Hematite	Fe_2O_3	5.2–5.3	Frequent cause of reddish-brown color in soil
Gypsum	$CaSO_4 \cdot 2H_2O$	2.35	Can lead to sulfate attack of concrete
Serpentine	$Mg_3Si_2O_5(OH)_4$	2.5–2.6	Silicate sheet or fibrous type mineral
Kaolinite	$Al_2Si_2O_5(OH)_4$	2.61–2.66	Silicate clay mineral, low activity
Illite	Complex	2.60–2.86	Silicate clay mineral, intermediate activity
Montmorillonite	Complex	2.74–2.78	Silicate clay mineral, highest activity

Note: Silicates are very common and account for about 80 percent of the minerals at the Earth's surface. Data accumulated from the following sources: Lambe and Whitman (1969) and Mottana et al. (1978).

cobbles that are retained on the $^3/_4$ in. sieve (Day, 1989). For these large particles, there may be internal rock fractures, voids, or moisture trapped within the gravel and cobbles. For these large particles, test procedures and calculations can include these features and the test result is referred to as the "bulk specific gravity," commonly designated G_b, see ASTM C 127-93 (2004), "Standard Test Method for Specific Gravity and Absorption of Coarse Aggregate." Because these large particles often have internal rock fractures and voids, the value of G_b is usually less than G_s.

3.2.4   Sieve Analysis

A basic element of a soil classification system is the determination of the amount and distribution of the particle sizes in the soil. Soil classification will be discussed in Sec. 4.2. The distribution of particle sizes larger than 0.075 mm (No. 200 sieve) is determined by sieving, while a sedimentation process (hydrometer test) is used to determine the distribution of particle sizes smaller than 0.075 mm.

A sieve is a piece of laboratory equipment that consists of a pan with a screen (square woven wire mesh) at the bottom. U.S. standard sieves are used to separate particles of a soil sample into various sizes. A sieve analysis is performed on dry soil particles that are larger than the No. 200 U.S. standard sieve (i.e., sand size, gravel size, and cobble size particles).

Sieves are designated in two different ways. The sieves having the largest openings are designated by their sieve openings in inches (such as the 4-in sieve, 3-in. sieve, 2-in. sieve, 1-in. sieve, $^3/_4$ in. sieve, and the $^3/_8$ in. sieve). For example, the 4-in. U.S. standard size sieve has square openings that are 4 in. (100 mm) wide. The second way that sieves are designated is by their U.S. sieve number. This identification is used for the finer sieves and it refers to the number of opening per inch. For example, a No. 4 sieve has four openings per inch, which are 0.19 in. (4.75 mm) wide. A common mistake is that the No. 4 sieve has openings 0.25 in. wide (i.e., 1.0 in. divided by 4), but because of the wire mesh, the openings are actually less than 0.25 inch. Commonly used U.S. standard sieve numbers versus their sieve opening are as follows:

No. 4 U.S. Standard Sieve	sieve opening = 4.75 mm
No. 10 U.S. Standard Sieve	sieve opening = 2.00 mm
No. 20 U.S. Standard Sieve	sieve opening = 0.85 mm
No. 40 U.S. Standard Sieve	sieve opening = 0.425 mm
No. 60 U.S. Standard Sieve	sieve opening = 0.25 mm
No. 100 U.S. standard Sieve	sieve opening = 0.15 mm
No. 140 U.S. Standard Sieve	sieve opening = 0.106 mm
No. 200 U.S. Standard Sieve	sieve opening = 0.075 mm

The laboratory test procedures for performing a sieve analysis are presented in ASTM D 422-02 (2004), "Standard Test Method for Particle-Size Analysis of Soils." The basic steps include first determining the initial dry mass of the soil (M_s). Then the soil is washed on the No. 200 sieve in order to remove all the fines (i.e., silt and clay size particles). Special No. 200 sieves are available that have high collars that facilitate the washing of the soil on the No. 200 sieve. The purpose of the washing of the soil on the No. 200 sieve is to ensue that all the fines and surface coatings are washed-off of the granular soil particles. Failure to use washed soil for the sieve analysis can lead to totally misleading sieve results (Rollins and Rollins, 1996).

The next step is to oven dry the soil retained on the No. 200 sieve. Then a stack of dry and clean sieves is assembled, such as illustrated in Fig. 3.3. The sieves are usually assembled so that the opening in any sieve screen is approximately double that of the next-finer (lower) screen. The top sieve should have openings that are large enough so that all of the soil particles will fall through the sieve. The dry soil is poured into the top of the stack of sieves, a top lid is installed, and the sieves are shaken. The stack of sieves can be shaken manually, although it is much easier to use a mechanical shaker, such as shown in Fig. 3.4. After shaking, a balance is used to determine the mass of soil retained on

Dry soil placed in top sieve

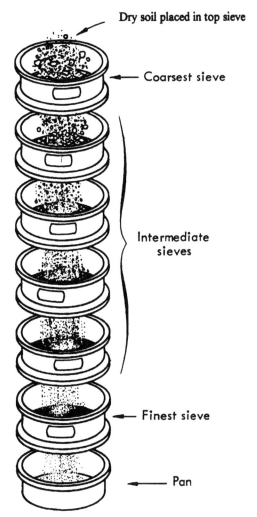

Coarsest sieve

Intermediate sieves

Finest sieve

Pan

FIGURE 3.3 Diagram illustrating the stacking of sieves. (*Adapted from The Asphalt Handbook, 1989.*)

each sieve. The percent finer F by dry weight, also known as the percent passing, is then calculated as follows:

$$F = 100 - 100 \left(\frac{R_{DS}}{M_S} \right)$$
(3.6)

where F = percent dry soil passing a particular sieve

R_{DS} = cumulative amount of dry soil retained on a given sieve (lb or g). This is calculated for a particular sieve by adding the mass of the soil retained on that sieve and the mass of the soil retained on all the coarser sieves

M_s = initial dry mass of the soil, obtained at the start of the test (lb or g)

FIGURE 3.4   Mechanical shaker.

3.2.5   Hydrometer Test

A sedimentation process is used to determine the particle distribution for fines (i.e., silt and clay size particles finer than the No. 200 sieve). A hydrometer is used to obtain the necessary data during the sedimentation process. The hydrometer test is based on Stokes law, which relates the diameter of a single sphere to the time required for the sphere to fall a certain distance in a liquid of known viscosity. The idea for the hydrometer test is that a larger, and hence heavier, soil particle will fall faster through distilled water than a smaller, and hence lighter, soil particle. The test procedure is approximate because many fine soil particles are not spheres, but are rather of a plate-like shape. Thus while the sieve analysis uses the size of a square sieve opening to define particle size, the hydrometer test uses the diameter of an equivalent sphere as the definition of particle size.

If the amount of fines is less than 5 percent (i.e., percent passing No. 200 sieve is less than 5 percent), typically a hydrometer test is not performed. Likewise, if the percent passing the No. 200 sieve is between 5 percent and 15 percent, the soil may be nonplastic and once again a hydrometer test may be unnecessary for classifying the soil. Usually if the percent passing the No. 200 sieve is greater than 15 percent, a hydrometer test could be performed. The test procedure is as follows:

Preparation of Soil.   The first step in the hydrometer test is to obtain a representative soil sample, i.e., the same soil that was used for the sieve analysis. Then the larger soil particles are removed (i.e., plus No. 40 sieve material). For the hydrometer test, it is desirable to have about 50 g of soil finer than the No. 40 sieve if it consists primarily of fines and about 100 g if it consists mostly of coarse-grained particles (i.e., sand particles and fines). A mass of 5.0 g of sodium hexametaphosphate is then added to the pan of soil and distilled water is added.

Mixing of Soil.   The water, soil, and sodium hexametaphosphate are thoroughly mixed and allowed to soak overnight. The purpose of the sodium hexametaphosphate is to act as a dispersing agent that prevents the clay size particles from forming flocs during the hydrometer test. At the end of the soaking period, a mechanical mixer is used to further disperse the soil-water-sodium hexametaphosphate slurry.

Sedimentation Process.   After the mixing is complete, all of the soil-water-sodium hexametaphosphate slurry is transferred to the 1000 mL glass sedimentation cylinder and distilled water is added to the 1000 mL mark. A rubber stopper is placed on the open end of the cylinder, and then the cylinder is shaken for a period of about 1 min to complete the soil dispersion process. As soon as the cylinder is set down, the hydrometer is inserted into the cylinder containing dispersed soil and readings are taken at 1, 2, 5, 15, 30, 60, 250, and 1440 min (see Fig. 3.5).

Calculations of Percent Finer.   A 152H hydrometer directly reads the mass of dissolved solids and soil particles in suspension. The calculations for percent finer are rather detailed because of temperature, specific gravity, and other required corrections. For the corrections and calculations as well as further details on the hydrometer test, see ASTM D 422-02 (2004), "Standard Test Method for Particle-Size Analysis of Soils."

As previously mentioned, the hydrometer test is based on Stokes law that assumes that the soil particles are spherical. Those soil particles that are smaller than about 0.005 mm are usually plate-shaped and can have a length or width that is from five to several hundred times their thickness. The fall of a plate-shaped soil particle through water has been described as somewhat like the downward drifting of a leaf from a tree (Lambe, 1951). Because the plate-shaped soil particles tend to stay in solution, the clay size fraction is typically overestimated by the hydrometer test.

Figure 3.6 shows the results of sieve and hydrometer tests performed on a soil. The plot shown in Fig. 3.6 is termed the "grain size curve," also known as the "particle size distribution." The sieve and hydrometer portions of the grain size curve are shown on the top of the graph. For the sieve analysis, the "percent finer" [Eq. (3.6)] is plotted for a corresponding sieve opening. For the hydrometer test, the percent finer is plotted for a corresponding soil grain size.

The grain size curve shown in Fig. 3.6 was actually obtained from a computer program (gINT, 1991). The raw data from the sieve and hydrometer tests were inputted into the computer program and the grain size curve and analysis of data were outputted. In Fig. 3.6, the particle sizes for *cobbles, gravel, sand,* and *silt or clay* are listed on the plot for easy reference.

At the bottom of Fig. 3.6, the computer program (gINT, 1991) has performed an analysis of the sieve and hydrometer tests. The percent gravel size particles (0 percent), sand size particles (58.1 percent), silt size particles (16.2 percent), and clay size particles (25.7 percent) have been calculated. The computer program also determines the D_{100}, D_{60}, D_{30}, and D_{10} particle sizes. In Fig. 3.6, the D_{100} is the largest particle size recorded (4.75 mm), the D_{60} is the particle size corresponding to

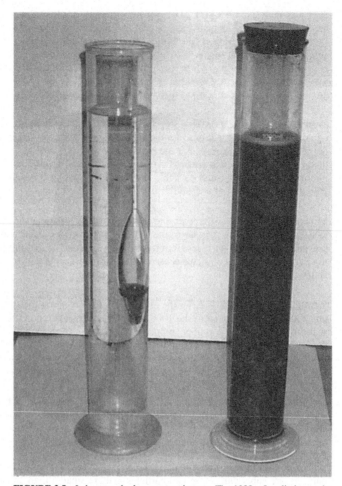

FIGURE 3.5   Laboratory hydrometer equipment. The 1000-mL cylinder on the left contains the hydrometer suspended in water; the 1000-mL cylinder on the right contains a dispersed soil specimen to be tested.

60 percent finer by dry weight (0.27 mm), D_{30} is the particle size corresponding to 30 percent finer by dry weight (0.006 mm), and D_{10} is the particle size corresponding to 10 percent finer by dry weight. Because of the presence of over 10 percent clay size particles for the soil data shown in Fig. 3.6, D_{10} could not be obtained for this soil.

If the sieve and the hydrometer tests are performed correctly, the portion of the grain size curve from the sieve analysis should merge smoothly into the portion of the curve from the hydrometer test, such as shown in Fig. 3.6. A large and abrupt jump in the grain size curve from the sieve to the hydrometer test indicates errors in the laboratory testing procedure.

3.2.6   Atterberg Limits Tests

The term plasticity is applied to silts and clays and indicates an ability to be rolled and molded without breaking apart. The Atterberg limits are defined as the water content corresponding to different behavior conditions of silts and clays. Although originally six limits were defined by Albert

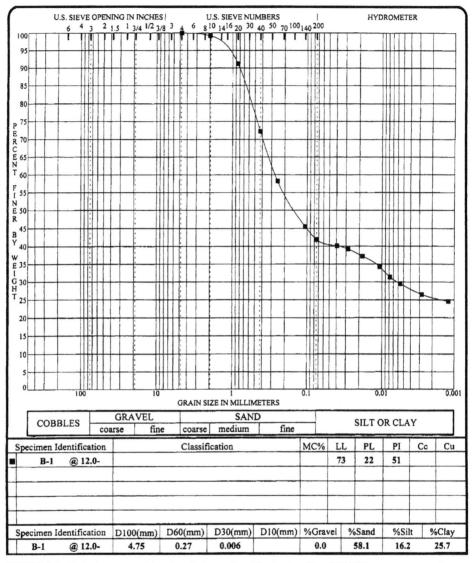

Specimen Identification	Classification	MC%	LL	PL	PI	Cc	Cu
■ B-1 @ 12.0-			73	22	51		

Specimen Identification	D100(mm)	D60(mm)	D30(mm)	D10(mm)	%Gravel	%Sand	%Silt	%Clay
B-1 @ 12.0-	4.75	0.27	0.006		0.0	58.1	16.2	25.7

FIGURE 3.6   Grain size curve and Atterberg limits test data (plot developed by gINT 1991 computer program).

Atterberg (1911), in geotechnical engineering, the term Atterberg limits only refers to the liquid limit (LL), plastic limit (PL), and shrinkage limit (SL), defined as follows:

Liquid Limit (LL).   The water content corresponding to the behavior change between the liquid and plastic state of a silt or clay. The liquid limit is determined by spreading a pat of soil in a brass cup, dividing it in two by use of a grooving tool, and then allowing it to flow together from the shock caused by repeatedly dropping the cup in a standard liquid limit device (see Fig. 3.7). In terms of specifics, the liquid limit is defined as the water content at which the pat of soil cut by the grooving tool will flow together for a distance of 0.5 in. (12.7 mm) under the impact of 25 blows in a standard

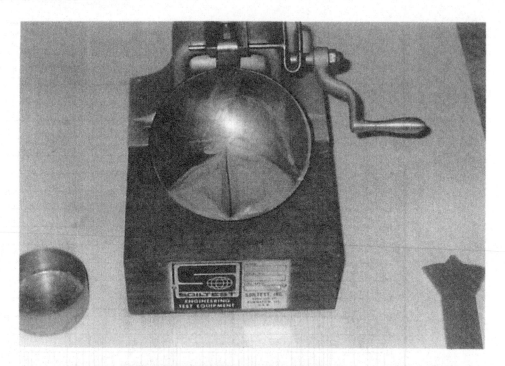

FIGURE 3.7 Liquid limit test. The upper photograph shows the liquid limit device containing soil with groove cut through the soil (the grooving tool is on the right side of the photograph). The lower photograph shows the soil after it has been tested (i.e., after the liquid limit cup has been raised and dropped).

liquid limit device. For laboratory testing details, see ASTM D 4318-00 (2004), "Standard Test Methods for Liquid Limit, Plastic Limit, and Plasticity Index of Soils."

Plastic Limit (PL).   The water content corresponding to the behavior change between the plastic and semi-solid state of a silt or clay. The plastic limit is determined by pressing together and rolling a small portion of the plastic soil so that its water content is slowly reduced with the end result that the thread of soil crumbles apart (see Fig. 3.8). In terms of specifics, the plastic limit is defined as the water content at which a silt or clay will just begin to crumble when rolled into a thread approximately 1/8 in. (3.2 mm) in diameter. For laboratory testing details, see ASTM D 4318-00 (2004), "Standard Test Methods for Liquid Limit, Plastic Limit, and Plasticity Index of Soils."

Shrinkage Limit (SL).   The water content corresponding to the behavior change between the semi-solid to solid state of a silt or clay. The shrinkage limit is also defined as the water content at which any further reduction in water content will not result in a decrease in volume of the soil mass. The shrinkage limit is rarely obtained in practice because of laboratory testing difficulties and limited use of the data. For testing details, see ASTM D 427-98 (2004), "Standard Test Method for Shrinkage Factors of Soils by the Mercury Method," or D 4943-02, 2004, "Standard Test Method for Shrinkage Factors of Soils by the Wax Method."

If the soil is nonplastic, the Atterberg limits tests are not performed. In many cases, it is evident that the soil is nonplastic because it cannot be rolled or molded. However, some soils may be incorrectly labeled as nonplastic because they are dry and crumbly. Water should be added to such soils to confirm that they cannot be rolled or molded at any water content. It is also possible that after completion of the Atterberg limits test, the final result is that the plastic limit (PL) is equal to or greater than the liquid limit (LL). This soil should also be classified as nonplastic.

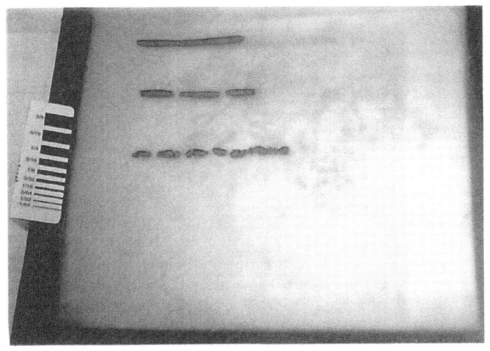

FIGURE 3.8 Plastic limit test. The upper thread of soil is still too wet, the middle thread of soil is approaching the plastic limit, and the lower thread of soil has been dried sufficiently and has reached the plastic limit.

According to ASTM, the LL and PL must be performed on that portion of the soil that passes the No. 40 sieve (0.425 mm). For many soils, a significant part of the soil specimen (i.e., those soil particles larger than the No. 40 sieve) will be excluded during testing. As indicated in Fig. 3.6, the LL and PL tests were performed for the portion of this soil passing the No. 40 sieve and the results are LL = 73 (i.e., the water content at 25 blows is 73 percent) and PL = 22 (i.e., the water content where the 1/8 in. thread of soil crumbles = 22 percent). The LL and PL are typically reported to the nearest whole number and the percent designation is omitted.

The plasticity index (PI) is defined as the liquid limit minus the plastic limit, or:

$$PI = LL - PL \qquad (3.7)$$

where PI = plasticity index of a cohesive soil
LL = liquid limit determined from the liquid limit test
PL = plastic limit determined from the plastic limit test

In Fig. 3.6, the PI is also indicated and is the LL minus the PL, or 51.

3.3 OEDOMETER TEST

The oedometer (also known as a consolidometer) is the primary laboratory equipment used to study the settlement and expansion behavior of soil. The oedometer test should only be performed on undisturbed soil specimens, or in the case of studies of fill behavior, on specimens compacted to anticipated field and moisture conditions.

The first step in the laboratory testing using the oedometer apparatus is to trim the soil specimen. In some cases, the undisturbed soil specimen can be trimmed directly from a block of soil by using a sharp cutting ring (arrow 1 in Fig. 3.9). In other cases, the undisturbed soil specimen can be obtained

FIGURE 3.9 Specimen preparation. Arrow 1 points to a cutting ring which can be used to trim a soil specimen from an undisturbed block of soil. Arrow 2 points to an undisturbed soil specimen extruded from a soil sampler directly into a confining ring.

by extruding the soil specimen from the soil sampler directly into a confining ring and then trimming the top and bottom of the specimen so that it is flush with the ring (arrow 2 in Fig. 3.9). The trimming of the soil specimen can be performed with a wire saw, sharp straight-edge knife, or a sharp putty knife. Soft soils are often effectively trimmed with a wire saw but stiffer soils usually require the use of a sharp straight-edge knife or sharp putty knife. Using the best cutting tool to minimize sample disturbance often requires considerable experience and judgment.

Once the soil specimen has been trimmed, the next step is to place the soil specimen in the oedometer apparatus. The equipment may vary, but in general, an oedometer consists of the following:

1. A metal ring that is used to laterally confine the soil specimen. For example, the undisturbed soil specimen in Fig. 3.9 is laterally confined by a metal ring. Figure 3.10 shows a diagram of the two basic types of soil specimen arrangements. For the fixed ring oedometer, the soil specimen will

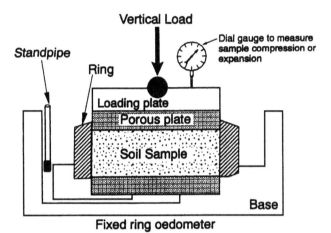

Fixed ring oedometer

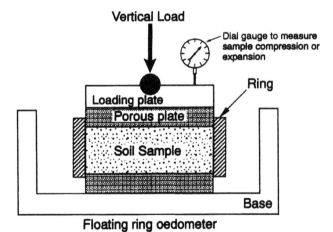

Floating ring oedometer

FIGURE 3.10   Fixed and floating ring oedometer apparatus. *Note*: Although not shown above, the dial gauge should be positioned at the center of the loading plate.

be compressed from the top downward as the load is applied, while for the floating ring oedometer, the soil specimen is compressed inward from the top and bottom. An advantage of the floating ring oedometer is that there is less friction between the confining ring and the soil. Disadvantages of the floating ring are that it is often more difficult to set up and soil may squeeze or fall out of the junction of the bottom porous plate and ring. Because of these disadvantages, the fixed ring oedometer is the most popular testing setup.

FIGURE 3.11 Loading device. Arrow 1 points to the soil specimen located at the center of the equipment, arrow 2 points to the loading arm located at the centerline of the equipment, arrow 3 points to a wheel that is rotated in order to level the loading arm, and arrow 4 points to the dial gauge that is used to measure the deformation of the soil specimen as it is loaded or unloaded.

2. Dry porous plates are placed on the top and bottom of the soil specimen (see Fig. 3.10). The porous plates must be clean (not clogged with soil) and the porous plates must have a high permeability to allow water to quickly flow through the plates and into or out of the soil specimen. Filter paper can be placed between the soil and porous plates in order to prevent intrusion of fines into the porous plates. The diameter of the porous plates should be 0.01 to 0.02 in. (0.2 to 0.5 mm) less than the inside diameter of the confining ring. If a floating ring oedometer is used, then the top and bottom plates should have the same diameter.

3. The soil specimen having porous plates on the top and bottom is then placed into a surrounding container. The purpose of the surrounding container is to allow the soil specimen to be submerged in distilled water during testing. A popular type of surrounding container is a Plexiglas dish.

4. A loading device is used to apply a concentric vertical load to the soil specimen. In general, a loading device must meet two criteria: (1) it must be able to apply a constant and concentric vertical load for a long period of time and (2) it must apply the vertical load quickly, but without inducing an impact load upon the soil. Figure 3.11 shows one type of loading device that meets these two criteria. In Fig. 3.11, arrow 1 points to the Plexiglas dish containing the laterally confined soil specimen with top and bottom porous plates (from step 3) that has been placed at the center of the testing apparatus. Arrow 2 in Fig. 3.11 points to a loading arm (located at the centerline of the equipment) upon which weights are gently placed in order to apply a vertical stress to the top of the soil specimen. Arrow 3 in Fig. 3.11 points to a wheel that is rotated in order to level the loading arm and maintain the load for a long period of time. The vertical stress σ_v applied to the soil is equal to P/A, where P = vertical concentric applied load and A = area of the soil specimen.

5. A dial gauge is used to measure the vertical deformation of the soil specimen as it is loaded or unloaded. Arrow 4 in Fig. 3.11 points to the dial gauge.

The oedometer test is popular because of its simplicity and it can be used to model and predict the behavior of the in situ soil. For example, a soil specimen can be placed in the oedometer and then subjected to an increase in pressure equivalent to the weight of the proposed structure. By analyzing the settlement versus load data, the geotechnical engineer can calculate the amount of expected settlement due to the weight of the proposed structure. The oedometer can be used to study the settlement behavior of collapsible soil (Chap. 7) and the consolidation of saturated clays (Chap. 8). As will be discussed in Chap. 9, the oedometer can also be used to predict the amount of heave of expansive soil.

3.4   SHEAR STRENGTH OF COHESIONLESS SOIL

3.4.1   Introduction

An understanding of the shear strength of soil is essential in foundation engineering. This is because most geotechnical failures involve a shear type failure of the soil. This is due to the nature of soil, which is composed of individual soil particles that slide (i.e., shear past each other) when the soil is loaded. The shear strength of soil is required for many different types of engineering analyses, such as the bearing capacity of shallow and deep foundations, slope stability analyses, and the design of retaining walls.

The mechanisms that control the shear strength of soil are complex, but in simple terms the shear strength of soils can be divided into two broad categories: (1) cohesionless soils, also known as nonplastic or granular soils, and (2) cohesive soils, also known as plastic soils. This section will discuss the shear strength of cohesionless soil and Sec. 3.5 will be devoted to the shear strength of cohesive soil.

The shear strength testing should be performed on saturated soil specimens. This is because the shear strength testing of partially saturated soil could overestimate the shear strength if the soil should become wetter. For example, it has been stated (Coduto, 1994):

The shear strength of a partially saturated soil is higher than if it was saturated. However, do not rely on this additional strength because of the possibility that the soil may become wetted sometime in the future. Therefore, it is usually best to soak all soil samples in the laboratory before testing them.

When conducting drained tests, this additional strength will be manifested as cohesion, so be cautious about using cohesive strength except in overconsolidated clays or in soils cemented with non–water-soluble agents. When conducting undrained tests, strength gains due to partial saturation are more insidious because they increase the measured s_u. Therefore, it is appropriate to determine the degree of saturation of laboratory test samples after they have been soaked to verify that they have been wetted sufficiently to match the worst-case condition that might appear in the field.

To fully understand shear strength testing, the concept of effective stress will first be introduced. The effective stress is defined as:

$$\sigma' = \sigma - u \tag{3.8}$$

where σ' = effective stress (psf or kPa)
σ = total stress (psf or kPa)
u = pore water pressure (psf or kPa)

In shear strength testing, the total stress acting on the soil specimen can be determined as the load divided by the area over which it acts. The pore water pressure in the soil is typically assumed to be equal to zero in the case of a saturated sand that is slowly sheared in direct shear apparatus or measured by a pore water pressure transducer in the case of a triaxial test on cohesive soil. The concept of effective stress is also used for field applications and will be further discussed in Sec. 4.4.

The shear strength of the soil can be defined as (Mohr-Coulomb failure law):

$$\tau_f = c' + \sigma'_n \tan \phi' \tag{3.9}$$

where τ_f = shear strength of the soil (psf or kPa)
c' = effective cohesion (psf or kPa)
σ'_n = effective normal stress acting on the shear surface (psf or kPa). The shear surface is also referred to as the slip surface or failure plane.
ϕ' = effective friction angle, also known as the angle of internal friction (degrees)

The effective cohesion c' and the effective friction angle ϕ' are known as the "shear strength parameters" of the soil. In essence, the shear strength parameters indicate how strong the soil will be when subjected to a shear stress. The higher the values of c' and ϕ', the higher the shear strength of the soil.

Nonplastic soils are known as cohesionless soils because there is no cohesion acting between the soil particles. Thus for cohesionless soil, $c' = 0$ and Eq. 3.9 reduces to:

$$\tau_f = \sigma'_n \tan \phi' \tag{3.10}$$

Cohesionless soils include gravels, sands, and nonplastic silt such as rock flour. A cohesionless soil develops its shear strength as a result of the frictional and interlocking resistance between the individual soil particles. Cohesionless soils can only be held together by confining pressures and will fall apart when the confining pressure is released.

Equation 3.10 indicates that in order to determine the shear strength τ_f of a cohesionless soil, two parameters need to be determined:

Effective Normal Stress.   The effective normal stress σ'_n is the effective stress that is acting perpendicular to the shear surface, i.e., it is the stress that is normal to the shear surface. The effective normal stress σ'_n can be determined from basic geotechnical engineering principles. For example, if it is desirable to determine the effective normal stress σ'_n on a horizontal shear plane in a soil deposit, then σ'_n would simply be equal to the vertical effective stress σ'_{vo} at that depth.

Effective Friction Angle.   The effective friction angle ϕ' is an intrinsic property of the soil and can be determined from laboratory or field testing.

Although $c' = 0$ is used for cohesionless soils, an exception is the testing of cohesionless soil at high normal pressures, where the shear strength envelope may actually be curved because of particle crushing (Holtz and Gibbs, 1956a). In this case, a straight line approximation at high normal stresses may indicate a cohesion intercept, but this value should be regarded as an extrapolated value that is not representative of the shear strength of cohesionless soils at low values of σ'_n.

During shear of cohesionless soils, those soils in a dense state will tend to dilate (increase in volume), while those soils in a loose state tend to contract (decrease in volume). Cohesionless soils have a high permeability and for the shear strength testing of saturated cohesionless soils, water usually flows quickly into the soil when it dilates or out of the soil when it contracts. Thus the effective shear strength, also known as the drained shear strength, is of most importance for cohesionless soils. An important exception is the liquefaction of saturated and loose cohesionless soils which will be discussed in Chap. 13.

The shear strength of cohesionless soils can be measured in the direct shear apparatus. There can be a small capillary tension in cohesionless soils and thus the soil specimen is saturated prior to shearing. Because of these test specifications which require the direct shear testing of soil in a saturated and drained state, the shear strength of the soil is expressed in terms of the effective friction angle ϕ'. Cohesionless soils can also be tested in a dry state and the shear strength of the soil is then expressed in terms of the friction angle ϕ. In a comparison of the effective friction angle ϕ' from drained direct shear tests on saturated cohesionless soil and the friction angle ϕ from direct shear tests on the same soil in a dry state, it has been determined that ϕ' is only $1°$ to $2°$ lower than ϕ (Terzaghi and Peck, 1967; Holtz and Kovacs, 1981). This slight difference is usually ignored and the friction angle ϕ and effective friction angle ϕ' are typically considered to mean the same thing for cohesionless soils.

Table 3.4 presents values of effective friction angles for different types of cohesionless soils. An exception to the values presented in Table 3.4 are cohesionless soils that contain appreciable mica flakes. A micaceous sand will often have a high void ratio and hence little interlocking and a lower friction angle (Horn and Deere, 1962).

For many projects, the effective friction angle ϕ' of a sand deposit is determined from laboratory testing, such as using the direct shear test which will be discussed in the next section. In other cases, indirect means, such as the standard penetration test (SPT) and the cone penetration test (CPT) can be used to estimate the effective friction angle of the soil. As indicated in Figs. 3.12 and 3.13, the effective friction angles ϕ' for clean quartz sand can be estimated from the results of SPT or CPT for various values of vertical effective stress σ'_{vo}.

Another useful chart is shown in Fig. 3.14. In this chart, the effective friction angle ϕ' can be estimated based on the soil type and using either the dry unit weight (γ_d, see Sec. 3.2.2) or the relative density (D_r, see Sec. 4.3).

TABLE 3.4 Typical Effective Friction Angles ϕ' for Different Cohesionless Soils

Soil types	Effective friction angles ϕ' at peak strength		Effective friction angle ϕ'_u at ultimate strength*
	Medium	Dense	
Silt (nonplastic)	28°–32°	30°–34°	26°–30°
Uniform fine to medium sand	30°–34°	32°–36°	26°–30°
Well-graded sand	34°–40°	38°–46°	30°–34°
Sand and gravel mixtures	36°–42°	40°–48°	32°–36°

*The effective friction angle ϕ'_u at the ultimate shear strength state could be considered to be the same as the friction angle ϕ' for the same soil in a loose state.

Source: Hough 1969.

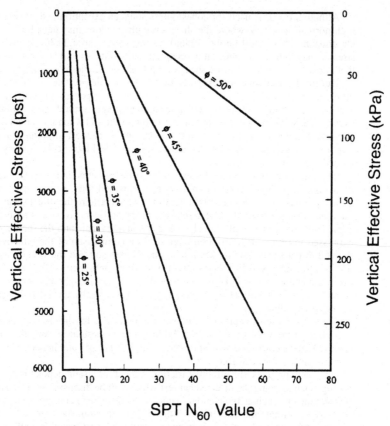

FIGURE 3.12   Empirical correlation between standard penetration test (SPT) N_{60} value, vertical effective stress, and friction angle for clean quartz sand deposits. *(Adapted from de Mello, 1970; reproduced from Coduto, 1994.)*

    In summary, for the shear strength of cohesionless soils, $c' = 0$ and the effective friction angle ϕ' depends on:

Soil type (Table 3.4).   Sand and gravel mixtures have a higher effective friction angle than non-plastic silts.

Soil density.   For a given cohesionless soil, the denser the soil, the higher the effective friction angle. This is due to the interlocking of soil particles, where at a denser state the soil particles are interlocked to a higher degree and hence the effective friction angle is greater than in a loose state. It has been observed that in the ultimate shear strength state, that the shear strength and density of a loose and dense sand tend to approach each other.

Grain size distribution.   A well graded cohesionless soil will usually have a higher friction angle than a uniform soil. With more soil particles to fill in the small spaces between soil particles, there is more interlocking and frictional resistance developed for a well graded than for a uniform cohesionless soil.

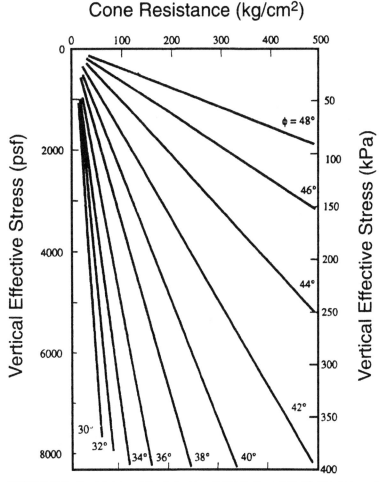

FIGURE 3.13 Empirical correlation between cone resistance, vertical effective stress, and friction angle for clean quartz sand deposits. *Note:* 1 kg/cm² approximately equals 1 tsf. *(Adapted from Robertson and Campanella, 1983; reproduced from Coduto, 1994.)*

Mineral type, angularity, and particle size.   Soil particles composed of quartz tend to have a higher friction angle than soil particles composed of weak carbonate. Angular soil particles tend to have rougher surfaces and better interlocking ability. Larger size particles, such as gravel size particles, typically have higher friction angles than sand.

Deposit variability.   Because of variations in soil types, gradations, particle arrangements, and dry density values, the effective friction angle is rarely uniform with depth. It takes considerable judgment and experience in selecting an effective friction angle.

Indirect methods.   For many projects, the effective friction angle of the sand is determined from indirect means, such as the SPT, Fig. 3.12 and the CPT, Fig. 3.13. Another useful chart is presented in Fig. 3.14, which correlates the effective friction angle with the soil type and dry unit weight or relative density.

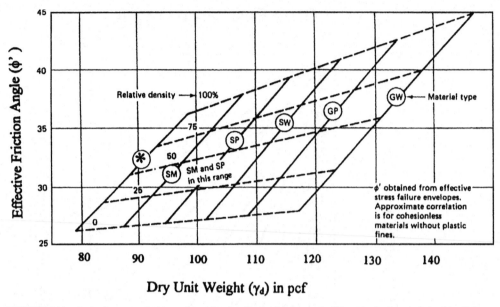

FIGURE 3.14 Approximate correlations to determine the effective friction angle ϕ' for cohesionless (nonplastic) soil. Enter the figure with the dry unit weight γ_d, intersect the soil type (Unified Soil Classification System) to determine ϕ'. As an alternative, enter the chart with the relative density and soil type to determine ϕ'. Asterisk indicates line for nonplastic silt, such as rock flour. (*Adapted from NAVFAC DM-7.1, 1982.*)

3.4.2 Direct Shear Test

The shear strength of cohesionless soil can be determined by using Eq. 3.10. The previous section presented typical values of effective friction angles for cohesionless soil (Table 3.4) and indirect methods for determining ϕ' from SPT, CPT, or based on soil classification and index properties. This section describes the determination of the shear strength of cohesionless soils based on laboratory testing. The most common laboratory test used to determine the effective friction angle is the direct shear test.

The direct shear test, first used by Coulomb in 1776, is the oldest type of shear testing equipment. The direct shear test is also the most common laboratory equipment used to obtain the drained shear strength (shear strength based on effective stress) of a cohesionless soil. Figure 3.15 presents an illustration the direct shear testing device. The purpose of the direct shear test is to literally shear the soil specimen in half along a horizontal failure surface.

As indicted in Fig. 3.15, the direct shear apparatus has the means for applying a vertical load, also known as the normal load. This vertical load can be converted into a pressure by taking the vertical load divided by the soil specimen area, and this pressure is often referred to as the normal stress or vertical pressure. As shown in Fig. 3.15, porous plates are also placed on both the top and bottom of the soil specimen to allow for migration of water into or out of the soil specimen. The direct shear box is usually circular and has two halves of equal thickness which are fitted together with alignment pins. The lower half of the direct shear box is firmly anchored, while the upper half of the direct shear box has the ability to be deformed laterally. By applying a horizontal force to the top half of the direct shear box, the soil specimen is sheared in half along a horizontal failure plane. Dial gauges are used to measure both the vertical and horizontal deformation of the soil specimen during the shearing process.

Although not shown in Fig. 3.15, the direct shear apparatus must also have provisions for allowing the soil specimen to be submerged in distilled water. To prevent corrosion of the apparatus, the

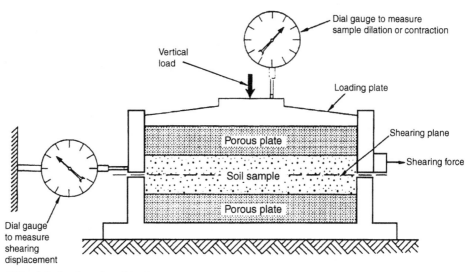

FIGURE 3.15   Illustration of the direct shear test.

equipment should consist of stainless steel, bronze, or aluminum and dissimilar metals are not permitted because they could lead to galvanic corrosive action.

In order to determine the effective friction angle ϕ' of the cohesionless soil, usually at least three soil specimens are sheared at different vertical pressures. It is important that the range in vertical pressures used during the direct shear testing is approximately the same as the range in pressure applicable for the field conditions. The test procedures for performing the direct shear test are listed in ASTM D 3080-03 (2004), "Standard Test Method for Direct Shear Test of Soils Under Consolidated Drained Conditions." The test procedures basically consist of the following:

1. *Direct shear box.*   The undisturbed soil specimen is placed in a direct shear box, such as shown in Fig. 3.16.
2. *Shearing apparatus.*   The direct shear box, containing the soil specimen, is then transferred to the shearing apparatus and locking screws are used to ensure that the bottom half of the direct shear box will not move when the shearing force is applied to the upper half of the box. Although the equipment shown in Fig. 3.17 has a removable direct shear box, for other types of direct shear equipment, the lower half of the direct shear box may be permanently attached to the base of the apparatus.
3. *Normal load.*   After the direct shear box is in place, a vertical load (i.e., normal load) is applied to the soil specimen, such as shown in Fig. 3.17. The vertical load is usually applied by placing weights on a hanger. Dial gauges are set-up to measure both the vertical and horizontal deformation of the soil specimen during the shearing operation. After the vertical load has been applied, the soil specimen is submerged in distilled water and allowed to equilibrate.
4. *Shearing.*   The final step is to apply a shear force to the upper half of the box, by either of the following methods:
   a. *Controlled strain test.*   This is the most common type of test where the shear force is applied so as to control the displacement rate. The objective is to shear the soil at a relatively uniform rate of horizontal displacement. Often this is achieved by using an electric motor and gear box arrangement. The shear force is determined by a load indicator device, such as a load cell or proving ring. The load cell or proving ring should be able to accurately measure the shear force to within 1 percent of the shear force at failure. The soil specimen should be sheared until the horizontal displacement is from 10 to 20 percent of the original specimen diameter. Thus for a typical diameter of 2.5 in. (6.4 cm), the soil specimen is sheared until the horizontal deformation is at least 0.25 in. (0.64 cm).

FIGURE 3.16   Soil specimen that has been inserted into the direct shear box. The arrow points to an alignment pin that holds together the two halves of the direct shear box.

b. Controlled stress test.   In this case, the shear force is applied in increments. A cable is attached to the upper half of the direct shear box. The cable is strung over a wheel and attached to a hanger. The shear force is then equal to the weight of the hanger and the loads applied to the hanger. The shear force is slowly increased by adding weights to the hanger. As the horizontal deformation of the soil specimen increases (i.e., approaching the shear strength of the soil), then progressively lighter weights should be applied to the hanger. When the shear strength of the soil is reached, the soil specimen often quite suddenly shears in half.

During the shearing of the soil specimen, the information recorded includes the dial gauge readings for the horizontal and vertical deformation of the soil specimen as well as the data from the proving ring, load cell, or hanger weights that can be used to calculate the shear force. Also, the shearing of the soil specimen must be slow enough that excess pore water pressures do not develop within the soil specimen. This means that if the soil specimen wants to dilate during shear (such as a dense soil specimen), then adequate time must be available to allow for the water to flow into the soil specimen. Likewise, if the soil specimen wants to contract during shear (such as a loose soil specimen), then adequate time must be available for water to flow out of the soil specimen. To enable this drained condition of the soil specimen during shearing, the total elapsed time to failure (i.e., the time from the start of shearing to the maximum applied shear force), should be as follows:

1. For clean sand having less than 5 percent nonplastic fines, the total elapsed time to failure should be at least 10 min.

2. For silty sand having greater than 5 percent nonplastic fines, the total elapsed time to failure should be at least 60 min.

3. For nonplastic silt, the total elapsed time to failure should be about 2 to 3 h.

FIGURE 3.17   Direct shear apparatus. Arrow 1 points to a locking screw which holds in place the lower half of the shear box. Arrow 2 points to the hanger which is used to apply a vertical load to the soil specimen. Arrow 3 points to a dial gauge used to record the vertical deformation of the soil specimen during shearing. Arrow 4 points to a loading ram used to apply a horizontal force to the upper half of the shear box.

Figure 3.18 presents the results of drained direct shear tests performed on undisturbed samples of silty sand. Undisturbed samples were obtained from two borings (B-1 and B-2) excavated in the silty sand deposit. A total of six direct shear tests were performed. For each direct shear test, the effective normal stress was based on the in situ vertical overburden pressure. The undisturbed soil specimen location and the vertical pressure (i.e., normal stress) during the direct shear tests were as follows:

Location	Vertical pressure (psf)
Boring B-1 at a depth of 3.5 ft	425
Boring B-1 at a depth of 6 ft	725
Boring B-2 at a depth of 2.5 ft	300
Boring B-2 at a depth of 4.5 ft	550
Boring B-2 at a depth of 6.5 ft	775
Boring B-2 at a depth of 10 ft	1250

During the shearing of the silty sand specimens in the direct shear apparatus, the shear stress (which equals the shearing force divided by the specimen area) versus lateral deformation was recorded and is plotted for two of the tests in Fig. 3.19. As shown in Fig. 3.19, the drained direct

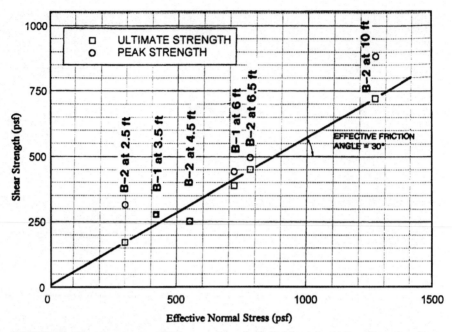

FIGURE 3.18 Shear strength versus effective normal stress for drained direct shear tests on silty sand specimens.

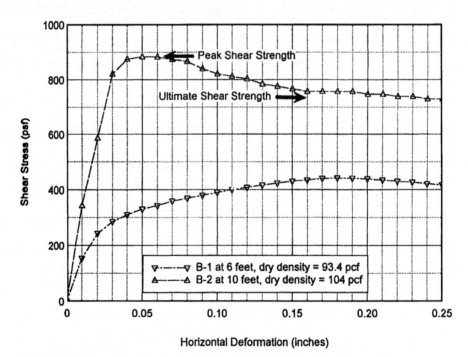

FIGURE 3.19 Shear stress versus horizontal deformation from the drained direct shear tests on silty sand specimens from a depth of 6 ft and 10 ft.

shear test performed on the silty sand at B-2 at 10 ft (3 m) has a distinct peak in the curve, while the drained direct shear test on the silty sand at B-1 at 6 ft (1.8 m) does not exhibit this distinctive peaking of the curve. This is due to the dry density of the soil specimens, where the specimen at 10 ft is in a denser state (γ_d = 104 pcf, 16.4 kN/m³) than the silty sand at a depth of 6 ft (γ_d = 93.4 pcf, 14.7 kN/m³). Both the peak (highest point) and ultimate (final value) from Fig. 3.19 have been plotted in Fig. 3.18 and are designated as the *ultimate shear strength* (squares) and the *peak shear strength* (circles). In those cases where the peak and ultimate values coincide, only one point is shown.

There is considerable scatter of the data in Fig. 3.18, but this is not unusual for soil which usually has variable engineering properties. The straight line drawn in Fig. 3.18 is the *shear strength envelope*, also known as the *failure envelope*, for the silty sand deposit. Considerable experience is needed in determining this envelope. For example, taking a conservative approach, the envelope was drawn by ignoring the two high peak shear strength points (B-2 at 2.5 ft and B-2 at 10 ft) and then a best-fit line was drawn through the rest of the data points. In addition, the angle of inclination of the failure envelope, which is known as the effective friction angle ϕ', is 30°, which is a typical value for silty sands. Thus the failure envelope drawn in Fig. 3.18 was based on engineering judgment and experience.

The failure envelope (straight line) shown in Fig. 3.18 is drawn through the origin of the plot. This means that the effective cohesion c' is zero. If the line had intersected the vertical axis, then that value would be defined as the effective cohesion value c'. In summary, based on the direct shear data presented in Fig. 3.18, the shear strength parameters for the silty sand are $c' = 0$ and $\phi' = 30°$.

The direct shear apparatus generally provides reasonably accurate values of the effective friction angle ϕ' for granular soils. The set-up and actual testing operation is relatively quick and simple, which is the main reason for the popularity of the test.

3.5 SHEAR STRENGTH OF COHESIVE SOIL

3.5.1 Introduction

This section deals with the laboratory shear strength testing of plastic soils, also known as cohesive soils. Cohesive soils have *fines*, which are silt and clay size particles that give the soil a plasticity or ability to be molded and rolled. Typical types of cohesive soils are silts and clays. The shear strength of cohesive soil is much more complicated than the shear strength of cohesionless soils. Also, in general the shear strength of cohesive soils tend to be lower than the shear strength of cohesionless soils. As a result, more shear induced failures occur in cohesive soils, such as clays, than in cohesionless soils.

The laboratory shear strength testing of cohesive soil can generally be divided into three broad groups:

Undrained Shear Strength.   Also known as the shear strength based on a total stress analysis. The purpose of these laboratory tests is to either obtain the undrained shear strength s_u of the soil or the failure envelope in terms of total stresses (total cohesion, c, and total friction angle, ϕ). These types of shear strength tests are often referred to as *undrained* shear strength tests because there is no change in water content of the soil during the shear portion of the test. Examples include the unconfined compression test, vane shear test, unconsolidated undrained triaxial compression test, and the consolidated undrained triaxial compression test. These types of laboratory tests are performed exclusively on cohesive soils.

Drained Shear Strength.   Also known as the shear strength based on an effective stress analysis. The purpose of these laboratory tests is to obtain the effective shear strength of the soil based on the failure envelope in terms of effective stress (effective cohesion, c', and effective friction angle, ϕ').

These types of shear strength tests are often referred to as "drained" shear strength tests because the water content of the soil is allowed to change during shearing. Examples of drained shear strength tests are the direct shear test and the consolidated drained triaxial compression test. Although technically not a drained shear strength test, the consolidated undrained triaxial compression test with pore water pressure measurements can also be used to obtain the effective shear strength parameters of the soil (c' and ϕ').

Drained Residual Shear Strength.   For some projects, it may be important to obtain the residual shear strength of cohesive soil, which is defined as the remaining (or residual) shear strength after a considerable amount of shear deformation has occurred. For example, a clay specimen could be placed in the direct shear box and then sheared back and forth several times to develop a well-defined shear failure surface. Once the shear surface is developed, the drained residual shear strength would be obtained by performing a final, slow shear of the specimen. The drained residual shear strength can be applicable to many types of soil conditions where a considerable amount of shear deformation has already occurred. For example, the stability analysis of ancient landslides, slopes in overconsolidated fissured clays, and slopes in fissured shales will often be based on the drained residual shear strength of the failure surface (Bjerrum, 1967a; Skempton and Hutchinson, 1969; Skempton, 1985; Hawkins and Privett, 1985; Ehlig, 1992).

In summary, the basic types of laboratory shear strength tests for cohesive soils are as follows:

1. *Unconsolidated undrained (UU).*   The unconsolidated undrained shear strength test is used to obtain the undrained shear strength s_u of the soil. Typical types of laboratory tests are the unconfined compression test and the vane shear test.

2. *Consolidated undrained (CU).*   The consolidated undrained shear strength test is used to obtain the failure envelope in terms of total stresses, i.e., total cohesion c and total friction angle ϕ. The triaxial apparatus is used to perform this test.

3. *Consolidated undrained with pore water pressure measurements (CU').*   The consolidated undrained shear strength test with pore pressure measurements is used to obtain the failure envelope in terms of effective stress, i.e., effective cohesion c' and effective friction angle ϕ'. The triaxial apparatus is used to perform this test.

4. *Consolidated drained (CD).*   The consolidated drained test is also used to obtain the failure envelope in terms of effective stress, i.e., effective cohesion c' and effective friction angle ϕ'. The consolidated drained test can be performed in the direct shear apparatus or in the triaxial apparatus.

5. *Drained residual shear strength.*   The drained residual shear strength is used to obtain the residual failure envelope in terms of effective stress, i.e., residual friction angle ϕ'_r. The drained residual shear strength can be obtained by shearing back and forth a specimen in the direct shear apparatus or by using the torsional ring shear apparatus.

3.5.2   Triaxial Test

Triaxial Apparatus.   In terms of understanding shear strength behavior, the triaxial test is probably the most important laboratory test and it is used extensively in the laboratory testing of cohesive soil. Because this test is so important, the test procedures and calculations required for this complex test will be discussed in detail.

The triaxial test procedure is to place a cylindrical specimen of cohesive soil in the center of the triaxial apparatus, seal the soil with a rubber membrane, subject the soil to a confining fluid pressure, and then the soil specimen is sheared by increasing the vertical pressure. The types of laboratory tests that require the triaxial apparatus are classified according to the soil specimen drainage conditions, and they are as follows:

1. Unconsolidated undrained triaxial compression test

2. Consolidated drained triaxial compression test

3. Consolidated undrained triaxial compression test

4. Consolidated undrained triaxial compression test with pore water pressure measurements

Because the data are of limited use, the unconsolidated undrained triaxial compression test is rarely used in practice. Likewise, the consolidated drained triaxial compression test is also rarely used in practice because the shearing of the cohesive soil in a drained state takes too long. The consolidated undrained triaxial compression test and the consolidated undrained triaxial compression test with pore water pressure measurements are identical, except that the latter test records the pore water pressure during shearing. The consolidated undrained triaxial compression test with pore water pressure measurements is by far the most popular of the four triaxial tests. The reason this triaxial test is so popular is because it can provide the shear strength in terms of total stresses (c and ϕ) and by measuring the pore water pressures during shearing, the shear strength in terms of effective stresses (c' and ϕ') can also be determined.

The left side of Fig. 3.20 presents an illustration of the triaxial apparatus. Figure 3.21 shows a photograph of the triaxial equipment and the loading apparatus. Arrow 1 in Fig. 3.21 points to the soil specimen which has been placed in the center of the triaxial apparatus and encased within a rubber membrane. The basic elements of the triaxial apparatus are as follows:

1. *Triaxial chamber.*   Arrow 2 in Fig. 3.21 points to the triaxial chamber. As indicated Fig. 3.20, the triaxial chamber consists of a top plate, base plate, and a chamber cylinder. Both the top plate and base plate are made of metal and have a round grove. A compressible O-ring and the chamber cylinder should fit snugly into each grove. Tie bars are used to clamp together the top plate, base plate, and chamber cylinder. Although only one tie bar is shown in Fig. 3.20, the triaxial chamber is

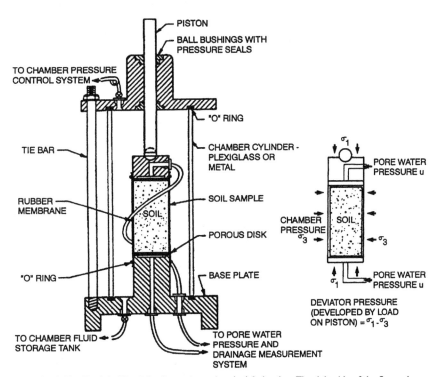

FIGURE 3.20   The left side of the figure shows the triaxial chamber. The right side of the figure shows the pressures that are exerted on the soil specimen during the triaxial test.

FIGURE 3.21 Photograph of the triaxial equipment and the loading apparatus. Arrow 1 points to the soil specimen which is encased in a rubber membrane, arrow 2 points to the triaxial chamber, arrow 3 points to the tube that leads to the chamber pressure control system, arrow 4 points to the tube that leads to the drainage measurement system and back pressure system, arrow 5 points to the loading piston, and arrow 6 points to a dial gauge used to measure the vertical deformation of the soil specimen.

typically fitted with at least three tie bars. The chamber cylinder can be made out of Plexiglas or metal and it must be able to safely resist the highest chamber pressure that will be used for the apparatus. A Plexiglas chamber, such as shown in Fig. 3.21, is more desirable than a metal cylinder because it enables the behavior of the soil specimen to be observed during testing.

The top plate is equipped with an axial load piston that is used to apply a vertical load to the top of the soil specimen. To reduce piston friction and the escape of chamber fluid along the sides of the piston, the top plate is equipped with ball bushings having pressure seals as shown in Fig. 3.20. Typically two ball bushings are used to guide the piston, minimize friction, and maintain vertical alignment. The top plate is also equipped with a vent valve such that air can be forced out of the chamber when it is filled with chamber fluid (usually water is used as the chamber fluid). One end of a tube is connected to this vent valve with the other end connected to a chamber pressure control system (discussed as Item No. 3).

The base plate is equipped with a central pedestal upon which the soil specimen is placed as shown in Fig. 3.20. The base plate also has an inlet through which the fluid is supplied to the chamber. Furthermore, the base plate has inlets leading to the soil specimen base and cap to allow saturation and drainage of the soil specimen when required (discussed as Item No. 4).

2. Rubber membrane. As indicated in Fig. 3.20, the soil specimen is sealed in a rubber membrane. Arrow 1 in Fig 3.21 points to the soil specimen encased in the rubber membrane. The rubber membrane must provide reliable protection against leakage of chamber fluid into the soil specimen. The membrane should be carefully inspected prior to use, and if any flaws or pinholes are discovered, then it should be discarded.

To offer minimum lateral restraint to the soil specimen, the unstressed membrane diameter should be between 90 to 95 percent of the specimen diameter. The membrane is sealed to the pedestal of the base plate and to the specimen cap by using O-rings (see Fig. 3.20).

3. *Chamber pressure control system.* The chamber fluid pressure is also known as the cell pressure. As indicated in Fig. 3.20, a tube connects the chamber pressure control system to the top valve in the triaxial chamber. Arrow 3 in Fig. 3.21 points to the tube that connects the chamber pressure control system to the triaxial chamber.

The purpose of the chamber pressure control system is to apply and maintain a pressure to the fluid contained within the triaxial chamber. This chamber fluid pressure applies a confining stress σ_c to the soil specimen. Several different types of chamber pressure control systems are available, such as self-compensating mercury pots, pneumatic pressure regulators, and combination pneumatic pressure and vacuum regulators. Because it may take several days to complete a triaxial test, chamber pressure systems having an air-water interface are not recommended by ASTM.

The chamber pressure is typically measured by using an electronic pressure transducer which directly measures the chamber fluid pressure. The chamber pressure transducer should be calibrated so that it records the chamber pressure that is exerted at the mid-height of the soil specimen.

4. *Pore water pressure and drainage measurement system.* As mentioned in Item 1, the base plate has inlets leading to the soil specimen base and cap to allow saturation and drainage of the soil specimen when required. For example, Fig. 3.20 shows the tubes that lead from the soil specimen to the pore water pressure and drainage measurement system. There are three components to the pore water pressure and drainage measurement system, as follows:

Pore water pressure. The purpose of the pore water pressure system is to measure the pore water pressure within the soil sample during testing. In order for this to be accomplished, the pore water measuring system must consist of non-flexible components. Thus the tubing between the specimen and the pore water measuring device must be short, thick walled, and have a small diameter. Likewise the pore water pressure measuring device must be very stiff, and this is often accomplished by using a very stiff electronic pressure transducer. The pore water pressure transducer should be calibrated so that it records the pore water pressure at the mid-height of the soil specimen. An air bubble that becomes trapped within the pore water pressure measuring system can cause significant inaccuracies in the pore water pressure readings.

The pore water pressure measurement system usually has the ability to be separated from the drainage measurement system. This enables the drainage measurement system to be shut-off, yet the pore water pressure within the soil specimen could still be monitored.

Drainage measurement system. The purpose of the drainage measurement system is to record the volume of water that enters or leaves the soil specimen. This system is needed for those soil specimens that are either consolidated or sheared in a drained state. A burette is usually used to measure the volume of water that enters or leaves the soil specimen.

Back pressure system. The triaxial equipment should also include the ability to apply a pore water pressure to the soil specimen. A pore water pressure that is applied to the soil specimen is known as the "back pressure." The system used to apply the back pressure is separate from the chamber pressure control system (Item 3). However, the device used to apply the back pressure could be the same type of device used to apply the chamber pressure.

Usually the drainage measurement system and the back pressure system are directly connected. Thus drainage of water from the soil specimen would flow through the drainage measurement system and into the back pressure system. This enables the flow of water into or out of the soil specimen to be measured for any change in the back pressure.

The pore water pressure measurement system, drainage measurement system, and back pressure system often have valves that are used to either open or close each individual system. These valves must be of a type that produce minimum volume changes during their operation. For example, ball valves have been found to produce minimum volume changes when opened and closed. In addition, the valves must be capable of withstanding the pore water pressure or back pressure without leaking.

5. *Loading apparatus.* Figure 3.21 shows the triaxial chamber that has been inserted into the loading apparatus. The purpose of the loading apparatus is to apply a vertical load to the piston. Arrow 5 in Fig. 3.21 points to the loading piston. The load from the piston is transferred to the specimen cap and ultimately to the top of the soil specimen. This axial load is used to shear the cylindrical specimen of soil.

The actual loading device can consist of weights applied to a hanger (controlled stress test) or a device used to control the displacement rate of the loading piston (controlled strain test) such as a screw jack driven by an electric motor acting through a geared transmission. The most commonly used loading device is the controlled strain test, and this is the only type of loading apparatus that will be further discussed in this chapter. For the controlled strain test, the load is determined by using a load indicator device, such as a load cell or proving ring. The load cell or proving ring should be able to accurately measure the axial force to within 1 percent of the axial force at failure.

During axial loading, a dial gauge is used to record the vertical deformation of the soil specimen. Arrow 6 in Fig. 3.21 points to the dial gauge that is used to measure the vertical deformation of the soil specimen. Other types of devices such as a linear variable differential transformer (LVDT) or an extensiometer can be used to measure the vertical deformation of the soil specimen.

The right side of Fig. 3.20 shows the pressures that are applied to the soil specimen during the triaxial compression test, as follows:

1. *Vertical total stress.* The major principal total stress σ_1 is equal to the vertical total stress σ_v. The vertical total stress σ_v is equal to the chamber fluid pressure plus the load induced by the piston divided by the area of the specimen, or:

$$\sigma_1 = \sigma_v = \sigma_c + (P/A) \tag{3.11}$$

where σ_1 = major principal total stress acting on the soil specimen (psi or kPa)
σ_v = vertical total stress acting on the top of the soil specimen (psi or kPa). It is equal to the sum of the cell pressure σ_c and the pressure exerted by the loading piston.
σ_c = chamber fluid pressure, also known as the cell pressure (psi or kPa)
P = load applied by the loading piston (lb or kN)
A = area of the soil specimen (in.^{2} or m^2). Because the area of the soil specimen changes as P is applied, an area correction is required

2. *Horizontal total stress.* The minor principal total stress σ_3 is equal to the horizontal total stress σ_h. The horizontal total stress is induced on the soil specimen by the chamber fluid pressure σ_c, or:

$$\sigma_3 = \sigma_h = \sigma_c \tag{3.12}$$

where: σ_3 = minor principal total stress acting on the soil specimen (psi or kPa)
σ_h = horizontal total stress (psi or kPa), which is equal to the cell pressure (σ_c)
σ_c = chamber fluid pressure, also known as the cell pressure (psi or kPa)

For the triaxial compression test, the intermediate principal total stress (σ_2) is equal to the minor principal total stress (i.e., $\sigma_2 = \sigma_3$).

3. *Pore water pressure.* The pore water pressure u that exists within the soil specimen can be applied to the soil specimen or it can be measured during testing. As previously mentioned, an applied pore water pressure is known as the back pressure.

4. *Deviatoric pressure.* The deviatoric pressure is also known as the *deviator stress* or *stress difference*. The deviatoric pressure is developed during the shearing of the soil specimen when the axial load is applied to the top of the soil specimen. The deviatoric pressure is defined as the major principal total stress minus the minor principal total stress, or deviatoric pressure = $\sigma_1 - \sigma_3 = \sigma_v - \sigma_h$.

By knowing the principal total stresses and the pore water pressure within the soil specimen, the principal effective stresses can be calculated as follows:

$$\sigma'_1 = \sigma_1 - u \qquad (3.13)$$

$$\sigma'_3 = \sigma_3 - u \qquad (3.14)$$

where σ'_1 and σ'_3 are the major and minor principal effective stresses.

Test Procedure.   As previously mentioned, by far the most popular triaxial test used in practice is the consolidated undrained triaxial compression test with pore water pressure measurements. The test procedures for this laboratory triaxial test are as follows:

1. *Soil specimen.*   An undisturbed soil specimen must be used for the triaxial test. For projects where it is anticipated that there will be fill placement, the fill specimens can be prepared by compacting them to the anticipated field as-compacted density and moisture condition. An undisturbed soil specimen can be trimmed from a block sample or extruded from a thin-walled sampler.

For the triaxial test, soil specimens must have a minimum diameter of 1.3 in. (33 mm). The specimen height divided by the specimen diameter should be between 2 and 2.5. The trimming process should be performed as quickly as possible to minimize the possibility of a change in water content of the cohesive soil. Often the soil specimen will be simply extruded from a Shelby tube and then the top and bottom of the specimen are trimmed flush. The top and bottom of the soil specimen must be trimmed so that they are perpendicular to the longitudinal axis of the specimen. In this case where the soil sample is extruded from a sampling tube, the initial diameter of the soil specimen is equal to the inside diameter of the sampling tube.

2. *Index properties.*   After the soil specimen has been trimmed, the index properties of the soil should be determined. Typically, the trimmings can be collected in order to determine the water content of the soil (see Sec. 3.2.1). In addition, a balance can be used to obtain the mass of the cylindrical soil specimen. By knowing the diameter of the soil specimen D_o and measuring the height of the soil specimen H_o, the total unit weight can be calculated (see Sec. 3.2.2). Using the water content, the dry unit weight can also be calculated.

3. *Porous disks and filter paper.*   Because the soil specimen will first be consolidated prior to shearing, filter paper strips should be attached to the sides of the soil specimen. The purpose of the filter paper strips is to enable the soil specimen to drain from its sides and hence speed-up the consolidation process. Rectangular strips are typically cut from a large sheet of filter paper, such as Whatman's No. 54 filter paper. Figure 3.22 shows a cohesive soil specimen with the filter paper strips attached to its perimeter. Usually by slightly moistening the filter paper strips, they will adhere to the cohesive soil. ASTM recommends that the filter paper cover no more than 50 percent of the specimen periphery. But for large diameter soil specimens, such as the 2.5 in. (6.35 cm) diameter specimen shown in Fig. 3.22, a larger area of filter paper may be needed in order to speed-up the consolidation process.

As shown in Fig. 3.22, porous disks are placed on the top and bottom of the soil specimen. The porous disks should have the same diameter as the soil specimen. The filter paper strips need to be long enough so that they are in contact with the sides of the porous disks. This direct contact will facilitate the transfer of water between the filter paper strips and the porous disks.

It is important to carefully handle the soil specimen in order to minimize disturbance and prevent a change in cross section. In addition, the filter paper and porous disks should be installed as quickly as possible to prevent a change of water content of the soil specimen.

4. *Rubber membrane.*   With the top plate, base plate, and chamber cylinder is a disassembled condition, the soil specimen containing top and bottom porous disks is placed on the base plate pedestal. A cap is then positioned on the top of the specimen. Check that the specimen cap, porous disks, and soil specimen are centered on the base plate pedestal. The specimen is then carefully encased in a rubber membrane. The membrane can be installed by rolling it up from the bottom or by using a membrane expander. Prior to installing the membrane, a thin coating of silicon grease can be installed on the vertical surfaces of the specimen cap and base plate pedestal and this grease will aid in sealing the membrane. The rubber membrane must extend beyond both the top and bottom porous disks and must be long enough to provide a seal around the perimeter of the base plate

FIGURE 3.22   Both photographs show the soil specimen with filter strips attached to its perimeter. In the upper photograph, the top porous disk has not yet been installed. Note in the upper photograph that the filter paper extends above the top of the soil specimen. The lower photograph shows the soil specimen with the top porous disk in place.

pedestal and the specimen cap. The rubber membrane is sealed to the pedestal of the base plate and to the specimen cap by using O-rings (the O-rings can be prepositioned on the specimen cap and base plate pedestal and then rolled into place). Figure 3.23 shows the completion of Step 4, i.e., the soil specimen and specimen cap have been placed on the base plate pedestal, the soil specimen has been sealed with a rubber membrane, and the O-rings are in place.

5. *Triaxial chamber.*   Once the soil specimen is in place and sealed in the rubber membrane, the triaxial chamber is assembled. The chamber cylinder is installed in the base plate groove and then the top plate is positioned on top of the chamber cylinder. The three components of the triaxial chamber (i.e., the base plate, chamber cylinder, and top plate) are clamped together by tightening the tie bars. Figure 3.24 shows the triaxial chamber in its assembled condition.

The specimen cap has a circular indentation at its center, and the alignment of the loading piston and specimen cap should be checked. This is accomplished by pushing down the loading piston and making sure it fits neatly into the center of the specimen cap. During this process, it is important that the soil specimen not be inadvertently loaded by the piston. With the piston in contact with the specimen cap, record the reading of the vertical dial indicator (e.g., the dial gauge). This dial reading is recorded as the initial dial reading for the vertical deformation of the soil specimen.

The chamber is then filled with fluid, usually distilled water. The fluid is allowed to flow into the bottom of the chamber and then up through the valve located in the top plate. The goal is to fill the chamber with fluid without trapping any air within the chamber.

6. *Saturation of soil specimen.*   The objective of the saturation process is to fill all the voids within the soil specimen with water without producing undesirable prestressing of the specimen or allowing the

FIGURE 3.23 The soil specimen and specimen cap have been placed on the base plate pedestal, the soil specimen has been sealed with a rubber membrane, and the O-rings are in place.

FIGURE 3.24   This photograph shows the triaxial chamber in its assembled condition.

soil to swell. There are two steps in this process: (1) achieving saturation of the pore water pressure and drainage measurement systems, and (2) ensuring saturation of the soil specimen itself.

There are many different methods of achieving saturation of the pore water pressure and drainage measurement system. One approach is to apply a vacuum to the top drainage tube and then slowly draw distilled water up through the lower porous disk, filter paper strips, and top porous disk and then out of the specimen cap. The purpose of this procedure is to remove as much air as possible and achieve saturation of the drainage tubes, porous disks, and filter paper strips. The partial vacuum applied to the system should never exceed the effective consolidation pressure used for the test.

After the pore water pressure and drainage measurement systems are saturated, the next step is to saturate the soil specimen. This is accomplished by using a back pressure which will force the air into solution. The procedure is to simultaneously increase both the chamber pressure and the back pressure, always keeping the chamber pressure greater than the back pressure. The difference between the chamber pressure and back pressure (also known as the effective confining pressure) should not exceed 5 psi (35 kPa). For most soil specimens that have a high degree of saturation, a back pressure of about 100 psi (700 kPa) is usually sufficient to ensure saturation (i.e., S = 100 percent) of the soil specimen. If initially the soil specimen has a low degree of saturation, such as a compacted clay specimen, then a much higher back pressure may be needed to ensure saturation.

In order to help with the saturation process, distilled water that has been deaired can be used in Step 6. Air remaining in the soil specimen and drainage system just prior to applying the back pressure will go into solution much more readily if deaired water is used for saturation. The use of deaired water may also decrease the time and back pressure required for saturation. There are many procedures that can create deaired water. For example, the distilled water can be placed in a flask and then a vacuum can be used to reduce the pressure in the flask. By agitating the flask, the distilled water can be deaired.

7. *Consolidation of soil specimen.*   The objective of this process is to consolidate the soil specimen in the triaxial apparatus. The fluid in the triaxial chamber applies the same pressure in all directions to the soil specimen, and thus this process is known as isotropic consolidation (i.e., $\sigma_h = \sigma_v$). The consolidation pressure σ'_c, also known as the effective confining pressure or effective consolidation pressure, is equal to the chamber pressure σ_c minus the back pressure u, or: $\sigma'_h = \sigma'_v = \sigma'_c = \sigma_c - u$.

At the completion of step 6, an initial burette reading is obtained. The initial effective confining pressure for the soil specimen should not exceed 5 psi (35 kPa). The burette is monitored, and if the soil specimen starts to swell (i.e., the burette indicates that water is flowing into the soil specimen), then the effective confining pressure should be immediately increased. If the burette indicates that the cohesive soil specimen is consolidating (i.e., water leaving the soil specimen), then the soil specimen should be allowed to equilibrate under this initial effective confining pressure.

After the soil specimen has equilibrated under the initial effective confining pressure, a burette reading is taken. The cell pressure is then increased and it is desirable to use a load increment ratio equal to 1.0. Thus the chamber fluid pressure is increased such that the effective confining pressure is doubled. At each effective confining pressure, burette readings can be taken at intervals of elapsed time of 0.1, 0.25, 0.5, 1, 2, 4, 8, 15, and 30 min and at 1, 2, 4, 8, and 24 h. As recorded by the burette, the volume of water leaving the soil specimen during consolidation can be plotted versus the log of time (see Chap. 8). The soil specimen should not be subjected to an increase in effective confining pressure until it has completed primary consolidation.

The soil specimen is incrementally loaded up to the effective confining pressure for which the shear strength will be determined. It is essential that at this effective confining pressure, the soil specimen be allowed to complete primary consolidation and enter secondary compression. The data can be plotted on a log of time plot to ensure that the soil specimen has completed primary consolidation.

In Fig. 3.25, the soil specimen has been subjected to a chamber pressure and back pressure and is in the process of being consolidated.

8. *Checking the B value.*   After consolidation is complete and just prior to shearing, the B value should be checked. The process consists of first recording the pore water pressure, which is equal to the back pressure u_o. Then the valve to the drainage measurement system and the back pressure system is closed (i.e., no water is allowed to enter or leave the soil specimen). Only the valve that leads to the measurement of the soil specimen pore water pressure is to remain open.

The next step is to increase the chamber pressure by about 10 psi (70 kPa), i.e., $\Delta\sigma_c = 10$ psi. After the increased chamber pressure has been applied to the soil specimen for about 2 min, the pore water pressure is measured u_f. The change in pore water pressure within the soil specimen Δu is equal to the final minus the initial pore water pressure, or $\Delta u = u_f - u_o$. The B value is calculated as the change in pore water pressure divided by the change in cell pressure, or B value = $\Delta u/\Delta\sigma_c$. The B value should be close to 1.0.

A large increase in Δu with time or a value of Δu greater than $\Delta\sigma_c$ usually indicates a leak of chamber fluid into the soil specimen. A decreasing value of Δu with time can indicate a leak in the pore water measurement system or a defective pore water pressure transducer.

After the B value has been checked, the chamber pressure is returned to its original pressure, i.e., return the chamber pressure to the same pressure that existed just prior to the B value check. Then the valve that leads to the drainage measurement system and the back pressure system is opened and the soil specimen is allowed to equilibrate.

9. *Loading device.*   The triaxial chamber containing the soil specimen is placed in the loading apparatus. Figure 3.21 shows the triaxial chamber in the loading apparatus and ready for testing. Since this is an undrained triaxial test, just prior to shearing, the valve to the drainage measurement system and back pressure system is closed. Once the valve has been closed, the back pressure can be reduced to zero pressure. For the undrained condition, the soil specimen is sheared without allowing water to enter or leave the soil specimen. Thus, during undrained shearing, the water content of the soil specimen is not allowed to change. The pore water pressures are monitored during the shearing portion of the triaxial test, which is why this test is known as a "consolidated undrained triaxial compression test with pore water pressure measurements."

10. *Shearing the soil specimen.*   After the completion of Step 9, the next step is to apply a load to the loading piston so that it just slowly moves down and barely comes in contact with the specimen

FIGURE 3.25 Saturation has been achieved of the pore water pressure and drainage measurement system, and of the soil specimen itself. The soil specimen is in the process of being consolidated at the effective confining pressure for which the shear strength will be determined.

cap. A load on the piston is required to overcome the upward thrust due to the chamber pressure and any piston friction induced by the ball bushings. The load-measuring device (such as the LVDT) can be adjusted to compensate for the chamber pressure thrust and piston friction. The loading piston should fit snugly into the indentation located at the top of the specimen cap.

Using the controlled strain loading apparatus, the soil specimen is sheared at a relatively constant rate of strain. The soil specimen should be sheared slow enough so that the excess pore water pressures are able to equalize throughout the soil specimen during the undrained shearing. The triaxial test is typically sheared at a strain rate of 0.5 percent axial strain per hour. Since the usual procedure is to shear the soil specimen to an axial strain of at least 15 percent, the shearing portion of the triaxial test should take at least 30 hours. ASTM D 4767-02 (2004) provides an alternate determination of the strain rate based on t_{50}, i.e., the time for 50 percent consolidation based on burette readings and using the log-of-time method.

11. *Recording of shear test data.* During the shearing of the soil specimen, the data can be recorded manually in tabular form. Some laboratories have equipment that electronically records the time, vertical load, vertical deformation, and pore water pressure during the shearing portion of the test.

12. *Failure of the soil specimen.* There are two different definitions of failure. The first definition of failure is based on total stresses where the shear failure of the soil specimen is considered to be the value of the maximum principal total stress difference (i.e., the maximum deviatoric pressure, or maximum value of $\sigma_1 - \sigma_3$). Since the minor principal total stress σ_3 never changes during the triaxial compression test, the maximum principal total stress difference also corresponds to the maximum vertical total pressure that the cohesive soil specimen can sustain. Thus failure could be said to occur at the maximum vertical total pressure (maximum σ_v) that the soil specimen is able to sustain. If the highest value of σ_v occurs after 15 percent axial strain, then σ_v corresponding to 15 percent axial strain should be considered to be the failure condition.

The second definition of failure is based on effective stresses. Similar to the earlier definition, failure could be assumed to occur at the maximum deviatoric effective stress (i.e., maximum value of $\sigma'_1 - \sigma'_3$). However, a more commonly used definition is that failure occurs at the maximum effective stress obliquity, or in other words, the maximum value of σ'_1/σ'_3.

13. *Removing the soil specimen.*   The shearing portion of the triaxial test is complete once the soil specimen has reached 15 percent axial strain. The controlled strain loading apparatus is turned off and the axial load is removed from the soil specimen. Then the chamber pressure is reduced to zero and the chamber fluid is drained from the triaxial chamber. The triaxial apparatus is then disassembled and the O-rings and rubber membrane are removed from the soil specimen. Figure 3.26 shows the soil specimen after the O-rings and rubber membrane have been removed. The type of failure for this cohesive soil specimen was a bulge type failure.

14. *Index properties at the end of test.*   The final step is to determine the water content of the soil specimen (see Sec. 3.2.1). The filter paper strips are first removed and any free water remaining on the soil specimen is blotted away. Often the entire cylindrical soil specimen is used for the water content test. When placing the soil specimen in the container for the water content test, it should be broken apart to determine if there are any large size particles within the soil specimen. Large size particles may restrict the shear of the soil specimen and lead to an overestimation of the in situ shear strength of the soil. If large particles are noted, the triaxial shear test may have to be repeated on a larger or different soil specimen.

FIGURE 3.26   Bulge-type failure condition of the soil specimen.

As previously mentioned, a bulge type shear failure of the cohesive soil can occur such as shown in Fig. 3.26. This is often the type of failure condition for soft cohesive soils. In other cases, there may be an actual shear failure of the soil specimen along an inclined shear surface.

In order to determine the shear strength parameters of the cohesive soil, usually at least three soil specimens are sheared at different effective confining pressures. It is important that the range in effective confining pressures used during the triaxial testing is approximately the same as the range in pressure applicable for the field conditions.

For further test details, see ASTM D 4767-02 (2004) "Standard Test Method for Consolidated Undrained Triaxial Compression Test for Cohesive Soils."

Mohr Circle, Stress Paths, and Pore Pressure Parameters.   Prior to presenting the calculations for the determination of the shear strength from the triaxial test, an understanding of the Mohr circle, stress paths, and the pore water pressure parameters A and B is required.

Mohr Circle.   The results of a triaxial test can be graphically represented by the Mohr circle, which is illustrated in Fig. 3.27. The upper diagram in Fig. 3.27 shows that the vertical total stress acting on the soil is the major principal total stress σ_1 and that the horizontal total stress acting on the soil is the minor principal total stress σ_3. This orientation of principal stresses is the same orientation of principal stresses as applied in the laboratory triaxial compression test.

In geotechnical engineering, compressive stresses are considered to be positive. As previously mentioned, in the triaxial compression test, the vertical total stress at failure is the major principal total stress σ_1 and is equal to the cell pressure plus the vertical load from the piston divided by the area of the specimen (using an area correction). The horizontal total stress is the minor principal total stress σ_3 and is equal to the cell pressure. By knowing the major σ_1 and minor σ_3 principal total stresses at failure, a Mohr circle can be drawn such as shown in Fig. 3.27. The major principal total stress minus the minor principal total stress (i.e., $\sigma_1 - \sigma_3$) is referred to as the deviatoric pressure, deviator stress, or stress difference. The maximum shear stress τ_{max} is equal to one-half the deviatoric pressure and equals the radius of the Mohr circle. In geotechnical engineering, usually only the upper half of the Mohr circle is used in the analysis. Although principal total stresses are shown in Fig. 3.27, the Mohr circle could also be plotted in terms of effective stresses, i.e., major principal effective stress σ'_1 and the minor principal effective stress σ'_3.

Stress Path.   A stress path is defined as a series of stress points that are connected together to form a line or curve. Figures 3.28 and 3.29 illustrate the construction of a stress path for a triaxial compression test. The state of stress for the triaxial test starts at Point A where the cohesive specimen is subjected to isotropic compression (i.e., $\sigma_c = \sigma_1 = \sigma_3$). Then during the shearing of the soil specimen as it is axially loaded, σ_1 increases while σ_3 remains constant. This results in a series of ever larger Mohr circles such as shown in Fig. 3.28. Mohr circle E represents stress conditions at failure.

Instead of drawing a series of Mohr circles to represent the change in stress during the triaxial shear test, a stress path can be drawn as shown in Fig. 3.29. The peak point of each Mohr circle (i.e., points A, B, C, D, and E) are used to draw the stress path on a *p-q* plot. The peak point of each Mohr circle is defined as:

In terms of total stresses:

$$p = \frac{\sigma_1 + \sigma_3}{2} \quad \text{and} \quad q = \frac{\sigma_1 - \sigma_3}{2} \tag{3.15}$$

In terms of effective stresses:

$$p' = \frac{\sigma'_1 + \sigma'_3}{2} \quad \text{and} \quad q = \frac{\sigma'_1 - \sigma'_3}{2} \tag{3.16}$$

Note that p represents the center of the Mohr circle and q is the radius of the Mohr circle. For a given stress state defined by σ_1, σ_3, and u, the value of q is the same for Eqs. 3.15 and 3.16. The shear strength parameters from the conventional plot (Fig. 3.28) are related to the shear strength parameters from the *p-q* plot (Fig. 3.29) by the following relationships:

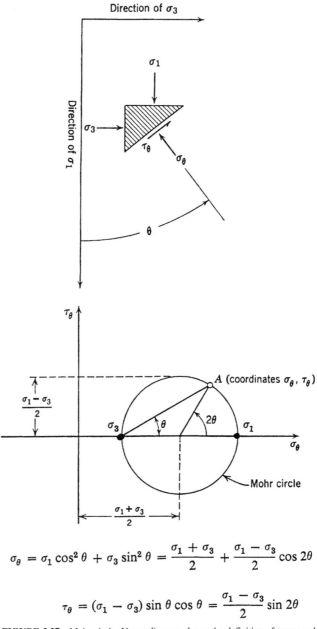

$$\sigma_\theta = \sigma_1 \cos^2 \theta + \sigma_3 \sin^2 \theta = \frac{\sigma_1 + \sigma_3}{2} + \frac{\sigma_1 - \sigma_3}{2} \cos 2\theta$$

$$\tau_\theta = (\sigma_1 - \sigma_3) \sin \theta \cos \theta = \frac{\sigma_1 - \sigma_3}{2} \sin 2\theta$$

FIGURE 3.27   Mohr circle. Upper diagram shows the definition of terms and the lower diagram shows the graphical construction of Mohr circle. The equations for the calculation of normal stress σ_θ and shear stress τ_θ on a given plane are also indicated in this figure. (*Adapted from Lambe and Whitman, 1969.*)

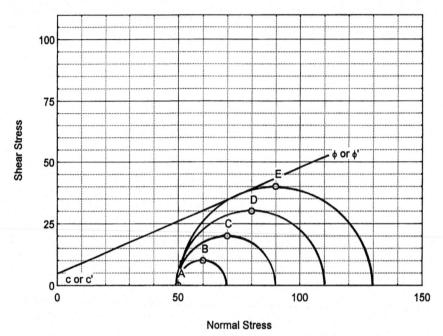

FIGURE 3.28   Series of Mohr circles representing the state of stress during a triaxial compression test (Mohr circle E represents the Mohr circle at failure). Figure 3.29 shows the stress path for these Mohr circles.

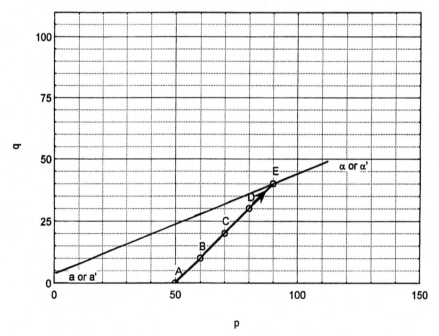

FIGURE 3.29   Peak point of the Mohr circles from Fig. 3.28 connected together to form a stress path.

In terms of total stresses:

$$\sin \phi = \tan \alpha \quad \text{and} \quad c \cos \phi = a \tag{3.17}$$

In terms of effective stresses:

$$\sin \phi' = \tan \alpha' \quad \text{and} \quad c' \cos \phi' = a' \tag{3.18}$$

The stress path shown in Fig. 3.29 can be based on the principal total stresses acting on the soil specimen and this is known as a "total stress path." Total stress paths are used to determine the shear strength parameters in terms of total stresses (i.e., c and ϕ) by using Eq. 3.17. A stress path can also be based on effective stresses, and this is known as an *effective stress path*. In order to draw an effective stress path, the pore water pressure during the shearing of the soil specimen must be measured, such as by using an electronic pressure transducer. Effective stress paths are used to determine the shear strength parameters in terms of effective stresses (c' and ϕ') by using Eq. 3.18.

Pore Water Pressure Parameters A and B.   The Skempton (1954) pore water pressure parameters relate the change in total stresses for an undrained loading. For an undrained triaxial test, the general expression is as follows:

$$\Delta u = B \left[\Delta \sigma_3 + A \left(\Delta \sigma_1 - \Delta \sigma_3 \right) \right] \tag{3.19}$$

where Δu = change in pore water pressure generated during undrained shear (psf or kPa)
$\quad \Delta \sigma_1$ = change in major principal total stress during undrained shear (psf or kPa)
$\quad \Delta \sigma_3$ = change in minor principal total stress during undrained shear (psf or kPa)
$\quad A, B$ = Skempton pore water pressure coefficients (dimensionless)

When performing a triaxial compression test on a saturated cohesive soil, the B value should be determined prior to shearing of the soil specimen. The process consists of preventing drainage of the cohesive soil and then applying a confining pressure ($\Delta \sigma_c$) to the specimen by increasing the cell pressure. In this case, $\Delta \sigma_c = \Delta \sigma_1 = \Delta \sigma_3$ and Eq. 3.19 reduces to:

$$B = \frac{\Delta u}{\Delta \sigma_3} \tag{3.20}$$

Based on the increase in confining pressure ($\Delta \sigma_c = \Delta \sigma_3$) and by measuring the increase in pore water pressure Δu during application of the confining pressure, the B value can be calculated from Eq. 3.20.

Table 3.5 lists values of measured B values for various types of clay. If the clay is saturated ($S = 100$ percent), the B value is essentially equal to 1.0. These data show that a stress applied to a saturated

TABLE 3.5   *B* Values for Various Clays

Types of soil	Degree of saturation %	B value
London clay (OC)	100	0.9981
London clay (NC)	100	0.9998
Vicksburg clay	100	0.9990
Kawasaki clay	100	0.999
Boulder clay	93	0.69
Boulder clay	87	0.33
Boulder clay	76	0.10

Note: OC = overconsolidated clay, NC = normally consolidated clay.
Source: Skempton (1954, 1961) and Lambe and Whitman (1969).

cohesive soil will be carried by an increase in pore water pressure and not the soil particle skeleton (i.e., $\Delta u = \Delta \sigma_c$). The measurement of the B value on saturated clay is confirmation that the shear strength of a saturated soil can not increase unless the effective stress of the soil first increases. A back pressure is used to achieve saturation prior to the consolidation and shearing of the soil specimen. The back pressure is defined as the pressure applied to the specimen pore water to cause any air in the pore space to compress and to pass into solution in the pore water thereby resulting in saturation of the soil specimen.

As indicated in Table 3.5, if the soil is not saturated, the B value is less than 1.0. The lower the degree of saturation, the lower the B value. Thus the lower the degree of saturation, the greater the portion of the load that will be immediately carried by the soil particle skeleton. In the case of a completely dry soil ($S = 0$ percent), the B value is zero and all of the load will be immediately carried by the soil particle skeleton.

While the B value is calculated before the triaxial shearing part of the test is performed, the A value is calculated during the actual shearing of the cohesive soil specimen. For the triaxial compression test, the confining pressure is held constant throughout shearing (i.e., $\Delta \sigma_3 = 0$), and if the degree of saturation is 100 percent ($B = 1.0$), then Eq. 3.19 reduces to the following:

$$A = \frac{\Delta u}{\Delta \sigma_1} \tag{3.21}$$

Based on triaxial compression tests and using Eq. 3.21, Skempton and Bjerrum (1957) provided typical A values at failure for various cohesive soils, as follows:

Very sensitive soft clay	A value at failure is greater than 1
Normally consolidated clay	A value at failure is 0.5–1
Overconsolidated clay	A value at failure is 0.25–0.5
Heavily overconsolidated clay	A value at failure is 0–0.25

The reason for high A values for very soft sensitive clay and normally consolidated clay is due to contraction of the soil structure during shear, and because it is an undrained loading, positive pore water pressures will develop. The reason for zero or even negative A values for highly overconsolidated soil is due to dilation of the soil structure during shear, and because it is an undrained loading, negative pore water pressures will develop. The A value can indicate a normally consolidated soil versus a heavily overconsolidated soil.

Calculations.   This part describes the calculations needed for the determination of the shear strength of cohesive soil. The calculations are for the triaxial test previously described, i.e., the consolidated undrained triaxial compression test with pore water pressure measurements.

Figure 3.30 presents actual laboratory test data, which were derived from the laboratory triaxial tests shown in Figs. 3.22 to 3.26. The soil specimen that was tested in the triaxial apparatus was classified as a silty clay of high plasticity (*CH*), having a liquid limit = 56 and a plasticity index = 41. The undisturbed soil specimen was obtained from a natural clay deposit, described as a dark brown silty clay, in a wet condition, and having a soft to medium consistency. The silty clay specimen was first saturated in the triaxial apparatus by using a back pressure equal to 85.9 psi (592 kPa). Then the soil specimen was allowed to consolidate under an effective confining pressure of 7.0 psi (48 kPa).

Just prior to starting the undrained shearing part of the triaxial test, the vertical dial indicator was reset so that it directly recorded the axial deformation and hence the values listed in Columns 1 and 2 are identical in Part *e*, Fig. 3.30. Also, the load cell was adjusted to compensate for the chamber pressure thrust and piston friction and thus directly measures the load (pounds) applied to the top of the soil specimen. Since the load cell directly measures the load applied to the top of the soil specimen, no conversion factor was required (i.e., the values listed in Columns 5 and 6 are identical in Part *e*, Fig. 3.30).

File no. 8316-03

Job name: TV Project

Sample location: TP-1 at 6–7 ft

Date: 12-15-99

(a) Initial conditions of soil specimen

Diameter (D_0): 2.493 in.   Height (H_0): 7.340 in.   Initial volume (V_0): 35.83 in.^{3}

Water content: 23.2%   Total unit weight: 116.7 pcf   Dry unit weight: 94.7 pcf

(b) Consolidation data

Initial dial reading: 1.2277 in.   Initial burette reading: 1.50 in.^{3}

Effective confining pressure, psi	Burette reading, in.^{3}	Change in volume (ΔV) of soil specimen, in.^{3}	Volumetric strain $\Delta V/V_0$
3.5	1.47	0.03	0.08
7.0	1.40	0.10	0.28

(c) Burette readings versus time at effective confining pressure (σ'_c) = 7.0 psi

Time, min	Burette reading	Time, min	Burette reading	Time, h	Burette reading
0	1.47	2	—	1	1.43
0.10	1.47	4	1.46	2	1.42
0.25	1.47	8	1.45	4	1.41
0.5	1.47	15	—	8	1.40
1	1.47	30	1.44	24	1.40

(d) B value check:

u_0 = 85.9 psi u_f = 96.0 psi $\Delta u = u_f - u_0$ = 10.1 psi

σ_{ci} = 92.9 psi σ_{cf} = 103.1 psi B value = $\Delta u / \Delta \sigma_c$ = 0.99

FIGURE 3.30 Laboratory test data: consolidated undrained triaxial compression test with pore water pressure measurements.

(e) Shear test data: $d_s = 1.2147$ in. $H_s = 7.327$ in. $V_s = 35.73$ in.³ $A_s = 4.877$ in.² $\sigma_c = 92.9$ pci $u_s = 85.9$ psi

| Axial deformation data | | | | Loading data | | | Pore water pressure | | | Failure | | Stress paths and A value | | | |
Dial	ΔH	$\epsilon = \Delta H/H_s$	A_c	Load cell	Axial load	σ_1	u	σ'_1	σ'_3	$\sigma_1 - \sigma_3$	σ'_1/σ'_3	p	p'	q	A value
0	0	0	4.877	0	0	92.90	85.9	7.00	7.0	0	1.000	92.9	7.00	0	0
0.050	0.050	0.68	4.910	20	20	96.97	86.2	10.77	6.7	4.07	1.607	94.9	8.74	2.04	0.074
0.100	0.100	1.36	4.944	36	36	100.18	87.0	13.18	5.9	7.28	2.234	96.5	9.54	3.64	0.151
0.150	0.150	2.04	4.979	48	48	102.54	88.6	13.94	4.3	9.64	3.242	97.7	9.12	4.82	0.280
0.200	0.200	2.73	5.014	59	59	104.67	89.4	15.27	3.5	11.77	4.363	98.8	9.39	5.89	0.297
0.250	0.250	3.41	5.049	67	67	106.17	89.7	16.47	3.2	13.27	5.147	99.5	9.84	6.64	0.286
0.300	0.300	4.09	5.085	71	71	106.86	89.8	17.06	3.1	13.96	5.504	99.9	10.08	6.98	0.279
0.350	0.350	4.78	5.122	73	73	107.15	89.4	17.75	3.5	14.25	5.072	100.0	10.63	7.13	0.246
0.400	0.400	5.46	5.159	74	74	107.24	89.3	17.94	3.6	14.34	4.985	100.1	10.77	7.17	0.237
0.450	0.450	6.14	5.196	76	76	107.53	89.2	18.33	3.7	14.63	4.953	100.2	11.01	7.31	0.226
0.500	0.500	6.82	5.234	78	78	107.80	89.1	18.70	3.8	14.90	4.922	100.4	11.25	7.45	0.215
0.550	0.550	7.51	5.273	78	78	107.69	88.9	18.79	4.0	14.79	4.698	100.3	11.40	7.40	0.203
0.600	0.600	8.19	5.312	80	80	107.96	88.8	19.16	4.1	15.06	4.673	100.4	11.63	7.53	0.193
0.650	0.650	8.87	5.352	81	81	108.04	88.8	19.24	4.1	15.14	4.692	100.5	11.67	7.57	0.192
0.700	0.700	9.55	5.392	83	83	108.29	88.8	19.49	4.1	15.39	4.754	100.6	11.80	7.70	0.188
0.750	0.750	10.24	5.433	84	84	108.36	88.6	19.76	4.3	15.46	4.596	100.6	12.03	7.73	0.175
0.800	0.800	10.92	5.475	86	86	108.61	88.6	20.01	4.3	15.71	4.653	100.6	12.15	7.85	0.172
0.850	0.850	11.60	5.517	88	88	108.85	88.5	20.35	4.4	15.95	4.625	100.9	12.38	7.98	0.163
0.900	0.900	12.28	5.560	89	89	108.91	88.4	20.51	4.5	16.01	4.557	100.9	12.50	8.00	0.156
0.950	0.950	12.97	5.604	89	89	108.78	88.3	20.48	4.6	15.88	4.453	100.8	12.54	7.94	0.151
1.000	1.000	13.65	5.648	89	89	108.66	88.2	20.46	4.7	15.76	4.353	100.8	12.58	7.88	0.146
1.050	1.050	14.33	5.693	90	90	108.71	88.1	20.61	4.8	15.81	4.294	100.8	12.70	7.90	0.139
1.100	1.100	15.01	5.739	91	91	108.76	88.0	20.76	4.9	15.86	4.236	100.8	12.83	7.93	0.132

(f) End of test: Water content: 27.1%   Start date: 12-18-99   End date: 12-19-99   Strain rate: 0.5%/h

FIGURE 3.30   (Continued)

The data recorded during the triaxial shearing include the dial reading corresponding to the axial deformation (Column 1, Part e, Fig. 3.30), the axial load as recorded by the load cell (Column 5, Part e, Fig. 3.30), and the pore water pressure (Column 8, Part e, Fig. 3.30). Prior to completing the shear test data (Part e), the following items must first be calculated:

Preshear Height of the Soil Specimen (H_s). The preshear height H_s of the soil specimen is calculated as the initial specimen height H_o minus the change in dial readings Δh, or:

$$H_s = H_o - \Delta h \qquad (3.22)$$

where H_s = preshear height of the soil specimen (in. or m)
$\quad H_o$ = initial height of the soil specimen (in. or m). This value is obtained from Part a of Fig. 3.30.
$\quad \Delta h$ = change in height of the soil specimen caused by the consolidation process (in. or m). The change in height of the soil specimen is calculated as the difference between the initial dial reading (from Part b, Fig. 3.30) and the preshear dial reading (d_s from Part e, Fig. 3.30).

Preshear Volume of the Soil Specimen V_s. The preshear volume V_s of the soil specimen is calculated as the initial specimen volume V_o minus the change in burette readings ΔV, or:

$$V_s = V_o - \Delta V \qquad (3.23)$$

where V_s = preshear volume of the soil specimen (in.^{3} or m^3)
$\quad V_o$ = initial volume of the soil specimen (in.^{3} or m^3). This value is obtained from Part a of Fig. 3.30.
$\quad \Delta V$ = change in volume of the soil specimen caused by the consolidation process (in.^{3} or m^3). The change in volume of the soil specimen is calculated as the difference between the initial burette reading (from Part b, Fig. 3.30) and the final end of consolidation burette reading (from Part c, Fig. 3.30).

Preshear Area of the Soil Specimen A_s. By knowing the preshear height of the soil specimen H_s and the preshear volume of the soil specimen V_s, the preshear area can be calculated as the volume divided by the height, or $A_s = V_s/H_s$.

Preshear Cell Pressure σ_c and Pore Water Pressure u_s. In Part e of Fig. 3.30, the preshear cell pressure and the preshear pore water pressure are also recorded. The preshear cell pressure σ_c must be identical to the highest chamber fluid pressure that the soil specimen was consolidated under. The preshear pore water pressure u_s should be approximately equal to the back pressure. The shearing portion consists of an undrained loading and thus just prior to shearing, the valve to the drainage measurement system and back pressure system is closed. But since the soil specimen should have been allowed to completely consolidate under the final effective confining pressure, there should be no change in pore water pressure within the soil specimen when the valve to the drainage measurement system and back pressure system is closed.

The shear test data table (i.e., see Part e, Fig. 3.30) has been divided into five different parts. Each part is individually discussed later:

1. *Axial deformation data.* During the shearing process, dial readings are taken that monitor the axial deformation of the soil specimen. These dial readings are recorded in Column 1, Part e, Fig. 3.30. If the initial dial reading is not zero, then the change in height of the soil specimen during the shearing process is defined as ΔH and is equal to the difference in the recorded dial reading and the initial dial reading.

The axial strain ε is calculated as the change in dial readings divided by the preshear height of the soil specimen, or $\varepsilon = \Delta H/H_s$. These calculations are listed in Column 3, Part e, Fig. 3.30 and the axial strain has been expressed as a percentage. As the soil specimen deforms during shearing, the height of the soil specimen decreases. But because it is an undrained triaxial test, the volume of the soil specimen does not change. It is common to assume that during shearing, the soil specimen retains its cylindrical shape, but this is often not the case as shown by the shear failure condition in Fig. 3.26. Nevertheless, assuming the soil specimen retains its cylindrical form during shearing, the corrected area A_c of the soil specimen is calculated as follows:

$$A_c = \frac{A_s}{1 - \varepsilon} \tag{3.24}$$

where A_c = corrected area of the soil specimen (in.^{2} or m^2)
A_s = preshear area of the soil specimen (in.^{2} or m^2)
ε = axial strain expressed in decimal form (dimensionless)

The corrected area determined from Eq. 3.24 is listed in Column 4, Part e, Fig. 3.30. During the triaxial shearing of the soil specimen, an axial load is induced by the piston to the top of the soil specimen. This axial load is measured by using either a load cell or proving ring. The data are listed in Column 5, Part e, Fig. 3.30. In this case, the load cell directly records the axial load. In other cases, such as for a proving ring, the data will need to be converted in order to obtain the axial load.

The major principal total stress σ_1 is equal to the vertical total stress σ_v. As defined by Eq. 3.11, the vertical total stress σ_v is equal to the chamber fluid pressure σ_c plus the axial load induced by the piston P divided by the corrected area of the specimen A_c. The calculations using Eq. 3.11 are listed in Column 7, Part e, Fig. 3.30. During the triaxial shearing of the soil specimen, it is possible that the filter paper strips and rubber membrane can offer some resistance to the axial load. Thus the vertical total stress acting on the soil specimen may need to be reduced to account for the axial resistance of the filter paper strips and rubber membrane. ASTM recommends that a correction for filter paper strips and the rubber membrane be applied only if the resistance exceeds 5 percent of the deviatoric pressure. As a practical matter, the axial resistance of the filter paper strips and rubber membrane can be neglected. However, in the case of very soft cohesive soils, these corrections may need to be included (see ASTM D 4767-02, 2004, for correction equations).

2. *Pore water pressure data.*   During the triaxial shearing, the pore water pressure (u) in the soil specimen is measured by an electronic pressure transducer. The soil specimen is sheared slow enough so that the excess pore water pressures are able to equalize within the soil specimen. Because both the soil specimen and the pore water pressure measurement system are saturated, there is a transfer of fluid pressure from the soil specimen, through the connecting tube, and to the transducer. The pore water pressures are listed in Column 8, Part e, Fig. 3.30.

The major principal effective stress σ'_1 and the minor principal effective stress σ'_3 are calculated by using Eqs. 3.13 and 3.14. For Eq. 3.13, the minor principal total stress σ_3 is equal to the cell pressure σ_c.

3. *Failure.*   The failure of the soil specimen can be based on total stresses or effective stresses. For total stresses, failure is defined as the maximum value of the deviatoric pressure (i.e., maximum value of $\sigma_1 - \sigma_3$). For effective stresses, failure is defined as the maximum effective stress obliquity, i.e., the maximum value of σ'_1/σ'_3.

4. *Stress paths and A value.*   The final columns in Part e, Fig. 3.30 are used to list various parameters that are used to plot the stress paths and determine the A value.

For the analysis of the laboratory data, usually several different plots are prepared, as follows:

Axial strain versus pore water pressure.   Figure 3.31 shows a plot of the axial strain ε versus the excess pore water pressure u_e generated during the shearing of the soil specimen. The axial strain ε was obtained from Column 3, Part e, Fig. 3.30, and the excess pore water pressure u_e was calculated by subtracting the back pressure (85.9 psi) from the pore water pressure u values listed in Column 8, Part e, Fig. 3.30. The excess pore water pressures initially increase, but then decrease once the axial strain reaches about 4 percent.

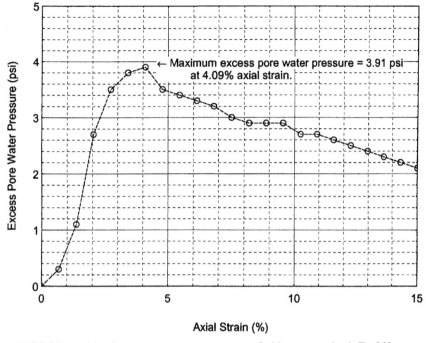

FIGURE 3.31 Axial strain versus excess pore water pressure for laboratory test data in Fig. 3.30.

Axial strain versus deviatoric pressure. Figure 3.32 shows a plot of the axial strain ε versus the deviatoric pressure ($\sigma_1 - \sigma_3$) generated during the shearing of the soil specimen. This plot is also commonly referred to as a stress-strain plot. The axial strain ε was obtained from Column 3, Part *e*, Fig. 3.30, and the deviatoric pressure represents the failure condition based on a total stress analysis and the values are listed in Column 11, Part *e*, Fig. 3.30. In terms of failure based on a total stress analysis, the maximum deviatoric pressure (i.e., maximum $\sigma_1 - \sigma_3$) is equal to 16.01 psi, which occurs at an axial strain of 12.28 percent. Thus in terms of a total stress analysis, the shear strength is mobilized at an axial strain of 12.28 percent.

Axial strain versus effective stress obliquity. Figure 3.33 shows a plot of the axial strain ε versus the effective stress obliquity (σ'_1/σ'_3) generated during the shearing of the soil specimen. The axial strain ε was obtained from Column 3, Part *e*, Fig. 3.30, and the effective stress obliquity represents the failure condition based on an effective stress analysis and the values are listed in Column 12, Part *e*, Fig. 3.30. In terms of failure based on an effective stress analysis, the maximum effective stress obliquity (i.e., maximum σ'_1/σ'_3) is equal to 5.504, which occurs at an axial strain of 4.09 percent. For the effective stress analysis, the shear strength is mobilized at an axial strain of 4.09 percent.

Effective stress path. Figure 3.34 shows the effective stress path, which is a plot of p' (Column 14, Part *e*, Fig. 3.30) versus q (Column 15, Part *e*, Fig. 3.30). The values of p' and q were calculated by using Eq. 3.16. The solid circle in Fig. 3.34 indicates the failure condition of the soil specimen in terms of the maximum effective stress obliquity (i.e., maximum value of σ'_1/σ'_3). The shear strength failure envelope must pass through this point. It has been observed that for continued strain after the maximum effective stress obliquity, that the effective stress path tends to progress along the shear strength envelope. In other words, the soil specimen continues to mobilize its shear strength with further straining beyond maximum obliquity. Thus the failure envelope

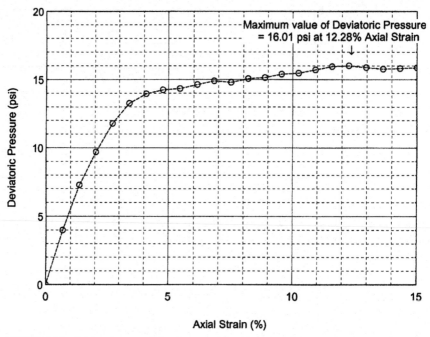

FIGURE 3.32   Axial strain versus deviatoric pressure for laboratory test data in Fig. 3.30.

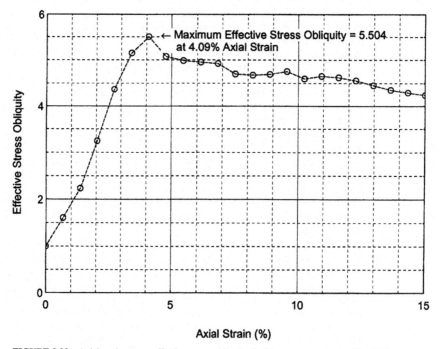

FIGURE 3.33   Axial strain versus effective stress obliquity for laboratory test data in Fig. 3.30.

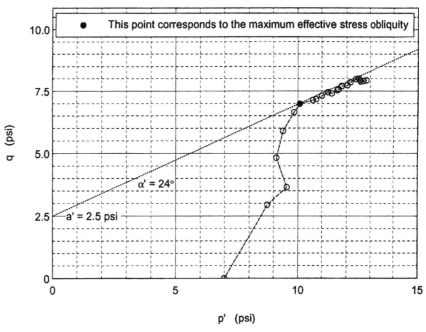

FIGURE 3.34   Effective stress path for laboratory test data in Fig. 3.30.

in Fig. 3.34 has been drawn through the data point corresponding to maximum effective stress obliquity as well as the data points representing axial strain beyond maximum obliquity.

The inclination of the straight line in Fig. 3.34 is defined as the angle α' and the intersection of the straight line on the vertical axis is defined as a'. For the straight line shown in Fig. 3.34, $\alpha' = 24°$ and $a' = 2.5$ psi (17 kPa). Using Eq. 3.18, the effective shear strength parameters can be calculated and they are $\phi' = 26°$ and $c' = 2.8$ psi (19 kPa).

A Values.   The last column in Part e, Fig. 3.30 presents the A values (calculated by using Eq. 3.21). At the maximum value of the effective stress obliquity, the A value is 0.279. Based on this A value, this clay when tested at an effective confining pressure of 7.0 psi (48 kPa) is considered to be an over-consolidated clay. The effective shear strength parameters determined for this test (i.e., $\phi' = 26°$ and $c' = 2.8$ psi) are indicative of an overconsolidated clay.

3.5.3 Unconfined Compression Test

The undrained shear strength has been introduced in Sec. 2.4.4, where it was discussed that the undrained shear strength s_u can be obtained from the vane shear test (VST) or by using a miniature vane. As the name implies, the undrained shear strength s_u refers to a shear condition where water does not enter or leave the cohesive soil during the shearing process. In essence, the water content of the soil must remain constant during the shearing process. There are many projects where the undrained shear strength is used in the design analysis. In general, these field situations must involve loading or unloading of the cohesive soil at a rate that is much faster than the shear induced pore water pressures can dissipate.

During rapid loading of saturated cohesive soils, the shear induced pore water pressures can only dissipate by the flow of water into (negative shear induced pore water pressures) or out of (positive shear induced pore water pressures) the soil. Cohesive soil has a low permeability, and if the load is applied quick enough, there will not be enough time for water to enter or leave the cohesive soil. For such a quick loading condition of a saturated cohesive soil, the undrained shear strength s_u should be used in the analysis.

Besides the VST, another type of test that can be used to determine the undrained shear strength is the unconfined compression test. The unconfined compression test is a very simple type of test that consists of applying a vertical compressive pressure to a cylinder of laterally unconfined cohesive soil. The unconfined compression test is also known as a *simple compression test*.

The unconfined compression test is most frequently performed on cohesive soils that are in a saturated condition, such as soil obtained from below the groundwater table. Because the soil specimen is laterally unconfined during testing (i.e., no lateral confining pressure), the soil specimen must be able to retain its plasticity during the application of the vertical pressure. In addition, the soil must not expel water (known as *bleed water*) during the compression test. For these reasons, the unconfined compression test is most frequently performed on saturated clays. Soils that tend to crumble, fall apart, or bleed water during the application of the vertical pressure should not be tested. Examples include fissured or varved clays, silts, peat, and all types of granular soils.

The test procedures for the unconfined compression test are as follows:

Soil Specimen.   An undisturbed soil specimen must be used for the unconfined compression test. The test is most frequently performed on saturated clays that retain their plasticity and do not bleed water during loading. The soil specimen must have a minimum diameter of 1.3 in. (33 mm) and the specimen height divided by the specimen diameter should be between 2 and 2.5. The trimming process should be performed as quick as possible to minimize the possibility of a change in water content of the cohesive soil. Often the soil specimen will be simply extruded from a Shelby tube and then the top and bottom of the specimen are trimmed flush. The top and bottom of the soil specimen must be trimmed so that they are perpendicular to the longitudinal axis of the specimen. In this case where the soil sample is extruded from a sampling tube, the initial diameter of the soil specimen is equal to the inside diameter of the sampling tube.

Loading Device.   The triaxial apparatus can be used as the loading device. In this sense, the unconfined compression test could be considered to be a special form of the triaxial test. Besides the triaxial apparatus, there are other types of loading devices that can accurately measure the vertical load applied to the soil specimen and record the vertical deformation during shearing. Examples include compression devices having a screw-jack activated load yoke or a hydraulic loading mechanism. The equipment must be capable of loading the soil specimen at a constant rate of strain.

Shearing the Soil Specimen.   The cylindrical soil specimen is placed in the loading apparatus so that it is centered on the bottom platen. It is important to carefully handle the soil specimen in order to minimize disturbance and prevent a change in cross section. The loading device is then adjusted so that the upper platen just makes contact with the top of the soil specimen. The dial gauge that measures the vertical deformation is then read or set to zero. Throughout the test, the soil specimen is unconfined in the horizontal direction.

Apply a Vertical Load to the Top of the Soil Specimen.   The soil specimen is typically sheared in a controlled strain apparatus at a strain rate of 0.5 to 2 percent axial strain per min. Since the usual procedure is to shear the soil specimen to an axial strain of 15 percent, the shearing portion of the test could take between 7.5 to 30 min. However, ASTM D 2166-00 (2004) recommends that the strain rate be chosen so that the time to failure (i.e., maximum σ_v) does not exceed 15 min. Softer cohesive soils will need a larger deformation to reach failure and should be tested at the higher strain rate. Stiff cohesive soils will need less deformation to reach failure and should be tested at the lower rate of strain.

Failure of the Soil Specimen.   For the unconfined compression test, failure is based on total stresses and occurs at the maximum principal total stress difference (i.e., the maximum deviatoric pressure, or maximum value of $\sigma_1 - \sigma_3$). Since the minor principal total stress σ_3 is equal to zero during the entire test procedure, failure corresponds to the maximum value of σ_1 which is equal to the maximum vertical total pressure (maximum σ_v) that the cohesive soil specimen can sustain. If the highest value of

σ_v occurs after 15 percent axial strain, then σ_v corresponding to 15 percent axial strain should be considered to be the failure condition.

Similar to the triaxial test, the change in height of the soil specimen during the shearing process is defined as ΔH and is equal to the difference in the recorded dial reading and the initial dial reading. The axial strain ε is calculated as the change in dial readings divided by the initial height of the soil specimen, or $\varepsilon = \Delta H/H_o$. As the soil specimen deforms during shearing, the height of the soil specimen decreases. But because it is an undrained triaxial test, the volume of the soil specimen does not change. It is common to assume that during shearing, the soil specimen retains its cylindrical shape and the corrected area A_c of the soil specimen is calculated using Eq. 3.24.

The axial load during shearing is measured by using either a load cell or proving ring. The major principal total stress σ_1 is equal to the vertical total stress σ_v. The vertical total stress σ_v is equal to the axial load P divided by the corrected area of the specimen (A_c), or:

$$\sigma_1 = \sigma_v = P/A_c \tag{3.25}$$

Shear failure for a total stress analysis is defined as the highest value of the vertical total stress (i.e., maximum value of σ_v). This highest value of σ_v is typically designated q_u and is known as the *unconfined compressive strength*. A Mohr circle in terms of total stresses could be drawn for the unconfined compression test. Since the lateral pressure is zero (i.e., no confining pressure), the minor principal total stress σ_3 is equal to zero. The major principal total stress σ_1 at failure is equal to the highest value of the vertical stress (i.e., maximum value of σ_v) which is designated as q_u. Thus the Mohr circle in terms of total stresses would have $\sigma_3 = 0$ and $\sigma_1 = q_u$.

For the unconfined compression test, the shear strength is defined as the peak point on the Mohr circle, which is equal to the maximum shear stress τ_{max}. The maximum shear stress is equal to the radius of the Mohr circle and thus the undrained shear strength s_u is equal to:

$$s_u = \tau_{max} = q_u/2 \tag{3.26}$$

where s_u = undrained shear strength of the cohesive soil (psf or kPa)

τ_{max} = maximum shear stress (psf or kPa) that the soil can withstand, which for this total stress analysis is assumed to be equal to the undrained shear strength

q_u = unconfined compressive strength of the soil (psf or kPa). The unconfined compressive strength of the soil is equal to the highest value of σ_v

A limitation of the unconfined compression test is the definition of the undrained shear strength s_u, which is assumed to be equal to the maximum shear stress (i.e., $s_u = \tau_{max}$). Using the Mohr circle (Fig. 3.27), the failure plane should always be inclined at an angle θ equal to 45°. But if a failure plane does develop during shear, it usually develops at an angle of inclination that is greater than 45°. Thus the actual undrained shear strength along the failure surface is less than τ_{max} and once again s_u could be overestimated.

Even with this limitation, the unconfined compression test is a popular shear strength test. Its popularity is due to the fact that it is one of the quickest and simplest laboratory tests used to measure the shear strength of cohesive soil. In addition, the unconfined compression test has the advantage that the failure surface will tend to develop in the weakest portion of the cohesive soil. In contrast, the VST forces the cohesive soil to shear along vertical and horizontal surfaces, which may not be the weakest surfaces.

For further details on the unconfined compression test, see ASTM D 2166-00 (2004), "Standard Test Method for Unconfined Compressive Strength of Cohesive Soil."

3.5.4 Unconsolidated Undrained Triaxial Compression Test

An alternate approach to determining the undrained shear strength s_u of the cohesive soil is to obtain the undrained shear strength in terms of the total shear strength parameters (c and ϕ). The usual process is to perform triaxial compression tests on specimens of the cohesive soil. The shear strength τ_f of cohesive soil for a total stress analysis can be expressed as follows (Mohr-Coulomb failure law):

$$\tau_f = c + \sigma_n \tan \phi \qquad (3.27)$$

where τ_f = shear strength of the soil (psf or kPa)
c = cohesion based on a total stress analysis (psf or kPa)
σ_n = total normal stress acting on the shear surface (psf or kPa)
ϕ = friction angle based on a total stress analysis (degrees)

Equation 3.27 is identical to Eq. 3.9, except that one equation is expressed in terms of effective stress (Eq. 3.9) and the other is expressed in terms of total stress (Eq. 3.27).

As previously mentioned, the unconsolidated undrained triaxial compression test is rarely used in engineering practice. It is more common to either use the undrained shear strength s_u from unconfined compression tests or vane shear tests, or to use the shear strength parameters (c and ϕ) from consolidated undrained triaxial compression tests.

The unconsolidated undrained triaxial compression test specifications are presented in ASTM D 2850-03 (2004), "Standard Test Method for Unconsolidated-Undrained Triaxial Compression Test on Cohesive Soils." The test procedure consists of placing the cylindrical specimen of cohesive soil in the center of the triaxial apparatus, sealing the soil specimen in a rubber membrane, and then subjecting the soil specimen to a confining pressure without allowing it to have a change in water content at any time during the triaxial testing. The soil is sheared in the triaxial apparatus by increasing the vertical stress on the soil specimen.

Figure 3.35 shows the results of four unconsolidated undrained triaxial compression tests (ASTM D 2850-03, 2004) performed on a cohesive soil (LL = 64, PI = 33) having a dry unit weight $\gamma_d = 70$ pcf (11.0 kN/m³), a water content $w = 49.3$ percent, and a degree of saturation (S) = 95 percent. Plots of the deviatoric pressure versus axial strain for these four tests are also shown on the left side of Fig. 3.35. The stress-strain plots shown on the left side of Fig. 3.35 do not have a peak point and are indicative of normally consolidated (OCR = 1) clays.

The middle plot in Fig. 3.35 shows the Mohr circles at failure for the four triaxial tests. These four triaxial tests are labeled 1 through 4 in Fig. 3.35 and the data are summarized below:

Test 1. Confining pressure = 0 (hence this is identical to an unconfined compression test). The vertical total stress at failure $\sigma_1 = 2.87$ psi (19.8 kPa). Thus the Mohr circle drawn in Fig. 3.35 has $\sigma_1 = 2.87$ psi, $\sigma_3 = 0$, and a radius of 1.44 psi.

Test 2. Confining pressure = 1.70 psi. The vertical total stress at failure $\sigma_1 = 5.87$ psi (40.5 kPa). Thus the Mohr circle drawn in Fig. 3.35 has $\sigma_1 = 5.87$ psi, $\sigma_3 = 1.70$ psi, and a radius of 2.1 psi.

Test 3. Confining pressure = 3.50 psi. The vertical total stress at failure $\sigma_1 = 7.72$ psi (53.2 kPa). Thus the Mohr circle drawn in Fig. 3.35 has $\sigma_1 = 7.72$ psi, $\sigma_3 = 3.50$ psi, and a radius of 2.1 psi.

Test 4. Confining pressure = 6.90 psi. The vertical total stress at failure $\sigma_1 = 10.93$ psi (75.4 kPa). Thus the Mohr circle drawn in Fig. 3.35 has $\sigma_1 = 10.93$ psi, $\sigma_3 = 6.90$ psi, and a radius of 2.0 psi.

For Test 1, the undrained shear strength s_u, which is the peak of the Mohr circle, is equal to 1.44 psi (9.9 kPa). Test 2 has a higher shear strength of 2.1 psi (14.5 kPa). This higher shear strength for test 2 can be attributed to the slight compression of the clay when a confining pressure of 1.70 psi (11.7 kPa) was applied to the soil specimen. Initially, the soil specimen was not saturated ($S = 95$ percent), but the confining pressure compressed the clay and created a saturated condition for Tests 2 through 4.

Tests 2 through 4 have essentially the same shear strength (peak point on the Mohr circle). The reason is because the increased cell pressure for tests 3 and 4 did not densify the soil specimen, but rather only increased the pore water pressure of the clay specimens. In essence, the effective stress did not change for tests 2 through 4 and thus the shear strength could not increase. This is an important concept in geotechnical engineering, that the shear strength of saturated soil can not increase unless the effective stress of the soil first increases.

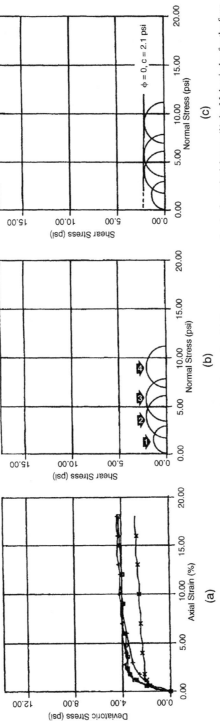

FIGURE 3.35 Unconsolidated undrained triaxial compression test data performed on a clay. (*a*) Stress-strain curves for the four triaxial tests; (*b*) the Mohr circles for the four triaxial tests; (*c*) the failure envelope for the clay.

For the plot on the right side Fig. 3.35, the failure envelope has been drawn for tests 2 through 4. Using Eq. 3.27 (Mohr-Coulomb failure law), the total stress parameters are $\phi = 0$ and $c = 2.1$ psi (14.5 kPa). These types of shear strength results have been termed the "$\phi = 0$ concept." It is important to recognize that this concept does not say that the clay has zero frictional resistance (i.e., $\phi = 0$). Instead, this concept is indicating that there is no shear strength increase in the saturated clay for the condition of a rapidly applied load. The saturated clay must first consolidate (flow of water from the clay) before it can increase its shear strength.

An example of the use of the shear strength data shown in Fig. 3.35 would a quick loading condition. Examples are the fast construction of a building or embankment fill on top of this clay. Because the construction is quick, there would be no drainage (i.e., consolidation) of the clay during construction. The most common method to model this quick loading condition would be to use the undrained shear strength s_u from tests described in Sec. 3.5.3. However, as an alternative, it would be appropriate to use the undrained shear strength parameters from Fig. 3.35, i.e., $\phi = 0$ and $c = 2.1$ psi (14.5 kPa).

3.5.5   Consolidated Undrained Triaxial Compression Test

As previously discussed, the consolidated undrained triaxial compression test and the consolidated undrained triaxial compression test with pore water pressure measurements are identical, except that the latter test measures the pore water pressure during shearing. In practice, the consolidated undrained triaxial compression test with pore water measurements (ASTM D 4767-02, 2004) is usually performed and then the pore water pressure data are ignored in order to determine the shear strength (c and ϕ) in terms of a total stress analysis.

In some cases, it may be appropriate to use the total stress shear strength parameters (c and ϕ) from consolidated undrained triaxial compression tests. For example, a structure (such as an oil tank or grain-elevator) could be constructed and then sufficient time elapses so that the saturated cohesive soil consolidates under this load. If the oil tank or grain-elevator is then quickly filled, the saturated cohesive soil would be subjected to an undrained loading. This condition can be modeled by performing consolidated undrained triaxial compression tests (ASTM D 4767-02, 2004) in order to determine the total stress parameters (c and ϕ).

As discussed in Sec. 3.5.2, the test procedure consists of placing a cylindrical specimen of cohesive soil in the center of the triaxial apparatus, sealing the specimen in a rubber membrane, applying a confining pressure, and then allowing enough time for the soil specimen to consolidate. At the completion of consolidation, an axial load is applied to the soil specimen without allowing a change in water content (i.e., undrained loading). Failure is based on total stresses and occurs at the maximum principal total stress difference (i.e., maximum deviatoric pressure, or maximum value of $\sigma_1 - \sigma_3$). At failure, the minor principal total stress $\sigma_3 =$ the confining pressure and the major principal total stress $\sigma_1 =$ the highest vertical pressure. By knowing σ_3 and σ_1, the Mohr circle at failure can be drawn.

Figure 3.36 presents laboratory data from three triaxial tests performed in accordance with ASTM D 4767-02 (2004) test specifications (Day and Marsh, 1995). The three specimens used for the triaxial tests were composed of silty clay, having a LL = 54, PI = 28, with a grain size distribution (based on dry weight) of 15 percent fine sand size particles, 33 percent silt size particles, and 52 percent clay size particles finer than 0.002 mm. In order to create the three triaxial specimens, the silty clay was compacted into a cylindrical mold to a relative compaction of 80 percent (i.e., dry unit weight = 85 pcf, 13.4 kN/m^3) at a water content of 20 percent. These three triaxial tests are labeled 1 through 3 in Fig. 3.36 and the data are summarized below:

Test 1.   The first silty clay specimen was saturated and allowed to consolidate at an effective confining pressure of 3.5 psi (24 kPa). The vertical total stress at failure (σ_1) = 9.4 psi (65 kPa). Thus the Mohr circle drawn in Fig. 3.36 has $\sigma_1 = 9.4$ psi, $\sigma_3 = 3.5$ psi, and a radius of 3.0 psi.

Test 2.   The second silty clay specimen was saturated and allowed to consolidate at an effective confining pressure of 13.9 psi (96 kPa). The vertical total stress at failure (σ_1) = 26.6 psi (183 kPa). Thus the Mohr circle drawn in Fig. 3.36 has $\sigma_1 = 26.6$ psi, $\sigma_3 = 13.9$ psi, and a radius of 6.4 psi.

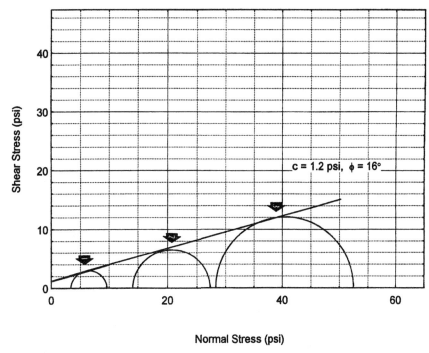

FIGURE 3.36   Consolidated undrained triaxial compression tests performed on a silty clay.

Test 3.   The third silty clay specimen was saturated and allowed to consolidate at an effective confining pressure of 27.8 psi (192 kPa). The vertical total stress at failure (σ_1) = 52.2 psi (360 kPa). Thus the Mohr circle drawn in Fig. 3.36 has σ_1 = 52.2 psi, σ_3 = 27.8 psi, and a radius of 12.2 psi.

The failure envelope (straight line) is drawn at a tangent to the three Mohr circles and the cohesion (*y* axis intercept) and friction angle (angle of inclination of the straight line) are *c* = 1.2 psi (8.3 kPa) and ϕ = 16°.

3.5.6   Consolidated Drained Triaxial Compression Test

The test procedures for the consolidated drained triaxial compression test are identical to the test procedures described in Sec. 3.5.2, except that during shearing the valve to the drainage measurement system and back pressure system remains open. For this test, the cohesive soil specimen must be sheared slow enough so that excess pore water pressures do not develop within the soil specimen. This means that if the cohesive soil wants to dilate during shear, then the shear process must be slow enough so that water can flow into the soil specimen. Likewise, if the cohesive soil wants to contract during shear, then the shear process must be slow enough so that water can flow out of the soil specimen.

Depending on the size of the soil specimen and the permeability of the soil, the requirement that the soil be sheared in a drained condition could require that the shearing portion of the test take several days or even weeks to complete. For this reason, the consolidated drained triaxial compression test is rarely performed in practice for cohesive soil.

3.5.7 Direct Shear Test

Cohesive soils usually have a very low permeability and it often takes a considerable amount of time to perform drained shear strength tests. The test procedure for performing a direct shear test on cohesive soil is identical to the procedure for cohesionless soil as outlined in Sec. 3.4.2. However, a major limitation is that cohesive soils must be sheared much slower than granular soils in order to obtain the drained shear strength. According to ASTM D 3080-03 (2004), the total elapsed time to failure t_f for a drained direct shear test should be:

$$t_f = 50\ t_{50} \tag{3.28}$$

where t_{50} is the time for the cohesive soil to achieve 50 percent consolidation under the vertical stress. This time t_{50} can be determined by using the log-of-time method (discussed in Chap. 8). For cohesive soil, Eq. 3.28 can require that the test be performed for a considerable amount of time. For example, for a clay that reaches 50 percent consolidation in 20 min, the total elapsed time to failure must be at least 1000 min (16.7 h). This may require the installation of special equipment (special gears to slow down the strain rate) to allow the direct shear device to shear very slowly. Probably the two most common problems with the direct shear testing of a cohesive soil are: (1) the soil is not saturated prior to shearing, and (2) the soil is sheared too quickly. Both of these conditions can result in the effective shear strength parameters (c' and ϕ') being overestimated.

3.5.8 Torsional Ring Shear Test

The drained residual shear strength ϕ'_r is defined as the remaining (or residual) shear strength of cohesive soil after a considerable amount of shear deformation has occurred. In essence, ϕ'_r represents the minimum shear resistance of a cohesive soil along a fully developed failure surface. The drained residual shear strength is primarily used to evaluate slope stability, where there is a preexisting shear surface. Examples include ancient landslides, slopes in overconsolidated fissured clays, and slopes in fissured shales. The drained residual shear strength may also be applicable for other types of preexisting shear surfaces, such as sheared bedding planes, joints, and faults. The residual shear strength ϕ'_r is independent of the original shear strength, water content, and liquidity index, and depends only on the size, shape, and mineralogical composition of the constituent particles (Skempton, 1964).

The drained residual friction angle ϕ'_r of cohesive soil could be determined by using the direct shear apparatus. The test procedure is similar to the standard direct shear testing, except that the soil specimen is sheared back and forth several times to develop a well-defined shear failure surface. By shearing the soil specimen back and forth, the clay particles become oriented parallel to the direction of shear. Once the shear surface is developed, the drained residual shear strength ϕ'_r would be determined by performing a final, slow shear of the specimen.

Besides the direct shear equipment, the drained residual shear strength can be determined by using the torsional ring shear apparatus, which is shown in Fig. 3.37 (Stark and Eid, 1994). Back calculations of landslide shear strength indicate that the drained residual shear strength from torsional ring shear tests are reasonably representative of the slip surface (Watry and Ehlig, 1995). The basic test procedures are as follows:

1. The torsional ring shear specimen is annular with a typical inside diameter of 2.8 in. (7 cm) and an outside diameter of 4 in. (10 cm). Drainage is provided by annular bronze porous plates secured to the bottom of the specimen container and the loading platen. Remolded specimens are often used for the ring shear testing.

2. The remolded specimen is typically obtained by air drying the soil (such as slide plane clay seam material), crushing it with a mortar and pestle, and processing it through the U.S. Standard Sieve No. 200.

3. Distilled water is added to the processed soil until a water content approximately equal to the liquid limit is obtained. The specimen is then allowed to rehydrate for one day in a high-humidity room.

FIGURE 3.37   Torsional ring shear device.

A spatula is used to place the remolded soil paste into the annular specimen container. The top porous plate and the top plate which is used to apply a torque to the soil specimen are installed, as shown in Fig. 3.38.

4. To measure the drained residual shear strength, the ring shear specimen is often consolidated at a high vertical pressure (such as 700 kPa) and then the specimen is unloaded to a much lower vertical pressure (such as 50 kPa). During both consolidation and shearing, the soil specimen is submerged in distilled water.

5. The specimen is then presheared by slowly rotating the ring shear base for one complete revolution using the hand wheel.

6. After preshearing, the specimen is sheared at a slow drained displacement rate (such as 0.02 mm/min). This slow shear displacement rate has been successfully used to test soils that are very plastic. Figure 3.39 shows a close-up view of the loading rams that are used to apply the torque to the ring shear soil specimen.

7. After a drained residual strength condition is obtained at the low vertical pressure, shearing is stopped and the normal stress is increased to a higher pressure (such as 100 kPa). After consolidation at this higher pressure, the specimen is sheared again until a drained residual condition is obtained.

FIGURE 3.38 Close-up view of the shear portion of the apparatus.

8. This procedure is also repeated for other effective normal stresses. The slow shear displacement rate (0.02 mm/min) should be used for all stages of the multistage test. Figure 3.40 shows the soil specimen at the completion of the torsional ring shear test.

For further details, see ASTM D 6467-99 (2004), "Standard Test Method for Torsional Ring Shear Test to Determine Drained Residual Shear Strength of Cohesive Soils."

An example of the type of data obtained from the torsional ring shear test is presented in Figs. 3.41 and 3.42. These data were obtained from the torsional ring shear testing of clay obtained from the slide plane of an actual landslide (Day and Thoeny, 1998). The index properties for the slide plane material are listed in Fig. 3.41. The landslide developed in the Friars Formation which consists primarily of montmorillonite clay particles (Kennedy, 1975). Figure 3.41 presents the drained residual failure envelope and Fig. 3.42 shows the stress-displacement plots for the torsional ring shear test on the slide plane sample. It can be seen in Fig. 3.41 that the failure envelope is nonlinear, which is a common occurrence for residual soil (Maksimovic, 1989a). If a linear failure envelope is assumed to pass through the origin and the shear stress at an effective normal stress of 2090 psf (100 kPa), the residual friction angle (ϕ'_r) is 8.2°. If a linear failure envelope is assumed to pass through the origin and the shear stress at an effective normal stress of 14,600 psf (700 kPa), the residual friction angle ϕ'_r is 6.2°. These drained residual friction

FIGURE 3.39 Close-up view of the shear portion of the apparatus showing the loading rams that are used to apply a torque to the top of the soil specimen.

angles are very low and are probably close to the lowest possible drained residual friction angles of soil.

Figure 3.43 presents a second example of data from laboratory tests performed on slide plane material from the Niguel Summit landslide. Figure 3.44 shows a photograph of the slide plane, which was exposed during the stabilization of the Niguel Summit landslide. The drained residual shear strength of the slide plane material was determined by using the torsional ring shear apparatus (solid line) and the direct shear apparatus (dashed line). As shown in Fig. 3.43, there is good correlation between the results of the torsional ring shear apparatus and the direct shear apparatus, with the results from the direct shear apparatus indicating a slightly higher residual shear strength.

FIGURE 3.40   Condition of the soil specimen at the end of the torsional ring shear test.

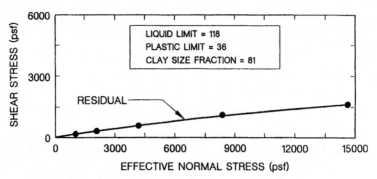

FIGURE 3.41   Drained residual shear strength envelope from torsional ring shear test on slide plane material.

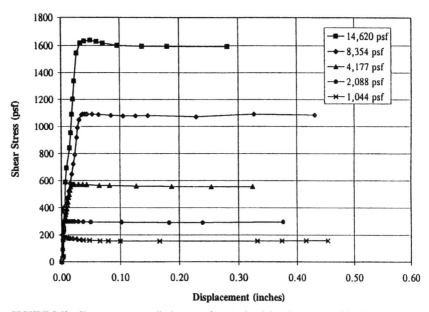

FIGURE 3.42 Shear stress versus displacement from torsional ring shear test on slide plane material.

3.6 LABORATORY COMPACTION TESTS

The laboratory compaction test consists of compacting a soil at a known water content into a mold of specific dimensions using a certain compaction energy. The procedure is repeated for various water contents to establish the compaction curve. The most common testing procedures (compaction energy, number of soil layers in the mold and the like.) are the Modified Proctor (ASTM D 1557-02, 2004) and the Standard Proctor (ASTM D 698-00, 2004). The term "Proctor" is in honor of R. R. Proctor, who in 1933 showed that the dry density of a soil for a given compactive effort depends on the amount of water the soil contains during compaction. This section will discuss the Modified Proctor laboratory compaction test and the Standard Proctor laboratory compaction test.

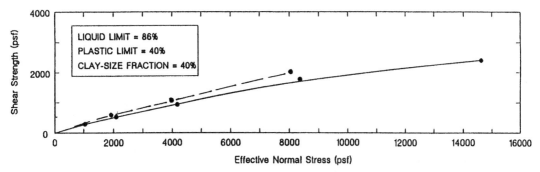

FIGURE 3.43 Drained residual shear strength envelope from torsional ring shear test (solid line) and direct shear test (dashed line) on slide plane material from the Niguel Summit landslide.

FIGURE 3.44 Photograph of the slide plane, which was exposed during the stabilization of the Niguel Summit landslide. Note that the direction of movement of the landslide can be inferred by the direction of striations in the slide plane.

3.6.1 Modified Proctor Compaction Test

In California, there is almost exclusive use of the Modified Proctor compaction specifications. For the Modified Proctor (ASTM D 1557-02, 2004, procedure A), the soil is compacted into a 4-in. (10.2 cm) diameter mold that has a volume of 1/30 ft³ (944 cm³), where five layers of soil are compacted into the mold with each layer receiving 25 blows from a 10-lbf (44.5-N) rammer that has a 18-in. (0.46-m) drop. The Modified Proctor has a compaction energy of 56,000 ft-lbf/ft³ (2700 kN-m/m³). The test procedure is to prepare soil at a certain water content, compact the soil into the mold, and then by recording the mass of soil within the mold, the wet density of the compacted soil is obtained. By knowing the water content of the compacted soil, the dry density can be calculated (i.e., divide the wet density by: $1 + w$, where w is in a decimal form). This compaction procedure is repeated for the soil at different water contents and then the dry density versus water content is plotted on a graph in order to obtain the compaction curve. The peak point of the compaction curve is known as the laboratory maximum dry density (pcf or Mg/m³) and is designated $\rho_{d\max}$. The water content corresponding to the peak point of the laboratory compaction curve is known as the *optimum moisture content*, which is designated w_{opt}. The specific steps in the Modified Proctor laboratory compaction test are as follows:

Sample Preparation.   The test procedure described in this section is applicable for soil that primarily passes the No. 4 sieve. If the soil has gravel and cobble size particles, then a different test procedure or even an oversize particle correction may be required (Day, 2001). It is important that during the compaction test, the soil not be reused. As such, a significant amount of soil is needed in order to complete the laboratory compaction test. Depending on the water content of the soil, about 20 to 40 lb (9 to 18 kg) of wet soil will be needed for the laboratory compaction test.

With experience and judgment, it is often possible to estimate the optimum moisture content w_{opt}. At least four soil specimens are prepared with water contents that bracket the estimated optimum moisture content. Two soil specimens should be prepared such that they have water contents below and two specimens should have water contents above the estimated optimum moisture content. The water contents should vary by about 2 percent, so for example, if the optimum moisture content is estimated to be 17 percent, then prepare soil specimens having water contents of 14 percent, 16 percent, 18 percent, and 20 percent. For some soils that have a very high optimum moisture content or have a flat compaction curve, the water contents of the soil specimens may need to vary by more than 2 percent. For each of the four soil specimens, about 5 lb (2.3 kg) of soil will be needed.

Compaction Mold.   Each of the four soil specimens will be compacted into the compaction mold. Figure 3.45 shows the compaction apparatus, which consists of three parts: base plate, compaction mold, and top collar. Figure 3.46 shows the compaction apparatus in its assembled condition, with the compaction mold fitting snugly into the base plate and then the top collar placed on top of the mold. The mold and collar are secured to the base plate by tightening the screws. Prior to compacting the soil specimens, a balance is used to determine the mass of the empty compaction mold M_o.

FIGURE 3.45   The compaction apparatus consisting of three parts: (1) base plate, (2) compaction mold, and (3) the top collar.

FIGURE 3.46 The compaction apparatus in an assembled condition.

The compaction apparatus must rest on a rigid foundation that has a mass of at least 200 lb (92 kg). The base plate is securely attached to the rigid foundation, as shown in Fig. 3.47. The purpose of this requirement is to enable the majority of the energy during compaction to be directed into the soil, with minimal loss of energy caused by the bouncing or shaking of the compaction apparatus. Chen (2000) describes an interesting case of a field technician who performed the laboratory compaction test on the open tailgate of his pickup truck instead of on a solid surface. Due to lost energy, the laboratory maximum dry density was as much as 5 pcf (0.08 Mg/m^3) less than the actual value. This example shows the importance of securely anchoring the base plate of the compaction apparatus to a rigid foundation.

Compacting the Soil Specimen. The soil specimen is compacted into the compaction mold in five layers, with each layer being of approximately equal thickness. For each layer, loose soil is placed into the compaction mold and spread-out so that it has a uniform thickness. Each of the five layers of soil receives 25 blows from a 10-lbf (44.5 N) rammer that has a 18 in. (0.46 m) drop. The rammer is contained within a guide sleeve, and during compaction, the guide sleeve is held in an vertical position and just above the soil layer surface. Holding the guide sleeve steady, the rammer is raised to the top of the guide sleeve. Then the rammer is released and allowed to free fall and impact the soil surface. Each layer of soil should be compacted so that the 25 blows are uniformly distributed with complete coverage of the soil layer surface. Figure 3.48 shows the rammer and the soil in the process of being compacted into the compaction mold.

FIGURE 3.47   The compaction apparatus attached to a rigid metal foundation.

After compaction of the fifth layer, the soil should extend no more than $1/4$ in. (6 mm) into the collar. If after compaction of the fifth layer, the soil is below the top of the compaction mold, then the specimen must be discarded and the test repeated on a fresh soil specimen. Figure 3.49 shows the completion of the compaction test with the collar having been removed.

The compaction mold is removed from the base plate and then both the top and bottom of the soil specimen are trimmed. The trimming process can be performed by scraping a metal straight edge across the top and bottom of the compaction mold in order to create flush soil surfaces. If small surface voids develop during the trimming process, then the surface voids should be filled-in with soil. Once the compacted soil has been trimmed flush with the top and bottom of the mold, a balance is used to obtain the mass of the mold and compacted soil M_F, such as shown in Fig. 3.50.

Water Content.   The soil specimen is next extruded from the compaction mold. Figure 3.51 shows a soil specimen in the process of being extruded from the compaction mold by using a hydraulic jack mechanism. After the soil specimen has been removed from the mold, the water content of the soil is determined. The entire soil specimen could be broken apart and used for the water content test, although it is more common to slice apart the extruded soil and obtain about 1 lb (450 g) of representative soil for the water content test. See Sec. 3.2.1 for the test procedure for determining the water content of the compacted soil.

Compacting the Other Three Soil Specimens.   After the test has been completed for the first soil specimen, the steps are repeated for the other three soil specimens.

FIGURE 3.48   Soil in the process of being compacted into the compaction mold using the Modified Proctor rammer.

The calculations consist of determining the wet density ($\rho_t = M/V$) and the dry density $\rho_d = \rho_t/(1 + w)$ of the compacted soil. For these calculations, M = mass of the compacted soil ($M = M_F - M_o$), V = volume of the compaction mold (1/30 ft^3), and w = water content of the compacted soil expressed as a decimal. In order to determine the laboratory maximum dry density ρ_{dmax} and the optimum moisture content w_{opt}, the dry density ρ_d versus water content w for the four tests are plotted. The peak point of the compaction curve is defined as the laboratory maximum dry density ρ_{dmax} and the corresponding water content is known as optimum moisture content w_{opt}.

An example of laboratory test data are presented in Fig. 3.52. Four soil specimens having water contents of about 8, 10, 12 and 14 percent were compacted. In Fig. 3.52, the water content versus dry density for each test has been plotted. The reason the dry density is plotted is because it is desirable to control the actual density of the soil particles, or in order words, to control how densely packed together are the soil solids. The four data points are connected in order to create the compaction curve. The peak point of the compaction curve is the laboratory maximum dry density ρ_{dmax}.

FIGURE 3.49 Completion of the compaction test with the top collar removed.

In Fig. 3.52, the laboratory maximum dry density is 122.5 pcf (1.96 Mg/m³). The water content corresponding to the laboratory maximum dry density is known as the optimum moisture content. In Fig. 3.52, the optimum moisture content w_{opt} for this soil is 11 percent. This information is important because it tells the grading contractor that the water content of the soil should be about 11 percent for the most efficient compaction of the soil.

The three lines to the right of the compaction curve are each known as a "zero air voids curve." These curves represent the relationship between water content and dry density for a condition of saturation ($S = 100$ percent) for a specified specific gravity. Usually the right side of the compaction curve will be approximately parallel to the zero air voids curve and can be used as a check on the laboratory test results. The following equation can be used to plot a "zero air voids curve."

$$\rho_z = \frac{G_s \rho_w}{1 + G_s w}$$

(3.29)

FIGURE 3.50   A balance is being used to determine the mass of the mold plus compacted soil.

Where ρ_z = vertical axis value of dry density of the soil for the plotting of the zero air voids curve (pcf or Mg/m³)

G_s = specific gravity of soil solids (dimensionless), see Sec. 3.2.3

ρ_w = density of water, which is equal to 62.4 pcf or 1.0 Mg/m³

w = water content of the soil, expressed as a decimal

Assuming a specific gravity G_s and water content w, the value of ρ_z can be calculated from Eq. 3.29. By plotting the ρ_z versus water content w, the zero air voids curve corresponding to the assumed specific gravity could be determined.

Figure 3.53 shows the compaction curves for several different soils. Well-graded soils, such as well-graded sand and gravel, have the highest values of laboratory maximum dry density ρ_{dmax} and the lowest optimum moisture contents w_{opt}. At the other extreme, uniform fine sands, silts, and fat clays have the lowest values of laboratory maximum dry density ρ_{dmax} and the highest values of

FIGURE 3.51   The compacted soil is in the process of being extruded from the compaction mold.

optimum moisture content w_{opt}. Another soil that usually has a very low laboratory maximum dry density $\rho_{d\max}$ and a very high optimum moisture content w_{opt} is diatomaceous earth.

During the compaction of soil, there can be the break-down of soft soil particles. For example, soft sedimentary particles, gypsum, and diatoms can be broken apart or crushed during the compaction process. Thus the compaction process will result in a change in the grain size distribution of the soil because the soil will have more finer soil size particles after compaction.

Another problem concerns the air-drying of soil specimens. In some cases, the soil specimens may be completely air-dried and then water is added to the soil specimen in order to achieve the desired water content. But the air-drying of the soil can affect the results. For example, it has been observed that the compaction tests made on air-dried halloysite samples give markedly different

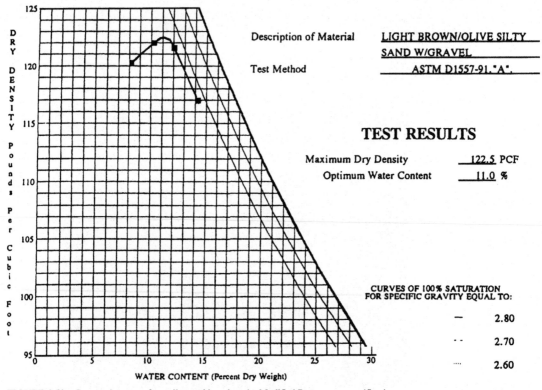

FIGURE 3.52 Compaction curve for a silty sand based on the Modified Proctor test specifications.

results than tests on samples at their natural water content (Holtz and Kovacs, 1981). Other soils that are sensitive to drying conditions include soils that contain diatoms, such as diatomaceous earth.

Clean granular soils may bleed water during compaction. For these highly permeable soils, water is often observed squirting out between the base plate and compaction mold during the compaction process. For such soils, a more representative method of controlling compaction may be to calculate the relative density (D_r, see Sec. 4.3) of the compacted soil.

Determining the water content of the soil is an important aspect of the compaction test. Certain soils, such as diatomaceous earth, are very sensitive to the drying temperature. Another type of soil that is sensitive to the drying temperature is *gypsiferous soil*, which is a soil that contains a high percentage of gypsum. In order to reduce the amount of dehydration of gypsum (Ca $SO_4 \cdot 2H_2O$), ASTM D 2216-98 (2004) indicates that it may be desirable for the water content test to use a drying temperature of 60°C, instead of the standard temperature of 110°C. Drying at a lower temperature can substantially effect the compaction curve. For example, Fig. 3.54 shows two compaction curves for a gypsiferous soil, where for the upper compaction curve, the water contents were determined by drying the soil at 60°C. For the lower compaction curve, the water contents were determined by drying the soil at 110°C. Using a lower temperature for determining the water content of the compacted soil resulted in a higher laboratory maximum dry density ρ_{dmax} and a lower optimum moisture content w_{opt} for the gypsiferous soil.

For further details on the modified Proctor laboratory compaction test, see ASTM D 1557-02 (2004), "Standard Test Methods for Laboratory Compaction Characteristics of Soil using Modified Effort [56,000 ft-lbf/ft³ (2700 kN-m/m³)]."

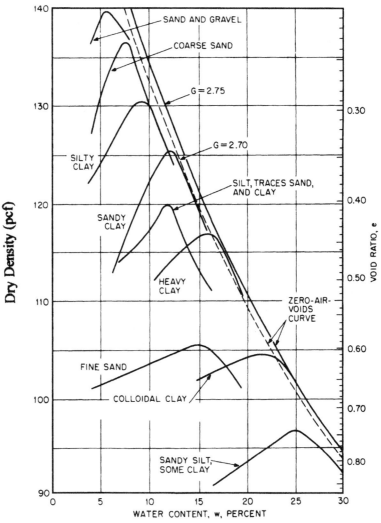

FIGURE 3.53 Compaction curves for different types of soil using the Modified Proctor compaction test. (*From Day, 2001b.*)

3.6.2 Standard Proctor Compaction Test

The test procedures for the Standard Proctor laboratory compaction test are identical to those procedures listed in Sec. 3.6.1, except that the Standard Proctor test has a lower compaction effort. For example, as indicated in ASTM D 698-00 (2004) "Standard Test Methods for Laboratory Compaction Characteristics of Soil Using Standard Effort [12,400 ft-lbf/ft^3 (600 kN-m/m^3)]," the compaction energy is only 12,400 ft-lbf/ft^3, or about 22 percent of the Modified Proctor compaction energy.

For the Standard Proctor (ASTM D 698-00, 2004, Procedure A), the soil is compacted into a 4 in. (10.2 cm) diameter mold that has a volume of 1/30 ft^3 (944 cm^3), where three layers of soil are compacted into the mold with each layer receiving 25 blows from a 5.5 lbf (24.4 N) rammer

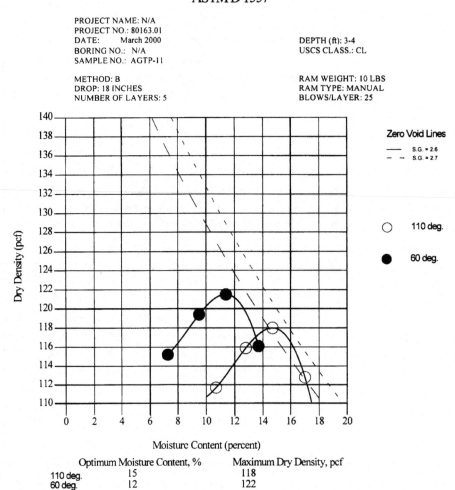

Modified Compaction Test
ASTM D 1557

PROJECT NAME: N/A
PROJECT NO.: 80163.01
DATE: March 2000
BORING NO.: N/A
SAMPLE NO.: AGTP-11

DEPTH (ft): 3-4
USCS CLASS.: CL

METHOD: B
DROP: 18 INCHES
NUMBER OF LAYERS: 5

RAM WEIGHT: 10 LBS
RAM TYPE: MANUAL
BLOWS/LAYER: 25

Zero Void Lines
——— S.G. = 2.6
– – S.G. = 2.7

○ 110 deg.

● 60 deg.

	Optimum Moisture Content, %	Maximum Dry Density, pcf
110 deg.	15	118
60 deg.	12	122

FIGURE 3.54 Compaction curves for a gypsiferous soil. The water content for the upper compaction curve was determined by using a drying temperature of 60°C. The water content for the lower compaction curve was determined by using a drying temperature of 110°C.

that has a 12 in. (0.31 m) drop. Thus a difference between the laboratory compaction tests is that the Modified Proctor has five layers of compacted soil, while the Standard Proctor has only three layers of compacted soil. For both laboratory tests, each layer is subjected to 25 blows, but the Modified Proctor rammer (i.e., 10 lbf rammer with a 18 in. drop) imparts much more energy as compared to the Standard Proctor test which uses a different rammer (5.5 lbf rammer with a 12 in. drop).

Since the Standard Proctor has much less compaction effort than the Modified Proctor, the laboratory maximum dry density ρ_{dmax} is lower and the optimum moisture content w_{opt} is higher for the Standard Proctor as compared to the Modified Proctor test on the same soil. Because of the modern

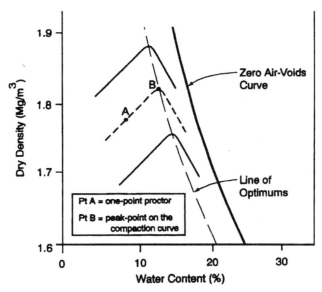

FIGURE 3.55 Procedure to estimate the laboratory maximum dry density from the one-point Proctor test.

use of heavier compaction equipment and the desirability of having a higher load-bearing fill, the Modified Proctor test is typically used more often than the Standard Proctor test.

3.6.3 One-Point Proctor Test

The one-point Proctor test is a quick method that can be used to obtain an estimate of the laboratory maximum dry density ρ_{dmax} and the optimum moisture content w_{opt}. Figure 3.55 illustrates the basic approach to the one-point Proctor test. In order to use the one-point Proctor test, the compaction curves on similar soils must already have been determined, such as the upper and lower compaction curves indicated as solid lines in Fig. 3.55. The one-point Proctor test consists of performing a single point maximum dry density test on soil that has a water content that is dry of optimum moisture content. The maximum dry density is then estimated based on the observation that compaction curves on similar soil types have the same basic shape. This procedure is illustrated in Fig. 3.55, where point A is the *one-point Proctor test* performed on the soil that has a water content dry of optimum and then the compaction curve is drawn. Note in Fig. 3.55 that the laboratory maximum dry density (Point B) is obtained by using the line of optimums, which is a line drawn through the peak point of the compaction curves. The one-point Proctor test is especially useful when there are several slightly different soil types at a site.

3.7 PERMEABILITY TESTS

Permeability is defined as the ability of water to flow through a saturated soil. A high permeability indicates that water flows rapidly through the void spaces, and vice versa. A measure of the soils permeability is the hydraulic conductivity, also known as the coefficient of permeability k. The hydraulic conductivity can be measured in the laboratory by using the constant head (ASTM D 2434-00, 2004) or falling head (ASTM D 5084-00, 2004) permeameter.

3.7.1 Constant Head Permeameter

Figure 3.56 shows a constant head permeameter apparatus. A saturated soil specimen is placed in the permeameter and then a head of water Δh is maintained. The hydraulic conductivity is based on Darcy's law, which states that the velocity v of flow in soil is proportional to the hydraulic gradient i, or:

$$v = k\,i \tag{3.30}$$

where k is the hydraulic conductivity (also known as coefficient of permeability), ft/sec or cm/sec and i is the hydraulic gradient (dimensionless) which is defined as the change in total head Δh divided by the length of the soil specimen L, or $i = \Delta h/L$.

Using Eq. 3.30, the coefficient of permeability k from the constant head permeameter can be calculated as follows:

$$k = \frac{QL}{(\Delta h)At} \tag{3.31}$$

where k = coefficient of permeability (ft/sec or cm/sec)
$\qquad Q$ = total discharged volume (ft^3 or cm^3) in a given time (t)
$\qquad L$ = length of the soil specimen (ft or cm)
$\qquad \Delta h$ = the total head loss for the constant head permeameter as defined in Fig. 3.56 (ft or cm)
$\qquad A$ = area of the soil specimen (ft^2 or cm^2)
$\qquad t$ = time (sec)

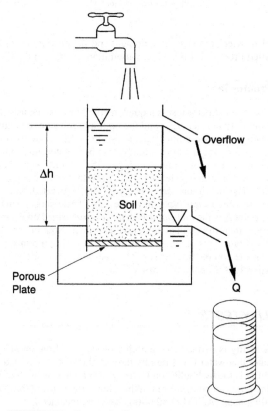

FIGURE 3.56   Constant head permeameter.

The constant head permeameter is often used for sandy soils that have a high permeability. It is important that the porous plate which supports the soil specimen (Fig. 3.56) has a very high permeability. As an alternative, a reinforced permeable screen can be used in place of the porous plate. Another important consideration is that the soil specimen diameter should be at least ten times larger than the size of the largest soil particle.

3.7.2 Falling Head Permeameter

Figure 3.57 (from Department of the Army, 1970) shows a laboratory falling head permeameter. This equipment is used to determine the hydraulic conductivity of a saturated silt or clay specimen. Filter paper is often placed over the porous plate to prevent the migration of soil fines through the porous plate. Also, a frequent cause of inaccurate results is the inability to obtain a seal between the soil specimen and the side of the permeameter. Because of these factors (migration of fines and inadequate sealing), a greater degree of skill is required to perform the falling head permeability test.

The objective of the falling head permeability test is to allow the water level in a small diameter tube to fall from an initial position h_o to a final position h_f. The amount of time it takes for the water level to fall from h_o to h_f is recorded. Based on Darcy's law, the equation to determine the hydraulic conductivity k for a falling head test is as follows:

$$k = 2.3 \frac{aL}{At} \log_{10} \frac{h_o}{h_f} \tag{3.32}$$

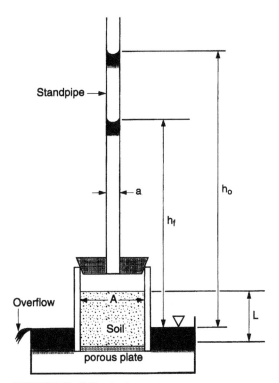

FIGURE 3.57 Falling head permeameter.

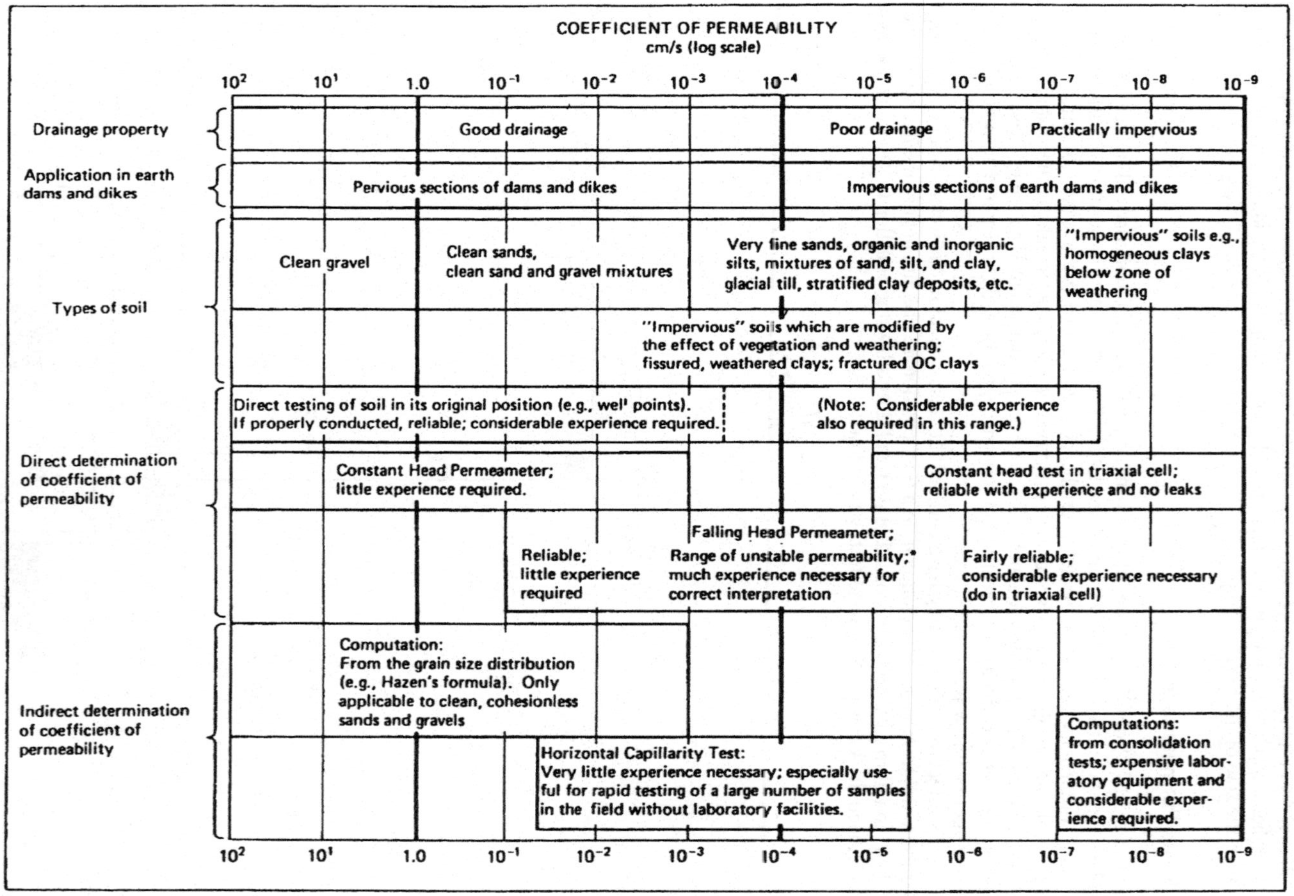

COEFFICIENT OF PERMEABILITY
cm/s (log scale)

	10^2	10^1	1.0	10^{-1}	10^{-2}	10^{-3}	10^{-4}	10^{-5}	10^{-6}	10^{-7}	10^{-8}	10^{-9}

Drainage property

Good drainage | Poor drainage | Practically impervious

Application in earth dams and dikes

Pervious sections of dams and dikes | Impervious sections of earth dams and dikes

Types of soil

Clean gravel | Clean sands, clean sand and gravel mixtures | Very fine sands, organic and inorganic silts, mixtures of sand, silt, and clay, glacial till, stratified clay deposits, etc. | "Impervious" soils e.g. homogeneous clays below zone of weathering

"Impervious" soils which are modified by the effect of vegetation and weathering; fissured, weathered clays; fractured OC clays

Direct determination of coefficient of permeability

Direct testing of soil in its original position (e.g., wel² points). If properly conducted, reliable; considerable experience required.

(Note: Considerable experience also required in this range.)

Constant Head Permeameter; little experience required.

Constant head test in triaxial cell; reliable with experience and no leaks

Reliable; little experience required

Falling Head Permeameter; Range of unstable permeability;* much experience necessary for correct interpretation

Fairly reliable; considerable experience necessary (do in triaxial cell)

Indirect determination of coefficient of permeability

Computation: From the grain size distribution (e.g., Hazen's formula). Only applicable to clean, cohesionless sands and gravels

Horizontal Capillarity Test: Very little experience necessary; especially useful for rapid testing of a large number of samples in the field without laboratory facilities.

Computations: from consolidation tests; expensive laboratory equipment and considerable experience required.

*Due to migration of fines, channels, and air in voids.

FIGURE 3.58 Coefficient of permeability versus drainage property, soil type, and method of determination. *(Developed by Casagrande, with minor additions by Holtz and Kovacs 1981; reproduced from Holtz and Kovacs 1981.)*

where k = coefficient of permeability (ft/sec or cm/sec)

$\quad a$ = area of the standpipe (ft^2 or cm^2)

$\quad L$ = length of the soil specimen (ft or cm)

$\quad h_o$ = initial height of water in the standpipe as defined in Fig. 3.57 (ft or cm)

$\quad h_f$ = final height of water in the standpipe as defined in Fig. 3.57 (ft or cm)

$\quad A$ = area of the soil specimen (ft^2 or cm^2)

$\quad t$ = time it takes for the water level in the standpipe to fall from h_o to h_f (sec)

3.7.3 Permeability Data

Figure 3.58 (adapted from Casagrande) presents a plot of the hydraulic conductivity (coefficient of permeability) versus drainage properties, type of soil, and type of permeameter apparatus best suited for the measurement of k. The bold lines in Fig. 3.58 indicate major divisions in the hydraulic conductivity. A hydraulic conductivity of about 1 cm/sec is the approximate boundary between laminar and turbulent flow. A hydraulic conductivity of about 1×10^{-4} cm/sec is the approximate dividing line between good drainage and poorly drained soils. In general, those soils that contain more fines, such as clays, will have the lowest hydraulic conductivity. This is because clay particles provide very small drainage paths, with a resultant large resistance to fluid flow, even though a clay will often have significantly more void space than a sand. According to Terzaghi and Peck (1967), the classification of soil according to hydraulic conductivity k is as follows:

High degree of permeability, k is over 0.1 cm/sec

Medium degree of permeability, k is between 0.1 and 0.001 cm/sec

Low permeability, k is between 0.001 and 1×10^{-5} cm/sec

Very low permeability, k is between 1×10^{-5} and 1×10^{-7} cm/sec

Practically impermeable, k is less than 1×10^{-7} cm/sec

In terms of engineering practice, the hydraulic conductivity is often reported to one or at most, two significant figures. This is because in many cases the laboratory permeability will not represent in situ conditions. For example, Tomlinson (1986) states:

> There is a difference between the horizontal and vertical permeability of natural soil deposits due to the effects of stratification with alternating beds of finer or coarser grained soils. Thus the results of laboratory tests on a few samples from a vertical borehole are of rather doubtful value in assessing the representative permeability of the soil for calculating the quantity of water to be pumped from a foundation excavation.

Other methods for determining the permeability of in situ soil will be described in Sec. 4.7.

NOTATION

The following notation is used in this chapter:

$\quad a$ = for a p-q plot, the y axis intercept of the failure envelope

$\quad a$ = area of standpipe (falling head permeameter)

$\quad a'$ = for a p'-q plot, the y axis intercept of the failure envelope

$\quad A$ = area of the soil specimen

$\quad A$ = cross sectional area (constant and falling head permeameters)

A, B = Skempton pore water pressure coefficients

$\quad A_c$ = corrected area of the soil specimen during shearing

$\quad A_s$ = preshear area of the soil specimen (triaxial test)

c = cohesion based on a total stress analysis

c' = cohesion based on an effective stress analysis

d_s = preshear dial reading for the height of the soil specimen

D_{10} = from grain size curve, particle size for 10 percent finer by dry weight

D_{30} = from grain size curve, particle size for 30 percent finer by dry weight

D_{60} = from grain size curve, particle size for 60 percent finer by dry weight

F = percent finer by dry weight (sieve analysis)

g = acceleration of gravity

G_b = bulk specific gravity (oversize particles)

G_s = specific gravity of soil solids

h_o, h_f = head measurements from a falling head permeameter test

Δh = change in height of the soil specimen during consolidation (triaxial test)

Δh = change in total head

H_o = initial height of the soil specimen (triaxial test)

H_s = preshear height of the soil specimen (triaxial test)

ΔH = change in height of the soil specimen during shearing

i = hydraulic gradient

k = hydraulic conductivity (also known as coefficient of permeability)

L = length of soil specimen (constant and falling head permeameters)

LL = liquid limit

M = total mass (soil solids plus water)

M = mass of the compacted soil (laboratory compaction test)

M_F = mass of the mold and compacted soil

M_o = mass of the empty compaction mold

M_s = mass of the soil solids (water content test)

M_s = mass of solids used for the specific gravity test (Sec. 3.2.3)

M_s = initial dry mass of the soil for sieve analysis (Sec. 3.2.4)

M_w = mass of water (water content test)

p = the x-coordinate for a total stress path

p' = the x-coordinate for an effective stress path

P = load applied by the loading piston

PI = plasticity index

PL = plastic limit

q = the y-coordinate for a total or effective stress path

q_u = unconfined compressive strength of the soil

Q = total discharge volume in a given time (constant head permeameter)

R_{DS} = cumulative amount of dry soil retained on a given sieve

s_u = undrained shear strength

S = degree of saturation

SL = shrinkage limit

t = time

t_f = time to failure (direct shear test)

t_{50} = time for the cohesive soil to achieve 50 percent consolidation

u = pore water pressure

u_e = excess pore water pressure

u_f = final pore water pressure

u_o = initial pore water pressure

u_s = preshear pore water pressure (triaxial test)

Δu = change in pore water pressure

v = velocity of flow (Darcy's law)

V = total volume (see Fig. 3.1)

V_o = initial volume of the soil specimen (triaxial test)

V_s = volume of the soil particles determined by using the pycnometer

V_s = preshear volume of the soil specimen (triaxial test)

ΔV = change in volume of the soil specimen due to consolidation (triaxial test)

w = water content of the soil (also known as the moisture content)

w_{opt} = optimum moisture content (laboratory compaction test)

α = on a p-q plot, the angle of inclination of the failure envelope

α' = on a p'-q plot, the angle of inclination of the failure envelope

ε = axial strain (triaxial test and unconfined compression test)

ϕ = friction angle based on a total stress analysis

ϕ' = friction angle based on an effective stress analysis

ϕ'_r = drained residual friction angle

ϕ'_u = effective friction angle at the ultimate shear strength state

γ_b = buoyant unit weight of the soil

γ_d = dry unit weight of the soil

γ_t = wet unit weight of soil (also known as total unit weight)

γ_w = unit weight of water (62.4 pcf or 9.81 kN/m^3)

ρ_{dmax} = laboratory maximum dry density (compaction test)

ρ_s = density of soil solids

ρ_t = wet density of soil (also known as total density)

ρ_w = density of water (62.4 pcf or 1.0 g/cm^3)

ρ_z = dry density for zero air voids curve (compaction test)

σ = total stress

σ' = effective stress

σ_c = chamber fluid pressure, also known as the cell pressure (triaxial test)

σ'_c = effective confining pressure, also known as the consolidation pressure

σ_h = horizontal total stress for the triaxial test

σ'_h = horizontal effective stress for the triaxial test

σ_n = total normal stress acting on the shear surface

σ'_n = effective normal stress

σ_v = vertical total stress for the triaxial test

σ'_v = vertical effective stress for the triaxial test

σ'_{vo} = vertical effective stress

σ_1 = major principal total stress

σ'_1 = major principal effective stress

σ_2 = intermediate principal total stress

σ_3 = minor principal total stress

σ'_3 = minor principal effective stress

$\Delta\sigma_c$ = change in cell pressure (triaxial test)

$\Delta\sigma_1$ = change in major principal total stress

$\Delta\sigma_3$ = change in minor principal total stress

τ_f = shear strength of the soil

τ_{max} = maximum shear stress

PROBLEMS

Solutions to the problems are presented in App. C of this book. The problems have been divided into basic categories as follows:

Water Content

3.1   A water content test was performed on a specimen of soil. The following data were obtained:

Mass of empty container = 105.6 g

Mass of the container plus wet soil = 530.8 g

Mass of the container plus dry soil = 483.7 g

Calculate the water content of the soil.

ANSWER: Water content = 12.5 percent.

Unit Weight

3.2   Using the data from Problem 3.1 and assuming that the initial volume of the wet soil specimen is equal to 225 cm^3, calculate the total unit weight and the dry unit weight.

ANSWER: Total unit weight = 18.5 kN/m^3 and dry unit weight = 16.5 kN/m^3.

3.3   A clay specimen has been obtained from below the groundwater table and the total unit weight γ_t of the soil specimen is 19.5 kN/m^3 (124 pcf). Calculate the buoyant unit weight of the clay.

ANSWER: γ_b = 9.7 kN/m^3 (61.6 pcf).

Specific Gravity

3.4   A specific gravity test was performed on a granular soil. The mass of dry soil used for the specific gravity test was equal to 102.2 g. The dry soil was placed in a pycnometer and it was determined that the volume of the soil particles was equal to 38.9 cm^3. Determine the specific gravity of solids G_s.

ANSWER: G_s = 2.63.

Sieve Analysis

3.5   Figure 3.59 presents the results of a sieve analysis for a soil specimen. Determine the particle size distribution for this soil.

ANSWER: See App. C for the solution.

A) Dry mass of the soil specimen:

Water content: 8.3 percent   Wet mass: 1386.9 g   Initial dry mass (M_s): _____

B) Dry mass of the soil specimen after washing on the No. 200 Sieve

Mass of empty evaporating dish: 234.8 g   Mass of dish plus dry soil: 1350.8 g

Mass of dry soil retained on the No. 200 Sieve (M_R): _____

C) Sieve analysis:

Sieve No.	Mass retained for each sieve (g)	Cumulative mass retained (R_{DS})	Percent finer (Eq. 3.6)
2-in.	0		
1½-in.	0		
1-in.	93.3		
¾-in.	71.9		
½-in.	114.3		
⅜-in.	135.7		
No. 4	182.2		
No. 10	150.1		
No. 20	142.2		
No. 40	112.8		
No. 60	47.8		
No. 100	29.6		
No. 200	35.9		
Pan	0.1		—

Check: Cumulative retained on Pan: 1115.9 g   Versus M_R: _____

FIGURE 3.59   Data for Problem 3.5.

Atterberg Limits

3.6   Figure 3.60 presents the results of Atterberg limits tests performed on a soil specimen. Determine the liquid limit, plastic limit, and plasticity index for this soil.

ANSWER: See App. C for the solution.

Shear Strength of Cohesionless Soil

3.7   The results of a standard penetration test indicate that a clean sand deposit has an N_{60} value equal to 5 at a depth where the vertical effective stress $\sigma'_v = 43$ kPa (910 psf). Using Fig. 3.12, determine the effective friction angle ϕ' of the sand.

ANSWER: $\phi' = 30°$.

3.8   For Problem 3.7, assume a cone penetration test was performed at the same depth and the cone resistance $q_c = 40$ kg/cm² (3900 kPa). Using Fig. 3.13, determine the effective friction angle ϕ' of the sand.

ANSWER: $\phi' = 40°$.

3.9   Use the test data from Fig. 3.19 for B-1 at a depth of 6 ft. Determine the effective friction angle ϕ' of the cohesionless silty sand (SM) by using Fig. 3.14. Compare this value with the effective friction angle obtained from the plot shown in Fig. 3.18.

ANSWER: $\phi' = 30°$, same as the value from Fig. 3.18

A) Water contents for liquid limit:			
Trial number	1	2	3
Container number	1	2	4
Container mass (M_c)	10.92	10.84	11.33
Container + wet soil (M_{wc})	20.89	22.90	24.07
Container + dry soil (M_{dc})	16.36	17.63	18.80
Mass of water (M_w)			
Mass of solids (M_s)			
Water content = M_w/M_s			
Number of blows	15	19	30
B) Water contents for plastic limit:			
Trial number	1	2	
Container number	3	5	
Container mass (M_c)	11.25	10.98	
Container + wet soil (M_{wc})	13.15	13.21	
Container + dry soil (M_{dc})	12.81	12.80	
Mass of water (M_w)			
Mass of Solids (M_s)			
Water content = M_w/M_s			
C) Summary			
Liquid limit (LL) = _____ Plastic limit (PL) = _____ Plasticity index (PI) = ____			

Note: All mass values are in g.

FIGURE 3.60   Data for Problem 3.6.

3.10   Use the test data from Fig. 3.19 for B-2 at a depth of 10 ft. Determine the effective friction angle ϕ' of the cohesionless silty sand (*SM*) by using Fig. 3.14.

ANSWER: $\phi' = 36°$.

3.11   A drained direct shear test was performed on a cohesionless soil. The specimen diameter = 6.35 cm (2.5 in.). At a vertical load of 150 N (34 lb), the peak shear force = 94 N (21 lb). For a second specimen tested at a vertical load of 300 N (68 lb), the peak shear force = 188 N (42 lb). Determine the effective friction angle ϕ' for the cohesionless soil.

ANSWER: $\phi' = 32°$.

Shear Strength Based on Total Stress Analyses for Cohesive Soil

3.12   An unconfined compression test was performed on an undisturbed specimen of saturated clay. The clay specimen had an initial diameter of 2.50 in. and an initial height of 6.0 in. The vertical force at the shear failure was 24.8 lb and the dial gauge recorded 0.80 in. of axial deformation from the initial condition to the shear failure condition. Calculate the undrained shear strength s_u of the clay.

ANSWER: $s_u = 315$ psf.

3.13   Use the triaxial test data from Fig. 3.30. Assuming that the cohesion (c) is zero, calculate the friction angle ϕ based on a total stress analysis.

ANSWER: $\phi = 32°$.

3.14   Use the triaxial test data from Fig. 3.30. Assuming that the cohesion (c) is 2.0 psi, calculate the friction angle ϕ based on a total stress analysis.

ANSWER: $\phi = 24°$.

3.15   Assume the data shown in Fig. 3.28 was obtained from a consolidated undrained triaxiaᴸ compression test with pore water pressure measurements performed on a saturated cohesive soil. The specimen was first consolidated at an effective confining pressure of 50 kPa (i.e., $\sigma'_1 = \sigma'_3$). After consolidation, the specimen height = 10.67 cm and the specimen diameter = 3.68 cm. The B value was checked by applying a cell pressure $\Delta\sigma'_c = 100$ kPa and then measuring the change in pore water pressure Δu which was equal to 99.8 kPa. During the undrained triaxial shearing of the specimen, the following data was recorded:

Point	Axial deformation (cm)	$\Delta\sigma_1$ (kPa)	Δu (kPa)
A	0	0	0
B	0.13	20	1.5
C	0.30	40	3.0
D	0.53	60	4.3
E	0.95	80	6.7

Note that point E represents the failure condition.

Calculate the friction angle ϕ based on a total stress analysis, assuming c = 5 kPa.

ANSWER: $\phi = 23°$.

3.16   For Problem 3.15, calculate the B value and A value at failure.

ANSWER: B value = 0.998, A value at failure = 0.08.

3.17   Calculate the undrained modulus E_u for the test data from Problem 3.15.

ANSWER: $E_u = 1600$ kPa.

Shear Strength Based on Effective Stress Analyses for Cohesive Soil

3.18   For Problem 3.15, determine α', assuming that $a' = 2$ kPa.

ANSWER: $\alpha' = 24.5°$.

3.19   Using the data from Problem 3.18, calculate the effective stress friction angle ϕ' and effective stress cohesion c'.

ANSWER: $\phi' = 27°$, $c' = 2.2$ kPa.

For Problems 3.20 through 3.25, use the following data from a consolidated undrained triaxial compression test with pore water pressure measurements performed on a saturated cohesive soil specimen. The soil specimen was first consolidated at an effective confining pressure of 100 kPa (i.e., $\sigma'_1 = \sigma'_3$). At the end of consolidation, the soil specimen had an area = 9.68 cm² (1.50 in.²) and a height = 11.7 cm (4.60 in.). The specimen was then subjected to an axial load, and at failure, the axial deformation = 1.48 cm (0.583 in.), the axial load = 48.4 N (10.9 lb), and the change in pore water pressure Δu = 45.6 kPa (950 psf).

3.20   Calculate the area of the specimen at failure.

ANSWER: Area at failure = 11.1 cm² (1.72 in.²).

3.21   Calculate the major principal effective stress at failure σ'_1 and the minor principal effective stress at failure σ'_3.

ANSWER: σ'_1 at failure = 98 kPa (2050 psf), σ'_3 at failure = 54.4 kPa (1140 psf).

3.22   Calculate q and p' (in terms of effective stresses).

ANSWER: $p' = 76.2$ kPa (1590 psf) and $q = 21.8$ kPa (455 psf).

3.23   Calculate the effective friction angle ϕ' assuming $c' = 0$.

ANSWER: $\phi' = 17°$.

3.24   Calculate the A value at failure.

ANSWER: A value at failure = 1.05.

3.25   Based on the results of Problems 3.20 to 3.24, what type of inorganic soil would most likely have this shear strength?

ANSWER: Normally consolidated clay of high plasticity (*CH*).

For Problems 3.26 and 3.27, data were obtained from consolidated undrained triaxial compression tests with pore water pressure measurements performed on saturated specimens of Atchafalaya clay (from Baton Rouge, Louisiana). The Atchafalaya clay has a liquid limit = 93 and a plasticity index = 68.

3.26   A specimen of Atchafalaya clay was consolidated at an effective confining pressure = 385 kPa, which is greater than its preconsolidation pressure. At failure (i.e., maximum value of σ'_1/σ'_3), $\sigma'_1 = 333$ kPa and $\sigma'_3 = 126$ kPa. Determine the effective friction angle ϕ'.

ANSWER: $\phi' = 27°$.

3.27   A second specimen of Atchafalaya clay was consolidated to a pressure in excess of the preconsolidation pressure and then unloaded and allowed to equilibrate. At failure (i.e., maximum value of σ'_1/σ'_3), $\sigma'_1 = 238$ kPa and $\sigma'_3 = 83$ kPa. Determine the effective cohesion c' for this overconsolidated specimen assuming that the Atchafalaya clay has the same effective friction angle for both the normally consolidated and overconsolidated state.

ANSWER: $c' = 7$ kPa.

For Problems 3.28 to 3.30, data were obtained from a consolidated undrained triaxial compression test with pore water pressure measurements performed on a saturated specimen of clay. The tabulated information in Fig. 3.61 presents the recorded data and calculations for the shearing portion of the triaxial test. For this tabulated data, S3 = σ'_3 and S1 = σ'_1.

3.28   Plot the effective stress path and determine a' and α'.

ANSWER: See Fig. 3.62 for the effective stress path, where $a' = 0.5$ psi and $\alpha' = 27°$.

3.29   Using the data from Fig. 3.62, calculate the effective shear strength parameters.

ANSWER: $c' = 0.6$ psi and $\phi' = 31°$.

3.30   Based on the A value at failure, determine the type of cohesive soil.

ANSWER: A value at failure = 0.018, therefore the soil type is a heavily overconsolidated clay.

```
STATIC TRIAXIAL TEST RESULTS
****************************
```

```
                                        **************************
                                        *  BY       *  DATE      *
                                        *           *            *
PROJECT NO. :                           *           *            *
BORING  NO. :  STP#1                    *  CHECK    *            *
SAMPLE  NO. :                           *           *            *
SAMPLE DEPTH: 4-4.5                     *           *            *
INTIAL  AREA:  4.90  SQ.INCH            *  CORR.    *            *
INITI.HEIGHT:  4.95     INCH            *           *            *
SOIL TYPE = CL   OLIVE GRAY CLAY        *  RECHECK  *            *
                                        *           *            *
                                        *           *            *
AREA AFTER CONSOLIDATION    =  4.901  SQ.INCH   *    *           *
HEIGHT AFTER CONSOLIDATION  =  4.949  INCH      *  ALPR.  *      *
                                        *           *            *
                                        **************************
```

```
CONSOLIDATED - UNDRAINED TEST
****************************
CELL PRESSURE  =   71.6 PSI
BACK PRESSURE  =   69.6 PSI
DRY DENSITY = 98.0 PCF
MOISTURE  =  19.0 (%)
```

PORE PR. (PSI)	LOAD READ	DEFORM (IN)	STRAIN (%)	AREA (SQ.IN)	S1-S3 (PSI)	POR.CH. (PSI)	- A	S3 (PSI)	S1 (PSI)	S1/S3	TAN MOD (PSI)	SEC MOD (PSI)	P' (PSI)	Q (PSI)
69.70	0.	.000	.00	4.901	.00	.00	.000	1.80	1.80	1.000	.0	.0	1.80	.00
70.10	17.	.025	.51	4.926	3.45	.40	.116	1.40	4.85	3.465	683.2	683.2	3.13	1.73
70.10	20.	.050	1.01	4.951	4.04	.40	.099	1.40	5.44	3.885	116.5	399.8	3.42	2.02
70.00	21.	.075	1.52	4.976	4.22	.30	.071	1.50	5.72	3.813	35.7	278.5	3.61	2.11
70.00	23.	.100	2.02	5.002	4.60	.30	.065	1.50	6.10	4.065	74.9	227.6	3.80	2.30
70.00	24.	.125	2.53	5.028	4.77	.30	.063	1.50	6.27	4.182	34.7	189.0	3.89	2.39
69.90	25.	.150	3.03	5.054	4.95	.20	.040	1.60,	6.55	4.091	34.3	163.2	4.07	2.47
69.80	26.	.175	3.54	5.081	5.12	.10	.020	1.70	6.82	4.010	33.9	144.7	4.26	2.56
69.80	27.	.200	4.04	5.107	5.29	.10	.019	1.70	6.99	4.110	33.5	130.8	4.34	2.64
69.80	28.	.225	4.55	5.134	5.45	.10	.018	1.70	7.15	4.208	33.0	120.0	4.43	2.73
69.80	29.	.250	5.05	5.162	5.62	.10	.018	1.70	7.32	4.305	32.6	111.2	4.51	2.81
69.70	29.	.275	5.56	5.189	5.59	.00	.000	1.80	7.39	4.105	-5.9	100.6	4.59	2.79
69.70	30.	.300	6.06	5.217	5.75	.00	.000	1.80	7.55	4.195	32.0	94.9	4.68	2.88
69.70	31.	.325	6.57	5.245	5.91	.00	.000	1.80	7.71	4.283	31.6	90.0	4.75	2.95
69.60	32.	.350	7.07	5.274	6.07	-.10	-.016	1.90	7.97	4.193	31.2	85.8	4.93	3.03
69.50	33.	.375	7.58	5.303	6.22	-.20	-.032	2.00	8.22	4.112	30.8	82.1	5.11	3.11
69.40	34.	.400	8.08	5.332	6.38	-.30	-.047	2.10	8.48	4.037	30.4	78.9	5.29	3.19
69.40	35.	.425	8.59	5.361	6.53	-.30	-.046	2.10	8.63	4.109	30.0	76.0	5.36	3.26
69.40	36.	.450	9.09	5.391	6.68	-.30	-.045	2.10	8.78	4.180	29.6	73.4	5.44	3.34
69.30	37.	.475	9.60	5.421	6.82	-.40	-.059	2.20	9.02	4.102	29.2	71.1	5.61	3.41
69.20	37.	.500	10.10	5.452	6.79	-.50	-.074	2.30	9.09	3.951	-7.5	67.2	5.69	3.39
69.10	38.	.525	10.61	5.483	6.93	-.60	-.087	2.40	9.33	3.888	28.6	65.3	5.87	3.47
69.20	39.	.550	11.11	5.514	7.07	-.50	-.071	2.30	9.37	4.075	28.1	63.6	5.84	3.54
69.10	40.	.575	11.62	5.545	7.21	-.60	-.083	2.40	9.61	4.006	27.7	62.1	6.01	3.61
69.00	41.	.600	12.12	5.577	7.35	-.70	-.095	2.50	9.85	3.941	27.3	60.6	6.18	3.68
68.90	42.	.625	12.63	5.609	7.49	-.80	-.107	2.60	10.09	3.880	26.9	59.3	6.34	3.74
68.90	43.	.650	13.13	5.642	7.62	-.80	-.105	2.60	10.22	3.931	26.5	58.0	6.41	3.81
68.90	43.	.675	13.64	5.675	7.58	-.80	-.106	2.60	10.18	3.914	-8.8	55.6	6.39	3.79
68.80	44.	.700	14.14	5.708	7.71	-.90	-.117	2.70	10.41	3.855	25.9	54.5	6.55	3.85
68.80	45.	.725	14.65	5.742	7.84	-.90	-.115	2.70	10.54	3.902	25.5	53.5	6.62	3.92
68.80	46.	.750	15.15	5.776	7.96	-.90	-.113	2.70	10.66	3.949	25.1	52.5	6.68	3.98
68.70	47.	.775	15.66	5.811	8.09	-1.00	-.124	2.80	10.89	3.889	24.7	51.6	6.84	4.04

FIGURE 3.61 Data for Problems 3.28 to 3.30.

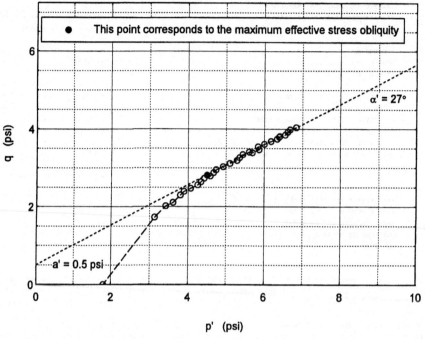

FIGURE 3.62   Solution for Problem 3.28.

3.31   Drained direct shear tests were performed on saturated specimens of remolded siltstone. The soil has a liquid limit = 48 and a plasticity index = 23. The diameter of the soil specimens were all equal to 2.50 in. The following data was obtained from the drained direct shear tests:

Test no.	Vertical load (*lb*)	Maximum shear force (*lb*)
1	1.7	5.8
2	8.5	12
3	15	15
4	32	27
5	70	41
6	133	64

Determine the effective shear strength parameters.

ANSWER: See Fig. 3.63 for the effective shear strength envelope. For $\sigma'_n > 700$ psf, $\phi' = 21°$ and $c' = 450$ psf. For $\sigma'_n < 700$ psf, the effective shear strength envelope is nonlinear.

3.32   For the torsional ring shear test data shown in Fig. 3.43, determine the drained residual friction angle ϕ'_r assuming a straight line through the origin and the data point at an effective normal stress of 14,600 psf.

ANSWER: $\phi'_r = 9.3°$.

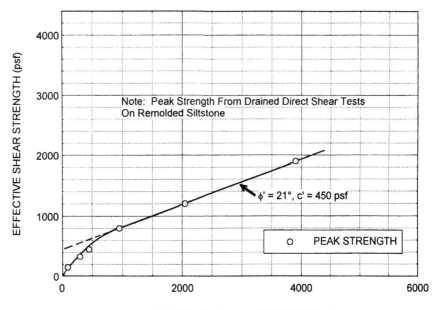

FIGURE 3.63   Solution for Problem 3.31.

Laboratory Compaction Tests

3.33   The following data were obtained from a laboratory compaction test performed in accordance with the Standard Proctor compaction specifications ($1/30$ ft^3 mold):

Test no.	Compacted wet soil (excluding mold), lb	Water content of soil,%
1	4.14	11.0
2	4.26	12.5
3	4.37	14.0
4	4.33	15.5

Determine the laboratory maximum dry density and optimum moisture content.

ANSWER: $\rho_{d\text{max}}$ = 115 pcf (1.84 Mg/m^3), w_{opt} = 14.0 percent.

3.34   If a higher compaction energy is to be used for soil pertaining to Problem 3.33, what will happen to the laboratory maximum dry density and optimum moisture content.

ANSWER: higher compaction energy results in a higher laboratory maximum dry density and a lower optimum moisture content.

3.35   For Problem 3.33, calculate the difference in water content between the optimum moisture content and the water content corresponding to the zero air voids curve. Assume the specific gravity of solids G_s = 2.65.

ANSWER: From Problem 3.33, w_{opt} = 14.0 percent. For ρ_{dmax} = 115 pcf, the water content corresponding to the zero air voids curve = 16.5 percent, and thus the difference in water content = 2.5 percent.

3.36   Use the data shown in Fig. 3.52 and assume that the line of optimums is parallel to the zero air voids curve. On a slightly different soil, a one point Proctor test was performed on soil that had a water content that was dry of optimum. The data recorded from this one point Proctor test was a dry density of 117 pcf at a water content of 8.0 percent. Estimate the laboratory maximum dry density of this slightly different soil.

ANSWER: ρ_{dmax} = 120 pcf (1.92 Mg/m³).

Laboratory Permeability Tests

3.37   In a constant head permeameter test, the outflow Q is equal to 782 mL in a measured time of 31 sec. The sand specimen has a diameter of 6.35 cm and a length L of 2.54 cm. The total head loss Δh for the permeameter is 2.0 m. Calculate the hydraulic conductivity (also known as the coefficient of permeability).

ANSWER: k = 0.01 cm/sec.

3.38   In a falling head permeameter test, the time required for the water in a standpipe to fall from h_o = 1.58 m to h_f = 1.35 m is 11.0 h. The clay specimen has a diameter of 6.35 cm and a length L of 2.54 cm. The diameter of the standpipe is 0.635 cm. Calculate the hydraulic conductivity.

ANSWER: k = 1 × 10⁻⁷ cm/sec.

CHAPTER 4
SOIL MECHANICS

4.1 INTRODUCTION

As discussed in Chap. 1, foundation engineering applies the knowledge of soil mechanics to the design and construction of foundations for buildings and other structures. The purpose of this chapter is to discuss some of the essential soil mechanics principles that are needed in foundation engineering. Topics covered in this chapter include the following:

1. Soil classification (Sec. 4.2)

2. Phase relationships (Sec. 4.3)

3. Effective stress (Sec. 4.4)

4. Stress distribution (Sec. 4.5)

5. Total stress and effective stress analyses (Sec. 4.6)

6. Permeability and seepage (Sec. 4.7)

4.2 SOIL CLASSIFICATION

Probably the most basic aspect of soil mechanics is soil classification. The purpose of a soil classification system is to provide the geotechnical engineer with a way to predict the behavior of the soil for engineering projects. In the United States, the most widely used soil classification system is the Unified Soil Classification System (U.S. Army Engineer Waterways Experiment Station 1960; Howard 1977). The Unified Soil Classification System is abbreviated to USCS and it should not be confused with the United States Customary System that has the same abbreviation. The USCS was initially developed by Casagrande (1948) and later modified by him in 1952.

4.2.1 Particle Size and Description

A basic element of the USCS is the determination of the amount and distribution of the particle sizes of the soil. The distribution of particle sizes larger than 0.075 mm (No. 200 sieve) is determined by sieving and the distribution of particle sizes smaller than 0.075 mm is determined by a sedimentation process (i.e., the hydrometer test). The sieve and hydrometer tests have been discussed in Sec. 3.2.4 and 3.2.5. For the USCS, the rock fragments or soil particles versus size are defined as follows (from largest to smallest particles sizes):

Boulders Rocks that have an average diameter greater than 12 in. (300 mm).

Cobbles Rocks that are smaller than 12 in. (300 mm) and are retained on the 3-in. (75 mm) U.S. standard sieve.

Gravel Size Particles   Rock fragments or soil particles that will pass a 3-in. (75 mm) sieve and be retained on a No. 4 (4.75 mm) U.S. standard sieve. Gravel size particles are subdivided into coarse gravel sizes or fine gravel sizes.

Sand Size Particles   Soil particles that will pass a No. 4 (4.75 mm) sieve and be retained on a No. 200 (0.075 mm) U.S. standard sieve. Sand size particles are subdivided into coarse sand size, medium sand size, or fine sand size.

Silt Size Particles   Fine soil particles that pass the No. 200 (0.075 mm) U.S. standard sieve and are larger than 0.002 mm.

Clay Size Particles   Fine soil particles that are smaller than 0.002 mm.

It is very important to distinguish between the size of a soil particle and the classification of the soil. For example, a soil could have a certain fraction of particles that are of *clay size*. The same soil could also be classified as *clay*. But the classification of *clay* does not necessarily mean that the majority of the soil particles are of clay size (smaller than 0.002 mm). In fact, it is not unusual for a soil to be classified as *clay* and have a larger mass of silt-size particles than clay-size particles. Throughout the book when reference is given to particle size, the terminology *clay-size particles* or *silt-size particles* will be used. When reference is given to a particular soil, then the terms such as *silt* or *clay* will be used.

Using the particle size dimension data, the coefficient of uniformity C_u and coefficient of curvature C_c can be calculated as follows:

$$C_u = \frac{D_{60}}{D_{10}} \tag{4.1}$$

$$C_c = \frac{(D_{30})^2}{(D_{10})(D_{60})} \tag{4.2}$$

As will be discussed in Sec. 4.2.3, these two parameters are used in the USCS to determine whether a soil is well graded (many different particle sizes) or poorly graded (many particles of about the same size).

As indicated earlier, the USCS has specific size dimensions for boulders, cobbles, gravel, and the like. There are many other classification systems that use different particle size dimensions and terminology. For example, the Modified Wentworth Scale is frequently used by geologists and includes such terms as *pebbles* and *mud* in the classification system, and uses different particle size dimensions to define sand, silt, and clay (Lane et al., 1947).

Soil scientists also use their own classification system that subdivides sand size particles into categories of very coarse, coarse, medium, fine, and very fine. Soil scientists also use different particle size dimension definitions (U.S. Department of Agriculture, 1975). The geotechnical engineer should be aware that terms or soil descriptions used by geologists or soil scientists may not match the USCS because they are using a different grain size scale.

4.2.2 Clay Mineralogy

Clay Minerals.   The amount and type of clay minerals present in a soil have a significant effect on soil engineering properties such as plasticity, swelling, shrinkage, shear strength, consolidation, and permeability. This is due in large part to their very small flat or plate-like shape that enables them to attract water to their surfaces, also known as the *double layer* effect. The double layer is a simplified description of the positively charged water layer, together with the negatively charged surface of the clay particle itself. Two reasons for the attraction of water to the clay particle (double layer) are:

1. The dipolar structure of the water molecule that causes it to be electrostatically attracted to the surface of the clay particle.

2. The clay particles attract cations that contribute to the attraction of water by the hydration process. Ion exchange can occur in the double layer, where under certain conditions, sodium, potassium, and calcium cations can be replaced by other cations present in the water. This property is known as *cation exchange capacity.*

In addition to the double layer, there is an *adsorbed water layer* that consists of water molecules that are tightly held to the clay particle face, such as by the process of hydrogen bonding. The presence of the very small clay particles surrounded by water helps explain their impact on the engineering properties of soil. For example, clays that have been deposited in lakes or marine environments often have a very high water content and are very compressible with a low shear strength because of this attracted and bonded water. Another example is desiccated clays, which have been dried, but have a strong desire for water and will swell significantly upon wetting.

There are many types of clay minerals. App. A (Glossary 2) presents definitions of three of the most common clay minerals, i.e., kaolinite, montmorillonite, and illite. Even within a clay mineral category, there can be different crystal components because of isomorphous substitution. This is the process where ions of approximately the same size are substituted in the crystalline framework.

Liquidity Index. A measure of a soil's plasticity is the plasticity index (PI), which has been defined in Sec. 3.2.6 as the liquid limit (LL) minus the plastic limit (PL) (i.e., PI = LL − PL). Another useful parameter is the liquidity index (LI), defined as:

$$LI = \frac{w - PL}{PI} \tag{4.3}$$

where w is the water content of the soil expressed as a percentage.

The liquidity index can be used to identify sensitive clays. For example, quick clays often have a water content w that is greater than the liquid limit, and thus the liquidity index is greater than 1.0. At the other extreme are clays that have liquidity index values that are zero or even negative. These liquidity index values indicate a soil that is desiccated and could have significant expansion potential. Per ASTM, the Atterberg limits are performed on soil that is finer than the No. 40 sieve, but the water content can be performed on soil containing larger soil particles and thus the liquidity index should only be calculated for soil that has all its particles finer than the No. 40 sieve.

Plasticity Chart. Using the Atterberg limits, the plasticity chart was developed by Casagrande (1932a) and is used in the USCS to classify soils. As shown in Fig. 4.1, the plasticity chart is a plot of LL versus PI. Casagrande (1932a) defined two basic dividing lines on the plasticity chart, as follows:

1. LL = 50 line.   This line is used to divide silts and clays into high-plasticity (LL > 50) and low-plasticity (LL < 50) categories.

2. A-line.   The A-line is defined as:

$$PI = 0.73 \, (LL - 20) \tag{4.4}$$

The A-line is used to separate clays, which plot above the A-line, from silts, which plot below the A-line. As shown in the lower diagram in Fig. 4.1, different types of soil tend to plot parallel to the A-line. An additional line has been added to the Casagrande plasticity chart, known as the U-line (see the upper diagram in Fig. 4.1). The U-line (or upper-limit line) is defined as:

$$PI = 0.90 \, (LL - 8) \tag{4.5}$$

The U-line is valuable because it represents the uppermost boundary of test data found thus far for natural soils. The U-line is a good check on erroneous data, and any test results that plot above the U-line should be rechecked. There have been other minor changes proposed for the original Casagrande plasticity chart. For example, at very low PI values, the A-line and U-line are sometimes defined differently than as shown in Fig. 4.1 (e.g., see Fig. 4, ASTM D 2487-00, 2004).

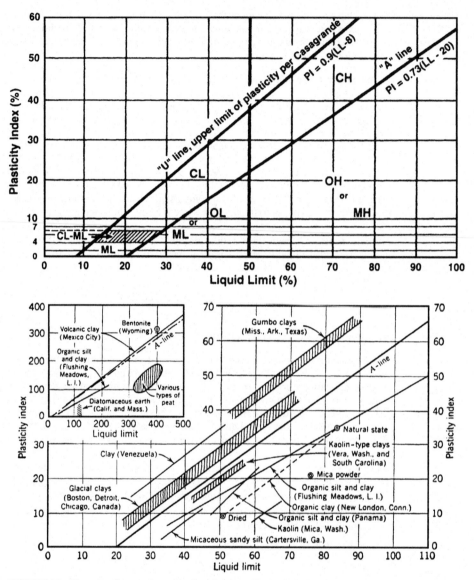

FIGURE 4.1   The upper diagram shows the plasticity chart. The lower diagram shows the location of various soils as they plot on the plasticity chart. Note in the lower diagram how the soils tend to plot parallel to the A-line. (*Reproduced from Casagrande, 1948; reprinted with permission of the American Society of Civil Engineers.*)

Clay minerals present in a soil can be identified by their x-ray diffraction patterns. This process is rather complicated, expensive, and involves special equipment that is not readily available to the geotechnical engineer. A more common approach is to use the location of clay particles as they plot on the plasticity chart to estimate the type of clay mineral in the soil (Fig. 4.2). This approach is often inaccurate, because soil can contain more than one type of clay mineral.

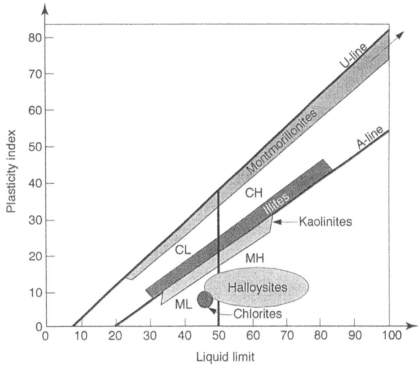

FIGURE 4.2   Plasticity characteristics of common clay minerals. (*Source: Mitchell, 1976; Holtz and Kovacs, 1981.*)

Also shown on Fig. 4.2 are two other less common clay minerals, chlorite and halloysite. Although not very common, halloysite is an interesting clay mineral because instead of the usual flat particle shape, it is tubular in shape, which can affect engineering properties in unusual ways. It has been observed that classification and compaction tests made on air-dried halloysite samples give markedly different result than tests on samples at their natural water content (Holtz and Kovacs, 1981).

Activity.   The activity A of clay is defined as (Skempton, 1953):

$$A = \frac{PI}{\text{clay fraction}} \qquad (4.6)$$

The clay fraction is that part of the soil specimen that is finer than 0.002 mm, based on dry weight. Clays that are *inactive* are defined as those clays that have an activity less than 0.75, *normal activity* is defined as those clays having an activity between 0.75 and 1.25, and *active* clay is defined as those clays having an activity greater than 1.25. Quartz has an activity of zero, while at the other extreme is sodium montmorillonite that can have an activity from 4 to 7.

Because the PI is determined from Atterberg limits that are performed on soil that passes the No. 40 sieve (0.425 mm), a correction is required for soils that contain a large fraction of particles coarser than the No. 40 sieve. For example, suppose a clayey gravel has 70 percent gravel particles (particles coarser than No. 40 sieve), 20 percent silt size particles, and 10 percent clay size particles. If the PI = 40 for the soil particles finer than the No. 40 sieve, then the activity for the clayey gravel would be 1.2 (i.e., 40/33.3).

4.2.3   Unified Soil Classification System

USCS separates soils into two main groups: coarse-grained soils and fine-grained soils. The basis of the USCS is that the engineering behavior of coarse-grained soils is based on their grain size distributions and the engineering behavior of fine-grained soil is related to their plasticity characteristics. Table 4.1 presents a summary of the USCS. As indicated in Table 4.1, the two main groups of soil are defined as follows:

1. *Coarse-grained soils.*   Defined as having more than 50 percent (by dry mass) of soil particles retained on the No. 200 sieve. Coarse-grained soils are divided into gravels and sands. Both gravels and sands are further subdivided into four secondary groups, depending on whether the soil is well-graded, poorly-graded, contains silt size particles, or clay size particles.

2. *Fine-grained soils.*   Defined as having 50 percent or more (by dry mass) of soil particles passing the No. 200 sieve. Fine-grained soils are divided into soils of low or high plasticity. The three secondary classifications are based on liquid limit and plasticity characteristics (PI).

Group symbols are used to identify different soil types. The group symbols consist of two capital letters and the first letter indicates the following:

G = gravel

S = sand

M = silt

C = clay

O = organic

The second letter indicates the following:

W = well-graded, indicating a coarse-grained soil has particles of all sizes

P = poorly-graded, indicating a coarse-grained soil has particles of the same size, or the soil is skip-graded or gap-graded

M = indicates a coarse-grained soil that has silt size particles

C = indicates a coarse-grained soil that has clay size particles

L = indicates a fine-grained soil of low plasticity

H = indicates a fine-grained soil of high plasticity

An exception is peat, where the group symbol is PT. As indicated in Table 4.1, certain soils require the use of dual symbols. In addition to the classification of a soil, other items should also be included in the field or laboratory description of a soil, such as:

Soil Color   Usually the standard primary color (red, orange, yellow) of the soil is listed. Although not frequently used in geotechnical engineering, color charts have been developed. For example, the Munsell Soil Color Charts (1975) display 199 different standard color chips systematically arranged according to their Munsell notation, on cards carried in a loose-leaf notebook. The arrangement is by the three variables that combine to describe all colors and are known in the Munsell system as hue, value, and chroma. Color can be very important in identifying different types of soil. For example, the Friars formation, which is a stiff-fissured clay and is a frequent cause of geotechnical problems such as landslides and expansive soil, can often be identified by its dark green color. Another example is the Sweetwater formation, which is also a stiff-fissured clay, and has a bright pink color due to the presence of montmorillonite.

Soil Structure   In some cases the structure of the soil may be evident. Definitions vary, but in general, the soil structure refers to both the geometric arrangement of the soil particles and the interparticle forces that may act between them (Holtz and Kovacs, 1981). There are many different types

TABLE 4.1 Unified Soil Classification System

Major divisions	Subdivisions	USCS symbol	Typical names	Laboratory classification criteria	
Coarse-grained soils (More than 50 percent retained on No. 200 sieve)	Gravels (More than 50 percent of coarse fraction retained on no. 4 sieve)	GW	Well-graded gravels or gravel-sand mixtures, little or no fines	Less than 5 percent fines*	$C_u \geq 4$ and $1 \leq C_c \leq 3$
		GP	Poorly graded gravels or gravelly sands, little or no fines	Less than 5 percent fines*	Does not meet C_u and/or C_c criteria listed above
		GM	Silty gravels, gravel-sand-silt mixtures	More than 12 percent fines*	Minus no. 40 soil plots below the A line
		GC	Clayey gravels, gravel-sand-clay mixtures	More than 12 percent fines*	Minus no. 40 soil plots on or above the A line
	Sands (50 percent of more of coarse fraction passes no. 4 sieve)	SW	Well-graded sands or gravelly sands, little or no fines	Less than 5 percent fines*	$C_u \geq 6$ and $1 \leq C_c \leq 3$
		SP	Poorly graded sands or gravelly sands, little or no fines	Less than 5 percent fines*	Does not meet C_u and/or C_c criteria listed above
		SM	Silty sands, sand-silt mixtures	More than 12 percent fines*	Minus no. 40 soil plots below the A line
		SC	Clayey sands, sand-clay mixtures	More than 12 percent fines*	Minus no. 40 soil plots on or above A line
Fine-grained soils (50 percent or more passes no. 200 sieve)	Silts and clays (Liquid limit less than 50)	ML	Inorganic silts, rock flour, silts of low plasticity	Inorganic soil	PI < 4 or plots below A line
		CL	Inorganic clays of low plasticity, gravelly clays, sandy clays, etc.	Inorganic soil	PI > 7 and plots on or above A line†
		OL	Organic silts and organic clays of low plasticity	Organic soil	LL (oven-dried)/LL (not dried) < 0.75
	Silts and clays (Liquid limit 50 or more)	MH	Inorganic silts, micaceous silts, silts of high plasticity	Inorganic soil	Plots below A line
		CH	Inorganic highly plastic clays, fat clays, silty clays, etc.	Inorganic soil	Plots on or above A line
		OH	Organic silts and organic clays of high plasticity	Organic soil	LL (oven-dried)/LL (not dried) < 0.75
Peat	Highly organic	PT	Peat and other highly organic soils	Primarily organic matter, dark in color, and organic odor	

*Fines are those soil particles that pass the no. 200 sieve. For gravels with 5 to 12 percent fines, use of dual symbols required (i.e., GW-GM, GW-GC, GP-GM, or GP-GC). For sands with 5 to 12 percent fines, use of dual symbols required (i.e., SW-SM, SW-SC, SP-SM, or SP-SC).

†If $4 \leq PI \leq 7$ and plots above A line, then dual symbol (i.e., CL-ML) is required.

of soil structure, such as cluster, dispersed, flocculated, honeycomb, single-grained, and skeleton (see App. A, Glossary 2, for definitions). In some cases, the soil structure may be visible under a magnifying glass, or in other cases the soil structure may be reasonably inferred from laboratory testing results.

Soil Texture   The texture of a soil refers to the degree of fineness of the soil. For example, terms such as smooth, gritty, or sharp can be used to describe the texture of the soil when it is rubbed between the fingers.

Soil Porosity   The soil classification should also include the in situ condition of the soil. For example, numerous small voids may be observed in the soil, and this is referred to as pinhole porosity.

Clay Consistency   For clays, the consistency (i.e., degree of firmness) should be listed. The consistency of clay varies from very soft to hard based on the undrained shear strength of the clay. The undrained shear strength can be determined from the unconfined compression test or from field or laboratory vane shear tests. Based on the undrained shear strength s_u of an undisturbed specimen, cohesive soils are deemed to have a very soft, soft, medium, stiff, very stiff, or hard consistency. The values of undrained shear strength versus consistency are listed below:

Clay consistency	Undrained shear strength (kPa)	Undrained shear strength (psf)
Very Soft	$s_u < 12$	$s_u < 250$
Soft	$12 \leq s_u < 25$	$250 \leq s_u < 500$
Medium	$25 \leq s_u < 50$	$500 \leq s_u < 1000$
Stiff	$50 \leq s_u < 100$	$1000 \leq s_u < 2000$
Very Stiff	$100 \leq s_u < 200$	$2000 \leq s_u < 4000$
Hard	$s_u \geq 200$	$s_u \geq 4000$

If the shear strength of the soil has not been determined, then the consistency of the clay can be estimated in the field or laboratory based on the following:

Very soft.   The clay is easily penetrated several centimeters by the thumb. The clay oozes out between the fingers when squeezed in the hand.

Soft.   The clay is easily penetrated about 1 in. (2 to 3 cm) by the thumb. The clay can be molded by slight finger pressure.

Medium.   The clay can be penetrated about 0.4 in. (1 cm) by the thumb with moderate effort. The clay can be molded by strong finger pressure.

Stiff.   The clay can be indented about 0.2 in. (0.5 cm) by the thumb with great effort.

Very stiff.   The clay cannot be indented by the thumb, but can be readily indented with the thumbnail.

Hard.   With great difficulty, the clay can only be indented with the thumbnail.

The estimates of the consistency of clay as listed above should only be used for classification purposes and never be used as the basis for engineering design parameters.

Sand Density Condition   For sands, the density state of the soil varies from very loose to very dense (see Table 2.6).

Soil Moisture Condition   The moisture condition of the soil should also be listed. Moisture conditions vary from a dry soil to a saturated soil. The moisture condition of a soil (i.e., the degree of saturation) will be further discussed in Sec. 4.3. Especially in the arid climate of the southwestern United States, soil may be in a dry and powdery state. Often these soils are misclassified as silts, when in fact they are highly plastic clays. It is always important to add water to dry or powdery soils in order to assess their plasticity characteristics.

Additional Descriptive Items   USCS is only applicable for soil and rock particles passing the 3-in. (75 mm) sieve. Cobbles and boulders are larger than 3 in. (75 mm), and if applicable, the words "with cobbles" or "with boulders" should be added to the soil classification. Other descriptive terminology includes the presence of rock fragments, such as crushed shale, claystone, sandstone, siltstone, or mudstone fragments, and unusual constituents such as shells, slag, glass fragments, and construction debris.

An example of a complete soil classification and description for the soil having the grain size distribution and Atterberg limits test data in Fig. 4.3 is as follows:

Clayey sand (SC).   Based on dry mass, the soil contains 13.8 percent gravel size particles, 48.7 percent sand size particles, 19.5 percent silt size particles, and 18.0 percent clay size particles. The gravel size particles are predominately hard and angular rock fragments. The sand size particles are predominately composed of angular quartz grains. Atterberg limits performed on the soil passing the No. 40 sieve indicate a LL = 28 and a PI = 15. The in situ soil has a reddish-brown color, soft consistency, gritty texture, and is wet.

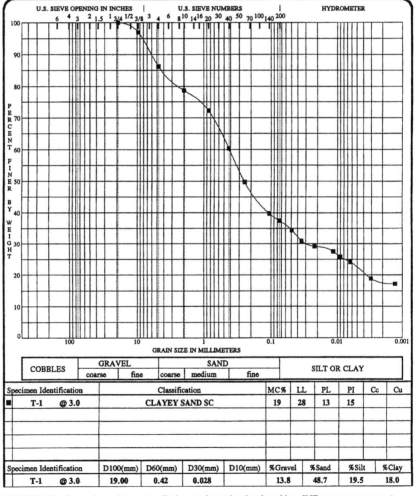

FIGURE 4.3   Grain size and Atterberg limits test data (plot developed by gINT computer program).

For further details on the USCS, see ASTM D 2487-00 (2004), "Standard Practice for Classification of Soils for Engineering Purposes (Unified Soil Classification System)."

4.2.4   AASHTO Soil Classification System

This classification system was developed by the American Association of State Highway and Transportation Officials (see Table 4.2). Inorganic soils are divided into seven groups (A-1 through A-7) with the eighth group (A-8) reserved for highly organic soils. Soil types A-1, A-2, and A-7 have subgroups as indicated in Table 4.2.

Those soils having plastic fines can be further categorized by using the group index (defined in Table 4.2). Groups A-1-a, A-1-b, A-3, A-2-4, and A-2-5 should be considered to have a group index equal to zero. According to AASHTO, the road supporting characteristics of a subgrade may be assumed as an inverse ratio to its group index. Thus a road subgrade having a group index of 0 indicates a good subgrade material that will often provide good drainage and adequate bearing when thoroughly compacted. A road subgrade material that has a group index of 20 or greater indicates a very poor subgrade material that will often be impervious and have a low bearing capacity.

Figure 4.4 shows the plasticity chart and a comparison of the classification of fine-grained soil per the USCS versus silt-clay materials per AASHTO. For further information on the AASHTO soil classification system, see ASTM D 3282-97 (2004), "Standard Practice for Classification of Soils and Soil-Aggregate Mixtures for Highway Construction Purposes."

4.2.5   USDA Textural Soil Classification System

Figure 4.5 presents the simple textural classification system per the U.S. Department of Agriculture (1975), where "percent sand" is defined as the percentage of soil particles between 2.0 and 0.050 mm in size, "percent silt" is defined as the percentage of soil particles between 0.050 and 0.002 mm, and "percent clay" is defined as the percentage of soil particles finer than 0.002 mm. Note that the definitions of sand and silt size particles differ from those used in the other classification systems. If a soil contains a significant soil fraction larger than 2.0 mm, the textural classification is preceded with the term *gravely* for soil fragments up to 76 mm in diameter, *cobbly* for rock fragments between 76 and 250 mm, and *stony* or *bouldery* for rock fragments larger than 250 mm.

4.2.6   Organic Soil Classification System

Table 4.3 presents a classification system for organic materials. As indicated in Table 4.3, there are four major divisions, as follows:

1. *Organic matter.*   These materials consist almost entirely of organic material. Examples include fibrous peat and fine-grained peat.

2. *Highly organic soils.*   These soils are composed of 30 to 75 percent organic matter mixed with mineral soil particles. Examples include silty peat and sandy peat.

3. *Organic soils.*   These soils are composed of 5 to 30 percent organic material. These soils are typically classified as organic soils of high plasticity (OH, i.e., LL > 50) or low plasticity (OL, i.e., LL < 50) and have a ratio of liquid limit (oven-dried soil) divided by liquid limit (not dried soil) that is less than 0.75 (see Table 4.1).

4. *Slightly organic soils.*   These soils typically have less than 5 percent organic matter. Per the USCS, they have a ratio of liquid limit (oven-dried soil) divided by liquid limit (not dried soil) that is greater than 0.75. Often a modifier, such as slightly organic soil, is used to indicate the presence of organic matter.

Also included in Table 4.3 is the typical range of laboratory test results for the four major divisions of organic material. As indicated in Table 4.3, the water content w increases and the total unit weight

TABLE 4.2 AASHTO Soil Classification System

Major divisions	Group	AASHTO symbol	Typical names	Sieve analysis (percent passing)	Atterberg limits	
Granular materials (35 percent or less passing no. 200 sieve)	A-1	A-1-a	Stone or gravel fragments	Percent passing: no. $10 \leq 50$ percent; no. $40 \leq 30$ percent; no. $200 \leq 15$ percent	$PI \leq 6$	
		A-1-b	Gravel and sand mixtures	No. $40 \leq 50$ percent; no. $200 \leq 25$ percent	$PI \leq 6$	
	A-3	A-3	Fine sand that is nonplastic	No. $40 > 50$ percent; no. $200 \leq 10$ percent	$PI = 0$ (nonplastic)	
	A-2	A-2-4	Silty gravel and sand	Percent passing no. 200 sieve ≤ 35 percent	$LL \leq 40$	$PI \leq 10$
		A-2-5	Silty gravel and sand	Percent passing no. 200 sieve ≤ 35 percent	$LL > 40$	$PI \leq 10$
		A-2-6	Clayey gravel and sand	Percent passing no. 200 sieve ≤ 35 percent	$LL \leq 40$	$PI > 10$
		A-2-7	Clayey gravel and sand	Percent passing no. 200 sieve ≤ 35 percent	$LL > 40$	$PI > 10$
Silt-clay materials (More than 35 percent passing no. 200 sieve)	A-4	A-4	Silty soils	Percent passing no. 200 sieve > 35 percent	$LL \leq 40$	$PI \leq 10$
	A-5	A-5	Silty soils	Percent passing no. 200 sieve > 35 percent	$LL > 40$	$PI \leq 10$
	A-6	A-6	Clayey soils	Percent passing no. 200 sieve > 35 percent	$LL \leq 40$	$PI > 10$
	A-7	A-7-5	Clayey soils	Percent passing no. 200 sieve > 35 percent	$LL > 40$ $PI > 10$	$PI \leq LL - 30$
		A-7-6	Clayey soils	Percent passing no. 200 sieve > 35 percent	$LL > 40$ $PI > 10$	$PI > LL - 30$
Highly organic	A-8	A-8	Peat and other highly organic soils	Primarily organic matter, dark in color, and organic odor		

Notes:

1. *Classification procedure.* First decide to which of the three main categories (granular materials, silt-clay materials, or highly organic) the soil belongs. Then proceed from the top to the bottom of the chart, and the first group that meets the particle size and Atterberg limits criteria is the correct classification.

2. Group index $= (F - 35)[0.2 + 0.005(LL - 40)] + 0.01(F - 15)(PI - 10)$, where F = percent passing no. 200 sieve, LL = liquid limit, and PI = plasticity index. Report group index to nearest whole number. For negative group index, report as 0. When working with A-2-6 and A-2-7 subgroups, use only the PI portion of the group index equation.

3. Atterberg limits tests are performed on soil passing the no. 40 sieve. LL = liquid limit, PL = plastic limit, and PI = plasticity index.

4. AASHTO definitions of particle sizes are as follows: Boulders: above 75 mm; gravel: 75 mm to no. 10 sieve; coarse sand: no. 10 to no. 40 sieve; fine sand: no. 40 to no. 200 sieve; and silt-clay size particles: material passing no. 200 sieve.

5. Example: An example of an AASHTO classification for a clay is A-7-6 (30), or group A-7, subgroup 6, group index 30.

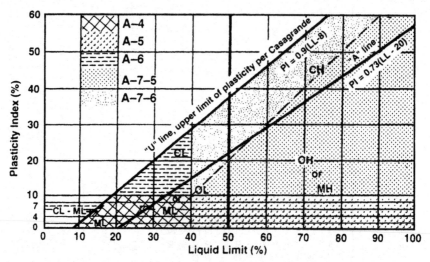

FIGURE 4.4 Plasticity chart showing the location of fine-grained soil per USCS and the location of silt-clay materials (shaded areas) per AASHTO.

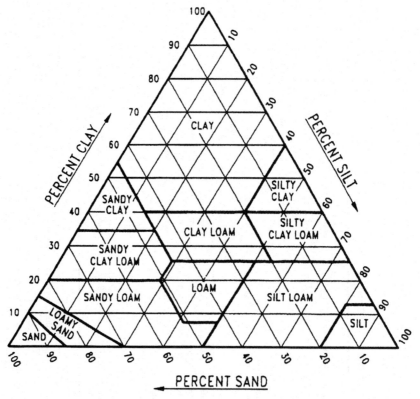

FIGURE 4.5 USDA textural classification system (*Note:* percent sand = 2.0 to 0.050 mm, percent silt = 0.050 to 0.002 mm, and percent clay is finer than 0.002 mm).

TABLE 4.3 Soil Classification for Organic Soil

Major divisions	Organic content	USCS symbol	Typical names	Distinguishing characteristics for visual identification	Typical range of laboratory test results
Organic matter	75–100 percent organics (either visible for inferred)	PT	Fibrous peat (woody, mats, and the like)	Lightweight and spongy. Shrinks considerably on air drying. Much water squeezes from sample	w = 500–1200 percent γ_t = 9.4–11 kN/m^3 (60–70 pcf) G = 1.2–1.8 $C_c/(1 + e_0) \geq 0.40$
		PT	Fine-grained peat (amorphous)	Lightweight and spongy. Shrinks considerably on air drying. Much water squeezes from sample	w = 400–800 percent PI = 200–500 γ_t = 9.4–11 kN/m^3 (60–70 pcf) G = 1.2–1.8 $C_c/(1 + e_0) \geq 0.35$
Highly organic soils	30–75 percent organics (either visible or inferred)	PT	Silty peat	Relatively lightweight, spongy. Shrinks on air drying. Usually can readily squeeze water from sample	w = 250–500 percent PI = 150–350 γ_t = 10–14 kN/m^3 (65 to 90 pcf) G = 1.8–2.3 $C_c/(1 + e_0)$ = 0.3–0.4
		PT	Sandy peat	Sand fraction visible. Shrinks on air drying. Often a "gritty" texture. Usually can squeeze water from sample	w = 100–400 percent PI = 50–150 γ_t = 11–16 kN/m^3 (70–100 pcf) G = 1.8–2.4 $C_c/(1 + e_0)$ = 0.2–0.3
Organic soils	5–30 percent organics (either visible or inferred)	OH	Clayey organic silt	Often has strong hydrogen sulfide (H$_2$S) odor. Medium dry strength and slow dilatancy.	w = 65–200 percent PI = 50–150 γ_t = 11–16 kN/m^3 (70–100 pcf) G = 2.3–2.6 $C_c/(1 + e_0)$ = 0.2–0.35
		OL	Organic sand or silt	Threads weak and friable near plastic limit, or will not roll at all. Low dry strength, medium to high dilatancy	w = 30–125 percent PI = NP–40 γ_t = 14–17 kN/m^3 (90 to 110 pcf) G = 2.4–2.6 $C_c/(1 + e_0)$ = 0.1–0.25
Slightly organic soils	Less than 5 percent organics	Use Table 4.1	Soil with slight organic fraction	Depends on characteristics of inorganic fraction	Depends on characteristics of inorganic fraction

Source: NAVFAC DM-7.1 (1982), based on unpublished work by Ayers and Plum.

Notes: w = in situ water content, PI = plasticity index, NP = nonplastic, γ_t = total unit weight, G = specific gravity (soil minerals plus organic matter), C_c = compression index, e_0 = initial void ratio, and $C_c/(1 + e_0)$ = modified compression index.

γ_t decreases as the organic content increases. The specific gravity G includes the organic matter, hence the low values for highly organic material. The compression index C_c is discussed in Chap. 8.

4.2.7 Description of Frozen Soil

Table 4.4 presents a description of frozen soil. The procedure is used in conjunction with the USCS. Thus a frozen soil would be provided with two group symbols, the first group symbol relating to the soil classification (Table 4.1) and the second group symbol delineating the frozen condition of the soil (Table 4.4). As an example, see the boring log for frozen soil presented in Fig. 2.33.

The process of describing frozen soil is based on a three-part identification process, as follows:

Part I: Description of the soil phase.   The soil phase is classified in accordance with the USCS (Table 4.1).

Part II: Description of the frozen soil.   Frozen soils where ice is not visible to the unaided eye are designed by the group symbol N and are divided into two subgroups as indicated in Table 4.4. Frozen soils in which significant segregated ice is visible to the unaided eye, but individual ice masses or layers are less than 1 in. (25 mm) in thickness, are designated by the group symbol V and are divided into five subgroups as shown in Table 4.4.

Part III: Description of substantial ice strata.   If the subsurface exploration reveals ice strata that are greater than 1 in. (25 mm) in thickness, they are designated by the group symbol ICE and divided into two subgroups as shown in Table 4.4.

For additional details concerning the classification system for frozen soil, see ASTM D 4083-01 (2004), "Standard Practice for Description of Frozen Soils (Visual-Manual Procedure)."

4.2.8 Rock Classification

The purpose of this section is to provide a brief introduction to rock classification. There are three basic types of rocks: igneous, sedimentary, and metamorphic. Because of their special education and training, usually the best person to classify rock is the engineering geologist. Table 4.5 presents a simplified rock classification and common rock types.

In addition to determining the type of rock, it is often important to determine the quality of the rock, which is related to its degree of weathering, defined as follows:

Fresh   No discoloration or oxidation.

Slightly Weathered   Discoloration or oxidation is limited to surface of, or short distance from, fractures; some feldspar crystals are dull.

Moderately Weathered   Discoloration or oxidation extends from fractures, usually throughout, Fe and Mg minerals are rusty and feldspar crystals are cloudy.

Intensely Weathered   Discoloration or oxidation throughout; all feldspars and Fe and Mg minerals are altered to clay to some extent; or chemical alteration produces in situ disaggregation.

Decomposed   Discolored or oxidized throughout, but resistant minerals such as quartz may be unaltered; all feldspars and Fe and Mg minerals are completely altered to clay.

Besides degree of weathering, a measure of the quality of rock is its hardness, which has been correlated with the unconfined compressive strength of rock specimens. Table 4.6 lists hardness of rock as a function of the unconfined compressive strength. Because the unconfined compressive strength is performed on small rock specimens, in most cases, it will not represent the actual condition of in situ rock. The reason is due to the presence of joints, fractures, fissures, and planes of weakness in the actual rock mass that govern its engineering properties, such as deformation characteristics, shear strength, and permeability. The unconfined compressive test also does not consider other rock quality factors, such as its resistance to weathering or behavior when submerged in water.

TABLE 4.4 Description of Frozen Soil

Description of frozen soil	Group symbol	Subgroup		Field identification
		Description	Symbol	
Ice not visible	N	Poorly bonded or friable	N_f	Identify by visual examination. To determine presence of excess ice, place some frozen soil in a small jar, allowing it to melt, and observing the quantity of supernatant water as a percentage of total volume. Also use hand-magnifying lens as necessary. For soil not fully saturated, estimate degree of ice saturation: medium, low. Note presence of crystals or of ice coatings around larger particles.
		Well-bonded		
		No excess ice	N_b	
		Excess ice	N_{bn}	
			N_{be}	
Visible ice less than 1 in. (25 mm) thick	V	Individual ice crystal or inclusions	V_x	For ice phase, record the following when applicable: location, orientation, thickness, length, spacing, hardness, structure, color, size, shape, and pattern.
		Ice coatings on particles	V_c	Estimate volume of visible segregated ice present as percentage of total sample volume.
		Random or irregularly oriented ice formations	V_r	
		Stratified or distinctly oriented ice formations	V_s	
		Uniformly distributed ice	V_u	
Visible ice strata greater than 1 in. (25 mm) thick	ICE	Ice with soil inclusions	ICE + soil type	Designate material as *ICE*. Where special forms of ice such as hoarfrost can be distinguished, more explicit description should be given. Use descriptive terms as follows (usually one item from each group, when applicable):
		Ice without soil inclusions	ICE	1. Hardness: hard, soft (of mass, not individual crystals).
				2. Color (examples): colorless, gray, blue.
				3. Structure: Clear, cloudy, porous, candled, granular, stratified
				4. Admixtures (example): contains few thin silt inclusions

Source: ASTM D 4083-01 (2004).

4.15

TABLE 4.5 Simplified Rock Classification

Common igneous rocks		
Major division	Secondary divisions	Rock types
Extrusive	Volcanic explosion debris (fragmental)	Tuff (lithified ash) and volcanic breccia
	Lava flows and hot siliceous clouds	Obsidian (glass), pumice, and scoria
	Lava flows (fine-grained texture)	Basalt, andesite, and rhyolite
Intrusive	Dark minerals dominant	Gabbro
	Intermediate (25–50 percent dark minerals)	Diorite
	Light color (quartz and feldspar)	Granite
Common sedimentary rocks		
Major division	Texture (grain size) or chemical composition	Rock types
Clastic rocks*	Grain sizes larger than 2 mm (pebbles, gravel, cobbles, and boulders)	Conglomerate (rounded cobbles) or breccia (angular rock fragments)
	Sand-size grains, 0.062–2 mm	Sandstone
	Silt-size grains, 0.004–0.062 mm	Siltstone
	Clay-size grains, less than 0.004 mm	Claystone and shale
Chemical and organic rocks	Carbonate minerals (e.g., calcite)	Limestone
	Halite minerals	Rock salt
	Sulfate minerals	Gypsum
	Iron-rich minerals	Hematite
	Siliceous minerals	Chert
	Organic products	Coal
Common metamorphic rocks		
Major division	Structure (foliated or massive)	Rock types
Coarse crystalline	Foliated	Gneiss
	Massive	Metaquartzite
Medium crystalline	Foliated	Schist
	Massive	Marble, quartzite, serpentine, soapstone
Fine to microscopic	Foliated	Phyllite, slate
	Massive	Hornfels, anthracite coal

*Grain sizes correspond to the Modified Wentworth scale.

For example, the author has observed complete disintegration (in only a few seconds) of the Friars formation claystone, which contains montmorillonite, when submerged in water.

Another measure of the quality of the rock is the rock quality designation (RQD), which is computed by summing the lengths of all pieces of the core (NX size) equal to or longer than 4 in. (10 cm) and dividing by the total length of the core run. The RQD is multiplied by 100 and expressed as a percentage. The mass rock quality can be defined as follows:

RQD = 0–25%, rock quality is defined as very poor

RQD = 25–50%, rock quality is defined as poor

RQD = 50–75%, rock quality is defined as fair

RQD = 75–90%, rock quality is defined as good

RQD = 90–100%, rock quality is defined as excellent

TABLE 4.6 Hardness of Rock Versus Unconfined Compressive Strength

Hardness	q_u	Rock description
Very soft	10–250 tsf	The rock can be readily indented, grooved, or gouged with fingernail, or carved with a knife. Breaks with light manual pressure. The rock disintegrates upon the single blow of a geologic hammer.
Soft	250–500 tsf	The rock can be grooved or gouged easily by a knife or sharp pick with light pressure. Can be scratched with a fingernail. Breaks with light to moderate manual pressure.
Hard	500–1000 tsf	The rock can be scratched with a knife or sharp pick with great difficulty (heavy pressure is needed). A heavy hammer blow is required to break the rock.
Very hard	1000–2000 tsf	The rock cannot be scratched with a knife or sharp pick. The rock can be broken with several solid blows of a geologic hammer.
Extremely hard	>2000 tsf	The rock cannot be scratched with a knife or sharp pick. The rock can only be chipped with repeated heavy hammer blows.

Notes:
1. One measure of the quality of rock is its hardness, which has been correlated with the unconfined compressive strength of rock specimens. This table lists hardness of rock as a function of the unconfined compressive strength q_u. Because the unconfined compressive strength is performed on small rock specimens, in most cases, it will not represent the actual condition of in situ rock. The reason is the presence of joints, fractures, fissures, and planes of weakness in the actual rock mass which govern its engineering properties, such as deformation characteristics, shear strength, and permeability. The unconfined compressive test also does not consider other rock quality factors, such as its resistance to weathering or behavior when submerged in water.
2. q_u = unconfined compressive strength (tsf) of the rock. 1 tsf is approximately equal to 100 kPa.
Sources: Basic Soils Engineering (Hough 1969) and *Engineering Geology Field Manual* (1987).

When calculating the RQD, only the natural fractures should be counted and any fresh fractures due to the sampling process should be ignored. RQD measurements can provide valuable data on the quality of the in situ rock mass, and can be used to locate zones of extensively fractured or weathered rock.

There are many excellent publications on the identification, sampling, and laboratory testing of rock. For example, the *Engineering Geology Field Manual* (U.S. Department of the Interior, 1987) provides guidelines on the performance of fieldwork leading to the development of geologic concepts and reports. This manual also provides a discussion of geologic mapping, discontinuity rock surveys, sampling and testing methods of rock, and provides instructions for core logging.

4.3   PHASE RELATIONSHIPS

Phase relationships are also known as weight-volume relationships. In essence, the phase relationships provide a mathematical description of the soil and they are used in engineering analyses. Certain phase relationships have been already described in Sec. 3.2, including water content, total unit weight, and specific gravity. This section describes those phase relationships that cannot be determined in a laboratory, but instead must be calculated, such as the volume of voids, void ratio, porosity, and degree of saturation.

Volume of Voids.   As indicated in Fig. 3.1, the volume of voids V_v is defined as:

$$V_v = V_g + V_w \tag{4.7}$$

where V_v = volume of voids (ft^3 or m^3)
V_g = volume of gas (air) as defined in Fig. 3.1 (ft^3 or m^3)
V_w = volume of water as defined in Fig. 3.1 (ft^3 or m^3)

Void Ratio and Porosity.   The void ratio e and porosity n are defined as:

$$e = \frac{V_v}{V_s} \tag{4.8}$$

$$n = \frac{V_v}{V} \tag{4.9}$$

where e = void ratio (dimensionless)
  n = porosity (dimensionless), sometimes expressed as a percentage
  V_s = volume of soil solids as defined in Fig. 3.1 (ft^3 or m^3)
  V = total volume of the soil as defined in Fig. 3.1 (ft^3 or m^3)

The void ratio e and porosity n are related as follows:

$$e = \frac{n}{1-n} \quad \text{and} \quad n = \frac{e}{1+e} \tag{4.10}$$

The void ratio and porosity indicate the relative amount of void space in a soil. The lower the void ratio and porosity, the denser the soil (and vice versa). The natural soil having the lowest void ratio is probably till. For example, a typical value of dry density for till is 146 pcf (2.34 Mg/m^3), which corresponds to a void ratio of 0.14 (NAVFAC DM-7.1, 1982). A typical till consists of a well-graded soil ranging in particle sizes from clay to gravel and boulders. The high density and low void ratio are due to the extremely high stress exerted by glaciers (Winterkorn and Fang, 1975). For compacted soil, the soil type with typically the lowest void ratio is well-graded decomposed granite (DG). A typical value of maximum dry density (Modified Proctor) for a well-graded DG is 137 pcf (2.20 Mg/m^3), which corresponds to a void ratio of 0.21. In general, the factors needed for a very low void ratio for compacted and naturally deposited soil are (Aberg, 1996; Day, 1997a):

1. A well-graded grain-size distribution
2. A high ratio of D_{100}/D_0 (ratio of the largest and smallest grain sizes)
3. Clay particles (having low activity) to fill in the smallest void spaces
4. A process, such as compaction or the weight of glaciers, to compress the soil particles into dense arrangements

At the other extreme are clays, such as sodium montmorillonite, which at low confining pressures can have a void ratio of more than 25. Highly organic soil, such as peat, can have an even higher void ratio.

Degree of Saturation   The degree of saturation S is defined as:

$$S(\%) = \frac{100\, V_w}{V_v} \tag{4.11}$$

where S = degree of saturation (dimensionless), usually expressed as a percentage
  V_w = volume of water as defined in Fig. 3.1 (ft^3 or m^3)
  V_v = volume of voids, from Eq. 4.7

The degree of saturation indicates the degree to which the soil voids are filled with water. A totally dry soil will have a degree of saturation of 0 percent, while a saturated soil, such as a soil below the groundwater table, will have a degree of saturation of 100 percent. Typical ranges of degree of saturation versus soil condition are as follows (Terzaghi and Peck, 1967):

Dry $S = 0\%$
Humid $S = 1$–25%

Damp	$S = 26\text{--}50\%$
Moist	$S = 51\text{--}75\%$
Wet	$S = 76\text{--}99\%$
Saturated	$S = 100\%$

If a soil is obtained below the groundwater table or after submergence in the laboratory, the degree of saturation is often assumed to be 100 percnt and then phase relationships (such as the void ratio) are back calculated. However, for soil below the groundwater table, a better approach is to use the degree of saturation as a final check on the accuracy on the laboratory test data (i.e., γ_t, w, and G_s).

Relative Density.   The relative density is a measure of the density state of a nonplastic soil. The relative density can only be used for soil that is nonplastic, such as sands and gravels. The relative density (D_r in percent) is defined as:

$$D_r(\%) = 100 \frac{e_{\max} - e}{e_{\max} - e_{\min}} \tag{4.12}$$

where $e_{\max}$ = void ratio corresponding to the loosest possible state of the soil, usually obtained by pouring the soil into a mold of known volume (ASTM D 4254-00, 2004)
$e_{\min}$ = void ratio corresponding to the densest possible state of the soil, usually obtained by vibrating the soil particles into a dense state (ASTM D 4253-00, 2004)
e = the natural void ratio of the soil

The density state of the natural soil can be described as follows (Lambe and Whitman, 1969):

$D_r = 0\text{--}15\%$	Very loose condition
$D_r = 15\text{--}35\%$	Loose condition
$D_r = 35\text{--}65\%$	Medium condition
$D_r = 65\text{--}85\%$	Dense condition
$D_r = 85$ to 100%	Very dense condition

The relative density D_r should not be confused with the relative compaction (RC), which will be discussed in Chap. 15. The relative density D_r of an in situ granular soil can be estimated from the results of standard penetration tests (see Table 2.6). The relative density D_r of sands and gravels is important because it is a primary factor in the amount of settlement due to applied foundation loads or the liquefaction potential of submerged soil.

Useful Relationship.   A frequently used method of solving phase relationships is to first fill in the phase diagram shown in Fig. 3.1. Once the different mass and volumes are known, then the various phase relationships can be determined. Another approach is to use equations that relate different parameters. A useful relationship is as follows:

$$G_s w = S e \tag{4.13}$$

where G_s = specific gravity of soil solids (dimensionless), see Sec. 3.2.3
w = water content of the soil (percent), see Sec. 3.2.1
S = degree of saturation from Eq. 4.11 (percent)
e = void ratio from Eq. 4.8 (dimensionless)

Other useful relationships are presented in Tables 4.7 and 4.8. Table 4.7 presents useful mass and volume relationships. Table 4.8 presents equations that can be used to calculate the total unit weight, dry unit weight, saturated unit weight, and buoyant unit weight.

Example Problem 4.1 A soil specimen has the following measured properties:

Total unit weight $\gamma_t = 121$ pcf $(19.0$ kN/m$^3)$
Water content $w = 18.0$ percent
Specific gravity $G_s = 2.65$

Calculate the parameters shown in Fig. 3.1 assuming that the total volume $V = 1$ ft^3 and 1 m^3. Also calculate the dry unit weight γ_t, void ratio e, porosity n, and degree of saturation S.

Solution For United States Customary System, $V = 1$ ft^3

$$\gamma_d = \frac{\gamma_t}{1+w} = \frac{121}{1+0.18} = 103 \text{ pcf}$$

$$\gamma_d = \frac{W_s}{V} \quad \text{or} \quad W_s = \gamma_d V = (103 \text{ pcf})(1 \text{ ft}^3) = 103 \text{ lb}$$

$$M_s = W_s = 103 \text{ lb}$$

$$M_w = M_s w = (103 \text{ lb})(0.18) = 18 \text{ lb}$$

$$V_s = \frac{W_s}{(G_s)(\gamma_w)} = \frac{103}{(2.65)(62.4)} = 0.62 \text{ ft}^3$$

$$V_w = \frac{M_w}{\rho_w} = \frac{18 \text{ lb}}{62.4 \text{ pcf}} = 0.30 \text{ ft}^3$$

$$V_a = V - V_s - V_w = 1 - 0.62 - 0.30 = 0.08 \text{ ft}^3$$

For SI, $V = 1$ m^3

$$\gamma_d = \frac{\gamma_t}{(1+w)} = \frac{19}{(1+0.18)} = 16.1 \text{ kN/m}^3$$

$$\gamma_d = \frac{W_s}{V} \quad \text{or} \quad W_s = \gamma_d V = (16.1 \text{ kN/m}^3)(1 \text{ m}^3) = 16.1 \text{ kN}$$

$$W_s = M_s a \quad \text{or} \quad M_s = \frac{W_s}{a} = \frac{16.1 \text{ kN}}{9.81 \text{ m/sec}^2} = 1.64 \text{ Mg}$$

$$M_w = M_s w = (1.64 \text{ Mg})(0.18) = 0.30 \text{ Mg}$$

$$V_s = \frac{W_s}{(G_s)(\gamma_w)} = \frac{16.1}{(2.65)(9.81)} = 0.62 \text{ m}^3$$

$$V_w = \frac{M_w}{\rho_w} = \frac{0.30 \text{ Mg}}{1.0 \text{ Mg/m}^3} = 0.30 \text{ m}^3$$

$$V_a = V - V_s - V_w = 1 - 0.62 - 0.30 = 0.08 \text{ m}^3$$

Void ratio, porosity, and degree of saturation:

$$e = \frac{V_v}{V_s} = \frac{(V_w + V_a)}{V_s} = \frac{(0.30 + 0.08)}{0.62} = 0.61$$

$$n = \frac{V_v}{V} = \frac{(0.30 + 0.08)}{1.0} = 0.38 \text{ or } 38\%$$

$$S = \frac{(G_s)(w)}{e} = \frac{(2.65)(0.180)}{0.61} = 0.78 \text{ or } 78\%$$

TABLE 4.7   Mass and Volume Relationships (see Figure 3.1 for definition of terms)

Parameter		Relationships
Mass	Mass of solids M_s	$M_s = \dfrac{M}{1+w} = \dfrac{M_w G_s}{eS} = G_s V \rho_w (1-n)$
	Mass of water M_w	$M_w = \dfrac{eM_s S}{G_s} = wM_s = S\rho_w V_v$
	Total mass M	$M = M_s + M_w = M_s(1 + w)$
Volume	Volume of solids V_s	$V_s = \dfrac{M_s}{G_s \rho_w} = \dfrac{V}{1+e} = \dfrac{V_v}{e} = V(1-n) = V - (V_g + V_w)$
	Volume of water V_w	$V_w = \dfrac{M_w}{\rho_w} = \dfrac{SVe}{1+e} = SV_s e = SV_v = V_v - V_g$
	Volume of gas V_g	$V_g = \dfrac{(1-S)Ve}{1+e} = (1-S)V_s e = V - (V_s + V_w) = V_v - V_w$
	Volume of voids V_v	$V_v = \dfrac{V_s n}{1-n} = V - \dfrac{M_s}{G_s \rho_w} = \dfrac{Ve}{1+e} = V_s e = V - V_s$
	Total volume V	$V = \dfrac{V_s}{1-n} = \dfrac{V_v(1+e)}{e} = V_s(1+e) = V_s + V_g + V_w$

4.4   EFFECTIVE STRESS

Having described the soil in terms of a written description (soil classification) and mathematical description (phase relationships), the next step in the analysis is often to determine the stresses acting on the soil. This is important because most of foundation engineering deals with a change in stress of the soil. For example, the construction of a building applies an additional stress onto the soil supporting the foundation, which results in settlement of the building.

Stress is defined as the load divided by the area over which it acts. In geotechnical engineering, a compressive stress is considered positive and tensile stress is negative. Stress and pressure are often used interchangeably in geotechnical engineering. When using the International System of Units (SI), the units for stress are kPa. In the United States Customary System, the units for stress are psf (pounds-force per square foot). Stress expressed in units of kg/cm^2 is used in some figures in this book (e.g., see Figs. 2.39 and 2.41). One kg/cm^2 is approximately equal to 1 ton per square foot (tsf).

Effective Stress Equation.   The effective stress equation ($\sigma' = \sigma - u$) has been previously defined in Eq. 3.8. Many engineering analyses use the vertical effective stress, also known as the effective overburden stress, which is designated σ'_v or σ'_{vo}. In this case, Eq. 3.8 becomes:

$$\sigma'_v = \sigma_v - u \qquad\qquad (4.14)$$

TABLE 4.8   Unit Weight Relationships (see Fig. 3.1 for definition of terms)

Parameter	Relationships
Total unit weight γ_t	$\gamma_t = \dfrac{W_s + W_w}{V} = \dfrac{G_s \gamma_w (1 + w)}{1 + e}$
Dry unit weight γ_d	$\gamma_d = \dfrac{W_s}{V} = \dfrac{G_s \gamma_w}{1 + e} = \dfrac{\gamma_t}{1 + w}$
Saturated unit weight γ_{sat}	$\gamma_{sat} = \dfrac{W_s + V_v \gamma_w}{V} = \dfrac{(G_s + e)\gamma_w}{1 + e} = \dfrac{G_s \gamma_w (1 + w)}{1 + G_s w}$

Note: The total unit weight γ_t is equal to the saturated unit weight γ_{sat} when all the void spaces are filled with water (that is, $S = 100$ percent).

Buoyant unit weight γ_b	$\gamma_b = \gamma_{sat} - \gamma_w$
	$\gamma_b = \dfrac{\gamma_w (G_s - 1)}{1 + e} = \dfrac{\gamma_w (G_s - 1)}{1 + G_s w}$

Note: The buoyant unit weight is also known as the submerged unit weight.

Notes:
1. In the equations listed here and in Table 4.7, water content w and degree of saturation S must be expressed as a decimal (not as a percentage).
2. ρ_w = density of water (1.0 Mg/m³, 62.4 pcf) and γ_w = unit weight of water (9.81 kN/m³, 62.4 pcf).
3. In the U.S. Customary System, it is common to concurrently use pounds to represent both a unit of mass M and a unit of force W. Thus the density of water (mass per unit volume: $\rho_w = 62.4$ pcf) has the same numerical value as the unit weight of water (weight per unit volume: $\gamma_w = 62.4$ pcf). In SI, the mass M is multiplied by the acceleration of gravity ($a = 9.81$ m/s²) in order to obtain the weight ($W = Ma$). Thus the density of water ρ_w equals 1.0 Mg/m³ or 1.0 g/cm³, and the unit weight of water γ_w equals 9.81 kN/m³.

where σ_v' = vertical effective stress (psf or kPa)
σ_v = vertical total stress (psf or kPa)
u = pore water pressure (psf or kPa)

Total Stress.   For the condition of a uniform soil and a level ground surface (geostatic condition), the vertical total stress at any depth below the ground surface is:

$$\sigma_v = \gamma_t z \tag{4.15}$$

where σ_v = vertical total stress (psf or kPa)
γ_t = total unit weight of the soil (pcf or kN/m³)
z = depth below ground surface (ft or m)

Suppose a uniform soil deposit has a total unit weight of 120 pcf (18.9 kN/m³). At a depth of 10 ft (3.0 m), the total vertical stress would be 120 pcf times 10 ft, or 1200 psf (57 kPa).

The more common situation is that there will be various soil layers, each having a different total unit weight. In order to calculate the vertical total stress, multiply the layer thickness times its corresponding total unit weight and then add together all the layers.

Pore Water Pressure.   For dry sand, the pore water pressure is zero. This is because there is no water in the soil pores and hence there is no water pressure. In this case with the pore water pressure u equal to zero, the total stress is equal to the effective stress per Eq. 3.8.

The usual case is that there will be water in the void spaces between the soil particles. For the condition of a soil below a groundwater table and for hydrostatic water pressure (i.e., no groundwater flow or excess pore water pressures), the pore water pressure is:

$$u = \gamma_w \, z_w \tag{4.16}$$

where u = pore water pressure (psf or kPa)
  γ_w = unit weight of water (62.4 pcf or 9.81 kN/m³)
  z_w = depth below the groundwater table (ft or m)

Suppose the uniform soil deposit previously discussed has a groundwater table corresponding to the ground surface. At a depth of 10 ft (3.0 m), the pore water pressure would be 62.4 pcf (9.81 kN/m³) times 10 ft, or 624 psf (29 kPa).

Calculation of Vertical Effective Stress. By knowing the vertical total stress (Eq. 4.15) and the pore water pressure (Eq. 4.16), the vertical effective stress σ'_v can be calculated at any depth below the ground surface by using Eq. 4.14. Using the prior example of a uniform soil deposit having a groundwater table at ground surface, the vertical effective stress at a depth of 10 ft (3.0 m) would be:

$$\sigma'_v = \sigma_v - u = 1200 - 624 = 576 \text{ psf (28 kPa)}$$

An alternative method is to use the buoyant unit weight γ_b to calculate the vertical effective stress. Using the prior example of a groundwater table at ground surface, the vertical effective stress σ'_v is simply the buoyant unit weight γ_b times the depth below the ground surface or $\gamma_b = 120 - 62.4 = 57.6$ pcf times 10 ft, which equals 576 psf (28 kPa). More often, the groundwater table is below the ground surface, in which case the vertical total stress of the soil layer above the groundwater table must be added to the buoyant unit weight calculations.

The vertical effective stress σ'_v is often plotted versus depth and included with the subsoil profile. For example, in Fig. 2.39, the vertical effective stress σ'_v and the pore water pressure (defined as u_s in Fig. 2.39) are plotted versus depth. Likewise in Figs. 2.40 and 2.41, the vertical effective stress, also known as effective overburden stress, has been plotted versus depth.

For cases where there is flowing groundwater or excess pore water pressure due to the consolidation of clay, the pore water will not be hydrostatic. Engineering analyses, such as seepage analyses (Sec. 4.7) or the theory of consolidation (Chap. 8) can be used to predict the pore water pressure. For some projects, piezometers can be installed to measure the pore water pressure u in the ground.

Capillarity. Soil above the groundwater table can be subjected to negative pore water pressure. This is known as capillarity, also known as capillary action, which is the rise of water through soil due to the fluid property known as surface tension. Due to capillarity, the pore water pressures are less than atmospheric values produced by the surface tension of pore water acting on the meniscus formed in the void spaces between the soil particles. The height of capillary rise (h_c) is related to the pore size of the soil, as follows (Hansbo, 1975):

Open graded gravel	$h_c = 0$
Coarse sand	$h_c = 0.1$–0.5 ft (0.03–0.15 m)
Medium sand	$h_c = 0.4$–3.6 ft (0.12–1.1 m)
Fine sand	$h_c = 1.0$–12 ft (0.3–3.5 m)
Silt	$h_c = 5$–40 ft (1.5–12 m)
Clay	$h_c > 33$ ft (> 10 m)

As this data indicates, there will be no capillary rise for open graded gravel because of the large void spaces between the individual gravel size particles. But for clay, which has very small void spaces, the capillary rise can be in excess of 33 ft (10 m). If the soil is saturated above the groundwater table, then the pore water pressure u is negative and can be calculated as the distance above the groundwater table h times the unit weight of water γ_w, or:

$$u = -\gamma_w h \tag{4.17}$$

Capillary is important in the understanding of soil behavior. Because of the negative value of pore water pressure due to capillarity, it essentially holds together the soil particles. For large size soil particles, such as gravel and coarse sand size particles, the effect of capillarity is negligible and the soil particles simply fall apart (hence they are cohesionless). Medium to fine sands do have some capillarity, which is enough to build a sand castle at the beach, but this small capillarity is lost when the sand becomes submerged in water or the sand completely dries.

Silt and clay size particles are so small that they are strongly influenced by capillarity. These fine soil size particles can be strongly held together by capillarity, which give the soil the ability to be remolded and rolled without falling apart (hence they are cohesive). Capillary is the mechanism that gives a silt or clay its plasticity. When the remolded silt or clay is submerged in water, the capillary tension is slowly eliminated and the soil particles will often disperse. The undrained shear strength of silts and clays are strongly influenced by capillarity.

Coefficient of Earth Pressure at Rest (k_o).   The preceding sections have discussed the vertical effective stress σ_v' for soil deposits. For many geotechnical projects, it may be important to determine the in situ horizontal effective stress σ_h'. For a level ground surface, the horizontal effective stress can be calculated as:

$$\sigma_h' = k_o\, \sigma_v' \tag{4.18}$$

where σ_h' = horizontal effective stress (psf or kPa)
    k_o = coefficient of lateral earth pressure at rest (dimensionless)
    σ_v' = vertical effective stress from Eq. 4.14 (psf or kPa)

The value of k_o is dependent on many factors, such as the soil type, density condition (loose versus dense), geological depositional environment (i.e., alluvial, glacial, and so forth), and the stress history of the site (Massarsch et al. 1975; Massarsch 1979). The value of k_o in natural soils can be as low as 0.4 for soils formed by sedimentation and never preloaded, up to 3.0 or greater for some heavily preloaded soil deposits (Holtz and Kovacs, 1981). For soil deposits that have not been significantly preloaded, a value of $k_o = 0.5$ is often assumed in practice, or the following equation is used (Jaky 1944, 1948; Brooker and Ireland, 1965):

$$k_o = 1 - \sin \phi' \tag{4.19}$$

where ϕ' is the effective friction angle of the soil (degrees).

As an approximation, the value of k_o for preloaded soil can be determined from the following equation (adapted from Alpan 1967; Schmertmann 1975; Ladd et al. 1977):

$$k_o = 0.5\,(OCR)^{0.5} \tag{4.20}$$

where OCR is the overconsolidation ratio, defined as the largest vertical effective stress ever experienced by the soil deposit σ_p' divided by the existing vertical effective stress σ_v'. The overconsolidation ratio (OCR) and the preconsolidation pressure σ_p' will be discussed in Chap. 8.

4.5   STRESS DISTRIBUTION

The previous section described methods that are used to determine the existing stresses within the soil mass. This section describes commonly used methods to determine the increase in stress in the soil deposit due to applied loads. This is naturally important in settlement analyses because the structure's weight results in an increase in stress in the underlying soil. In most cases, it is the increase in vertical stress that is of most importance in settlement analyses. The symbol σ_z is often used to denote an increase in vertical stress in the soil, although $\Delta\sigma_v$ (change in total vertical stress) is also used.

When dealing with stress distribution, a distinction must be made between one-dimensional and two- or three-dimensional loading. A one-dimensional loading applies a stress increase at depth that is 100 percent of the applied surface stress. An example of a one-dimensional loading would be the placement of a fill layer of uniform thickness and large area extent at ground surface. Beneath the center of the uniform fill, the in situ soil is subjected to an increase in vertical stress σ_z that equals the following:

$$\sigma_z = \Delta\sigma_v = h\,\gamma_t \qquad (4.21)$$

where $\sigma_z = \Delta\sigma_v$ = increase in vertical stress (psf or kPa)
h = thickness of the uniform fill layer (ft or m)
γ_t = total unit weight of the fill (pcf or kN/m^3)

In this case of one-dimensional loading, the soil would only be compressed in the vertical direction (i.e., strain only in the vertical direction). Another example of one-dimensional loading is the uniform lowering of a groundwater table. If the total unit weight of the soil does not change as the groundwater table is lowered, then the one-dimensional increase in vertical stress for the in situ soil located below the groundwater table would equal the following:

$$\sigma_z = \Delta\sigma_v = h\,\gamma_w \qquad (4.22)$$

where h = vertical distance that the groundwater table is uniformly lowered (ft or m) and γ_w = unit weight of water (62.4 pcf or 9.81 kN/m^3)

4.5.1 Shallow Foundations

Surface loadings can cause both vertical and horizontal strains, and this is referred to as two- or three-dimensional loading. Common examples of two-dimensional loading are from strip footings or long embankments (i.e., plane strain conditions). Examples of three-dimensional loading would be square and rectangular footings (spread footings) and round storage tanks. This section describes methods that can be used to determine the change in vertical stress for two-dimensional (strip footings and long embankments) and three-dimensional (spread footings and round storage tanks) loading conditions. In these cases, the load usually dissipates rapidly with depth.

2:1 Approximation.   A simple method to determine the increase in vertical stress with depth is the 2:1 *approximation* (also known as the 2:1 *method*). Figure 4.6 illustrates the basic principle of the 2:1 approximation. This method assumes that the stress dissipates with depth in the form of a trapezoid that has 2:1 (vertical:horizontal) inclined sides as shown in Fig. 4.6. The purpose of this method is to approximate the actual *pressure bulb* stress increase beneath a foundation.

If a strip footing of width B has a concentric vertical load P per unit length of footing, then as indicated in Fig. 4.6, the stress applied by the footing onto the soil σ_o would be:

$$\sigma_o = \frac{P}{B} \qquad (4.23)$$

where σ_o = vertical stress applied by the footing (psf or kPa)
P = concentric vertical load per unit length of footing (lb per ft or kN per m)
B = width of the strip footing (ft or m)

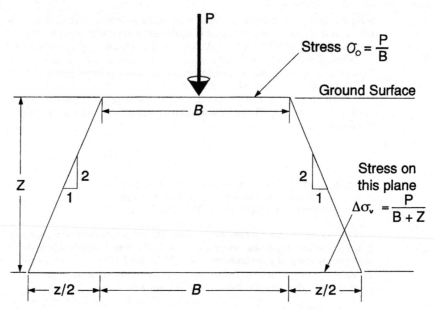

FIGURE 4.6   2:1 approximation for the calculation of the increase in vertical stress at depth due to an applied load P.

As indicated in Fig. 4.6, at a depth z below the footing, the vertical stress increase σ_z due to the strip footing load would be:

$$\sigma_z = \Delta\sigma_v = \frac{P}{B+z} \tag{4.24}$$

For a rectangular spread footing, the applied footing stress σ_o would be:

$$\sigma_o = \frac{P}{BL} \tag{4.25}$$

where σ_o = vertical stress applied by the footing (psf or kPa)
P = total concentric vertical load on the footing (lb or kN)
B = width of the spread footing (ft or m)
L = length of the spread footing (ft or m)

Based on the 2:1 approximation, the vertical stress increase σ_z at a depth of z below the rectangular spread footing would be:

$$\sigma_z = \Delta\sigma_v = \frac{P}{(B+z)(L+z)} \tag{4.26}$$

For a circular foundation subjected to a total concentric vertical force of P and having a diameter D, the vertical stress applied by the foundation σ_o and the vertical stress increase σ_z at a depth of z below the circular footing would be:

$$\sigma_o = \frac{P}{{}^{1}/_{4}\pi D^2} \tag{4.27}$$

$$\sigma_z = \Delta\sigma_v = \frac{P}{^1\!/_4\,\pi(D+z)^2}$$ (4.28)

A major advantage of the 2:1 approximation is its simplicity, and for this reason, it is probably used more often than any other type of stress distribution method. The main disadvantage with the 2:1 approximation is that the stress increase under the center of a uniformly loaded area does not equal the stress increase under the corner or side of the loaded area. The actual situation is that the soil underlying the center of the uniform loaded area is subjected to higher vertical stress increase than the soil underneath a corner or edge of the loaded area. Thus the 2:1 approximation is often only used to estimate the average settlement of the loaded area. Different methods, such as stress distribution based on the theory of elasticity, can be used to calculate the change in vertical stress between the center and corner of the loaded area.

Equations and Charts Based on the Theory of Elasticity.   Equations and charts have been developed to determine the change in stress due to applied loads based on the theory of elasticity. The solutions assume an elastic and homogeneous soil that is continuous and in static equilibrium. The elastic solutions also use a specific type of applied load, such as a point load, uniform load, or linearly increasing load (triangular distribution). For loads where the length of the footing is greater than five times the width, such as for strip footings, the stress distribution is considered to be plane strain. This means that the horizontal strain of the elastic soil only occurs in the direction perpendicular to the long axis of the footing.

Although equations and charts based on the theory of elasticity are often used to determine the change in soil stress, soil is not an elastic material. For example, if a heavy foundation load is applied to a soil deposit, there will be vertical deformation of the soil in response to this load. If this heavy load is removed, the soil will rebound but not return to its original height because soil is not elastic. However, it has been stated that as long as the factor of safety against shear failure exceeds about 3, then stresses imposed by the foundation load are roughly equal to the values computed from elastic theory (NAVFAC DM-7.2, 1982).

In 1885, Boussinesq published equations based on the theory of elasticity. For a surface point load Q applied at the ground surface such as shown in Fig. 4.7, the vertical stress increase at any depth z and distance r from the point load, can be calculated by using the following Boussinesq (1885) equation:

$$\sigma_z = \Delta\sigma_v = \frac{3Q\,z^3}{2\pi(r^2 + z^2)^{5/2}}$$ (4.29)

If there were a uniform line load Q (force per unit length), the vertical stress increase at a depth z and distance r from the line load would be:

$$\sigma_z = \Delta\sigma_v = \frac{2Q\,z^3}{\pi(r^2 + z^2)^2}$$ (4.30)

In 1935, Newmark performed an integration of Eq. 4.29 and derived an equation to determine the vertical stress increase σ_z under the corner of a loaded area. Convenient charts have been developed based on the Newmark (1935) equation. For example, the chart shown in Fig. 4.8 is easy to use and consists of first calculating m and n. The value m is defined as the width of the loaded area x divided by the depth to where the vertical stress increase σ_z is to be calculated. The value n is defined as the length of the loaded area y divided by the depth z. The chart is entered with the value of n and upon intersecting the desired m curve, the influence value I is then obtained from the vertical axis. As indicated in Fig. 4.8, vertical stress increase σ_z is then calculated as the loaded area pressure q_o times the influence value I.

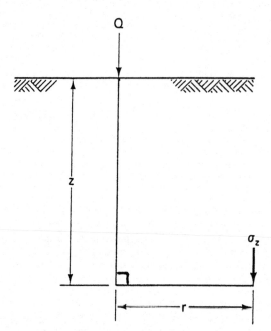

FIGURE 4.7   Definition of terms for Eqs. 4.29 and 4.30.

Figure 4.8 can also be used to determine the vertical stress increase σ_z below the center of a rectangular loaded area. In this case, the rectangular loaded area would be divided into four parts and then Fig. 4.8 would be used to find the stress increase below the corner of one of the parts. By multiplying this stress by 4 (i.e., four parts), the vertical stress increase σ_z below the center of the total loaded area is obtained. This type of analysis is possible because of the principle of superposition for elastic materials. To find the vertical stress increase σ_z outside the loaded area, additional rectangular areas can be added and subtracted as needed in order to model the loading condition.

Figures 4.9 to 4.14 present additional charts for different types of loading conditions based on the integration of the original Boussinesq equations, as follows:

Square and continuous footings.   Figure 4.9 can be used to find the vertical stress increase q beneath square or continuous footings for a footing stress equal to q_o. For the same footing stress q_o, the continuous footing induces a higher vertical stress at any depth than the square footing.

Circular foundation.   Figure 4.10 presents a chart that can be used to calculate the stress increase σ_z beneath a circular foundation exerting a uniform stress equal to q_o. An example of such a loading would be from an oil tank.

Triangular load.   Figure 4.11 can be used to determine the vertical stress increase σ_z for a rectangular footing subjected to a triangular load. An example would be a footing subjected to an eccentric load.

Long embankment.   Figure 4.12 can be used to determine the vertical stress increase σ_z beneath the center of a long embankment by splitting the embankment down the middle and then multiplying the final result by 2 (i.e., two parts of the embankment).

Loaded area of any shape.   Figure 4.13 presents a Newmark (1942) chart that can be used to determine the vertical stress increase σ_z beneath a uniformly loaded area of any shape. There are numerous influence charts, each having a different influence value. Figure 4.13 has an influence

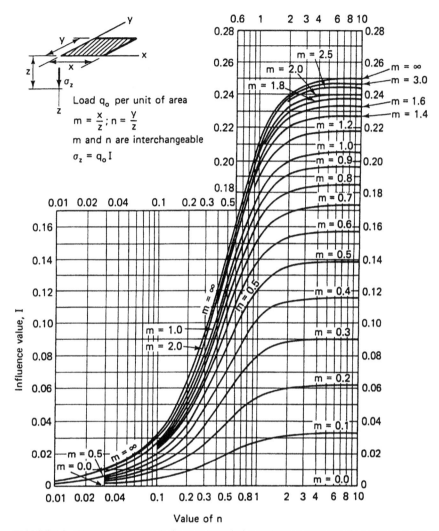

FIGURE 4.8 Chart for calculating the increase in vertical stress beneath the corner of a uniformly loaded rectangular area. (*From NAVFAC DM-7.1, 1982; reproduced from Holtz and Kovacs, 1981.*)

value I of 0.005. The first step is to draw the loaded area onto the chart, using a scale where AB equals the depth z. The center of the chart must correspond to the point where the increase in vertical stress σ_z is desired. The increase in vertical stress σ_z is then calculated as:

$$\sigma_z = q_o I N \qquad (4.31)$$

where q_o = applied uniform stress from the irregular area (psf or kPa)
$\quad I$ = influence value (0.005 for Fig. 4.13)
$\quad N$ = number of blocks within the irregular shaped area plotted on Fig. 4.13

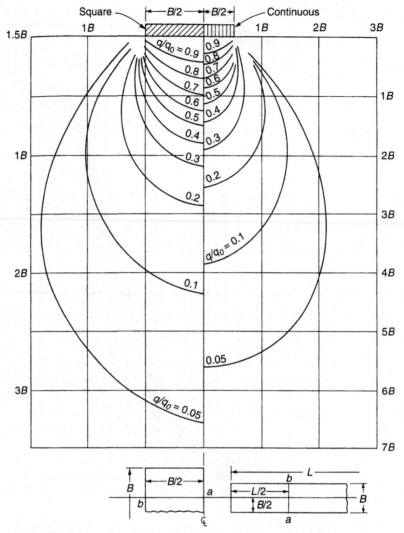

FIGURE 4.9 Pressure isobars based on the Boussinesq equation for square and strip footings. Applicable only along line *ab* as shown. (*From Bowles, 1982; reproduced with permission of McGraw-Hill.*)

When obtaining the value of *N*, portions of blocks are also counted. Note that the entire procedure must be repeated if the increase in vertical stress σ_z is needed at a different depth.

Irregular loads.   Figure 4.14 presents a modification of the Newmark method that can be used for loaded areas of irregular shape as well as irregular loads. The example shown in Fig. 4.14 is for a uniform load of 2.0 tsf (192 kPa). For irregular loads, the process is to multiply the load at given radii by the product of *A* times *I*.

In addition to the equations and charts described earlier, the theory of elasticity has been applied to many other types of loading conditions (e.g., see Poulos and Davis, 1974).

Example Problem 4.2   Assume a round storage tank will impose a surface load of 1000 psf (48 kPa) and the tank has a diameter of 100 ft (30.5 m). Determine the increase in vertical stress σ_z at a depth z of 100 ft (30.5 m) directly beneath the center of the round storage tank.

Solution   In this simple case, the diameter of the tank equals the depth z and thus a circle having a diameter of AB is drawn on Fig. 4.13 with the center of the circle corresponding to the center of Fig. 4.13. The number of blocks (N) within the circle drawn on Fig. 4.13 is 60, and thus the increase in vertical stress σ_z at a depth of 100 ft (30.5 m) below the center of the tank is equal to:

$$\sigma_z = (1000 \text{ psf})(0.005)(60) = 300 \text{ psf or } 14 \text{ kPa}$$

Figure 4.10 can be used as a check on these calculations. For Fig. 4.10, $z/r = 2$ and the offset distance $= 0$. Using these values, the influence value (I) from Fig. 4.10 is 30 and by using the equation listed in Fig. 4.10, the increase in vertical stress (σ_z) is equal to:

$$\sigma_z = (30)(1000 \text{ psf})/100 = 300 \text{ psf or } 14 \text{ kPa}$$

Using the 2:1 approximation (Eq. 4.28), the vertical stress σ_z is equal to 250 psf (12 kPa). The 2:1 approximation value is lower because it is an average value of the stress increase at a depth of 100 feet (30.5 m).

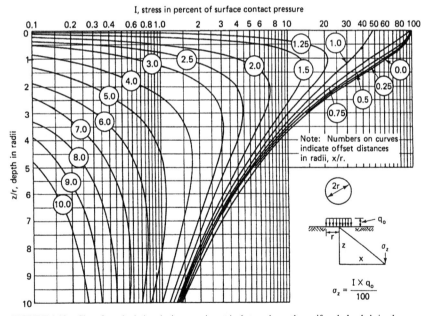

FIGURE 4.10   Chart for calculating the increase in vertical stress beneath a uniformly loaded circular area. (*From NAVFAC DM-7.1, 1982, and Foster and Ahlvin, 1954; reproduced from Holtz and Kovacs, 1981.*)

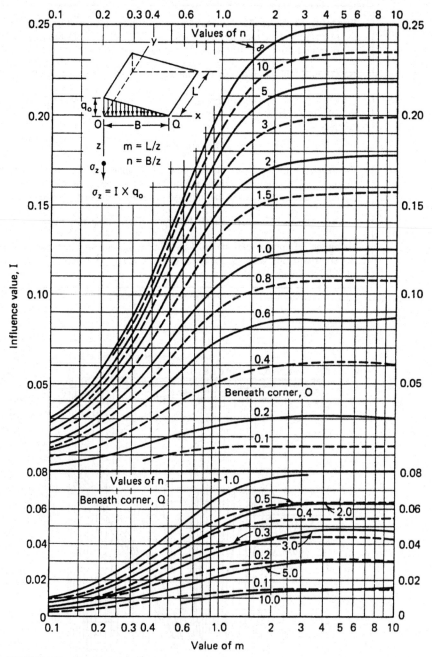

FIGURE 4.11 Chart for calculating the increase in vertical stress beneath the corner of a rectangular area that has a triangular load. (*From NAVFAC DM-7.1, 1982; reproduced from Holtz and Kovacs, 1981.*)

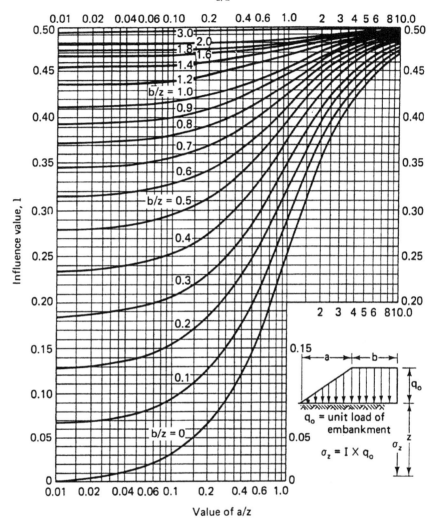

FIGURE 4.12 Chart for calculating the increase in vertical stress beneath the center of a very long embankment. (*From NAVFAC DM-7.1, 1982 and Osterberg, 1957; reproduced from Holtz and Kovacs, 1981.*)

Layered Soil. Figure 4.15 presents the Westergaard (1938) analysis for a soft elastic material reinforced by numerous strong horizontal sheets. This chart was based on an elastic material that contains numerous thin and perfectly rigid layers that allow for only vertical strain but not horizontal strain. This chart may represent a better model for the increase in vertical stress σ_z for layered soils, such as a soft clay deposit that contains numerous horizontal layers of sand.

Other useful charts are shown in Fig. 4.16. These charts can be used when there is a uniform circular loading of radius r underlain by two different layers. Both layers are assumed to have a Poisson's ratio of 0.25 (i.e., $\mu_1 = \mu_2 = 0.25$). The steps in using these charts are as follows:

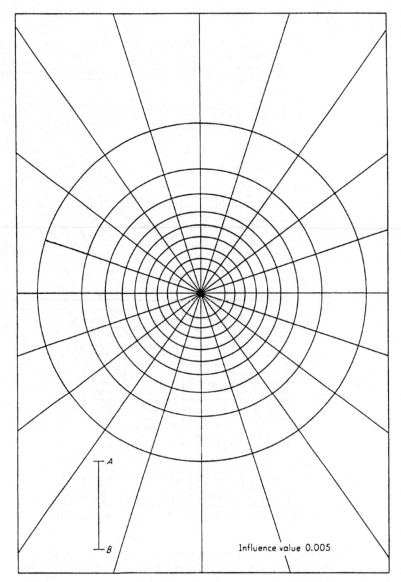

FIGURE 4.13 Newmark chart for calculating the increase in vertical stress beneath a uniformly loaded area of any shape. (*From Newmark, 1942; reproduced from Bowles, 1982.*)

1. *Needed data.* The modulus of elasticity must be estimated for both soil layers (i.e., E_1 and E_2). Often the modulus of elasticity can be estimated from triaxial compression tests. The value of k is calculated as the ratio of E_1 divided by E_2 (i.e., $k = E_1/E_2$)

2. *Upper portion in Fig. 4.16.* Calculate α, which is defined as $\alpha = r/H$, where r is the radius of the uniformly loaded circular area and H is the thickness of the upper soil layer. The upper chart in Fig. 4.16 determines the vertical stress increase σ_z beneath the center of the circular area at the location of the interface between the upper and lower soil layers. Enter the upper chart in Fig. 4.16

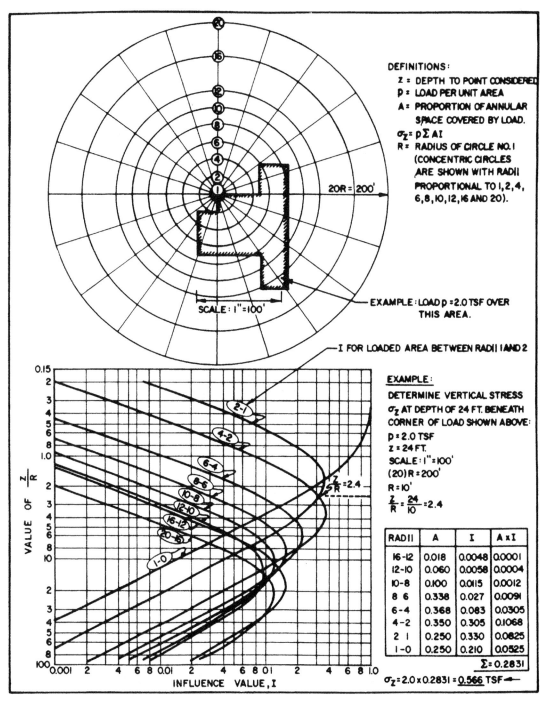

FIGURE 4.14   Modification of the Newmark chart for calculating the increase in vertical stress beneath a loaded area of any shape. (*From Jimenez Salas, 1948; reproduced from NAVFAC DM-7.1, 1982.*)

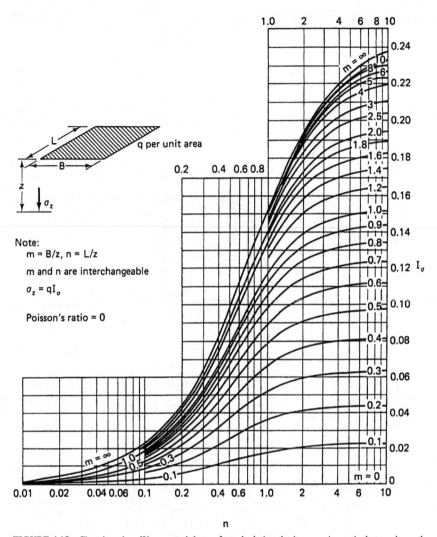

FIGURE 4.15   Chart based on Westergaard theory for calculating the increase in vertical stress beneath the corner of a uniformly loaded rectangular area. (*From Duncan and Buchignani, 1976; reproduced from Holtz and Kovacs, 1981.*)

with the value of k, intersect the appropriate α-curve, and then determine the influence value I from the vertical axis. Then σ_z is equal to I times p, where p is the magnitude of the uniform surface stress applied by the circular area.

3. *Lower portion in Fig. 4.16.*   The lower three charts in Fig. 4.16 can be used to determine the influence values at the center ($r = 0$), edge ($r = 1$), and middle point ($r = 0.5$) of the circular loaded area. The three lower charts in Fig. 4.16 can only be used to determine the influence value I for the case where $r = H$ (i.e., $\alpha = 1.0$). The value of σ_z is equal to I times p, where p is the magnitude of the uniform surface stress applied by the circular area.

Figure 4.16 can also be used for square footings by calculating the area of the square footing and then using this area to determine the radius of an equivalent circular loading.

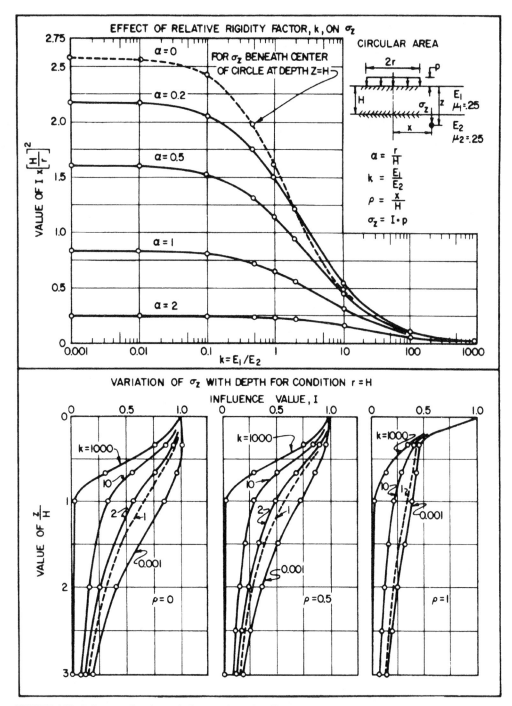

FIGURE 4.16 Influence values for vertical stresses beneath uniformly loaded circular area underlain by two layers. (*From NAVFAC, DM-7.1, 1982.*)

Example Problem 4.3   A site consists of two layers of soil, where the upper layer is 10 ft (3 m) thick. For both soil layers, assume $\mu_1 = \mu_2 = 0.25$. Also assume that the upper layer is very soft and that $E_1/E_2 = 0.001$ (i.e., $k = 0.001$). A proposed oil tank, 20 ft (6 m) in diameter, will be situated at the site and the uniform loading from the oil tank is 2000 psf (96 kPa). Determine the stress increase σ_z beneath the center of the tank Compare these values to Fig. 4.10 (i.e., uniform soil deposit) and the 2:1 approximation.

Solution   The radius of the tank $r = 20/2 = 10$ ft and $\alpha = r/H = 1.0$. Since $r = H$, the lower charts in Fig. 4.16 can be used. Since $\rho = x/H = 0/10 = 0$, use the lower left chart in Fig. 4.16.

Depth (ft)	z/H	Influence value (I)	$\sigma_z = I\rho$	Figure 4.10	2:1 Method
5	0.5	1.00	2000 psf	1750 psf	1280 psf
10	1.0	0.85	1700 psf	1300 psf	890 psf
15	1.5	0.62	1240 psf	900 psf	650 psf
20	2.0	0.40	800 psf	550 psf	500 psf
25	2.5	0.28	560 psf	420 psf	395 psf
30	3.0	0.20	400 psf	300 psf	320 psf

As can be seen by the above values, the situation of a soft layer over a denser layer will produce higher values of σ_z as compared to a uniform soil deposit. Also the 2:1 approximation significantly underestimates the value of σ_z at the center of the loaded area for the upper soft layer.

Guidelines.   Use the following guidelines for stress distribution beneath shallow foundations:

1. If the soil beneath the shallow foundation is somewhat homogeneous, then use the Boussinesq equations or charts. Usually the bottom of the foundation will be at a shallow depth below ground surface. For this case, neglect the shallow embedment of the foundation and use the charts and equations assuming the ground surface is at the bottom of the foundation. Also use the net foundation pressure in the analysis.

2. If the soil beneath the shallow foundation is highly anistropic, or if there is a distinct horizontal stratification, then use the Westergaard method (Fig. 4.15). For two layer materials, use Fig. 4.16. Be aware that a softer layer over a denser layer will produce higher values of σ_z.

3. For quick results, use the 2:1 approximation. At any given depth, the 2:1 approximation provides average values of σ_z (i.e., the value of σ_z is the same at the center and edge of a uniformly loaded area). Hence if the foundation is very flexible, the settlement at the center of the foundation could be much greater than as predicted by using the 2:1 approximation.

4.5.2   Deep Foundations

Similar to solutions for shallow foundations, charts have been developed for estimating the distribution of stress beneath deep foundations. For example, Fig. 4.17 shows the pressure distributions for pile foundations for four different soil conditions. These charts are valid for relatively rigid pile caps, where the rigidity of the foundation system tends to restrict differential settlement across the pile cap. In this case, the 2:1 approximation is ideal because with a relatively rigid foundation system, it is desirable to calculate the average settlement of the pile cap. The four conditions in Fig. 4.17 are individually discussed later:

Friction Piles in Clay.   The upper left figure shows the pressure distribution for a pile group embedded in clay. The load P that the pile cap supports is first turned into an applied stress q by using the following equation:

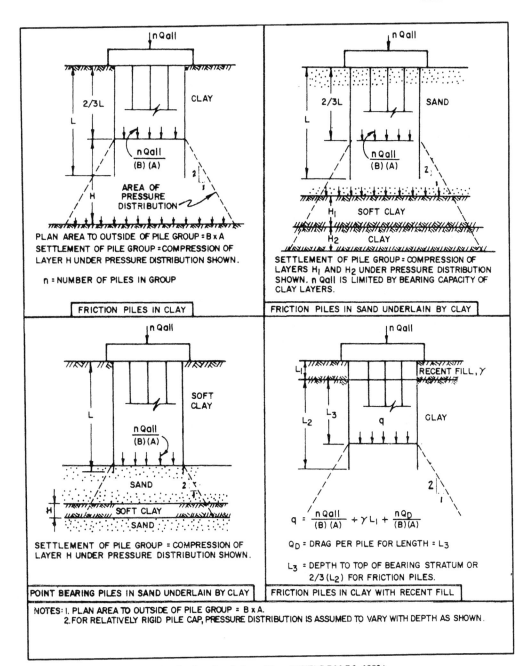

FIGURE 4.17 Pressure distribution for deep foundations. (*From NAVFAC DM-7.2, 1982.*)

$$q = \sigma_o = \frac{P}{BA} = \frac{nQ_{all}}{BA} \tag{4.32}$$

where $q = \sigma_o$ = vertical stress (psf or kPa) applied by the deep foundation at a depth of $(2/3)L$
   below ground surface, where L = length (ft or m) of the piles
   P = total concentric vertical load (lb or kN) on the pile cap. The maximum load the pile cap
   can support is n times Q_{all}
   n = number of piles that support the pile cap
   Q_{all} = allowable vertical load for each pile (lb or kN)
   B = width of the pile group, taken to the outside edge of the group (ft or m)
   A = length of the pile group, taken to the outside edge of the group (ft or m)

In essence, the 2:1 approximation starts at a depth of $(2/3)L$ below ground surface. In Fig. 4.17, it is assumed that there is a hard layer located below the clay. In this case, the compression or consolidation of the clay would be calculated for the thickness of H as defined in the upper left diagram of Fig. 4.17.

Friction Piles in Sand Underlain by Clay.   The upper right diagram in Fig. 4.17 shows the pile group embedded in sand with two underlying clay layers. For this situation, two conditions would have to be evaluated. The first is a bearing capacity failure of the pile group where it punches through the sand and into the upper soft clay layer. The second condition is the compression or consolidation of the two clay layers located below the pile group. Similar to the case outlined above, the 2:1 approximation is used to calculate the increase in vertical stress σ_z due to the pile cap loading for the two clay layers.

Point Bearing Piles in Sand Underlain By Clay.   The lower left diagram in Fig. 4.17 shows the condition of a soft clay layer underlain by a sand stratum. The sand stratum provides most of the vertical resistance and hence the pile group is considered to be point bearing (also known as end bearing). For this case, the 2:1 approximation is assumed to start at the top of the sand layer and is used to calculate the increase in vertical stress σ_z due to the pile cap loading for the soft clay layer.

Friction Piles in Clay With Recent Fill.   The lower right diagram in Fig. 4.17 shows the condition of piles embedded in clay with the recent placement of a fill layer at ground surface. In this case, the piles will be subjected to a downdrag load due to placement of the fill layer. The 2:1 approximation is assumed to start at a distance of L_3 below the top of the clay layer, where $L_3 = (2/3)L_2$. The value of q applied at this depth includes two additional terms, the first is the total unit weight of fill γ_t times the thickness of the fill L_1 and the second is the downdrag load converted to a stress, defined as $(n\ Q_D)/(B\ A)$. The calculation of the downdrag load Q_D will be discussed in Chap. 6.

If the pile caps are spaced close together, there could be additional settlement as the pressure distribution from one pile cap overlaps with the pressure distribution from a second nearby pile cap.

Example Problem 4.4   A 4 by 4 pile group supports a pile cap that is subjected to a concentric vertical load of 500 kips (2200 kN). The piles have a diameter of 1.0 ft (0.3 m) and are spaced apart at a distance of 4 pile diameters in both directions. Assume the design situation is as shown in the lower left of Fig. 4.17 (i.e., point bearing piles in sand underlain by clay). The site consists of an upper 40 ft (12 m) thick layer of soft clay, underlain by a sand layer that is 10 ft (3 m) thick, which is in turn underlain by a 4 ft (1.2 m) thick soft clay layer. Determine the vertical stress increase σ_z at the center of the 4 ft (1.2 m) thick soft clay layer.

(Continued)

Solution

$B = A$ = distance to outside edge of pile group

$B = A$ = three times spacing of piles + 2 r = (3)(4)(1.0 ft) + (2)(0.5 ft) = 13 ft (4 m)

z = thickness of sand layer + $1/2$ thickness of soft clay layer = 10 + (1/2) (4) = 12 ft (3.7 m)

From Eq. 4.26:

$$\sigma_z = \frac{P}{(B+z)(A+z)} = \frac{(500)(1000)}{(13+12)(13+12)} = 800 \text{ psf} (38 \text{ kPa})$$

4.6 TOTAL STRESS AND EFFECTIVE STRESS ANALYSES

4.6.1 Introduction

Sections 3.4 and 3.5 have presented a discussion of shear strength tests performed in the laboratory. As discussed in those sections, the shear strength of cohesionless soil is relatively simple and straightforward, while the shear strength for cohesive soil is complex. The laboratory tests discussed in Secs. 3.4 and 3.5 can generally be divided into two main categories:

1. *Shear strength tests based on total stress.*   The purpose of these laboratory tests is to obtain the undrained shear strength s_u of the soil or the failure envelope in terms of total stresses (total cohesion, c, and total friction angle, ϕ). These types of shear strength tests are often referred to as *undrained* shear strength tests. Examples include the vane shear test, unconfined compression test, unconsolidated undrained triaxial test, and the consolidated undrained triaxial test. These types of laboratory tests are performed exclusively on cohesive soils, such as silts and clays. Examples of the types of laboratory data obtained from these types of tests are shown in Fig. 4.18, parts (a) and (b). This type of shear strength is used in total stress analyses.

2. *Shear strength tests based on effective stress.*   The purpose of these laboratory tests is to obtain the effective shear strength of the soil based on the failure envelope in terms of effective stress (effective cohesion, c', and effective friction angle, ϕ'). These types of shear strength tests are often referred to as drained shear strength tests. Examples include the direct shear test, consolidated drained triaxial test, and the consolidated undrained triaxial test with pore water pressure measurements. These types of tests can be performed on cohesive and cohesionless soils. Examples of the types of laboratory data obtained from these types of tests on cohesive soil are shown in Fig. 4.18, parts (c) and (d).

The drained residual shear strength is also classified as a shear strength test based on effective stress, but it is reserved for cohesive soil. Shear strength based on effective stress is used in effective stress analyses.

An understanding of total stress and effective stress analyses is essential in geotechnical engineering and these two methods are used to evaluate all types of earth projects, such as slope stability, earth pressure calculations, and foundation bearing capacity. The total stress and effective stress analyses are defined as follows:

1. *Total stress analyses.*   Total stress analyses use the undrained shear strength of the soil. Total stress analyses are often used for the evaluation of foundations and embankments to be supported by cohesive soil. The actual analyses are performed for rapid loading or unloading conditions often encountered during the construction phase or just at the end of construction. These analyses are applicable to field situations where there is a change in shear stress which occurs quickly enough so

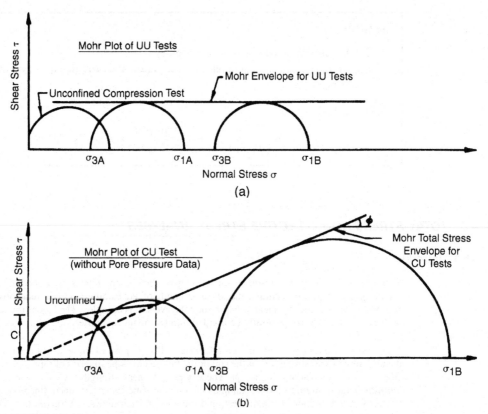

FIGURE 4.18 Summary of triaxial testing for cohesive soil. (*a*) Data from unconsolidated undrained triaxial compression test; (*b*) data from consolidated undrained triaxial compression test; (*c*) data from consolidated undrained triaxial compression test with pore water pressure measurements, (*d*) stress paths. (*Reproduced from NAVFAC DM-7.1, 1982.*)

that the cohesive soil does not have time to consolidate, or in the case of heavily overconsolidated cohesive soils, the negative pore water pressures do not have time to dissipate. For this reason, total stress analyses are often termed short-term analyses. Total stress analyses use the total unit weight γ_t of the soil and the pore water pressures, such as from a groundwater table, are not included in the analyses.

Total stress analyses are also applicable when there is a sudden change in loading condition of a cohesive soil. Examples include wind or earthquake loadings that exert compression or tension forces on piles supported by clay strata, or the rapid draw-down of a reservoir that induces shear stresses upon the clay core of the dam. Total stress analyses use the undrained shear strength s_u of the cohesive soil obtained from unconfined compression tests or vane shear tests. An alternative approach is to use the total stress parameters (c and ϕ) from unconsolidated undrained triaxial compression tests or consolidated undrained triaxial tests.

An advantage of total stress analyses is that the undrained shear strength is obtained from tests, such as the unconfined compression test or vane shear test, that are easy to perform. A major disadvantage of this approach is that the accuracy of the undrained shear strength is always in doubt because it depends on the shear induced pore water pressures, which are not measured or included in the analyses. Also, the undrained shear strength can be significantly influenced by such factors as sample disturbance, strain rate effects, and soil anisotropy (Lambe and Whitman, 1969).

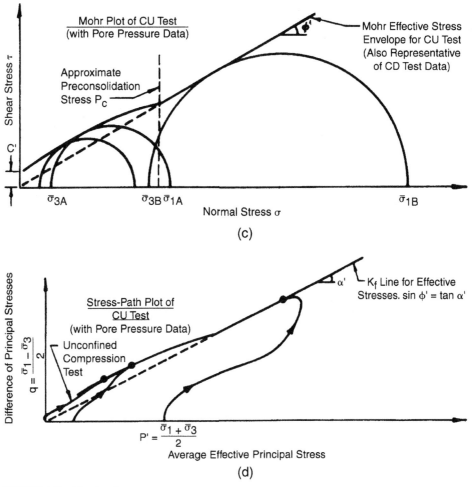

FIGURE 4.18   (*Continued*)

2. *Effective stress analyses.*   Effective stress analyses use the drained shear strength parameters (c' and ϕ'). Except for earthquake loading (i.e., liquefaction), essentially all analyses of cohesionless soils are made using effective stress analyses. For cohesionless soil, $c' = 0$ and the effective friction angle ϕ' is often obtained from drained direct shear tests, drained triaxial tests, or from empirical correlations, such as shown in Figs. 3.12 to 3.14.

For cohesive soils, the effective shear strength could be obtained from drained direct shear tests or drained triaxial tests, although a more common procedure is to perform consolidated undrained triaxial compression tests with pore water measurements performed on saturated cohesive soil in order to determine the effective shear strength parameters, c' and ϕ'. An example of this test data has been presented in Sec. 3.5.2, see Figs. 3.30 to 3.34.

Effective stress shear strength parameters are used for all long-term analyses where conditions are relatively constant. Examples include the long-term stability of slopes, embankments, earth supporting structures, and bearing capacity of foundations for cohesionless and cohesive soil. Effective stress analyses can also be used for any situation where the pore water pressures induced by loading can be estimated or measured. Effective stress analyses can use either one of the following approaches: (1) total unit weight of the soil γ_t and the boundary pore water pressures, or (2) buoyant unit weight of the soil γ_b and the seepage forces.

An advantage of effective stress analyses are that they more fundamentally model the shear strength of the soil, because shear strength is directly related to effective stress. A major disadvantage of effective stress analyses is that the pore water pressures must be included in the analyses. The accuracy of the pore water pressure is often in doubt because of the many factors that affect the magnitude of pore water pressure changes, such as the determination of changes in pore water pressure resulting from changes in external loads (Lambe and Whitman, 1969). For effective stress analyses, assumptions are frequently required concerning the pore water pressures that will be used in the analyses.

Under no circumstances can a total stress analysis and an effective stress analysis be combined. For example, suppose a slope stability analysis is needed for a slope consisting of alternating sand and clay layers. Either the factor of safety of the slope must be determined by using a total stress analysis or by using an effective stress analysis. For example, you could not perform a slope stability analysis that uses the undrained shear strength s_u for one soil layer and effective shear strength parameters (c' and ϕ') and a groundwater table for the second soil layer.

4.6.2 Undrained Shear Strength of Cohesive Soil

The undrained shear strength has been introduced in Sec. 3.5 and is used for total stress analyses. As the name implies, the undrained shear strength s_u refers to a shear condition where water does not enter or leave the cohesive soil during the shearing process. In essence, the water content of the soil must remain constant during the shearing process. Those design situations using the undrained shear strength involve loading or unloading of the plastic soil at a rate that is much faster than the shear induced pore water pressures can dissipate. During rapid loading of saturated plastic soils, the shear induced pore water pressures can only dissipate by the flow of water into (negative shear induced pore water pressures) or out of (positive shear induced pore water pressures) the soil. But cohesive soils, such as clays, have a low permeability, and if the load is applied quickly enough, there will not be enough time for water to enter or leave the plastic soil.

Normalized Undrained Shear Strength.   Figure 4.19 presents examples of the undrained shear strength s_u versus depth for borings E1 and F1 excavated in an offshore deposit of Orinoco clay, which was created by sediments from the Orinoco River, Venezuela. This data was needed for the design of offshore oil platforms to be constructed off the coast of Venezuela. For this project, the undrained shear strength s_u of the clay deposit was needed for the design of the oil platforms because of the critical storm wave loading condition, which is a quick loading condition compared to the drainage ability of 140 ft (43 m) of clay.

The Orinoco clay can be generally classified as clay of high plasticity (CH). This offshore clay deposit can be considered to be a relatively uniform soil deposit. The undrained shear strength s_u was obtained from the Torvane device, the laboratory miniature vane, and the unconfined compression test (UUC). In Fig. 4.19, there is a distinct discontinuity in the undrained shear strength s_u at a depth of 60 ft (18 m) for boring E1 and 40 ft (12 m) for boring F1. This discontinuity was due to different sampling procedures. Above a depth of 60 ft (18 m) at boring E1 and 40 ft (12 m) at boring F1, samplers where hammered into the clay deposit causing sample disturbance and a lower shear strength value for the upper zone of clay. For the deeper zone of clay, a *WIP* sampling procedure was utilized which produced less sample disturbance and hence a higher undrained shear strength.

There is a considerable amount of scatter in the undrained shear strength s_u shown in Fig. 4.19. This is not unusual for soil deposits and nonuniform deposits often have much larger scatter in data than as shown in Fig. 4.19. It takes considerable experience and judgment to decide what value of undrained shear strength s_u to use based on the data shown in Fig. 4.19. One approach is to use the normalized undrained shear strength in order to evaluate shear strength data. For the data shown in Fig. 4.19, an average line can be drawn through the data points, such as:

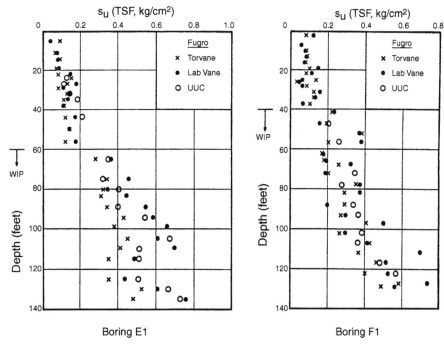

FIGURE 4.19 Undrained shear strength s_u versus depth for Orinoco clay at borings E1 and F1. (*From Day 1980 and Ladd et al. 1980.*)

$$s_u = 0.0043\ z \qquad (4.33)$$

where s_u is the undrained shear strength (tons per square foot or kg/cm^2) and z is the depth in feet.

For the Orinoco clay deposit, the average total unit weight γ_t of the Orinoco clay is 105 pcf (16.5 kN/m^3) at boring E1 and 101 pcf (15.9 kN/m^3) at boring F1. This offshore clay is submerged, in a saturated state, and has hydrostatic pore water pressures u. The vertical effective stress σ'_v would be:

$$\sigma'_v\ (\text{psf}) = \gamma_b\ z = (\gamma_t - \gamma_w)\ z = (105 - 62.4)\ z = 43\ z \quad \text{or} \quad \text{in tsf, } \sigma'_v = 0.022\ z \qquad \text{(for boring E1)}$$
$$(4.34)$$

$$\sigma'_v\ (\text{psf}) = \gamma_b\ z = (\gamma_t - \gamma_w)\ z = (101 - 62.4)\ z = 39\ z \quad \text{or} \quad \text{in tsf, } \sigma'_v = 0.020\ z \qquad \text{(for boring F1)}$$
$$(4.35)$$

Substituting Eqs. 4.34 and 4.35 into Eq. 4.33, the result is:

$$\frac{s_u}{\sigma'_v} = 0.20 \quad \text{(for boring E1)} \qquad (4.36)$$

$$\frac{s_u}{\sigma'_v} = 0.22 \quad \text{(for boring F1)} \qquad (4.37)$$

The parameter s_u/σ'_v is known as the normalized undrained shear strength. It is a valuable parameter for the analysis of the undrained shear strength of saturated clay deposits. For example, in Fig. 2.40, the values of s_u/σ'_v (labeled c/p in Fig. 2.40) have been listed for the various clay deposits.

As a check on the undrained shear strength tests, the value of s_u/σ_v' can be compared with published results, such as (Jamiolkowski et al., 1985):

$$\frac{s_u}{\sigma_v'} = (0.23 \pm 0.04)\text{OCR}^{0.8} \tag{4.38}$$

where OCR is the overconsolidation ratio (dimensionless), defined as the largest vertical effective stress ever experienced by the soil deposit σ_p' divided by the existing vertical effective stress σ_v'.

The overconsolidation ratio (OCR) and the preconsolidation pressure σ_p' will be discussed in Chap. 8. The Orinoco clay deposit is essentially normally consolidated (OCR = 1) and thus Eq. 4.38 indicates that s_u/σ_v' should be approximately equal to 0.23 ± 0.04, which is very close to the values indicated in Eqs. 4.36 and 4.37. Thus assuming an average undrained shear strength line through the data points in Fig. 4.19 seems to be a reasonable approach based on a comparison of s_u/σ_v' with published results.

Sensitivity.   The sensitivity of cohesive soil has been briefly introduced in Sec. 2.4.4. The unconfined compressive test and the vane shear test can be performed on completely remolded soil specimens in order to determine the sensitivity S_t of the cohesive soil. The sensitivity S_t is defined as the undrained shear strength s_u of an undisturbed soil specimen divided by the undrained shear strength s_{ur} of a remolded soil specimen. Based on the sensitivity, the cohesive soil can be classified as follows (Holtz and Kovacs, 1981):

S_t less than 4 indicates a *low* sensitivity

S_t from 4 to 8 indicates a *medium* sensitivity

S_t from 8 to 16 indicates a *high* sensitivity

S_t greater than 16 indicates a *quick* clay

When testing a remolded soil specimen, it is important to retain the same water content of the undisturbed soil. To accomplish this objective in the laboratory, the soil can be placed in a plastic bag and then thoroughly remolded by continuously squeezing and deforming the soil. If the soil specimen bleeds water during this process, then the sensitivity cannot be determined for the soil. After remolding, the soil is carefully pressed down into a mold, without trapping any air within the soil specimen. Once extruded from the mold, the remolded soil is ready for vane shear or unconfined compression testing.

In the last column of Fig. 2.41, sensitivity values have been listed, with S_t values up to 150. This Canadian clay is definitely a *quick* clay. Quick clay indicates an unstable soil structure, where the bonds between clay particles are destroyed upon remolding. Such unstable soil can be formed by the leaching of soft glacial clays deposited in salt water and subsequently uplifted, or by the creation of unstable soil due to weathering of volcanic ash (Terzaghi and Peck, 1967).

A remolded *quick* clay may actually regain some of its lost shear strength, because the soil particles begin to reform bonds, and this increase in shear strength of a remolded soil is known as thrixotropy.

4.6.3   Shear Strength Based on Effective Stress

Effective stress analyses must use either the drained shear strength or the effective shear strength parameters (c' and ϕ') from triaxial tests with pore water pressure measurements. Values of typical effective friction angles ϕ' for cohesionless soil (sand and gravel) have been presented in Table 3.4. The remainder of this section will discuss the effective shear strength of cohesive soil.

For undisturbed natural clays, the values of ϕ' range from around 20° for normally consolidated highly plastic clays up to 30° or more for other types of cohesive soil (Holtz and Kovacs, 1981). For example, Fig. 4.20 shows the relationship between ϕ' and plasticity index (PI) for normally consolidated cohesive soil. For compacted clay, the value of ϕ' is typically in the range of 25° to 30° and occasionally as high as 35°.

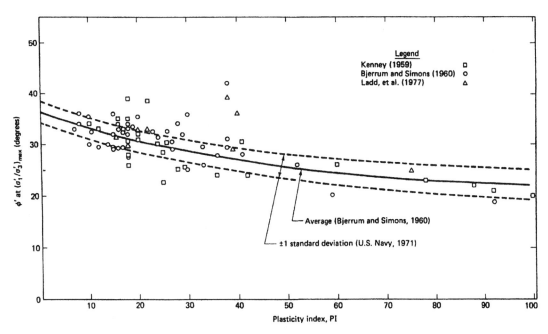

FIGURE 4.20 Empirical correlation between ϕ' and PI from triaxial compression tests on normally consolidated undisturbed clays. (*From NAVFAC DM-7, 1971; Ladd et al. 1977; reproduced from Holtz and Kovacs 1981.*)

The effective cohesion c' for normally consolidated noncemented clays is very small and can be assumed to be zero for practical work (Holtz and Kovacs, 1981). If the cohesive soil is overconsolidated (such as due to compaction), then there may be an effective cohesion intercept c'. Often the higher the overconsolidation ratio (i.e., the higher the preloaded condition of the cohesive soil), the higher the effective cohesion intercept. For overconsolidated cohesive soil, the failure envelope at low effective stresses is often curved and passes through the origin. Thus the effective cohesion c' is usually an extrapolated value that does not represent the actual effective shear strength at low effective stresses (Maksimovic, 1989b). For example, Fig. 4.21 shows the effective stress failure envelope for compacted London Clay. Note the nonlinear nature of the shear strength envelope. For an effective stress of about 2100 to 6300 psf (100 to 300 kPa), the failure envelope is relatively linear and has effective shear strength parameters of $\phi' = 16°$ and $c' = 520$ psf (25 kPa). But below about 2100 psf (100 kPa), the failure envelope is curved (see Fig. 4.21) and the shear strength is less than the extrapolated line from high effective stresses. Because of the nonlinear nature of the shear strength envelope, the shear strength parameters (c' and ϕ') obtained at high effective stresses can overestimate the shear strength of the soil and should not be used in engineering analysis that use low effective stresses (such as surficial slope stability).

4.6.4 Factors that Affect Shear Strength of Cohesive Soil

There can be many factors that affect the shear strength of cohesive soil, such as:

Sample Disturbance.   As previously discussed in Sec. 2.4.2, a soil sample cannot be taken from the ground and be in a perfectly undisturbed state. Disturbance causes a decrease in effective stress, a reduction in the interparticle bonds, and a rearrangement of the soil particles. Sample disturbance can cause the greatest reduction in shear strength compared to any other factor. As an example of the effects

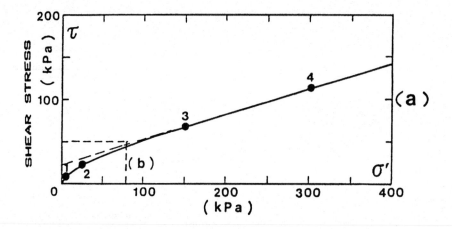

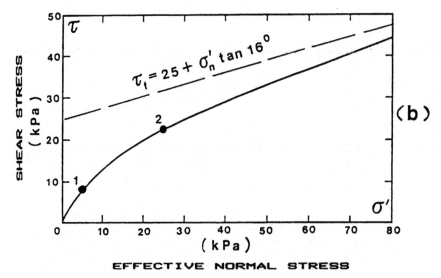

FIGURE 4.21   Failure envelope for compacted London Clay. (*a*) Investigated stress range with detail, (*b*) low stress range. (*From Maksimovic 1989b; reprinted with permission from the American Society of Civil Engineers.*)

of sample disturbance, Fig. 4.22 shows the undrained shear strength s_u of undisturbed and remolded Orinoco clay and indicates about a 75 percent reduction in the undrained shear strength (Ladd et al., 1980; Day, 1980).

For clays having a high sensitivity, such as quick clays, disturbance can severely affect the shear strength. In instances of severe disturbance, quick clays can be disturbed to such an extent that they even become liquid and have essentially no shear strength.

Strain Rate.   The faster a soil specimen is sheared, i.e., a fast strain rate, the higher the value of the undrained shear strength s_u. For the unconfined compression test and vane shear test, the strain rate is very fast with failure occurring in only a few minutes or less. In Fig. 4.22, vane shear tests were performed on the Orinoco clay with the tests having different times to failure. Note in this figure that,

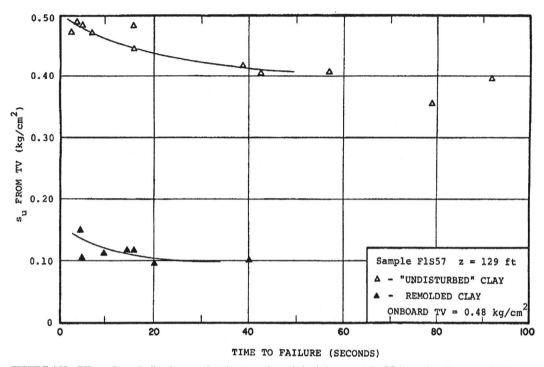

FIGURE 4.22   Effects of sample disturbance and strain rate on the undrained shear strength of Orinoco clay. (*From Day, 1980.*)

for both the undisturbed specimens and the remolded specimens, a slower strain rate results in a lower undrained shear strength. This is because a slower strain rate allows the soil particles more time to slide, deform, and creep around each other resulting in lower undrained shear strength.

Anisotropy.   Soil, especially clays, have a natural strength variation where s_u depends on the orientation of the failure plane, thus s_u along a horizontal failure plane will not equal s_u along a vertical failure plane. For example, the very fine layering of the Orinoco clay is visible from the radiograph shown in Fig. 2.18. The laboratory miniature vane and Torvane tests have simultaneous horizontal and vertical failure planes with an undrained shear strength that rarely equals s_u from the unconfined compression test which has an oblique failure plane.

These three factors listed earlier can affect soil in different ways. For example, it has been stated that for the vane shear test, the values of the undrained shear strength from field vane tests are likely to be higher than can be mobilized in practice (Bjerrum 1972, 1973). This has been attributed to a combination of anisotropy of the soil and the fast rate of shearing involved with the vane shear test. Because of these factors, Bjerrum (1972) has proposed that when vane shear tests are performed on saturated normally consolidated clays, the undrained shear strength s_u be reduced based on the plasticity index of the clay. Figure 2.26 shows Bjerrum's (1972) recommendation, where the in situ undrained shear strength s_u is equal to the shear strength determined from the field vane test times the correction factor determined from Fig. 2.26.

Similar to the vane shear test, the unconfined compression test can provide values of undrained shear strength that are too high because the saturated cohesive soil is sheared too quickly. In many cases, the unconfined compression test has provided reasonable values of undrained shear strength because of a compensation of factors: too high an undrained shear strength due to the fast strain rate that is compensated by a reduction in undrained shear strength due to sample disturbance (Ladd, 1971).

Because of all of these factors (sample disturbance, strain rate, and anisotropy), considerable experience and judgment are needed for the selection of the undrained shear strength parameters to be used in total stress analyses for cohesive soil.

4.6.5   Examples of Total Stress and Effective Stress Analyses

Figure 4.23 presents five examples where either a total stress analysis or an effective stress analysis would be utilized for slope stability. The reason one or the other type of analysis is selected is because of the soil type and/or the analysis provides for the lowest factor of safety of the slope.

1. *Slope in coarse-grained soil with some cohesion.*   For both the low groundwater and high groundwater conditions, effective stress analyses are utilized to determine the factor of safety of the slope.

2. *Slope in coarse-grained, cohesionless soil.*   For both the low groundwater and high groundwater conditions, effective stress analyses are utilized to determine the factor of safety of the slope.

3. *Slope in normally consolidated or slightly preconsolidated clay.*   Total stress analyses are used to determine the factor of safety of the slope. The location of the groundwater table is ignored. However, a fluctuation in the groundwater table could change the undrained shear strength of the clay. This change in s_u should be considered in the slope stability analysis.

4. *Slope in stratified soil.*   A wedge type failure based on an effective stress analysis is used to determine the factor of safety of the slope.

5. *Old slide mass.*   An effective stress analysis would be used to determine the factor of safety of the slope. The slope stability analysis would use steady-state pore water pressures from the groundwater table and the drained residual friction angle ϕ_r' determined from laboratory tests, such as the torsional ring shear (Sec. 3.5.8).

There are many slope stability or foundation construction cases where both a total stress analysis and an effective stress analysis are required. The reason that both types of analyses are needed is because the shear strength and/or pore water pressures change with time from the short-term to long-term condition. Figure 4.24 presents three examples, as follows:

Failure of fill embankment on soft clay having sand drains.   For this situation, both total stress and effective stress analyses would be performed. Both methods would be performed based on conditions existing just after the placement of the fill embankment (i.e., short-term condition). Whichever method gives the lowest factor of safety would be the governing condition. For the effective stress analysis, use effective shear strength parameters for the soft clay (c' and ϕ') and estimate pore water pressures from piezometers. For the total stress analysis, ignore pore water pressures and use the undrained shear strength s_u from the unconfined compression test or the unconsolidated undrained triaxial compression test.

Failure of fill embankment on soft clay (no sand drains).   Similar to case 1, both total stress and effective stress analyses would be performed. The first analysis is based on the end of construction conditions (i.e., short-term stability), which usually produces the lowest factor of safety. Use a total stress analysis, ignore pore water pressures, neglect the shear strength of the fill embankment, and use the undrained shear strength s_u for the clay layer. The second analysis considers the embankment stability at a time well after construction is complete (i.e., long-term stability). For this condition, use an effective stress analysis (i.e., c' and ϕ' for the soft clay) and use steady-state pore water pressures due to the groundwater table.

Cut in stiff fissured clay.   The third example shows a slope cut into a stiff-fissured clay. For the short-term stability, i.e., just after the cut is made, use a total stress analysis in order to determine the factor of safety. Analyze long-term stability based on an effective stress analysis using the drained residual shear strength parameters (c_r' and ϕ_r') and steady-state pore water pressures due to the groundwater table.

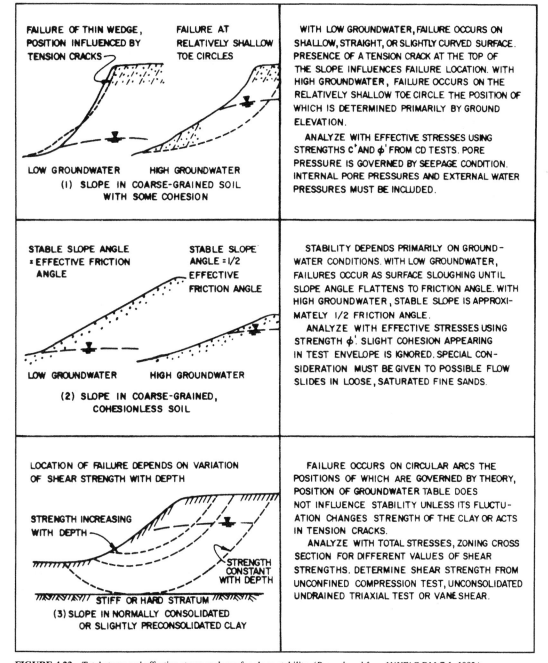

FIGURE 4.23   Total stress and effective stress analyses for slope stability. (*Reproduced from NAVFAC DM-7.1, 1982.*)

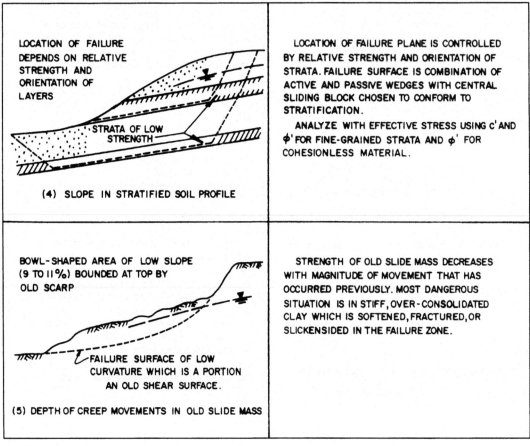

LOCATION OF FAILURE DEPENDS ON RELATIVE STRENGTH AND ORIENTATION OF LAYERS

STRATA OF LOW STRENGTH

(4) SLOPE IN STRATIFIED SOIL PROFILE

LOCATION OF FAILURE PLANE IS CONTROLLED BY RELATIVE STRENGTH AND ORIENTATION OF STRATA. FAILURE SURFACE IS COMBINATION OF ACTIVE AND PASSIVE WEDGES WITH CENTRAL SLIDING BLOCK CHOSEN TO CONFORM TO STRATIFICATION.

ANALYZE WITH EFFECTIVE STRESS USING c' AND ϕ' FOR FINE-GRAINED STRATA AND ϕ' FOR COHESIONLESS MATERIAL.

BOWL-SHAPED AREA OF LOW SLOPE (9 TO 11%) BOUNDED AT TOP BY OLD SCARP

FAILURE SURFACE OF LOW CURVATURE WHICH IS A PORTION AN OLD SHEAR SURFACE.

(5) DEPTH OF CREEP MOVEMENTS IN OLD SLIDE MASS

STRENGTH OF OLD SLIDE MASS DECREASES WITH MAGNITUDE OF MOVEMENT THAT HAS OCCURRED PREVIOUSLY. MOST DANGEROUS SITUATION IS IN STIFF, OVER-CONSOLIDATED CLAY WHICH IS SOFTENED, FRACTURED, OR SLICKENSIDED IN THE FAILURE ZONE.

FIGURE 4.23   (*Continued*)

As these examples illustrate, the use of total stress and effective stress analyses are often complex. For some situations, both a short-term condition and a long-term condition may need to be analyzed. The key to using the total stress and effective stress analyses is to determine which analysis will provide the lowest factor of safety or lead to the lowest allowable design value. Many slope and foundation failures occur because the wrong analysis was selected and the critical design situation was not analyzed. Throughout this book, foundation design situations that require using either total stress analyses or effective stress analyses will be discussed.

4.6.6 Geotechnical Earthquake Engineering

For geotechnical earthquake engineering, the total stress and effective stress analyses are based on the following:

1. Total stress analyses
   a. Use total stress shear strength parameters (s_u or c and ϕ)

LOCATION OF FAILURE DEPENDS ON GEOMETRY AND STRENGTH OF CROSS SECTION.

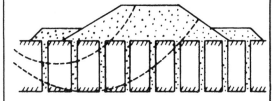

(1) FAILURE OF FILL ON SOFT COHESIVE FOUNDATION WITH SAND DRAINS

USUALLY MINIMUM STABILITY OCCURS DURING PLACING OF FILL. IF RATE OF CONSTRUCTION IS CONTROLLED, ALLOW FOR GAIN IN STRENGTH WITH CONSOLIDATION FROM DRAINAGE.

ANALYZE WITH EFFECTIVE STRESS USING c' AND ϕ' FROM CU TEST WITH PORE PRESSURE MEASUREMENT. APPLY ESTIMATED PORE PRESSURES OR PIEZOMETRIC PRESSURES. ANALYZE WITH TOTAL STRESS FOR RAPID CONSTRUCTION WITHOUT OBSERVATION OF PORE PRESSURES; USE SHEAR STRENGTH FROM UNCONFINED COMPRESSION OR UNCONSOLIDATED UNDRAINED TRIAXIAL.

FAILURE SURFACE MAY BE ROTATION ON CIRCULAR ARC OR TRANSLATION WITH ACTIVE AND PASSIVE WEDGES.

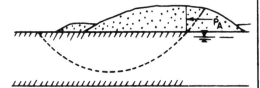

(2) FAILURE OF STIFF COMPACTED FILL ON SOFT COHESIVE FOUNDATION

USUALLY, MINIMUM STABILITY OBTAINED AT END OF CONSTRUCTION. FAILURE MAY BE IN THE FORM OF ROTATION OR TRANSLATION, AND BOTH SHOULD BE CONSIDERED.

FOR RAPID CONSTRUCTION IGNORE CONSOLIDATION FROM DRAINAGE AND UTILIZE SHEAR STRENGTHS DETERMINED FROM U OR UU TESTS OR VANE SHEAR IN TOTAL STRESS ANALYSIS. IF FAILURE STRAIN OF FILL AND FOUNDATION MATERIALS DIFFER GREATLY, SAFETY FACTOR SHOULD EXCEED ONE, IGNORING SHEAR STRENGTH OF FILL. ANALYZE LONG-TERM STABILITY USING c' AND ϕ' FROM CU TESTS WITH EFFECTIVE STRESS ANALYSIS, APPLYING PORE PRESSURES OF GROUNDWATER ONLY.

ORIGINAL GROUND LINE

CUT AT TOE

FAILURE SURFACE DEPENDS ON PATTERN OF FISSURES OR DEPTH OF SOFTENING.

(3) FAILURE FOLLOWING CUT IN STIFF FISSURED CLAY

RELEASE OF HORIZONTAL STRESSES BY EXCAVATION CAUSES EXPANSION OF CLAY AND OPENING OF FISSURES, RESULTING IN LOSS OF COHESIVE STRENGTH.

ANALYZE FOR SHORT TERM STABILITY USING c AND ϕ WITH TOTAL STRESS ANALYSIS. ANALYZE FOR LONG TERM STABILITY WITH c'_r AND ϕ'_m BASED ON RESIDUAL STRENGTH MEASURED IN CONSOLIDATED DRAINED TESTS.

FIGURE 4.24   Total stress and effective stress analyses for short-term and long-term conditions. (*Revised from NAVFAC DM-7.1, 1982.*)

      b. Use total unit weight of the soil γ_t

      c. Ignore pore water pressures

   2. Effective tress analyses

      a. Use effective stress shear strength parameters (c' and ϕ')

      b. Determine the earthquake induced pore water pressures u_e

Depending on the type of soil, the type of analyses would be as follows:

Cohesionless Soil.   For the earthquake analysis of cohesionless soil (i.e., sands and gravels), it is often easier to perform an effective stress analysis, as follows:

   1. *Cohesionless soil above the groundwater table.*   Often the cohesionless soil above the groundwater table will have negative pore water pressures due to capillary tension of pore water fluid. The capillary tension tends to hold together the soil particles and provide additional shear strength to the soil. For geotechnical engineering analyses, it is commonly assumed that the pore water pressures are equal to zero, which ignores the capillary tension. This conservative assumption is also utilized for earthquake analyses. Thus the shear strength of soil above the groundwater table is assumed to be equal to the effective friction angle ϕ' from empirical correlations (such as Figs. 3.12 and 3.13) or it is equal to the effective friction angle ϕ' from drained direct shear tests performed on saturated soil (ASTM D 3080-03, 2004).

   2. *Dense cohesionless soil below the groundwater table.*   Dense cohesionless soil tends to dilate during the earthquake shaking. This causes the excess pore water pressures to become negative, and the shear strength of the soil is actually momentarily increased. Thus for dense cohesionless soil below the groundwater table, the shear strength is assumed to be equal to the effective friction angle ϕ' from empirical correlations (such as Figs. 3.12 and 3.13) or it is equal to the effective friction angle ϕ' from drained direct shear tests performed on saturated soil (ASTM D 3080-03, 2004). In the effective stress analysis, the negative excess pore water pressures are ignored and the pore water pressure is assumed to be hydrostatic. Once again, this is a conservative approach.

   3. *Loose cohesionless soil below the groundwater table.*   Loose cohesionless soil tends to contract during the earthquake shaking. This causes the development of pore water pressures, and the shear strength of the soil is decreased. If liquefaction occurs, the shear strength of the soil can be decreased to essentially zero. For any cohesionless soil that is likely to liquefy during the earthquake, one approach is to assume that ϕ' is equal to zero (i.e., no shear strength). For those loose cohesionless soils that have a factor of safety against liquefaction greater than 1.0, the analysis will usually need to take into account the reduction in shear strength due to the increase in pore water pressure as the soil contracts. One approach is to use the effective friction angle ϕ' from empirical correlations (such as Figs. 3.12 and 3.13) or the effective friction angle ϕ' from drained direct shear tests performed on saturated soil (ASTM D 3080-03, 2004). In addition, the earthquake-induced pore water pressures must be used in the effective stress analysis. The disadvantage of this approach is that it is very difficult to estimate the pore water pressures generated by the earthquake-induced contraction of the soil. One option is to use Fig. 4.25, which presents a plot of the factor of safety against liquefaction FS_L versus pore water pressure ratio r_u, defined as:

$$r_u = \frac{u}{\gamma_t h} \tag{4.39}$$

where r_u = pore water pressure ratio (dimensionless)

      u = pore water pressure (psf or kPa)

      γ_t = total unit weight of the soil (pcf or kN/m^3)

      h = depth below the ground surface (ft or m)

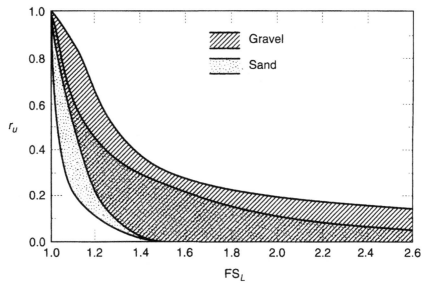

FIGURE 4.25   Factor of safety against liquefaction FS_L versus the pore water pressure ratio r_u for gravel and sand. (*Developed by Marcuson and Hynes, 1990; reproduced from Kramer 1996.*)

As indicated in Fig. 4.25, at a factor of safety against liquefaction FS_L equal to 1.0 (i.e., liquefied soil), the value of $r_u = 1.0$. Using a value of $r_u = 1.0$ in Eq. 4.39, then $r_u = 1.0 = u/(\gamma_t h)$. This means that the pore water pressure u must be equal to the total stress ($\sigma = \gamma_t h$), and hence the effective stress σ' is equal to zero ($\sigma' = \sigma - u$). For a granular soil, an effective stress equal to zero means that the soil will not possess any shear strength (i.e., it has liquefied). Chapter 13 will present the analyses that are used to determine the factor of safety against liquefaction.

4. *Flow failures in cohesionless soil.*   As indicated earlier, the earthquake analyses for cohesionless soil will often be performed using an effective stress analysis, using ϕ' and assumptions concerning the earthquake-induced pore water pressure. Flow failures are also often analyzed using an effective stress analysis with a value of the pore water pressure ratio = 1.0, or by using a shear strength of the liquefied soil equal to zero (i.e., $\phi' = 0$ and $c' = 0$). This will be further discussed in Chap. 13.

Cohesive Soil.   For cohesive soils, such as silts and clays, it is often easier to perform total stress analyses, as follows:

1. *Cohesive soil above the groundwater table.*   Often the cohesive soil above the groundwater table will have negative pore water pressures due to capillary tension of the pore water fluid. In some cases, the cohesive soil may even be dry and desiccated. The capillary tension tends to hold together the soil particles and provide additional shear strength to the soil. For a total stress analysis, the undrained shear strength s_u of the cohesive soil could be determined from unconfined compression tests or vane shear tests. As an alternative, total stress parameters (c and ϕ) could be determined from triaxial tests (e.g., ASTM D 2850-03 and ASTM D 4767-02, 2004).

Because of the negative pore water pressures, a future increase in water content would tend to decrease the undrained shear strength s_u of partially saturated cohesive soil above the groundwater table. Thus a possible change in water content in the future should be considered. In addition, a triaxial test performed on a partially saturated cohesive soil often has a stress-strain curve that exhibits a peak shear strength that then reduces to an ultimate value. If there is a significant drop-off in shear strength with strain, it may be prudent to use the ultimate value in earthquake analyses.

2. *Cohesive soil below the groundwater table having low sensitivity.* The sensitivity represents the loss of undrained shear strength as a cohesive soil specimen is remolded. An earthquake also tends to shear a cohesive soil back and forth, much like the remolding process. For cohesive soil having low sensitivity ($S_t < 4$), the reduction in the undrained shear strength during the earthquake should be small.

3. *Cohesive soil below the groundwater table having high sensitivity.* For highly sensitive and quick clays ($S_t > 8$), there could be a significant shear strength loss during the earthquake shaking. An example is the Turnagain Heights Landslide in Alaska.

The stress-strain curve from a triaxial test performed on a highly sensitive or quick clay often exhibits a peak shear strength that develops at a low vertical strain, followed by a dramatic drop-off in strength with continued straining of the soil specimen. The analysis will need to include the estimated reduction in undrained shear strength due to the earthquake shaking. In general, the most critical conditions exist when the highly sensitive or quick clay is subjected to a high static shear stress (such as the Turnagain Heights Landslide). If during the earthquake, the sum of the static shear stress and the seismic induced shear stress exceeds the undrained shear strength of the soil, then a significant reduction in shear strength would be expected to occur. Cohesive soils having a medium sensitivity ($4 < S_t < 8$) would tend to be an intermediate case.

4. *Drained residual shear strength ϕ_r' for cohesive soil.* As indicated earlier, the earthquake analyses for cohesive soil will often be performed using a total stress analysis (i.e., s_u from unconfined compression tests and vane shear tests, or c and ϕ from triaxial tests).

An exception is cohesive slopes that have been subjected to a significant amount of shear deformation. For example, the stability analysis of ancient landslides, slopes in overconsolidated fissured clays, and slopes in fissured shales will often be based on the drained residual shear strength of the failure surface (Bjerrum, 1967; Skempton and Hutchinson, 1969; Skempton, 1985; Hawkins and Privett, 1985; Ehlig, 1992). When the stability of such a slope is to be evaluated for earthquake shaking, then the drained residual shear strength ϕ_r' should be used in the analysis. The drained residual shear strength can be determined from laboratory tests by using the torsional ring shear or direct shear apparatus.

In order to perform effective stress analyses, the pore water pressures are usually assumed to be unchanged during the earthquake shaking. The slope or landslide mass will also be subjected to additional destabilizing forces due to the earthquake shaking. These destabilizing forces can be included in the effective stress slope stability analysis and this approach is termed the pseudostatic method (see Chap. 13).

Analyses For Subsoil Profiles Consisting of Cohesionless and Cohesive Soil. For earthquake analysis where both cohesionless and cohesive soil must be considered, either a total stress analysis or an effective stress analysis must be performed. As indicated earlier, usually the effective shear strength parameters will be known for the cohesionless soil. Thus subsoil profiles having layers of sand and clay are often analyzed using effective stress analyses (c' and ϕ') with an estimation of the earthquake induced pore water pressures.

If the sand layers were to liquefy during the anticipated earthquake, then a total stress analysis could be performed using the undrained shear strength s_u for the clay and assuming the undrained shear strength of the liquefied sand layer is equal to zero (i.e., $s_u = 0$). Bearing capacity or slope stability analyses using total stress parameters would then be performed so that the circular or planar slip surfaces passes through or along the liquefied sand layer.

Summary of Shear Strength Analyses for Geotechnical Earthquake Engineering. Table 4.9 presents a summary of the soil type versus type of analysis and shear strength that should be used for earthquake analyses.

TABLE 4.9 Soil Type versus Type of Analysis and Shear Strength for Earthquake Engineering

Soil type	Type of analysis	Field condition	Shear strength
Cohesionless soil	Use an effective stress analysis	Cohesionless soil above the groundwater table	Assume pore water pressures are equal to zero, which ignores the capillary tension. Use ϕ' from empirical correlations or from laboratory tests such as drained direct shear tests.
		Dense cohesionless soil below the groundwater table	Dense cohesionless soil dilates during the earthquake shaking (hence negative excess pore water pressure). Assume earthquake-induced negative excess pore water pressures are zero, and use ϕ' from empirical correlations or from laboratory tests such as drained direct shear tests.
		Loose cohesionless soil below the groundwater table	Excess pore water pressures u_e generated during the contraction of soil structure. For $FS_L \leq 1.0$, use $\phi' = 0$ or $r_u = 1.0$. For $FS_L > 1$, use r_u from Fig. 4.25 and ϕ' from empirical correlations or from laboratory tests such as drained direct shear tests.
		Flow failures	Flow failures are also often analyzed using an effective stress analysis with a value of the pore water pressure ratio = 1.0, or by using a shear strength of the liquefied soil equal to zero ($\phi' = 0$ and $c' = 0$).
Cohesive soil	Use a total stress analysis	Cohesive soil above the groundwater table	Determine s_u from unconfined compression tests or vane shear tests. As an alternative, use total stress parameters (c and ϕ) from triaxial tests. Consider shear strength decrease due to increase in water content. For a significant drop-off in strength with strain, consider using ultimate shear strength for earthquake analysis.
		Cohesive soil below the groundwater table with $S_t \leq 4$	Determine s_u from unconfined compression tests or vane shear tests. As an alternative, use total stress parameters (c and ϕ) from triaxial tests.
		Cohesive soil below the groundwater table with $S_t > 8$	Include an estimated reduction in undrained shear strength due to earthquake shaking. Most significant strength loss occurs when the sum of the static shear stress and the seismic-induced shear stress exceeds the undrained shear strength of the soil. Cohesive soils having a medium sensitivity ($4 < S_t \leq 8$) are an intermediate case.
	Possible exception	Existing landslides	Use an effective stress analysis and the drained residual shear strength (ϕ'_r) for the slide plane. Assume pore water pressures are unchanged during earthquake shaking. Include destabilizing earthquake forces in slope stability analyses (pseudostatic method).

4.7 PERMEABILITY AND SEEPAGE

4.7.1 Introduction

This last section of Chap. 4 presents the basic principles of permeability and groundwater seepage through soil. In the previous sections, the discussion of groundwater has included the following:

- *Subsurface exploration (Sec. 2.3).*   One of the main purposes of subsurface exploration is to determine the location of the groundwater table.
- *Laboratory testing (Sec. 3.7).*   The rate of flow through soil is dependent on its permeability. The constant head permeameter (Fig. 3.56) or the falling head permeameter (Fig. 3.57) can be used to determine the coefficient of permeability (also known as the hydraulic conductivity). Equations to calculate the coefficient of permeability in the laboratory are based on Darcy's law ($v = k\ i$). Figure 3.58 shows the coefficient of permeability versus drainage properties and soil type.
- *Pore water pressure (Sec. 4.4).*   The pore water pressure u for a level groundwater table with no flow and hydrostatic conditions is calculated by using Eq. 4.16 for soil below the groundwater table and Eq. 4.17 for soil above the groundwater table that is saturated due to capillary rise.
- *Shear strength (Sec. 3.5).*   The shear strength of saturated soil is directly related to the water that fills the soil pores. Normally consolidated clays have an increase in pore water pressure as the soil structure contracts, while heavily overconsolidated clays have a decrease in pore water pressure as the soil structure dilates during undrained shear.

The main items that will be discussed in this section are permeability, seepage velocity, seepage forces, and two-dimensional flow nets. In the following chapters, these principles will be used for all types of engineering analyses, such as consolidation, slope stability, and foundation excavation dewatering. Chapter 16 will specifically discuss foundation excavation dewatering.

4.7.2 Permeability

As indicted in Fig. 3.58, cohesive soils such as clays have very low values of the coefficient of permeability k, while cohesionless soils such as clean sands and gravels can have a coefficient of permeability k that is a billion times larger than cohesive soils. The reason is because of the size of the drainage paths through the soil. Open graded gravel has very large, interconnected void spaces. At the other extreme are clay size particles, which produce minuscule void spaces with a large resistance to flow and hence a very low permeability.

In many cases, the determination of the coefficient of permeability k from laboratory tests on small soil specimens may not be representative of the overall field condition. A common method of determining the field coefficient of permeability is through the measurements of the change in water levels in open standpipes. Figures 4.26 and 4.27 presents various standpipe conditions and the equations used to calculate the field permeability k. A simple equation is presented in Fig. 4.28.

Although the type of soil is the most important parameter that governs the coefficient of permeability k of saturated soil, there are many other factors that affect k, such as:

1. *Void ratio.*   For a given soil, the higher the void ratio, the higher the coefficient of permeability. For example, Fig. 4.29 presents data from various soils and shows how the coefficient of permeability k increases as the void ratio increases.
2. *Particle size distribution.*   A well-graded soil will tend to have a lower coefficient of permeability k than a uniform soil. This is because a well-graded soil will have more particles to fill in the void spaces and make the flow paths smaller.
3. *Soil structure.*   Some types of soil structure, such as the honeycomb structure or cardhouse structure, have larger interconnected void spaces with a corresponding higher permeability.
4. *Layering of soil.*   Natural soils are often stratified and have layers or lenses of permeable soil, resulting in a much higher horizontal permeability than vertical permeability. This is important because oftentimes it is only the vertical permeability that is measured in the falling head and constant head permeameters.
5. *Soil imperfections or discontinuities.* Natural soil has numerous imperfections, such as root holes, animal burrows, joints, fissures, seams, and soil cracks, that significantly increase the permeability of the soil mass. For compacted clay liners, there may be incomplete bonding between fill lifts, which allows water to infiltrate through the fill mass.

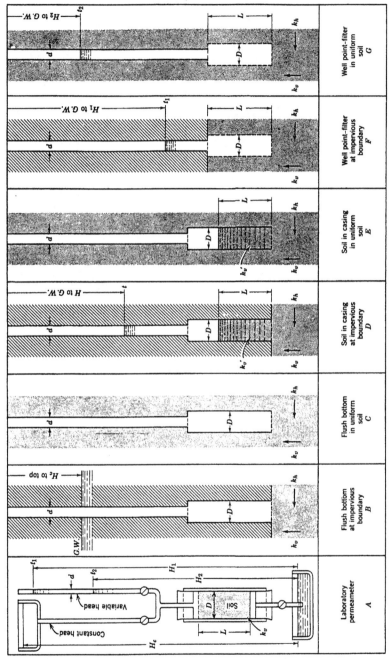

FIGURE 4.26 Methods for determining permeability. See Figure 4.27 for the formulas for determining permeability. *(From Hvorslev, 1951; reproduced from Lambe and Whitman, 1969.)*

Case	Constant Head	Variable Head
A	$k_v = \dfrac{4 \cdot q \cdot L}{\pi \cdot D^2 \cdot H_c}$	$k_v = \dfrac{d^2 \cdot L}{D^2 \cdot (t_2 - t_1)} \ln \dfrac{H_1}{H_2}$ $k_v = \dfrac{L}{t_2 - t_1} \ln \dfrac{H_1}{H_2} \quad \text{for } d = D$
B	$k_m = \dfrac{q}{2 \cdot D \cdot H_c}$	$k_m = \dfrac{\pi \cdot d^2}{8 \cdot D \cdot (t_2 - t_1)} \ln \dfrac{H_1}{H_2}$ $k_m = \dfrac{\pi \cdot D}{8 \cdot (t_2 - t_1)} \ln \dfrac{H_1}{H_2} \quad \text{for } d = D$
C	$k_m = \dfrac{q}{2.75 \cdot D \cdot H_c}$	$k_m = \dfrac{\pi \cdot d^2}{11 \cdot D \cdot (t_2 - t_1)} \ln \dfrac{H_1}{H_2}$ $k_m = \dfrac{\pi \cdot D}{11 \cdot (t_2 - t_1)} \ln \dfrac{H_1}{H_2} \quad \text{for } d = D$
D	$k_v' = \dfrac{4 \cdot q \left(\dfrac{\pi}{8} \cdot \dfrac{k_v'}{k_v} \cdot \dfrac{D}{m} + L \right)}{\pi \cdot D^2 \cdot H_c}$	$k_v' = \dfrac{d^2 \cdot \left(\dfrac{\pi}{8} \cdot \dfrac{k_v'}{k_v} \cdot \dfrac{D}{m} + L \right)}{D^2 \cdot (t_2 - t_1)} \ln \dfrac{H_1}{H_2}$ $k_v = \dfrac{\dfrac{\pi}{8} \cdot \dfrac{D}{m} + L}{t_2 - t_1} \ln \dfrac{H_1}{H_2} \quad \text{for } \begin{cases} k_v' = k_v \\ d = D \end{cases}$
E	$k_v' = \dfrac{4 \cdot q \cdot \left(\dfrac{\pi}{11} \cdot \dfrac{k_v'}{k_v} \cdot \dfrac{D}{m} + L \right)}{\pi \cdot D^2 \cdot H_c}$	$k_v' = \dfrac{d^2 \cdot \left(\dfrac{\pi}{11} \cdot \dfrac{k_v'}{k_v} \cdot \dfrac{D}{m} + L \right)}{D^2 \cdot (t_2 - t_1)} \ln \dfrac{H_1}{H_2}$ $k_v = \dfrac{\dfrac{\pi}{11} \cdot \dfrac{D}{m} + L}{t_2 - t_1} \ln \dfrac{H_1}{H_2} \quad \text{for } \begin{cases} k_v' = k_v \\ d = D \end{cases}$
F	$k_h = \dfrac{q \cdot \ln \left[\dfrac{2mL}{D} + \sqrt{1 + \left(\dfrac{2mL}{D} \right)^2} \right]}{2 \cdot \pi \cdot L \cdot H_c}$	$k_h = \dfrac{d^2 \cdot \ln \left[\dfrac{2mL}{D} + \sqrt{1 + \left(\dfrac{2mL}{D} \right)^2} \right]}{8 \cdot L \cdot (t_2 - t_1)} \ln \dfrac{H_1}{H_2}$ $k_h = \dfrac{d^2 \cdot \ln \left(\dfrac{4mL}{D} \right)}{8 \cdot L \cdot (t_2 - t_1)} \ln \dfrac{H_1}{H_2} \quad \text{for } \dfrac{2mL}{D} > 4$
G	$k_h = \dfrac{q \cdot \ln \left[\dfrac{mL}{D} + \sqrt{1 + \left(\dfrac{mL}{D} \right)^2} \right]}{2 \cdot \pi \cdot L \cdot H_c}$	$k_h = \dfrac{d^2 \cdot \ln \left[\dfrac{mL}{D} + \sqrt{1 + \left(\dfrac{mL}{D} \right)^2} \right]}{8 \cdot L \cdot (t_2 - t_1)} \ln \dfrac{H_1}{H_2}$ $k_h = \dfrac{d^2 \cdot \ln \left(\dfrac{2mL}{D} \right)}{8 \cdot L \cdot (t_2 - t_1)} \ln \dfrac{H_1}{H_2} \quad \text{for } \dfrac{mL}{D} > 4$

ASSUMPTIONS

Soil at intake, infinite depth, and directional isotropy (k_v and k_n constant). No disturbance, segregation, swelling, or consolidation of soil. No sedimentation or leakage. No air or gas in soil, well point, or pipe. Hydraulic losses in pipes, well point, or filter negligible.

FIGURE 4.27 Formulas for determination of permeability. See Figure 4.26 for identification of cases *A* through *G*. (*From Hvorslev, 1951; reproduced from Lambe and Whitman, 1969.*)

Basic Time Lag	Notation
$$k_v = \frac{d^2 \cdot L}{D^2 \cdot T}$$ $$k_v = \frac{L}{T} \quad \text{for} \quad d = D$$	D = Diam, intake, sample (cm) d = Diameter, standpipe (cm) L = Length, intake, sample (cm) H_c = Constant piez. head (cm) H_1 = Piez. head for $t = t_1$ (cm) H_2 = Piez. head for $t = t_2$ (cm) q = Flow of water (cm³/sec) t = Time (sec) T = Basic time lag (sec) k_v' = Vert. perm. casing (cm/sec)
$$k_m = \frac{\pi d^2}{8 \cdot D \cdot T}$$ $$k_m = \frac{\pi \cdot D}{8 \cdot T} \quad \text{for} \quad d = D$$	
$$k_m = \frac{\pi \cdot d^2}{11 \cdot D \cdot T}$$ $$k_m = \frac{\pi \cdot D}{11 \cdot T} \quad \text{for} \quad d = D$$	
$$k_v' = \frac{d^2 \cdot \left(\frac{\pi}{8} \cdot \frac{k_v'}{k_v} \cdot \frac{D}{m} \right) + L}{D^2 \cdot T}$$ $$k_v = \frac{\frac{\pi}{8} \cdot \frac{D}{m} + L}{T} \quad \text{for} \quad \begin{cases} k_v' = k_v \\ d = D \end{cases}$$	k_v = Vert. perm. ground (cm/sec) k_h = Horz. perm. ground (cm/sec) k_m = Mean coeff. perm. (cm/sec) m = Transformation ratio $k_m = \sqrt{k_h \cdot k_v} \qquad m = \sqrt{k_h / k_v}$ $\ln = \log_e = 2.3 \log_{10}$
$$k_v' = \frac{d^2 \cdot \left(\frac{\pi}{11} \cdot \frac{k_v'}{k_v} \cdot \frac{D}{m} + L \right)}{D^2 \cdot T}$$ $$k_v' = \frac{\frac{\pi}{11} \cdot \frac{D}{m} + L}{T} \quad \text{for} \quad \begin{cases} k_v' = k_v \\ d = D \end{cases}$$	
$$k_h = \frac{d^2 \cdot \ln \left[\frac{2mL}{D} + \sqrt{1 + \left(\frac{2mL}{D} \right)^2} \right]}{8 \cdot L \cdot T}$$ $$k_h = \frac{d^2 \cdot \ln \left(\frac{4mL}{D} \right)}{8 \cdot L \cdot T} \quad \text{for} \quad \frac{2mL}{D} > 4$$	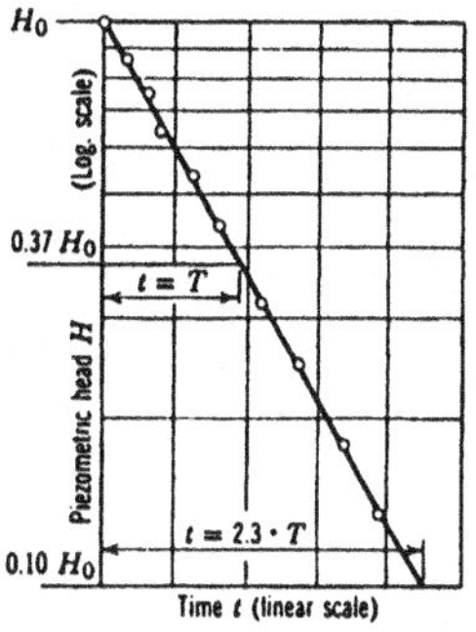
$$k_h = \frac{d^2 \cdot \ln \left[\frac{mL}{D} + \sqrt{1 + \left(\frac{mL}{D} \right)^2} \right]}{8 \cdot L \cdot T}$$ $$k_h = \frac{d^2 \cdot \ln \left(\frac{2mL}{D} \right)}{8 \cdot L \cdot T} \quad \text{for} \quad \frac{mL}{D} > 4$$	Determination basic time lag T

FIGURE 4.27 *(Continued)*

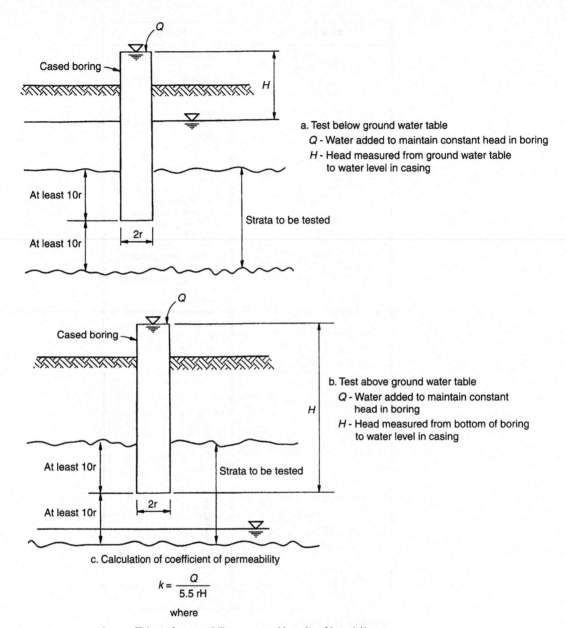

a. Test below ground water table

   Q - Water added to maintain constant head in boring

   H - Head measured from ground water table to water level in casing

b. Test above ground water table

   Q - Water added to maintain constant head in boring

   H - Head measured from bottom of boring to water level in casing

c. Calculation of coefficient of permeability

$$k = \frac{Q}{5.5\,rH}$$

where

k = coefficient of permeability measured in units of length/time

Q = flow of water required to maintain constant head in units of volume/time

H = head in units of length

r = radius of casing in units of length

FIGURE 4.28   Simple equation for field determination of permeability. (*Adapted from Cedergren, 1989; reproduced from Rollings and Rollings, 1996.*)

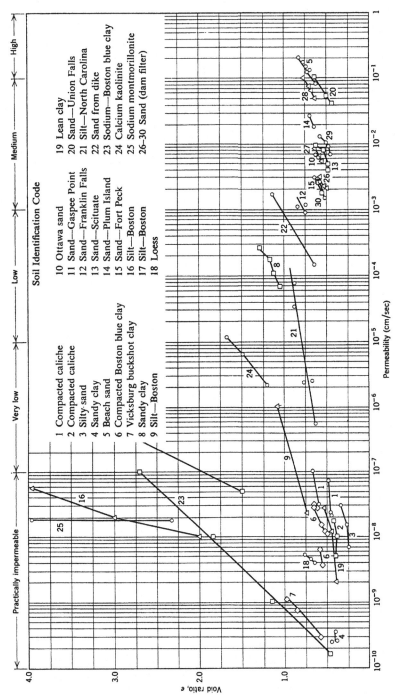

FIGURE 4.29 Coefficient of permeability k versus void ratio e. (*From Lambe and Whitman 1969; reproduced with permission from John Wiley & Sons.*)

Soil Identification Code

1 Compacted caliche
2 Compacted caliche
3 Silty sand
4 Sandy clay
5 Beach sand
6 Compacted Boston blue clay
7 Vicksburg buckshot clay
8 Sandy clay
9 Silt—Boston

10 Ottawa sand
11 Sand—Gaspee Point
12 Sand—Franklin Falls
13 Sand—Scituate
14 Sand—Plum Island
15 Sand—Fort Peck
16 Silt—Boston
17 Silt—Boston
18 Loess

19 Lean clay
20 Sand—Union Falls
21 Silt—North Carolina
22 Sand from dike
23 Sodium—Boston blue clay
24 Calcium kaolinite
25 Sodium montmorillonite
26–30 Sand (dam filter)

4.7.3 Seepage Velocity

From Bernoulli's energy equation, the total head h is equal to the sum of the velocity head, pressure head, and elevation head. Head has units of length. For seepage problems in soil, the velocity head is usually small enough to be neglected and thus for laminar flow in soil, the total head h is equal to the sum of the pressure head h_p and the elevation head h_e.

In order for water to flow through soil, there must be a change in total head Δh. For example, given a level groundwater table with hydrostatic pore water pressures, the total head will be identical at all points below the groundwater table. Since there is no change in total head ($\Delta h = 0$), there can be no flow of water.

Darcy's law ($v = k\,i$) is used for the analysis of laminar flow in soil and it relates the velocity v to the change in total head ($i = \Delta h/L$, where $L =$ length between the total head differences). Since the coefficient of permeability k is measured in a permeameter where the quantity of water is measured flowing out of (constant head permeameter) or into (falling head permeameter) the soil, the velocity v is actually a *superficial velocity* and not the actual velocity of flow in the soil voids v_s. The *seepage velocity* v_s is related to the *superficial velocity* v, by the following equation:

$$v_s = \frac{v}{n} = \frac{ki}{n} = \frac{k\,\Delta h}{nL} \tag{4.40}$$

where $v_s =$ seepage velocity, i.e., velocity of flow of water in the soil (ft/day or cm/sec)
$v =$ superficial velocity, i.e., velocity of flow of water into or out of the soil (ft/day or cm/sec)
$k =$ the coefficient of permeability, also known as the hydraulic conductivity (ft/day or cm/sec)
$i =$ hydraulic gradient ($i = \Delta h/L$), which is a dimensionless parameter
$n =$ porosity of the soil, Eq. 4.9, expressed as a decimal (dimensionless)
$\Delta h =$ change in total head between two points in the soil mass (ft or m)
$L =$ length between the two points in the soil mass (ft or m)

Between any two points in a saturated soil mass below a sloping groundwater table or below a level groundwater table that does not have hydrostatic pore water pressure (such as an artesian condition), Eq. 4.40 can be used to determine the velocity of flow v_s in the soil.

4.7.4 Seepage Force

When water flows through a soil, it exerts a drag force on the individual soil particles. This force has been termed the *seepage force*. For example, an artesian condition can cause an upward flow of water to the ground surface. If this upward flow of water exerts enough of a seepage force upon the individual soil particles, they can actually lose contact with each other. This fluid condition of a cohesionless soil has been termed *quicksand*. In order to create this fluid condition, the vertical effective stress σ'_v in the soil must equal zero.

There are two basic approaches to evaluating a seepage condition, and each approach gives the same answer. The first approach is to use the total unit weight of the soil γ_t and the boundary pore water pressures. The second approach is to use the buoyant unit weight of the soil γ_b and the *seepage force j*, which is equal to the unit weight of water γ_w times the hydraulic gradient i, or $j = \gamma_w\,i$. The seepage force j has units of force per unit volume of soil (pcf or kN/m³).

Because it is often easier to measure or predict the pore water pressure, the first approach which uses the total unit weight of the soil and the boundary pore water pressures is used more often in engineering analyses. The most common method used to predict the pore water pressures in the ground due to flowing groundwater is to use a flow net. An example of an engineering analysis that may need to include the seepage effects of groundwater is the slope stability of an earth dam.

4.7.5 Flow Nets

As shown in Fig. 3.56 and 3.57, the constant head and falling head permeameters have a one-dimensional flow condition where the water only flows vertically through the soil. For one-dimensional flow conditions, the quantity of water Q that exits the soil can be obtained by multiplying the superficial velocity v times the cross sectional area A and time t, or:

$$Q = v A t = k i A t \tag{4.41}$$

For two-dimensional flow conditions, where for example the groundwater is flowing horizontally and vertically, the situation is more complex. The most common method of estimating the quantity of water Q and pore water conditions in the soil for a two-dimensional case is to construct a flow net. A flow net is defined as a graphical representation used to study the flow of water through soil. A flow net is composed of two types of lines: (1) flow lines, which are lines that indicate the path of flow of water through the soil, and (2) equipotential lines, which are lines that connect points of equal total head. All along an equipotential line, the numerical value of the total head (h) is constant.

Figure 4.30 presents a cross section depicting the seepage of water underneath a sheet pile wall and into a foundation pit. On one side of the sheet pile wall, the level of water is at a height $= h_w$. On the other side of the sheet pile wall in the foundation pit, the water has been pumped out and the groundwater table corresponds to the ground surface. In this case, the change in total head (Δh) between the outside water level and the inside foundation pit water level is equal to h_w (i.e., $\Delta h = h_w$). Because there is a difference in total head, there will be the flow of water from the outside to the inside of the foundation pit. As shown in Fig. 4.30, the sheet pile wall is embedded a depth of 12 m (39 ft) into the soil. The site consists of 25 m (82 ft) of pervious sand overlying a relatively impervious clay stratum.

Assuming isotropic and homogeneous soil, the flow net has been drawn as shown in Fig. 4.30. The flow lines are those lines that have arrows, which indicate the direction of flow. The equipotential lines are those lines that are generally perpendicular to the flow lines. In Fig. 4.30, the flow lines and equipotential lines intersect at right angles and tend to form squares. By using the flow net, the quantity of seepage (Q) entering the foundation pit, per unit length of sheet pile wall, can be estimated by using Eq. 4.41, or:

$$Q = k i A t = k \Delta h t (A/L) = k \Delta h t (n_f/n_d) \tag{4.42}$$

where Q = quantity of water that enters the foundation pit per unit length of sheet pile wall (ft³ for a 1 foot length of wall or m³ for a 1 meter length of wall)
  k = coefficient of permeability of the pervious stratum (ft/day or m/day)
  Δh = change in total head (ft or m)
  t = time (days)
  A/L = cross sectional area of flow divided by the length of flow, which is equal to n_f/n_d, i.e., the number of flow channels (n_f) divided by the number of equipotential drops (n_d)

Example Problem 4.5 Using the flow net shown in Fig. 4.30, assume the coefficient of permeability $k = 0.1$ m/day (0.33 ft/day), the height of the water level outside the foundation pit $h_w = 10$ m (33 ft), and the length of the sheet pile wall is 100 m (328 ft). Calculate the quantity of water that enters the foundation pit per week ($t = 7$ days).

Solution The solution is obtained by using Eq. 4.42 with the following values:

  $k = 0.1$ m/day (0.33 ft/day)

  $\Delta h = h_w = 10$ m (33 ft)

  $T = 7$ days

  n_f = number of flow channels from the flow net = 9 (see Fig. 4.30)

(Continued)

n_d = number of equipotential drops from the flow net = 18 (see Fig. 4.30). Inserting these values into Eq. 4.42:

$$Q = k\,\Delta h\,t\,(n_f/n_d) = (0.1)(10)(7)(9/18) = 3.5 \text{ m}^3 \text{ per day}$$

The quantity of water Q that enters the foundation pit = 3.5 m³ per linear meter of wall length (38 ft³ per linear foot of wall length). For a 100 m (328 ft) long sheet pile wall, the total amount of water that enters the foundation pit per week = 350 m³ (12,500 ft³). This is a considerable amount of water and extensive pumping would be required to keep the foundation pit dry.

In addition to determining the quantity of water Q that enters the foundation pit, the flow net can be used to determine the pore water pressures u in the ground. The equipotential lines are used to determine the pore water pressure in the soil. In Fig. 4.30, the value of h' is shown. This value represents the drop in total head between each successive equipotential line. Because there are 18 equipotential drops for the flow net shown in Fig. 4.30, $h' = \Delta h/18 = h_w/18$. Thus each successive equipotential line represents an additional drop in total head = h'. By establishing a datum to obtain the elevation head h_e, and knowing the value of the total head h along each equipotential line, the pressure head ($h_p = h - h_e$) and hence pore water pressure can be calculated ($u = h_p\gamma_w$).

Example Problem 4.6 Assume the same conditions as the previous example. Calculate the pore water pressure u at Point A in Fig. 4.30, if Point A is located 2 m (6.6 ft) above the impervious stratum.

Solution For convenience, let the datum be at the top of the impervious stratum. Then the total head h corresponding to the ground surface on the left side of the sheet pile wall is equal to the elevation head (h_e = 25 m) plus the pressure head ($h_p = h_w$ = 10 m) or a total head h of 35 m (115 ft). In Fig. 4.30, the value of $h' = \Delta h/n_d = h_w/18 = 0.556$ m (1.82 ft). Since there are 9 equipotential drops for the flow of water to Point A (Fig. 4.30), the reduction in total head = 9 times 0.556 m, or 5.0 m (16.5 ft).

The total head along the equipotential line that contains Point A can now be calculated. The total head h at Point A would equal 35 m (115 ft) minus the reduction in total head due to the 9 equipotential drops (5.0 m), or 30 m (98.5 ft). Since the elevation head at Point A = 2 m (6.6 ft), and recognizing that the total head h is equal to the sum of the pressure head h_p and the elevation head h_e, the pressure head is equal to 28 m (i.e., $h_p = h - h_e$ = 30 m – 2 m = 28 m). Converting head to pore water pressure ($u = h_p\gamma_w$), the pore water pressure u at Point A = 275 kPa (5730 psf).

Check: Because Point A is located at exactly the midpoint in the flow net, a quick check can be performed on the earlier calculations. At the mid-point, the drop in head of water would equal $\frac{1}{2}\,h_w$, or 5 m. Installing a piezometer at Point A, the water level would rise 5 m above ground surface, or a total of 28 m (23 m plus 5 m).

In a sense, the reduction in total head (i.e., loss of energy) as the water flows underneath the sheet pile wall is absorbed by the frictional drag resistance of the soil particles. If this frictional drag resistance of the soil particles becomes too great, piping may occur. Piping is the physical transport of soil particles by the groundwater seepage. Piping can result in the opening of underground voids that could lead to the sudden failure of the sheet pile wall.

The exit hydraulic gradient i_e can be used to assess the piping potential of soil into the foundation pit. The most likely location for a piping failure is at Point B (Fig. 4.30), because this is where the flow of groundwater is directly upward and the distance between equipotential lines is least. If the distance between the two equipotential lines at Point B is 2 m, then the exit hydraulic gradient at Point B (i_e) would equal h'/L = 0.556 m/2 m = 0.28.

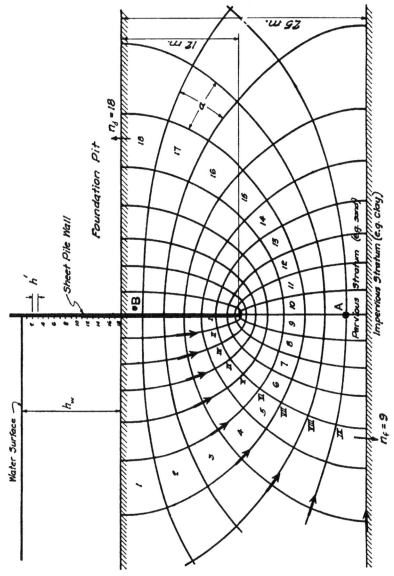

FIGURE 4.30 Flow net depicting the flow of water underneath a sheet pile wall and into a foundation pit. (*Adapted from Casagrande, 1940.*)

It has been determined that for a *quicksand* condition (also known as a *blowout* or *boiling* condition), the critical hydraulic gradient i_c is equal to the buoyant unit weight γ_b divided by the unit weight of water γ_w, or $i_c = \gamma_b/\gamma_w$. The buoyant unit weight is often about equal to the unit weight of water, and thus the critical hydraulic gradient i_c is about equal to 1. Certainly if the soil at Point B in Fig. 4.30 should turn to quicksand, then there would be significant piping (loss of soil particles) at this location. Thus it is often useful to compare the exit hydraulic gradient i_e to the critical hydraulic gradient which will cause quicksand i_c, or:

$$F = \frac{i_c}{i_e} \tag{4.43}$$

where F = the factor of safety for the development of a quicksand condition
i_c = critical hydraulic gradient that will cause quicksand (dimensionless)
i_e = exit hydraulic gradient (dimensionless)

For the earlier example, i_e (Point B) = 0.28 and using i_c = 1.0, the factor of safety = 3.6. Because of the catastrophic effects of a piping failure, usually a high factor of safety is required for the exit hydraulic gradient. Cedergren (1989) indicates that the factor of safety for the exit hydraulic gradient should be at least 2 to 2.5, while Chen (1995) states that factors of safety of 4 to 5 are generally considered reasonable when using flow nets.

Figure 4.31 shows an earth dam with the seepage of water through the dam and into a longitudinal drainage filter. In Fig. 4.31, the flow lines are those lines that have arrows that indicate the direction of flow. The equipotential lines are those lines that are generally perpendicular to the flow lines. The vertical difference between the water level on the upstream side of the earth dam and the level of water in the drainage filter (i.e., dashed line) is the change in total head Δh. In Fig. 4.31, a series of horizontal lines have been drawn adjacent the line representing the groundwater table in the earth dam. The distance between these horizontal lines are equal and represent equal drops in total head for successive equipotential lines.

The quantity of flow Q collected by the drainage filter can be determined by using Eq. 4.42. Another use for the flow net shown in Fig. 4.31 would be to determine the stability of earth dam slopes. Based on the flow net, the pore water pressures u could be determined and then the effective shear strength could be used in conjunction with a slope stability analysis to determine the stability of the earth dam.

The usual procedure in the preparation of a flow net is trial and error sketching. A flow net is first sketched in pencil with a selected number of flow channels and then the equipotential lines are inserted. If the flow net does not have flow lines and equipotential lines intersecting at right angles, or if the flow net is not composed of approximate squares, then the lines are adjusted. The rules for flow net construction are as follows (NAVFAC DM-7.2, 1982):

1. When the soil is isotropic with respect to permeability, the pattern of flow lines and equipotential lines intersect at right angles. Draw a pattern in which squares are formed between flow lines and equipotential lines.

2. Usually it is expedient to start with an integer number of equipotential drops, dividing the total head loss by a whole number and drawing flow lines to conform to these equipotential drops. In a general sense, the outer flow path will form a rectangle rather than a square. The shape of these rectangles must be constant (i.e., constant ratio of B/L).

3. The upper boundary of the flow net that is at atmospheric pressure is a *free water surface*. Equipotential lines intersect the free water surface at points spaced at equal vertical intervals (for example, see Fig. 4.31).

4. A discharge face through which seepage passes is an equipotential line if the discharge is submerged, or a free water surface if the discharge is not submerged. If it is a free water surface, the flow net directly adjacent the discharge face will not be composed of squares.

5. In a stratified soil profile where the ratio of permeability of different layers exceeds a factor of 10, the flow in the more permeable layer controls. In this case, the flow net may be drawn for the more permeable layer assuming the less permeable layer to be impervious.

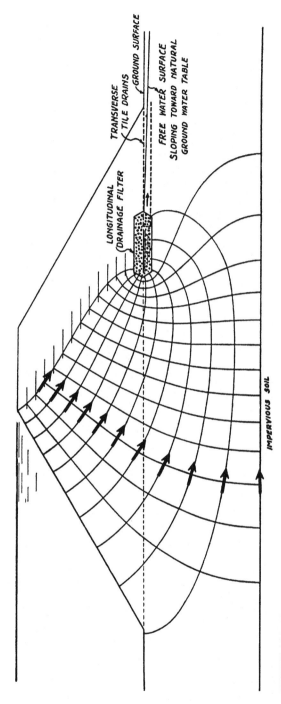

FIGURE 4.31 Flow net depicting the flow of water through an earth dam. (*Adapted from Casagrande, 1940.*)

4.69

6. Where only the quantity of seepage is to be determined, an approximate flow net suffices. If pore water pressures are to be determined, the flow net must be accurately drawn.

The flow nets presented in this section have been based on the assumption of isotropic and homogeneous soil. Flow nets can also be drawn for soils having different coefficients of permeability or for anisotropic soil where the coefficient of permeability is higher in the horizontal direction than the vertical direction (Casagrande, 1940). When materials are anisotropic with respect to permeability, the cross section may be transformed by changing the scale as indicated in Fig. 4.32 and the flow net is then drawn as for an isotropic soil. In computing the quantity of seepage Q, the total head difference Δh is not changed for the transformation. Fig. 4.32 also shows the flow net for seepage into a dry dock and the transfer conditions for the flow of water from one soil stratum to another.

Computer programs have been developed to aid in the construction of a flow net. For example, the SEEP/W (Geo-Slope, 1992) computer program can be used to determine the total head h distributions and the flow vectors (i.e., flow lines) for both simple and highly complex seepage problems.

4.7.6   Groundwater Lowering by Pumping Wells

The final part of this section deals with groundwater lowering by pumping from wells. This is often an important aspect of foundation dewatering. One commonly used equation to determine the amount of drawdown of the groundwater table for pumping from a well is shown in Fig. 4.33. This equation is

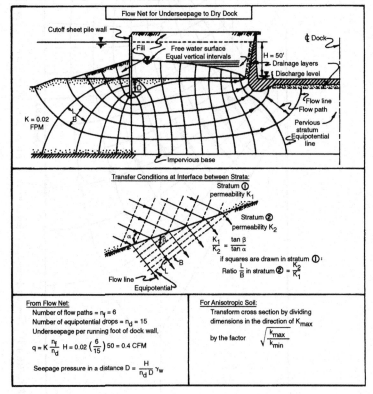

FIGURE 4.32   Flow net for seepage into a dry dock. Also shown are the transfer conditions for the flow of water from one soil stratum to another and the procedure for construction of a flow net for anisotropic soil. (*From NAVFAC DM-7.1, 1982.*)

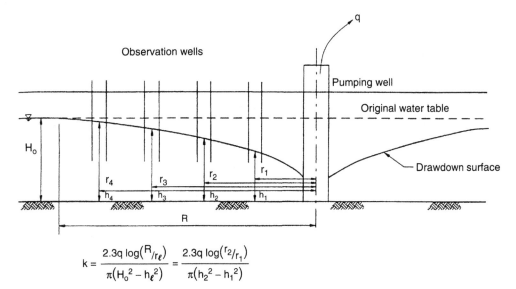

$$k = \frac{2.3q \, \log\left(\tfrac{R}{r_\ell}\right)}{\pi\left(H_o^2 - h_\ell^2\right)} = \frac{2.3q \, \log\left(\tfrac{r_2}{r_1}\right)}{\pi\left(h_2^2 - h_1^2\right)}$$

k - Coefficient of permeability in units of length/time
H_o - Original height to ground water table in units of length
h - Height of drawdown surface in observation well in units of length
R - Radius of influence of pumping well in units of length
r - Distance from pumping well to observation well in units of length
q - Steady state pumping rate in units of volume/time

FIGURE 4.33   Schematic drawing of well pumping test. (*Adapted from Bureau of Reclamation, 1990; reproduced from Rollings and Rollings, 1996.*)

applicable for a homogeneous unconfined aquifer (i.e., permeable soil) underlain by an impervious soil layer. The pumping well penetrates all the way to the top surface of the impervious soil layer. The coefficient of permeability k can be calculated by knowing the pumping rate q and the height h of the groundwater table above the impervious soil layer from an observation well. Once the coefficient of permeability k is known for the unconfined aquifer, the drawdown at any distance from the pumping well can be determined. Groundwater lowering by pumping wells will be further discussed in Chap. 16.

Example Problem 4.7   Assume the conditions shown in Fig. 4.33, with $q = 1000$ ft³/day (28 m³/day), $H_o = 50$ ft (15.2 m), $R = 200$ ft (61 m), $h_1 = 45$ ft (13.7 m), and $r_1 = 80$ ft (24 m). Determine the coefficient of permeability k of the soil comprising the homogenous and unconfined aquifer.

Solution   Using the dewatering well equation in Fig. 4.33:

$$k = 2.3q \frac{\log(R/r_1)}{\left[\pi\left(H_o^2 - h_1^2\right)\right]}$$

$$k = 2.3 \frac{(1000 \text{ ft}^3/\text{day}) \log(200/80)}{[\pi(50^2 - 45^2)]} = 0.6 \text{ ft/day} (0.2 \text{ m/day})$$

NOTATION

The following notation is used in this chapter:

A = activity of clay, Eq. 4.6

A = cross–sectional area (Sec. 4.7.5)

A = length of the pile group (Sec. 4.5.2)

B = width of the footing (Sec. 4.5.1)

B = width of the pile group (Sec. 4.5.2)

c = cohesion based on a total stress analysis

c' = cohesion based on an effective stress analysis

c'_r = drained residual cohesion (effective stress analysis)

C_c = coefficient of curvature

C_c = compression index for consolidation

C_u = coefficient of uniformity

D = diameter of circular loaded area

D_r = relative density

D_0 = from grain size curve, smallest particle size

D_{10} = from grain size curve, particle size for 10 percent finer by dry weight

D_{30} = from grain size curve, particle size for 30 percent finer by dry weight

D_{60} = from grain size curve, particle size for 60 pecent finer by dry weight

D_{100} = from grain size curve, largest particle size

e = void ratio of the soil

e_{max} = void ratio corresponding to the loosest state of the soil

e_{min} = void ratio corresponding to the densest state of the soil

E = modulus of elasticity

FS_L = factor of safety against liquefaction

G = specific gravity of soil (including organic matter)

G_s = specific gravity of soil solids

h = distance above the groundwater table, Eq. 4.17

h = thickness of fill layer, Eq. 4.21

h = depth below the ground surface, Eq. 4.39

h = total head (Sec. 4.7.3)

h = height of the groundwater table (Sec. 4.7.6)

h' = equipotential drop, i.e., change in total head (Sec. 4.7.5)

h_c = height of capillary rise

h_e = elevation head

h_p = pressure head

h_w = height of water behind the sheet pile wall (Fig. 4.30)

h_1 = height of drawdown in observation well (Sec. 4.7.6)

Δh = change in total head (Sec. 4.7.3)

H = thickness of the upper soil layer (Fig. 4.16)

H_o = original height of groundwater table (Sec. 4.7.6)

i = hydraulic gradient ($i = \Delta h/L$)

i_c = critical hydraulic gradient

i_e = exit hydraulic gradient

I = influence value

j = seepage force per unit volume of soil

k = ratio used for Fig. 4.16

k = coefficient of permeability (also known as hydraulic conductivity)

k_o = coefficient of lateral earth pressure at rest

L = length of the footing for Eqs. 4.25 and 4.26

L = length between the total head differences (Sec. 4.7.3)

L = various lengths used in Fig. 4.17

LI = liquidity index

LL = liquid limit

m, n = parameters used to determine σ_z below the corner of a loaded area

M = mass of water plus mass of soil solids

M_s = dry mass of the soil solids

M_w = mass of water in the soil

n = porosity

n = number of piles that support the pile cap (Sec. 4.5.2)

n_d = number of equipotential drops

n_f = number of flow channels

N = number of blocks for Eq. 4.31

OCR = overconsolidation ratio

p = magnitude of the stress applied by the circular area (Fig. 4.16)

P = vertical footing load

PI = plasticity index

PL = plastic limit

q = increase in vertical stress in the soil due to the applied stress (Fig. 4.9)

q = pumping rate (Sec. 4.7.6)

Q = surface point or line load for Eqs. 4.29 and 4.30

Q = quantity of water that exits or enters the soil (Sec. 4.7.5)

Q_{all} = allowable vertical load for each pile

Q_D = pile downdrag load (Fig. 4.17)

r = horizontal distance from the point load (Fig. 4.7)

r = radius for circular loaded area (Sec. 4.5.1)

r_1 = distance from pumping well to observation well (Sec. 4.7.6)

r_u = pore water pressure ratio (Eq. 4.39)

R = radius of influence of pumping well (Sec. 4.7.6)

s_u = undrained shear strength of the soil

s_{ur} = undrained shear strength of remolded soil

S = degree of saturation of the soil

S_t = sensitivity of the cohesive soil

t = time

u = pore water pressure in the soil

u_e = excess pore water pressure in the soil

u_s = hydrostatic pore water pressure in the soil

v = superficial velocity, which equals v in Darcy's equation ($v = k\,i$)

v_s = seepage velocity

V = total volume of the soil

V_g = volume of gas (air) in the soil

V_s = volume of soil solids

V_v = volume of voids in the soil, which equals $V_g + V_w$

V_w = volume of water in the soil

w = water content (also known as moisture content)

x = width of the loaded area (Fig. 4.8)

y = length of the loaded area (Fig. 4.8)

z = depth below ground surface

z_w = depth below groundwater table

α = ratio used in Fig. 4.16

ϕ = friction angle based on a total stress analysis

ϕ' = friction angle based on an effective stress analysis

ϕ'_r = drained residual friction angle (effective stress analysis)

γ_b = buoyant unit weight of saturated soil below the groundwater table

γ_d = dry unit weight of soil

γ_t = total unit weight of soil

γ_w = unit weight of water

μ = poisson's ratio

ρ_s = density of the soil solids

ρ_t = total density of the soil

ρ_w = density of water

σ = total stress

σ_o, q_o = uniform pressure applied by the footing

σ_v = vertical total stress

$\sigma_z, \Delta\sigma_v$ = increase in vertical stress in the soil due to the applied stress

σ' = effective stress

σ'_h = horizontal effective stress

σ'_p = preconsolidation pressure (largest σ'_v ever experienced by the soil)

σ'_v, σ'_{vo} = vertical effective stress

PROBLEMS

Solutions to the problems are presented in App. C of this book. The problems have been divided into basic categories as follows:

Soil Classification

4.1 A soil has the following particle size gradation based on dry mass:

Gravel and sand size particles coarser than No. 40 sieve = 65%	
Sand size particles finer than No. 40 sieve	= 10%
Silt size particles	= 5%
Clay size particles (finer than 0.002 mm)	= 20%
Total	= 100%

In accordance with ASTM test procedures, the Atterberg limits were performed on the soil finer than the No. 40 sieve and the results are LL = 93 and PL = 18. Calculate the activity A of this soil.

ANSWER: A = 1.3.

4.2 A clay has a liquid limit = 60 and a plastic limit = 20. Using Fig. 4.2, determine the predominate clay mineral in the soil.

ANSWER: Montmorillonite.

4.3 Based on a particle size analysis, the following values were obtained: $D_{60} = 15$ mm, $D_{50} = 12$ mm, $D_{30} = 2.5$ mm, and $D_{10} = 0.075$ mm. Calculate C_u and C_c.

ANSWER: $C_u = 200$ and $C_c = 5.6$.

4.4 For the soil in Problem 4.3, assume that 18 percent of the soil passes the No. 40 sieve and the LL = 68 and PI = 34 for this soil fraction. What is the USCS group symbol?

ANSWER: GP-GM.

4.5 A sand has a $C_u = 7$ and a $C_c = 1.5$. The sand contains 4 percent nonplastic fines (by dry mass). What is the USCS group symbol?

ANSWER: SW.

4.6 An inorganic soil has 100 percent passing (by dry mass) the No. 200 sieve. The LL = 43 and the PI = 16. What is the USCS group symbol?

ANSWER: ML.

4.7 An inorganic clay has a LL = 60 and a PL = 20. What is the USCS group symbol?

ANSWER: CH.

4.8 A fine-grained soil has a black color and organic odor. The LL = 65 on non-oven dried soil and LL = 40 on oven-dried soil. What is the USCS group symbol?

ANSWER: OH.

4.9 Based on dry mass, a soil has 20 percent gravel size particles, 40 percent sand size particles, and 40 percent fines. Based on dry mass, 48 percent of the soil particles pass the No. 40 sieve and the LL = 85 and PL = 18 for soil passing the No. 40 sieve. What is the USCS group symbol?

ANSWER: SC.

For Problems 4.10 through 4.14, use the laboratory testing data summarized in Fig. 4.34. Consider all five soils to consist of inorganic soil particles with the Atterberg limits performed on

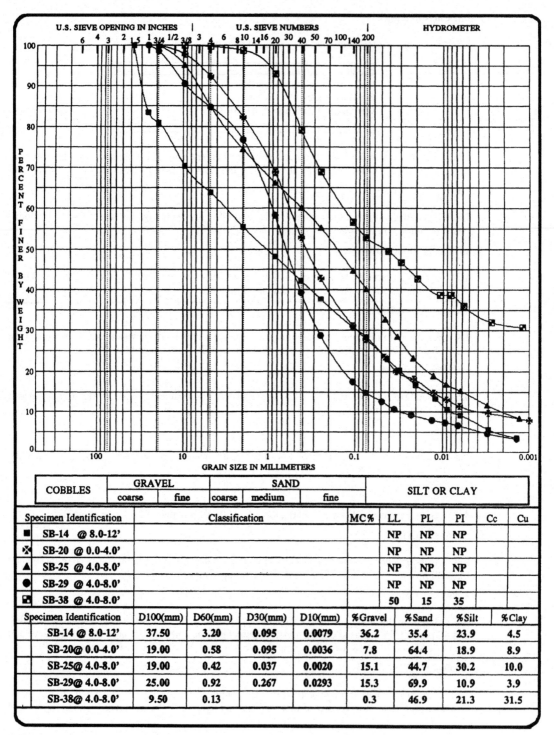

COBBLES	GRAVEL		SAND			SILT OR CLAY	
	coarse	fine	coarse	medium	fine		

Specimen Identification	Classification					MC%	LL	PL	PI	Cc	Cu
■ SB-14 @ 8.0-12'							NP	NP	NP		
✖ SB-20 @ 0.0-4.0'							NP	NP	NP		
▲ SB-25 @ 4.0-8.0'							NP	NP	NP		
● SB-29 @ 4.0-8.0'							NP	NP	NP		
▣ SB-38 @ 4.0-8.0'							50	15	35		

Specimen Identification	D100(mm)	D60(mm)	D30(mm)	D10(mm)	%Gravel	%Sand	%Silt	%Clay
SB-14 @ 8.0-12'	37.50	3.20	0.095	0.0079	36.2	35.4	23.9	4.5
SB-20@ 0.0-4.0'	19.00	0.58	0.095	0.0036	7.8	64.4	18.9	8.9
SB-25@ 4.0-8.0'	19.00	0.42	0.037	0.0020	15.1	44.7	30.2	10.0
SB-29@ 4.0-8.0'	25.00	0.92	0.267	0.0293	15.3	69.9	10.9	3.9
SB-38@ 4.0-8.0'	9.50	0.13			0.3	46.9	21.3	31.5

FIGURE 4.34   Particle size distributions for Problems 4.10 to 4.14 (plot developed by gINT computer program).

soil passing the No. 40 sieve (per ASTM D 4318-00, 2004). Note in Fig. 4.34 that LL is the liquid limit, PL is the plastic limit, PI is the plasticity index, and NP indicates that the soil is nonplastic.

4.10 For SB-14 at 8 to 12 ft, determine the soil classification per the USCS and AASHTO soil classification systems.

ANSWER: GM per USCS and A-2-4 per AASHTO.

4.11 For SB-20 at 0 to 4 ft, determine the soil classification per the USCS and AASHTO soil classification systems.

ANSWER: SM per USCS and A-2-4 per AASHTO.

4.12 For SB-25 at 4 to 8 ft, determine the soil classification per the USCS and AASHTO soil classification systems.

ANSWER: SM per USCS and A-4 (0) per AASHTO.

4.13 For SB-29 at 4 to 8 ft, determine the soil classification per the USCS and AASHTO soil classification systems.

ANSWER: SM per USCS and A-1-b per AASHTO.

4.14 For SB-38 at 4 to 8 ft, determine the soil classification per the USCS, AASHTO, and USDA soil classification systems.

ANSWER: CH per USCS, A-7-6 (14) per AASHTO, and sandy clay loam per USDA.

For Problems 4.15 through 4.18, use the laboratory testing data summarized in Fig. 4.35. Consider all four soils to consist of inorganic soil particles with the Atterberg limits performed on soil passing the No. 40 sieve (per ASTM D 4318-00, 2004). Note in Fig. 4.35 that LL is the liquid limit, PL is the plastic limit, PI is the plasticity index, and NP indicates that the soil is nonplastic.

4.15 For SB-39 at 4 to 8 ft, determine the soil classification per the USCS, AASHTO, and USDA soil classification systems.

ANSWER: SC per USCS, A-7-6 (9) per AASHTO, and sandy clay loam per USDA.

4.16 For SB-42 at 4 to 8 ft, determine the soil classification per the USCS, AASHTO, and USDA soil classification systems.

ANSWER: CL per USCS, A-6 (8) per AASHTO, and loam per USDA.

4.17 For SB-44 at 4 to 8 ft, determine the soil classification per the USCS and AASHTO soil classification systems.

ANSWER: SM per USCS and A-1-b per AASHTO.

4.18 For SB-45 at 4 to 8 ft, determine the soil classification per the USCS and AASHTO soil classification systems.

ANSWER: SC per USCS, and A-2-6 (1) per AASHTO.

For Problems 4.19 through 4.21, use the laboratory testing data summarized in Fig. 4.36. Consider the three soils to consist of inorganic soil particles with the Atterberg limits performed on soil passing the No. 40 sieve (per ASTM D 4318-00, 2004). Note in Fig. 4.36 that LL is the liquid limit, PL is the plastic limit, and PI is the plasticity index.

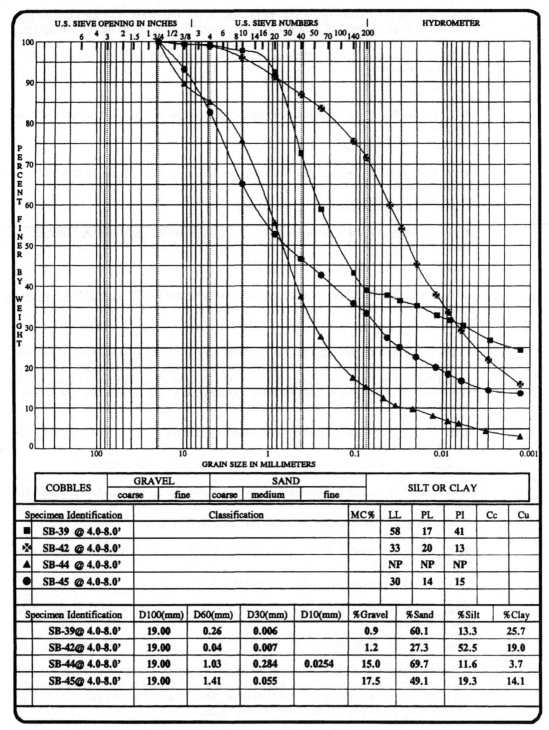

COBBLES	GRAVEL		SAND			SILT OR CLAY		
	coarse	fine	coarse	medium	fine			

Specimen Identification	Classification				MC%	LL	PL	PI	Cc	Cu
■ SB-39 @ 4.0-8.0'						58	17	41		
✕ SB-42 @ 4.0-8.0'						33	20	13		
▲ SB-44 @ 4.0-8.0'						NP	NP	NP		
● SB-45 @ 4.0-8.0'						30	14	15		

Specimen Identification	D100(mm)	D60(mm)	D30(mm)	D10(mm)	%Gravel	%Sand	%Silt	%Clay
SB-39@ 4.0-8.0'	19.00	0.26	0.006		0.9	60.1	13.3	25.7
SB-42@ 4.0-8.0'	19.00	0.04	0.007		1.2	27.3	52.5	19.0
SB-44@ 4.0-8.0'	19.00	1.03	0.284	0.0254	15.0	69.7	11.6	3.7
SB-45@ 4.0-8.0'	19.00	1.41	0.055		17.5	49.1	19.3	14.1

FIGURE 4.35   Particle size distributions for Problems 4.15 to 4.18 (plot developed by gINT computer program).

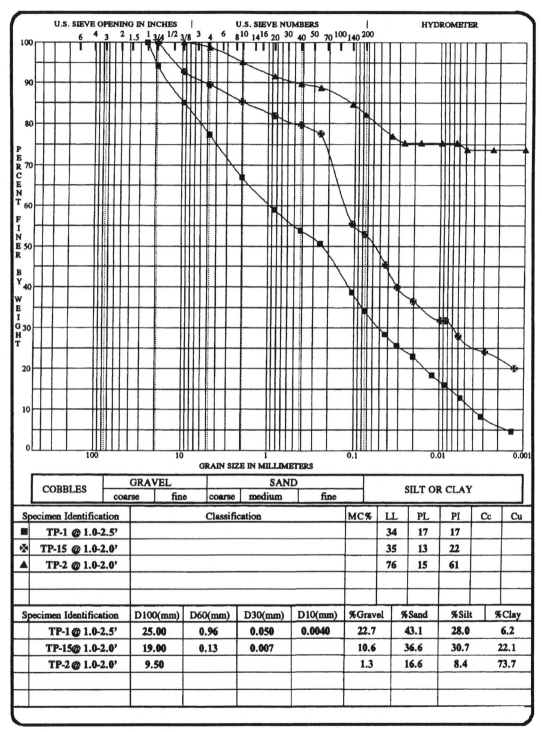

FIGURE 4.36 Particle size distributions for Problems 4.19 to 4.21 (plot developed by gINT computer program).

4.19 For TP-1 at 1 to 2.5 ft, determine the soil classification per the USCS and AASHTO soil classification systems.

ANSWER: SC per USCS and A-2-6 (1) per AASHTO.

4.20 For TP-15 at 1 to 2 ft, determine the soil classification per the USCS and AASHTO soil classification systems.

ANSWER: CL per USCS and A-6 (8) per AASHTO.

4.21 For TP-2 at 1 to 2 ft, determine the soil classification per the USCS, AASHTO, and USDA soil classification systems.

ANSWER: CH per USCS, A-7-6 (52) per AASHTO, and clay per USDA.

4.22 Use the laboratory testing data summarized in Fig. 4.37. Consider the soils to consist of inorganic soil particles with the Atterberg limits performed on the entire soil. Note in Fig. 4.37 that LL is the liquid limit, PL is the plastic limit, and PI is the plasticity index. Determine the soil classification per the USCS, AASHTO, and USDA soil classification systems.

ANSWER: CL per USCS, A-7-6 (22) per AASHTO, and clay loam per USDA.

For Problems 4.23 and 4.24, use the laboratory testing data summarized in Fig. 4.38. Consider the soils to consist of inorganic soil particles with the Atterberg limits performed on soil passing the No. 40 sieve (per ASTM D 4318-00, 2004). Note in Fig. 4.38 that LL is the liquid limit, PL is the plastic limit, and PI is the plasticity index.

4.23 For AGC-2 at 0.6 to 0.8 ft, determine the soil classification per the USCS, AASHTO, and USDA soil classification systems.

ANSWER: SM per USCS, A-2-4 per AASHTO, and sand per USDA.

4.24 For AGC-2 at 0.8 to 1.4 ft, determine the soil classification per the USCS and AASHTO soil classification systems. Assume that there are 12 percent clay size particles (i.e., 12 percent finer than 0.002 mm).

ANSWER: SC per USCS and A-2-7 (1) per AASHTO.

For Problems 4.25 through 4.33, use the laboratory testing data summarized in Fig. 4.39. Consider all eight soils to consist of inorganic soil particles with the Atterberg limits performed on soil passing the No. 40 sieve (per ASTM D 4318-00, 2004). Note in Fig. 4.39 that W_l is the liquid limit, W_p is the plastic limit, and PI is the plasticity index.

4.25 For soil number 1, determine the soil classification per the USCS and AASHTO soil classification systems.

ANSWER: GP-GC per USCS and A-2-4 per AASHTO.

4.26 For soil number 2, determine the soil classification per the USCS and AASHTO soil classification systems.

ANSWER: GW-GC per USCS and A-1-a per AASHTO.

4.27 For soil number 3, determine the soil classification per the USCS and AASHTO soil classification systems.

ANSWER: SP per USCS and A-1-b per AASHTO.

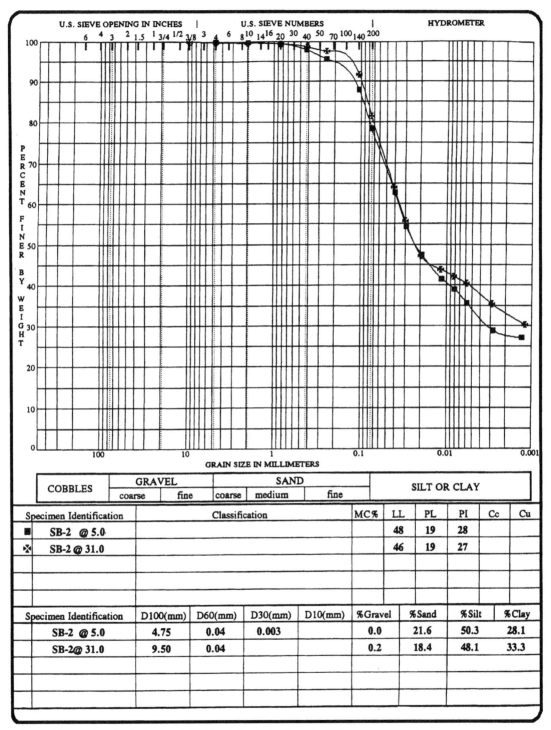

FIGURE 4.37 Particle size distributions for Problem 4.22 (plot developed by gINT computer program).

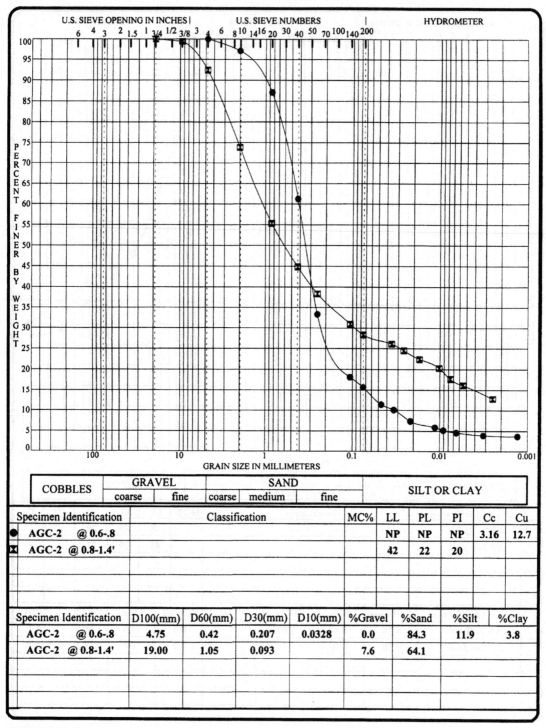

Specimen Identification	Classification	MC%	LL	PL	PI	Cc	Cu
● AGC-2 @ 0.6-.8			NP	NP	NP	3.16	12.7
⊠ AGC-2 @ 0.8-1.4'		42	22	20			

Specimen Identification	D100(mm)	D60(mm)	D30(mm)	D10(mm)	%Gravel	%Sand	%Silt	%Clay
AGC-2 @ 0.6-.8	4.75	0.42	0.207	0.0328	0.0	84.3	11.9	3.8
AGC-2 @ 0.8-1.4'	19.00	1.05	0.093		7.6	64.1		

FIGURE 4.38 Particle size distributions for Problems 4.23 and 4.24 (plot developed by gINT computer program).

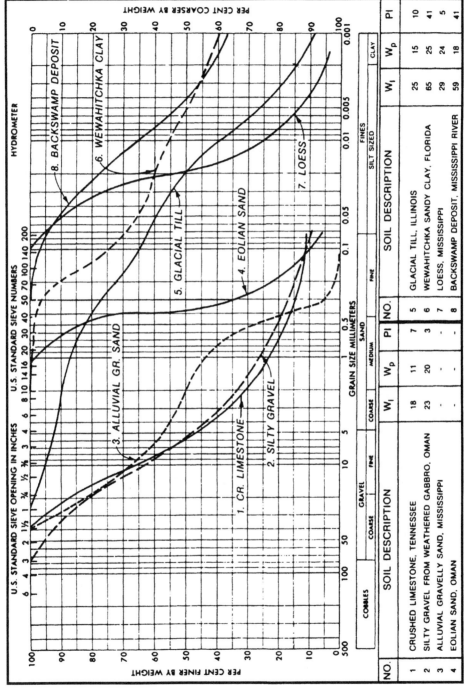

FIGURE 4.39 Grain size curves and Atterberg limits test data for eight different soils. (*From Rollings and Rollings, 1996; reproduced with permission of McGraw-Hill, Inc.*)

4.28   For soil number 4, determine the soil classification per the USCS, AASHTO, and USDA soil classification systems.

ANSWER: SP-SM per USCS, A-3 per AASHTO, and sand per USDA.

4.29   For soil number 5, determine the soil classification per the USCS, AASHTO, and USDA soil classification systems.

ANSWER: CL per USCS, A-4 (3) per AASHTO, and loam per USDA.

4.30   For soil number 6, determine the soil classification per the USCS, AASHTO, and USDA soil classification systems.

ANSWER: CH per USCS, A-7-6 (28) per AASHTO, and clay per USDA.

4.31   For soil number 7, determine the soil classification per the USCS, AASHTO, and USDA soil classification systems.

ANSWER: ML per USCS, A-4 (5) per AASHTO, and silt per USDA.

4.32   For soil number 8, determine the soil classification per the USCS, AASHTO, and USDA soil classification systems.

ANSWER: CH per USCS, A-7-6 (43) per AASHTO, and silty clay per USDA.

4.33   For soil numbers 5 to 8, calculate the activity of the clay size particles.

ANSWER: Soil no. 5, activity = 0.53; soil no. 6, activity = 0.93; soil no. 7, activity = 1.0; and soil no. 8, activity = 0.98.

4.34   A frozen soil has visible ice that is about $1/2$-inch (12-mm) thick. It was also observed that there are ice coatings on the individual soil particles. Using Table 4.4, determine the group system for this frozen soil.

ANSWER: V_c.

Phase Relationships

4.35   A soil specimen obtained from below the groundwater table has the following measured properties:

   Total unit weight $\gamma_t = 121.4$ pcf (19.1 kN/m^3)

   Water content $w = 29.5$ percent

   Specific gravity $G_s = 2.70$

   Calculate the parameters shown in Fig. 3.1 assuming that the total volume $(V) = 1$ ft^3 and 1 m^3. Also calculate the dry unit weight γ_t, void ratio e, porosity n, and degree of saturation (S).

ANSWER: See App. C for mass and volume calculations. The dry unit weight = 93.7 pcf (14.7 kN/m^3), void ratio = 0.79, porosity = 44 percent, and degree of saturation = 100 percent.

4.36   Assume the soil specimen in Problem 4.35 is sand having a maximum void ratio = 0.85 and a minimum void ratio = 0.30. Calculate the relative density and indicate the density state of the sand.

ANSWER: Relative density = 11 percent, very loose condition.

4.37   Using the data in Part (*a*) of Fig. 3.30, calculate the initial void ratio, porosity, and degree of saturation for the clay specimen. Assume the clay has a specific gravity = 2.70.

ANSWER: $e = 0.79$, $n = 44$ percent, and $S = 79$ percent.

4.38 A specimen of clay has a void ratio $e = 1.16$, specific gravity $G_s = 2.72$, and water content $w = 42.7$ percent. Calculate the total unit weight γ_t, dry unit weight γ_d, saturated unit weight γ_{sat}, and buoyant unit weight γ_b.

ANSWER: $\gamma_t = 112$ pcf, $\gamma_d = 78.6$ pcf, $\gamma_{sat} = 112$ pcf, and $\gamma_b = 49.6$.

Total Stress, Effective Stress, and Pore Water Pressure

4.39 Assume a uniform soil deposit and the groundwater table coincides with the ground surface. The total unit weight γ_t of the soil is equal to 19.5 kN/m³ (124 pcf). For a level ground surface and hydrostatic pore water pressures in the ground, calculate the total vertical stress σ_v, pore water pressure u, and vertical effective stress σ_v' at a depth of 6 m (20 ft).

ANSWER: $\sigma_v = 117$ kPa (2480 psf), $u = 59$ kPa (1250 psf), and $\sigma_v' = 58$ kPa (1230 psf).

4.40 Assume a uniform soil deposit with a level ground surface and the groundwater table is at a depth of 1.5 m (5 ft) below the ground surface. Above the groundwater table, assume the soil is saturated. The total unit weight γ_t of the soil is equal 19.5 kN/m³ (124 pcf). For hydrostatic pore water pressures in the ground, calculate the total vertical stress σ_v, pore water pressure u, and vertical effective stress σ_v' at a depth of 6 m (20 ft).

ANSWER: $\sigma_v = 117$ kPa (2480 psf), $u = 44$ kPa (940 psf), and $\sigma_v' = 73$ kPa (1540 psf).

4.41 Assume a uniform soil deposit and the site is a lake where the water is 3 m (10 ft) deep. The total unit weight γ_t of the soil is equal 19.5 kN/m³ (124 pcf). For hydrostatic pore water pressures in the ground and a level lake bottom, calculate the total vertical stress σ_v, pore water pressure u, and vertical effective stress σ_v' at a depth of 6 m (20 ft) below the lake bottom.

ANSWER: $\sigma_v = 146$ kPa (3100 psf), $u = 88$ kPa (1870 psf), and $\sigma_v' = 58$ kPa (1230 psf). Note that the effective stress for Problems 4.39 and 4.41 are identical.

4.42 Assume uniform clay stratum with a total unit weight γ_t equal to 19.5 kN/m³ (124 pcf), a level ground surface, and the groundwater table is at a depth of 3 m (10 ft) below the ground surface. Above the groundwater table, assume the clay is saturated. For the condition of capillary rise in the clay, calculate the total vertical stress σ_v, pore water pressure u, and vertical effective stress σ_v' at a depth of 1.5 m (5 ft) below the ground surface.

ANSWER: $\sigma_v = 29$ kPa (620 psf), $u = -15$ kPa (−310 psf), and $\sigma_v' = 44$ kPa (930 psf).

4.43 Assume a piezometer is installed in a clay deposit and the piezometer records a pressure head of 3 m (10 ft) of water at a depth of 6 m (20 ft) below ground surface. If the total vertical stress $\sigma_v = 117$ kPa (2480 psf) at this depth, calculate the vertical effective stress σ_v'.

ANSWER: $\sigma_v' = 88$ kPa (1860 psf).

4.44 For the soil profile shown in Fig. 2.39, calculate the total vertical stress σ_v, pore water pressure u, and vertical effective stress σ_v' at elevation −90 ft.

ANSWER: $\sigma_v = 620$ kPa (13,000 psf or 6.3 kg/cm²), $u = 310$ kPa (6400 psf or 3.1 kg/cm²), and $\sigma_v' = 320$ kPa (6600 psf or 3.2 kg/cm²).

Stress Distribution

4.45 A fill of uniform thickness that has a large area will be placed at the ground surface. The thickness of the fill layer will be 3 m (10 ft). The total unit weight of the fill = 18.7 kN/m³ (119 pcf). If the groundwater table is below the ground surface, calculate the increase in vertical stress $\Delta\sigma_v$ beneath the center of the constructed fill mass.

ANSWER: $\Delta\sigma_v = 56$ kPa (1190 psf).

4.46   Assume the same conditions as Problem 4.45 except that the fill mass has a width of 6 m (20 ft) and a length of 10 m (33 ft). If the groundwater table is below the ground surface, calculate the increase in vertical stress $\Delta\sigma_v$ beneath the center of the constructed fill mass at a depth of 12 m (39 ft) below original ground surface using the 2:1 approximation.

ANSWER: $\Delta\sigma_v = 8.5$ kPa (180 psf).

4.47   Solve Problem 4.46 by using Eq. 4.29 and assuming the fill is a concentrated load (Q).

ANSWER: $\Delta\sigma_v = 11$ kPa (230 psf).

4.48   Solve Problem 4.46 by using the Newmark chart (Fig. 4.8).

ANSWER: $\Delta\sigma_v = 9.9$ kPa (210 psf).

4.49   Solve Problem 4.46 by using the Westergaard chart (Fig. 4.15).

ANSWER: $\Delta\sigma_v = 6.1$ kPa (130 psf).

4.50   Solve Problem 4.46 by using Fig. 4.13.

ANSWER: Number of blocks = 34, $\Delta\sigma_v = 9.5$ kPa (200 psf).

4.51   For the proposed fill mass in Problem 4.46, subsurface exploration will be performed at the center of the fill mass. If it is proposed to excavate the boring to a depth where the change in vertical stress $\Delta\sigma_v$ is 10 percent of the applied stress, determine the depth of subsurface exploration below the original ground surface.

ANSWER: Depth of subsurface = 17 m (54 ft) based on the 2:1 approximation.

Permeability and Seepage

4.52   In a uniform soil deposit having a level ground surface, an open standpipe with a constant interior diameter of 5.0 cm is installed to a depth of 10 m. The standpipe has a flush bottom (case C, Fig. 4.26). The groundwater table is located 4.0 m below the ground surface. The standpipe is filled with water and the water level is maintained at ground surface by adding 1.0 L per every 33 sec. Calculate the mean coefficient of permeability k for the soil deposit.

ANSWER: $k = 0.006$ cm/sec.

4.53   Assume the same conditions as Problem 4.52, except that after the standpipe is filled with water to ground surface, the water level is allowed to drop without adding any water to the standpipe. If the water level in the standpipe drops from ground surface to a depth of 0.9 m below ground surface in 60 sec, calculate the mean coefficient of permeability (k) of the soil deposit.

ANSWER: $k = 0.006$ cm/sec.

4.54   For the flow net shown in Fig. 4.30, what would happen to the quantity of water (Q) per unit time that enters the foundation pit if h_w were doubled.

ANSWER: A doubling of h_w (i.e.,Δh), doubles Q per Eq. 4.42.

4.55   At the middle of the flow net square labeled 15 in Fig. 4.30, calculate the pore water pressure u. Assume that $h_w = 10$ m and the middle of the flow net square labeled 15 in Fig. 4.30 is 13 m above the impervious stratum.

ANSWER: $u = 137$ kPa.

4.56   Assume that the sand strata shown in Fig. 4.30 has a total unit weight $\gamma_t = 19.8$ kN/m^3. Using the data from Problem 4.55, calculate the vertical effective stress σ_v' at the middle of the flow net square labeled 15 in Fig. 4.30.

ANSWER: $\sigma_v' = 101$ kPa.

4.57   Using the data from Problems 4.55 and 4.56, calculate the exit gradient i_e and the factor of safety for piping failure F for the flow net square labeled 18 in Fig. 4.30. To estimate the length L, scale from the distances shown on the right side of Fig. 4.30.

ANSWER: $i_e = 0.14$, factor of safety = 7.

4.58   Using the data from Problem 4.57, determine the seepage velocity v_s and direction of seepage at the flow net square labeled 18 in Fig. 4.30. Assume the coefficient of permeability $k = 0.1$ m/day.

ANSWER: $v_s = 0.037$ m/day in an approximately upward direction.

4.59   For the flow net shown in Fig. 4.30, where does the seepage velocity v_s have the highest value?

ANSWER: Since the equipotential drops are all equal, the highest seepage velocity occurs where the length of the flow net squares are the smallest, or at the sheet pile tip.

4.60   In Fig. 4.31, assume the height of water behind the earth dam $h_w = 20$ m, the soil comprising the earth dam has a coefficient of permeability $k = 1 \times 10^{-6}$ cm/sec, and the earth dam is 200 m long. Determine the quantity of water Q that will be collected by the drainage system each day. Assume water from the lowest flow channel also enters the drainage system.

ANSWER: $Q = 2.5$ m^3 of water per day.

4.61   Using the data from Problem 4.60, determine the number of equipotential drops and the head drop h' for each equipotential drop.

ANSWER: Number of equipotential head drops = 14 and $h' = 1.43$ m (Note: For the uppermost flow channel, one of the equipotential drops occurs in the drainage filter).

4.62   Based on the flow net in Fig. 4.31, at what location will the seepage velocity (v_s) be the highest?

ANSWER: Since the equipotential drops are all equal, the highest seepage velocity occurs where the length of the flow net squares are the smallest, or in the soil that is located in front of the longitudinal drainage filter.

4.63   Based on the flow net in Fig. 4.31, at what location will there most likely be piping of soil from the earth dam?

ANSWER: The soil located in front of the longitudinal drainage filter has the highest seepage velocity v_s and this is the most likely location for piping of soil into the drainage filter.

4.64   Assuming the same conditions as Problem 4.60, calculate the pore water pressure u at a point located at the centerline of the dam and 20 m below the dam top.

ANSWER: $u = 110$ kPa.

4.65   For Problem 4.64, assume $\gamma_t = 20.0$ kN/m^3 for the earth dam soil. Determine the vertical effective stress σ_v' at the same point referred to in Problem 4.64.

ANSWER: $\sigma_v' = 290$ kPa.

4.66   Assume the effective shear strength parameters are $\phi' = 28°$ and $c' = 2$ kPa for the soil comprising the earth dam. Calculate the shear strength τ_f on a horizontal plane at the point referred to in Problem 4.65.

ANSWER: $\tau_f = 156$ kPa.

4.67   A slope stability analysis is to be performed for the earth dam shown in Fig. 4.31. Assume the shear strength calculated in Problem 4.66 represents the average shear strength along the critical slip surface. If the average existing shear stress in the soil along the same slip surface = 83 kPa, calculate the factor of safety F for slope stability of the earth dam.

ANSWER: Factor of safety $= 1.88$.

FOUNDATION DESIGN

CHAPTER 5
SHALLOW AND DEEP FOUNDATIONS

5.1 INTRODUCTION

Part 2 of the book will deal with geotechnical engineering aspects of foundation design. Specific items covered in Part 2 include the following:

1. Bearing capacity of the foundation (Chap. 6)
2. Settlement of the foundation (Chap. 7)
3. Consolidation (Chap. 8)
4. Foundations on expansive soil (Chap. 9)
5. Slope stability (Chap. 10)
6. Retaining walls (Chap. 11)
7. Deterioration and moisture intrusion of foundations (Chap. 12)
8. Geotechnical earthquake engineering for soils and foundations (Chaps. 13 and 14)

The purpose of this chapter is to provide an introduction to the selection of a foundation type and a discussion of the characteristics and uses of shallow and deep foundations. It is important to recognize that without adequate and meaningful data from the subsurface exploration (Chap. 2) and laboratory testing (Chap. 3), the analyses presented in the next nine chapters will be of doubtful value and may even lead to erroneous conclusions.

5.2 SELECTION OF FOUNDATION TYPE

Table 1.1 has presented a brief introduction of shallow and deep foundations. This section deals with the selection of the type of foundation. The selection of a particular type of foundation is often based on a number of factors, such as:

1. *Adequate depth.* The foundation must have an adequate depth to prevent frost damage. For such foundations as bridge piers, the depth of the foundation must be sufficient to prevent undermining by scour.
2. *Bearing capacity failure.* The foundation must be safe against a bearing capacity failure.
3. *Settlement.* The foundation must not settle to such an extent that it damages the structure.
4. *Quality.* The foundation must be of adequate quality so that it is not subjected to deterioration, such as from sulfate attack.

5. *Adequate strength.*   The foundation must be designed with sufficient strength that it does not fracture or break apart under the applied superstructure loads. The foundation must also be properly constructed in conformance with the design specifications.

6. *Adverse soil changes.*   The foundation must be able to resist long-term adverse soil changes. An example is expansive soil, which could expand or shrink causing movement of the foundation and damage to the structure.

7. *Seismic forces.*   The foundation must be able to support the structure during an earthquake without excessive settlement or lateral movement.

Based on an analysis of all of the factors listed above, a specific type of foundation (i.e., shallow versus deep) would be recommended by the geotechnical engineer. The following sections discuss various types of shallow and deep foundations.

5.3   SHALLOW FOUNDATIONS

A shallow foundation is often selected when the structural load will not cause excessive settlement of the underlying soil layers. In general, shallow foundations are more economical to construct than deep foundations. Common types of shallow foundations are listed in Table 1.1 and described later:

5.3.1   Spread Footings, Combined Footings, and Strip Footings

These types of shallow foundations are probably the most common types of building foundations. Figure 5.1 shows various types of shallow foundations.

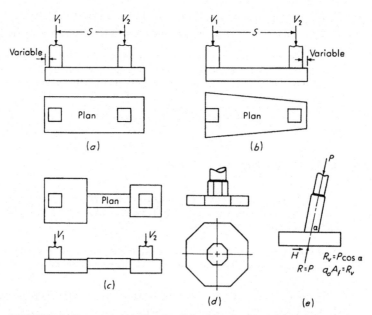

FIGURE 5.1   Examples of shallow foundations. (*a*) Combined footing; (*b*) combined trapezoidal footing; (*c*) cantilever or strap footing; (*d*) octagonal footing; (*e*) eccentric loaded footing with resultant coincident with area so soil pressure is uniform. (*Reproduced from Bowles, 1982; with permission of McGraw-Hill, Inc.*)

5.3.2 Mat Foundation

Figure 5.2 shows various types of mat foundations. Based on economic considerations, mat foundations are often constructed for the following reasons (NAVFAC DM-7.2, 1982):

1. *Large individual footings.*   A mat foundation is often constructed when the sum of individual footing areas exceeds about one-half of the total foundation area.

2. *Cavities or compressible lenses.*   A mat foundation can be used when the subsurface exploration indicates that there will be unequal settlement caused by small cavities or compressible lenses below the foundation. A mat foundation would tend to span over the small cavities or weak lenses and create a more uniform settlement condition.

3. *Shallow settlements.*   A mat foundation can be recommended when shallow settlements predominate and the mat foundation would minimize differential settlements.

4. *Unequal distribution of loads.*   For some structures, there can be a large difference in building loads acting on different areas of the foundation. Conventional spread footings could be subjected to excessive differential settlement, but a mat foundation would tend to distribute the unequal building loads and reduce the differential settlements.

5. *Hydrostatic uplift.*   When the foundation will be subjected to hydrostatic uplift due to a high groundwater table, a mat foundation could be used to resist the uplift forces.

5.3.3 Posttensioned Slab-on-Grade

Posttensioned slab-on-grade is common in southern California and other parts of the United States. The most common uses of posttensioned slab-on-grade are to resist expansive soil forces or when the projected differential settlement exceeds the tolerable value for a conventional (lightly reinforced)

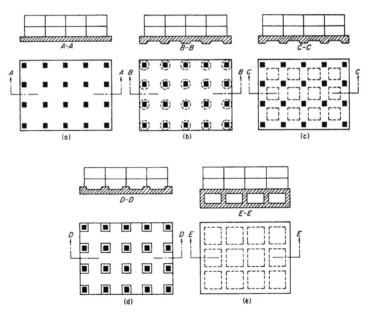

FIGURE 5.2   Examples of mat foundations. (*a*) Flat plate; (*b*) plate thickened under columns; (*c*) beam-and-slab; (*d*) plate with pedestals; (*e*) basement walls as part of mat. (*Reproduced from Bowles, 1982; with permission of McGraw-Hill, Inc.*)

slab-on-grade. For example, posttensioned slabs-on-grade are frequently recommended if the projected differential settlement is expected to exceed 0.75 in. (2 cm).

The Post-Tensioning Institute (1996) has prepared installation and field inspection procedures for posttensioned slab-on-grade. Posttensioned slab-on-grade consists of concrete with embedded steel tendons that are encased in thick plastic sheaths. The plastic sheath prevents the tendon from coming in contact with the concrete and permits the tendon to slide within the hardened concrete during the tensioning operations. Usually tendons have a dead end (anchoring plate) in the perimeter (edge) beam and a stressing end at the opposite perimeter beam to enable the tendons to be stressed from one end. However, it is often recommend that the tendons in excess of 100 ft (30 m) be stressed from both ends. For typical anchorage details for the tendons, see the Post-Tensioning Institute (1996).

Because posttensioned slab-on-grade perform better (i.e., less shrinkage related concrete cracking) than conventional slab-on-grade, they are more popular even for situations where low levels of settlement are expected. Posttensioned slab-on-grade has become common for situations where it is desirable to limit the amount and width of concrete shrinkage cracks.

5.3.4 Shallow Foundation Alternatives

If the expected settlement for a proposed shallow foundation is too large, then other options for foundation support or soil stabilization must be evaluated. Some commonly used alternatives are as follows:

1. *Grading.*   Grading operations can be used to remove the compressible soil layer and replace it with structural fill. Usually the grading option is only economical if the compressible soil layer is near the ground surface and the groundwater table is below the compressible soil layer or the groundwater table can be economically lowered. Grading will be discussed in Chap. 15.

2. *Surcharge.*   If the site contains an underlying compressible cohesive soil layer, the site can be surcharged with a fill layer placed at the ground surface. Vertical drains (such as wick drains or sand drains) can be installed in the compressible soil layer to reduce the drainage paths and speed-up the consolidation process. Once the compressible cohesive soil layer has had sufficient consolidation, the fill surcharge layer is removed and the building is constructed.

3. *Densification of soil.*   There are many different methods that can be used to densify loose or soft soil. For example, vibro-flotation and dynamic compaction are often effective at increasing the density of loose sand deposits. Another option is compaction grouting, which consists of intruding a mass of very thick consistency grout into the soil, which both displaces and compacts the loose soil. These soil improvement options will be discussed in Chap. 15.

4. *Floating foundation.*   A floating foundation is a special type of deep foundation where the weight of the structure is balanced by the removal of soil and construction of an underground basement.

5.4   DEEP FOUNDATIONS

5.4.1 Pile Foundations

Probably the most common type of deep foundation is the pile foundation. Piles are defined as relatively long, slender, column-like members. Piles are usually driven into specific arrangements and are used to support reinforced concrete pile caps or a mat foundation. For example, the building load from a steel column may be supported by a concrete pile cap that is in turn supported by four piles located near the corners of the concrete pile cap. Typical pile configurations are shown in Fig. 5.3.

Piles can consist of wood (timber), steel H-sections, precast concrete, cast-in-place concrete, pressure injected concrete, concrete filled steel pipe piles, and composite type piles. Examples of

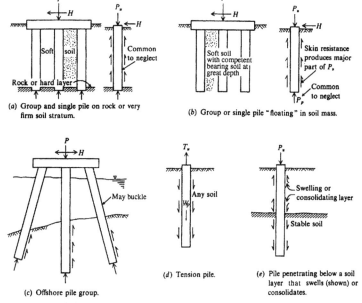

FIGURE 5.3   Typical pile configurations. (*Reproduced from Bowles, 1982; with permission of McGraw-Hill, Inc.*)

cast-in-place piles are shown in Fig. 5.4. Various types of prestressed piles are depicted in Fig. 5.5. A discussion of the typical pile characteristics and uses are presented in Table 5.1.

Piles are either driven into place or installed in predrilled holes. Piles that are driven into place are generally considered to be low displacement or high displacement depending on the amount of soil that must be pushed out of the way as the pile is driven. Examples of low displacement piles are steel H-sections and open-ended steel pipe piles that do not form a soil plug at the end. Examples of high-displacement piles are solid section piles, such as round timber piles or square precast concrete piles, and steel pipe piles with a closed end.

Various types of piles, in terms of their support capacity, are as follows:

- *End-bearing pile.*   A pile the support capacity of which is derived principally from the resistance of the foundation material on which the pile tip rests. End-bearing piles are often used when dense or hard strata underlie a soft upper layer. If the upper soft layer should settle, the pile could be subjected to downdrag forces, and the pile must be designed to resist these soil-induced forces (see Sec. 6.3.4).

- *Friction pile.*   A pile the support capacity of which is derived principally from the resistance of the soil friction and/or adhesion mobilized along the side of the pile. Friction piles are often used in soft clays where the end-bearing resistance is small because of punching shear at the pile tip. A pile that resists upward loads (i.e., tension forces) would also be considered to be a friction pile.

- *Combined end-bearing and friction pile.*   A pile that derives its support capacity from combined end-bearing resistance developed at the pile tip and frictional and/or adhesion resistance on the pile perimeter.

- *Batter pile.*   A pile driven in at an angle inclined to the vertical to provide high resistance to lateral loads.

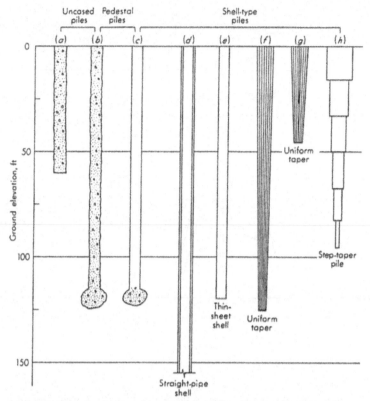

FIGURE 5.4 Common Types of Cast-in-Place Concrete Piles. (*a*) uncased pile; (*b*) Franki uncased-pedestal pile; (*c*) Franki cased-pedestal pile; (*d*) welded or seamless pipe pile; (*e*) cased pile using a thin sheet shell; (*f*) monotube pile; (*g*) uniform tapered pile; (*h*) step-tapered pile. (*Reproduced from Bowles, 1982; with permission of McGraw-Hill, Inc.*)

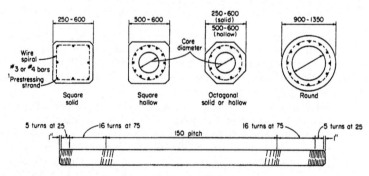

FIGURE 5.5 Typical prestressed concrete piles; dimensions in millimeters. (*Reproduced from Bowles, 1982; with permission of McGraw-Hill, Inc.*)

TABLE 5.1 Typical Pile Characteristics and Uses

Pile type	Timber	Steel	Cast-in-place concrete piles (shells driven without mandrel)	Cast-in-place concrete piles (shells withdrawn)
Maximum length	35 m	Practically unlimited	45 m	36 m
Optimum length	9–20 m	12–50 m	9–25 m	8–12 m
Applicable material specifications	ASTM-D25 for piles; PI-54 for quality of creosote; CI-60 for creosote treatment (standards of American Wood Preservers Assoc.)	ASTM-A36 for structural sections ASTM-A1 for rail sections	ACI	ACI[†]
Recommended maximum stresses	Measured at midpoint of length: 4–6 MPa for cedar, western hem lock, Norway pine, spruce, and depending on code 5–8 MPa for southern pine. Douglas fir, oak cypress, hickory	$f_s = 65\text{–}140$ MPa $f_s = 0.35\text{–}0.5 f_y$	$0.33 f'_c$; $0.4 f'_c$ if shell gage ≤ 14; shell stress $= 0.35 f_y$ if thickness of shell ≥ 3 mm	$0.25\text{–}0.33 f'_c$
Maximum load for usual conditions	270 kN	Maximum allowable stress × cross section	900 kN	1300 kN
Optimum-load range	130–225 kN	350–1050 kN	450–700 kN	350–900 kN
Disadvantages	Difficult to splice Vulnerable to damage in hard driving Vulnerable to decay unless treated, when piles are intermittently submerged	Vulnerable to corrosion HP section may be damaged or deflected by major obstructions	Hard to splice after concreting Considerable displacement	Concrete should be placed in dry hole More than average dependence on quality of workmanship

(*Continued*)

TABLE 5.1 Typical Pile Characteristics and Uses (*Continued*)

Pile type	Timber	Steel	Cast-in-place concrete piles (shells driven without mandrel)	Cast-in-place concrete piles (shells withdrawn)
Advantages	Comparatively low initial cost Permanently submerged piles are resistant to decay Easy to handle	Easy to splice High capacity Small displacement Able to penetrate through light obstructions	Can be redriven Shell not easily damaged	Initial economy
Remarks	Best suited for friction pile in granular material	Best suited for end bearing on rock Reduce allowable capacity for corrosive locations	Best suited for friction piles of medium length	Allowable load on pedestal pile is controlled by bearing capacity of stratum immediately below pile
Typical illustrations				

Notes: Stresses given for steel piles and shells are for noncorrosive locations. For corrosive locations estimate possible reduction in steel cross section or provide protection from corrosion.

Pile type	Concrete filled steel pipe piles	Composite piles	Precast concrete (including prestressed)	Cast in place (thin shell driven with mandrels)	Auger placed pressure-injected concrete (grout) piles
Maximum length	Practically unlimited	55 m	30 m for precast 60 m for prestressed	30 m for straight sections 12 m for tapered sections	9–25 m
Optimum length	12–36 m	18–36 m	12–15 m for precast 18–30 m for prestressed	12–18 m for straight 5–12 m for tapered	12–18 m
Applicable material specifications	ASTM A36 for core ASTM A252 for pipe ACI Code 318 for concrete	ACI Code 318 for concrete ASTM A36 for structural section ASTM A252 for steel pipe ASTM D25 for timber	ASTM A15 reinforcing steel ASTM A82 cold-drawn wire ACI Code 318 for concrete	ACI	See ACI†
Recommended maximum stresses	$0.40 f_y$ reinforcement < 205 MPa $0.50 f_y$ for core < 175 MPa $0.33 f'_c$ for concrete	Same as concrete in other piles Same as steel in other piles Same as timber piles for wood composite	$0.33 f'_c$ unless local building code is less; $0.4 f_y$ for reinforced unless prestressed	$0.33 f'_c$; $f_s = 0.4 f_y$ if shell gauge is ≤ 14; use $f_s = 0.35 f_y$ if shell thickness ≥ 3 mm	0.225–$0.40 f_c$
Maximum load for usual conditions	1800 kN without cores 18,000 kN for large sections with steel cores	1800 kN	8500 kN for prestressed 900 kN for precast	675 kN	700 kN
Optimum-load range	700–1100 kN without cores 4500–14,000 kN with cores	250–725 kN	350–3500 kN	250–550 kN	350–550 kN
Disadvantages	High initial cost Displacement for closed-end pipe	Difficult to attain good joint between two materials	Difficult to handle unless prestressed High initial cost Considerable displacement Prestressed difficult to splice	Difficult to splice after concreting Redriving not recommended Thin shell vulnerable during driving Considerable displacement	Dependence on workmanship Not suitable in compressible soil

(Continued)

5.11

TABLE 5.1 Typical Pile Characteristics and Uses (*Continued*)

Pile type	Concrete filled steel pipe piles	Composite piles	Precast concrete (including prestressed)	Cast in place (thin shell driven with mandrels)	Auger placed pressure-injected concrete (grout) piles
Advantages	Best control during installation No displacement for open-end installation Open-end pipe best against obstructions High load capacities Easy to splice	Considerable length can be provided at comparatively low cost	High load capacities Corrosion resistance can be attained Hard driving possible	Initial economy Taped sections provide higher bearing resistance in granular stratum	Freedom from noise and vibration Economy High skin friction No splicing
Remarks	Provides high bending resistance where unsupported length is loaded laterally	The weakest of any material used shall govern allowable stresses and capacity	Cylinder piles in particular are suited for bending resistance	Best suited for medium-load friction piles in granular materials	Patented method
Typical illustrations					

†ACI Committee 543: "Recommendations for Design, Manufacture, and Installation of Concrete Piles," *JACI*, August 1973, October 1974.
Sources: NAVFAC DM-7.2, 1982 and Bowles (1982).

An important consideration is to design and construct the foundation so that it can span unsupported between the piles if the underlying soil is expected to settle. Figure 5.6 shows an example of floor slab settlement due to consolidation and secondary compression of peat at the "Meadowlands," which is a marshy area in New Jersey. Piles have been used to support bearing walls, but as shown in Fig. 5.6, there is often cracking and deformation of the floor slabs around the piles (Whitlock and Moosa, 1996). The reason for the settlement and damage to the floor slab shown in Fig. 5.6 is because the floor slab was not designed to span unsupported between the piles. The construction of

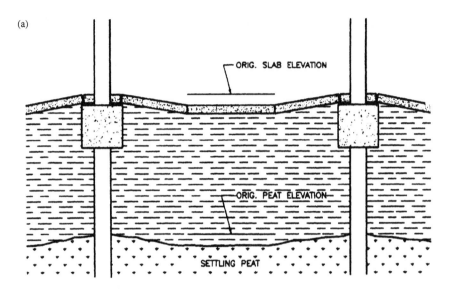

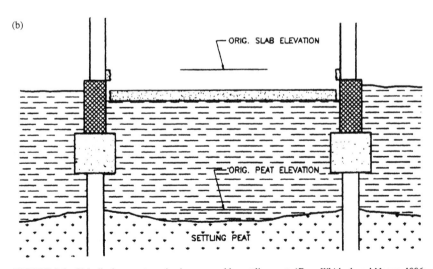

FIGURE 5.6 Slab displacement mechanisms caused by settling peat. (*From Whitlock and Moosa, 1996; reprinted with permission of the American Society of Civil Engineers*).

a structural floor slab that can transfer loads to the piles is an important design feature for sites having settling compressible soil, such as the peat layer shown in Fig. 5.6.

5.4.2 Pier Foundations

A pier is defined as a deep foundation system, similar to a cast-in-place pile that consists of a column-like reinforced concrete member. Piers are often of large enough diameter to enable down-hole inspection. Piers are also commonly referred to as drilled shafts, bored piles, or drilled caissons. Figure 5.7 shows the typical steps in the construction of a drilled pier.

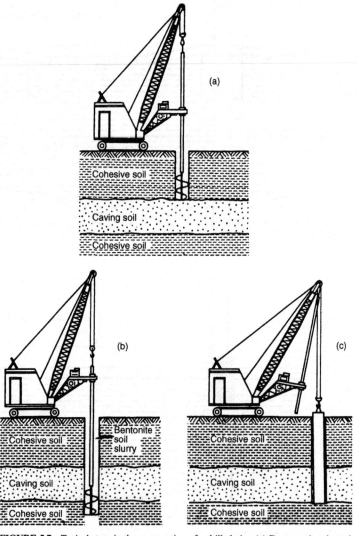

FIGURE 5.7   Typical steps in the construction of a drilled pier. (*a*) Dry augering through self-supporting cohesive soil; (*b*) augering through water-bearing cohesionless soil with aid of slurry; (*c*) setting the casing; (*d*) dry augering into cohesive soil after sealing; (*e*) forming a bell. (*After O'Neill and Reese 1970; reproduced from Peck, Hanson, and Thornburn 1974.*)

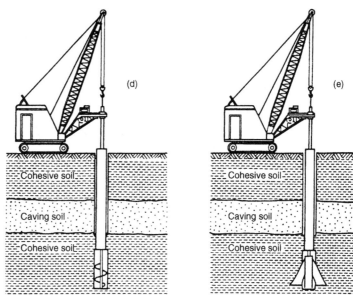

FIGURE 5.7 *(Continued)*

5.4.3 Other Types of Deep Foundation Elements

There are many other methods available for forming deep foundation elements. Examples include earth stabilization columns, such as (NAVFAC DM-7.2, 1982):

Mixed-in-place piles.   A mixed-in-place soil-cement or soil-lime pile. The pile is created by forcing a grout through a hollow shaft in the ground. As the grout is forced into the soil, an auger-like head (that is attached to the hollow shaft) mixes the soil to create the mixed-in-place pile.

Vibro-replacement stone columns.   Vibro-flotation or other method is used to make a cylindrical, vertical hole that is filled with compacted gravel or crushed rock.

Grouted stone columns.   This is similar to the above but includes filling voids with bentonite-cement or water-sand-bentonite cement mixtures.

Concrete vibro columns.   Similar to stone columns, but concrete is used instead of gravel.

Another type of deep foundation element is the helical anchor. The principle behind a helical anchor is the same as a corkscrew, except that the helical anchor has fewer blades. The advantages of helical anchors are that they are usually lower cost then other types of deep foundations, they are quick and easy to install, and because they are rotated into place, they tend to produce much less vibrations than driven piles, which could be important if there are nearby settlement sensitive structures.

Helical anchors can be used for new construction, where the helical anchors are attached to the footings or grade beams. It is usually desirable that heavy vertical building loads coincide with the location of the helical anchors. In addition, the tops of the helical anchors should be securely attached to the grade beams or footings. Embedding the tops of the helical anchors within the grade beams and wrapping the steel in the grade beam around the top of the helical anchor can accomplish this. With settling soil, the grade beams, which support a structural slab, can be designed to span unsupported between the helical anchors.

Besides supporting bearing loads for new construction, helical anchors can also be used to resist uplift loads (i.e., tie-down anchors), for foundation underpinning, and as tieback anchors for retaining walls. Figure 5.8 shows an example of helical anchors being used for foundation underpinning. Helical anchors will be further discussed in Sec. 6.3.4.

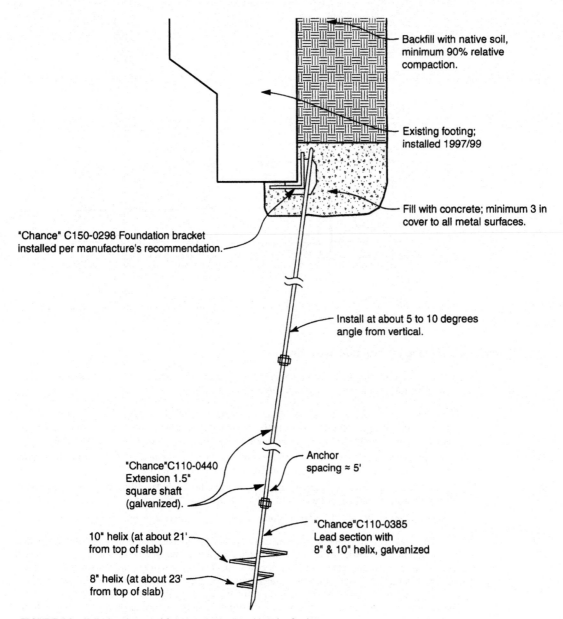

Backfill with native soil,
minimum 90% relative
compaction.

Existing footing;
installed 1997/99

Fill with concrete; minimum 3 in
cover to all metal surfaces.

"Chance" C150-0298 Foundation bracket
installed per manufacture's recommendation.

Install at about 5 to 10 degrees
angle from vertical.

Anchor
spacing ≈ 5'

"Chance"C110-0440
Extension 1.5"
square shaft
(galvanized).

"Chance"C110-0385
Lead section with
8" & 10" helix, galvanized

10" helix (at about 21'
from top of slab)

8" helix (at about 23'
from top of slab)

FIGURE 5.8   Helical anchors used for the underpinning of exterior footings.

PROBLEMS

5.1 Assume that the subsoil conditions at a site are as shown in Fig. 2.38. It is proposed to build a state highway bridge (essential facility) at this location. It is anticipated that during major flood conditions, there could be scour that could remove sediments down to an elevation of –10 m. What type of bridge foundation should be installed for the highway bridge?

ANSWER: A deep foundation system consisting of piles or piers embedded in the sandstone.

5.2 Assume that the subsoil conditions at a site are as shown in Fig. 2.39. It is proposed to construct a two-story research facility that will also contain an underground basement to be used as the laboratory. Based on these requirements, what type of foundation should be installed for the research facility?

ANSWER: Assuming the weight of soil excavated for the basement is approximately equal to the weight of the two-story structure, a floating foundation would be desirable.

5.3 Assume that the subsoil conditions at a site are as shown in Fig. 2.40. It is proposed to construct single-family detached housing at this site. What type of foundation should be used?

ANSWER: Because the upper 10 ft of the site consists of overconsolidated clay, it would be desirable to use a shallow foundation system based on the assumption of light building loads.

5.4 Assume that the subsoil conditions at a site are as shown in Fig. 2.41. The structural engineer is recommending that high displacement piles be driven to a depth of 50 ft in the clay deposit. Are these types of pile appropriate?

ANSWER: No, because of the very high sensitivity of the clay, high-displacement piles will remold the clay and result in a loss of shear strength. The preferred option is to install low-displacement piles or use predrilled, cast-in-place concrete piles.

CHAPTER 6
BEARING CAPACITY
OF FOUNDATIONS

6.1 INTRODUCTION

A bearing capacity failure is defined as a foundation failure that occurs when the shear stresses in the soil exceed the shear strength of the soil. Bearing capacity failures of foundations can be grouped into three categories, as follows (Vesic, 1963, 1975):

1. *General shear.* As shown in Fig. 6.1, a general shear failure involves total rupture of the underlying soil. There is a continuous shear failure of the soil (solid lines) from below the footing to the ground surface. When the load is plotted versus settlement of the footing, there is a distinct load at which the foundation fails (solid circle), and this is designated Q_{ult}. The value of Q_{ult} divided by the width B and length L of the footing is considered to be the *ultimate bearing capacity* (q_{ult}) of the footing. The ultimate bearing capacity has been defined as the bearing stress that causes a sudden catastrophic failure of the foundation (Lambe and Whitman, 1969).

As shown in Fig. 6.1, a general shear failure ruptures and pushes up the soil on both sides of the footing. For actual failures in the field, the soil is often pushed up on only one side of the footing with subsequent tilting of the structure. A general shear failure occurs for soils that are in a dense or hard state.

2. *Local shear failure.* As shown in Fig. 6.2, local shear failure involves rupture of the soil only immediately below the footing. There is soil bulging on both sides of the footing, but the bulging is not as significant as in general shear. Local shear failure can be considered as a transitional phase between general shear and punching shear. Because of the transitional nature of local shear failure, the bearing capacity could be defined as the first major nonlinearity in the load-settlement curve (open circle) or at the point where the settlement rapidly increases (solid circle). A local shear failure occurs for soils that are in a medium dense or firm state.

3. *Punching shear.* As shown in Fig. 6.3, a punching shear failure does not develop the distinct shear surfaces associated with a general shear failure. For punching shear, the soil outside the loaded area remains relatively uninvolved and there is minimal movement of soil on both sides of the footing. The process of deformation of the footing involves compression of soil directly below the footing as well as the vertical shearing of soil around the footing perimeter. As shown in Fig. 6.3, the load settlement curve does not have a dramatic break and for punching shear, the bearing capacity is often defined as the first major nonlinearity in the load-settlement curve (open circle). A punching shear failure occurs for soils that are in a loose or soft state.

Table 6.1 presents a summary of the type of bearing capacity failure that would most likely develop based on soil type and soil properties. As compared to the number of structures damaged by settlement, there are far fewer structures that have bearing capacity failures. This is because of the following factors:

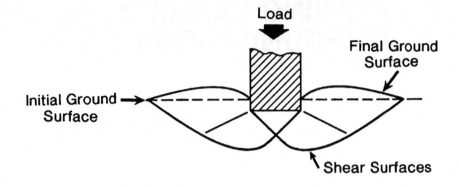

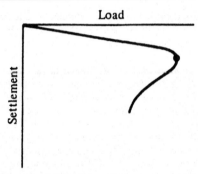

FIGURE 6.1 General shear foundation failure. (*After Vesic, 1963.*)

1. *Settlement governs.* The foundation design is based on several requirements (see Sec. 5.2) and two of the main considerations are: (1) settlement due to the building loads must not exceed tolerable values, and (2) there must be an adequate factor of safety against a bearing capacity failure. In most cases, settlement governs and the foundation bearing pressures recommended by the geotechnical engineer are based on limiting the amount of settlement.

2. *Extensive studies.* There have been extensive studies of bearing capacity failures, which have led to the development of bearing capacity equations that are routinely used in practice to determine the ultimate bearing capacity of the foundation.

3. *Factor of safety.* In order to determine the allowable bearing pressure q_{all}, the ultimate bearing capacity q_{ult} is divided by a factor of safety. The normal factor of safety used for bearing capacity analyses is 3. This is a high factor of safety as compared to other factors of safety, such as only 1.5 for slope stability analyses.

4. *Minimum footing sizes.* Building codes often require minimum footing sizes and embedment depths (e.g., see Chap. 18).

5. *Allowable bearing pressures.* In addition, building codes often have maximum allowable bearing pressures for different soil and rock conditions (e.g., see Table 18.4). Especially in the case of dense or stiff soils, these allowable bearing pressures often have adequate factors of safety.

6. *Footing dimensions.* Usually the structural engineer will determine the size of the footings by dividing the maximum footing load (dead load plus live load) by the allowable bearing pressure. Typically the structural engineer uses values of dead and live loads that also contain factors of

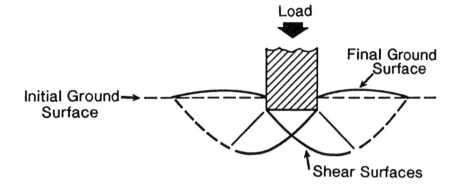

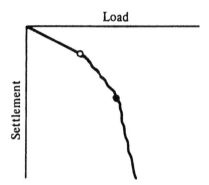

FIGURE 6.2   Local shear foundation failure. *(After Vesic, 1963.)*

safety. For example, the live load may be from the local building code, which specifies minimum live load requirements for specific building uses. Such building code values often contain a factor of safety, which is in addition to the factor of safety of 3 that was used to determine the allowable bearing pressure.

Because the bearing capacity failure involves a shear failure of the underlying soil (Figs. 6.1 to 6.3), the analysis will naturally include the shear strength of the soil. As indicated in Figs. 6.1 to 6.3, the depth of the bearing capacity failure is rather shallow. It is often assumed that the soil involved in the bearing capacity failure can extend to a depth equal to B (footing width) below the bottom of the footing. Thus for bearing capacity analysis, this zone of soil should be evaluated for its shear strength properties.

The documented cases of bearing capacity failures indicate that usually the following three factors (separately or in combination) are the cause of the failure: (1) there was an overestimation of the shear strength of the underlying soil, (2) the actual structural load at the time of the bearing capacity failure was greater than that assumed during the design phase, or (3) the site was altered, such as the construction of an adjacent excavation, which resulted in a reduction in support and a bearing capacity failure. Bearing capacity failures also can occur during earthquakes and these are discussed in Sec. 6.5.

A famous case of a bearing capacity failure is the Transcona Grain-Elevator, located at Transcona, near Winnipeg, Canada. Figure 6.4 shows the October 1913 failure of the grain elevator. At the time of failure, the grain elevator was essentially fully loaded. The foundation had been constructed on clay, which was described as stiff clay. As shown in Fig. 6.4, the soil has been pushed up on only one side of the foundation with subsequent tilting of the structure.

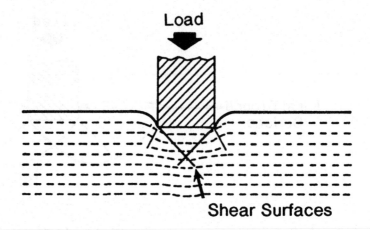

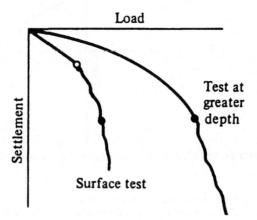

FIGURE 6.3 Punching shear foundation failure. (*After Vesic, 1963.*)

TABLE 6.1 Summary of Type of Bearing Capacity Failure versus Soil Properties

Type of bearing capacity failure	Cohesionless soil (e.g., sands)			Cohesive soil (e.g., clays)	
	Density condition	Relative density (D_r)	$(N_1)_{60}$	Consistency	Undrained shear strength (s_u)
General shear failure (Fig. 6.1)	Dense to very dense	65–100%	> 20	Very stiff to hard	> 2000 psf >100 kPa
Local shear failure (Fig. 6.2)	Medium	35–65%	5–20	Medium to stiff	500–2000 psf 25–100 kPa
Punching shear failure (Fig. 6.3)	Loose to very loose	0–35%	< 5	Soft to very soft	< 500 psf < 25 kPa

FIGURE 6.4   Transcona grain elevator failure.

6.2   BEARING CAPACITY FOR SHALLOW FOUNDATIONS

As indicated in Table 1.1, common types of shallow foundation include spread footings for isolated columns, combined footings for supporting the load from more than one structural unit, strip footings for walls, and mats or raft foundations constructed at or near ground surface. Shallow footings often have an embedment that is less than the footing width.

6.2.1   Bearing Capacity Equation

The most commonly used bearing capacity equation is that equation developed by Terzaghi (1943). For a uniform vertical loading of a strip footing, Terzaghi (1943) assumed a general shear failure (Fig. 6.1) in order to develop the following bearing capacity equation:

$$q_{ult} = \frac{Q_{ult}}{BL} = cN_c + \tfrac{1}{2}\gamma_t\, BN_\gamma + \gamma_t\, D_f N_q \qquad (6.1)$$

where q_{ult} = ultimate bearing capacity for a strip footing (psf or kPa)
Q_{ult} = vertical concentric load causing a general shear failure of the underlying soil (Fig. 6.1), lb or kN
B = width of the strip footing (ft or m)
L = length of the strip footing (ft or m)
γ_t = total unit weight of the soil (pcf or kN/m³)
D_f = vertical distance from ground surface to bottom of strip footing (ft or m)
c = cohesion of the soil underlying the strip footing (psf or kPa)
N_c, N_γ, and N_q = dimensionless bearing capacity factors

As indicated in Eq. 6.1, there are three terms that are added together to obtain the ultimate bearing capacity of the strip footing. These terms represent the following:

$c\,N_c$ The first term accounts for the cohesive shear strength of the soil located below the strip footing. If the soil below the footing is cohesionless (i.e., $c = 0$), then this term is zero.

$\tfrac{1}{2}\,\gamma_t\,BN_\gamma$ This second term accounts for the frictional shear strength of the soil located below the strip footing. The friction angle ϕ is not included in this term, but is accounted for by the bearing capacity factor N_γ. Note that γ_t represents the total unit weight of the soil located below the footing.

$\gamma_t\,D_f N_q$ This third term accounts for the soil located above the bottom of the footing. The value of γ_t times D_f represents a surcharge pressure that helps to increase the bearing capacity of the footing. If the footing were constructed at ground surface (i.e., $D_f = 0$), then this term would equal zero. This term indicates that the deeper the footing, the greater the ultimate bearing capacity of the footing. In this term, γ_t represents the total unit weight of the soil located above the bottom of the footing. The total unit weight above and below the footing bottom may be different, in which case different values are used in the second and third terms of Eq. 6.1.

In order to calculate the allowable bearing pressure q_{all}, which is used to determine the size of the footings, the following equation is used:

$$q_{\text{all}} = \frac{q_{\text{ult}}}{F} \tag{6.2}$$

where q_{all} = allowable bearing pressure, also known as the allowable bearing capacity (psf or kPa)
  q_{ult} = ultimate bearing capacity from Eq. 6.1 (psf or kPa)
  F = factor of safety, where the commonly used factor of safety is equal to 3

There are many charts, graphs, and figures that present bearing capacity factors (N_c, N_γ, and N_q) developed by engineers and researchers based on varying assumptions (Terzaghi, 1943; Meyerhof, 1951, 1963; Peck, Hanson, and Thornburn, 1974; Vesic, 1975; Myslivec and Kysela, 1978). Some bearing capacity charts have presented bearing capacity factors up to five significant figures. These bearing capacity factors imply an accuracy that simply does not exist because the bearing capacity equation is only an approximation of the actual bearing failure. Typically two significant figures are more than adequate for the dimensionless bearing capacity factors.

Terzaghi (1943) originally developed the bearing capacity equation (Eq. 6.1) for a general shear bearing capacity failure. This type of bearing capacity failure is shown in Fig. 6.1 and will develop for dense or stiff soil. For loose or soft soil, there will be a punching shear failure as shown in Fig. 6.3. The bearing capacity factors must be reduced to account for a punching shear failure. Two commonly used charts that present bearing capacity factors and account for punching shear failure are as follows:

1. *Cohesionless soil.*   Figure 6.5 presents bearing capacity factors (N_γ and N_q) that automatically incorporate allowance for punching and local shear failure of cohesionless soils. Simply enter the figure with the friction angle of the cohesionless soil (ϕ) to obtain N_γ and N_q. Another option is to enter the chart with the standard penetration test (SPT) N_{60} value in order to obtain N_γ and N_q.

2. *All soil types.*   Figure 6.6 presents bearing capacity factors (N_c, N_γ, and N_q) that do not include allowance for punching shear failure. If the soil is loose enough or soft enough that the footing will punch into the soil, then adjust the value of ϕ and c as indicted in the figure title.

In Figs. 6.5 and 6.6, the bearing capacity factors rapidly increase at high friction angles (ϕ). These bearing capacity factors should be used with caution, because natural soils are not homogeneous and the natural variability of such soil will result in weaker layers that will be exploited during a bearing capacity failure.

Example Problem 6.1   A proposed strip footing that is 4 ft (1.2 m) wide will be located 2 ft (0.6 m) below ground surface. The soil type is uniform dense sand that has a friction angle ϕ of 35°. The total unit weight of the soil is equal to 125 pcf (19.7 kN/m³). The groundwater table is well below the bottom of the footing and will not be a factor in the bearing capacity analysis. Using a factor of safety of 3, calculate the allowable bearing pressure using Figs. 6.5 and 6.6.

Solution   From Fig. 6.5, for $\phi = 35°$, $N_\gamma = 37$ and $N_q = 33$. Using Eq. 6.1 with $c = 0$:

$$q_{ult} = \frac{1}{2}\gamma_t\, BN_\gamma + \gamma_t\, D_f N_q = \frac{1}{2}\,(125)(4)(37) + (125)(2)(33) = 17,500 \text{ psf } (840 \text{ kPa})$$

Using $F = 3$, $q_{all} = \dfrac{q_{ult}}{3} = \dfrac{17,500}{3} = 5800 \text{ psf} (280 \text{ kPa})$

From Fig. 6.6, for $\phi = 35°$, $N_\gamma = 40$ and $N_q = 36$. Using Eq. 6.1 with $c = 0$:

$$q_{ult} = \frac{1}{2}\gamma_t\, BN_\gamma + \gamma_t D_f N_q = \frac{1}{2}\,(125)(4)(40) + (125)(2)(36) = 19,000 \text{ psf } (910 \text{ kPa})$$

Using $F = 3$, $q_{all} = \dfrac{q_{ult}}{3} = \dfrac{19,000}{3} = 6300 \text{ psf} (300 \text{ kPa})$

For this example problem, the allowable bearing capacity is about 6000 psf (290 kPa).

Example Problem 6.2   Use the same data as the above example problem, but assume silty sand with a friction angle ϕ of 30°. Also assume the footing will not punch into the soil.

Solution   From Fig. 6.5, for $\phi = 30°$, $N_\gamma = 15$ and $N_q = 19$. Using Eq. 6.1 with $c = 0$:

$$q_{ult} = \frac{1}{2}\gamma_t BN_\gamma + \gamma_t D_f N_q = \frac{1}{2}\,(125)(4)(15) + (125)(2)(19) = 8500 \text{ psf } (400 \text{ kPa})$$

Using $F = 3$, $q_{all} = \dfrac{q_{ult}}{3} = \dfrac{8500}{3} = 2800 \text{ psf} (130 \text{ kPa})$

From Fig. 6.6, for $\phi = 30°$, $N_\gamma = 17$ and $N_q = 20$. Using Eq. 6.1 with $c = 0$:

$$q_{ult} = \frac{1}{2}\gamma_t BN_\gamma + \gamma_t\, D_f N_q = \frac{1}{2}\,(125)(4)(17) + (125)(2)(20) = 9250 \text{ psf } (440 \text{ kPa})$$

Using $F = 3$, $q_{all} = \dfrac{q_{ult}}{3} = \dfrac{9250}{3} = 3100 \text{ psf} (150 \text{ kPa})$

For this example problem, the allowable bearing capacity is about 3000 psf (140 kPa).

In addition to Eq. 6.1 for strip footings, Terzaghi also developed equations for the general shear failure of square and circular footings, as follows:

$$\text{Square footings: } q_{ult} = 1.3cN_c + 0.4\gamma_t BN_\gamma + \gamma_t D_f N_q \tag{6.3}$$

$$\text{Circular footings: } q_{ult} = 1.3cN_c + 0.3\gamma_t\, BN_\gamma + \gamma_t D_f N_q \tag{6.4}$$

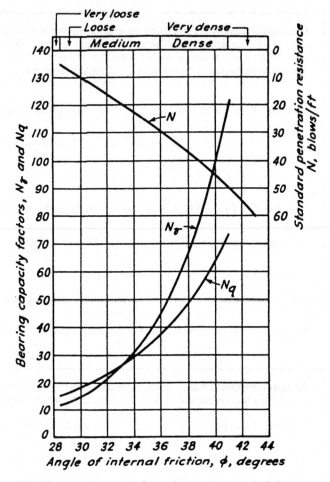

FIGURE 6.5 Bearing capacity factors N_γ and N_q that automatically incorporate allowance for punching and local shear failure. *Note:* For analysis using the standard penetration test, assume N refers to the N_{60} value from Eq. 2.4. (*From Peck, Hanson, and Thornburn, 1974, reproduced with permission of John Wiley & Sons.*)

In essence, Eq. 6.1 has been slightly altered to account for footing shape factors. The Terzaghi bearing capacity equation is often presented in a form that includes shape factors, as follows:

$$q_{ult} = s_c c N_c + \tfrac{1}{2} s_\gamma \gamma_t B N_\gamma + \gamma_t s_q D_f N_q \tag{6.5}$$

where s_c, s_γ, s_q are the shape factors (dimensionless).

For a strip footing, all of the shape factors equal 1.0, and Eq. 6.5 is the same as Eq. 6.1. For square footings, the Terzaghi shape factors are $s_c = 1.3$, $s_\gamma = 0.8$, and $s_q = 1.0$, in which case Eq. 6.5 is the same as Eq. 6.3. For rectangular footings of width B and length L, commonly used shape factors are:

$$s_c = [1 + 0.3(B/L)]$$

$$s_\gamma = 0.8$$

$$s_q = 1.0$$

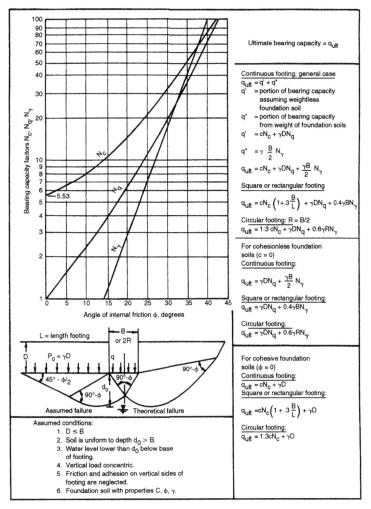

FIGURE 6.6 Bearing capacity factors N_γ, N_q, and N_c that do not include allowance for punching shear failure. [*Note:* For punching shear of loose sands or soft clays, the value of ϕ to be used in this figure = $\tan^{-1}(0.67\tan\phi)$ and the cohesion used in the bearing capacity equation = 0.67 c]. (*Reproduced from NAVFAC DM-7.2, 1982.*)

or for rectangular footings, Eq. 6.5 becomes

$$q_{\text{ult}} = cN_c\,[1 + 0.3\,(B/L)] + 0.4\gamma_t BN_\gamma + \gamma_t D_f N_q \qquad (6.6)$$

Similar to shape factors, various researchers have proposed factors for load inclination, depth, and footing inclination (e.g., Meyerhof, 1951, 1953, 1963, 1965; Hansen, 1961). In addition, other bearing capacity equations or alterations to the original Terzaghi bearing capacity equation have been proposed. However, the extra effort involved in using these equations is usually not justified because of the large factor of safety applied to the ultimate result and the fact that settlement, not bearing capacity, often dictates the allowable bearing pressure.

An important consideration is that for the strip footing, the shear strength is actually based on a plane strain condition (i.e., the soil is confined along the long axis of the footing). It has been stated that the friction angle ϕ is about 10 percent higher in the plane strain condition as compared to the friction angle ϕ measured in the triaxial apparatus (Meyerhof, 1961; Perloff and Baron, 1976). Ladd et al. (1977) indicated that the friction angle ϕ in plane strain is larger than ϕ in triaxial shear by 4° to 9° for dense sands. A difference in friction angle of 4° to 9° has a significant impact on the bearing capacity factors. In practice, plane strain shear strength tests are not performed and thus there is an added factor of safety for the strip footing as compared to the analysis for spread or combined footings.

6.2.2 Bearing Capacity for Cohesionless Soil

As discussed in Sec. 3.4, cohesionless soil is nonplastic and includes gravels, sands, and nonplastic silt such as rock flour. A cohesionless soil develops its shear strength as a result of the frictional and interlocking resistance between the soil particles. For cohesionless soil, $c = 0$ and the first term in Eqs. 6.1 and 6.5 is zero.

For cohesionless soil, the location of the groundwater table can affect the ultimate bearing capacity. As mentioned in Sec. 3.4, the saturation of sand usually does not have much effect on the friction angle (i.e., $\phi' \cong \phi$). But a groundwater table creates a buoyant condition in the soil which results in less resistance of the soil and hence a lower ultimate bearing capacity. The depth of the bearing capacity failure is rather shallow (Fig. 6.1 to 6.3) and it is often assumed that the soil involved in the bearing capacity failure extends to a depth equal to B (footing width) below the bottom of the footing. Thus for a groundwater table located in this zone, an adjustment to the ultimate bearing capacity is performed by adjusting the unit weight of the second term of Eqs. 6.1 and 6.5 by using the following equation (Myslivec and Kysela, 1978):

$$\gamma_a = \gamma_b + \frac{h' - D_f}{B}(\gamma_t - \gamma_b) \tag{6.7}$$

where γ_a = adjusted unit weight that is used in place of γ_t for the second term in Eqs. 6.1 and 6.5 (pcf or kN/m³)

γ_b = buoyant unit weight of the soil (pcf or kN/m³)
γ_t = total unit weight of the soil (pcf or kN/m³)
h' = depth of the groundwater table below ground surface (ft or m)
D_f = depth from ground surface to the bottom of the footing (ft or m)
B = width of the footing (ft or m)

Equation 6.7 is valid for $0 \le (h' - D_f) \le B$

A correction to the unit weight of the soil is not required if the groundwater table is at a depth h' that is equal to or greater than $D_f + B$. The following example illustrates the use of the bearing capacity equation with the groundwater correction.

Example Problem 6.3   A strip footing will be constructed on a nonplastic silty sand deposit that has the shear strength properties as shown in Fig. 3.18 (i.e., $c' = 0$ and $\phi' = 30°$) and a saturated unit weight of 125 pcf (19.7 kN/m³). The proposed strip footing will be 4 ft (1.2 m) wide and embedded 2 ft (0.6 m) below the ground surface. Use a factor of safety of 3 and use Fig. 6.5 for the bearing capacity factors. Assume the groundwater table is located 4 ft (1.2 m) below ground surface. Determine the allowable bearing pressure q_{all} and the maximum vertical concentric load the strip footing can support for the nonplastic silty sand.

(Continued)

Solution

$$\gamma_b = \gamma_{sat} - \gamma_w = 125 - 62.4 = 62.6 \text{ pcf } (9.89 \text{ kN/m}^3)$$

$$\gamma_a = \gamma_b + [(h' - D_f)/B](\gamma_t - \gamma_b)$$

$$= 62.6 + [(4 - 2)/4](125 - 62.6) = 93.9 \text{ pcf } (14.8 \text{ kN/m}^3)$$

From Fig. 6.5, for $\phi = 30°$, $N_\gamma = 15$ and $N_q = 19$. Using Eq. 6.1 with $c' = 0$:

$$q_{ult} = \tfrac{1}{2}\gamma_a B N_\gamma + \gamma_t D_f N_q = \tfrac{1}{2}(93.9)(4)(15) + (125)(2)(19) = 7600 \text{ psf } (360 \text{ kPa})$$

Using $F = 3$, $q_{all} = q_{ult}/3 = 7600/3 = 2500 \text{ psf } (120 \text{ kPa})$

This 4 ft (1.2 m) wide strip footing could carry a maximum vertical load Q_{all} of 10 kips per linear foot (140 kN per linear meter).

6.2.3 Bearing Capacity for Cohesive Soil

The bearing capacity of cohesive (plastic) soil, such as silts and clays, is more complicated than the bearing capacity of cohesionless (nonplastic) soil. In the southwestern United States, surface deposits of clay are common and can have a hard and rocklike appearance when they become dried-out during the summer. Instead of being susceptible to bearing capacity failure, desiccated clays can cause heave (upward movement) of lightly loaded foundations when the clays get wet during the rainy season. Expansion of clay will be discussed in Chap. 9.

Plastic saturated soils (silts and clays) usually have a lower shear strength than nonplastic cohesionless soil, and are more susceptible to bearing capacity failure such as the Transcona Grain-Elevator shown in Fig. 6.4. For saturated plastic soils, the bearing capacity often has to be calculated for two different conditions:

1. Total stress analyses (short-term condition) that use the undrained shear strength of the plastic soil

2. Effective stress analyses (long-term condition) that use the drained shear strength parameters (c' and ϕ') of the plastic soil

Total Stress Analysis.   The total stress analysis uses the undrained shear strength of the plastic soil. The undrained shear strength s_u could be determined from field tests, such as the vane shear test (VST), or in the laboratory from unconfined compression tests. If the undrained shear strength is approximately constant with depth, then $s_u = c$ and $\phi = 0$ for Eqs. 6.1 and 6.5.

If unconsolidated undrained triaxial compression tests (ASTM D 2850-03, 2004) are performed, then an envelope similar to Fig. 3.35 should be obtained for the saturated plastic soil and the $\phi = 0$ concept should be utilized (Sec. 3.5.4). For example, as shown in Fig. 3.35, the undrained shear strength parameters are $c = 2.1 \text{ psi } (14.5 \text{ kPa})$ and $\phi = 0$.

In Fig. 6.6, for $\phi = 0$, the bearing capacity factors are $N_c = 5.5$, $N_\gamma = 0$, and $N_q = 1$. For both of these cases of an undrained shear strength s_u and shear strength from unconsolidated undrained triaxial compression tests on saturated cohesive soil ($\phi = 0$ concept), the bearing capacity Eqs. 6.1 and 6.6 reduce to the following:

For strip footings:

$$q_{ult} = 5.5c + \gamma_t D_f = 5.5s_u + \gamma_t D_f \tag{6.8}$$

For spread footings:

$$q_{ult} = 5.5c[1 + 0.3\,(B/L)] + \gamma_t D_f = 5.5s_u\,[1 + 0.3\,(B/L)] + \gamma_t D_f \tag{6.9}$$

Because of the use of total stress parameters, the groundwater table does not affect Eqs. 6.8 and 6.9. An example of the use of the bearing capacity equation for cohesive soil is presented below:

Example Problem 6.4   Assume the same conditions as the prior example, except that the strip footing will be constructed in a clay deposit having the shear strength parameters shown in Fig. 3.35. Neglect possible punching shear and perform a total stress analysis (i.e., short-term condition).

Solution   For the strip footing, use Eq. 6.8 where $c = 2.1$ psi (300 psf, 14.5 kPa).

$$q_{ult} = 5.5c + \gamma_t D_f = (5.5)(300) + (125)(2) = 1900 \text{ psf (91 kPa)}$$

Using $F = 3$, $q_{all} = q_{ult}/3 = 1900/3 = 630$ psf (30 kPa)

This 4 ft (1.2 m) wide strip footing could carry a maximum concentric vertical load Q_{all} of 2.5 kips per liner foot (36 kN per linear meter).

As indicated by the above examples, the ultimate bearing capacity of plastic soil is often much less than the ultimate bearing capacity of cohesionless soil. This is the reason that building codes (such as Table 18.4) allow higher allowable bearing pressures for cohesionless soil (such as sand) than plastic soil (such as clay). Also, because the ultimate bearing capacity does not increase with footing width for saturated plastic soils, there is often no increase allowed for an increase in footing width.

In some cases, it may be appropriate to use total stress parameters c and ϕ in order to calculate the ultimate bearing capacity. For example, a structure (such as an oil tank or grain-elevator) could be constructed and then sufficient time elapses so that the saturated plastic soil consolidates under this load. If an oil tank or grain elevator were then quickly filled, the saturated plastic soil would be subjected to an undrained loading. This condition can be modeled by performing consolidated undrained triaxial compression tests (ASTM D 4767-02, 2004) in order to determine the total stress parameters (c and ϕ). Based on the ϕ value, the bearing capacity factors would be obtained from Fig. 6.6 and then the ultimate bearing capacity would be calculated from Eqs. 6.1 or 6.5.

If the site consists of two layers of cohesive soil having different shear strength properties, then Fig. 6.7 can be utilized. In order to use this chart, perform the following steps:

1. Calculate the ratio of the undrained shear strength of layer 2 divided by the undrained shear strength of layer 1, i.e., $c_2/c_1 = s_{u2}/s_{u1}$.
2. Determine the ratio T/B, where T = vertical distance from the bottom of the foundation to the top of the layer 2 and B = width of the foundation.
3. Enter the chart with the value of c_2/c_1, intersect the appropriate T/B curve, and determine the value of N_c.
4. For strip footings, use Eq. 6.1 with $\phi = 0$ (i.e., $N_\gamma = 0$ and $N_q = 1$), or:

$$q_{ult} = c_1 N_c + \gamma_t D = s_{u1} N_c + \gamma_t D \tag{6.10}$$

5. For spread footings, use Eq. 6.6 with $\phi = 0$ (i.e., $N_\gamma = 0$ and $N_q = 1$), or:

$$q_{ult} = N_c c_1 [1 + 0.3 (B/L)] + \gamma_t D = N_c s_{u1} [1 + 0.3 (B/L)] + \gamma_t D \tag{6.11}$$

where N_c = bearing capacity factor from Fig. 6.7 (dimensionless)
$\quad c_1 = s_{u1}$ = undrained shear strength of layer 1 (psf or kPa)
$\quad D$ = depth of the foundation below ground surface (ft or m)

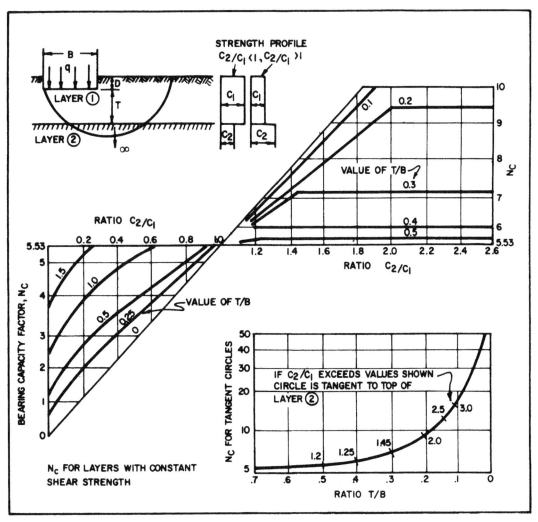

FIGURE 6.7   Ultimate bearing capacity of a shallow foundation constructed on two layers of cohesive soil. (*From NAVFAC DM-7.2, 1982.*)

Effective Stress Analysis.   The effective stress analysis uses the drained shear strength (c' and ϕ') of the plastic soil. The drained shear strength could be obtained from triaxial compression tests with pore water pressure measurements performed on saturated specimens of the plastic soil. This analysis is termed a *long-term analysis* because the shear induced pore water pressures (positive or negative) from the loading have dissipated and the hydrostatic pore water conditions now prevail in the field. Because an effective stress analysis is being performed, the location of the groundwater table must be considered in the analysis (use Eq. 6.7).

The first step to perform the bearing capacity analysis would be to obtain the bearing capacity factors (N_c, N_γ, and N_q) from Fig. 6.6 using the value of ϕ'. An adjustment to the total unit weight may be required depending on the location of the groundwater table (Eq. 6.7). Then Eq. 6.1 or 6.5 would be utilized (with c' substituted for c) to obtain the ultimate bearing capacity, with a factor of safety of 3 applied in order to calculate the allowable bearing pressure per Eq. 6.2.

Governing Case.   Usually the total stress analysis will provide a lower allowable bearing capacity for soft or very soft saturated plastic soils. This is because the foundation load will consolidate the plastic soil leading to an increase in the shear strength as time passes. For the long-term case (effective stress analysis), the shear strength of the plastic soil is higher with a resulting higher bearing capacity.

Usually the effective stress analysis will provide a lower allowable bearing capacity for very stiff or hard saturated plastic soils. This is because such plastic soils are usually heavily overconsolidated and they tend to dilate (increase in volume) during undrained shear deformation. A portion of the undrained shear strength is due to the development of negative pore water pressures during shear deformation. As these negative pore water pressures dissipate with time, the shear strength of the heavily overconsolidated plastic soil decreases. For the long-term case (effective stress analysis), the shear strength will be lower resulting in a lower bearing capacity.

Firm to stiff saturated plastic soils are intermediate conditions. The overconsolidation ratio (OCR) and the tendency of the saturated plastic soil to consolidate (gain shear strength) will determine whether the short-term condition (total stress parameters) or the long-term condition (effective stress parameters) provides the lower bearing capacity.

6.2.4   Other Bearing Capacity Considerations

There are many other possible considerations in the evaluation of bearing capacity for shallow foundations. Some common items are as follows:

Inclined Loads.   In addition to the vertical load acting on the footing, it may also be subjected to a lateral load; hence the resultant of the load will be inclined. One possible method as proposed by Merehof is to reduce the allowable bearing capacity based on the inclination of the load. However, this approach has a drawback in that the geotechnical engineer usually does not know the inclination of the various loads when preparing the foundation report. And if the inclinations were known, then numerous allowable bearing capacities would be needed for the various inclinations of the load.

A more commonly used procedure is to treat lateral loads separately and resist the lateral loads by using the soil pressure acting on the sides of the footing (allowable passive pressure) and by using the frictional resistance along the bottom of the footing. This is the approach used in the analysis of retaining wall foundations.

Moments and Eccentric Loads.   It is always desirable to design and construct shallow footings so that the vertical load is concentric, i.e., the vertical load is applied at the center of gravity of the footing. For combined footings that carry more than one vertical load, the combined footing should be designed and constructed so that the vertical loads are symmetric.

There may be design situations where the footing is subjected to a moment, such as where there is a fixed end connection between the building frame and the footing. This moment can be represented by a load P that is offset a certain distance (known as the eccentricity) from the center of gravity of the footing. For other projects, there may be property line constraints and the load must be offset a certain distance (eccentricity) from the center of gravity of the footing.

There are several different methods used to evaluate eccentrically loaded footings. Because an eccentrically loaded footing will create a higher bearing pressure under one side as compared to the opposite side, one approach is to evaluate the actual pressure distribution beneath the footing. The usual procedure is to assume a rigid footing (hence linear pressure distribution) and use the section modulus ($\frac{1}{6} B^2$) in order to calculate the largest and lowest bearing pressure. For a footing having a width B, the largest (q') and lowest (q'') bearing pressures are as follows:

$$q' = \frac{Q(B+6e)}{B^2} \tag{6.12}$$

$$q'' = \frac{Q(B-6e)}{B^2} \tag{6.13}$$

where q' = largest bearing pressure underneath the footing, which is located along the same side of the footing as the eccentricity (psf or kPa)

q'' = lowest bearing pressure underneath the footing, which is located at the opposite side of the footing (psf or kPa)

$Q = P$ = vertical load applied to the footing (pounds per linear foot of footing width or kN per linear meter of footing width)

e = Eccentricity of the load Q, i.e., the lateral distance from Q to the center of gravity of the footing (ft or m)

B = width of the footing (ft or m)

A usual requirement is that the load Q must be located within the middle one-third of the footing and the above equations are only valid for this condition. The value of q' must not exceed the allowable bearing pressure q_{all}.

Figure 6.8 presents another approach for footings subjected to moments. As indicted in Part *a* of Fig. 6.8, the moment M is converted to a load Q that is offset from the center of gravity of the footing by an eccentricity e. This approach is identical to the procedure outlined for Eqs. 6.12 and 6.13.

The next step is to calculate a reduced area of the footing. As indicated in Part *b* of Fig. 6.8, the new footing dimensions are calculated as $L' = L - 2e_1$ and $B' = B - 2e_2$. A reduction in footing dimensions in both directions would only be applicable for the case where the footing is subjected to two moments, one moment in the long direction of the footing (hence e_1) and the other moment across the footing (hence e_2). If the footing were subjected to only one moment in either the long or short direction of the footing, then the footing would be reduced in only one direction. Similar to Eqs. 6.12

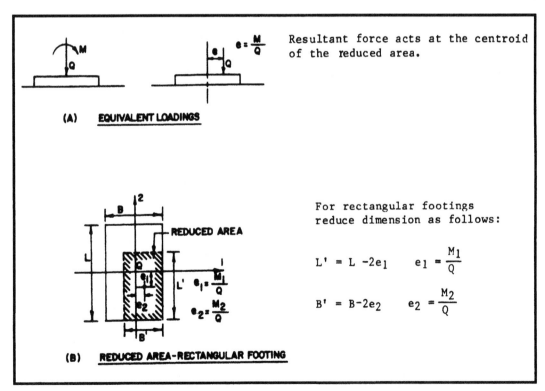

FIGURE 6.8   Reduced area method for a footing subjected to a moment. (*Reproduced from NAVFAC DM-7.2, 1982.*)

and 6.13, this method should only be utilized if the load Q is located within the middle one-third of the footing.

Once the new dimensions of the footing have been calculated (i.e., L' and B'), then Eq. 6.1 can be used by substituting L' for L and B' for B.

Footings at the Top of Slopes. Although methods have been developed to determine the allowable bearing capacity of foundations at the top of slopes (e.g., NAVFAC DM-7.2, 1982), these methods should be used with caution when dealing with plastic (cohesive) soil. This is because the outer face of slopes composed of plastic soils may creep downslope, leading to a loss in support for the footing constructed at the top of such slopes. Structures constructed at the top of clayey slopes will be further discussed in Chap. 10.

Inclined Base of Footing. Charts have been developed to determine the bearing capacity factors for footings having inclined bottoms. However, it has been stated that inclined bases should never be constructed for footings (AASHTO, 1996). Sometimes a sloping contact of underlying hard material will be encountered during excavation of the footing. Instead of using an inclined footing base along the sloping contact, the hard material should be excavated in order to construct a level footing that is entirely founded within the hard material.

6.3 BEARING CAPACITY FOR DEEP FOUNDATIONS

6.3.1 Introduction

Deep foundations are used when the upper soil stratum is too soft, weak, or compressible to support the static and earthquake-induced foundation loads. Deep foundations are also used when there is a possibility of the undermining of the foundation. For example, bridge piers are often founded on deep foundations to prevent a loss of support due to flood conditions that could cause river bottom scour. In addition, deep foundations are used when the expected settlement is excessive, to prevent ground surface damage of the structure, or to prevent a bearing capacity failure caused by the liquefaction of an underlying soil deposit. The types of deep foundations have already been discussed in Sec. 5.4.

Except for when a single pile or pier is used, a pile cap usually supports the vertical and horizontal structural loads and transfers these loads to the piles. The pile cap is often of reinforced concrete, the pile ends are embedded within the cap, and the cap is cast directly on the ground. It is assumed that each pile in the group carries the same load for the situation where the piles are all vertical and symmetrical, with the vertical load located at the center of gravity of the pile cap. This will not be the case if the pile cap is subjected to a moment or an eccentric load. These cases will be discussed in Sec. 6.3.4.

There are several different methods that are used in the design and construction of piles, as follows:

1. *Engineering analysis.* Based on the results of subsurface exploration and laboratory testing, the bearing capacity of the deep foundation can be calculated in a similar manner to the previous section on shallow foundations. The next two sections (Secs. 6.3.2 and 6.3.3) will describe the engineering analyses for deep foundations in cohesionless and cohesive soil.

2. *Field load tests.* Prior to the construction of the foundation, a pile or pier could be load tested in the field to determine its carrying capacity. Because of the uncertainties in the design of piles based on engineering analyses, pile load tests are common. The pile load test can often result in a more economical foundation than one based solely on engineering analyses. Pile load tests will be further discussed in Chap. 16.

3. *Application of pile driving resistance.* In the past, the pile capacity was estimated based on the driving resistance during the installation of the pile. Pile driving equations, such as the *Engineering News Formula* (Wellington, 1888), were developed that related the pile capacity to the energy of the pile driving hammer and the average net penetration of the pile per blow of the pile hammer. But studies have shown that there is no satisfactory relationship between the pile capacity from pile driving equations as compared to the pile capacity measured from load tests. Based on these studies, it has been concluded that use of pile driving equations are no longer justified (Terzaghi and Peck, 1967).

Often the pile driving resistance (i.e., blows per foot) is recorded as the pile is driven into place. When the anticipated bearing layer is encountered, the driving resistance (blows per foot) should substantially increase, assuming the same energy is applied per blow from the pile driver.

4. *Wave equation.* This method, first put into a practical form by Smith (1962), is based on using the stress wave from the hammer impact in a finite-element analysis. The actual analysis is rather complex and usually requires a computer program. Various parameters are input into the computer program, such as the energy induced by the hammer (fall height, weight, and the like), modulus of elasticity of the pile, cushion spring constant, and soil properties such as the side and point damping. The computer program output can include the pile capacity and the stresses in the pile due to driving. For additional details on the computer program, see Bowles (1974, 1982).

The main problem with this approach is the uncertainty associated with the input data for the computer program. For example, pile driving equipment operators usually do not apply the same energy per blow throughout the driving of a pile. Initially, as the pile starts to penetrate the soil, the equipment operator applies less energy, such as by adjusting the hammer free fall height, in order to slowly drive the pile into the ground. Likewise, if the pile is penetrating too quickly into the soil, they will back off the pile driving energy. They do this so that they can more accurately control the seating and driving of the pile at the designated location. When the pile is seated properly and the soil resistance starts to increase, the equipment operator typically increases the energy delivered to the pile.

Another important parameter is the cushion spring constant, which refers to the cushion material that is used between the hammer and the pile so that the top of the pile will not be damaged from the impact of the hammer. The cushion material often consists of wood blocks placed between the pile top and hammer, such as shown in Fig. 6.9. Initially, the wood blocks absorb a considerable amount of the impact energy, but as driving progresses, the wood blocks become compressed and transfer more of the energy to the pile (see Figs. 6.10 and 6.11). Hence the amount of energy transmitted to the piles can be highly variable both due to equipment operator performance and due to compression of the cushion material. These and other parameters that are constantly changing or hard to predict have led to the conclusion from Bowles (1982) that for the wave equation "any comparison between the computer output and predicted pile capacity within a 30 percent deviation is likely to be a happy coincidence of input data."

5. *Specifications and experience.* Other factors that should be considered in the deep foundation design include governing building code or agency requirements. In addition, local experience in terms of what has worked best in the past for the local soil conditions may prove valuable in the design and construction of pile foundations.

6.3.2 Bearing Capacity for Cohesionless Soil

End Bearing Pile or Pier. For an end bearing pile or pier, the Terzaghi bearing capacity equation can be used to determine the ultimate bearing capacity q_{ult}. Assuming a pile that has a square cross-section (i.e., $B = L$) and since $c = 0$ (cohesionless soil), Eq. 6.3 reduces to

$$q_{ult} = \frac{Q_p}{B^2} = 0.4\gamma_t BN_\gamma + \gamma_t D_f N_q \qquad (6.14)$$

In comparing the second and third term in Eq. 6.14, the value of B (width of pile) is much less than the embedment depth D_f of the pile. Therefore, the first term in Eq. 6.14 can be neglected.

The value of $\gamma_t D_f$ in the second term in Eq. 6.14 is equivalent to the total vertical stress σ_v at the pile tip. For cohesionless soil and using an effective stress analysis, the groundwater table must be included and hence the vertical effective stress σ'_v can be substituted for σ_v. Equation 6.14 reduces to

FIGURE 6.9 A prestressed concrete pile is being hoisted into place for driving. Wood blocks have been placed at the top of the pile to cushion the impact of the drive hammer and prevent damage to the top of the pile.

FIGURE 6.10 The prestressed concrete pile at the center of the photograph is the same pile as shown in Fig. 6.9 and has just been driven into place. Note that the cushion blocks of wood have been compressed during the pile driving. The steam emanating from the wood is due to heat generated during the impact with the hammer.

FIGURE 6.11   Close-up view of Fig. 6.10. The wood cushion blocks have been crushed together during the pile driving operation.

For end-bearing piles having a square cross-section:

$$q_{\text{ult}} = \frac{Q_p}{B^2} = \sigma'_v N_q \qquad (6.15)$$

For end-bearing piles or piers having a circular cross-section:

$$q_{\text{ult}} = \frac{Q_p}{\pi r^2} = \sigma'_v N_q \qquad (6.16)$$

where q_{ult} = ultimate bearing capacity of the end-bearing pile or pier (psf or kPa)
  Q_p = ultimate point resistance force (pounds or kN)
  B = width of the piles having a square cross-section (ft or m)
  r = radius of the piles or piers having a round cross-section (ft or m)
  σ'_v = vertical effective stress at the pile tip (psf or kPa)
  N_q = bearing capacity factor (dimensionless)

For drilled piers or piles placed in predrilled holes, the value of N_q can be obtained from Figs. 6.5 and 6.6 based on the friction angle ϕ of the cohesionless soil located at the pile tip. However, for driven piles, the N_q values from Figs. 6.5 and 6.6 are generally too conservative. Figure 6.12 presents a chart prepared by Vesic (1967) that provides the bearing capacity factor N_q from several different sources. As shown in Fig. 6.12, for $\phi = 30°$, N_q varies from about 30 to 150, while for $\phi = 40°$, N_q varies from about 100 to 1000. This is a tremendous variation in N_q values and is related to the different approaches used by the various researchers, where in some cases the basis of the relationship shown in Fig. 6.12 is theoretical, while in other cases the relationship is based on analysis of field data such as pile load tests.

There is a general belief that the bearing capacity factor N_q is higher for driven piles than for shallow foundations. One reason for a higher N_q value is the effect of driving the pile, which displaces and densifies the cohesionless soil at the bottom of the pile. The densification could be due to both the physical process of displacing the soil and the driving vibrations. These actions would tend to increase the friction angle of the cohesionless soil in the vicinity of the driven pile. High displacement piles, such as large diameter solid piles, would tend to displace and densify more soil than low displacement piles, such as hollow piles that do not form a soil plug.

Example Problem 6.5 Assume a pile having a diameter of 0.3 m (1 ft) and a length of 6 m (20 ft) will be driven into a nonplastic silty sand deposit which has the shear strength properties shown in Fig. 3.18 (i.e., $c = 0$ and $\phi' = 30°$). Assume that the location of the groundwater table is located 3 m (10 ft) below the ground surface and the total unit weight above the groundwater table is 19 kN/m³ (120 pcf) and the buoyant unit weight (γ_b) below the groundwater table is 9.9 kN/m³ (63 pcf). Using the Terzaghi correlation shown in Fig. 6.12, calculate the allowable end-bearing capacity using a factor of safety of 3.

Solution The vertical effective stress σ'_v at the pile tip equals:

$$\sigma'_v = (3 \text{ m}) (19 \text{ kN/m}^3) + (3 \text{ m}) (9.9 \text{ kN/m}^3) = 87 \text{ kPa} (1,800 \text{ psf})$$

From Fig. 6.12, using the Terzaghi relationship, $N_q = 30$
Using Eq. 6.16:
Ultimate end-bearing capacity $q_{ult} = (87 \text{ kPa})(30) = 2600 \text{ kPa} (54,000 \text{ psf})$
Multiplying q_{ult} by the area of the pile tip (πr^2), or:

$$Q_p = (2600)(\pi)(0.15)^2 = 180 \text{ kN} (42 \text{ kips})$$

Using a factor of safety of 3, the allowable pile tip resistance = 60 kN (14 kips)

Friction Pile. As the name implies, a friction pile develops its load carrying capacity due to the frictional resistance between the cohesionless soil and the pile perimeter. Piles subjected to vertical uplift forces would be designed as friction piles because there would be no end-bearing resistance as the pile is pulled from the ground. Based on a linear increase in frictional resistance with confining pressure, the average ultimate frictional capacity (q_{ult}) can be calculated as follows:
For piles having a square cross-section:

$$q_{ult} = \frac{Q_s}{4BL} = \sigma'_h \tan \phi_w = \sigma'_v k \tan \phi_w \tag{6.17}$$

For piles or piers having a circular cross-section:

$$q_{ult} = \frac{Q_s}{2\pi rL} = \sigma'_h \tan \phi_w = \sigma'_v k \tan \phi_w \tag{6.18}$$

where q_{ult} = average ultimate frictional capacity for the pile or pier (psf or kPa)
  Q_s = ultimate skin friction resistance force (lb or kN)
  B = width of the piles having a square cross-section (ft or m)
  r = radius of the piles or piers having a round cross-section (ft or m)
  L = length of the pile or pier (ft or m)
  σ'_h = average horizontal effective stress over the length of the pile or pier (psf or kPa)
  σ'_v = average vertical effective stress over the length of the pile or pier (psf or kPa)
  k = dimensionless parameter equal to σ'_h divided by σ'_v (i.e., similar to Eq. 4.18). Equation 4.19 can be used to estimate the value of k for loose sand deposits. Because of the densification of the cohesionless soil associated with driven displacement piles, values of k between 1 and 2 are often assumed.
  ϕ_w = friction angle between the cohesionless soil and the perimeter of the pile or pier (degrees). Commonly used friction angles are $\phi_w = \frac{3}{4} \phi$ for wood and concrete piles and $\phi_w = 20°$ for steel piles.

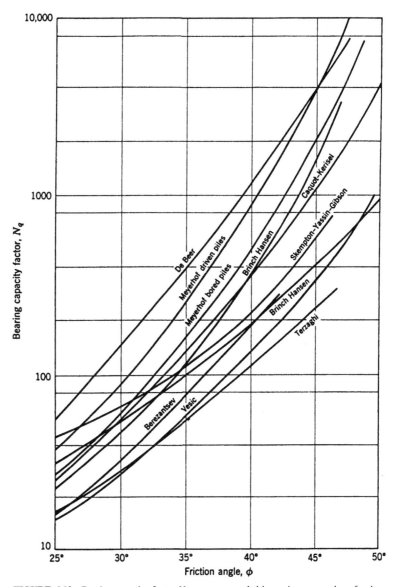

FIGURE 6.12   Bearing capacity factor N_q as recommended by various researchers for deep foundations. (*From Vesic, 1967; reproduced from Lambe and Whitman, 1969.*)

In Eqs. 6.17 and 6.18, the terms $4BL$ (for square piles) and $2\pi r L$ (for circular piles) are the perimeter surface area of the pile or pier. In Eqs. 6.17 and 6.18, the term: σ_h' tan ϕ_w equals the shear strength τ_f between the pile or pier surface and the cohesionless soil. This term is identical to Eq. 3.9 (with $c' = 0$), i.e., $\tau_f = \sigma_n'$ tan ϕ'. Thus the frictional resistance force Q_s in Eqs. 6.17 and 6.18 is equal to the perimeter surface area times the shear strength of the soil at the pile or pier surface.

Example Problem 6.6   Assume the same conditions as the previous example and that the pile is made of concrete. For driven displacement piles and using $k = 1$, calculate the allowable frictional capacity of the pile.

Solution   The easiest solution consists of dividing the pile into two sections. The first section is located above the groundwater table ($z = 0$ to 3 m) and the second section is that part of the pile below the groundwater table ($z = 3$ to 6 m). The average vertical stress will be at the midpoint of these two sections, or:

$$\sigma_v \text{ (at } z = 1.5 \text{ m)} = (1.5 \text{ m}) (19 \text{ kN/m}^3) = 29 \text{ kPa (600 psf)}$$

$$\sigma'_v \text{ (at } z = 4.5 \text{ m)} = (3 \text{ m}) (19 \text{ kN/m}^3) + (1.5 \text{ m}) (9.9 \text{ kN/m}^3) = 72 \text{ kPa (1500 psf)}$$

For a concrete pile: $\phi_w = 3/4 \ \phi = 3/4 \ (30°) = 22.5°$

Substituting values into Eq. 6.18: For $z = 0$ to 3 m, $L = 3$ m and Q_s equals:

$$Q_s = (2\pi r L)(\sigma'_v k \tan \phi_w) = (2\pi)(0.15 \text{ m})(3 \text{ m})(29 \text{ kPa})(1)(\tan 22.5°) = 34 \text{ kN (7.6 kips)}$$

For $z = 3$ to 6 m, $L = 3$ m and Q_s equals:

$$Q_s = (2\pi r L)(\sigma'_v k \tan \phi_w) = (2\pi)(0.15 \text{ m})(3 \text{ m})(72 \text{ kPa})(1)(\tan 22.5°) = 84 \text{ kN (19 kips)}$$

Adding together both values of Q_s, the total frictional resistance force = 34 kN + 84 kN = 118 kN (26.6 kips). Applying a factor of safety of 3, the allowable frictional capacity of the pile is approximately equal to 40 kN (9 kips).

Combined End-Bearing and Friction Pile.   For piles and piers subjected to vertical compressive loads and embedded in a deposit of cohesionless soil, they are usually treated in the design analysis as combined end-bearing and friction piles or piers. This is because the pile or pier can develop substantial load carrying capacity from both end-bearing and frictional resistance. To calculate the ultimate pile or pier capacity for a condition of combined end-bearing and friction, the value of Q_p is added to the value of Q_s.

In the previous example of a 6 m (20 ft) long pile embedded in a silty sand deposit, the allowable tip resistance (60 kN) and the allowable frictional capacity (40 kN) are added together to obtain the allowable combined end-bearing and frictional resistance capacity of the pile of 100 kN (23 kips).

Pile Groups.   The previous discussion has dealt with the load capacity of a single pile in cohesionless soil. Usually pile groups are used to support the foundation elements, such as a group of piles supporting a pile cap or a mat slab. In loose sand and gravel deposits, the load carrying capacity of each pile in the group may be greater than a single pile because of the densification effect due to driving the piles. Because of this densification effect, the load capacity of the group is often taken as the load capacity of a single pile times the number of piles in the group. An exception would be a situation where a weak layer underlies the cohesionless soil. In this case, group action of the piles could cause them to punch through the cohesionless soil and into the weaker layer or cause excessive settlement of the weak layer located below the pile tips (see Fig. 4.17).

In order to determine the settlement of the strata underlying the pile group, the 2:1 approximation can be used to determine the increase in vertical stress ($\Delta \sigma_v$) for those soil layers located below the pile tip. If the piles in the group are principally end-bearing, then the 2:1 approximation starts at the tip of the piles ($L = $ length of the pile group, $B = $ width of the pile group, and $z = $ depth below the tip of the piles). If the pile group develops its load carrying capacity principally through side friction, then the 2:1 approximation starts at a depth of two-thirds D, where $D = $ depth of the pile group. See Fig. 4.17 for examples.

6.3.3   Bearing Capacity for Cohesive Soil

The load carrying capacity of piles and piers in cohesive soil is more complex than the analysis for cohesionless soil. Some of the factors that may need to be considered in the analysis are as follows (AASHTO, 1996):

- A lower load carrying capacity of a pile in a pile group as compared to that of a single pile.
- The settlement of the underlying cohesive soil due to the load of the pile group.
- The effects of driving piles on adjacent structures or slopes. The ground will often heave around piles driven into soft and saturated cohesive soil.
- The increase in load on the pile due to negative skin friction (i.e., downdrag loads) from consolidating soil.
- The effects of uplift loads from expansive and swelling clays.
- The reduction in shear strength of the cohesive soil due to construction techniques, such as the disturbance of sensitive clays or development of excess pore water pressures during the driving of the pile. There is often an increase in load carrying capacity of a pile after it has been driven into a soft and saturated clay deposit. This increase in load carrying capacity with time is known as *freeze* or *setup* and is caused primarily by the dissipation of excess pore water pressures.
- The influence of fluctuations in the elevation of the groundwater table on the load carrying capacity when analyzed in terms of effective stresses.

Total Stress Analyses.   The ultimate load capacity of a single pile or pier in cohesive soil is often determined by performing a total stress analysis. This is because the critical load on the pile, such as from wind or earthquake loads, is a short-term loading condition and thus the undrained shear strength of the cohesive soil will govern. The total stress analysis for a single pile or pier in cohesive soil typically is based on the undrained shear strength s_u of the cohesive soil or the value of cohesion c determined from unconsolidated undrained triaxial compression tests (i.e., $\phi = 0$ analysis, see Part a, Fig. 4.18).

The ultimate load capacity of the pile or pier in cohesive soil would equal the sum of the ultimate end-bearing and ultimate side adhesion components. In order to determine the ultimate end-bearing capacity, Eq. 6.6 can be utilized with $B = L$, and $\phi = 0$, in which case $N_c = 5.5$, $N_\gamma = 0$, and $N_q = 1$ (Fig. 6.6). The term $\gamma_t D_f N_q = \gamma_t D_f$ (for $\phi = 0$) is the weight of overburden, which is often assumed to be balanced by the pile weight and thus this term is not included in the analysis. Note in Eq. 6.6 that the term $1.3\, cN_c = 1.3c\, (5.5) = 7.2c$ (or $N_c = 7.2$). However, N_c is commonly assumed to be equal to 9 for deep foundations (Mabsout et al., 1995). Thus the ultimate load capacity (Q_{ult}) of a single pile or pier in cohesive soil equals:

$$Q_{ult} = \text{end bearing} + \text{side adhesion} = cN_c \text{ (area of tip)} + c_A \text{ (surface area)},$$

or

$$Q_{ult} = c9\,(\pi R^2) + c_A\,(2\pi Rz) = 9\pi cR^2 + 2\pi c_A Rz \qquad (6.19)$$

where Q_{ult} = ultimate load capacity of the pile or pier (lb or kN)
c = cohesion of the cohesive soil at the pile tip (psf or kPa). Because it is a total stress analysis, the undrained shear strength $(s_u = c)$ is used, or the undrained shear strength is obtained from unconsolidated undrained triaxial compression tests (i.e., $\phi = 0$ analysis, c = peak point of Mohr circles, see Part a, Fig. 4.18).
R = radius of the uniform pile or pier (ft or m). If the pier bottom is belled or a tapered pile is used, then R at the tip would be different than the radius of the shaft.
z = embedment depth of the pile (ft or m)
c_A = adhesion between the cohesive soil and pile or pier perimeter (psf or kPa). Figure 6.13 can be used to determine the value of the adhesion (c_A) for different types of piles and cohesive soil conditions.

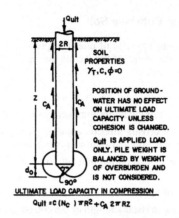

ULTIMATE LOAD CAPACITY IN COMPRESSION

$$Q_{ult} = C(N_C)\pi R^2 + C_A 2\pi RZ$$

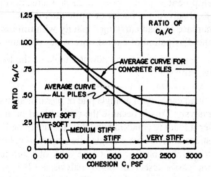

RECOMMENDED VALUES OF ADHESION

PILE TYPE	CONSISTENCY OF SOIL	COHESION, C PSF	ADHESION, C_A PSF
	VERY SOFT	0 - 250	0 - 250
TIMBER AND CONCRETE	SOFT	250 - 500	250 - 480
	MED. STIFF	500 - 1000	480 - 750
	STIFF	1000 - 2000	750 - 950
	VERY STIFF	2000 - 4000	950 - 1300
	VERY SOFT	0 - 250	0 - 250
	SOFT	250 - 500	250 - 460
STEEL	MED. STIFF	500 - 1000	460 - 700
	STIFF	1000 - 2000	700 - 720
	VERY STIFF	2000 - 4000	720 - 750

ULTIMATE LOAD CAPACITY IN TENSION

$$T_{ult} = C_A 2\pi RZ$$

FIGURE 6.13   Ultimate capacity for a single pile or pier in cohesive soil. (*Reproduced from NAVFAC DM-7.2, 1982.*)

If the pile or pier is subjected to an uplift force, then the ultimate capacity (T_{ult}) for the pile or pier in tension is equal to

$$T_{ult} = 2\pi c_A Rz \tag{6.20}$$

where c_A is the adhesion between the cohesive soil and the pile or pier perimeter (Fig. 6.13). In order to determine the allowable capacity of a pile or pier in cohesive soil, the values calculated from the above equations would be divided by a factor of safety. A commonly used factor of safety equals 3.

Effective Stress Analyses.   For long-term loading conditions of piles or piers, effective stress analyses could be performed. In this case, the effective cohesion c' and effective friction angle ϕ' would be used in the analysis for end bearing. The location of the groundwater table would also have to be considered in the analysis. Along the pile perimeter, the ultimate resistance could be based on the effective shear strength between the pile or pier perimeter and the cohesive soil.

Pile Groups.   The bearing capacity of pile groups in cohesive soils is normally less than the sum of individual piles in the group and this reduction in the group capacity must be considered in the analysis. The *group efficiency* is defined as the ratio of the ultimate load capacity of each pile in the group to the ultimate load capacity of a single isolated pile. If the spacing between piles in the group are at a distance that is greater than about seven times the pile diameter, then the group efficiency is equal to 1 (i.e., no reduction in pile capacity for group action). The group efficiency decreases as the piles become closer together in the pile group. For example, a 9×9 pile group with

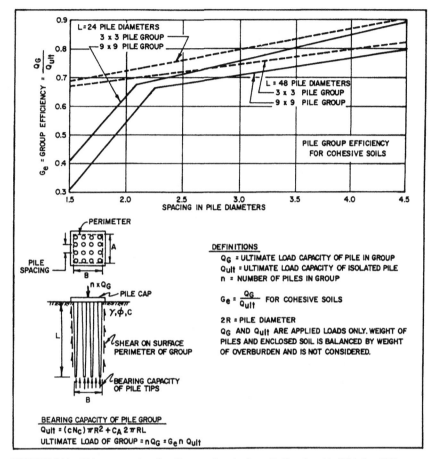

FIGURE 6.14   Ultimate capacity of a pile group in cohesive soil. (*Developed by Whitaker 1957, reproduced from NAVFAC DM-7.2, 1982.*)

a pile spacing equal to 1.5 times the pile diameter has a group efficiency of only 0.3. Figure 6.14 can be used to determine the ultimate load capacity of a pile group in cohesive soil.

Similar to pile groups in cohesionless soil, the settlement of the strata underlying the pile group can be evaluated by using the 2:1 approximation to calculate the increase in vertical stress ($\Delta\sigma_v$) for those soil layers located below the pile tip. If the piles in the group develop their load carrying capacity principally by end-bearing in cohesive soil, then the 2:1 approximation starts at the tip of the piles (L = length of the pile group, B = width of the pile group, and z = depth below the tip of the piles). If the pile group develops its load carrying capacity principally through cohesive soil adhesion along the pile perimeter, then the 2:1 approximation starts at a depth of $^2/_3\, D$, where D = depth of the pile group. See Fig. 4.17 for examples.

6.3.4   Other Deep Foundation Considerations

There are many other possible considerations in the evaluation of bearing capacity for deep foundations. Some common items are as follows:

Belled Piers.   For some projects, deep foundations may need to carry very high vertical loads. In this case, piers are usually constructed such as shown in Fig. 6.15. By installing a bell, the

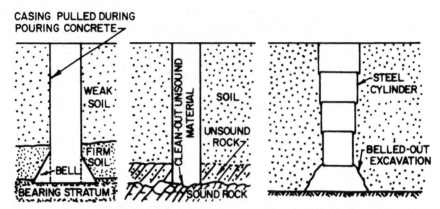

FIGURE 6.15   Illustrations of the construction of piers and bells. (*Reproduced from NAVFAC DM-7.2, 1982.*)

end-bearing capacity of the pier can be significantly increased. Some of the advantages of a belled pier are as follows:

- The pier and bell can be visually inspected when the pier diameter is about 30 in. (0.76 m) or larger.
- Pile caps are usually not needed since most loads can be carried on a single belled pier.
- With belling, the uplift capacity of the pier can be significantly increased.
- The design depth and diameter of the pier and bell can be modified based on observed field conditions.
- The pier and bell can be installed in sound bedrock to carry very high loads.

  Some of the disadvantages of belled piers are as follows:

- Loose granular soil below the groundwater table can cause installation problems. The bell usually cannot be formed in granular soil below the groundwater table.
- Small diameter piers and bells cannot be easily inspected to confirm bearing and are particularly susceptible to necking problems.
- If the rock is too hard, it may not be possible to excavate the bell, in which case a rock socket may need to be drilled.

Belled piers can be designed in the same manner as conventional deep foundations. They are usually designed as end-bearing members with the allowable end-bearing capacity determined by the geotechnical engineer. For sound rock, the allowable end-bearing capacity could be based on allowable bearing code values (e.g., Table 18.4), field load tests, and local experience. For the ultimate end-bearing capacity in soil, Eqs. 6.15 and 6.16 can be used for cohesionless soil and the end-bearing part of Eq. 6.19 can be used for cohesive soil.

Inclined Loads.   In addition to the vertical load acting on the pile cap or pier, they may also be subjected to a lateral load; hence the resultant of the load will be inclined. Lateral loads on deep foundations are more complex than lateral loads on shallow foundations. Section 6.4 will be devoted to lateral loads on deep foundations.

Moments and Eccentric Loads.   It is always desirable to design and construct deep foundations so that the vertical load is concentric, i.e., the vertical load is applied at the center of gravity of the pile cap or the center of gravity of the pier. For a pile cap that carries more than one vertical load, the pile cap should be designed and constructed so that the vertical loads are symmetric.

There may be design situations where the pile cap is subjected to a moment, such as where there is a fixed end connection between the building frame and the cap. This moment can be represented by a load Q that is offset a certain distance (known as the eccentricity) from the center of gravity of the pile cap. For other projects, there may be property line constraints and the load must be offset a certain distance (eccentricity) from the center of gravity of the pile cap.

One method commonly used to evaluate an eccentrically loaded pile cap is to use Eqs. 6.12 and 6.13. It is assumed that the eccentrically loaded cap will cause the piles located under one side to carry a higher load than as compared to the piles located on the opposite side of the pile cap. The first step in the procedure is to ignore the piles, assume a rigid pile cap, and use a linear pressure distribution such as shown in Fig. 6.16. The values of q' and q'' are then calculated by using Eqs. 6.12 and 6.13. The final step is to proportion the pressure to an individual pile, such as shown in Fig. 6.16. A usual requirement is that the load Q must be located within the middle 1/3 of the pile cap and Eqs. 6.12 and 6.13 are only valid for this condition. The pile carrying the largest load (i.e., number 1 in Fig. 6.16) must not exceed its allowable load capacity. Equations 6.12 and 6.13 can be used for an eccentricity in one direction (i.e., across the width or length of the pile cap) or for an eccentricity in both pile cap directions.

Example Problem 6.7   A 4 by 4 pile group will support a pile cap that is subjected to a vertical load of 500 kips that is offset from the centerline of the pile cap by a distance of 2.0 ft (i.e., $e = 2.0$ ft in only one direction). The piles have a diameter of 1.0 ft (0.3 m), are spaced at four pile diameters (centerline to centerline) in both directions, and the pile cap is 16 ft by 16 ft (4.9 m by 4.9 m). Using Fig. 6.16, determine the load that each pile supports.

Solution   The first step is to check that the eccentricity is within the middle one-third of the pile cap. The middle third of the pile cap is a distance of 2.7 ft (0.8 m) from both sides of the centerline of the pile cap. Thus the eccentricity is within the middle one-third of the pile cap. Since there are four rows of piles, each row will carry a $Q = 500/4 = 125$ kips per row or 125,000 lb per row.

Using Eq. 6.12, $q' = Q (B + 6e)/B^2 = (125,000)[16 + (6)(2)]/(16)^2 = 13,700$ lb/ft

Using Eq. 6.13, $q'' = Q (B - 6e)/B^2 = (125,000)[16 - (6)(2)]/(16)^2 = 1,950$ lb/ft

Proportioning for each pile per Fig. 6.16:

Change in bearing per foot = $(13,700 - 1,950)/16 = 730$ psf

Vertical load for Pile No. 1 = (4 ft)[13,700 − (730)(2)] = 48,900 lb = 48.9 kips

Vertical load for Pile No. 2 = (4 ft)[13,700 − (730)(6)] = 37,300 lb = 37.3 kips

Vertical load for Pile No. 3 = (4 ft)[13,700 − (730)(10)] = 25,600 lb = 25.6 kips

Vertical load for Pile No. 4 = (4 ft)[13,700 − (730)(14)] = 13,900 lb = 13.9 kips

---

Check:

4 No. 1 piles = (4)(48.9) = 195 kips

4 No. 2 piles = (4)(37.3) = 149 kips

4 No. 3 piles = (4)(25.6) = 101 kips

4 No. 4 piles = (4)(13.9) = 55 kips

Total   = 500 kips

For this example problem, the allowable pile capacity would have to be at least 48.9 kips.

Downdrag Loads on a Single Pile or Pier.   A downdrag load is defined as a force induced on deep foundations resulting from the downward movement of adjacent soil relative to the foundation element. Downdrag is also referred to as negative skin friction. Conditions that could cause downdrag loads are as follows:

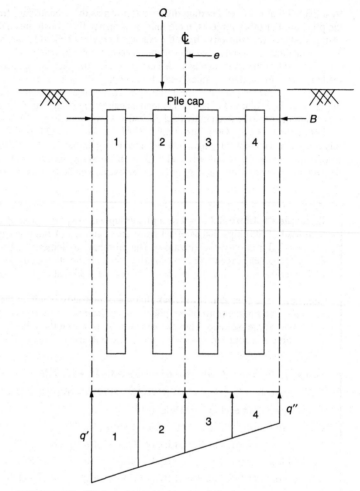

FIGURE 6.16   Pile cap subjected to an eccentric load.

- Above the end-bearing strata, there is a soil layer that is in the process of settling due to consolidation or collapse.
- Above the end-bearing strata, the original groundwater table is permanently lowered causing an increase in effective stress and resulting in settlement of the soil strata.
- A fill layer is placed at ground surface that causes settlement of the soil strata, such as shown in Fig. 4.17.

The following equations can be used to calculate the downdrag load Q_D for a single isolated pile or pier:

For a pile or pier in cohesionless soil subjected to downdrag:

$$Q_D = 2\pi R L_1 \sigma_v' k \tan \phi_w \tag{6.21}$$

For a pile or pier in cohesive soil subjected to downdrag:

$$Q_D = 2\pi R L_1 c_A \tag{6.22}$$

where Q_D = downdrag load acting on the pile or pier (lb or kN)

R = radius of the pile or pier (m or ft)

L_1 = vertical distance over which the pile or pier is subjected to the downdrag load (ft or m)

σ'_v = average vertical effective stress over the portion of the pile or pier subjected to the downdrag load (psf or kPa)

k = dimensionless parameter equal to σ'_h divided by σ'_v (i.e., similar to Eq. 4.18). Equation 4.19 can be used to estimate the value of k for loose sand deposits. Because of the densification of the cohesionless soil associated with driven displacement piles, values of k between 1 and 2 are often assumed

ϕ_w = friction angle between the cohesionless soil and the perimeter of the pile or pier (degrees). Commonly used friction angles are $\phi_w = \frac{3}{4} \phi$ for wood and concrete piles and $\phi_w = 20°$ for steel piles

c_A = adhesion between the cohesive soil and pile or pier perimeter (psf or kPa). Figure 6.13 can be used to determine the value of the adhesion (c_A) for different types of piles and cohesive soil conditions

---

Example Problem 6.8   Assume a site consists of an end-bearing pier founded on solid rock. The diameter of the pier is 3 ft (0.9 m) and a 40 ft (12 m) long zone of the pier is embedded in clay that is in the process of consolidating. Assuming the clay has a uniform undrained shear strength s_u of 500 psf (24 kPa), calculate the downdrag load on the pier.

Solution   Using a total stress analysis, therefore, s_u = cohesion (c). From Fig. 6.13, the value of c_A = 480 psf (23 kPa) for a concrete pier embedded in clay having c = 500 psf (24 kPa). Using Eq. 6.22:

$$Q_D = 2\pi R L_1 c_A = 2\pi (1.5 \text{ ft})(40 \text{ ft})(480 \text{ psf}) = 181{,}000 \text{ lb (800 kN)}$$

Downdrag Loads on Pile Groups.   The downdrag load on a pile group or pier group is more complicated. Two possible downdrag loading conditions for the group must be calculated, as follows:

1. *Large spacing of piles.*   If the piles are spaced far apart, then it is likely that the downdrag will act on each pile separately. Equations 6.21 and 6.22 can be used to determine the downdrag load for each pile. The downdrag of each pile in the group would then be summed up in order to calculate the total downdrag for a given pile group.

2. *Small spacing of piles.*   When the pile spacing is small, the downdrag load may act effectively on the entire perimeter of the group, in which case the following equations are utilized:

For square pile groups in cohesionless soil subjected to downdrag:

$$Q_D = 4BL_1 \sigma'_v k_o \tan \phi' \tag{6.23}$$

For square pile groups in cohesive soil subjected to downdrag:

$$Q_D = 4BL_1 s_u \tag{6.24}$$

where Q_D = downdrag load acting on the pile group (lb or kN)

B = width of the pile group measured to the outside edge of the perimeter piles (m or ft)

L_1 = vertical distance over which the pile group is subjected to the downdrag load (ft or m)

σ'_v = average vertical effective stress at the portion of the pile group subjected to the downdrag load (psf or kPa)

k_o = coefficient of earth pressure at rest, see Eq. 4.18 (dimensionless). Equation 4.19 can be used to estimate the value of k_o for loose sand deposits.

ϕ' = effective friction angle of the cohesionless soil (degrees)

s_u = average undrained shear strength along the portion of the pile group subjected to the downdrag load (psf or kPa)

As a conservative approach, the downdrag load should be calculated for both cases outlined above and the maximum value used for design.

Example Problem 6.9   Consider the same situation as the previous example, except that there is a 2 by 2 pier group. The piers are spaced 3 pier diameters (centerline to centerline). Determine the downdrag load on the pier group.

Solution   Since there are four piers in the 2 by 2 group and assuming each pier is separately subjected to a downdrag load, or:

$$Q_D = (4)(181,000) = 724,000 \text{ lb (3200 kN)}$$

The second downdrag condition is to use Eq. 6.24, or:

$$B = 3 \text{ times the diameter} + 2r = (3)(3) + (2)(1.5) = 12 \text{ ft (3.7 m)}$$

$$Q_D = 4BL_1 s_u = (4)(12)(40)(500) = 960,000 \text{ lb (4300 kN)}$$

Hence the governing condition would be assuming the downdrag over the entire perimeter of the pier group, or $Q_D = 960,000$ lb (4300 kN).

Helical Anchors.   Helical anchors have been introduced in Sec. 5.4.3. There are two possible methods that can be used to determine the ultimate bearing capacity of a helical anchor, as follows (A. B. Chance Co., 1989):

1. *Bearing plus cylindrical shear method.*   This method is based on determining the ultimate bearing capacity of the projected area of the lowest helix. In addition, there is frictional resistance of a cylinder of soil with a diameter equal to the average diameter of the remaining helices and a length equal to the distance from the top helix to the bottom helix.

2. *Bearing capacity method.*   This method assumes that the ultimate bearing capacity of the anchor is equal to the sum of the capacities of individual helices. Thus each helix bearing capacity is calculated and then summed up in order to obtain the total ultimate bearing capacity.

It has been stated that when the helices are spaced quite close (such as 6 in. apart), the Bearing Plus Cylindrical Shear Method is applicable. However, when the helix spacing is great (such as 10 ft apart) and all the helices are in the bearing strata, then the Bearing Capacity Method should be used (A. B. Chance Co., 1989). The allowable axial load capacity is typically determined by applying a factor of safety of 3 to the ultimate capacity calculated by using either method outlined above. The allowable axial load capacity must not exceed the "safe limit" specified for a particular type of helical anchor that is selected for use. In addition, the shaft of the helical anchor must be strong enough to resist the applied axial load and withstand the estimated torque required to install the anchor. Because helical anchors are rather flexible, they typically have a much lower capacity when resisting lateral loads.

6.4   LATERAL LOAD CAPACITY OF DEEP FOUNDATIONS

6.4.1   Introduction

This section will discuss deep foundations subjected to lateral loads. If the soil is unable to provide sufficient lateral support, then failure of the structure could occur. In terms of the amount of lateral resistance, the most important factor is the shear strength of the soil or rock. Obviously solid rock can impart much more lateral resistance than a soft clay. Other important factors in the design of deep foundations subjected to lateral loads are as follows:

1. *Magnitude of load.*   In many cases, the magnitude of the lateral load will be small and will not significantly affect the design of the deep foundation. An extensive investigation or field load tests are usually not required and the allowable lateral bearing of the soil or rock could be based on building code values. For example, Table 18.4 presents allowable lateral bearing values for soil and rock per the *International Building Code* (2009). The values listed in Table 18.4 can be increased for each additional foot (0.3 m) of depth up to a maximum of 15 times. However, the allowable lateral bearing values should only be applied for the upper portion of the pile that laterally deflects to the extent that passive soil pressure is mobilized. This will be further discussed in Item 3.

For those projects where the lateral load to be carried by the deep foundation is low, the allowable lateral bearing values from Table 18.4 can be recommended for design. In most cases, the values in Table 18.4 will be conservative and it is often recommended that these values be applied over the pile or pier diameter. However, there may be cases where the allowable lateral bearing values in Table 18.4 are too high, such as when the rock is highly fractured or weathered. Likewise, the allowable lateral bearing values in Table 18.4 may be too high for soil that is in a loose or soft state.

2. *Periodic or constant load.*   In some cases, the deep foundation may need to only resist periodic lateral loads, such as a wind-induced or earthquake-induced lateral loads. For periodic lateral loads, the allowable lateral bearing capacity for a deep foundation embedded in cohesive soil would be a short-term case and hence a total stress analysis using the undrained shear strength of the soil would be used.

In other cases, the lateral load may be constant. An example of a constant lateral load would be a retaining wall footing supported by a deep foundation. The backfill soil would exert a constant lateral load onto the retaining wall and this lateral load would be transmitted to the deep foundation. For constant lateral loads, the long-term condition could govern and an effective stress analysis would be needed. In addition, for constant lateral loads on a deep foundation in cohesive soil, the possibility of long-term creep of the soil must be evaluated. For example, Fig. 6.17 shows a picture of soil movement around a concrete pier. The soil shown in Fig. 6.17 was classified as silty clay, having a liquid limit of 56 and a plasticity index of 32. At this site, the soil was plastic enough

FIGURE 6.17   Soil movement around a concrete pier.

to simply flow around the pier. Thus if the cohesive soil is plastic enough, the constant lateral load may cause the deep foundation to slowly deflect laterally with the clay creeping around the pile or pier and opening a gap such as shown in Fig. 6.17.

3. *Allowable lateral movement.*   The allowable lateral deflection of the deep foundation will govern the amount of lateral resistance exerted by the soil. This is because, especially for cohesive soil, a considerable amount of lateral deformation is required in order to achieve passive pressure. If the deep foundation is very flexible and can sustain a large lateral deflection, then a greater depth of soil will be subjected to passive pressure. On the other hand, if the acceptable amount of lateral deflection is low, then only the upper few feet of soil may mobilize passive soil resistance.

6.4.2 Design and Construction Options

Because piles are usually rather slender members, they are not efficient in resisting lateral loads. Hence the pile cap is often equipped with additional components, such as batter piles, lateral tieback anchors, or deadman, which are used to resist the lateral load. The lateral resistance can also be increased by placing and compacting sand or gravel near ground surface to provide a surface zone of high lateral resistance material and by increasing the diameter of the pile near ground surface by using a concrete collar, pile wings, deepened concrete cap, or constructing short piers adjacent to and in contact with the laterally loaded pile (Broms, 1972).

Batter Piles.   A cost-effective method to resist lateral loads on pile foundations is to use batter piles. If the lateral load were transmitted to the pile cap in only one direction, then batter piles would be installed in line with the lateral load. For lateral loads in both directions, batter piles would need to be installed on all sides of the pile cap. Inclinations of batter piles typically range from 1:12 to 5:12 (horizontal:vertical).

Early analyses of batter piles were based on the assumption that the vertical piles would only carry vertical loads and the batter piles would only carry lateral loads. This made the analyses easy in the sense that standard pile design from Secs. 6.3.2 and 6.3.3 could be used to determine the allowable load capacity of the vertical piles. The batter piles were designed as compression or tension members with their resultant horizontal component carrying the lateral loads. However, studies have shown that when a pile group with batter piles is subjected to lateral load, the batter piles do indeed resist most of the lateral load, but the vertical piles also carry some lateral load. This results in bending moments being introduced into the vertical piles, which were only designed to resist compressive stresses. The bending moment led to failure of the vertical piles due to lack of tensile reinforcement.

Since the simple analysis is not appropriate for lateral loading of pile groups having batter piles, more complicated methods have been developed. These methods often use complex computer programs that consider the relative stiffness of the entire pile system (i.e., vertical and batter piles) as well as the soil lateral resistance.

Pressuremeter Test.   Section 2.4.4 has presented a discussion of the pressuremeter test. In order to determine the stress-strain curve for horizontal loading, the equipment is either inserted in a predrilled borehole or a self-boring pressuremeter is utilized. Since this field test measures the in situ horizontal response to lateral pressure, it is ideally suited for the design of piles subjected to lateral loads. For information on the design methods, see "Laterally Loaded Piles and the Pressuremeter: Comparison of Existing Methods," (Briaud et. al., 1984).

Lateral Load Tests.   Lateral load tests can be performed on an individual vertical pile, an individual batter pile, pile groups having only vertical piles, and pile groups with both vertical and batter piles. Lateral load tests are considered to be the most accurate method of determining the lateral load capacity of a pile or pile group. Unfortunately, a lateral load test can be very expensive and time consuming. Such a test measures the lateral load versus lateral deflection of the pile or pile group. For lateral load tests on pile groups, a pile cap must be constructed or the piles must be interconnected so that they act together. In a general sense, the type of piles, depth of embedment, and pile cap conditions should be constructed so that they simulate in-service conditions.

Often the pile cap is designed so that it also resists the lateral load due to passive soil pressure along its side. To simulate this condition, the cap can be cast against the soil or fill can be compacted against the side of the pile cap.

The usual method of applying the lateral load is to construct a rigid reaction system and then install hydraulic jacks between the reaction system and the test piles. As shown in Fig. 6.18, rigid reaction systems can consist of the following:

1. *Reaction piles.* A series of vertical and batter piles are anchored to a pile cap. The lateral resistance of the reaction piles must be greater than the test piles.

2. *Deadman.* This can consist of timbers or steel supports bearing against the sides of an embankment or slope so as to provide the necessary solid end support for the hydraulic jacks.

3. *Weighted platform.* A platform can be constructed and then weights added to the platform in order to provide the necessary resistance against the maximum lateral load to be applied to the test piles.

Once the test setup is constructed, the test pile or group is subjected to an incremental increase in lateral load. Depending on the type of lateral load that will be exerted on the pile during its actual use, different types of loading schedules have been developed. For example, ASTM D 3966-95 (2004) provides loading schedules for standard loading, excess loading, cyclic loading, and surge loading. The ideal situation would be to model anticipated field lateral-loading conditions. Thus, for example, if the pile is subjected to cyclic lateral loading for in-service conditions, then a cyclic loading sequence should be used. The pile is usually loaded to a value that is at least twice the lateral design load. During the loading schedule, the horizontal deflection of the pile or pile group is measured. As shown in Fig. 6.19, the lateral load versus pile head deflection can be plotted and used for the design of the pile or to check that the design load will not cause lateral movement in excess of the allowable value.

The allowable lateral load is usually based on the maximum allowable lateral defection of the pile head. For some projects, the structural engineer may specific this value. In other cases, the building code may dictate the allowable lateral load capacity from load tests. For example, the *International Building Code* (2009) in Sec. 1810.3.3.2 states: "The resulting allowable load shall not be more than one-half of that test load that produces a gross lateral movement of 1 inch (25 mm) at the ground surface."

For further information on the lateral load testing of a pile or pile group, see ASTM D 3966-95 (2004), "Standard Test Method for Piles Under Lateral Load."

Example Problem 6.10   Figure 6.19 presents a plot of the lateral load kN versus lateral pile head deflection (mm). The project specifications state that the maximum allowable lateral movement of the pile is 10 mm (0.4 in.). It is also a requirement that a factor of safety of at least 2 be applied to the ultimate lateral load capacity when determined from pile load tests. Based on Fig. 6.19, determine the maximum allowable lateral load.

Solution   From Fig. 6.19, at a pile head lateral deflection of 10 mm (0.4 in.), the corresponding lateral load is 30 kN (6.7 kips).

Also from Fig. 6.19, the ultimate lateral load capacity is 50 kN (11.2 kips). Applying a factor of safety equal to 2, or:

Allowable lateral load = 50/2 = 25 kN (5.6 kips)

The lower value is the governing criteria, or the allowable lateral load is 25 kN (5.6 kips).

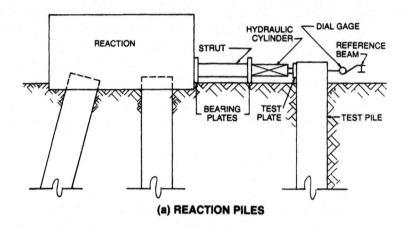

(a) REACTION PILES

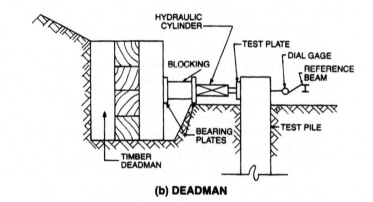

(b) DEADMAN

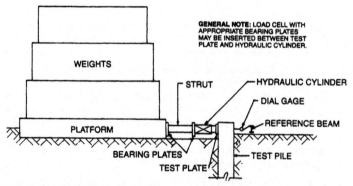

FIGURE 6.18 Typical set ups for applying lateral load with conventional hydraulic jack. (*Reproduced from ASTM D 3966-95, 2004; with permission from the American Society for Testing and Materials.*)

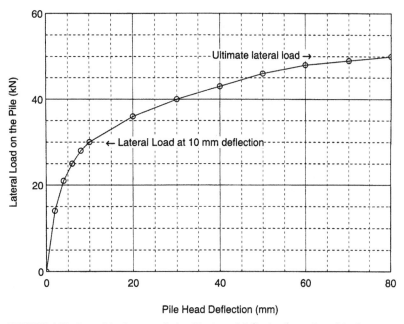

FIGURE 6.19 Lateral load versus pile-head horizontal deflection from a lateral load test on a single pile.

p-y Curves. This method predicts lateral pile response by using a finite-difference model along with horizontal nonlinear springs, with each spring representing the lateral soil resistance as defined by its *p-y* curve. Each *p-y* curve represents the relationship between *p*, which is the horizontal soil resistance (expressed in units of force per length) and *y*, which is the horizontal displacement. The *p-y* curves depend on many factors, such as the soil type, type of loading (i.e., periodic or constant), pile characteristics (e.g., diameter), depth below ground surface, and group interaction effects. Generic *p-y* curves, based on lateral load tests of piles, have been developed for soft clays, stiff clays, and sands and have been incorporated into computer programs that are used to obtain the pile defections as well as the shear and bending moments in the piles (Matlock, 1970; Reese et al., 1974, 1975). For closely spaced piles in a group, it has been proposed that the *p-y* curves be reduced by using "*p*-multipliers" to reduce all the *p*-values on a given *p-y* curve (Brown et al., 1987). The use of reduction factors for pile groups subjected to lateral loads will be discussed in Secs. 6.4.3 and 6.4.4. For more information on *p-y* curves, see Coduto (1994).

Passive Earth Pressure Theory. Since the pile or pier deforms laterally into the soil, passive earth pressure theory can be used to obtain the lateral bearing capacity. Passive earth pressure theory is also used for the design of retaining walls, where for example, the retaining wall footing moves laterally into the soil due to the lateral load induced by the wall backfill soil (Chap. 11).

Table 11.1 presents magnitudes of rotation needed to develop passive pressure. Assuming uniform rotation of the upper 5 ft (1.5 m) of the pile, the amount of lateral movement for the soil to reach passive pressure is about 0.4 in. (1 cm) for a pile embedded in loose sand (i.e., 0.006 times 60 in. = 0.4 in.) and about $2^1/_2$ in. (6 cm) for a pile embedded in soft cohesive soil (i.e., 0.04 times 60 in. = $2^1/_2$ in.). Thus for a loose sand deposit, a pile head that has a maximum allowable lateral deflection of 0.4 in. (1 cm) would only develop passive resistance in approximately the upper 5 ft (1.5 m) of soil.

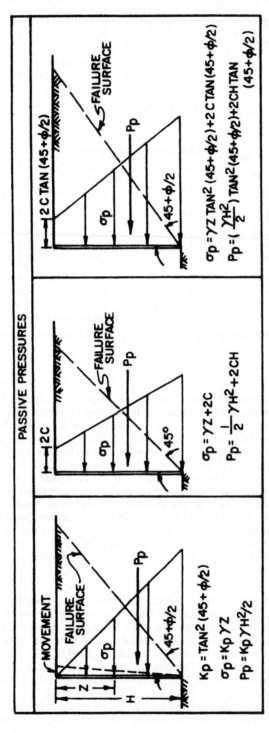

FIGURE 6.20 Passive earth pressure theory. Left diagram presents passive earth pressure theory for cohesionless soil ($c' = 0$). Middle diagram presents passive earth pressure theory for the undrained shear strength of cohesive soil ($\phi = 0$, $c = s_u$). The diagram on the right side is used for combined cohesion and friction, e.g., consolidated undrained shear strength parameters (c and ϕ) for cohesive soil such as shown in Fig. 3.36. *(Reproduced from NAVFAC DM-7.2, 1982.)*

This limited depth of passive resistance has been recognized by the *International Building Code* (2009), which states: "piles driven into firm ground can be considered fixed and laterally supported at 5 ft (1.5 m) below the ground surface and in soft material at 10 ft (3 m) below the ground surface" (Sec. 1810.2.1). The next two sections will present passive earth pressure theory for cohesionless and cohesive soil.

6.4.3 Deep Foundations in Cohesionless Soil

Passive Earth Pressure Theory (Single Pile or Pier). A simple approach to determining the lateral resistance of cohesionless soil is to determine the passive resistance of the soil acting on the side of the pile. The passive resistance of soil can be calculated by using the passive earth pressure coefficient k_p, which is defined on the left side of Fig. 6.20. As indicated in this diagram, for cohesionless soil $c' = 0$, the passive resistance exerted on a rigid foundation element that is laterally pushed into the soil is equal to:

$$P_p = \tfrac{1}{2}k_p\gamma_t H^2 \qquad (6.25)$$

where P_p = passive resistance per unit width of pile (lb/ft or kN/m).
  k_p = passive earth pressure coefficient [$k_p = \tan^2(45° + \tfrac{1}{2}\,\phi')$], dimensionless.
  ϕ' = effective friction angle of the cohesionless soil. The effective friction angle could be determined from drained direct shear tests or from correlations such as shown in Figs. 3.12 to 3.14.
  γ_t = total unit weight of the soil (pcf or kN/m³). The buoyant unit weight is required if the pile is below the groundwater table.
  H = length of the pile that is laterally pushed into the uniform cohesionless soil and subjects the pile to passive pressure (ft or m). Because of restrictions on lateral deflections, typically only the upper few feet of the pile will be subjected to passive pressure.

As the pile is pushed laterally into the soil, the soil resistance tends to develop in the form of a wedge. Hence, it is often recommended that the passive resistance be applied over the distance of 3 pile diameters (Broms, 1965; Coduto, 1994). Multiplying Eq. 6.25 by $3D$, where D diameter of the pile, the final result is as follows:

$$P_p = 1.5k_p\gamma_t DH^2 \qquad (6.26)$$

Equation 6.26 should only be used when the pile spacing is equal to or greater then $3D$. Using a value of $3D$ to account for the anticipated wedge-type failure of the soil as the pile laterally deflects is probably an acceptable assumption for medium to dense sand. For loose sand, the pile may simply laterally punch into the sand, much like a punching shear failure for the bearing capacity of shallow foundations on loose sands. Hence for nondisplacement (i.e., cast-in-place piles) and low-displacement piles in loose sands, it may be appropriate to only multiply the result from Eq. 6.25 by the width of the pile, or:

$$P_p = \tfrac{1}{2}k_p\gamma_t DH^2 \qquad (6.27)$$

The final step would be to divide the passive resistance P_p by a factor of safety. The factor of safety should be at least 3.

Example Problem 6.11   A pile that has a diameter of 1.0 ft (0.3 m) and that is 20 ft (6 m) long is driven into a uniform sand deposit that has a medium density. The sand has an effective friction angle $\phi' = 34°$ and a total unit weight = 125 pcf (20 kN/m³). The groundwater table is well below the bottom of the pile. Assume that only the upper 5 ft (1.5 m) of the pile deflects sufficiently to develop passive pressure. Determine the amount of lateral deflection of the pile head assuming the values in Table 11.1 are applicable. Also determine the passive resistance of this upper 5 ft (1.5 m) of soil.

(Continued)

Solution From Table 11.1, for sand having a medium density, use a value of horizontal displacement divided by height of pile subjected to passive resistance of:

$$\frac{1}{2}(0.002 + 0.006) = 0.004$$

Horizontal deflection of pile head for passive resistance in the upper 5 ft (1.5 m) of soil = (0.004)(60 in.) = 1/4 in. (0.6 cm)

To determine passive resistance, use Eq. (6.26):

$$P_p = 1.5k_p\gamma_t DH^2 = 1.5\ [\tan^2(45° + 34°/2)](125)(1.0)(5)^2 = 16,600\ \text{lb}\ (72\ \text{kN})$$

Passive Earth Pressure Theory (Pile Groups). When considering the passive resistance of a closely spaced group of piles, the total lateral resistance cannot be calculated as the value of Eq. 6.26 or 6.27 times the number of piles. The reasons are because of overlapping passive soil wedges and soil gap formation. The formation of a gap, such as shown in Fig. 6.17, results in a reduction in passive resistance for the trailing row.

Shadowing is the process where the passive soil wedge in the trailing row of piles overlaps into the leading row of piles, resulting in a reduction of passive resistance. The effects of shadowing have led to the development of row reduction factors. Hence the leading pile row may have no or a very slight reduction in lateral capacity, while the pile rows that trail behind the leading row could have reduction factors of 0.3 to 0.4.

Example Problem 6.12 Use the data from the previous example. Assume the pile group consists of a 2 by 2 pile group with no row reduction for the leading row and a row reduction factor of 0.4 for the trailing row. Determine the passive resistance for the pile group.

Solution The pile group consists of four piles, two in the leading row and two in the trailing row. Passive resistance = (2)(16,600) + (2)(0.4)(16,600) = 46,500 lb (202 kN)

Study of Pile Group Behavior. There have been experimental lateral loadings of full-scale pile groups embedded in cohesionless soil. For example, results from a study by Brown et al. (1988) are as follows:

- For the 3 by 3 pile group tested, results clearly showed the effect of shadowing, in which the soil resistance of a pile in a trailing row was greatly reduced because of the presence of the pile ahead of it. The soil resistance of the piles in the leading row was only slightly less than that of a laterally loaded isolated single pile. The key in the design of pile groups subjected to lateral loads is to account for the loss of soil resistance for the piles in the trailing rows. One method to account for this loss of soil resistance is to use a "*p*-multiplier," which is a constant used to modify the *p-y* curves of a single isolated pile.

- When the pile group was subjected to two-way cyclic lateral loading (i.e., back and forth lateral loading) at the pile head, the shadowing effect was not appreciably diminished.

- Although the two-way cyclic lateral loading did not influence the shadowing effect, there was densification of the soil adjacent the piles related to sand falling into the voids around the pile as it was pushed back and forth. For 100 cycles of loading, about 9 in. (23 cm) of ground surface settlement directly around the pile was observed due to this effect of sand falling into the voids created by pushing the pile back and forth. This densification effect of the sand caused by the back and forth lateral loads appeared to improve the soil resistance at subsequently larger lateral loads.

- Cyclic lateral loading in only one direction would not produce as much densification as the two-way cyclic lateral loading and would result in a greater loss of soil resistance with increasing cycles of load. In this respect, the results of the experiment show the beneficial effects of back and forth cyclic loading but the worst case would be lateral loading in only one direction. The results of the study clearly show the importance of the nature of the loading.

6.4.4   Deep Foundations in Cohesive Soil

Passive Earth Pressure Theory for Short-Term Condition (Single Pile or Pier).   A simple approach to determining the lateral resistance of cohesive soil for a quick lateral loading condition (i.e., short-term condition) is to use a total stress analysis, assume large deformations, and determine the passive resistance of the soil acting on the side of the pile. The passive resistance of the cohesive soil can be calculated by using the middle diagram in Fig. 6.20 (i.e., $\phi = 0$, $s_u = c$). As indicated in this diagram, for cohesive soil, the passive resistance exerted on a rigid foundation element that is laterally pushed into the soil is equal to:

$$P_p = \frac{1}{2}\gamma_t H^2 + 2cH = \frac{1}{2}\gamma_t H^2 + 2s_u H \tag{6.28}$$

where P_p = passive resistance per unit width of pile (lb/ft or kN/m)
γ_t = total unit weight of the soil (pcf or kN/m³)
H = length of the pile that is laterally pushed into the uniform cohesive soil and subjects the pile to passive pressure (ft or m). Because of restrictions on lateral deflections, typically only the upper few feet of the pile will be subjected to passive pressure.
$c = s_u$ = undrained shear strength (total stress analysis), such as obtained from vane shear strength tests or unconfined compression tests (psf or kPa)

For cohesive soils, the passive resistance is usually only applied over the diameter of the pile or pier. Multiplying Eq. 6.28 by D, where D is the diameter of the pile, the final result is as follows:

$$P_p = \frac{1}{2}\gamma_t DH^2 + 2cDH = \frac{1}{2}\gamma_t DH^2 + 2s_u DH \tag{6.29}$$

The diagram on the right side of Fig. 6.20 can be used for combined cohesion and friction. These undrained shear strength parameters are often obtained from triaxial tests, such as the unconsolidated undrained triaxial compression test (ASTM D 2850-03, 2004) or the consolidated undrained triaxial compression test (ASTM D 4767-02, 2004, see Fig. 3.36).

The final step would be to divide the passive resistance P_p by a factor of safety. The factor of safety should be at least 3.

Example Problem 6.13   A pile that has a diameter of 1.0 ft (0.3 m) and that is 20 ft (6 m) long is driven into a uniform clay deposit that has a medium consistency. The clay has a uniform undrained shear strength $s_u = 750$ psf (36 kPa) and a total unit weight = 125 pcf (20 kN/m³). Assume that only the upper 5 ft (1.5 m) of the pile deflects sufficiently to develop passive pressure. Determine the amount of lateral deflection of the pile head assuming the values in Table 11.1 are applicable. Also determine the passive resistance of this upper 5 ft (1.5 m) of soil.

Solution   From Table 11.1, for clay having a medium consistency, use a value of horizontal displacement divided by height of pile subjected to passive resistance of

$$\frac{1}{2}(0.02 + 0.04) = 0.03$$

Horizontal deflection of pile head for passive resistance in the upper 5 ft (1.5 m) of soil is $(0.03)(60 \text{ in.}) = 1.8$ in. (4.6 cm)

To determine passive resistance, use Eq. (6.29):

$$P_p = \frac{1}{2}\gamma_t DH^2 + 2s_u DH = \frac{1}{2}(125)(1.0)(5)^2 + (2)(750)(1)(5) = 9,100 \text{ lb} (40 \text{ kN})$$

Passive Earth Pressure Theory for Long-Term Condition (Single Pile or Pier).   The above discussion is for the short-term lateral loading of a pile or pier in cohesive soil, using a total stress analysis. A much more complex case would be a pile embedded in cohesive soil and subjected to a constant lateral load. Both the short-term (total stress analysis, Eq. 6.29) and long-term (effective stress analysis) would need to be evaluated.

For the effective stress analysis, the effective shear strength parameters c' and ϕ' could be obtained from consolidated undrained triaxial compression tests with pore pressure measurements (e.g., see test data in Sec. 3.5.2). If the groundwater table is shallow, its effect may need to be included in the analysis (i.e., use buoyant unit weight). Ignoring the effective cohesion of the soil, Eq. 6.27 could be used to determine the passive resistance of the pile or pier. Depending on the amount of allowable lateral movement, usually only the upper few feet of soil will be subjected to passive pressure. For soft cohesive soil that is highly plastic and creeps around the pile such as shown in Fig. 6.17, there could be additional lateral deflection of the pile head.

Example Problem 6.14   A pile that has a diameter of 1.0 ft (0.3 m) and that is 20 ft (6 m) long is driven into a uniform clay deposit that has a medium consistency. The clay has a uniform undrained shear strength s_u = 750 psf (36 kPa) and a total unit weight = 125 pcf (20 kN/m³). Consolidated undrained triaxial compression tests with pore water pressure measurements were also performed on specimens of the clay and indicate effective stress parameters of c' = 100 psf and ϕ' = 27°. Assume that only the upper 5 ft (1.5 m) of the pile deflects sufficiently to develop passive pressure. Determine the passive resistance of this upper 5 ft (1.5 m) of soil for both the short-term and long-term conditions. For the long-term condition, assume the groundwater table is at the ground surface and neglect possible creep of the soil around the pile.

Solution   The short-term condition was analyzed in the prior example (using Eq. 6.29), or:

$$P_p = \tfrac{1}{2}\gamma_t DH^2 + 2s_u DH = \tfrac{1}{2}(125)(1.0)(5)^2 + (2)(750)(1)(5) = 9100 \text{ lb (40 kN)}$$

For the long-term condition, use Eq. 6.27 by assuming c' = 0. Since the groundwater table is at ground surface, use buoyant unit weight (125 – 62.4 = 62.6 pcf), or:

$$P_p = \tfrac{1}{2}k_p\gamma_b DH^2 = \tfrac{1}{2}[\tan^2(45° + 27°/2)](62.6)(1.0)(5)^2 = 2100 \text{ lb (9.3 kN)}$$

For this example, the long-term condition governs.

Passive Earth Pressure Theory (Pile Groups).   When considering the passive resistance of a closely spaced group of piles in cohesive soil, the total resistance cannot be calculated as the value of Eq. 6.29 times the number of piles. The reasons are because of the overlapping of passive soil wedges, soil gap formation, and creep of clay around the pile (long-term condition). As previously mentioned, shadowing is the process where the passive soil wedge in the trailing row of piles overlaps into the leading row of piles, resulting in a reduction of passive resistance.

Especially for the leading row, the row reduction factors tend to be lower for cohesive soil than cohesionless soil. Typical values of row reduction factors are 0.6 for the leading row and 0.3 to 0.4 for the trailing row.

Example Problem 6.15   Use the data from the previous example. Assume the pile group consists of a 2 by 2 pile group with row reduction factors 0.6 for the leading row and 0.3 for the trailing row. Determine the passive resistance of the pile group.

Solution   The pile group consists of four piles, two in the leading row and two in the trailing row. The long-term conditions govern and hence the passive resistance equals

$$(2)(0.6)(2,100) + (2)(0.3)(2,100) = 3,800 \text{ lb (17 kN)}$$

Study of Pile Group Behavior. There have been experimental lateral loadings of full-scale pile groups embedded in cohesive soil. For example, results from a study by Rollins et al. (1998) are as follows:

• Lateral load capacity in the pile group was a function of row position. For a given deflection, piles in trailing rows carried significantly less load than piles in the leading row due to shadowing.

• For the 3 by 3 pile group tested, with all piles spaced at three pile diameters, the results clearly showed the effect of shadowing, in which the soil resistance of a pile in a trailing row was greatly reduced because of the presence of the pile ahead of it. The *p*-multipliers were found to be 0.6, 0.38, and 0.43 for the front, middle, and back rows respectively.

• Based on the study by Cox et al. (1984) using 5.4 mm diameter piles tested in the laboratory, Rollins et al. (1998) recommended that no lateral load reduction is required when the pile spacing is equal to or greater than six pile diameters.

In summary, for pile groups in cohesive soil, a reduction in vertical load capacity is based on pile spacing (Fig. 6.14), while a reduction in lateral load capacity for closely spaced piles is based on row location (i.e., leading or trailing rows).

6.5 GEOTECHNICAL EARTHQUAKE ENGINEERING

6.5.1 Introduction

The documented cases of bearing capacity failures during earthquakes indicate that usually the following three factors (separately or in combination) are the cause of the failure:

1. *Soil shear strength.* Common problems include an overestimation of the shear strength of the underlying soil. Another common situation leading to a bearing capacity failure is the loss of shear strength during the earthquake, because of the liquefaction of the soil or the loss of shear strength for sensitive clays.

2. *Structural load.* Another common problem is that the structural load at the time of the bearing capacity failure was greater than that assumed during the design phase. This can often be the case when the earthquake causes rocking of the structure, and the resulting structural overturning moments produce significant cyclic vertical thrusts on the foundation elements and underlying soil.

3. *Change in site conditions.* An altered site can produce a bearing capacity failure. For example, if the groundwater table rises, then the potential for liquefaction is increased. Another example would be the construction of an adjacent excavation, which could result in a reduction in support and a bearing capacity failure.

The most common cause of a seismic bearing capacity failure or excessive settlement is due to liquefaction of the underlying soil. Section 13.4 will present the analyses used to determine if a soil will liquefy during the design earthquake.

When presenting the recommendations for the allowable bearing pressures at a site, it is common practice for the geotechnical engineer to recommend that the allowable bearing pressure be increased by a factor of one-third when performing seismic analyses. For example, the *International Building Code* (2009) states:

> An increase [in allowable bearing capacity] of one-third is permitted when using the alternate load combinations in Sec. 1605.3.2 that include wind or earthquake.

In soil reports, it is commonly recommended that for the analysis of earthquake loading, the allowable bearing pressure may be increased by a factor of $1/_3$. The rational behind this recommendation is that the allowable bearing pressure has an ample factor of safety and thus for seismic analyses, a lower factor of safety would be acceptable. Usually the above recommendation is appropriate for the following materials:

1. Massive crystalline bedrock and sedimentary rock that remains intact during the earthquake
2. Dense to very dense granular soil
3. Heavily overconsolidated cohesive soil, such as very stiff to hard clays

These materials do not lose shear strength during the seismic shaking and therefore an increase in bearing pressure is appropriate.

A one-third increase in allowable bearing pressure should not be recommended for the following materials:

1. Foliated or friable rock that fractures apart during the earthquake
2. Loose soil subjected to liquefaction or a substantial increase in excess pore water pressure
3. Sensitive clays that lose shear strength during the earthquake
4. Soft clays and organic soils that are overloaded and subjected to plastic flow

These materials have a reduction in shear strength during the earthquake. Since the seismic shaking weakens them, the static values of allowable bearing pressures should not be increased for the earthquake analyses. In fact, the allowable bearing pressure may actually have to be reduced to account for the weakening of the soil during the earthquake. The remainder of this section will deal with the determination of the bearing capacity of soils that are weakened by the seismic shaking.

6.5.2 Bearing Capacity Analyses for Liquefied Soil

For cases involving earthquake-induced liquefaction failures or punching shear failures, the depth of soil involvement could exceed the footing width. For buildings with numerous spread footings that occupy a large portion of the building area, the individual pressure bulbs from each footing may combine and thus the entire width of the building could be involved in a bearing capacity failure. Either a total stress analysis or an effective stress analysis must be used in order to determine the bearing capacity of a foundation. Table 4.9 presents a summary of the type of analyses and the shear strength parameters that should be used for the bearing capacity calculations.

Figure 6.21 illustrates the earthquake-induced punching shear analysis. The soil layer portrayed by dashed lines represents unliquefiable soil that is underlain by a liquefied soil layer. For the punching

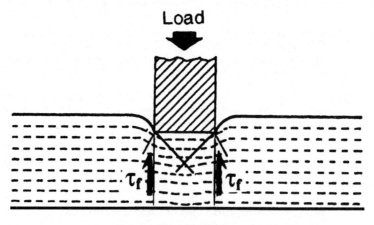

FIGURE 6.21 Illustration of a punching shear analysis. The dashed lines represent unliquefiable soil that is underlain by a liquefied soil layer. In the analysis, the footing will punch vertically downward and into the liquefied soil.

shear analysis, it is assumed that the load will cause the foundation to punch straight downward through the upper unliquefiable soil layer and into the liquefied soil layer. As shown in Fig. 6.21, this assumption means that there will be vertical shear surfaces in the soil that start at the sides of the footing and extend straight downward to the liquefied soil layer. It is also assumed that the liquefied soil has no shear strength.

Using the assumptions outlined above, the factor of safety F can be calculated as follows:

For strip footings:

$$F = \frac{R}{P} = \frac{2T\tau_f}{P}$$ (6.30)

For spread footings:

$$F = \frac{R}{P} = \frac{2\,(B+L)(T\tau_f)}{P}$$ (6.31)

where F = factor of safety. For seismic analyses, usually the minimum acceptable factor of safety is 5.

R = shear resistance of the soil. For strip footings, R is the shear resistance per unit length of the footing (lb/ft or kN/m). For spread footings, R is the shear resistance beneath the entire footing perimeter (lb or kN).

P = footing load. For strip footings, P is the load per unit length of the footing (lb/ft or kN/m). For spread footings, P is the total load of the footing (lb or kN). The footing load includes dead, live, and seismic loads acting on the footing as well as the weight of the footing itself. Typically the structural engineer would provide the value of P.

T = vertical distance from the bottom of the footing to the top of the liquefied soil layer (ft or m)

τ_f = shear strength of the unliquefiable soil layer (psf or kPa)

B = width of the footing (ft or m)

L = length of the footing (ft or m)

In Eq. 6.31, the term $2(B + L)$ represents the entire perimeter of the spread footing. When this term is multiplied by T, it represents the total perimeter area that the footing must push through in order to reach the liquefied soil layer. For an assumed footing size and given loading condition, the only unknowns in Eqs. 6.30 and 6.31 are the vertical distance from the bottom of the footing to the top of the liquefied soil layer T and the shear strength of the unliquefiable soil layer τ_f. The value of T would be based on the liquefaction analysis (Sec. 13.4) and the proposed depth of the footing. The shear strength of the unliquefiable soil layer τ_f can be calculated as follows:

1. For an unliquefiable soil layer consisting of cohesive soil (e.g., clays) use a total stress analysis:

$$\tau_f = s_u$$ (6.32)

or

$$\tau_f = c + \sigma_h \tan \phi$$ (6.33)

where s_u = undrained shear strength of the cohesive soil (total stress analysis), psf or kPa. The undrained shear strength can be obtained from unconfined compression tests or vane shear tests.

c, ϕ = undrained shear strength parameters (total stress analysis). These undrained shear strength parameters are often obtained from triaxial tests, such as the unconsolidated undrained triaxial compression test (ASTM D 2850-03, 2004) or the consolidated undrained triaxial compression test (ASTM D 4767-02, 2004, see Fig. 3.36).

σ_h = horizontal total stress (psf or kPa). Since vertical shear surfaces are assumed (see Fig. 6.21), then the normal stress acting on the shear surfaces will be the horizontal total stress. For cohesive soil, σ_h is often assumed to be equal to $\frac{1}{2}\,\sigma_v$.

2. For an unliquefiable soil layer consisting of cohesionless soil (e.g., sands) use an effective stress analysis:

$$\tau_f = \sigma'_h \tan \phi' = k_o \sigma'_{vo} \tan \phi' \qquad (6.34)$$

where σ'_h = horizontal effective stress (psf or kPa). Since vertical shear surfaces are assumed (see Fig. 6.21), then the normal stress acting on the shear surface will be the horizontal effective stress. The horizontal effective stress σ'_h is equal to the coefficient of earth pressure at rest k_o times the vertical effective stress σ'_{vo} or $\sigma'_h = k_o \sigma'_{vo}$.
ϕ' = effective friction angle of the cohesionless soil (effective stress analysis). The effective friction angle could be determined from drained direct shear tests or from empirical correlations such as shown in Figs. 3.12 and 3.14.

The following example problems illustrate the use of Eqs. 6.30 and 6.31:

Example Problem 6.16   For cohesive surface layer (total stress analysis)—Assume that based on a liquefaction analysis (Sec. 13.4), there will be a zone of liquefaction that extends from a depth of 3 to 8.5 m below ground surface. Assume the surface soil (upper 3 m) consists of an unliquefiable cohesive soil. It is proposed to construct a sewage disposal plant. The structural engineer would like to use shallow strip footings to support exterior walls and interior spread footings to support isolated columns. It is proposed that the bottom of the footings will be at a depth of 0.5 m below ground surface. The structural engineer has also indicated that the maximum total loads (including the weight of the footing and the dynamic loads) are 50 kN/m for the strip footings and 500 kN for the spread footings. It is desirable to use 1-m wide strip footings and square spread footings that are 2 m wide.

For the 3-m thick unliquefiable cohesive soil layer, assume that the undrained shear strength s_u of the soil is equal to 50 kPa. Calculate the factor of safety of the footings using Eqs. 6.30 and 6.31.

Solution   In order to calculate the factor of safety in terms of a bearing capacity failure for the strip and spread footings, the following values are used:

$P = 50$ kN/m for the strip footing and 500 kN for the spread footing.

$T = 2.5$ m (i.e., total thickness of the unliquefiable soil layer minus footing embedment depth = 3 m − 0.5 m = 2.5 m)

$$\tau_f = s_u = 50 \text{ kPa} = 50 \text{ kN/m}^2$$

$$B = L = 2 \text{ m}$$

Substituting the above values into Eqs. 6.30 and 6.31:

For the strip footing:

$$F = 2T\tau_f/P = (2)(2.5 \text{ m})(50 \text{ kN/m}^2)/(50 \text{ kN/m}) = 5.0$$

For the spread footing:

$$F = 2 (B + L) T\tau_f/P = (2)(2 \text{ m} + 2 \text{ m})(2.5 \text{ m})(50 \text{ kN/m}^2)/(500 \text{ kN}) = 2.0$$

For a seismic analysis, a factor of safety of 5.0 would be acceptable, but the factor of safety of 2.0 would probably be too low.

Example Problem 6.17   For cohesionless surface layer (effective tress analysis)—Use the same data, but assume the 3 m thick unliquefiable surface layer is sand with an effective friction angle ϕ' equal to 32°, a coefficient of earth pressure at rest (k_o) equal to 0.5, and a total unit weight = 18.3 kN/m^3. Also assume that the groundwater table is at a depth of 3 m below the existing ground surface. Calculate the factor of safety of the footings using Eqs 6.30 and 6.31.

Solution   In order to calculate the factor of safety in terms of a bearing capacity failure for the strip and spread footings, the following values are used:

$P = 50$ kN/m for the strip footing and 500 kN for the spread footing.

$T = 2.5$ m (i.e., total thickness of the unliquefiable soil layer minus footing embedment depth = 3 m − 0.5 m = 2.5 m)

$$\sigma'_{vo} = \sigma_v - u$$

Since the soil is above the groundwater table, assume $u = 0$. Using a total unit weight = 18.3 kN/m^3 and an average depth of $(0.5 + 3.0)/2 = 1.75$ m

$$\sigma'_{vo} = (18.3)(1.75) = 32 \text{ kPa}$$

$$\tau_f = k_o \sigma'_{vo} \tan \phi' = (0.5)(32 \text{ kPa})(\tan 32°) = 10 \text{ kPa} = 10 \text{ kN/m}^2 \text{ (Eq. 6.34)}$$

$$B = L = 2 \text{ m}$$

Substituting the above values into Eqs. 6.30 and 6.31:

For the strip footing:

$$F = 2T\tau_f/P = (2)(2.5 \text{ m})(10 \text{ kN/m}^2)/(50 \text{ kN/m}) = 1.0$$

For the spread footing:

$$F = 2 (B + L) T\tau_f/P = (2)(2 \text{ m} + 2 \text{ m})(2.5 \text{ m})(10 \text{ kN/m}^2)/(500 \text{ kN}) = 0.4$$

For the seismic bearing capacity analyses, these factors of safety would indicate that both the strip and spread footings would punch down through the upper sand layer and into the liquefied soil layer.

As a final check, the factor of safety (F) must be calculated for the static bearing capacity condition (i.e., non-earthquake analysis, Sec. 6.2). The reason is because the factor of safety for the static condition could be the governing criteria. This often occurs when the liquefied soil layer is at a significant depth below the bottom of the footing, or in other words, at high values of T/B. In any event, the factor of safety for the non-earthquake bearing capacity condition (Sec. 6.2) should be at least 3 and the factor of safety for the earthquake punching shear analysis (Eqs. 6.30 and 6.31) should be at least 5.

6.5.3 Bearing Capacity Analyses for Granular Soil with Earthquake-Induced Pore Water Pressures

The previous section has dealt with soil that is weakened during the earthquake due to liquefaction. This section deals with granular soil that does not liquefy, but rather there is a reduction in shear strength due to an increase in pore water pressure. Examples include sands and gravels that are below the groundwater table and have a factor of safety against liquefaction that is greater than 1.0, but less than 2.0. If the factor of safety against liquefaction is greater than 2.0, the earthquake-induced excess pore water pressures will typically be small enough that their effect can be neglected.

Using the Terzaghi bearing capacity equation and an effective stress analysis, and recognizing that sands and gravels are cohesionless (i.e., $c' = 0$), Eq. 6.1 reduces to following:

$$q_{ult} = \frac{1}{2}\gamma BN_\gamma + \gamma_t D_f N_q \tag{6.35}$$

For shallow foundations, it is best to neglect the second term (i.e., $\gamma_t D_f N_q$) in Eq. 6.35. This is because this term represents the resistance of the soil located above the bottom of the footing, which may not be mobilized for a punching shear failure into the underlying weakened granular soil layer. Thus neglecting the second term in Eq. 6.35, the result is as follows:

$$q_{ult} = \frac{1}{2}\gamma_t BN_\gamma \tag{6.36}$$

Assuming that the location of the groundwater table is close to the bottom of the footing, the buoyant unit weight γ_b is used in place of the total unit weight γ_t in Eq. 6.36. In addition, since this is an effective stress analysis, the increase in excess pore water pressures that are generated during the design earthquake must be accounted for in Eq. 6.36. Using Fig. 4.25 can accomplish this, which is a plot of the pore water pressure ratio (i.e., $r_u = u_e/\sigma'$) versus the factor of safety against liquefaction. Using the buoyant unit weight γ_b in place of the total unit weight γ_t and inserting the term $1 - r_u$ to account for the excess pore water pressures that are generated by the design earthquake, the final result for the ultimate bearing capacity q_{ult} is as follows:

For strip footings:

$$q_{ult} = \frac{1}{2}(1 - r_u)\gamma_b BN_\gamma \tag{6.37}$$

For spread footings based on Eq. 6.6:

$$q_{ult} = 0.4(1 - r_u)\gamma_b BN_\gamma \tag{6.38}$$

where r_u = pore water pressure ratio from Fig. 4.25 (dimensionless). In order to determine r_u, the factor of safety against liquefaction of the soil located below the bottom of the footing must be determined (i.e., see Sec. 13.4). Equations 6.37 and 6.38 are only valid if the factor of safety against liquefaction is greater than 1.0. When the factor of safety against liquefaction is greater than 2.0, the Terzaghi bearing capacity equation can be utilized, taking into account the location of the groundwater table (see Sec. 6.2.2).

γ_b = buoyant unit weight of the soil below the footing (pcf or kN/m^3). Equations 6.37 and 6.38 were developed based on an assumption that the groundwater table is located near the bottom of the footing or it is anticipated that the groundwater table could rise so that it is near the bottom of the footing.

B = width of the footing (ft or m)

N_γ = bearing capacity factor (dimensionless). Figure 6.5 can be used to determine the value of N_γ based on the effective friction angle ϕ' of the cohesionless soil.

The final step would be to divide the ultimate bearing capacity q_{ult} by a factor of safety in order to calculate the allowable bearing capacity q_{all}. For seismic analyses, the factor of safety is usually at least 5.

Example Problem 6.18 A site consists of a sand deposit with a fluctuating groundwater table. The proposed development will consist of buildings having shallow strip footings to support bearing walls and interior spread footings to support isolated columns. The expected depth of the footings will be between 0.5 to 1.0 m. Assume that the groundwater table could periodically rise to a level that is close to the bottom of the footings. Also assume the following parameters: buoyant unit weight of the sand = 9.7 kN/m^3, the sand below the groundwater table has a factor of safety against liquefaction = 1.3 for the design earthquake, the effective friction angle of the sand (ϕ') = 32°, and the footings will have a minimum width of 1.5 and 2.5 m for the strip and spread footings, respectively. Using a factor of safety of 5, determine the allowable bearing capacity of the footings.

(Continued)

Solution   Using the following values:

$$\gamma_b = 9.7 \text{ kN/m}^3$$

$N_\gamma = 21$ (entering Fig. 6.5 with $\phi' = 32°$, and intersecting the N_γ curve, the value of N_γ from the vertical axis is equal to 21)

$B = 1.5$ m for the strip footings and 2.5 m for the spread footings

$r_u = 0.20$ (entering Fig. 4.25 with a factor of safety against liquefaction = 1.3, the value of r_u for sand varies from 0.05 to 0.35. Using an average value, $r_u = 0.20$)

Inserting the above values into Eqs. 6.37 and 6.38, or:

For the strip footings:

$$q_{ult} = \tfrac{1}{2} (1 - r_u) \, \gamma_b BN_\gamma = \tfrac{1}{2} (1 - 0.20)(9.7 \text{ kN/m}^3)(1.5 \text{ m})(21) = 120 \text{ kPa}$$

And using a factor of safety = 5.0

$$q_{all} = q_{ult}/F = (120 \text{ kPa})/5.0 = 24 \text{ kPa}$$

For the spread footings:

$$q_{ult} = 0.4 (1 - r_u) \, \gamma_b BN_\gamma = 0.4 (1 - 0.20)(9.7 \text{ kN/m}^3)(2.5 \text{ m})(21) = 160 \text{ kPa}$$

And using a factor of safety = 5.0

$$q_{all} = q_{ult}/F = (160 \text{ kPa})/5.0 = 32 \text{ kPa}$$

Thus provided the strip and spread footings are at least 1.5 and 2.5 m wide, respectively, the allowable bearing capacity is equal to 24 kPa (500 psf) for the strip footings and 32 kPa (670 psf) for the spread footings. These allowable bearing pressures would be used to determine the size of the footings based on the anticipated dead, live, and seismic loads.

6.5.4   Bearing Capacity Analyses for Cohesive Soil Weakened by the Earthquake

Cohesive soils and organic soils can also be susceptible to a loss of shear strength during the earthquake. Examples include sensitive clays, which lose shear strength when they are strained back and forth. When dealing with such soils, it is often desirable to limit the stress exerted by the footing during the earthquake so that it is less than the maximum past pressure σ'_{vm} of the cohesive or organic soils. As discussed in Chap. 8, the maximum past pressure is the highest vertical effective stress that the clay was subjected to and completely consolidated under. The goal in limiting the imposed footing stress to a value of σ'_{vm} is to prevent the soil from squeezing out or deforming laterally from underneath the footing.

It is often very difficult to predict the amount of earthquake-induced settlement for foundations bearing on cohesive and organic soils. One approach is to ensure that the foundation has an adequate factor of safety in terms of a bearing capacity failure. In order to perform a bearing capacity analysis, a total stress analysis can be performed by assuming that $c = s_u$. Neglecting the third term in Eq. 6.1 and assuming relatively constant undrained shear strength versus depth below the footing, the ultimate bearing capacity is as follows:

For strip footings:

$$q_{ult} = cN_c = 5.5s_u \tag{6.39}$$

For spread footings:

$$q_{ult} = cN_c\,[1 + 0.3\,(B/L)] = 5.5s_u\,[1 + 0.3\,(B/L)] \tag{6.40}$$

For a given footing size, the only unknown in Eqs. 6.39 and 6.40 is the undrained shear strength s_u. Table 4.9 presents guidelines in terms of the undrained shear strength that should be utilized for earthquake engineering analyses. These guidelines for the selection of the undrained shear strength s_u as applied to bearing capacity analyses are as follows:

1. *Earthquake parameters.* These parameters define the nature of the design earthquake, such as the peak ground acceleration a_{max} and earthquake magnitude. The higher the peak ground acceleration and the higher the magnitude of the earthquake, the greater the tendency for the cohesive soil to be strained and remolded by the earthquake shaking.

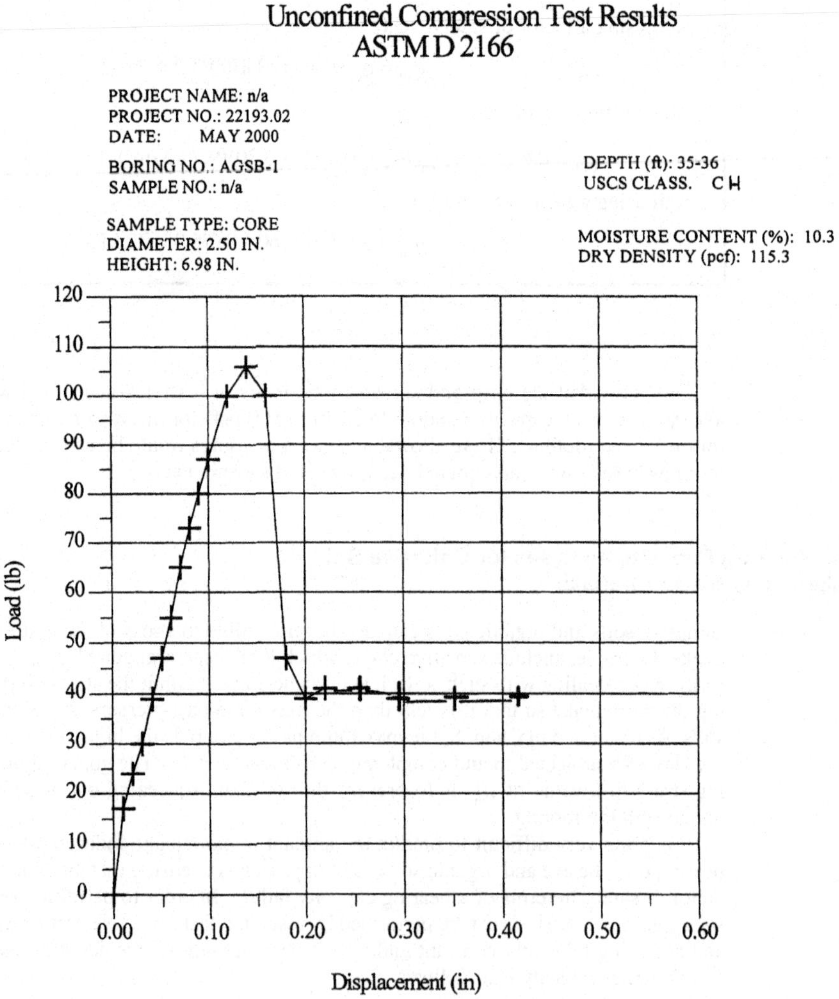

Unconfined Compression Test Results
ASTM D 2166
PROJECT NAME: n/a
PROJECT NO.: 22193.02
DATE: MAY 2000
BORING NO.: AGSB-1
SAMPLE NO.: n/a
DEPTH (ft): 35-36
USCS CLASS. CH
SAMPLE TYPE: CORE
DIAMETER: 2.50 IN.
HEIGHT: 6.98 IN.
MOISTURE CONTENT (%): 10.3
DRY DENSITY (pcf): 115.3
Load (lb)
0
10
20
30
40
50
60
70
80
90
100
110
120
0.00
0.10
0.20
0.30
0.40
0.50
0.60
Displacement (in)

FIGURE 6.22 Stress-strain curve from an unconfined compression test on a clay of high plasticity.

2. *Soil behavior.* The important soil properties for the bearing capacity analysis are the undrained shear strength s_u, sensitivity S_t, stress-strain behavior of the soil, and the maximum past pressure (σ'_{vm}, Chap. 8), as follows:

   a. Undrained shear strength: As the cohesive soil is strained back and forth during the earthquake, there could be a weakening of the soil. Consider a reduction in undrained shear strength based on the soils sensitivity and stress-strain behavior.

   b. Sensitivity: Highly sensitive and quick clays could be significantly weakened during the earthquake. On the other hand, clays of low sensitivity are less likely to lose shear strength upon straining during the earthquake.

   c. Stress-strain behavior: Figure 6.22 shows the stress-strain curve for an unconfined compression test on clay of high plasticity. The clay reached the peak undrained shear strength at a relatively low displacement and then with further displacement, the shear strength was reduced to an ultimate value. If the project is likely to be subjected to severe ground shaking, one approach may be to use the ultimate undrained shear strength in the bearing capacity analyses.

   d. Maximum past pressure σ'_{vm}: As previously mentioned, the goal in limiting the imposed footing stress to a value of σ'_{vm} is to prevent the soil from squeezing out or deforming laterally from underneath the footing.

3. *Rocking.* The increase in shear stress caused by the dynamic loads acting on the foundation must be considered in the analysis. Lightly loaded foundations would tend to produce the lowest dynamic loads, while heavy and tall buildings would subject the foundation to high dynamic loads due to rocking.

Given the many variables as outlined previously, it takes considerable experience and judgment in the selection of the undrained shear strength s_u to be used in Eqs. 6.39 and 6.40. The value of s_u should be adjusted for the anticipated loss in shear strength as the cohesive soil is strained back and forth during the earthquake. This weakening of the soil during the earthquake will be further discussed in Sec. 13.5.

NOTATION

The following notation is used in this chapter:

a_{max} = peak ground acceleration

B = width of the footing (Secs. 6.1, 6.2, and 6.5)

B = width of the pile (Sec. 6.3)

B = width of the pile group measured to the outside edge of the perimeter piles (for down-drag load)

B' = reduced footing width to account for eccentricity of load

c = cohesion based on a total stress analysis

c' = cohesion based on an effective stress analysis

c_A = adhesion between the cohesive soil and the pile or pier perimeter

c_1, c_2 = cohesion based on a total stress analysis (two soil layers)

D = depth of the pile group

D = diameter of the pile (Sec. 6.4)

D_f = depth below ground surface to the bottom of the footing

D_r = relative density

e = eccentricity of the vertical load Q

e_1, e_2 = eccentricities along and across the footing (Fig. 6.8)

F = factor of safety

h' = depth of the groundwater table below ground surface (Eq. 6.7)

H = length of the pile that is laterally pushed into the soil and subjects the pile to passive pressure

k = dimensionless parameter equal to σ'_h divided by σ'_v

k_o = coefficient of earth pressure at rest

k_p = passive earth pressure coefficient

L = length of the footing (Secs. 6.1, 6.2, and 6.5)

L = length of the pile or pier (Sec. 6.3)

L' = reduced footing length to account for eccentricity of load

L_1 = vertical distance over which the pile or pier is subjected to the downdrag load

N_c, N_γ, N_q = dimensionless bearing capacity factors

$(N_1)_{60}$ = N value corrected for field testing procedures and overburden pressure

P = applied load to the footing

P_p = passive soil resistance

q_{all} = allowable bearing pressure

q_{ult} = ultimate bearing capacity

q' = largest bearing pressure exerted by an eccentrically loaded footing

q'' = lowest bearing pressure exerted by an eccentrically loaded footing

Q = applied load to the footing

Q_{all} = maximum allowable footing, pile, or pier load

Q_p = ultimate pile tip or pier tip resistance force

Q_s = ultimate skin friction resistance force for the pile or pier

Q_{ult} = load causing a bearing capacity failure

Q_{ult} = ultimate load capacity of a pile or pier (Sec. 6.3)

Q_D = downdrag load acting on a pile or pile group

r_u = pore water pressure ratio

r, R = radius of the pile or pier

R = shear resistance of the soil (Sec. 6.5)

s_u = undrained shear strength of the soil

s_{u1}, s_{u2} = undrained shear strength of the soil (two soil layers)

S_t = sensitivity of the clay

T = vertical distance from foundation base to top of layer 2 (Sec. 6.2.3)

T = vertical distance from the bottom of the footing to the top of the liquefied soil layer (Sec. 6.5)

T_{ult} = ultimate uplift load of the pile or pier

u = pore water pressure

u_e = excess pore water pressure

z = depth below the pile group or embedment depth of the pile

ϕ = friction angle based on a total stress analysis

ϕ' = friction angle based on an effective stress analysis

ϕ_w = friction angle between the cohesionless soil and the perimeter of the pile

γ_a = adjusted unit weight for a groundwater table (Eq. 6.7)

γ_b = buoyant unit weight of saturated soil below the groundwater table

γ_t = total unit weight of the soil

σ' = effective stress

σ_h = horizontal total stress

σ_h' = horizontal effective stress

σ_n' = effective normal stress on the shear surface

σ_v = vertical total stress

σ_v', σ_{vo}' = vertical effective stress

σ_{vm}' = maximum past pressure, also known as the *preconsolidation* pressure

$\Delta\sigma_v$ = increase in vertical stress

τ_f = shear strength of the soil

PROBLEMS

Solutions to the problems are presented in App. C of this book. The problems have been divided into basic categories as indicated below:

Bearing Capacity of Shallow Foundations on Cohesionless Soil

6.1   Use the data from Example Problem 6.3 (i.e., the 4-ft-wide strip footing). Assume the ground-water table is located at the bottom of the footing. Also assume that the total unit weight γ_t is the same above and below the groundwater table ($\gamma_t = 125$ pcf). Calculate the ultimate bearing capacity q_{ult} and the allowable load Q_{all}.

ANSWER: $q_{ult} = 2200$ psf, $Q_{all} = 8.8$ kips/ft.

6.2   Use the data from Example Problem 6.3. Instead of a strip footing, assume it is a square footing (4 ft by 4 ft). Calculate the ultimate bearing capacity q_{ult} and the allowable load Q_{all}.

ANSWER: $q_{ult} = 2300$ psf, $Q_{all} = 37$ kips.

6.3   Use the data from Example Problem 6.3. Assume the strip footing must support a vertical load of 150 kN per linear meter of wall. Also assume the groundwater table is well below the bottom of the footing. Determine the minimum width of the strip footing based on a factor of safety of 3.

ANSWER: $B = 1.14$ m.

6.4   A site consists of a sand deposit that has a relative density D_r of 65 percent. The groundwater table is well below the bottom of the footing and the total unit weight γ_t of the sand = 120 pcf. For a square spread footing that is 10 ft wide and at a depth of 5 ft below ground surface, calculate the ultimate bearing capacity q_{ult} and the allowable load Q_{all}.

ANSWER: $q_{ult} = 44{,}400$ psf, $Q_{all} = 1480$ kips.

Bearing Capacity for Shallow Foundations on Cohesive Soil

6.5   Use the data from Example Problem 6.4 and assume the subsoil consists of clay that has an undrained shear strength (s_u) = 20 kPa. Based on a total stress analysis, calculate the ultimate bearing capacity q_{ult} and the allowable load Q_{all}.

ANSWER: $q_{ult} = 122$ kPa and $Q_{all} = 49$ kN/m.

6.6 Use the data from Example Problem 6.4. Assume the strip footing must support a vertical load of 50 kN per linear meter of wall. Determine the minimum width B of the strip footing based on a factor of safety of 3.

ANSWER: $B = 1.64$ m.

6.7 Use the data from Example Problem 6.4, except that the site is underlain by heavily overconsolidated clay that has an undrained shear strength $s_u = 200$ kPa and a drained shear strength of $\phi' = 28°$ and $c' = 5$ kPa. Assume the groundwater table is located at a depth of 0.6 m and a factor of safety of 3. Performing both a total stress analysis and an effective stress analysis, determine the allowable load Q_{all}.

ANSWER: The effective stress analysis governs and $Q_{all} = 180$ kN/m.

Eccentrically Loaded Footings

6.8 Use the data from Example Problem 6.3. Assume the vertical load exerted by the strip footing = 100 kN per linear meter of wall length and that this load is offset from the centerline of the strip footing by 0.15 m (i.e., $e = 0.15$ m). Determine the largest bearing pressure q' and the least bearing pressure q'' exerted by the eccentrically loaded footing. Is q' acceptable from an allowable bearing capacity standpoint?

ANSWER: $q' = 146$ kPa and $q'' = 21$ kPa and since $q' > q_{all}$, then q' is unacceptable.

6.9 Solve Problem 6.8 using the reduced area method (Fig. 6.8).

ANSWER: $q = 111$ kPa and $q_{all} = 120$ kPa, then $q < q_{all}$ and q is acceptable by this method.

6.10 Use the data from Problem 6.9, but assume the footing is a square footing (1.2 m by 1.2 m) and the load exerted by the square footing = 100 kN that is offset from the center of the footing by 0.15 m (i.e., $e = 0.15$ m). Determine the largest bearing pressure q' and the least bearing pressure q'' exerted by the eccentrically loaded footing. Is q' acceptable from an allowable bearing capacity standpoint?

ANSWER: $q' = 122$ kPa and $q'' = 17$ kPa and since $q' > q_{all}$, then q' is unacceptable.

6.11 Solve Problem 6.10 using the reduced area method (Fig. 6.8).

ANSWER: $q = 93$ kPa and $q_{all} = 120$ kPa, then $q < q_{all}$ and q is acceptable by this method.

Bearing Capacity for Deep Foundations (Cohesionless Soil)

6.12 Use the data from Example Problem 6.5 (i.e., the 0.3-m diameter pile with a length of 6 m). A field load test on this pile indicates an ultimate load capacity Q_{ult} of 250 kN. Assume all of the load capacity is developed by end-bearing. Also assume at the time of construction that 0.4-m diameter piles are used instead of 0.3-m diameter piles. Calculate the ultimate end-bearing capacity Q_p and the allowable pile capacity Q_{all} for the 0.4-m diameter pile using a factor of safety of 3.

ANSWER: $Q_p = 444$ kN and $Q_{all} = 148$ kN.

6.13 Use the data from Problem 6.12, but instead of end-bearing, assume all of the 250 kN is carried by skin friction. Calculate the ultimate frictional capacity Q_s and the allowable frictional capacity Q_{all} for the 0.4-m diameter pile using a factor of safety of 3.

ANSWER: $Q_s = 333$ kN and $Q_{all} = 111$ kN.

6.14 Use the data from Problem 6.12, but assume that 60 percent of the 250 kN load is carried by end-bearing and 40 percent of the 250 kN load is carried by skin friction. Calculate the ultimate load capacity Q_{ult} and the allowable load capacity Q_{all} for the 0.4-m diameter pile using a factor of safety of 3.

ANSWER: $Q_{ult} = 400$ kN and $Q_{all} = 133$ kN.

6.15   Assume a site has the subsoil profile as shown in Fig. 2.38 and a bridge pier foundation consisting of piers will be installed at the location of B-25. Assume that each pier will be embedded 3 m into the sandstone. For the sediments above the sandstone, use an average buoyant unit weight $\gamma_b = 9.2$ kN/m³. For the sandstone, use an average buoyant unit weight $\gamma_b = 11.7$ kN/m³ and the effective shear strength parameters are $c' = 50$ kPa and $\phi' = 40°$. Based on scour conditions (scour to elevation −10 m) and considering only end-bearing, determine the allowable pier capacity (Q_{all}) if the piers are 1.0 m in diameter and using a factor of safety of 3.

ANSWER: $Q_{all} = 3,770$ kN.

6.16   In Fig. 2.38, assume at Boring B-25 that the uppermost soil layer (silt, silty sand, and peat layer) extends from elevation −2 to −5 m. Use the data from Problem 6.15 and assume that after construction of the bridge pier, a 0.5 m thick sand layer ($\gamma_b = 9.2$ kN/m³) is deposited on the river bottom. Also assume that $k = 0.5$ and $\phi_w = 20°$ for the 0.5 m thick sand sediment layer, and $k = 0.4$ and $\phi_w = 15°$ for the 3-m-thick silt-peat layer. Determine the downdrag load on each pier due to consolidation and compression of the silt-peat layer.

ANSWER: Downdrag load = 19.3 kN.

Bearing Capacity of Deep Foundations (Cohesive Soil)

6.17   Assume the subsoil conditions at a site are as shown in Fig. 2.39. A 9 by 9 pile group, having spacing in pile diameters of 3, supports a pile cap at ground surface (elevation +20 ft). Assume each pile in the group has a diameter of 1.5 ft and the piles are 70 ft long. Neglecting skin friction resistance in the upper 30 ft, determine the load capacity of the pile group using a total stress analysis and a factor of safety = 3.

ANSWER: Using an average $s_u = 0.6$ kg/cm² from elevation −10 to −50 ft, $Q_s = 140$ kips. Using an average $s_u = 0.4$ kg/cm² at elevation −50 ft, $Q_p = 13$ kips. The allowable load capacity of the pile group = 2900 kips.

6.18   Assume that the subsoil conditions at a site are as shown in Fig. 2.40. Also assume that 1.5-ft diameter piles are installed so that their tip is at a depth of 35 ft below ground surface. Using a total stress analysis, determine the ultimate skin adhesion capacity Q_s neglecting the shear strength in the upper 5 ft, the ultimate end-bearing capacity Q_p, and the allowable load capacity Q_{all} using a factor of safety of 3.

ANSWER: Using an average undrained shear strength $s_u = 350$ psf, $Q_s = 50$ kips. Using an average undrained shear strength $s_u = 400$ psf at a depth of 35 ft, $Q_p = 6$ kips. $Q_{all} = 19$ kips.

6.19   Use the data from Problem 6.18. For a 9 by 9 pile group with a spacing in pile diameter = 1.5, determine the allowable load capacity of the pile group using a factor of safety of 3.

ANSWER: Allowable load capacity of pile group = 630 kips.

6.20   Assume that the subsoil conditions at a site are as shown in Fig. 2.41. Predrilled cast-in-place concrete piles that have a diameter of 1 ft will be installed to a depth of 50 ft. Ignoring the soil adhesion in the upper 10 ft of the pile, calculate the allowable load capacity Q_{all} of the pile using a total stress analysis and a factor of safety of 3.

ANSWER: Using an average $s_u = 0.6$ kg/cm² (1200 psf) from 10 to 50 ft depth, $Q_s = 100$ kips and $Q_p = 8$ kips. $Q_{all} = 36$ kips.

Deep Foundations Subjected to Lateral Loads

6.21 As discussed in Sec. 6.4.2, the *International Building Code* (2006) states: "The resulting allowable load shall not be more than one-half of that test load that produces a gross lateral movement of 1 in. (25 mm) at the ground surface." Using the test data shown in Fig. 6.19, determine the allowable lateral load based on these building code requirements.

ANSWER: 19 kN.

Bearing Capacity Analyses for Geotechnical Earthquake Engineering

6.22 Use the data from Example Problem 6.16 for the cohesive surface layer (total stress analysis). Calculate the spread footing size so that the factor of safety is equal to 5.

ANSWER: 5 m by 5 m spread footing.

6.23 Use the data from Example Problem 6.16 for the cohesive surface layer (total stress analysis). Calculate the maximum concentric load that can be exerted to the 2 m by 2 m spread footing such that the factor of safety is equal to 5.

ANSWER: $P = 200$ kN.

6.24 Use the data from Example Problem 6.17 for the cohesionless surface layer (effective stress analysis). Calculate the maximum concentric load that can be exerted to the strip footing such that the factor of safety is equal to 5.

ANSWER: $P = 10$ kN/m.

6.25 Use the data from Example Problem 6.17 for the cohesionless surface layer (effective stress analysis). Calculate the maximum concentric load that can be exerted to the 2 m by 2 m spread footing such that the factor of safety is equal to 5.

ANSWER: $P = 40$ kN.

6.26 A site consists of an upper 3-m-thick sand layer that is above the groundwater table and hence will not liquefy during the design earthquake. However, the groundwater table is located at a depth of 3 m and the sand located below this depth is expected to liquefy during the design earthquake. For the upper 3 m of sand at the site, use the following values: total unit weight = 18.3 kN/m^3, effective friction angle $\phi' = 33°$, and $k_o = 0.5$. Further assume that the foundation will consist of shallow strip and spread footings that are 0.3 m deep. Using a factor of safety equal to 5 and footing widths of 1 m, determine the allowable bearing pressure for the strip and spread footings.

ANSWER: $q_{all} = 10$ kPa for the 1-m-wide strip footings and $q_{all} = 21$ kPa for the 1 m by 1 m spread footings.

6.27 Solve Problem 6.26, except assume that the soil above the groundwater table is cohesive soil that has undrained shear strength of 20 kPa.

ANSWER: $q_{all} = 20$ kPa for the 1-m-wide strip footings and $q_{all} = 40$ kPa for the 1 m by 1 m spread footings.

6.28 Solve Problem 6.26 using the Terzaghi bearing capacity equation. What values should be used in the design of the footings?

ANSWER: $q_{all} = 48$ kPa for the 1-m-wide strip footings and $q_{all} = 38$ kPa for the 1 m by 1 m spread footings. For the design of the footings, use the lower values calculated in Problem 6.26.

6.29 Solve Problem 6.27 using the Terzaghi bearing capacity equation. What values should be used in the design of the footings?

ANSWER: $q_{all} = 22$ kPa for the 1-m-wide strip footings and $q_{all} = 29$ kPa for the 1 m by 1 m spread footings. For the design of the strip footings, use the value from Problem 6.27 (i.e., $q_{all} = 20$ kPa). For the design of the spread footings, use the lower value calculated in this problem (i.e., $q_{all} = 29$ kPa).

6.30 Solve Problem 6.27 assuming that the spread footing is 3 m by 3 m. Use Eq. 6.31 and Fig. 6.7 to solve the problem. What bearing capacity values should be used in the design of the footings?

ANSWER: Using Eq. 6.31, $q_{all} = 14$ kPa. Based on Fig. 6.7, $q_{all} = 12$ kPa. Use the lower value of 12 kPa for the design of the 3 m by 3 m spread footing.

6.31 Assume that a proposed 20 m by 20 m mat foundation will be supported by piles, with the tip of the piles located at a depth of 15 m. The piles are evenly spaced along the perimeter and interior portion of the mat. The structural engineer has determined that the critical design load (sum of live, dead, and seismic loads) is equal to 50 MN, which can be assumed to act at the center of the mat and will be transferred to the pile tips. The soil at a depth of 15 to 17 m below ground surface is sand having an effective friction angle $\phi' = 34°$, $k_o = 0.60$, and the average vertical effective stress $\sigma'_{vo} = 168$ kPa. During the design earthquake, it is anticipated that the sand at a depth of 17 to 20 m below ground surface will liquefy. Calculate the factor of safety (using Eq. 6.31) for an earthquake-induced punching shear failure of the pile foundation into the liquefied soil layer.

ANSWER: $F = 0.22$ and therefore the pile foundation will punch down into the liquefied soil layer located at a depth of 17 to 20 m below ground surface.

6.32 Use the data from Problem 6.31, but assume that high-displacement friction piles are used to support the mat. Further assume the friction piles will primarily resist the 50 MN load by soil friction along the pile perimeters. Using the 2:1 approximation and assuming it starts at a depth of $\frac{2}{3} L$ (where L = pile length), determine the factor of safety (using Eq. 6.31) for an earthquake-induced punching shear failure into the liquefied soil located at a depth of 17 to 20 m below ground surface.

ANSWER: $F = 0.27$ and therefore the pile foundation will punch down into the liquefied soil layer located at a depth of 17 to 20 m below ground surface.

6.33 Use the data from Example Problem 6.17 for the cohesive surface layer (total stress analysis). Assume that the eccentricity e is 0.10 m for the strip footing and 0.3 m for the spread footing. Determine the values of the allowable load Q_{all} and allowable moment M_{all} using a factor of safety of 5. Solve the problem by using Eq. 6.12 and the method outlined in Fig. 6.8.

ANSWER: For the strip footing, $Q_{all} = 34$ kN/m and $M_{all} = 3.4$ kN-m/m. For the spread footing, $Q_{all} = 88$ kN and $M_{all} = 26$ kN-m.

6.34 Use the data from Example Problem 6.17 for the cohesive surface layer (total stress analysis). Assume that there are 150 kN-m moments acting in both the B and L directions. Using Fig. 6.8, calculate the factor of safety F in terms of a bearing capacity failure.

ANSWER: $F = 0.82$.

CHAPTER 7
SETTLEMENT OF FOUNDATIONS

7.1 INTRODUCTION

Settlement can be defined as the permanent downward displacement of the foundation. There are two basic types of settlement:

1. *Settlement due directly to the weight of the structure.* The first type of settlement is directly caused by the weight of the structure. For example, the weight of a building may cause compression of an underlying sand deposit (Sec. 7.3) or consolidation of an underlying clay layer (Chap. 8). The settlement analysis must include the actual dead load of the structure. The dead load is defined as the structural weight due to beams, columns, floors, roofs, and other fixed members.

Live loads are defined as the weight of nonstructural members, such as furniture, occupants, inventory, and snow. Live loads can also result in settlement of the structure. For example, if the proposed structure is a library, then the actual weight of the books (a live load) will contribute to the settlement of the structure. Likewise, for a proposed warehouse, it may be appropriate to include the actual weight of anticipated stored items in the settlement analyses. In other projects where the live loads represent a significant part of the loading, such as large electrical transmission towers that will be subjected to wind loads, the live load (wind) should also be considered in the settlement analysis.

In summary, the load used for settlement analyses must consist of the actual dead weight of the structure, and in many cases, will also include live loads. Considerable experience and judgment are required to determine the load that is to be used in the settlement analyses.

2. *Settlement due to secondary influences.* The second basic type of settlement of a building is caused by secondary influence, which may develop at a time long after the completion of the structure. This type of settlement is not directly caused by the weight of the structure. For example, the foundation may settle as water infiltrates the ground and causes unstable soils to collapse (i.e., collapsible soil, Sec. 7.2). The foundation may also settle due to yielding of adjacent excavations or the collapse of limestone cavities or underground mines and tunnels (Sec. 7.4). Other causes of settlement that would be included in this category are natural disasters, such as settlement caused by earthquakes (Chap. 13) or undermining of the foundation from floods.

Subsidence is usually defined as a sinking down of a large area of the ground surface. Subsidence could be caused by the extraction of oil or groundwater that leads to a compression of the underlying porous soil or rock structure. Since subsidence is due to a secondary influence (extraction of oil or groundwater), its effect on the structure would be included in the second basic type of settlement described earlier.

A special case is the downward displacement of the foundation due to the drying of underlying wet clays, which will be discussed in Chap. 9. Often this downward displacement of the foundation caused by the desiccation of clays is referred to as settlement. But upon the introduction of moisture, such as during the rainy season, the desiccated clay will swell and the downward displacement will be reversed. The foundation could even heave more than it initially settled. When dealing with expansive clays, it is best to consider the downward displacement of the foundation as part of the cyclic heave and shrinkage of expansive soil and not permanent settlement.

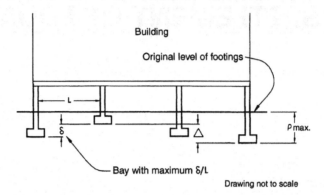

FIGURE 7.1   Diagram illustrating the definitions of total settlement ρ_{max}, maximum differential settlement Δ, and maximum angular distortion δ/L.

Determining the settlement behavior of the proposed structure is one of the primary obligations of the geotechnical engineer. The following parameters are often required:

1. *Total settlement ρ_{max}.*   Also known as the maximum settlement, it is the largest amount of settlement experienced by any part of the foundation, such as shown in Fig. 7.1.

2. *Maximum differential settlement Δ.*   The maximum differential settlement is the largest difference in settlement between two different foundation locations, such as shown in Fig. 7.1. The maximum differential settlement does not necessarily occur at the same location as the total settlement.

3. *Rate of settlement.*   It is often desirable to know if the settlement will occur during construction as the dead load is applied to the soil, or if the settlement will occur over the life of the project, in which case there may be on-going cracking to the structure.

4. *Maximum angular distortion δ/L.*   The angular distortion is defined as the differential settlement between two points divided by the distance between them less the tilt, where tilt equals rotation of the entire building (Skempton and MacDonald, 1956). As shown in Fig. 7.1, the highest value of δ/L would be the maximum angular distortion. The location of the maximum angular distortion does not necessarily occur at the location of the total settlement or maximum differential settlement.

One approach for the design of the foundation is to first calculate the total and maximum differential settlement of the proposed structure for the site soil conditions (Secs. 7.2 to 7.5). Then these values can be compared with allowable values (Sec. 7.6). The foundation type may have to be changed or mitigation measures adopted if the calculated total settlement of the structure exceeds the allowable settlement.

7.2   COLLAPSIBLE SOIL

7.2.1   Introduction

In the southwestern United States, a common cause of foundation settlement is collapsible soil, which can be broadly classified as soil that is susceptible to a large and sudden reduction in volume upon wetting. Collapsible soil usually has a low dry density and low moisture content. Such soil can withstand a large applied vertical stress with a small compression, but then experience much larger settlements after wetting, with no increase in vertical pressure (Jennings and Knight, 1957). As such, collapsible soil falls within the second basic category of settlement, which is settlement of the structure due to secondary influences.

For collapsible soil, compression will often occur as the overburden pressure increases due to the placement of overlying fill or the construction of a building on top of the soil. The compression due to this increase in overburden pressure involves a decrease in void ratio of the soil due to expulsion of air. The compression usually occurs at constant moisture content. After completion of the project, water may infiltrate the soil due to irrigation, rainfall, or leaky water pipes.

In general, there has been an increase in damage due to collapsible soil, probably because of the lack of available land in many urban areas. This causes development of marginal land, which may contain deposits of dumped fill or deposits of natural collapsible soil. Also, substantial grading can be required to develop level building pads, which results in more areas having deep fill.

There are two common types of collapsible soil: (1) fill, such as debris fill, uncontrolled fill, and deep fill, and (2) natural soil, such as alluvium or colluvium, as discussed below:

1. *Fill.*   Uncontrolled fills include fills that were not documented with compaction testing as they were placed: these include dumped fills, fills dumped under water, hydraulically placed fills, and fills that may have been compacted but there is no documentation of testing or the amount of effort that was used to perform the compaction (Greenfield and Shen, 1992). These conditions may exist in rural areas where inspections are lax or for structures built many years ago when the standards for fill compaction were less rigorous.

The mechanism that usually causes the collapse of a loose soil structure is a decrease in negative pore water pressure (capillary tension) as the fill becomes wet. For a fill specimen submerged in distilled water, the main variables that govern the amount of one-dimensional collapse are the soil type, compacted moisture content, compacted dry density, and the vertical pressure (Dudley, 1970; Johnpeer, 1986; Lawton et al., 1989, 1991, 1992; Tadepalli and Fredlund, 1991; Day, 1994a). In general, the one-dimensional collapse of fill will increase as the dry density decreases, the moisture content decreases, or the vertical pressure increases. For a constant dry density and moisture content, the one-dimensional collapse will decrease as the clay fraction increases once the optimum clay content (usually a low percentage) is exceeded (Rollins et al., 1994).

2. *Natural soil.*   For natural deposits of collapsible soil in the arid climate of the southwest, the common collapse mechanism entails breaking of bonds at coarse particle contacts by weakening of fine-grained materials brought there by surface tension in evaporating water. In other cases, the alluvium or colluvium may have an unstable soil structure, such as pin hole porosity, which collapses as the wetting front passes through the soil.

7.2.2 Laboratory Testing

If the results of field exploration indicate the possible presence of collapsible soil at the site, then soil specimens should be obtained and tested in the laboratory. One-dimensional collapse is typically measured in the oedometer apparatus using standard test procedures such as outlined in ASTM D 5333-03 (2004), "Standard Test Method for Measurement of Collapse Potential of Soils." The test procedures are as follows:

1. *Soil specimen.*   An undisturbed soil specimen must be used for the collapse test. For projects where it is anticipated that there will be fill placement, the fill specimens can be prepared by compacting them to the anticipated field as compacted density and moisture condition. An undisturbed soil specimen can be trimmed from a block sample or extruded from the sampler directly into a confining ring. The specimen diameter divided by the specimen height should be equal to or greater than 2.5. A common diameter of soil specimen is 2.5 in. (6.4 cm) and height of 1.0 in. (2.5 cm). The trimming process should be performed as quickly as possible to minimize the possibility of a change in water content of the soil. Because collapsible soil often has a low water content, specimens should not be stored or trimmed in a high humidity moisture room. Typically the soil specimen is trimmed in the confining ring and the initial thickness of the soil specimen h_o is then equal to the height of the confining ring.

2. *Index properties.*   After the soil specimen has been trimmed, the index properties of the soil should be determined. Typically the trimmings can be collected in order to determine the water content of the soil. In addition, a balance can be used to obtain the mass of the confining ring and soil specimen. By subtracting the mass of the confining ring, the mass of the soil specimen can be calculated. Knowing the volume of the confining ring, the total unit weight can be calculated. Using the water content, the dry unit weight can also be calculated.

3. *Loading device.*   Dry and clean porous plates are placed on the top and bottom of the soil specimen and it is then placed in a surrounding container, such as a Plexiglas dish. The Plexiglas dish containing the soil specimen is placed at the center of the oedometer (see Sec. 3.3). A seating load equivalent to a vertical pressure of 125 psf (6 kPa) is then applied to the soil specimen and a dial reading is taken and recorded on the data sheet. Within 5 min of applying the seating load, the load is increased. Typical loading increments are 250, 500, 1000, 2000 psf, and the like (12, 25, 50, 100 kPa, and so on) and each load increment should remain on the soil specimen for less than 1 h to prevent excessive evaporation of moisture from the specimen. The soil specimen can be surrounded with a loose-fitting plastic membrane or aluminum foil to reduce moisture loss during the loading process. Prior to applying the next load increment, a dial reading should be taken and recorded on a data sheet in order to determine the deformation versus loading behavior. The soil should be loaded to a vertical pressure that approximately equals the anticipated overburden pressure after completion of the structure. After the deformation has ceased at this vertical pressure, a dial reading d_o is taken.

4. *Soil specimen wetting.*   The soil specimen is then submerged in distilled water by filling the Plexiglas dish with distilled water. The distilled water should be quickly poured into the bottom of the Plexiglas dish. The water should be poured into the Plexiglas dish and allowed to flow upward through the soil specimen, which will reduce the amount of entrapped air. As soon as the distilled water has been poured into the Plexiglas dish, dial readings versus time at approximately 0.1, 0.25, 0.5, 1, 2, 4, 8, 15, and 30 min, and 1, 2, 4, 8, and 24 h should be recorded. Often the collapse process occurs very quickly and the first few seconds after inundation are very hectic and consist of filling the Plexiglas dish with distilled water, releveling the loading arm as the soil specimen collapses, and recording dial readings versus time. The collapse process is usually complete after 24 h of inundation with distilled water (dial reading at 24 hours = d_F).

5. *Continued loading of soil specimen.*   Additional vertical stress can be placed on the soil specimen to determine the soil behavior after collapse. Each vertical stress should remain on the soil specimen for about 24 h. Before each additional vertical stress is applied to the soil specimen, a dial reading should be taken.

6. *Unloading of the soil specimen.*   After loading to the desired highest vertical pressure, the soil specimen is unloaded by reducing the load upon the soil specimen. Once again, at each reduction in load upon the soil specimen, it should be allowed to equilibrate for 24 h and a dial reading should be taken prior to the next reduction in load upon the soil specimen.

7. *Index properties at end of test.*   Once the soil specimen has been completely unloaded, it is removed from the oedometer. The water content of the soil specimen can then be determined. When placing the soil specimen in the container for the water content test, it should be broken apart to determine if there are any large size particles within the soil specimen. Large size particles, such as gravel size particles, may restrict the collapse of the soil specimen and lead to an underestimation of the in situ collapse potential of the soil. If large particles are noted, the collapse test may have to be repeated on a larger size soil specimen.

7.2.3 Percent Collapse

The percent collapse (%C) can be calculated as follows:
In terms of vertical deformation:

$$\%C = \frac{100\Delta h}{h_o} = \frac{100(d_o - d_F)}{h_o}$$ (7.1)

In terms of void ratio:

$$\%C = \frac{100\Delta e}{1 + e_o}$$ (7.2)

where Δh = change in height of the soil specimen upon wetting (in. or cm)
  h_o = initial thickness of the soil specimen (in. or cm)
  d_o = dial reading taken just before the soil specimen is wetted (in. or cm)
  d_F = final dial reading after the soil specimen has completed its collapse (in. or cm)
  Δe = change in void ratio upon wetting (dimensionless)
  e_o = initial void ratio (dimensionless)

Figure 7.2 presents the results of a one-dimensional collapse test performed on a soil specimen. The soil specimen contained approximately 60 percent sand size particles, 30 percent silt size particles, and 10 percent clay size particles and was classified as silty sand (SM). The soil specimen had an initial dry unit weight γ_t of 14.5 kN/m^3 (92.4 pcf) and a water content w of 14.8 percent. Other specifics of the collapse test are as follows:

1. *Figure 7.2.* The silty sand specimen, having an initial height of 25.4 mm (1.0 in.), was incrementally loaded to a vertical stress of 144 kPa (3000 psf) and then inundated with distilled water. After the collapse was complete, the soil specimen was loaded to a vertical stress of 288 kPa and then unloaded. In Fig. 7.2, the percent collapse (10.3 percent) was calculated using Eq. 7.1.

2. *Figure 7.3.* Using the data from Fig. 7.2, the void ratio (vertical axis) was plotted versus vertical pressure (horizontal axis). The arrows indicate the loading, collapse, and unloading portions of the test. First, the soil specimen was incrementally loaded to a vertical pressure of 144 kPa (Point A). Then the specimen was inundated with distilled water and the soil collapsed (decrease in void ratio with no change in vertical pressure). Following the completion of collapse, the soil specimen was loaded to a vertical stress of 288 kPa and then incrementally unloaded to a vertical stress of 18 kPa. Note that Fig. 7.3 is a semilog plot. This is the standard way to present the data because the loading and unloading portions of the plot are often approximately straight lines when using a semilog plot.

3. *Figure 7.4.* Using the data from Part c of Fig. 7.2, the amount of vertical deformation (collapse) as a function of time after inundation was plotted. As shown in Fig. 7.4, the collapse occurs very quickly, with most of the collapse occurring within about 1 min after inundation with distilled water.

Once the laboratory tests are complete and the data have been analyzed, the settlement analysis can be performed and the procedure is usually as follows:

1. The amount of collapse for the different soil layers underlying the site are obtained from the laboratory oedometer tests on undisturbed samples. The vertical stress used in the laboratory testing should equal the overburden pressure plus the weight of the structure (based on stress distribution theory).

2. To obtain the settlement of each collapsible soil layer, the percent collapse is multiplied by the thickness of the soil layer.

3. The total settlement is the sum of the collapse value from the different soil layers.

For example, suppose the data shown in Fig. 7.2 represents a collapse test on an undisturbed soil specimen taken from the middle of a 3 m (10 ft) uniform layer of collapsible soil. Then the estimated settlement for this layer of collapsible soil would be 0.103 (10.3 percent) times 3 m (10 ft), or a settlement of 0.31 m (1.0 ft).

This analysis could be performed for each boring excavated within the proposed footprint of the building. The settlement could then be calculated for each boring and the largest value would be the total settlement ρ_{max}. The maximum differential settlement Δ is more difficult to determine because it depends to a large extent on how water infiltrates the collapsible soil. For example, for the wetting of the collapsible soil only under a corner or portion of a building, the maximum differential settlement Δ could approach in value the total settlement ρ_{max}. Usually because of the unknown wetting conditions that will prevail in the field, the maximum differential settlement Δ is assumed to be from 50 to 75 percent of the total settlement ρ_{max}. The amount of time for this settlement to occur depends on the availability of water and the rate of infiltration of the water. Usually the collapse process in the field is a slow process and may take many years to complete.

File no. 6329.01   Job name: Office Plaza

Sample location: B-1 @ 4.6 m   Date: June 10, 1999

(a) Initial conditions of soil specimen:

Specimen dia: 6.35 cm   Specimen height (h_0): 2.54 cm   Inundation load: 144 kPa

Water content: 14.8%   Wet unit weight: 16.6 kN/m^3 Dry unit weight: 14.53 kN/m^3

(b) Loading data

Vertical pressure, kPa	Dial reading, mm	Specimen height, mm	Void ratio (G_s = 2.72)
4.5	7.29	25.40	0.837
9	7.28	25.39	0.836
18	7.22	25.33	0.832
36	7.10	25.21	0.823
72	6.90	25.01	0.809
144	6.74	24.85	0.797
144	4.12	22.23	0.607
288	3.48	21.59	0.561
72	3.74	21.85	0.580
18	3.99	22.10	0.599

(c) Dial readings at inundation with distilled water:

Time, min	Dial reading	Time, min	Dial reading	Time, h	Dial reading
0 (d_0)	6.74	2	4.24	1.3	4.15
0.10	—	4	4.22	3	4.14
0.25	4.74	8	4.19	3.8	4.14
0.5	4.42	15	4.17	18	4.13
1	4.31	40	4.16	30 (d_F)	4.12

Percent collapse: $100(d_0 - d_F) / h_0 =$ 100(6.74 – 4.12) / 25.4 = 10.3%

(d) End of test: Water content 20.9%

NOTE: Dial gauge measures deformation in inches, valves converted to millimeters.

FIGURE 7.2   Collapse test data for a silty sand (see Figs. 7.3 and 7.4 for data plots).

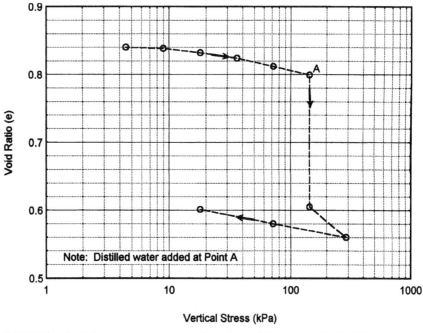

FIGURE 7.3   Vertical pressure versus void ratio for a silty sand. Data presented in Fig. 7.2.

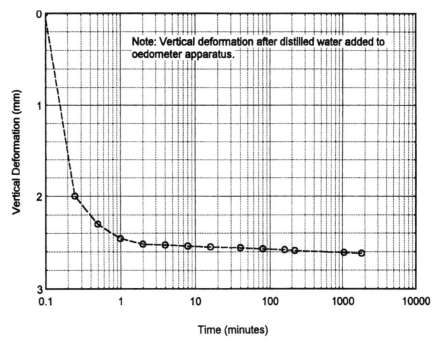

FIGURE 7.4   Vertical deformation versus time for a silty sand (after inundation). Also see Fig. 7.3. Data presented in Fig. 7.2.

TABLE 7.1 Degree of Specimen Collapse Versus Collapse Index

Degree of specimen collapse	Collapse index I_e %
None	0
Slight	0.1–2
Moderate	2.1–6
Moderately severe	6.1–10
Severe	>10

Source: ASTM D 5333-03 (2004).

7.2.4 Collapse Index

The collapse index I_e of a soil can be determined by applying a vertical stress of 200 kPa (2 tsf) to the soil specimen, submerging it in distilled water, and then determining the percent collapse by using (Eq. 7.1 or 7.2), which is designated the collapse index I_e. This collapse index I_e can be considered as an index test to compare the susceptibility of collapse for different soils. The collapse index I_e versus degree of specimen collapse has been listed in Table 7.1.

7.2.5 Design and Construction for Collapsible Soil

There are many different methods to deal with collapsible soil. If there is a shallow deposit of collapsible soil, then the deposit can be removed and recompacted during the grading of the site. In some cases, the soil can be densified, such as by compaction grouting, to reduce the collapse potential of the soil. Another method to deal with collapsible soil is to flood the building footprint or

FIGURE 7.5 Damage to a pool shell due to collapse of underlying uncompacted fill (arrows point to cracks in the pool shell).

force water into the collapsible soil stratum by using wells. As the wetting front moves through the ground, the collapsible soil will densify and reach an equilibrium state. Flooding or forcing water into collapsible soil should not be performed if there are adjacent buildings because of the possibility of damaging these structures. Also, after the completion of the flooding process, subsurface exploration and laboratory testing should be performed to evaluate the effectiveness of the process.

There are also foundation options that can be used for sites containing collapsible soil. A deep foundation system that derives support from strata below the collapsible soil could be constructed. Also, post-tensioned foundations or mat slabs can be designed and installed to resist the anticipated settlement from the collapsible soil.

As previously mentioned, the triggering mechanism for the collapse of fill or natural soil is the introduction of moisture. Common reasons for the infiltration of moisture include water from irrigation, broken or leaky water lines, and an altering of surface drainage that allows rainwater to pond near the foundation. Another source of moisture infiltration is from leaking pools. For example, Fig. 7.5 shows a photograph of a severely damaged pool shell that cracked due to the collapse of an underlying uncompacted debris fill. For sites with collapsible soil, it is good practice to emphasize the importance of positive drainage (no ponding of water at the site) and an immediate repair of any leaking utilities.

7.3 SETTLEMENT OF COHESIONLESS SOIL

7.3.1 Introduction

Chapter 8 will be devoted to the settlement of cohesive soil. This section deals with the settlement of cohesionless (nonplastic) soil, such as sands and gravels, caused by the structural load. The settlement of granular soil subjected to foundation loads occurs primarily from the compression of the soil skeleton due to the rearrangement of soil particles into denser arrangements. Hence a very loose sand or gravel subjected to a foundation load will have much more rearrangement of soil particles and resulting settlement than the same soil in a dense or very dense state.

A major difference between saturated cohesive soil and cohesionless soil is that the settlement of cohesionless soil is usually not time dependent. Because of the generally high permeability of cohesionless soil, the settlement of a saturated cohesionless soil usually occurs as the dead load is applied during the construction of the building. Cohesionless soils typically do not have long-term settlement (i.e., such as consolidation). Exceptions are as follows:

1. *Collapsible cohesionless soil.* As indicated in Sec. 7.2, such soils could be subjected to long-term settlement as water infiltrates the soil. If the cohesionless soil is collapsible, then the procedures as outlined in Sec. 7.2 should be followed.

2. *Seismic loading.* Seismic shaking from the earthquake can cause densification of loose cohesionless soil resulting in foundation settlement. This will be covered in Chap. 14.

3. *Vibrations.* An example of vibrations would be from machinery, such as a printing press, that is supported by the foundation. Especially for loose sands, the printing press vibrations can slowly densify the soil, resulting in long-term foundation settlement.

4. *Fluctuating loads.* Similar to vibrations, fluctuating loads can also cause long-term foundation settlement. An example would be an oil storage tank that is filled and emptied many times over its life. If loose sand supports the oil tank foundation, then the loading-unloading cycles could slowly densify the sand, resulting in long-term settlement. It has been suggested that for structures subjected to slight fluctuating loads, the settlement of footings on sand 30 years after construction might be as much as 1.5 times the immediate post-construction settlement, while for heavy fluctuating loads, the settlement might be as much as 2.5 times the immediate post-construction settlement (Burland and Burbidge, 1985). Actual values of post-construction settlement would be dependent on the number of loading-unloading cycles, the magnitude of the fluctuating load, and the density of the sand (loose versus dense).

Figure 7.6 presents a plot of maximum settlement (i.e., total settlement, ρ_{max}) versus maximum differential settlement Δ from case studies. Bjerrum (1963) considered only those structures having comparably loaded footings on sands of uniform thickness. For the case studies used to develop Fig. 7.6, the maximum differential settlement Δ is that which occurred between comparable footings designed for the same total settlement. As shown in Fig. 7.6, when the total settlement was less than about 20 mm (0.8 in.), there was considerable scatter in data with no discernable correlation between total settlement and maximum differential settlement. However, for those structures where the recorded settlement exceeded about 20 mm (0.8 in.), the differential settlement typically was close in value to the total settlement. The large values of maximum differential settlement are attributed to the nonuniformities in natural sand deposits. The differential settlement is often difficult to determine because it depends on the rigidity of the foundation and the nonuniformities in sand deposits, and hence the maximum differential settlement is often assumed to be 50 to 75 percent of the total settlement (i.e., $\Delta = 0.5$ to 0.75 of ρ_{max}) for cohesionless soil.

For cohesionless soil, there are many different methods that can be used to determine the settlement S of a footing due to structural loading. The methods described below will underestimate the settlement if the footing pressure bulbs were to overlap (i.e., closely spaced footings). In addition, the methods described in this section do not consider contributions to settlement from collapsible soil or from foundation vibrations.

7.3.2 Field Testing

Various field tests can be used to estimate the amount of settlement, as follows:

Plate Load Test.   The plate load test can be used to directly estimate the settlement of the footing. The plate load test is performed using standard test procedures such as outlined in ASTM D 1194-94

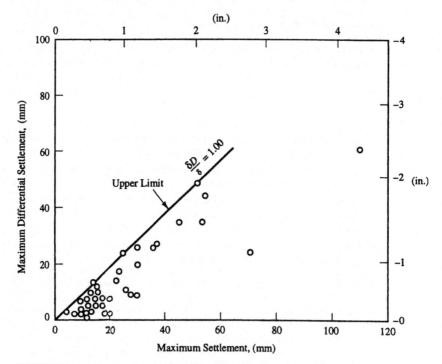

FIGURE 7.6   Maximum settlement (i.e., total settlement ρ_{max}) versus maximum differential settlement (Δ) from case studies. (*Adapted from Bjerrum, 1963; reproduced from Coduto, 1994.*)

(2003), "Standard Test Method for Bearing Capacity of Soil for Static Load and Spread Footings." The test procedures are as follows:

1. *Bearing plate.* The ASTM test procedures state: "Three circular steel bearing plates, not less than 1 in. (25 mm) in thickness and varying in diameter from 12 to 30 in. (305 to 762 mm), including the minimum and maximum diameter specified or square steel bearing plates of equivalent area." As an alternative, small concrete footings can be cast and tested in place of the steel plate.

2. *Test location.* Ideally, the plate load test should be at the same location and elevation as the bottom of the proposed footing. The plate load test could be performed at the bottom of test pits or trenches and, in general, it is desirable to perform at least three plate load tests. Also the site conditions should be similar to the in-service conditions. Thus if it is expected that the soil will become wetted in the future, then the soil in the area to be tested should be prewetted.

3. *Loading of plate.* Loads are applied in increments such that each load is not more than one-tenth of the bearing capacity of the soil in the area being tested. The load must be accurately measured and applied in such a manner that the load reaches the soil as a static load, without impact, fluctuation, or eccentricity. After the application of each load increment, it should be maintained for a time interval of not less than 15 min. The load is increased until there is a definite failure of the soil or until the settlement of the plate reaches 10 percent of the plate width. The load is then removed in increments and the rebound deflections are also recorded.

4. *Pressure versus settlement curve.* The bearing pressure is calculated as the load divided by the area of the plate. The pressure versus settlement from the plate load test can then be plotted.

The plate load test can also be used to directly estimate the settlement of a footing. For settlement of medium to dense sands caused by an applied surface loading, an empirical equation that relates the depth of penetration of the steel plate S_1 to the settlement of the actual footing S is as follows (Terzaghi and Peck, 1967):

$$S = \frac{4S_1}{(1 + D_1/D)^2} \tag{7.3}$$

where S = settlement of the footing (ft or m)
 S_1 = depth of penetration of the steel plate (ft or m)
 D_1 = smallest dimension of the steel plate (ft or m)
 D = smallest dimension of the actual footing (ft or m)

In order to use Eq. 7.3, the pressure exerted by the steel plate corresponding to S_1 must be the same pressure as exerted on the sand by the actual footing. Assuming independent footings with no overlap of pressure bulbs, the following conditions apply:

1. *Footings with the same pressure.* If all the footings subject the sand to about the same pressure, then the total settlement ρ_{max} would occur for the footing having the largest value of D. The maximum differential settlement Δ would be the difference in settlement for this footing minus the settlement for the footing having the smallest value of D.

2. *Footing of same size.* If all the footings have variable bearing pressures, but have about the same size, then the total settlement ρ_{max} would occur for the footing that exerts the highest pressure on the sand. The maximum differential settlement Δ would be the difference in settlement for this footing minus the settlement for the footing that exerts the lowest pressure onto the sand.

The calculated settlement from Eq. 7.3 can significantly underestimate the actual settlement at the site in cases where there is settlement due to secondary influences (such as collapsible soil) or in cases where there is a deep looser or softer layer that is not affected by the small plate, but is loaded by the much larger footing. For example, the plate load test only gives information on the soil to a depth of about two times the bearing plate width or diameter. In addition to this limitation, the test is time consuming and expensive to perform and hence it is used less frequently than the other methods described below. Probably because of its infrequent use, ASTM withdrew this test from publication in December 2002.

Modulus of Subgrade Reaction.   The plate load test can also be used to calculate the modulus of subgrade reaction, also known as the subgrade modulus K_v. The procedure (per NAVFAC DM-7.1, 1982) is to plot the stress exerted by the plate versus the penetration of the plate. Using this plot, the yield point at which the penetration rapidly increases is estimated. Then the stress q and depth of penetration of the plate (δ) corresponding to half the yield point are determined from the plot and the subgrade modulus K_v is defined as:

$$K_v = \frac{q}{\delta} \tag{7.4}$$

Assuming the modulus of elasticity increases linearly with depth, the settlement S caused by a uniform vertical footing pressure q can be estimated from the following equations (NAVFAC DM-7.1, 1982):
For shallow foundations with $D \leq B$ and $B \leq 20$ ft:

$$S = \frac{4qB^2}{K_v(B+1)^2} \tag{7.5}$$

Using SI units, for shallow foundations with $D \leq B$ and $B \leq 6$ m:

$$S = \frac{4qB^2}{K_v(B+0.3)^2} \tag{7.6}$$

For shallow foundations with $D \leq B$ and $B \geq 40$ ft:

$$S = \frac{2qB^2}{K_v(B+1)^2} \tag{7.7}$$

Using SI units, for shallow foundations with $D \leq B$ and $B \geq 12$ m:

$$S = \frac{2qB^2}{K_v(B+0.3)^2} \tag{7.8}$$

where S = settlement of the footing (ft or m)
$\quad q$ = vertical footing pressure (psf or kPa)
$\quad B$ = footing width (ft or m)
$\quad K_v$ = subgrade modulus from the plate load test (pcf or kN/m³)
$\quad D$ = depth of the footing below ground surface (ft or m)

It is necessary to interpolate between Eqs. 7.5 and 7.7 for shallow foundations having a foundation width B between 20 and 40 ft. Eqs. 7.7 and 7.8 can also be used for deep foundations having $D \geq 5B$ and $B \leq 20$ ft (6 m).
Besides the plate load test, the subgrade modulus can also be obtained from Fig. 7.7. The values of K_v from Fig. 7.7 apply to dry or moist cohesionless (granular) soil with the groundwater table at a depth of at least 1.5 B below the bottom of the footing. If a groundwater table is at the base of the footing, then $\frac{1}{2} K_v$ should be used in computing the settlement. For continuous footings, the settlement calculated earlier should be multiplied by a factor of 2. The relative density D_r can be estimated from the standard penetration test N value (see Table 2.6). Equations 7.5 to 7.8 may underestimate the settlement in cases of large footings where soil deformation properties vary significantly with depth or where the thickness of granular soil is only a fraction of the width of the loaded area.

Example Problem 7.1   A site consists of a sand deposit where the standard penetration test N value corrected for field testing procedures and overburden pressure $[(N_1)_{60}]$ is equal to 20. The groundwater table is well below the bottom of the footing and the total unit weight γ_t of the sand = 120 pcf (18.9 kN/m³). For a square spread footing that is 10 ft (3 m) wide and at a depth of 5 ft (1.5) below ground surface, determine the settlement if the footing exerts a vertical pressure of 6600 psf (320 kPa) onto the sand.

Solution   From Table 2.6, for $(N_1)_{60}$ = 20, the relative density D_r of sand is at the junction between medium and dense and hence D_r is approximately 65 percent.

Entering Fig. 7.7 at D_r = 65 percent, the value of K_v = 190 tons/ft³ (380,000 pcf or 60,000 kN/m³).

It is the net pressure exerted by the footing on the soil that will cause settlement, or:

$$q = 6600 \text{ psf} - (5 \text{ ft})(120 \text{ pcf}) = 6000 \text{ psf } (290 \text{ kPa})$$

Since $D \leq B$ and $B \leq 20$ ft, use Eq. 7.5:

$$S = (4qB^2)/[K_v (B + 1)^2]$$

$$= [(4)(6,000)(10)^2]/[(380,000)(10 + 1)^2] = 0.052 \text{ ft} = 0.63 \text{ in.}$$

Using Eq. 7.6:

$$S = (4qB^2)/[K_v (B + 0.3)^2]$$

$$= [(4)(290)(3)^2]/[(60,000)(3 + 0.3)^2] = 0.016 \text{ m} = 1.6 \text{ cm}$$

Terzaghi and Peck Empirical Chart.   Figure 7.8 shows a chart by Terzaghi and Peck (1967) that presents an empirical correlation between the standard penetration test N value and the allowable soil pressure (tsf) that will produce a settlement of the footing of 1.0 in. (2.5 cm). Terzaghi and Peck (1967) developed this chart specifically for the standard penetration test (i.e., E_m = 60, and C_b = 1.0).

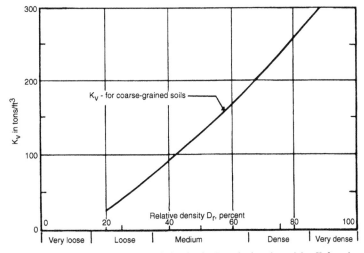

FIGURE 7.7   Correlation between relative density D_r and subgrade modulus K_v for cohesionless soil. (_Adapted from NAVFAC DM-7.1, 1982._)

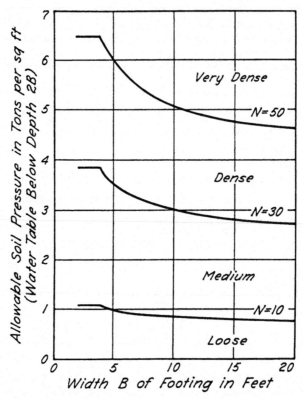

FIGURE 7.8 Allowable soil bearing pressures for footings on sand based on the standard penetration test. *Note:* assume N refers to the N_{60} value from Eq. 2.4. (*From Terzaghi and Peck, 1967; reprinted with permission of John Wiley & Sons.*)

Hence as a practical matter, the N value referred to in Fig. 7.8 can be assumed to be essentially equivalent to the N_{60} value calculated from Eq. 2.4. Terzaghi and Peck (1967) defined the soil density states versus N_{60} as follows:

Soil density	N_{60} value
Very loose	Less than 4
Loose	4–10
Medium	10–30
Dense	30–50
Very dense	Greater than 50

Thus the curves in Fig. 7.8 are at the boundaries between different soil density states. For N_{60} values other than those for which the curves are drawn in Fig. 7.8, the allowable soil pressure can be obtained by linear interpolation between curves. According to Terzaghi and Peck (1967), if all of the footings are proportioned in accordance with the allowable soil pressure corresponding to Fig. 7.8, then the total settlement ρ_{max} of the foundation should not exceed 1.0 in. (2.5 cm) and the maximum differential settlement Δ should not exceed 3/4 in. (2 cm). Figure 7.8 was developed for the groundwater table located at a depth equal to or greater than a depth of 2B below the bottom of the footing.

For conditions of a high groundwater table close to the bottom of the shallow foundation, the allowable soil pressures obtained from Fig. 7.8 should be reduced by 50 percent.

Figure 7.8 shows the importance of the density of the soil on the amount of settlement. For example, a footing that is 5 ft (1.5 m) wide will settle 1.0 in. (2.5 cm) if the soil density is at the boundary between loose and medium density (i.e., $N_{60} = 10$) and the footing pressure is 1 tsf (2000 psf, 96 kPa). For the same conditions except that the sand has a soil density at the boundary between dense and very dense (i.e., $N_{60} = 50$), the footing must exert 6 times the pressure, or 6 tsf (12,000 psf, 580 kPa) to produce the same 1.0 in. (2.5 cm) settlement.

Building codes limit the allowable bearing capacity to restrict the amount of settlement. For example, in Table 18.4 (reproduced from the *International Building Code*, 2009), the allowable bearing capacity for sands is 2000 psf (96 kPa). Using this value for sands that are in a medium, dense, or very dense state will generally produce a settlement of 1.0 in. (2.5 cm) or less per Fig. 7.8. However, if the sand is in a very loose to loose state, then the settlement of the footing will exceed 1.0 in. (2.5 cm) for a footing bearing pressure of 2000 psf (96 kPa). Hence building code values, such as those listed in Table 18.4, are very conservative for dense to very dense soil, but can result in unacceptable settlement for loose to very loose soil.

Example Problem 7.2   Using the same conditions as the prior example, determine the footing settlement from Fig. 7.8.

Solution   As per Terzaghi and Peck (1967), the boundary between medium and dense sand occurs when $N_{60} = 30$. Entering Fig. 7.8 with $B = 10$ ft (3 m) and intersecting the $N_{60} = 30$ curve, the foundation pressure to cause 1.0 in. (2.5 cm) of settlement is 3 tsf (6000 psf, 290 kPa). Since the net footing pressure is also 6000 psf (290 kPa), then the expected settlement $S = 1.0$ in. (2.5 cm).

7.3.3   Theory of Elasticity

The theory of elasticity can be used to estimate the settlement of cohesionless soil. The settlement would be based on Poisson's ratio μ and the modulus of elasticity E. For sands, Poisson's ratio μ typically varies from about 0.2 to 0.4, with a value of $1/3$ often assumed for settlement analyses.

The total settlement is highly dependent on the modulus of elasticity E of the soil. Sometimes an undisturbed specimen can be obtained for granular soil and then a consolidated drained triaxial compression test can be used to obtain the stress-strain curve. There are different methods of determining the modulus of elasticity E from the stress-strain curve, such as illustrated in Fig. 7.9. Depending on the initial shape of the stress-strain curve, the modulus of elasticity would be calculated as either the initial tangent modulus or the initial secant modulus. The modulus of elasticity for granular soil is often referred to as the drained modulus and is designated E_s.

The modulus of elasticity E_s of the cohesionless soil can also be obtained from empirical correlations with the standard penetration test (SPT) or the cone penetration test (CPT). Approximate correlations are as follows (Schmertmann, 1970):

Soil type	E_s/N_{60}
Silts, sandy silts, slightly cohesive silt-sand mixtures	4
Clean, fine to medium sands and slightly silty sands	7
Coarse sands and sands with little gravel	10
Sandy gravel and gravel	12

The units for E_s/N_{60} are tons per square foot (tsf), with N_{60} obtained from Eq. 2.4. For example, if a clean medium sand has a N_{60} value = 10, then $E_s = 70$ tsf (140,000 psf or 6.7 MPa). For the cone penetration test (CPT), an approximate correlation (Schmertmann, 1970) is $E_s = 2q_c$, where $q_c =$ cone

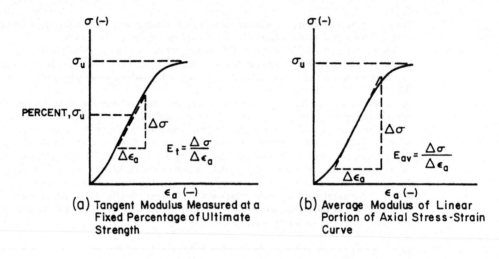

(a) Tangent Modulus Measured at a
Fixed Percentage of Ultimate
Strength

(b) Average Modulus of Linear
Portion of Axial Stress-Strain
Curve

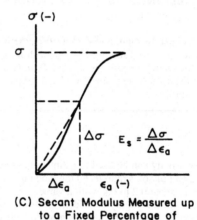

(C) Secant Modulus Measured up
to a Fixed Percentage of
Ultimate Strength

FIGURE 7.9 Methods for calculating the modulus of elasticity from a stress-strain curve. (*From ASTM D 3148-02, reprinted with permission from the American Society for Testing and Materials, 2004.*)

resistance in kg/cm². Once the modulus of elasticity E_s and Poisson's ratio μ are known, the settlement can be calculated as follows:

$$S = qBI \frac{1-\mu^2}{E_s} \tag{7.9}$$

where S = settlement of the footing (ft or m)
q = vertical footing pressure (psf or kPa)
B = footing width (ft or m)
I = shape and rigidity factor (dimensionless). This factor is derived from the theory of
elasticity to account for the thickness of the soil layer, shape of the foundation, and
flexibility of the foundation. Theoretical values of I can be obtained for various situations as shown in Fig. 7.10.

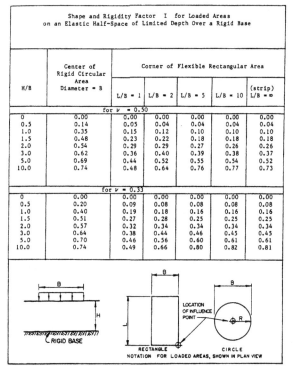

Shape and Rigidity Factor I for Loaded Areas on an Elastic Half-Space of Infinite Depth				
Shape and Rigidity	Center	Corner	Edge/Middle of Long Side	Average
Circle (flexible)	1.00		0.64	0.85
Circle (rigid)	0.79		0.79	0.79
Square (flexible)	1.12	0.56	0.76	0.95
Square (rigid)	0.82	0.82	0.82	0.82
Rectangle: (flexible) length/width				
2	1.53	0.76	1.12	1.30
5	2.10	1.05	1.68	1.82
10	2.56	1.28	2.10	2.24
Rectangle: (rigid) length/width				
2	1.12	1.12	1.12	1.12
5	1.6	1.6	1.6	1.6
10	2.0	2.0	2.0	2.0

Shape and Rigidity Factor I for Loaded Areas on an Elastic Half-Space of Limited Depth Over a Rigid Base						
H/B	Center of Rigid Circular Area Diameter = B	Corner of Flexible Rectangular Area				
		L/B = 1	L/B = 2	L/B = 5	L/B = 10	(strip) L/B = ∞
for ν = 0.50						
0	0.00	0.00	0.00	0.00	0.00	0.00
0.5	0.14	0.05	0.04	0.04	0.04	0.04
1.0	0.35	0.15	0.12	0.10	0.10	0.10
1.5	0.48	0.23	0.22	0.18	0.18	0.18
2.0	0.54	0.29	0.29	0.27	0.26	0.26
3.0	0.62	0.36	0.40	0.39	0.38	0.37
5.0	0.69	0.44	0.52	0.55	0.54	0.52
10.0	0.74	0.48	0.64	0.76	0.77	0.73
for ν = 0.33						
0	0.00	0.00	0.00	0.00	0.00	0.00
0.5	0.20	0.09	0.08	0.08	0.08	0.08
1.0	0.40	0.19	0.18	0.16	0.16	0.16
1.5	0.51	0.27	0.28	0.25	0.25	0.25
2.0	0.57	0.32	0.34	0.34	0.34	0.34
3.0	0.64	0.38	0.44	0.46	0.45	0.45
5.0	0.70	0.46	0.56	0.60	0.61	0.61
10.0	0.74	0.49	0.66	0.80	0.82	0.81

FIGURE 7.10 Shape and rigidity factors I for calculating settlement from the theory of elasticity. (*From NAVFAC DM-7.1, 1982.*)

μ = Poisson's ratio (dimensionless), which is often assumed to be $\frac{1}{3}$ for granular soil

E_s = modulus of elasticity of the soil (psf or kPa)

For complex foundation problems, computer programs have been developed to estimate the settlement based on the finite-element analysis. For example, the SIGMA/W (Geo-Slope, 1993) computer program can determine the settlement assuming the soil is elastic or inelastic.

Example Problem 7.3   Using the same conditions as the prior example, determine the settlement of the footing from Eq. 7.9. Assume a Poisson's ratio = $1/3$, an elastic half space of infinite depth, and that the square footing is rigid.

Solution   From the prior example, the following values are obtained:

$$q = 6000 \text{ psf (290 kPa)}$$
$$B = 10 \text{ ft (3 m)}$$
$$N_{60} = 30$$

Using Fig. 7.10, $I = 0.82$ (elastic half space of infinite depth and rigid square footing)

$$E_s = 7N_{60} = (7)(30) = 210 \text{ tsf} = 420{,}000 \text{ psf (20,000 kPa) for fine to medium sands}$$
$$E_s = 10N_{60} = (10)(30) = 300 \text{ tsf} = 600{,}000 \text{ psf (29,000 kPa) for coarse sands}$$

Using Eq. 7.9, for fine to medium sands:

$$S = qBI \, [(1 - \mu^2)/E_s]$$
$$= (6000)(10)(0.82)[(1 - 0.33^2)/420{,}000] = 0.10 \text{ ft} = 1.2 \text{ in. (3.0 cm)}$$

Using Eq. 7.9, for coarse sands:

$$S = qBI \, [(1 - \mu^2)/E_s]$$
$$= (6000)(10)(0.82)[(1 - 0.33^2)/600{,}000] = 0.072 \text{ ft} = 0.87 \text{ in. (2.2 cm)}$$

7.3.4   Schmertmann's Method

In many cases, the soil modulus E_s will vary with depth. Schmertmann's method (Schmertmann, 1970; Schmertmann et al., 1978) is ideally suited to determining the settlement of footings when the modulus of elasticity of the soil (E_s) varies with depth. Schmertmann's method is outlined in Fig. 7.11 and an example problem is also included in Fig. 7.11. The basic concept is to model the settlement of the footing by using a triangular distribution of vertical strain, known as the "2B-0.6 distribution." As shown in Fig. 7.11, the strain influence factor I_z linearly increases from zero at the footing bottom to a value of 0.6 at a depth of $1/2 B$ below the footing bottom. The strain influence value I_z then linearly decreases to zero at a depth of $2B$ below the footing bottom. In essence, Schmertmann's method considers settlement of soil to a depth of $2B$ below the bottom of the footing.

The equation for Schmertmann's method is as follows:

$$S = C_1 C_2 \Delta P \sum \frac{\Delta z I_z}{E_S} \tag{7.10}$$

where $S = \Delta H$ = settlement of the footing (ft or m)

Δz = thickness of the various soil layers located below the footing (ft or m)

I_z = strain influence value (dimensionless), which is a triangular shape as previously described

E_s = modulus of elasticity of the soil (psf or kPa)

C_1 = embedment correction factor (dimensionless). It is recommended that $C_1 \geq 0.5$. The embedment correction factor is defined as follows:

$$C_1 = 1 - \frac{P_o}{2\Delta P} \tag{7.11}$$

DATA REQUIRED:

1. A profile of standard penetration resistance N (blows/ft) versus depth, from the proposed foundation level to a depth of 2B, or to boundary of an incompressible layer, whichever occurs first. Value of soil modulus E_s is established using the following relationships.

Soil Type	E_s/N
Silts, sands silts, slightly cohesive silt-sand mixtures	4
Clean, fine to med, sands & slightly silty sands	7
Coarse sands & sands with little gravel	10
Sandy gravels and gravel	12

2. Least width of foundation = B, depth of embedment = D, and proposed average contact pressure = P.

3. Approximate unit weights of surcharge soils, and position of water table if within D.

4. If the static cone bearing value q_c is measured compute E_s based on $E_s = 2\ q_c$.

ANALYSIS PROCEDURE:

Refer to table in example problem for column numbers referred to by parenthesis:

1. Divide the subsurface soil profile into a convenient number of layers of any thickness, each with constant N over the depth interval 0 to 2B below the foundation.

2. Prepare a table as illustrated in the example problem, using the indicated column headings. Fill in columns 1, 2, 3 and 4 with the layering assigned in Step 1.

3. Multiply N values in column 3 by the appropriate factor E_s/N (col. 4) to obtain values of E_s; place values in column 5.

4. Draw an assumed 2B-0.6 triangular distribution for the strain influence factor I_z, along a scaled depth of 0 to 2B below the foundation. Locate the depth of the mid-height of each of the layers assumed in Step 2, and place in column 6. From this construction, determine the I_z value at the mid-height of each layer, and place in column 7.

FIGURE 7.11 Settlement of footings on granular soil: example computation using Schmertmann's method. (*From NAVFAC DM-7.1, 1982.*)

5. Calculate $(I_z/E_s) \Delta Z$, and place in column 8. Determine the sum of all values in column 8.

6. Total settlement = ΔH = $C_1 C_2 \Delta p \sum_0^{2B} (\dfrac{I_z}{E_s}) \Delta Z$,

    where $C_1 = 1 - 0.5 (p_o/\Delta p)$; $C_1 \gtrless 0.5$ embedment correction factor

    $C_2 = 1 + 0.2 \log(10t)$ creep correction factor

    p_o = overburden pressure at foundation level

    Δp = net foundation pressure increase

    t = elaspsed time in years.

EXAMPLE PROBLEM:

GIVEN THE FOLLOWING SOIL SYSTEM AND CORRESPONDING STANDARD PENETRATION TEST (SPT) DATA, DETERMINE THE AMOUNT OF ULTIMATE SETTLEMENT UNDER A GIVEN FOOTING AND FOOTING LOAD:

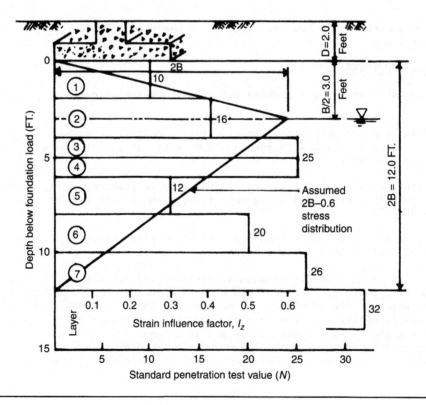

FIGURE 7.11   (Continued)

Footing Details:

Footing width: 6.0 ft. (min.) by 8.0 ft. (max.)

Depth of Embedment: 2.0 ft. Load (Dead + Live): 120 tons

Soil Properties:

Depth Below Base of Footing (ft.)	Unit Wt. (pcf) Moist	Sat.	Soil Description
<5	95	105	Fine sandy silt
5 - 10	105	120	Fine to medium sand
10 - 15	120	130	Coarse sand

Solution:

Layer (1)	ΔZ (in.) (2)	N (3)	E_s/N (4)	E_s (tsf) (5)	Z_c (in.) (6)	I_z (7)	$\dfrac{I_z \Delta Z}{E_s}$ (in./tsf) (8)
1	24	10	4	40	12	.20	0.120
2	24	16	4	64	36	.60	0.225
3	12	25	4	100	54	.50	0.060
4	12	25	7	175	66	.43	0.029
5	24	12	7	84	84	.33	0.094
6	24	20	7	140	108	.20	0.034
7	24	26	10	260	132	.07	0.006

$$\Sigma = 0.568$$

$$p_0 = (2.0 \text{ ft.})(95 \text{ pcf}) = 190 \text{ psf} = 0.095 \text{ tsf}$$
$$\Delta p = 120 \text{ tons}/(6 \text{ ft.})(8 \text{ ft.}) = 2.50 \text{ tsf}$$

At t = 1 yr,

$$C_1 = 1 - 0.5(.095/2.50) = 0.981$$

$$C_2 = 1 + 0.2 \log (10)(1) = 1.20$$

$$\Delta H = (0.981)(1.20)(2.50)(0.568) = \underline{1.67 \text{ in.}}$$

FIGURE 7.11 (*Continued*)

p_o = overburden soil pressure at foundation level (psf or kPa)
ΔP = net foundation pressure (psf or kPa)
C_2 = creep correction factor (dimensionless), defined as follows:

$$C_2 = 1 + 0.2 \log(10t) \tag{7.12}$$

t = time in years

Schmertmann (1970) observed long-term settlement of the foundations on sand that were studied. Thus, a creep correction factor C_2 was incorporated into Eq. 7.10. Perhaps this observed long-term settlement was due to slight fluctuating loads. As previously mentioned, for structures subjected to slight fluctuating loads, the settlement of footings on sand 30 years after construction might be as much as 1.5 times the immediate postconstruction settlement, while for heavy fluctuating loads, the settlement might be as much as 2.5 times the immediate postconstruction settlement (Burland and Burbidge, 1985).

Example Problem 7.4   Using the same conditions as the prior examples, determine the settlement of the footing from Eq. 7.10. Neglect creep effects (i.e., $C_2 = 1$).

Solution

$$p_o = (120)(5) = 600 \text{ psf (30 kPa)}$$
$$\Delta P = 6600 - 600 = 6000 \text{ psf (290 kPa)}$$
$$C_1 = 1 - \tfrac{1}{2}\,(p_o/\Delta P) = 1 - \tfrac{1}{2}\,(600/6000) = 0.95$$

For this simple case of a constant modulus of elasticity with depth, only one soil layer is required:

Layer 1: From the bottom of the footing to a depth of $2B$ below the footing:

$$I_z = 0.3$$
$$\Delta z = 2B = 20 \text{ ft (6 m)}$$
$$E_s = 7N_{60} = (7)(30) = 210 \text{ tsf} = 420,000 \text{ psf (20,000 kPa) for fine to medium sands}$$
$$\Sigma \Delta z I_z / E_s = (20)(0.3)/(420,000) = 1.4 \times 10^{-5} \text{ ft}^3/\text{lb} \ (9.0 \times 10^{-5} \text{ m}^3/\text{kN})$$

Using Eq. 7.10:

$$S = C_1 C_2 \Delta P \Sigma \, (\Delta z\, I_z / E_s)$$
$$= (0.95)(1.0)(6000 \text{ psf})(1.4 \times 10^{-5} \text{ ft}^3/\text{lb}) = 0.080 \text{ ft} = 0.98 \text{ in. (2.5 cm)}$$

Repeating the above steps for coarse sands:

$$E_s = 10N_{60} = (10)(30) = 300 \text{ tsf} = 600,000 \text{ psf (29,000 kPa)}$$
$$\Sigma \Delta z I_z / E_s = (20)(0.3)/(600,000) = 1.0 \times 10^{-5} \text{ ft}^3/\text{lb} \ (6.2 \times 10^{-5} \text{ m}^3/\text{kN})$$
$$S = C_1 C_2 \Delta P \Sigma \, (\Delta z I_z / E_s)$$
$$= (0.95)(1.0)(6000 \text{ psf})(1.0 \times 10^{-5} \text{ ft}^3/\text{lb}) = 0.057 \text{ ft} = 0.68 \text{ in. (1.7 cm)}$$

Summary   In summary, for this example problem of a 10 ft (3 m) wide square footing that subjects the medium-dense sand ($D_r = 65$ percent) to a net pressure of 6000 psf (290 kPa), the calculated settlements from the various methods are as follows:

(Continued)

Method	Calculated settlement S
Modulus of subgrade reaction (Eq. 7.5)	0.63 in. (1.6 cm)
Terzaghi and Peck chart (Fig. 7.8)	1.0 in. (2.5 cm)
Theory of elasticity (Eq. 7.9)	
Fine to medium sand	1.2 in. (3.0 cm)
Coarse sand	0.87 in. (2.2 cm)
Schmertmann's method (Eq. 7.10)	
Fine to medium sand	0.98 in. (2.5 cm)
Coarse sand	0.68 in. (1.7 cm)

Thus the calculated settlement of the footing varies from 0.63 in. (1.6 cm) to 1.2 in. (3.0 cm), or almost a doubling in values. When dealing with settlement calculations, it is not unusual to have a range in values that are double from one method to another. Hence, it is always best to perform the settlement analysis using different methods and compare the final result. The final decision on the expected amount of settlement would be based on experience and judgment.

7.3.5   Laboratory Testing

It is sometimes possible to obtain undisturbed soil specimens of granular soil. For soil above the groundwater table, capillary tension may hold the soil particles together. In other cases there might be a slight cementation that allows for undisturbed sampling of sands. If undisturbed soil specimens of the cohesionless soil can be obtained, such as from test pits or Shelby tube samplers, then the oedometer apparatus can be used to measure the settlement versus deformation behavior of the soil (see Sec. 3.3).

Figure 7.12 shows laboratory test results on an undisturbed specimen of silty sand that was tested in the oedometer apparatus (i.e., one-dimensional compression test). The soil was classified as a residual soil that was held together by capillary action and a slight cementation. The capillary action and slight cementation allowed for the undisturbed sampling of the soil. The soil is classified as nonplastic silty sand (SM) having 79 percent sand size particles and 21 percent silt size particles. The initial void ratio e_o of the silty sand specimen was 0.41 and the initial dry unit weight was 117.4 pcf (18.5 kN/m³). The silty sand was considered to be in a medium to dense state.

The appearance of the compression curve shown in Fig. 7.12 would suggest highly compressible silty sand. However, the silty sand actually has a relatively low compressibility since it experienced less than 3 percent vertical strain when loaded to a vertical effective stress of 16,000 psf (770 kPa). Because of the low compressibility of the soil specimen, the apparatus compressibility (i.e., compressibility of the porous stones, Plexiglas dish, and the like) was determined and then subtracted from the deformation readings in order to obtain the compressibility of only the silty sand specimen.

For data plotted on a void ratio e versus σ'_{vc} plot, the settlement is calculated from the following equation:

$$S = \frac{\Delta e H_o}{1 + e_o} \tag{7.13}$$

For data plotted on a vertical strain ε_v versus σ'_{vc} plot, such as shown in Fig. 7.12, the settlement is calculated from the following equation:

$$S = \Delta \varepsilon_v H_o \tag{7.14}$$

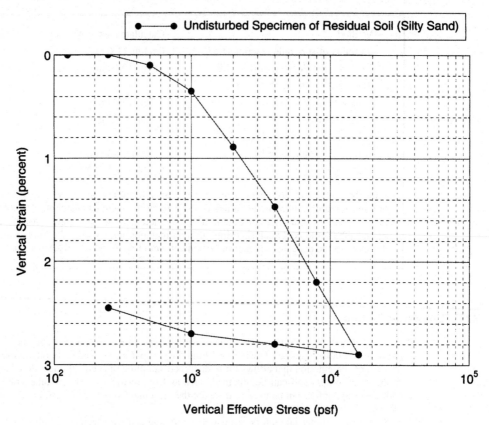

FIGURE 7.12   Laboratory one-dimensional compression test performed on a silty sand.

where S = settlement of the footing (ft or m)

Δe = change in void ratio (dimensionless) from the void ratio e versus σ'_{vc} plot. Enter the plot at the vertical effective stress σ'_{vo} and upon intersecting the compression curve, determine the initial void ratio e_i. Based on the pressure applied to the soil by the footing, determine the increase in vertical stress $\Delta\sigma_v$ at the center of the sand layer by using stress distribution theory from Sec. 4.5. Then enter the plot at the final vertical effective stress (i.e., $\sigma'_{vo} + \Delta\sigma_v$) and upon intersecting the compression curve, determine the final void ratio e_f. The change in void ratio $\Delta e = e_i - e_f$

H_o = initial height of the in situ cohesionless soil layer (ft or m). Often several undisturbed soil specimens are obtained and tested in the laboratory and H_o represents the thickness of the in situ layer that is represented by an individual oedometer test. For the analysis of several soil layers, Eq. 7.13 or 7.14 is used for each layer and then the total settlement is the sum of the settlements calculated for each soil layer

e_o = initial void ratio of the cohesionless soil layer (dimensionless)

$\Delta\varepsilon_v$ = change in vertical strain (dimensionless) from the vertical strain ε_v versus σ'_{vc} plot (Fig. 7.12). Enter the plot at the vertical effective stress σ'_{vo} and upon intersecting the compression curve, determine the initial vertical strain ε_i. Also enter the plot at the final vertical effective stress (i.e., $\sigma'_{vo} + \Delta\sigma_v$) and upon intersecting the compression curve, determine the final vertical strain ε_f and then $\Delta\varepsilon = \varepsilon_f - \varepsilon_i$

Example Problem 7.5 Use the same data as the prior example problems. Assume the laboratory test data shown in Fig. 7.12 is applicable for a zone of silty sand located from ground surface to a depth of 25 ft (7.6 m). At a depth of 25 ft (7.6 m), assume hard rock is encountered. Calculate the settlement of the footing using the 2:1 approximation and using one soil layer that is 20 ft (6.1 m) thick. Assume the groundwater table is located at a depth of 5 ft (1.5 m) below ground surface and the buoyant unit weight $\gamma_b = 63$ pcf (9.9 kN/m³)

Solution From the prior example, the following values are obtained:

Net foundation pressure $q = 6000$ psf (290 kPa)

Square footing, width $B = 10$ ft (3 m)

Depth of embedment of the footing $D = 5$ ft (1.5 m)

Total unit weight of the sand (depth of 0 to 5 ft) = 120 pcf (18.9 kN/m³)

Below a depth of 5 ft, use $\gamma_b = 63$ pcf (9.9 kN/m³)

Layer No. 1:

Extends from a depth of 5 to 25 ft (1.5 to 7.6 m) below ground surface. Calculating the vertical effective stress at the center of layer 1, or at a depth of 15 ft below ground surface:

$$\sigma'_{vo} = \sigma_{vo} - u = (5 \text{ ft})(120 \text{ pcf}) + (10 \text{ ft})(63 \text{ pcf}) = 1200 \text{ psf (59 kPa)}$$

In terms of the depth below the bottom of the footing to the center of the sand layer:

$$z = 10 \text{ ft (3 m)}$$

Using the 2:1 approximation (Eq. 4.26), with $B = L = 10$ ft:

$$\sigma_z = \Delta\sigma_v = P/[(B + z)(L + z)]$$

$$= [(6000 \text{ psf})(10)(10)]/[(10 + 10)(10 + 10)] = 1500 \text{ psf (74 kPa)}$$

Final vertical effective stress = $\sigma'_{vo} + \Delta\sigma_v = 1200 + 1500 = 2700$ psf

Entering Fig. 7.12 at $\sigma'_{vo} = 1200$ psf, $\varepsilon_i = 0.50$ percent = 0.0050

Entering Fig. 7.12 at the final vertical effective stress = 2700 psf:

$$\varepsilon_f = 1.15 \text{ percent} = 0.0115$$

$$\Delta\varepsilon_v = \varepsilon_f - \varepsilon_i = 0.0115 - 0.0050 = 0.0065$$

Using Eq. 7.14:

$$S = \Delta\varepsilon_v H_o = (0.0065)(20 \text{ ft}) = 0.13 \text{ ft} = 1.6 \text{ in. (4 cm)}$$

Example Problem 7.6 Solve the prior example, but calculate the settlement using three soil layers, the upper two are 5 ft (1.5 m) thick and the third layer is 10 ft (3 m) thick.

Layer No. 1:

Extends from a depth of 5 to 10 ft (1.5 to 3 m) below ground surface. Calculating the vertical effective stress at the center of layer 1, or at a depth of 7.5 ft below ground surface:

$$\sigma'_{vo} = \sigma_{vo} - u = (5 \text{ ft})(120 \text{ pcf}) + (2.5 \text{ ft})(63 \text{ pcf}) = 750 \text{ psf (36 kPa)}$$

(Continued)

In terms of the depth below the bottom of the footing to the center of the first sand layer: $z = 2.5$ ft (0.8 m)

Using the 2:1 approximation (Eq. 4.26), with $B = L = 10$ ft:

$$\sigma_z = \Delta\sigma_v = P/[(B + z)\,(L + z)]$$

$$= [(6000 \text{ psf})(10)(10)]/[(10 + 2.5)(10 + 2.5)] = 3840 \text{ psf } (180 \text{ kPa})$$

Final vertical effective stress $= \sigma'_{vo} + \Delta\sigma_v = 750 + 3840 = 4600$ psf
Entering Fig. 7.12 at $\sigma'_{vo} = 750$ psf, $\varepsilon_i = 0.25$ percent $= 0.0025$
Entering Fig. 7.12 at the final vertical effective stress $= 4600$ psf:

$$\varepsilon_f = 1.63 \text{ percent} = 0.0163$$

$$\Delta\varepsilon_v = \varepsilon_f - \varepsilon_i = 0.0163 - 0.0025 = 0.0138$$

Using Eq. 7.14:

$$S = \Delta\varepsilon_v\,H_o = (0.0138)(5 \text{ ft}) = 0.069 \text{ ft} = 0.83 \text{ in. } (2.1 \text{ cm})$$

Layer No. 2:

Extends from a depth of 10 to 15 ft (3 to 4.6 m) below ground surface. Calculating the vertical effective stress at the center of layer 2, or at a depth of 12.5 ft below ground surface:

$$\sigma'_{vo} = \sigma_{vo} - u = (5 \text{ ft})(120 \text{ pcf}) + (7.5 \text{ ft})(63 \text{ pcf}) = 1100 \text{ psf } (51 \text{ kPa})$$

In terms of the depth below the bottom of the footing to the center of the second sand layer: $z = 7.5$ ft (2.3 m)

Using the 2:1 approximation (Eq. 4.26), with $B = L = 10$ ft:

$$\sigma_z = \Delta\sigma_v = P/[(B + z)\,(L + z)]$$

$$= [(6000 \text{ psf})(10)(10)]/[(10 + 7.5)(10 + 7.5)] = 2000 \text{ psf } (94 \text{ kPa})$$

Final vertical effective stress $= \sigma'_{vo} + \Delta\sigma_v = 1100 + 2000 = 3100$ psf
Entering Fig. 7.12 at $\sigma'_{vo} = 1100$ psf, $\varepsilon_i = 0.40$ percent $= 0.0040$
Entering Fig. 7.12 at the final vertical effective stress $= 3100$ psf:

$$\varepsilon_f = 1.28 \text{ percent} = 0.0128$$

$$\Delta\varepsilon_v = \varepsilon_f - \varepsilon_i = 0.0128 - 0.0040 = 0.0088$$

Using Eq. 7.14:

$$S = \Delta\varepsilon_v\,H_o = (0.0088)(5 \text{ ft}) = 0.044 \text{ ft} = 0.53 \text{ in. } (1.3 \text{ cm})$$

Layer No. 3:

Extends from a depth of 15 to 25 ft (4.6 to 7.6 m) below ground surface. Calculating the vertical effective stress at the center of layer 3, or at a depth of 20 ft below ground surface:

$$\sigma'_{vo} = \sigma_{vo} - u = (5 \text{ ft})(120 \text{ pcf}) + (15 \text{ ft})(63 \text{ pcf}) = 1500 \text{ psf } (74 \text{ kPa})$$

In terms of the depth below the bottom of the footing to the center of the third sand layer: $z = 15$ ft (4.6 m)

(Continued)

Using the 2:1 approximation (Eq. 4.26), with $B = L = 10$ ft:

$$\sigma_z = \Delta\sigma_v = P/[(B + z)(L + z)]$$

$$= [(6000 \text{ psf})(10)(10)]/[(10 + 15)(10 + 15)] = 960 \text{ psf (46 kPa)}$$

Final vertical effective stress $= \sigma'_{vo} + \Delta\sigma_v = 1500 + 960 = 2500$ psf

Entering Fig. 7.12 at $\sigma'_{vo} = 1500$ psf, $\varepsilon_i = 0.70$ percent $= 0.0070$

Entering Fig. 7.12 at the final vertical effective stress $= 2500$ psf:

$$\varepsilon_f = 1.10 \text{ percent} = 0.0110$$

$$\Delta\varepsilon_v = \varepsilon_f - \varepsilon_i = 0.0110 - 0.0070 = 0.0040$$

Using Eq. 7.14:

$$S = \Delta\varepsilon_v \, H_o = (0.0040)(10 \text{ ft}) = 0.040 \text{ ft} = 0.48 \text{ in. (1.2 cm)}$$

Adding together the settlement from the three layers:

$$S = 0.83 + 0.53 + 0.48 = 1.8 \text{ in. (4.6 cm)}$$

Comparing the two example problems, with just one layer, the calculated settlement is 1.6 in. (4 cm). But using three layers, the calculated settlement is 1.8 in. (4.6 cm). Using more layers would produce a more accurate answer, although in this case the difference is small.

7.4 OTHER COMMON CAUSES OF SETTLEMENT

There are many other causes of settlement of structures. Some of the more common causes are discussed in this section. The types of settlement discussed in this section are within the second basic category of settlement (see Sec. 7.1), settlement of the structure primarily due to secondary influences. An exception is the construction of a structure atop a landfill, where the weight of the structure could directly cause compression of the underlying loose material.

7.4.1 Limestone Cavities or Sinkholes

Settlement related to limestone cavities or sinkholes will usually be limited to areas having karst topography. Karst topography is a type of landform developed in a region of easily soluble limestone bedrock. It is characterized by vast numbers of depressions of all sizes, sometimes by great outcrops of limestone, sinks and other solution passages, an almost total lack of surface streams, and larger springs in the deeper valleys (Stokes and Varnes, 1955).

Identification techniques and foundations constructed on karst topography are discussed by Sowers (1997). Methods to investigate the presence of sinkholes are geophysical techniques and the cone penetration device (Foshee and Bixler, 1994). A low cone penetration resistance could indicate the presence of raveling, which is a slow process where granular soil particles migrate into the underlying porous limestone. An advanced state of raveling will result in subsidence of the ground, commonly referred to as sinkhole activity. Figure 7.13 shows the settlement process starting with an initial condition of underground caverns (dark areas), followed by the development

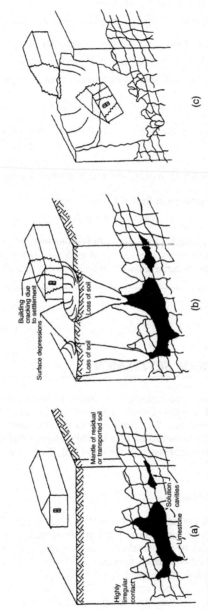

FIGURE 7.13 Development of sinkholes. (*a*) Initial condition; (*b*) subsidence or raveling sinkhole; (*c*) collapse sinkhole. (*From Rollings and Rollings, 1996; reprinted with permission of McGraw-Hill Book Co.*)

of surface depressions due to raveling of soil into the caverns, and the final condition of a collapsed sinkhole.

7.4.2 Underground Mines and Tunnels

According to Gray (1988), damage to residential structures in the United States caused by the collapse of underground mines is estimated to be between $25 million and $35 million each year, with another $3 million to $4 million in damage to roads, utilities, and services. There are approximately 2 million hectares of abandoned or inactive coal mines, with 200,000 of these hectares in populated urban areas (Dyni and Burnett, 1993).

It has been stated that ground subsidence associated with long-wall mining can be predicted fairly well with respect to magnitude, time, and area position (Lin et al., 1995). Once the amount of ground subsidence has been estimated, there are measures that can be taken to mitigate the effects of mine-related subsidence (National Coal Board, 1975; Kratzsch, 1983; Peng, 1986, 1992). For example, in a study of different foundations subjected to mining induced subsidence, it was concluded that post-tensioning of the foundation was most effective, because it prevented the footings from cracking (Lin et al., 1995).

The collapse of underground mines and tunnels can produce tension and compression type features within the buildings. The location of the compression zone will be in the center of the subsided area as shown in Fig. 7.14. The tension zone is located along the perimeter of the subsided area.

Besides the collapse of underground mines and tunnels, there can be settlement of buildings constructed on spoil extracted from the mines. Mine operators often dispose of other debris, such as trees, scrap metal, and tires, within the mine spoil. In many cases, the mine spoil is dumped (no compaction) and can be susceptible to large amounts of settlement. For example, Cheeks (1996) describes an interesting case of a motel unknowingly built on spoil that had been used to fill in a strip-mining operation. The motel building experienced about 3 ft (1 m) of settlement within the 5-year monitoring period. The settlement and damage for this building actually started during construction and the motel owners could never place the building into service. A lesson from this case study was the importance of subsurface exploration. In many cases, the borings may encounter

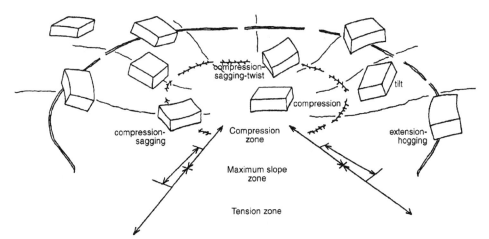

FIGURE 7.14 Location of tension and compression zones due to collapse of underground mines. (*From Marino et al., 1988; reprinted with permission from the American Society of Civil Engineers.*)

refusal on boulders generated from the mining operation and the geotechnical engineer may believe that the depth to solid rock is shallow. It is important when dealing with spoil generated from mining operations that the thickness of the spoil and the compression or collapse behavior of the material be adequately investigated. This may require use of rock coring or geophysical techniques to accurately define the limits and depth of the spoil pile.

7.4.3 Subsidence Due to Extraction of Oil or Groundwater

Large scale pumping of water or oil from the ground can cause settlement of the ground surface over a large area. The pumping can cause a lowering of the groundwater table, which then increases the overburden pressure on underlying sediments. This can cause consolidation of soft clay deposits. In other cases, the removal of water or oil can lead to compression of the soil or porous rock structure, which then results in ground subsidence.

Lambe and Whitman (1969) describe two famous cases of ground surface subsidence due to oil or groundwater extraction. The first is oil pumping from Long Beach, California that affected a 25-mi^2 (65-km^2) area and caused 25 ft (8 m) of ground surface subsidence. Because of this ground surface subsidence, the Long Beach Naval Shipyard had to construct sea walls to keep the ocean from flooding the facilities. A second famous example is ground surface subsidence caused by pumping of water for domestic and industrial use in Mexico City. Rutledge (1944) shows that the underlying Mexico City clay, which contains a porous structure of microfossils and diatoms, has a very high void ratio (up to $e = 14$) and is very compressible. Ground surface subsidence in Mexico City has been reported to be 30 ft (9 m) since the beginning of the twentieth century.

The theory of consolidation (Chap. 8) can often be used to estimate the ground surface subsidence caused by the pumping of shallow groundwater. In this case, the lowering of the groundwater table will increase the overburden pressure acting on the clay layer. For example, if pumping causes the groundwater table to be permanently lowered, then the increase in overburden stress $\Delta\sigma_v$ will be equal to the difference between the final σ'_{vf} and initial σ'_{vo} vertical effective stress in the clay layer (i.e., $\Delta\sigma_v = \sigma'_{vf} - \sigma'_{vo}$). In other cases, pumping of shallow groundwater may not cause a significant lowering of the groundwater table, but instead the pumping can cause a decrease in the pore water pressure in a permeable layer located below the clay layer. In this case, the reduction in pore water pressure will also lead to consolidation of the clay, and once again the increase in overburden stress $\Delta\sigma_v$ will be equal to the difference between the final σ'_{vf} and initial σ'_{vo} vertical effective stress in the clay layer.

7.4.4 Decomposition of Organic Matter and Landfills

Organic Matter.   Organic matter consists of a mixture of plant and animal products in various stages of decomposition, of substances formed biologically and/or chemically from the breakdown of products, and of microorganisms and small animals and their decaying remains. For this very complex system, organic matter is generally divided into two groups: nonhumic and humic substances (Schnitzer and Khan, 1972). Nonhumic substances include numerous compounds such as carbohydrates, proteins, and fats that are easily attacked by microorganisms in the soil and normally have a short existence. Most organic matter in soils consists of humic substances, defined as amorphous, hydrophilic, acidic, and polydisperse substances (Schnitzer and Khan, 1972). An example of humic substances are the brown or black part of organic matter that are so well decomposed that the original source cannot be identified. The important characteristics exhibited by humic substances are a resistance to microbic deterioration, and the ability to form stable compounds (Kononova, 1966).

The geotechnical engineer can identify organic matter by its brown or black color, pungent odor, spongy feel, and in some cases its fibrous texture. The change in character of organic matter due to decomposition has been studied by Al-Khafaji and Andersland (1981). They present visual evidence of the changes in pulp fibers due to microbic activity, using a scanning electron microscope (SEM).

Decomposition of pulp fibers includes a reduction in length, diameter, and the development of rougher surface features.

The ignition test is commonly used to determine the percent organics (ASTM D 2974-00, 2004, "Standard Test Methods for Moisture, Ash, and Organic Matter of Peat and Other Organic Soils"). In this test the humic and nonhumic substances are destroyed using high ignition temperatures. The main source of error in the ignition test is the loss of surface hydration water from the clay minerals. Franklin et al. (1973) indicate that large errors can be produced if certain minerals, such as montmorillonite, are present in quantity. The ignition test also does not distinguish between the humic and nonhumic fraction of organics.

The problem with decomposing organics is the development of voids and corresponding settlement. The rate of settlement will depend on how fast the nonhumic substances decompose, and the compression characteristics of the organics. For large mass graded projects, it is difficult to keep all organic matter out of the fill. Most engineers recognize the detrimental effects of organic matter. Common organic matter inadvertently placed in fill includes branches, shrubs, leaves, grass, and construction debris such as pieces of wood and paper. Figure 7.15 shows a photograph of decomposing organics placed in a structural fill.

If possible, the ideal situation would be to remove the debris from the site and dispose of it off-site. In some cases, the debris can be incorporated into the fill. For example, if deep fill is to be placed at the site, then the concrete, brick, or glass debris can be mixed-in with the fill and compacted as deep structural fill. It is important that nesting of concrete debris not be allowed and that no degradable material (such as organic waste products) or compressible (such as plastic containers) be mixed in with the fill.

Landfills.   Landfills can settle due to compression of the loose waste products and decomposition of organic matter that was placed within the landfill. Landfills are a special case of settlement because they could have both basic types of settlement, for example, settlement due directly to the weight of the structure that causes an initial compression of the loose waste products and settlement due to secondary effects, such as when the organic matter in the landfill slowly decomposes.

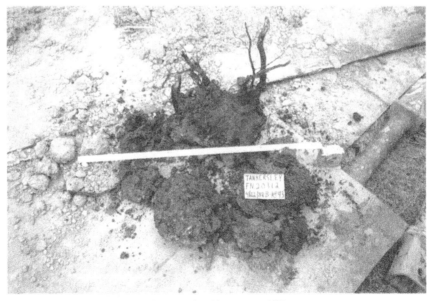

FIGURE 7.15   Decomposing organic matter placed in a structural fill.

The study of landfills provides data on the rate of decomposition of large volumes of organic matter. An excavation at the Mallard North Landfill in Hanover Park, Illinois, unearthed a 15-year-old steak (dated by a legible newspaper buried nearby) that had bone, fat, meat, everything (Rathje and Psihoyos, 1991). The lack of decomposition of these nonhumic substances can be attributed to the fact that a typical landfill admits no light and little air or moisture, so the organic matter in the trash decomposes slowly. The nonhumic substances will decompose eventually, but because little air (and hence oxygen) circulates around the organics, the slower working anaerobic microorganisms thrive.

Rathje and Psihoyos (1991) have found that 20 to 50 percent of food and yard waste biodegrades in the first 15 years. There can be exceptions such as the Fresh Kills landfill, opened by New York City in 1948, on Staten Island's tidal marshland. Below a certain level in the Fresh Kills landfill, there are no food debris or yard wastes, and practically no paper. The reason is, apparently, water from the tidal wetlands has seeped into the landfill, causing the anaerobic microorganisms to flourish. The study by Rathje and Psihoyos (1991) indicates that the type of environment is very important in terms of how fast the nonhumic substances decompose.

There are many old abandoned landfills throughout the United States. In some cases it may be a large municipal landfill that has been abandoned, but a more frequent case is that the site contains old debris and other waste products that have been dumped at a site. Settlement of structures constructed atop municipal landfill could be due to compression of the underlying loose waste products or the decomposition of any organic matter remaining in the landfill. During decomposition of the organic matter, there will also be the generation of methane (a flammable gas), which must be safely vented to the atmosphere.

7.4.5 Soluble Soil

Especially in arid parts of the world, the soil may contain soluble soil particles, such as halite (salt). Deposits of halite can form in salt playas, sabkhas (coastal salt marshes), and salinas (Bell, 1983). Besides halite, the soil may contain other minerals that are soluble, such as magnesium or calcium carbonate (caliche) and gypsum (gypsiferous soil).

These soluble soil particles are dense and hard enough to carry the overburden pressure. But after the site is developed, there can be infiltration of water into the ground from irrigation or leaky water pipes. As this water penetrates the soil containing soluble minerals, two types of settlement can occur: (1) the collapse of the soil structure due to weakening of salt cemented bonds at particle contacts, and (2) the water can dissolve away the soluble minerals (i.e., a loss of solids) resulting in ground surface settlement.

A simple method to determine the presence of soluble minerals is to perform a permeameter test. A specimen of the soil, having a dry mass M_o of about 100 g, is placed in the permeameter apparatus. Filter paper should be used to prevent the loss of fines during the test. Usually about 2 L of distilled water is slowly flushed through the soil specimen. After flushing, the soil is dried and the percent soluble soil particles (% soluble) is determined as the initial (M_o) minus final (M_f) dry mass divided by the initial dry mass of soil (M_o), expressed as a percentage, or:

$$\text{\% soluble } (S_L) = \frac{100(M_o - M_f)}{M_o} \tag{7.15}$$

As a check on the amount of soluble minerals, the water flushed through the soil specimen can be collected, placed in a sedimentation cylinder (1000 mL), and a hydrometer can be used to determine the amount (grams) of dissolved minerals. As an alternative, the water flushed through the soil specimen can be boiled and the residue collected and weighed.

In order to help identify the type of dissolved minerals, a chemical analysis could be performed on the water that was flushed through the soil. For example, the data shown in Fig. 7.16 presents an analysis of the water that was flushed through a soil specimen. The soil specimen was obtained from Las Vegas, Nevada, and was suspected of being a gypsiferous soil. This was confirmed by the

Analysis of Water Samples

Component Analyzed	Method	Unit	PQL	Analysis Result	
				AGTP-4 00-02175-2	AGTP-5 00-02175-3
General minerals					
Alkalinity	310.1	mg/L	2	36	41
Bicarbonate	SM2320B	mg/L	2	36	41
Carbonate	SM2320B	mg $CaCO_3$/L	2	ND	ND
Hydroxide	SM2320B	mg $CaCO_3$/L	2	ND	ND
Chloride Cl^-	325.3	mg/L	1	14	11
Hardness	130.2	mg/L	2	1900	1830
Surfactants (MBAS)	425.1	mg/L	0.1	0.05J	0.04J
pH	9040	pH unit	0.01	6.78	6.72
Electric conductivity	120.1/9050	µS/cm	1	3080	2980
Sulfate (SO_4^{2-})	375.4	mg/L	2	1820	1650
Solids, total dissolved (TDS)	160.1	mg/L	10	2610	2720
Nitrate (NO_3^-) as N	SM4500N03D	mg/L	1	0.8J	0.8J
Calcium, Ca	6010	mg/L	0.4	798	783
Copper, Cu	6010	mg/L	0.02	0.15	0.17
Iron, Fe	6100	mg/L	0.1	0.38	1.8
Magnesium, Mg	6010	mg/L	0.2	17.8	15.1
Manganese, Mn	6010	mg/L	0.01	0.041	0.066
Potassium, K	6010	mg/L	0.8	7.4	6.9
Sodium, Na	6010	mg/L	4	16.3	12.7
Zinc, Zn	6010	mg/L	0.02	0.079	0.11

PQL: Practical quantitation limit. MDL: Method detection limit. CRDL: Contract-required detection limit.
N.D.: Not detected or less than the practical limit.   "—": Analysis is not required.
J: Reported between PQL and MDL.

FIGURE 7.16 Laboratory test results on water flushed through a soil specimen. (*Laboratory tests performed by Applied P & Ch Laboratory, Chino, California.*)

laboratory analysis summarized in Fig. 7.16, which indicated a high concentration of total dissolved solids (TDS of 2610 and 2720 mg/L), with the highest fraction of dissolved solids being sulfate (SO_4 = 1820 and 1650 mg/L) and calcium (Ca = 798 and 783 mg/L). These two compounds are the primary components of gypsum ($CaSO_4 \cdot 2H_2O$).

Ground surface settlement due to the dissolution of soluble soil particles can be calculated from the following equation:

$$S = S_L H_o \frac{G_s}{G_{sol}} \qquad (7.16)$$

where S = settlement of the soil layer due to loss of soluble soil particles (ft or m)
$\quad S_L$ = soluble soil particles in the soil, expressed in decimal form (from Eq. 7.15)
$\quad H_o$ = initial thickness of the soluble soil layer (ft or m)
$\quad G_s$ = specific gravity of the insoluble soil minerals (dimensionless)
$\quad G_{sol}$ = specific gravity of the soluble soil minerals (dimensionless)

The calculated settlement S from Eq. 7.16 should be added to the collapsed settlement determined from oedometer tests (see Sec. 7.3).

> **Example Problem 7.7**   As an example of the use of Eq. 7.16, suppose a 2 ft (0.6 m) thick soil layer has an average value of 10 percent soluble gypsum soil particles ($G_{sol} = 2.35$) and a specific gravity G_s of the insoluble soil minerals = 2.70. Determine the amount of settlement due to loss of soluble soil particles.
>
> **Solution**   Using Eq. 7.16:
>
> $$S = S_L H (G_s/G_{sol}) = (0.10)(2 \text{ ft})(2.70/2.35) = 0.23 \text{ ft} = 2.8 \text{ in. (7 cm)}$$

If a soil contains above 6 percent soluble soil particles, then it cannot be used as structural fill (Converse Consultants Southwest, Inc. 1990). Soil having between 2 to 6 percent soluble minerals can be blended with nonsoluble soil in accordance with the following ratios (per Converse Consultants Southwest, Inc. 1990):

Percent soluble soil particles	Blending proportions(nonsoluble soil: soluble soil)
2–4	1:1
4–6	2:1
above 6	Cannot be used as structural fill

7.4.6   Ground Fissures

Besides ground surface subsidence, the extraction of groundwater or oil can also cause the opening of ground fissures. For example, there has been up to about 5 ft (1.5 m) of ground surface subsidence in Las Vegas valley between 1963 and 1987 due primarily to groundwater extraction. It has been stated that the subsidence has been focused on preexisting geologic faults, which serve as points of weakness for ground movement (Purkey et al., 1994). Figures 7.17 to 7.21 show five views of damage caused by ground fissures at a housing development in Las Vegas, Nevada, as follows:

1. *Figure 7.17.*   This photograph shows an open ground fissure. In many cases the ground fissures have an upper soil plug that hides the presence of the fissure. Water from irrigation or leaky pipes can soften the soil plug and cause the fissure to open at ground surface.
2. *Figures 7.18 and 7.19.*   These two photographs show foundation damage. The house structure was so badly damaged that it was demolished and the only remaining part is the foundation. In Fig. 7.18, the fissure can be seen running directly underneath the foundation.
3. *Figures 7.20 and 7.21.*   These last two photographs show damage to the roads and sidewalks. The houses adjacent the roads were badly damaged and demolished. Although the fissures are not exposed at ground surface, the settlement of the street and sidewalk caused by the underground fissures are clearly evident in these photographs.

Ground fissures can also be caused by other geologic mechanisms. For example, strike-slip fault rupture can create open fissures at ground surface. Another possibility is landslide movement, where fissures may open up at the crown, flanks, or main body of the slide.

7.5   FOUNDATIONS ON ROCK

7.5.1   Lightly Loaded Foundations on Rock

For lightly loaded foundations supported by hard and sound rock, the total settlement ρ_{max} and maximum differential settlement Δ may be essentially zero. In some cases, the rock may actually be stronger than the foundation concrete. When dealing with lightly loaded foundations to be supported

FIGURE 7.17 Ground fissure in Las Vegas, Nevada.

by hard and sound rock, an extensive investigation is usually not justified and the allowable bearing capacity from the building code is often recommended (e.g., see Table 18.4, reproduced from the *International Building Code*, 2009). For hard and sound rock, usually the allowable bearing capacity from the building code will have an ample factor of safety. Exceptions include the following:

1. *Weathered or fractured rock.* Rock may become so weathered that its behavior is closer to a soil than intact rock. The allowable bearing capacity from the building code, such as the values listed

FIGURE 7.18 Ground fissure running underneath a foundation, Las Vegas, Nevada.

in Table 18.4, may also be too high for foliated, friable, weakly cemented, highly jointed, or other conditions that result in weak rock.

2. *Expansive rock.* Some types of rock, such as claystone and shale, may be expansive. Instead of settlement problems, these types of rock may expand upon moisture intrusion. Chap. 9 will discuss expansive rock.

FIGURE 7.19 Foundation damage caused by ground fissures, Las Vegas, Nevada.

FIGURE 7.20   Settlement of the street and sidewalk due to ground fissures, Las Vegas, Nevada.

3. *Secondary influences.*   Even for lightly loaded foundations on sound rock, they could still experience significant settlement due to secondary influences, such as the opening of limestone cavities or sinkholes (Sec. 7.4.1) or settlement related to the collapse of underground mines and tunnels (Sec. 7.4.2).

FIGURE 7.21   Settlement of the street and sidewalk due to ground fissures, Las Vegas, Nevada. Note that the Las Vegas strip is visible in the background of this photograph.

4. *Cut-fill transition.* A cut-fill transition occurs when a building pad has some rock removed (the cut portion), with a level building pad being created by filling in with soil the remaining portion. If the cut side of the building pad contains non-expansive rock that is hard and sound, then very little settlement would be expected for that part of the building on cut. But the fill portion could settle under its own weight and cause damage. For example, a slab crack will typically open at the location of the cut-fill transition as illustrated in Fig. 7.22. Damage is caused by the vertical foundation movement (settlement of fill) and the horizontal movement that manifests itself as a slab crack and drag effect. One option is to use deepened footings or deep foundations to underpin that portion of the structure located on fill such that the entire foundation is anchored in rock. This option would be the preferred method, especially for heavily loaded foundations.

In summary, for lightly loaded foundations bearing on rock, an extensive investigation is usually not justified and the recommended allowable bearing capacity is often based on code values, such those values listed in Table 18.4.

7.5.2 Heavily Loaded Foundations on Rock

For heavily loaded foundations, using the allowable bearing capacity from Table 18.4 may be too conservative for hard and sound rock, which will result in an uneconomical foundation. For example, Table 18.4 specifies a maximum allowable bearing capacity of 12,000 psf (570 kPa) for crystalline bedrock. But other sources have recommended much higher allowable bearing values, such as 160,000 psf (7.7 MPa) for massive crystalline bedrock (NAVFAC DM-7.2, 1982). Local experience may even dictate higher allowable bearing values, but the allowable bearing capacity should never exceed the compressive strength of the foundation concrete.

It has been stated that there is no reliable method of predicting the overall strength and deformation behavior of a rock mass from the results of laboratory tests, such as the unconfined compression test on small rock specimens. This is because the settlement behavior of rock is strongly influenced by large-scale in situ properties, such as joints, fractures, faults, inhomogeneities, weakness planes, and other factors. Hence using small core specimens to predict settlement behavior will often produce highly inaccurate results, with the inaccuracy increasing as the RQD decreases (see Sec. 4.2.8 for definition of RQD). Because of this limitation, field tests are preferable, with some options for determining the settlement of heavily loaded foundations on rock as follows:

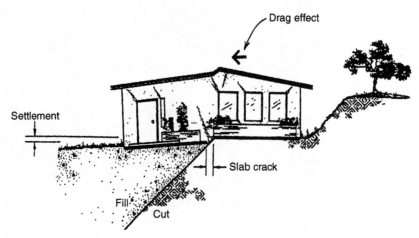

FIGURE 7.22 Cut-fill transition lot.

1. *Foundation load test.* If the proposed foundation were to consist of piers founded on rock, then test piers could be constructed and load tested. Performing load tests would be the most accurate method of determining the load-settlement behavior of a foundation element supported on rock (i.e., see ASTM D 1143-94, 2004, "Standard Test Method for Piles Under Static Axial Compressive Load"). Unfortunately, such a test would be very expensive and time consuming and would normally only be justified for essential facilities or very high foundation loads.

2. *Plate loading method.* Similar to the plate load test described in Sec. 7.3.2, the rock could be subjected to a plate loading test in order to determine the load-deformation behavior of the rock. For further details, see ASTM D 4394-98 (2004), "Standard Test Method for Determining the In Situ Modulus of Deformation of Rock Mass Using the Rigid Plate Loading Method" and ASTM D 4395-98 (2004), "Standard Test Method for Determining the In Situ Modulus of Deformation of Rock Mass Using the Flexible Plate Loading Method."

3. *Rock mass strength.* The allowable bearing value q_{all} could be based on the rock mass strength. The approach would be to determine the unconfined compressive strength q_c and then apply a factor or safety, or $q_{all} = q_c/F$. The goal is to take into account the effect of both intact material behavior and the behavior of discontinuities contained within the specimen block. The rock mass strength could be determined by performing in situ unconfined compressive tests as described in ASTM D 4555-01 (2004), "Standard Test Method for Determining Deformability and Strength of Weak Rock by an In Situ Uniaxial Compressive Test." The procedure is to perform in situ compression tests on specimens of rock that are large enough so that the rock mass unconfined compressive strength q_c is obtained, such as illustrated in Fig. 7.23. Similar to the load tests described above, this approach is very expensive and time consuming.

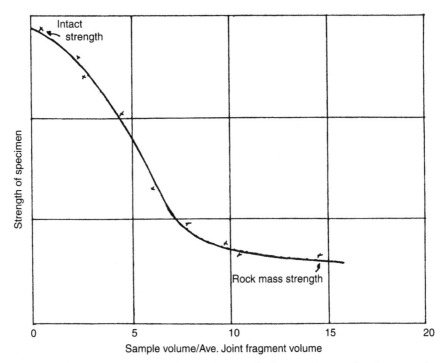

FIGURE 7.23 Example showing how the rock mass unconfined compressive strength q_c decreases as the specimen size increases. The curve eventually flattens out and the rock mass unconfined compressive strength q_c can be determined as illustrated in the figure. (*From ASTM D 4555-01, reprinted with permission from the American Society for Testing and Materials, 2004.*)

4. *Laboratory testing.*   Especially for soft rock, such as weakly cemented sedimentary rock, undisturbed specimens of the material can be obtained from Shelby tube samplers. The specimens could then be set-up in an oedometer apparatus and the vertical stress versus vertical stain curve can be obtained (i.e., similar to Fig. 7.12). Several specimens of rock would be obtained at various depths and the settlement calculations would be identical to the procedure as outlined in Sec. 7.3.5.

Two other approaches for determining the settlement of heavily loaded foundations on rock are the elastic method (i.e., Eq. 7.9) and the finite element method. Unfortunately, both of these methods require that the modulus of elasticity E and Poisson's ratio μ be determined for the rock mass. A major limitation of these two methods is that the value of E is often obtained from an unconfined compression test on a small rock specimen (i.e., ASTM D 3148-02, 2004, "Standard Test Method for Elastic Moduli of Intact Rock Core Specimens in Uniaxial Compression."). Using this approach will undoubtedly overestimate the value of E, with the overestimation increasing as the RQD decreases. The methods will have much greater accuracy if the value of E is obtained from in situ tests on large rock specimens (i.e., ASTM D 4555-01, 2004). For large scale in situ tests on rock, the value of E is obtained from the vertical stress versus vertical strain curve and is often defined as the tangent modulus at 50 percent of the maximum strength and is then referred to as E_t (see item a, Fig. 7.9).

When using high allowable bearing values for rock, it is essential that the footing or deep foundation excavation be inspected and cleaned of all loose debris so that the foundation bears directly on intact rock. Even a thin layer of disturbed and loose material left at the bottom of the foundation excavation can lead to settlement that is greatly in excess of calculated values.

7.6 ALLOWABLE SETTLEMENT

The allowable settlement is defined as the acceptable amount of settlement and it usually includes a factor of safety. In many cases the structural engineer or architect will determine the allowable settlement of the proposed structure. At the start of a project, the structural engineer or architect should be consulted about the anticipated loading conditions, allowable settlement of the structure, and preliminary thoughts on the type of foundation.

If this information is unavailable, then the geotechnical engineer will have to estimate the allowable settlement of the structure in order to determine an appropriate type of foundation. There is a considerable amount of data available on the allowable settlement of structures (e.g., Leonards, 1962; ASCE, 1964; Feld, 1965; Peck et al., 1974; Bromhead, 1984; Wahls, 1994). For example, it has been stated that the allowable differential and total settlement should depend on the flexibility and complexity of the structure including the construction materials and type of connections (Winterkorn and Fang, 1975). Brief discussions of other studies are as follows:

Coduto (1994).   Coduto (1994) states that the allowable settlement depends on many factors, including the following:

1. *The type of construction.*   For example, wood-frame buildings with wood siding would be much more tolerant than unreinforced brick buildings.

2. *The use of the structure.*   Even small cracks in a house might be considered unacceptable, whereas much larger cracks in an industrial building might not even be noticed.

3. *The presence of sensitive finishes.*   Tile or other sensitive finishes are much less tolerant of movements.

4. *The rigidity of the structure.*   If a footing beneath part of a very rigid structure settles more than the others, the structure will transfer some of the load away from the footing. However, footings beneath flexible structures must settle much more before any significant load transfer occurs. Therefore, a rigid structure will have less differential settlement than a flexible one.

Coduto (1994) also states that the allowable settlement for most structures, especially buildings, will be governed by aesthetic and serviceability requirements, not structural requirements. Unsightly cracks, jamming doors and windows, and other similar problems will develop long before the integrity of the structure is in danger.

Skempton and MacDonald (1956).   Because the determination of the allowable settlement is so complex, engineers often rely on correlations between observed behavior of structures and the settlement that results in damage. A major reference for the allowable settlement of structures based on such correlations is the paper by Skempton and MacDonald (1956) titled "The Allowable Settlements of Buildings." Skempton and MacDonald (1956) studied 98 buildings, where 58 had suffered no damage and 40 had been damaged in varying degrees as a consequence of settlement. From a study of these 98 buildings, Skempton and MacDonald (1956) in part concluded the following:

1. The cracking of the brick panels in frame buildings or load-bearing brick walls is likely to occur if the angular distortion of the foundation exceeds 1/300. Structural damage to columns and beams is likely to occur if the angular distortion of the foundation exceeds 1/150.

2. By plotting the maximum angular distortion (δ/L, see Fig. 7.1) versus the maximum differential settlement Δ such as shown in Fig. 7.24, a correlation was obtained that is defined as $\Delta = 350\ \delta/L$ (note: Δ is in inches). Using this relationship and a maximum angular distortion δ/L of 1/300, cracking of brick panels in frame buildings or load-bearing brick walls is likely to occur if the maximum differential settlement Δ exceeds $1\frac{1}{4}$ in. (32 mm).

3. The angular distortion criteria of 1/150 and 1/300 were derived from an observational study of buildings of load-bearing-wall construction, and steel and reinforced-concrete-frame buildings with conventional brick panel walls but without diagonal bracing. The criteria are intended as no

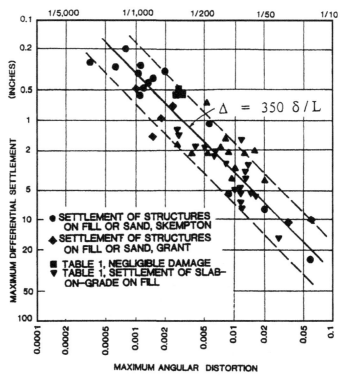

FIGURE 7.24   Maximum differential settlement Δ versus maximum angular distortion δ/L. (*Initial data from Skempton and MacDonald, 1956; Table 1 in Day, 1990a.*)

more than a guide for day-to-day work in designing typical foundations for such buildings. In certain cases they may be overruled by visual or other considerations.

Data concerning the behavior of lightly reinforced, conventional slab-on-grade foundations have also been included in Fig. 7.24. This data indicates that cracking of gypsum wallboard panels is likely to occur if the angular distortion of the slab-on-grade foundation exceeds 1/300 (Day, 1990a). The ratio of 1/300 appears to be useful for both wood-frame gypsum wallboard panels and the brick panels studied by Skempton and MacDonald (1956). The data plotted in Fig. 7.24 would indicate that the relationship $\Delta = 350 \; \delta/L$ can also be used for buildings supported by lightly reinforced slab-on-grade foundations. Using $\delta/L = 1/300$ as the boundary where cracking of panels in wood-frame residences supported by concrete slab-on-grade is likely to occur and substituting this value into the relationship $\Delta = 350 \; \delta/L$ (Fig. 7.24), the calculated differential slab displacement is $1^{1}/_{4}$ in. (32 mm). For buildings on lightly reinforced slabs-on-grade, cracking of gypsum wallboard panels is likely to occur when the maximum slab differential exceeds $1^{1}/_{4}$ in. (32 mm).

Grant et al. (1974).   The paper by Grant et al. (1974) updated the Skempton and MacDonald data pool and also evaluated the rate of settlement with respect to the amount of damage incurred. Grant et al. (1974) in part concluded the following:

1. A building foundation that experiences a maximum value of deflection slope δ/L greater than 1/300 will probably suffer some damage. However, damage does not necessarily occur at the point where the local deflection slope exceeds 1/300.
2. For any type of foundation on sand or fill, new data tend to support Skempton and MacDonalds's suggested correlation of $\Delta = 350 \; \delta/L$ (see Fig. 7.24).
3. Consideration of the rate of settlement is important only for the extreme situations of either very slow or very rapid settlement. Based on the limited data available, the values of maximum δ/L corresponding to building damage appear to be essentially the same for cases involving slow and fast settlements.

Terzaghi (1938).   Concerning allowable settlement, Terzaghi (1938) stated:

> Differential settlement must be considered inevitable for every foundation, unless the foundation is supported by solid rock. The effect of the differential settlement on the building depends to a large extent on the type of construction.

Terzaghi (1938) summarized his studies on several buildings in Europe where he found that walls 60 ft (18 m) and 75 ft (23 m) long with differential settlements over 1 in. (2.5 cm) were all cracked, but four buildings with walls 40 ft (12 m) to 100 ft (30 m) long were undamaged when the differential settlement was $^{3}/_{4}$ in. (2 cm) or less. This is probably the basis for the general design guide that building foundations should be designed so that the differential settlement is $^{3}/_{4}$ in. (2 cm) or less.

Sowers (1962).   Another example of allowable settlements for buildings is Table 7.2 (from Sowers, 1962). In this table, the allowable foundation displacement has been divided into three categories: total settlement, tilting, and differential movement. Table 7.2 indicates that structures that are more flexible, such as simple steel frame buildings, or have more rigid foundations, such as mat foundations, can sustain larger values of total settlement and differential movement.

Bjerrum (1963).   Figure 7.25 presents data from Bjerrum (1963). Similar to the studies previously mentioned, this figure indicates that cracking of panel walls is to be expected at a maximum angular distortion δ/L of 1/300 and that structural damage of buildings is to be expected at a maximum angular distortion δ/L of 1/150. This figure also provides other limiting values of maximum angular distortion δ/L, such as for buildings containing sensitive machinery or overhead cranes.

Settlement Versus Cracking Damage.   Table 7.3 summarizes the severity of cracking damage versus approximate crack widths, typical values of maximum differential movement (Δ), and maximum

TABLE 7.2   Allowable Settlement

Type of movement	Limiting factor	Maximum settlement
Total settlement	Drainage	15–30 cm (6–12 in.)
	Access	30–60 cm (12–24 in.)
	Probability of nonuniform settlement:	
	Masonry walled structure	2.5–5 cm (1–2 in.)
	Framed structures	5–10 cm (2–4 in.)
	Smokestacks, silos, mats	8–30 cm (3–12 in.)
Tilting	Stability against overturning	Depends on H and W
	Tilting of smokestacks, towers	$0.004L$
	Rolling of trucks, etc.	$0.01L$
	Stacking of goods	$0.01L$
	Machine operation—cotton loom	$0.003L$
	Machine operation—turbogenerator	$0.0002L$
	Crane rails	$0.003L$
	Drainage of floors	0.01–$0.02L$
Differential movement	High continuous brick walls	0.0005-$0.001L$
	One-story brick mill building, wall cracking	0.001–$0.002L$
	Plaster cracking (gypsum)	$0.001L$
	Reinforced-concrete-building frame	0.0025–$0.004L$
	Reinforced-concrete-building curtain walls	$0.003L$
	Steel frame, continuous	$0.002L$
	Simple steel frame	$0.005L$

Source: Sowers (1962).
Notes: L = distance between adjacent columns that settle different amounts, or between any two points that settle differently. Higher values are for regular settlements and more tolerant structures. Lower values are for irregular settlement and critical structures. H = height and W = width of structure.

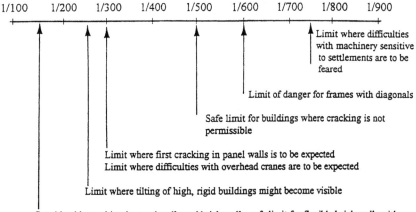

Angular Distortion (δ / L)

FIGURE 7.25   Damage criteria. (*After Bjerrum, 1963.*)

TABLE 7.3 Severity of Cracking Damage

Damage category	Description of typical damage	Approx. crack width	Δ	δ/L
Negligible	Hairline cracks	< 0.1 mm	< 3 cm (<1.2 in.)	< 1/300
Very slight	Very slight damage includes fine cracks that can be easily treated during normal decoration, perhaps an isolated slight fracture in building, and cracks in external brickwork visible on close inspection	1 mm	3–4 cm (1.2–1.5 in.)	1/300 to 1/240
Slight	Slight damage includes cracks that can be easily filled and redecoration would probably be required; several slight fractures may appear showing on the inside of the building; cracks that are visible externally and some repointing may be required; doors and windows may stick	3 mm	4–5 cm (1.5–2.0 in.)	1/240 to 1/175
Moderate	Moderate damage includes cracks that require some opening up and can be patched by a mason; recurrent cracks that can be masked by suitable linings; repointing of external brickwork and possibly a small amount of brickwork replacement may be required; doors and windows stick; service pipes may fracture; weather-tightness is often impaired	5 to 15 mm or a number of cracks > 3 mm	5–8 cm (2.0–3.0 in.)	1/175 to 1/120
Severe	Severe damage includes large cracks requiring extensive repair work involving breaking out and replacing sections of walls (especially over doors and windows); distorted windows and door frames; noticeably sloping floors; leaning or bulging walls; some loss of bearing in beams; and disrupted service pipes	15 to 25 mm but also depends on number of cracks	8–13 cm (3.0–5.0 in.)	1/120 to 1/70
Very severe	Very severe damage often requires a major repair job involving partial or complete rebuilding; beams lose bearing; walls lean and require shoring; windows are broken with distortion; and there is danger of structural instability	Usually > 25 mm but also depends on number of cracks	> 13 cm (> 5 in.)	>1/70

angular distortion δ/L of the foundation (Burland et al., 1977; Boone, 1996; Day, 1998). The relationship between differential settlement Δ and maximum angular distortion δ/L was based on the equation $\Delta = 350 \, \delta/L$ (from Fig. 7.24).

When assessing the severity of damage for an existing structure, the damage category (Table 7.3) should be based on multiple factors, including crack widths, differential settlement, and the maximum angular distortion of the foundation. Relying on only one parameter, such as crack width, can be inaccurate in cases where cracking has been hidden or patched, or in cases where other factors (such as concrete shrinkage) contribute to crack widths. For example, wall cracks can be hidden with wallpaper, but the area will recrack with additional foundation movement, such as shown in Fig. 7.26.

Foundations subjected to settlement can be damaged by a combination of both vertical and horizontal movements. For example, a common cause of foundation damage is fill settlement. Figure 7.27 shows an illustration of the settlement of fill in a canyon environment. Over the sidewalls of the canyon, there tends to be a pulling or stretching of the ground surface (tensional features), with compression effects near the canyon centerline. This type of damage is due to two-dimensional settlement, where the fill compresses or collapses in both the vertical and horizontal directions (Lawton et al., 1991; Day, 1991a). Another common situation where both vertical and horizontal foundation displacement occurs is at a cut-fill transition, such as shown in Fig. 7.22. In these cases, the lateral movement is a secondary result of the primary vertical movement due to settlement of the foundation. Table 7.3 can therefore be

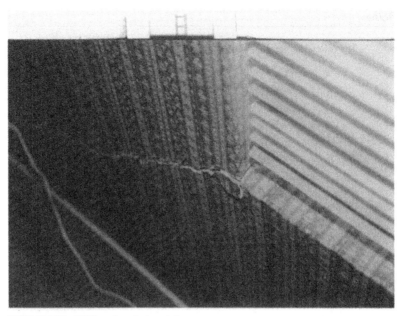

FIGURE 7.26   Wallboard crack at a window corner.

used as a guide to correlate damage category with Δ and δ/L. In cases where lateral movement is the most predominant or critical mode of foundation displacement, Table 7.3 may underestimate the severity of cracking damage for values of Δ and δ/L (Day, 1998; Boone, 1998).

Building Code Regulations.   The geotechnical engineer should always check the local building code or other governing regulations for allowable settlement. An example of a very restrictive code regulation is as follows (Southern Nevada Building Code Amendments, 1997):

> Section 1806.12, Maximum Design Total and Differential Settlements: Total settlements shall not exceed the tolerance for the materials and type of construction used. Slab deflection shall not exceed the length of the slab divided by 600 [i.e., $\delta/L < 1/600$] nor more than $^1/_2$ inch (1.3 cm) over the length of the slab.

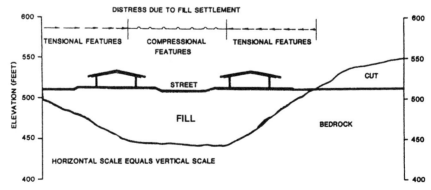

FIGURE 7.27   Fill settlement in a canyon environment.

NOTATION

The following notation is used in this chapter:

B = footing width

$\%C$ = percent collapse

C_b = borehole diameter correction for the SPT

C_1 = embedment correction factor

C_2 = creep correction factor

d_o = dial reading taken just before the soil specimen is wetted (Sec. 7.2)

d_F = final dial reading after the soil specimen has completed its collapse

D = depth of the footing below ground surface (Sec. 7.3)

D = smallest dimension of the actual footing (plate load test)

D_r = relative density

D_1 = smallest dimension of the steel plate (plate load test)

e_i, e_f = void ratio obtained from the compression curve (Sec. 7.3.5)

e_o = initial void ratio

Δe = change in void ratio

E, E_s = modulus of elasticity of the soil

E_m = hammer efficiency for the SPT

E_t = tangent modulus of elasticity (Fig. 7.9)

F = factor of safety

G_s = specific gravity of the insoluble soil minerals

G_{sol} = specific gravity of the soluble soil minerals

h_o = initial thickness of soil specimen (Sec. 7.2)

Δh = change in height of the soil

H_o = initial thickness of the in situ soil layer

ΔH = settlement of the footing

I = shape and rigidity factor

I_e = collapse index

I_z = strain influence value

K_v = modulus of subgrade reaction, also known as the subgrade modulus

L = footing length (Sec. 7.3.5)

L = horizontal distance for calculation of maximum angular distortion

M_o = initial dry mass of soil (solubility test)

M_f = final dry mass of soil (solubility test)

N = standard penetration test N value

N_{60} = N value corrected for field testing procedures

$(N_1)_{60}$ = N value corrected for field testing procedures and overburden pressure

p_o = overburden soil pressure at foundation level

P = foundation load

ΔP = net foundation pressure

q = vertical footing pressure

q_{all} = allowable bearing value

q_c = rock mass unconfined compressive strength (Sec. 7.5)

q_c = cone resistance (CPT)

S = settlement of the footing

S_L = soluble soil particles in the soil

S_1 = depth of penetration of the steel plate (plate load test)

t = time

u = pore water pressure

z = depth below the foundation bottom

Δz = thickness of the various soil layers located below the footing

Δ = maximum differential settlement of the foundation

δ = vertical displacement for calculation of maximum angular distortion

δ = depth of penetration of the plate (plate load test)

δ/L = maximum angular distortion of the foundation

$\varepsilon_i, \varepsilon_f$ = strain obtained from the compression curve (Sec. 7.3.5)

ε_v = vertical strain

$\Delta\varepsilon_v$ = change in vertical strain

γ_b = buoyant unit weight of the soil

μ = Poisson's ratio

ρ_{max} = total settlement of the foundation, also known as the maximum settlement

σ'_{vc} = vertical effective stress (compression test)

σ'_{vf} = final vertical effective stress

σ_{vo} = vertical total stress

σ'_{vo} = vertical effective stress

$\Delta\sigma_v, \sigma_z$ = change in vertical stress

PROBLEMS

Solutions to the problems are presented in App. C of this book. The problems have been divided into basic categories as follows:

Loading for Settlement Calculations

7.1 A proposed project will consist of an industrial building (30 m wide and 42 m long) that will be used by a furniture moving company to store household items. The structural engineer indicates that preliminary plans call for exterior tilt-up walls, interior isolated columns on spread footings, and a floor slab. Perimeter tilt-up wall dead loads are 60 kN per linear meter. Interior columns will be spaced 6 m on center in both directions and each column will support 900 kN of dead load. The interior floor slab dead load is 6 kPa. The structural engineer also indicates that it is likely that the industrial building could be full of household items for a considerable length of time and this anticipated live load is 30 kPa. Assume the subsurface exploration has discovered the presence of a compressible soil layer and for the purposes of the settlement analysis, the weight of the industrial building can be assumed to be a uniform stress σ_o applied at ground surface. Determine the uniform stress σ_o that should be used in the settlement analysis.

ANSWER: $\sigma_o = 60$ kPa (1250 psf).

Settlement Analyses for Collapsible Soil

7.2   A canyon contains collapsible alluvium. One side of the proposed building will be constructed over the canyon centerline, where the thickness of collapsible alluvium = 5 ft. Along the opposite side of the building, the thickness of the alluvium = 1 ft. If undisturbed specimens of the alluvium experience 5 percent collapse under the overburden pressure and proposed building loads, determine the maximum differential settlement Δ of the building after saturation of the alluvium.

ANSWER: $\Delta = 2.4$ in.

7.3   It is proposed to construct a building on top of an existing fill mass. Undisturbed specimens of the fill were obtained and tested in the laboratory oedometer apparatus in order to determine the percent collapse. The data is summarized below:

Depth of fill layer below ground surface, ft	Laboratory oedometer testing (vertical pressure, psf)	Percent collapse
0–4	250	0
4–8	750	0.10
8–12	1250	0.25
12–16	1750	0.63
16–20	2250	1.14

Assume the fill weight = 125 pcf and there is 20 ft of fill underneath the building. Neglecting the weight of the building, determine the maximum settlement ρ_{max} assuming the entire fill mass becomes saturated.

ANSWER: $\rho_{max} = 1.0$ in.

7.4   An undisturbed soil specimen, having an initial height = 25.4 mm, is placed in an oedometer apparatus. During loading to a vertical pressure of 200 kPa, the soil specimen compresses 0.3 mm. When submerged in distilled water, the soil structure collapses and an additional 2.8 mm of vertical deformation occurs. Determine the collapse index and degree of specimen collapse.

ANSWER: Collapse index = 11 percent, degree of specimen collapse = severe.

Settlement of Cohesionless Soil

7.5   Use the data from Example Problem 7.2 (i.e., 10 ft square spread footing on sand having D_r = 65 percent). Assume that the maximum allowable settlement of this footing is 1.0 in. based on the Terzaghi and Peck design chart (Fig. 7.8). Determine what governs the design of this footing, settlement analysis or bearing capacity analysis (using a factor of safety = 3).

ANSWER: Settlement governs and hence the allowable net bearing pressure is 6000 psf.

7.6   Use the data from Problem 7.5. Determine the maximum allowable bearing pressure per the *International Building Code* (see Table 18.4 of this book).

ANSWER: As per the *International Building Code*, the allowable bearing pressure is 2000 psf.

7.7   A proposed building will have a square mat foundation with $B = 20$ m. The shallow mat foundation will exert a uniform pressure $q = 30$ kPa. Assume the site consists of clean coarse sand having $N_{60} = 20$ and Poisson's ratio $\mu = 0.3$. Calculate the maximum total settlement ρ_{max} and the maximum differential settlement Δ. For the analyses, assume a flexible loaded area on an elastic half-space of infinite depth.

ANSWER: $\rho_{max} = 3.1$ cm (center) and $\Delta = 1.6$ cm.

7.8   A square footing (6 ft by 6 ft) will be subjected to a vertical load of 230 kips, which includes the weight of the footing. For the underlying clean fine to medium sand deposit, assume $N_{60} = 30$

and Poisson's ratio $\mu = 0.3$. Calculate the maximum total settlement ρ_{max}. For the analysis, assume a rigid loaded area on an elastic half-space of infinite depth.

ANSWER: $\rho_{max} = 0.82$ in.

7.9   Assume the same conditions as Problem 7.8. Using the Terzaghi and Peck chart (Fig. 7.8), calculate the maximum total settlement ρ_{max}.

ANSWER: From Fig. 7.8, for $B = 6$ ft and $N_{60} = 30$, $q = 3.4$ tsf for 1 in. settlement. Since actual $q = 3.2$ tsf, ρ_{max} will be slightly less than 1 in. (Note: for this problem, the theory of elasticity and Fig. 7.8 provide similar answers).

7.10   Assume the same conditions as Problems 7.8 and 7.9, except that the groundwater table rises to near the bottom of the footing. Using the Terzaghi and Peck chart (Fig. 7.8), calculate the size of the square footing so that the maximum total settlement ρ_{max} is approximately 1 in.

ANSWER: By trial and error, the size of the footing should be 8.5 ft by 8.5 ft.

7.11   Subsurface exploration indicates that a level site has 20 m of sand overlying dense rock. From ground surface to a depth of 4 m, the sand is loose with a total unit weight $\gamma_t = 19.3$ kN/m³. From a depth of 4 to 12 m below ground surface, the sand has a medium density with a total unit weight $\gamma_t = 20.6$ kN/m³. From a depth of 12 to 20 m, the sand is dense with a total unit weight $\gamma_t = 21.4$ kN/m³. The groundwater table is located 5 m below ground surface. It is proposed to construct a large building at the site that will be supported by a mat foundation (30 m by 40 m). The bottom of the mat foundation will be located 4 m below ground surface and the structural engineer has indicated that the building load can be approximated as a uniform pressure of 200 kPa that the bottom of the mat applies to the sand at a depth of 4 m. Undisturbed soil specimens of the sand were obtained at various depths and the results of one-dimensional compression tests performed on saturated sand specimens are presented in Fig. 7.28. Using the data from Fig. 7.28, calculate the net stress applied by

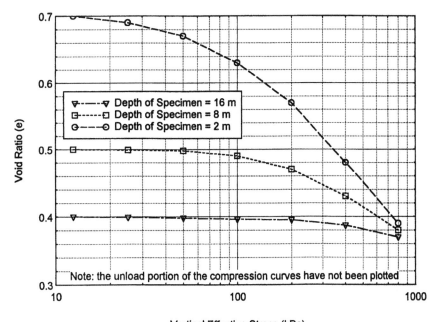

FIGURE 7.28   Compression curves for Problem 7.11.

the foundation at a depth of 4 m (σ_o), the increase in stress $\Delta\sigma_v$ at the center of the sand layers using the 2:1 approximation, and the total settlement ρ_{max} of the mat foundation.

ANSWER: $\sigma_o = 123$ kPa, $\Delta\sigma_v = 98.5$ kPa at the center of the medium density sand layer and 67.5 kPa at the center of the dense sand layer, and $\rho_{max} = 12$ cm.

Allowable Settlement

7.12   A building will be constructed with a continuous steel frame. If the length L between adjacent columns will be 6 m (20 ft), determine the maximum allowable differential settlement Δ.

ANSWER: As per Table 7.2, $\Delta = 1.2$ cm (0.5 in.).

7.13   A building will be constructed with sensitive machinery. Determine the maximum allowable angular distortion δ/L and maximum differential settlement Δ if the supporting columns are spaced 6 m (20 ft) apart.

ANSWER: As per Fig. 7.25, $\delta/L = 1/750$ and thus $\Delta = 0.8$ cm (0.3 in.).

7.14   It is proposed to construct a house with a conventional slab-on-grade foundation. It is desirable that the house should have a factor of safety of 2.5 against gypsum wallboard cracking. What is the maximum allowable differential settlement?

ANSWER: $\Delta = 0.5$ in.

7.15   It is proposed to construct a flexible oil storage tank. If the tank has a diameter of 30 m and the maximum allowable angular distortion $\delta/L = 1/200$, determine the maximum allowable differential settlement Δ.

ANSWER: $\Delta = 7.5$ cm (3 in.).

CHAPTER 8
CONSOLIDATION

8.1 INTRODUCTION

Saturated cohesive and organic soil can be susceptible to a large amount of settlement from structural loads. It is usually the direct weight of the structure that causes settlement of the cohesive or organic soil (i.e., first basic category of settlement, see Sec. 7.1). However, secondary influences such as the lowering of the groundwater table can also lead to settlement of cohesive or organic soils. The settlement of saturated clay or organic soil can have three different components: immediate settlement, primary consolidation, and secondary compression.

1. *Immediate settlement* (*also known as initial settlement, s_i*).   In most situations, surface loadings causes both vertical and horizontal strains, and this is referred to as two- or three-dimensional loading. Immediate settlement is due to undrained shear deformations, or in some cases contained plastic flow, caused by the two- or three-dimensional loading (Ladd et al., 1977; Foott and Ladd, 1981). Section 8.3 will be devoted to immediate settlement.

2. *Primary consolidation s_c.*   The increase in vertical pressure due to the weight of the structure constructed on top of saturated soft clays and organic soil will initially be carried by the pore water in the soil. This increase in pore water pressure is known as an *excess pore water pressure* (u_e). The excess pore water pressure will decrease with time, as water slowly flows out of the cohesive soil. This flow of water from cohesive soil (which has a low permeability) as the excess pore water pressures slowly dissipate is known as *primary consolidation*, or simply *consolidation*. As the water slowly flows from the cohesive soil, the structure settles as the load is transferred to the soil particle skeleton, thereby increasing the effective stress of the soil. Consolidation is a time-dependent process that may take many years to complete. Primary consolidation will be discussed in Secs. 8.4 and 8.5.

The typical one-dimensional case of consolidation involves strain in only the vertical direction. Common examples of one-dimensional loading include the lowering of the groundwater table or a uniform fill surcharge applied over a very large area. In the case of a one-dimensional loading, both the strain and flow of water from the cohesive soil as it consolidates will only be in the vertical direction. The oedometer apparatus is ideally suited for the study of one-dimensional consolidation behavior.

3. *Secondary compression s_s.*   The final component of settlement is due to secondary compression, which is that part of the settlement that occurs after essentially all of the excess pore water pressures have dissipated (i.e., settlement that occurs at constant effective stress). The usual assumption is that secondary compression does not start until after primary consolidation is complete. The amount of secondary compression is often neglected because it is rather small as compared to the primary consolidation settlement. However, secondary compression can constitute a major part of the total settlement for peat or other highly organic soil (Holtz and Kovacs, 1981).

Secondary compression has been described as a process where the particle contacts are still rather unstable at the end of primary consolidation and the particles will continue to move until finding a stable arrangement. This would explain why the rate of secondary compression often increases with compressibility. The more compressible the soil, the greater the tendency for a larger number of particles to be unstable at the end of primary consolidation (Ladd, 1973). Because there are no excess pore water pressures during secondary compression, it is often described as *drained creep*. Similar to the analysis of consolidation, the oedometer apparatus is ideally suited for the study of secondary compression. Secondary compression will be further discussed in Sec. 8.6.

In order to calculate the total settlement ρ_{max} of the in situ cohesive or organic soil layer due to a structural loading, the three components of settlement are added together, or:

$$\rho_{max} = s_i + s_c + s_s \tag{8.1}$$

where ρ_{max} = total settlement, also known as the maximum settlement
$\quad\quad s_i$ = immediate settlement (Sec. 8.3)
$\quad\quad s_c$ = primary consolidation settlement (Secs. 8.4 and 8.5)
$\quad\quad s_s$ = secondary compression settlement (Sec. 8.6)

Figure 8.1 presents a plot of maximum settlement (i.e., total settlement ρ_{max}) versus maximum differential settlement Δ from case studies of footings on clay. Bjerrum (1963) divided the data into two categories, as follows:

1. Rigid Structures: such as those structures having load-bearing brick walls

2. Flexible Structures: such as those structures having open steel or concrete frames

As shown in Fig. 8.1, when the total settlement was less than about 20 mm (0.8 in.), there was considerable scatter in data with no discernable correlation between total settlement and maximum differential settlement. This conclusion is similar to the study of foundations on sand (i.e., see Fig. 7.6). However, for those structures where the recorded settlement exceeded about 20 mm (0.8 in.), the data were significantly different for sands and clays (e.g., compare Figs. 7.6 and 8.1).

In Fig. 8.1, four lines have been drawn that represent ratios of Δ/ρ_{max} of 1.00, 0.75, 0.50, and 0.25. Also shown in Fig. 8.1 are two dashed lines that represent the upper limit boundaries of total settlement ρ_{max} versus maximum differential settlement Δ for flexible structures and rigid structures. As the total settlement of footings on clay increases, the ratio of Δ/ρ_{max} tends to decrease, as shown by the dashed lines in Fig. 8.1. This means that the structure tends to settle more as a unit due to the underlying soil consolidation.

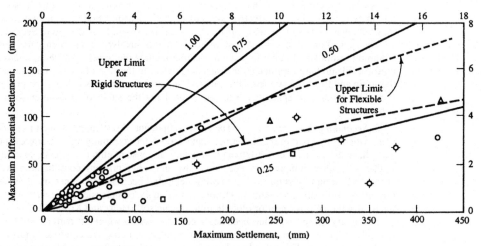

FIGURE 8.1   Maximum settlement (i.e., total settlement ρ_{max}) versus maximum differential settlement Δ from case studies of spread footings on clay. (*Adapted from Bjerrum, 1963; reproduced from Coduto, 1994.*)

8.2 *LABORATORY CONSOLIDATION TEST*

8.2.1 Test Procedures

The purpose of this section is to discuss the laboratory test procedures for determining the consolidation properties of soil. If the results of field exploration indicate the presence of saturated cohesive soils (such as silts or clays) or organic soils (such as those listed in Table 4.3), then undisturbed soil specimens should be obtained and tested in the laboratory. One-dimensional consolidation is typically measured in the oedometer apparatus using standard test procedures such as outlined in ASTM D 2435-03 (2004), "Standard Test Methods for One-Dimensional Consolidation Properties of Soils Using Incremental Loading." The test procedures are as follows:

1. *Soil specimen.*   An undisturbed soil specimen must be used for the consolidation test. An undisturbed soil specimen can be trimmed from a block sample or obtained from an undisturbed soil sampler, such as a Shelby tube. The specimen diameter divided by the specimen height should be equal to or greater than 2.5. A common diameter of soil specimen is 2.5 in. (6.4 cm) and height of 1.0 in. (2.5 cm). The trimming process should be performed as quickly as possible to minimize the possibility of a change in water content of the soil. Because saturated cohesive soils and organic soils often have a high water content, specimens can be stored and trimmed in a high-humidity moisture room (if it is available). Typically the soil specimen is trimmed in the confining ring and the initial height of the soil specimen (h_o) is then equal to the height of the confining ring. Because the soil specimen is confined laterally by the metal-confining ring, only one-dimensional vertical settlement will be allowed for the soil specimen during the laboratory testing.

2. *Index properties.*   After the soil specimen has been trimmed, the index properties of the soil should be determined. Typically, the trimmings can be collected in order to determine the water content of the soil. In addition, a balance can be used to obtain the mass of the confining ring and soil specimen. By subtracting the mass of the confining ring, the mass of the soil specimen can be calculated. Knowing the volume of the confining ring, the total unit weight can be calculated. Using the water content, the initial dry unit weight γ_d can also be calculated using basic phase relationships.

3. *Loading device.*   Dry and clean porous plates are placed on the top and bottom of the soil specimen and it is then placed in a surrounding container, such as a Plexiglas dish. The Plexiglas dish containing the soil specimen is placed at the center of the oedometer (see Fig. 3.11) and a dial gauge is set-up to measure vertical deformation. A seating load equivalent to a vertical pressure of 125 psf (6 kPa) is then applied to the soil specimen and a dial gauge reading is taken and recorded on the data sheet. Within 5 min of applying the seating load, the soil specimen is submerged in distilled water. If the soil specimen starts to swell (increase in height), the load must be immediately increased in order to prevent the soil specimen from swelling.

4. *Soil specimen loading.*   Typical loading increments are 250, 500, 1000, 2000 psf and so on (12, 25, 50, 100 kPa and so on) and each load increment should remain on the soil specimen for 24 h. Prior to increasing the load on the soil specimen, a dial gauge reading is taken. The vertical effective stress σ'_{vc} that equals the effective overburden pressure σ'_{vo} plus the increase in vertical stress caused by the proposed building load $(\Delta\sigma_v)$ should be estimated (i.e., $\sigma'_{vc} = \sigma'_{vo} + \Delta\sigma_v$). When the loading of the soil specimen approaches this estimated value of σ'_{vc}, dial gauge readings versus time at approximately 0.1, 0.25, 0.5, 1, 2, 4, 8, 15, and 30 min, and 1, 2, 4, 8, and 24 h should be recorded. The consolidation process is usually complete after 24 h.

5. *Continued loading of soil specimen.*   Additional vertical stress should be placed on the soil specimen to determine the soil behavior at higher pressures. In general, at least two additional load increments should be applied to the soil specimen (i.e., two load increments in excess of σ'_{vc} should be applied to the soil specimen). Each vertical stress should remain on the soil specimen for about 24 h. Before each additional vertical stress is applied to the soil specimen, a dial gauge reading should be taken.

6. *Unloading of the soil specimen.*   After loading to the desired highest vertical pressure, the soil specimen is unloaded by reducing the load upon the soil specimen. Once again, at each reduction in load upon the soil specimen, it should be allowed to equilibrate for 24 h and a dial gauge reading should be taken prior to the next reduction in load upon the soil specimen.

7. *Index properties at end of test.*   Once the soil specimen has been completely unloaded, it is removed from the oedometer. The water content of the soil specimen can then be determined. When placing the soil specimen in the container for the water content test, it should be broken apart to determine if there are any large size particles within the soil specimen. Large size particles, such as gravel size particles, may restrict the consolidation of the soil specimen and lead to an underestimation of the in situ consolidation potential of the soil. If large particles are noted, the consolidation test may have to be repeated on a larger size soil specimen.

Common laboratory errors associated with the consolidation test are as follows (Rollings and Rollings, 1996):

1. The soil specimen is disturbed or there is excessive disturbance during trimming.

2. The soil specimen does not fit snugly into the oedometer ring (i.e., the specimen does not fill the oedometer ring).

3. The permeability of the porous stones is too low.

4. There is excessive friction between the soil specimen and the confining ring.

5. An inappropriate load is applied during inundation.

6. The soil specimen does not have an appropriate height.

Another common error is misuse of the loading device. For example, the soil specimen may not be concentrically loaded within the equipment. Other problems could be that incorrect weights are placed on the loading hanger, leading to a vertical stress that is greater than or less than the desired vertical stress at inundation. Furthermore, the dial gauge measuring the vertical deformation may be inaccurate or incorrectly read, leading to inaccuracies in the calculated deformation.

The results of the consolidation test are highly sensitive to sample disturbance. Sample disturbance must be minimized during the sampling and trimming operation. The soil specimens should be trimmed from block samples or obtained from thin-walled sampling tubes. In addition, considerable experience is required in order to minimize sample disturbance during the trimming and set-up operations.

For the test procedures described in this section, apparatus compressibility has not been included. It has been stated that apparatus compressibility should be included in the calculations if the apparatus compressibility exceeds 5 percent of the measured deformation and in all tests where filter paper are used (ASTM D 2435-03, 2004). As a practical matter, apparatus compressibility is often neglected based on the assumption that the additional deformation associated with apparatus compressibility is compensated by a reduced soil deformation due to friction between the soil specimen and confining ring.

8.2.2   Laboratory Test Data

The laboratory test data can either be presented in the form of percent strain ε_v or void ratio e. The calculations are as follows:

$$\varepsilon_v = \frac{100\Delta h}{h_o} \tag{8.2}$$

$$e = \frac{V_v}{V_s} = \frac{1-(\Delta h/h_o)-V_s}{V_s} \tag{8.3}$$

where ε_v = vertical strain expressed as a percentage

  e = void ratio (dimensionless)

  Δh = change in height of the soil specimen (ft or m)

  h_o = initial height of the soil specimen (ft or m)

  V_v = volume of voids (ft^3 or m^3)

  V_s = volume of the solids (ft^3 or m^3), which is equal to the following:

$$V_s = \frac{\gamma_d}{G_s \gamma_w} \tag{8.4}$$

  γ_d = initial dry unit weight (pcf or kN/m^3), calculated in Step 2, Sec. 8.2.1

  G_s = specific gravity of soil solids (dimensionless)

  γ_w = unit weight of water (62.4 pcf or 9.81 kN/m^3)

Laboratory Consolidation Curve.   After the percent vertical strain (Eq. 8.2) or the void ratio (Eq. 8.3) has been calculated, the data are typically plotted on a semilog graph with the strain or void ratio on the vertical axis and the vertical effective stress σ'_v on the horizontal axis which is a logarithm scale. The vertical total stress σ_v is applied to the soil specimen by the oedometer apparatus. Once primary consolidation is complete, i.e., the excess pore water pressures are zero, the vertical stress σ_v from the loading device is equal to the vertical effective stress (σ'_v also known as σ'_{vc}).

This plot of void ratio e versus vertical effective stress σ'_{vc} is often referred to as the *consolidation curve*. A primary objective of the laboratory consolidation test is to obtain this consolidation curve and Fig. 8.2 illustrates the type of data obtained from the consolidation curve. Note in Fig. 8.2 that the loading portion of the consolidation curve can often be approximated as two straight line segments: (1) the recompression curve, and (2) the virgin consolidation curve. The recompression curve represents the reloading of the saturated cohesive soil, and the virgin consolidation curve represents the loading of the saturated cohesive soil beyond the maximum past pressure σ'_{vm}. The reason for the relatively steep virgin consolidation curve, as compared to the recompression curve, is because the soil has never experienced such a vertical stress and thus there is a tendency for the soil structure to break down and contract once the pressure exceeds the maximum past pressure σ'_{vm}.

The values of the recompression index C_r and the compression index C_c are simply the slope of the recompression curve and the virgin consolidation curve, respectively. These indices can be calculated as a change in void ratio Δe divided by the corresponding change in the effective pressures ($\log \sigma'_{vc2} - \log \sigma'_{vc1}$), or for the recompression curve:

$$C_r = \frac{\Delta e}{\log(\sigma'_{vc2}/\sigma'_{ve1})} \tag{8.5}$$

For the virgin consolidation curve:

$$C_c = \frac{\Delta e}{\log(\sigma'_{vc2}/\sigma'_{ve1})} \tag{8.6}$$

An easier method to obtain C_r and C_c is to determine Δe over one log cycle: or for example, if $\sigma'_{vc2} = 100$ and $\sigma'_{vc1} = 10$, then the log ($\sigma'_{vc2}/\sigma'_{vc1}$) = log (100/10) = 1. By using one log cycle, the values of $C_r = \Delta e$ for the recompression curve and $C_c = \Delta e$ for the virgin consolidation curve.

If the consolidation data are plotted on a graph of vertical strain ε_v versus consolidation stress σ'_{vc}, then the recompression curve and virgin consolidation curve can also be approximated as straight lines. When using a vertical strain ε_v versus consolidation stress σ'_{vc} plot, the slope of the recompression curve is designated the *modified recompression index* C_{re} and the slope of the virgin

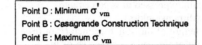

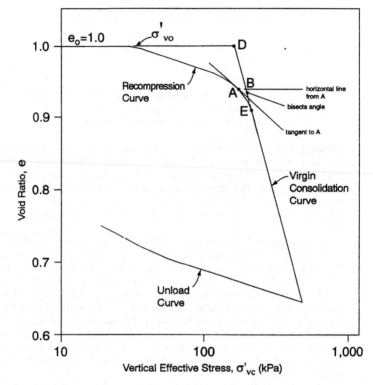

FIGURE 8.2   Illustration of the consolidation curve and the Casagrande (1936) technique for determining the maximum past pressure.

consolidation curve is designated the *modified compression index* $C_{c\varepsilon}$. These indices are related as follows:

$$C_{r\varepsilon} = \frac{C_r}{1+e_o} \tag{8.7}$$

$$C_{c\varepsilon} = \frac{C_c}{1+e_o} \tag{8.8}$$

As shown in Fig. 8.2, the final segment of the consolidation curve is the *unload curve*. Often the unloading curve is approximately parallel to the recompression curve.

Maximum Past Pressure.   The maximum past pressure σ'_{vm} is also known as the *preconsolidation pressure* σ'_p. The laboratory consolidation curve (Fig. 8.2) can also be used to determine the maximum past pressure σ'_{vm}. The most commonly used procedure for determining the maximum past pressure σ'_{vm} is to use the Casagrande construction technique (1936). Figure 8.2 illustrates this empirical procedure, which is performed as follows:

1. Locate the point of minimum radius (i.e., maximum curvature) of the consolidation curve (Point A, Fig. 8.2).

2. Draw a line tangent to the consolidation curve at Point A.

3. Draw a horizontal line from Point A.

4. Bisect the angle made by steps 2 and 3.

5. Extend the straight line portion of the virgin consolidation curve up to where it meets the bisect line. The point of intersection (Point B) of these two lines is the maximum past pressure σ'_{vm}, also known as preconsolidation pressure σ'_p.

This construction procedure will give the same results if the vertical axis is in terms of vertical strain ε_v or void ratio e. Since the Casagrande technique is only an approximate method of determining the maximum past pressure, it is often useful to obtain a range in values. As shown in Fig. 8.2, there is a range in possible values of σ'_{vm} varying from Point D to Point E.

Example of Laboratory Test Data.   Figure 8.3 presents results of consolidation tests performed on two specimens of Orinoco clay at a depth of 128 ft (40 m) (Ladd et al., 1980; Day, 1980). The upper plot is known as the consolidation curve for the two Orinoco clay soil specimens. The horizontal axis is the effective consolidation stress (i.e., the applied vertical pressure) which is designated σ'_{vc}, with the subscripts vc referring to vertical consolidation pressure and the prime mark indicating that it is an effective stress. Note that the horizontal axis is a logarithm scale. In Fig. 8.3, the vertical axis is percent vertical strain ε_v. The vertical axis could also be in terms of the void ratio e.

The two consolidation tests shown in Fig. 8.3 start out at zero vertical strain. Then as the vertical pressure is increased in increments, the soil consolidates and the vertical strain increases. The usual procedure is to apply a vertical pressure that is double the previously applied pressure (load increment ratio = 1.0). However, as shown in Fig. 8.3, in order to better define the consolidation curve, the loading increments can be adjusted to better define the breaking point of the consolidation curve. For the two consolidation tests performed on Orinoco clay (Fig. 8.3), the specimens were loaded up to a vertical pressure of 12 kg/cm² (24,600 psf, 1180 kPa) and then unloaded.

In Fig. 8.3, the arrows indicate the numerical values of the vertical effective stress (σ'_{vo}) and the maximum past pressure σ'_{vm} for both consolidation tests. Based on the Casagrande construction technique, the maximum past pressure σ'_{vm} was determined for the two consolidation tests on Orinoco clay. These two tests have significantly different values of maximum past pressure (2.75 versus 1.35 kg/cm²). The reason for the difference was due to sample disturbance. Test No. 12 (solid circles) was disturbed during sampling and only had a vane shear strength (TV) of 0.30 tsf (29 kPa), as compared to the undisturbed specimen (Test No. 18) which had a vane shear strength (TV) of 0.51 tsf (49 kPa). Thus sample disturbance can significantly lower the value of the maximum past pressure of saturated cohesive soil.

In Fig. 8.3, the vertical effective stress σ'_{vo} has a numerical value that is close to the maximum past pressure σ'_{vm} for the undisturbed soil specimen (open circles). This means that the Orinoco clay is essentially normally consolidated. In Fig. 8.4, all of the laboratory consolidation test data for the Orinoco clay are summarized on one sheet. The vertical axis is depth (feet) and the horizontal axis is effective stress (kg/cm²). In Fig. 8.4, the vertical effective stress σ'_{vo} versus depth for the Orinoco clay at borings E1 and F1 has been plotted. The maximum past pressure σ'_{vm} for each consolidation test has also been plotted with possible ranges in values based on the procedure shown in Fig. 8.2. The open symbols represent consolidation tests performed on undisturbed specimens, while the solid symbols represent consolidation tests performed on disturbed specimens. In general, the maximum past pressure data points σ'_{vm} are close to the vertical effective stress σ'_{vo} and for this offshore deposit of clay, it could be concluded that the clay is essentially normally consolidated (OCR = 1) to very slightly overconsolidated. The concepts of normally consolidated and overconsolidated clay will be further discussed in Sec. 8.4.

Log-of-Time Method.   As previously discussed in Step 4, Sec. 8.2.1, during the laboratory consolidation test, when the load is applied to the clay specimen, dial readings versus time can be recorded. These data are used to determine the coefficient of consolidation, which will be discussed in Sec. 8.5.

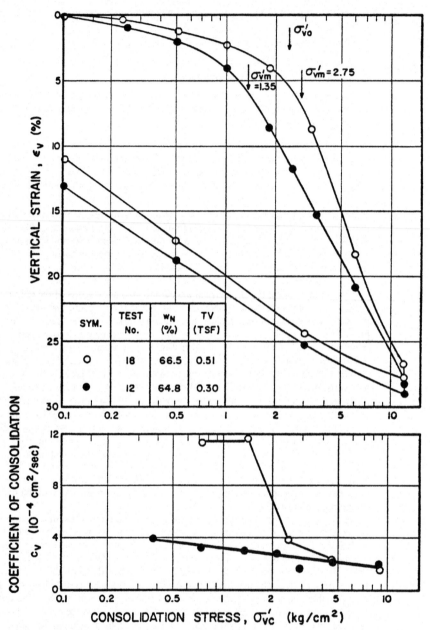

FIGURE 8.3   Consolidation test data for the Orinoco clay at a depth of 40 m. (*From Ladd et al., 1980; Day, 1980.*)

Figure 8.5 shows an example of the log-of-time method where the vertical deformation (mm) has been plotted versus time (on a log scale). This vertical deformation versus time data was recorded from a laboratory consolidation test performed on a highly plastic saturated soil (LL = 73, PI = 51, $e_o = 1.22$). The vertical deformation versus time data was recorded when the vertical pressure on the

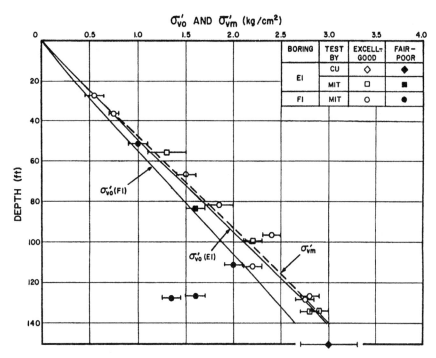

FIGURE 8.4 Stress history of the Orinoco clay. (*From Ladd et al., 1980; Day, 1980.*)

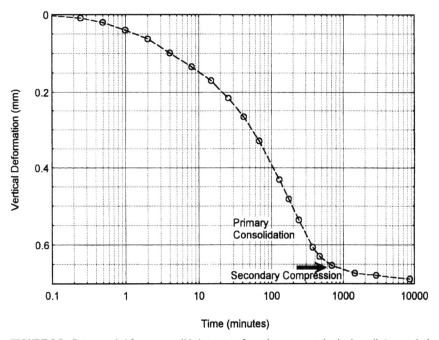

FIGURE 8.5 Data recorded from a consolidation test performed on a saturated cohesive soil. At a vertical pressure σ_y of 200 kPa, the vertical deformation as a function of time after loading was recorded and plotted above (log-of-time method). The arrow indicates the end of primary consolidation.

soil (σ'_{vc}) = 200 kPa. The purpose of the log-of-time method is to determine the time for 50 percent consolidation (i.e., t_{50}), as follows:

Determine End of Primary Consolidation.   The end of primary consolidation (location of arrow) is estimated as the intersection of the two straight line segments. The value of d_{100} = the end of primary consolidation (d_{100} = 0.66 mm).

Determine d_o.   As an approximation, d_o can be assumed to be the dial reading at t = 0.1 min (i.e., the curve is extended back to t = 0.1 min), or d_o = 0 mm. ASTM D 2435-03 (2004) presents an alternative method for the determination of d_o.

Determine d_{50} and t_{50}.   The value of d_{50} is equal to the d_o plus d_{100} divided by two [i.e., $d_{50} = \frac{1}{2} (d_o + d_{100}) = \frac{1}{2} (0 + 0.66) = 0.33$]. Using the curve shown in Fig. 8.5, at a vertical deformation of 0.33 mm, the corresponding time (t_{50}) = 65 min. The value of t_{50} (65 min) is the length of time it took for the saturated cohesive soil specimen to experience 50 percent of its primary consolidation when subjected to a vertical pressure of 200 kPa. The calculations for the coefficient of consolidation using the log time method will be presented in Sec. 8.5.

Square-Root-of-Time Method.   Figure 8.6 shows an example of the square-root-of-time method. The data used to develop both Figs. 8.5 and 8.6 are identical, except that for Fig. 8.5 time has been plotted on a logarithm scale, while in Fig. 8.6 time has been plotted as the square root of time. The laboratory data shown in Fig. 8.6 can also be used to determine the coefficient of consolidation. The procedure is as follows:

Determine the Average Degree of Consolidation U_{avg} = 90 percent.   Usually a straight line can be drawn through the initial vertical deformation versus square root of time data points. Then a second straight line with all x-axis values 1.15 times larger than the corresponding actual test data is drawn (i.e., see the dashed line in Fig. 8.6). The intersection of the dashed line and the actual test data curve

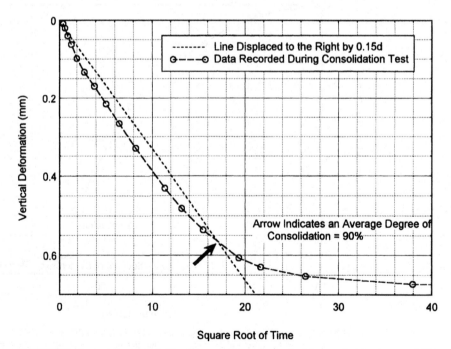

FIGURE 8.6   Same data as used in Fig. 8.5, except the square-root-of-time method has been used. The arrow indicates an average degree of consolidation U_{avg} = 90 percent.

is assumed to be the vertical deformation corresponding to an average degree of consolidation = 90 percent. The arrow in Fig. 8.6 points to the intersection of the dashed line and the actual test data curve.

Determine Time Corresponding to U_{avg} = 90 percent. The next step is to determine the time corresponding to an average degree of consolidation = 90 percent. In Fig. 8.6, the x-axis value corresponding to U_{avg} = 90 percent is 17. Since the x-axis is square root of time, the value of 17 must be squared, or the time it takes to achieve a U_{avg} of 90 percent is $(17)^2$ = 290 min. The calculations for the coefficient of consolidation using the square-root-of-time method will be presented in Sec. 8.5.

8.3 IMMEDIATE SETTLEMENT

8.3.1 Introduction

Immediate settlement, also known as *initial settlement*, is the first component of settlement of saturated cohesive soil and it occurs as the load is applied to the soil. In order to visualize the concept of immediate settlement, the difference between one-dimensional and two- or three-dimensional loading will be first introduced.

1. *One-dimensional loading.* A common example of a one-dimensional loading is a uniform fill surcharge applied over a very large area. In this case of a one-dimensional loading, the soil strain will only be in the vertical direction. From a theoretical standpoint, a one-dimensional loading on a saturated cohesive soil will not cause any immediate settlement ($s_i = 0$). This is because the one-dimensional loading does not induce any horizontal strain into the soil.

2. *Two- or three-dimensional loading.* A common example of a two-dimensional loading is a strip footing. In the case of the loading of a strip footing, there will be strain in the vertical direction and in the direction perpendicular to the long axis of the footing (i.e., strain in two directions). For strip footings, it is typically assumed that there will be no strain in the long direction of the footing. Examples of three-dimensional loading are from square footings and round storage tanks. For three-dimensional loading, there can be strain of the soil in all three directions (x, y, and z).

Immediate settlement can most easily be visualized by assuming an instantaneous two- or three-dimensional loading. In this case, the saturated cohesive soil is suddenly subjected to a load which induces both shear stresses and increased pore water pressures in the soil. The shear stresses tend to cause the soil to deform laterally. If the load were high enough, the cohesive soil would deform laterally to such an extent that there would be a bearing capacity failure.

Because the load is assumed to act instantaneous, there is not enough time for the pore water pressures to dissipate. Thus immediate settlement is assumed to be caused by undrained shear deformations of the saturated cohesive soil. This means that there will be no change in volume of the soil and the amount of settlement of the structure must be compensated by an equivalent amount of lateral deformation or squeezing of soil from beneath the foundation. Hence, immediate settlement is due to undrained shear deformations, or in some cases contained plastic flow of the soil from beneath the foundation, caused by the two- or three-dimensional loading (Ladd et al., 1977; Foott and Ladd, 1981).

Soil and Loading Conditions Susceptible to Immediate Settlement. For most field situations, immediate settlement will not be significant. For example, Foott and Ladd (1981) state:

> Most field evidence indicates that s_i (immediate settlement) is usually not an important design consideration and that conventional predictions of s_c (consolidation settlement) using the one-dimensional model will generally yield an adequate estimate of the total settlement that will occur. Therefore, design practice does not need to be changed with those situations. The objective is rather to try and detect special cases wherein undrained settlements might become troublesome and to make allowance for potential problems before they occur.

In this regard, Foott and Ladd (1981) indicate that those troublesome cases where initial settlements may become very significant are highly plastic and organic soils, especially when the foundation is loaded quickly and the factor of safety for a bearing capacity failure is low.

Maximum Past Pressure.   Laboratory tests were performed in order to observe other troublesome conditions that might produce significant immediate settlement (Day, 1995a). The procedure first consisted of consolidating large diameter specimens of highly plastic (LL = 93, PI = 71) and saturated clay to a vertical effective stress of 64 kPa (1300 psf). After the soil specimens had fully consolidated under this stress of 64 kPa (1300 psf), a much smaller diameter circular loading disk was placed on the top of the large samples and loaded. Test results are shown in Fig. 8.7, which is a plot of vertical strain versus vertical pressure induced by the loaded area. Three conditions were tested in the laboratory:

1. A one-dimensional loading (i.e., a standard laboratory consolidation test)
2. A three-dimensional loading where the ratio of the height of the soil specimen to the width of the loaded area (*H/D* ratio) equals 0.10
3. A three-dimensional loading where *H/D* = 0.30

The arrow in Fig. 8.7 indicates the maximum past pressure σ'_{vm}, i.e., the largest vertical effective stress the soil specimens were ever subjected to and completely consolidated under, which is equal to 64 kPa. Note in Fig. 8.7 the substantial increase in vertical strain when the vertical stress exceeds the maximum past pressure. For the three-dimensional loadings (*H/D* = 0.10 and 0.30), the plastic soil squeezed out from underneath the load (i.e., plastic flow) and there was much more vertical strain as compared to the one-dimensional loading. In Fig. 8.7, the amount of immediate settlement is approximately equal to the difference between the one-dimensional loading curve and the three-dimensional loading curves. As Fig. 8.7 shows, the amount of immediate settlement can even exceed the amount of one-dimensional consolidation. Thus, these laboratory tests show that the maximum

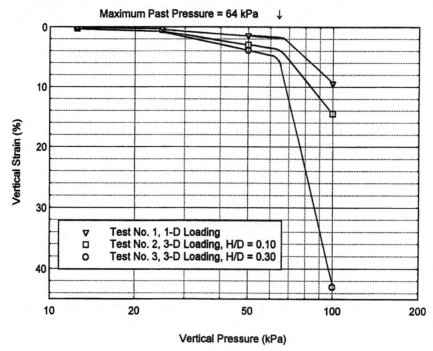

FIGURE 8.7   Vertical strain versus vertical pressure for one-dimensional and three-dimensional loading of a highly plastic saturated soil.

past pressure σ'_{vm} is an important parameter when considering the amount of immediate settlement. If the structural loading of the soil is greater than its maximum past pressure σ'_{vm}, the amount of immediate settlement could be very large.

In summary, this laboratory testing program indicated that troublesome cases occur when there is a near surface highly plastic and saturated clay layer, the foundation is constructed at or near ground surface, the load is applied very quickly, and the load causes the maximum past pressure of the clay to be exceeded. A field loading that satisfies these conditions leading to a substantial amount of immediate settlement is a thick near- or at-surface clay layer that is normally consolidated (OCR = 1) or is subjected to a quick loading $\Delta\sigma_v$ that results in a vertical stress exceeding the maximum past pressure σ'_p. An example would be an oil tank constructed at ground surface as illustrated in Fig. 8.8, with the quick loading of the clay occurring when the oil tank is filled for the first time.

As previously mentioned, in most cases immediate settlement is assumed to be zero. Examples where immediate settlements are usually ignored are as follows:

1. Deep foundation.   Provided the deep foundation has an adequate factor of safety in terms of a bearing capacity failure, the overburden pressure limits the ability of the saturated clay to squeeze out from beneath the bottom of the piles.

2. Shallow foundations with deep clay layer.   Another example is where there is an isolated clay layer at depth. Because the clay is not in direct contact with the bottom of the foundation and because of the overburden pressure, the clay is once again restricted in its ability to squeeze laterally from underneath the foundation.

3. Alternating sand and clay layers.   For this situation, the sand layers tend to provide lateral rigidity to the soil mass, reducing the ability of the clay to squeeze or flow laterally.

According to Leonards (1976), the assumption of one-dimensional consolidation and hence zero initial settlement is valid for the following cases:

1. Width of the loaded area exceeds four times the thickness of the clay strata (i.e., $B > 4H$, where B = width of the loaded area and H = thickness of the clay strata).

2. Depth to the top of the clay stratum exceeds twice the loaded area (i.e., $D > 2B$, where D = depth from the bottom of the foundation to the top of the clay stratum and B = width of the loaded area).

3. Compressible soil lies between two stiffer soil strata whose presence tends to reduce the magnitude of horizontal strain.

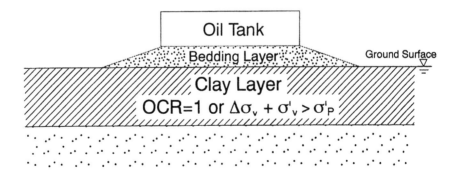

Note: Thickness of Bedding Layer Has Been Exaggerated For Viewing Purposes

FIGURE 8.8   Example of a condition causing significant immediate settlement.

8.3.2 Theory of Elasticity

There are several different methods that can be used to estimate the immediate settlement (s_i). One approach is to use the modulus of elasticity E of the soil from consolidated undrained triaxial compression tests with pore water pressure measurements (ASTM D 4767-02, 2004) in order to estimate the immediate settlement based on the theory of elasticity. Because the modulus of elasticity is obtained from an undrained triaxial test, it is often referred to as the *undrained modulus* and designated E_u. The equation is the same as Eq. 7.9, or:

$$s_i = qBI \frac{1-\mu^2}{E_u}$$

(8.9)

where s_i = immediate settlement of the footing (ft or m)
$\quad q$ = vertical footing pressure (psf or kPa)
$\quad B$ = footing width (ft or m)
$\quad I$ = shape and rigidity factor (dimensionless). This factor is derived from the theory of elasticity to account for the thickness of the soil layer, shape of the foundation, and flexibility of the foundation. Theoretical values of I can be obtained for various situations as shown in Fig. 7.10.
$\quad \mu$ = Poisson's ratio (dimensionless), which is usually assumed to be 0.5 for saturated plastic soil subjected to undrained loading
$\quad E_u$ = modulus of elasticity of the soil (psf or kPa). This is often obtained from an undrained triaxial compression test performed on an undisturbed specimen where the stress-strain curve is plotted (i.e., see Fig. 3.32). The initial tangent modulus to the stress-strain curve is often assumed to be the value of E_u.

This approach based on the theory of elasticity will provide approximate results provided the soil deposit underlying the foundation is overconsolidated and the load plus the existing vertical effective stress does not exceed the maximum past pressure (i.e., $\Delta\sigma_v + \sigma'_{vo} < \sigma_{vm}$). However, using the theory of elasticity could significantly underestimate the amount of immediate settlement for situations where there is plastic flow of the soil (i.e., such as the example shown in Fig. 8.8).

Example Problem 8.1   An oil tank is to be constructed at a site that contains a thick deposit of saturated and overconsolidated clay. Assume that when the cohesive soil is subjected to the loading, it will still be in an overconsolidated state. Further assume that the clay beneath the oil tank can be represented by the triaxial compression test data shown in Fig. 3.30. The oil tank will have a diameter of 30 ft (9.1 m) and when the oil tank if full of oil, it will subject the ground surface to a stress of 1000 psf (48 kPa). Calculate the initial settlement s_i of the oil tank assuming a relatively flexible foundation constructed on an elastic half space of infinite depth.

Solution   From the upper portion of Fig. 7.10, for an elastic half space of infinite depth and a circular flexible foundation, the shape and rigidity factor (I) = 1.0 at the center of the loaded area.

Also use the following values:

$$\mu = 0.5$$
$$q = 1000 \text{ psf (48 kPa)}$$
$$B = 30 \text{ ft (9.1 m)}$$

(*Continued*)

Using the first two data points from Fig. 3.30, Part c, the initial tangent modulus is equal to:

$$E_u = \Delta\sigma_1/\Delta\varepsilon = (4.07 \text{ psi} - 0)/(0.0068 - 0) = 600 \text{ psi} = 86,000 \text{ psf} (4100 \text{ kPa})$$

Using Eq. 8.9:

$$s_i = qBI \left[(1 - \mu^2)/E_u\right]$$

$$= (1000 \text{ psf})(30 \text{ ft})(1.0) \left[(1 - 0.5^2)/86,000 \text{ psf}\right] = 0.26 \text{ ft} = 3.1 \text{ in.} (8.0 \text{ cm})$$

Example Problem 8.2   A site consists of 5 ft (1.5 m) of sand ($\gamma_t = 120$ pcf, 18.9 kN/m³) overlying clay that has the same properties as described in the prior example and is 15 ft (4.6 m) thick. Assume that when the cohesive soil is subjected to the loading, it will still be in an over-consolidated state. Further assume that a hard strata underlies the 15 ft (4.6 m) thick clay layer. For a square relatively rigid foundation that is 27 ft (8.2 m) wide and at a depth of 5 ft (1.5 m) below ground surface, determine the immediate settlement if the foundation exerts a vertical pressure of 1600 psf (77 kPa) onto the clay.

Solution   From the prior example, use the following values:

$$\mu = 0.5$$

$$E_u = 86,000 \text{ psf} (4100 \text{ kPa})$$

Also use the following values:

Net pressure exerted by the foundation:

$$q = 1600 \text{ psf} - (5 \text{ ft})(120 \text{ pcf}) = 1000 \text{ psf} (48 \text{ kPa})$$

Turning the square foundation into an equivalent round foundation, or:

$$\text{Area} = (B)(L) = (27 \text{ ft})(27 \text{ ft}) = 729 \text{ ft}^2$$

$$\tfrac{1}{4} \pi B^2 = 729 \text{ ft}^2 \quad \text{or} \quad B = 30 \text{ ft} (9.1 \text{ m})$$

Assuming the hard strata represents a rigid base and for the 15 ft thick clay layer:

$$H/B = 15/30 = 0.5$$

Using the lower part of Fig. 7.10, which is applicable for an elastic half space over a rigid base, with $H/B = 0.5$ and $\mu = 0.5$, the shape and rigidity factor $(I) = 0.14$ at the center of a rigid circular area of diameter B.

Using Eq. 8.9:

$$s_i = qBI \left[(1 - \mu^2)/E_u\right]$$

$$= (1000 \text{ psf})(30 \text{ ft})(0.14) \left[(1 - 0.5^2)/86,000 \text{ psf}\right] = 0.036 \text{ ft} = 0.43 \text{ in.} (1.1 \text{ cm})$$

For these two examples, which both have essentially the same size foundation and loading conditions, the 15 ft (4.6 m) thick clay layer has only 0.43 in. (1.1 cm) of initial settlement as compared to 3.1 in. (8.0 cm) of initial settlement for the clay layer of infinite depth.

8.3.3   Other Methods

Plate Load Tests.   Another approach is to perform field plate load tests (see Sec. 7.3.2) to measure the amount of immediate settlement due to an applied load. The plate load test could significantly underestimate the immediate settlement if the test is performed on a near surface sandy layer or surface crust of clay that is heavily overconsolidated. Low values of immediate settlement would be recorded because the pressure bulb of the plate load test is very small. But when a large structure is built, the pressure bulb is much larger which could result in significant plastic flow if there is a normally consolidated clay layer underlying the stiff surface layer. Because of these limitations, the plate load test may not be applicable for many construction sites.

Stress-Path Method.   The amount of immediate settlement can also be determined from actual laboratory tests that model the field loading conditions. An undisturbed soil specimen could be set-up in the triaxial apparatus and then the specimen could be subjected to vertical and horizontal stresses that are equivalent to the anticipated loading condition. The undrained vertical deformation (i.e., immediate settlement) due to the applied loading could then be measured. By measuring the amount of vertical deformation from a series of specimens taken from various depths below the proposed structure, the total amount of immediate settlement could be calculated. This approach has been termed the *stress-path method* (Lambe, 1967).

8.3.4   Mitigation Measures

If the site is one of those troublesome cases where the immediate settlement is expected to be large, a deep foundation could be constructed. Another option would be to add a surcharge fill in order to increase the maximum past pressure of the clay. Other mitigation measures to limit the amount of immediate settlement are as follows (Foott and Ladd, 1981):

> The safest way of preventing substantial (immediate) settlements is to ensure that the clay consolidates quite rapidly, thus increasing its stiffness and decreasing the shear stress levels in the foundation clay. If necessary, the installation of vertical drains to accelerate consolidation in portions of the foundation soils, in order to limit lateral movements, should be considered.
>
> Another possible approach is the use of incremental loading via stage construction, allowing stiffening of the clay soils through consolidation before adding the next load increment. This will reduce both the immediate and undrained creep components of settlement below the values that would occur with immediate full loading. However, the soil's modulus (of elasticity) may be so low that even these reduced undrained settlements are large enough to be a significant problem.

8.4   PRIMARY CONSOLIDATION

8.4.1   Introduction

This section will deal with one-dimensional primary consolidation of saturated cohesive soil, such as clays, where there is strain only in the vertical direction. Part 1 of Fig. 8.9 shows the conditions for a saturated cohesive soil that is in equilibrium with hydrostatic pore water pressures. In this figure, σ_v is the vertical total stress and σ_v' is the vertical effective stress, which is equal to the total stress minus the pore water pressure u, or: $\sigma_v' = \sigma_v - u$ (Eq. 3.8).

Examples of one-dimensional loading are as follows:

1. *Lowering of groundwater level (Fig. 8.9, Part 2).*   For a lowering of the groundwater table, the total stress σ_v remains essentially unchanged. But because the pore water pressure decreases due to the groundwater table lowering, the effective stress must increase, resulting in primary consolidation. The amount of increase in effective stress as the clay consolidates is the difference between the two parallel dashed lines in Fig. 8.9 (Part 2). The line labeled effective stress in Fig. 8.9 (Part 2) represents the effective stress when about half of the primary consolidation settlement is complete.

2. *Added fill (Fig. 8.9, Part 3).*   An added fill surcharge applied over a very large area causes an increase in vertical pressure acting on the saturated clay layer, resulting in primary consolidation. Initially the fill surcharge will be carried by the pore water in the soil and this increase in pore water pressure is known as the initial excess pore water pressure u_o. The excess pore water pressure will dissipate over time, as water flows in a vertical direction out of the cohesive soil. Because of the low permeability of cohesive soil, water will slowly flow out of the cohesive soil, causing the excess pore

Stress condition	Diagram of vertical stresses	Description
(1) Simple overburden pressure	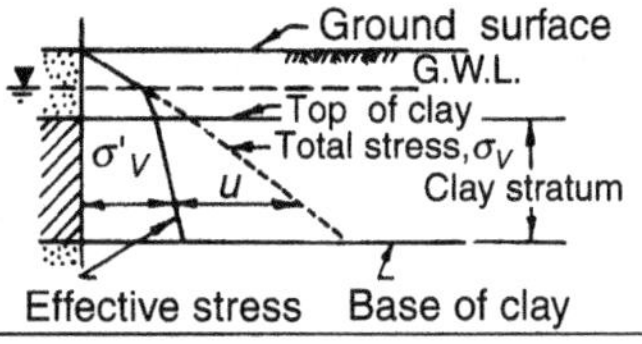	Total stress: It is computed using total unit weight γ_T above and below the G.W.L. Pore water pressure u is due to G.W.L.
(2) Lowering of ground water level	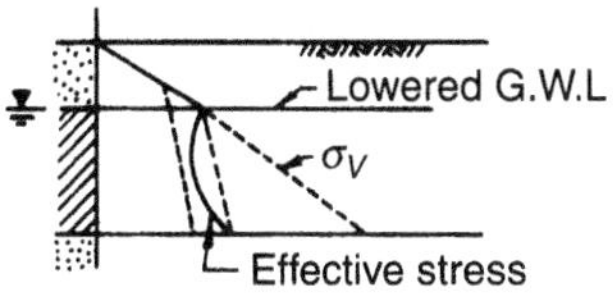	Immediately after lowering of the groundwater, total stress in top sand layer remains practically unchanged, but the effective stresses increase. Since the water escapes slowly from the clay layer, the effective stress requires a long time to reach the new equilibrium value.
(3) Partial consolidation under weight of initial fill	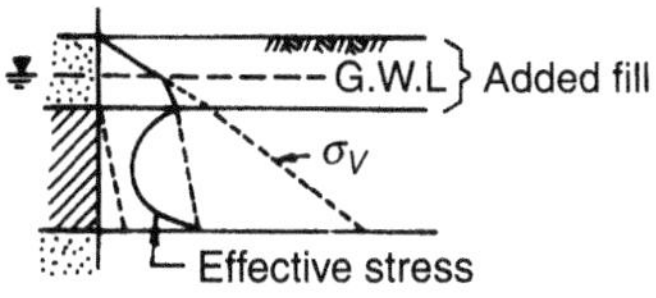	Total stresses on a clay layer increased by the addition of surcharge load. Initially this load is carried by pore water in the form of excess pore pressure. As the settlement progresses in the clay layer, the effective stress increases to correspond to the stress from surcharge load.
(4) Rise of ground water level	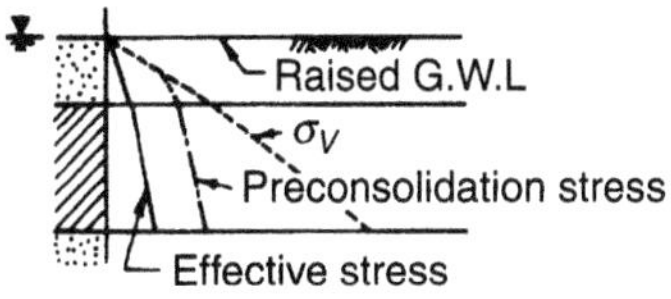	Rise of ground water level decreases effective pressure of overburden. Effective stress line moves to left. Then pre-consolidation stress equals original effective stress overburden. Total stress practically unchanged.
(5) Excavation	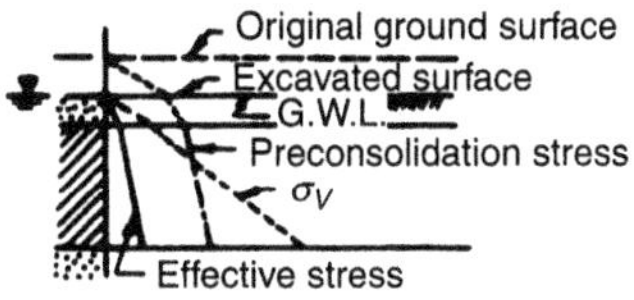	Excavation of overburden material unloads clay layer. Effective stress line moves to the left. Then preconsolidation stress equals original effective stress of overburden.
(6) Preconsolidation from loading in the past	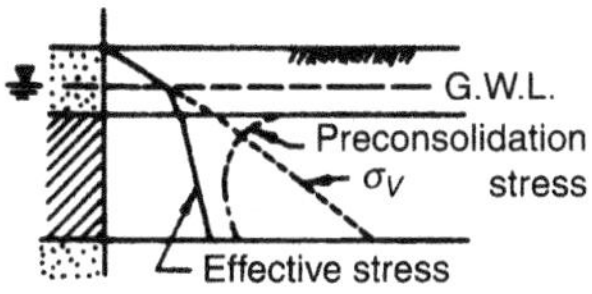	Preconsolidation from past loadings greater than the existing overburden may have been caused by weight of glacial ice, erosion of former overburden, lower ground water level plus desiccation, or removal of former structures.
(7) Artesian pressure	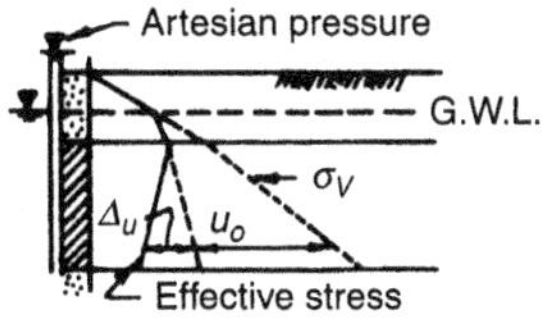	Sand stratum below the clay may be subject to artesian hydraulic pressures that decrease effective stress at base of clay. Total stress remains unchanged.

The left side of the table is labeled "Consolidation" spanning rows (1)–(3) and "Preconsolidated conditions" spanning rows (4)–(7).

FIGURE 8.9 Primary consolidation of saturated clay: stress condition, diagram of vertical stresses, and description. Note: G.W.L. groundwater level. (*Adapted from NAVFAC DM-7.1, 1982.*)

water pressures to slowly dissipate. As the water slowly flows from the cohesive soil, the soil settles as the load is transferred to the soil particle skeleton, thereby increasing the effective stress of the soil. The line labeled effective stress in Fig. 8.9 (Part 3) represents the effective stress when about half of the primary consolidation settlement is complete.

From a theoretical standpoint, the one-dimensional loading on a saturated cohesive soil will not cause any immediate settlement ($s_i = 0$). This is because the one-dimensional loading does not induce any horizontal strain into the soil. In addition, in this section it will also be assumed that an instantaneous load is applied to the saturated cohesive soil.

Geologists also use the term *consolidation*, but it has a totally different meaning. In geology, consolidation is defined as the processes, such as cementation and crystallization, that transforms a soil into a rock. The term consolidation can also be used to describe the change of lava or magma into firm rock (Stokes and Varnes, 1955).

8.4.2   Primary Consolidation Settlement

The standard method to determine the primary consolidation settlement (s_c) is to use the results of laboratory consolidation tests as described in Sec. 8.2. In particular, the objective of the laboratory consolidation test is to obtain the laboratory consolidation curve, which is shown in Figs. 8.2 and 8.3.

Overconsolidation Ratio.   Based on the stress history of saturated cohesive soils, they are considered to be either underconsolidated, normally consolidated, or overconsolidated. The overconsolidation ratio (OCR) is used to describe the stress history of cohesive soil, and it is defined as:

$$\text{OCR} = \frac{\sigma'_{vm}}{\sigma'_{vo}} \tag{8.10}$$

where OCR = overconsolidation ratio (dimensionless)

$\sigma'_{vm} = \sigma'_p$ = maximum past pressure σ'_{vm}, also known as the preconsolidation pressure σ'_p, which is equal to the highest previous vertical effective stress that the cohesive soil was subjected to and completely consolidated under (psf or kPa). This value is often estimated from the Casagrande construction technique (see Fig. 8.2).

$\sigma'_{vo} = \sigma'_v$ = existing vertical effective stress (psf or kPa)

In terms of the stress history of a cohesive soil, there are three possible conditions, as follows:

1. *Underconsolidated (OCR < 1)*   A saturated cohesive soil is considered underconsolidated if the soil is not fully consolidated under the existing overburden pressure and excess pore water pressures u_e exist within the soil. Underconsolidation occurs in areas where a cohesive soil is being deposited very rapidly and not enough time has elapsed for the soil to consolidate under its own weight.
2. *Normally consolidated (OCR = 1)*   A saturated cohesive soil is considered normally consolidated if it has never been subjected to a vertical effective stress greater than the existing overburden pressure and if the deposit is completely consolidated under the existing overburden pressure.
3. *Overconsolidated or preconsolidated (OCR > 1).*   A saturated cohesive soil is considered overconsolidated if it has been subjected in the past to a vertical effective stress greater than the existing vertical effective stress. Table 8.1 indicates various mechanisms that can cause a cohesive soil to become overconsolidated as follows:
   a. Rise of groundwater level: As illustrated in Part 4 of Fig. 8.9, a permanent rise in the groundwater table can create an overconsolidated soil. Once the groundwater rises, it subjects soil to buoyant conditions and the load on the clay is reduced. An overconsolidated soil is created because the vertical effective stress is now lower with the permanent rise in the groundwater table. In Fig. 8.9 (Part 4), the line labeled preconsolidation stress σ'_p represents the highest previous vertical effective stress that the cohesive soil was subjected to and completely consolidated under.
   b. Excavation: An overconsolidated soil can be created when soil is excavated, such as shown in Part 5, Fig. 8.9. The line labeled preconsolidation stress σ'_p equals the original vertical effective stress before the excavation was started.

TABLE 8.1 Mechanisms that Can Create an Overconsolidated Soil

Main mechanism	Item creating overconsolidated soil	Remarks or references
Change in total stress	Removal of overburden Past structures Weight of past glaciers	Soil erosion Human-induced factors Melting of glaciers

Change in pore water pressure	Change in groundwater
	  table elevation
	Artesian pressures
	Deep pumping of groundwater
	Desiccation due to drying
	Desiccation due to plant life
		Sea-level changes
		  (Kenney, 1964)
		Common in glaciated areas
		Common in many cities
		May have occurred during
		  deposition

Change in soil structure	Secondary compression (aging)
	Changes in environment, such
	  as pH, temperature, and salt
	  concentration
	Chemical alterations due to
	  weathering, precipitation of
	  cementing agents, and
	  ion exchange
		Lambe, 1958a, 1958b;
		Bjerrum, 1967b; Leonards
		and Altschaeffl, 1964
		Bjerrum, 1967b; Cox, 1968

Source: Ladd (1973).

c. *Loading in the past:* Another common process that creates an overconsolidated soil is past loading. For example, the weight of an overlying glacier will subject the clay to a high vertical overburden pressure and when the glacier melts, the clay becomes overconsolidated (see Fig. 8.9, Part 6). Another common situation that creates an overconsolidated soil is where a thick overburden layer of soil has been removed by erosion over time.

d. *Artesian pressure:* If a site initially had hydrostatic pore water pressure and then a permanent artesian pressure condition develops, the clay will become overconsolidated. As illustrated in Fig. 8.9 (Part 7), the artesian condition increases the water pressure at the bottom of the clay layer causing water to flow upward through the clay layer. The increase in pore water pressure will decrease the effective stress, resulting in an overconsolidated soil at the bottom of the clay layer. The amount of overconsolidation is essentially equal to Δu, or the clay at the bottom of the layer is overconsolidated, but normally consolidated at the top of the clay layer (where conditions are unchanged).

e. *Clay desiccation:* The upper portion of a clay layer may become desiccated during hot and dry periods. The process of desiccation causes the clay to shrink with a corresponding increase in dry density. This densification of the clay creates an overconsolidated state. Two examples of near surface overconsolidated clays due to desiccation are shown in Figs. 2.40 and 2.41.

f. *Aging:* As indicated in Table 8.1, secondary compression (aging) can create a slight overconsolidation of a cohesive soil, even though the effective stress has not changed. While the site history may indicate a normally consolidated soil, it is possible that the soil may actually be slightly overconsolidated due to aging. At these sites, a slight load can be applied to the cohesive soil without triggering virgin consolidation because of this secondary compression aging effect that creates a slightly overconsolidated soil.

For structures constructed on top of saturated cohesive soil, determining the OCR of the soil is very important in the settlement analysis. For example, if the cohesive soil is underconsolidated, then considerable settlement due to continued consolidation by the soil's own weight as well as the applied structural load would be expected. On the other hand, if the cohesive soil is highly overconsolidated, then a load can often be applied to the cohesive soil without significant settlement.

This information on the stress history of saturated clay deposits is very important and it is often added to the subsoil profile. For example, as shown in Figs. 2.39 to 2.41, the maximum past pressures (i.e., preconsolidation pressures) obtained from laboratory consolidation tests have been included on these subsoil profiles.

Settlement Calculations.   In addition to using the consolidation curve to determine the maximum past pressure σ'_{vm} of a cohesive soil, the consolidation curve obtained from the laboratory test (Sec. 8.2) can also be used to estimate the primary consolidation settlement s_c due to one-dimensional loading of the cohesive soil. As shown in Fig. 8.2, the consolidation curve can often be approximated as two straight line segments. The recompression index C_r represents the reloading of the saturated cohesive soil, and the compression index C_c represents the loading of the saturated cohesive soil beyond the maximum past pressure σ'_{vm}. The compression index is often referred to as the slope of the virgin consolidation curve because the in situ saturated cohesive soil has never experienced this loading condition. The reason for the relatively steep virgin consolidation curve, as compared to the recompression curve, is because there is a tendency for the soil structure to breakdown and contract once the pressure exceeds the maximum past pressure σ'_{vm}.

Using the calculated values or C_r and C_c and the maximum past pressure σ'_{vm} from the laboratory consolidation curve, the primary consolidation settlement s_c due to an increase in load $\Delta\sigma_v$ can be determined from the following equations:

A) For underconsolidated soil (OCR < 1)

$$s_c = C_c \frac{H_o}{1+e_o}\left[\log\frac{\sigma'_{vo}+\Delta\sigma_v}{\sigma'_{vo}}+\log\frac{\sigma'_{vo}}{\sigma'_{vo}-\Delta\sigma'_v}\right] \tag{8.11}$$

B) For normally consolidated soil (OCR = 1)

$$s_c = C_c \frac{H_o}{1+e_o}\log\frac{\sigma'_{vo}+\Delta\sigma_v}{\sigma'_{vo}} \tag{8.12}$$

C) For overconsolidated soil (OCR > 1)
Case I: $\sigma'_{vo}+\Delta\sigma_v \leq \sigma'_{vm}$

$$s_c = C_r \frac{H_o}{1+e_o}\log\frac{\sigma'_{vo}+\Delta\sigma_v}{\sigma'_{vo}} \tag{8.13}$$

Case II: $\sigma'_{vo}+\Delta\sigma_v > \sigma'_{vm}$

$$s_c = C_r \frac{H_o}{1+e_o}\log\frac{\sigma'_{vm}}{\sigma'_{vo}}+C_c\frac{H_o}{1+e_o}\log\frac{\sigma'_{vo}+\Delta\sigma_v}{\sigma'_{vm}} \tag{8.14}$$

where s_c = settlement due to primary consolidation caused by an increase in load (ft or m)
C_c = compression index, obtained from the virgin consolidation curve, see Fig. 8.2 and Eq. 8.6 (dimensionless)
C_r = recompression index, obtained from the recompression portion of the laboratory consolidation curve, see Fig. 8.2 and Eq. 8.5 (dimensionless)
H_o = initial thickness of the in situ saturated cohesive soil layer (ft or m). The value of H_o in Eqs. 8.11 to 8.14 represents the initial thickness of the in situ cohesive soil layer. Because of changing parameters versus depth, such as a change in the maximum past pressure σ'_{vm}, the cohesive soil layer may need to be broken into several horizontal layers in order to obtain an accurate value of the primary consolidation settlement (s_c).

e_o = initial void ratio of the in situ saturated cohesive soil layer (dimensionless)

σ'_{vo} = initial vertical effective stress of the in situ soil (psf or kPa)

$\Delta\sigma'_v$ = for an underconsolidated soil, this represents the increase in vertical effective stress that will occur as the cohesive soil consolidates under its own weight (psf or kPa)

$\Delta\sigma_v$ = increase in stress due to the one-dimensional loading, such as the construction of a fill layer at ground surface (psf or kPa). Two- and three-dimensional loadings will be discussed in Secs. 8.7 and 8.8. As previously mentioned, a drop in the groundwater table or a reduction in pore water pressure can also result in an increase in load on the cohesive soil.

σ'_{vm} = maximum past pressure (psf or kPa), also known as the preconsolidation pressure σ'_p. It is often obtained from the consolidation curve using the Casagrande (1936) construction technique (see Fig. 8.2).

For overconsolidated soil, there are two possible cases that can be used to calculate the amount of settlement.

1. *Overconsolidated soil, case I.* The first case occurs when the existing vertical effective stress σ'_{vo} plus the increase in vertical stress $\Delta\sigma_v$ due to the proposed building weight does not exceed the maximum past pressure σ'_{vm}. For this first case, there will only be recompression of the cohesive soil.

2. *Overconsolidated soil, case II.* For the second case, the sum of the existing vertical effective stress σ'_{vo} plus the increase in vertical stress $\Delta\sigma_v$ due to the proposed building weight exceeds the maximum past pressure σ'_{vm}. For the second case, there will be virgin consolidation of the cohesive soil. Given the same cohesive soil and identical field conditions, the settlement due to the second case will be significantly more than the first case.

Example Problem 8.3 A site has a level ground surface and a level groundwater table located 5 m below the ground surface. As shown in Fig. 8.10, subsurface exploration has discovered that the site is underlain with sand, except for a uniform and continuous clay layer that is located at a depth of 10 to 12 m below ground surface. Below the groundwater table, the pore water pressures are hydrostatic in the sand layers (i.e., no artesian pressures). The average void ratio e_o of the clay layer is 1.10 and the buoyant unit weight γ_b of the clay layer = 7.9 kN/m³. The total unit weight γ_t of the sand above the groundwater table = 18.7 kN/m³ and the total unit weight γ_t of the sand below the groundwater table = 19.7 kN/m³. A laboratory consolidation test performed on an undisturbed specimen obtained from the center of the clay layer (Point A, Fig. 8.10) indicates the maximum past pressure σ'_{vm} = 100 kPa and the compression index C_c = 0.83. Determine the primary consolidation settlement s_c of the 2-m-thick clay layer if a uniform fill surcharge of 50 kPa is applied over a very large area at ground surface.

Solution The first step is to determine the vertical effective stress σ'_{vo} at the center of the clay layer (Point A, Fig. 8.10), or:

$$\sigma'_{vo} = (5 \text{ m})(18.7 \text{ kN/m}^3) + (5 \text{ m})(19.7 - 9.81 \text{ kN/m}^3) + (1 \text{ m})(7.9 \text{ kN/m}^3) = 150 \text{ kPa}$$

Since σ'_{vo} is greater than σ'_{vm}, the clay is underconsolidated (OCR < 1) by an amount of 50 kPa at the center of the clay layer. The greatest underconsolidation is at the center of the clay layer, with the clay adjacent to the sand layers essentially normally consolidated. Hence assuming a linear distribution of underconsolidation increasing from zero at the top and bottom of the clay layer to a maximum value of 50 kPa, the average underconsolidation for the

(Continued)

entire 2-m-thick layer is equal to:

$$\Delta\sigma_v' = (^1/_2)(50 \text{ kPa}) = 25 \text{ kPa}$$

As indicated in the problem statement, use the following values:

$$C_c = 0.83$$
$$H_o = 2 \text{ m}$$
$$e_o = 1.10$$
$$\Delta\sigma_v = 50 \text{ kPa}$$

For underconsolidated soil (OCR < 1), use Eq. 8.11:

$$s_c = C_c [H_o/(1 + e_o)] [\log (\sigma_{vo}' + \Delta\sigma_v)/\sigma_{vo}' + \log \sigma_{vo}'/(\sigma_{vo}' - \Delta\sigma_v')]$$
$$s_c = (0.83)[(2 \text{ m})/(1 + 1.1)] [\log (150 + 50)/150 + \log 150/(150 - 25)]$$
$$s_c = 0.16 \text{ m } (6.4 \text{ in.})$$

Example Problem 8.4   Use the same data as the previous example problem, except that the laboratory consolidation test performed on an undisturbed specimen obtained from the center of the clay layer indicates a maximum past pressure $\sigma_{vm}' = 150$ kPa.

Solution   From the prior example problem, at the center of the clay layer:

$$\sigma_{vo}' = 150 \text{ kPa}$$

Since σ_{vo}' is equal to σ_{vm}', the clay layer is normally consolidated (OCR = 1). From the prior example, use the following values:

$$C_c = 0.83$$
$$H_o = 2 \text{ m}$$
$$e_o = 1.10$$
$$\Delta\sigma_v = 50 \text{ kPa}$$

Since OCR = 1, use Eq. 8.12:

$$s_c = C_c [H_o/(1 + e_o)] \log [(\sigma_{vo}' + \Delta\sigma_v)/\sigma_{vo}']$$
$$s_c = (0.83)[(2 \text{ m})/(1 + 1.1)] \log [(150 \text{ kPa} + 50 \text{ kPa})/150 \text{ kPa}] = 0.10 \text{ m } (3.9 \text{ in.})$$

Example Problem 8.5   Use the same data as the previous example problem, except that the laboratory consolidation test performed on an undisturbed specimen obtained from the center of the clay layer indicates a maximum past pressure $\sigma_{vm}' = 175$ kPa. Also use a recompression index $C_r = 0.05$.

(*Continued*)

Solution From the prior example problems, at the center of the clay layer:

$$\sigma'_{vo} = 150 \text{ kPa}$$

Since σ'_{vo} is less than σ'_{vm}, the clay layer is overconsolidated (OCR > 1). The OCR can be calculated as follows:

$$\text{OCR} = \sigma'_{vm}/\sigma'_{vo} = 175 \text{ kPa}/150 \text{ kPa} = 1.17$$

From the prior examples, use the following values:

$$C_c = 0.83$$
$$H_o = 2 \text{ m}$$
$$e_o = 1.10$$
$$\Delta\sigma_v = 50 \text{ kPa}$$

Since $\sigma'_{vo} + \Delta\sigma_v = 200 \text{ kPa} > \sigma'_{vm}$
Use Case II (Eq. 8.14):

$$s_c = C_r \, [H_o/(1 + e_o)] \log (\sigma'_{vm}/\sigma'_{vo}) + C_c \, [H_o/(1 + e_o)] \log [(\sigma'_{vo} + \Delta\sigma_v)/\sigma'_{vm}]$$
$$s_c = (0.05)[(2 \text{ m})/(1 + 1.1)] \log (175 \text{ kPa}/150 \text{ kPa})$$
$$+ (0.83)[(2 \text{ m})/(1 + 1.1)] \log [(150 \text{ kPa} + 50 \text{ kPa})/175 \text{ kPa}] = 0.049 \text{ m} (1.9 \text{ in.})$$

Example Problem 8.6 Use the same data as the previous example problem, except that the laboratory consolidation test performed on an undisturbed specimen obtained from the center of the clay layer indicates a maximum past pressure $\sigma'_{vm} = 250$ kPa. Also use a recompression index $C_r = 0.05$.

Solution From the prior example problem, at the center of the clay layer:

$$\sigma'_{vo} = 150 \text{ kPa}$$

Since σ'_{vo} is less than σ'_{vm}, the clay layer is overconsolidated (OCR > 1). The OCR can be calculated as follows:

$$\text{OCR} = \sigma'_{vm}/\sigma'_{vo} = 250 \text{ kPa}/150 \text{ kPa} = 1.67$$

From the prior examples, use the following values:

$$H_o = 2 \text{ m}$$
$$e_o = 1.10$$
$$\Delta\sigma_v = 50 \text{ kPa}$$

Since $\sigma'_{vo} + \Delta\sigma_v = 200 \text{ kPa} < \sigma'_{vm}$
Use Case I (Eq. 8.13):

$$s_c = C_r \, [H_o/(1 + e_o)] \log [(\sigma'_{vo} + \Delta\sigma_v)/\sigma'_{vo}]$$
$$s_c = (0.05)[(2 \text{ m})/(1 + 1.1)] \log [(150 \text{ kPa} + 50 \text{ kPa})/150 \text{ kPa}] = 0.006 \text{ m} (0.23 \text{ in.})$$

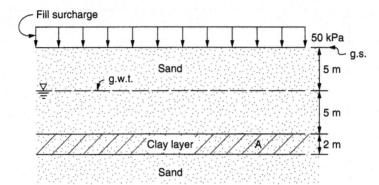

FIGURE 8.10   Example problem.

In summary, for the prior four example problems (Fig. 8.10), which are identical except for the value of the maximum past pressure σ'_{vm}, the calculated primary consolidation settlements are as follows:

Stress history	Calculated primary consolidation settlement (s_c)
Underconsolidated (OCR < 1)	16 cm (6.4 in.)
Normally consolidated (OCR = 1)	10 cm (3.9 in.)
Overconsolidated (OCR = 1.17), Case II	4.9 cm (1.9 in.)
Overconsolidated (OCR = 1.67), Case I	0.6 cm (0.23 in.)

Thus for these example problems, the primary consolidation settlement (s_c) varies from 16 to 0.6 cm depending on the value of the maximum past pressure.

8.5   RATE OF PRIMARY CONSOLIDATION

8.5.1   Terzaghi Equation

When a saturated cohesive or organic soil is loaded, there is an increase in the pore water pressure, known as the initial excess pore water pressure u_o. As water flows out of the soil, this excess pore water pressure slowly dissipates. In order to determine how fast this consolidation process will take, Terzaghi (1925) developed the one-dimensional consolidation equation to describe the time-dependent settlement behavior of clays. The purpose of the Terzaghi theory of consolidation is to estimate the settlement versus time relationship after loading of the saturated cohesive or organic soil.

The Terzaghi equation is a form of the diffusion equation from mathematical physics. There are other phenomena that can be described by the diffusion equation, such as the heat flow through solids. In order to derive Terzaghi's diffusion equation, the following assumptions are made (Terzaghi, 1925; Taylor, 1948; Holtz and Kovacs, 1981):

1. The clay is homogeneous and the degree of saturation is 100 percent (saturated soil).

2. Drainage is provided at the top and/or bottom of the compressible layer.

3. Darcy's law ($v = ki$) is valid.

4. The soil grains and pore water are incompressible.

5. Both the compression and flow of water are one-dimensional.

6. The load increment results in small strains so that the coefficient of permeability k and the coefficient of compressibility a_v remain constant.

7. There is no secondary compression.

Terzaghi's diffusion equation is derived by considering the continuity of flow out of a soil element, and is:

$$c_v \frac{\partial^2 u_e}{\partial z^2} = \frac{\partial u_e}{\partial t} \qquad (8.15)$$

where u_e = excess pore water pressure
z = vertical height to the nearest drainage boundary
t = time
c_v = coefficient of consolidation (ft²/day or m²/day), usually determined from a laboratory consolidation test, defined as:

$$c_v = \frac{k(1+e_o)}{\gamma_w a_v} \qquad (8.16)$$

k = coefficient of permeability, also known as the hydraulic conductivity (ft/day or m/day)
e_o = initial void ratio (dimensionless)
γ_w = unit weight of water (62.4 pcf or 9.81 kN/m³)
a_v = coefficient of compressibility (ft²/lb or m²/kN), defined as:

$$a_v = \frac{\Delta e}{\Delta \sigma_v'} \qquad (8.17)$$

where Δe is the change in void ratio corresponding to the change in vertical effective stress ($\Delta \sigma_v'$).

The mathematical solution to Eq. 8.15 can be expressed in graphical form as shown in Fig. 8.11. The vertical axis is defined as Z, which equals:

$$Z = \frac{z}{H} \qquad (8.18)$$

where z is the vertical depth below the top of the clay layer (ft or m) and H is one-half the initial thickness of the clay layer (double drainage), ft or m.

The horizontal axis in Fig. 8.11 is the consolidation ratio U_z defined as:

$$U_z = 1 - \frac{u_e}{u_o} \qquad (8.19)$$

where u_e is the excess pore water pressures in the cohesive soil (psf or kPa) and u_o is the initial excess pore water pressure (psf or kPa), which is equal to the applied load on the saturated cohesive soil (i.e., $u_o = \Delta \sigma_v$).

In Fig. 8.11, the value of U_z can be determined for different time factors T, defined as:

$$T = \frac{c_v t}{H_{dr}^2} \qquad (8.20)$$

where T = time factor from Table 8.2 (dimensionless)
t = time since the application of the load to the in situ saturated cohesive or organic soil layer. The load is assumed to be applied instantaneously.
c_v = coefficient of consolidation from laboratory testing (ft²/day or m²/day)
H_{dr} = height of the drainage path (ft or m). If water can drain out through the top and bottom of the cohesive soil layer (double drainage), then $H_{dr} = \frac{1}{2} H_o$, where H_o = initial thickness of the cohesive soil layer. An example of double drainage would be if there are sand layers located on top

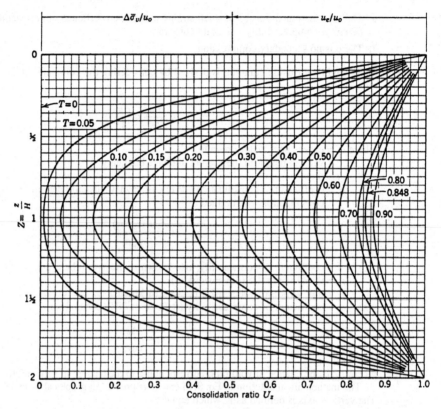

FIGURE 8.11 Consolidation ratio as a function of depth and time factor for uniform initial excess pore water pressure. (*From Lambe and Whitman, 1969; reproduced with permission of John Wiley & Sons.*)

and bottom of the cohesive soil layer. If water can only drain out of the top or bottom of the cohesive soil layer (single drainage), then $H_{dr} = H_o$. An example of single drainage would be if the clay layer is underlain by dense shale that is essentially impervious.

An important part of the mathematical solution to the Terzaghi theory of consolidation is the relationship between the average degree of consolidation U_{avg} and the time factor T. The average degree of consolidation U_{avg} is defined as:

$$U_{avg} = \frac{100\, s_t}{s_c} \tag{8.21}$$

where s_t is the settlement at any given time t, ft or m, and s_c is the primary consolidation settlement (ft or m) calculated from Eqs. 8.11, 8.12, 8.13, or 8.14.

The purpose of the Terzaghi theory of consolidation is to estimate the settlement versus time relationship after loading of the cohesive soil. The amount of settlement, in terms of the average degree of consolidation settlement U_{avg}, is related to the time factor T as follows:

$$U_{avg} = 100 - \frac{100\,(\text{area within a } T \text{ curve})}{(\text{total area of Fig. 8.11})} \tag{8.22}$$

Using Fig. 8.11 and Eq. 8.22, the relationship between the average degree of consolidation settlement U_{avg} and the time factor T has been calculated and is summarized in Table 8.2. When the

TABLE 8.2 Average Degree of Consolidation (U_{avg}) versus Time Factor (T)

Average degree of consolidation U_{avg}, %	Time factor T
0	0
5	0.002
10	0.008
15	0.017
20	0.031
25	0.049
30	0.071
35	0.092
40	0.126
45	0.159
50	0.197
55	0.238
60	0.286
65	0.340
70	0.403
75	0.477
80	0.567
85	0.683
90	0.848
95	1.13
100	∞

Assumptions: Terzaghi theory of consolidation, linear initial excess pore water pressures, and instantaneous loading.
Note: The above values are based on the following approximate equations by Casagrande (unpublished notes) and Taylor (1984):

For $U_{avg} < 60\%$, the time factor $T = \frac{1}{4}\pi(U_{avg}/100)^2$

For $U_{avg} \geq 60\%$, the time factor $T = 1.781 - 0.933 \log(100 - U_{avg})$

one-dimensional load is applied, the average degree of consolidation U_{avg} is zero (no settlement) and the time factor T is also zero. When one-half of the consolidation settlement has occurred (i.e., $U_{avg} = 50$ percent), the time factor $T = 0.197$. The Terzaghi theory of consolidation predicts that the time t for complete consolidation ($U_{avg} = 100$ percent, i.e., total settlement) is infinity, but as a practical matter, a time factor T equal to 1.0 is often assumed for $U_{avg} = 100$ percent.

Figure 8.12 shows the average degree of consolidation U_{avg} versus time factor T for various values of α. For this chart, α is defined as:

$$\alpha = \frac{u_o\,(\text{top})}{u_o\,(\text{bottom})} \tag{8.23}$$

where u_o (top) is the initial excess pore water pressure at the top of the clay layer (psf or kPa) and u_o (bottom) is the initial excess pore water pressure at the bottom of the clay layer (psf or kPa).

For one-dimensional instantaneous loading, the value of $\alpha = 1.0$. For this case of $\alpha = 1.0$, the values of average degree of consolidation U_{avg} versus time factor T are identical for Fig. 8.12 and Table 8.2. Examples where α may not be equal to 1.0 include two- or three-dimensional loadings and pore water pressure changes caused by pumping of groundwater from wells.

In all cases where there is two-way drainage (drainage at the top and bottom of the clay layer), a variation in initial excess pore water pressure from the top to bottom of the clay layer (i.e., $\alpha \neq 1.0$) will not effect the U_{avg} versus T values. However, for one way drainage, U_{avg} versus T is altered in the early stage of consolidation. For example, if a clay layer has single drainage and $\alpha = 0$, the time factors are higher during the initial stage of consolidation indicating that it takes longer for consolidation

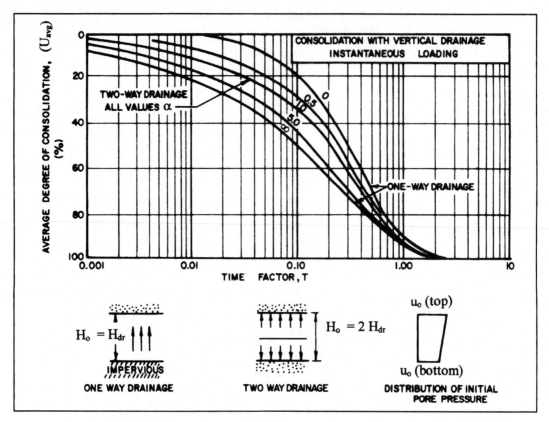

FIGURE 8.12 Average degree of consolidation U_{avg} versus time factors T for various values of α. (*Adapted from NAVFAC DM-7.1, 1982.*)

to get started. At the other extreme is $\alpha = \infty$, where the initial consolidation occurs very quickly (lower values of T) because the highest initial excess pore water pressures are at the drainage boundary. Note in Fig. 8.12 that no matter what the value of α, all of the curves eventually approach each other at high values of U_{avg}, indicating the time to complete primary consolidation is not really dependent on the magnitude or variation of initial excess pore water pressure across the clay layer.

8.5.2 Coefficient of Consolidation from Laboratory Tests

In Fig. 8.13, the data from Table 8.2 have been plotted on a semi-log plot. This curve has a characteristic shape, which has been termed a Type 1 curve (Leonards and Altschaeffl, 1964). In Fig. 8.13, the two straight line segments have been extended and the intersection point is at $U_{avg} = 100$ percent, i.e., when the primary consolidation settlement is complete. The plot in Fig. 8.13 is known as the log-of-time method, which has been introduced in Sec. 8.2.2.

In Fig. 8.14, the data from Table 8.2 have been plotted on a square root of time factor plot. Note in Fig. 8.14 that the time factor is essentially a straight line for an average degree of consolidation from 0 to 60 percent. This is because the equation for U_{avg} versus time factor T (listed at the bottom of Table 8.2) can be rearranged such that U_{avg} is equal to the square root of T. If a line is drawn such that it is placed at a distance of $0.15d$ (as illustrated in Fig. 8.14), it will intersect the curve at an average degree of consolidation equal to 90 percent. This plot in Fig. 8.14 is known as the square-root-of-time method, which was also introduced in Sec. 8.2.2.

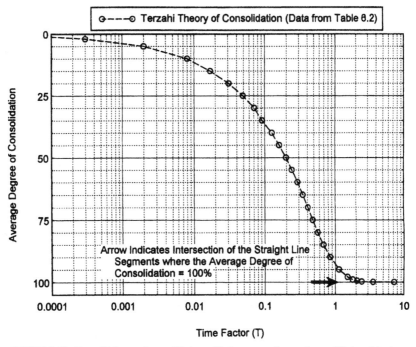

FIGURE 8.13 Terzaghi theory of consolidation, with the average degree of consolidation plotted versus time factor (log scale). The data are from Table 8.2.

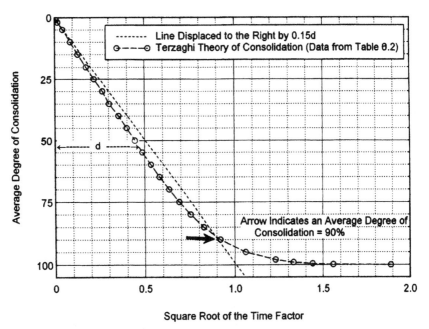

FIGURE 8.14 Terzaghi theory of consolidation, with the average degree of consolidation plotted versus square root of time factor. The data are from Table 8.2.

The usual procedure is to calculate c_v from the results of laboratory consolidation tests. During the incremental loading of the saturated soil in the oedometer apparatus, dial readings of vertical deformation versus time can be recorded (i.e., see Step 4, Sec. 8.2.1). As discussed in Sec. 8.2.2, there are two different possible plots that can be developed with these data, the first based on the log-of-time method (Fig. 8.5) and the second based on the square-root-of-time method (Fig. 8.6).

Note that the shape of the curve in Fig. 8.5 is very similar to the shape of the Terzaghi log of time theoretical curve in Fig. 8.13. Likewise, the shape of the curve in Fig. 8.6 is very similar to the Terzaghi square root of time theoretical curve in Fig. 8.14. Thus the laboratory data shown in Figs. 8.5 and 8.6 can be used to determine the coefficient of consolidation. The procedures are as follows:

1. *Log-of-time method.*   Equation 8.20 is used to calculate the coefficient of consolidation (c_v) for the log-of-time method. As per Table 8.2, the time factor T for an average degree of consolidation U_{avg} of 50 percent is 0.197. For the data shown in Fig. 8.5, the height H of the specimen at $d_{50} = 7.84$ mm, and since the specimen has double drainage in the laboratory oedometer apparatus, $H_{dr} = 3.92$ mm. Inserting $t_{50} = 65$ min (from Sec. 8.2.2), $H_{dr} = 3.92$ mm, and $T = 0.197$ into Eq. 8.20, the coefficient of consolidation $c_v = 0.047$ mm^2/min (0.025 m^2/year). This very low coefficient of consolidation is due to the clay particles (montmorillonite) in the soil (i.e., PI = 51).

2. *Square-root-of-time method.*   Equation 8.20 is also used to calculate the coefficient of consolidation c_v for the square-root-of-time method. As per Table 8.2, the time factor T for an average degree of consolidation U_{avg} of 90 percent is 0.848. The average height H of the specimen = 7.84 mm, and since the specimen has double drainage in the laboratory oedometer apparatus, $H_{dr} = 3.92$ mm. Inserting $t_{90} = 290$ min (from Sec. 8.2.2), $H_{dr} = 3.92$ mm, and $T = 0.848$ into Eq. 8.20, the coefficient of consolidation $c_v = 0.045$ mm^2/min (0.024 m^2/year).

There is a slight difference in the calculated value of the coefficient of consolidation c_v based on the log-of-time method ($c_v = 0.047$ mm^2/min) versus the square-root-of-time method ($c_v = 0.045$ mm^2/min). This is because both of the curve fitting techniques are approximations of the Terzaghi theory presented in Figs. 8.13 and 8.14.

The coefficient of consolidation c_v for laboratory specimens under different sample heights, load increment ratios, and load duration has been determined for numerous cohesive soil (Taylor, 1948; Leonards and Ramiah, 1959). Ladd (1973) states that it can be assumed that variations in sample height, load increment ratio, and load duration will generally lead to insignificant difference in c_v provided that the resulting dial versus time reading have the characteristic shape shown in Fig. 8.5 (i.e., a Type 1 curve; Leonards and Altschaeffl, 1964) when plotted on a log-time scale. This means that there must be appreciable primary consolidation during the loading increment.

A factor that can significantly affect the coefficient of consolidation is sample disturbance. For example, in Fig. 8.3 (lower plot) the coefficient of consolidation is shown for both the disturbed Orinoco clay specimen (solid circles) and the undisturbed Orinoco clay specimen (open circles). Especially for that part of the consolidation test that involves recompression of the cohesive soil, sample disturbance can significantly reduce the values of the coefficient of consolidation.

In summary, the four steps in using the Terzaghi theory of consolidation are as follows:

Step 1. Coefficient of consolidation: The first step is to determine the coefficient of consolidation c_v of the saturated cohesive or organic soil. Either the log-of-time method or the square-root-of-time method can be used to obtain the coefficient of consolidation.

Step 2. Determine the drainage height H_{dr} of the in situ cohesive soil: After the coefficient of consolidation c_v has been determined from laboratory testing of undisturbed soil specimens, the next step is to determine the drainage height of the in situ cohesive soil layer. As previously mentioned, if water can drain from the cohesive soil layer at both the top and bottom of the clay layer (double drainage), then $H_{dr} = \frac{1}{2} H_o$, where H_o = thickness of the cohesive soil layer. If water can only drain from the top or bottom of the cohesive soil layer (single drainage), then $H_{dr} = H_o$.

Step 3. Determine the time factor T: Based on an average degree of consolidation U_{avg}, the time factor T can be obtained from Table 8.2 or Fig. 8.12.

Step 4. Use Eq. 8.20 to determine the time t: Once c_v, T, and H_{dr} are known, the time t corresponding to a certain amount of settlement U_{avg} can be calculated from Eq. 8.20.

In summary, the steps listed above basically consist of determining the coefficient of consolidation c_v from laboratory consolidation tests performed on undisturbed soil specimens and then these data are used to predict the time-settlement behavior of the in situ cohesive soil layer. Terzaghi's consolidation equation is one of the most widely taught and applied theories in geotechnical engineering. Some of the limitations of this theory are as follows (Duncan, 1993):

1. c_v is commonly assumed to be constant, which is often not the case in the field.
2. The stress-strain behavior of the soil skeleton is assumed to be linear and elastic.
3. The strains are assumed to be uniform.
4. There are often permeable sand lenses within the in situ cohesive soil that can significantly decrease the length of time for primary consolidation as predicted by assuming a constant c_v.
5. The strain often decreases with depth because the stress increase caused by surface loads decreases with depth, or the clay compressibility decreases with depth, or both.
6. A final factor is that the theory is based on the vertical flow of water from the cohesive soil, but many cases involve two or three-dimensional loading which would allow the clay to drain partially from its sides.

Another factor is temperature, where the temperature in the laboratory can be significantly greater than the temperature of the in situ soil. It has been stated that temperature can significantly influence the coefficient of consolidation c_v, where higher values of c_v are recorded as the temperature increases (Lambe, 1951). This is because temperature affects the permeability of the soil (k) which in turn affects the rate at which the water can flow from the soil.

Because of all these factors, the Terzaghi theory of consolidation often does not accurately predict the time required to reach a certain average degree of consolidation. Frequently the in situ clay layer will contain layer or lenses of permeable material that decrease the drainage paths and cause the Terzaghi theory of consolidation to overpredict the time required to reach a certain average degree of consolidation. When applied to field situations, the theory should only be considered as an approximation of the time-settlement behavior for loading of saturated cohesive or organic soil. Field measurements (such as settlement monuments and piezometers) are often essential in comparing the actual time-settlement behavior with the predicted behavior.

Example Problem 8.7   Use the example problem shown in Fig. 8.10 and assume the clay layer is normally consolidated (OCR = 1.0). Assuming quick construction of the fill surcharge and using the Terzaghi theory of consolidation, predict how long after construction it will take for 50 and 90 percent of the primary consolidation settlement to occur. Based on laboratory oedometer testing of an undisturbed clay specimen, where dial readings versus time were recorded at a vertical pressure of 200 kPa, the coefficient of consolidation c_v was calculated to be = 0.32 m^2/year.

Solution   Since there is sand both on the top and bottom of the 2-m-thick clay layer, the clay layer has double drainage and $H_{dr} = 1$ m.

For U_{avg} = 50 percent, the time factor $T = 0.197$ (Table 8.2)

For U_{avg} = 90 percent, the time factor $T = 0.848$ (Table 8.2)

(Continued)

For 50 percent consolidation:

$$T = c_v t/H_{dr}^2 \quad \text{or} \quad t = TH_{dr}^2/c_v = (0.197)(1 \text{ m})^2/0.32 \text{ m}^2/\text{year} = 0.61 \text{ years}$$

For 90 percent consolidation:

$$t = TH_{dr}^2/c_v = (0.848)(1 \text{ m})^2/0.32 \text{ m}^2/\text{year} = 2.65 \text{ years}$$

With a value of $c_v = 0.32$ m²/year, it will take 0.61 years for 50 percent of the primary consolidation to occur (i.e., 0.61 years for 5 cm of settlement) and it will take 2.65 years for 90 percent of the primary consolidation to occur (i.e., 2.65 years for 9 cm of settlement). Using other values of U_{avg}, the primary consolidation settlement versus time plot can be obtained, as shown in Fig. 8.15.

Example Problem 8.8   Solve the prior example problem, but assume that at the bottom of the clay layer there is impervious rock.

Solution   In this case, the clay layer can only drain from the top. Hence the clay layer has single drainage and $H_{dr} = 2$ m.

For 50 percent consolidation:

$$t = TH_{dr}^2/c_v = (0.197)(2 \text{ m})^2/0.32 \text{ m}^2/\text{year} = 2.46 \text{ years}$$

For 90 percent consolidation:

$$t = TH_{dr}^2/c_v = (0.848)(2 \text{ m})^2/0.32 \text{ m}^2/\text{year} = 10.6 \text{ years}$$

Note that to reach a specific amount of settlement, it takes four times as long for a clay to consolidate for single drainage as compared to double drainage.

Example Problem 8.9   Use the data from the previous example problem and assume the clay can only drain from the top of the layer. Determine the time it takes for the clay to consolidate 50 and 90 percent if the initial distribution of excess pore water pressure u_o varies from 100 kPa at the top of the clay layer to 0 kPa at the bottom of the clay layer.

Solution   Using Eq. 8.23:

$$\alpha = (u_o \text{ at top of clay layer})/(u_o \text{ at bottom of clay layer}) = 100 \text{ kPa/0 kPa} = \infty$$

Entering the chart in Fig. 8.12 at $U_{avg} = 50$ percent and intersecting the one-way drainage $\alpha = \infty$ curve, the value of T from the horizontal axis = 0.10

$$t = TH_{dr}^2/c_v = (0.10)(2 \text{ m})^2/0.32 \text{ m}^2/\text{year} = 1.25 \text{ years}$$

Entering the chart in Fig. 8.12 at $U_{avg} = 90$ percent and intersecting the one-way drainage $\alpha = \infty$ curve, the value of T from the horizontal axis = 0.80

$$t = TH_{dr}^2/c_v = (0.80)(2 \text{ m})^2/0.32 \text{ m}^2/\text{year} = 10 \text{ years}$$

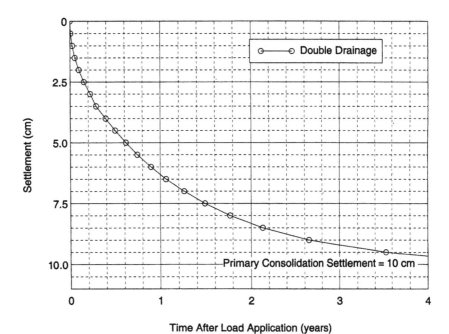

FIGURE 8.15 Primary consolidation settlement versus time for the example problem.

The time for 50 percent of the primary consolidation settlement is only 1.25 years for single drainage with $\alpha = \infty$ as compared to the usual situation of 2.46 years for $\alpha = 1$. Hence the closer the excess pore water pressures are to a drainage boundary, the quicker the time for primary consolidation. This is why vertical drains are often installed in the clay layer in order to speed up the consolidation process.

8.6 SECONDARY COMPRESSION

The final component of settlement is due to secondary compression, which is that part of the settlement that occurs after essentially all of the excess pore water pressures have dissipated (i.e., settlement that occurs at constant effective stress). The usual assumption is that secondary compression does not start until after primary consolidation is complete. Secondary compression has been described as a process where the particle contacts are still rather unstable at the end of primary consolidation and the particles will continue to move until finding a stable arrangement. In Fig. 8.5, that part of the vertical deformation above the arrow is due to primary consolidation settlement, while that part of the vertical deformation below the arrow is due to secondary compression. For laboratory test data such as shown in Fig. 8.5, the secondary compression is a small portion of the total settlement and it is often ignored. But as previously mentioned, for other soils such as organic soils or soft highly plastic clays, the secondary compression may constitute a significant portion of the total settlement and it must be included in the analysis.

The amount of settlement s_s due to secondary compression can be calculated as follows:

$$s_s = C_\alpha H_o \Delta \log t \tag{8.24}$$

where s_s = settlement due to secondary compression (occurs after the end of primary consolidation), ft or m.

C_α = secondary compression ratio (dimensionless) which is defined as the slope of the secondary compression curve. For example, the secondary compression curve shown in

Fig. 8.5 (portion of curve below the arrow) can be approximated as a straight line. If the vertical axis in Fig. 8.5 is converted to strain ε_v, then C_α equals the change in strain $\Delta\varepsilon_v$ divided by the change in time (log scale), or $C_\alpha = \Delta\varepsilon_v/\Delta \log t$. Using one log cycle of time (i.e., $\Delta \log t = 1$), then the value of secondary compression $C_\alpha = \Delta\varepsilon_v$.

H_o = initial thickness of the in situ cohesive or organic soil layer (ft or m).

$\Delta \log t$ = change in log of time from the end of primary consolidation to the end of the design life of the structure. For example, if primary consolidation is complete at a time = 10 years, and the design life of the structure is 100 years, then $\Delta \log t = \log 100 - \log 10 = 1$.

The final calculation for estimating the total settlement ρ_{max} of the in situ cohesive soil would be to add together the three components of settlement using Eq. 8.1, or: $\rho_{max} = s_i + s_c + s_s$.

Example Problem 8.10   Use the example problem shown in Fig. 8.10 and assume the clay layer is normally consolidated (OCR = 1.0) with double drainage. Determine the total settlement ρ_{max} of the in situ cohesive soil 50 years after application of the 50 kPa fill surcharge. Use the data in Fig. 8.5 to calculate the secondary compression ratio C_α.

Solution   From Sec. 8.5.2, the value of $H = 7.84$ mm at t_{50} from the consolidation test. The height of the soil specimen at the start of secondary compression is equal to:

$$H_o = 7.84 \text{ mm} - (0.66 \text{ mm} - 0.33 \text{ mm}) = 7.51 \text{ mm}$$

Using the data in Fig. 8.5, from 1000 to 10,000 min:

Change in height = 0.69 mm – 0.66 mm = 0.03 mm

$$\Delta\varepsilon_v = \Delta H/H_o = (0.03 \text{ mm})/(7.51 \text{ mm}) = 0.004$$

$$C_\alpha = \Delta\varepsilon_v/\Delta \log t = (0.004)/(\log 10{,}000 - \log 1000) = 0.004$$

Using $T = 1$ as the end of primary consolidation:

$$t = TH_{dr}^2/c_v = (1.0)(1 \text{ m})^2/0.32 \text{ m}^2/\text{year} = 3.1 \text{ years}$$

Using Eq. 8.24, with the in situ $H_o = 2$ m

$$s_s = C_\alpha H_o \Delta \log t = (0.004)(2 \text{ m})(\log 50 - \log 3.1) = 0.01 \text{ m} = 1 \text{ cm}$$

Thus the total settlement at 50 years is equal to:

$$\rho_{max} = s_i + s_c + s_s = 0 + 10 \text{ cm} + 1 \text{ cm} = 11 \text{ cm} \ (4.3 \text{ in.})$$

8.7   CONSOLIDATION OF SOIL BENEATH SHALLOW FOUNDATIONS

Shallow foundations will usually induce two- or three-dimensional loading onto the underlying soil. As discussed in Sec. 8.3, there may be some troublesome cases where the two- or three-dimensional loading causes a significant amount of immediate settlement. However, for many cases, the immediate settlement can be neglected ($s_i = 0$) and only the primary consolidation settlement s_c and secondary compression s_s are calculated for the shallow foundation. In order to determine the amount of primary consolidation settlement and secondary compression, it is assumed that the one-dimensional laboratory consolidation curve (Fig. 8.2) is applicable for the two- or three-dimensional shallow foundation loading. Thus the primary consolidation settlement is calculated using Eqs. 8.11 to 8.14 with the increase in vertical stress due to the shallow foundation load $\Delta\sigma_v$ determined from stress distribution theory (Sec. 4.5).

For two- and three-dimensional loading of saturated plastic soil, the water will not only be able to drain in a vertical direction, but it will also be able to drain in a horizontal direction. This will result in a faster rate of consolidation as predicted by the Terzaghi theory of consolidation (Sec. 8.5). To account for lateral drainage, the time factor T can be adjusted based on the ratio of H/B, where H = thickness of the clay layer and B = width of a circular loaded area. As shown in Fig. 8.16, when the ratio of H/B decreases, the time factors T approach the values listed in Table 8.2.

Example Problem 8.11   Use the example problem shown in Fig. 8.10 and assume the clay layer is normally consolidated (OCR = 1.0) with double drainage. Instead of a fill surcharge, assume that a building having a width = 20 m and a length = 30 m is proposed for the site and will be constructed at ground surface. The structural engineer has determined that the weight of the building can be approximated as a uniform pressure applied at ground surface equivalent to 50 kPa. Determine the following:

Using the Newmark chart (Fig. 4.8), calculate the primary consolidation settlement (s_c) due to the building load. Neglect possible settlement of the sand layers and ignore immediate settlement and secondary compression (i.e., $s_i = s_s = 0$).

Assuming quick construction of the building and using the Terzaghi theory of consolidation, predict how long after construction it will take for 50 and 90 percent of the primary consolidation settlement to occur.

Solution   From the prior example problems, use the following values:

$\sigma'_{vo} = \sigma'_{vm} = 150$ kPa (normally consolidated clay layer, OCR = 1).

$C_c = 0.83$

$H_o = 2$ m

$e_o = 1.10$

$c_v = 0.32$ m²/year

Part 1)

The maximum settlement will occur at the center of the loaded area. For the building, $B = 20$ m, $L = 30$ m, and $z = 11$ m. Dividing the rectangular loaded area into four equal parts, or:

$$x = B/2 = 20/2 = 10 \text{ m}$$
$$y = L/2 = 30/2 = 15 \text{ m}$$
$$m = x/z = 10/11 = 0.91$$
$$n = y/z = 15/11 = 1.4$$

Entering Fig. 4.8 with $n = 1.4$, intersecting the $m = 0.9$ curve, $I = 0.185$

Finding the increase in stress below the corner of the loaded area:

$$\sigma_z = \Delta\sigma_v = q_o I = (50 \text{ kPa})(0.185) = 9.25 \text{ kPa}$$

For center of the loaded area, multiply the above result by 4:

$$\sigma_z = \Delta\sigma_v = (4)(9.25 \text{ kPa}) = 37 \text{ kPa}$$

Since OCR = 1, use Eq. 8.12:

(Continued)

$$s_c = C_c \left[H_o/(1 + e_o)\right] \log \left[(\sigma'_{vo} + \Delta\sigma_v)/\sigma'_{vo}\right]$$

$$s_c = (0.83)[(2 \text{ m})/(1 + 1.1)] \log \left[(150 \text{ kPa} + 37 \text{ kPa})/150 \text{ kPa}\right] = 0.076 \text{ m} = 7.6 \text{ cm}$$

Part 2)

Using the following value for $H = H_o = 2$ m

Turning the rectangular foundation into an equivalent round foundation, or:

$$\text{Area} = (B)(L) = (20 \text{ m})(30 \text{ m}) = 600 \text{ m}^2$$

$$\tfrac{1}{4} \pi B^2 = 600 \text{ m}^2 \quad \text{or} \quad B = 28 \text{ m}$$

If the clay layer was directly below the loaded area:

$$H/B = 2/28 = 0.07$$

Based on Fig. 8.16, this value is so low that essentially one-dimensional flow conditions can be assumed, or:

For 50 percent consolidation, $T = 0.197$ from Table 8.2:

$$t = TH_{dr}^2/c_v = (0.197)(1 \text{ m})^2/0.32 \text{ m}^2/\text{year} = 0.61 \text{ years}$$

For 90 percent consolidation, $T = 0.848$ from Table 8.2:

$$t = TH_{dr}^2/c_v = (0.848)(1 \text{ m})^2/0.32 \text{ m}^2/\text{year} = 2.65 \text{ years}$$

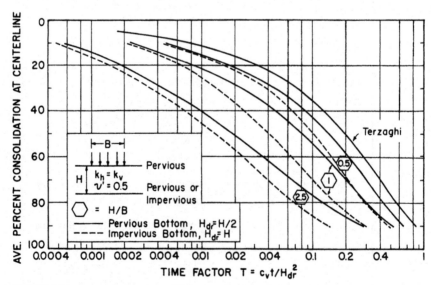

FIGURE 8.16 Average degree of consolidation U_{avg} versus time factors T beneath the centerline of a uniform circular load on a clay foundation with isotropic permeability. (*From Ladd et al., 1977; redrawn from Davis and Poulos, 1972.*)

Example Problem 8.12   Solve Part 1 of the prior example problem by using the 2:1 approximation.

Solution   Using the 2:1 approximation (Eq. 4.26), with $B = 20$ m, $L = 30$ m, and $z = 11$ m:

$$\sigma_z = \Delta\sigma_v = P/[(B + z)(L + z)]$$

$$= [(50 \text{ kPa})(20)(30)]/[(20 + 11)(30 + 11)] = 24 \text{ kPa}$$

Since OCR = 1, use Eq. 8.12:

$$s_c = C_c [H_o/(1 + e_o)] \log [(\sigma'_{vo} + \Delta\sigma_v)/\sigma'_{vo}]$$

$$s_c = (0.83)[(2 \text{ m})/(1 + 1.1)] \log [(150 \text{ kPa} + 24 \text{ kPa})/150 \text{ kPa}] = 0.051 \text{ m} = 5.1 \text{ cm}$$

Example Problem 8.13   Solve the above example problem by using the layered soil chart (Fig. 4.16). Assume the sand layer ($z = 0$ to 10 m) and the clay layer ($z = 10$ to 12 m) have a Poisson's ratio of 0.25 (i.e., $\mu_1 = \mu_2 = 0.25$) and the value of $k = E_1/E_2 = 10$.

Solution   From Example Problem 8.11, turning the rectangular foundation into an equivalent round foundation:

$$B = 2r = 28 \text{ m} \quad \text{or} \quad r = 14 \text{ m}$$

Calculating the increase in stress to the top of the clay layer, use $H = 10$ m

$$\alpha = r/H = 14/10 = 1.4$$

Entering the upper part of Fig. 4.16 with $k = 10$, extrapolating between the $\alpha = 1$ and $\alpha = 2$ curves, the value of $(I)(H/r)^2 = 0.24$

$$I = (0.24)/(H/r)^2 = (0.24)/(10/14)^2 = 0.47$$

$$\sigma_z = \Delta\sigma_v = Ip = (0.47)(50 \text{ kPa}) = 23 \text{ kPa}$$

The stress increase of 23 kPa is only applicable at the top of the clay layer. However, using this value for the center of the clay layer and since OCR = 1, from Eq. 8.12:

$$s_c = C_c [H_o/(1 + e_o)] \log [(\sigma'_{vo} + \Delta\sigma_v)/\sigma'_{vo}]$$

$$s_c = (0.83)[(2 \text{ m})/(1 + 1.1)] \log [(150 \text{ kPa} + 23 \text{ kPa})/150 \text{ kPa}] = 0.049 \text{ m} = 4.9 \text{ cm}$$

Comparing the settlement for these three prior example problems of a normally consolidated clay (OCR = 1.0):

Method	Calculated primary consolidation settlement (s_c)
Section 8.4.2 (one-dimensional loading)	10 cm (3.9 in.)
Newmark chart (Fig. 4.8)	7.6 cm (3.0 in.)
2:1 Approximation	5.1 cm (2.0 in.)
Layered soil (Fig. 4.16)	4.9 cm (1.9 in.)

As expected, all three methods have a primary consolidation settlement that is less than the one-dimensional loading case. The 2:1 approximation and Fig. 4.16 (layered soil) provide about the same answer, with the Newmark chart having the highest value of primary consolidation settlement. The amount of foundation settlement would probably vary from about 5 to 7.5 cm (2 to 3 in.), depending on the rigidity of the foundation and the actual variation of stress increase with depth.

8.8 CONSOLIDATION OF SOIL BENEATH DEEP FOUNDATIONS

The analysis for the load carrying capacity and settlement behavior of piles and piers in cohesive soil is more complex than shallow foundations. Some of the factors that may need to be considered in the analysis are as follows:

1. *Immediate settlement.* Provided there is an adequate factor of safety for the bearing capacity of the piles and pile groups, the immediate settlement s_i is often ignored. This is because the load is usually transferred to a great depth and the overburden pressure limits the ability of the saturated clay to squeeze out from beneath the bottom of the piles.

2. *Pile load tests.* Prior or during the construction of the foundation, a pile or pile group could be load tested in the field to determine its carrying capacity. Because of the uncertainties in the design of piles based on engineering analyses, pile load tests are common. However, the load test only determines the short-term load-deformation behavior of the pile. In addition, the zone of influence of a single pile will be much less than the zone of influence of the entire building.

   Thus long-term settlement of the entire structure due to consolidation of the cohesive soil must still be evaluated. For example, the pile load test may demonstrate a high capacity, but once constructed, the structure experiences excessive settlement due to consolidation of clay layers located well below the pile tips.

3. *Downdrag loads.* Section 6.3.4 has presented a discussion and the equations needed to determine downdrag loads for single piles and pile groups. Piles that penetrate soft clay or organic soil will often be subjected to downdrag loads as the soil consolidates.

4. *Bearing capacity.* The bearing capacity of clay layers underlying the piles must be evaluated. The danger is that the piles will punch through the bearing layer and into underlying softer clay strata.

Similar to solutions for shallow foundations, charts have been developed for estimating the distribution of stress beneath deep foundations. For example, Fig. 4.17 shows the pressure distributions for pile foundations for four different soil conditions. For easy reference, Fig. 4.17 has been reproduced as Fig. 8.17, with an asterisk included that shows the zone of clay that would need to be evaluated for primary consolidation settlement and secondary compression. These charts are valid for relatively rigid pile caps, where the rigidity of the foundation system tends to restrict differential settlement across the pile cap. In this case, the 2:1 approximation is ideal because with a relatively rigid foundation system, it is desirable to calculate the average settlement of the pile cap. The four conditions in Fig. 8.17 are individually discussed below:

Friction Piles in Clay. The upper left figure shows the pressure distribution for a pile group embedded in clay. The load P that the pile cap supports is first turned into an applied stress q by using Eq. 4.32.

In essence, the 2:1 approximation starts at a depth of $2/3 L$ below ground surface. In Fig. 8.17, it is assumed that there is a hard layer located below the clay. In this case, the primary consolidation and secondary compression would be calculated for the zone of clay defined as H in the upper left diagram of Fig. 8.17. The clay of thickness H may have to be divided into several layers in order to calculate the primary consolidation and secondary compression.

Friction Piles in Sand Underlain by Clay. The upper right diagram in Fig. 8.17 shows the pile group embedded in sand with two underlying clay layers. For this situation, two conditions would have to be evaluated. The first is a bearing capacity failure of the pile group where it punches through the sand and into the upper soft clay layer. The second condition is the primary consolidation and secondary compression of the two clay layers located below the pile group. Similar to the case outlined above, the 2:1 approximation is used to calculate the increase in vertical stress σ_z due to the pile cap loading for the two clay layers.

Point Bearing Piles in Sand Underlain by Clay.   The lower left diagram in Fig. 8.17 shows the condition of a soft clay layer underlain by a sand stratum. The sand stratum provides most of the vertical resistance and hence the pile group is considered to be end bearing. For this case, the upper sand layer is assumed to be thick enough that a punching failure is not a concern. As shown in Fig. 8.17, the 2:1 approximation is assumed to start at the top of the sand layer and is used to calculate the increase in vertical stress ($\sigma_z = \Delta\sigma_v$) and hence the primary consolidation settlement and secondary compression of the soft clay layer having a thickness defined as H.

Friction Piles in Clay with Recent Fill.   The lower right diagram in Fig. 8.17 shows the condition of piles embedded in clay with the recent placement of a fill layer at ground surface. In this case, the piles will be subjected to a downdrag load due to placement of the fill layer. Section 6.3.4 presents a discussion and the equations needed to determine downdrag loads for single piles and pile groups.

As shown in Fig. 8.17, the 2:1 approximation is assumed to start at a distance of L_3 below the top of the clay layer, where $L_3 = 2/3\ L_2$. The value of q applied at this depth includes two additional terms, the first is the total unit weight of fill γ_t times the thickness of the fill L_1 and the second is the downdrag load converted to a stress. As discussed in Sec. 6.3.4, two cases should be evaluated, the first where the downdrag load is determined for each pile and summed, or $(nQ_D)/(BA)$, and the second is the downdrag load along the entire perimeter of the pile group. The higher downdrag load would be used in the analysis. Once q is known, the primary consolidation and secondary compression can be calculated for the zone of clay located below a depth defined as L_3 in Fig. 8.17.

If the pile caps are spaced close together, there could be additional primary consolidation settlement as the pressure distribution from one pile cap overlaps with the pressure distribution from a second near-by pile cap.

Example Problem 8.14   Use the example problem shown in Fig. 8.10 and assume the clay layer is normally consolidated (OCR = 1.0) with double drainage. Instead of a fill surcharge, assume that a building having a width = 20 m and a length = 30 m is proposed for the site and will be constructed at ground surface, with piles that are 7.5 m long. The structural engineer has determined that the weight of the building can be approximated as a uniform pressure applied at ground surface equivalent to 50 kPa. Using the 2:1 approximation (Fig. 8.17), calculate the primary consolidation settlement s_c due to the building load. Neglect possible settlement of the sand layers, ignore immediate settlement and secondary compression (i.e., $s_i = s_s = 0$), and assume there is an adequate factor of safety so that the piles do not punch into the clay layer.

Solution   From the prior example problems, use the following values:

$$\sigma'_{vo} = \sigma'_{vm} = 150 \text{ kPa (normally consolidated clay layer, OCR = 1)}$$

$$C_c = 0.83$$

$$H_o = 2 \text{ m}$$

$$e_o = 1.10$$

(Continued)

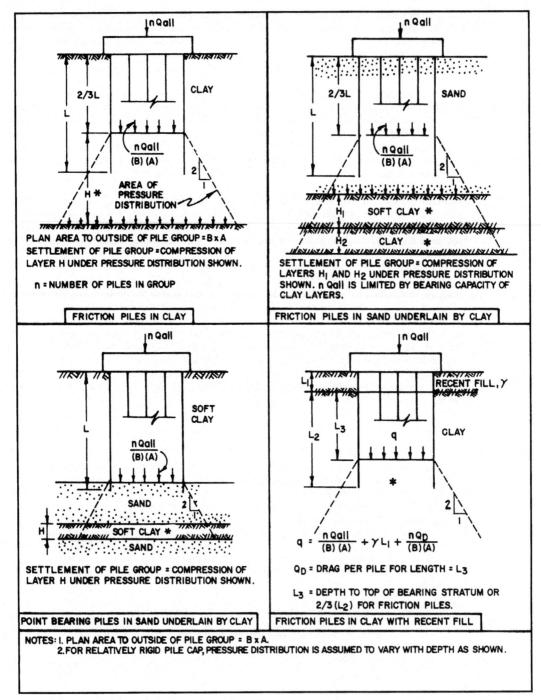

FIGURE 8.17 Pressure distribution for deep foundations. The asterisk shows the zone of clay that must be evaluated for primary consolidation settlement and secondary compression. (*Adapted from NAVFAC DM-7.2, 1982.*)

The analysis should be performed as shown in the upper right diagram in Fig. 8.17, with only one clay layer. As shown in Fig. 8.17, the 2:1 approximation (Eq. 4.26) starts at a depth of 2/3 L, or: $(2/3)(7.5 \text{ m}) = 5$ m. Hence using the following values:

$$B = 20 \text{ m}, L = 30 \text{ m}, \text{ and } z = 6 \text{ m}:$$

$$\sigma_z = \Delta\sigma_v = P/[(B + z)(L + z)]$$

$$= [(50 \text{ kPa})(20)(30)]/[(20 + 6)(30 + 6)] = 32 \text{ kPa}$$

Since OCR = 1, use Eq. 8.12:

$$s_c = C_c [H_o/(1 + e_o)] \log [(\sigma'_{vo} + \Delta\sigma_v)/\sigma'_{vo}]$$

$$s_c = (0.83)[(2 \text{ m})/(1 + 1.1)] \log [(150 \text{ kPa} + 32 \text{ kPa})/150 \text{ kPa}] = 0.066 \text{ m} = 6.6 \text{ cm}$$

This answer is close to the calculated values from Sec. 8.7 because for this example problem where the piles terminate above the clay layer, they have little effect on the stress increase experienced by the clay layer. The settlement of the foundation due to primary consolidation could be eliminated if the piles were long enough so that they penetrated through the clay layer.

8.9 SETTLEMENT OF UNSATURATED COHESIVE SOIL

Unsaturated cohesive soil can also be subjected to a significant amount of settlement. If the cohesive soil is near ground surface, it may become desiccated and hence susceptible to expansion and this will be covered in Chap. 9. Usually the compression curve for unsaturated clay does not have linear recompression and virgin compression indices such as shown in Fig. 8.2. As an alternative to using Eqs. 8.11 to 8.14, Eqs. 7.13 and 7.14 can be used to estimate the settlement due to a building load.

An example of laboratory test data for an unsaturated cohesive soil is presented in Figs. 8.18 to 8.20, as follows:

1. *Figure 8.18, soil classification data.*   This figure summarizes the grain size distribution and Atterberg limits for soil obtained at a depth of 10 to 13 ft (3 to 4 m). The test pit was excavated using a backhoe in order to obtain undisturbed block samples of the soil. As indicated in Fig. 8.18, the soil is classified as a sandy clay of high plasticity (CH). During the subsurface exploration, it was observed that the sandy clay was located above the groundwater table. Based on laboratory testing of an undisturbed specimen from the block sample, the sandy clay has a total unit weight = 117.6 pcf (18.5 kN/m³), dry unit weight = 94.1 pcf (14.8 kN/m³), water content = 25.0 percent, and degree of saturation = 85 percent.

2. *Figure 8.19, consolidation curve.*   An undisturbed specimen of the sandy clay was trimmed from the block sample, placed in an oedometer (see Sec. 3.3), and subjected to a seating pressure of 125 psf (6 kPa). The sandy clay specimen was then submerged in distilled water. For many soils, such as open graded gravel, submergence in water will usually cause the large interconnected void spaces to fill with water. But for the sandy clay, which has very small void spaces, the air is typically entrapped within the soil and will not easily be removed by submerging the soil in distilled water. Hence for this test, because of the entrapped air within the soil, the pore water pressures are unknown and it is best to plot the data as the applied vertical total stress versus void ratio.

After submergence in distilled water, the soil specimen was subjected to an increase in vertical stress, with each vertical stress increment sustained for a minimum of 24 h. Note that the consolidation curve shown in Fig. 8.19 does not have a distinct maximum past pressure σ'_{vm}. This could be due to several factors, such as the lack of saturation of the sandy clay as well as prior site factors, such as drying or capillary effects, that have created an overconsolidated soil.

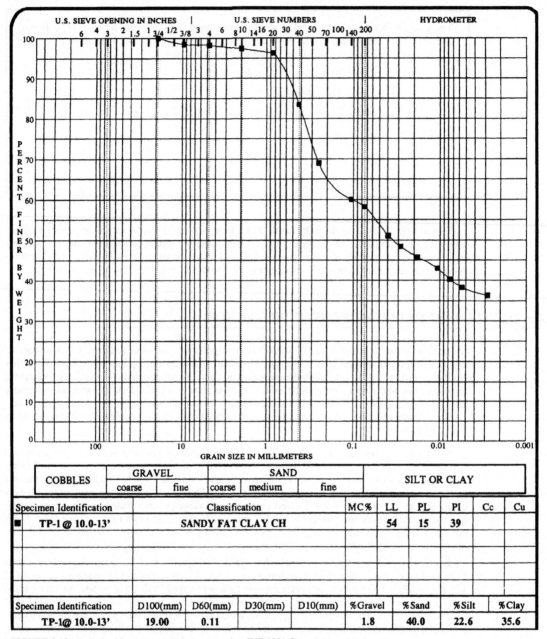

FIGURE 8.18 Soil classification data (Plot developed by gINT 1991 Computer program).

3. *Figure 8.20, time versus deformation readings.* Figure 8.20 shows a plot of the vertical deformation versus time (on a logarithm scale) when the sandy clay specimen was subjected to a vertical stress of 8000 psf (380 kPa). Comparing Fig. 8.20 with Figs. 8.5 and 8.13, it is evident that the deformation behavior does not have the classic deformation versus log-time shape

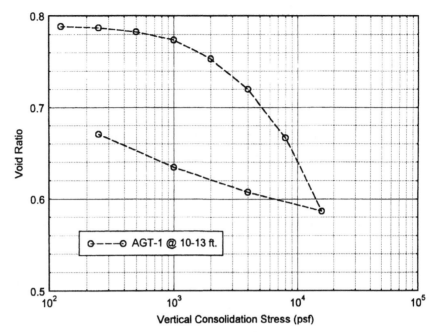

FIGURE 8.19   Consolidation curve for an unsaturated sandy clay.

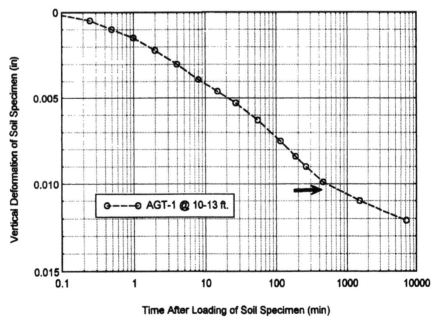

FIGURE 8.20   Log-of-time method, with data recorded for an unsaturated sandy clay subjected to a vertical pressure σ_v of 8000 psf. The arrow indicates the estimated end of primary consolidation.

(i.e., Type 1 curve). This could again be due to the lack of saturation of the sandy clay. Even assuming that only air is initially expelled from the sandy clay, it must achieve a void ratio of 0.675 before it becomes saturated (i.e., $e = G_s w/S = 2.7 \times 0.25/1.0 = 0.675$). Thus at a vertical stress of 8000 psf (380 kPa), the deformation is probably due to the expulsion of both air and water from the void spaces in the soil.

In summary, the laboratory results for the unsaturated sandy clay indicated that the consolidation curve did not have a distinct maximum past pressure σ'_{vm}, and the recompression index C_r and compression index C_c were not linear on a semi-log plot (Fig. 8.19). Also the time versus deformation plot did not have the characteristic shape when using the log-of-time method (Fig. 8.20). These laboratory data make the in situ settlement of the unsaturated sandy clay layer more difficult to predict.

Example Problem 8.15   Use the example problem shown in Fig. 8.10 and assume the clay layer has the consolidation curve shown in Fig. 8.19. Also assume that the groundwater table is located below the clay layer. Calculate the settlement of the 2-m thick unsaturated clay for the fill surcharge of 50 kPa applied over a large area at ground surface. Ignore secondary compression.

Solution   Use the following values from the prior examples:

$$H_o = 2 \text{ m}$$

Total unit weight (γ_t) of the sand = 18.7 kN/m³

Total unit weight γ_t of the clay = $\gamma_b + \gamma_w$ = 7.9 kN/m³ + 9.81 kN/m³ = 17.7 kN/m³

The first step is to determine the vertical stress σ_v at the center of the clay layer (i.e., 11 m below ground surface), or:

$$\sigma_v = (10 \text{ m})(18.7 \text{ kN/m}^3) + (1 \text{ m})(17.7 \text{ kN/m}^3) = 205 \text{ kPa (4300 psf)}$$

The increase in vertical total stress due to the fill surcharge (one-dimensional condition) is equal to:

$$\Delta\sigma_v = 50 \text{ kPa}$$

Final vertical total stress = $\sigma_v + \Delta\sigma_v$ = 205 + 50 = 255 kPa (5300 psf)

Entering Fig. 8.19 at the initial σ_v = 4300 psf, e_i = 0.712

Entering Fig. 8.19 at the final vertical total stress = 5300 psf:

$$e_f = 0.697$$

$$\Delta e = e_i - e_f = 0.712 - 0.697 = 0.015$$

Using Eq. 7.13 and the same e_o as the prior example problems (for comparison purposes), or $e_o = 1.10$:

$$S = \Delta e H_o/(1 + e_o) = (0.015)(2 \text{ m})/(1 + 1.10) = 0.014 \text{ m} = 1.4 \text{ cm (0.56 in.)}$$

Comparing this settlement to the four example problems in Sec. 8.4.2:

Stress history	Calculated primary consolidation settlement (s_c)
Underconsolidated (OCR < 1)	16 cm (6.4 in.)
Normally consolidated (OCR = 1)	10 cm (3.9 in.)
Overconsolidated (OCR = 1.17), Case II	4.9 cm (1.9 in.)
Overconsolidated (OCR = 1.67), Case I	0.6 cm (0.23 in.)
Unsaturated clay (Fig. 8.19)	1.4 cm (0.56 in.)

The settlement of the unsaturated clay is closest in value to the Case I overconsolidated clay. This often occurs because a partially saturated clay becomes overconsolidated due to drying or capillary effects.

NOTATION

The following notation is used in this chapter:

a_v = coefficient of compressibility
A = length to the outside edge of the pile group (Fig. 8.17)
B = footing width
B = width of a circular loaded area (Fig. 8.16)
B = distance to the outside edge of the pile group (Fig. 8.17)
c_v = coefficient of consolidation
C_c = compression index
C_r = recompression index
$C_{c\varepsilon}$ = modified compression index
$C_{r\varepsilon}$ = modified recompression index
C_α = secondary compression ratio
d = distance as defined in Fig. 8.14
d_o = deformation at the start of primary consolidation
d_{100} = deformation at the end of primary consolidation
D = diameter of the specimen loaded area (Sec. 8.3.1)
e = void ratio
e_i, e_f = initial and final void ratio (consolidation curve)
e_o = initial void ratio of the in situ soil layer
Δe = change in void ratio
E, E_u = modulus of elasticity of the soil
G_s = specific gravity of soil solids
h_o = height of the soil specimen
Δh = change in height of the soil specimen
H = thickness of the soil specimen (Sec. 8.3.1)
H = one-half the clay layer thickness for double drainage (Fig. 8.11)
H = thickness of the clay layer (Figs. 8.16 and 8.17)
H_{dr} = height of the drainage path
H_o = initial thickness of the in situ soil layer
i = hydraulic gradient
I = shape and rigidity factor
k = coefficient of permeability, also known as the hydraulic conductivity
k = parameter defined in Fig. 4.16
L, L_2, L_3 = various lengths defined in Fig. 8.17
LL = liquid limit
m, n = defined in Fig. 4.8

n = number of piles in a pile group (Fig. 8.17)

OCR = overconsolidation ratio

P = foundation load

PI = plasticity index

q = vertical footing or pile cap pressure

Q_D = downdrag load on a pile or pile group

s_c = primary consolidation settlement

s_i = immediate settlement, also known as initial settlement

s_s = settlement due to secondary compression

s_t = settlement at any given time

S = settlement of unsaturated clay

t = time

t_{50} = time needed to achieve 50 percent primary consolidation

t_{90} = time needed to achieve 90 percent primary consolidation

T = time factor

u = pore water pressure

u_e = excess pore water pressure

u_o = initial excess pore water pressure

U_{avg} = average degree of consolidation settlement

U_z = consolidation ratio

v = superficial velocity, i.e., velocity of flow of water into or out of the soil

V_v = volume of voids

V_s = volume of soil solids

w = water content

z = vertical depth below the top of the clay layer (Eq. 8.18)

z = depth below the foundation

α = defined in Eq. 8.23

Δ = maximum differential settlement of the foundation

ε_v = vertical strain

$\Delta\varepsilon$ = change in strain

γ_d = dry unit weight of the soil

γ_t = total unit weight of the soil

γ_w = unit weight of water

μ = Poisson's ratio

ρ_{max} = total settlement, also known as maximum settlement of the foundation

$\sigma'_v, \sigma'_{vc}, \sigma'_{vo}$ = vertical effective stress

$\sigma'_{vc1}, \sigma'_{vc2}$ = vertical effective stress used to calculate C_c and C_r

σ_v = vertical total stress

σ'_{vm}, σ'_p = maximum past pressure, also known as the preconsolidation pressure

$\Delta\sigma_1$ = change in principal total stress (i.e., vertical total stress, triaxial test)

$\Delta\sigma_v$ = change in vertical stress, such as due to a building load

$\Delta\sigma'_v$ = for an underconsolidated soil, change in vertical effective stress

$\Delta\sigma'_v$ = change in effective stress (Terzaghi consolidation theory)

PROBLEMS

Solutions to the problems are presented in App. C of this book. The problems have been divided into basic categories as indicated below:

Immediate Settlement

8.1   A building site consists of a thick and saturated, heavily overconsolidated, cohesive soil layer. Assume that when the cohesive soil is subjected to the building load, it will still be in an overconsolidated state. Based on triaxial compression tests, the average undrained modulus E_u of the cohesive soil = 20,000 kPa. If the applied surface stress $q = 30$ kPa and if the square building has a width $B = 20$ m, calculate the immediate settlement s_i beneath the center and corner of the building. For the analyses, assume a flexible loaded area on an elastic half-space of infinite depth.

ANSWER: $s_i = 2.5$ cm (center) and $s_i = 1.3$ cm (corner).

8.2   A site consists of a thick, unsaturated cohesive fill layer. Assume that an oil tank, having a diameter of 10 m, will be constructed on top of the fill. Based on triaxial compression tests performed on the unsaturated cohesive fill, the average undrained modulus E_u of the cohesive fill = 40,000 kPa. When the tank is filled, the applied surface stress $q = 50$ kPa. Calculate the immediate settlement s_i beneath the center of the oil tank assuming Poisson's ratio = 0.4. For the analysis, assume a flexible loaded area on an elastic half-space of infinite depth.

ANSWER: $s_i = 1.0$ cm.

8.3   An oil tank is constructed at a site that contains a thick deposit of soft saturated clay. Figure 8.21 shows the time versus projected settlement of the oil tank, determined by the design engineer. The actual time versus settlement behavior of the tank (once it is filled with oil) is also plotted in Fig. 8.21. There is a substantial difference between the predicted and actual settlement behavior of the oil tank. What is the most likely cause of the substantial difference between the predicted and measured settlement behavior of the oil tank?

ANSWER: Immediate settlement s_i due to undrained creep of the soft saturated clay was not included in the original settlement analysis by the design engineer.

Primary Consolidation Settlement

8.4   Figure 8.22 shows the consolidation curves for two tests performed on saturated cohesive soil specimens. Using the Casagrande construction technique, determine the maximum past pressure σ'_{vm} for both tests.

ANSWER: $\sigma'_{vm} = 20$ kPa (solid line) and $\sigma'_{vm} = 30$ kPa (dashed line).

8.5   For the two consolidation curves shown in Fig. 8.22, calculate the compression index (C_c).

ANSWER: $C_c = 2.8$ (solid line), $C_c = 1.8$ (dashed line).

8.6   For the Orinoco clay data shown in Fig. 8.3, calculate the modified compression index $C_{c\varepsilon}$ for the undisturbed and disturbed soil specimens.

ANSWER: $C_{c\varepsilon} = 0.36$ for the undisturbed Orinoco clay specimen and $C_{c\varepsilon} = 0.24$ for the disturbed Orinoco clay specimen.

8.7   Using the one-dimensional compression curve shown in Fig. 8.7, calculate the modified recompression index $C_{r\varepsilon}$ and the modified compression index $C_{c\varepsilon}$.

ANSWER: $C_{r\varepsilon} = 0.04$, $C_{c\varepsilon} = 0.45$.

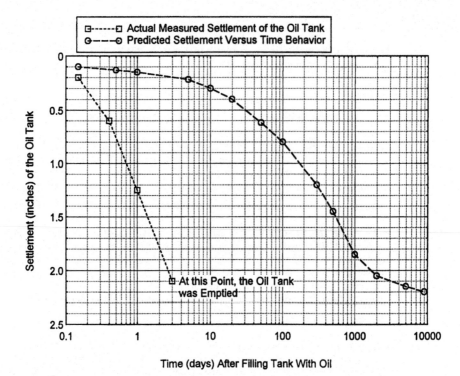

FIGURE 8.21   Settlement data for Problem 8.3.

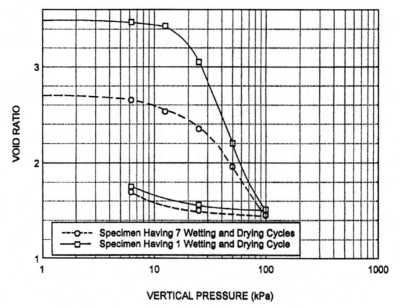

FIGURE 8.22   Consolidation curves for Problem 8.4.

8.8   Use the data from Example Problem 8.3., i.e., the clay layer that is underconsolidated. Determine the primary consolidation settlement s_c of the 2-m thick clay layer if the stress exerted by the uniform fill surcharge is doubled (i.e., $\Delta\sigma_v = 100$ kPa).

ANSWER: $s_c = 0.24$ m, and note that a doubling of the surcharge load does not double the primary consolidation settlement.

8.9   Use the data from Example Problem 8.4., i.e., the clay layer that is normally consolidated. Determine the primary consolidation settlement s_c of the 2-m thick clay layer if the stress exerted by the uniform fill surcharge is doubled (i.e., $\Delta\sigma_v = 100$ kPa).

ANSWER: $s_c = 0.18$ m, and note that a doubling of the surcharge load does not double the primary consolidation settlement.

8.10   Use the data from Example Problem 8.6., i.e., the clay layer that is overconsolidated (case I). Determine the primary consolidation settlement s_c of the 2-m thick clay layer if the stress exerted by the uniform fill surcharge is doubled (i.e., $\Delta\sigma_v = 100$ kPa).

ANSWER: $s_c = 1.1$ cm, and note that a doubling of the surcharge load does not double the primary consolidation settlement.

8.11   Use the data from Example Problem 8.3., i.e., the clay layer that is normally consolidated. Assume that the clay layer completely consolidates under the uniform fill surcharge and then the groundwater table permanently rises such that it is now located at a depth of 2 m below original ground surface. After the clay layer has equilibrated under the change in effective stress, determine the OCR at the center of the clay layer.

ANSWER: OCR = 1.15.

8.12   Use the data from Example Problem 8.3., i.e., the clay layer that is normally consolidated. Assume that the clay layer completely consolidates under the uniform fill surcharge and then the entire fill surcharge is removed. After the clay layer has equilibrated under the change in effective stress, determine the OCR at the center of the clay layer.

ANSWER: OCR = 1.33.

8.13   Use the data from Example Problem 8.12., i.e., the settlement of the 20 m by 30 m building based on the 2:1 approximation. Assume that enough time has elapsed so that primary consolidation due to the building weight is complete. At this time, an adjacent property owner initiates pumping of the groundwater table and the groundwater table is permanently lowered by 4 m (i.e., the level of the groundwater table is now 9 m below ground surface). Determine the primary consolidation settlement (s_c) due to the permanent lowering of the groundwater table.

ANSWER: $s_c = 6.4$ cm.

8.14   Use the data from Example Problems 8.12., i.e., the settlement of the 20 m by 30 m building based on the 2:1 approximation. Assume that enough time has elapsed so that primary consolidation due to the building weight is complete. At this time, an adjacent property owner initiates pumping of the sand layer below the clay layer and the pore water pressure is permanently lowered by an amount of 40 kPa. Assume that at the top of the clay layer, the pore water pressures are still hydrostatic. Using two layers (each 1 m thick), determine the primary consolidation settlement s_c due to the permanent reduction in pore water pressure of the sand layer located below the clay layer.

ANSWER: 3.7 cm.

8.15   Assume the subsoil conditions at a site are as shown in Fig. 2.39. Suppose a large building, 400 ft long and 200 ft wide, will be constructed at ground surface (elevation +21 ft). The weight of the building can be represented as a uniform pressure = 1000 psf applied at ground surface. Consider just the clay from elevation −50 to −90 ft and for the analysis, divide this soil into four 10-ft thick layers. Determine the primary consolidation settlement s_c of the proposed building based on $C_c = 0.35$ and $e_o = 0.9$.

ANSWER: The clay from elevation −50 to −90 ft is essentially normally consolidated and using the 2:1 approximation, $s_c = 3.9$ in.

8.16   Assume the subsoil conditions at a site are as shown in Fig. 2.39 and assume that a pile group (37.5 ft by 37.5 ft) supports a load of 2900 kips and this load is resisted by pile side friction and end bearing of the soil from elevation −10 to −50 ft. By dividing the clay from elevation −50 to −90 ft into four 10-ft thick layers, calculate the primary consolidation settlement s_c due to the load of 2900 kips. Use $C_c = 0.35$ and $e_o = 0.9$. Determine the settlement based on the 2:1 approximation and assume it starts at an elevation of −37 ft (i.e., $^2/_3 D$, where D = depth of that portion of the pile group supported by soil adhesion).

ANSWER: The clay from elevation −50 to −90 ft is essentially normally consolidated and thus $s_c = 4.4$ in.

Rate of One-Dimensional Consolidation

8.17   A consolidation test is performed on a highly plastic, saturated cohesive soil specimen by using the oedometer apparatus (drainage provided on the top and bottom of the soil specimen). At a vertical stress of 50 kPa, the vertical deformation of the soil specimen as a function of time after the start of loading was recorded and the data have been plotted on a semi-log plot as shown in Fig. 8.23. Assume that at zero vertical deformation in Fig. 8.23, that the specimen height = 10.05 mm. Calculate the coefficient of consolidation c_v.

ANSWER: $c_v = 4.4 \times 10^{-6}$ cm^2/sec.

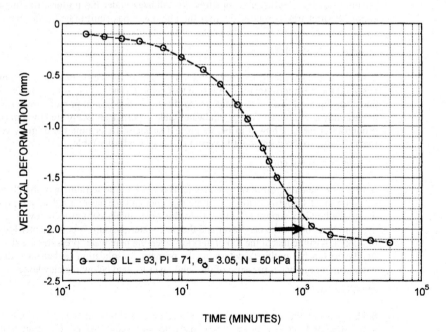

FIGURE 8.23   Data for Problem 8.17.

8.18 Use the data from Example Problem 8.7., i.e., the 2-m-thick normally consolidated clay layer that has double drainage. At a time of 180 days after the placement of the fill surcharge, it is determined that the clay layer has experienced 5 cm of primary consolidation. Assuming all of this 5 cm of settlement is due to one-dimensional primary consolidation, determine the field coefficient of consolidation c_v.

ANSWER: $c_v = 0.4$ m²/year.

8.19 Use the data from Example Problem 8.7., i.e., the 2-m-thick normally consolidated clay layer that has double drainage. At a time of 180 days after the placement of the fill surcharge, a piezometer located in the middle of the clay layer records a pressure head h_p of 9.92 m. Based on this pressure head h_p measurement, determine the field coefficient of consolidation c_v.

ANSWER: $c_v = 0.4$ m²/year.

Secondary Compression

8.20 Using the data from Problem 8.17, calculate the secondary compression ratio C_α.

ANSWER: $C_\alpha = 0.007$.

8.21 Use the data from Example Problem 8.12., i.e., the settlement of the 20 m by 30 m building based on the 2:1 approximation. Assume the design life of the structure is 50 years, $C_\alpha = 0.01$, and $c_v = 0.32$ m²/year. Calculate the secondary compression settlement s_s over the design life of the structure, using a time factor T equal to 1.0 to determine the time corresponding to the end of primary consolidation. Also calculate the total settlement ρ_{max} assuming the initial settlement s_i is zero.

ANSWER: $s_s = 2.4$ cm, $\rho_{max} = 7.5$ cm.

CHAPTER 9
FOUNDATIONS ON EXPANSIVE SOIL

9.1 INTRODUCTION

Expansive soils are a worldwide problem, causing extensive damage to civil engineering structures. Jones and Holtz estimated in 1973 that the annual cost of damage in the United States due to expansive soil movement was $2.3 billion (Jones and Holtz, 1973). A more up-to-date figure is about $9 billion in damages annually to buildings, roads, airports, pipelines, and other facilities (Jones and Jones, 1987).

Chapter 8 discussed the consolidation of clay, which is basically the compression of soft clays that have a high water content. Expansive clays are different in that the near-surface clay often varies in density and moisture condition from the wet season to the dry season. For example, near- or at-surface clays often dry out during periods of drought but then expand during the rainy season or when they get water from irrigation or leaky pipes.

There are many factors that govern the expansion behavior of soil. The primary factors are a change in water content and the amount and type of clay size particles in the soil. Other important factors affecting the expansion behavior include the type of soil (natural or fill), condition of the soil in terms of dry density and moisture content, magnitude of the surcharge pressure, and amount of nonexpansive material such as gravel or cobble size particles (Ladd and Lambe, 1961; Kassiff and Baker, 1971; Chen, 1988; Day, 1991b, 1992a). These main factors are individually discussed below:

Change in Water Content. An important factor for expansive soils is a change in water content. For example, expansive soils cause extensive damage to structures located in the desert southwest of the United States. Because of the lack of rain, the near-surface clays are often in a desiccated or dry powdery state. After a structure has been built on desiccated clay, water is often introduced through irrigation or leaky pipes, and the clay absorbs the water and expands causing extensive damage.

Other areas may have significant surface deposits of clays, but there is enough yearly precipitation so that the clays stay in a permanently wet condition. Since there is always enough rainfall to keep the clays in a wet state, they tend to remain relatively dormant and they neither swell nor shrink. There can be exceptions to this rule, such as an area that normally has a wet climate, but a severe drought occurs which causes the clays to become desiccated and shrunken, resulting in expansive soil movement.

Although most states have expansive soil, Chen (1988) reported that certain areas of the United States, such as Colorado, Texas, Wyoming, and California, are more susceptible to damage from expansive soils than others. These areas have both large surface deposits of clay and have climates characterized by alternating periods of rainfall and drought.

Amount of Clay Size Particles. The more clay size particles of a particular type a soil has, the more swell there will be (all other factors being the same). Clay size particles attract water to their particle faces due to the double layer effect. Water is also drawn into the soil due to the negative pore water pressures associated with dried clay. Thus, the more clay size particles in a dry soil, the greater the need for water to be drawn into the soil and hence the higher the swell potential of the soil.

Type of Clay Size Particles.   The type of clay size particles significantly affects swell potential. Given the same dry weight, kaolinite clay particles (activity between 0.3 and 0.5) are much less expansive than sodium montmorillonite clay particles (activity between 4 and 7) (Holtz and Kovacs, 1981). Montmorillonite is a much smaller and more active clay mineral than kaolinite, and this results in much more attracted water per unit dry mass of clay particles. Once again, the need for more water results in a greater amount of water drawn into the dry soil and hence higher swell potential of the soil.

Density and Water Content.   The dry density and water content are important factors in the amount of expansion of a soil. In general, expansion potential increases as the dry density increases and the water content decreases. The most expansive condition is when the soil has a high dry density and a low moisture condition. This often occurs when near-surface clays becomes desiccated, such as during a hot and dry summer season.

Clays that have a low dry density and high water content may not have additional swell, but they could still cause the structure to experience downward movement if they should dry out. Thus it is not unusual for near-surface clays to experience changes in dry density and water content throughout the season, depending on whether they have absorbed water and swelled-up during the rainy season, or shrunk and dried-out during the dry season.

Surcharge Pressure.   Laboratory and field studies have shown that the amount of swell will decrease as the confining pressure increases. The effect of surcharge is important because it is usually the lightly loaded structures such as concrete flatwork, pavements, slab-on-grade foundations, or concrete canal liners that are often impacted by expansive soil.

9.2   EXPANSION POTENTIAL

An important aspect of the laboratory testing of expansive soil is to classify them according to their degree of potential expansiveness. The most commonly used system is to classify soils as having either a very low, low, medium, high, or very high expansion potential. There are many different ways to classify expansive soils and some of the more commonly used methods are discussed in the following subsections.

9.2.1   Index Properties

The first method for the classification of expansive soils is to use their index properties. For example, Table 9.1 lists typical soil properties versus the expansion potential (Holtz and Gibbs, 1956b; United States Department of Housing and Urban Development, 1971; Holtz and Kovacs, 1981; Meehan and Karp, 1994; ASTM D 4829-03, 2004). Commonly used approaches to determining the expansion potential of a soil are based on the clay content and plasticity index, as follows:

Clay Content.   The clay content is defined as the percentage of soil particles that are finer than 0.002 mm, based on dry weight. In essence, the clay content is simply the percent clay in the soil. The percent clay in the soil is determined from a particle size analysis. For example, in Fig. 8.18 the clay content in the whole soil is 35.6 percent, and this soil would be classified as having a very high expansion potential per Table 9.1.

Plasticity Index.   As previously discussed, the plasticity index (PI) is defined as the liquid limit (LL) minus the plastic limit (PL). Per ASTM, the LL and PL are performed on soil that is finer than the No. 40 sieve. Thus when correlating the PI and expansion potential, the PI to be used in Table 9.1 should be the PI of the whole sample. The PI of the whole sample is equal the PI from the Atterberg limits times the fraction of soil passing the No. 40 sieve.

TABLE 9.1 Typical Soil Properties versus Expansion Potential

Expansion potential	Very low	Low	Medium or moderate	High	Very high or critical
Clay content (<2 μm)	0–10%	10–15%	15–25%	25–35%	35–100%
Plasticity index (see note)	0–10	10–15	15–25	25–35	>35
Expansion index test	0–20	21–50	51–90	91–130	>130
HUD criteria (% swell at $\sigma_v = 0$ psf)	—	0–10	10–20	20–30	>30
HUD criteria (% swell at $\sigma_v = 60$ psf)	—	0–4	4–8	8–12	>12
HUD criteria (% swell at $\sigma_v = 144$ psf)	—	0–2	2–6	6–10	>10
HUD criteria (% swell at $\sigma_v = 650$ psf)	—	0–0.5	0.5–1.5	1.5–3.5	>3.5

Note: The plasticity index is a more reliable indicator of expansion potential than clay content. For standard Atterberg limits test data (i.e., tests performed on minus no. 40 soil), the PI to be used for this chart is the PI of the whole sample, which is equal to the PI times the fraction of soil passing the no. 40 sieve.

For example, using the test data shown in Fig. 8.18, the PI for the soil passing the No. 40 sieve is equal to 39. Since there is 84 percent passing the No. 40 sieve, the PI for the whole sample is equal to 39 times 0.84, or 33. As indicated in Table 7.1, for a PI of the whole sample is 33, the soil is classified as having a high expansion potential.

For many soils, the PI is usually a more reliable indicator of expansion potential than clay content (percent clay size particles). This is because the type of clay mineral (i.e., kaolinite versus sodium montmorillonite) has such a large effect on expansion potential. For example, a soil may have a very high clay content consisting of kaolinite, but because kaolinite is usually a relatively inactive mineral, the expansion potential could nevertheless be low.

Those soils classified as clays of high plasticity (CH) will usually have a high to very high expansion potential. At the other extreme, those soils that are classified as nonplastic (PI = 0), such as clean sands and gravels, are nonexpansive (no expansion potential).

Expansive Soil Classification Chart. The expansion potential can also be estimated from expansive soil classification charts. For example, Seed et al. (1962) developed a classification chart based solely on the amount and type (activity) of clay size particles (see Fig. 9.1). When using this chart, the percent clay size refers to the clay fraction of the whole sample and the activity A is defined in Eq. 4.6. As previously mentioned in Sec. 4.2.2, the activity must be adjusted if the soil has particles that are coarser than the No. 40 sieve.

As an example of the use of Fig. 9.1, consider the test data shown in Fig. 8.18, where the percent clay size of the whole sample is equal to 35.6 percent. The activity is more difficult to calculate because only the soil passing the No. 40 sieve was used to determine the plasticity index. The activity is equal to the plasticity index (PI = 39) divided by the clay fraction of the soil passing the No. 40 sieve. The clay fraction of the soil passing the No. 40 sieve is equal to 42.4 percent (i.e., 35.6/0.84 = 42.4 percent). Thus the activity is equal to the plasticity index (PI = 39) divided by the adjusted clay fraction (42.4 percent), or 0.92. The chart in Fig. 9.1 is then entered with a percent clay size of the whole sample equal to 35.6 percent and an activity of the clay particles equal to 0.92. This plots just barely within the high expansive soil classification.

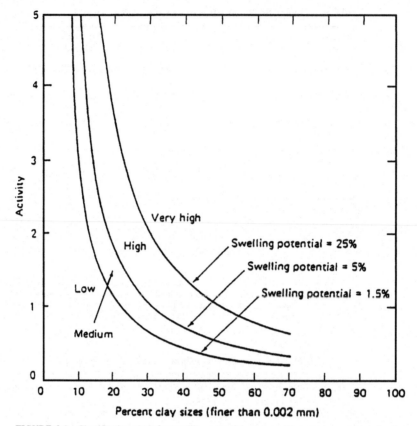

FIGURE 9.1 Classification chart for swelling potential. *Note:* soil specimens compacted using standard Proctor energy and tested with a normal stress of 6.9 kPa (1 psi). (*From Seed et al., 1962; reprinted with permission from the American Society of Civil Engineers.*)

Expansive Soil Classification Chart. Figure 9.2 presents another example of an expansive soil classification chart. For this chart, two of the following three items are required: plasticity index of the whole sample, percent clay of the whole sample, and activity of the clay size particles. Using the data shown in Fig. 8.18, the percent clay of the whole sample = 35.6 percent, the plasticity index of the whole sample is equal to 33 (that is, $39 \times 0.84 = 33$), and the activity of the clay particles is 0.92. Entering the chart with these values, the data plots just barely within the very high expansive soil classification.

In summary, for the data shown in Fig. 8.18, the results are relatively consistent and are summarized as follows:

1. *Method based on clay fraction.* Clay fraction of the whole sample is 35.6 percent, therefore it is very high expansion potential per Table 9.1.

2. *Method based on plasticity index.* The Atterberg limits were performed on minus No. 40 soil and the plasticity index is 39. The plasticity index of the whole sample is 33 (i.e., $39 \times 0.84 = 33$), and therefore the soil has a high expansion potential per Table 9.1.

3. *Method based on Fig. 9.1.* Using a clay fraction of the whole sample of 35.6 percent and the activity of the clay particles of 0.92, as per Fig. 9.1 the soil has a high expansion potential.

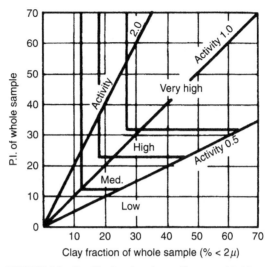

FIGURE 9.2   Classification chart for swelling potential. (*From Van der Merwe, 1964.*)

4. *Method based on Fig. 9.2.*   Using a clay fraction of the whole sample of 35.6 percent, a plasticity index of the whole sample of 33, and an activity of the clay particles of 0.92, as per Fig. 9.2 the soil has a very high expansion potential.

For this soil, two of the methods indicate a high expansion potential while the other two methods indicate very high expansion potential. It should be recognized that this soil classification is only for potential expansiveness. For example, the soil data shown in Fig. 8.18 indices a high to very high expansion potential, but actual consolidation test data (Fig. 8.19) indicates that this clay is relatively compressible in its in situ state. Therefore this classification of expansion potential should only be considered as an index test of its possible expansiveness as related to other soils.

9.2.2   Expansion Index Test

Another method for determining the expansion potential of a soil is to perform an expansion index test. The laboratory test procedures are stated in ASTM D 4829-03 (2004), "Standard Test Method for Expansion Index of Soils." The purpose of this laboratory test is to determine the expansion index, which is then used to classify the soil as having a very low, low, medium, high, or very high expansion potential (see Table 9.1). The expansion index test basically consists of compacting a soil specimen so that it has a degree of saturation of approximately 50 percent and then placing the soil specimen in an oedometer apparatus and allowing it to swell. The test procedures are as follows:

Soil Specimen.   A disturbed soil specimen or a bulk sample can be used for this test. The first step is to sieve the soil on the No. 4 sieve (4.75 mm). If the particles retained on the No. 4 sieve are possibly expansive, such as fragments of claystone, shale, or weathered volcanic rock, then these particles should be broken down so that they pass the No. 4 sieve. By using a balance, the mass of nonexpansive material retained on the No. 4 sieve is determined and then the particles retained on the No. 4 sieve are discarded. The soil passing the No. 4 sieve should be weighed and then thoroughly mixed up. A portion of this soil should be used for a water content test and the remainder of the soil should be sealed in an airtight container and allowed to equilibrate overnight. Ideally about 1 kg (2 lb) of wet soil will be needed for the expansion index test. By knowing the water content of the soil passing the No. 4 sieve, the fraction of soil passing the No. 4 sieve (based on dry weight) can be calculated.

Compaction of the Soil.   After the soil is allowed to equilibrate overnight, the soil specimen is compacted into a ring having an internal diameter of 4.0 in. (10.2 cm) and a height of 1.0 in. (2.54 cm). The compaction of the soil is performed in a special mold (see Fig. 9.3). To prevent loss of energy during compaction of the soil within the mold, the mold can be firmly attached to a rigid anchor block such as shown in Fig 9.4.

The soil is placed in two equal layers within the mold. Each layer of soil receives 15 uniformly distributed blows from a 5.5-lb (2.5-kg) tamper having a 12-in. (30.5-cm) drop. At the end of the compaction process, the depth of soil in the mold should be about 2 in. (5 cm).

The compaction energy used for the expansion index test is close to the compaction energy used for the standard Proctor compaction test (Sec. 3.6.2). Usually the optimum moisture content for compacted soil equates to a degree of saturation of about 80 percent, and thus, the final result of a compacted soil specimen at a degree of saturation of 50 percent means that the soil will have a relative compaction that is less than 100 percent of the standard Proctor maximum dry density.

Trimming of the Soil Specimen.   After the soil has been compacted in the mold, it is disassembled by removing the upper and lower portions of the mold in order to extract the inner ring that contains

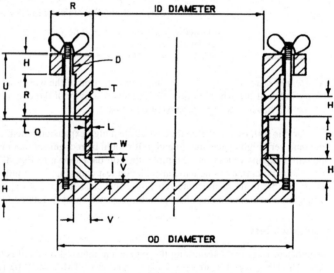

Letter	in.	mm
ID	4.01	101.9
OD	5½	139.7
H	½	12.7
D	7/32 Hole	5.6 Hole
U	15/8	41.3
T	3/8	9.5
O	1/8	3.2
R	1	25.4
W	7/16	11.1
V	9/16	14.3
L	0.120	3.05

FIGURE 9.3   Mold, containing an inner ring, that is used for the compaction of the specimen for the expansion index test. (*Reproduced with permission of ASTM.*)

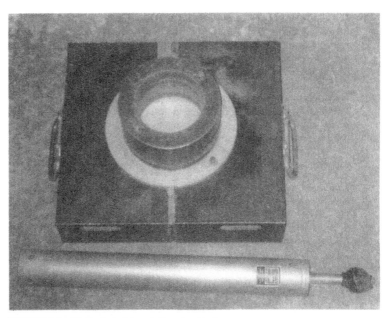

FIGURE 9.4 Mold attached to a rigid anchor block.

the soil specimen. The soil specimen is then carefully trimmed by using a metal straight edge so that both the top and bottom of the soil specimen are flush with the confining ring. If any surface voids develop during the trimming process, they should be filled in with soil. Figure 9.5 shows the soil specimen after the top and bottom have been trimmed. The initial height of the soil specimen h_o is then equal to the height of the confining ring (1.0 in., 2.54 cm).

FIGURE 9.5 Trimmed soil specimen.

Calculation of Degree of Saturation.   The water content of the soil specimen was determined from Step 1. A balance is used to obtain the mass of the ring plus the wet soil specimen. The mass of the wet soil and dry soil within the confining ring can then be readily calculated. If the specific gravity of the soil has been determined, then this value is used or otherwise a value of 2.7 is typically assumed. The degree of saturation of the soil specimen is then calculated by using basic phase relationships.

If the calculated degree of saturation is between 49 and 51 percent, the specimen is ready for testing, or else the test specimen is discarded and the preparation procedure is repeated after adjustment of the initial water content. If the degree of saturation is below 49 percent, then distilled water will need to be added to the soil. If the degree of saturation is above 51 percent, the soil specimen will need to be air dried to reduce its water content. Having an initial degree of saturation between 49 and 51 percent will provide the most accurate test results, but ASTM D 4829-03 (2004) does allow the test to be performed for a degree of saturation between 40 and 60 percent, provided a correction is applied to the final result.

Loading Device.   The laterally confined soil specimen that has a degree of saturation between 49 and 51 percent is ready for testing. Dry and clean porous plates are placed on the top and bottom of the soil specimen and it is then placed in a surrounding container (such as a Plexiglas dish). The Plexiglas dish containing the soil specimen is placed at the center of the oedometer or equivalent loading apparatus, such as shown in Fig. 9.6. A dial gauge is set up in order to measure the vertical

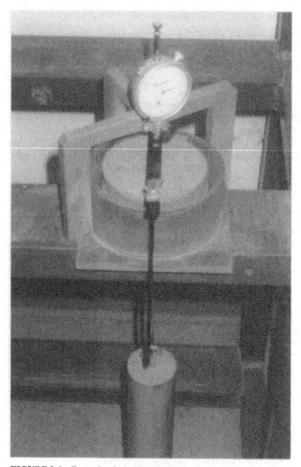

FIGURE 9.6   Example of a loading device for the expansion index test.

swell of the soil specimen. A load equivalent to a vertical pressure of 144 psf (6.9 kPa) is then applied to the soil specimen and a dial gauge reading is taken.

Because the soil specimen has been compacted, there should be negligible vertical deformation when the vertical stress of 144 psf (a very light load) is applied to the soil specimen. If deformation does occur under this load, it may indicate that the specimen was not properly compacted and there are loose zones of soil, in which case the specimen preparation procedure should be repeated. Within 10 min of applying the load, the soil specimen is submerged in distilled water. After inundation, time versus dial gauge readings can be recorded.

End of Test. After the soil specimen has swelled, it is removed from the apparatus and the final water content of the soil is determined. The final saturation of the soil specimen can also be calculated using basic phase relationships.

The final step in the expansion index test is to calculate the expansion index. In the calculation of the expansion index, ASTM recommends that the soil specimen should be allowed to swell for 24 h or until the rate of expansion becomes less than 0.005 mm/h, whichever occurs first (but never less than 3 h). But as shown in Fig. 9.7, there can be significant swell beyond the 24-h time period. Another approach is to calculate the expansion index (EI) based on the end of primary swell, defined as (Day, 1993a):

$$EI = \frac{1000(h_p - h_o)}{h_o} = (10)(\% \text{ primary swell}) \tag{9.1}$$

where EI = expansion index (dimensionless)
  h_p = height of the soil specimen at the end of primary swell (in. or mm)
  h_o = initial height of the soil specimen (1.0 in., 25.4 mm)

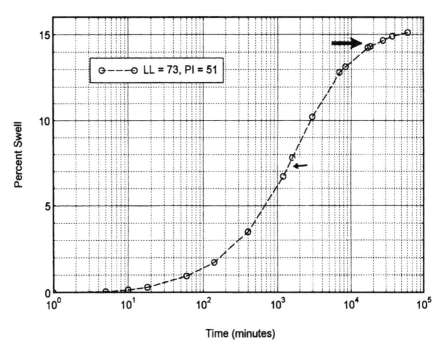

FIGURE 9.7   Swell versus time for an expansion index test. The small arrow indicates the time corresponding to 24 h after inundation (7.4 percent swell). Note that the soil specimen continues to experience significant swell after 24 h of inundation. The large arrow points to the end of primary swell (14.5 percent).

Based on the expansion index, the expansion potential of the soil is determined as indicated in Table 9.1. In order to obtain the specimen height at the end of primary swell h_p, dial readings can be converted to percent swell and plotted versus time on a semi-log plot. Figure 9.8 presents an expansion index test performed on clay of low plasticity (CL). The shape of the swelling versus time curve (Fig. 9.8) is similar to the consolidation of a saturated clay (Fig. 8.5), except that swelling has been plotted as positive values and consolidation as negative values. Based on this similarity, the end of primary swell can be determined as the intersection of the straight line portions of the swell curve. The arrow in Fig. 9.8 indicates a primary swell of 6.2 percent (or EI = 62), indicating a medium expansion potential (Table 9.1).

During the expansion index test, if there is no change or a decrease in height of the soil specimen, then the expansion index is zero (i.e., EI = 0 indicating a nonexpansive soil). Even for soil classified as having a very low expansion potential, the swell versus time plot will have the characteristic shape enabling the end of primary swell to be calculated (see Fig. 9.9).

In some cases, the expansion index test can yield misleading results for soils classified as clayey gravels (GC). The reason is because clayey gravels will have a significant fraction retained on the No. 4 sieve, but the expansion index test only uses those particles passing the No. 4 sieve. There is no correction in the test procedures to account for nonexpansive (i.e., hard rock) gravel particles retained on the No. 4 sieve. One approach used in practice is to reduce the expansion index based on the percentage of particles by dry mass that pass the No. 4 sieve. This correction is as follows (Day, 1993a):

$$\text{EI (corrected)} = \frac{(\text{EI})(\% \text{ passing No. 4 sieve})}{100} \tag{9.2}$$

Suppose a clayey gravel has 40 percent by dry mass passing the No. 4 sieve and for the particles passing the No. 4 sieve, the EI is 100. Then according to Eq. 9.2, the corrected EI would be 100 times 0.4, or 40 (low expansion potential).

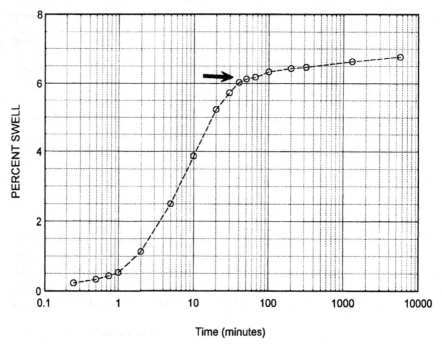

FIGURE 9.8 Percent swell versus time for an expansion index test of a clay of low plasticity. *Note:* The arrow indicates the end of primary swell.

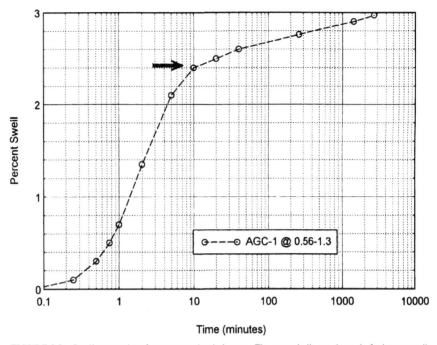

FIGURE 9.9   Swell versus time for an expansion index test. The arrow indicates the end of primary swell, with test results indicating a low expansion potential.

Figure 9.10 presents the results of an expansion index test performed on soil that has the soil classification shown in Fig. 8.18. As previously mentioned in Sec. 9.2.1, this soil has a high to very high expansion potential based on the results of index tests. The test data shown in Fig. 9.10 indicates an end of primary swell of 13.8 percent, and using Eq. 9.1, the expansion index is equal to 138. As per Table 9.1, an expansion index of 138 indicates a very high expansion potential. Thus the final result from the expansion index test is consistent with the results from the various index classifications described in Sec. 9.2.1.

Figure 9.11 presents a plot of the plasticity index versus the expansion index for different tested soils. As expected, there is a correlation between PI and EI, where as the PI increases, the EI also increases. The line shown in Fig. 9.11 represents the best fit line through the data points. The scatter of data in Fig. 9.11 is partly due to the fact that the plasticity index and the expansion index are not performed on the same fraction of soil. In particular, the expansion index is performed on soil that passes the No. 4 sieve (4.75 mm), while the Atterberg limits and hence plasticity index are determined on soil that passes the No. 40 sieve (0.425 mm). Nevertheless, the best fit line in Fig. 9.11 is close to the correlation between plasticity index and expansion index shown in Table 9.1.

The expansion index test is often used to separate those soils that will require a special design due to the swell and shrinkage of the soil versus soils that will not cause any expansion soil problems. For example, Sec. 1803.5.3 of the *International Building Code* (2009) states that expansive soils are those soils that have an expansion index greater than 20. Special soil treatment options or special foundations would be required if the soil has an expansion index greater than 20.

9.2.3   HUD Swell Test

Another laboratory test that is commonly used to determine the expansion potential is the test developed by the U.S. Department of Housing and Urban Development (1971) and it is often referred to

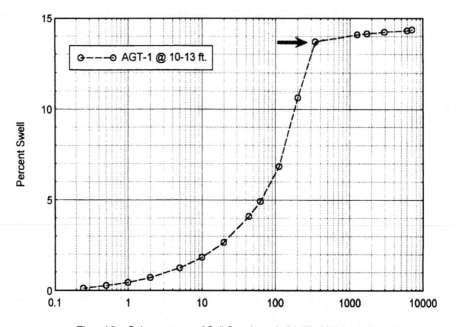

FIGURE 9.10 Swell versus time for an expansion index test. Arrow indicates the end of primary swell.

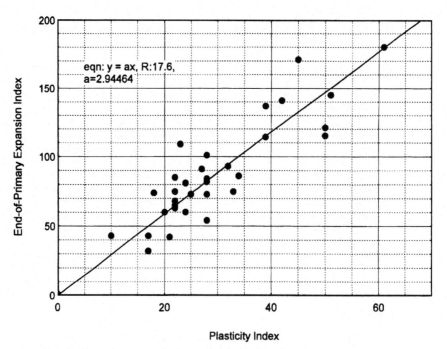

FIGURE 9.11 Expansion index versus plasticity index for different soils.

as the HUD swell test. This test is similar to the expansion index test in that the soil specimen is laterally confined by a metal ring, a loading device (such as shown in Fig. 9.6) is used to apply a vertical pressure to the soil specimen, and then the soil specimen is inundated with distilled water and a dial gauge is used to record the amount of vertical swell of the soil. The original HUD specifications are as follows (U.S. Department of Housing and Urban Development, 1971):

> One inch thick specimen—swell measured from air dried, shrinkage limit to saturated condition on a remolded sample compacted to field placement density and moisture.

Although the HUD swell test can be performed at a vertical pressure of either 0, 60, 144, or 650 psf, the most commonly used surcharge pressures are 60 and 144 psf. The bottom four rows of Table 9.1 list values of percent swell versus expansion potential. Note in Table 9.1 the importance of surcharge pressure on percent swell. At a surcharge pressure of 650 psf (31 kPa) the percent swell is much less than at a surcharge pressure of 0 psf. For example, for highly expansive soil, the percent swell for a surcharge pressure of 0 psf is 20 to 30 percent, while at a surcharge pressure of 650 psf, the swell is only 1.5 to 3.5 percent.

As indicated in Table 9.1, the original HUD specifications did not have a very low expansion potential category. In addition, the HUD specifications used the terminology of moderate instead of medium expansion potential, and critical instead of very high expansion potential. Over the years, the original HUD specifications have been modified and adapted to local conditions. The commonly used test procedures are as follows:

Soil Specimen.   As mentioned earlier, originally the specifications required that a soil specimen be utilized that was compacted to the same dry density and water content conditions that existed in the field. The soil specimen was typically compacted into a confining ring having a height of 1.0 in. (2.54 cm) and a diameter of 2.5 in. (6.35 cm). As an alternative to the earlier requirement, in many cases an undisturbed soil specimen will be utilized. It is often easier and simpler to trim an undisturbed soil specimen into a confining ring, rather than try to duplicate the field dry density and water content conditions.

Typically, the trimmings can be collected in order to determine the water content of the soil. In addition, a balance can be used to obtain the mass of the confining ring and soil specimen. By subtracting the mass of the confining ring, the mass of the soil specimen can be calculated. Knowing the volume of the confining ring, the total unit weight can be calculated. Using the water content, the dry unit weight can also be calculated from basic phase relationships.

Air-Drying.   The next step is to air-dry the soil specimen. As mentioned earlier, the original HUD specifications required that the specimen be air-dried to the shrinkage limit of the soil. But the shrinkage limit test is not routinely performed on soil and in addition, it is often very difficult to air-dry the soil to exactly the shrinkage limit.

The more commonly used test procedures require the soil specimen to be air-dried for a certain period of time, such as a minimum of 24 or 48 h. The objective is to air-dry the soil specimen so that it has a water content that is equal to or less than the shrinkage limit. The HUD specifications provided typical values of the shrinkage limit (SL) for various categories, as follows:

1. Low expansion potential: SL = 15 percent
2. Moderate expansion potential: SL = 10 to 15 percent
3. High expansion potential: SL = 7 to 12 percent
4. Critical expansion potential: SL = 11 percent

Thus as a guide for clayey soils, the water content should be less than 7 percent to ensure that it has a water content that is less than or equal to the shrinkage limit. Note that if the water content decreases below the shrinkage limit, the soil specimen will not experience any additional shrinkage. Thus once below the shrinkage limit, any additional air-drying will not change the size of the soil specimen.

After the soil specimen has been air-dried to a water content equal to or less than the estimated shrinkage limit, the water content of the soil should be calculated. Since the original water content of the soil specimen is known (Step 1), the dry mass of the soil specimen can be calculated. By weighting the soil specimen and ring after air-drying, the wet mass and hence water content at the end of the air-drying period can be calculated. If the soil specimen has a water content that is deemed to be above the estimated shrinkage limit, then the soil specimen should be subjected to additional air-drying.

Loading Device.   Once the soil specimen has completed its air-drying, it is ready for testing. Figure 9.12 shows a specimen of soil ready for testing. Soil desiccation cracks will often develop as the soil specimen dries, such as shown in Fig. 9.12, especially if it initially has a high water content.

Porous plates are placed on the top and bottom of the soil specimen. In some cases, during the shrinkage of the soil, the soil may stick to the inside of the confining ring or shrink unevenly. It is important to press the top porous plate down upon the soil specimen to ensure a solid contact with the desiccated soil. Soil particles may need to be moved aside in order to get a good seating contact between the porous plates and the soil specimen. Care should be taken during this process to prevent the loss of soil.

After dry and clean porous plates have been properly seated on the top and bottom of the soil specimen, it is placed in a surrounding container (such as a Plexiglas dish). The Plexiglas dish containing the soil specimen is placed at the center of the oedometer or equivalent loading apparatus such as shown in Fig. 9.6. A dial gauge is set up to measure the vertical swell of the soil specimen. As previously mentioned, the vertical pressure during testing can be 0, 60, 144, or 650 psf, although a load equivalent to a vertical pressure of 60 psf (2.9 kPa) or 144 psf (6.9 kPa) is most commonly utilized. After the surcharge pressure has been applied to the soil specimen, a dial gauge reading is taken. Because the soil specimen has been air-dried, there should be negligible vertical deformation when the vertical stress of 60 or 144 psf (very light loads) is applied to the soil specimen. If deformation does occur under this load, it may indicate that the porous plates were not properly seated on the soil specimen, in which case the apparatus should be disassembled and porous plates reseated on the specimen. Within 10 min of applying the load, the soil

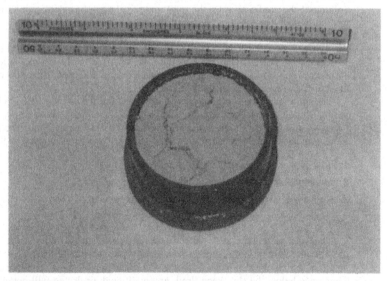

FIGURE 9.12   Soil specimen that has been air dried to a water content below the estimated shrinkage limit and is ready for testing.

specimen is submerged in distilled water. After inundation, time versus dial gauge readings can be recorded.

End of Test.   After the soil specimen has swelled, it is removed from the apparatus and the final water content of the soil is then determined. The final saturation of the soil specimen can also be calculated using basic phase relationships.

The final step in the HUD swell test is to calculate the percent swell. Percent swell is defined as the change in height of the soil specimen divided by the initial height of the soil specimen (i.e., initial height = 1.0 in.). Similar to the expansion index test, the percent swell can be plotted versus time on a semilog plot. As a consistent approach, the percent swell for the HUD test could be that swell corresponding to the end of primary swell (i.e., the arrow in Fig. 9.13). This value of percent swell is then compared with the values in Table 9.1 to determine the expansion potential of the soil. For example, for the test data shown in Fig. 9.13, the percent swell at the end of primary swell is equal to 6.6 percent, and using Table 9.1 (vertical surcharge pressure = 144 psf), the soil has a high expansive potential.

During the HUD test, if there is no change or a decrease in height of the soil specimen upon inundation with distilled water, then the percent swell is zero (i.e., swell = 0, indicating a nonexpansive soil).

9.2.4 Standard 60 psf Swell Test

In southern Nevada, a variation of the HUD swell test is used. The test was initially known as the City of North Las Vegas Expansion Test, but the name was changed when other southern Nevada building departments adopted this test. The test procedures are as follows (Southern Nevada Amendments to the *2000 International Building Code*, 2003.):

> When the standard 60 pounds per square foot swell test is performed on any soil with a swell greater than 4 percent, it shall be considered expansive. When soils are determined to be expansive, special design consideration is required. In the event that expansive soil properties vary with depth, the variation shall be included in the engineering analysis of the expansive soil effect on the structure. The foundation design and special inspection for grading/foundations shall be based upon the results obtained from the standard 60 pounds per square foot swell test.

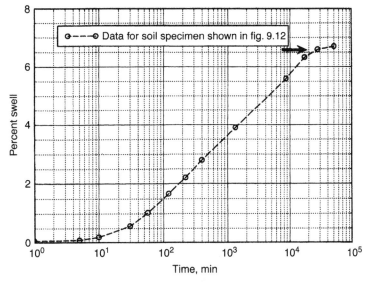

FIGURE 9.13   Swell versus time for the soil specimen shown in Fig. 9.12. The vertical pressure during testing was 144 psf. The arrow indicates the end of primary swell.

Section 1802.2.3, Standard 60 Pounds Per Square Foot Swell Test: The swell test samples may be remolded to the in-place density required for the particular soil type as called for in the geotechnical report, or it may be an in situ undisturbed sample. The test samples shall be one inch thick and laterally confined by placing them in a retaining ring constructed in accordance with ASTM D 2435. The swell test sample shall be oven dried at 60°C, and the sample shall be dried a minimum of eight hours. The test sample shall be inundated with water and kept in a saturated moisture condition until measurable swelling or vertical movement ceases. The swell test shall use a 60 pounds per square foot surcharge load. The balance of the swell test will be per ASTM D 2435. Swell test results shall be interpreted using Table 1805.8 or as permitted in Sec. 1805.8.

Table 1805.8 has been reproduced as Table 9.2. As described earlier, this swell test is nearly identical to the HUD Swell Test described in Sec. 9.2.3. Differences include drying the soil in an oven at a temperature of 60°C and requiring the use of a 60 psf (2.9 kPa) surcharge pressure during testing. Comparing Tables 9.1 (HUD for 60 psf) and Table 9.2, the percent swell range versus expansion potential (low, moderate, high, and critical) is identical.

9.2.5 Summary

In summary, all of the various methods described in this section only evaluate potential expansion. These methods include correlations with soil index properties, classification charts (Figs. 9.1 and 9.2), expansion index test, HUD swell test, and the standard 60 psf swell test. All of these methods do not predict the actual expansive soil movement beneath the foundation, but rather are used to classify the soil as having a potential degree of expansion, with the expansion potential varying from nonexpansive (0 percent swell) up to very high or critical. While all of these methods do not predict the actual expansive soil movement beneath the foundation, they can be used to set minimum design requirements, such as shown in Table 9.2.

Of the three methods where actual soil specimens are tested in the laboratory, the expansion index test will often classify the soil as having a lower expansion potential than the other two methods. The reasons are because the expansion index test has a lower initial dry density, higher initial water content, and higher surcharge pressure than the other two methods (see Table 9.3). All of these factors will provide a lower percent swell for the expansion index test as compared to the percent swell for the HUD swell test and the standard 60 psf swell test. In addition, the percent swell versus expansion potential is less restrictive for the expansion index test. For example, a percent swell of 4.5 percent would only be classified as low expansion potential per the expansion index test, but would be classified as moderate for both the HUD swell test and standard 60 psf swell test.

TABLE 9.2   Expansion Potential versus Percent Swell and Minimum Foundation Design Criteria

Expansion potential	Percent swell under 60 psf surcharge	Minimum thickened edge or footing depth (in.)	Minimum design values y_m (in.) for posttensioned slabs	
			Edge lift	Center lift
Low	0–4	12	$1/8$–$1/4$	—
Moderate	4–8	12	$1/4$–$1/2$	$1/8$–$3/8$
High	8–12	18	$1/2$–1	$3/8$–1
Critical	12–16	24	See note below	
	16–20	30		
	>20	36		

Note: Specific recommendations from geotechnical engineer required. Design value y_m shall be minimum of 1 in. (2.5 cm).
Source: Table 1805.8 of Southern Nevada Amendments to the 2000 *International Building Code* (2003). Additional notes included with Table 1805.8 have not been reproduced. See Sec. 9.5.3 for a discussion of y_m.

TABLE 9.3   Summary of Methods

Method	Sample type	Initial moisture condition	Surcharge pressure	Percent swell versus expansion potential	Ranking*
Expansion index test	Compacted (close to standard Proctor energy)	Water content corresponding to a degree of saturation of 50%	144 psf (6.9 kPa)	0–2 Very low 2–5 Low 5–9 Medium 9–13 High >13 Very high	1
HUD swell test	Compacted or undisturbed specimens	Air-dry	Usually 60 psf (2.9 kPa) or 144 psf (6.9 kPa)	For 60 psf: 0–4 Low 4–8 Moderate 8–12 High >12 Critical	2
Standard 60 psf swell test	Compacted or undisturbed specimens	Oven-dry (60°C)	60 psf (2.9 kPa)	0–4 Low 4–8 Moderate 8–12 High >12 Critical	3

*Ranking: 1 indicates the least percent swell and 3 indicates the highest percent swell when comparing the initial dry density, initial moisture condition, and surcharge pressure for the three methods.

9.3   BASIC EXPANSIVE SOIL PRINCIPLES

The purpose of this section is to introduce basic expansive soil principles. Items that will be discussed include the depth of seasonal moisture change, soil suction, Thornthwaite moisture index, identification and swelling of desiccated clay, types of expansive soil foundation movement, and effects of vegetation.

9.3.1   Depth of Seasonal Moisture Change

Near surface clay deposits will often have different values of water content depending on the time of year. During a hot and dry period, the water content will be significantly lower than during a rainy period. The greatest variation in water content occurs at ground surface, with the variation deceasing with depth.

Figure 9.14 shows the water content versus depth for two clay deposits located in Irbid, Jordan (Al-Homoud et al., 1997). Soil deposit A has a liquid limit of 35 and a plasticity index of 22, while soil deposit B has a liquid limit of 79 and a plasticity index of 27 (Al-Homoud et al., 1995). Note in Fig. 9.14 that during the hot and dry summer, the water content of the soil is significantly lower than during the wet winter. During the summer, the lowest water contents are recorded near ground surface, and the water contents are below the shrinkage limit (SL). A near-surface water content below the shrinkage limit (SL) is indicative of severe desiccation of the clay. Below a depth of about 3.2 m (10 ft) for soil deposit A and a depth of about 4.5 m (15 ft) for soil deposit B, the water content is relatively unchanged between the summer and winter monitoring period, and this depth is commonly known as the *depth of seasonal moisture change*. As shown in Fig. 9.14, soil deposit B has a greater variation in water content from the dry summer to wet winter and a greater depth of seasonal moisture change. This is probably because soil deposit B has a higher clay content than soil deposit A.

The depth of seasonal moisture change is also sometimes referred to as the *depth of the active zone* or simply the *active depth*. The depth of seasonal moisture change would depend on many different factors, as follows:

1. *Climate.*   Such as the duration of the dry and wet seasons as well as the temperature and humidity during the dry season and the amount of rainfall during the wet season.

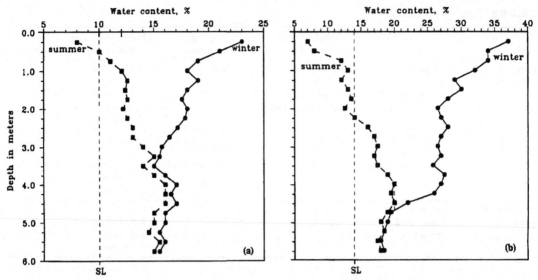

FIGURE 9.14 Water content versus depth for (*a*) soil A and (*b*) soil B. (*From Al-Homoud et al., 1997; reprinted with permission from the American Society of Civil Engineers.*)

2. *Soil characteristics.* Such as the nature of the soil in terms of clay content, clay mineralogy, and density.

3. *Site variables.* Such as the presence of vegetation that can extract soil moisture and the depth of the groundwater table.

The typical depth of seasonal moisture change has been reported to be in the range of 1 to 4.6 m (3 to 15 ft) depending on the soil and the climate (Sorensen and Tasker, 1976). However, deeper zones of seasonal moisture change have been reported. For example, Nelson and Miller (1992) state:

> Along the Front Range of Colorado, the active zone appears to be generally about 15 to 20 ft (4.6 to 6.1 m) deep. Estimates of active zone depths less than 10 ft (3 m) should be considered suspect.

The depth of seasonal moisture change is important because it defines the minimum zone of soil that will have changes in seasonal moisture and subsequent heave or shrinkage. This is the minimum depth of soil that should be sampled and then tested in the laboratory to determine its expansion behavior. However, the depth of soil expansion could be deeper, if for example, once the structure is completed, a steady source of water is introduced into the ground from irrigation or leaky water pipes.

9.3.2 Soil Suction

The total suction of an unsaturated soil is the sum of the matric suction s_m and the osmotic suction s_o, as follows (Fredlund and Rahardjo, 1993):

$$s_T = s_m + s_o = (u_a - u_w) + s_o \tag{9.3}$$

where s_T = total suction of the soil (psf or kPa)
$s_m = u_a - u_w$ = matric suction (psf or kPa)
u_a = air pressure in the soil voids (psf or kPa)
u_w = pore water pressure acting between the soil particles (psf or kPa)
s_o = osmotic suction (psf or kPa)

The matric suction and osmotic suction are due to the following:

1. *Matric suction* $(s_m = u_a - u_w)$.   For unsaturated soil, the air pressure in the soil voids is usually atmospheric and hence $u_a = 0$. But soils can have positive or negative pore water pressures u. For example, a clay can have excess pore water pressures u_e, such as during consolidation, which causes water to flow out of the soil. The same clay when unsaturated can have negative pore water pressures $-u_w$, due to capillarity, which will cause water to be drawn into the clay. In fact, a dried clay can have negative pore water pressures as low as minus 7000 kPa (Olson and Langfelder, 1965). Note that using a negative pore water pressure in Eq. 9.3 results in a positive value of total suction.

In cases where the soil is saturated and the pore water pressures are positive, such as soil located below the groundwater table, the matric suction is assumed to be equal to zero. For soils above the groundwater table that are saturated due to capillary rise, the matric suction can be calculated from Eq. 4.17, or:

$$s_m = u_a - u_w = 0 - (-\gamma_w h) = \gamma_w h$$

where γ_w is the unit weight of water and h is distance above the groundwater table.

The concept of capillary rise can be used to distinguish those soil types that will have low versus high matric suction. For example, as indicated in Sec. 4.4, open graded gravel has a capillary rise equal to zero $(h_c = 0)$ and hence it will always have zero matric suction. At the other extreme are clays that can have high values of capillary rise because of the small pore sizes and hence such soils can also have high values of matric suction.

2. *Osmotic suction* s_o.   The role of osmotic suction is equally applicable to both unsaturated and saturated soils. In a general sense, osmotic suction is due to the salt content of the pore-water which is present in both saturated and unsaturated soils. For example, the pore-water may contain cations that are attracted to the negatively charged clay particle face (i.e., double layer). Water wants to dilute the double layer that contains cations, resulting in an osmotic pressure.

As the water content of clay decreases, the total suction increases. For example, Fig. 9.15 presents laboratory test data that shows values of matric suction and osmotic suction versus the water

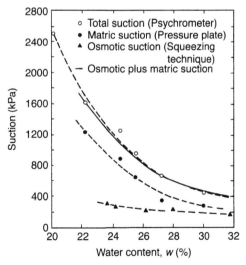

FIGURE 9.15   Total suction, matric suction, and osmotic suction measured in the laboratory on compacted Regina clay. (*Originally developed by Krahn and Fredlund, 1972; reproduced from Fredlund and Rahardjo, 1993.*)

content of the soil. The data shown in Fig. 9.15 was obtained from Regina clay specimens compacted at various initial water contents. Each component of soil suction and the total suction were independently measured. This figure shows the following:

1. *Osmotic suction.* There is a slight increase in osmotic suction as the water content of the soil decreases. But as a practical matter, this change in osmotic suction is generally insignificant. This data would indicate that over the range of water contents tested, the cations in the double layer are fully hydrated and hence the osmotic pressure is fairly constant.

2. *Matric suction.* As the water content of the clay decreases, the matric suction rapidly increases. This is in response to the negative pore water pressures that develop as the clay is dried. As shown in Fig. 9.15, the matric suction greatly exceeds in value the osmotic suction at low water content values. As an approximation, a change in total suction is essentially equivalent to a change in matric suction. Thus the absorption of water and subsequent swelling of a clay can be thought of as primarily in response to the negative pore water pressures $-u_w$ in a soil.

3. *Total suction.* The solid line in Fig. 9.15 represents the sum of the measured osmotic suction and measured matric suction. In addition, the total suction was measured in the laboratory using a psychrometer. As shown in the figure, the calculated value and measured value of total suction are essentially identical.

There are many different methods that can be used to determine the matric, osmotic, and total suction of a soil [see Fredlund and Rahardjo (1993) and Nelson and Miller (1992) for a discussion of the methods]. One commonly used laboratory technique is the filter paper method, which is described in ASTM D 5298-03 (2004), "Standard Test Method for Measurement of Soil Potential (Suction) Using Filter Paper."

Many publications, especially those in soil science, use the *pF* unit for soil suction. The *pF* has been defined as the logarithm of the height of a column of water in centimeters. Hence a *pF* = 2 is equivalent to the pressure exerted at the base of a column of water that is 100 cm high (i.e., 9.8 kPa). Likewise a *pF* = 3 is equivalent to the pressure exerted at the base of a column of water that is 1000 cm high (i.e., 98 kPa). Other publications use standard SI units of pressure (kPa) for soil suction, as shown in Fig. 9.15.

Driscoll (1983) has suggested relationships between soil suction and the level of desiccation, where *pF* = 2 would indicate the onset of desiccation and *pF* = 3 would indicate when desiccation becomes significant.

9.3.3 Thornthwaite Moisture Index

The Thornthwaite moisture index (Thornthwaite, 1948) is a measure of the long-term severity of the climate. It is determined from a calculation of the water balance and employs estimates of rainfall and evaporation obtained from climate records. Values can range from +100 to −100 and areas can be contoured such as shown in Fig. 9.16. A positive Thornthwaite moisture index represents a net long-term increase in soil moisture due to rainfall, with the higher the positive value, generally the wetter the climate. A negative Thornthwaite moisture index represents a net long-term decrease in soil moisture due to a lack of rainfall or harsh climatic conditions. From a theoretical standpoint, a Thornthwaite moisture index of zero would indicate a perfectly balanced soil moisture condition, with just enough rainfall to compensate for loss of soil moisture due to evaporation.

If the Thornthwaite moisture index is a high positive value, then expansive soil problems are unlikely. This does not mean that there is an absence of expansive soils, but rather there is a mild climate and/or enough rainfall to keep these soils in a permanently wetted condition. As shown in Fig. 9.16, the northwest coastal region of the United States, which has a cool and wet climate, has a Thornthwaite moisture index of +100.

Given the same soil type, the more negative the Thornthwaite moisture index, the lower the soil moisture, and the greater the potential for expansive soil problems. As shown in Fig. 9.16, areas having the largest negative values include large sections of Arizona, Nevada, and the southern California desert areas. Some of the most challenging expansive soil problems are encountered in the northern

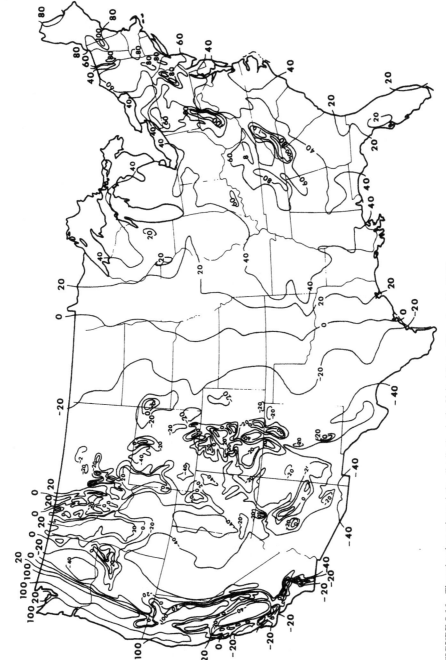

FIGURE 9.16   Thornthwaite moisture index distribution in the United States (Thornthwaite, 1948).

9.21

Las Vegas metropolitan area, which has both critically expansive clays and Thornthwaite moisture index values of about −40.

It should be mentioned that areas that have positive values of Thornthwaite moisture index can still experience expansive soil related damage. For example, London, England, which normally has a cool and wet climate, has experienced significant expansive soil related foundation damage during periods of drought when large trees have extracted moisture from beneath the foundations and caused the clay to shrink, resulting in downward movement of the foundation. This will be further discussed in Sec. 9.3.6.

9.3.4 Identification and Swelling of Desiccated Clay

Areas having surface deposits of clay and high negative values of the Thornthwaite moisture index or periods of prolonged drought can have desiccated clay. Structures constructed on top of desiccated clay can be severely damaged due to expansive soil heave (Jennings, 1953). For example, Chen (1988) states: "Very dry clays with natural moisture content below 15 percent usually indicate danger. Such clays will easily absorb moisture to as high as 35 percent with resultant damaging expansion to structures." There can also be desiccation and damage to final clay cover systems for landfills and site remediation projects, and for shallow clay landfill liners (Boardman and Daniel, 1996). The geotechnical engineer can often visually identify desiccated clay because of the numerous ground surface cracks, such as shown in Fig. 9.17.

Figure 9.18 shows a specimen of desiccated natural clay obtained from Otay Mesa, California. The near-surface deposit of natural clay has caused extensive damage to structures, pavements, and flatwork constructed in this area. The clay particles in this soil are almost exclusively montmorillonite (Kennedy and Tan, 1977; Cleveland, 1960). Figure 9.19 presents the results of a swell test (lower half of Fig. 9.19) and a falling head permeameter test (upper half of Fig. 9.19) performed on a specimen of desiccated Otay Mesa clay (Day, 1997b). At time zero, the desiccated clay specimen was inundated with distilled water. The data in Fig. 9.19 indicates three separate phases of swelling of the clay, as follows:

1. *Primary swell.*   The first phase of swelling of the desiccated clay was primary swell. The primary swell occurs from time equals zero (start of wetting) to about 100 min. The end of primary

FIGURE 9.17   View of a deposit of desiccated clay located in Death Valley, California. Note that the hat in the center of the photograph provides a scale for the size of the desiccation cracks.

FIGURE 9.18   Specimen of desiccated Otay Mesa clay.

swell (100 min) was estimated from the log-of-time method, which as previously mentioned, can also be applied to the swelling of clays.

Figure 9.19 shows that during primary swell, there was a rapid decrease in the hydraulic conductivity (also known as permeability) of the clay. The rapid decrease in hydraulic conductivity was due to the closing of soil cracks as the clay swells. At the end of primary swell, the main soil cracks have probably closed and the hydraulic conductivity was about 7×10^{-7} cm/sec.

2. *Secondary swell.*   The second phase of swelling was secondary swell. The secondary swell occurs from a time of about 100 to 20,000 min after wetting. Figure 9.19 shows that during secondary swell, the hydraulic conductivity continues to decrease as the clay continues to swell and the micro-cracks close-up. The lowest hydraulic conductivity of about 1.5×10^{-8} cm/sec. occurs at a time of about 5000 min, when most of the micro-cracks have probably sealed-up. From a time of 5000 to 20,000 min after wetting, there was a very slight increase in hydraulic conductivity. This is probably due to a combination of additional secondary swell which increases the void ratio and a reduction in entrapped air.

3. *Steady-state.*   The third phase started when the clay stopped swelling. This occurred at about 20,000 min after inundation with distilled water. No swell was recorded from a time of 20,000 min after wetting to the end of the test (50,000 min). As shown in Fig. 9.19, the hydraulic conductivity is constant once the clay has stopped swelling. From a time of 20,000 min after wetting to the end of the test (50,000 min), the hydraulic conductivity of the clay was constant at about 3×10^{-8} cm/sec.

The rate of swelling is important because it governs how fast water will enter the soil and cause foundation heave. Chen (1988) states that the permeability (and hence coefficient of swell) is an important factor because the higher the permeability, the greater the probability of differential movement of the foundation. This is because the water could quickly penetrate underneath one portion of the structure, resulting in damaging differential movement. A slower moisture migration into the soil could result in a more gradual and uniform foundation heave.

There appear to be three factors that govern the permeability and rate of swelling of desiccated clay: the development of cracks as the clay dries, the increased suction at a lower water content, and the process of slaking.

    1. *Development of desiccation cracks.*   The amount and distribution of desiccation cracks, such as shown in Fig. 9.17, are probably the greatest factors in the rate of swelling. Clays will shrink until the shrinkage limit (usually a low water content) is reached. Even as the moisture

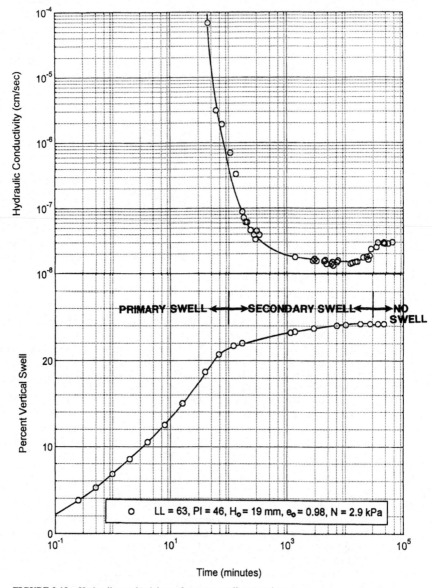

FIGURE 9.19 Hydraulic conductivity and percent swell versus time.

content decreases below the shrinkage limit, there is probably still the development of additional microcracks as the clay dries. The more cracks in the clay, the greater the pathways for water to penetrate the soil, and the quicker the rate of swelling.

2. *Increased suction at a lower water content.*   The second factor that governs the rate of swelling of a desiccated clay is suction pressure. As shown in Fig. 9.15, the total suction increases as the water content decreases. At low water contents, the water is drawn into the clay by the suction pressures. The combination of both shrinkage cracks and high suction pressures allows water to be quickly sucked into the clay, resulting in a higher rate of swell.

3. *Slaking.* The third reason is the process of slaking. Slaking is defined as the breaking of dried clay when submerged in water, due either to compression of entrapped air by inwardly migrating capillary water or to the progressive swelling and sloughing off of the outer layers (Stokes and Varnes, 1955). Slaking breaks apart the dried clay clods and allows water to quickly penetrate all portions of the desiccated clay. The process of slaking is quicker and more disruptive for clays having the most drying time and lowest initial water content.

It would be desirable to use laboratory test data to predict how long it will take for the short-term and long-term expansive soil conditions to develop. For example, there is a similarity of the shape of time versus deformation curves for consolidation and swell of clay in the oedometer apparatus (i.e., compare Figs. 8.5 and 9.8). The rate of swell can be estimated from Terzaghi's diffusion equation where the coefficient of consolidation (c_v, see Eq. 8.15) is replaced by the coefficient of swell c_s, or (Blight, 1965):

$$c_s = \frac{T H_{dr}^2}{t} \tag{9.4}$$

where c_s = coefficient of swell from laboratory testing (ft²/day or m²/day)
T = time factor from Table 8.2 (dimensionless)
t = time since water enters the expansive soil (days)
H_{dr} = height of the swell path (ft or m). If water enters at the top and bottom of the cohesive soil layer, then $H_{dr} = \frac{1}{2} H_o$, where H_o = initial thickness of the expansive soil layer. If water only enters at the top or bottom of the expansive soil layer, then $H_{dr} = H_o$.

The coefficient of swell (c_s) can be determined from laboratory time versus swell data by using the log-of-time method or the square-root-of-time method (see Sec. 8.2.2). For example, for the data shown in Fig. 9.8, the time for 50 percent primary swell t_{50} is about 6.5 min. The specimen height corresponding to 50 percent primary swell is 1.03 in. (2.62 cm) and recognizing that water can be drawn into the soil from the top and bottom, the coefficient of swell (c_s) from Eq. 9.4 is 8×10^{-3} in.²/min (0.05 cm²/min). By using the coefficient of swell, the length of time it will take the in situ soil to achieve a certain amount of heave can be estimated.

From a theoretical standpoint, the approach for predicting the rate of swell of a expansive soil should be similar in approach to the rate of consolidation (Chap. 8). Unfortunately, in practice, this is not the case because expansive soil typically does not have continuous access to water, but rather the clay will shrink during the dry season or when irrigation water is reduced and then swell during the wet season. In addition, because of soil desiccation during the dry season, the permeability of the expansive soil can be quite variable (see Fig. 9.19). Hence applying the principles of the coefficient of swell is usually not performed in practice and determining the rate of swell of an expansive soil is difficult to predict.

9.3.5 Expansive Soil Foundation Movement

Two Types of Expansive Soil Movement. If a shallow foundation, such as a slab-on-grade, is constructed on top of a clay, there are usually two main types of expansive soil movement. The first is the cyclic heave and shrinkage around the perimeter of the foundation and the second is the long-term progressive swelling beneath the center of the structure, as follows:

1. *Cyclic heave and shrinkage.* Clays are characterized by their moisture sensitivity, they will expand when given access to water and shrink when they are dried out. A soil classified as having a very high expansion potential will swell or shrink much more than a soil classified as having a very low expansion potential (Table 9.1). For example, the perimeter of a slab-on-grade foundation could heave during the rainy season and then deform downward during the drought if the clay dries out. This causes cycles of up and down movement, causing cracking and damage to the structure. Field measurements of this up-and-down cyclic movement have been recorded by Johnson (1980).

The amount of cyclic heave and shrinkage depends on the change in moisture content of the clays below the perimeter of the foundation. The moisture change in turn depends on the severity of the drought and rainy seasons, the influence of drainage and irrigation, and the presence of live tree roots, which can extract moisture and cause clays to shrink. The cyclic heave and shrinkage around the perimeter of a structure is generally described as a seasonal or short-term condition.

2. *Progressive swelling beneath the center of the foundation (center lift).* There are several ways that moisture can accumulate underneath the center of a slab-on-grade. Probably the most important is the foundation barrier effect. Because of capillary action, moisture can move upward through soil, where it will evaporate at the ground surface. But when a slab-on-grade is constructed, it acts as a ground surface barrier, reducing or preventing the evaporation of moisture. Thus slowly with time water will tend to accumulate underneath the center of the foundation, resulting in the swelling of the clay and uplift of the foundation, and this process is known as center lift.

Similar to capillary action, water can also be drawn into the soil due to suction pressures. Hence water could also move laterally through the soil by the action of soil suction and again accumulate beneath the center of the foundation resulting in center lift.

A third possible reason for moisture migration to the center of a slab-on-grade is due to thermal gradients (Sowers, 1979). It has been stated that water at a higher temperature than its surroundings will migrate in the soil towards the cooler area to equalize the thermal energy of the two areas (Chen, 1988; Nelson and Miller, 1992; Day, 1996a). Especially during the summer months, the temperature under the center of a foundation tends to be much cooler than at the exterior ground surface.

The progressive heave of the center of the structure is generally described as a long-term condition, because the maximum value may not be reached until many years after construction. Figure 9.20 illustrates center lift beneath a house foundation and Fig. 9.21 shows the typical crack pattern in the concrete slab-on-grade due to expansive soil center lift.

The above descriptions of the two main types of expansive soil movement are simplified and many different variations are possible. For example, in some cases there may be no cyclic heave and shrinkage, but instead just an edge lift condition (Fig. 9.22). In the early years of the project, water migrates under the perimeter of the foundation with the source of water often due to the irrigation of landscaping installed by the owners. As water migrates under the perimeter of the foundation, the edge of the foundation is uplifted (see middle diagram, Fig. 9.22). In cases where a slab-on-grade is constructed on a dry clay, there may only be edge lift of the perimeter of the slab during the early years of the project and not the seasonal heave and shrinkage as described earlier.

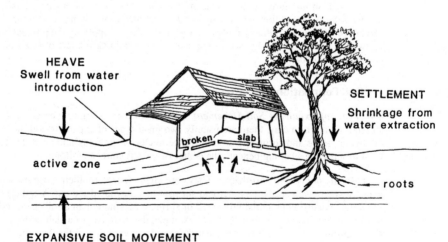

EXPANSIVE SOIL MOVEMENT

FIGURE 9.20 Illustration of center lift of the foundation.

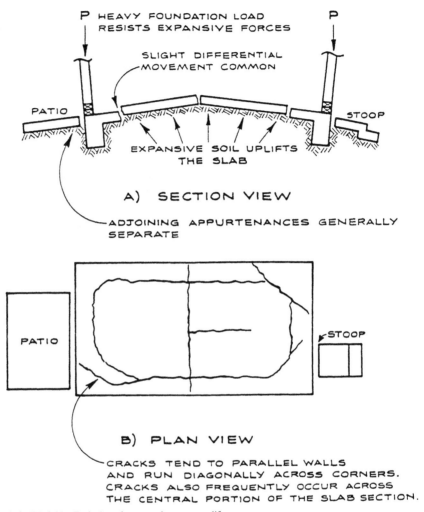

FIGURE 9.21   Typical crack pattern due to center lift.

As the project matures, a center lift condition may eventually be reached as shown in the bottom diagram in Fig. 9.22.

Rate of Foundation Movement.   It has been stated that damages due to expansive soils typically occur within the first two to three years after house construction (Meehan and Karp, 1994). This is likely because owners often install landscaping, which is then heavily irrigated in order to establish the ground cover. The sudden influx of water beneath the edge of the foundation causes the perimeter of the foundation to heave upward, especially if the foundation was constructed on a dry and desiccated clay.

For an expansive soil test site near Newcastle, Australia, where a flexible cover with 0.5 m (1.6 ft) deep edge beams was constructed at ground surface over a relatively dry clay of high plasticity and monitored over several years, the following was observed (Fityus, et al., 2004):

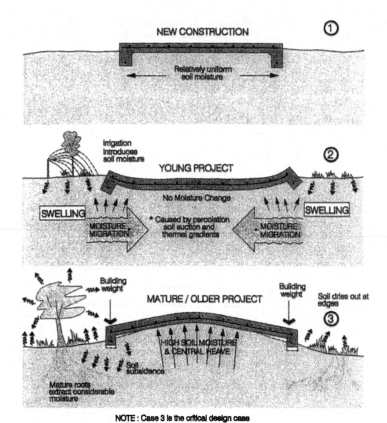

FIGURE 9.22 Expansive soil damage progression. (1) New construction of the foundation on expansive soil, which often has relatively uniform soil moisture. (2) Water from irrigation and rainfall migrates underneath the perimeter of the foundation, resulting in edge lift. (3) For older projects, the water will eventually migrate beneath the center of the foundation, resulting in center lift.

1. *Open ground areas.* In the open areas away from the flexible cover, the ground surface increased and decreased in elevation in response to the seasonal wetting and drying cycles.

2. *Edge lift of flexible cover.* Movement beneath the flexible cover has been consistently upward (i.e., heaving), as the initially dry clay slowly becomes wet and approaches an equilibrium water content. The edges of the flexible cover rose rather quickly and attained their final heaved level in 5 to 6 years. Little or no cyclic heave and shrinkage was observed around the perimeter. This is because the flexible cover with edge beams was constructed over relatively dry clay and only edge lift has been documented so far.

3. *Center lift of flexible cover.* The center of the flexible cover rose more slowly and is still rising slowly after 7 years of monitoring, although the majority of the heave occurred within the first 4 years. The amount of center lift (after 7 years) is about 5.5 cm (2.2 in.). Fityus et al. (2004) state that the center lift condition is likely to be most severe after a long time when the covered soil has attained equilibrium wetness and can swell no further.

It should be mentioned that other factors can trigger expansive soil movement many years, even decades, after construction. For example, although the soil may have reached an equilibrium state, pipe leaks can introduce large quantities of water that reactivates expansive soil movement. An example of this condition is shown in Fig. 9.23. Another possible trigger for expansive soil movement could be the installation and heavy irrigation of new landscaping adjacent the foundation. A final factor could be the effect of trees, which can cause expansive soil related movement at any time after original construction, as discussed in the next section.

Maximum Angular Distortion and Differential Movement.   As discussed in Sec. 7.6, by plotting the maximum angular distortion δ/L versus the maximum differential settlement Δ such as shown in Fig. 7.24, Skempton and MacDonald (1956) obtained a correlation for the settlement of foundations that is defined as $\Delta = 350\ \delta/L$ (note: Δ is in inches). Using this relationship and a maximum angular distortion δ/L of 1/300, cracking of brick panels in frame buildings or load-bearing brick walls is likely to occur if the maximum differential settlement Δ exceeds $1^1/_4$ in. (32 mm).

However, when plotting maximum differential foundation movement Δ versus maximum angular distortion δ/L, a different relationship is often obtained for foundations on expansive soil, such as $\Delta = 300\ \delta/L$ (note: Δ is in inches), Marsh and Thoeny (1999). This difference in relationships between settlement and expansive soil seems to be due to the interaction of the foundation and underlying soil. For settlement, the movement is often distributed across the entire foundation, i.e., the entire foundation settles. But for expansive soil, the movement is often concentrated at one or more foundation locations, such as the uplift of the corners or sides of the building. This different interaction between the soil and foundation probably accounts for the different relationships for foundation settlement as compared to expansive soil movement.

9.3.6   Effects of Vegetation

One frequent cause of expansive soil damage is heave of the foundation (center lift) as illustrated in Fig. 9.20. But foundation movement can also be caused by the shrinkage of clay. For example, tree

FIGURE 9.23   Expansive soil damage of a raised wood floor foundation. Note the wet condition of the soil in the crawl space.

roots and rootlets can extract moisture from the ground, which can cause the near surface clay to shrink and the foundation to deform downward.

There are cases (Cheney and Burford, 1975) where the opposite can also occur, where large trees have been removed and the clay has expanded as the soil moisture increases to its natural state. In the United States, Holtz (1984) states that large, broad-leaf, deciduous trees located near the structure cause the greatest changes in moisture and the greatest resulting damage in both arid and humid areas. However, the most dramatic effects are during periods of drought, such as the severe drought in Britain from 1975 to 1976, when the amount of water used by the trees during transpiration greatly exceeded the amount of rainfall within the area containing tree roots. Biddle (1979, 1983) investigated 36 different trees, covering a range of tree species and clay types and concluded that Poplars have much greater effects than other species and that the amount of soil movement will depend on the clay shrinkage characteristics. Ravina (1984) states that it is the nonuniform moisture changes and soil heterogeneities that cause the uneven soil movements that damage shallow foundations, structures, and pavements.

Driscoll (1983) has suggested relationships between water content w of the soil and the liquid limit (LL), which indicate the level of desiccation, where

$$w = 0.5\,LL \tag{9.5}$$

would indicate the onset of desiccation, and

$$w = 0.4\,LL \tag{9.6}$$

would indicate when desiccation becomes significant. As mentioned in Sec. 9.3.1, a near-surface water content w below the shrinkage limit (SL) would be indicative of severe desiccation of the clay.

Tucker and Poor (1978) indicated that trees located at distances closer than their heights to structures caused significantly larger movements due to clay shrinkage than those trees located at greater distances. Hammer and Thompson (1966) stated that trees should not be planted at a minimum of one-half of their anticipated mature height from a shallow foundation and slow-growing, shallow-rooted varieties of trees were preferred. However, Cutler and Richardson (1989) indicated that total crown volume (hence leaf area) is generally more important than absolute height in relation to water demand. Cutler and Richardson (1989) used case histories to determine the frequency of damage as a function of tree-trunk distance from the structure for different species. Biddle (1983) recommended that for very high shrinkage clay, the perimeter footings should be at least 5 ft (1.5 m) deep and this would be sufficient to accommodate most tree-planting designs.

In summary, when dealing with expansive clays, it is important for the geotechnical engineer to consider the possibility of damage due to clay shrinkage caused by the drying of wet clay around the perimeter of the foundation and due to extraction of moisture by tree roots. Shallow foundations, such as slab-on-grade or raised wood floor foundations having shallow perimeter footings are especially vulnerable to damage due to clay shrinkage.

9.4 METHODS USED TO PREDICT FOUNDATION MOVEMENT

9.4.1 Soil Suction

Soil suction has been introduced in Sec. 9.3.2. As previously described, expansive soil movements occur in response to changes in total suction s_T. As the soil dries, the total suction increases, with subsequent shrinkage of the soil (e.g., see Fig. 9.15). Likewise, if the soil is wetted, the total suction decreases, and the soil expands. If the initial soil suction of the clay can be measured and the final condition can also be estimated, then the amount of heave or shrinkage of the soil beneath the foundation can be calculated, as follows (Wray, 1989, 1997):

$$\Delta H = (H)(\gamma_h)(\Delta \log s_T - \Delta \log \sigma_v) \tag{9.7}$$

where ΔH = change in elevation (heave or shrinkage), ft or m

H = thickness of the soil layer experiencing heave or shrinkage (ft or m)

γ_h = suction compressibility index (Lytton, 1977), sometimes referred to as the coefficient of suction compressibility (dimensionless)

$\Delta \log s_T$ = change in total soil suction for the soil layer of H thickness (psf or kPa)

$\Delta \log \sigma_v$ = change in vertical total stress, from the top to the bottom of the soil layer of H thickness (psf or kPa)

Values of the suction compressibility index γ_h are often between 0.01 and 0.10, although values up to 0.22 have been reported (Nelson and Miller, 1992). It is defined as:

$$\gamma_h = \frac{\varepsilon_v}{\Delta \log s_T} \qquad (9.8)$$

where ε_v is the vertical strain = $\Delta H/H$. The suction compressibility index γ_h can be determined from laboratory testing or from empirical correlations. The term $\Delta \log \sigma_v$ was included in Eq. 9.7 to account for the change in total stress for the layer of H thickness because the heave or shrinkage depends on both a change in soil suction and the overburden pressure. The suction compressibility index is assumed to apply equally to the total suction term and the overburden pressure term.

The soil suction method has provided accurate predictions of heave and shrinkage (Wray, 1997; Masia et al., 2004). But it has also been stated (Fityus et al., 2004):

> An important outcome of the field monitoring work is confirmation that in situ measurement of soil suction is a difficult task that has a strong likelihood of unreliable results for the methods assessed. This outcome is consistent with the recently published findings of Harrison and Blight (2000).

Because of the difficult task of measuring soil suction as well as predicting the change in soil suction, other methods, such as described in the following sections, are more commonly utilized for the determination of the movement of foundations on expansive soil.

Example Problem 9.1   A site has a surface deposit of clay that is 2 m thick. Assume the clay layer is underlain by rock. Based on laboratory testing, the suction compressibility index $\gamma_h = 0.05$ and the total unit weight = 18.0 kN/m³. Assume a slab-on-grade foundation will be built in the dry season and that the foundation exerts a dead load pressure of 5 kPa at ground surface. Determine the amount of center lift of the slab-on-grade for the change in conditions from the dry season to the wet season using the following soil suction values:

Depth below ground surface	Dry conditions	Wet conditions
0–0.5 m	30,000 kPa	40 kPa
0.5–1 m	10,000 kPa	100 kPa
1–2 m	3,000 kPa	250 kPa

Solution   Layer No. 1 (0 to 0.5 m):

$$\sigma_v = 5 \text{ kPa (top of soil layer)}$$

Assuming one-dimensional conditions:

$$\sigma_v = 5 \text{ kPa} + (18.0 \text{ kN/m}^3)(0.5 \text{ m}) = 14 \text{ kPa (bottom of soil layer)}$$

$$\Delta \log \sigma_v = \log 14 - \log 5 = 0.45$$

$$\Delta \log s_T = \log 30{,}000 - \log 40 = 2.88$$

(Continued)

Using Eq. 9.7:

$$\Delta H = (H)(\gamma_h)(\Delta\log s_T - \Delta\log \sigma_v) = (0.5 \text{ m})(0.05)(2.88 - 0.45) = 0.060 \text{ m}$$

Layer No. 2 (0.5 to 1 m):

$$\sigma_v = 14 \text{ kPa (top of soil layer)}$$

Assuming one-dimensional conditions:

$$\sigma_v = 5 \text{ kPa} + (18.0 \text{ kN/m}^3)(1 \text{ m}) = 23 \text{ kPa (bottom of soil layer)}$$

$$\Delta\log \sigma_v = \log 23 - \log 14 = 0.22$$

$$\Delta\log s_T = \log 10{,}000 - \log 100 = 2.0$$

Using Eq. 9.7:

$$\Delta H = (H)(\gamma_h)(\Delta\log s_T - \Delta\log \sigma_v) = (0.5 \text{ m})(0.05)(2.0 - 0.22) = 0.044 \text{ m}$$

Layer No. 3 (1 to 2 m):

$$\sigma_v = 23 \text{ kPa (top of soil layer)}$$

Assuming one-dimensional conditions:

$$\sigma_v = 5 \text{ kPa} + (18.0 \text{ kN/m}^3)(2 \text{ m}) = 41 \text{ kPa (bottom of soil layer)}$$

$$\Delta\log \sigma_v = \log 41 - \log 23 = 0.25$$

$$\Delta\log s_T = \log 3000 - \log 250 = 1.1$$

Using Eq. 9.7:

$$\Delta H = (H)(\gamma_h)(\Delta\log s_T - \Delta\log s_v) = (1 \text{ m})(0.05)(1.1 - 0.25) = 0.043 \text{ m}$$

Total center lift = 0.060 + 0.044 + 0.043 = 0.15 m (5.8 in.)

9.4.2 Swell Tests

This section will discuss the determination of expansive soil movement by testing undisturbed soil specimens. If it is proposed to place fill at the site, then the amount of expansion of the fill can be determined by testing laboratory specimens that have been compacted to anticipated field dry density and water content conditions. These tests are commonly referred to as *intact swell tests* or simply *swell tests*. These tests are different than the expansion index test, HUD swell test, and standard 60 psf swell test in that the goal is to predict the actual amount of soil movement for field situations, instead of simply trying to determine the expansion potential of the soil.

The test procedures are listed in ASTM D 4546-03 (2004), "Standard Test Methods for One-Dimensional Swell or Settlement Potential of Cohesive Soils." The commonly used test procedures are similar to Method B of ASTM D 4546-03 (2004), as follows:

1. *Soil specimen.* For projects where it is anticipated that there will be fill placement, fill specimens can be prepared by compacting them to the anticipated field as-compacted density and moisture condition. For other situations, undisturbed soil specimens are required. An undisturbed soil specimen can be trimmed from a block sample or obtained from a Shelby tube sampler. The specimen diameter divided by the specimen height should be equal to or greater than 2.5 and a common diameter of soil specimen is 2.5 in. (6.4 cm) and height is 1.0 in. (2.5 cm). The trimming process should be performed as quickly as possible to minimize a change in water content of the soil. Because expansive soil often has a low water content, specimens should not be stored or trimmed in a high humidity moisture room. Typically the soil specimen is trimmed in the confining ring and the initial height of the soil specimen (h_o) is then equal to the height of the confining ring.

2. *Index properties.* After the soil specimen has been trimmed, the index properties of the soil should be determined. Typically, the trimmings can be collected in order to determine the water content of the soil. In addition, a balance can be used to obtain the mass of the confining ring and soil specimen. By subtracting the mass of the confining ring, the mass of the soil specimen can be calculated. Knowing the volume of the confining ring, the total unit weight can be calculated. Using the water content, the dry unit weight can also be calculated from basic phase relationships.

3. *Loading device.* Dry and clean porous plates are placed on the top and bottom of the soil specimen and it is then placed in a surrounding container (such as a Plexiglas dish). The Plexiglas dish containing the soil specimen is placed at the center of the oedometer (such as shown in Fig. 3.11). A seating load equivalent to a vertical pressure of 125 psf (6 kPa) is then applied to the soil specimen and a dial gauge is set-up to record vertical expansion of the soil specimen. Within 5 min of applying the seating load, the load is increased. Typical loading increments are 250, 500, 1000, 2000 psf, and so on (12, 25, 50, 100 kPa, and so on) and each load increment should remain on the soil specimen for less than 5 min to prevent excessive evaporation of moisture from the specimen. The soil specimen can be surrounded with a loose-fitting plastic membrane or aluminum foil to reduce moisture loss during the loading process. Prior to applying the next load increment, a dial reading should be taken and recorded in order to determine the deformation versus loading behavior. The soil should be loaded to a vertical pressure that approximately equals the anticipated overburden pressure after completion of the structure. After the deformation has ceased at this vertical pressure, a dial gauge reading d_o is taken.

4. *Soil specimen wetting.* The soil specimen is then submerged in distilled water by filling the Plexiglas dish with distilled water. The distilled water should be quickly poured into the bottom of the Plexiglas dish. The water should not be poured onto the soil specimen, but rather the water is poured into the Plexiglas dish and allowed to flow upward through the soil specimen, which may reduce the amount of entrapped air. As soon as the distilled water has been poured into the Plexiglas dish, dial readings versus time at approximately 0.1, 0.25, 0.5, 1, 2, 4, 8, 15, and 30 min, and 1, 2, 4, 8, and 24 h should be recorded. The dial gauge readings or percent swell can be plotted versus time on a semi-log plot. The soil specimen must be allowed to swell for a long enough period of time so that it swells beyond primary swell and into secondary swell. A final dial gauge reading is obtained (d_f) in order to calculate the percent swell of the soil specimen, defined as follows:

$$\text{percent swell} = \frac{100(d_f - d_o)}{h_o} \tag{9.9}$$

where d_f = final dial reading (in. or mm)
$\quad d_o$ = initial dial reading (in. or mm)
$\quad h_o$ = initial specimen height determined from Step 1 (in. or mm)

5. *Unloading of the soil specimen.* After swelling is complete, the soil specimen is quickly unloaded by removing the load upon the soil specimen. Once the soil specimen has been completely unloaded, it is removed from the oedometer. The water content of the soil specimen can then be determined. When placing the soil specimen in the container for the water content test, it should be broken apart to determine if there are any large size particles within the soil specimen. Large size particles, such as gravel size particles, may restrict the swell of the soil specimen and lead to an underestimation of the in situ swell potential of the soil. If large particles are noted, the swell test may have to be repeated on a larger size soil specimen.

Concerning the limitations of the swell test, ASTM states:

> Estimates of the swell of soil determined by this test method (ASTM D 4546-03) are often of key importance in design of floor slabs on grade and evaluation of their performance. However, when using these estimates, it is recognized that swell parameters determined from these test methods for the purpose of estimating in situ heave of foundations and compacted soils may not be representative of many field conditions because:

1. Lateral swell and lateral confining pressure are not simulated.
2. Swell in the field usually occurs under constant overburden pressure, depending on the availability of water. Swell in the laboratory is evaluated by observing changes in volume due to changes in applied pressure while the specimen is inundated with water. Method B is designed to avoid this limitation.
3. Rates of swell indicated by swell test are not always reliable indicators of field rates of heave due to fissures in the in situ soil mass and inadequate simulation of the actual availability of water to the soil. The actual availability of water to the foundation may be cyclic, intermittent, or depend on in-place situations, such as pervious soil-filled trenches and broken water and drain lines.
4. Secondary or long-term swell may be significant for some soils and should be added to primary swell.
5. Chemical content of the inundating water affects volume changes and swell pressure; that is, field water containing large concentrations of calcium ions will produce less swelling than field water containing large concentrations of sodium ions or even rain water.
6. Disturbance of naturally occurring soil samples greatly diminishes the meaningfulness of the results.

Another major limitation of the swell test is that the water content of the tested soil may not be the same water content at the time of construction of the foundation. For example, if a drought occurs and the clay dries out, the percent expansion will be underestimated. On the other hand, if a rainy season occurs prior to construction and the clay absorbs moisture, the percent swell will be overestimated. One approach is to perform laboratory swell tests on soil conditions most likely to exist at the time of construction. A conservative option is to test the soil under the worst-case conditions and then the foundation is designed accordingly. Even with all of these limitations and assumptions, swell tests on undisturbed soil specimens are generally considered to be the most reliable method of predicting the future potential heave of the foundation.

Figure 9.24 presents the commonly used method for predicting the foundation heave based on laboratory swell test data. The percent swell versus depth is plotted as shown in Fig. 9.24 and the foundation heave is equal to the area under the percent swell versus depth curve. Figure 9.24 also illustrates the procedure for estimating the depth of undercut (and then replacement with nonexpansive soil) necessary to reduce foundation heave to an allowable value.

Example Problem 9.2   Use the data from Example Problem 9.1. Determine the amount of center lift of the slab-on-grade for the change in conditions from the dry season to the wet season using the following swell test data performed on undisturbed soil specimens. Assume the percent swell was measured under a surcharge pressure equivalent to the overburden pressure plus the foundation dead load pressure of 5 kPa.

Depth below ground surface	Percent swell
0.25 m	9.8
0.75 m	6.2
1.5 m	2.9

Solution   Layer No. 1 (0 to 0.5 m):

$$\Delta H = (H)(\text{swell}) = (0.5 \text{ m})(0.098) = 0.049 \text{ m}$$

Layer No. 2 (0.5 to 1 m):

$$\Delta H = (H)(\text{swell}) = (0.5 \text{ m})(0.062) = 0.031 \text{ m}$$

Layer No. 3 (1 to 2 m):

$$\Delta H = (H)(\text{swell}) = (1.0 \text{ m})(0.029) = 0.029 \text{ m}$$

Total center lift = 0.049 + 0.031 + 0.029 = 0.11 m (4.3 in.)

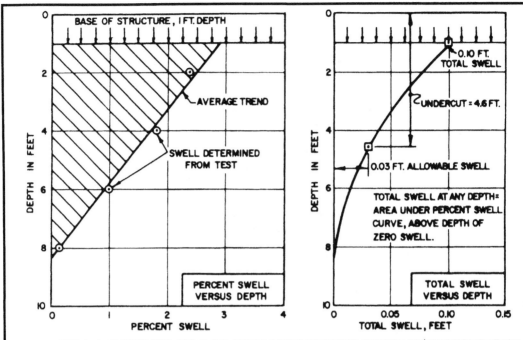

MATERIALS INVESTIGATED ARE CLAYS, HIGHLY OVERCONSOLIDATED BY CAPILLARY STRESSES THAT ARE
EFFECTIVE PRIOR TO THE CONSTRUCTION OF THE STRUCTURE UPON THEM.

PROCEDURE FOR ESTIMATING TOTAL SWELL UNDER STRUCTURE LOAD.

1. OBTAIN REPRESENTATIVE UNDISTURBED SAMPLES OF THE SHALLOW CLAY STRATUM AT
   A TIME WHEN CAPILLARY STRESSES ARE EFFECTIVE ; I.E., WHEN NOT FLOODED OR
   SUBJECTED TO HEAVY RAIN.
2. LOAD SPECIMENS (AT NATURAL WATER CONTENT) IN CONSOLIDOMETER UNDER A
   PRESSURE EQUAL TO THE ULTIMATE VALUE OF OVERBURDEN FOR HIGH GROUND WATER,
   PLUS WEIGHT OF STRUCTURE. ADD WATER TO SATURATE AND MEASURE SWELL.
3. COMPUTE FINAL SWELL IN TERMS OF PERCENT OF ORIGINAL SAMPLE HEIGHT AND
   PLOT SWELL VERSUS DEPTH, AS IN THE LEFT PANEL.
4. COMPUTE TOTAL SWELL WHICH IS EQUAL TO THE AREA UNDER THE PERCENT SWELL
   VERSUS DEPTH CURVE. FOR THE ABOVE EXAMPLE:
   TOTAL SWELL = 1/2 (8.2 −1.0) x 2.8/100 = 0.10 FT.

PROCEDURE FOR ESTIMATING UNDERCUT NECESSARY TO REDUCE SWELL TO AN ALLOWABLE VALUE.

1. FROM PERCENT SWELL VERSUS DEPTH CURVE PLOT RELATIONSHIP OF TOTAL SWELL
   VERSUS DEPTH AT THE RIGHT. TOTAL SWELL AT ANY DEPTH EQUALS AREA UNDER
   THE CURVE AT LEFT, INTEGRATED UPWARD FROM THE DEPTH OF ZERO SWELL.
2. FOR A GIVEN ALLOWABLE VALUE OF SWELL, READ THE AMOUNT OF UNDERCUT
   NECESSARY FROM THE TOTAL SWELL VERSUS DEPTH CURVE. FOR EXAMPLE, FOR AN
   ALLOWABLE SWELL OF 0.03 FT, UNDERCUT REQUIRED ≈ 4.6 FT. UNDERCUT CLAY IS
   REPLACED BY AN EQUAL THICKNESS OF NONSWELLING COMPACTED FILL.

FIGURE 9.24 Computation of foundation heave for expansive soils. (*Reproduced from NAVFAC DM-7.1, 1982.*)

9.4.3 Swelling Index

Another approach for the calculation of foundation heave is to use the swelling index. This test is different than the swell test in that the soil specimen is actually loaded in the oedometer apparatus much like a consolidation test. The test procedures are similar to Method C of ASTM D 4546-03, (2004), "Standard Test Methods for One-Dimensional Swell or Settlement Potential of Cohesive Soils," as follows:

1. *Soil specimen.* An undisturbed soil specimen must be used for the this test. For projects where it is anticipated that there will be fill placement, the fill specimens can be prepared by compacting them to the anticipated field as-compacted density and moisture condition. An undisturbed soil specimen can be trimmed from a block sample or extruded from the sampler directly into a confining ring. The specimen diameter divided by the specimen height should be equal to or greater than 2.5 and a common diameter of soil specimen is 2.5 in. (6.4 cm) and height is 1.0 in. (2.5 cm). The trimming process should be performed as quickly as possible to minimize a change in water content of the soil. Because expansive soil often has a low water content, specimens should not be stored or trimmed in a high humidity moisture room. Typically the soil specimen is trimmed in the confining ring and the initial height of the soil specimen (h_o) is then equal to the height of the confining ring.

2. *Index properties.* After the soil specimen has been trimmed, the index properties of the soil should be determined. Typically, the trimmings can be collected in order to determine the water content of the soil. In addition, a balance can be used to obtain the mass of the confining ring and soil specimen. By subtracting the mass of the confining ring, the mass of the soil specimen can be calculated. Knowing the volume of the confining ring, the total unit weight can be calculated. Using the water content, the dry unit weight can also be calculated using basic phase relationships.

3. *Loading device.* Dry and clean porous plates are placed on the top and bottom of the soil specimen and it is then placed in a surrounding container (such as a Plexiglas dish). The Plexiglas dish containing the soil specimen is placed at the center of the oedometer apparatus (see Fig. 3.11). A seating load equivalent to a vertical pressure of 125 psf (6 kPa) is then applied to the soil specimen and a dial reading is taken and recorded on the data sheet.

4. *Soil specimen wetting.* Within 5 min of applying the seating load, the soil specimen is inundated by filling the Plexiglas dish with distilled water. The distilled water should be quickly poured into the bottom of the Plexiglas dish. The water should not be poured onto the soil specimen, but rather the water should be poured into the Plexiglas dish and allowed to flow upward through the soil specimen, which may reduce the amount of entrapped air. As soon as the distilled water has been poured into the Plexiglas dish, the dial gauge measuring vertical movement should be monitored. When the soil specimen begins to swell, the vertical pressure on the soil specimen must be immediately increased. Typical loading increments are 250, 500, 1000, 2000 psf, and so on (12, 25, 50, 100 kPa, and so on) with the next load increment applied if the soil should continue to swell. This process of preventing the soil specimen from swelling by increasing the vertical stress on the soil usually takes continuous observation of the dial gauge that records vertical deformation. As soon as a vertical pressure is reached where the soil specimen begins to permanently compress together (i.e., no longer a tendency for swell), then this vertical stress is maintained on the soil specimen for a time period of 24 h.

5. *Continued loading of soil specimen.* Additional vertical stress is placed on the soil specimen to enable the determination of the swelling pressure P'_s. The soil specimen should be subjected to at least two additional incremental loads. Each vertical stress should remain on the soil specimen for about 24 h. Before each vertical stress is applied to the soil specimen, a dial reading should be taken and recorded on the data sheet.

6. *Unloading of the soil specimen.* After loading to the largest vertical pressure, the soil specimen is unloaded by reducing the load upon the soil specimen. The purpose of the unloading of the soil specimen is to obtain the swelling index C_s. At each reduction in load upon the soil specimen, it should be allowed to equilibrate for at least 24 h and a dial reading should be taken prior to the next reduction in load upon the soil specimen. During unloading, time versus dial readings at approximately 0.1, 0.25, 0.5, 1, 2, 4, 8, 15, and 30 min, and 1, 2, 4, 8, and 24 h can be recorded. The time versus dial readings are plotted by using the log-of-time method in order to ensure that primary swell is complete before the next reduction in vertical pressure.

7. *Index properties at the end of test.*   Once the soil specimen has been completely unloaded, it is removed from the oedometer. The water content of the soil specimen can then be determined. When placing the soil specimen in the container for the water content test, it should be broken apart to determine if there are any large size particles within the soil specimen. Large size particles, such as gravel size particles, may restrict the compression and swell of the soil specimen and lead to an underestimation of the in situ expansion potential of the soil. If large particles are noted, the test may have to be repeated on a larger size soil specimen.

The apparatus compressibility should be included in the analysis. This is because the swelling pressure P_s' is strongly influenced by the initial specimen height. In order to determine the apparatus compressibility, the loading apparatus is assembled with a metal disk of approximately the same height as the soil specimen. It is assumed that the metal disk is relatively incompressible. In the loading device, the metal disk is substituted for the soil specimen, and the top and bottom porous plates, the Plexiglas container, and filter paper (if used during the testing of the soil) are also included. By subjecting the metal disk to an increasing vertical pressure, the vertical deformation versus vertical pressure relationship for only the apparatus can be determined. The soil specimen height h is then adjusted by the amount of apparatus compressibility, as follows:

$$h = h_o + (d_v - d_s) + d_c \tag{9.10}$$

where h = soil specimen height at a specific vertical stress (in. or mm)
$\quad h_o$ = initial height of the soil specimen (in. or mm)
$\quad d_s$ = dial reading corresponding to the seating load (125 psf) acting on the soil specimen (in. or mm)
$\quad d_v$ = dial reading at a specific vertical stress during the loading of the specimen (in. or mm). If the soil has swelled, then d_v has a larger value than d_s and if the soil specimen has compressed, then vice versa.
$\quad d_c$ = amount of the apparatus compressibility for the same vertical stress corresponding to the dial reading of d_v (in. or mm)

An example of laboratory test results are shown in Figs. 9.25 to 9.27, as follows:

Test data (Fig. 9.25).   An undisturbed specimen of clay that has liquid limit of 73 and a plasticity index of 51 was placed in an oedometer apparatus and subjected to a seating pressure of 125 psf (6 kPa). When inundated with water at this seating pressure, the soil specimen stated to swell and the vertical pressure was increased to prevent the swelling. The vertical pressure had to be increased to 4000 psf (192 kPa) before the specimen began to permanently compress. The specimen was then loaded to a vertical pressure of 8000 psf (383 kPa), with this pressure remaining on the specimen for one week. In order to determine the swell index, the soil specimen was unloaded and dial versus time readings were taken.

Vertical pressure versus void ratio (Fig. 9.26).   This figure presents a plot of the void ratio versus vertical pressure (semilog plot with data from Part *b*, Fig. 9.25). This plot is used to obtain two parameters: (1) the swelling pressure, and (2) the swelling index, as follows:

a. *Swelling pressure P_s'.*   There are many different definitions of the swelling pressure of a soil. In this case, the swelling pressure P_s' is determined as the point of intersection of a horizontal line corresponding to the initial void ratio e_o and the compression curve which is extended back to the initial void ratio line. For the laboratory test data shown in Fig. 9.26, the swelling pressure P_s' is approximately equal to 4000 psf (200 kPa). The swelling pressure is important because it indicates the vertical pressure that can be applied by a footing or pier which will prevent expansion of the soil.

b. *Swelling index C_s.*   The swelling index is defined as the slope of the swelling curve (i.e., the slope of the unload curve). The swelling index C_s can be approximated as a straight line in Fig. 9.26 and is defined as follows:

File no. 4328	Job name: Royal Tern

Sample location: TP-3 @ 4 ft   Date: February 3, 2000

(a) Initial conditions of soil specimen:

Specimen diameter: 2.50 in.   Specimen height (h_0): 1.00 in.

Water content: 11.6%   Wet unit weight: 116.8 pcf   Dry unit weight: 104.7 pcf

(b) Loading data

Vertical pressure, psf	Dial reading, in.	Apparatus Compressibility d_c, in.	Specimen height, in.	Void ratio ($G_s = 2.70$)
125	0.4935 (d_s)	0	1.0000	0.6092
250	0.4930	0.0005	1.0000	0.6092
500	0.4917	0.0018	1.0000	0.6092
1000	0.4897	0.0038	1.0000	0.6092
2000	0.4880	0.0055	1.0000	0.6092
4000	0.4861	0.0070	0.9996	0.6085
8000	0.4465	0.0090	0.9620	0.5480
2000	0.4650	0.0065	0.9780	0.5738
500	0.5020	0.0040	1.0125	0.6293
125	0.5590	0.0035	1.0690	0.7202

(c) Dial readings during unloading: Vertical pressure = 2000 lb/ft^2

Time, min	Dial reading	Time, min	Dial reading	Time, h	Dial reading
0	0.4465	2	0.4515	3.2	0.4555
0.10	0.4504	5	0.4520	8.1	0.4574
0.25	0.4506	14	0.4527	2 days	0.4622
0.5	0.4509	52	0.4538	14 days	0.4647
1	0.4512	93	0.4545	28 days	0.4650

NOTE: Plot in Fig. 9.27 based on percent swell = 0 at time = 0.1 min.

(d) End of test: Water content: 26.8%

FIGURE 9.25   Laboratory test data for an expansive clay (see Fig. 9.26 and 9.27 for data plots.)

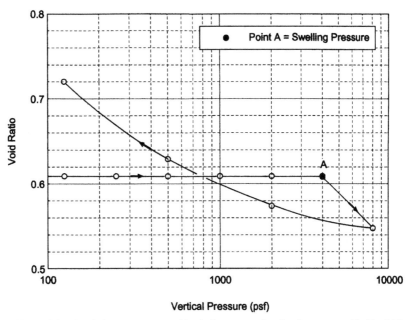

FIGURE 9.26   Vertical pressure versus void ratio for an expansive clay. Data presented in Fig. 9.25. Arrows indicate the sequence of loading and unloading. Soil has a LL = 73 and a PI = 51.

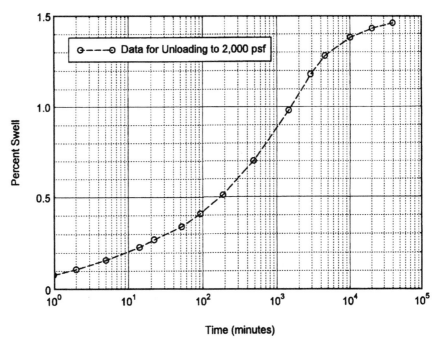

FIGURE 9.27   Vertical swell versus time for an expansive clay. Data recorded during the unloading of the soil specimen. Also see Fig. 9.26. Data presented in Fig. 9.25.

$$C_s = \frac{\Delta e}{\log(\sigma'_{vc2}/\sigma'_{vc1})} \tag{9.11}$$

An easier method to obtain C_s is to determine Δe over one log cycle: or for example, if $\sigma'_{vc2} = 100$ and $\sigma'_{vc1} = 10$, then the log $(\sigma'_{vc2}/\sigma'_{vc1}) = \log (100/10) = 1$. By using one log cycle, the values of $C_s = \Delta e$ for the swelling curve. In Fig. 9.26, using $e = 0.5738$ at 2000 psf and $e = 0.7202$ at 125 psf, the swelling index C_s is approximately equal to 0.12.

Vertical swell versus time (Fig. 9.27). Figure 9.27 presents a plot of the time versus dial readings from Part c of Fig. 9.25. Using the log-of-time method, the end of primary swell has been estimated to be about 5000 min (3.5 days). Thus, during unloading, the soil specimen was allowed to swell for one week after each reduction in vertical pressure.

Figure 9.28 presents the commonly used method for predicting the foundation heave based on the swell index C_s. The first step is to determine the depth of swelling soil, which as previously mentioned, is at least as deep as the depth of seasonal moisture change (see Sec. 9.3.1). The depth of expansive soil can also be governed by the depth to nonexpansive material. For the example shown in Fig. 9.28, the depth of swelling clay has been determined to be 2 m (6.6 ft). An undisturbed soil sample has been obtained and tested in the oedometer apparatus using the procedure outlined earlier. Similar to Fig. 9.26, the results of the laboratory test are shown in the lower left corner of Fig. 9.28 and indicate a swelling pressure $P'_s = 200$ kPa and a coefficient of swell $C_s = 0.10$.

The equation used to calculate the heave of the foundation is shown in Fig. 9.28. Note that this equation used to calculate the heave of the foundation is similar to Eq. 8.12, with h_i = the initial thickness of the in situ soil layer ($h_i = H_o$), the swelling index C_s used in place of the compression index C_c, the value of P_f = the final vertical effective stress in the soil after swelling has occurred, and P_o = swelling pressure (i.e., $P_o = P'_s$). Note that in Fig. 9.28 the swelling soil was divided into three layers and that the swell was calculated for each layer and then the total heave was calculated as the sum of the swell from the three layers.

Example Problem 9.3 Use the data from Example Problem 9.1. Determine the amount of center lift of the slab-on-grade for the change in conditions from the dry season to the wet season using the test data shown in Fig. 9.28.

Solution The example problem shown in Fig. 9.28 does not include the foundation dead load pressure of 5 kPa. Including the 5 kPa in the analysis, the results are as follows:

Layer No. 1 (0 to 0.5 m):

$$\Delta h_i = C_s [h_i/(1 + e_o)] \log(P_f/P_o)$$

$$\Delta h_i = (0.10)[(0.5 \text{ m})/(1 + 1.0)] \log[(4.5 + 5)/200] = 0.033 \text{ m}$$

Layer No. 2 (0.5 to 1 m):

$$\Delta h_i = C_s [h_i/(1 + e_o)] \log(P_f/P_o)$$

$$\Delta h_i = (0.10)[(0.5 \text{ m})/(1 + 1.0)] \log[(13.5 + 5)/200] = 0.026 \text{ m}$$

Layer No. 3 (1 to 2 m):

$$\Delta h_i = C_s [h_i/(1 + e_o)] \log(P_f/P_o)$$

$$\Delta h_i = (0.10)[(1.0 \text{ m})/(1 + 1.0)] \log[(27 + 5)/200] = 0.040 \text{ m}$$

Total center lift = 0.033 + 0.026 + 0.040 = 0.10 m (3.9 in.)

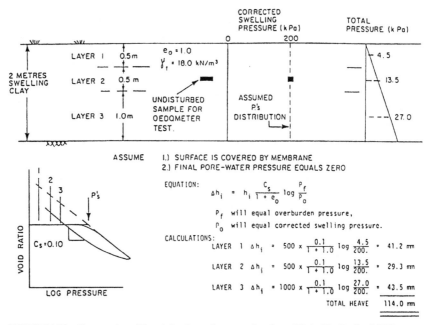

FIGURE 9.28 Computation of foundation heave for expansive clays. (*Method by Fredlund 1983; reproduced from Chen 1988.*)

9.5 *FOUNDATION DESIGN FOR EXPANSIVE SOIL*

There are many different methods used to deal with expansive soil. Options include removal of the expansive soil during grading operations, soil treatment or stabilization, deep foundation systems, and posttensioned slab-on-grade foundations.

9.5.1 Soil Treatment or Stabilization

There are many different soil treatment or stabilization methods that are used in practice. Table 9.4 presents a summary of these methods and the advantages and disadvantages of each method. Some of the more commonly used methods are as follows:

1. *Removal and replacement.*   During the grading of a project, it may be possible to remove the expansive soil and replace it with nonexpansive or less expansive soil.

2. *Remolding and compaction.*   This method is also commonly known as compaction control (Gromko, 1974). This process is based on the observation that by compacting a clay at a water content that is wet of the optimum moisture content, the initial percent swell will be less than for the same soil compacted dry of optimum (Holtz and Gibbs, 1956b; Holtz, 1959). Thus by compacting a clay wet of optimum, the swell potential of the soil will be reduced. However, if the clay should dry out prior to placement of the structure, then the beneficial effect of compacting the clay wet of optimum will be destroyed, causing the soil to become significantly more expansive (Day, 1994b).

3. *Surcharge loading.*   This method basically consists of applying sufficient pressure to the expansive soil in order to reduce the amount of swell. For example, a layer of nonexpansive soil or less

TABLE 9.4 Expansive Soil Treatment Alternatives

Method	Salient points
Removal and replacement	Nonexpansive, impermeable fill must be available and economical.
	Nonexpansive soils can be compacted at higher densities than expansive clay, producing high bearing capacities.
	If granular fill is used, special precautions must be taken to control drainage away from the fill, so water does not collect in the pervious material.
	Replacement can provide safe slab-on-grade construction.
	Expansive material may be subexcavated to a reasonable depth, then protected by a vertical and/or a horizontal membrane. Sprayed asphalt membranes are effectively used in highway construction.
Remolding and compaction	Beneficial for soils having low potential for expansion, high dry density, low natural water content, and in a fractured condition.
	Soils having a high swell potential may be treated with hydrated lime, thoroughly broken up, and compacted— if they are lime reactive.
	If lime is not used, the bearing capacity of the remolded soil is usually lower since the soil is generally compacted wet of optimum at a moderate density.
	Quality control is essential.
	If the active zone is deep, drainage control is especially important.
	The specific moisture-density conditions should be maintained until construction begins and checked prior to construction.
Surcharge loading	If swell pressures are low and some deformation can be tolerated, a surcharge load may be effective.
	A program of soil testing is necessary to determine the depth of the active zone and the maximum swell pressures to be counteracted.
	Drainage control is important when using a surcharge. Moisture migration can be both vertical and horizontal.
Prewetting	Time periods up to as long as a year or more may be necessary to increase moisture contents in the active zone.
	Vertical sand drains drilled in a grid pattern can decrease the wetting time.
	Highly fissured, desiccated soils respond more favorably to prewetting.
	Moisture contents should be increased to at least 2–3% above the plastic limit.
	Surfactants may increase the percolation rate.
	The time needed to produce the expected swelling may be significantly longer than the time to increase moisture contents.
	It is almost impossible to adequately prewett dense unfissured clays.
	Excess water left in the upper soil can cause swelling in deeper layers at a later date.
	Economics of prewetting can compare favorably to other methods, but funds must be available at an early date in the project.
	Lime treatment of the surface soil following prewetting can provide a working table for equipment and increase soil strength.
	Without lime treatment soil strength can be significantly reduced, and the wet surface may make equipment operation difficult.
	The surface should be protected against evaporation and surface slaking.
	Quality control improves performance.
Lime treatment	Sustained temperatures over 70°F for a minimum of 10 to 14 days is necessary for the soil to gain strength. Higher temperatures over a longer time produce higher strength gains.
	Organics, sulfates, and some iron compounds retard the pozzolanic reaction of lime.
	Gypsum and ammonium fertilizers can increase the soil's lime requirements.
	Calcareous and alkaline soil have good reactivity.
	Poorly drained soils have higher reactivities than well-drained soils.
	Usually 2–10% lime stabilizes reactive soil.
	Soil should be tested for lime reactivity and percentage of lime needed.
	The mixing depth is usually limited to 12–18 in., but large tractors with ripper blades have successfully allowed in-place mixing of 2 ft of soil.
	Lime can be applied dry or in a slurry, but excess water must be present.
	Some delay between application and final mixing improves workability and compaction.
	Quality control is especially important during pulverization, mixing, and compaction.

(Continued)

Lime treatment (*Continued*)	Lime-treated soils should be protected from surface and groundwater. The lime can be leached out and the soil can lose strength if saturated. Dispersion of the lime from drill holes is generally ineffective unless the soil has an extensive network of fissures. Stress relief from drill holes may be a factor in reducing heave. Smaller-diameter drill holes provide less surface area to contact the slurry. Penetration of pressure-injected lime is limited by the slow diffusion rate of the lime, the amount of fracturing in the soil, and the small pore size of clay. Pressure injection of lime may be useful to treat layers deeper than possible with the mixed-in-place technique.
Cement treatment	Portland cement (4–6%) reduces the potential for volume change. Results are similar to lime, but shrinkage may be less with cement. Method of application is similar to mix-in-place lime treatment, but there is significantly less time delay between application and final placement. Portland cement may not be as effective as lime in treating highly plastic clays. Portland cement may be more effective in treating soils that are not lime-reactive. Higher strength gains may result with cement. Cement-stabilized material may be prone to cracking and should be evaluated prior to use.
Salt treatment	There is no evidence that use of salts other than NaCl or $CaCl_2$ is economically justified. Salts may be leached easily. Lack of permanence of treatment may make salt treatment uneconomical. The relative humidity must be a least 30% before $CaCl_2$ can be used. Calcium and sodium chloride can reduce frost heave by lowering the freeing point of water. $CaCl_2$ may be useful to stabilize soils having a high sulfur content.
Fly ash	Fly ash can increase the pozzolanic reaction of silty soils. The gradation of granular soils can be improved.
Organic compounds	Spraying and injection are not very effective because of the low rate of diffusion in expansion soil. Many compounds are not water-soluble and react quickly and irreversibly. Organic compounds do not appear to be more effective than lime. None is as economical and effective as lime.
Horizontal barriers	Barrier should extend far enough from the roadway or foundation to prevent horizontal moisture movement into the foundation soils. Extreme care should be taken to securely attach barrier to foundation, seal the joints, and slope the barrier down and away from the structure. Barrier material must be durable and nondegradable. Seams and joints attaching the membrane to a structure should be carefully secured and made waterproof. Shrubbery and large plants should be planted well away from the barrier. Adequate slope should be provided to direct surface drainage away from the edges of the membranes.
Asphalt	When used in highway construction, a continuous membrane should be placed over subgrade and ditches. Remedial repair may be less complex than for concrete pavement. Strength of pavement is improved over untreated granular base. Can be effective when used in slab-on-grade construction.
Rigid barrier	Concrete sidewalks should be reinforced. A flexible joint should connect sidewalk and foundation. Barriers should be regularly inspected to check for cracks and leaks.
Vertical barrier	Placement should extend as deep as possible, but equipment limitations often restrict the depth. A minimum of half of the active zone should be used. Backfill material in the trench should be impervious. Types of barriers that have provided control of moisture content are capillary barrier (coarse limestone), lean concrete, asphalt and ground-up tires, polyethylene, and semihardening slurries. A trenching machine is more effective than a backhoe for digging the trench.
Membrane-encapsulated soil layers	Joints must be carefully sealed. Barrier material must be durable to withstand placement. Placement of the first layer of soil over the bottom barrier must be controlled to prevent barrier damage.

Source: Nelson and Miller (1992).

expansive soil can be placed on top of the clay. This soil cap increases the pressure on the clay and as shown in Table 9.1, the greater the surcharge pressure, the lower the percent swell.

Another possibility is to use the dead load of the structure to provide the surcharge loading. This approach is basically a delicate balancing act of providing enough dead load to prevent damaging uplift, but not too much dead load that settlement becomes excessive. If the foundation is constructed after the rainy season when the clay has increased in moisture and softened, the heavy dead load could actually cause excessive settlement due to consolidation of the clay.

4. *Prewetting.* This method is also commonly known as *presaturation* or *presoaking*. The idea is to flood the expansive soil and allow it to absorb water and swell. The foundation, such as a slab-on-grade, is then constructed on top of the swelled clay. Usually the perimeter footings and interior bearing wall footings are deepened in order to provide bearing beneath the softened clay. In addition, the deepened perimeter footings tend to act as a cut-off wall that traps the moisture beneath the slab-on-grade. This method has many disadvantages including the possibility that excess water left in the upper soil can cause swelling in deeper layers at a later date. Another disadvantage is that if tree roots grow underneath the slab-on-grade, they can extract moisture from the wetted clay resulting in shrinkage of the clay and downward displacement of the foundation.

5. *Soil cementation.* Many different compounds can be added to the expansive soil in order to cement the soil particles together or reduce the expansiveness of the soil. As discussed in Table 9.4, these methods include lime treatment and cement treatment.

6. *Barriers.* Horizontal and vertical barriers can be constructed around the perimeter of the foundation in order to reduce the potential for cyclic heave and shrinkage.

9.5.2 Deep Foundations

The prior section has discussed methods of dealing with expansive soils by soil treatment and stabilization. The next two sections discuss foundation types that are used to resist the expansive soil forces. For example, one possible foundation type for expansive soil is a deep foundation system, such as pier and grade beam support. The basic principle is to construct the piers such that they are below the depth of seasonal moisture change. The piers can be belled at the bottom in order to increase their uplift resistance. Grade beams and structural floor systems, that are free of the ground, are supported by the piers.

Chen (1988) provides several examples of proper construction details for pier and grade beam foundations. The design considerations for the pier and grade beam support are as follows (Woodward et al., 1972; Nadjer and Werno, 1973; Jubenville and Hepworth, 1981; Chen, 1988):

1. *Sufficient pier length.* A model test for pier uplift indicated that the expansive soil uplift of a pier is similar to the extraction of a pile (i.e., uplift capacity) from the ground (Chen, 1988). Based on this test data, the uplift force T_u on the pier due to swelling of soil in the active zone can be estimated as follows:

$$T_u = c_A \, 2\pi R Z_a \tag{9.12}$$

where T_u = uplift force on the pier (lb or kN)
    c_A = adhesion between the clay and pier (psf or kPa)
    R = radius of the pier (ft or m)
    Z_a = depth of wetting which usually corresponds to the depth of seasonal moisture
        change, i.e., the active zone (ft or m). Based on the type of pier and the undrained
        shear strength of the clay, the value for c_A can be obtained from Fig. 6.13.

There are methods to reduce the uplift force T_u on the pier. For example, an air gap can be provided around the pier or the pier hole can be enlarged and an easily deformable material placed between the pier and expansive clay.

The total resisting force T_r for straight concrete piers can be calculated as follows:

$$T_r = P + c_A \, 2\pi R Z_{na} \tag{9.13}$$

where T_r = total resisting force for straight concrete piers (lb or kN)

P = resisting force of the pier, equal to the weight of the pier and the dead load applied to the top of the pier (lb or kN)

Z_{na} = portion of the pier below the active zone (ft or m)

Note that the total length of the pier = $Z_a + Z_{na}$. If the ends of the piers are belled, then there would be additional resisting forces. By equating Eqs. 9.12 and 9.13, the value of Z_{na} can be obtained. By multiplying Z_{na} by an appropriate factor of safety (at least 1.5), the total depth of the pier can be determined.

As an alternate to the earlier calculations, it has been stated that the depth of piles or piers should be at least 1.5 times the depth where the swelling pressure is equal to the overburden pressure (David and Komornik, 1980).

2. *Pier diameter.*   In terms of pier diameter, Chen (1988) states that to exert enough dead load pressure on the piers, it is necessary to use small diameter piers in combination with long spans of the grade beams. Piers for expansive soil typically have a diameter of 1 ft (0.3 m).

3. *Pier reinforcement.*   Because a large uplift force T_u can be developed on the pier as the clay swells in the active zone, steel reinforcement is required to prevent the piers from failing in tension. Reinforcement for the full length of the pier is essential to avoid tensile failures.

4. *Construction process.*   It is important to use proper construction procedures when constructing the piers. For example, because the piers are often heavily loaded, the bottom of the piers must be cleaned of all loose debris or slough. After the pier hole is drilled and the concrete has been placed, excess concrete is often not removed from the top of the pier. This excess concrete has a mushroom shape and because of its large area, the expansive soil can exert a substantial and unanticipated uplift force onto the top of the pier (Chen, 1988).

5. *Void space below grade beams.*   A common procedure is to use a void-forming material (such as cardboard) to create a void space below the grade beam. It is important to create a complete and open void below the grade beam so that when the soil expands, it will not come in contact with the grade beam and cause damaging uplift forces.

9.5.3 Posttensioned Slab-on-Grade

Another type of foundation for expansive soils is the posttensioned slab-on-grade. There can be many different types of posttensioned designs. In Texas and Louisiana, the early posttensioned foundations consisted of a uniform thickness slab with stiffening beams in both directions, which became known as the ribbed foundation. In California, a commonly used type of posttensioned slab consists of a uniform thickness slab with an edge beam at the entire perimeter, but no or minimal interior stiffening beams or ribs. This type of posttensioned slab has been termed the California slab (Post-Tensioning Institute, 1996).

In *Design and Construction of Post-Tensioned Slabs-on-Ground*, prepared by the Post-Tensioning Institute (1996), the design moments, shears, and differential deflections under the action of soil loading resulting from changes in water contents of expansive soils are predicted using equations developed from empirical data and a computer study of a plate on an elastic foundation. To get the required rigidity to reduce foundation deflections, the stiffening beams (perimeter and interior footings) can be deepened. Although the differential movement to be expected for a given expansive soil is supplied by a soil engineer, the actual design of the foundation is usually by the structural engineer.

The posttensioned slab-on-grade should be designed for two conditions: (1) center-lift (also called center heave or doming) and (2) edge-lift (also called edge heave or dishing). These two conditions are illustrated in Fig. 9.29. Center-lift is the long-term progressive swelling beneath the center of the slab, or because the soil around the perimeter of the slab dries and shrinks (causing the perimeter to deform downward), or a combination of both. Edge-lift is the cyclic heave beneath the perimeter of the foundation. In order to complete the design, the soils engineer usually provides the maximum anticipated vertical differential soil movement y_m and the horizontal distance of moisture variation

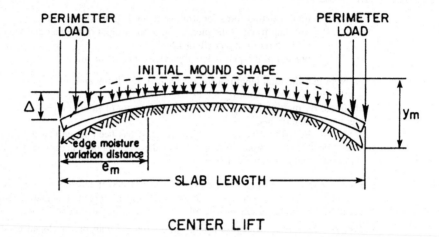

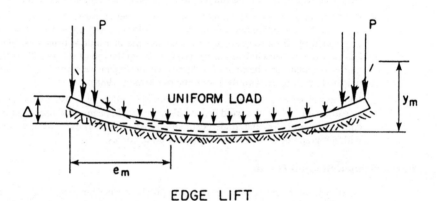

FIGURE 9.29 Depiction of center lift and edge lift for posttensioned slab-on-grade. (*Reproduced with permission from the Post-Tensioning Institute, 1996.*)

from the slab perimeter e_m for both the center-lift and edge-lift conditions (see Fig. 9.29). Values of moisture variation from the slab perimeter e_m are usually obtained from the Post-Tensioning Institute (1996), with typical values of e_m equal to 3 to 6 ft (0.9–1.8 m) for center lift and 2 to 6 ft (0.6–1.8 m) for edge lift. Other methods used to determine the edge moisture variation distance e_m are presented by El-Garhy and Wray (2004).

There are many design considerations for posttensioned slabs-on-grade. Four important considerations are as follows (Day, 1994c):

1. The design of the foundation is based on static values of y_m, but the actual movement is cyclic. To mitigate leaks from utilities, flexible utility lines that enter the slab should be used. Also, the posttensioned slab-on-grade must be rigid enough so that cracks do not develop in interior wallboard due to cyclic movement.

2. The values of e_m are difficult to determine because they are dependent on climatic conditions (i.e., nature of the wetting and drying cycles) and soil-structure interaction. The structural parameters that govern e_m include the magnitude and distribution of dead loads, rigidity of the foundation, slab length, and depth of the perimeter footings. The soil parameters include the permeability, suction values, and the heave-shrinkage characteristics of the soil.

3. The geotechnical engineer might test the expansive clays during the rainy season, but the foundation might be built during a drought. In this case, by using the methods outlined in Sec. 9.4, the values of y_m for center-lift could be considerably underestimated. The geotechnical engineer must test the clay under the conditions anticipated at the time of construction or use the worst-case scenario.

4. The Post-Tensioning Institute (1996) provides tables that can be used to determine y_m for center-lift and edge-lift conditions. However, these values of y_m are based solely on climate considerations. The values of y_m provided by the Post-Tensioning Institute (1996) do not include possible movement related to vegetation, cut-fill transitions, surface drainage, time of construction (i.e., wet versus dry season), and post-construction effects.

9.6   FLATWORK

Flatwork can be defined as appurtenant structures that surround a building or house such as concrete walkways, patios, driveways, and pool decks. It is the lightly loaded structures, such as pavements or lightly loaded foundations, that are commonly damaged by expansive soil (Williams, 1965; Van der Merwe and Ahronovitz, 1973; Snethen, 1979). Because flatwork usually only supports its own weight, it can be especially susceptible to expansive soil related damage such as shown in Fig. 9.30. The arrows in Fig. 9.30 indicate the amount of uplift of the lightly loaded exterior sidewalk relative to the heavily loaded exterior wall of the tilt-up building. The sidewalk heaved to such an extent that the door could not be opened and the door had to be rehung so that it opened into the building.

Another example is Fig. 9.31, which shows cracking to a concrete driveway due to expansive soil uplift. The expansive soil uplift tends to produce a distinct crack pattern, which has been termed a *spider* or *x-type* crack pattern.

Besides the concrete itself, utilities can also be damaged due to the upward movement of the flatwork. For example, Fig. 9.32 shows a photograph of concrete flatwork that was uplifted due to expansive soil heave. The upward movement of the flatwork bent the utility line and chipped-off the stucco as shown in Fig. 9.32.

FIGURE 9.30   Sidewalk uplift due to expansive soil.

FIGURE 9.31   Driveway cracking due to expansive soil.

Besides differential movement of the flatwork, there can also be progressive movement of flatwork away from the structure. This lateral movement is known as *walking*. Figure 9.33 shows an example of walking of flatwork where a backyard concrete patio slab was built atop highly expansive soils. The patio slab originally abutted the wall of the house, but is now separated from the house by about 1.5 in. (4 cm). At one time, the gap was filled in with concrete, but as Fig. 9.33 shows, the patio slab has continued to walk away from the house. In many cases, there will be appurtenant structures

FIGURE 9.32   Concrete flatwork uplifted by expansive soil.

FIGURE 9.33   Walking of flatwork on expansive soil.

(such as a patio shade cover) that are both attached to the house and also derive support from the flat-work. As the flatwork walks away from the house, these appurtenant structures are pulled laterally and frequently damaged.

The results of a field experiment indicated that most walking occurs during the wet period (Day, 1992c). The expansion of the clay causes the flatwork to move up and away from the structure. During the dry period, the flatwork does not return to its original position. Then during the next wet period, the expansion of the clay again causes an upward and outward movement of the flatwork. The cycles of wetting and drying cause a progressive movement of flatwork away from the building. An important factor in the amount of walking is the moisture condition of the clay prior to construction of the flatwork. If the clay is dry, then more initial upward and outward movement can occur during the first wet cycle.

9.7 EXPANSIVE ROCK

There are several different mechanisms that can cause the expansion of rock. Some rock types, such as shale, slate, mudstone, siltstone, and claystone, can be especially susceptible to expansion. Common mechanisms that can cause rock to expand are as follows:

1. *Rebound.*   For cut areas, where the overburden has been removed by erosion or during mass-grading operations, the rock will rebound due to the release in overburden pressure. The rebound can cause the opening of cracks and joints. Usually rebound of rock occurs during the rock excavation and because it is a relatively rapid process, it is not included in the engineering analyses.

2. *Expansion due to physical factors.*   Rock, especially soft sedimentary and fractured rock, can expand due to the physical growth of plants roots or by the freezing of water in the fractures. Studies have also shown that the precipitation of gypsum in rock pores, cracks, and joints can cause rock expansion and disintegration. Such conditions occur in arid climates where subsurface moisture evaporates at ground surface, precipitating the minerals in the rock pores. Gypsum crystals have been observed to grow in rock fractures and are believed to exert the most force at their

growing end (Hawkins and Pinches, 1987). Gypsum growth has even been observed in massive sandstone, which resulted in significant heave of the rock (Hollingsworth and Grover, 1992).

3. *Expansion due to weathering of rock.*   Probably the most frequent cause of heave of rock and resulting damage to foundations is due to the weathering of rock. Weathering of rock can occur by physical and chemical methods. Typical types of chemical weathering include oxidation, hydration of clay minerals, and the chemical alteration of the silt size particles to clay. Factors affecting oxidation include the presence of moisture and oxygen (aerobic conditions), biological activity, acidic environment, and temperature (Hollingsworth and Grover, 1992). An example of expansive rock is pyritic shale, which often expands upon exposure to air and moisture. Another example is bentonite, which is a rock that is composed of montmorillonite clay minerals. This type of rock will also rapidly weather and greatly expand when exposed to air and moisture.

It is often difficult to predict the amount of heave that will occur due to expansive rock. One approach is similar to the swell testing of soil (Sec. 9.4.2), where a rock specimen is placed in an oedometer, subjected to the anticipated overburden pressure, submerged in distilled water, and then the expansion of the rock is measured. However, this approach could considerably underestimate the expansion potential of the rock because in an intact (unweathered) state, it is much less expansive than in a fractured and weathered state.

Another approach is to break apart the rock and then subject the fragments to wetting and drying cycles. The cycles of wetting and drying are often very effective in rapidly weathering expansive rock (Day, 1994d). Once the rock fragments have sufficiently weathered, an expansion index test (Sec. 9.2.2) could be performed on this material. Based on the expansion index test results, the type of foundation could then be selected.

NOTATION

The following notation is used in this chapter:

A = activity (Eq. 4.6)
c_A = adhesion between clay and pier
c_s = coefficient of swell
c_v = coefficient of consolidation
C_c = compression index
C_s = swelling index (Eq. 9.11)
d_c = amount of apparatus compressibility
d_f = final dial reading
d_o = initial dial reading
d_s = dial reading corresponding to the seating load
d_v = dial reading at a specific vertical stress
e = void ratio
e_m = edge moisture variation distance
e_o = initial void ratio
Δe = change in void ratio
EI = expansion index
h = distance above the groundwater table (Sec. 9.3.2)

h = soil specimen height (Sec. 9.4.3)

h_c = height of capillary rise

h_i = initial thickness of the in situ soil layer (Fig. 9.28)

h_o = initial height of specimen

h_p = height of specimen at the end of primary swell

H = thickness of the soil layer experiencing heave or shrinkage (Sec. 9.4.1)

H_{dr} = height of the swell path

H_o = initial height of the in situ soil layer

ΔH = change in ground surface elevation due to soil swelling or shrinkage

LL = liquid limit

N = vertical pressure for one-dimensional swell test (Fig. 9.19)

P = pier weight plus dead load applied at the top of the pier

PI = plasticity index

PL = plastic limit

P_f = final overburden pressure (Fig. 9.28)

P_o = equivalent to P_s' (Fig. 9.28)

P_s' = swelling pressure (Fig. 9.28)

R = radius of pier

s_m = matric suction

s_o = osmotic suction

s_T = total suction of the soil

SL = shrinkage limit

T = time

T = time factor from Table 8.2

T_r = total resisting force for straight concrete piers

T_u = uplift force acting on the pier

u = pore water pressure

u_a = air pressure in the soil voids

u_e = excess pore water pressure

u_w = pore water pressure acting between the soil particles

w = water content

y_m = maximum anticipated vertical differential movement

Z_a = depth of wetting of the clay (active zone)

Z_{na} = portion of pier below the active zone

Δ = maximum differential movement

δ/L = maximum angular distortion

ε_v = vertical strain

γ_h = suction compressibility index

γ_w = unit weight of water

σ_v = vertical total stress

$\sigma_{vc2}', \sigma_{vc1}'$ = vertical effective stress from the swell curve (Fig. 9.26)

PROBLEMS

Solutions to the problems are presented in App. C of this book. The problems have been divided into basic categories as indicated below:

Expansion Potential. For Problems 9.1 through 9.10, consider the soils to consist of inorganic soil particles with the Atterberg limits performed on soil passing the No. 40 sieve (per ASTM D 4318-00, 2004). Note that LL is the liquid limit, PL is the plastic limit, and PI is the plasticity index.

9.1 Using the laboratory test data shown in Fig. 3.6, determine the expansion potential of the soil based on clay content, plasticity index, and the expansion soil classification charts shown in Figs. 9.1 and 9.2.

ANSWER: Based on clay content: high expansion potential. Based on plasticity index: very high expansion potential. Based on Fig. 9.1: high expansion potential. Based on Fig. 9.2: high expansion potential.

9.2 Using the laboratory test data shown in Fig. 4.34 for the soil at SB-38 at 4 to 8 ft, determine the expansion potential of the soil based on clay content, plasticity index, and the expansion soil classification charts shown in Figs. 9.1 and 9.2.

ANSWER: Based on clay content: high expansion potential. Based on plasticity index: high expansion potential. Based on Fig. 9.1: medium expansion potential. Based on Fig. 9.2: high expansion potential.

9.3 Using the laboratory test data shown in Fig. 4.35 for the soil at SB-39 at 4 to 8 ft, determine the expansion potential of the soil based on clay content, plasticity index, and the expansion soil classification charts shown in Figs. 9.1 and 9.2.

ANSWER: Based on clay content: high expansion potential. Based on plasticity index: high expansion potential. Based on Fig. 9.1: medium expansion potential. Based on Fig. 9.2: high expansion potential.

9.4 Using the laboratory test data shown in Fig. 4.35 for the soil at SB-42 at 4 to 8 ft, determine the expansion potential of the soil based on clay content, plasticity index, and the expansion soil classification charts shown in Figs. 9.1 and 9.2.

ANSWER: Based on clay content: medium expansion potential. Based on plasticity index: low expansion potential. Based on Fig. 9.1: low expansion potential. Based on Fig. 9.2: low expansion potential.

9.5 Using the laboratory test data shown in Fig. 4.35 for the soil at SB-45 at 4 to 8 ft, determine the expansion potential of the soil based on clay content, plasticity index, and the expansion soil classification charts shown in Figs 9.1 and 9.2.

ANSWER: Based on clay content: low expansion potential. Based on plasticity index: very low expansion potential. Based on Fig. 9.1: low expansion potential. Based on Fig. 9.2: low expansion potential.

9.6 Using the laboratory test data shown in Fig. 4.36 for the soil at TP-1 at 1 to 2.5 ft, determine the expansion potential of the soil based on clay content, plasticity index, and the expansion soil classification charts shown in Figs. 9.1 and 9.2.

ANSWER: Based on clay content: very low expansion potential. Based on plasticity index: very low expansion potential. Based on Fig. 9.1: low expansion potential. Based on Fig. 9.2: low expansion potential.

9.7 Using the laboratory test data shown in Fig. 4.36 for the soil at TP-15 at 1 to 2 ft, determine the expansion potential of the soil based on clay content, plasticity index, and the expansion soil classification charts shown in Figs. 9.1 and 9.2.

ANSWER: Based on clay content: medium expansion potential. Based on plasticity index: medium expansion potential. Based on Fig. 9.1: low expansion potential. Based on Fig. 9.2: medium expansion potential.

9.8   Using the laboratory test data shown in Fig. 4.36 for the soil at TP-2 at 1 to 2 ft, determine the expansion potential of the soil based on clay content, plasticity index, and the expansion soil classification charts shown in Figs. 9.1 and 9.2.

ANSWER: Based on clay content: very high expansion potential. Based on plasticity index: very high expansion potential. Based on Fig. 9.1: very high expansion potential. Based on Fig. 9.2: very high expansion potential.

9.9   Using the laboratory test data shown in Fig. 4.37 for the soil at SB-2 at 5 ft, determine the expansion potential of the soil based on clay content, plasticity index, and the expansion soil classification charts shown in Figs. 9.1 and 9.2.

ANSWER: Based on clay content: high expansion potential. Based on plasticity index: high expansion potential. Based on Fig. 9.1: medium expansion potential. Based on Fig. 9.2: high expansion potential.

9.10   Use the laboratory test data shown in Fig. 4.38 for the soil at AGC-2 at 0.8 to 1.4 ft and assume that the soil has 12 percent clay size particles (i.e., 12 percent finer than 0.002 mm). Determine the expansion potential of the soil based on clay content, plasticity index, and the expansion soil classification charts shown in Figs. 9.1 and 9.2.

ANSWER: Based on clay content: low expansion potential. Based on plasticity index: very low expansion potential. Based on Fig. 9.1: low expansion potential. Based on Fig. 9.2: low expansion potential.

For Problems 9.11 through 9.14, use the laboratory testing data summarized in Fig. 4.39. Consider the soils to consist of inorganic soil particles with the Atterberg limits performed on soil passing the No. 40 sieve (per ASTM D 4318-00, 2004). Note in Fig. 4.39 that W_l is the liquid limit, W_p is the plastic limit, and PI is the plasticity index.

9.11   Using the laboratory test data shown in Fig. 4.39 for soil number 5 (glacial till from Illinois), determine the expansion potential of the soil based on clay content, plasticity index, and the expansion soil classification charts shown in Figs. 9.1 and 9.2.

ANSWER: Based on clay content: low expansion potential. Based on plasticity index: very low expansion potential. Based on Fig. 9.1: low expansion potential. Based on Fig. 9.2: low expansion potential.

9.12   Using the laboratory test data shown in Fig. 4.39 for soil number 6 (Wewahitchka sandy clay from Florida), determine the expansion potential of the soil based on clay content, plasticity index, and the expansion soil classification charts shown in Figs. 9.1 and 9.2.

ANSWER: Based on clay content: very high expansion potential. Based on plasticity index: very high expansion potential. Based on Fig. 9.1: high expansion potential. Based on Fig. 9.2: very high expansion potential.

9.13   Using the laboratory test data shown in Fig. 4.39 for soil number 7 (loess from Mississippi), determine the expansion potential of the soil based on clay content, plasticity index, and the expansion soil classification charts shown in Figs. 9.1 and 9.2.

ANSWER: Based on clay content: very low expansion potential. Based on plasticity index: very low expansion potential. Based on Fig. 9.1: low expansion potential. Based on Fig. 9.2: low expansion potential.

9.14   Using the laboratory test data shown in Fig. 4.39 for soil number 8 (backswamp deposit from Mississippi river), determine the expansion potential of the soil based on clay content, plasticity index, and the expansion soil classification charts shown in Figs. 9.1 and 9.2.

ANSWER: Based on clay content: very high expansion potential. Based on plasticity index: very high expansion potential. Based on Fig. 9.1: high expansion potential. Based on Fig. 9.2: very high expansion potential.

9.15   Figure 9.34 shows the results of an expansion index test performed on soil taken from TP-1 at a depth of 5 to 7 ft. Determine the end of primary expansion index and the expansion potential.

ANSWER: End of primary expansion index = 140 and per Table 9.1, very high expansion potential.

Rate of Swell

9.16   Assume the upper 2 m of a clay deposit has the swell behavior shown in Fig. 9.8. Assume the ground surface is deliberately flooded. How long will it take for 90 percent of primary swell to occur?

ANSWER: 1.2 year.

9.17   For Problem 9.16, assume that in addition to the ground surface being flooded, the groundwater table rises to a depth of 2 m below ground surface. How long will it take for 90 percent of primary swell to occur.

ANSWER: 0.3 year.

9.18   Using the data shown in Fig. 9.8, determine the secondary swell ratio $C_{\alpha s}$.

ANSWER: The equation used to calculate the secondary compression ratio (Sec. 8.6) can be used to calculate $C_{\alpha s} = 0.003$.

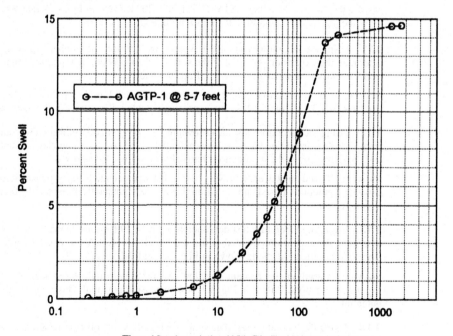

FIGURE 9.34   Data for Problem 9.15.

Foundation Heave

9.19   Use the data shown in Fig. 9.24. If the allowable foundation heave = 0.9 in., what depth of undercut and replacement with nonexpansive soil would be required?

ANSWER: Depth of undercut and replacement = 2 ft.

9.20   Use the data shown in Fig. 9.24, except assume that the swell tests performed on undisturbed soil specimens indicate the following:

Depth below ground surface of soil specimen (ft)	Percent swell
2	4
4	3
6	2
8	1
10	0

Using the above data, calculate the total heave of the foundation.

ANSWER: Total heave = 2.4 in.

9.21   Use the data shown in Fig. 9.28. Assume that the swell pressure $P'_s = 100$ kPa. Determine the total heave of the foundation assuming the foundation is constructed at ground surface and its weight is neglected.

ANSWER: Total heave of the foundation = 84 mm.

9.22   Use the data shown in Fig. 9.28. Assume that the swell index $C_s = 0.20$. Determine the total heave of the foundation assuming the foundation is constructed at ground surface and its weight is neglected.

ANSWER: Total heave of the foundation = 228 mm.

9.23   Use the data shown in Fig. 9.28. Assume that the final condition P_f will be a groundwater table at a depth of 0.5 m with hydrostatic pore water pressures below the groundwater table and zero pore water pressures above the groundwater table. Also assume the total unit weight γ_t of 18 kN/m^3 can be used for the clay above and below the groundwater table. Determine the total heave of the foundation assuming the foundation is constructed at ground surface and its weight is neglected.

ANSWER: Total heave of the foundation = 126 mm.

9.24   Use the data from Problem 9.23. Also assume that there is a mat foundation with the bottom of the mat located at a depth of 0.5 m and the mat foundation exerts a vertical pressure = 25 kPa on the soil at this level. If the mat foundation is large enough so that one-dimensional conditions exist beneath the center of the mat foundation, determine the total heave of the center of the mat foundation.

ANSWER: Total heave of the center of the mat foundation = 61 mm.

9.25   Use the data from Problem 9.23. Also assume that there is a square footing (1.2 m by 1.2 m) with the bottom of the square footing located at a depth of 0.5 m and the square footing exerts a vertical pressure = 50 kPa on the soil at this level. Using the 2:1 approximation, determine the total heave of the square footing.

ANSWER: Total heave of the square footing = 59 mm.

9.26   Solve Problem 9.25, but assume that an undisturbed soil specimen was obtained from layer no. 3 (see Fig. 9.28) and testing in the oedometer apparatus indicates a swell pressure (P'_s) of 100 kPa. Using the 2:1 approximation, determine the total heave of the square footing.

ANSWER: Total heave of the square footing = 44 mm.

9.27   Use the data shown in Fig. 9.28. Assume the 2 m of swelling clay has an undrained shear strength s_u of 50 kPa. It is proposed to construct a concrete pier and grade beam foundation such that the bottom of the grade beams are located at a depth of 0.4 m below ground surface. The piers will have a diameter of 0.3 m. Also assume that there is unweathered shale (adhesion value, $c_A = 80$ kPa) below the 2 m of swelling clay. If each pier supports a dead load (including the weight of the pier) of 20 kN, determine the depth of the piers below ground surface (include a factor of safety = 2.0). Also determine the thickness of the air gap that should be provided beneath the grade beams (include a factor of safety = 1.5).

ANSWER: Depth of piers below ground surface = 3.0 m, thickness of air gap = 110 mm.

CHAPTER 10
SLOPE STABILITY

10.1 INTRODUCTION

The design of the foundation must consider the consequences of slope movement. For example, if the foundation is too close to the top of the slope, then any slope movement could cause settlement and lateral deformation of the structure. To mitigate this effect, building codes often require that the foundation be setback from the top of slope. For example, the *International Building Code* (2009) requires that the foundation be setback from the face of the slope a minimum distance of $H/3$, where H = height of the slope, but the distance need not exceed 40 ft (12 m).

This chapter will be devoted to the slope stability analyses for static conditions. Slope stability analyses for earthquakes will be covered in Chap. 13. Slope movement can generally be divided into six basic categories, as follows:

1. *Rockfalls or topples.* This is usually an extremely rapid movement that includes the free fall of rocks, movement of rocks by leaps and bounds down the slope face, and/or the rolling of rocks or fragments of rocks down the slope face (Varnes, 1978). A rock topple is similar to a rockfall, except that there is a turning moment about the center of gravity of the rock, which results in an initial rotational type movement and detachment from the slope face. Rockfalls are discussed in Sec. 10.2.

2. *Surficial slope stability.* Surficial slope instability involves shear displacement along a distinct failure (or slip) surface. As the name implies, surficial slope stability analyses deals with the outer face of the slope, generally up to 4 ft (1.2 m) deep. The typical surficial slope stability analysis assumes the failure surface is parallel to the slope face. Surficial slope stability analyses are discussed in Sec. 10.3.

3. *Gross slope stability.* As contrasted with a shallow (surficial) stability analysis, gross slope stability involves an analysis of the entire slope. In the stability analysis of slopes, planar or circular slip surfaces are often assumed. Terms, such as fill slope stability analyses and earth or rock slump analyses, have been used to identify similar processes. Gross slope stability analyses are discussed in Sec. 10.4.

4. *Landslides.* Gross slope stability could be referred to as *landslide analysis.* However, landslides in some cases may be so large that they involve several different slopes. Landslides are discussed in Sec. 10.5.

5. *Debris flow.* A debris flow is commonly defined as soil with entrained water and air that moves readily as a fluid on low slopes. Debris flow can include a wide variety of soil-particle sizes (including boulders) as well as logs, branches, tires, and automobiles. Other terms, such as mud flow, debris slide, mud slide, and earth flow, have been used to identify similar processes. While categorizing flows based on rate of movement or the percentage of clay particles may be important, the mechanisms of all these flows are essentially the same (Johnson and Rodine, 1984). Debris flows are discussed in Sec. 10.6.

6. *Creep.* Creep is generally defined as an imperceptibly slow and more or less continuous downward and outward movement of slope-forming soil or rock (Stokes and Varnes, 1955). Creep can

affect both the near-surface (surficial) soil or deep-seated (gross) materials. The process of creep is frequently described as viscous shear that produces permanent deformations, but not failure as in landslide movement. Creep is discussed in Sec. 10.7.

In order to determine the stability of a slope, the factor of safety is calculated. A factor of safety of 1.0 indicates a failure condition, while a factor of safety greater than 1.0 indicates that the slope is stable. The higher the factor of safety, the higher the stability of a slope. If the factor of safety of the slope is deemed to be too low, then remedial treatments will be needed.

Table 10.1 presents a checklist for slope stability and landslide analysis. This table provides a comprehensive list of the factors that may need to be considered by the geotechnical engineer when designing slopes.

The calculation of the factor of safety of slopes is based on the *limit equilibrium method*. This is an approximate method that considers a state of equilibrium between the driving forces that cause failure and the forces resisting failure that are due to the mobilized shear stresses of the soil. The primary assumption of the limit equilibrium method is that the shear strength can be simultaneously mobilized along the entire failure surface. This assumption would be applicable to soil that has a uniform shear strength and identical stress-strain behavior throughout the slope. However, for many slope stability analyses, there may be several layers of soil, each having different shear strength parameters and stress-strain curves. Thus the primary assumption of the limit equilibrium method would not be satisfied, i.e., the shear strength will not be mobilized simultaneously for the different layers comprising the slope. This will be further discussed in Sec. 10.4.3.

10.1.1 Groundwater

A very important factor in slope stability is the presence of or anticipated rise of groundwater within the slope. Groundwater can affect slopes in many different ways and Table 10.2 presents common examples and the influence of groundwater on slope failures. The main destabilizing factors of groundwater on slope stability are as follows (Harr, 1962; Cedergren, 1989):

1. Reducing or eliminating cohesive strength

2. Producing pore water pressures that reduce effective stresses, thereby lowering shear strength

3. Causing horizontally inclined seepage forces that increase the driving forces and reduce the factor of safety of the slope

4. Providing for the lubrication of slip surfaces

5. Trapping of groundwater in soil pores during earthquakes or other severe shocks that leads to liquefaction failures

There are many different construction methods to mitigate the effects of groundwater on slopes. During construction of slopes, built-in drainage systems can be installed. For existing slopes, drainage devices such as trenches or galleries, relief wells, or horizontal drains can be installed. Another common slope stabilization method is the construction of a drainage buttress at the toe of a slope. In its simplest form, a drainage buttress can consist of cobbles or crushed rock placed at the toe of a slope. The objective of the drainage buttress is to be as heavy as possible to stabilize the toe of the slope and also have a high permeability so that seepage is not trapped in the underlying soil.

10.1.2 Allowable Lateral Movement of Foundations

As compared to the settlement of buildings, there is less work available on the allowable lateral movement of foundations. To evaluate the lateral movement of buildings, a useful parameter is the horizontal strain ε_h, defined as the change in length divided by the original length of the foundation. Figure 10.1 shows a correlation between horizontal strain ε_h and severity of damage (Boone, 1996; originally from Boscardin and Cording, 1989). Assuming a 20 ft (6 m) wide zone of the foundation subjected to lateral movement, Fig. 10.1 indicates that a building can be damaged by as little as

TABLE 10.1   Checklist for the Study of Slope Stability and Landslides

Main topics	Relevant items
Topography	Contour map, consider land form and anomalous patterns (jumbled, scarps, bulges). Surface drainage, evaluate conditions such as continuous or intermittent drainage. Profiles of slope, to be evaluated along with geology and the contour map. Topographic changes, such as the rate of change by time and correlate with groundwater, weather, and vibrations.
Geology	Formations at site, consider the sequence of formations, colluvium (bedrock contact and residual soil), formations with bad experience, and rock minerals susceptible to alteration. Structure: evaluate three-dimensional geometry, stratification, folding, strike and dip of bedding or foliation (changes in strike and dip and relation to slope and slide), and strike and dip of joints with relation to slope. Also investigate faults, breccia, and shear zones with relation to slope and slide. Weathering, consider the character (chemical, mechanical, and solution) and depth (uniform or variable).
Groundwater	Piezometric levels within slope, such as normal, perched levels, or artesian pressures with relation to formations and structure. Variations in piezometric levels due to weather, vibration, and history of slope changes. Other factors include response to rainfall, seasonal fluctuations, year-to-year changes, and effect of snowmelt. Ground surface indication of subsurface water, such as springs, seeps, damp areas, and vegetation differences. Effect of human activity on groundwater, such as groundwater utilization, groundwater flow restriction, impoundment, additions to groundwater, changes in ground cover, infiltration opportunity, and surface water changes. Groundwater chemistry, such as dissolved salts and gases and changes in radioactive gases.
Weather	Precipitation from rain or snow. Also consider hourly, daily, monthly, or annual rates. Temperature, such as hourly and daily means or extremes, cumulative degree-day deficit (freezing index), and sudden thaws. Barometric changes.
Vibration	Seismicity, such as seismic events, microseismic intensity, and microseismic changes. Human induced from blasting, heavy machinery, or transportation (trucks, trains, etc.)
History of slope changes	Natural processes, such as long-term geologic changes, erosion, evidence of past movement, submergence, or emergence. Human activities, including cutting, filling, clearing, excavation, cultivation, paving, flooding, and sudden drawdown of reservoirs. Also consider changes caused by human activities, such as changes in surface water, groundwater, and vegetation cover. Rate of movement from visual accounts, evidence in vegetation, evidence in topography, or photographs (oblique, aerial, stereoptical data, and spectral changes). Also consider instrumental data, such as vertical changes, horizontal changes, and internal strains and tilt, including time history. Correlate movements with groundwater, weather, vibration, and human activity.

Source: Adapted from Sowers and Royster, 1978. Reprinted with permission from *Landslides: Analysis and Control*, Special Report 176. Copyright 1978 by National Academy of Sciences. Courtesy of the National Academy Press, Washington, D.C.

TABLE 10.2   Common Examples of Slope Failures

Kind of slope	Conditions leading to failure	Type of failure and its consequences
Natural earth slopes above developed land areas (homes, industrial)	Earthquake shocks, heavy rains, snow, freezing and thawing, undercutting at toe, mining excavations	Mud flows, avalanches, landslides; destroying property, burying villages, damming rivers
Natural earth slopes within developed land areas	Undercutting of slopes, heaping fill on unstable slopes, leaky sewers and water lines, lawn sprinkling	Usually slow creep type of failure; breaking water mains, sewers, destroying buildings, roads
Reservoir slopes	Increased soil and rock saturation, raised water table, increased buoyancy, rapid drawdown	Rapid or slow landslides; damaging highways, railways, blocking spillways, leading to overtopping of dams, causing flood damage with serious loss of life
Highway or railway cut or fill slopes	Excessive rain, snow, freezing, thawing, heaping fill on unstable slopes, undercutting, trapping groundwater	Cut slope failures blocking roadways, foundation slipouts removing roadbeds or tracks; property damage, some loss of life
Earth dams and levees, reservoir ridges	High seepage levels, earthquake shocks; poor drainage	Sudden slumps leading to total failure and floods downstream; much loss of life, property damage
Excavations	High groundwater level, insufficient groundwater control, breakdown of dewatering systems	Slope failures or heave of bottoms of excavations; largely delays in construction, equipment loss, property damage

Source: From Cedergren, 1989; reprinted with permission from John Wiley & Sons.

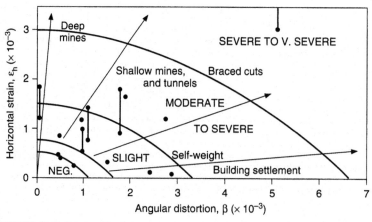

FIGURE 10.1   Relationship of damage to angular distortion and horizontal extension strain. (*From Boscardin and Cording 1989; reprinted with permission from the American Society of Civil Engineers.*)

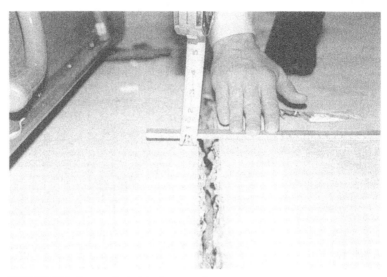

FIGURE 10.2   Damage due to lateral movement.

0.1 in. (3 mm) of lateral movement. Figure 10.1 also indicates that a lateral movement of 1 in. (25 mm) would cause *severe* to *very severe* building damage.

It should be mentioned that in Fig. 10.1, Boscardin and Cording (1989) used a *distortion factor* in their calculation of angular distortion β for foundations subjected to settlement from mines, tunnels, and braced cuts. Because of this *distortion factor*, the angular distortion β by Boscardin and Cording (1989) in Fig. 10.1 is different than the definition (δ/L) as used in Chap. 7.

The ability of the foundation to resist lateral movement will depend on the tensile strength of the foundation. Those foundations that cannot resist the tensile forces imposed by lateral movement will be the most severely damaged. For example, Figs. 10.2 and 10.3 show damage to a tilt-up building. For a tilt-up building, the exterior walls are cast in segments upon the concrete floor slab and then once they gain sufficient strength, they are tilted up into position. The severe damage shown in Figs. 10.2 and 10.3 was caused by slope movement, which affected the tilt-up building because it was constructed near the top of the slope. Figure 10.2 shows lateral separation of the concrete floor slab at the location of a floor joint. Figure 10.3 shows separation at the junction of two tilt-up panels. Because of the presence of joints between tilt-up panels and joints in the concrete floor slab, the building was especially susceptible to slope movement, which literally pulled apart the tilt-up building.

Those foundations that have joints or planes of weakness, such as the tilt-up building shown in Figs. 10.2 and 10.3, will be most susceptible to damage from lateral movement. Buildings having a mat foundation or a posttensioned slab would be less susceptible to damage because of the high tensile resistance of these foundations.

10.2   ROCKFALL

A *rockfall* is defined as a relatively free falling rock or rocks that have detached themselves from a cliff, steep slope, cave, arch, or tunnel (Stokes and Varnes, 1955). The movement may be by the process of a vertical fall, by a series of bounces, or by rolling down the slope face. The free fall nature of the rocks and the lack of movement along a well-defined slip surface differentiate a rockfall from a rockslide.

FIGURE 10.3   Joint separation between wall panels.

A rock slope is characterized by a heterogeneous and discontinuous medium of solid rocks that are separated by discontinuities. The rocks comprising a rockfall tend to detach themselves from these preexisting discontinuities in the slope or tunnel walls. The sizes of the individual rocks in a rockfall are governed by the attitude, geometry, and spatial distribution of the rock discontinuities. The basic factors governing the potential for a rockfall include (Piteau and Peckover, 1978):

1. The geometry of the slope
2. The system of joints and other discontinuities and the relation of these systems to possible failure surfaces
3. The shear strength of the joints and discontinuities
4. Destabilizing forces such as water pressure in the joints, freezing water, or vibrations

Rockfall can cause significant damage to structures in their path. For example, Fig. 10.4 shows a large rock that detached itself from the slope and landed on the roof of a house. The house is a one-story structure, having typical wood frame construction, and a stucco exterior. Figure 10.5 shows the location where the rock detached itself from the slope. The rock that impacted the house was part of the Santiago Peak Volcanics, which consists of an elongated belt of mildly metamorphosed volcanic rock. The rock is predominantly dacite and andesite and can be classified as hard and extremely resistant to weathering and erosion (Kennedy, 1975). The part of the house below the area of impact of the rock was severely damaged as shown in Figs. 10.6 and 10.7. The oven, refrigerator, and other

FIGURE 10.4 Rockfall in Santiago Peak volcanics.

FIGURE 10.5 Location of rockfall.

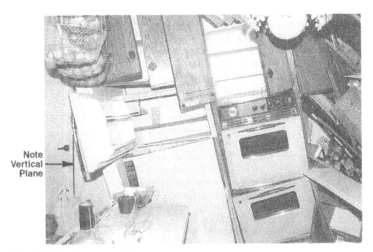

FIGURE 10.6 Damage to kitchen area.

FIGURE 10.7   Damage to bathroom.

items are turned over at an angle in Fig. 10.6. The hanging baskets in the upper left corner of Fig. 10.6 provide a vertical plane of reference. The damages shown in Figs. 10.6 and 10.7 were due to the force of impact that crushed the roof and interior walls that were beneath the rock. There was also damage in the house at other locations away from the area of rock impact. For example, Fig. 10.8 shows interior wallboard cracking and distortion of the door frames. Note that the door frames in Fig. 10.8 were originally rectangular, but are now highly distorted.

The investigation and analysis of rockfall will often be jointly performed by the engineering geologist and geotechnical engineer. If the analysis indicates that rockfall are likely to impact the site, then remedial measures can be implemented. For slopes, the main measures to prevent damage due to a rockfall are as follows (Peckover, 1975; Piteau and Peckover, 1978):

FIGURE 10.8   Damage to interior wallboard and distortion of door frames.

1. *Alter the slope configuration.*   Altering the slope can include such measures as removing the unstable or potentially unstable rocks, flattening the slope, or incorporating benches into the slope.

2. *Retain the rocks on the slope face.*   Measures to retain the rocks on the slope face include the use of anchoring systems (such as bolts, rods, or dowels), shotcrete applied to the rock slope face, or retaining walls.

3. *Intercept the falling rocks before they reach the structure.*   Intercepting or deflecting falling rocks around the structure can be accomplished by using toe-of-slope ditches, wire mesh catch fences, and catch walls. Recommendations for the width and depth of toe-of-slope ditches have been presented by Ritchie (1963) and Piteau and Peckover (1978) and are shown in Fig. 10.9.

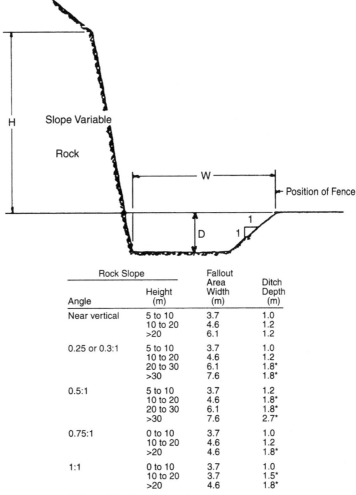

Rock Slope		Fallout Area Width (m)	Ditch Depth (m)
Angle	Height (m)		
Near vertical	5 to 10	3.7	1.0
	10 to 20	4.6	1.2
	>20	6.1	1.2
0.25 or 0.3:1	5 to 10	3.7	1.0
	10 to 20	4.6	1.2
	20 to 30	6.1	1.8*
	>30	7.6	1.8*
0.5:1	5 to 10	3.7	1.2
	10 to 20	4.6	1.8*
	20 to 30	6.1	1.8*
	>30	7.6	2.7*
0.75:1	0 to 10	3.7	1.0
	10 to 20	4.6	1.2
	>20	4.6	1.8*
1:1	0 to 10	3.7	1.0
	10 to 20	3.7	1.5*
	>20	4.6	1.8*

* May be 1.2 m if catch fence is used.

FIGURE 10.9 Design criteria for ditches at the base of rock slopes. (*From Piteau and Peckover, 1978.*)

4. *Direct the falling rocks around the structure.* Walls, fences, or ditches can be constructed in order to deflect the falling rocks around the structure.

10.3 SURFICIAL SLOPE STABILITY

Surficial failures of slopes are quite common throughout the United States and are often referred to as *shallow slides* or *shallow surface slips* (Day and Axten, 1989; Wu et al., 1993; Aubeny and Lytton, 2004). In southern California, surficial failures usually occur during the winter rainy season, after a prolonged rainfall or during a heavy rainstorm, and are estimated to account for more than 95 percent of the problems associated with slope movement upon developed properties (Gill, 1967).

10.3.1 Surficial Stability Equation

Figure 10.10 illustrates a typical surficial slope failure. The surficial failure by definition is shallow with the failure surface parallel to the slope face and usually at a depth of 4 ft (1.2 m) or less (Evans, 1972). The common surficial failure mechanism for clay slopes in southern California is as follows:

1. During the hot and dry summer period, the slope face can become desiccated and shrunken. The extent and depth of the shrinkage cracks depends on many factors, such as the temperature and humidity, the plasticity of the clay, and the extraction of moisture by plant roots.

2. When the winter rains occur, water percolates into the shrinkage cracks causing the slope surface to swell and saturate with a corresponding reduction in shear strength. Initially, water percolates downward into the slope through desiccation cracks and in response to the suction pressures of the dried clay.

3. As the outer face of the slope swells and saturates, the permeability parallel to the slope face increases. With continued rainfall, seepage develops parallel to the slope face.

4. Because of a reduction in shear strength due to saturation and swell coupled with the condition of seepage parallel to the slope face, failure occurs.

To determine the factor of safety (F) for surficial stability, an effective stress slope stability analysis is performed and the following equation is used:

$$F = \frac{\text{shear strength of soil}}{\text{shear stress in the soil}} = \frac{c' + \sigma'_n \tan \phi'}{\tau} \tag{10.1}$$

where c' = cohesion based on an effective stress analysis (psf or kPa)
σ'_n = effective normal stress on the assumed failure surface (psf or kPa)
ϕ' = friction angle based on an effective stress analysis (degrees)
τ = shear stress (psf or kPa)

In order to derive the surficial stability equation, it is assumed that there is an infinite slope with steady-state seepage of groundwater parallel to the slope face and that this seepage extends from the slope face to a depth of d. Based on these assumptions, the values of σ'_n and τ can be determined and the final result is as follows (Lambe and Whitman, 1969):

$$F = \frac{c' + \gamma_b d \cos^2 \alpha \tan \phi'}{\gamma_t d \cos \alpha \sin \alpha} \tag{10.2}$$

Terraced Lots Above

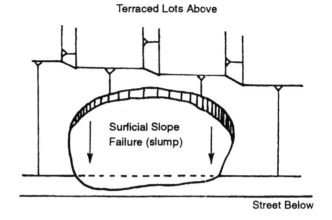

A) PLAN VIEW

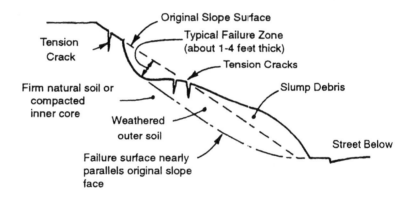

B) SECTION VIEW

FIGURE 10.10   Illustration of typical surficial slope failure: (*a*) plan view; (*b*) cross-sectional view.

where γ_b = buoyant unit weight of the soil (pcf or kN/m^3)
γ_t = total unit weight of the soil for a saturated state (pcf or kN/m^3)
α = slope inclination (degrees)
d = vertical zone of steady state seepage (ft or m)

Since an effective stress analysis is being performed with steady-state flow conditions, effective shear strength soil parameters (ϕ' = effective friction angle; c' = effective cohesion) must be used in the analysis. In Eq. 10.2, the parameter d is also the depth of the failure surface where the factor of safety is computed.

The factor of safety for surficial stability (Eq. 10.2) is highly dependent on the effective cohesion value of the soil. Automatically assuming $c' = 0$ for Eq. 10.2 may be overly conservative. Because of the shallow nature of surficial failures, the effective normal stress on the failure surface is very low. Studies have shown that the effective shear strength envelope for soil can be nonlinear at low effective stresses. For example, Fig. 10.11(from Maksimovic, 1989b) shows the effective stress failure envelope for compacted London clay. Note the nonlinear nature of the shear strength envelope.

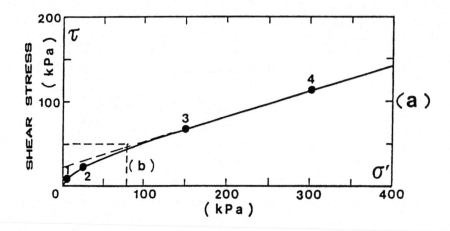

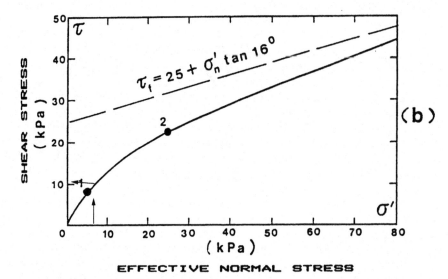

EFFECTIVE NORMAL STRESS

FIGURE 10.11   Failure envelope for compacted London clay: (a) Investigated stress range with detail; (b) low stress range. For Table 10.3 calculations (F = 0.98), enter lower chart at σ'_n = 8.3 kPa to obtain shear strength = 10.7 kPa as shown above. (*Adapted from Maksimovic, 1989b.*)

For an effective stress of about 100 to 300 kPa (2100 to 6300 psf), the failure envelope is relatively linear and has effective shear strength parameters of ϕ' = 16° and c' = 25 kPa (520 psf). But below about 100 kPa (2100 psf), the failure envelope is curved (see Fig. 10.11b) and the shear strength is less than the extrapolated line from high effective stresses. Because of the nonlinear nature of the shear strength envelope, the shear strength parameters (c' and ϕ') obtained at high normal stresses can overestimate the shear strength of the soil and should not be used in Eq. 10.2 (Day, 1994e). As an example, suppose a fill slope having an inclination of 1.5:1 (α = 33.7°) was constructed using compacted London clay having the shear strength envelope shown in Fig. 10.11. Table 10.3 lists the parameters used for the calculations and shows that using the linear portion of the shear strength envelope (Fig. 10.11a), the factor of safety = 2.5; but using the actual nonlinear shear strength envelope (Fig. 10.11b), the factor of safety = 0.98, indicating a failure condition.

TABLE 10.3 Example of Surficial Stability Calculations

Shear strength	Factor of safety
Linear portion of shear strength envelope, $\phi' = 16°$ and $c' = 25$ kPa (520 psf), i.e., Fig. 10.11a	2.5
Nonlinear portion of shear strength envelope at low effective stress, i.e., Fig. 10.11b	0.98

Note: Both surficial stability analyses use Eq. 10.2, with slope inclination $\alpha = 33.7°$, total unit weight $\gamma_t = 19.8$ kN/m³ (126 pcf), buoyant unit weight $\gamma_b = 10.0$ kN/m³ (63.6 pcf), and depth of seepage and failure plane $d = 1.2$ m (4 ft).

10.3.2 Surficial Failures

There can be surficial slope failures of cut slopes, natural slopes, and fill slopes, as follows:

Cut Slopes. Figure 10.12 shows a picture of a surficial slope failure in a cut slope. The slope is located in Poway, California and was created in 1991 by cutting down the hillside during the construction of the adjacent road. The cut slope has an area of about 4000 ft² (400 m²), a maximum

FIGURE 10.12 Surficial failure, cut slope for road.

height of 20 ft (6 m), and a slope inclination that varies from 1.5:1 (34°) to 1:1 (45°). These are rather steep slope inclinations, but are not uncommon for cut slopes in rock. The failure mechanism is a series of thin surficial failures, about 0.5 ft (0.15 m) thick. The depth to length ratio D/L of a single failure mass is around 5 to 6 percent. Using this ratio, the slides shown in Fig. 10.12 fall within the classification range (3 to 6 percent) that Hansen (1984) has defined as *shallow surface slips*.

The type of rock exposed in the cut slope is the Friars formation. This rock is of middle to late Eocene and has thick layers of nonmarine lagoonal sandstone and claystone (Kennedy, 1975). The clay minerals are montmorillonite and kaolinite. The Friars formation is common in San Diego and Poway, California, and is a frequent source of geotechnical problems such as landslides and heave of foundations.

The cause of the surficial slope failures shown in Fig. 10.12 was weathering of the Friars formation. Weathering breaks down the rock and reduces the effective shear strength of the material. The weathering process also opens up fissures and cracks, which increases the permeability of the near surface rock and promotes seepage of water parallel to the slope face (Ortigao et al., 1997). Failure will eventually occur when the material has weathered to such a point that the effective cohesion approaches zero.

Surficial failures are most common for cut slopes in soft sedimentary rocks, such as claystones or weakly cemented sandstones. As mentioned, the most common reason for the surficial failure is because a relatively steep slope (such as 1.5:1 or 1:1) is excavated into the sedimentary rock. Then with time, as vegetation is established on the slope face and the face of the cut slope weathers, the probability of surficial failures increases. The surficial failure usually develops during or after a period of heavy and prolonged rainfall.

Natural Slopes. A similar process as described above can cause surficial failures in natural slopes. There is often an upper weathered zone of soil that slowly grades into solid rock with depth. The upper weathered zone of soil is often much more permeable because of its loose soil structure and hence higher porosity than the underlying dense rock. Water that seeps into the natural slope from rainstorms will tend to flow in the outer layer of the slope which has the much higher permeability than the deeper unweathered rock. Thus the seepage condition and eventual failure condition for natural slopes is often similar to that as described earlier for cut slopes in rock.

Fill Slopes. Studies by Pradel and Raad (1993) indicate that in southern California, fill slopes made of clayey or silty soils are more prone to develop the conditions for surficial instability than slopes made of sand or gravel. This is probably because water tends to migrate downward in sand or gravel fill slopes, rather than parallel to the slope face.

Figure 10.13 shows a surficial failure in a fill slope. Such failures can cause extensive damage to landscaping and can even carry large trees downslope. Besides the landscaping, there can be damage to the irrigation and drainage lines. The surficial failure can also damage appurtenant structures, such as fences, walls, or patios.

A particularly dangerous condition occurs when the surficial failure mobilizes itself into a debris flow. In such cases, severe damage can occur to any structure located in the path of the debris flow. Figure 10.14 shows partial mobilization of the surficial failure, which flowed over the sidewalk and into the street. Surficial failures, such as the failure shown in Figs. 10.13 and 10.14, can be sudden and unexpected, without any warning of potential failure. Other surficial failures, especially in clays, may have characteristic signs of imminent failure. For example, Fig. 10.15 shows a clay slope having a series of nearly continuous semicircular ground cracks. After this picture was taken (during the rainy season), this slope failed in a surficial failure mode.

Surficial failures can also develop on the downstream face of earth dams. For example, Sherard et al. (1963) state:

> Shallow slides, most of which follow heavy rainstorms, do not as a rule extend into the embankment in a direction normal to the slope more than 4 or 5 ft (1.2 – 1.5 m). Some take place soon after construction, while other occur after many years of reservoir operation . . . . Shallow surface slips involving only the upper few inches of the embankment have sometimes occurred when the embankment slopes have been poorly compacted. This is a frequent difficulty in small, cheaply constructed

FIGURE 10.13 Surficial slope failure in a fill slope.

FIGURE 10.14 Partial mobilization of a surficial failure.

FIGURE 10.15   Ground cracks associated with incipient surficial instability.

dams, where the construction forces often do not make the determined effort necessary to prevent a loose condition in the outer slope. The outer few feet then soften during the first rainy season, and shallow slides result.

A contributing factor in the instability of the slope shown in Fig. 10.15 was the loss of vegetation due to fire. Also notice in Figs. 10.13 and that the surficial failure appeared to have developed just beneath the bottom of the grass roots. Roots can provide a large resistance to shearing and the shear resistance of root-permeated homogeneous and stratified soil has been studied by Waldron (1977) and Merfield (1992). The increase in shear strength due to plant roots is due directly to mechanical reinforcement of the soil and indirectly by removal of soil moisture by transpiration. Even grass roots can provide an increase in shear resistance of the soil equivalent to an effective cohesion of 60 to 100 psf (3 to 5 kPa), Day 1993b. When the vegetation is damaged or destroyed due to fire, the slope can be much more susceptible to surficial failure such as shown in Fig. 10.15.

10.3.3   Surficial Stability Analysis

In order to calculate the factor of safety for surficial stability of cut, natural, or fill slopes, the method of analysis should be as follows:

1. *Determine if a surficial failure is possible.*   For cut slopes, determine if the rock is likely to weather. Local experience can be of use in identifying sedimentary rocks, such as claystones or shales that are known to quickly weather.

For fill slopes, determine if the soil type is likely to experience surficial failures. Sands and gravels are unlikely to develop surficial failures while clay is most susceptible to such failures. Also, Eq. 10.2 is based on the assumption of an infinite slope and thus slopes of small height are unlikely to develop infinite slope seepage conditions.

2. *Determine the shear strength parameters.*   Obtain undisturbed samples of the fill or weathered rock and perform shear strength tests, such as drained direct shear tests, at confining pressures as low as possible to obtain the effective shear strength parameters (ϕ' and c'). The factor of safety for surficial stability (Eq. 10.2) is very dependent on the value of effective cohesion c'. If the shear strength tests indicate a large effective cohesion intercept, then perform additional drained shear strength tests to verify this cohesion value.

3. *Calculate the factor of safety.*   Use Eq. 10.2 to calculate the factor of safety. The parameters in Eq. 10.2 can be determined as follows:

a. Inclination α.   The slope inclination can be measured for natural slopes or based on the anticipated constructed condition for cut and fill slopes.

b. Total unit weight γ_t.   For fill slopes, the total unit weight γ_t can be based on anticipated unit weight conditions at the end of grading. For cut or natural slopes, the total unit weight γ_t can be obtained from the laboratory testing of undisturbed samples. Note that the total unit weight used in Eq. 10.2 must be based on a saturated soil condition ($S = 100$ percent).

c. Buoyant unit weight γ_b.   By knowing the total unit weight for saturated soil, the buoyant unit weight γ_b can be obtained from Eq. 3.4.

d. Depth of failure surface d.   In Eq. 10.2, the depth of the failure surface (d) is also the depth of seepage parallel to the slope face. In southern California, the value of d is typically assumed to equal to 4 ft (1.2 m) for fill slopes. This may be overly conservative for cut or natural slopes, and different values of d may be appropriate given local rainfall and weathering conditions.

The acceptable minimum factor of safety for surficial stability is often 1.5. However, as previously mentioned, root reinforcement can significantly increase the surficial stability of a slope. It may be appropriate to accept a lower factor of safety in cases where deep rooting plants will be quickly established on the slope face.

10.3.4 Design and Construction

If the factor of safety for surficial stability is deemed to be too low, there are many different methods that can be used to increase the factor of safety, as follows:

1. *Flatten the slope.*   The surficial stability can be increased by building a flatter slope (i.e., decrease α = slope inclination). It has been observed that 1.5:1 (horizontal:vertical) or steeper slope inclinations are often most susceptible to surficial instability, while 2:1 or flatter slope inclinations have significantly fewer surficial failures.

2. *Use soil having a higher shear strength.*   The surficial stability could also be increased by facing the slope with soil that has a higher shear strength (i.e., increase c' and/or ϕ'). This process can be performed during the grading of the site and is similar to the installation of a stabilization fill.

3. *Increase the shear strength of the soil.*   There are many different methods that can be used to increase the shear strength of the soil, such as adding lime or cement to the soil in order to increase c' and ϕ'. The slope face could also be constructed with layers of geogrid, which will provide tensile reinforcement and increase the soils resistance to slope movement.

4. *Reduce infiltration of water.*   Maintaining adequate vegetation and irrigating the slope during the dry season will promote root reinforcement and reduce desiccation cracks. Surface vegetation and the absence of desiccation cracks will reduce the infiltration of water into the slope face.

Another option is to face the slope with gunite. The gunite facing will not allow rainwater to infiltrate the slope, which should prevent surficial instability from developing.

5. *Mitigate infinite slope conditions.*   A tall slope will have a long slope face that can promote the development of infinite slope conditions and seepage parallel to the slope face. This condition can be mitigated by adding a ditch or berm at the top of slope to prevent water from flowing over the top of slope as well as gunite terrace ditches that effectively make a series of slopes of smaller height. The terrace ditches will also intercept water flowing on the slope face.

10.4   GROSS SLOPE STABILITY

10.4.1   Introduction

As contrasted with a shallow (surficial) slope failure discussed in the previous section, a gross slope failure often involves shear displacement of the entire slope. The shear displacement can occur on a slip surface that is planar, circular, or irregular. Different terms, such as slides or slumps, are often used to identify gross slope instability. Gross slope stability analyses are often divided into two general categories, as follows:

1. *Translational slope stability.*   A translational slope failure can develop on a steeply inclined weak layer. A wedge analyses is often used to determine the factor of safety for translational slope failures, as illustrated in Fig. 10.16.

2. *Rotational slope stability.*   If the soil is relatively homogeneous, then a rotational slope failure could occur such as illustrated in Fig. 10.17. In a rotational slope failure, the slide mass moves downward and outward. The method of slices is often used to determine the factor of safety for rotational slope failures.

Similar to surficial stability analysis, the objective of a gross slope stability analysis is to determine the factor of safety of an existing or proposed slope. For permanent slopes, the minimum factor of safety is usually 1.5. This is a relatively low factor of safety, especially considering other minimum factors of safety in foundation engineering, such as a factor of safety of 3 for bearing capacity analyses.

Gross slope stability analyses can either be performed in terms of a total stress analysis (e.g., short-term condition using the undrained shear strength), or an effective stress analysis (e.g., long-term condition using the drained shear strength). Figure 4.23 presents slope stability examples where the total stress or effective stress analysis should be performed. For a total stress slope stability analysis, the total unit weight of the soil is utilized and the groundwater table is not considered in the analysis. For an effective stress analysis, the pore water pressures (such as based on a groundwater table)

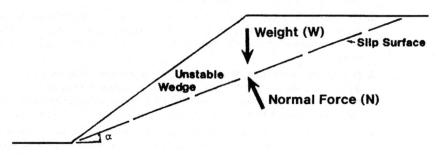

FIGURE 10.16   Example of translational slope movement (wedge method).

FIGURE 10.17   Example of rotational slide movement.

must be included in the analysis. In some cases, the slope may need to be analyzed for both a total stress analysis (short-term condition) and an effective stress analysis (long-term condition), as discussed in Sec. 4.6.5 and illustrated in Fig. 4.24. Table 10.4 provides additional guidelines.

10.4.2   Wedge Method

The simplest type of gross slope stability analysis uses a free body diagram such as illustrated in Fig. 10.16, where there is a planar slip surface inclined at an angle α to the horizontal. The wedge method is a two-dimensional analysis based on a unit length of slope. The assumption in this slope

TABLE 10.4   Effective Stress and Total Stress Analyses

Situation	Preferred method	Comments
Slope stability at the end of construction: saturated cohesive soil and construction period short compared to consolidation time	Total stress analysis: use undrained shear strength (i.e., $c = s_u$ and $\phi = 0°$)	As a check, use an effective stress analysis during construction with c' and ϕ' and measured pore water pressures
Slope stability at the end of construction: unsaturated cohesive soil and construction period short compared to consolidation time	Either method: use total stress analysis with c and ϕ from UU tests or use effective stress analysis with c' and ϕ' with estimated pore water pressures	As a check, the effective stress analysis could also be performed using the actual measured pore water pressures
Slope stability at intermediate times	Effective stress analysis: use with c' and ϕ' and estimated pore water pressures	Actual pore water pressures should be checked in the field
Long term slope stability	Effective stress analysis: use with c' and ϕ' and steady-state pore water pressures	An example of steady-state pore water pressures would be those determined from a flow net

Note: UU = unconsolidated undrained triaxial tests.
Source: Lambe and Whitman, 1969.

stability analysis is that there will be a wedge type failure of the slope along a planer slip surface. The factor of safety (F) of the slope can be derived by summing forces parallel to the slip surface, and is:

Total stress analysis:

$$F = \frac{\text{resisting force}}{\text{driving force}} = \frac{cL + N \tan \phi}{W \sin \alpha} = \frac{cL + W \cos \alpha \tan \phi}{W \sin \alpha} \tag{10.3}$$

Effective stress analysis:

$$F = \frac{\text{resisting force}}{\text{driving force}} = \frac{c'L + N' \tan \phi'}{W \sin \alpha} = \frac{c'L + (W \cos \alpha - uL) \tan \phi'}{W \sin \alpha} \tag{10.4}$$

where F = factor of safety for gross slope stability (dimensionless)

c, ϕ = shear strength parameters based on total stress, e.g., undrained shear strength parameters of the slide plane

c', ϕ' = shear strength parameters based on effective stress, e.g., drained shear strength parameters of the slide plane

L = length of the slip surface (ft or m)

N = normal force, i.e., the force acting perpendicular to the slip surface (lbs per unit length of slope or kN per unit length of slope)

N' = effective normal force, i.e., the effective force acting perpendicular to the slip surface (lbs per unit length of slope or kN per unit length of slope)

W = total weight of the failure wedge (i.e., $W = \gamma_t A$, where γ_t = total unit weight of the soil and A = area of the wedge). Units are lbs per unit length of slope or kN for a unit length of slope.

u = average pore water pressure along the slip surface (psf or kPa)

α = slip surface inclination (degrees)

Because the wedge method is a two-dimensional analysis based on a unit length of slope (i.e., length = 1 ft or 1 m), the numerator and denominator of Eqs. 10.3 and 10.4 are in lb/ft or kN/m. Similar to the surficial stability analysis (Eq. 10.1), the resisting force in Eqs. 10.3 and 10.4 is equal to the shear strength (in terms of total stress or effective stress) of the soil along the slip surface. The driving force is caused by the pull of gravity and it is equal to the component of the weight of the wedge parallel to the slip surface.

The total stress analysis would be applicable for the quick construction of the slope or a condition where the slope is loaded very quickly. A total stress analysis could be performed by using the consolidated undrained shear strength (c and ϕ) or the undrained shear strength s_u of the slip surface material. When using the undrained shear strength, $s_u = c$ and $\phi = 0$ are substituted into Eq. 10.3.

The purpose of the effective stress slope stability analysis is often to model the long-term condition of the slope and effective stresses must be utilized along the slip surface. The term $W \cos \alpha - uL$ (Eq. 10.4) is equal to the effective normal force on the slip surface. Because the analysis uses the boundary pore water pressures and total unit weight of the soil, the weight W in the numerator and denominator of Eq. 10.4 is equal to the total weight of the wedge.

For a constant slope inclination with the slope consisting of a uniform nonplastic cohesionless soil (i.e., $c' = 0$) and no pore water pressures ($u = 0$), Eq. 10.4 reduces to:

$$F = \frac{W \cos \alpha \tan \phi'}{W \sin \alpha} = \frac{\tan \phi'}{\sin \alpha / \cos \alpha} = \frac{\tan \phi'}{\tan \alpha} \tag{10.5}$$

Note that the above equation is independent of the total unit weight γ_t of the cohesionless soil and the height of the slope H. Based on the above equation, if the slope inclination of a cohesionless soil is slowly increased until the slope just begins to fail, this maximum slope inclination will equal the friction angle of the cohesionless soil. This maximum slope inclination at which the soil is barely stable has been termed the *angle of repose*.

Example Problem 10.1   A slope has an height of 30 ft (9.1 m) and the slope face is inclined at a 2:1 (horizontal:vertical) ratio. Assume a wedge type analysis where the slip surface is planar through the toe of the slope and is inclined at a 3:1 (horizontal:vertical) ratio. The total unit weight of the slope material (γ_t) = 126 pcf (19.8 kN/m³). Using the undrained shear strength parameters of c = 70 psf (3.4 kPa) and ϕ = 29°, calculate the factor of safety.

Solution   The area of the wedge is first determined from simple geometry and is equal to 450 ft² (41.4 m²). For a unit length of the slope, the total weight W of the wedge equals the area times total unit weight, or:

W = (126 pcf)(450 ft²) = 56,700 lb/ft of slope length (820 kN/m of slope length). Using Eq. 10.3 and the following values:

$$c = 70 \text{ psf (3.4 kPa)}$$

$$\phi = 29°$$

Length of slip surface (L) = 95 ft (29 m)

Slip surface inclination α = 18.4°

Total weight of wedge (W) = 56,700 lb/ft (820 kN/m)

$$F = (cL + W\cos\alpha\tan\phi)/(W\sin\alpha)$$

$$F = [(70)(95) + (56,700)(\cos 18.4°)(\tan 29°)]/[(56,700)(\sin 18.4°)] = 2.04$$

Example Problem 10.2   Use the same situation as the previous example except that the slip surface has the effective shear strength parameters: c' = 70 psf (3.4 kPa) and ϕ' = 29°. Also assume that piezometers have been installed along the slip surface and the average measured steady-state pore water pressure u = 50 psf (2.4 kPa). Calculate the factor of safety of the failure wedge based on an effective stress analysis.

Solution   Using Eq. 10.4 and the following values:

$$c' = 70 \text{ psf (3.4 kPa)}$$

$$\phi' = 29°$$

Length of slip surface (L) = 95 ft (29 m)

Slip surface inclination α = 18.4°

Average pore water pressure acting on the slip surface (u) = 50 psf (2.4 kPa)

Total weight of wedge (W) = 56,700 lbs/ft (820 kN/m)

$$F = [c'L + (W\cos\alpha - uL)\tan\phi']/(W\sin\alpha)$$

$$F = \{(70)(95) + [(56,700)(\cos 18.4°) - (50)(95)](\tan 29°)\}/[(56,700)(\sin 18.4°)] = 1.89$$

10.4.3   Method of Slices

The most commonly used method of gross slope stability analysis is the *method of slices*, where the failure mass is subdivided into vertical slices and the factor of safety is calculated based on force equilibrium equations. A circular arc slip surface and rotational type of failure mode is often used for the method of slices, and for homogeneous soil, a circular arc slip surface provides a lower factor of safety than assuming a planar slip surface. The slope failure shown in Fig. 10.17 is an example of a rotational failure.

Figure 10.18 shows an example of a slope stability analysis using a circular arc slip surface. The failure mass has then been divided into 30 vertical slices. The calculations are similar to the wedge type analysis, except that the resisting and driving forces are calculated for each slice and then summed up in order to obtain the factor of safety of the slope. For the *ordinary method of slices* (also known as the *Swedish circle method* or *Fellenius method*; Fellenius, 1936), the equations used to calculate the factor of safety are identical to Eqs. 10.3 and 10.4, with the resisting and driving forces calculated for each slice and then summed up in order to obtain the factor of safety.

Commonly used method of slices to obtain the factor of safety are listed in Table 10.5. The method of slices is not an exact method because there are more unknowns than equilibrium equations. This requires that an assumption be made concerning the interslice forces. Table 10.5 presents a summary of the assumptions for the various methods. For example, Fig. 10.19 shows that for the ordinary method of slices (Fellenius, 1936), it is assumed that the resultant of the interslice forces are parallel to the average inclination of the slice (α). It has been determined that because of this interslice assumption for the ordinary method of slices, this method provides a factor of safety that is too low for some situations (Whitman and Bailey, 1967). As a result, the other methods listed in Table 10.5 are used more often than the ordinary method of slices.

Because of the tedious nature of the calculations, computer programs are routinely used to perform the analysis. Duncan (1996) states that the nearly universal availability of computers and much improved understanding of the mechanics of slope stability analyses have brought about considerable change in the computational aspects of slope stability analysis. Analyses can be done much

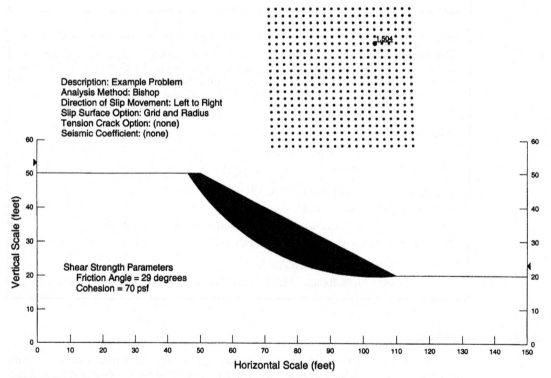

FIGURE 10.18   Example of a slope stability analysis using the Geo-Slope computer program.

TABLE 10.5 Assumptions Concerning Interslice Forces for Different Methods of Slices

Type of Method of slices	Assumption concerning interslice forces	Reference
Ordinary method of slices	Resultant of the interslice forces is parallel to the average inclination of the slice	Fellenius (1936)
Bishop simplified method	Resultant of the interslice forces is horizontal (no interslice shear forces)	Bishop (1955)
Janbu simplified method	Resultant of the interslice forces is horizontal (a correction factor is used to account for interslice shear forces)	Janbu (1968)
Janbu generalized method	Location of the interslice normal force is defined by an assumed line of thrust	Janbu (1957)
Spencer method	Resultant of the interslice forces is of constant slope throughout the sliding mass	Spencer (1967, 1968)
Morgenstern-Price method	Direction of the resultant of interslice forces is determined by using a selected function	Morgenstern and Price (1965)

Sources: Lambe and Whitman (1969); Geo-Slope (1991).

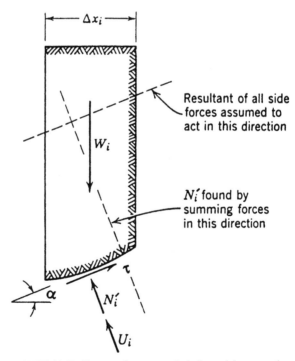

FIGURE 10.19 Forces acting on a vertical slice and the assumption concerning interslice forces for the ordinary method of slices. (*Adapted from Lambe and Whitman, 1969.*)

more thoroughly, and, from the point of view of mechanics, more accurately than was possible previously. However, problems can develop because of a lack of understanding of soil mechanics, soil strength, and the computer programs themselves, as well as the inability to analyze the results in order to avoid mistakes and misuse (Duncan, 1996).

Most slope stability computer programs can perform both total stress and effective stress slope stability analyses. For an effective stress analysis, the effective shear strength parameters (c' and ϕ') and the pore water pressures must be inputted into the computer program. In most cases, the pore water pressures have a significant impact on slope stability and they are often very difficult to estimate. There are several different options that can be used concerning the pore water pressures, as follows:

1. *Zero pore water pressure.* A common assumption for slopes that have or will be constructed with drainage devices is to use a pore water pressure equal to zero.

2. *Groundwater table.* A second option is to specify a groundwater table. For the soil above the groundwater table, it is common to assume zero pore water pressures. If the groundwater table is horizontal, then the pore water pressures below the groundwater table are typically assumed to be hydrostatic (i.e., Eq. 4.16). For the condition of seepage through the slope (i.e., a sloping groundwater table), the computer program can develop a flow net in order to estimate the pore water pressures below the groundwater table (Sec. 4.7.5). Figure 10.20 shows the pore water pressures along the failure surface due to a sloping groundwater table.

3. *Pore water pressure ratio (r_u).* The pore water pressure ratio has been defined in Eq. 4.39 and is reproduced below:

$$r_u = \frac{u}{\gamma_t h} \tag{10.6}$$

where r_u = pore water pressure ratio (dimensionless)
u = pore water pressure (psf or kPa)
γ_t = total unit weight of the soil (pcf or kN/m^3)
h = depth below the ground surface (ft or m)

If a value of $r_u = 0$ is selected, then the pore water pressures u are assumed to be equal to zero in the slope. Suppose an r_u value is used for the entire slope. In many cases the total unit weight is about

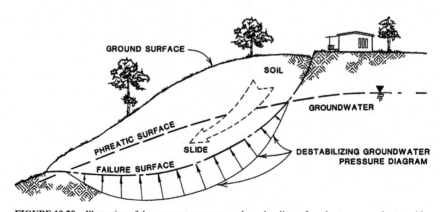

FIGURE 10.20 Illustration of the pore water pressures along the slip surface due to a groundwater table.

equal to two times the unit weight of water (i.e., $\gamma_t = 2\gamma_w$), and thus a value of $r_u = 0.25$ is similar to the effect of a groundwater table at mid-height of the slope. A value of $r_u = 0.5$ would be similar to the effect of a groundwater table corresponding to the ground surface. The pore water pressure ratio r_u can be used for existing slopes where the pore water pressures have been measured in the field, or for the design of proposed slopes where it is desirable to obtain a quick estimate of the effect of pore water pressures on the stability of the slope.

The object of the slope stability analysis is to accurately model the existing or design conditions of the slope. Some of the important factors that may need to be considered in a slope stability analysis are as follows:

1. *Effective stress and total stress analyses.* Figures 4.23 and 4.24 have provided examples where a total stress analysis and/or an effective stress analysis should be performed. Additional guidelines have been provided in Table 10.4.

2. *Different soil layers.* If a proposed slope or existing slope contains layers of different soil types with different engineering properties, then these layers must be input into the slope stability computer program. Most slope stability computer programs have this capability.

    As discussed in Sec. 10.1, the primary assumption of the limit equilibrium method is that the shear strength is simultaneously mobilized along the entire failure surface. When performing a slope stability analysis for a slope containing layers of different soil types, the stress-strain behavior of each soil must be taken into consideration. For example, if each layer of soil reaches its peak shear strength at a different value of shear strain, then it may be appropriate to use the ultimate values of shear strength in the slope stability analysis.

3. *Slip surfaces.* In some cases, a composite type slip surface may need to be included in the analysis. This option is discussed in Sec. 10.5.

4. *Tension cracks.* It has been stated that tension cracks at the top of the slope can reduce the factor of safety of a slope by as much as 20 percent and are usually regarded as an early and important warning sign of impending failure in cohesive soil (Cernica, 1995a). Slope stability programs often have the capability to model or input tension crack zones. The destabilizing effects of water in tension cracks or even the expansive forces caused by freezing water can also be modeled by some slope stability computer programs.

5. *Surcharge loads.* There may be surcharge loads (such as a building load) at the top of the slope or even on the slope face. Most slope stability computer programs have the capability of including surcharge loads. In some computer programs, other types of loads, such as due to tieback anchors, can also be included in the analysis.

6. *Nonlinear shear strength envelope.* In some cases, the shear strength envelope is nonlinear (for example, see Fig. 10.11). If the shear strength envelope is nonlinear, then a slope stability computer program that has the capability of using a nonlinear shear strength envelope in the analysis should be used.

7. *Plane strain condition.* Similar to strip footings, long uniform slopes will be in a plane strain condition. The friction angle ϕ is about 10 percent higher in the plane strain condition as compared to the friction angle ϕ measured in the triaxial apparatus (Meyerhof, 1961; Perloff and Baron, 1976). Since plane strain shear strength tests are not performed in practice, there will be an additional factor of safety associated with the plane strain condition. For uniform fill slopes that have a low factor of safety, it is often observed that the end slopes (slopes that make a 90° turn) are the first to show indications of slope movement. This is because the end slope is not subjected to a plane strain condition and the shear strength is actually lower than in the center of a long, continuous slope.

8. *Progressive failure.* For the method of slices, the factor of safety is an average value of all the slices. Some slices, such as at the toe of the slope, may have a lower factor of safety which is balanced by other slices which have a higher factor of safety. For those slices that have a low factor of safety, the shear stress and strain may exceed the peak shear strength. For some soils, such as

stiff-fissured clays, there may be a significant drop in shear strength as the soil deforms beyond the peak values. This reduction in shear strength will then transfer the load to an adjacent slice, which will cause it to experience the same condition. Thus the movement and reduction of shear strength will progress along the slip surface, eventually leading to failure of the slope. The progressive nature of the failure may even reduce the shear strength of the soil to its residual value ϕ'_r.

9. *Pseudo-static analysis.*   Slope stability analysis can also be adapted to perform earthquake analysis. This option will be discussed in Chap. 13.

10. *Other structures.*   Slope stability analysis can be used for other types of engineering structures. For example, the stability of a retaining wall is often analyzed by considering a slip surface beneath the foundation of the wall.

Example Problem 10.3   For Example Problem 10.1 (i.e., total stress analysis example problem), use the method of slices and assume the shear stress parameters are applicable for the entire slope. Calculate the factor of safety of the slope.

Solution   The slope stability analysis shown in Fig. 10.18 was performed by using the SLOPE/W (Geo-Slope, 1991) computer program. The following data were inputted into the computer program:

Slope cross section: The slope has a height of 30 ft (9.1 m) and a 2:1 (horizontal:vertical) slope inclination.

Type of slope stability analysis: A total stress slope stability analysis was selected.

The shear strength parameters of $c = 70$ psf (3.4 kPa) and $\phi = 29°$ were used in the slope stability analysis.

Total unit weight of soil: The total unit weight $\gamma_t = 126$ pcf (19.8 kN/m³).

Critical slip surface: For this analysis, the computer program was requested to perform a trial and error search for the critical slip surface (i.e., the slip surface having the lowest factor of safety). Note the grid of points that has been produced above the slope. Each one of these points represents the center of rotation of a circular arc slip surface. The computer program has actually performed 2646 slope stability analyses.

In Fig. 10.18, the dot with the number 1.504 indicates the center of rotation of the circular arc slip surface with the lowest factor of safety (i.e., lowest factor of safety = 1.504, Bishop method of slices). Although not shown, other slope stability methods were used to solve the problem using the Geo-slope (1991) computer program and the results are as follows:

$$\text{Ordinary method of slices: } F = 1.42$$

$$\text{Janbu method of slices: } F = 1.41$$

The difference in the methods is due to the assumptions used for each method (i.e., see Table 10.5).

Example Problem 10.4   For Example Problem 10.2 (i.e., effective stress analysis example problem), use the method of slices and assume the shear stress parameters are applicable for the entire slope. Also assume a pore water pressure ratio $r_u = 0.25$. Calculate the factor of safety of the slope.

Solution   The slope stability analysis was performed by using the SLOPE/W (Geo-Slope, 1991) computer program. The following data were inputted into the computer program:

(*Continued*)

Slope cross section: The slope has a height of 30 ft (9.1 m) and a 2:1 (horizontal:vertical) slope inclination.

Type of slope stability analysis: An effective stress slope stability analysis was selected.

The shear strength parameters of $c' = 70$ psf (3.4 kPa) and $\phi' = 29°$ were used in the slope stability analysis.

Total unit weight of soil: The total unit weight $\gamma_t = 126$ pcf (19.8 kN/m³).

A pore water pressure ratio $r_u = 0.25$ was used in the slope stability analysis.

Critical slip surface: For this analysis, the computer program was requested to perform a trial and error search for the critical slip surface (i.e., the slip surface having the lowest factor of safety). The computer program actually performed 2646 slope stability analyses and the factor of safety = 1.12 (Bishop method)

Although not shown, other slope stability methods were used to solve the problem using the Geo-Slope (1991) computer program and the results are as follows:

Ordinary method of slices: $F = 1.02$

Janbu method of slices: $F = 1.04$

Once again, the difference in the methods is due to the assumptions used for each method (i.e., see Table 10.5).

10.4.4 Slope Stability Charts

Slope stability analyses can also be performed by using slope stability charts. Many different charts have been developed that can be used to determine the factor of safety of slopes having constant slope inclination and composed of a single soil where the soil properties are approximately constant. Stability charts are useful because they can provide a quick check on the factor of safety of a slope for different design conditions.

Taylor Chart.   The Taylor (1937) chart is presented in Fig. 10.21. This chart can only be used for a total stress analysis of plastic soils using the undrained shear strength (i.e., $c = s_u$). Note that Fig. 10.21 cannot be used for cases where the consolidated undrained shear strength (i.e., $\phi > 0°$) has been obtained. The steps in using this chart are as follows:

1. *Depth to firm base.*   The parameter d is determined as the vertical distance below the toe of the slope to a firm base (D) divided by the height of the slope (H), i.e., $d = D/H$. For slope inclinations greater than 53°, the slope failure occurs through the toe of the slope and the value of d is not required.

2. *Stability number N_o.*   The chart in Fig. 10.21 is entered with the slope inclination β and upon intersecting the curve for a specific value of d, the stability number N_o is determined from the vertical axis.

3. *Factor of safety.*   The factor of safety F is calculated from the following equation:

$$F = \frac{N_o c}{\gamma_t H} \tag{10.7}$$

where F = factor of safety (dimensionless)
    N_o = stability number from Step 2 (dimensionless)
    $c = s_u$ = undrained shear strength of the cohesive soil (psf or kPa)
    γ_t = total unit weight of the cohesive soil (pcf or kN/m³)
    H = height of the slope (ft or m)

Example Problem 10.5   Solve the problem as illustrated in Fig. 10.21, except calculate the factor of safety for failure through the toe of the slope.

Solution   Enter the chart (Fig. 10.21) at $\beta = 35°$ and intersect the solid line (toe circles). From the vertical axis, $N_o = 6.2$

The factor of safety = $[(6.2)(600)]/[(115)(25)] = 1.29$

Stability Charts by Cousins.   Figures 10.22 through 10.24 present stability charts prepared by Cousins (1978). These three charts were developed for failure through the toe of the slope. The pore water pressure ratio r_u has been discussed in Sec. 10.4.3. Each chart has been developed for a different pore water pressure ratio r_u, as follows:

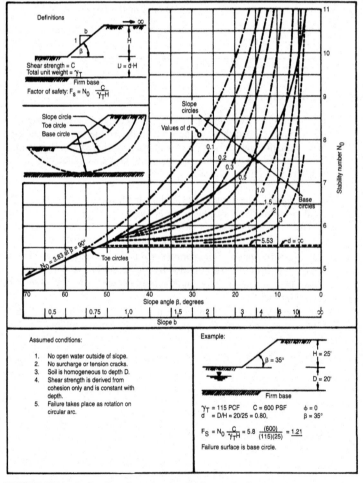

FIGURE 10.21   Taylor chart for estimating the factor of safety of a slope using a total stress analysis. (*Reproduced from NAVFAC DM-7.1, 1982.*)

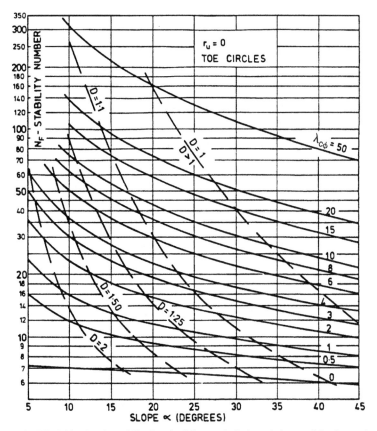

FIGURE 10.22 Cousins (1978) chart for failure analysis through the toe of the slope and zero pore water pressures in the slope (i.e., $r_u = 0$). (*Reprinted with permission of the American Society of Civil Engineers.*)

1. Figure 10.22: $r_u = 0$. For this chart, the pore water pressure equals zero for the entire slope.

2. Figure 10.23: $r_u = 0.25$. This chart is similar to the effect of a groundwater table at mid-height of the slope.

3. Figure 10.24: $r_u = 0.50$. This chart is similar to the effect of a groundwater table corresponding to the ground surface.

Note that in each chart there are lines that are labeled with various D values. The value of $D = H'/H$, where H' = vertical distance from the top of slope to the lowest point on the slip surface and H = height of the slope.

The Cousins charts can only be used if the soil has a cohesion value. The steps to be used in the analysis are presented below:

Effective stress analysis:

1. Based on the existing groundwater table or the anticipated groundwater level in the slope, select a value of r_u. Use either Fig. 10.22, 10.23, or 10.24 depending on the value of r_u.

2. Calculate the value of $\lambda_{c\phi}$ which is defined as follows:

$$\lambda_{c\phi} = \frac{\gamma_t H \tan \phi'}{c'} \tag{10.8}$$

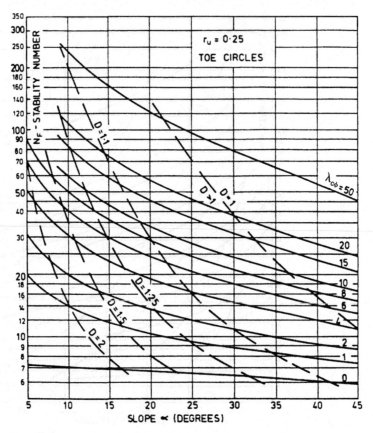

FIGURE 10.23 Cousins (1978) chart for failure analysis through the toe of the slope and a pore water pressure ratio $r_u = 0.25$. (*Reprinted with permission of the American Society of Civil Engineers.*)

where γ_t = total unit weight of the soil (pcf or kN/m³)
H = height of the slope (ft or m)
ϕ' = effective friction angle of the soil (degrees)
c' = effective cohesion of the soil (psf or kPa)

3. Enter the chart along the horizontal axis at the value of the slope inclination α. Select the appropriate curve based on the value of $\lambda_{c\phi}$ and then determine the stability number N_F from the vertical axis.

4. Calculate the factor of safety F from the following equation:

$$F = \frac{N_F \, c'}{\gamma_t \, H} \tag{10.9}$$

Total stress analysis: For a total stress analysis, use the undrained shear strength from consolidated undrained triaxial compression tests (i.e., ϕ and c). Figure 10.22 must be used ($r_u = 0$) and the four steps outlined previously would be performed using ϕ and c (in place of ϕ' and c').

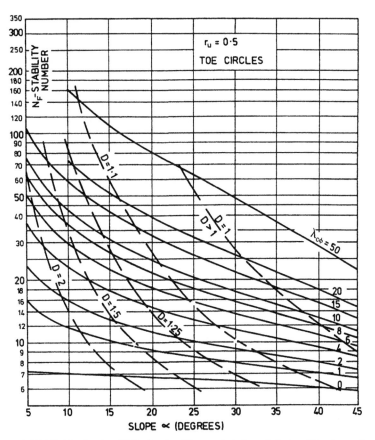

FIGURE 10.24 Cousins (1978) chart for failure analysis through the toe of the slope and a pore water pressure ratio $r_u = 0.50$. (*Reprinted with permission of the American Society of Civil Engineers.*)

Example Problem 10.6 Use the data from Example Problem 10.3. Determine the factor of safety using a total stress analysis and the Cousins charts.

Solution Slope cross section: The slope has a height of 30 ft (9.1 m) and a 2:1 (horizontal:vertical) slope inclination ($\alpha = 26.6°$).

Type of slope stability analysis: For a total stress slope stability analysis, $r_u = 0$ (i.e., pore water pressures are ignored in a total stress analysis). The shear strength parameters are $c = 70$ psf (3.4 kPa) and $\phi = 29°$.

Total unit weight of soil: The total unit weight $\gamma_t = 126$ pcf (19.8 kN/m³).

Using Eq. 10.8:

$$\lambda_{c\phi} = (\gamma_t H \tan \phi)/c = [(126)(30)(\tan 29°)]/(70) = 30$$

Entering Fig. 10.22 ($r_u = 0$ plot) with $\alpha = 26.6°$ and for $\lambda_{c\phi} = 30$ (interpolating between the 20 and 50 curves), the value of $N_F = 75$. Using Eq. 10.9:

$$F = (N_F c)/(\gamma_t H) = [(75)(70)]/[(126)(30)] = 1.39$$

Example Problem 10.7   Use the data from Example Problem 10.4. Determine the factor of safety using an effective stress analysis ($r_u = 0.25$) and the Cousins charts.

Solution   Slope cross section: The slope has a height of 30 ft (9.1 m) and a 2:1 (horizontal: vertical) slope inclination ($\alpha = 26.6°$).

Type of slope stability analysis: For the effective stress slope stability analysis, use $r_u = 0.25$. The shear strength parameters are $c' = 70$ psf (3.4 kPa) and $\phi' = 29°$.

Total unit weight of soil: The total unit weight $\gamma_t = 126$ pcf (19.8 kN/m³).

Using Eq. 10.8:

$$\lambda_{c\phi} = (\gamma_t H \tan \phi')/c' = [(126)(30)(\tan 29°)]/(70) = 30$$

Entering Fig. 10.23 ($r_u = 0.25$ plot) with $\alpha = 26.6°$ and for $\lambda_{c\phi} = 30$ (interpolating between the 20 and 50 curves), the value of $N_F = 55$. Using Eq. 10.9:

$$F = (N_F c')/(\gamma_t H) = [(55)(70)]/[(126)(30)] = 1.02$$

In summary, the results of the gross slope stability example problems from Secs. 10.4.3 and 10.4.4 are as follows:

A) Total Stress Analysis: $c = 70$ psf and $\phi = 29°$	
Method	Factor of safety
Bishop Method of slices	1.50
Ordinary method of slices	1.42
Janbu method of slices	1.41
Cousins chart ($r_u = 0$)	1.39

B) Effective Stress Analysis: $c' = 70$ psf, $\phi' = 29°$, and $r_u = 0.25$	
Method	Factor of safety
Bishop method of slices	1.12
Ordinary method of slices	1.02
Janbu method of slices	1.04
Cousins chart	1.02

The effective stress analyses include pore water pressures that result in a lower factor of safety as compared to the total stress analyses. For the total stress analyses, the factor of safety varies from 1.39 to 1.50. For the effective stress analyses, the factor of safety varies from 1.02 to 1.12. For this example problem, all of the different method of slices and the Cousins charts provide about the same answer, with the Bishop method having the highest value of the factor of safety. The Bishop method often has the highest factor of safety and it is always a good idea to check several different methods of slices and then use an average value of the factor of safety.

10.5 LANDSLIDES

10.5.1 Introduction

Landslides can be some of the most challenging projects worked on by geotechnical engineers and engineering geologists. Landslides can cause extensive damage to structures and may be very expensive to stabilize when they impact developed property. In terms of damage, the National Research Council (1985a) states:

> Landsliding in the United States causes at least $1 to $2 billion in economic losses and 25 to 50 deaths each year. Despite a growing geologic understanding of landslide processes and a rapidly developing engineering capability for landslide control, losses from landslides are continuing to increase. This is largely a consequence of residential and commercial development that continues to expand onto the steeply sloping terrain that is most prone to landsliding.

Figure 10.25 shows an example of a landslide and Table 10.6 presents common nomenclature used to describe landslide features (Varnes, 1978). Landslides are described as mass movement of soil or rock that involves shear displacement along one or several rupture surfaces, which are either visible or may be reasonably inferred (Varnes, 1978). As previously mentioned, it is the shear displacement along a distinct rupture surface that distinguishes landslides from other types of soil or rock movement such as falls, topples, or flows.

Similar to gross slope failures, landslides are generally classified as either rotational or translational. Rotational landslides are due to forces that cause a turning movement about a point above the center of gravity of the failure mass, which results in a curved or circular surface of rupture. Translational landslides occur on a more or less planar or gently undulatory surface of rupture. Translational landslides are frequently controlled by weak layers, such as faults, joints, or bedding planes; examples include the variations in shear strength between layers of tilted bedded deposits or the contact between firm bedrock and weathered overlying material.

Active landslides are those that are either currently moving or that are only temporarily suspended, which means that they are not moving at present but have moved within the last cycle of seasons (Varnes, 1978). Active landslides have fresh features, such as a main scarp, transverse ridges and cracks, and a distinct main body of movement. The fresh features of an active landslide enable the limits of movement to be easily recognized. Generally active landslides are not significantly modified by the processes of weathering or erosion.

Landslides that have long since stopped moving are typically modified by erosion and weathering; or may be covered with vegetation so that the evidence of movement is obscure. The main scarp

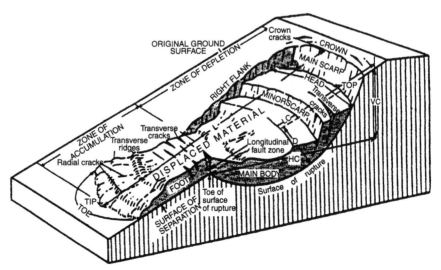

FIGURE 10.25   Landslide illustration. [Reprinted with permission from Landslides: Analysis and Control, Special Report 176 (From Varnes 1978). Copyright 1978 by the National Academy of Sciences, Courtesy of the National Academy Press, Washington, D.C.]

TABLE 10.6 Common Landslide Nomenclature

Terms	Definitions
Main scarp	A steep surface on the undisturbed ground around the periphery of the slide, caused by the movement of slide material away from the undisturbed ground. The projection of the scarp surface under the displaced material becomes the surface of rupture.
Minor scarp	A steep surface on the displaced material produced by differential movements within the sliding mass.
Head	The upper parts of the slide material along the contact between the displaced material and the main scarp.
Top	The highest point of contact between the displaced material and the main scarp.
Toe, surface of rupture	The intersection (sometimes buried) between the lower part of the surface of rupture and the original ground surface.
Toe	The margin of displaced material most distant from the main scarp.
Tip	The point on the toe most distant from the top of the slide.
Foot	That portion of the displaced material that lies downslope from the toe of the surface of rupture.
Main body	That part of the displaced material that overlies the surface of rupture between the main scarp and toe of the surface of rupture.
Flank	The side of the landslide.
Crown	The material that is still in place, practically undisplaced and adjacent to the highest parts of the main scarp.
Original ground surface	The slope that existed before the movement which is being considered took place. If this is the surface of an older landslide, that fact should be stated.
Left and right	Compass directions are preferable in describing a slide, but if right and left are used they refer to the slide as viewed from the crown.
Surface of separation	The surface separating displaced material from stable material but not known to have been a surface on which failure occurred.
Displaced material	The material that has moved away from its original position on the slope. It may be in a deformed or undeformed state.
Zone of depletion	The area within which the displaced material lies below the original ground surface.
Zone of accumulation	The area within which the displaced material lies above the original ground surface.

Source: Reprinted with permission from *Landslides: Analysis and Control, Special Report 176* (from Varnes, 1978). Copyright 1978 by the National Academy of Sciences. Courtesy of the National Academy Press, Washington, D.C.

and transverse cracks will have been eroded or filled in with debris. Such landslides are generally referred to as ancient or fossil landslides (Zaruba and Mencl, 1969; Day 1995b). These landslides have commonly developed under different climatic conditions thousands or more years ago.

Many different conditions can trigger a landslide. Landslides can be triggered by an increase in shear stress or a reduction in shear strength. The following factors contribute to an increase in shear stress:

1. Removing lateral support, such as erosion of the toe of the landslide by streams or rivers
2. Applying a surcharge at the head of the landslide, such as the construction of a fill mass for a road
3. Applying a lateral pressure, such as the raising of the groundwater table
4. Applying vibration forces, such as an earthquake or construction activities

Factors that result in a reduction in shear strength include:

1. Natural weathering of soil or rock
2. Development of discontinuities, such as faults or bedding planes
3. Increase in moisture content or pore water pressure of the slide plane material

10.5.2   Landslide Example

The purpose of this section is to describe the Laguna Niguel landslide, located in Laguna Niguel, California that was not discovered during the design and construction of the project. The Laguna Niguel landslide occurred during the *El Niño* winter of 1997–1998. The landslide was triggered by the heavy California rainfall due to the El Niño weather pattern. Rainfall can infiltrate into the ground where it can trigger landslides by lubricating slide planes and by raising the groundwater table, which increases driving forces due to seepage pressures and decreases resisting forces due to the buoyancy effect.

The mass movement of the landslide occurred on March 19, 1998 (see Fig. 10.26 for a site plan and Figs. 10.27and 10.28 for overviews of the landslide). This landslide was very large and deep and

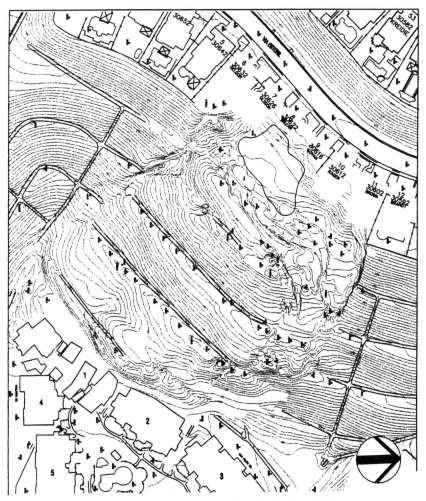

FIGURE 10.26   Laguna Niguel site plan. (*Note*: Approximate Scale: 1 in. = 150 ft.)

FIGURE 10.27   Overview of the Laguna Niguel landslide.

caused extensive damage. Figure 10.29 shows a cross section through the landslide and shows the original predevelopment topography (dashed line), the final as-graded topography, and the failure condition. As shown in Fig. 10.29, the mass movement of the landslide caused a dropping down at the top (head) and a bulging upward of the ground at the base (toe) of the landslide.

Figures 10.30 to 10.40 show pictures of the damage caused by the Laguna Niguel landslide. The figures show the following:

Figures 10.30 to 10.33.   These photographs show damage to condominiums at the toe of the land-slide. The area shown in Fig. 10.30 was originally relatively level, but the toe of the landslide has

FIGURE 10.28   Overview of the Laguna Niguel landslide.

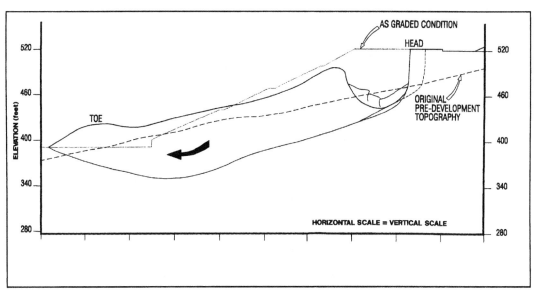

FIGURE 10.29   Laguna Niguel landslide: cross-section through the center of the landslide.

uplifted both the road and the condominiums. The arrow in Fig. 10.30 indicates the original level of the road and condominiums. Note also in Fig. 10.30 that the upward thrust caused by the toe of the landslide has literally ripped the building in half. At other locations, the toe of the landslide uplifted the road by about 20 ft (6 m) as shown in Figs. 10.31 and 10.32. Many buildings located at the toe of the landslide were completely crushed by the force of the landslide movement (Fig. 10.33).

Figures 10.34 to 10.39.   These photographs show damage at the head of the landslide. The main scarp, which is shown in Fig. 10.34, is about 40 ft (12 m) in height. The vertical distance between the two arrows in Fig. 10.34 represents the down-dropping of the top of the landslide. Figure 10.35 shows a corner of a house that is suspended in mid-air because the landslide has dropped down and away. Figure 10.36 shows a different house where the rear is also suspended in mid-air because of the landslide movement. A common feature of all the houses located at the top of the landslide was the vertical and lateral movement caused by the landslide as it dropped down and away. For example, Fig. 10.37 shows one house that has dropped down relative to the driveway, Fig. 10.38 shows a second house where a gap opened up in the driveway as a house was pulled downslope, and Fig. 10.39 shows a third house (visible in the background) that was pulled downslope, leaving behind a groove in the ground which was caused by the house footings being dragged across the ground surface.

Figure 10.40.   This photograph shows the graben area that opened up at the head of the landslide.
Except in cases where the landslide fails because of a sudden loading or unloading (such as during an earthquake, Chap. 13), the usual procedure is to perform an effective stress slope stability analysis. Figure 10.41 shows the effective stress analysis for the Laguna Niguel landslide. The actual slope stability analysis was performed by using the SLOPE/W (Geo-Slope, 1991) computer program. Although the method of slices was originally developed for circular slip surfaces, the analyses can be readily adapted to planar or composite slip surfaces.
In order to calculate the factor of safety for the landslide based on an effective stress analysis, the following parameters were inputted into the SLOPE/W (Geo-Slope, 1991) computer program:

1. *Landslide cross section.*   The landslide cross section was developed by the engineering geologist. The different layers in Fig. 10.41 have been identified as *ef* (engineered fill), Q_{al} (alluvium), T_m (fractured formational rock), and slide plane material.

FIGURE 10.30 Laguna Niguel landslide: view of the toe of the landslide (note that the arrow indicates the original level of this area, but the landslide uplifted both a portion of the road and the condominium).

2. *Failure mass.* As mentioned in Sec. 10.4.3, the computer program can efficiently search for the critical failure surface that has the lowest factor of safety. Another option is to specify the location of the slip surface. For the Laguna Niguel landslide, the location of the slip surface was determined from inclinometer monitoring. This location of the slip surface was then input into the computer program (i.e., the slip surface is fully specified). Note in Fig. 10.41 that for the stability analysis of the inputted failure mass, the computer program has divided the failure mass into 64 vertical slices.

3. *Groundwater table.* The location of the groundwater table was determined from piezometer readings and inputted into the slope stability program. The dashed line in Fig. 10.41 is the location of the groundwater table.

4. *Drained residual shear strength.* The drained residual shear strength (Sec. 3.5.8) of the slide plane material was obtained from laboratory tests using the modified Bromhead ring shear apparatus (Stark and Eid, 1994). Figure 3.43 presents the drained residual failure envelope from ring shear tests on the slide plane specimen. It can be seen in Fig. 3.43 that the failure envelope is nonlinear.

FIGURE 10.31   Laguna Niguel landslide: view of the toe of the landslide where the road has been uplifted 20 ft.

Given the great depth of the slide plane (i.e., a high effective normal stress), a drained residual friction angle ϕ'_r of 12° was selected for the slope stability analysis.

5. *Shear strength of other soil layers.*   The effective shear strength parameters (c' and ϕ') were determined from laboratory shear strength tests and the data are summarized in Fig. 10.41.

6. *Total unit weight.*   Based on laboratory testing of soil and rock specimens, total unit weights of the various strata were determined and the values are summarized in Fig. 10.41.

FIGURE 10.32   Laguna Niguel landslide: another view of the toe of the landslide where the road has been uplifted 20 ft.

FIGURE 10.33 Laguna Niguel landslide: view of damage at the toe of the landslide. The landslide has crushed the buildings located in this area.

FIGURE 10.34 Laguna Niguel landslide: head of the landslide. (*Note*: The vertical distance between the two arrows is the distance that the landslide dropped downward.)

FIGURE 10.35 Laguna Niguel landslide: head of the landslide. (*Note*: The smaller arrow points to the corner of a house that is suspended in mid-air and the larger arrow points to the main scarp.)

FIGURE 10.36 Laguna Niguel landslide: head of the landslide showing another house with a portion suspended in mid-air. (*Note*: The arrow points to a column and footing that are suspended in midair.)

FIGURE 10.37   Laguna Niguel landslide: head of the landslide where a house has dropped downward. (*Note*: The arrow points to the area where the house foundation has punched through the driveway on its way down.)

Based on the input parameters, the factor of safety using the Spencer method of slices was calculated by the SLOPE/W (Geo-Slope, 1991) computer program and the value is 0.995. This value is consistent with the actual failure (factor of safety = 1.0) of the slope.

The repair of the landslide consisted of the construction of a buttress at the toe of the slope, removal of material from the head of the landslide, and the construction of a retaining wall with tieback anchors at the head of the landslide as shown in Fig. 10.42.

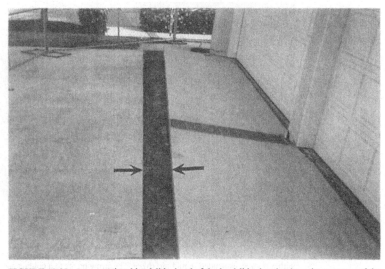

FIGURE 10.38   Laguna Niguel landslide: head of the landslide showing lateral movement of the entire house. (*Note*: The distance between the arrows indicates the amount of lateral movement.)

FIGURE 10.39 Laguna Niguel landslide: head of the landslide showing a house that has been pulled downslope.

FIGURE 10.40 Laguna Niguel landslide: graben area created as the landslide moved downslope.

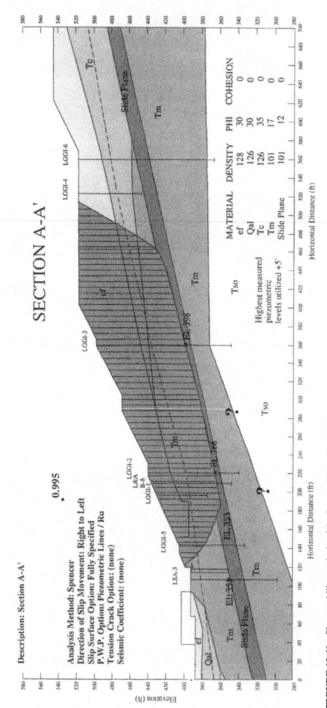

FIGURE 10.41 Slope stability analysis of the Laguna Niguel landslide.

FIGURE 10.42 Retaining wall constructed at the head of the Laguna Niguel landslide.

As this example illustrates, landslides can be very destructive. It is especially important that the geotechnical engineer and engineering geologist identify such hazards at the site prior to design and construction of the project.

10.5.3 Stabilization of Slopes and Landslides

If an ancient landslide were discovered during the design stage of a new project, the method of analysis would be similar to that as described in Sec. 10.5.2. Extensive subsurface exploration would be required so that the engineering geologist could develop cross sections of the ancient landslide based on subsurface exploration. The drained residual friction angle could be obtained from laboratory shear strength tests. The unit weight of the landslide could be estimated from laboratory testing of undisturbed specimens of the landslide materials.

A major unknown in the effective stress slope stability analysis of the ancient landslide would be the groundwater table. It is not uncommon that the groundwater table will rise once the project has been completed. This is because there will be additional infiltration of water into the landslide mass from irrigation or leaky pipes. The approximate location of the future (long-term) groundwater table could be based on the local topography and presence of drainage facilities that are to be installed during development of the site. Once the cross section of the landslide, location of the slip surface, estimated location of the long-term groundwater table, residual shear strength parameters, and unit weight are determined, the factor of safety of the ancient landslide mass would be calculated by a slope stability program such as shown in Fig. 10.41. The standard requirement is that a landslide must have a factor of safety of at least 1.5.

If the analysis shows that the landslide has too low of a factor of safety, there are three basic approaches that can be used to increase the factor of safety of a slope or landslide: (1) increase the resisting forces, (2) decrease the driving forces, or (3) rebuild the slope, as follows:

1. *Increase the resisting forces.*   Methods to increase the resisting forces include the construction of a buttress at the toe of the slope or the installation of piles or reinforced concrete pier walls which provide added resistance to the slope. Pier walls can also be used to stabilize slopes, excavations, or landslides deemed to have an unacceptable factor of safety (Abramson et al. 1996; Ehlig, 1986).

Another technique is soil nailing which is a practical and proven system used to stabilize slopes by reinforcing the slope with relatively short, fully bonded inclusions such as steel bars (Bruce and Jewell, 1987).

2. *Decrease the driving forces.*   Methods to decrease the driving forces include the lowering of the groundwater table by improving surface drainage facilities, installing underground drains, or pumping groundwater from wells. Other methods to decrease the driving forces consist of removing soil from the head of the landslide or regrading the slope in order to decrease its height or slope inclination.

3. *Rebuild the slope.*   The slope failure could also be rebuilt and strengthened by using geogrids or other soil reinforcement techniques (Rogers, 1992). Other techniques could be employed during the grading of the site, such as the construction of a shear key. A *shear key* is defined as a deep and wide trench cut through the landslide mass and into intact material below the slide. The shear key is excavated from ground surface to below the basal rupture surface, and then backfilled with soil that provides a higher shear strength than the original rupture surface. By interrupting the original weak rupture surface with a higher strength soil, the factor of safety of the landslide is increased. During construction, the shear key is also normally provided with a drainage system. It is generally recognized that during the excavation of a shear key, there is a risk of the failure of the landslide. To reduce this risk, a shear key is usually constructed in several sections with only a portion of the slide plane exposed at any given time. Figure 10.43 shows a cross-section through an ancient landslide where a shear key was installed during grading of the site. At this project, the shear key was not successful in stabilizing the landslide and a portion of the landslide became reactivated (Day and Thoeny, 1998).

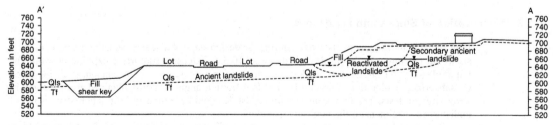

FIGURE 10.43   Cross section through an ancient landslide showing the construction of a shear key and reactivated portion of the landslide.

10.6   DEBRIS FLOW

10.6.1   Introduction

Debris flows cause a tremendous amount of damage and loss of life throughout the world. An example is the loss of 6000 lives from the devastating flows that occurred in Leyte, Philippines, on November 5, 1991, due to deforestation and torrential rains from tropical storm Thelma. Due to continued population growth, deforestation, and poor-land development practices, it is expected that debris flows will increase in frequency and devastation.

A *debris flow* is commonly defined as soil with entrained water and air that moves readily as a fluid on low slopes. As shown in Fig. 10.14, in many cases there is an initial surficial slope failure that transforms itself into a debris flow (Ellen and Fleming, 1987; Anderson and Sitar, 1995, 1996). Figure 10.44 shows two views of a debris flow. The upper photograph shows the source area of the debris flow and the lower photograph shows how the debris flow forced its way into the house.

Debris flow can include a wide variety of soil-particle sizes (including boulders) as well as logs, branches, tires, and automobiles. Other terms, such as *mudflow, debris slide, mudslide,* and *earth flow* have been used to identify similar processes. While categorizing flows based on rate of movement or the percentage of clay particles may be important, the mechanisms of all these flows are essentially the same (Johnson and Rodine, 1984).

There are generally three segments of a debris flow: the source area, main track, and depositional area (Baldwin et al., 1987). The source area is the region where a soil mass becomes detached and transforms itself into a debris flow. The main track is the path over which the debris flow descends the slope and increases in velocity depending on the slope steepness, obstructions, channel configuration, and the viscosity of the flowing mass. When the debris flow encounters a marked decrease in slope gradient and deposition begins, this is called the *depositional area.*

10.6.2   Predication and Mitigation Measures

It can be very difficult to predict the potential for a debris flow at a particular site. The historical method is one means of predicting debris-flow activity in a particular area. For example, as Johnson and Rodine (1984) indicated, many alluvial fans in southern California contain previous debris-flow deposits, which in the future will likely again experience debris flow. However, using the historical method for predicting debris flow is not always reliable. For example, the residences of Los Altos Hills experienced an unexpected debris flow mobilization from a road fill after several days of intense rainfall (Johnson and Hampton, 1969). Using the historical method to predict debris flow is not always reliable, because the area can be changed which could increase or decrease the potential for a debris flow.

Johnson and Rodine (1984) stated that a single parameter should not be used to predict either the potential or actual initiation of a debris flow. Several parameters appear to be of prime importance. Two such parameters, which numerous investigators have studied, are rainfall amount and rainfall intensity. For example, Neary and Swift (1987) stated that hourly rainfall intensity of 3.5 to 4 in./h (90 to 100 mm/h) was the key to triggering debris flows in the southern Appalachians. Other important factors include the type and thickness of soil in the source area, the steepness and length of the slope in the source area, the destruction of vegetation due to fire or logging, and other factors such as the cutting of roads.

The engineering geologist is usually the best individual to investigate the possibility of a debris flow impacting the site. Based on this analysis, measures can be taken to prevent a debris flow from damaging the site. For example, grading could be performed to create a raised building pad so that the structure is elevated above the main debris flow path. Another possibility is the construction of deflection walls, such as shown in Fig. 10.45. Other options include the construction of retention basins or channels to control or direct the debris flow away from the structures.

10.6.3   Debris Flow Example

This section deals with a debris flow adjacent to the Pauma Indian Reservation, in San Diego County, California. The debris flow occurred in January 1980, during heavy and intense winter rains. A review

FIGURE 10.44   Two views of a debris flow. Upper photograph shows the area of detachment and the lower photograph shows how the flow forced its way into the house.

of aerial photographs indicates that alluvial fans are being built at the mouths of the canyons in this area due to past debris flows.

The debris flow in January 1980 hit a house, which resulted in a lawsuit being filed. The author was retained in June of 1990 as an expert for one of the cross-defendants, a lumber company. The case settled out of court in July, 1990.

The lumber company had cut down trees on the Pauma Indian Reservation and it was alleged that the construction of haul roads and the lack of tree cover contributed to the debris flow. These factors were observed to be contributing factors in the debris flow. For example, after the debris flow, deep

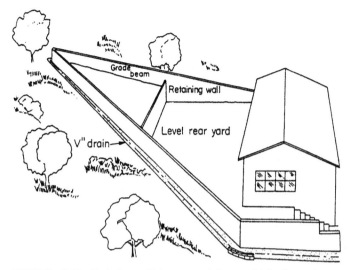

FIGURE 10.45 Typical A-wall layout to deflect a debris flow. (*Source: Hollingsworth and Kovacs, 1981.*)

erosional channels were discovered in the haul roads, indicating that at least a portion of the soil in the flow came from the road subgrade.

Figure 10.46 presents a topographic map of the area, showing the location of the house and the path of the debris flow. The debris flow traveled down a narrow canyon (top of Fig. 10.46) that had an estimated drainage basin of 200 acres (0.8 km²). The complete extent of the source area could not be determined because of restricted access. The source area and main track probably extended from an elevation of about 3200 ft (980 m) to 1300 ft (400 m) and the slope inclination of the canyon varied from about 34° to 20°. At an elevation of about 1300 ft (400 m), the slope inclination changes to about 7°, corresponding to the beginning of the depositional area.

Figure 10.47 presents a photograph of the house, which is a one-story single-family structure, having typical wood-frame construction and stucco exterior. It was observed in the area of the site that the debris flow was uniform with a thickness of about 2 ft (0.6 m). There was no damage to the structural frame of the house due to two factors. The first factor was that the debris flow impacted the house and moved around the sides of the house, rather than through the house, because of the lack of any opening on the impact side of the house (Fig. 10.47). The second factor was that the debris flow had traveled about 1200 ft (370 m) in the depositional area before striking the house and then proceeded only about 50 ft (15 m) beyond the house. By the time the debris flow reached the house, its energy had been nearly spent.

A sample of the debris-flow material was tested for its grain-size distribution. The soil comprising the debris flow was classified as a nonplastic silty sand (SM), with 16 percent gravel, 69 percent sand, and 15 percent silt and clay. The material in the debris flow was derived from weathering of rock from the source area. Geologic maps of the area indicated that the watershed is composed of metamorphic rocks, probably the Julian schist.

Based on the historical method, for the site shown in Fig. 10.46, it seems probable that there will be future debris flows. Preventative solutions include the restriction of houses in the depositional area or the construction of houses on elevated building pads. Notice in the center of Fig. 10.46 that two houses were built very close to the path of the debris flow, but they are situated at an elevation of about 50 ft (15 m) above the depositional area and, hence, were unaffected by the debris flow.

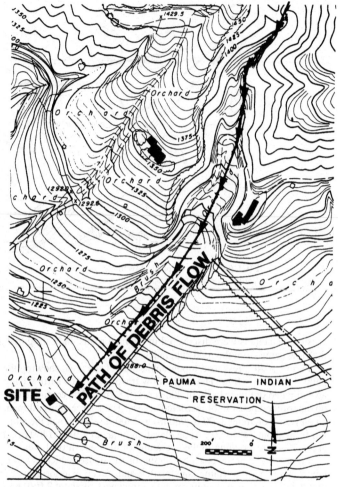

FIGURE 10.46   Path of the debris flow.

10.7   SLOPE SOFTENING AND CREEP

This last section describes the process of slope softening and creep. Both of these processes cause lateral movement of the slopes. As a practical matter, slope softening and creep need only be considered for plastic (cohesive) soil. Slopes composed of cohesionless soil, such as sands and gravels, do not need to be evaluated for slope softening or creep.

10.7.1   Slope Softening

In many urban areas, there is a tendency toward small lot sizes because of the high cost of the land, with most of the lot occupied by building structures. Fill in slope areas is generally placed and compacted near optimum moisture content, which is often well below saturation. After construction of

FIGURE 10.47 Photograph of site. Arrows indicate height of the debris flow.

the slope, additional moisture is introduced into the fill by irrigation, rainfall, groundwater sources, and leaking water pipes. At optimum moisture content, a compacted clay fill can have high shear strength because of negative pore water pressures. As water infiltrates the clay, the slope softens as the pore spaces fill with water and the pore water pressures tend toward zero. If a groundwater table then develops, the pore water pressures will become positive. The elimination of negative pore water pressure results in a decrease in effective stress and deformation of the slope in order to mobilize the needed shear stress to maintain stability.

This process of moisture infiltration into a compacted clay slope that results in slope deformation has been termed *slope softening* (Day and Axten, 1990). The moisture migration into a cohesive fill slope that leads to slope deformation has also been termed *lateral fill extension*.

Some indications of slope softening include the rear patio pulling away from the structure, pool decking pulling away from the coping, tilting of improvements near the top of the slope, stair-step cracking in walls perpendicular to the slope, and downward deformation of that part of the building near the top of the slope. In addition to the slope movement caused by the slope softening process, there can be additional movement due to the process of creep.

10.7.2 Creep

Creep is generally defined as an imperceptibly slow and more or less continuous downward and outward movement of slope-forming soil or rock (Stokes and Varnes, 1955). Creep can affect both the near-surface soil and deep-seated materials. The process of creep is frequently described as viscous shear that produces permanent deformations, but not failure as in landslide movement. Typically the amount of movement is governed by the following factors: shear strength of the clay, slope angle, slope height, elapsed time, moisture conditions, and thickness of the active creep zone (Lytton and Dyke, 1980).

The process of creep is often divided into three different stages: primary or transient, steady state or secondary, and tertiary that could lead to failure of the slope. These three stages of creep are illustrated in Fig. 10.48. Because of its relatively short duration, the primary (transient) creep is often ignored in slope deformation analysis. The secondary phase of creep often produces a relatively constant rate of strain, and depending on the slope conditions, it could continue for a

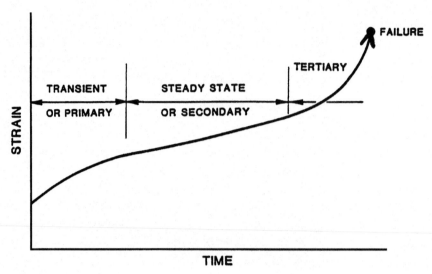

FIGURE 10.48   Three stages of creep. (*After Price, 1966.*)

considerable number of years. If the tertiary phase of creep is reached, the strain rate accelerates, and the slope could ultimately be subjected to a shear failure.

10.7.3   Example of Slope Creep

The purpose of this section is to present an example of a project having secondary creep (slope creep at a relatively constant rate). Figure 10.49 presents a cross-section through the site that was experiencing creep of a compacted clay slope. Typical damage included differential foundation displacement of the building located near the top of slope and cracking and separation of the patios located at the top of slope. A boring was excavated at the top of slope (Boring B-1) and an inclinometer casing was installed in the boring (see Fig. 10.49 for location of Boring B-1).

An inclinometer is a device used to measure the amount of lateral movement of slopes. The procedure is to install a flexible casing into a vertical borehole. Then the inclinometer survey is performed by lowering an inclinometer probe into the flexible vertical casing. As the slope deforms, the flexible casing moves with the slope. By performing successive surveys, the shape and position of the flexible casing can be measured and the lateral deformation of the slope can be determined.

Figure 10.50 shows the results of the inclinometer monitoring. In Figure 10.50, the vertical axis is depth below the ground surface and the horizontal axis is the lateral (out-of-slope) movement. The initial survey (base reading) was obtained on February 11, 1987. Additional successive surveys were performed from 1987 to 2004. As shown in Fig. 10.50, the inclinometer readings indicate a slow, continuous lateral movement (steady state creep) of the slope. The inclinometer data do not show movement on a discrete failure surface, but rather indicates a progressive creep of the entire fill slope. The inclinometer recorded the greatest amount of lateral movement [1.7 in. (43 mm)] at a depth of 1 ft (0.3 m). The amount of slope movement decreases with depth.

Figure 10.51 shows a plot of the amount of lateral slope movement versus time for readings at a depth of 1 ft (0.3 m). This plot shows a steady state creep of about 0.16 in./year (4.1 mm/year) from 1987 to 1995 and a steady state creep of about 0.05 in./year (1.3 mm/year) from 1995 to 2004. The reduction in the rate of creep is believed to be due to a reduction in irrigation and drainage improvements that prevent water from flowing over the top of slope.

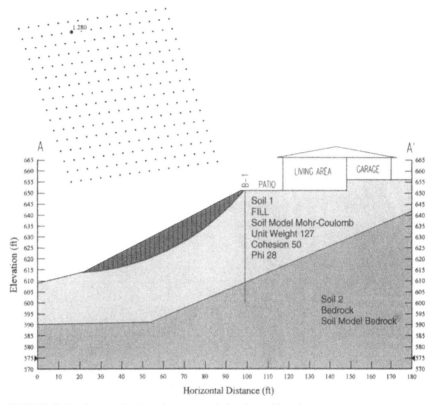

FIGURE 10.49 Cross-section through a compacted clay slope subjected to creep.

An effective stress slope stability analysis was performed by using the SLOPE/W (Geo-Slope, 1991) computer program. In Figure 10.49, the cross-section was developed from the pre- and post-grading topographic maps and the results of the boring. The effective shear strength parameters of the compacted clay fill were obtained from drained direct shear strength tests and $c' = 50$ psf (2.4 kPa) and $\phi' = 28°$ were inputted into the computer program. Also used in the analysis was a total unit weight of the compacted clay fill $(\gamma_t) = 127$ pcf (20 kN/m^3) and the pore water pressures were assumed to be equal to zero.

According to the SLOPE/W (Geo-Slope, 1991) computer program, the minimum factor of safety of the compacted fill slope based on an effective stress analysis and using the Janbu method of slices is 1.28. The results of the slope stability analysis are shown in Fig. 10.49. The stability analysis indicates that the fill closest to the slope face has the lowest factor of safety, which is consistent with the greatest amount of movement recorded from the inclinometer. In addition to the low factor of safety of the slope, there are apparently two additional factors that contributed to the creep of the slope:

1. *Loss of peak shear strength.* In the stability analysis, the peak effective shear strength parameters were used [$c' = 50$ psf (2.4 kPa) and $\phi' = 28$]. Figure 10.52 shows the shear strength versus horizontal deformation for the drained direct shear test specimen having a normal pressure during testing of 400 psf (19 kPa). The peak strength is identified in this figure. Note that with continued deformation, the shear strength decreases and reaches an ultimate value (which is less than the peak). It has been stated (Lambe and Whitman, 1969) that overconsolidated soil, such as compacted clay, has a peak drained shear strength which it loses with further strain, and thus the overconsolidated

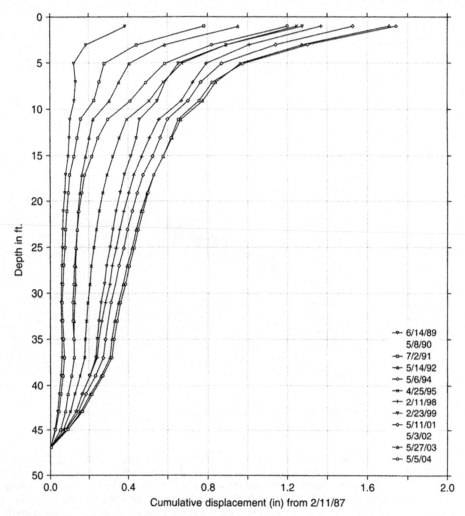

FIGURE 10.50 Inclinometer monitoring.

and normally consolidated strengths approach each other at large strains. This reduction in shear strength of the compacted clay, as it deforms during drained loading, would result in an even lower factor safety (lower than 1.28), which would further promote creep of the slope.

2. Seasonal moisture changes.   It has been stated that seasonal moisture changes can cause creep of fill slopes. For example, Bromhead (1984) states:

> These strains reveal themselves (usually at the surface) as deformations or ground movement. Some movement, particularly close to the surface, occurs even in slopes considered stable, as a result of seasonal moisture content variation; creep rates of a centimeter or so per annum are easily reached. Because of the shallow nature it is only likely to affect poorly-founded structures.

At this site, there would be seasonal moisture changes that probably contributed to the near-surface creep of the compacted clay fill slope.

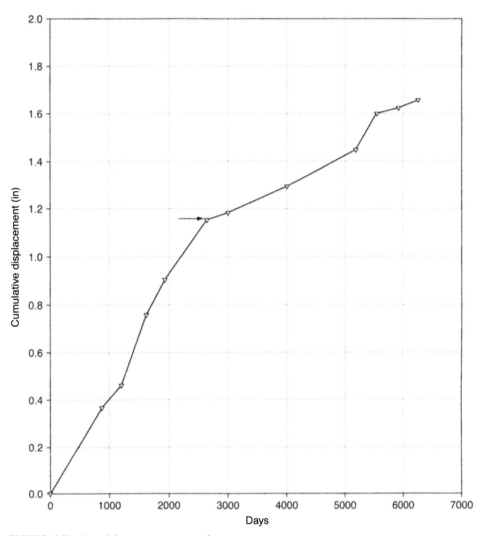

FIGURE 10.51   Lateral slope movement versus time.

10.7.4 Method of Analysis

The engineering geologist will often be involved with the studies of creep of natural slopes. In many cases, subsurface exploration such as test pits or trenches excavated into the slope face will reveal the depth of active creep, which can be identified by the lateral offset of rock strata or soil layers.

The zone of a compacted clay slope subjected to slope softening and creep is often difficult to estimate because it depends on many different factors. In general, the higher the plasticity index of the clay, the larger the zone of slope softening and creep. Also, the lower the factor of safety of the slope face, the more active the creep zone. In some cases grading options such as facing the slope with a cohesionless soil stabilization fill can be used to mitigate the effects of slope creep.

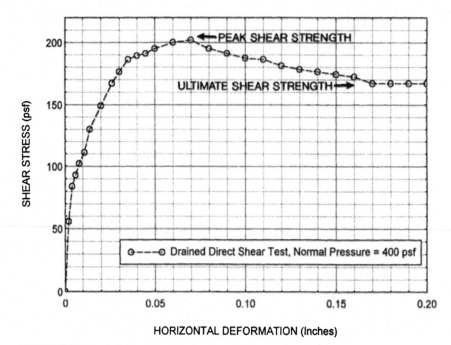

FIGURE 10.52   Drained direct shear test, silty clay fill.

For projects where clay slopes cannot be avoided, the depth of slope softening and creep will be at least as deep as the depth of seasonal moisture changes in the clay. Structures should not be founded on the slope face or near the top of slope that corresponds to the zone of clay having seasonal moisture changes. Also, this zone should not be relied upon for support, such as passive pressure support for retaining wall footings.

The clay slope could creep at a depth that is deeper than the depth of seasonal moisture changes. One design approach is to test the clay in the laboratory and model its slope softening and creep behavior. For example, a clay specimen could be prepared at anticipated field conditions and then placed in the triaxial apparatus, sealed in a rubber membrane, and then subjected to the anticipated horizontal and vertical total stresses based on the slope configuration (stress path method). The measurement of the deformation upon saturation and deformation measurements versus time during the loading could be used to estimate the depth of the creep zone and the amount of anticipated slope movement.

NOTATION

The following notation is used in this chapter:

A = area of the failure wedge

c = cohesion based on a total stress analysis

c' = cohesion based on an effective stress analysis

d = equal to D/H for Fig. 10.21

d, D = depth of the slip surface

D = vertical distance below the toe of the slope to a firm base (Fig. 10.21)

D = equal to H'/H for Cousins stability charts

F = factor of safety for slope stability

h = depth below the ground surface (for calculation of r_u)

H = height of the slope

H' = vertical distance from top of slope to lowest point on the slip surface

L = length of the slip surface

N = normal force on the slip surface

N' = effective normal force on the slip surface

N_F = stability number from Cousins stability charts

N_o = stability number from Fig. 10.21

r_u = pore water pressure ratio

s_u = undrained shear strength of the soil

S = degree of saturation

T = shear force along the slip surface

u = pore water pressure

U = pore water force acting on the slip surface

W = total weight of the failure wedge or failure slice

α = slope inclination or slip surface inclination

β = angular distortion as defined by Boscardin and Cording (1989)

β = slope inclination (Fig. 10.21)

δ/L = maximum angular distortion of the foundation

ε_h = horizontal strain of the foundation

ϕ = friction angle based on a total stress analysis

ϕ' = friction angle based on an effective stress analysis

ϕ'_r = drained residual friction angle

γ_b = buoyant unit weight of the soil

γ_t = total unit weight of the soil

γ_w = unit weight of water

$\lambda_{c\phi}$ = parameter used for Cousins stability charts (see Eq. 10.8)

σ'_n = effective normal stress on the slip surface

τ = shear stress along the slip surface

PROBLEMS

Solutions to the problems are presented in App. C of this book. The problems have been divided into basic categories as indicated below:

Surficial Slope Stability

10.1   For the surficial stability analysis summarized in Table 10.3, calculate the shear strength τ_f along the slip surface.

ANSWER: Using the linear shear strength envelope, $\tau_f = 27.4$ kPa. For the nonlinear portion of the shear strength envelop at low effective stress, $\tau_f = 10.7$ kPa.

10.2   Perform the surficial stability calculations summarized in Table 10.3 for a slope inclination α of 2:1 (horizontal:vertical).

ANSWER: Using the linear shear strength envelope, $F = 2.9$. For the nonlinear portion of the shear strength envelope at low effective stress, $F = 1.37$.

Gross Slope Stability

10.3   Revise Eqs. 10.3 and 10.4 so that it includes a vertical surcharge load Q applied to the failure wedge.

ANSWER: see App. C.

10.4   Revise Eqs. 10.3 and 10.4 so that it includes a horizontal earthquake force = bW, where b = a constant and W = weight of the wedge.

ANSWER: see App. C.

10.5   For Example Problems 10.1 and 10.2, assume there is a uniform vertical surcharge pressure of 200 psf applied to the entire top of the slope. Calculate the factors of safety.

ANSWER: For the total stress example problem, $F = 2.00$. For the effective stress example problem, $F = 1.87$.

10.6   For Example Problems 10.1 and 10.2, assume that an earthquake causes a horizontal force = 0.1 W (i.e., $b = 0.1$). Calculate the factors of safety (pseudostatic slope stability analysis).

ANSWER: For the total stress example problem, $F = 1.52$. For the effective stress example problem, $F = 1.41$.

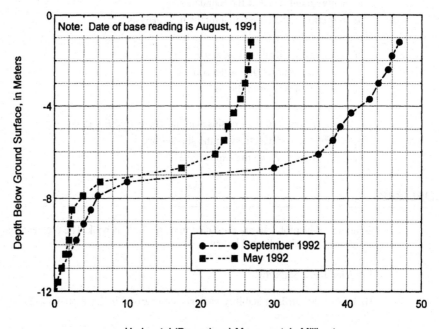

FIGURE 10.53   Results from inclinometer monitoring.

10.7   A computer program is used to determine the factor of safety of a slope. Assume that the computer program uses the ordinary method of slices. For a particular slice located within the assumed failure mass, the total weight W of the slice = 100 kN, the average slope inclination of the slice (α) = 26°, the pore water pressure u at the bottom of the slice is zero, the width of the slice (Δx) equals 1.0 m, and the peak effective shear strength parameters along the assumed slip surface are $\phi' = 25°$ and $c' = 2$ kPa. Assuming an effective stress slope stability analysis is being performed, calculate the shear stress τ and the shear strength τ_f along the base of the slice.

ANSWER: $\tau = 39.4$ kPa and $\tau_f = 39.7$ kPa.

10.8   For Problem 10.7, calculate the factor of safety of the slice. If the factor of safety for the entire failure mass = 1.2 and the slope is composed of stiff-fissured clay that has a substantial reduction in shear strength for strain past the peak values, is progressive failure likely for this slope?

ANSWER: $F = \tau_f/\tau = 1.01$, and yes, progressive failure is likely for this slope.

10.9   Figure 10.53 presents a plot of the lateral downslope movement of a slope. The data were obtained from inclinometer monitoring of a casing that was installed at the top of slope. The base reading for the inclinometer monitoring is August 1991, with readings taken through September 1992. Based on the inclinometer results (Fig. 10.53), what can be concluded concerning the stability of the slope.

ANSWER: The slope is deforming laterally on a slip surface located about 7 m below ground surface.

10.10   Assume a 2:1 (horizontal:vertical) slope is composed of Soil B (Fig. 9.14). Also assume the slope height is 20 m. Based on the depth of seasonal moisture changes and assuming slope creep will occur in this zone, how far back from the top of slope should the structure be located?

ANSWER: 9 m.

CHAPTER 11
RETAINING WALLS

11.1 INTRODUCTION

A *retaining wall* is defined as a structure whose primary purpose is to provide lateral support for soil or rock. In some cases, the retaining wall may also support vertical loads. Examples include basement walls and certain types of bridge abutments.

Some of the more common types of retaining walls are gravity walls, counterfort walls, cantilevered walls, and crib walls. Gravity retaining walls are routinely built of plane concrete or stone and the wall depends primarily on its massive weight to resist failure from overturning and sliding. Counterfort walls consist of a footing, a wall stem, and intermittent vertical ribs (called counterforts) that tie the footing and wall stem together. Crib walls consist of interlocking concrete members that form cells, which are then filled with compacted soil. Common types of retaining walls are shown in Fig. 11.1.

Although mechanically stabilized earth retaining walls have become more popular in the past decade, cantilever retaining walls are still probably the most common type of retaining structure. There are many different types of cantilevered walls, with the common features being a footing that supports the vertical wall stem. Typical cantilevered walls are T-shaped, L-shaped, or reverse L-shaped (Cernica, 1995a, 1995b).

Clean granular material (no silt or clay) is the standard recommendation for backfill material. There are several reasons for this recommendation:

1. *Predictable behavior.* Import granular backfill generally has a more predictable behavior in terms of earth pressure exerted on the wall. Also, expansive soil related forces (Chap. 9) would not be generated by clean granular soil. If clay backfill should be used, the seepage of water into the clay backfill could cause horizontal swelling pressures well in excess of at-rest values. For example, Fourie (1989) measured the swell pressure of compacted clay for zero lateral strain to be 8800 psf (420 kPa). Besides the swelling pressure induced by the expansive soil, there can also be groundwater or perched water pressure on the retaining or basement wall because of the poor drainage of clayey soils.

2. *Drainage system.* To prevent the buildup of hydrostatic water pressure on the retaining wall, a drainage system is often constructed at the heel of the wall. The drainage system will be more effective if highly permeable soil, such as clean granular soil, is used as backfill.

3. *Frost action.* In cold climates, frost action has caused many retaining walls to move so much that they have become unusable. If freezing temperatures prevail, the backfill soil can be susceptible to frost action, where ice lenses will form parallel to the wall and cause horizontal movements of up to 2 to 3 ft (0.6 to 0.9 m) in a single season (Sowers and Sowers, 1970). Backfill soil consisting of clean granular soil and the installation of a drainage system at the heel of the wall will help to protect the wall from frost action.

Movement of retaining walls (i.e., active condition) involves the shear failure of the wall backfill and the analysis will naturally include the shear strength of the backfill soil. Similar to the analysis of

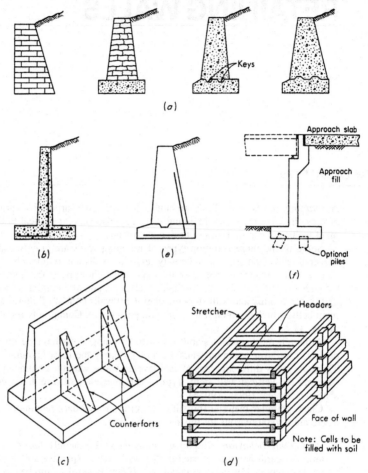

FIGURE 11.1   Common types of retaining walls. (*a*) Gravity walls of stone, brick, or plain concrete. Weight provides overturning and sliding stability. (*b*) Cantilevered wall. (*c*) Counterfort retaining wall or buttressed retaining wall. If backfill covers counterforts, the wall is termed a counterfort retaining wall. (*d*) Crib wall. (*e*) Semigravity wall (often steel reinforcement is used). (*f*) Bridge abutment. (*Reproduced from Bowles, 1982; with permission of McGraw-Hill, Inc.*)

strip footings and slope stability, for most field situations involving retaining structures, the backfill soil is in a plane strain condition (i.e., the soil is confined along the long axis of the wall). As previously mentioned, the friction angle ϕ is about 10 percent higher in the plane strain condition as compared to the friction angle ϕ measured in the triaxial apparatus. In practice, plane strain shear strength tests are not performed, which often results in an additional factor of safety for retaining wall analyses.

The next two sections (11.2 and 11.3) of this chapter will discuss the basic retaining wall equations for a simple retaining wall without and with wall friction. The following sections will then discuss in more detail the design and construction of retaining walls (11.4), restrained retaining walls (11.5), mechanically stabilized earth retaining walls (11.6), sheet pile walls (11.7), and temporary retaining walls (11.8), which are often needed to support foundation excavations. The final section (11.9) is devoted to moisture migration through retaining walls. Geotechnical earthquake engineering analyses for retaining walls will be covered in Chap. 14.

11.2   SIMPLE RETAINING WALL WITHOUT WALL FRICTION

Figure 11.2 shows a reverse L-shaped cantilever retaining wall. This type of simple retaining wall will be used to introduce the basic types of retaining wall design analyses. There are three pressures acting on the retaining wall, as follows:

1. *Active earth pressure.*   The pressure exerted on the back of the wall is the active earth pressure.

2. *Bearing pressure.*   The vertical bearing pressure of the soil or rock supports the retaining wall footing.

3. *Passive earth pressure.*   Lateral movement of the wall is resisted by passive earth pressure and slide friction between the footing and bearing material.

The following discussion of the design analyses for retaining walls are divided into two categories, (1) simple retaining wall without wall friction and (2) simple retaining wall with wall friction.

11.2.1   Active Earth Pressure

As shown in Fig. 11.2, the active earth pressure is often assumed to be horizontal by neglecting the friction developed between the vertical wall stem and the backfill. This friction force has a stabilizing effect on the wall and therefore it is usually a safe assumption to ignore friction. However, if the wall should settle more than the backfill, such as due to high vertical loads imposed on the top of the wall, then a negative skin friction can develop between the wall and backfill, which has a destabilizing effect on the wall.

In the evaluation of the active earth pressure, it is common for the soil engineer to recommend clean granular soil as backfill material. In order to calculate the active earth pressure resultant force P_A, in pounds per linear foot of wall or kilonewtons per linear meter of wall, the following equation is used for clean granular backfill:

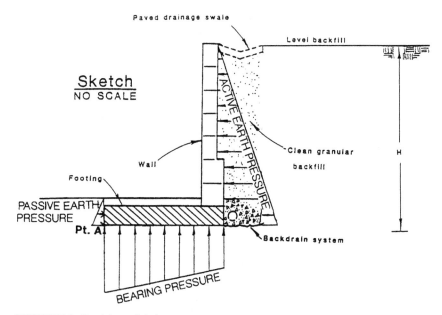

FIGURE 11.2   Retaining wall design pressures.

$$P_A = \frac{1}{2} k_A \gamma_t H^2 \tag{11.1}$$

where k_A = active earth pressure coefficient (dimensionless)
γ_t = total unit weight of the backfill (pcf or kN/m³)
H = height over which the active earth pressure acts as defined in Fig. 11.2 (ft or m)

The active earth pressure coefficient k_A is equal to:

$$k_A = \tan^2 (45° - \frac{1}{2}\phi) \tag{11.2}$$

where ϕ is the friction angle of the clean granular backfill (degrees).

Equation 11.2 is known as the *active Rankine state*, after the British engineer Rankine, who in 1857 obtained this relationship.

In Eq. 11.1, the product of k_A times γ_t is referred to as the *equivalent fluid pressure* (even though the product is actually a unit weight). In the design analysis, the soil engineer usually assumes a total unit weight γ_t of 120 pcf (18.9 kN/m³) and a friction angle ϕ of 30° for the granular backfill. Using Eq. 11.2 and $\phi = 30°$, the active earth pressure coefficient k_A is 0.333. Multiplying 0.333 times the total unit weight γ_t of backfill results in an equivalent fluid pressure of 40 pcf (6.3 kN/m³). This is a common recommendation for equivalent fluid pressure from soil engineers. It is valid for the conditions of clean granular backfill, a level ground surface behind the wall, a backdrain system, and no surcharge loads. Note that this recommended value of equivalent fluid pressure of 40 pcf (6.3 kN/m³) does not include a factor of safety and is the actual pressure that would be exerted on a smooth wall when the friction angle ϕ of the granular backfill equals 30°. When designing the vertical wall stem in terms of wall thickness and size and location of steel reinforcement, a factor of safety (F) can be applied to the active earth pressure in Eq. 11.1. A factor of safety may be prudent because higher wall pressures will most likely be generated during compaction of the backfill or when translation of the footing is restricted (Goh, 1993).

Additional important details concerning the active earth pressure are as follows:

1. *Sufficient movement.*   There must be sufficient movement of the retaining wall in order to develop the active earth pressure of the backfill. For example, Table 11.1 indicates the amount of wall rotation that must occur for different backfill soils in order to reach the active earth pressure state.

2. *Triangular distribution.*   As shown in Fig. 11.2, the active earth pressure is a triangular distribution and thus the active earth pressure resultant force P_A is located at a distance equal to $\frac{1}{3} H$ above the base of the wall.

3. *Active wedge.*   The *active wedge* is defined as that zone of soil involved in the development of the active earth pressures upon the wall. This active wedge must move laterally in order to develop the active earth pressures. It is important that building footings or other load carrying members are not supported by the active wedge, or else they will be subjected to lateral movement. Figure 11.3 shows an illustration of the active wedge of soil behind the retaining wall. As indicated in Fig. 11.3, the active wedge is inclined at an angle of $45° + \phi/2$ from the horizontal.

TABLE 11.1   Magnitudes of Wall Rotation to Reach Active and Passive States

Soil type and condition	Rotation (Y/H) for active state	Rotation (Y/H) for passive state
Dense cohesionless	0.0005	0.002
Loose cohesionless	0.002	0.006
Stiff cohesive	0.01	0.02
Soft cohesive	0.02	0.04

Note: Y = wall displacement and H = height of the wall.
Source: From NAVFAC DM-7.2, 1982.

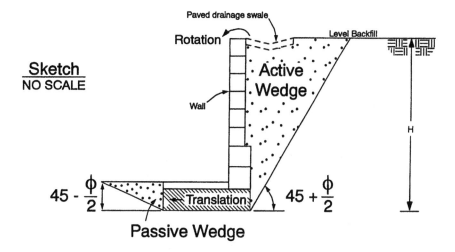

Note: For active and passive wedge development there must be movement of the retaining wall as illustrated above.

FIGURE 11.3   Active wedge behind retaining wall.

4. *Surcharge pressure.*   If there is a uniform surcharge pressure Q acting upon the entire ground surface behind the wall, then there would be an additional horizontal pressure exerted upon the retaining wall equal to the product of k_A times Q. Thus the resultant force P_Q, in pounds per linear foot of wall or kilonewtons per linear meter of wall, acting on the retaining wall due to the surcharge Q is equal to:

$$P_Q = QHk_A \tag{11.3}$$

where Q = uniform vertical surcharge (psf or kPa) acting upon the entire ground surface behind the retaining wall
k_A = active earth pressure coefficient (Eq. 11.2)
H = height of the retaining wall (ft or m)

Because this pressure acting upon the retaining wall is uniform, the resultant force P_Q is located at mid-height of the retaining wall.

11.2.2 Passive Earth Pressure

As shown in Fig. 11.2, the passive earth pressure is developed along the front side of the footing. Passive pressure is developed when the wall footing moves laterally into the soil and a passive wedge is developed such as shown in Fig. 11.3. In order to calculate the passive resultant force P_p, the following equation is used assuming that there is cohesionless soil in front of the wall footing:

$$P_p = \tfrac{1}{2}k_p\gamma_t D^2 \tag{11.4}$$

where P_p = passive resultant force (lb per linear ft or kN per linear m of wall length)
k_p = passive earth pressure coefficient (dimensionless)
γ_t = total unit weight of the soil located in front of the wall footing (pcf or kN/m³)
D = depth of the wall footing, i.e., vertical distance from the ground surface in front of the retaining wall to the bottom of the footing (ft or m)

The passive earth pressure coefficient k_p is equal to:

$$k_p = \tan^2 (45° + \tfrac{1}{2}\phi) \qquad (11.5)$$

where ϕ is the friction angle of the soil in front of the wall footing (degrees).

Equation 11.5 is known as the passive Rankine state. In order to develop passive pressure, the retaining wall foundation must move laterally into the soil. As indicated in Table 11.1, the wall translation to reach the passive state is at least twice that required to reach the active earth pressure state.

Usually it is desirable to limit the amount of foundation translation by applying a reduction factor to the passive pressure. A commonly used reduction factor is 2.0 (Lambe and Whitman, 1969). The soil engineer routinely reduces the passive pressure by $\frac{1}{2}$ (reduction factor = 2.0) and then refers to the value as the allowable passive pressure. To limit wall translation, the structural engineer should use the allowable passive pressure for design of the retaining wall. The passive pressure may also be limited by building codes (e.g., see Table 18.4).

If the soil in front of the retaining wall is a plastic (clayey) soil, then usually the long-term effective stress analysis will govern. For the effective stress analysis, the effective cohesion c' and effective friction angle ϕ' can be determined from laboratory tests. As a conservative approach, often the effective cohesion c' is ignored and the effective friction angle ϕ' is used in Eq. 11.5 in order to determine the allowable passive resistance.

11.2.3 Foundation Bearing Pressure

In order to calculate the foundation bearing pressure, the first step is to sum the vertical loads, such as the wall and footing weights. The vertical loads can be represented by a single resultant vertical force, per linear foot or meter of wall length, which is offset by a distance from the toe of the retaining wall foundation. This can then be converted to a pressure distribution as shown in Fig. 11.2. The largest bearing pressure is routinely at the toe of the foundation (Point A, Fig. 11.2). The largest bearing pressure should not exceed the allowable bearing pressure, which is usually provided by the geotechnical engineer.

11.2.4 Foundation Sliding Analysis

The *factor of safety* (F) for sliding of the retaining wall foundation is often defined as the resisting forces divided by the driving force. The forces are per linear meter or foot of wall length, or:

$$F = \frac{W \tan \delta + P_p}{P_A} \qquad (11.6)$$

where δ = friction angle between the bottom of the concrete foundation and bearing soil (degrees)
W = weight of the retaining wall and foundation (lb per linear ft or kN per linear m of wall length)
P_p = allowable passive resultant force, i.e., P_p from Eq. 11.4 divided by a reduction factor (lb per linear ft or kN per linear m of wall length)
P_A = active earth pressure resultant force from Eq. 11.1 (lb per linear ft or kN per linear m of wall length)

The typical recommendation for minimum factor of safety for sliding of the retaining wall foundation is 1.5 to 2.0 (Cernica, 1995a).

In some situations, there may be adhesion between the bottom of the retaining wall foundation and the bearing soil. This adhesion is often neglected because the wall is designed for active pressures, which typically develop when there is translation of the footing. Translation of the foundation will break the adhesive forces between the bottom of the footing and the bearing soil and therefore adhesion is often neglected for the factor of safety of sliding.

11.2.5 Overturning Analysis

The factor of safety (F) for overturning of the retaining wall is calculated by taking moments about the toe of the foundation (Point A, Fig. 11.2), and is:

$$F = \frac{Wa}{\frac{1}{3} P_A H} \tag{11.7}$$

where W = weight of the retaining wall and foundation (lb per linear ft or kN per linear m of wall length)

a = lateral distance from W to the toe of the foundation (ft or m)

P_A = active earth pressure resultant force, i.e., Eq. 11.1 (lb per linear ft or kN per linear m of wall length)

H = height of the retaining wall (ft or m)

In Eq. 11.7, the moment due to passive pressure is neglected. The reason is because with a rotation type failure mode, the wall may not move enough laterally to induce passive earth pressures. The typical recommendation for minimum factor of safety for overturning is 1.5 to 2.0 (Cernica, 1995a).

11.3 SIMPLE RETAINING WALL WITH WALL FRICTION

In some cases, the geotechnical engineer may want to include the friction between the soil and the rear side of the retaining wall. For this situation, the design analysis is more complicated as discussed below:

11.3.1 Active Earth Pressure

A common equation that is used to calculate the active earth pressure coefficient k_A for the case of wall friction is the Coulomb equation, which is shown in Fig. 11.4. The Coulomb equation can also be used if the back face of the wall is sloping or if there is a sloping backfill behind the wall. Once the active earth pressure coefficient k_A is calculated, the active earth pressure resultant force P_A can be calculated by using Eq. 11.1.

11.3.2 Example Problem

Figure 11.5 (from Lambe and Whitman, 1969) presents an example of a proposed concrete retaining wall that will have a height of 20 ft (6.1 m) and a base width of 7 ft (2.1 m). The wall will be backfilled with sand that has a total unit weight γ_t of 110 pcf (17.3 kN/m^3), friction angle ϕ of 30°, and an assumed wall friction ϕ_w of 30°. Although $\phi_w = 30°$ will be used for this example problem, more typical values of wall friction are $\phi_w = \frac{3}{4}\phi$ for the wall friction between granular soil and wood or concrete walls, and $\phi_w = 20°$ for the wall friction between granular soil and steel walls such as sheet-pile walls.

For the example problem shown in Fig. 11.5, the value of the active earth pressure coefficient k_A can be calculated by using Coulomb's equation (Fig. 11.4) and inserting the following values:

- Slope inclination $\beta = 0°$ (no slope inclination)
- Back face of the retaining wall $\theta = 0°$ (vertical back face of the wall)
- Friction between the back face of the wall and the soil backfill $\delta = \phi_w = 30°$
- Friction angle of backfill sand $\phi = 30°$

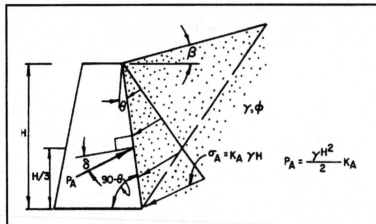

A) Coulomb's Equation (Static Condition):

$$K_A = \frac{\cos^2(\phi - \theta)}{\cos^2\theta \, \cos(\delta + \theta)\left[1 + \sqrt{\dfrac{\sin(\delta + \phi)\sin(\phi - \beta)}{\cos(\delta + \theta)\cos(\beta - \theta)}}\right]^2}$$

B) Mononobe-Okabe Equation (Earthquake Condition):

$$K_{AE} = \frac{\cos^2(\phi - \theta - \psi)}{\cos\psi \, \cos^2\theta \, \cos(\delta + \theta + \psi)\left[1 + \sqrt{\dfrac{\sin(\delta + \phi)\sin(\phi - \beta - \psi)}{\cos(\delta + \theta + \psi)\cos(\beta - \theta)}}\right]^2}$$

FIGURE 11.4 Coulomb's earth pressure k_A equation. Part A) presents Coulomb's equation for static conditions and Part B) presents the modified Coulomb's equation for earthquake conditions. (*Figure reproduced from NAVFAC DM-7.2, 1982; with equations from Kramer 1996.*)

Inputting the above values into Coulomb's equation (Fig. 11.4), the value of the active earth pressure coefficient $k_A = 0.297$.

By using Eq. 11.1 with $k_A = 0.297$, total unit weight $\gamma_t = 110$ pcf (17.3 kN/m³), and the height of the retaining wall $H = 20$ ft (6.1 m, see Fig. 11.5A), the active earth pressure resultant force $P_A = 6540$ pounds per linear foot of wall (95.4 kilonewtons per linear meter of wall). As indicated Fig. 11.5A, the active earth pressure resultant force ($P_A = 6540$ lb/ft) is inclined at an angle of 30° due to the wall friction assumptions. The vertical ($P_v = 3270$ lb/ft) and horizontal ($P_H = 5660$ lb/ft) resultants of P_A are also shown in Fig. 11.5A. Note in Fig. 11.4 that even with wall friction, the active earth pressure is still a triangular distribution acting upon the retaining wall and thus the location of the active earth pressure resultant force P_A is at a distance of ⅓H above the base of the wall, or 6.7 ft (2.0 m).

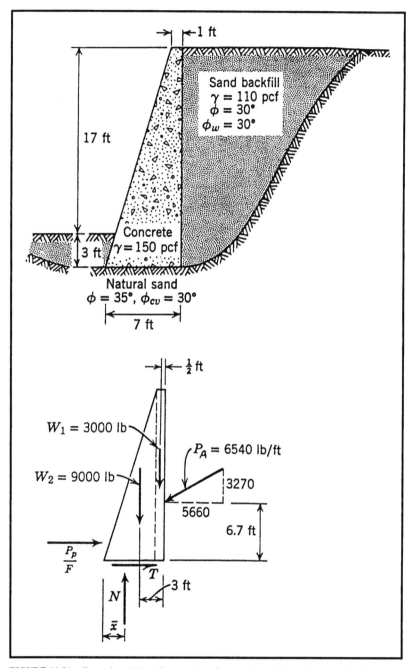

FIGURE 11.5A   Example problem: Cross section of proposed retaining wall and resultant forces acting on the retaining wall. (*From Lambe and Whitman, 1969; reproduced with permission of John Wiley & Sons.*)

Find.   Adequacy of wall.

Solution.   The first step is to determine the active thrust;
The next step is to compute the weights:

$$W_1 = (1)(20)(150) = 3000 \text{ lb/ft}$$

$$W_2 = \tfrac{1}{2}(6)(20)(150) = 9000 \text{ lb/ft}$$

Next N and $\bar{x}$ are computed:

$$N = 9000 + 3000 + 3270 = 15{,}270 \text{ lb/ft}$$

$$\text{Overturning moment} = 5660(6.67) - 3270(7) = 37{,}800 - 22{,}900 = 14{,}900$$

$$\text{Moment of weight} = (6.5)(3000) + (4)(9000) = 19{,}500 + 36{,}000 = 55{,}500$$

$$\text{Ratio} = 3.73 \quad \underline{\underline{\text{OK}}}$$

$$\bar{x} = \frac{55{,}500 - 14{,}900}{15{,}270} = \frac{40{,}600}{15{,}270} = 2.66 \text{ ft} \quad \underline{\underline{\text{OK}}}$$

The location of N

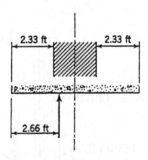

FIGURE 11.5B   Example problem: Calculation of the factor of safety of overturning and the location of the resultant force N. (*From Lambe and Whitman, 1969; reproduced with permission of John Wiley & Sons.*)

Passive Earth Pressure.   As shown in Fig. 11.5A, the passive earth pressure is developed by the soil located at the front of the retaining wall. Usually wall friction is ignored for the passive earth pressure calculations. For the example problem shown in Fig. 11.5, the passive resultant force P_p was calculated by using Eqs. 11.4 and 11.5 and neglecting wall friction and the slight slope of the front of the retaining wall (see Fig. 11.5C for passive earth pressure calculations).

Footing Bearing Pressure.   The procedures for the calculations of the footing bearing pressures are as follows:

1. *Calculate N.*   As indicated in Fig. 11.5B, the first step is to calculate N (15,270 lb/ft), which equals the sum of the weight of the wall, footing, and vertical component of the active earth pressure resultant force (i.e., $N = W + P_A \sin \phi_w$).

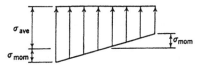

Next the bearing stress is computed. The average bearing stress is 15,270/7 = 2180 psf. Assuming that the bearing stress is distributed linearly, the maximum stress can be found

$$\sigma_{mom} = \frac{M}{S}$$

where

$$M = \text{moment about } \mathcal{L} = 15{,}270(3.5 - 2.66) = 12{,}820 \text{ lb-ft/ft}$$

$$S = \text{section modulus} = \tfrac{1}{6}B^2 = \tfrac{1}{6}(7)^2 = 8.17 \text{ ft}^2$$

where B is width of base

$$\sigma_{mom} = \frac{12{,}820}{8.17} = 1570 \text{ psf}$$

$$\text{Maximum stress} = 2180 + 1570 = 3750 \text{ psf}$$

Finally, the resistance to horizontal sliding is checked. Assuming passive resistance without wall friction,

$$K_p = 3$$

$$P_p = \tfrac{1}{2}(110)(3^2)(3) = 1500 \text{ lb/ft}$$

With reduction factor of 2,

$$\frac{P_p}{F} = 750 \text{ lb/ft}$$

$$T = 5660 - 750 = 4910 \text{ lb/ft}$$

$$N \tan 30° = 8810 \text{ lb/ft}$$

$$\frac{N \tan \phi_{cv}}{T} = 1.79 < 2 \qquad \text{not OK}$$

Ignoring passive resistance

$$T = 5660 \text{ lb/ft}$$

$$\frac{N \tan \phi_{cv}}{T} = 1.55 > 1.5 \qquad \text{OK}$$

FIGURE 11.5C Example problem: Calculation of the maximum bearing stress and the factor of safety of sliding. (*From Lambe and Whitman, 1969; reproduced with permission of John Wiley & Sons.*)

2. *Determine $\bar{x}$.* The value of $\bar{x}$ (2.66 ft) is calculated as shown in Fig. 11.5B. The moments are determined about the toe of the retaining wall. Then $\bar{x}$ equals the difference in the opposing moments divided by N.

3. *Determine average bearing pressure.* The average bearing pressure (2180 psf) is calculated in Fig. 11.5C as N divided by the width of the footing (7 ft).

4. *Calculate moment about the centerline of the footing.* The moment about the centerline of the footing is calculated as N times the eccentricity (0.84 ft).

5. *Section modulus.* The section modulus of the footing is calculated as shown in Fig. 11.5C.

6. *Portion of bearing pressure due to moment.* The portion of the bearing pressure due to the moment σ_{mom} is determined as the moment divided by the section modulus.

7. *Maximum bearing pressure.* The maximum bearing pressure is then calculated as the sum of the average pressure ($\sigma_{avg} = 2180$ psf) plus the bearing pressure due to the moment ($\sigma_{mom} = 1570$ psf).

As indicated in Fig. 11.5C, the maximum bearing pressure is 3750 psf (180 kPa). This maximum bearing pressure must be less than the allowable bearing pressure (Chap. 6). It is also a standard requirement that the resultant normal force N be located within the middle third of the footing such as illustrated in Fig. 11.5B.

An alternate method to determine the maximum q' and minimum q'' bearing pressure is to use Eqs. 6.12 and 6.13.

Sliding Analysis. The factor of safety (F) for sliding of the retaining wall is often defined as the resisting forces divided by the driving force. The forces are per linear foot or meter of wall length, or:

$$F = \frac{N \tan \delta + P_p}{P_H} \qquad (11.8)$$

where $\delta = \phi_{cv}$ = friction angle between the bottom of the concrete foundation and bearing soil (degrees)

N = sum of the weight of the wall, footing, and vertical component of the active earth pressure resultant force, i.e., $N = W + P_A \sin \phi_w$ (lb per linear ft or kN per linear m of wall length)

P_p = allowable passive resultant force, i.e., P_p from Eq. 11.4 divided by a reduction factor (lb per linear ft or kN per linear m of wall length)

P_H = horizontal component of the active earth pressure resultant force, i.e., $P_H = P_A \cos \phi_w$ (lb per linear ft or kN per linear m of wall length)

There are variations of Eq. 11.8 that are used in practice. For example, as illustrated in Fig. 11.5C, the value of P_p is subtracted from P_H in the denominator of Eq. of 11.8, instead of P_p being used in the numerator. For the example problem shown in Fig. 11.5, the factor of safety for sliding (F) = 1.79 when passive pressure is included and $F = 1.55$ when passive pressure is excluded. As previously mentioned, the typical recommendation for minimum factor of safety for sliding is 1.5 to 2.0 (Cernica, 1995a).

Overturning Analysis. The factor of safety (F) for overturning of the retaining wall is calculated by taking moments about the toe of the footing (Point A, Fig. 11.2), and is:

$$F = \frac{Wa}{\frac{1}{3}P_H H - P_v e} \qquad (11.9)$$

where a = lateral distance from the resultant weight of the wall and footing (W) to the toe of the footing (ft or m)

P_H = horizontal component of the active earth pressure resultant force (lb per linear ft or kN per linear m of wall length)

H = height of the retaining wall (ft or m)

P_v = vertical component of the active earth pressure resultant force (lb per linear ft or kN per linear m of wall length)

e = lateral distance from the location of P_v to the toe of the wall (ft or m)

In Fig. 11.5B, the factor of safety (ratio) for overturning is calculated to be 3.73. As previously mentioned, the typical recommendation for minimum factor of safety for overturning is 1.5 to 2.0 (Cernica, 1995a).

11.4 DESIGN AND CONSTRUCTION OF RETAINING WALLS

The previous section dealt with simple retaining walls. This section will provide an additional discussion of the design and construction of retaining walls. Figure 11.6 (from NAVFAC DM-7.2, 1982) shows several examples of different types of retaining walls. Figure 11.6A shows gravity and semigravity retaining walls, Fig. 11.6B shows cantilever and counterfort retaining walls, and the

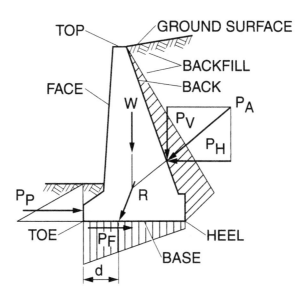

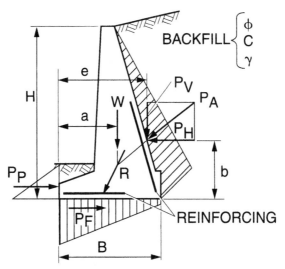

FIGURE 11.6A Gravity and semigravity retaining walls. (*From NAVFAC DM-7.2, 1982.*)

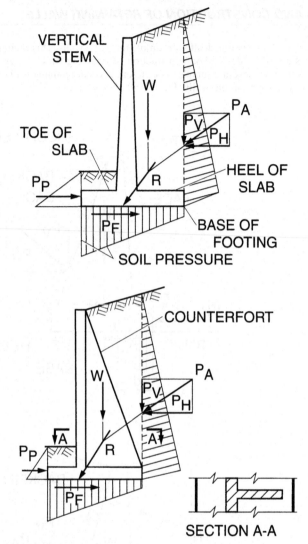

FIGURE 11.6B Cantilever and counterfort retaining walls. (*From NAVFAC DM-7.2, 1982.*)

design analyses are shown in Fig. 11.6C. Although the equations in Fig. 11.6C include an adhesion value c_a, as previously mentioned, the adhesion is often neglected. This is because active pressures develop when there is translation (movement) of the footing that would tend to break the adhesive resistance.

11.4.1 Retaining Walls at the Top of Slopes

Retaining walls are sometimes constructed at the top of slopes. In this case, there is a descending ground surface in front of the retaining wall and Eq. 11.5 cannot be used to determine the passive earth pressure coefficient. For either a descending slope ($-\beta$) or ascending slope ($+\beta$) in front of

LOCATION OF RESULTANT

MOMENTS ABOUT TOE:

$$d = \frac{Wa + P_ve - P_Hb}{W + P_v}$$

ASSUMING $P_P = 0$

OVERTURNING

MOMENTS ABOUT TOE:

$$F = \frac{W_a}{P_Hb - P_ve} \geqq 1.5$$

IGNORE OVERTURNING IF R IS WITHIN MIDDLE
THIRD (SOIL), MIDDLE HALF (ROCK).
CHECK R AT DIFFERENT HORIZONTAL PLANES
FOR GRAVITY WALLS.

RESISTANCE AGAINST SLIDING

$$F = \frac{(W + P_v) \text{ TAN } \delta + C_aB}{P_H} \geqq 1.5$$

$$F = \frac{(W + P_v) \text{ TAN } \delta + C_aB + P_P}{P_H} \geqq 2.0$$

$$P_F = (W + P_v) \text{ TAN } \delta + C_aB$$

C_a = ADHESION BETWEEN SOIL AND BASE

TAN δ = FRICTION FACTOR BETWEEN SOIL
AND BASE

W = INCLUDES WEIGHT OF WALL AND SOIL IN FRONT
FOR GRAVITY AND SEMIGRAVITY WALLS.
INCLUDES WEIGHT OF WALL AND SOIL ABOVE
FOOTING, FOR CANTILEVER AND COUNTERFORT
WALLS.

FIGURE 11.6C   Design analysis for retaining walls shown in Figs. 11.6A and 11.6B. (*From NAVFAC DM-7.2, 1982.*)

the retaining wall, the following equation can be used to determine the passive earth pressure coefficient k_p:

$$k_p = \frac{\cos^2 \phi}{[1 - (\sin^2 \phi + \sin \phi \cos \phi \tan \beta)^{0.5}]^2} \tag{11.10}$$

where: ϕ is the friction angle of the soil in front of the retaining wall (degrees) and β is the slope inclination measured from a horizontal plane, where a descending slope in front of the retaining wall has a negative β value (degrees).

Although not readily apparent, if $\beta = 0$, Eq. 11.10 will give exactly the same values of k_p as Eq. 11.5. For example, substituting $\beta = 0°$ and $\phi = 30°$ into Eq. 11.10, $k_p = 3$ which is exactly the value obtained from Eq. 11.5.

Suppose that a retaining wall is constructed at the top of a 2:1 (horizontal:vertical) slope. In this case, the slope inclination is $\beta = -26.6°$. Assuming $\phi = 30°$ for the soil comprising the slope, then according to Eq. 11.10, the value of $k_p = 1.12$. Thus for the case of a 2:1 descending slope in front of the retaining wall, the passive resistance will be significantly reduced (k_p decreases from 3 to 1.12). Many retaining walls tilt or deform downslope because the condition of a sloping ground surface in front of the retaining wall significantly reduces the passive resistance for the wall foundation.

There can be additional factors that contribute to the failure of retaining walls constructed at the top of slopes. In some cases, there may be a low factor of safety of the entire slope that is exacerbated by the construction of a top of slope wall. In other cases, where the soil in front of the retaining wall is clayey, there can be slope creep that causes a soil gap to open up at the front of the retaining wall footing. In these cases of a descending clayey slope in front of the retaining wall, the depth of creep of the outer face of the slope must be estimated and then the passive pressure should only be utilized below the depth of surface slope creep.

11.4.2 Retaining Wall Failures

There are many construction factors that can result in excessive lateral movement, bearing capacity failures, sliding failures, or failure by overturning of the retaining wall. Common causes can include inadequate subsurface exploration or laboratory testing, incorrect design, improper construction, or unanticipated loadings. Typical construction related problems are discussed below:

1. *Clay backfill.*   A frequent cause of failure is because the wall was backfilled with clay. As previously mentioned, clean granular sand or gravel is usually recommended as backfill material. This is because of the undesirable effects of using clay or silt as a backfill material. When clay is used as backfill material, the clay backfill can exert swelling pressures on the wall (Fourie, 1989; Marsh and Walsh, 1996). The highest swelling pressures develop when water infiltrates a backfill consisting of clay that was compacted to a high dry density at low moisture content. The type of clay particle that will exert the highest swelling pressures is montmorillonite. Because the clay backfill is not free draining, there could also be additional hydrostatic forces or ice-related forces that substantially increase the thrust on the wall. Fig. 11.7 shows the collapse of retaining wall that has a clay backfill.

2. *Inferior backfill soil.*   To reduce construction costs, soil available on-site is sometimes used for backfill. This soil may not have the properties, such as being a clean granular soil with a high shear strength, that were assumed during the design stage. Using on-site available soil such as clays, rather than importing granular material, is probably the most common reason for retaining wall failures.

3. *Compaction induced pressures.*   As previously mentioned, one reason for applying a factor of safety (F) to the active earth pressure (Eq. 11.1) is because larger wall pressures will typically be generated during compaction of the backfill. By using heavy compaction equipment in close proximity to the wall, excessive pressures can be developed that damage the wall. The best compaction equipment, in terms of exerting the least compaction induced pressures on the wall, are small vibrator plate (hand-operated) compactors such as models VPG 160B and BP 19/75 (Duncan et al., 1991). The vibrator plates effectively densify the granular backfill, but do not induce high lateral loads because of their light weight. Besides hand-operated compactors, other types of relatively light-weight equipment (such as a bobcat) can be used to compact the backfill.

4. *Failure of the back-cut.*   There could also be the failure of the back-cut for the retaining wall. A vertical back-cut is often used when the retaining wall is less than 5 ft (1.5 m) high. In other cases, the back-cut is usually sloped. The back-cut can fail if it is excavated too steeply and does not have an adequate factor of safety.

FIGURE 11.7   Collapse of a retaining wall that has a clay backfill.

Retaining wall movement is often gradual and the wall deforms by intermittently tilting or moving laterally. It is also possible that a failure can occur suddenly, such as when there is a slope-type failure beneath the wall or when the foundation of the wall fails due to inadequate bearing capacity. These rapid failure conditions could develop if the wall is supported by soft clay and there is an undrained shear failure beneath the foundation.

11.5   RESTRAINED RETAINING WALLS

As mentioned in the previous section, in order for the active wedge (Fig. 11.3) to be developed, there must be sufficient movement of the retaining wall (Table 11.1). There are many cases where movement of the retaining wall is restricted. Examples include massive bridge abutments, rigid basement walls, and retaining walls that are anchored in non-yielding rock. These cases are often described as *restrained retaining walls*.

In order to determine the earth pressure acting on a restrained retaining wall, Eq. 11.1 can be utilized where the coefficient of earth pressure at rest (k_o) is substituted for k_A. The value of k_o can be estimated from Eqs. 4.19 and 4.20. A common value of k_o that is used for restrained retaining walls is 0.5. Restrained retaining walls are especially susceptible to higher earth pressures induced by heavy compaction equipment and extra care must be taken during the compaction of backfill for restrained retaining walls.

For either a line load or point load acting on the ground surface behind the restrained retaining wall, Fig. 11.8 can be used to determine the additional horizontal force P_H acting on the restrained retaining wall and the location of this horizontal force above the base of the wall. The upper plots in Fig. 11.8 can be used to determine the horizontal pressure distribution acting on the retaining wall due to the line load or point load.

If there is a uniform surcharge Q acting upon the entire ground surface behind the wall, then there would be an additional horizontal pressure exerted upon the restrained retaining wall equal to the product of k_o times Q. Thus the resultant force P_Q, in pounds per linear foot of wall or kilonewtons per linear meter of wall, could be calculated by using Eq. 11.3 and substituting k_o for k_A.

Example Problem 11.1   For the example problem shown in Fig. 11.5, assume that the retaining wall is part of a bridge abutment and that the lateral deformation of the wall will be restricted. Considering the retaining wall to be a restrained wall and using Eq. 4.19 to calculate the coefficient of earth pressure at rest, determine the resultant of the lateral pressure exerted on the retaining wall by the backfill soil.

Solution   From Eq. 4.19:

$$k_o = 1 - \sin \phi' = 1 - \sin 30° = 0.5$$

$$P = \tfrac{1}{2} k_o \gamma_t H^2 = \tfrac{1}{2}(0.5)(110)(20)^2 = 11{,}000 \text{ lb/ft}$$

Resultant force = 11,000 lb/ft acting at a distance of 6.7 ft above the base of the wall

---

Example Problem 11.2   Using the previous example problem, assume that the ground surface behind the retaining wall will be subjected to a line load of 1000 pounds per foot (i.e., $Q_L =$ 1000 lb/ft) located at a distance of 6 feet from the wall (i.e., $x = 6$ ft). Determine the horizontal force P_H acting on the restrained retaining wall and the location of P_H above the base of the wall using Fig. 11.8.

Solution   Using the left side of Fig. 11.8:

$$x = (m)(H) \text{ with } x = 6 \text{ ft and using } H = 20 \text{ ft, therefore, } m = 0.30$$

Using the lower left corner of Fig. 11.8 and since $m \leq 0.4$:

$$P_H = 0.55 \, Q_L = (0.55)(1000 \text{ lb/ft}) = 550 \text{ lb/ft}$$

In the upper left plot, there is a box that provides values of m versus R. For m = 0.3, the value of $R = 0.60H = (0.60)(20 \text{ ft}) = 12$ ft

Resultant force = 550 lb/ft acting at a distance of 12 ft above the base of the wall

---

Example Problem 11.3   Using the previous example problem, assume that the ground surface behind the retaining wall will be subjected to a uniform pressure of 100 psf (i.e., $Q =$ 100 psf). Determine the horizontal force P_Q acting on the restrained retaining wall due to this surcharge pressure and the location of P_Q above the base of the wall.

Solution   From Eq. 4.19:

$$k_o = 1 - \sin \phi' = 1 - \sin 30° = 0.5$$

Using Eq. 11.3 and substituting k_o for k_A:

$$P_Q = QHk_o = (100 \text{ psf})(20 \text{ ft})(0.5) = 1000 \text{ lb/ft}$$

Because this pressure acting upon the retaining wall is uniform, the resultant force P_Q is located at mid-height of the retaining wall = $\tfrac{1}{2} H = \tfrac{1}{2}$ (20 ft) = 10 ft

Resultant force = 1000 lb/ft acting at a distance of 10 ft above the base of the wall

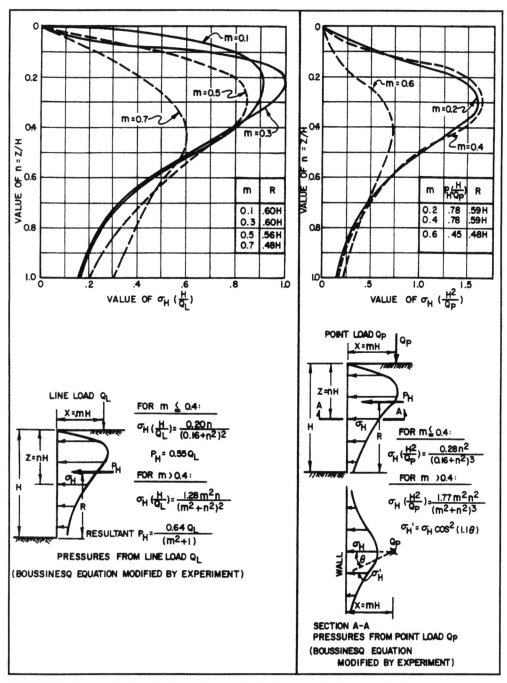

FIGURE 11.8   Horizontal resultant load P_H and pressure distribution due to a line load or point load acting on the ground surface behind a restrained retaining wall. (*From NAVFAC DM-7.2, 1982.*)

In summary, for the retaining wall shown in Fig. 11.5, the values of the earth pressure acting on the wall are as follows:

Condition	Value of earth pressure
Active earth pressure, wall friction, no surcharge	$P_A = 6540$ lb/ft
At-rest earth pressure, no surcharge	$P = 11,000$ lb/ft
Additional pressure due to line load ($Q_L = 1000$ lb/ft)	$P_H = 550$ lb/ft
Additional pressure due to uniform surcharge ($Q = 100$ psf)	$P_Q = 1000$ lb/ft

11.6   MECHANICALLY STABILIZED EARTH RETAINING WALLS

Mechanically stabilized earth retaining walls (also known as MSE retaining walls) are typically composed of strip- or grid-type (geosynthetic) reinforcement (Koerner, 1998). Because they are often more economical to construct than conventional concrete retaining walls, mechanically stabilized earth retaining walls have become very popular.

11.6.1   Construction of MSE Retaining Walls

The construction of a mechanically stabilized earth retaining wall is shown in Fig. 11.9 to 11.11, as follows:

1. *Excavation of key and installation of drainage system (Fig. 11.9).*   This photograph shows the first step in the construction of a mechanically stabilized earth retaining wall. A key (slot) has been excavated into the natural ground and a drainage system is being constructed at the back of the key.
2. *Construction of mechanically stabilized earth retaining wall (Fig. 11.10).*   A mechanically stabilized earth retaining wall is composed of three elements: (1) wall facing material, (2) soil reinforcement, such as strip- or grid-type reinforcement, and (3) compacted fill between the soil reinforcement, as follows:

   a. Wall facing element.   The wall facing elements, which are precast concrete members, are being installed as shown in Fig. 11.10. Other types of wall facing elements are often utilized, such as wood planks or concrete interlocking panels. The wall facing members are first aligned and then connected together by using steel dowels.
   b. Geogrid.   Figure 11.10 shows the black geogrid, which is a type of geosynthetic. Geogrids are usually composed of a high strength polymer used to create an open grid pattern. The large arrows in Fig. 11.10 indicate the width of the geogrid, which is considered to be the zone of mechanically stabilized earth. Geogrid is transported to the site in rolls, which are then spread out and cut as needed. Geogrid has different tensile strengths along the roll versus perpendicular to the roll and it is important that the highest strength of the geogrid be placed in a direction that is perpendicular to the wall face. The small arrow in Fig. 11.10 points to a splice, where the geogrid is overlapped. A smaller width of geogrid is used at the end of the wall (location of small arrow) because the required resistance is less at the wall perimeter. The geogrid is often attached to the wall facing elements and the vertical spacing of the geogrid is equal to the thickness of the wall facing elements.
   c. Compaction of fill.   The final step is to place fill on top of the geogrid and the fill is then compacted. Since this mechanically stabilized earth retaining wall only has a drain in the key, granular (permeable) soil is being used to prevent the buildup of pore water pressure behind or in the mechanically stabilized earth zone. For geogrid installed in cohesive (plastic) soil, a backdrain system is often installed.

FIGURE 11.9 Excavation of key and installation of drainage system for a mechanically stabilized earth retaining wall.

The geogrid and compacted fill derive frictional resistance and interlocking resistance between each other. When the mechanically stabilized soil mass is subjected to shear stress, the soil tends to transfer the shear stress to the stronger geogrid. Also, if high stress concentrations develop in the mechanically stabilized soil mass, such as along a potential slip surface, then the geogrid tends to redistribute stresses away from areas of high stress.

Lightweight equipment must be used to compact the fill that is placed on top of each layer of geogrid. Heavy compaction equipment could push the wall facing elements out of alignment.

FIGURE 11.10   Two views of the installation of the geogrid for a mechanically stabilized earth retaining wall. The large arrows indicate the width of the mechanically stabilized zone and the small arrow points to a geogrid splice.

3. *Final constructed condition (Fig.11.11).*   This photograph shows the final constructed condition of the mechanically stabilized earth retaining wall. The arrow in Fig. 11.11 points to a gap in the wall facing element, which can be used to install plants which will grow over the wall facing elements and eventually blend the wall into the surrounding vegetation.

The design analysis for a mechanically stabilized earth retaining wall is more complex than for a cantilevered retaining wall. For a mechanically stabilized retaining wall, both the internal and external stability must be checked as discussed in the next two sections.

FIGURE 11.11 Final constructed condition of the mechanically stabilized earth retaining wall. The arrow points to an opening in the wall facing element that can be used to establish vegetation on the wall face.

11.6.2 Design Analysis for MSE Walls, External Stability

The analysis for the external stability is similar to a gravity retaining wall. For example, Figs. 11.12 and 11.13 (adapted from AASHTO, 1996) present the design analysis for external stability for a level backfill condition and a sloping backfill condition. In both Figs. 11.12 and 11.13, the zone of mechanically stabilized earth mass is treated in a similar fashion as a massive gravity retaining wall. The following analyses must be performed:

1. *Allowable bearing pressure.* The bearing pressure due to the reinforced soil mass must not exceed the allowable bearing pressure.

2. *Factor of safety of sliding.* The reinforced soil mass must have an adequate factor of safety for sliding ($F = 1.5$ to 2).

3. *Factor of safety of overturning.* The reinforced soil mass must have an adequate factor of safety ($F = 2$) for overturning about Point O.

4. *Resultant of vertical forces N.* The resultant of the vertical forces N should be within the middle $\frac{1}{3}$ of the base of the reinforced soil mass.

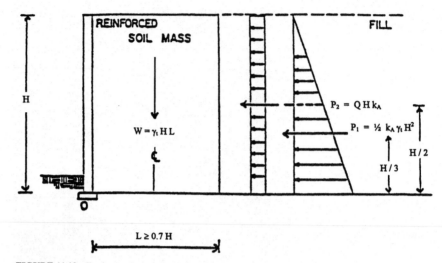

FIGURE 11.12 Design analysis for mechanically stabilized earth retaining wall having horizontal backfill. (*Adapted from AASHTO, 1996.*)

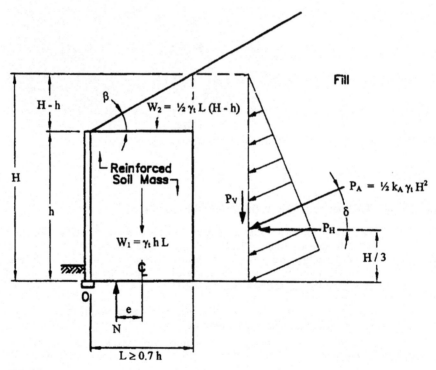

FIGURE 11.13 Design analysis for mechanically stabilized earth retaining wall having sloping backfill. (*Adapted from AASHTO, 1996.*)

5. *Stability of reinforced soil mass.*   The stability of the entire reinforced soil mass (i.e., shear failure below the bottom of the wall) would have to be checked.

Note in Fig. 11.12 that two forces (P_1 and P_2) are shown acting on the reinforced soil mass. The first force P_1 is determined from the standard active earth pressure resultant equation (i.e., Eq. 11.1). The second force P_2 is due to a uniform surcharge Q applied to the entire ground surface behind the mechanically stabilized earth retaining wall (i.e., Eq. 11.3). If the wall does not have a surcharge, then P_2 is equal to zero.

Figure 11.13 presents the active earth pressure force for an inclined slope behind the retaining wall. Note in Fig. 11.13 that the friction δ of the soil along the back side of the reinforced soil mass has been included in the analysis. The value of k_A would be obtained from Coulomb's earth pressure equation (Fig. 11.4). As a conservative approach, the friction angle δ can be assumed to be equal to zero and then $P_H = P_A$. Note in both Figs. 11.12 and 11.13 that the minimum width of the reinforced soil mass must be at least 7/10 the height of the reinforced soil mass.

11.6.3 Design Analysis for MSE Walls, Internal Stability

In terms of the internal stability of MSE walls, Zornberg et al. (1998) have shown that there is very good agreement between the actual location of the slip surface and the location of the critical slip surface predicted by slope stability analyses (i.e., method of slices). To check the stability of the mechanically stabilized zone, a slope stability analysis can be performed where the soil reinforcement is modeled as horizontal forces equivalent to the allowable tensile resistance of the geogrid. In addition to calculating the factor of safety, the pullout resistance of the reinforcement along the slip surface should also be checked.

The analysis of mechanically stabilized earth retaining walls is based on active earth pressures. It is assumed that the wall will move enough to develop the active wedge. Similar to concrete retaining walls, it is important that building footings or other load carrying members are not supported by the mechanically stabilized earth retaining wall and the active wedge, or else they will be subjected to lateral movement.

11.7   SHEET PILE WALLS

Sheet pile retaining walls are widely used for waterfront construction and consist of interlocking members that are driven into place. Sheet pile walls are also frequently used as temporary retaining walls for the construction of deep foundations. Individual sheet piles come in many different sizes and shapes. In addition, sheet piles have an interlocking joint that enables the individual segments to be connected together to form a solid wall.

11.7.1 Earth Pressures Acting on Sheet Pile Walls

There are many different types of design methods that are used for sheet pile walls. Figure 11.14 (from NAVFAC DM-7.2, 1982) shows the most common type of design method. In Fig. 11.14, the term H represents the unsupported face of the sheet pile wall. As indicated in Fig. 11.14, this sheet pile wall is being used as a waterfront retaining structure and the level of water in front of the wall is at the same elevation as the groundwater table elevation behind the wall. For highly permeable soil, such as clean sand and gravel, this often occurs because the water can quickly flow underneath the wall in order to equalize the water levels.

In Fig. 11.14, the term D represents that portion of the sheet pile wall that is anchored in soil. Also shown in Fig. 11.14 is a force designated as A_p. This represents a restraining force on the sheet pile wall due to the construction of a tieback, such as by using a rod that has a grouted end or is attached to an anchor block. Tieback anchors are often used in sheet pile wall construction in order to reduce the bending moments in the sheet pile. When tieback anchors are used, the sheet pile wall

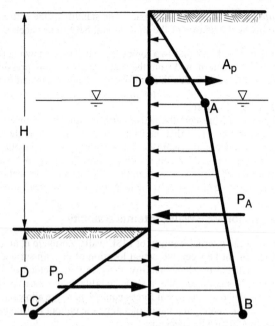

FIGURE 11.14   Earth pressure diagram for design of sheet pile wall. (*Reproduced from NAVFAC DM-7.2, 1982.*)

is typically referred to as an anchored bulkhead, while if no tiebacks are utilized, the wall is called a cantilevered sheet pile wall.

Sheet pile walls tend to be relatively flexible. Thus, as indicated in Fig. 11.14, the design is based on active and passive earth pressures. The soil behind the wall is assumed to exert an active earth pressure on the sheet pile wall. At the groundwater table (Point A), the active earth pressure is equal to:

$$\text{Active earth pressure at Point A (psf or kPa)} = k_A \gamma_t d_1 \tag{11.11}$$

where k_A = active earth pressure coefficient from Eq. 11.2 (dimensionless). The friction between the sheet pile wall and the soil is usually neglected in the design analysis.
   γ_t = total unit weight of the soil above the groundwater table (pcf or kN/m^3)
   d_1 = depth from the ground surface to the groundwater table (ft or m)

In using Eq. 11.11, a unit length (1 ft or 1 m) of sheet pile wall is assumed. At Point B in Fig. 11.14, the active earth pressure equals:

$$\text{Active earth pressure at Point B (psf or kPa)} = k_A \gamma_t d_1 + k_A \gamma_b d_2 \tag{11.12}$$

where γ_b is the buoyant unit weight of the soil below the groundwater table (pcf or kN/m^3) and d_2 is the depth from the groundwater table to the bottom of the sheet pile wall (ft or m)

For a sheet pile wall having assumed values of H and D (see Fig. 11.14), and using the calculated values of active earth pressure at Points A and B, the active earth pressure resultant force P_A, in pounds per linear foot of wall length or kilonewtons per linear meter of wall length, can be calculated.

The soil in front of the wall is assumed to exert a passive earth pressure on the sheet pile wall. The passive earth pressure at Point C in Fig. 11.14 is equal to:

$$\text{Passive earth pressure at Point C (psf or kPa)} = k_p \gamma_b D \qquad (11.13)$$

The passive earth pressure coefficient k_p can be calculated from Eq. 11.5. Similar to the analysis of cantilever retaining walls, if it is desirable to limit the amount of sheet pile wall translation, then a reduction factor can be applied to the passive pressure. Once the allowable passive pressure is known at Point C, the passive resultant force P_p can be readily calculated.

As an alternative solution for the passive pressure, Eq. 11.4 can be used to calculate P_p with the buoyant unit weight γ_b substituted for the total unit weight γ_t and the depth D as shown in Fig. 11.14. Note that a water pressure has not been included in the analysis. This is because the water level is the same on both sides of the wall and because the water pressure cancels out, it should not be included in the analysis.

11.7.2 Anchored Bulkhead

The design of sheet pile walls requires the following analyses:

1. Evaluation of the earth pressures that act on the wall, such as shown in Fig. 11.14

2. Determination of the required depth D of piling penetration

3. Calculation of the maximum bending moment M_{max}, which is used to determine the maximum stress in the sheet pile

4. Selection of the appropriate piling type, size, and construction details

A typical design process is to assume a depth D (Fig. 11.14) and then calculate the factor of safety for toe failure (i.e., toe kick-out) by the summation of moments at the tieback anchor (Point D). The factor of safety is defined as the moment due to the passive force divided by the moment due to the active force. Values of acceptable factor of safety for toe failure are 2 to 3. An alternative solution is to first select the factor of safety and then develop the active and passive resultant forces and moment arms in terms of D. By solving the equation, the value of D for a specific factor of safety can be directly calculated.

Once the depth D of the sheet pile wall is known, the anchor pull A_p must be calculated. The anchor pull is determined by the summation of forces in the horizontal direction, or:

$$A_p = P_A - P_p/F \qquad (11.14)$$

where P_A = active resultant force, Fig. 11.14 (lb/ft or kN/m)
P_p = the resultant passive force, Fig. 11.14 (lb/ft or kN/m)
F = factor of safety that was obtained from the toe failure analysis (dimensionless)

Based on the earth pressure diagram (Fig. 11.14) and the calculated value of A_p, elementary structural mechanics can be used to determine the maximum moment in the sheet pile wall. The maximum moment divided by the section modulus can then be compared with the allowable design stresses of a particular type of sheet pile.

For cohesive soil, the long-term condition often governs the design. In this case, an effective stress analysis can be performed (c' and ϕ') using the estimated location of the groundwater table. For convenience, the effective cohesion is neglected and the analysis is performed by using only ϕ'. Therefore, the long-term effective stress condition for cohesive soil is analyzed as described in the preceding discussion for sheet pile walls having granular soil. Since ϕ' for a cohesive soil is usually less than ϕ' for granular soil, the active earth pressure will be higher and the passive resistance lower for cohesive soil.

Some other important design considerations for sheet pile walls include the following:

1. *Soil layers.*   The active and passive earth pressures should be adjusted for soil layers having different engineering properties.

2. *Penetration depth.*   The penetration depth D of the sheet pile wall should be increased by at least an additional 20 percent to allow for the possibility of dredging and scour. Deeper penetration depths may be required based on a scour analysis.

3. *Surcharge loads.*   The ground surface behind the sheet pile wall is often subjected to surcharge loads. Equation 11.3 can be used to determine the active earth pressure resultant force due to a uniform surcharge pressure applied to the ground surface behind the wall. Note in Eq. 11.3 that the entire height of the sheet pile wall (i.e., $H + D$, see Fig. 11.14) must be used in place of H. Typical surcharge pressures exerted on sheet pile walls are caused by adjacent foundation loads, railroads, highways, dock loading facilities and merchandise, ore piles, and cranes.

4. *Unbalanced hydrostatic and seepage forces.*   The previous discussion has assumed that the water levels on both sides of the sheet pile wall are at the same elevation. Depending on factors such as the water tightness of the sheet pile wall and the backfill permeability, it is possible that the groundwater level could be higher than the water level in front of the wall, in which case the wall would be subjected to water pressures. This condition could develop when there is a receding tide or a heavy rainstorm that causes a high groundwater table. A flow net can be used to determine the unbalanced hydrostatic and upward seepage forces in the soil in front of the sheet pile wall. Adjustments to the active and passive resultant forces for seepage underneath the sheet pile wall can be determined by using Fig. 11.15.

5. *Other loading conditions.*   The sheet pile wall may have to be designed to resist the lateral loads due to ice thrust, wave forces, ship impact, mooring pull, and earthquake forces. If granular soil behind or in front of the sheet pile wall is in a loose state, it could be susceptible to liquefaction during an earthquake.

6. *Factors increasing the stability.*   A factor usually not considered in the design analysis is the densification of loose sand during driving of the sheet piles. However, since sheet piles are relatively thin, the densification effect would be less for a sheet pile than for a comparable round pile, because the sheet pile displaces less soil.

Another beneficial effect in the design analysis is that many sheet pile walls used for waterfront construction or for foundation excavations are relatively long and hence the soil is in a plane strain condition. As previously discussed, the plane strain shear strength is higher than the shear strength determined from conventional shear strength tests.

11.7.3   Cantilevered Sheet Pile Wall

In the case of cantilevered sheet pile walls, the sheet piling is driven to an adequate depth to become fixed as a vertical cantilever in resisting the lateral active earth pressure. Cantilevered walls usually undergo large lateral deflections and are readily affected by scour and erosion in front to the wall. Because of these factors, penetration depths can be quite high, resulting in excess stresses and severe yield. According to the *USS Steel Sheet Piling Design Manual* (1984), cantilevered walls using steel sheet piling are restricted to a maximum height H of approximately 15 ft (4.6 m).

Design charts have been developed for the analysis of cantilevered sheet pile walls. For example, Fig. 11.16 is a design chart reproduced from the *USS Steel Sheet Piling Design Manual* (1984). The value of α is based on the location of the water level as indicated in Fig. 11.16. The chart is entered with the ratio of the passive earth pressure coefficient k_p divided by the active earth pressure coefficient k_A. By intersecting the depth ratio curves or moment ratio curves, the depth of the sheet pile wall (D) or maximum moment M_{max} can be calculated. Note that the depth D from Fig. 11.16 corresponds to a factor of safety of 1.0. The *USS Steel Sheet Piling Design Manual* (1984) states that increasing the depth of embedment (D) by 20 to 40 percent will provide for a factor of safety of toe failure of approximately 1.5 to 2.0.

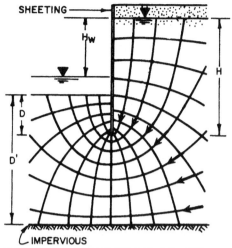

SHEETING

H_W

H

D

D'

IMPERVIOUS

FLOW NET FOR SEEPAGE BENEATH WALL FOR
CONSTANT DIFFERENTIAL HEAD $= H_W$

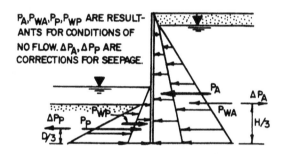

P_A, P_{WA}, P_P, P_{WP} ARE RESULT-
ANTS FOR CONDITIONS OF
NO FLOW. $\Delta P_A, \Delta P_P$ ARE
CORRECTIONS FOR SEEPAGE.

P_A

ΔP_A

P_{WP}

P_{WA}

$H/3$

ΔP_P P_P

$D/3$

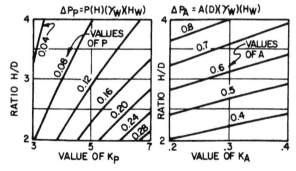

$\Delta P_P = P(H)(\gamma_W)(H_W)$

$\Delta P_A = A(D)(\gamma_W)(H_W)$

VALUES OF P

0.04   0.08   0.12   0.16   0.20   0.24   0.28

RATIO H/D

VALUE OF K_P

VALUES OF A

0.8   0.7   0.6   0.5   0.4

RATIO H/D

VALUE OF K_A

FIGURE 11.15   Effect of seepage beneath sheet pile wall. (*From NAV-FAC DM-7.2, 1982; based on the work by Richart and Schmertmann.*)

Example Problem 11.4 Assume a cantilevered sheet pile wall with $k_A = 0.2$, $k_p = 8.65$, $\alpha = 1.0$, $H = 13.8$ ft (4.2 m), and $\gamma_b = 57$ pcf (9.0 kN/m^3). Determine the total length and the maximum moment in the sheet pile wall.

Solution Total length $(H + D)$:

Entering Fig. 11.16 at $k_p/k_A = 43$, and intersecting the depth ratio curve of $\alpha = 1$, $D/H = 0.6$. Therefore, using $H = 13.8$ ft (4.2 m), the value of D for a factor of safety of 1 is:

$$D/H = 0.6 \text{ or: } D = (0.6)(13.8 \text{ ft}) = 8.3 \text{ ft (2.5 m)}$$

Using a 30 percent increase in embedment depth, the required embedment depth $D = 11$ ft (3.4 m). Thus, the total length of the sheet pile wall is 25 ft (7.6 m).

Maximum moment M_{max}:

Entering Fig. 11.16 at $k_p/k_A = 43$, and intersecting the moment ratio curve of $\alpha = 1$, $M_{max}/(\gamma_b k_A H^3) = 0.53$

$$M_{max} = (0.53)(\gamma_b k_A H^3) = (0.53)(57 \text{ pcf})(0.2)(13.8 \text{ ft})^3 = 15{,}900 \text{ ft-lb/ft}$$

Thus the maximum moment in the sheet pile wall = 15,900 ft-lb/ft (70.7 kN-m/m)

11.7.4 Cofferdams

Steel sheet piling is widely used for the construction of a cofferdam, which is defined as a temporary structure designed to support the sides of an excavation and to exclude water from the excavation. Tomlinson (1986) presents an in-depth discussion of the different types of cofferdams and

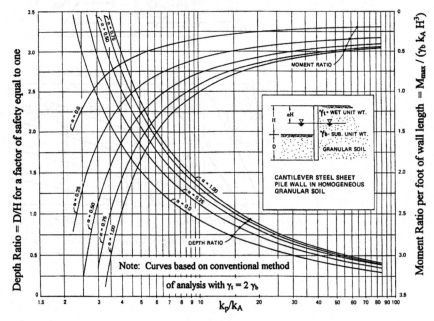

FIGURE 11.16 Design chart for cantilever sheet pile wall. (*From USS Sheet Piling Design Manual, 1984.*)

construction techniques. For example, the first step in the construction of a bridge foundation could be the installation of a large circular cofferdam. Then the river water would be pumped out from inside the cofferdam and the river bottom would be excavated to the desired bearing strata for the bridge foundation.

In general, the following loads could be exerted on the cofferdam:

1. Hydrostatic groundwater pressures located outside the cofferdam. If the cofferdam is constructed in a river or other body of water, then the hydrostatic water pressure should be based on the anticipated high water level.

2. Hydrostatic pressure of water inside the cofferdam if the water level outside falls below the interior level during initial pumping of water from the cofferdam. This condition could cause bursting of the cofferdam by interlock tension.

3. Earth pressure outside the cofferdam.

4. Other loads, such as surcharge loads, wave pressures, and earthquake loading.

Based on these loading conditions, the cofferdam must have an adequate factor of safety for the following conditions:

1. Heave of the soil located at the bottom of the cofferdam.

2. Sliding of the cofferdam along its base.

3. Overturning and tilting of the cofferdam.

4. The inward or outward yielding of the cofferdam (i.e., maximum moment and shear in the sheet piling and interlock tension).

5. Piping of soil that could undermine the cofferdam or lead to sudden flooding of the interior of the cofferdam.

An important consideration in the design of cofferdams is the ability of the sheet pile wall to deflect. If struts are used that brace the opposite sides of the excavation (i.e., braced excavation), the deflection of the sheet pile wall will be restricted. The next section will discuss the earth pressures exerted on braced excavations. If the sheet pile wall is able to deform, such as by using raking braces, then the deflections of the wall during construction will permit mobilization of active pressures in accordance with the previous discussion. The *USS Steel Sheet Piling Design Manual* (1984) and NAVFAC DM-7.2 (1982) present a further discussion of the design of cofferdams.

11.8 TEMPORARY RETAINING WALLS

Temporary retaining walls are often used during construction, such as for the support of the sides of an excavation that is made below-grade in order to construct the building foundation. If the temporary retaining wall has the ability to develop the active wedge (Fig. 11.3), then the basic active earth pressure principles described in the previous sections can be used for the design of the temporary retaining walls.

Especially in urban areas, movement of the temporary retaining wall may have to be restricted to prevent damage to adjacent property. If movement of the retaining wall is restricted, the earth pressures will typically be between the active k_A and at-rest k_o values. Common types of temporary retaining systems and braced excavations are shown in Fig. 11.17.

11.8.1 Braced Excavations

For some projects, temporary retaining walls may be constructed of sheeting (such as sheet piles) that are supported by horizontal braces, also known as *struts* (see upper right diagram of Fig. 11.17).

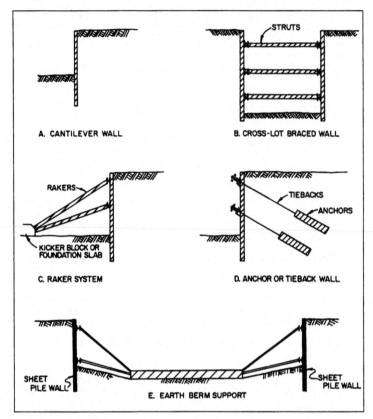

FIGURE 11.17 Common types of retaining systems and braced excavations. (*Reproduced from NAVFAC DM-7.2, 1982.*)

Near or at the top of the temporary retaining wall, the struts restrict movement of the retaining wall and prevent the development of the active wedge. Because of this inability of the retaining wall to deform at the top, earth pressures near the top of the wall are in excess of the active k_A pressures (see Fig. 11.18). At the bottom of the wall, the soil is usually able to deform into the excavation, which results in a reduction in earth pressure, and the earth pressures at the bottom of the excavation tend to be constant or even decrease as shown in Fig. 11.18.

The earth pressure distributions shown in Fig. 11.18 were developed from actual measurements of the forces in struts during the construction of braced excavations (Terzaghi and Peck, 1967). In Fig. 11.18, case a shows the earth pressure distribution for braced excavations in sand and cases b and c show the earth pressure distribution for clays. In Fig. 11.18, the distance H represents the depth of the excavation (i.e., the height of the exposed wall surface). The earth pressure distribution is applied over the exposed height H of the wall surface with the earth pressures transferred from the wall sheeting to the struts (the struts are labeled with the forces F_1, F_2, and so on).

Any surcharge pressures, such as surcharge pressures on the ground surface adjacent the excavation, must be added to the pressure distributions shown in Fig. 11.18. In addition, if the sand deposit has a groundwater table that is above the level of the bottom of the excavation, then water pressures must be added to the case a pressure distribution shown in Fig. 11.18.

Because the excavations are temporary (i.e., short-term condition), a total stress analysis is used with the earth pressure distributions for clay based on the undrained shear strength ($s_u = c$). The earth

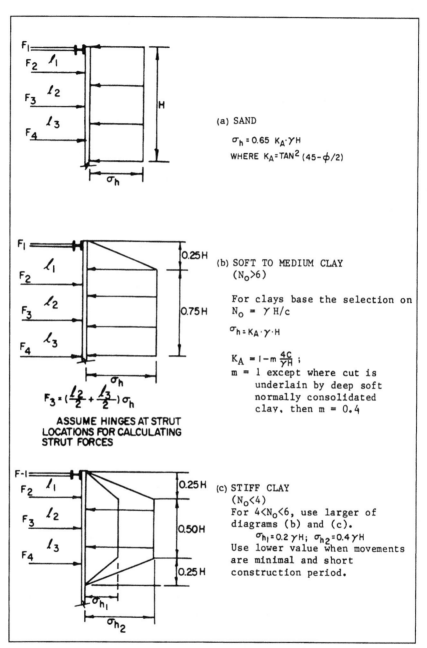

(a) SAND

$\sigma_h = 0.65\ K_A \cdot \gamma H$

WHERE $K_A = TAN^2\ (45 - \phi/2)$

(b) SOFT TO MEDIUM CLAY ($N_o > 6$)

For clays base the selection on $N_o = \gamma H/c$

$\sigma_h = K_A \cdot \gamma \cdot H$

$K_A = 1 - m\ \dfrac{4c}{\gamma H}$;
m = 1 except where cut is underlain by deep soft normally consolidated clay, then m = 0.4

$F_3 = (\dfrac{\ell_2}{2} + \dfrac{\ell_3}{2})\ \sigma_h$

ASSUME HINGES AT STRUT LOCATIONS FOR CALCULATING STRUT FORCES

(c) STIFF CLAY ($N_o < 4$)
For $4 < N_o < 6$, use larger of diagrams (b) and (c).
$\sigma_{h1} = 0.2\ \gamma H$; $\sigma_{h2} = 0.4\ \gamma H$
Use lower value when movements are minimal and short construction period.

FIGURE 11.18 Earth pressure distribution on temporary braced walls. (*By Terzaghi and Peck, 1967; reproduced from NAVFAC DM-7.2, 1982.*)

pressure distributions for clay (i.e., cases *b* and *c*) are not valid for permanent walls or for walls where the groundwater table is above the bottom of the excavation.

11.8.2   Steel I-beams and Wood Lagging

Figure 11.19 shows a picture of a temporary retaining wall that consists of steel I-beams, wood lagging, and tieback anchors. The temporary retaining wall is being used to support the excavation during the construction of the foundation. This type of temporary retaining wall consists of steel I-beams that are either driven into place or installed in pre-drilled holes with the bottom portion cemented into place. During the excavation of the interior area, wood lagging is installed between the steel I-beams and the earth pressure exerted on the wood lagging is transferred to the steel I-beam flanges. In order to reduce bending moments in the steel I-beams and to reduce lateral deflections, tieback anchors are often installed as the excavation proceeds.

In some cases, the steel I-beams will be braced with struts and the earth pressure distributions would be similar to those as shown in Fig. 11.18. For cantilevered steel I-beam and wood lagging retaining walls, there will be sufficient movement of the top of the wall to develop the active earth pressure (i.e., Eq. 11.1).

11.8.3   Tieback Anchors

Tieback anchors, such as shown in Fig. 11.20, can be used for all types of retaining walls. They are often installed during the construction of steel I-beams and wood lagging systems in order to reduce the bending moments of the steel I-beams for deep excavations. Tieback anchors can also be used for many other purposes, in which case they are often referred to as ground anchors. For example, ground anchors have been used to stabilize slopes and tunnels and as tie-down support for foundations.

Tieback anchors consist of the following items:

1. *Borehole.*   Tiebacks are installed in boreholes drilled by specially adapted equipment. Auger boring, percussion drilling, or rotary coring can be used to excavate the borehole.

FIGURE 11.19   Temporary steel I-beam and wood lagging retaining wall having tieback anchors. The temporary retaining wall is being used to support the excavation during the construction of the foundation.

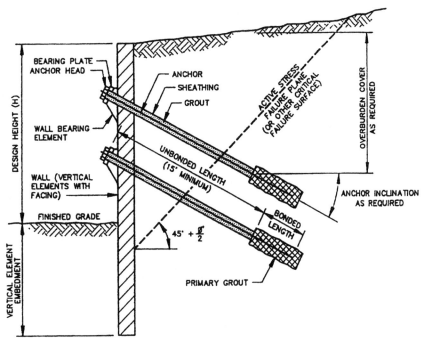

FIGURE 11.20 Cross section showing tieback anchors for retaining walls. (*Reproduced with permission from AASHTO, 1996.*)

2. *Tendon.* The tendon is usually made of prestressing steel wires, strands, or bars. The tendon includes the following:

a. *Bonded length.* The bonded length is that part of the tendon that is fixed in the primary grout and transfers the tension force to the surrounding soil or rock. The anchor bond length is designed so that it can resist the required pullout load of the anchor. The bonded length is often created by the pressure injection of a Portland cement-based mixture. As shown in Fig. 11.20, the bonded length should be located well behind the active wedge or other critical failure surface.

Especially for cohesionless soil, the tieback anchors will need an adequate overburden pressure to increase the bond stress at the grouted end. To accomplish this, the tieback anchors are often installed at a downward angle as shown in Fig. 11.20.

b. *Unbonded length.* This is the part of the tendon that is able to elongate and hence transfers the tension force to the bonded length. As shown in Fig. 11.20, the unbonded length is often filled with grout, but because the tendon is contained within a sheath, the tension force is transferred to the bonded length. Grouting of the unbonded length will prevent cave-in of this portion of the borehole and it will protect the tendon from corrosion.

3. *Anchorage.* The anchorage consists of a bearing plate and anchor head that permits stressing of the tendon. Because tiebacks are often inclined, the anchorage must resist both horizontal and vertical forces. If the anchorage is not adequately designed to resist these forces, deformation can substantially reduce the effectiveness of the tieback anchor. For example, if the anchorage should slide downward, the tensioning force will be reduced, allowing the retaining wall to deflect or fall into the excavation.

In order to determine the bonded length, the bond stress between the soil and rock and the grout must be known. Some of the variables that govern the soil-grout and rock-grout bond stress are the method of drilling and cleaning of the tieback drill hole, hole diameter, shear strength of the soil or rock, overburden pressure, and the method of grout installation such as gravity grouting versus pressure grouting (Post-Tensioning Institute, 1996). In terms of minimum bond lengths, it has been recommended that the bonded length be at least 15 ft (4.6 m) long (Post-Tensioning Institute, 1996). Provided there is sufficient overburden pressure, usually cohesionless soil will have a higher bond stress then cohesive soil.

The following equation can be used to determine the bonded length of the tieback anchor:

$$L_b = \frac{PF}{\pi d \tau_f} \qquad (11.15)$$

where L_b = bonded length of the tieback anchor (ft or m)
  P = design load for the tieback anchor (lb or kN)
  F = factor of safety (typically 3 to 5)
  d = diameter of the drill hole (ft or m)
  τ_f = ultimate soil-grout or rock-grout bond stress from Table 11.2 (psf or kPa)

Given the uncertainties in the design of tieback anchors, they should always be load tested in the field. Table 11.3 presents test procedures and acceptance criteria for the load testing of tieback anchors. The performance test is the most rigorous procedure (lasting 24 h) and it is usually only performed on a selected few tieback anchors. The proof test is a much quicker test and it is often performed on all of the remaining tieback anchors. As indicated in Table 11.3, it is important to analyze the tieback anchors in terms of acceptance criteria.

TABLE 11.2   Ultimate Bond Stress for Tieback Anchors

	Soil or rock type	Ultimate bond stress*	
		psi	MPa
Cohesive soil	Soft silty clay	5–10	0.03–0.07
	Silty clay	5–10	0.03–0.07
	Stiff clay, medium to high plasticity	5–15	0.03–0.10
	Very stiff clay, medium to high plasticity	10–25	0.07–0.17
	Stiff clay, medium plasticity	15–35	0.10–0.25
	Very stiff clay, medium plasticity	20–50	0.14–0.35
	Very stiff sandy silt, medium plasticity	40–55	0.28–0.38
Cohesionless soil	Fine to medium sand, medium dense to dense	12–55	0.08–0.38
	Medium coarse sand with gravel, medium dense	16–95	0.11–0.66
	Medium coarse sand with gravel, dense to very dense	35–140	0.25–0.97
	Silty sands	25–60	0.17–0.41
	Dense glacial till	43–75	0.30–0.52
	Sandy gravel, medium dense to dense	31–200	0.21–1.38
	Sandy gravel, dense to very dense	40–200	0.28–1.38
Rock	Limestone	100–250	0.70–1.7
	Slates and hard shales	100–200	0.70–1.4
	Soft shales	35–100	0.25–0.70
	Sandstone	100–250	0.70–1.7

*Ultimate bond stress for anchors that are pressure grouted and have a straight shaft. Actual values for pressure grouted anchors depend on the ability to develop high pressures in each soil or rock type. For cohesionless soils, ultimate bond stress is for small diameter (3 – 6 in., 7.5 – 15 cm) holes and an overburden pressure of at least 15 ft (4.6 m) of soil.
  Sources: Post-Tensioning Institute (1996) and Hayward Baker.

TABLE 11.3 Test Procedures and Acceptance Criteria for Each Tieback

A. Stages and observation periods for 24-hour test (performance test)	
Load Level	Recommended period of observation (in min)
Seating or alignment load = 0.1DL	None
0.25DL	10
0.50DL	10
0.75DL	10
1.00DL	10
1.25DL	10
1.50DL	10
1.75DL	30
2.00DL	480

DL = design load

After each stage of loading, the load should be reduced to seating or alignment load and replaced after 2 minutes. Deflection should be measured immediately after application of each load level and at 5-minute intervals thereafter. Measurements should also be taken immediately prior to any unloading and reloading steps.

B. Load stages and observation periods for proof test	
Load Level	Recommended period of observation (in min)
Seating or alignment load = 0.1DL	None
0.50DL	10
1.00DL	10
1.25DL	10
1.50DL	30

DL = design load

After each stage of loading, the load should be reduced to seating or alignment load and replaced after 2 minutes. Deflection should be measured before and after application of each loading increment.

Acceptance criteria:

1. Total deflection during the long term testing should not exceed 12 inches.
2. Creep deflection at 200% design load should not exceed 0.1 in. during the 4-hour period.
3. Total deflection during short-term testing should not exceed 12 inches.
4. Creep deflection at 150% design load should not exceed 0.1 in. during the 15-minute period.
5. All the tiebacks should be locked off at 120% of the design load.
6. Following final tensioning, the nongrouted portion of the anchor boring should be grouted.

It should be mentioned that tieback anchors in soft or medium cohesive soils could slowly lose their tensioning effect over time due to long-term creep of the soil. For permanent tieback anchors, long-term creep testing of the tieback anchors is required to determine the loss of tensioning with time.

11.9 MOISTURE MIGRATION THROUGH BASEMENT WALLS

11.9.1 Introduction

Water can penetrate basement walls due to hydrostatic pressure, capillary action, and water vapor. If a groundwater table exists behind a basement wall, then the wall will be subjected to hydrostatic pressure, which can force large quantities of water through wall cracks or joints. Figures 11.21 and 11.22

FIGURE 11.21   Photograph of a basement wall subjected to hydrostatic water pressure.

show two photographs of a basement wall subjected to hydrostatic water pressure that forced large quantities of water through the wall. A subdrain is usually placed behind the basement wall to prevent the buildup of hydrostatic pressure. For the drains to be effective, the backfill material should consist of granular (i.e., permeable) soil.

Another way for moisture to penetrate basement walls is by capillary action in the soil or the wall itself. By capillary action, water can travel from a lower to higher elevation in the soil or wall. Capillary rise in walls is related to the porosity of the wall and the fine cracks in both the masonry

FIGURE 11.22   Overview of the same area shown in Fig. 11.21.

and, especially, the mortar. To prevent moisture migration through basement walls, an internal or surface waterproofing agent is used. Chemicals can be added to cement mixes to act as internal waterproofs. More common are the exterior applied waterproof membranes. Oliver (1988) lists and describes various types of surface-applied waterproof membranes.

A third way that moisture can penetrate through basement walls is by water vapor. Similar to concrete floor slabs, water vapor can penetrate a basement wall whenever there is a difference in vapor pressure between the two areas.

11.9.2 Types of Damage

Moisture that travels through basement walls can damage wall coverings, such as wood paneling. Moisture traveling through the basement walls can also cause musty odors or mildew growth in the basement areas. If the wall should freeze, then the expansion of freezing water in any cracks or joints may cause deterioration of the wall.

The moisture that is passing through the basement wall will usually contain dissolved salts. The penetrating water may often contain salts originating from the ground or mineral salts naturally present in the wall materials. As the water evaporates at the interior wall surface, the salts form white crystalline deposits (efflorescence) on the basement walls. Figure 11.23 shows photographs of the build-up of salt deposits on the interior surface of basement walls.

The salt crystals can accumulate in cracks or wall pores, where they can cause erosion, flaking, or ultimate deterioration of the contaminated surface. This is because the process of crystallization often involves swelling and considerable forces are generated. Another problem is if the penetrating water contains dissolved sulfates, because they can accumulate and thus increase their concentration on the exposed wall surface, resulting in chemical deterioration of the concrete (ACI, 1990).

The effects of dampness, freezing, and salt deposition are major contributors to the weathering and deterioration of basement walls. Some of the other common deficiencies that contribute to moisture migration through basement walls are as follows (Diaz et al., 1994; Day, 1994f):

1. The wall is poorly constructed (for example, joints are not constructed to be watertight), or poor-quality concrete that is highly porous or shrinks excessively is used.

2. There is no waterproofing membrane or there is a lack of waterproofing on the basement-wall exterior.

3. There is improper installation, such as a lack of bond between the membrane and the wall, or deterioration with time of the waterproofing membrane.

4. There is no drain, the drain is clogged, or there is improper installation of the drain behind the basement wall. Clay, rather than granular backfill, is used.

5. There is settlement of the wall, which causes cracking or opening of joints in the basement wall.

6. There is no protection board over the waterproofing membrane. During compaction of the backfill, the waterproofing membrane is damaged.

7. The waterproofing membrane is compromised. This happens, for example, when a hole is drilled through the basement wall.

8. There is poor surface drainage, or downspouts empty adjacent the basement wall.

11.9.3 Design and Construction Details

The main structural design and construction details to prevent moisture migration through basement walls is a drainage system at the base of the wall to prevent the build-up of hydrostatic water pressure and a waterproofing system applied to the exterior wall surface. Such a system for basement walls is shown in Fig. 11.24. A perforated drain is installed at the bottom of the basement wall footing. Open graded gravel, wrapped in a geofabric, is used to convey water down to the perforated drain. The drain outlet could be tied to a storm drain system.

The waterproofing system frequently consists of a high strength membrane, a primer for wall preparation, liquid membrane for difficult to reach areas, and mastic to seal holes in the wall. The

FIGURE 11.23 Two views of groundwater migration through basement walls at a condominium complex in Los Angeles, California.

primer is used to prepare the concrete wall surface for the initial application of the membrane and to provide long-term adhesion of the membrane. A protection board is commonly placed on top of the waterproofing membrane to protect it from damage during compaction of the backfill soil. Self-adhering waterproofing systems have been developed to make the membrane easier and quicker to install. For example, Fig. 11.25 shows a self-adhering waterproofing system that has been installed to the back side of a basement wall. In Fig. 11.25, the waterproofing system is located on the right side of the photograph and the perforated drain is visible in the center of the photograph.

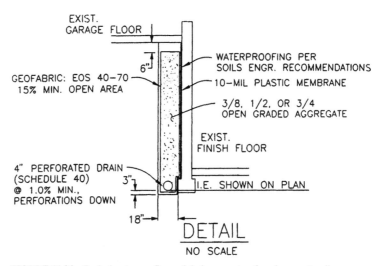

FIGURE 11.24   Typical waterproofing and drainage system for a basement wall.

NOTATION

The following notation is used in this chapter:

a = horizontal distance from W to the toe of the footing

A_p = anchor pull force (sheet pile wall)

c = cohesion based on a total stress analysis

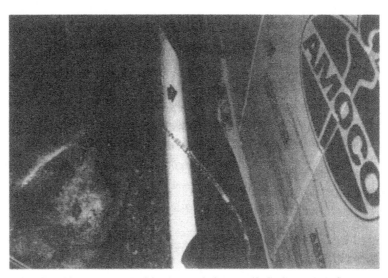

FIGURE 11.25   The right side of the photograph shows a self-adhering waterproofing system that has been installed on the back side of a basement wall. The perforated drain line is visible in the center of the photograph.

c' = cohesion based on an effective stress analysis

c_a = adhesion between the bottom of the footing and the underlying soil

d = diameter of the drill hole

d_1 = depth from ground surface to the groundwater table

d_2 = depth from the groundwater table to the bottom of the sheet pile wall

D = depth of the retaining wall footing

D = portion of the sheet pile wall anchored in soil (Fig. 11.14)

e = lateral distance from P_v to the toe of the retaining wall

F = factor of safety

F_1, F_2 = braced excavation forces (Fig. 11.18)

H = height of the retaining wall

H = unsupported face of the sheet pile wall (Fig. 11.14)

k_A = active earth pressure coefficient

k_o = coefficient of earth pressure at rest

k_p = passive earth pressure coefficient

L_b = bonded length of the tieback anchor

M_{max} = maximum moment in the sheet pile wall

N = sum of the wall weights W plus P_v

P = design load for the tieback anchor

P_A = active earth pressure resultant force

P_H = horizontal component of the active earth pressure resultant force

P_H = horizontal force acting on a restrained retaining wall (Fig. 11.8)

P_p = passive resultant force

P_Q = resultant force due to a uniform surcharge on top of the wall backfill

P_v = vertical component of the active earth pressure resultant force

P_1 = active earth pressure resultant force (i.e., $P_1 = P_A$, Fig. 11.12)

P_2 = resultant force due to a uniform surcharge (i.e., $P_2 = P_Q$, Fig. 11.12)

Q = uniform vertical surcharge pressure acting on the wall backfill

Q_L = line load acting on the backfill of a restrained retaining wall (Fig. 11.8)

R = location of the resultant force above the base of the wall (Fig. 11.8)

W = resultant of the vertical retaining wall loads

x = defined in Fig. 11.8

$\bar{x}$ = location of N from the toe of the footing

Y = horizontal displacement of the retaining wall (Table 11.1)

α = groundwater location parameter (Fig. 11.16)

β = slope inclination behind or in front of the retaining wall

δ, ϕ_{cv} = friction angle between bottom of wall footing and underlying soil

δ, ϕ_w = friction angle between the back face of the wall and the soil backfill

ϕ = friction angle based on a total stress analysis

ϕ' = friction angle based on an effective stress analysis

γ_b = buoyant unit weight of the soil

γ_t = total unit weight of the soil

θ = back face inclination of the retaining wall

σ_{avg} = average bearing pressure of the retaining wall foundation

σ_{mom} = the portion of the bearing pressure due to the eccentricity of N

τ_f = ultimate soil-grout or rock-grout bond stress

PROBLEMS

Solutions to the problems are presented in App. C of this book. The problems have been divided into basic categories as indicated below:

Retaining Wall Analyses with no Wall Friction

11.1   Using the retaining wall shown in Fig. 11.2, assume $H = 4$ m, the thickness of the reinforced concrete wall stem = 0.4 m, the reinforced concrete wall footing is 3 m wide by 0.5 m thick, the ground surface in front of the wall is level with the top of the wall footing, and the unit weight of concrete = 23.5 kN/m³. The wall backfill will consist of sand having $\phi = 32°$ and $\gamma_t = 20$ kN/m³. Also assume that there is sand in front of the footing with these same soil properties. The friction angle between the bottom of the footing and the bearing soil (δ) = 24°. For the condition of a level backfill and neglecting the wall friction on the back side of the wall and the front side of the footing, calculate the active earth pressure coefficient k_A, the passive earth pressure coefficient k_p, the active earth pressure resultant force P_A, and the passive resultant force P_p using a reduction factor = 2.0.

ANSWER: $k_A = 0.307$, $k_p = 3.25$, $P_A = 49.2$ kN/m, and $P_p = 4.07$ kN/m.

11.2   For Problem 11.1, determine the amount of wall displacement needed to develop the active wedge if the sand will be compacted into a dense state.

ANSWER: 0.2 cm.

11.3   For Problem 11.1, determine the width of the active wedge at the top of the retaining wall.

ANSWER: 2.2 m.

11.4   For Problem 11.1, determine the resultant normal force N and the distance of N from the toe of the footing ($\bar{x}$).

ANSWER: $N = 68.2$ kN/m and $\bar{x} = 1.165$ m.

11.5   Using the data from Problems 11.1 and 11.4, determine the maximum bearing pressure q' and the minimum bearing pressure q'' exerted by the retaining wall foundation.

ANSWER: $q' = 37.9$ kPa and $q'' = 7.5$ kPa.

11.6   For Problem 11.1, determine the allowable bearing capacity q_{all} of the retaining wall foundation using Eq. 6.1, assuming the soil beneath the footing is cohesionless sand ($\phi = 32°$), and using a factor of safety = 3. Assume the groundwater table is well below the bottom of the retaining wall foundation. Based on a comparison of the allowable bearing pressure q_{all} and the largest bearing pressure exerted by the retaining wall foundation (q'), is the design acceptable in terms of the foundation bearing pressures?

ANSWER: q_{all} = 290 kPa and since $q_{all} > q'$, the design is acceptable in terms of bearing pressures.

11.7 Using the data from Problem 11.1, calculate the factor of safety (F) for sliding of the retaining wall.

ANSWER: $F = 0.70$.

11.8 Using the data from Problems 11.1 and 11.4, calculate the factor of safety (F) for overturning of the retaining wall.

ANSWER: $F = 2.2$.

11.9 For the retaining wall described in Problem 11.1, it is proposed to increase the factor of safety for sliding by installing a concrete key at the bottom of the retaining wall foundation. Determine the depth of the key (measured from the ground surface in front of the retaining wall) that is needed to increase the factor of safety of sliding to 1.5. Neglect the weight of the concrete key in the analysis and include a reduction factor of 2 for P_p.

ANSWER: $D - 1.64$ m.

Retaining Wall Analyses with Wall Friction

11.10 Solve Problem 11.1, but include wall friction in the analysis (use Coulomb's earth pressure equation, Fig. 11.4). Assume the friction angle between the back side of the retaining wall and the backfill is equal to $^3/_4$ of ϕ (i.e., $\phi_w = ^3/_4\phi = 24°$).

ANSWER: $k_A = 0.275$, $P_A = 43.9$ kN/m, $P_H = 40.1$ kN/m, and $P_v = 17.9$ kN/m.

11.11 For Problem 11.10, determine the resultant normal force N and the distance of N from the toe of the footing ($\overline{x}$).

ANSWER: $N = 86.1$ kN/m and $\overline{x} = 1.69$ m.

11.12 Using the data from Problems 11.10 and 11.11, determine the maximum bearing pressure q' and the minimum bearing pressure q'' exerted by the retaining wall foundation.

ANSWER: $q' = 39.6$ kPa and $q'' = 17.8$ kPa.

11.13 For Problem 11.10, determine the allowable bearing capacity q_{all} of the retaining wall foundation using Eq. 6.1, assuming the soil beneath the footing is cohesionless sand ($\phi = 32°$), and using a factor of safety = 3. Assume the groundwater table is well below the bottom of the retaining wall foundation. Based on a comparison of the allowable bearing pressure q_{all} and the largest bearing pressure exerted by the retaining wall foundation (q'), is the design acceptable in terms of the foundation bearing pressures?

ANSWER: q_{all} = 290 kPa and since $q_{all} > q'$, the design is acceptable in terms of bearing pressures.

11.14 Using the data from Problem 11.10, calculate the factor of safety (F) for sliding of the retaining wall.

ANSWER: $F = 1.06$.

11.15 Using the data from Problem 11.10, calculate the factor of safety (F) for overturning of the retaining wall.

ANSWER: $F = \infty$.

11.16 For the retaining wall described in Problem 11.10, it is proposed to increase the factor of safety for sliding by installing a concrete key at the bottom of the retaining wall foundation. Determine the depth of the key (measured from the ground surface in front of the retaining wall) that is needed to increase the factor of safety of sliding to 1.5. Neglect the weight of the concrete key in the analysis and include a reduction factor of 2 for P_p.

ANSWER: $D = 1.16$ m.

11.17 Using the retaining wall shown at the top of Fig. 11.6B (i.e., a cantilevered retaining wall), assume $H = 4$ m, the thickness of the reinforced concrete wall stem $= 0.4$ m and the wall stem is located at the centerline of the footing, the reinforced concrete wall footing is 2 m wide by 0.5 m thick, the ground surface in front of the wall is level with the top of the wall footing, and the unit weight of concrete $= 23.5$ kN/m³. The wall backfill will consist of sand having $\phi = 32°$ and $\gamma_t = 20$ kN/m³. Also assume that there is sand in front of the footing with these same soil properties. The friction angle between the bottom of the footing and the bearing soil $(\delta) = 24°$. For the condition of a level backfill and assuming total mobilization of the shear strength along the vertical plane at the heel of the wall, calculate the active earth pressure coefficient k_A, the passive earth pressure coefficient k_p, the active earth pressure resultant force P_A, the vertical P_v and horizontal components P_H of P_A, and the passive resultant force P_p using a reduction factor $= 2.0$.

ANSWER: $k_A = 0.277$, $k_p = 3.25$, $P_A = 44.3$ kN/m, $P_v = 23.5$ kN/m, $P_H = 37.6$ kN/m, and $P_p = 4.07$ kN/m.

11.18 For Problem 11.17, determine the resultant normal force N and the distance of N from the toe of the footing $(\bar{x})$.

ANSWER: $N = 135.9$ kN/m and $\bar{x} = 1.05$ m.

11.19 Using the data from Problems 11.17 and 11.18, determine the maximum bearing pressure q' and the minimum bearing pressure q'' exerted by the retaining wall foundation.

ANSWER: $q' = 78.1$ kPa and $q'' = 57.8$ kPa.

11.20 For Problem 11.17, determine the allowable bearing capacity q_{all} of the retaining wall foundation using Eq. 6.1, assuming the soil beneath the footing is cohesionless sand ($\phi = 32°$), and using a factor of safety $= 3$. Assume the groundwater table is well below the bottom of the retaining wall foundation. Based on a comparison of the allowable bearing pressure q_{all} and the largest bearing pressure exerted by the retaining wall foundation (q'), is the design acceptable in terms of the foundation bearing pressures?

ANSWER: $q_{all} = 220$ kPa and since $q_{all} > q'$, the design is acceptable in terms of bearing pressures.

11.21 Using the data from Problem 11.17, calculate the factor of safety (F) for sliding of the retaining wall.

ANSWER: $F = 1.72$.

11.22 Using the data from Problems 11.17 and 11.18, calculate the factor of safety (F) for overturning of the retaining wall.

ANSWER: $F = 47$.

11.23 For the retaining wall described in Problem 11.17, it is proposed to increase the factor of safety for sliding by installing a concrete key at the bottom of the retaining wall foundation. Determine the depth of the key (measured from the ground surface in front of the retaining wall) that is needed to increase the factor of safety of sliding to 2.0. Neglect the weight of the concrete key in the analysis and include a reduction factor of 2 for P_p.

ANSWER: $D = 0.95$ m.

Additional Design Considerations for Retaining Walls

11.24   For the example problem shown in Fig. 11.5, assume that there is a vertical surcharge pressure of 200 psf located at ground surface behind the retaining wall. Calculate the factor of safety for sliding, factor of safety for overturning, and determine if N is within the middle one-third of the retaining wall foundation.

ANSWER: Factor of safety for sliding = 1.48, factor of safety for overturning = 2.64, and N is not within the middle one-third of the retaining wall foundation.

11.25   For the example problem shown in Fig. 11.5, assume that the ground surface behind the retaining wall slopes upward at a 3:1 (horizontal:vertical) slope inclination. Calculate the factor of safety for sliding, factor of safety for overturning, and determine if N is within the middle one-third of the retaining wall foundation.

ANSWER: Factor of safety for sliding = 1.32, factor of safety for overturning = 2.73, and N is not within the middle one-third of the retaining wall foundation.

11.26   Assume a 2.5:1 (horizontal:vertical) slope is composed of soil B (Fig. 9.14) and the slope height is 20 m. Also assume that the depth of slope creep will correspond to the depth of seasonal moisture changes and that the effective shear strength parameters for this clay are $\phi' = 28°$ and $c' = 4$ kPa and $\gamma_t = 19$ kN/m^3. It is proposed to construct a retaining wall at the top of the fill slope. If the required passive resistance force P_p for the proposed retaining wall is equal to 10 kN/m, determine the required depth of the retaining foundation (measured from the top of the fill slope). Use a reduction factor = 2 for P_p.

ANSWER: $D = 5.6$ m.

11.27   A cantilevered retaining wall (3 m in height) has a granular backfill with $\phi = 30°$ and $\gamma_t = 20$ kN/m^3. Neglect wall friction and assume the drainage system fails and the water level rises 1.5 m above the bottom of the retaining wall. Determine the initial active earth pressure resultant force P_A, the resultant force (due to earth plus water pressure) on the wall due to the rise in water level, and the percent increase in force against the wall due to the rise in water level.

ANSWER: $P_A = 30$ kN/m (initial condition). With a rise in water level, the force acting on the wall = 37.3 kN/m, representing a 24 percent increase in force acting on the retaining wall.

Restrained Retaining Wall

11.28   Use the data from Problem 11.1, but assume the wall is restrained. Determine the resultant of the lateral pressure exerted on the retaining wall by the backfill soil. Also determine the percent increase in resultant force for the restrained condition versus the unrestrained case.

ANSWER: Resultant force = 75.2 kN/m, representing a 53 percent increase in force acting on the retaining wall.

Mechanically Stabilized Earth Retaining Walls

11.29   Using the mechanically stabilized earth retaining wall shown in Fig. 11.12, assume $H = 20$ ft, the width of the mechanically stabilized earth retaining wall = 14 ft, the depth of embedment at the front of the mechanically stabilized zone = 3 ft, the soil behind the mechanically stabilized zone is a clean sand with a friction angle $\phi = 30°$, and the total unit weight $\gamma_t = 110$ pcf. Also assume that there is sand in front of the wall with these same properties and there is a level backfill with no surcharge pressures (i.e., $P_2 = 0$). Further assume that the mechanically stabilized zone will have a total unit weight $\gamma_t = 120$ pcf, there will be no shear stress (i.e., $\delta = 0°$) along the vertical back side of the mechanically stabilized zone, and $\delta = 23°$ along the bottom of the mechanically stabilized zone. Calculate the active earth pressure coefficient k_A, the passive earth pressure coefficient k_p, the active earth pressure resultant force P_A, and the passive resultant force P_p using a reduction factor = 2.0.

ANSWER: $k_A = 0.333$, $k_p = 3.0$, $P_A = 7330$ lb/ft, and $P_p = 740$ lb/ft.

11.30   For Problem 11.29, determine the width of the active wedge at the top of the mechanically stabilized earth retaining wall.

ANSWER: 25.5 ft measured from the upper front corner of the mechanically stabilized earth retaining wall (Fig. 11.12).

11.31   For Problem 11.29, determine the resultant normal force N and the distance of N from the toe of the mechanically stabilized zone, i.e., point O.

ANSWER: $N = 33,600$ lb/ft and $\bar{x} = 5.55$ ft.

11.32   Using the data from Problems 11.29 and 11.31, determine the maximum bearing pressure q' and the minimum bearing pressure q'' exerted by the mechanically stabilized earth retaining wall.

ANSWER: $q' = 3890$ psf and $q'' = 910$ psf.

11.33   For Problem 11.29, determine the allowable bearing capacity q_{all} of the base of the mechanically stabilized earth retaining wall using Eq. 6.1, assuming the soil beneath the wall is cohesionless sand ($\phi = 30°$ and $\gamma_t = 120$ pcf), and using a factor of safety = 3. Assume the groundwater table is well below the bottom of the mechanically stabilized zone. Based on a comparison of the allowable bearing pressure q_{all} and the largest bearing pressure exerted at the base of the mechanically stabilized earth retaining wall (q'), is the design acceptable in terms of the bearing pressures?

ANSWER: $q_{all} = 6300$ psf and since $q_{all} > q'$, the design is acceptable in terms of bearing pressures.

11.34   Using the data from Problem 11.29, calculate the factor of safety (F) for sliding of the mechanically stabilized earth retaining wall.

ANSWER: $F = 2.05$.

11.35   Using the data from Problems 11.29 and 11.31, calculate the factor of safety (F) for overturning of the mechanically stabilized earth retaining wall.

ANSWER: $F = 4.81$.

11.36   Use the data from Problem 11.29 and assume that there is a vertical surcharge pressure of 200 psf located at ground surface behind the mechanically stabilized earth retaining wall. Calculate the factor of safety for sliding, factor of safety for overturning, and the maximum pressure exerted by the base of the mechanically stabilized earth retaining wall.

ANSWER: Factor of safety for sliding = 1.73, factor of safety for overturning = 3.78, and maximum pressure $q' = 4300$ psf.

11.37   Use the data from Problem 11.29 and assume that the ground surface behind the mechanically stabilized earth retaining wall slopes upward at a 3:1 (horizontal:vertical) slope inclination. Also assume that the 3:1 slope does not start at the upper front corner of the rectangular reinforced soil mass (such as shown in Fig. 11.13), but instead the 3:1 slope starts at the upper back corner of the rectangular reinforced soil mass. Calculate the factor of safety for sliding, factor of safety for overturning, and the maximum pressure exerted by the retaining wall foundation.

ANSWER: Factor of safety for sliding = 1.60, factor of safety for overturning = 3.76, and maximum pressure $q' = 4,310$ psf.

11.38   For Problem 11.29, the internal stability of the mechanically stabilized zone is to be checked by using the wedge analysis. Assume a planar slip surface that is inclined at an angle of 61° (i.e., $\alpha = 61°$) and passes through the toe of the mechanically stabilized zone. Also assume that the mechanically stabilized zone contains 40 horizontal layers of Tensar SS2 geogrid which has an allowable tensile strength = 300 lb per ft of wall length for each geogrid. In the wedge analysis, these 40 layers of geogrid can be represented as an allowable horizontal resistance force = 12,000 lb per ft of wall length (i.e., 40 layers times 300 lb). If the friction angle ϕ of the sand = 32° in the mechanically stabilized zone, calculate the factor of safety for internal stability of the mechanically stabilized zone using the wedge analysis.

ANSWER: $F = 1.82$.

Sheet Pile Walls

11.39   Using the sheet pile wall diagram shown in Fig. 11.14, assume that the soil behind and in front of the sheet wall is uniform sand with a friction angle $\phi' = 33°$, buoyant unit weight $\gamma_b = 64$ pcf, and above the groundwater table, the total unit weight $\gamma_t = 120$ pcf. Also assume that the sheet pile wall has $H = 30$ ft, $D = 20$ ft, the water level in front of the wall is at the same elevation as the groundwater table which is located 5 ft below the ground surface, and the tieback anchor is located 4 ft below the ground surface. Neglecting wall friction, calculate the active earth pressure coefficient k_A, the passive earth pressure coefficient k_p, the active earth pressure resultant force P_A, and the passive resultant force P_p using no reduction factor.

ANSWER: $k_A = 0.295$, $k_p = 3.39$, $P_A = 27,500$ lb/ft, and $P_p = 43,400$ lb/ft.

11.40   For Problem 11.39, calculate the factor of safety for toe failure (i.e., toe kick-out).

ANSWER: $F = 2.19$.

11.41   For Problem 11.39, calculate the anchor pull force A_p assuming that the anchors will be spaced 10 ft on-center.

ANSWER: $A_p = 76.8$ kips.

11.42   For Problem 11.39, calculate the depth D that will provide a factor of safety for toe failure (i.e., toe kick-out) equal to 1.5.

ANSWER: $D = 14.6$ ft.

11.43   For Problem 11.39, assume that there is a uniform vertical surcharge pressure = 200 psf applied to the ground surface behind the sheet pile wall. Calculate the factor of safety for toe failure.

ANSWER: $F = 2.03$.

11.44   For Problem 11.39, assume that the ground surface slopes upward at a 2:1 (horizontal:vertical) slope ratio behind the sheet pile wall. Calculate the factor of safety for toe failure.

ANSWER: $F = 1.46$.

11.45   For Problem 11.39, assume that the ground in front of the sheet pile wall (i.e., the passive earth zone) slopes downward at a 3:1 (horizontal:vertical) slope ratio. Calculate the factor of safety for toe failure.

ANSWER: $F = 1.18$.

11.46   For Problem 11.39, assume that there are two different horizontal layers of sand at the site. The first layer of sand is located only behind the sheet pile wall and extends from the ground surface to a depth of 30 ft. This sand layer has the engineering properties as stated in Problem 11.39. The second horizontal layer of sand is located on both sides of the sheet pile wall. All of the sand in front of the sheet pile wall is composed of this second layer and the soil behind the sheet pile wall below a depth of 30 ft also consists of this second layer. The second sand layer is a clean sand, in a loose state, and has a friction angle $\phi' = 30°$ and a buoyant unit weight $\gamma_b = 60$ pcf. Neglecting wall friction, calculate the active earth pressure coefficients k_A, the passive earth pressure coefficient k_p, the active earth pressure resultant force P_A, and the passive resultant force P_p using no reduction factor.

ANSWER: $k_A = 0.295$ (upper sand layer), $k_A = 0.333$ (lower sand layer), $k_p = 3.0$, $P_A = 29,400$ lb/ft, and $P_p = 36,000$ lb/ft.

11.47   For Problem 11.46, calculate the factor of safety for toe failure (i.e., toe kick-out).

ANSWER: $F = 1.67$.

11.48   For Problem 11.46, calculate the anchor pull force A_p assuming that the anchors will be spaced 10 ft on-center.

ANSWER: $A_p = 78.4$ kips.

11.49   For Problem 11.46, assume that there is a uniform vertical surcharge pressure = 200 psf applied to the ground surface behind the sheet pile wall. Calculate the factor of safety for toe failure.

ANSWER: $F = 1.55$.

11.50   Use the data from Problem 11.46, except assume that the second soil layer is a clay that has effective shear strength parameters of $c' = 2$ kPa and $\phi' = 25°$. Neglecting the effective cohesion value and using an effective stress analysis, calculate the factor of safety for toe failure.

ANSWER: $F = 1.16$.

11.51   For Example Problem 11.4, calculate the depth D of the sheet pile wall and the maximum moment M_{max} in the sheet pile wall if the sand has a friction angle ϕ' of 30°. Neglect friction between the sheet pile wall and the sand.

ANSWER: $D = 27$ ft and $M_{max} = 55,000$ ft-lb/ft.

11.52   For Example Problem 11.4, calculate the depth D of the sheet pile wall and the maximum moment M_{max} in the sheet pile wall if the water level on both sides of wall is located at the top of the sheet pile wall (i.e., $\alpha = 0$).

ANSWER: $D = 8$ ft and $M_{max} = 6000$ ft-lb/ft.

Braced Excavations

11.53   A braced excavation will be used to support the vertical sides of a 20 ft deep excavation (i.e., $H = 20$ ft in Fig. 11.18). If the site consists of a sand with a friction angle $\phi = 32°$ and a total unit weight $\gamma_t = 120$ pcf, calculate the earth pressure σ_h and the resultant earth pressure force acting on the braced excavation. Assume the groundwater table is well below the bottom of the excavation.

ANSWER: $\sigma_h = 480$ psf, and the resultant force = 9600 lb per linear ft of wall length.

11.54   Solve Problem 11.53, but assume the site consists of a soft clay having an undrained shear strength $s_u = 300$ psf (i.e., $c = s_u = 300$ psf).

ANSWER: $\sigma_h = 1200$ psf, and the resultant force = 21,000 lb per linear ft of wall length.

11.55   Solve Problem 11.53, but assume the site consists of a stiff clay having an undrained shear strength $s_u = 1200$ psf and use the higher earth pressure condition (i.e., σ_{h2}).

ANSWER: $\sigma_{h2} = 960$ psf, and resultant force = 14,400 lb per linear ft of wall length.

Steel I-beams and Wood Lagging

11.56   A cantilevered steel I-beam and wood lagging system is installed for a 20-ft-deep excavation. Assume the site consists of a sand with a friction angle $\phi = 32°$ and a total unit weight $\gamma_t = 120$ pcf, and the temporary retaining wall deforms enough to develop active earth pressures. Also assume the groundwater table is well below the bottom of the excavation. If the steel I-beams are spaced 7 ft on-center, calculate the active earth pressure resultant force acting on each steel I-beam.

ANSWER: The resultant force = 51,600 lb acting on each steel I-beam.

11.57   For Problem 11.56, assume that that each steel I-beam is braced (i.e., a braced excavation). For the braced excavation, calculate the earth pressure resultant force acting on each steel I-beam.

ANSWER: The resultant force = 67,200 lb acting on each steel I-beam.

11.58   Assume that for a steel I-beam and wood lagging temporary retaining wall, tieback anchors will be used to reduce the bending moments in the steel I-beams. Each tieback will have a diameter of 6 in. and the allowable frictional resistance between the grouted tieback anchor and the soil is 10 psi. If each tieback must resist a load of 20 kips, determine the bonded length L_b.

ANSWER: $L_b = 8.8$ ft.

CHAPTER 12
FOUNDATION DETERIORATION AND CRACKING

12.1 INTRODUCTION

Prior chapters have presented a discussion of bearing capacity analyses, settlement analyses, foundations on expansive soil, slope stability, and retaining walls. The purpose of this chapter is to discuss foundation deterioration and cracking, as follows:

1. *Deterioration.* All man-made and natural materials are susceptible to deterioration. In terms of deterioration, the National Science Foundation (1992) states:

> The infrastructure deteriorates with time, due to aging of the materials, excessive use, overloading, climatic conditions, lack of sufficient maintenance, and difficulties encountered in proper inspection methods. All of these factors contribute to the obsolescence of the structural system as a whole. As a result, repair, retrofit, rehabilitation, and replacement become necessary actions to be taken to insure the safety of the public.

This topic is so broad that it is not possible to cover every type of geotechnical and foundation element susceptible to deterioration. Instead, this chapter will discuss three of the more common types of deterioration: timber decay, sulfate attack of concrete foundations, and frost action. These three topics will be covered in Secs. 12.2 to 12.4. In addition, deterioration of the foundations for historic structures will be discussed in Sec. 12.5.

2. *Shrinkage cracking.* Foundation cracks can develop due to soil movement, such as settlement or expansive soil. Foundation cracks can also develop due to the shrinkage of concrete and excessive shrinkage can cause damage to floor coverings, such as brittle tile. Shrinkage of concrete foundations will be discussed in Sec. 12.6.

3. *Moisture intrusion.* Moisture can penetrate the foundation at the location of the shrinkage cracks. Moisture can also penetrate the slab in the form of water vapor and cause damage to floor coverings, such as wood and linoleum. Moisture intrusion of slab-on-grade concrete foundations will be covered in Sec. 12.7.

12.2 TIMBER DECAY

12.2.1 Introduction

The decay of timber can be caused by several different factors, such as the following (Singh, 1994; Singh and White, 1997):

1. *Fungal decay.* Common types of fungal decay are due to dry rot, wet rot, and mold.

   a. *Dry rot.* A fungus belonging to the same group as common mushrooms and toadstools causes dry rot. Reproduction of the fungus is by spores, which can be produced in enormous numbers, but reproduction requires favorable timber moisture content of 20 to 30 percent. The fungus produce an enzyme called *cellulase*, which digests the wood cellulose, but is unable to attack the rigidifying polymer (in the cell walls), called *lignin*. The lignin remains as a brittle matrix that cracks into cubical pieces. The fungal strands have the ability to spread beyond the initial area of attack and these strands are known as *rhizomorphs*. These strands are able to transport nutrients and water and may be up to $1/4$ in. (6 mm) in diameter. If the timber moisture content falls below about 20 percent, the fungus will become dormant and eventually die within 9 months to 1 year.

In terms of the appearance of dry rot, Singh and White (1997) state:

> Wood thoroughly rotted with dry-rot fungus is light in weight, dull brown in color, crumbles under the fingers, and loses its fresh resinous smell. It is also called brown rot, a term relating to the manner in which it destroys the cellulose, leaving the lignin largely unaltered, so that the wood acquires a distinctive brown color. As a result of this, the structural strength is almost entirely lost.

    b. *Wet rot.*   Wet-rot fungus develops when the timber is in a persistently wet condition with moisture content of 50 to 60 percent. Wet rot decay is responsible for up to 90 percent of wood decay in buildings.

    c. *Molds.*   Molds usually develop on the surface of the wood. They increase the porosity of the wood and allow the wood to get wet more easily and to stay wet, which increases wet-rot decay. Often the presence of surface deposits of mold will indicate that the wood has been subjected to excessive moisture.

2. *Insect decay.*   There are many different types of insects that will attack timber. Some insects, such as beetles, are wood boring and directly consume the wood. Other insects prefer damp wood or wood that has fungal decay. Some insects, such as termites, have complex colonies of winged adults, workers, and soldiers. Termites live within the timber, often hollowing out the interior, but leaving the outer shell for protection.

3. *Other factors.*   There are other factors that can cause timber decay. Examples include chemical decay, mechanical wear, and decomposition by physical agents, such as prolonged heating, fire, and moisture.

12.2.2   Timber Piles

Timber piles can be especially susceptible to deterioration because they can be attacked by a variety of organisms. For example, the immersed portions of the wood piles in marine or river environments are liable to severe attack by marine organisms (marine borers, and the like). Timber piles above the groundwater table can also experience decay due to fungal growth and insect attack. To reduce the deterioration due to fungal decay and insect attack, timber piles should be treated with a preserving chemical. This process consists of placing the timber piles in a pressurized tank filled with creosote or some other preserving chemical. The pressure treatment forces the creosote into the wood pores and creates a thick coating on the pile perimeter. Creosote timber piles normally last as long as the design life of the structure. An exception is the case where they are subjected to prolonged high temperatures (such as supporting blast furnaces) because studies show that they lose strength with time in such an environment (Coduto, 1994).

12.2.3   Timber Foundations

With the exception of timber piles, most foundations in the United States are made of concrete. However, there may be older foundations of historic structures that consist of wood and are susceptible to decay. In addition, Sections 1810.3.2.4 and 1810.3.2.5 of the *International Building Code* (2009) states that timber can be used for footings and foundations provided the wood is treated to resist decay.

12.3 SULFATE ATTACK OF CONCRETE

12.3.1 Introduction

Sulfate attack of concrete is defined as a chemical and/or physical reaction between sulfates (usually in the soil or groundwater) and concrete or mortar, primarily with hydrated calcium aluminate in the cement-paste matrix, often causing deterioration (ACI, 1990). Sulfate attack of concrete occurs throughout the world, especially in arid areas, such as the southwestern United States. In arid regions, the salts are drawn up into the concrete and then deposited on the concrete surface as the groundwater evaporates. Sulfate attack of concrete can cause a physical loss of concrete and unusual cracking and discoloration of the concrete. Sulfate attack can also cause scaling, pitting, and etching of the concrete surface (Detwiler et al., 2000). Figures 12.1 and 12.2 show sulfate attack of a concrete patio.

Typically the geotechnical engineer obtains the representative soil or groundwater samples to be tested for sulfate content. The geotechnical engineer can analyze the soil samples or groundwater in-house, but a more common procedure is to send the samples to a chemical laboratory for testing. There are different methods to determine the soluble sulfate content in soil or groundwater. One method is to precipitate out and then weigh the sulfate compounds. A faster and easier method is to add barium chloride to the solution and then compare the turbidity (relative cloudiness) of barium sulfate with known concentration standards.

The geotechnical engineer is often required to provide soluble sulfate specifications and recommendations for concrete foundations. In addition to concrete foundation deterioration due to sulfate attack, there could also be deterioration of concrete or the steel reinforcement due to chloride attack and acid attack. According to AASHTO (1996), laboratory testing of soil and groundwater samples for sulfates, chloride, and pH should be sufficient to assess deterioration potential. When chemical wastes are suspected, a more thorough chemical analysis of soil and groundwater samples would be required.

FIGURE 12.1   Sulfate attack of a concrete patio.

FIGURE 12.2   Another example of sulfate attack of a concrete patio.

12.3.2   Mechanisms of Sulfate Attack of Concrete

There has been considerable research, testing, and chemical analysis of sulfate attack. Two different mechanisms of sulfate attack have been discovered: chemical reactions and the physical growth of crystals.

Chemical Reactions.   The chemical reactions involving sulfate attack of concrete are complex and studies have discovered two main chemical reactions (Lea, 1971; Mehta, 1976). The first is a chemical reaction of sulfate and calcium hydroxide (which was generated during the hydration of the cement) to form calcium sulfate, commonly known as *gypsum.* The second is a chemical reaction of gypsum and hydrated calcium aluminate to form calcium sulfoaluminate, commonly called *ettringite* (ACI, 1990).

As with many chemical reactions, the final product of ettringite causes an increase in volume of the concrete. Hurst (1968) indicates that the chemical reactions produce a compound of twice the volume of the original tricalcium aluminate compound. Concrete has a low tensile strength and thus the increase in volume fractures the concrete, allowing for more sulfates to penetrate the concrete, resulting in accelerated deterioration.

Physical Growth of Crystals.   It has been shown that there can be crystallization of the sulfate salts in the pores of the concrete. The growth of crystals exerts expansive forces within the concrete, causing flaking and spalling of the outer concrete surface. Besides sulfate, the concrete, if porous enough, can be disintegrated by the expansive force exerted by the crystallization of almost any salt in its pores (Tuthill, 1966; Reading, 1975).

Damage due to crystallization of salt is commonly observed in areas where water is migrating through the concrete and then evaporating at the concrete surface. Examples include the surfaces of concrete dams, basement and retaining walls that lack proper waterproofing, and concrete structures that are partially immersed in salt-bearing water (such as seawater) or soils. Figures 12.3 and 12.4 show two examples of concrete patio deterioration due to the crystallization of salts.

12.3.3   Sulfate Resistance of Concrete

The sulfate resistance of concrete depends on many different factors. In general, the degree of sulfate attack of concrete will depend on the type of cement used, quality of the concrete, soluble

FIGURE 12.3 Concrete patio deterioration, Mojave Desert. The arrows point to deterioration of the concrete surface.

sulfate concentration that is in contact with the concrete, and the surface preparation of the concrete (Mather, 1968).

1. *Type of cement.* There is a correlation between the sulfate resistance of cement and its tricalcium aluminate content. As previously discussed, it is the chemical reaction of hydrated calcium aluminate and gypsum that forms ettringite. Therefore, limiting the tricalcium aluminate content

FIGURE 12.4 Concrete patio deterioration, Mojave Desert (note the salt deposits and deterioration of the concrete surface).

of cement reduces the potential for the formation of ettringite. It has been stated that the tricalcium aluminate content of the cement is the greatest single factor that influences the resistance of concrete to sulfate attack, where in general, the lower the tricalcium aluminate content, the greater the sulfate resistance (Bellport, 1968).

2. *Quality of concrete.*   In general, the more impermeable the concrete, the more difficult for the waterborne sulfate to penetrate the concrete surface. To have a low permeability, the concrete must be dense, have high cement content, and a low water-cementitious materials ratio. Using a low water-cementitious materials ratio decreases the permeability of mature concrete (PCA, 1994). A low water-cementitious materials ratio is a requirement of ACI (1990) for concrete subjected to soluble sulfate in the soil or groundwater. For example, the maximum water-cementitious materials ratio must be equal to or less than 0.50 for concrete exposed to moderate sulfate exposure and 0.45 for concrete exposed to severe or very severe sulfate exposure (see Table 18.2).

There are many other conditions that can affect the quality of the concrete. For example, a lack of proper consolidation of the concrete can result in excessive voids. Another condition is the corrosion of reinforcement, which may crack the concrete and increase its permeability. Cracking of concrete may also occur when structural members are subjected to bending stresses. For example, the tensile stress due to a bending moment in a footing may cause the development of microcracks that increases the permeability of the concrete.

3. *Concentration of sulfates.*   As previously mentioned, it is often the geotechnical engineer who obtains the soil or water specimens that will be tested for soluble sulfate concentration. In some cases after construction is complete, the sulfate may become concentrated on crack faces. For example, water evaporating through cracks in concrete flatwork will deposit the sulfate on the crack faces. This concentration of sulfate may cause accelerated deterioration of the concrete.

4. *Surface preparation of concrete.*   An important factor in concrete resistance is the surface preparation, such as the amount of curing of the concrete. Curing results in a stronger and more impermeable concrete, which is better able to resist the effects of salt intrusion (PCA, 1994).

12.3.4   Design and Construction

The soil or groundwater samples should be obtained after the site has been graded and the location of the proposed building is known. For the planned construction of shallow foundations, near surface soil samples should be obtained for the sulfate testing. For the planned construction of deep foundations that consist of concrete piles or piers, soil and groundwater samples for sulfate testing should be taken at various depths that encompass the entire length of the concrete foundation elements.

The geotechnical engineer should select representative soil samples for testing. If the site contains clean sand or gravel, then these types of soil often have low sulfate contents because of their low capillary rise and high permeability that enables any sulfate to be washed from the soil. But clays can have a higher sulfate content because of their high capillary rise that enables them to draw up sulfate bearing groundwater and their low permeability that prevents the sulfate from being washed from the soil.

Sulfate concentrations in the soil and groundwater can vary throughout the year. For example, the highest concentration of sulfates in the soil and groundwater will tend to occur at the end of the dry season or after a long dry spell. Likewise, the lowest concentration of sulfates will occur at the end of the rainy season, when the sulfate has been diluted or partially flushed out of the soil. The geotechnical engineer should recognize that sampling of soil or groundwater at the end of a heavy rainfall season will most likely be unrepresentative of the most severe condition.

Soil that contains gypsum should always be tested for sulfate content. This is because soluble sulfate (SO_4) is released as gypsum ($Ca\,SO_4 \cdot 2H_2O$) weathers. Gypsum is an evaporate and can rapidly weather upon exposure to air and water.

There are many environments that could lead to the chemical attack of concrete. For example, if the site had been previously used as a farm, there may be fertilizers or animal wastes in the soil that will have a detrimental effect on concrete. A corrosive environment, such as drainage of acid mine water, will also lead to the deterioration of concrete. Another corrosive environment is seawater, which has both moderate soluble sulfate content and a high salt content that can attack concrete through the process of chemical reactions involving sulfate attack and deterioration of concrete due to the physical growth of salt crystals.

As indicated in Table 18.2, of the types of Portland cements, the most resistant cement is type V, in which the tricalcium aluminate content must be less than 5 percent. Depending on the percentage of soluble sulfate (SO_4) in the soil or groundwater, a certain cement type is required as indicated in Table 18.2. In addition, minimum water-cementitious ratios and minimum unconfined compressive strengths f'_c are required depending on the amount of sulfate in the soil or groundwater (see Table 18.2).

12.4 FROST

There have been extensive studies on the detrimental effects of frost, which can impact both pavements and foundations (Casagrande, 1932b; Kaplar, 1970; Yong and Warkentin, 1975; Reed et al., 1979). Two common types of damage related to frost are: (1) freezing of water in cracks, and (2) formation of ice lenses. In many cases, deterioration or damage is not evident until the frost has melted. In these instances, it may be difficult for the geotechnical engineer to conclude that frost was the primary cause of the deterioration.

There is about a 10 percent increase in volume of water when it freezes and this volumetric expansion of water upon freezing can cause deterioration or damage to many different types of materials. Common examples include rock slopes and concrete, as discussed below.

1. *Rock slopes.*   The expansive forces of freezing water results in a deterioration of the rock mass, additional fractures, and added driving (destabilizing) forces. Feld and Carper (1997) describe several rock slope failures caused by freezing water, such as the February 1957 failure where 1000 tons (900 Mg) of rock fell out of the slope along the New York State thruway, closing all three southbound lanes north of Yonkers.

2. *Concrete foundations.*   The American Concrete Institute (ACI, 1982) defines durability as the ability of concrete to resist weathering, chemical attack, abrasion, or any other type of deterioration. Durability is affected by strength, but also by density, permeability, air entrainment, dimensional stability, characteristics and proportions of constituent materials, and construction quality (Feld and Carper, 1997). Durability is harmed by freezing and thawing, sulfate attack, corrosion of reinforcing steel, and reactions between the various constituents of the cements and aggregates. Damage to concrete caused by freezing could occur during the original placement of the concrete or after it has hardened. To prevent damage during placement, it is important that the fresh concrete not be allowed to freeze. Air-entraining admixtures can be added to the concrete mixture to help protect the hardened concrete from freeze-thaw deterioration.

Frost penetration and the formation of ice lenses in the soil can damage shallow foundations and pavements. It is well known that silty soils are more likely to form ice lenses because of their high capillarity and sufficient permeability that enables them to draw up moisture to the ice lenses. Feld and Carper (1997) describe several interesting cases of foundation damage due to frost action. At Fredonia, New York, the frost from a deep-freeze storage facility froze the soil and heaved the foundations upward 4 in. (100 mm). A system of electrical wire heating was installed to maintain soil volume stability.

Another case involved an extremely cold winter in Chicago, where frost penetrated below an underground garage and broke a buried sprinkler line. This caused an ice buildup that heaved the structure above the street level and sheared-off several supporting columns.

As these cases show, it is important that the foundation be constructed below the depth of frost action. Usually there will be local building requirements on the minimum depth of the foundation to prevent damage caused by the formation of ice lenses. There have also been studies to determine the annual maximum frost depths for various site conditions and 50-year or 100-year return periods (e.g., DeGaetano, et al., 1997).

12.5 *HISTORIC STRUCTURES*

12.5.1 Introduction

The repair or maintenance of historic structures presents unique challenges to the geotechnical and foundation engineer. The geotechnical engineer could be involved in many different types of problems with historic structures. Common problems include structural weakening and deterioration due to age or environmental conditions, original poor construction practices, inadequate design, or faulty maintenance. The purpose of this section is to describe the deterioration of foundations for historic structures. Two examples will be discussed, the Bunker Hill Monument and the Guajome Ranch House.

12.5.2 Bunker Hill Monument

Figure 12.5 shows the exposed foundation of the Bunker Hill Monument, which was erected at the top of Bunker Hill in 1825 in tribute to the courageous Minutemen who died at Breed's Hill. Figure 12.6 shows the mortar between the hornblende granite blocks that comprise the foundation. In some cases the mortar was completely disintegrated, while in other cases, the mortar could be easily penetrated such as shown in Fig. 12.6 where a pen has been used to pierce the mortar. The causes of the disintegration were chemical and physical weathering, such as cycles of freezing and thawing of the foundation.

12.5.3 Adobe Construction

As compared to New England where many of the foundations for historic structures are made of brick or stone, the historic structures in the southwestern United States are commonly constructed of adobe. *Adobe* is a sun-dried brick made of soil mixed with straw. The preferred soil type is a clayey material that shrinks upon drying to form a hard, rocklike brick. Straw is added to the adobe to provide tensile reinforcement. The concept is similar to the addition of steel fibers to modern day concrete.

FIGURE 12.5   Exposed foundation of the Bunker Hill Monument.

FIGURE 12.6 Deteriorated mortar between the granite blocks of the Bunker Hill Monument foundation.

Adobe construction was developed thousands of years ago by the indigenous people, and traditional adobe brick construction was commonly used prior to the twentieth century. The use of adobe as a building material was due in part to the lack of abundant alternate construction materials, such as trees. The arid climate of the southwestern United States also helped to preserve the adobe bricks.

Adobe structures in the form of multifamily dwellings were typically constructed in low-lying valleys, adjacent to streams, or near springs. In an arid environment, a year-round water source was essential for survival. Mud was scooped out of streambeds and used to make the adobe bricks. As a matter of practicality, many structures were constructed close to the source of the mud. These locations were frequent in floodplains, which were periodically flooded, or in areas underlain by a shallow groundwater table.

Adobe construction does not last forever and many historic adobe structures have simply melted away. The reasons for deterioration of adobe with time include the uncemented or weakly cemented nature of adobe that makes it susceptible to erosion or disintegration from rainfall, periodic flooding, or water infiltration due to the presence of a shallow groundwater table. For many historic adobe structures and typical of archeological sites throughout the world, all that remains is a mound of clay.

12.5.4 Guajome Ranch House

The purpose of this section is to describe the performance of a historic adobe structure known as the *Guajome Ranch House*, located in Vista, California. The Guajome Ranch House, a one-story adobe structure, is considered to be one of the finest large Mexican colonial ranch houses remaining in southern California (Guajome, 1986). The name Guajome means *frog pond* (Engstrand and Ward, unpublished report, 1991).

The main adobe structure was built during the period 1852–1854. It was constructed with the rooms surrounding and enclosing a main inner courtyard, which is typical Mexican architecture (Fig. 12.7). The main living quarters were situated in the front of the house. Sleeping quarters were located in another wing, and the kitchen and bake house in a third wing. At the time of initial construction, the

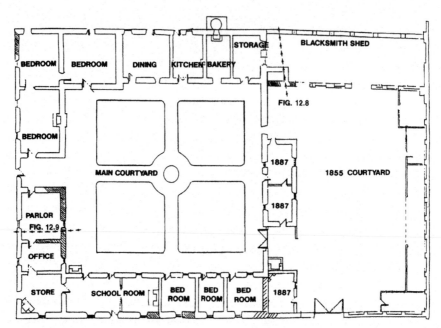

FIGURE 12.7   Site plan of the Guajome Ranch House.

house was sited in a vast rural territory, and was self-sufficient with water from a nearby stream and farming and livestock provided the food supply for its habitants.

Additional rooms were added around 1855, enclosing a second courtyard (Fig. 12.7). Several additional rooms were added in 1887 as indicated in Fig. 12.7. Contemporary additions (not shown in Fig. 12.7) include a garage and gable roof sewing room (built near the parlor).

The County of San Diego purchased the Guajome Ranch from the Couts family in 1973. The Guajome Ranch has been declared a National Historic Landmark and the County of San Diego performed the restoration. The author was a member of the restoration team and investigated the damage to the foundation caused by poor surface drainage at the site.

Deteriorated Condition.   The original exterior adobe walls, which are 2 to 4 ft (0.6 to 1.2 m) thick, have been covered with stucco as a twentieth century modification. Portions of the original low-pitched tile roof have been replaced with corrugated metal.

The Guajome Ranch House was built on gently sloping topography. In the 1855 courtyard, water drained toward one corner and then passed beneath the foundation. The arrows in Fig. 12.7 indicate the path of the surface water. At the location where the water comes in contact with the adobe, there was considerable deterioration, as shown in Fig. 12.8. Note in this photograph that the individual adobe blocks are visible.

Both the interior and exterior of the structure had considerable deterioration due to a lack of maintenance. The adobe was further eroded when the gable-roofed sewing room burned in 1974 and water from the fire hoses severely eroded the adobe. The hatch walls in Fig. 12.7 indicate those adobe walls having the most severe damage.

The main courtyard drains toward the front of the house, as indicated by the arrows in Fig. 12.7. There are no drains located underneath the front of the house and water is removed by simply letting it flow in ditches beneath the floorboards. Figure 12.9 shows the drainage beneath the front of the house and the moisture contributed to the deterioration of the wood floorboards. Because the water is in contact with the adobe foundation, the parlor had some of the most badly damaged adobe walls in the structure.

FIGURE 12.8 Deterioration of adobe. The lower photograph is a close-up view of the upper photograph. Note that the individual adobe blocks are visible in the lower photograph.

Laboratory Testing of Original Adobe Materials. Classification tests performed on remolded samples of the original adobe bricks indicate the soil can be classified as a silty sand to clayey sand (SM-SC), the plasticity index varies from about 2 to 4, and the liquid limit is about 20. Based on dry weight, the original adobe bricks contain about 50 percent sand-size particles, 40 percent silt-size particles, and 10 percent clay-size particles smaller than 0.002 mm.

The resistance of the original adobe bricks to moisture is an important factor in their preservation. To determine the resistance of the adobe to moisture infiltration, an index test for the erosion potential of an adobe brick was performed (Day, 1990b). The test consisted of trimming an original adobe brick to a diameter of 2.5 in. (6.35 cm) and a height of 1.0 in. (2.54 cm). Porous plates having a

FIGURE 12.9   Drainage beneath parlor.

diameter of 2.5 in. (6.35 cm) were placed on the top and bottom of the specimen. The specimen of adobe was then subjected to a vertical stress of 60 psf (2.9 kPa) and was unconfined in the horizontal direction. A dial gauge measured vertical deformation.

After obtaining an initial dial gauge reading, the adobe specimen was submerged in distilled water. Time-versus-dial readings were then recorded. The dial readings were converted to percent strain and plotted versus time, as shown in Fig. 12.10. When initially submerged in distilled water, some soil particles sloughed off in the horizontal (unconfined) direction, but the specimen remained essentially intact. This indicates that there is a weak bond between the soil particles.

After submergence in distilled water for eight days, the water was removed from the apparatus and the adobe specimen was placed outside and allowed to dry in the summer sun. After seven days of drying, the adobe specimen showed some shrinkage and corresponding cracking.

The adobe specimen was again submerged in distilled water and after only 18 min, the specimen had completely disintegrated as soil particles sloughed off in the horizontal (unconfined) direction. The experiment demonstrated the rapid disintegration of the adobe when it is subjected to wetting-drying cycles. Initially, the adobe is weakly cemented and resistant to submergence, but when the soaked adobe is dried, it shrinks and cracks, allowing for an accelerated deterioration when again submerged.

Drainage Repair.   Drainage was probably not a major design consideration when the Guajome Ranch House was built. However, the deterioration of the adobe where water is present indicates the importance of proper drainage. The drainage repair consisted of regrading of the courtyards so that surface water flows to box and grate inlets connected to storm-water piping. Surface drainage water is removed from the site through underground pipes. The new drainage system prevents surface water from coming in contact with the adobe foundation.

In summary, the interior and exterior of the Guajome Ranch house had severe adobe and wood floor decay in areas where water from surface drainage came in contact with the foundation (Figs. 12.8 and 12.9). The results of laboratory testing demonstrated the rapid disintegration of the adobe when it is subjected to cycles of wetting and drying. The repair for damage related to surface drainage consisted of a new drainage system that prevents surface water from coming in contact with the adobe foundation and wood floorboards.

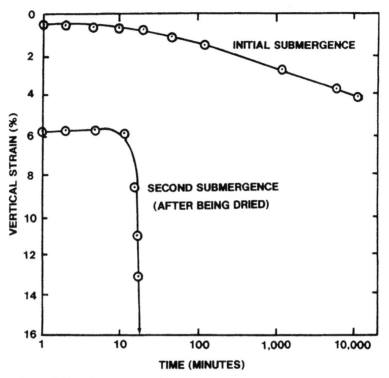

FIGURE 12.10 Laboratory test results.

12.6 *SHRINKAGE CRACKING*

12.6.1 Drying Shrinkage

Concrete expands slightly with a gain in moisture and contracts with a loss in moisture. This contraction of the concrete with a loss of moisture has been termed *drying shrinkage*. The United States Department of the Interior (1947) performed drying shrinkage tests on small concrete beams (4 × 4 × 40 in.) and they measured the drying shrinkage for a period of 38 months. The small concrete beams were initially moist-cured for 14 days at 70°F and then stored for 38 months in air at the same temperature and 50 percent relative humidity. The results of the study showed that 34 percent of the drying shrinkage occurred within the first month and 90 percent of the drying shrinkage occurred within 11 months. The remaining 10 percent of the shrinkage occurred from 11 months to the end of the test (38 months). Hence the results of this study on small concrete specimens showed that most drying shrinkage occurs within the first year after placement, with the amount of shrinkage decreasing with time.

It has also been observed that concrete near the surface dries and shrinks faster than the inner concrete core (Hanson, 1968). Other studies have shown that the volume-to-surface ratio is very important in the amount of drying shrinkage (Hansen and Mattock, 1966). Concrete having a low volume-to-surface ratio will shrink much more than the same concrete having a high volume-to-surface ratio. In addition, it has been observed that drying shrinkage continues longer for larger masses of concrete.

For a constant volume-to-surface ratio of concrete, the factors governing the amount of drying shrinkage are as follows (PCA, 1994):

1. *Water per unit volume of concrete.*   The amount of water per unit volume of concrete is by far the most important factor affecting the amount of drying shrinkage. The more water in the concrete mix, the greater the amount of drying shrinkage. One study at M.I.T. showed that for each 1 percent increase in mixing water, concrete drying shrinkage was increased by about 2 percent (Carlson, 1938). Another study showed that doubling the water in the concrete mix from 200 to 400 lb/yd^3 caused a five times increase in the amount of drying shrinkage.

2. *Increased water requirement.*   Any alteration of the mix design that increases the water requirement of the cement paste will lead to an increase in drying shrinkage. For example, additional mix water may be needed to increase the slump or to compensate for high temperatures of the freshly mixed concrete. Likewise, more water per unit volume will be required for those concrete mixes that have a high proportion of fine aggregates.

3. *Coarse aggregates.*   The coarse aggregates physically restrain the drying shrinkage of the cement paste. Hence dense and relatively incompressible coarse aggregates will be more difficult to compress and will reduce the amount of drying shrinkage as compared to soft and deformable aggregates. Examples of dense and relatively incompressible aggregates that reduce the amount of drying shrinkage include aggregates composed of quartz, granite, and feldspar (ACI, 1980).

4. *Curing.*   Curing of concrete is accomplished by maintaining a satisfactory moisture content and temperature in the concrete for a specified period of time immediately following the placement and finishing of the concrete. In essence, the objective of curing is to prevent the loss of moisture from the concrete. Curing can improve many desirable concrete properties, such as strength, durability, watertightness, abrasion resistance, and resistance to freezing and thawing. In addition, because the loss of water from concrete causes it to shrink, the process of curing will reduce the amount of drying shrinkage.

5. *Admixtures.*   Some admixtures require an increase in the water content of the concrete mix and hence they will cause an increase in drying shrinkage. In addition, studies have shown that accelerators, such as calcium chloride, can substantially increase the drying shrinkage of concrete.

6. *Cement.*   The type of cement, cement fineness and composition, and the cement contents may have some effect on the amount of drying shrinkage, but the effect is usually small.

The basic reason for the development of shrinkage cracks in concrete slab-on-grade foundations is that drying shrinkage produces tensile stresses that are in excess of the concrete's tensile strength. There can be many different reasons for the development of excessive concrete shrinkage. For example, excessive shrinkage cracking can develop in concrete if the mix was inadequately prepared with a water-cementitious materials ratio that is too high (i.e., too much water in the mix). Another possibility is that water may be added to the concrete at the job site in order to make the concrete easier to place and consolidate, in which case the extra water will cause excessive drying shrinkage. Other possibilities are that the concrete slab-on-grade is too thin (i.e., low volume-to-surface ratio) or it is not adequately cured. Figures 12.11 and 12.12 show two examples of excessive concrete shrinkage cracks in a slab-on-grade foundation.

There are three basic methods that can be used to limit or prevent the development of shrinkage cracks in concrete slab-on-grade foundations, as follows (PCA, 1994):

1. *Post-tensioned slabs.*   A concrete slab-on-grade foundation can be built with post-tensioning cables, which basically consist of steel strands in ducts that are tensioned after the concrete hardens. By tensioning the cables, the concrete is subjected to a compressive stress that will counteract the development of tensile stress due to concrete shrinkage. If the compressive stress is high enough, an essentially crack-free foundation can be created. Especially for a long slab-on-grade foundation, special effort may be needed to reduce subgrade friction.

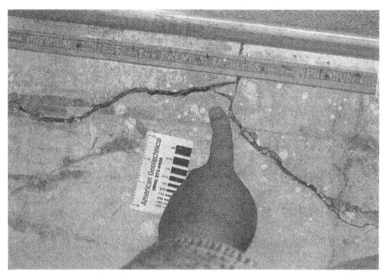

FIGURE 12.11 Excessive shrinkage cracks in a concrete slab-on-grade foundation.

2. *Steel reinforcement.* It has been stated that steel reinforcement equal to 0.5 percent of the cross-sectional area of the slab-on-grade foundation will be sufficient to essentially eliminate visible shrinkage cracking.

3. *Expansive cements.* Expansive cement is defined as *hydraulic cement* that expands slightly during the early hardening period after setting. Concrete that contains expansive cement can be

FIGURE 12.12 Excessive shrinkage crack in a concrete slab-on-grade foundation. The concrete has been cored and the crack is observed to penetrate the entire thickness of the concrete. Note that a visqueen moisture barrier is visible at the location of the core.

used to offset the amount of anticipated drying shrinkage. However, steel reinforcement is still needed in order to produce compressive stresses during and after the expansion period of the cement.

12.6.2 Contraction Joints

One option to control the location of shrinkage cracking is to use *contraction joints*. A contraction joint, also known as a *control joint*, is usually formed as a continuous straight slot in the top of the concrete slab. The slot forms a plane of weakness in which the shrinkage crack will develop. A common method of constructing the slot is to saw-cut the concrete after it is strong enough to resist tearing or other damage by the saw blade. Normally the concrete slab-on-grade foundation is saw-cut 4 to 12 h after the concrete begins to harden. It is often recommended that the concrete slab-on-grade be saw-cut to a depth equal to $\frac{1}{4}$ the thickness of the slab. If the slab is not saw-cut to an appropriate depth, then the drying shrinkage crack will deviate from the saw-cut, such as shown in Fig. 12.13.

There are many other methods for installing contraction joints in concrete slab-on-grade foundations. For example, during the finishing of the top surface of the concrete, grooves can be formed in the concrete. Another possible option is to install plastic strips in the top of the slab before the concrete has set, such as shown in Fig. 12.14.

The goals of contraction joints are to force the shrinkage cracking to develop at specific areas and to minimize the width of the shrinkage cracks. By limiting the width of the shrinkage crack, vertical loads are transmitted across the contraction joint by aggregate interlock between the opposite faces of the crack. In order to increase load transfer across contraction joints, steel reinforcement or steel dowels can be placed across the joint.

Design and Control of Concrete Mixtures (PCA, 1994) provides two tables that can be useful for the design of contraction joints, the first (Table 9-1) provides dowel sizes and spacing at contraction joints and the second (Table 9-2) provides maximum spacing of contraction joints as a function of floor slab thickness, slump, and maximum-size aggregate. As this document indicates, contraction joints should not be random, rather they should consist of a standard square pattern with the spacing of contraction joints as recommended by the text. The joint spacing should be decreased if the concrete is suspected of having high shrinkage characteristics and the resulting panels should be approximately square.

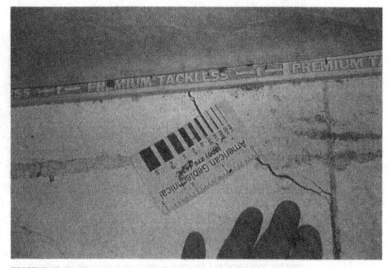

FIGURE 12.13   Concrete slab-on-grade having a contraction joint that was not saw-cut deep enough. Note how the slab crack deviates from the shallow saw-cut joint.

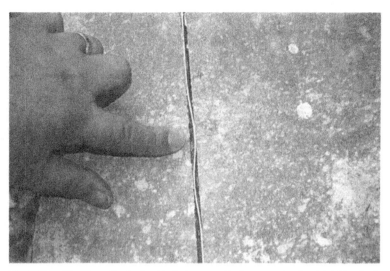

FIGURE 12.14   View of a plastic strip that was used to form a contraction joint. The plastic strip is sometimes referred to as a *zip-strip*. The gaps that have opened-up on both sides of the plastic strip were caused by drying shrinkage.

Contraction joints are commonly used for buildings such as warehouses that have large concrete slab-on-grade foundations. It should be recognized that if there is no steel reinforcement across the contraction joints, they become planes of weakness in the foundation. For example, Fig. 12.15 shows about 1 in. (25 mm) opening of a contraction joint due to slope movement that pulled apart the foundation at the contraction joints. Contraction joints placed at the living areas of slab-on-grade

FIGURE 12.15   About 1 in. (25 mm) opening of a contraction joint due to slope movement that pulled apart the foundation at the contraction joints. A crack monitoring device has been placed across the contraction joint.

FIGURE 12.16   The opening of a contraction joint (lower arrow) has resulted in the cracking of shower tiles (upper arrow) that span the contraction joint.

foundations can also result in damage to brittle floor coverings placed over the contraction joints. For example, Fig. 12.16 shows the opening of a contraction joint (lower arrow) that has resulted in the cracking of shower tiles (upper arrow) that span the contraction joint.

As previously mentioned, concrete having a low volume-to-surface ratio will produce more drying shrinkage. This is important because excessive shrinkage is often observed for slab-on-grade foundations that are too thin.

12.6.3   Plastic Shrinkage

Plastic shrinkage cracking is defined as the cracking of the surface of concrete that develops in freshly mixed concrete soon after it has been placed and while it is still being finished. Because the concrete is still plastic, this type of shrinkage cracking is termed *plastic shrinkage* to distinguish it from drying shrinkage described in the previous sections. Plastic shrinkage cracking is most often associated with climate conditions, such as high temperatures, low humidity, and high winds that result in rapid evaporation of moisture from the concrete surface. Especially in the desert areas of the southwestern United States, the water may be quickly drawn from the concrete resulting in plastic shrinkage.

Plastic shrinkage cracks develop because of the tensile stresses generated in the concrete due to the evaporating water. The concrete cracks often form in distinct patterns of short and irregular cracks. The lengths are generally from a few inches to a few feet in length with spacing similar to crack lengths. For a discussion of the measures to prevent plastic shrinkage cracking, see *Design and Control of Concrete Mixtures* (PCA, 1994).

12.7   MOISTURE MIGRATION THROUGH SLAB-ON-GRADE FOUNDATIONS

12.7.1   Introduction

This last section of this chapter will discuss the moisture migration through concrete slab-on-grade foundations. Moisture migration into buildings is one of the major challenges faced by engineers, architects, and contractors. Many times, the project architect provides waterproofing recommendations, but in other instances, the geotechnical engineer may need to provide recommendations.

For example, in southern California, it is common practice for the geotechnical engineer to provide the moisture barrier requirements located below the slab-on-grade foundation.

Moisture can migrate into the structure through the foundation, exterior walls, and through the roof. Four ways that moisture can penetrate a concrete slab-on-grade foundation are as follows (WFCA, 1984):

1. *Water vapor.*   Water vapor acts in accordance with the physical laws of gases, where water vapor will travel from one area to another whenever there is a difference in vapor pressure between the two areas.

2. *Hydrostatic pressure.*   Hydrostatic pressure is the buildup of water pressure beneath the foundation that can force large quantities of water through cracks or joints.

3. *Leakage.*   Leakage refers to water traveling from a higher to a lower elevation due solely to the force of gravity, and such water can surround or flood the area below the slab.

4. *Capillary action.*   Capillary action is different from leakage in that water can travel from a lower to a higher elevation. The controlling factor in the height of capillary rise in soils is the pore size (Holtz and Kovacs, 1981). Open-graded gravel has large pore spaces and hence very low capillary rise. This is why open-graded gravel is frequently placed below the floor slab to act as a capillary break (Butt, 1992).

Moisture that travels through the concrete slab-on-grade foundation can damage floor coverings such as carpet, hardwood, and vinyl. When a concrete slab-on-grade foundation has floor coverings, the moisture can collect at the top of the slab, where it weakens the floor-covering adhesive. Hardwood floors can be severely affected by moisture migration through concrete slab-on-grade foundations because they can warp or swell from the moisture. Moisture that penetrates the slab-on-grade foundation can also cause musty odors or mildew growth in the space above the slab. Some people are allergic to mold and mildew spores and they can develop health problems from the continuous exposure to such allergens. In most cases, the moisture that passes through the concrete slab-on-grade contains dissolved salts. As the water evaporates at the slab surface, the salts form a white crystalline deposit, commonly called *efflorescence*. The salt can build up underneath the floor covering, where it attacks the adhesive as well as the flooring material itself.

As an example of damage due to moisture migration through a concrete slab-on-grade foundation, Fig. 12.17 shows damage to wood flooring installed on top of the concrete slab. Only 6 months

FIGURE 12.17   Damage to wood flooring. The arrow points to a moisture stain and the asterisk indicates upward warping of the wood floor.

after completion of the house, the wood flooring developed surface moisture stains and warped upward in some places as much as 0.5 ft (15 cm). Most of the moisture stains developed at the joints where the wood planking had been spliced together. The joints would be the locations where most of the moisture penetrates the wood flooring. In Fig. 12.17, the arrow points to one of the moisture stains. The asterisk in Fig. 12.17 shows the location of the upward warping of the wood floor.

Oliver (1988) believes that it is the shrinkage cracks in concrete that provide the major pathways for rising dampness. Figures 12.18 and 12.19 show examples of damage due to moisture intrusion through shrinkage cracks in slab-on-grade foundations. Sealing of slab cracks may be necessary in situations of rising dampness affecting sensitive floor coverings. In other cases, the concrete may be porous or there may be an inadequate moisture barrier beneath the foundation, in which case a floor seal may be required in order to reduce the migration of moisture through the concrete slab-on-grade foundation (Floor Seal Technology, 1998).

12.7.2 Design and Construction Details

As previously mentioned, the geotechnical engineer often provides the design and construction specifications that are used to prevent both water vapor and capillary rise through floor slabs. An example of moisture barrier recommendations below the slab-on-grade foundation is as follows (WFCA, 1984):

> Over the subgrade, place 4 to 8 in. (10 to 20 cm) of washed and graded gravel. Place a leveling bed of 1 to 2 in. (2.5 to 5 cm) of sand over the gravel to prevent moisture barrier puncture. Place a moisture barrier over the sand leveling bed and seal the joints to prevent moisture penetration. Place a 2 in. (5 cm) sand layer over the moisture barrier.

The gravel layer should consist of an open graded gravel. This means that the gravel should not contain any fines and that all of the soil particles are retained on the No. 4 (4.75-mm) sieve. This will provide for large void spaces between the gravel particles that will help prevent capillary rise of water through the gravel (Day, 1992c). In addition to a gravel layer, the installation of a moisture barrier (such as visqueen) will further reduce the moisture migration through concrete (Brewer, 1965).

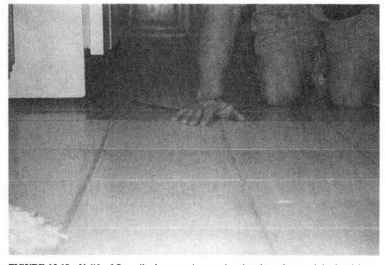

FIGURE 12.18   Uplift of floor tile due to moisture migration through a crack in the slab-on-grade foundation. The floor tile absorbed moisture and expanded, resulting in the uplift shown in the photograph.

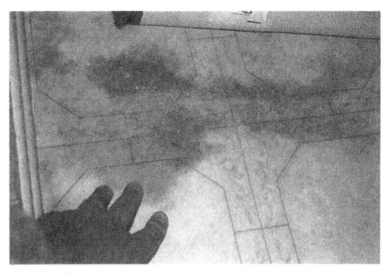

FIGURE 12.19   Linoleum discoloration due to moisture migration through a crack in the slab-on-grade foundation. The moisture accumulated beneath the linoleum at the location of the slab crack and caused the darkening of the linoleum as shown in the photograph.

In some areas, there may be local building requirements for the construction of moisture barriers. For example, Fig. 12.20 indicates the County of San Diego requirements for below slab moisture barriers when constructing on clays. Note in this figure that 4 in. (10 cm) of open-graded gravel or rock is required below the concrete slab-on-grade foundation. A 6-mil. visqueen moisture barrier is

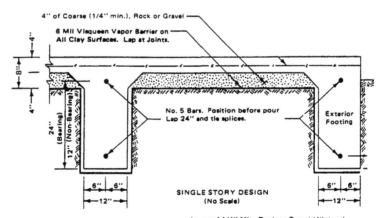

FIGURE 12.20   Moisture barrier specifications below a concrete slab-on-grade foundation according to the County of San Diego (1983).

FIGURE 12.21   Poor surface drainage at a condominium complex.

required below the open-graded gravel. These recommendations are similar to the example listed earlier, except that the sand layers have been omitted. The purpose of the sand layer is to prevent puncture of the visqueen moisture barrier. When using specifications similar to Fig. 12.20, it is best to use rounded gravel or add a geotextile so that the visqueen is not punctured during placement and densification of the gravel.

Inadequate surface drainage can be an important factor in moisture migration through concrete slab-on-grade foundations. For example, water ponding adjacent to a foundation can weep in through the perimeter footing and/or seep underneath the perimeter footing and up through the slab. The infiltration of ponding water can also raise the groundwater table, which could subject the foundation to hydrostatic water pressures. Figure 12.21 shows an example of poor surface drainage and ponding of water at a condominium complex. In Fig. 12.21, the condominium is visible along the right side of the photograph. In order to prevent such ponding of water at the site, it is important to have proper drainage gradients around the structure. In addition, the surface runoff must be transferred to suitable disposal systems, such as storm drain lines.

Using a below slab moisture barrier system as described previously and providing positive surface drainage around the foundation will not be enough to resist a naturally occurring high groundwater table or artesian groundwater condition. In these cases, a more extensive waterproofing system will need to be installed. One common approach is to provide a below foundation waterproofing system and a sump equipped with a pump to catch and dispose of any water that seeps through the waterproofing system. A *sump* is defined as a small pit excavated in the ground or through the basement floor to serve as a collection basin for groundwater. A sump pump is used to periodically drain the pit when it fills with water. When the water level reaches a certain level in the collector box, the submerged pump will be activated, and the collection box will be emptied of water.

CHAPTER 13
GEOTECHNICAL EARTHQUAKE ENGINEERING FOR SOILS

13.1 INTRODUCTION

Geotechnical earthquake engineering can be defined as that subspecialty within the field of geotechnical engineering that deals with the design and construction of projects in order to resist the effects of earthquakes. Geotechnical earthquake engineering requires an understanding of basic geotechnical principles as well as an understanding of geology, seismology, and earthquake engineering. In a broad sense, *seismology* can be defined as the study of earthquakes. This would include the internal behavior of the earth and the nature of seismic waves generated by the earthquake.

For geotechnical earthquake engineering of soils and foundations, the types of activities that may need to be performed by the geotechnical engineer include the following:

- Subsurface exploration for geotechnical earthquake engineering, i.e., the screening investigation and the quantitative evaluation (see Sec. 2.8)
- Total stress and effective stress analyses for geotechnical earthquake engineering (see Sec. 4.6.6)
- Bearing capacity analyses for geotechnical earthquake engineering (see Sec. 6.5)
- For the design earthquake, determining the peak ground acceleration and magnitude of the earthquake (Sec. 13.3)
- Investigating the possibility of liquefaction at the site (Sec. 13.4)
- Investigating the stability of slopes for the additional forces imposed during the design earthquake (Sec. 13.5)
- Calculating the settlement of the foundation caused by the design earthquake (Sec. 14.3)
- Performing calculations to assess the effect of the earthquake on retaining walls (Sec. 14.4)
- Considering foundation alternatives to mitigate the effects of earthquakes (Sec. 14.5)

A list of terms and definitions as applied to geotechnical earthquake engineering is presented in App. A, Glossary 5.

In terms of the accuracy of the calculations used to determine the earthquake induced soil movement, Tokimatsu and Seed (1984) conclude:

> It should be recognized that, even under static loading conditions, the error associated with the estimation of settlement is on the order of ±25 to 50%. It is therefore reasonable to expect less accuracy in predicting settlements for the more complicated conditions associated with earthquake loading.... In the application of the methods, it is essential to check that the final results are reasonable in the light of available experience.

13.2 BASIC EARTHQUAKE PRINCIPLES

13.2.1 Plate Tectonics

The 1960s theory of plate tectonics has helped immeasurably in the understanding of earthquakes. According to the plate tectonic theory, the earth's surface contains tectonic plates, also known as *lithosphere plates*, with each plate consisting of the crust and the more rigid part of the upper mantle. Figure 13.1 shows the locations of the major tectonic plates and the arrows indicate the relative directions of plate movement. Figure 13.2 shows the locations of the epicenters of major earthquakes. In comparing Figs. 13.1 and 13.2, it is evident that the locations of the great majority of earthquakes correspond to the boundaries between plates. Depending on the direction of movement of the plates, there are three types of plate boundaries: divergent boundary, convergent boundary, and transform boundary.

Divergent Boundary. This occurs when the relative movement of two plates is away from each other. The upwelling of hot magma that cools and solidifies as the tectonic plates move away from each other forms spreading ridges. An example of a spreading ridge is the mid-Atlantic ridge (see Fig. 13.1). Earthquakes on spreading ridges are limited to the ridge crest, where new crust is being formed. These earthquakes tend to be relatively small and occur at shallow depths (Yeats et al., 1997).

When a divergent boundary occurs within a continent, it is called *rifting*. Molten rock from the asthenosphere rises to the surface, forcing the continent to break and separate. With enough movement, the rift valley may fill with water and eventually form a mid-ocean ridge.

Convergent Boundary. This occurs when the relative movement of the two plates is toward each other. The amount of crust on the earth's surface remains relatively constant and therefore when a divergent boundary occurs in one area, a convergent boundary must occur in another area. There are three types of convergent boundaries: oceanic-continental subduction zone, oceanic-oceanic subduction zone, and continent-continent collision zone.

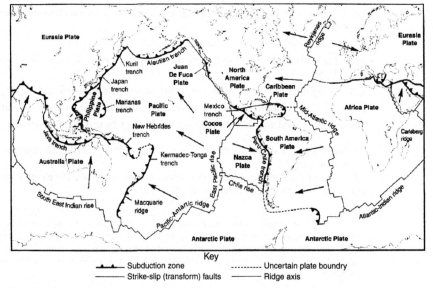

FIGURE 13.1 The major tectonic plates, mid-oceanic ridges, trenches, and transform faults of the earth. Arrows indicate directions of plate movement. (*Developed by Fowler, 1990, reproduced from Kramer, 1996*).

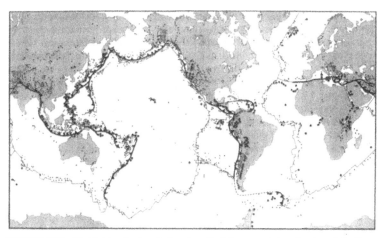

FIGURE 13.2   Worldwide seismic activity, where the dots represent the epicenters of significant earthquakes. In comparing Figs. 13.1 and 13.2, the great majority of the earthquakes are located at the boundaries between plates. (Developed by Bolt, 1988, reproduced from Kramer, 1996).

Oceanic-Continental Subduction Zone.   In this case, one tectonic plate is forced beneath the other. For an oceanic subduction zone, it is usually the denser oceanic plate that will subduct beneath the less dense continental plate. A deep-sea trench forms at the location where one plate is forced beneath the other. Once the subducting oceanic crust reaches a depth of about 60 mi (100 km), the crust begins to melt and some of this magma is pushed to the surface. An example of an oceanic-continental subduction zone is at the Peru-Chile trench (see Fig. 13.1).

Oceanic-Oceanic Subduction Zone.   An oceanic-oceanic subduction zone often results in the formation of an island arc system. As the subducting oceanic crust meets with the asthenosphere, the newly created magma rises to the surface and forms volcanoes. The volcanoes may eventually grow tall enough to form a chain of islands. An example of an oceanic-oceanic subjection zone is the Aleutian Island chain (see Fig. 13.1).

The earthquakes related to subduction zones have been attributed to four different conditions (Christensen and Ruff, 1988):

1. Shallow interplate thrust events caused by failure of the interface between the down-going plate and the overriding plate

2. Shallow earthquakes caused by deformation within the upper plate

3. Earthquakes at depths from 25 to 430 mi (40 to 700 km) within the down-going plate

4. Earthquakes that are seaward of the trench, caused mainly by the flexing of the down-going plate, but also by compression of the plate

In terms of the seismic energy released at subduction zones, it has been determined that the largest earthquakes and the majority of the total seismic energy released during the past century have occurred as shallow earthquakes at subduction-zone plate boundaries (Pacheco and Sykes, 1992).

Continent-Continent Collision Zone.   The third type of convergent boundary is the continent-continent collision zone. This condition occurs when two continental plates collide into each other, causing the two masses to squeeze, fold, deform, and thrust upward. According to Yeats et al. (1997), the Himalaya Mountains mark the largest active continent-continent collision zone on earth. They indicate that the collision between the Indian subcontinent and the Eurasia plate began in early Tertiary time, when the northern edge of the Indian plate was thrust back onto itself, with the subsequent uplifting of the Himalaya Mountains.

Transform Boundary.   A transform boundary, or transform fault, involves the plates sliding past each other, without the construction or destruction of the earth's crust. When the relative movement of two plates is parallel to each other, strike-slip fault zones can develop at the plate boundaries. *Strike-slip faults* are defined as faults on which the movement is parallel to the strike of the fault, or in other words, there is horizontal movement that is parallel to the direction of the fault.

California has numerous strike-slip faults, with the most prominent being the San Andreas Fault. Figure 13.3 presents an example of the horizontal movement along this fault (1906 San Francisco earthquake). Since a boundary between two plates occurs in California, it has numerous earthquakes and the highest seismic hazard in the continental United States.

The theory of plate tectonics is summarized in Table 13.1. This theory helps to explain the location and nature of earthquakes. Once a fault has formed at a plate boundary, the shearing resistance for continued movement of the fault is less than the shearing resistance required to fracture new intact rock. Thus faults at the plate boundaries that have generated earthquakes in the recent past are likely to produce earthquakes in the future. This principle is the basis for the development of seismic hazard maps.

The theory of plate tectonics also helps explain such geologic features as the islands of Hawaii. The islands are essentially large volcanoes that have risen from the ocean floor. The volcanoes are believed to be the result of a thermal plume or *hotspot* within the mantle, which forces magna to the surface and creates the islands. The thermal plume is believed to be relatively stationary with respect to the center of the earth, but the Pacific plate is moving to the northwest. Thus the islands of the

FIGURE 13.3   San Francisco earthquake, 1906. The fence has been offset 8.5 ft (2.6 m) by the San Andreas Fault displacement. The location is 0.5 mi (0.8 km) northwest of Woodville, Marin County, California. (*Photograph courtesy of USGS*).

TABLE 13.1   Summary of Plate Tectonics Theory

Plate boundary type	Type of plate movement	Categories	Types of earthquakes	Examples
Divergent boundary	Relative movement of the two plates is away from each other.	Sea-floor spreading ridge	Earthquakes on spreading ridges are limited to the ridge crest, where new crust is being formed. These earthquakes tend to be relatively small and occur at shallow depths.	Mid-Atlantic Ridge
		Continental rift valley	Earthquakes generated along normal faults in the rift valley.	East-African Rift
Convergent boundary	Relative movement of the two plates is toward each other.	Oceanic-continental subduction zone and oceanic-oceanic subduction zone	Shallow interplate thrust events caused by failure of the interface between the down-going plate and the overriding plate. Shallow earthquakes caused by deformation within the upper plate. Earthquakes at depths from 25 to 430 mi (40 to 700 km) within the down-going plate. Earthquakes that are seaward of the trench, caused mainly by the flexing of the down-going plate, but also by compression of the plate.	Peru-Chile Trench (Oceanic-continental) Aleutian Island chain (Oceanic-oceanic)
		Continent-continent collision zone	Earthquakes generated at the collision zone, such as at reverse faults and thrust faults.	Himalaya Mountains
Transform boundary	Plates sliding past each other.	Strike-slip fault zones	Earthquakes often generated on strike-slip faults.	San Andreas Fault

Hawaiian chain to the northwest are progressively older and contain dormant volcanoes that have weathered away. Yeats et al. (1997) use an analogy of the former locations of the Pacific plate with respect to the plume as being much like a piece of paper passed over the flame of a stationary candle, which shows a linear pattern of scorch marks.

13.2.2   Fault Rupture

Most earthquakes will not create ground surface fault rupture. For example, there is typically an absence of surface rupture for small earthquakes, earthquakes generated at great depths at subduction zones, and earthquakes generated on blind faults. Krinitzsky et al. (1993) state that fault ruptures commonly occur in the deep subsurface with no ground breakage at the surface. They further state that such behavior is widespread, accounting for all earthquakes in the central and eastern United States.

On the other hand, large earthquakes at transform boundaries will usually be accompanied by ground surface fault rupture on strike slip faults. An example of ground surface fault rupture of the San Andreas fault is shown in Fig. 13.3. Another example of ground surface rupture is shown in Fig. 13.4.

Fault displacement is defined as the relative movement of the two sides of a fault, measured in a specific direction (Bonilla, 1970). Examples of very large surface fault rupture are the 35 ft (11 m) of vertical displacement in the Assam earthquake of 1897 (Oldham, 1899) and the 29 ft (9 m) of horizontal movement during the Gobi-Altai earthquake of 1957 (Florensov and Solonenko, 1965). The length of

FIGURE 13.4   Surface fault rupture associated with the Kocaeli (Turkey) earthquake on August 17, 1999 (*Photograph by Tom Fumal, USGS.*)

the fault rupture can be quite significant. For example, the estimated length of surface faulting in the 1964 Alaskan earthquake varied from 600 to 720 km (Savage and Hastie, 1966; Housner, 1970).

Surface fault rupture associated with earthquakes is important because it has caused severe damage to buildings, bridges, dams, tunnels, canals, and underground utilities (Lawson et al., 1908; Ambraseys, 1960; Duke, 1960; California Department of Water Resources, 1967; Bonilla, 1970; Steinbrugge, 1970). For example, there were spectacular examples of surface fault rupture associated with the Chi-chi (Taiwan) earthquake on September 21, 1999. An example is as follows:

Collapsed Bridge North of Fengyuen.   Figures 13.5 to 13.7 show three photographs of the collapse of a bridge just north of Fengyuen, Taiwan. The bridge generally runs in a north-south direction, with the collapse occurring at the southern portion of the bridge. The bridge was originally straight and level. The surface fault rupture passes underneath the bridge and apparently caused the bridge to shorten such that the southern spans were shoved off of their supports. In addition, the fault rupture developed beneath one of the piers, resulting in its collapse. Note in Fig. 13.7 that there is a waterfall to the east of the bridge. The fault rupture that runs underneath the bridge caused this displacement and development of the waterfall. The waterfall is estimated to have a height of about 30 to 33 ft (9 to 10 m).

Figure 13.8 shows a close-up view of the new waterfall created by the surface fault rupture. This photograph shows the area to the east of the bridge. Apparently the dark rocks located in front of the waterfall are from the crumpling of the leading edge of the thrust fault movement.

As indicated by the photographs in this section, structures and foundations are simply unable to resist the shear movement associated with surface faulting. One design approach is to simply restrict

FIGURE 13.5   Collapsed bridge north of Fengyuen caused by surface fault rupture associated with the Chi-chi (Taiwan) earthquake on September 21, 1999. (*Photograph from the USGS Earthquake Hazards Program, NEIC, Denver.*)

FIGURE 13.6   Another view of the collapsed bridge north of Fengyuen caused by surface fault rupture associated with the Chi-chi (Taiwan) earthquake on September 21, 1999. (*Photograph from the USGS Earthquake Hazards Program, NEIC, Denver.*)

FIGURE 13.7   Another view of the collapsed bridge north of Fengyuen caused by surface fault rupture associated with the Chi-chi (Taiwan) earthquake on September 21, 1999. Note that the surface faulting has created the waterfall on the right side of the bridge. (*Photograph from the USGS Earthquake Hazards Program, NEIC, Denver.*)

construction in the active fault shear zone. Often the local building code will restrict the construction in fault zones. For example, the *Southern Nevada Building Code Amendments* (1997) state the following:

Minimum Distances to Ground Faulting:

1. No portion of the foundation system of any habitable space shall be located less than five feet to a fault.
2. When the geotechnical report establishes that neither a fault nor a fault zone exists on the project, no fault zone set back requirements shall be imposed.
3. If through exploration, the fault location is defined, the fault and/or the no build zone shall be clearly shown to scale on grading and plot plan(s).
4. When the fault location is not fully defined by explorations but a no build zone of potential fault impact is established by the geotechnical report, no portion of the foundation system of any habitable space shall be constructed to allow any portion of the foundation system to be located within that zone. The no build zone shall be clearly shown to scale on grading and plot plan(s).
5. For single lot, single-family residences, the fault location may be approximated by historical research as indicated in the geotechnical report. A no build zone of at least 50 feet each side of the historically approximated fault edge shall be established. The no build zone shall be clearly shown to scale on grading and plot plan(s).

In many cases, structures will have to be constructed in the surface rupture zone. For example, transportation routes may need to cross the active shear fault zones. One approach is to construct the roads such that they cross the fault in a perpendicular direction. In addition, it is desirable to cross the surface rupture zone at a level ground location so that bridges or overpasses need not be constructed in the surface rupture zone.

Pipelines also must often pass through surface rupture zones. Similar to pavements, it is best to cross the fault rupture zone in a perpendicular direction and at a level ground site. There are many different types of design alternatives for pipelines that cross the rupture zone. For example, a large tunnel can be constructed with the pipeline suspended within the center of the tunnel. The amount of open space between the tunnel wall and the pipeline would be based on the expected amount of surface rupture. Another option is to install automatic shut-off valves that will close the pipeline

FIGURE 13.8 Close-up view of the waterfall shown in Fig. 13.7. The waterfall was created by the surface fault rupture associated with the Chi-chi (Taiwan) earthquake on September 21, 1999 and has an estimated height of 9 to 10 m. (*Photograph from the USGS Earthquake Hazards Program, NEIC, Denver.*)

if there is a drop in pressure. With additional segments of the pipeline stored nearby, the pipe can then be quickly repaired.

13.2.3 Regional Subsidence

In addition to the surface fault rupture, another tectonic effect associated with the earthquake could be uplifting or regional subsidence. For example, at continent-continent collision zones, the plates collide into each other, causing the ground surface to squeeze, fold, deform, and thrust upward.

Besides uplifting, there could also be regional subsidence associated with the earthquake. There was extensive damage due to regional subsidence during the August 17, 1999 Kocaeli earthquake in Turkey along the North Anatolian Fault, which is predominantly a strike-slip fault due to the Anatolian Plate shearing past the Eurasian Plate. But to the west of Izmit, there is a localized extension zone where the crust is being stretched apart and has formed the Gulf of Izmit. An extension zone is similar to a rift valley and it occurs when a portion of the earth's crust is stretched apart and a graben develops. A *graben* is defined as a crustal block that has dropped down relative to adjacent rocks along bounding faults. The down-dropping block is usually much longer than its width, creating a long and narrow valley.

The city of Golcuk is located on the south shore of the Gulf of Izmit. It has been reported that during the earthquake, 2 mi (3 km) of land along the Gulf of Izmit subsided at least 10 ft (3 m).

Water from the Gulf of Izmit flooded inland and several thousand people drowned or were crushed as buildings collapsed in Golcuk.

13.2.4   Earthquake Magnitude

There are two basic ways to measure the strength of an earthquake: (1) based on the earthquake magnitude, and (2) based on the intensity of damage. Magnitude measures the amount of energy released from the earthquake and intensity is based on the damage to buildings and reactions of people. This section will discuss earthquake magnitude and the next section will discuss the intensity of the earthquake.

There are many different earthquake magnitude scales used by seismologists. This section will discuss two of the more commonly used magnitude scales.

Local Magnitude Scale (M_L).   In 1935, Professor Charles Richter, from the California Institute of Technology, developed an earthquake magnitude scale for shallow and local earthquakes in southern California. This magnitude scale has often been referred to as the *Richter magnitude scale*. Because this magnitude scale was developed for shallow and local earthquakes, it is also known as the *local magnitude scale M_L*. This magnitude scale is the best known and most commonly used magnitude scale. The magnitude is calculated as follows (Richter 1935, 1958):

$$M_L = \log_{10}A - \log_{10}A_o = \log_{10}(A/A_o) \qquad (13.1)$$

where M_L = local magnitude (also often referred to as the Richter magnitude scale)
$\quad A_o$ = 0.001 mm. The zero of the local magnitude scale was arbitrarily fixed as an amplitude
$\qquad$ of 0.001 mm, which corresponded to the smallest earthquakes then being recorded.
$\quad A$ = maximum trace amplitude (mm)

The *maximum trace amplitude* is defined as that amplitude recorded by a standard Wood-Anderson seismograph that has a natural period of 0.8 sec, a damping factor of 80 percent, and a static magnification of 2800. The maximum trace amplitude must be that amplitude that would be recorded if a Wood-Anderson seismograph was located on firm ground at a distance of exactly 100 km (62 mi) from the epicenter of the earthquake. Charts and tables are available to adjust the maximum trace amplitude for the usual case where the seismograph is not located exactly 100 km (62 mi) from the epicenter.

As indicated above, Richter (1935) designed the magnitude scale so that a magnitude of 0 corresponds to approximately the smallest earthquakes then being recorded. There is no upper limit to the Richter magnitude scale, although earthquakes over M_L of 8 are rare. Often the data from Wood-Anderson seismographs located at different distances from the epicenter provide different values of the Richter magnitude. This is to be expected because of the different soil and rock conditions that the seismic waves travel through and because the fault rupture will not release the same amount of energy in all directions.

Since the Richter magnitude scale is based on the logarithm of the maximum trace amplitude, there is a 10 times increase in the amplitude for an increase in one unit of magnitude. In terms of the energy released during the earthquake, Yeats et al. (1997) indicate that the increase in energy for an increase of one unit of magnitude is roughly 30-fold and is different for different magnitude intervals.

For the case of small earthquakes (i.e., $M_L < 6$), the center of energy release and the point where the fault rupture begins is not far apart. But in the case of large earthquakes, these points may be very far apart. For example, the Chilean earthquake of 1960 had a fault rupture length of about 600 mi (970 km) and the epicenter was at the northern end of the ruptured zone that was about 300 mi (480 km) from the center of the energy release (Housner, 1963, 1970). This increased release of energy over a longer rupture distance resulted in both a higher peak ground acceleration a_{max} and a longer duration of shaking. For example, Table 13.2 presents approximate correlations between the local magnitude M_L and the peak ground acceleration a_{max}, duration of shaking, and modified Mercalli intensity level (discussed in Sec. 13.2.5) near the vicinity of the fault rupture. At distances further away from the epicenter or location of fault rupture, the intensity will decrease but the duration of ground shaking will increase.

TABLE 13.2 Approximate Correlations between the Local Magnitude M_L and the Peak Ground Acceleration a_{max}, Duration of Shaking, and Modified Mercalli Level of Damage near the Vicinity of the Fault Rupture

Local magnitude M_L	Typical peak ground acceleration a_{max} near the vicinity of the fault rupture	Typical duration of ground shaking near the vicinity of the fault rupture	Modified Mercalli intensity level near the vicinity of the fault rupture (see Table 13.3)
≤2	—	—	I – II
3	—	—	III
4	—	—	IV – V
5	0.09g	2 sec	VI – VII
6	0.22g	12 sec	VII – VIII
7	0.37g	24 sec	IX – X
≥8	≥0.50g	≥34 sec	XI – XII

Sources: Yeats et al. (1997), Gere and Shah (1984), and Housner (1970).

Moment Magnitude Scale (M_w).   The moment magnitude scale has become the more commonly used method for determining the magnitude of large earthquakes. This is because it tends to take into account the entire size of the earthquake. The first step in the calculation of the moment magnitude is to calculate the seismic moment M_o. The seismic moment can be determined from a seismogram using very long period waves for which even a fault with a very large rupture area appears as a point source (Yeats et al., 1997). The seismic moment can also be estimated from the fault displacement, as follows (Idriss, 1985):

$$M_o = \mu A_f D \tag{13.2}$$

where M_o = seismic moment (N·m)

  μ = shear modulus of the material along the fault plane (N/m²). The shear modulus is often assumed to be 3×10^{10} N/m² for surface crust and 7×10^{12} N/m² for the mantle.

  A_f = area of fault plane undergoing slip (m²). This can be estimated as the length of surface rupture times the depth of the aftershocks.

  D = average displacement of the ruptured segment of the fault (m). Determining the seismic moment works best for strike-slip faults where the lateral displacement on one side of the fault relative to the other side can be readily measured.

In essence, to determine the seismic moment requires taking the entire area of the fault rupture surface (A_f) times the shear modulus (μ) in order to calculate the seismic force (newtons). This force is converted to a moment by multiplying the seismic force (newtons) by the average slip (meters), in order to calculate the seismic moment (N·m).

Kanamori (1977) and Hanks and Kanamori (1979) introduced the moment magnitude (M_w) scale, in which the magnitude is calculated from the seismic moment by using the following equation:

$$M_w = -6.0 + 0.67 \log_{10} M_o \tag{13.3}$$

where M_w = moment magnitude of the earthquake

  M_o = seismic moment of the earthquake (N·m). The seismic moment is calculated from Eq. 13.2.

Comparison of Magnitude Scales.   Figure 13.9 shows the approximate relationships between several different earthquake magnitude scales. When viewing the data shown in Fig. 13.9, it would appear that there is an exact relationship between the moment magnitude M_w and the other various magnitude scales. But in comparing Eqs. 13.1 and 13.3, it is evident that these two equations cannot be equated. Therefore, there is not an exact and unique relationship between the maximum trace amplitude from a standard Wood-Anderson seismograph (Eq. 13.1) and the seismic moment (Eq. 13.3).

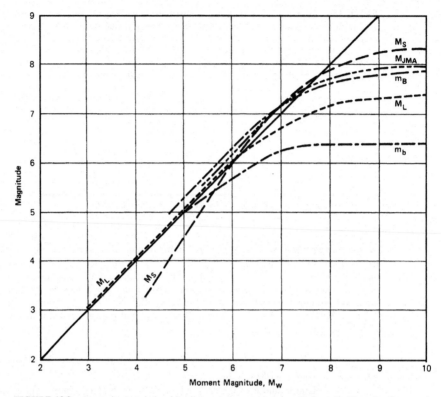

FIGURE 13.9 Approximate relationships between the moment magnitude scale M_w and other magnitude scales. Shown are the short-period body-wave magnitude scale m_b, the local magnitude scale M_L, the long-period body-wave magnitude scale m_B, the Japan Meteorological Agency magnitude scale M_{JMA} and the surface-wave magnitude scale M_S. (*Developed by Heaton et al. 1982, reproduced from Idriss 1985.*)

The lines drawn in Fig. 13.9 should only be considered as approximate relationships, representing a possible wide range in values.

Recognizing the limitations of Fig. 13.9, it could still be concluded that the local magnitude M_L and moment magnitude M_w scales are reasonably close to each other below a value of about 7. At high magnitude values, the moment magnitude M_w tends to significantly deviate from the other magnitude scales.

13.2.5 Earthquake Intensity

The intensity of an earthquake is based on the observations of damaged structures and the presence of secondary effects, such as earthquake induced landslides, liquefaction, and ground cracking. The intensity of an earthquake is also based on the degree to which individuals felt the earthquake, which is determined through interviews.

The intensity of the earthquake may be easy to determine in an urban area where there is a considerable amount of damage, but could be very difficult to evaluate in rural areas. The most commonly used scale for the determination of the intensity of an earthquake is the modified Mercalli intensity scale (Table 13.3), where the intensity ranges from an earthquake that is not felt (I) up to an earthquake that results in total destruction (XII). In general, the larger the magnitude of the earthquake, the greater the area affected by the earthquake and the higher the intensity level.

TABLE 13.3 Modified Mercalli Intensity Scale

Intensity level	Reaction of observers and types of damage
I	Reactions: Not felt except by a very few people under especially favorable circumstances.
II	Reactions: Felt only by a few persons at rest, especially on upper floors of buildings. Many people do not recognize it as an earthquake. Damage: No damage. Delicately suspended objects may swing.
III	Reactions: Felt quite noticeably indoors, especially on upper floors of buildings. The vibration is like the passing of a truck, and the duration of the earthquake may be estimated. However, many people do not recognize it as an earthquake. Damage: No damage. Standing motor cars may rock slightly.
IV	Reactions: During the day, felt indoors by many, outdoors by a few. At night, some people are awakened. The sensation is like a heavy truck striking the building. Damage: Dishes, windows, and doors are disturbed. Walls make a creaking sound. Standing motor cars rock noticeably.
V	Reactions: Felt by nearly everyone, many awakened. Damage: Some dishes, window, etc., broken. A few instances of cracked plaster and unstable objects overturned. Disturbances of trees, poles, and other tall objects sometimes noticed. Pendulum clocks may stop.
VI	Reactions: Felt by everyone. Many people are frightened and run outdoors. Damage: There is slight structural damage. Some heavy furniture is moved, and there are a few instances of fallen plaster or damaged chimneys.
VII	Reactions: Everyone runs outdoors. Noticed by persons driving motor cars. Damage: Negligible damage in buildings of good design and construction, slight to moderate damage in well-built ordinary structures, and considerable damage in poorly built or badly designed structures. Some chimneys are broken.
VIII	Reactions: Persons driving motor cars are disturbed. Damage: Slight damage in specially designed structures. Considerable damage in ordinary substantial buildings, with partial collapse. Great damage in poorly built structures. Panel walls are thrown out of frame structures. There is the fall of chimneys, factory stacks, columns, monuments, and walls. Heavy furniture is overturned. Sand and mud are ejected in small amounts, and there are changes in well-water levels.
IX	Damage: Considerable damage in specially designed structures. Well-designed frame structures are thrown out of plumb. There is great damage in substantial buildings with partial collapse. Buildings are shifted off their foundations. The ground is conspicuously cracked, and underground pipes are broken.
X	Damage: Some well-built wooden structures are destroyed. Most masonry and frame structures are destroyed, including the foundations. The ground is badly cracked. There are bent train rails, a considerable number of landslides at river banks and steep slopes, shifted sand and mud, and water is splashed over their banks.
XI	Damage: Few, if any, masonry structures remain standing. Bridges are destroyed, and train rails are greatly bent. There are broad fissures in the ground, and underground pipelines are completely out of service. There are earth slumps and land slips in soft ground.
XII	Reactions: Waves are seen on the ground surface. The lines of sight and level are distorted. Damage: Total damage with practically all works of construction greatly damaged or destroyed. Objects are thrown upward into the air.

13.3 *PEAK GROUND ACCELERATION*

13.3.1 Introduction

The ground motion caused by earthquakes is generally characterized in terms of ground surface displacement, velocity, and acceleration. Geotechnical engineers traditionally use acceleration, rather than velocity or displacement, because acceleration is directly related to the dynamic forces that earthquakes induce on the soil mass. For geotechnical analyses, the measure of the cyclic ground motion is represented by the maximum horizontal acceleration at the ground surface (a_{max}). The maximum horizontal acceleration at ground surface is also known as the *peak horizontal ground acceleration*.

For most earthquakes, the horizontal acceleration is greater than the vertical acceleration, and thus the peak horizontal ground acceleration also turns out to be the peak ground acceleration (PGA).

For earthquake engineering analyses, the peak ground acceleration a_{max} is one of the most difficult parameters to determine. It represents an acceleration that will be induced sometime in the future by an earthquake. Since it is not possible to predict earthquakes, the value of the peak ground acceleration must be based on prior earthquakes and fault studies.

Often attenuation relationships are used in the determination of the peak ground acceleration. An *attenuation relationship* is defined as a mathematical relationship that is used to estimate the peak ground acceleration at a specified distance from the earthquake. Numerous attenuation relationships have been developed and they relate the peak ground acceleration to: (1) the earthquake magnitude and (2) the distance between the site and the seismic source (the causative fault). The increasingly larger pool of seismic data recorded in the world, and particularly in the western United States, has allowed researchers to develop reliable empirical attenuation equations that are used to model the ground motions generated during an earthquake (Federal Emergency Management Agency, 1994).

13.3.2   Methods Used to Determine the Peak Ground Acceleration

The engineering geologist is often the best individual to determine the peak ground acceleration a_{max} at the site based on fault, seismicity, and attenuation relationships. Some of the more commonly used methods to determine the peak ground acceleration a_{max} at a site are as follows:

Historical Earthquake.   One approach is to consider the past earthquake history of the site. For the more recent earthquakes, data from seismographs can be used to determine the peak ground acceleration. For older earthquakes, the location of the earthquake and its magnitude are based on historical accounts of damage. Computer programs, such as EQSEARCH (Blake, 2000), have been developed that incorporate past earthquake data. By inputting the location of the site, the peak ground acceleration a_{max} could be determined.

The peak horizontal ground acceleration a_{max} should never be based solely on the past history of seismic activity in an area. The reason is because the historical time frame of recorded earthquakes is usually too small. Thus the value of a_{max} determined from historical studies should be compared with the value of a_{max} as determined from the other methods described below.

Maximum Credible Earthquake.   The maximum credible earthquake (MCE) is often considered to be the largest earthquake that can reasonably be expected to occur based on known geologic and seismologic data. In essence, the maximum credible earthquake is the maximum earthquake that an active fault can produce, considering the geologic evidence of past movement and recorded seismic history of the area.

According to Kramer (1996), other terms that have been used to describe similar worst-case levels of shaking include *safe shutdown earthquake* (used in the design of nuclear power plants), *maximum capable earthquake, maximum design earthquake, contingency level earthquake, safe level earthquake, credible design earthquake*, and *contingency design earthquake*. In general, these terms are used to describe the upper-most level of earthquake forces in the design of essential facilities.

The maximum credible earthquake is determined for particular earthquakes or levels of ground shaking. As such, the analysis used to determine the maximum credible earthquake is typically referred to as a *deterministic method*.

Maximum Probable Earthquake.   There are many different definitions of the maximum probable earthquake. The maximum probable earthquake is based on a study of nearby active faults. By using attenuation relationships, the maximum probable earthquake magnitude and maximum probable peak ground acceleration can be determined.

A commonly used definition of maximum probable earthquake is the largest predicted earthquake that a fault is capable of generating within a specified time period, such as 50 or 100 years. Maximum probable earthquakes are most likely to occur within the design life of the project, and therefore, have been commonly used in assessing seismic risk (Federal Emergency Management Agency, 1994).

Another commonly used definition of a maximum probable earthquake is an earthquake that will produce a peak ground acceleration a_{max} with a 50 percent probability of exceedance in 50 years (USCOLD, 1985).

According to Kramer (1996), other terms that have been used to describe earthquakes of similar size are *operating basis earthquake, operating level earthquake, probable design earthquake*, and *strength level earthquake*.

USGS Earthquake Maps. Another method for determining the peak ground acceleration is to determine the value of a_{max} that has a certain probability of exceedance in a specific number of years. The design basis ground motion can often be determined by a site-specific hazard analysis or it may be determined from a hazard map. Various USGS maps are available that show peak ground acceleration with a 10, 5, and 2 percent probability of being exceeded in 50 years and they provide the user with the choice of the appropriate level of hazard or risk. Such an approach is termed a *probabilistic method*, with the choice of the peak ground acceleration based on the concept of acceptable risk.

Code or Other Regulatory Requirements. There may be local building code or other regulatory requirements that specify design values of peak ground acceleration.

A typical ranking of the value of peak ground acceleration a_{max} obtained from the different methods described earlier, from the least to greatest value, is as follows:

1. Maximum probable earthquake (deterministic method)
2. USGS earthquake map: 10 percent probability of exceedance in 50 years (probabilistic method)
3. USGS earthquake map: 5 percent probability of exceedance in 50 years (probabilistic method)
4. USGS earthquake map: 2 percent probability of exceedance in 50 years (probabilistic method)
5. Maximum credible earthquake (deterministic method)

13.3.3 Example of the Determination of Peak Ground Acceleration

This example deals with the proposed W.C.H. Medical Library in La Mesa, California. The different methods used to determine the peak ground acceleration for this project were as follows:

1. *Historical earthquake.* The purpose of the EQSEARCH (Blake, 2000b) computer program is to perform a historical search of earthquakes. For this computer program, the input data included the site coordinates in terms of latitude and longitude, search parameters, attenuation relationship, and other earthquake parameters. The output data indicated that the largest earthquake site acceleration from 1800 to 1999 was $a_{max} = 0.189g$.

2. *Largest maximum earthquake.* The EQFAULT computer program (Blake, 2000a) was developed to determine the largest maximum earthquake site acceleration. For this computer program, the input data also included the site coordinates in terms of latitude and longitude, search radius, attenuation relationship, and other earthquake parameters. The output data indicated that the largest maximum earthquake site acceleration (a_{max}) is $0.420g$.

3. *USGS earthquake maps.* Instead of using seismic maps, the USGS also enables the Internet user to obtain the peak ground acceleration for a specific zip-code location (see Fig. 13.10). In Fig. 13.10, PGA is the peak ground acceleration, PE is the probability of exceedance, and SA is the spectral acceleration.

For this project (i.e., the W.C.H. Medical Library), a summary of the different values of peak ground acceleration a_{max} are provided below:

$a_{max} = 0.189g$ (historical earthquakes, EQSEARCH computer program)

$a_{max} = 0.212g$ (10 percent probability of exceedance in 50 years, see Fig. 13.10)

$a_{max} = 0.280g$ (5 percent probability of exceedance in 50 years, see Fig. 13.10)

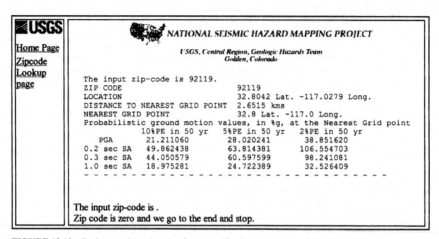

FIGURE 13.10   Peak ground acceleration for a specific zip code location. (*USGS, 1996*).

$a_{max} = 0.389g$ (2 percent probability of exceedance in 50 years, see Fig. 13.10)

$a_{max} = 0.420g$ (largest maximum earthquake, EQFAULT computer program)

There is a considerable variation in values for a_{max} as indicated above, from a low of $0.189g$ to a high of $0.420g$. The geotechnical engineer should work with the engineering geologist in selecting the most appropriate value of a_{max}. For the earlier data, based on a design life of 50 years and recognizing that the library is not an essential facility, an appropriate range of a_{max} to be used for the earthquake analyses is 0.189 to $0.212g$. Using a probabilistic approach, a value of $0.21g$ would seem appropriate.

If the project was an essential facility or had a design life in excess of 50 years, then a higher peak ground acceleration should be selected. For example, if the project required a 2 percent probability of exceedance in 50 years, then a peak ground acceleration a_{max} of about $0.39g$ should be used in the earthquake analyses. On the other hand, if the project is an essential facility that must be able to resist the largest maximum earthquake, then an appropriate value of peak ground acceleration a_{max} would be $0.42g$. As these examples illustrate, it takes considerable experience and judgment in selecting the value of a_{max} to be used for the earthquake analyses.

13.3.4   Local Soil and Geologic Conditions

For the determination of the peak ground acceleration a_{max} as discussed in the previous sections, local soil and geologic conditions were not included in the analysis. USGS recommends that the final step in the determination of a_{max} for a particular site is to adjust the value (if needed) for such factors as:

1. Directivity of ground motion that can cause stronger shaking in certain directions.
2. Basin effects, such as the conversion into surface waves and reverberation experienced by sites in an alluvial basin.
3. Soft soils, which can increase the peak ground acceleration. Often a site that may be susceptible to liquefaction will also contain thick deposits of soft soil. The local soil condition of a thick deposit of soft clay is the most common reason for increasing the peak ground acceleration a_{max}. If the site is underlain by soft ground, such as a soft and saturated clay deposit, then there could be an increased peak ground acceleration a_{max} and a longer period of vibration of the ground. The following two examples illustrate the effect of soft clay deposits.

Michoacan Earthquake in Mexico on September 19, 1985.   There was extensive damage to Mexico City caused by the September 19, 1985 Michoacan earthquake. The greatest damage in

Mexico City occurred to those buildings underlain by 125 to 164 ft (39 to 50 m) of soft clays, which are within the part of the city known as the Lake Zone (Stone et. al., 1987). Because the epicenter of the earthquake was so far from Mexico City, the peak ground acceleration a_{max} recorded in the foothills of Mexico City (rock site) was about 0.04g. However, at the Lake Zone, the peak ground accelerations a_{max} were up to five times greater than the rock site (Kramer, 1996). In addition, the characteristic site periods were estimated to be 1.9 to 2.8 sec (Stone et al., 1987).

This longer period of vibration of the ground tended to coincide with the natural period of vibration of the taller buildings in the 5- to 20-story range. The increased peak ground acceleration and the effect of resonance caused either collapse or severe damage of these taller buildings. To explain this condition of an increased peak ground acceleration and a longer period of surface vibration, an analogy is often made between the shaking of these soft clays and the shaking of a bowl of jelly.

Loma Prieta Earthquake in San Francisco Bay Area on October 17, 1989. A second example of soft ground effects is the Loma Prieta earthquake on October 17, 1989. Figure 13.11 presents the ground accelerations (East-West direction) at Yerba Buena Island and at Treasure Island (Seed et al., 1990). Both sites are about the same distance from the epicenter of the Loma Prieta earthquake. However, the Yerba Buena Island seismograph is located directly on a rock outcrop, while the Treasure Island seismograph is underlain by 45 ft (13.7 m) of loose sandy soil over 55 ft (16.8 m) of San Francisco Bay Mud (a normally consolidated silty clay). Note the significantly different ground acceleration plots for these two sites. The peak ground acceleration in the E-W direction at Yerba Buena Island was only 0.06g, while at Treasure Island, the peak ground acceleration in the East-West

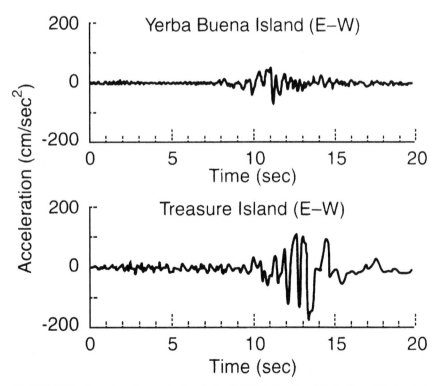

FIGURE 13.11 Ground surface acceleration in the East-West direction at Yerba Buena Island and at Treasure Island for the Loma Prieta earthquake in California on October 17, 1989. (*From Seed et al., 1990.*)

FIGURE 13.12   Overview of the collapse of the Cypress Street Viaduct caused by the Loma Prieta earthquake in California on October 17, 1989. (*From USGS.*)

direction was 0.16*g* (Kramer, 1996). Thus the soft clay site had a peak ground acceleration that was 2.7 times the hard rock site.

The amplification of the peak ground acceleration by soft clay also contributed to damage of structures throughout the San Francisco Bay Area. For example, the northern portion of the Interstate 880 highway (Cypress Street Viaduct) that collapsed was underlain by the San Francisco Bay Mud (see Figs. 13.12 to 13.14). The bay mud did not underlie the southern portion of the Interstate 880 highway and it did not collapse.

FIGURE 13.13   Close-up view of the collapse of the Cypress Street Viaduct caused by the Loma Prieta earthquake in California on October 17, 1989. (*From USGS.*)

FIGURE 13.14 Another close-up view of the collapse of the Cypress Street Viaduct caused by the Loma Prieta earthquake in California on October 17, 1989. (*From USGS.*)

As these two examples illustrate, local soft ground conditions can significantly increase the peak ground acceleration a_{max} by a factor of 3 to 5 times. The soft ground can also increase the period of ground surface shaking, leading to resonance of taller structures. The geotechnical engineer and engineering geologist will need to evaluate the possibility of increasing the peak ground acceleration a_{max} and increasing the period of ground shaking for sites that contain thick deposits of soft clay.

13.4 LIQUEFACTION

13.4.1 Introduction

Casagrande first introduced the concept of liquefaction in the late 1930s (also see Casagrande, 1975). The typical subsurface soil condition that is susceptible to liquefaction is loose sand that has been newly deposited or placed, with a groundwater table near ground surface. During an earthquake, the application of cyclic shear stresses induced by the propagation of shear waves causes the loose sand to contract, resulting in an increase in pore water pressure. Because the seismic shaking occurs so quickly, the cohesionless soil is subjected to an undrained loading (total stress analysis). The increase in pore water pressure causes an upward flow of water to the ground surface, where it emerges in the form of mud spouts or sand boils. The development of high pore water pressures due to the ground shaking and the upward flow of water may turn the sand into a liquefied condition, which has been termed *liquefaction*. For this state of liquefaction, the effective stress is zero and the individual soil particles are released from any confinement, as if the soil particles were floating in water (Ishihara, 1985).

Because liquefaction typically occurs in soil with a high groundwater table, its effects are most commonly observed in low-lying areas or adjacent rivers, lakes, bays, and oceans. When liquefaction occurs, the soil can become a liquid and thus the shear strength of the soil can be decreased to essentially zero. Without any shear strength, the liquefied soil will be unable to support the foundations for buildings and bridges. For near surface liquefaction, buried tanks will float to the surface and buildings will sink or fall over (Seed, 1970). Sand boils often develop when there has been liquefaction at a site.

After the soil has liquefied, the excess pore water pressure will start to dissipate. The length of time that the soil will remain in a liquefied state depends on two main factors: (1) the duration of the seismic shaking from the earthquake and (2) the drainage conditions of the liquefied soil. The longer and the stronger the cyclic shear stress application from the earthquake, the longer the state of liquefaction persists. Likewise, if the liquefied soil is confined by an upper and lower clay layer, then it will take longer for the excess pore water pressures to dissipate by the flow of water from the liquefied soil. After the liquefaction process is complete, the soil will be in a somewhat denser state.

This chapter will be devoted solely to *level-ground liquefaction*. Liquefaction can result in foundation settlement (Sec. 14.3) or even a bearing capacity failure of the foundation (Sec. 6.5). Liquefaction can also cause or contribute to lateral movement of slopes, which will be discussed in Sec. 13.5.

13.4.2   Factors that Govern Liquefaction

There are many factors that govern the liquefaction process for in situ soil. Based on the results of laboratory tests as well as field observations and studies, the most important factors that govern liquefaction are as follows:

1. *Earthquake intensity and duration.*   In order to have liquefaction of soil, there must be ground shaking. The character of the ground motion, such as acceleration and duration of shaking, determines the shear strains that cause the contraction of the soil particles and the development of excess pore water pressures leading to liquefaction.

   The most common cause of liquefaction is due to the seismic energy released during an earthquake. The potential for liquefaction increases as the earthquake intensity and duration of shaking increases. Those earthquakes that have the highest magnitude will both produce the largest ground acceleration and longest duration of ground shaking (see Table 13.2). Although data are sparse, there would appear to be a shaking threshold that is needed in order to produce liquefaction. These threshold values are a peak ground acceleration a_{max} of about $0.10g$ and local magnitude M_L of about 5 (National Research Council, 1985b; Ishihara, 1985). Thus, a liquefaction analysis would typically not be needed for those sites having a peak ground acceleration a_{max} less than $0.10g$ or a local magnitude M_L less than 5.

   Besides earthquakes, other conditions can cause liquefaction, such as subsurface blasting, pile driving, and vibrations from train traffic.

2. *Groundwater table.*   The condition most conducive to liquefaction is a near surface groundwater table. Unsaturated soil located above the groundwater table will not liquefy. If it can be demonstrated that the soils are currently above the groundwater table and are highly unlikely to become saturated for given foreseeable changes in the hydrologic regime, then such soils generally do not need to be evaluated for liquefaction potential.

   At sites where the groundwater table significantly fluctuates, the liquefaction potential will also fluctuate. Generally, the historic high groundwater level should be used in the liquefaction analysis unless other information indicates a higher or lower level is appropriate (Division of Mines and Geology, 1997).

   Poulos et al. (1985) state that liquefaction can also occur in very large masses of sands or silts that are dry and loose and loaded so rapidly that the escape of air from the voids is restricted. Such movement of dry and loose sands is often referred to as *running soil* or *running ground*. Although such soil may flow like liquefied soil, in this text, such soil deformation will not be termed liquefaction. It is best to consider that liquefaction only occurs for soils that are located below the groundwater table.

3. *Soil type.*   In terms of the soil types most susceptible to liquefaction, Ishihara (1985) states:

   > The hazard associated with soil liquefaction during earthquakes has been known to be encountered in deposits consisting of fine to medium sand and sands containing low-plasticity fines. Occasionally, however, cases are reported where liquefaction apparently occurred in gravely soils.

   Thus, the soil types susceptible to liquefaction are nonplastic (cohesionless) soils. An approximate listing of cohesionless soils from least to most resistant to liquefaction is clean sands, nonplastic silty sands, nonplastic silt, and gravels. There could be numerous exceptions to this sequence. For example, Ishihara (1985, 1993) describes the case of tailings derived from the mining industry that were essentially composed of ground-up rocks and were classified as rock flour. Ishihara (1985, 1993) states that the rock flour in a water-saturated state did not possess significant cohesion and behaved as if it were clean sand. These tailings were shown to exhibit as low a resistance to liquefaction as clean sand.

Based on laboratory testing and field performance, the majority of cohesive soils will not liquefy during earthquakes (Seed et al., 1983). Based on various studies, cohesive soils that are susceptible to liquefaction should meet the following two criteria (Seed and Idriss, 1982; Youd and Gilstrap, 1999; Bray et al., 2004):

a. The soil must have a liquid limit (LL) that is less than about 35 (i.e., LL < 35).

b. The water content w of the soil must be greater than about 90 percent of the liquid limit (i.e., $w > 0.9$ LL).

Hence only those clays or silts that have a low plasticity (i.e., CL or ML) and high water content will be susceptible to liquefaction. However, even if the cohesive soil does not liquefy, there can still be the possibility of a significant undrained shear strength loss due to the seismic shaking.

4. *Soil relative density (D$_r$).* Based on field studies, cohesionless soils in a loose relative density state are susceptible to liquefaction. Loose nonplastic soils will contract during the seismic shaking which will cause the development of excess pore water pressures. Upon reaching initial liquefaction, there will be a sudden and dramatic increase in shear displacement for loose sands.

For dense sands, the state of initial liquefaction does not produce large deformations because of the dilation tendency of the sand upon reversal of the cyclic shear stress. Poulos et al. (1985) state that if the in situ soil can be shown to be dilative, then it need not be evaluated because it will not be susceptible to liquefaction. In essence, dilative soils are not susceptible to liquefaction because their undrained shear strength is greater than their drained shear strength.

5. *Particle size gradation.* Uniformly graded nonplastic soils tend to form unstable particle arrangements and are more susceptible to liquefaction than well-graded soils. Well-graded soils will also have smaller size particles that fill in the void spaces between the larger particles. This tends to reduce the potential contraction of the soil, resulting in less excess pore water pressures being generated during the earthquake. Kramer (1996) states that field evidence indicates that most liquefaction failures have involved uniformly graded granular soils.

6. *Placement conditions or depositional environment.* Hydraulic fills (fill placed under water) tend to be more susceptible to liquefaction because of the loose and segregated soil structure created by the soil particles falling through water. Natural soil deposits formed in lakes, rivers, or in the ocean also tend to form a loose and segregated soil structure and are more susceptible to liquefaction. Soils that are especially susceptible to liquefaction are formed in lacustrine, alluvial, and marine depositional environments.

7. *Drainage conditions.* If the excess pore water pressure can quickly dissipate, the soil may not liquefy. Thus highly permeable gravel drains or gravel layers can reduce the liquefaction potential of adjacent soil.

8. *Confining pressures.* The greater the confining pressure, the less susceptible the soil is to liquefaction. Conditions that can create a higher confining pressure are a deeper groundwater table, soil that is located at a deeper depth below ground surface, and a surcharge pressure applied at ground surface. Case studies have shown that the possible zone of liquefaction usually extends from the ground surface to a maximum depth of about 50 ft (15 m). Deeper soils generally do not liquefy because of the higher confining pressures.

This does not mean that a liquefaction analysis should not be performed for soil that is below a depth of 50 ft (15 m). In many cases, it may be appropriate to perform a liquefaction analysis for soil that is deeper than 50 ft (15 m). An example would be sloping ground, such as a sloping berm in front of a water front structure or the sloping shell of an earth dam. In addition, a liquefaction analysis should be performed for any soil deposit that has been loosely dumped in water (i.e., the liquefaction analysis should be performed for the entire thickness of loosely dumped fill in water, even if it exceeds 50 ft in thickness). Likewise, a site where alluvium is being rapidly deposited may also need a liquefaction investigation below a depth of 50 ft (15 m).

Considerable experience and judgment are required in the determination of the proper depth to terminate a liquefaction analysis.

9. *Particle shape.* The soil particle shape can also influence liquefaction potential. For example, soils having rounded particles tend to densify more easily than angular shaped soil particles. Hence a soil containing rounded soil particles is more susceptible to liquefaction than a soil containing angular soil particles.

10. *Aging and cementation.* Newly deposited soils tend to be more susceptible to liquefaction than older deposits of soil. It has been shown that the longer a soil is subjected to a confining pressure, the greater the liquefaction resistance (Ohsaki, 1969; Seed, 1979a; Yoshimi et al., 1989). Table 13.4 presents the estimated susceptibility of sedimentary deposits to liquefaction versus the geologic age of the deposit.

The increase in liquefaction resistance with time could be due to the deformation or compression of soil particles into more stable arrangements. With time, there may also be the development of bonds due to cementation at particle contacts.

TABLE 13.4   Estimated Susceptibility of Sedimentary Deposits to Liquefaction during Strong Seismic Shaking Based on Geologic Age and Depositional Environment

Type of deposit	General distribution of cohesionless sediments in deposits	Likelihood that cohesionless sediments, when saturated, would be susceptible to liquefaction (by age of deposit)			
		<500 years	Holocene	Pleistocene	Pre-Pleistocene
(a) Continental deposits					
Alluvial fan and plain	Widespread	Moderate	Low	Low	Very low
Delta and fan-delta	Widespread	High	Moderate	Low	Very low
Dunes	Widespread	High	Moderate	Low	Very low
Marine terrace/plain	Widespread	Unknown	Low	Very low	Very low
Talus	Widespread	Low	Low	Very low	Very low
Tephra	Widespread	High	High	Unknown	Unknown
Colluvium	Variable	High	Moderate	Low	Very low
Glacial till	Variable	Low	Low	Very low	Very low
Lacustrine and playa	Variable	High	Moderate	Low	Very low
Loess	Variable	High	High	High	Unknown
Floodplain	Locally variable	High	Moderate	Low	Very low
River channel	Locally variable	Very high	High	Low	Very low
Sebka	Locally variable	High	Moderate	Low	Very low
Residual soils	Rare	Low	Low	Very low	Very low
Tuff	Rare	Low	Low	Very low	Very low
(b) Coastal zone					
Beach—large waves	Widespread	Moderate	Low	Very Low	Very low
Beach—small waves	Widespread	High	Moderate	Low	Very low
Delta	Widespread	Very high	High	Low	Very low
Estuarine	Locally variable	High	Moderate	Low	Very low
Foreshore	Locally variable	High	Moderate	Low	Very low
Lagoonal	Locally variable	High	Moderate	Low	Very low
(c) Artificial					
Compacted fill	Variable	Low	Unknown	Unknown	Unknown
Uncompacted fill	Variable	Very high	Unknown	Unknown	Unknown

Source: Data from Youd and Hoose (1978), reproduced from R. B. Seed (1991).

11. *Historical environment.*   It has also been determined that the historical environment of the soil can affect liquefaction potential. For example, older soil deposits that have already been subjected to seismic shaking have an increased liquefaction resistance as compared to a newly formed specimen of the same soil having an identical density (Finn et al., 1970; Seed et al., 1975).

   Liquefaction resistance also increases with an increase in the overconsolidation ratio (OCR) and the coefficient of lateral earth pressure at rest (k_o) (Seed and Peacock, 1971; Ishihara et al., 1978). An example would be the removal of an upper layer of soil due to erosion. Because the underlying soil has been preloaded, it will have a higher OCR and it will have a higher coefficient of lateral earth pressure at rest (k_o). Such a soil that has been preloaded will be more resistant to liquefaction than the same soil that has not been preloaded.

12. *Building load.*   The construction of a heavy building on top of a sand deposit can decrease the liquefaction resistance of the soil. For example, suppose a mat slab at ground surface supports a heavy building. The soil underlying the mat slab will be subjected to shear stresses caused by the building load. These shear stresses induced into the soil by the building load can make the soil more susceptible to liquefaction. The reason is because a smaller additional shear stress will be required from the earthquake in order to cause contraction and hence liquefaction of the soil. For level-ground liquefaction discussed in this chapter, the effect of the building load is ignored. Although building loads will not be considered in the liquefaction analysis in this chapter, the building loads must be included in all liquefaction-induced settlement, bearing capacity, and stability analyses.

In summary, the site conditions and soil type most susceptible to liquefaction are as follows:

Site conditions:

- Site that is close to the epicenter or location of fault rupture of a major earthquake
- Site that has a groundwater table close to ground surface

Soil type most susceptible to liquefaction for given site conditions: Sand that has uniform gradation, rounded soil particles, very loose or loose density state, recently deposited with no cementation between soil grains, and no prior preloading or seismic shaking.

13.4.3 Liquefaction Analysis

Introduction.   The first step in the liquefaction analysis is to determine if the soil has the ability to liquefy during an earthquake. As discussed in Sec. 13.4.2 (item number 3), the majority of soils that are susceptible to liquefaction are cohesionless soils. Cohesive soils should not be considered susceptible to liquefaction unless they meet the two criteria listed in Sec. 13.4.2 (see item number 3, soil type).

   The most common type of analysis to determine the liquefaction potential is to use the standard penetration test (SPT) (Seed et al., 1985; Stark and Olson, 1995). The analysis is based on the simplified method proposed by Seed and Idriss (1971). This method of liquefaction analysis proposed by Seed and Idriss (1971) is often termed the *simplified procedure* and is the most commonly used method to evaluate the liquefaction potential of a site. The steps are as follows:

1. *Appropriate soil type.*   As discussed earlier, the first step is to determine if the soil has the ability to liquefy during an earthquake. The soil must meet the two requirements listed in Sec. 13.4.2 (item number 3).

2. *Groundwater table.*   The soil must be below the groundwater table. The liquefaction analysis could also be performed if it is anticipated that the groundwater table will rise in the future and thus the soil will eventually be below the groundwater table.

3. *CSR induced by earthquake.*   If the soil meets the above two requirements, then the simplified procedure can be performed. The first step in the simplified procedure is to determine the cyclic stress ratio (CSR) that will be induced by the earthquake.

   A major unknown in the calculation of the CSR induced by the earthquake is the peak horizontal ground acceleration a_{max} that should be used in the analysis. The peak horizontal ground acceleration has been discussed in Sec. 13.3. Threshold values needed to produce liquefaction have been discussed in Sec. 13.4.2 (item number 1). As previously mentioned, a liquefaction analysis would typically not be needed for those sites having a peak ground acceleration a_{max} less than $0.10g$ or a local magnitude M_L less than 5.

4. *CRR from standard penetration test.* By using the standard penetration test, the cyclic resistance ratio (CRR) of the in situ soil is then determined. If the CSR induced by the earthquake is greater than the CRR determined from the standard penetration test, then it is likely that liquefaction will occur during the earthquake, and vice versa.

5. *Factor of safety (FS).* The final step is to determine the factor of safety against liquefaction, which is defined as FS = CRR/CSR.

Cyclic Stress Ratio Caused by the Earthquake. If it is determined that the soil has the ability to liquefy during an earthquake and the soil is below or will be below the groundwater table, then the liquefaction analysis is performed. The first step in the simplified procedure is to calculate the CSR, also commonly referred to as the *seismic stress ratio* (SSR), which is caused by the earthquake.

In order to develop the CSR earthquake equation, it is assumed that there is a level ground surface, a soil column of unit width and length, and that the soil column will move horizontally as a rigid body in response to the maximum horizontal acceleration a_{max} exerted by the earthquake at ground surface. Figure 13.15 shows a diagram of these assumed conditions. Given these assumptions, the weight W of the soil column is equal to γ_t times z, where γ_t = total unit weight of the soil and z = depth below ground surface. The horizontal earthquake force F acting on the soil column (which has a unit width and length) would be equal to

$$F = ma = (W/g)a = (\gamma_t z/g)\, a_{max} = \sigma_{vo}\,(a_{max}/g) \tag{13.4}$$

where F = horizontal earthquake force acting on the soil column that has a unit width and length (lb or kN)
m = total mass of the soil column (lb or kg), which is equal to W/g
W = total weight of the soil column (lb or kN). For the assumed unit width and length of soil column, the total weight of the soil column = $\gamma_t z$
γ_t = total unit weight of the soil (pcf or kN/m³)
z = depth below ground surface of the soil column as shown in Fig. 13.15

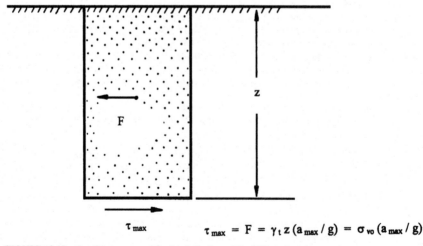

FIGURE 13.15 Conditions assumed for the derivation of the CSR earthquake equation.

a = acceleration, which in this case is the maximum horizontal acceleration at ground surface caused by the earthquake ($a = a_{max}$) (ft/sec^2 or m/sec^2)

a_{max} = maximum horizontal acceleration at ground surface that is induced by the earthquake (ft/sec^2 or m/sec^2). The maximum horizontal acceleration is also commonly referred to as the PGA (see Sec. 13.3).

σ_{vo} = total vertical stress at the bottom of the soil column (psf or kPa). The total vertical stress = $\gamma_t z$

As shown in Fig. 13.15, by summing forces in the horizontal direction, the force F acting on the rigid soil element is equal to the maximum shear force at the base of the soil element. Since the soil element is assumed to have a unit base width and length, the maximum shear force F is equal to the maximum shear stress τ_{max}, or from Eq. 13.4:

$$\tau_{max} = F = \sigma_{vo} \, (a_{max}/g) \tag{13.5}$$

Dividing both sides of the equation by the vertical effective stress σ'_{vo}, or:

$$\tau_{max}/\sigma'_{vo} = (\sigma_{vo}/\sigma'_{vo}) \, (a_{max}/g) \tag{13.6}$$

Since the soil column does not act as a rigid body during the earthquake, but rather the soil is deformable, Seed and Idriss (1971) incorporated a depth reduction factor r_d into the right side of Eq. 13.6, or:

$$\tau_{max}/\sigma'_{vo} = r_d(\sigma_{vo}/\sigma'_{vo}) \, (a_{max}/g) \tag{13.7}$$

For the simplified method, Seed et al. (1975) converted the typical irregular earthquake record into an equivalent series of uniform stress cycles by assuming the following:

$$\tau_{cyc} = 0.65 \, \tau_{max} \tag{13.8}$$

where τ_{cyc} is the uniform cyclic shear stress amplitude of the earthquake (psf or kPa).

In essence, the erratic earthquake motion was converted into an equivalent series of uniform cycles of shear stress, referred to as τ_{cyc}. Substituting Eq. 13.8 into Eq. 13.7, the earthquake induced CSR is obtained, or:

$$\text{CSR} = \tau_{cyc}/\sigma'_{vo} = 0.65 \, r_d \, (\sigma_{vo}/\sigma'_{vo}) \, (a_{max}/g) \tag{13.9}$$

where CSR = cyclic stress ratio (dimensionless). The CSR is also commonly referred to as the SSR.

a_{max} = maximum horizontal acceleration at ground surface that is induced by the earthquake (ft/sec^2 or m/sec^2), which is also commonly referred to as the PGA (see Sec. 13.3).

g = acceleration of gravity (32.2 ft/sec^2 or 9.81 m/sec^2).

σ_{vo} = total vertical stress at a particular depth where the liquefaction analysis is being performed (psf or kPa). In order to calculate the total vertical stress, the total unit weight γ_t of the soil layer(s) must be known.

σ'_{vo} = vertical effective stress at that same depth in the soil deposit where σ_{vo} was calculated (psf or kPa). In order to calculate the vertical effective stress, the location of the groundwater table must be known.

r_d = depth reduction factor, also known as the *stress reduction coefficient* (dimensionless)

As previously mentioned, the depth reduction factor was introduced to account for the fact that the soil column shown in Fig. 13.15 does not behave as a rigid body during the earthquake. Figure 13.16 presents the range in values for the depth reduction factor r_d versus depth below ground surface. Note that with depth, the depth reduction factor decreases to account for the fact that the soil is not a rigid body, but is rather deformable. As indicated in Fig. 13.16, Idriss (1999) indicates that the values

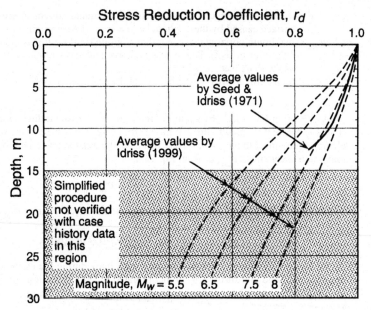

FIGURE 13.16   Reduction factor r_d versus depth below level or gently sloping ground surfaces. (*From Andrus and Stokoe 2000, reproduced with permission from the American Society of Civil Engineers.*)

of r_d depend on the magnitude of the earthquake. As a practical matter, the r_d values are usually obtained from the curve labeled *Average values by Seed & Idriss (1971)* in Fig. 13.16.

Another option is to assume a linear relationship of r_d versus depth and use the following equation (Kayen et al., 1992):

$$r_d = 1 - (0.012)\,(z) \tag{13.10}$$

where z is the depth in meters below the ground surface where the liquefaction analysis is being performed (i.e., the same depth used to calculate σ_{vo} and σ'_{vo}).

For Eq. 13.9, the vertical total stress σ_{vo} and vertical effective stress σ'_{vo} can be readily calculated using basic geotechnical principles. Equation 13.10 or Fig. 13.16 could be used to determine the depth reduction factor r_d. Thus all parameters in Eq. 13.9 can be readily calculated, except for the peak ground acceleration a_{max}, which has been discussed in Sec. 13.3.

Cyclic Resistance Ratio from the Standard Penetration Test.   The second step in the simplified procedure is to determine the cyclic resistance ratio (CRR) of the in situ soil. The CRR represents the liquefaction resistance of the in situ soil. The most commonly used method for determining the liquefaction resistance is to use the data obtained from the standard penetration test (SPT). The SPT has been discussed in Sec. 2.4.3. The advantages of using the standard penetration test to evaluate liquefaction potential are as follows:

1. *Groundwater table.*   A boring must be excavated in order to perform the standard penetration test. The location of the groundwater table can be measured in the borehole. Another option is to install a piezometer in the borehole, which can then be used to monitor the groundwater level over time.

2. *Soil type.*   In clean sand, the SPT sampler may not be able to retain a soil sample. But for most other types of soil, the SPT sampler will be able to retrieve a soil sample. The soil sample

retrieved in the SPT sampler can be used to visually classify the soil and estimate the percent fines in the soil. In addition, the soil specimen can be returned to the laboratory and classification tests can be performed in order to further assess the liquefaction susceptibility of the soil (see item number 3, Sec. 13.4.2).

3. *Relationship between N-value and liquefaction potential.*   In general, the factors that increase the liquefaction resistance of a soil will also increase the $(N_1)_{60}$ from the standard penetration test [see Sec. 2.4.3 for the procedure to calculate $(N_1)_{60}$]. For example, a well-graded dense soil that has been preloaded or aged will be resistant to liquefaction and will have high values of $(N_1)_{60}$. Likewise, a uniformly graded soil with a loose and segregated soil structure will be more susceptible to liquefaction and will have much lower values of $(N_1)_{60}$.

Based on the standard penetration test and field performance data, Seed et al. (1985) concluded that there are three approximate potential damage ranges that can be identified, as follows:

$(N_1)_{60}$	Potential damage
0–20	High
20–30	Intermediate
>30	No significant damage

As indicated in Table 2.6, an $(N_1)_{60}$ value of 20 is the approximate boundary between the medium and dense states of the sand. Above an $(N_1)_{60}$ of 30, the sand is either in a dense or very dense state. For this condition, initial liquefaction does not produce large deformations because of the dilation tendency of the sand upon reversal of the cyclic shear stress. This is the reason that such soils produce no significant damage as indicated by the above table.

Figure 13.17 presents a chart that can be used to determine the cyclic resistance ratio of the in situ soil. This figure was developed from investigations of numerous sites that had liquefied or did not liquefy during earthquakes. For most of the data used in Fig. 13.17, the earthquake magnitude was close to 7.5 (Seed et al., 1985). The three lines shown in Fig. 13.17 are for soil that contains 35, 15, or ≤5 percent fines. The lines shown in Fig. 13.17 represent approximate dividing lines, where data to the left of each individual line indicate field liquefaction, while data to the right of the line indicate sites that generally did not liquefy during the earthquake.

In order to use Fig. 13.17 to determine the CRR of the in situ soil, the following four steps are performed:

1. *Standard penetration test $(N_1)_{60}$ value.*   Note in Fig. 13.17 that the horizontal axis represents data from the SPT that must be expressed in terms of the $(N_1)_{60}$ value. In the liquefaction analysis, the standard penetration test N_{60} value (Eq. 2.4) is corrected for the overburden pressure (see Eq. 2.5). As discussed in Sec. 2.4.3, when a correction is applied to the N_{60} value to account for the effect of overburden pressure, this value is referred to as $(N_1)_{60}$.

2. *Percent fines.*   Once the $(N_1)_{60}$ value has been calculated, the next step is to determine or estimate the percent fines in the soil. For a given $(N_1)_{60}$ value, soils with more fines have a higher liquefaction resistance. Figure 13.17 is applicable for nonplastic silty sands or for plastic silty sands that meet the criteria for cohesive soils listed in Sec. 13.4.2 (see item number 3, soil type).

3. *Cyclic resistance ratio (CRR) for an anticipated magnitude 7.5 earthquake.*   Once the $(N_1)_{60}$ value and the percent fines in the soil have been determined, then Fig. 13.17 can be used to obtain the CRR of the soil. In order to use Fig. 13.17, the figure is entered with the corrected standard penetration test $(N_1)_{60}$ value from Eq. 2.5, and then by intersecting the appropriate fines content curve, the CRR is obtained.

   As shown in Fig. 13.17, for a magnitude 7.5 earthquake, clean sand will not liquefy if the $(N_1)_{60}$ value exceeds 30. For an $(N_1)_{60}$ value of 30, the sand is either in a dense or very dense state (see Table 2.6). As previously mentioned, dense sands will not liquefy because they tend to dilate during shearing.

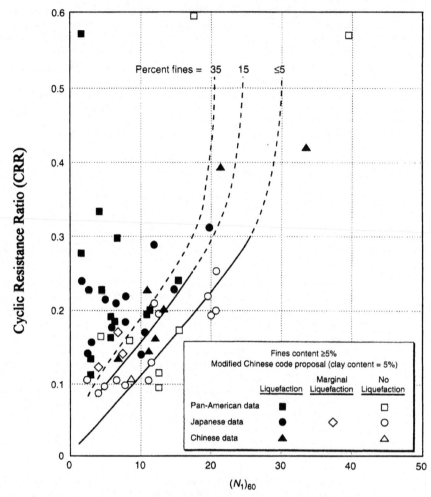

FIGURE 13.17 Plot used to determine the CRR for clean and silty sands for $M = 7.5$ earthquakes. (*After Seed et al. 1985, reprinted with permission of the American Society of Civil Engineers.*)

4. *Correction for other magnitude earthquakes.* Figure 13.17 is for a projected earthquake that has a magnitude of 7.5. The final factor that must be included in the analysis is the magnitude of the earthquake. As indicated in Table 13.2, the higher the magnitude of the earthquake, the longer the duration of ground shaking. A higher magnitude will thus result in a higher number of applications of cyclic shear strain, which will decrease the liquefaction resistance of the soil. Figure 13.17 was developed for an earthquake magnitude of 7.5, and for other different magnitudes, the CRR values from Fig. 13.17 would be multiplied by the magnitude scaling factor indicated in Table 13.5. Figure 13.18 presents other suggested magnitude scaling factors.

As shown in Fig. 13.9, it could be concluded that the local magnitude M_L, the surface wave magnitude M_s, and moment magnitude M_w scales are reasonably close to each other below a value of about 7. Thus for a magnitude of 7 or below, any one of these magnitude scales can be used to determine the magnitude scaling factor. At high magnitude values, the moment magnitude M_w tends to

TABLE 13.5 Magnitude Scaling Factors

Anticipated earthquake magnitude	Magnitude scaling factor (MSF)
$8\frac{1}{2}$	0.89
$7\frac{1}{2}$	1.00
$6\frac{3}{4}$	1.13
6	1.32
$5\frac{1}{4}$	1.50

Note: In order to determine the CRR of the in situ soil, multiply the magnitude scaling factor indicated above by the cyclic resistance ratio determined from Fig. 13.17.
Source: Seed et al. (1985).

significantly deviate from the other magnitude scales, and the moment magnitude M_w should be used to determine the magnitude scaling factor from Table 13.5 or Fig. 13.18.

Two additional correction factors may need to be included in the analysis. The first possible correction factor is for deep soil layers (i.e., depths where $\sigma'_{vo} > 100$ kPa) because the Seed and Idriss simplified procedure has not verified liquefaction potential for such a condition (see Youd and

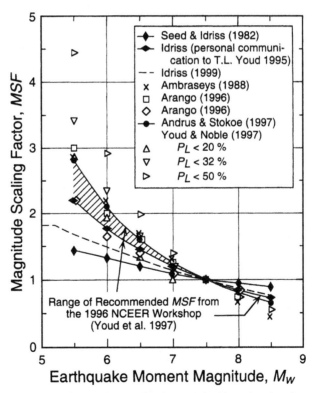

FIGURE 13.18 Magnitude scaling factors derived by various investigators. (*From Andrus and Stokoe 2000, reprinted with permission of the American Society of Civil Engineers.*)

Idriss, 2001). The second possible correction factor is for sloping ground conditions, which will be discussed in Sec. 13.5.

As indicated in Sec. 13.3, both the peak ground acceleration a_{max} and the length of ground shaking increase for sites having soft, thick, and submerged soils. In a sense, the earthquake magnitude accounts for the increased shaking at a site, i.e., the higher the magnitude, the longer the ground is subjected to shaking. Thus for sites having soft, thick, and submerged soils, it may be prudent to both increase the peak ground acceleration a_{max} and the earthquake magnitude to account for local site effects.

Factor of Safety Against Liquefaction.   The final step in the liquefaction analysis is to calculate the factor of safety against liquefaction. If the CSR caused by the anticipated earthquake (Eq. 13.9) is greater than the CRR of the in situ soil (Fig. 13.17), then liquefaction could occur during the earthquake, and vice versa. The factor of safety against liquefaction FS is defined as follows:

$$FS = \frac{CRR}{CSR} \tag{13.11}$$

The higher the factor of safety, the more resistant the soil is to liquefaction. However, soil that has a factor of safety slightly greater than 1.0 may still liquefy during an earthquake. For example, if a lower layer liquefies, then the upward flow of water could induce liquefaction of the layer that has a factor of safety that is slightly greater than 1.0.

In the above liquefaction analysis, there are many different equations and corrections that are applied to both the CSR induced by the anticipated earthquake and the CRR of the in situ soil. For example, there are four different corrections (i.e., E_m, C_b, C_r, and σ'_{vo}) that are applied to the standard penetration test N value in order to calculate the $(N_1)_{60}$ value. All of these different equations and various corrections may provide the engineer with a sense of high accuracy, when in fact, the entire analysis is only a gross approximation. The analysis should be treated as such and engineering experience and judgment are essential in the final determination of whether or not a site has liquefaction potential.

As an alternative to using the standard penetration test, the cone penetration test can be used to determine the CRR of the in situ soil. The shear wave velocity of the soil can also be used to determine the CRR of the in situ soil (see Day, 2002).

The following example problem illustrates the procedure that is used to determine the factor of safety against liquefaction:

Example Problem 13.1   It is planned to construct a building on a cohesionless sand deposit (fines < 5 percent). There is a nearby major active fault and the engineering geologist has determined that for the anticipated earthquake, the peak ground acceleration a_{max} will be equal to $0.40g$. Assume a level ground surface with the groundwater table located 1.5 m below ground surface, total unit weight of sand above the groundwater table is 18.9 kN/m³, buoyant unit weight of sand below the groundwater table is 9.81 kN/m³, and the standard penetration test was performed at a depth of 3 m where $(N_1)_{60} = 7.7$. Assuming an anticipated earthquake magnitude of 7.5, calculate the factor of safety against liquefaction for the saturated clean sand located at a depth of 3 m below ground surface.

Solution   At a depth of 3 m:

$$\sigma'_{vo} = (18.9 \text{ kN/m}^3)(1.5 \text{ m}) + (9.81 \text{ kN/m}^3)(1.5 \text{ m}) = 43 \text{ kPa}$$

$$\sigma_{vo} = (18.9 \text{ kN/m}^3)(1.5 \text{ m}) + (9.81 + 9.81 \text{ kN/m}^3)(1.5 \text{ m}) = 58 \text{ kPa}$$

(Continued)

Using Eq. 13.10 with $z = 3$ m, then $r_d = 0.96$

Using Eq. 13.9:

$$\text{CSR} = 0.65\ r_d(\sigma_{vo}/\sigma'_{vo})\ (a_{max}/g) = (0.65)(0.96)(58/43)(0.40) = 0.34$$

The next step is to determine the CRR of the in situ soil. Entering Fig. 13.17 with $(N_1)_{60} = 7.7$ and intersecting the curve labeled less than 5 percent fines, the CRR of the in situ soil at a depth of 3 m = 0.09.

The final step is to calculate the factor of safety against liquefaction FS by using Eq. 13.11, or:

$$\text{FS} = \text{CRR/CSR} = 0.09/0.34 = 0.26$$

Based on the factor of safety against liquefaction, it is probable that during the anticipated earthquake the in situ sand located at a depth of 3 m below ground surface will liquefy.

13.5 SLOPE STABILITY

13.5.1 Introduction

All types of slopes can fail during earthquakes, from small road embankments to massive landslides. Tables 13.6 and 13.7 present the different types of slope movement in soil and rock that can be triggered by an earthquake.

There would appear to be a shaking threshold that is needed to produce earthquake-induced slope movement. For example, those sites having a peak ground acceleration a_{max} less than 0.10g or a local magnitude M_L less than 5 would typically not require analyses of potential liquefaction related slope failures (i.e., flow slides or lateral spreading). Other threshold values for different types of slope movement are summarized in Tables 13.6 and 13.7.

Tables 13.6 and 13.7 also indicate the relative abundance of earthquake-induced slope failures based on a historical study of 40 earthquakes by Keefer (1984). In general, the most abundant types of slope failures during earthquakes tend to have the lowest threshold values and can involve both small and large size masses

Those slope failures listed as *uncommon* in Tables 13.6 and 13.7 tend to have higher threshold values and also typically involve larger masses of soil and rock. Because of their large volume, they tend to be less common. For example, in comparing rock slides and rock block slides in Table 13.6, the rock block slides would tend to involve massive blocks of rock that remain relatively intact during the earthquake-induced slope movement. Another example is a rock avalanche, which by definition implies a large mass of displaced material.

The seismic evaluation of slope stability can be grouped into two general categories: weakening slope stability analysis and inertia slope stability analysis.

Weakening Slope Stability Analysis. The weakening slope stability analysis is preferred for those materials that will experience a significant reduction in shear strength during the earthquake. Examples of these types of soil and rock are as follows:

- Foliated or friable rock that fractures apart during the earthquake resulting in rockfalls, rock slides, and rock slumps.

- Sensitive clays that lose shear strength during the earthquake. An example of a weakening landslide is the Turnagain Heights Landslide that will be described in Sec. 13.5.3.

- Soft clays and organic soils that are overloaded and subjected to plastic flow during the earthquake. The type of slope movement involving these soils is often termed *slow earth flows*.

TABLE 13.6 Earthquake-Induced Slope Movement in Rock

Main type of slope movement	Subdivisions	Material type	Minimum slope inclination	Threshold values	Relative abundance
Falls	Rockfalls	Rocks weakly cemented, intensely fractured, or weathered; contain conspicuous planes of weakness dipping out of slope or contain boulders in a weak matrix	40° (1.2:1)	$M_L = 4.0$	Very abundant (more than 100,000 in the 40 earthquakes)
Slides	Rock slides	Rocks weakly cemented, intensely fractured, or weathered; contain conspicuous planes of weakness dipping out of slope or contain boulders in a weak matrix	35° (1.4:1)	$M_L = 4.0$	Very abundant (more than 100,000 in the 40 earthquakes)
	Rock avalanches	Rocks intensely fractured and exhibiting one of the following properties: significant weathering, planes of weakness dipping out of slope, weak cementation, or evidence of previous landsliding	25° (2.1:1)	$M_L = 6.0$	Uncommon (100 to 1000 in the 40 earthquakes)
	Rock slumps	Intensely fractured rocks, preexisting rock slump deposits, shale, and other rocks containing layers of weakly cemented or intensely weathered material	15° (3.7:1)	$M_L = 5.0$	Moderately common (1000 to 10,000 in the 40 earthquakes)
	Rock block slides	Rocks having conspicuous bedding planes or similar planes of weakness dipping out of slopes	15° (3.7:1)	$M_L = 5.0$	Uncommon (100 to 1000 in the 40 earthquakes)

Sources: Keefer (1984) and Division of Mines and Geology (1997).

- Loose soils located below the groundwater table and subjected to liquefaction or a substantial increase in excess pore water pressure. There are two cases of weakening slope stability analyses involving the liquefaction of soil, as follows:

  1. *Flow slide.* Flow slides develop when the static driving forces exceed the shear strength of the soil along the slip surface and thus the factor of safety is less than 1.0.

  2. *Lateral spreading.* In this case, the static driving forces do not exceed the shear strength of the soil along the slip surface, and thus the ground is not subjected to a flow slide. Instead, the driving forces only exceed the resisting forces during those portions of the earthquake that impart net inertial forces in the downslope direction.

Weakening slope stability analyses will be discussed in Secs. 13.5.2 and 13.5.3.

TABLE 13.7 Earthquake-Induced Slope Movement in Soil

Main type of slope movement	Subdivisions	Material type	Minimum slope inclination	Threshold values	Relative abundance
Falls	Soil falls	Granular soils that are slightly cemented or contain clay binder	40° (1.2:1)	$M_L = 4.0$	Moderately common (1000 to 10,000 in the 40 earthquakes)
Slides	Soil avalanches	Loose, unsaturated sands	25° (2.1:1)	$M_L = 6.5$	Abundant (10,000 to 100,000 in the 40 earthquakes)
	Disrupted soil slides	Loose, unsaturated sands	15° (3.7:1)	$M_L = 4.0$	Very abundant (more than 100,000 in the 40 earthquakes)
	Soil slumps	Loose, partly to completely saturated sand or silt; uncompacted or poorly compacted artificial fill composed of sand, silt, or clay, preexisting soil slump deposits	10° (5.7:1)	$M_L = 4.5$	Abundant (10,000 to 100,000 in the 40 earthquakes)
	Soil block slides	Loose, partly or completely saturated sand or silt; uncompacted or slightly compacted artificial fill composed of sand or silt, bluffs containing horizontal or subhorizontal layers of loose, saturated sand or silt	5° (11:1)	$M_L = 4.5$	Abundant (10,000 to 100,000 in the 40 earthquakes)
Flow slides and lateral spreading	Slow earth flows	Stiff, partly to completely saturated clay and preexisting earth flow deposits	10° (5.7:1)	$M_L = 5.0$	Uncommon (100 to 1000 in the 40 earthquakes)
	Flow slides	Saturated, uncompacted or slightly compacted artificial fill composed of sand or sandy silt (including hydraulic fill earth dams and tailings dams); loose, saturated granular soils	2.3° (25:1)	$M_L = 5.0$ $a_{max} = 0.10g$	Moderately common (1000 to 10,000 in the 40 earthquakes)
	Subaqueous flows	Loose, saturated granular soils	0.5° (110:1)	$M_L = 5.0$ $a_{max} = 0.10g$	Uncommon (100 to 1000 in the 40 earthquakes)
	Lateral spreading	Loose, partly or completely saturated silt or sand, uncompacted or slightly compared artificial fill composed of sand	0.3° (190:1)	$M_L = 5.0$ $a_{max} = 0.10g$	Abundant (10,000 to 100,000 in the 40 earthquakes)

Sources: Keefer (1984) and Division of Mines and Geology (1997).

Inertia Slope Stability Analysis.   The inertia slope stability analysis is preferred for those materials that retain their shear strength during the earthquake. Examples of these types of soil and rock are as follows:

• Massive crystalline bedrock and sedimentary rock that remains intact during the earthquake, such as earthquake-induced rock block slide.

• Soils that tend to dilate during the seismic shaking, or for example, dense to very dense granular soil and heavily overconsolidated cohesive soil such as very stiff to hard clays.

• Soils that have a stress-strain curve that does not exhibit a significant reduction in shear strength with strain. Earthquake-induced slope movement in these soils often takes the form of soil slumps or soil block slides.

• Clay that has a low sensitivity.

• Soils located above the groundwater table. These soils often have negative pore water pressure due to capillary action.

• Landslides that have a distinct rupture surface where the shear strength along the rupture surface is equal to the drained residual shear strength ϕ'_r.

There are different types of inertia slope stability analyses and two of the most commonly used are the pseudostatic approach and the Newmark method (1965). These two methods will be described in Sec. 13.5.4.

13.5.2   Weakening Slope Stability—Flow Slides and Lateral Spreading

A weakening slope instability analysis is used for slopes that contain soil that is likely to liquefy during the earthquake. As mentioned in the previous section, there are two cases of weakening slope stability analyses involving the liquefaction of soil: flow slides and lateral spreading.

Flow Slides.   For flow slides, Seed (1970) states:

> If liquefaction occurs in or under a sloping soil mass, the entire mass will flow or translate laterally to the unsupported side in a phenomenon termed a flow slide. Such slides also develop in loose, saturated, cohesionless materials during earthquakes and are reported at Chile (1960), Alaska (1964), and Niigata (1964).

There are three general types of flow slides, as follows:

1. *Mass liquefaction.*   This type of flow slide occurs when nearly the entire sloping mass is susceptible to liquefaction. These types of failures often occur to partially or completely submerged slopes, such as shoreline embankments. For example, Figure 13.19 shows damage to a marine facility at Redondo Beach King Harbor. The California Northridge earthquake on January 17, 1994 caused this damage. The 18 ft (5.5 m) of horizontal displacement was due to the liquefaction of the offshore sloping fill mass that was constructed as part of the marine facility.

For design conditions, the first step in the analysis would be to determine the factor of safety against liquefaction. If it is determined that the entire sloping mass, or a significant portion of the sloping mass, will be subjected to liquefaction during the design earthquake, then the slope will be susceptible to a flow slide.

2. *Zonal liquefaction.*   This second type of flow slide develops because there is a specific zone of liquefaction within the slope. A classic example of zonal liquefaction resulting in a flow slide is the failure of the Lower San Fernando Dam caused by the San Fernando earthquake, also known as the *Sylmar* earthquake, on February 9, 1971. Seismographs located on the abutment and on the crest of the dam recorded peak ground accelerations a_{max} of about 0.5 to 0.55g. These high peak ground accelerations caused the liquefaction of a zone of hydraulic sand fill near the base of the upstream shell. Figure 13.20 shows a cross section through the earthen dam and the location of the zone of

FIGURE 13.19 Damage to a marine facility caused by the California Northridge earthquake on January 17, 1994. (*From Kerwin and Stone, 1997; reprinted with permission from the American Society of Civil Engineers.*)

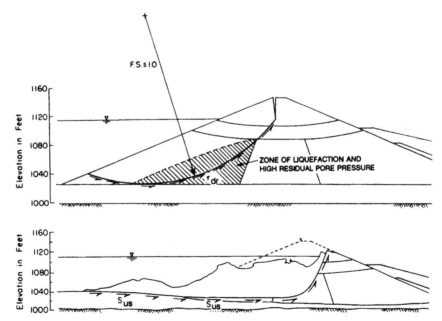

FIGURE 13.20 Cross section through the Lower San Fernando Dam. The upper diagram shows the condition immediately prior to the flow slide caused by the San Fernando earthquake on February 9, 1971. The lower diagram shows the configuration after the flow slide of the upstream slope and crest of the dam. (*From Castro et al. 1992, reproduced with permission of the American Society of Civil Engineers.*)

material that was believed to have liquefied during the earthquake. Once liquefied, the upstream portion of the dam was subjected to a flow slide.

The upper part of Fig. 13.20 indicates the portion of the dam and the slip surface along which the flow slide is believed to have initially developed. As indicated in the upper part of Fig. 13.20, the flow slide developed when the driving forces exceeded the shear strength along the slip surface and the factor of safety became 1.0 or less. The lower part of Fig. 13.20 depicts the final condition of the dam after the flow slide. The flow slide caused the upstream toe of the dam to move about 150 ft (46 m) into the reservoir.

3. *Landslide movement caused by liquefaction of soil layers or seams.*   The third type of flow slide develops because of liquefaction of horizontal soil layers or seams of soil. For example, there can be liquefaction of seams of loose saturated sands within a slope. This can cause the entire slope to move laterally along the liquefied layer at the base. These types of landslides caused by liquefied seams of soil caused extensive damage during the 1964 Alaskan earthquake (Shannon and Wilson, Inc., 1964; Hansen, 1965). Buildings located in the graben area are subjected to large differential settlements and are often completely destroyed by this type of liquefaction-induced landslide movement (Seed, 1970).

The steps in determining the potential for a flow slide are as follows:

1. *Extent of liquefaction.*   The first step is to determine the extent of soil that will liquefy during the design earthquake. This can be accomplished by using the liquefaction analysis presented in Sec. 13.4.3. However, this liquefaction analysis was developed for level-ground sites. For sites that have a level-ground surface and geostatic soil conditions, the horizontal shear stress is equal to zero. But for sites with sloping ground, there will be a horizontal static shear stress that is induced into the soil. The presence of this horizontal static shear stress makes loose soil more susceptible to liquefaction. The reason is because less earthquake-induced shear stress is required to cause contraction and hence liquefaction when the soil is already subjected to a static horizontal shear stress. Hence, especially for loose soil, a correction may need to be applied to Eq. 13.9 (see Day, 2002).

If it is determined that the entire sloping mass, or a significant portion of the sloping mass, will be subjected to liquefaction during the design earthquake, then the slope will be susceptible to a flow slide. No further analyses would be required for the *mass liquefaction* case. On the other hand, if only zones or thin layers of soil will liquefy during the design earthquake, then slope stability analyses are required.

2. *Slope stability analyses.*   For the conditions of a zone of potentially liquefiable soil or for potential liquefaction of soil layers or seams, a slope stability analysis using the method of slices (see Sec. 10.4.3) is required, as described below:

a. *Zonal liquefaction.*   A slope stability analysis is performed by using various circular arc slip surfaces that pass through the zone of expected liquefaction. The slope stability analysis is often performed using an effective stress analysis with the shear strength of the liquefied soil zones equal to zero, i.e., $\phi' = 0$ and $c' = 0$ (see Sec. 4.6.6 and Table 4.9). If the factor of safety of the slope is equal to or less than 1.0, then a flow slide is likely to occur during the earthquake.

b. *Liquefaction of soil layers or seams.*   It can be difficult to evaluate the possibility of flow slides due to liquefaction of soil layers or seams. This is because the potentially liquefiable soil layers or seams can be rather thin and may be hard to discover during the subsurface exploration. When performing the slope stability analysis, the slip surface must pass through these horizontal layers or seams of liquefied soil. Thus a slope stability analysis is often performed using a block type failure mode (rather than using circular arc slip surfaces). The slope stability analysis is typically performed using an effective stress analysis with the shear strength of the liquefied soil layers or seams equal to zero, i.e., $\phi' = 0$ and $c' = 0$ (see Sec. 4.6.6 and Table 4.9). If the factor of safety of the slope is equal to or less than 1.0, then a flow slide is likely to occur during the earthquake. An example of a slope stability analysis for a 5 ft (1.5 m) thick layer of liquefiable soil is shown in Figure 13.21. Because the factor of safety is less than 1.0, a flow slide would be expected for this slope during the earthquake.

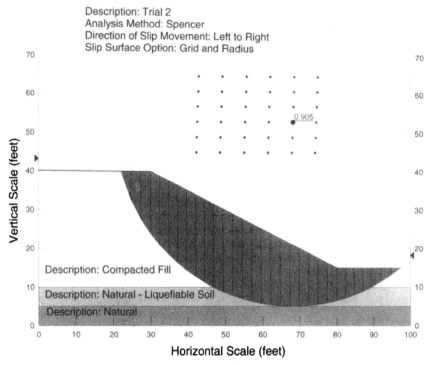

FIGURE 13.21 Slope stability analysis for the weakening condition based on zero shear strength of a 5-ft-thick layer of the liquefied soil. The SLOPE/W computer program was used to perform the stability analysis (Geo-Slope, 1991).

It has been stated that the shear strength of liquefied soil may not necessarily be equal to zero. For example, even though the soil liquefies, there may still be a small value of undrained shear strength caused by the individual soil particles trying to shear past each other as the flow slide develops. This undrained shear strength of liquefied soil has been termed the *liquefied shear strength* (Seed and Harder, 1990; Stark and Mesri, 1992; Olson et al., 2000). However, this liquefied shear strength tends to be very small, especially if the soil has low values of $(N_1)_{60}$. As a practical matter, this small value of liquefied shear strength is often ignored in practice and as described previously, the liquefied zones or liquefied layers are assumed to have zero shear strength for the slope stability analysis of flow slides.

Lateral Spreading. The concept of cyclic mobility is used to describe large-scale lateral spreading. Because the ground is gently sloping or flat, the static driving forces do not exceed the resistance of the soil along the slip surface, and thus the ground is not subjected to a flow slide (i.e., the factor of safety is greater than 1.0). Instead, the driving forces only exceed the resisting forces during those portions of the earthquake that impart net inertial forces in the downslope direction. Each cycle of net inertial forces in the downslope direction causes the driving forces to exceed the resisting forces along the slip surface, resulting in progressive and incremental lateral movement. Often the lateral movement and ground surface cracks first develop at the unconfined toe and then the slope movement and ground cracks progressively move upslope.

A commonly used approach for predicting the amount of horizontal ground displacement resulting from liquefaction-induced lateral spreading is to use empirical equations (Bartlett and Youd, 1995; Youd et al., 2002). Both U.S. and Japanese case histories of lateral spreading of liquefied sand were used to develop the empirical equations. Based on the regression analysis, two different equations were developed: (1) for lateral spreading toward a free face, such as a river bank, and (2) for lateral spreading of gently sloping ground where a free face is absent. By using these empirical equations, the amount of lateral spreading during the earthquake can be estimated (see Bartlet and Youd, 1995; Youd et al., 2002).

Summary.   As discussed in this section, the liquefaction of soil can cause flow failures or lateral spreading. It is also possible that even with a factor of safety against liquefaction greater than 1.0, there could be still be significant weakening of the soil and deformation of the slope. In summary, the type of analysis should be based on the factor of safety against liquefaction FS, as follows:

1. *FS ≤ 1.0.*   In this case, the soil is expected to liquefy during the design earthquake and thus a flow slide analysis and/or a lateral spreading analysis would be performed.

2. *FS > 2.0.*   If the factor of safety against liquefaction is greater than about 2.0, the pore water pressures generated by the earthquake-induced contraction of the soil are usually small enough so that they can be neglected. In this case, it could be assumed that the soil is not weakened by the earthquake and thus the inertia slope stability analyses (Sec. 13.5.4) could be performed.

3. *1.0 < FS ≤ 2.0.*   For this case, the soil is not anticipated to liquefy during the earthquake. However, as the loose granular soil contracts during the earthquake, there could still be a substantial increase in pore water pressure and hence weakening of the soil. Figure 4.25 can be used to estimate the pore water pressure ratio for various values of the factor of safety against liquefaction FS. Using the estimated pore water pressure ratio from Fig. 4.25, an effective stress slope stability analysis could be performed. If the results of the effective stress slope stability analysis indicate a factor of safety less than 1.0, then failure of the slope would be expected during the earthquake.

Even with a slope stability factor of safety greater than 1.0, there could still be substantial deformation of the slope. There could be two different types of slope deformation. The first type of deformation would occur as the earthquake-induced pore water pressures dissipate and the soil contracts. The second type of deformation would occur when the earthquake imparts net inertial forces that cause the driving forces to exceed the resisting forces. Each cycle of net inertial forces in the downslope direction that cause the driving forces to exceed the resisting forces along the slip surface would result in progressive and incremental lateral movement. If the factor of safety from the slope stability analysis is only marginally in excess of 1.0, then the lateral spreading approach could be used to obtain a rough estimate of the lateral deformation of the slope.

13.5.3   Weakening Slope Stability—Strain-Softening Soil

In addition to soil that liquefies during the earthquake, other types of soil that will be weakened during the earthquake are as follows:

• Sensitive clays, which are strained back and forth during the earthquake and lose shear strength.

• Soft clays and organic soils that are overloaded and subjected to plastic flow during the earthquake. The type of slope movement involving these soils is often termed *slow earth flows.*

These types of plastic soils are often characterized as strain softening soils because there is a substantial reduction in shear strength once the peak shear strength is exceeded. During the earthquake, failure often first occurs at the toe of the slope and then the ground cracks and displacement of the slope progresses upslope. Blocks of soil are often observed to move laterally during the earthquake.

It is very difficult to evaluate the amount of lateral movement of slopes containing strain-softening soil. The most important factors are the level of static shear stress versus the peak shear stress of the soil and the amount of additional shear stress that will be induced into the soil by the earthquake. If the existing static shear stress is close to the peak shear stress, then only a small additional earthquake-induced shear stress will be needed to exceed the peak shear strength. Once this happens, the shear strength will significantly decrease with strain, resulting in substantial lateral movement of the slope. If it is anticipated that this will occur during the design earthquake, then one approach is to use the ultimate (i.e., softened) shear strength of the soil.

Turnagain Heights Landslide.   The March 27, 1964 Prince William Sound earthquake in Alaska was the largest earthquake in North America and the second largest this past century (largest occurred in Chile in 1960). Some details concerning this earthquake are as follows (Pflaker, 1972; Christensen, 2000; Sokolowski, 2000):

- The epicenter was in the northern Prince William Sound about 75 mi (120 km) east of Anchorage and about 55 mi (90 km) west of Valdez. The local magnitude M_L for this earthquake is estimated to be from 8.4 to 8.6. The moment magnitude M_w is reported as 9.2.
- The depth of the main shock was approximately 15 mi (25 km).
- The duration of shaking as reported in the Anchorage area lasted about 4 to 5 min.
- In terms of plate tectonics, the northwestward motion of the Pacific plate at about 2 to 3 in. (5 to 7 cm) per year causes the crust of southern Alaska to be compressed and warped, with some areas along the coast being depressed and other areas inland being uplifted. After periods of tens to hundreds of years, the sudden southeastward motion of portions of coastal Alaska relieves this compression as they move back over the subducting Pacific plate.
- There was both uplifting and regional subsidence. For example, some areas east of Kodiak were raised about 30 ft (9 m) and areas near Portage experienced regional subsidence of about 8 ft (2.4 m).
- The maximum intensity per the modified Mercalli Intensity scale was XI.
- There were 115 deaths in Alaska and about $300 to $400 million in damages (1964 dollars). The death toll was extremely small for a quake of this size, due to low population density, time of day (holiday), and type of material used to construct many buildings (wood).

During the strong ground shaking from this earthquake, seams of loose saturated sands and sensitive clays suffered a loss of shear strength. This caused entire slopes to move laterally along these weakened seams of soil. These types of landslides devastated the Turnagain Heights residential development and many downtown areas in Anchorage. It has been estimated that 56 percent of the total cost of damage was caused by earthquake-induced landslides (Shannon and Wilson, Inc., 1964; Hansen, 1965; Youd, 1978; Wilson and Keefer, 1985).

The cross sections shown in Figure 13.22 illustrate the sequence of movement of this landslide during the earthquake. The landslide movement has been described as follows (Nelson, 2000):

> During the Good Friday earthquake on March 27, 1964, a suburb of Anchorage, Alaska, known as Turnagain Heights broke into a series of slump blocks that slid toward the ocean. This area was built on sands and gravels overlying marine clay. The upper clay layers were relatively stiff, but the lower layers consisted of sensitive clay. The slide moved about 2,000 ft (610 m) toward the ocean, breaking up into a series of blocks. It began at the sea cliffs on the ocean after about 1.5 minutes of shaking caused by the earthquake, when the lower clay layer became liquefied. As the slide moved into the ocean, clays were extruded from the toe of the slide. The blocks rotating near the front of the slide, eventually sealed off the sensitive clay layer preventing further extrusion. This led to pull-apart basins being formed near the rear of the slide and the oozing upward of the sensitive clays into the space created by the extension [see Figure 13.22]. The movement of the mass of material toward the ocean destroyed 75 homes on the top of the slide.

13.5.4 Inertia Slope Stability

Pseudostatic Method.   As previously mentioned, the inertial slope stability analysis is preferred for those materials that retain their shear strength during the earthquake. The most commonly used inertial slope stability analysis is the pseudostatic approach. The advantages of this method are that it is

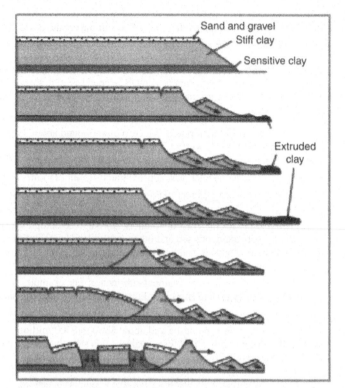

FIGURE 13.22   The above cross sections illustrate the sequence of movement of the Turnagain Heights landslide during the Prince William Sound earthquake in Alaska on March 27, 1964. (*Reproduced from Nelson 2000, based on work by Abbott 1996, with original version by USGS.*)

easy to understand and apply, and the method is applicable for both total stress and effective stress slope stability analyses.

The original application of the pseudostatic method has been credited to Terzaghi (1950). This method ignores the cyclic nature of the earthquake and treats it as if it applies an additional static force upon the slope. In particular, the pseudostatic approach is to apply a lateral force acting through the centroid of the sliding mass, acting in an out-of-slope direction. The pseudostatic lateral force F_h is calculated by using Eq. 13.4, or:

$$F_h = ma = (W/g)a = (W)(a_{max}/g) = k_h W \qquad (13.12)$$

where F_h = horizontal pseudostatic force acting through the centroid of the sliding mass, in an out-of-slope direction (lb or kN). For the slope stability analysis, the slope is usually assumed to have a unit length (i.e., two-dimensional analysis).

m = total mass of the slide material (lb or kg), which is equal to W/g

W = total weight of the slide material (lb or kN)

a = acceleration, which in this case is the maximum horizontal acceleration at ground surface caused by the earthquake ($a = a_{max}$) (ft/sec² or m/sec²)

a_{max} = maximum horizontal acceleration at ground surface that is induced by the earthquake (ft/sec² or m/sec²). The maximum horizontal acceleration is also commonly referred to as the *PGA* (see Sec. 13.3).

$a_{max}/g = k_h$ = seismic coefficient, also known as the pseudostatic coefficient (dimensionless)

Note that an earthquake could subject the sliding mass to both vertical and horizontal pseudostatic forces. However, the vertical force is usually ignored in the standard pseudostatic analysis. This is because the vertical pseudostatic force acting on the sliding mass usually has much less effect on the stability of a slope. In addition, most earthquakes produce a peak vertical acceleration that is less than the peak horizontal acceleration, and hence k_v is smaller than k_h.

As indicated in Eq. 13.12, the only unknowns in the pseudostatic method are the weight of the sliding mass (W) and the seismic coefficient k_h. Based on the results of subsurface exploration and laboratory testing, the unit weight of the soil or rock can be determined and then the weight of the sliding mass (W) can be readily calculated. The other unknown is the seismic coefficient k_h, which is much more difficult to determine. The selection of the seismic coefficient k_h takes considerable experience and judgment. Guidelines for the selection of k_h are as follows:

1. *Peak ground acceleration.* Section 13.3 has presented an in-depth discussion of the determination of the peak ground acceleration a_{max} for a given site. The higher the value of the peak ground acceleration a_{max}, the higher the value of k_h that should be used in the pseudostatic analysis.

2. *Earthquake magnitude.* The higher the magnitude of the earthquake, the longer the ground will shake (see Table 13.2) and consequently the higher the value of k_h that should be used in the pseudostatic analysis.

3. *Maximum value of k_h.* When considering items 1 and 2 as outlined earlier, the value of k_h should never be greater than the value of a_{max}/g.

4. *Minimum value of k_h.* There may be agency requirements that require a specific seismic coefficient. For example, a common requirement by many local agencies in California is the use of a minimum seismic coefficient $k_h = 0.15$ (Division of Mines and Geology, 1997).

5. *Size of the sliding mass.* A lower seismic coefficient should be used as the size of the slope failure mass increases. The larger the slope failure mass, the less likely that during the earthquake, the entire slope mass will be subjected to a destabilizing seismic force acting in the out-of-slope direction. Suggested guidelines are as follows:

   a. Small slide mass. A value of $k_h = a_{max}/g$ is often used for a small slope failure mass. Examples would include small rock falls or surficial stability analyses.

   b. Intermediate slide mass. A value of $k_h = 0.65 \, a_{max}/g$ is often used for slopes of moderate size (Krinitzsky et al., 1993; Taniguchi and Sasaki, 1986). Note that this value of 0.65 was used in the liquefaction analysis (i.e., see Eq. 13.9).

   c. Large slide mass. The lowest values of k_h are used for large failure masses, such as large embankments, dams, and landslides. Seed (1979b) recommended the following: $k_h = 0.10$ for sites near faults capable of generating magnitude 6.5 earthquakes with an acceptable pseudostatic factor of safety of 1.15 or greater and $k_h = 0.15$ for sites near faults capable of generating magnitude 8.5 earthquakes with an acceptable pseudostatic factor of safety of 1.15 or greater.

Other guidelines for the selection of the value of k_h include the following:

- Terzaghi (1950). Suggested the following values: $k_h = 0.10$ for *severe* earthquakes, $k_h = 0.20$ for *violent and destructive* earthquakes, and $k_h = 0.50$ for *catastrophic* earthquakes.

- Seed and Martin (1966) and Dakoulas and Gazetas (1986). Using shear beam models, they showed that the value of k_h for earth dams depends on the size of the failure mass. In particular, the value of k_h for a deep failure surface is substantially less than the value of k_h for a failure surface that does not extend far below the dam crest. This conclusion is identical to item 5 (size of sliding mass) as outlined earlier.

- Marcuson (1981). Suggested that for dams, $k_h = 0.33 \, a_{max}/g$ to $0.50 \, a_{max}/g$, and consideration should be given to possible amplification or deamplification of the seismic shaking due to the dam configuration.

- Hynes-Griffin and Franklin (1984). Based on a study of the earthquake records from over 350 accelerograms, they recommended $k_h = 0.50 \, a_{max}/g$ for earth dams. By using this seismic coefficient and having a pseudostatic factor of safety greater than 1.0, it was concluded that earth dams will not be subjected to *dangerously large* earthquake deformations.

• Kramer (1996).   Stated that the study on earth dams by Hynes-Griffin and Franklin (1984) would be appropriate for most slopes. Also Kramer indicated that there are no hard and fast rules for the selection of the pseudostatic coefficient for slope design, but that it should be based on the actual anticipated level of acceleration in the failure mass (including any amplification or deamplification effects).

Because of the tedious nature of the calculations, computer programs are routinely used to perform the pseudostatic analysis. Most slope stability computer programs have the ability to perform pseudostatic slope stability analyses and the only additional data that need to be inputted is the seismic coefficient k_h. In southern California, an acceptable minimum factor of safety of the slope is 1.1 to 1.15 for a pseudostatic slope stability analysis (Day and Poland, 1996).

Newmark Method.   The purpose of the Newmark (1965) method is to estimate the slope deformation for those cases where the pseudostatic factor of safety is less than 1.0 (i.e., the failure condition). The Newmark method assumes that the slope will only deform during those portions of the earthquake when the out-of-slope earthquake forces cause the pseudostatic factor of safety to drop below 1.0. When this occurs, the slope will no longer be stable and it will be accelerated downslope. The longer the time that the slope is subjected to a pseudostatic factor of safety below 1.0, the greater the amount of slope deformation. On the other hand, if the pseudostatic factor of safety drops below 1.0 for a mere fraction of a second, then the slope deformation will be limited.

Figure 13.23 can be used to illustrate the basic premise of the Newmark method. Part (*a*) in this figure shows the horizontal acceleration of the slope during an earthquake. Those accelerations that plot above the zero line are considered to be out-of-slope accelerations, while those accelerations that plot below the zero line are considered to be into-the-slope accelerations. It is only the out-of-slope accelerations that cause downslope movement, and thus only the acceleration that plots above the zero line will be considered in the analysis. In Part (*a*) of Fig. 13.23, a dashed line has been drawn that corresponds to the horizontal yield acceleration, which is designated a_y. This horizontal yield acceleration a_y is considered to be the horizontal earthquake acceleration that results in a pseudostatic factor of safety that is exactly equal to 1.0. The potions of the two acceleration pulses that plot above a_y have been darkened. According to the Newmark method, it is these darkened portions of the acceleration pulses that will cause lateral movement of the slope.

Parts (*b*) and (*c*) in Fig. 13.23 present the corresponding horizontal velocity and slope displacement that occurs in response to the darkened portions of the two acceleration pulses. Note that the slope displacement is incremental and only occurs when the horizontal acceleration from the earthquake exceeds the horizontal yield acceleration a_y. The magnitude of the slope displacement would depend on the following factors:

1. *Horizontal yield acceleration (a_y).*   The higher the horizontal yield acceleration a_y, the more stable the slope is for any given earthquake.

2. *Peak ground acceleration (a_{max}).*   The peak ground acceleration a_{max} represents the highest value of the horizontal ground acceleration. In essence, this is the amplitude of the maximum acceleration pulse. The greater the difference between the peak ground acceleration a_{max} and the horizontal yield acceleration a_y, the larger the downslope movement.

3. *Length of time.*   The longer the length of time the earthquake acceleration exceeds the horizontal yield acceleration a_y, the larger the downslope deformation. Considering the combined effects of items 2 and 3, it can be concluded that the larger the shaded area shown in Part (*a*) of Fig. 13.23, the greater the downslope movement.

4. *Number of acceleration pulses.*   The larger the number of acceleration pulses that exceed the horizontal yield acceleration a_y, the greater the cumulative downslope movement during the earthquake.

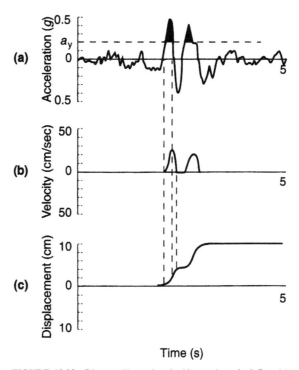

FIGURE 13.23   Diagram illustrating the Newmark method. Part (*a*) shows the acceleration versus time, Part (*b*) shows the velocity versus time for the darkened portions of the acceleration pulses, and Part (*c*) shows the corresponding downslope displacement versus time in response to the velocity pulses. (*After Wilson and Keefer, 1985.*)

Many different equations have been developed utilizing the basic Newmark method as outlined earlier. One simple equation that is based on the use of two of the four main parameters discussed earlier, is as follows (Ambraseys and Menu, 1988):

$$\log_{10}d = 0.90 + \log_{10}[(1 - a_y/a_{max})^{2.53}\,(a_y/a_{max})^{-1.09}]\tag{13.13}$$

where d = estimated downslope movement caused by the earthquake (cm)

a_y = yield acceleration, defined as the horizontal earthquake acceleration that results in a pseudostatic factor of safety that is exactly equal to 1.0 (ft/sec^2 or m/sec^2)

a_{max} = PGA of the design earthquake (ft/sec^2 or m/sec^2)

Based on the Newmark method, Eq. 13.13 is only valid for those cases where the pseudostatic factor of safety is less than 1.0. In essence, the peak ground acceleration a_{max} must be greater then the horizontal yield acceleration a_y. In order to use Eq. 13.13, the first step would be to determine the pseudostatic factor of safety. Provided the pseudostatic factor of safety is less than 1.0, the next step would be to reduce the value of the seismic coefficient k_h until a factor of safety exactly equal to 1.0 is obtained. This can usually be quickly accomplished when using a slope stability computer program. The value of k_h that corresponds to a pseudostatic factor of safety equal to 1.0 multiplied by the acceleration of gravity g is equal to the horizontal yield acceleration a_y. Substituting the values of the peak ground acceleration a_{max} and the yield acceleration a_y into Eq. 13.13, the slope deformation in centimeters can then be determined.

Because Eq. 13.13 uses the peak ground acceleration a_{max} from the earthquake, the analysis would tend to be more accurate for small or medium sized failure masses where the seismic coefficient k_h is approximately equal to a_{max}/g.

Example Problem 13.2   Assume that for a slope the pseudostatic factor of safety = 0.734 for a peak ground acceleration a_{max} = 0.40 g (i.e., the seismic coefficient k_h is equal to 0.40). Since the pseudostatic factor of safety is less than 1.0, the Newmark method can be used to estimate the slope deformation. Further assume that a value of k_h = 0.22 corresponds to a pseudostatic factor of safety of 1.0. Determine the lateral movement of this slope using Eq. 13.13.

Solution   For a psuedostatic factor of safety = 1.0, the value of k_h is equal to 0.22 and thus the yield acceleration a_y is equal to 0.22g. Substituting the ratio of a_y/a_{max} = 0.22g/0.40g = 0.55 into Eq. 13.13, the result is as follows:

$$\log_{10} d = 0.90 + \log_{10}[(1 - 0.55)^{2.53} \, (0.55)^{-1.09}]$$

or:

$$\log_{10} d = 0.90 + \log_{10}(0.254) = 0.306$$

Solving the above equation, the slope deformation d is equal to about 2 cm. Thus, although the pseudostatic factor of safety is well below 1.0 (i.e., pseudostatic factor of safety = 0.734), Eq. 13.13 predicts that only about 2 cm of downslope movement will occur during the earthquake.

Limitation of Newmark Method.   The major assumption of the Newmark method is that the slope will only deform when the peak ground acceleration a_{max} exceeds the yield acceleration a_y. This type of analysis would be most appropriate for a slope that deforms as a single massive block, such as a wedge type failure. In fact, Newmark (1965) used the analogy of a sliding block on an inclined plane in order to develop the displacement equations.

A limitation of the Newmark method is that it may prove unreliable for those slopes that do not tend to deform as a single massive block. An example would be a slope composed of dry and loose granular soil (i.e., sands and gravels). The individual soil grains that compose a dry and loose granular soil will tend to individually deform, rather than the entire slope deforming as one massive block. Thus a slope composed of dry and loose sand could both settle and deform laterally even if the pseudostatic factor of safety is greater than 1.0. The Newmark (1965) method should only be used for slopes that will deform as an intact massive block, and not for those cases of individual soil particle movement (such as a dry and loose granular soil).

13.5.5 Mitigation of Slope Hazards

In order to evaluate the effect of the earthquake-induced slope movement upon the structure, the first step is to estimate the stability of the slope or the amount of lateral movement. If the factor of safety of the slope is too low or if the anticipated earthquake-induced lateral movement exceeds the allowable lateral movement, then slope stabilization options will be required.

In general, mitigation options can be divided into three basic categories, as follows (Division of Mines and Geology, 1997):

1. *Avoid the failure hazard.*   Where the potential for failure is beyond the acceptable level and not preventable by practical means, as in mountainous terrain subject to massive planar slides or rock and debris avalanches, the hazard should be avoided. Developments should be built sufficiently

far away from the threat that they will not be affected even if the slope does fail. Planned development areas on the slope or near its base should be avoided and relocated to areas where stabilization is feasible.

2. *Protect the site from the failure.*   While it is not always possible to prevent slope failures occurring above a project site, it is sometimes possible to protect the site from the runout of failed slope materials. This is particularly true for sites located at or near the base of steep slopes that can receive large amounts of material from shallow disaggregated landslides or debris flows. Methods include catchment and/or protective structures such as basins, embankments, diversion or barrier walls, and fences. Diversion methods should only be employed where the diverted landslide materials will not affect other sites.

3. *Reduce the hazard to an acceptable level.*   Unstable slopes affecting a project can be rendered stable (i.e., by increasing the factor of safety to >1.5 for static and >1.1 for dynamic loads) by eliminating the slope, removing the unstable soil and rock materials, or applying one or more appropriate slope stabilization methods (such as buttress fills, subdrains, soil nailing, crib walls, and the like). For deep-seated slope instability, strengthening the design of the structure (e.g., reinforced foundations) is generally not by itself an adequate mitigation measure.

NOTATION

The following notation is used in this chapter:

a = acceleration

a_{max} = peak ground acceleration

a_y = horizontal yield acceleration

A = maximum trace amplitude recorded by a Wood-Anderson seismograph

A_f = area of the fault plane

A_o = maximum trace amplitude for the smallest recorded earthquake

c' = cohesion based on an effective stress analysis

C_b = borehole diameter correction

C_r = rod length correction

CRR = cyclic resistance ratio

CSR = cyclic stress ratio

d = downslope movement caused by the earthquake

D = average displacement of the ruptured segment of the fault

D_r = relative density

E_m = hammer efficiency

F = horizontal earthquake force

FS, FS_L = factor of safety against liquefaction

F_h = horizontal pseudostatic force

g = acceleration of gravity

k_h = seismic coefficient, also known as the *pseudostatic* coefficient

k_o = coefficient of lateral earth pressure at rest

LL = liquid limit

m = mass of the soil column

m = total mass of the slide material (Sec. 13.5.4)

$$m_b = \text{body-wave magnitude scale}$$
$$M_{\text{JMA}} = \text{Japanese meteorological agency magnitude scale}$$
$$M_L = \text{local magnitude of the earthquake}$$
$$M_o = \text{seismic moment of the earthquake}$$
$$M_S = \text{surface wave magnitude of the earthquake}$$
$$M_w = \text{moment magnitude of the earthquake}$$
$$\text{MSF} = \text{magnitude scaling factor}$$
$$(N_1)_{60} = N \text{ value corrected for field testing procedures and overburden pressure}$$
$$\text{OCR} = \text{overconsolidation ratio}$$
$$r_d = \text{depth reduction factor}$$
$$\text{SPT} = \text{standard penetration test}$$
$$w = \text{water content}$$
$$W = \text{weight of the soil column}$$
$$W = \text{total weight of the slide material (Sec. 13.5.4)}$$
$$z = \text{depth below ground surface}$$
$$\phi' = \text{friction angle based on an effective stress analysis}$$
$$\gamma_t = \text{total unit weight of the soil}$$
$$\mu = \text{shear modulus of the material along the fault plane}$$
$$\sigma_{vo} = \text{total vertical stress}$$
$$\sigma'_{vo} = \text{vertical effective stress}$$
$$\tau_{\text{cyc}} = \text{uniform cyclic shear stress amplitude of the earthquake}$$
$$\tau_{\text{max}} = \text{maximum shear stress}$$

PROBLEMS

Solutions to the problems are presented in App. C of this book. The problems have been divided into basic categories as indicated below:

Soil Liquefaction

13.1 Figure 13.24 shows the subsoil profile at Kawagishi-cho in Niigata. Assume a level-ground site with the groundwater table at a depth of 1.5 m below ground surface, the medium sand and medium-fine sand have less than 5 percent fines, the total unit weight γ_t of the soil above the groundwater table is 18.3 kN/m³, and the buoyant unit weight γ_b of the soil below the groundwater table is 9.7 kN/m³. The standard penetration data shown in Fig. 13.24 are uncorrected N values. Assume a hammer efficiency E_m of 60 percent, boring diameter of 100 mm, and the length of drill rods is equal to the depth of the SPT test below ground surface. The earthquake conditions are a peak ground acceleration a_{max} of 0.16g and a magnitude of 7.5. Using the standard penetration test data, determine the factor of safety against liquefaction versus depth.

ANSWER: See App. C for the solution.

13.2 In Fig. 13.24, assume the cyclic resistance ratio (labeled *cyclic strength* in Fig. 13.24) for the soil was determined by modeling the earthquake conditions in the laboratory (i.e., the amplitude and number of cycles of the sinusoidal load are equivalent to $a_{\text{max}} = 0.16g$ and magnitude = 7.5). Using the

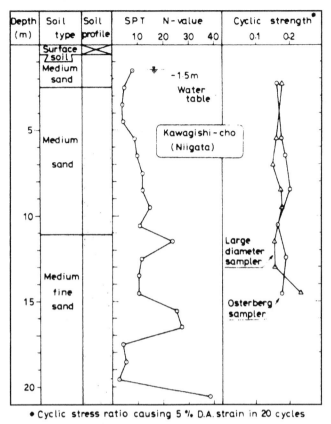

FIGURE 13.24 Subsoil profile, Kawagishi-cho, Niigata. (*Reproduced from Ishihara, 1985.*)

laboratory cyclic strength tests performed on large diameter samples, determine the factor of safety against liquefaction versus depth.

ANSWER: See App. C for the solution.

13.3 Based on the results from Problems 13.1 and 13.2, what zones of soil will liquefy during the earthquake?

ANSWER: The standard penetration test data indicate that there are three zones of liquefaction from about 2 to 11 m, 12 to 15 m, and 17 to 20 m below ground surface. The laboratory cyclic strength tests indicate that there are two zones of liquefaction from about 6 to 8 m and 10 to 14 m below ground surface.

13.4 Figure 13.25 shows the subsoil profile at a sewage disposal site in Niigata. Assume a level-ground site with the groundwater table at a depth of 0.4 m below ground surface, the medium to coarse sand has less than 5 percent fines, the total unit weight γ_t of the soil above the groundwater table is 18.3 kN/m³, and the buoyant unit weight γ_b of the soil below the groundwater table is 9.7 kN/m³. The standard penetration data shown in Fig. 13.25 are uncorrected N values. Assume a hammer efficiency E_m of 60 percent, boring diameter of 100 mm, and the length of drill rods is equal to the depth

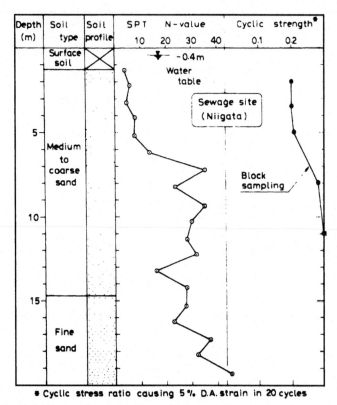

FIGURE 13.25 Subsoil profile, sewage site, Niigata. (*Reproduced from Ishihara, 1985.*)

of the SPT test below ground surface. The earthquake conditions are a peak ground acceleration a_{max} of 0.16g and a magnitude of 7.5. Using the standard penetration test data, determine the factor of safety against liquefaction versus depth.

ANSWER: See App. C for the solution.

13.5 In Fig. 13.25, assume the cyclic resistance ratio (labeled *cyclic strength* in Fig. 13.25) for the soil was determined by modeling the earthquake conditions in the laboratory (i.e., the amplitude and number of cycles of the sinusoidal load are equivalent to a_{max} = 0.16g and magnitude = 7.5). Using the laboratory cyclic strength tests performed on block samples, determine the factor of safety against liquefaction versus depth.

ANSWER: See App. C for the solution.

13.6 Based on the results from Problems 13.4 and 13.5, what zones of soil would be most likely to liquefy?

ANSWER: The standard penetration test data indicate that there are two zones of liquefaction from about 1.2 to 6.7 m and 12.7 to 13.7 m below ground surface. The laboratory cyclic strength tests indicate that the soil has a factor of safety against liquefaction in excess of 1.0.

13.7 Figure 13.26 presents *before improvement* and *after improvement* standard penetration resistance profiles at a warehouse site. Assume a level-ground site with the groundwater table at a depth

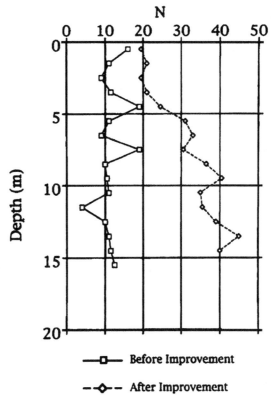

FIGURE 13.26 Example of pre- and post- treatment standard penetration resistance profiles at a site.

of 0.5 m below ground surface, the soil type is a silty sand with an average of 15 percent fines, the total unit weight γ_t of the soil above the groundwater table is 18.9 kN/m³, and the buoyant unit weight γ_b of the soil below the groundwater table is 9.8 kN/m³. Neglect any increase in unit weight of the soil due to the improvement process. The standard penetration data shown in Fig. 13.26 are uncorrected N values. Assume a hammer efficiency E_m of 60 percent, boring diameter of 100 mm, and the length of drill rods is equal to the depth of the SPT test below ground surface. The design earthquake conditions are a peak ground acceleration a_{max} of 0.40g and moment magnitude M_w of 8.5. Determine the factor of safety against liquefaction for the before improvement and after improvement conditions. Was the improvement process effective in reducing the potential for liquefaction at the warehouse site?

ANSWER: See App. C for the solution. Since the after improvement factor of safety against liquefaction exceeds 1.0 for the design earthquake, the improvement process was effective in eliminating liquefaction potential at the site.

Slope Stability

13.8 A slope has a height of 10 m and the slope face is inclined at a 2:1 (horizontal:vertical) ratio. Assume a wedge type analysis where the slip surface is planar through the toe of the slope and is inclined at a 3:1 (horizontal:vertical) ratio. The total unit weight of the slope material $\gamma_t = 18$ kN/m³.

Using the undrained shear strength parameters of $c = 15$ kPa and $\phi = 0$, calculate the factor of safety (F) for the static case and an earthquake condition of $k_h = 0.3$. Assume that the shear strength does not decrease with strain (i.e., not a weakening type soil).

ANSWER: Static $F = 1.67$, pseudostatic $F = 0.88$.

13.9   Use the data from Problem 13.8, except assume that the slip surface has an effective shear strength of $c' = 4$ kPa and $\phi' = 29°$. Also assume that piezometers have been installed along the slip surface and the average measured steady-state pore water pressure $u = 5$ kPa. Calculate the factor of safety (F) of the failure wedge based on an effective stress analysis for the static case and an earthquake condition of $k_h = 0.2$. Assume that the shear strength does not decrease with strain (i.e., not a weakening type soil) and the pore water pressures will not increase during the earthquake.

ANSWER: Static $F = 1.80$, pseudostatic $F = 1.06$.

13.10   A near vertical rock slope has a continuous horizontal joint through the toe of the slope and another continuous vertical joint located 10 ft back from the top of the slope. The height of the rock slope is 20 ft and the unit weight of the rock is 140 pcf. The shear strength parameters for the horizontal joint are $c' = 0$ and $\phi' = 40°$ and the pore water pressure u is equal to zero. For the vertical joint, assume zero shear strength. Neglecting possible rotation of the rock block and considering only a sliding failure, calculate the pseudostatic factor of safety (F) if $k_h = 0.50$.

ANSWER: Pseudostatic $F = 1.68$.

13.11   Use the data from Problem 13.8 and calculate the slope deformation based on the Newmark method (i.e., Eq. 13.13).

ANSWER: $d = 0.4$ cm

13.12   Use the data from Problem 13.9 and calculate the slope deformation based on the Newmark method (i.e., Eq. 13.13).

ANSWER: Since pseudostatic $F > 1.0$, $d = 0$.

13.13   Use the data from Problem 13.1. Assume the subsoil profile shown in Fig. 13.24 pertains to a level ground site that is adjacent to a river bank. The river bank has a 3:1 (horizontal:vertical) slope inclination and assume that the average level of water in the river corresponds to the depth of the groundwater table (i.e., 1.5 m below ground surface). Further assume that the depth of water in the river is 9 m. Will the river bank experience a flow failure during the design earthquake? What type of flow failure is expected? Assume that a correction need not be applied to Eq. 13.9 for the sloping ground condition (i.e., the factor of safety against liquefaction is the same for the level ground and sloping ground).

ANSWER: A mass liquefaction flow failure would develop during the earthquake.

CHAPTER 14
GEOTECHNICAL EARTHQUAKE ENGINEERING FOR FOUNDATIONS AND RETAINING WALLS

14.1 INTRODUCTION

As discussed in Chap. 13, the actual rupture of the ground due to fault movement could damage a structure. Secondary effects, such as the liquefaction of loose granular soil and slope movement or failure could also cause structural damage. This chapter will discuss some of the other earthquake-induced effects or structural conditions that can result in damage to foundations and retaining walls. Topics will include earthquake-induced settlement and foundation alternatives to mitigate earthquake effects.

Earthquakes throughout the world cause a considerable amount of death and destruction. Earthquake damage can be classified as being either structural or nonstructural. For example, the Federal Emergency Management Agency (1994) states:

> Damage to buildings is commonly classified as either structural or non-structural. Structural damage means the building's structural support has been impaired. Structural support includes any vertical and lateral force resisting systems, such as the building frames, walls, and columns. Non-structural damage does not affect the integrity of the structural support system. Examples of non-structural damage include broken windows, collapsed or rotated chimneys, and fallen ceilings. During an earthquake, buildings get thrown from side to side, and up and down. Heavier buildings are subjected to higher forces than light-weight buildings, given the same acceleration. Damage occurs when structural members are overloaded, or differential movements between different parts of the structure strain the structural components. Larger earthquakes and longer shaking durations tend to damage structures more. The level of damage resulting from a major earthquake can be predicted only in general terms, since no two buildings undergo the exact same motions during a seismic event. Past earthquakes have shown us, however, that some buildings are likely to perform more poorly than others.

There are four main factors that cause structural damage during an earthquake, as follows:

1. *Strength of shaking.* For small earthquakes (magnitude less than 6), the strength of shaking decreases rapidly with distance from the epicenter of the earthquake. According to the United States Geological Survey (2000b), the strong shaking along the fault segment that slips during an earthquake becomes about half as strong at a distance of 8 mi, a quarter as strong at a distance of 17 mi, an eighth as strong at a distance of 30 mi, and a sixteenth as strong at a distance of 50 mi.

   In the case of a small earthquake, the center of energy release and the point where slip begins is not far apart. But in the case of large earthquakes, which have a significant length of fault rupture, these two points may be hundreds of miles apart. Thus for big earthquakes, the strength of shaking decreases in a direction away from the fault rupture.

2. *Length of shaking.*   The length of shaking depends on how the fault breaks during the earthquake. For example, the maximum shaking during the Loma Prieta earthquake lasted only 10 to 15 sec But during other magnitude earthquakes in the San Francisco bay area, the shaking may last 30 to 40 sec. The longer the ground shakes, the greater the potential for structural damage. In general, the higher the magnitude of an earthquake, the longer the duration of the shaking ground (see Table 13.2).

3. *Type of subsurface conditions.*   Ground shaking can be increased if the site has a thick deposit of soil that is soft and submerged. Many other subsurface conditions can cause or contribute to structural damage. For example, as discussed in Sec. 13.4, there could be structural damage due to liquefaction of loose submerged sands.

4. *Type of building.*   Certain types of buildings and other structures are especially susceptible to the side-to-side shaking common during earthquakes. For example, sites located within approximately 10 mi (16 km) of the epicenter or location of fault rupture are generally subjected to rough, jerky, and high frequency seismic waves that are often more capable of causing short buildings to vibrate vigorously. For sites located at greater distance, the seismic waves often develop into longer period waves that are more capable of causing high-rise buildings and buildings with large floor areas to vibrate vigorously (Federal Emergency Management Agency, 1994).

Much like diseases will attack the weak and infirm, earthquakes damage those structures that have inherent weaknesses or age-related deterioration. Those buildings that are nonreinforced, poorly constructed, weakened from age or rot, or underlain by soft or unstable soil are most susceptible to damage. The next section will discuss some of these susceptible structures.

14.2   EARTHQUAKE STRUCTURAL DAMAGE

The purpose of this section is to present examples of structural damage caused by earthquakes. A lot of this damage is related to foundation conditions. For example, shear walls, which will be discussed in Sec. 14.2.4, can be ineffective and severely damaged if they are not adequately attached to the foundation.

14.2.1   Torsion

Torsional problems develop when the center of mass of the structure is not located at the center of its lateral resistance, which is also known as the *center of rigidity*. A common example is a tall building that has a first floor area consisting of a space that is open and supports the upper floors by the use of isolated columns, while the remainder of the first floor area contains solid load-bearing walls that are interconnected. The open area having isolated columns will typically have much less lateral resistance than that part of the floor containing the interconnected load-bearing walls. While the center of mass of the building may be located at the mid-point of the first floor area, the center of rigidity is offset toward the area containing the interconnected load-bearing walls. During the earthquake, the center of mass will twist about the center of rigidity, causing torsional forces to be induced into the building frame.

14.2.2   Soft Story

A *soft story*, also known as a *weak story*, is defined as a story in a building that has substantially less resistance, or stiffness, than the stories above or below it. In essence, a soft story has inadequate shear resistance or inadequate ductility (energy absorption capacity) to resist the earthquake-induced building stresses. Although not always the case, the usual location of the soft story is at the ground floor of the building. This is because many buildings are designed to have an open first-floor area that is easily accessible to the public. Thus the first floor may contain large open areas between columns, without adequate shear resistance. The earthquake-induced building movement also causes the first floor to be subjected to the greatest stress, which compounds the problem of a soft story on the ground floor.

Concerning soft stories, the National Information Service for Earthquake Engineering (2000) states:

> In shaking a building, an earthquake ground motion will search for every structural weakness. These weaknesses are usually created by sharp changes in stiffness, strength and/or ductility, and the effects of these weaknesses are accentuated by poor distribution of reactive masses. Severe structural damage suffered by several modern buildings during recent earthquakes illustrates the importance of avoiding sudden changes in lateral stiffness and strength. A typical example of the detrimental effects that these discontinuities can induce is in the case of buildings with a 'soft story.' Inspections of earthquake damage as well as the results of analytical studies have shown that structural systems with a soft story can lead to serious problems during severe earthquake ground shaking. [Numerous examples] illustrate such damage and therefore emphasize the need for avoiding the soft story by using an even distribution of flexibility, strength, and mass.

Three examples of buildings having a soft story on the ground floor are as follows:

1. *Chi-chi earthquake in Taiwan on September 21, 1999.*   In Taiwan, it is common practice to have an open first floor area by using columns to support the upper floors. In some cases, the spaces between the columns are filled-in with plate-glass windows in order to create ground-floor shops. Figure 14.1 shows an example of this type of construction and the resulting damage caused by the Chi-chi earthquake.

2. *Northridge earthquake in California on January 17, 1994.*   Many apartment buildings in southern California contain a parking garage on the ground floor. In order to provide an open area for

FIGURE 14.1   Damage due to a soft story at the ground floor. The damage occurred during the Chi-chi earthquake in Taiwan on September 21, 1999. (*Photograph from the USGS Earthquake Hazards Program, NEIC, Denver.*)

the ground floor parking area, isolated columns are used to support the upper floors. These isolated columns often do not have adequate shear resistance and are susceptible to collapse during an earthquake. For example, Figs. 14.2 and 14.3 show the collapse of an apartment building during the Northridge earthquake caused by the weak shear resistance of the first floor garage area.

3. *Kocaeli earthquake in Turkey on August 17, 1999.*   In terms of building conditions during this earthquake, it has been stated (Bruneau, 1999):

> A typical reinforced concrete frame building in Turkey consists of a regular, symmetric floor plan, with square or rectangular columns and connecting beams. The exterior enclosures as well as interior partitioning are of non-bearing unreinforced brick masonry infill walls. These walls contributed significantly to the lateral stiffness of buildings during the earthquake and, in many instances, controlled the lateral drift and resisted seismic forces elastically. This was especially true in low-rise buildings, older buildings where the ratio of wall to floor area was very high, and buildings located on firm soil. Once the brick infills failed, the lateral strength and stiffness had to be provided by the frames alone, which then experienced significant inelasticity in the critical regions. At this stage, the ability of reinforced concrete columns, beams, and beam-column joints to sustain deformation demands depended on how well the seismic design and detailing requirements were followed both in design and in construction.
>
> A large number of residential and commercial buildings were built with soft stories at the first-floor level. First stories are often used as stores and commercial areas, especially in the central part of cities. These areas are enclosed with glass windows, and sometimes with a single masonry infill at the back. Heavy masonry infills start immediately above the commercial floor. During the earthquake, the presence of a soft story increased deformation demands very significantly, and put the burden of energy dissipation on the first-story columns. Many failures and collapses can be attributed to the increased deformation demands caused by soft stories, coupled with lack of deformability of poorly designed columns. This was particularly evident on a commercial street where nearly all buildings collapsed towards the street.

Examples of this soft story condition are shown in Figs. 14.4 and 14.5.

Concerning the retrofitting of a structure that has a soft story, the National Information Service for Earthquake Engineering (2000) states:

> There are many existing buildings in regions of high seismic risk that, because of their structural systems and/or of the interaction with non-structural components, have soft stories with either inadequate shear resistance or inadequate ductility (energy absorption capacity) in the event of being subjected to

FIGURE 14.2   Building collapse caused by a soft story due to the parking garage on the first floor. The building collapse occurred during the Northridge earthquake in California on January 17, 1994.

FIGURE 14.3 View inside the collapsed first floor parking garage (the arrows point to the columns). The building collapse occurred during the Northridge earthquake in California on January 17, 1994.

FIGURE 14.4 Damage caused by a soft story at the first-floor level. The damage occurred during the Kocaeli earthquake in Turkey on August 17, 1999. (*Photograph by Mehmet Celebi, USGS.*)

FIGURE 14.5   Building collapse caused by a soft story at the first-floor level. The damage occurred during the Kocaeli earthquake in Turkey on August 17, 1999. (*Photograph by Mehmet Celebi, USGS.*)

severe earthquake ground shaking. Hence they need to be retrofitted. Usually the most economical way of retrofitting such a building is by adding proper shear walls or bracing to the soft stories.

14.2.3   Pancaking

Pancaking occurs when the earthquake shaking causes a soft story to collapse, leading to total failure of the overlying floors. These floors crush and compress together such that the final collapsed condition of the building consists of one floor stacked on top of another, much like a stack of pancakes.

Pancaking of reinforced concrete multistory buildings was common throughout the earthquake-stricken region of Turkey due to the Kocaeli earthquake on August 17, 1999. Concerning the damage caused by the Kocaeli earthquake, Bruneau (1999) states:

> Pancaking is attributed to the presence of 'soft' lower stories and insufficiently reinforced connections at the column-beam joints. Most of these buildings had a 'soft' story—a story with most of its space unenclosed—and a shallow foundation and offered little or no lateral resistance to ground shaking. As many as 115,000 of these buildings—some engineered, some not—were unable to withstand the strong ground shaking and were either badly damaged or collapsed outright, entombing sleeping occupants beneath the rubble. Partial collapses involved the first two stories. The sobering fact is that Turkey still has an existing inventory of several hundred thousand of these highly vulnerable buildings. Some will need to undergo major seismic retrofits; others will be demolished.

An example of pancaking caused by the Kocaeli earthquake is shown in Fig. 14.6.

14.2.4   Shear Walls

There are many different types of structural systems that can be used to resist the inertia forces in a building that are induced by the earthquake ground motion. For example, the structural engineer

FIGURE 14.6   Pancaking of a building during the Kocaeli earthquake in Turkey on August 17, 1999. (*Photograph by Mehmet Celebi, USGS.*)

could use braced frames, moment resisting frames, and shear walls to resist the lateral earthquake-induced forces. Shear walls are designed to hold adjacent columns or vertical support members in place and then transfer the lateral forces to the foundation. The forces resisted by shear walls are predominantly shear forces, although a slender shear wall could also be subjected to significant bending (Arnold and Reitherman, 1982).

Common problems with shear walls are that they have inadequate strength to resist the lateral forces and that they are inadequately attached to the foundation. For example, having inadequate shear walls on a particular building level can create a soft story. A soft story could also be created if there is a discontinuity in the shear walls from one floor to the other, such as a floor where its shear walls are not aligned with the shear walls on the upper or lower floors.

14.2.5   Wood-Frame Structures

It is generally recognized that single-family wood frame structures that include shear walls in their construction are very resistant to collapse from earthquake shaking. This is due to several factors, such as their flexibility, strength, and light dead loads, which produce low earthquake-induced inertia loads. These factors make the wood-frame construction much better at resisting shear forces and hence more resistant to collapse.

There are exceptions to the general rule that wood frame structures are resistant to collapse. For example, in the 1995 Kobe earthquake, the vast majority of deaths were due to the collapse of one- and two-story residential and commercial wood-frame structures. More than 200,000 houses, about 10 percent of all houses in the Hyogo prefecture, were damaged, including over 80,000 collapsed houses, 70,000 severely damaged, and 7000 consumed by fire. The collapse of the houses has been attributed to several factors, such as (EQE Summary Report, 1995):

* Age-related deterioration, such as wood rot, that weakened structural members.

* Post and beam construction that often included open first floor areas (i.e., a soft first floor), with few interior partitions that were able to resist lateral earthquake loads.

- Weak connections between the walls and the foundation.
- Inadequate foundations that often consisted of stones or concrete blocks.
- Poor soil conditions consisting of thick deposits of soft or liquefiable soil that settled during the earthquake. Because of the inadequate foundations, the wood-frame structures were unable to accommodate the settlement.
- Inertia loads from heavy roofs that exceeded the lateral earthquake load-resisting capacity of the supporting walls. The heavy roofs were created by using thick mud or heavy tile and were used to resist the winds from typhoons. However, when the heavy roofs collapsed during the earthquake, they crushed the underlying structure.

14.2.6 Pounding Damage

Pounding damage can occur when two buildings are constructed close to each other and as they rock back-and-forth during the earthquake, they collide into each other. Even when two buildings having dissimilar construction materials or different heights are constructed adjacent to each other, it does not necessarily mean that they will be subjected to pounding damage.

The common situation for pounding damage is when a much taller building, which has a higher period and larger amplitude of vibration, is constructed against a squat and short building that has a lower period and smaller amplitude of vibration. Thus during the earthquake, the buildings will vibrate at different frequencies and amplitudes, and they can collide into each other. The effects of pounding can be especially severe if the floors of one building impact the other building at different elevations, so that, for example, the floor of one building hits a supporting column of an adjacent building.

It is very difficult to model the pounding effects of two structures and hence design structures to resist such damage. As a practical matter, the best design approach to prevent pounding damage is to provide sufficient space between the structures in order to avoid the problem. If two buildings must be constructed adjacent to each other, then one design feature should be to have the floors of both buildings at the same elevations, so that the floor of one building does not hit a supporting column of an adjacent building.

14.2.7 Tsunami and Seiche

Tsunami is a Japanese word that when translated into English means *harbor wave* and it is defined as an ocean wave that is created by a disturbance that vertically displaces a column of seawater. Tsunami can be generated by an oceanic meteorite impact, submarine landslide, or earthquake if the sea floor abruptly deforms and vertically displaces the overlying water. Earthquakes generated at sea-floor subduction zones are particularly effective in generating tsunamis. Waves are formed as the displaced water mass, which acts under the influence of gravity, attempts to regain its equilibrium.

Tsunami is different from a normal ocean wave in that it has a long period and wavelength. For example, tsunami can have a wavelength in excess of 60 mi (100 km) and a period on the order of 1 h. In the Pacific Ocean, where the typical water depth is about 13,000 ft (4000 m), tsunami travels at about 650 ft/sec (200 m/sec). Because the rate at which a wave loses its energy is inversely related to its wavelength, tsunami not only propagates at high speeds, it also can travel long transoceanic distances with limited energy losses.

The tsunami is transformed as it leaves the deep water of the ocean and travels into the shallower water near the coast. The tsunami's speed diminishes as it travels into the shallower coastal water and its height grows. While the tsunami may be imperceptible at sea, the shoaling effect near the coast causes the tsunami to grow to be several meters or more in height. When it finally reaches the coast, the tsunami may develop into a rapidly rising or falling tide, a series of breaking waves, or a tidal bore.

Seiche is identical to tsunami, except that it occurs in an inland body of water, such as a lake. It can be caused by lake-bottom earthquake movements or by volcanic eruptions and landslides within the lake. Seiche has been described as being similar to the sloshing of water in a bathtub.

Because of the tremendous destructive forces, options to mitigate structural damage caused by tsunami or seiche are often limited. Some possibilities include the construction of walls to deflect the surging water or the use of buildings having weak lower-floor partitions that will allow the water to flow through the building, rather than knocking it down.

14.3 FOUNDATION SETTLEMENT

14.3.1 Introduction

Previous sections have discussed various ways that an earthquake can damage foundations. Examples are as follows:

Tectonic Surface Effects

- Surface fault rupture, which can cause a foundation that straddles the fault to be displaced vertically and laterally (Sec. 13.2.2).
- Regional uplifting or subsidence associated with the tectonic movement (Sec. 13.2.3).

Liquefaction

- Liquefaction-induced bearing capacity failure of the foundation (Sec. 6.5). Localized liquefaction could also cause limited punching type failure of individual footings.
- Liquefaction-induced flow slides that can pull apart the foundation (Sec. 13.5.2).
- Liquefaction-induced localized or large-scale lateral spreading that can also pull apart the foundation (Sec. 13.5.2).

Seismic-Induced Slope Movement

- Seismic-induced slope movement or failure that could damage structures located at the top or toe of the slope (Sec. 13.5).

Those buildings founded on solid rock are least likely to experience earthquake-induced differential settlement. However, foundations on soil could be subjected to many different types of earthquake-induced settlement. Three conditions that can cause settlement of a structure founded on soil that will be discussed in this section are as follows:

1. *Liquefaction-induced settlement.* There could be foundation settlement associated with liquefaction of soil located well below the base of the foundation. In addition, there could also be settlement of the foundation due to a loss of soil through the development of ground surface sand boils (Sec. 14.3.3).
2. *Volumetric compression, also known as cyclic soil densification.* This type of settlement is due to ground shaking that causes the soil to compress together, which is often described as *volumetric compression* or *cyclic soil densification*. An example would be the settlement of dry and loose sands that densify during the earthquake, resulting in ground surface settlement (Sec. 14.3.4).
3. *Settlement due to dynamic loads caused by rocking.* This type of settlement is due to dynamic structural loads that momentarily increase the foundation pressure acting on the soil. The soil will deform in response to the dynamic structural load, resulting in settlement of the building. This settlement due to dynamic loads is often a result of the structure rocking back and forth (Sec. 14.3.5). Volumetric compression and rocking can also work in combination and cause settlement of the foundation.

14.3.2 Liquefaction-Induced Settlement

Table 14.1 summarizes the requirements and analyses for soil susceptible to liquefaction.
    The steps are as follows:

1. *Requirements.*   The first step is to determine if the two requirements listed in Table 14.1 are met. If either of these two requirements is not met, then the foundation is susceptible to failure during

TABLE 14.1   Requirements and Analyses for Soil Susceptible to Liquefaction

Requirements and analyses	Design conditions
Requirements	Bearing location of foundation. The foundation must not bear on soil that will liquefy during the design earthquake. Even lightly loaded foundations will sink into the liquefied soil. Surface layer H_1. There must be an adequate thickness of an unliquefiable soil surface layer H_1 in order to prevent damage due to sand boils and surface fissuring (see Fig. 14.10). Without this layer, there could be damage to shallow foundations, pavements, flatwork, and utilities.
Settlement analysis	Use Figs. 14.7 and 14.8 for the following conditions: Lightweight structures. Settlement of lightweight structures, such as wood frame buildings bearing on shallow foundations. Low net bearing stress. Settlement of any other type of structure that imparts a low net bearing pressure onto the soil. Floating foundation. Settlement of floating foundations, provided the zone of liquefaction is below the bottom of the foundation and the floating foundation does not impart a significant net stress upon the soil. Heavy structures with deep liquefaction. Settlement of heavy structures, such as massive buildings founded on shallow foundations, provided the zone of liquefaction is deep enough that the stress increase caused by the structural load is relatively low. Differential settlement. Differential movement between a structure and adjacent appurtenances, where the structure contains a deep foundation that is supported by strata below the zone of liquefaction.
Bearing capacity analysis	Use the analyses presented in Sec. 6.5 for the following conditions: Heavy buildings with underlying liquefied soil. Use a bearing capacity analysis when there is a soil layer below the bottom of the foundation that will be susceptible to liquefaction during the design earthquake. In this case, once the soil has liquefied, the foundation load could cause it to punch or sink into the liquefied soil, resulting in a bearing capacity failure (see Sec. 6.5.2). Check bearing capacity. Perform a bearing capacity analysis whenever the footing imposes a net pressure onto the soil and there is an underlying soil layer that will be susceptible to liquefaction during the design earthquake. Positive induced pore water pressures. For cases where the soil will not liquefy during the design earthquake, but there will be the development of excess pore water pressures, perform a bearing capacity analysis (see Sec. 6.5.3).
Special considerations	Buoyancy effects. Consider possible buoyancy effects. Examples include buried storage tanks or large pipelines that are within the zone of liquefied soil. Instead of settling, the buried storage tanks and pipelines may actually float to the surface when the ground liquefies. Sloping ground condition. Determine if the site is susceptible to liquefaction-induced flow slide or lateral spreading (see Sec. 13.5.2).

the design earthquake and special design considerations, such as the use of deep foundations or soil improvement, are required.

2. *Settlement analysis.*   Provided that the two design requirements are met, the next step is to perform a settlement analysis (to be discussed below). Note that in some cases, the settlement analysis is unreliable (e.g., heavy buildings with an underlying liquefied soil layer close to the bottom of the foundation).

3. *Bearing capacity analysis.*   A bearing capacity analysis is required when there is a possibility that the footing will punch into the liquefied soil layer (Sec. 6.5.2). For cases where the soil will not liquefy during the design earthquake, but there will be the development of excess pore water pressures, a bearing capacity analysis is also required (see Sec. 6.5.3).

4. *Special considerations.*   Special considerations may be required if the structure is subjected to buoyancy or if there is a sloping ground condition.

Method of Analysis by Ishihara and Yoshimine (1992). The settlement of the ground will occur as water drains from the liquefied soil due to the excess pore water pressures generated during the earthquake shaking. Even for a factor of safety against liquefaction that is greater than 1.0, there could still be the generation of excess pore water pressures and hence settlement of the soil. However, the amount of settlement will be much more for the liquefaction condition as compared to the nonliquefied state.

Figure 14.7 shows a chart developed by Ishihara and Yoshimine (1992) that can be used to estimate the ground surface settlement of saturated clean sands for a given factor of safety against liquefaction (FS_L). The procedure used in Fig. 14.7 is as follows:

1. *Calculate the factor of safety against liquefaction* (FS_L). The first step is to calculate the factor of safety against liquefaction using the procedure outlined in Sec. 13.4.

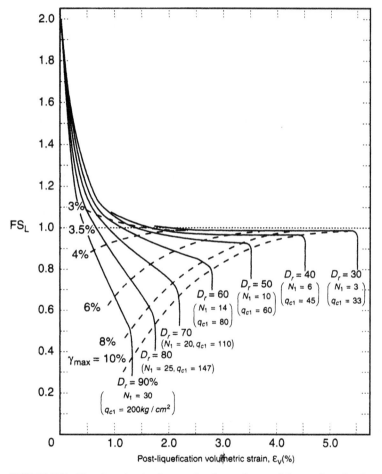

FIGURE 14.7 Chart for estimating the ground surface settlement of clean sand as a function of the factor of safety against liquefaction (FS_L). In order to use this figure, one of the following properties must be determined: relative density D_r of the in situ soil, maximum shear strain to be induced by the design earthquake (γ_{max}), corrected cone penetration resistance (q_{c1} in kg/cm^2), or Japanese standard penetration test N_1 value. For practical purposes, assume Japanese standard penetration test N_1 value is equal to the $(N_1)_{60}$ value from Eq. 2.5. (*Reproduced from Kramer, 1996; originally developed by Ishihara and Yoshimine, 1992.*)

2. *Soil properties.* The second step is to determine one of the following properties: relative density D_r of the in situ soil, maximum shear strain to be induced by the design earthquake (γ_{max}), corrected cone penetration resistance (q_{c1} in kg/cm^2), or Japanese standard penetration test N_1 value.

Kramer (1996) indicates that the Japanese standard penetration test typically transmits about 20 percent more energy to the SPT sampler and the equation: $N_1 = 0.83\ (N_1)_{60}$ can be used to convert the $(N_1)_{60}$ value into the Japanese N_1 value. However, Seed (1991) states that Japanese SPT results require corrections for blow frequency effects and hammer release and that these corrections are equivalent to an overall effective energy ratio E_m of 55 percent (versus $E_m = 60$ percent for U.S. safety hammer). Thus Seed (1991) states that the $(N_1)_{60}$ values should be increased by about 10 percent (i.e., 60/55) when using Fig. 14.7 to estimate volumetric compression, or $N_1 = 1.10\ (N_1)_{60}$. As a practical matter, it can be assumed that the Japanese N_1 value is approximately equivalent to the $(N_1)_{60}$ value calculated from Eq. 2.5 (Sec. 2.4.3).

3. *Volumetric strain.* In Fig. 14.7, enter the vertical axis with the factor of safety against liquefaction (FS_L), intersect the appropriate curve corresponding to the Japanese N_1 value [assume Japanese $N_1 = (N_1)_{60}$ from Eq. 2.5], and then determine the volumetric strain ε_v from the horizontal axis. Note in Fig. 14.7 that each N_1 curve can be extended straight downward in order to obtain the volumetric strain for very low values of the factor of safety against liquefaction.

4. *Settlement.* The settlement of the soil is calculated as the volumetric strain, expressed as a decimal, times the thickness of the liquefied soil layer.

Note in Fig. 14.7 that the volumetric strain can also be calculated for clean sand that has a factor of safety against liquefaction (FS_L) in excess of 1.0. For an FS_L greater than 1.0 but less than 2.0, the contraction of the soil structure during the earthquake shaking results in excess pore water pressures that will dissipate and cause a smaller amount of settlement. At an FS_L equal to or greater than 2.0, Fig. 14.7 indicates that the volumetric strain will be essentially equal to zero. This is because for an FS_L higher than 2.0, only small values of excess pore water pressures u_e will be generated during the earthquake shaking (i.e., see Fig. 4.25).

Method of Analysis by Tokimatsu and Seed (1984, 1987). Figure 14.8 shows a chart developed by Tokimatsu and Seed (1984, 1987) that can be used to estimate the ground surface settlement of saturated clean sands. The solid lines in Fig. 14.8 represent the volumetric strain for liquefied soil (i.e., factor of safety against liquefaction less than or equal to 1.0). Note that the solid line labeled 1 percent volumetric strain in Fig. 14.8 is similar to the dividing line in Fig. 13.17 between liquefiable and nonliquefiable clean sand.

The dashed lines in Fig. 14.8 represent the volumetric strain for a condition where excess pore water pressures are generated during the earthquake, but the ground shaking is not sufficient to cause liquefaction (i.e., $FS_L > 1.0$). This is similar to the data in Fig. 14.7, in that the contraction of the soil structure during the earthquake shaking could cause excess pore water pressures that will dissipate and result in smaller amounts of settlement. Thus by using the dashed lines in Fig. 14.8, the settlement of clean sands having a factor of safety against liquefaction in excess of 1.0 can also be calculated.

The procedure used in Fig. 14.8 is as follows:

1. *Calculate the cyclic stress ratio.* The first step is to calculate the cyclic stress ratio (CSR) by using Eq. 13.9. Usually a liquefaction analysis (Sec. 13.4) is first performed, and thus the value of CSR should have already been calculated.

2. *Adjusted CSR value.* Figure 14.8 was developed for a magnitude 7.5 earthquake. Tokimatsu and Seed (1987) suggest that the CSR calculated from Eq. 13.9 be adjusted if the magnitude of the anticipated earthquake is different from 7.5. The corrected CSR value is obtained by dividing the CSR value from Eq. 13.9 by the magnitude scaling factor from Table 13.5. The chart in Fig. 14.8 is entered on the vertical axis by using this corrected CSR value.

3. *$(N_1)_{60}$ value.* The next step is to calculate the $(N_1)_{60}$ value (Eq. 2.5, see Sec. 2.4.3). Usually a liquefaction analysis (Sec. 13.4) is first performed, and thus the value of $(N_1)_{60}$ should have already been calculated.

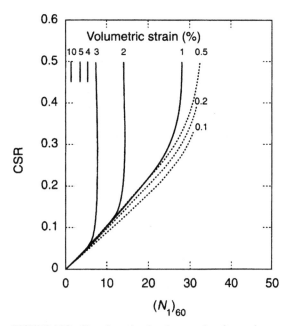

FIGURE 14.8 Chart for estimating the ground surface settlement of clean sand for factor of safety against liquefaction less than or equal to 1.0 (solid lines) and greater than 1.0 (dashed lines). In order to use this figure, the CSR from Eq. 13.9 and the $(N_1)_{60}$ value from Eq. 2.5 must be determined. (*Reproduced from Kramer, 1996; originally developed by Tokimatsu and Seed 1984.*)

4. *Volumetric strain.* In Fig. 14.8, the volumetric strain is determined by entering the vertical axis with the CSR from Eq. 13.9 and entering the horizontal axis with the $(N_1)_{60}$ value from Eq. 2.5.

5. *Settlement.* The settlement of the soil is calculated as the volumetric strain, expressed as a decimal, times the thickness of the liquefied soil layer.

Silty Soils. Figures 14.7 and 14.8 were developed for clean sand deposits (fines ≤ 5 percent). For silty soils, Seed (1991) suggests that the most appropriate adjustment is to increase the $(N_1)_{60}$ values by adding the values of N_{corr} indicated below:

Percent fines	N_{corr}
≤5	0
10	1
25	2
50	4
75	5

Limitations. The methods presented in Figs. 14.7 and 14.8 can only be used for the following cases:

• *Lightweight structures.* Settlement of lightweight structures, such as wood frame buildings bearing on shallow foundations.

- *Low net bearing stress.*   Settlement of any other type of structure that imparts a low net bearing pressure onto the soil.

- *Floating foundation.*   Settlement of floating foundations, provided the zone of liquefaction is below the bottom of the foundation and the floating foundation does not impart a significant net stress upon the soil.

- *Heavy structures with deep liquefaction.*   Settlement of heavy structures, such as massive buildings founded on shallow foundations, provided the zone of liquefaction is deep enough that the stress increase caused by the structural load is relatively low.

- *Differential settlement.*   Differential movement between a structure and adjacent appurtenances, where the structure contains a deep foundation that is supported by strata below the zone of liquefaction.

The methods presented in Figs. 14.7 and 14.8 cannot be used for the following cases:

- *Foundations bearing on liquefiable soil.*   Do not use Figs. 14.7 and 14.8 when the foundation is bearing on soil that will liquefy during the design earthquake. Even lightly loaded foundations will sink into the liquefied soil.

- *Heavy buildings with underlying liquefiable soil.*   Do not use Figs. 14.7 and 14.8 when the liquefied soil is close to the bottom of the foundation and the foundation applies a large net load onto the soil. In this case, once the soil has liquefied, the foundation load will cause it to punch or sink into the liquefied soil. There could even be a bearing capacity type failure. Obviously these cases will lead to settlement well in excess of the values obtained from Figs. 14.7 and 14.8. It is usually very difficult to determine the settlement for these conditions and the best engineering solution is to provide a sufficiently high static factor of safety so that there is ample resistance against a bearing capacity failure (see Sec. 6.5).

- *Buoyancy effects.*   Consider possible buoyancy effects. Examples include buried storage tanks or large pipelines that are within the zone of liquefied soil. Instead of settling, the buried storage tanks and pipelines may actually float to the surface when the ground liquefies.

- *Sloping ground condition.*   Do not use Figs. 14.7 and 14.8 when there is a sloping ground condition. If the site is susceptible to liquefaction-induced flow slide or lateral spreading, the settlement of the building could be well in excess of the values obtained from Figs. 14.7 and 14.8.

- *Liquefaction induced ground damage.*   The calculations using Figs. 14.7 and 14.8 do not include settlement that is related to the loss of soil through the development of ground surface sand boils or the settlement of shallow foundations caused by the development of ground surface fissures. These types of settlement are discussed in the next section.

Example Problem 14.1   Use the data from Example Problem 13.1. Assume that the liquefied soil layer is 1.0 m thick. As indicated in Example Problem 13.1, the factor of safety against liquefaction $FS_L = 0.26$ and the calculated value of $(N_1)_{60}$ determined at a depth of 3 m below ground surface is equal to 7.7. Calculate the ground surface settlement of the liquefied soil using Figs. 14.7 and 14.8.

Solution   Figure 14.7: Assume that the Japanese N_1 value is approximately equal to the $(N_1)_{60}$ value from Eq. 2.5, or use Japanese $N_1 = 7.7$. The Japanese N_1 curves labeled 6 and 10 are extended straight downward to a $FS_L = 0.26$ and then extrapolating between the curves for an N_1 value of 7.7, the volumetric strain is equal to 4.1 percent. Since the in situ liquefied soil layer is 1.0 m thick, the ground surface settlement of the liquefied soil is equal to 1.0 m times 0.041, or a settlement of 4.1 cm.

(Continued)

Figure 14.8: Per Example Problem 13.1, the CSR from Eq. 13.9 is equal to 0.34 and the calculated value of $(N_1)_{60}$ determined at a depth of 3 m below ground surface is equal to 7.7. Entering Fig. 14.8 with CSR = 0.34 and $(N_1)_{60}$ = 7.7, the volumetric strain is equal to 3.0 percent. Since the in situ liquefied soil layer is 1.0 m thick, the ground surface settlement of the liquefied soil is equal to 1.0 m times 0.030, or a settlement of 3.0 cm.

Based on the two methods, the ground surface settlement of the 1.0 m thick liquefied sand layer would be expected to be on the order of 3 to 4 cm.

Suppose instead of assuming the earthquake will have a magnitude of 7.5, this example problem is repeated for a magnitude $5\frac{1}{4}$ earthquake. As indicated in Table 13.5, the magnitude scaling factor = 1.5 and thus the corrected CSR is equal to 0.34 divided by 1.5, or 0.23. Entering Fig. 14.8 with the modified CSR = 0.23 and $(N_1)_{60}$ = 7.7, the volumetric strain is still equal to 3.0 percent. Thus, provided the sand liquefies for both the magnitude $5\frac{1}{4}$ and 7.5 earthquakes, the settlement of the liquefied soil is the same in this case.

14.3.3 Liquefaction-Induced Ground Damage

In addition to ground surface settlement caused by the liquefaction of soil, there could also be liquefaction induced ground damage that is illustrated in Fig. 14.9. As shown in this figure, there are two main aspects to the ground surface damage, as follows:

1. *Sand boils.* There could be liquefaction-induced ground loss below the structure, such as the loss of soil through the development of ground surface sand boils. Often a line of sand boils will be observed at ground surface. A row of sand boils will often develop at the location of cracks or fissures in the ground.

2. *Surface fissures.* The liquefied soil could also cause the development of ground surface fissures which break the overlying soil into blocks that open and close during the earthquake.

The liquefaction-induced ground conditions illustrated in Fig. 14.9 can damage all types of structures, such as buildings supported on shallow foundations, pavements, flatwork, and utilities. In terms of the main factor influencing the liquefaction-induced ground damage, Ishihara (1985) states:

> One of the factors influencing the surface manifestation of liquefaction would be the thickness of a mantle of unliquefied soils overlying the deposit of sand which is prone to liquefaction. Should the mantle near the ground surface be thin, the pore water pressure from the underlying liquefied sand deposit will be able to easily break through the surface soil layer, thereby bringing about the ground rupture such as sand boiling and fissuring. On the other hand, if the mantle of the subsurface soil is sufficiently thick, the uplift force due to the excess water pressure will not be strong enough to cause a breach in the surface layer, and hence, there will be no surface manifestation of liquefaction even if it occurs deep in the deposit.

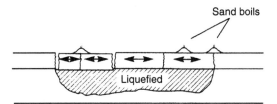

FIGURE 14.9 Ground damage caused by the liquefaction of an underlying soil layer. (*Reproduced from Kramer, 1996; originally developed by Youd, 1984.*)

Based on numerous case studies, Ishihara (1985) developed a chart [Part (*a*), Fig. 14.10] that can be used to determine the thickness of the unliquefiable soil surface layer (H_1) in order to prevent damage due to sand boils and surface fissuring. Three different situations were used by Ishihara (1985) in the development of the chart and they are shown in Part (*b*) of Fig. 14.10.

Since it is very difficult to determine the amount of settlement due to liquefaction-induced ground damage, one approach is to ensure that the site has an adequate surface layer of unliquefiable soil by using Fig. 14.10. If the site has an inadequate surface layer of unliquefiable soil, then mitigation measures such as the placement of fill at ground surface, soil improvement, or the construction of deep foundations may be needed.

In order to use Fig. 14.10, the thickness of layers H_1 and H_2 must be determined. Guidelines are as follows:

1. *Thickness of the unliquefiable soil layer (H_1).*   For two of the three situations in Part (*b*) of Fig. 14.10, the unliquefiable soil layer is defined as that thickness of soil located above the groundwater table. As previously mentioned, soil located above the groundwater table will not liquefy.

   One situation in Part (*b*) of Fig. 14.10 is for a portion of the unliquefiable soil below the groundwater table. Based on the case studies, this soil was identified as unliquefiable cohesive soil (Ishihara, 1985). As a practical matter, it would seem the unliquefiable soil below the groundwater table that is used to define the layer thickness H_1 would be applicable for any soil that has a factor of safety against liquefaction that is in excess of 1.0. However, if the factor of safety against liquefaction is only slightly in excess of 1.0, it could still liquefy due to the upward flow of water from layer H_2. Considerable experience and judgment are required when determining the thickness H_1 of the unliquefiable soil when a portion of this layer is below the groundwater table.

2. *Thickness of the liquefied soil layer (H_2).*   Note in Part (*b*) of Fig. 14.10 that for all three situations, the liquefied sand layer H_2 has an uncorrected N value that is less than or equal to 10. This N value data were applicable for the case studies evaluated by Ishihara (1985). It would seem that

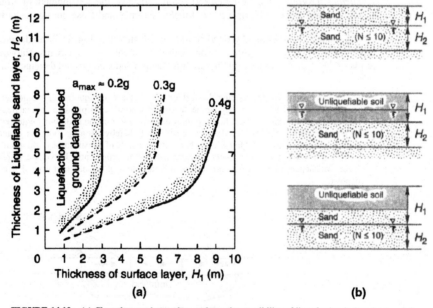

FIGURE 14.10   (a) Chart that can be used to evaluate the possibility of liquefaction induced ground damage based on H_1, H_2, and the peak ground acceleration a_{max}. (b) Three situations used for the development of the chart, where H_1 = thickness of the surface layer that will not liquefy during the earthquake and H_2 = thickness of the liquefiable soil layer. (*Reproduced from Kramer, 1996; originally developed by Ishihara, 1985.*)

irrespective of the N value, H_2 could be the thickness of the soil layer that has a factor of safety against liquefaction that is less than or equal to 1.0.

Example Problem 14.2 Based on standard penetration test data, a zone of liquefaction extends from a depth of 1.2 to 6.7 m below ground surface at a level ground site. Assume the surface soil (upper 1.2 m) consists of an unliquefiable soil. Using a peak ground acceleration a_{max} of 0.20g, will there be liquefaction-induced ground damage at this site?

Solution Since the zone of liquefaction extends from a depth of 1.2 to 6.7 m, the thickness of the liquefiable sand layer (H_2) is equal to 5.5 m. Entering Fig. 14.10 with H_2 = 5.5 m and intersecting the a_{max} = 0.2g curve, the minimum thickness of the surface layer H_1 needed to prevent surface damage is 3 m. Since the surface layer of unliquefiable soil is only 1.2 m thick, then there will be liquefaction induced ground damage.

Some appropriate solutions would be as follows: (1) at ground surface, add a fill layer that is at least 1.8 m thick, (2) densify the soil and hence improve the liquefaction resistance of the upper portion of the liquefiable layer, or (3) use a deep foundation supported by soil below the zone of liquefaction.

14.3.4 Volumetric Compression

Volumetric compression is also known as *soil densification*. This type of settlement is due to earthquake-induced ground shaking that causes the soil particles to compress together. Noncemented cohesionless soils, such as dry and loose sands or gravels, are susceptible to this type of settlement. Volumetric compression can result in a large amount of ground surface settlement. For example, Grantz et al. (1964) describe an interesting case of ground vibrations from the 1964 Alaskan Earthquake that caused 0.8 m (2.6 ft) of alluvium settlement.

Silver and Seed (1971) state that the earthquake-induced settlement of dry cohesionless soil depends on three main factors, as follows:

1. *Relative density D_r of the soil.* The looser the soil, the more susceptible it is to volumetric compression. Those cohesionless soils that have the lowest relative densities will be most susceptible to soil densification. Often the standard penetration test is used to assess the density condition of the soil.

2. *Maximum shear strain γ_{max} induced by the design earthquake.* The larger the shear strain induced by the earthquake, the greater the tendency for a loose cohesionless soil to compress together. The amount of shear strain will depend on the peak ground acceleration a_{max}. A higher value of a_{max} will lead to a greater shear strain of the soil.

3. *Number of shear strain cycles.* The more cycles of shear strain, the greater the tendency for the loose soil structure to compress. For example, it is often observed that the longer a loose sand is vibrated, the denser the soil. The number of shear strain cycles can be related to the earthquake magnitude. As indicated in Table 13.2, the higher the earthquake magnitude, the longer the duration of ground shaking.

In summary, the three main factors that govern the settlement of loose and dry cohesionless soil are the relative density, amount of shear strain, and number of shear strain cycles. These three factors can be accounted for by using the standard penetration test, peak ground acceleration, and earthquake magnitude. For example, Fig. 14.11 presents a simple chart that can be used to estimate the settlement of dry sand (Krinitzsky et al., 1993). The figure uses the standard penetration test N value and the peak ground acceleration (i.e., a_p) in order to calculate the earthquake-induced volumetric

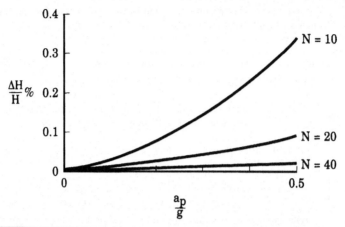

FIGURE 14.11   Simple chart that can be used to determine the settlement of dry sand. In this figure, use the peak ground acceleration a_p and assume that the N in the figure refers to $(N_1)_{60}$ values from Eq. 2.5. (*Reproduced from Krinitzsky et al., 1993; with permission from John Wiley & Sons.*)

strain (i.e., $\Delta H/H$, expressed as a percentage). Figure 14.11 accounts for two of the three main factors causing volumetric compression, i.e., (1) the looseness of the soil based on the standard penetration test, and (2) the amount of shear strain based on the peak ground acceleration a_p.

Note in Fig. 14.11 that the curves are labeled in terms of the uncorrected N values. As a practical matter, the curves should be in terms of the standard penetration test $(N_1)_{60}$ values (i.e., Eq. 2.5, Sec. 2.4.3). This is because the $(N_1)_{60}$ value more accurately represents the density condition of the sand. For example, given two sand layers having the same uncorrected N value, the near surface sand layer will be in a much denser state than the sand layer located at a great depth.

In order to use Fig. 14.11, both the $(N_1)_{60}$ value of the sand and the peak ground acceleration a_p must be known. Then by entering the chart with the a_p/g value and intersecting the desired $(N_1)_{60}$ curve, the volumetric strain ($\Delta H/H$, expressed as a percentage) can be determined. The volumetric compression (i.e., settlement) is then calculated by multiplying the volumetric strain, expressed as a decimal, times the thickness of the soil layer (H).

A much more complicated method for estimating the settlement of dry sand has been proposed by Tokimatsu and Seed (1987), based on the prior work by Seed and Silver (1972) and Pyke et al. (1975).

These methods for the calculation of volumetric compression can only be used for the following cases:

- *Lightweight structures.*   Settlement of lightweight structures, such as wood frame buildings bearing on shallow foundations.

- *Low net bearing stress.*   Settlement of any other type of structure that imparts a low net bearing pressure onto the soil.

- *Floating foundation.*   Settlement of floating foundations, provided the floating foundation does not impart a significant net stress upon the soil.

- *Heavy structures with deep settlement.*   Settlement of heavy structures, such as massive buildings founded on shallow foundations, provided the zone of settlement is deep enough that the stress increase caused by the structural load is relatively low.

- *Differential settlement.*   Differential movement between a structure and adjacent appurtenances, where the structure contains a deep foundation that is supported by strata below the zone of volumetric compression.

These methods cannot be used for the following cases:

- *Heavy buildings bearing on loose soil.*   Do not use the methods when the foundation applies a large net load onto the loose soil. In this case, the heavy foundation will punch downward into the loose soil during the earthquake. It is usually very difficult to determine the settlement for these conditions and the best engineering solution is to provide a sufficiently high static factor of safety so that there is ample resistance against a bearing capacity failure.

- *Sloping ground condition.*   Do not use these methods when there is a sloping ground condition. The loose sand may deform laterally during the earthquake and the settlement of the building could be well in excess of the calculated values.

14.3.5 Settlement Due to Dynamic Loads Caused by Rocking

This type of settlement is due to dynamic structural loads that momentarily increase the foundation pressure acting on the soil, such as illustrated in Fig. 14.12. The soil will deform in response to the dynamic structural load, resulting in settlement of the building. This settlement due to dynamic loads is often a result of the structure rocking back and forth.

Both cohesionless and cohesive soils are susceptible to rocking settlement. For cohesionless soils, loose sands and gravels are prone to rocking settlement. In addition, rocking settlement and volumetric compression (Sec. 14.3.4) often work in combination to cause settlement of the structure.

Cohesive soils can also be susceptible to rocking settlement. The types of cohesive soils most vulnerable are normally consolidated soils (OCR = 1.0), such as soft clays and organic soils. There can be significant settlement of foundations on soft saturated clays and organic soils because of undrained plastic flow when the foundations are overloaded during the seismic shaking. Large settlement can also occur if the existing vertical effective stress σ'_{vo} plus the dynamic load $\Delta\sigma_v$ exceeds the maximum past pressure σ'_{vm} of the cohesive soil, or $\sigma'_{vo} + \Delta\sigma_v > \sigma'_{vm}$. Another type of cohesive soil that can be especially vulnerable to rocking settlement is sensitive clay. These soils can lose a

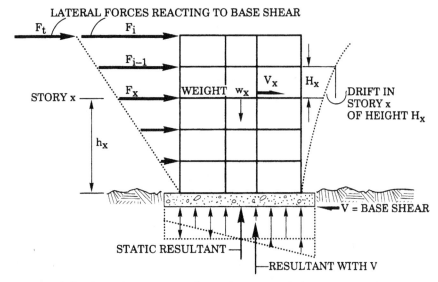

FIGURE 14.12 Diagram illustrating lateral forces *F* in response to the base shear *V* caused by the earthquake. Note that the uniform static bearing pressure is altered by the earthquake such that the pressure is increased along one side of the foundation. (*Reproduced from Krinitzsky et al., 1993; with permission from John Wiley & Sons.*)

portion of their shear strength during the cyclic loading. The higher the sensitivity, the greater the loss of shear strength for a given shear strain.

Lightly loaded structures would be least susceptible to rocking settlement. On the other hand, tall and heavy buildings that have shallow foundations bearing on vulnerable soils would be most susceptible to this type of settlement. In terms of the analysis for rocking settlement, Seed (1991) states:

> Vertical accelerations during earthquake seldom produce sufficient vertical thrust to cause significant foundation settlements. Horizontal accelerations, on the other hand, can cause 'rocking' of a structure, and the resulting structural overturning moments can produce significant cyclic vertical thrusts on the foundation elements. These can, in turn, result in cumulative settlements, with or without soil liquefaction or other strength loss. This is generally a potentially serious concern only for massive, relatively tall structures. Structures on deep foundations are not necessarily immune to this hazard; structures founded on 'friction piles' (as opposed to more solidly-based end-bearing piles) may undergo settlements of up to several inches or more in some cases. It should be noted that the best engineering solution is generally simply to provide a sufficiently high static factor of safety in bearing in order to allow for ample resistance to potential transient seismic loading.

14.4 RETAINING WALLS

14.4.1 Introduction

Chapter 11 has presented an in-depth discussion of retaining walls. The purpose of this section is to describe retaining wall design considerations for earthquakes. The performance of retaining walls during earthquakes is very complex. As stated by Kramer (1996), laboratory tests and analyses of gravity walls subjected to seismic forces have indicated the following:

1. Walls can move by translation and/or rotation. The relative amounts of translation and rotation depend on the design of the wall; one or the other may predominate for some walls, and both may occur for others (Nadim and Whitman, 1984; Siddharthan et al., 1992).
2. The magnitude and distribution of dynamic wall pressures are influenced by the mode of wall movement, e.g., translation, rotation about the base, or rotation about the top (Sherif et al., 1982; Sherif and Fang, 1984a,b).
3. The maximum soil thrust acting on a wall generally occurs when the wall has translated or rotated toward the backfill (i.e., when the inertial force on the wall is directed toward the backfill). The minimum soil thrust occurs when the wall has translated or rotated away from the backfill.
4. The shape of the earthquake pressure distribution on the back of the wall changes as the wall moves. The point of application of the soil thrust therefore moves up and down along the back of the wall. The position of the soil thrust is highest when the wall has moved toward the soil and lowers when the wall moves outward.
5. Dynamic wall pressures are influenced by the dynamic response of the wall and backfill and can increase significantly near the natural frequency of the wall-backfill system (Steedman and Zeng, 1990). Permanent wall displacements also increase at frequencies near the natural frequency of the wall-backfill system (Nadim, 1982). Dynamic response effects can also cause deflections of different parts of the wall to be out of phase. This effect can be particularly significant for walls that penetrate into the foundation soils when the backfill soils move out of phase with the foundation soils.
6. Increase residual pressures may remain on the wall after an episode of strong shaking has ended (Whitman, 1990).

Because of the complex soil-structure interaction during the earthquake, the most commonly used method for the design of retaining walls is the pseudostatic method.

14.4.2 Pseudostatic Method

The pseudostatic method has been previously discussed in Sec. 13.5.4 and it is applicable for slope stability and retaining wall analyses. The advantages of this method are that it is easy to understand and apply. Similar to earthquake slope stability analyses, this method ignores the cyclic nature of the earthquake and treats it as if it applies an additional static force upon the retaining wall. In particular, the pseudostatic approach is to apply a lateral earthquake force upon the retaining wall. To derive the lateral force, it can be assumed that the force acts through the centroid of the active wedge. The pseudostatic lateral force P_E is calculated by using Eq. 13.12 (i.e., $P_E = k_h W$) and it has units of pounds per linear foot of wall length or kilonewtons per linear meter of wall length.

Note that an earthquake could subject the active wedge to both vertical and horizontal pseudostatic forces. However, the vertical force is usually ignored in the standard pseudostatic analysis. This is because the vertical pseudostatic force acting on the active wedge usually has much less effect on the design of the retaining wall. In addition, most earthquakes produce a peak vertical acceleration that is less than the peak horizontal acceleration, and hence k_v is smaller than k_h.

The only unknowns in the pseudostatic method are the weight of the active wedge (W) and the seismic coefficient k_h. Because of the usual relatively small size of the active wedge, the seismic coefficient k_h can be assumed to be equal to a_{max}/g. Using Fig. 11.3, the weight of the active wedge can be calculated as follows:

$$W = \tfrac{1}{2} H L \gamma_t = \tfrac{1}{2} H \, [H \tan (45° - \tfrac{1}{2}\phi)] \gamma_t = \tfrac{1}{2} (k_A)^{1/2} H^2 \gamma_t \qquad (14.1)$$

where W = weight of the active wedge (lb per linear ft of wall length or kN per linear m of wall length)
H = height of the retaining wall (ft or m)
L = length of the active wedge at the top of the retaining wall. Note in Fig. 11.3 that the active wedge is inclined at an angle equal to: $45° + \tfrac{1}{2}\phi$. Therefore the internal angle of the active wedge is equal to: $90° - (45° + \tfrac{1}{2}\phi) = 45° - \tfrac{1}{2}\phi$. The length L can then be calculated as $L = H \tan (45° - \tfrac{1}{2}\phi) = H(k_A)^{1/2}$.
k_A = active earth pressure coefficient (dimensionless)
γ_t = total unit weight of the backfill soil (i.e., the unit weight of the soil comprising the active wedge), pcf or kN/m³

Using Eq. 14.1, the final result is as follows:

$$P_E = k_h W = \tfrac{1}{2} k_h (k_A)^{1/2} H^2 \gamma_t = \tfrac{1}{2} (k_A)^{1/2} (a_{max}/g) H^2 \gamma_t \qquad (14.2)$$

Note that since the pseudostatic force is applied to the centroid of the active wedge, the location of the force P_E is at a distance of $\tfrac{2}{3}H$ above the base of the retaining wall. Seed and Whitman (1970) developed a similar equation that can be used to determine the horizontal pseudostatic force acting on the retaining wall, as follows:

$$P_E = (\tfrac{3}{8}) \, (a_{max}/g) \, H^2 \gamma_t \qquad (14.3)$$

Note that the terms in Eq 14.3 have the same definitions as the terms in Eq. 14.2. In comparing Eqs. 14.2 and 14.3, the two equations are identical for the case where $\tfrac{1}{2} (k_A)^{1/2} = \tfrac{3}{8}$. According to Seed and Whitman (1970), the location of the pseudostatic force from Eq. 14.3 can be assumed to act at a distance of $0.6 H$ above the base of the wall.

Mononobe and Matsu (1929) and Okabe (1926) also developed an equation that can be used to determine the horizontal pseudostatic force acting on the retaining wall. This method is often referred to as the *Mononobe-Okabe* method. The equation is an extension of the Coulomb approach and is as follows:

$$P_{AE} = P_A + P_E = \tfrac{1}{2} k_{AE} H^2 \gamma_t \qquad (14.4)$$

where P_{AE} is the sum of the static (P_A) and the pseudostatic earthquake force (P_E). The equation for k_{AE} is shown in Fig. 11.4. Note that in Fig. 11.4, the term ψ is defined as follows:

$$\psi = \tan^{-1} k_h = \tan^{-1} (a_{max}/g) \qquad (14.5)$$

The original approach by Mononobe and Okabe was to assume that the force P_{AE} from Eq. 14.4 acts at a distance of $\tfrac{1}{3} H$ above the base of the wall.

Example Problem 14.3   Figure 11.5 (from Lambe and Whitman, 1969) presents an example of a proposed concrete retaining wall that will have a height of 20 ft (6.1 m) and a base width of 7 ft (2.1 m). The wall will be backfilled with sand that has a total unit weight γ_t of 110 pcf (17.3 kN/m³) and a friction angle ϕ of 30°. The retaining wall has been previously analyzed for the static case in Sec. 11.3.2. For this problem, analyze the retaining wall for an earthquake condition of $k_h = 0.2$ and use Eq. 14.2, the method by Seed and Whitman (Eq. 14.3), and the method by Mononobe-Okabe (Eq. 14.4). Also assumed that the backfill soil, bearing soil, and soil located in the passive wedge will not be weakened by the earthquake.

Solution

a) Equation 14.2:

From Eq. 11.2, and neglecting the wall friction for the determination of the earthquake force:
$k_A = \tan^2(45° - \frac{1}{2}\phi) = \tan^2(45° - \frac{1}{2}30°) = 0.333$

Substituting into Eq. 14.2, or:

$$P_E = \frac{1}{2}(k_A)^{1/2}(a_{max}/g)\,H^2\,\gamma_t$$

$P_E = \frac{1}{2}(0.333)^{1/2}(0.2)(20\text{ ft})^2(110\text{ pcf}) = 2540$ lb per linear foot of wall length

This pseudostatic force acts at a distance of $\frac{2}{3}H$ above the base of the wall, or $\frac{2}{3}H = \frac{2}{3}(20\text{ ft}) = 13.3$ ft. Similar to Eq. 11.6, the factor of safety (F) for sliding is equal to:

$$F = \frac{N\tan\delta + P_P}{P_H + P_E} \tag{14.6}$$

Substituting values into Eq. 14.6:

$$F = \frac{15{,}270\tan 30° + 750}{5660 + 2540} = 1.17$$

Based on Eq. 11.7, the factor of safety for overturning is as follows:

$$F = \frac{Wa}{\frac{1}{3}P_H H - P_v e + \frac{2}{3}HP_E} \tag{14.7}$$

Inserting values into Eq. 14.7:

$$F = \frac{55{,}500}{\frac{1}{3}(5660)(20) - (3270)(7) + \frac{2}{3}(20)(2540)} = 1.14$$

b) Method by Seed and Whitman (1970):

Using Eq. 14.3, and neglecting the wall friction for the determination of the earthquake force:

$$P_E = (\tfrac{3}{8})(a_{max}/g)\,H^2\,\gamma_t$$

$P_E = \frac{3}{8}(0.2)(20\text{ ft})^2(110\text{ pcf}) = 3300$ lb per linear foot of wall length

This pseudostatic force acts at a distance of $0.6\,H$ above the base of the wall, or $0.6\,H = (0.6)(20\text{ ft}) = 12$ ft. Using Eq. 14.6, the factor of safety (F) for sliding is equal to:

$$F = \frac{N\tan\delta + P_P}{P_H + P_E} = \frac{15{,}270\tan 30° + 750}{5660 + 3300} = 1.07$$

(Continued)

Similar to Eq. 14.7, the factor of safety for overturning is as follows:

$$F = \frac{Wa}{\frac{1}{3}\,P_H H - P_v e + 0.6\,HP_E} \tag{14.8}$$

Substituting values into Eq. 14.8:

$$F = \frac{55{,}500}{\frac{1}{3}\,(5660)(20) - (3270)(7) + (0.6)(20)(3300)} = 1.02$$

c) Mononobe-Okabe method:

Using the following values:

θ (wall inclination) = 0°

ϕ (friction angle of backfill soil) = 30°

β (backfill slope inclination) = 0°

$\delta = \phi_w$ (friction angle between the backfill and wall) = 30°

$$\psi = \tan^{-1} k_h = \tan^{-1}(a_{max}/g) = \tan^{-1}(0.2) = 11.3°$$

Inserting the above values into the K_{AE} equation in Fig. 11.4, the value of $K_{AE} = 0.471$. Therefore, using Eq. 14.4:

$$P_{AE} = P_A + P_E = \frac{1}{2} k_{AE} H^2 \gamma_t$$

$$P_{AE} = \frac{1}{2}(0.471)(20)^2\,(110) = 10{,}400 \text{ lb per linear foot of wall length}$$

This force P_{AE} is inclined at an angle of 30° and acts at a distance of $\frac{1}{3} H$ above the base of the wall, or $\frac{1}{3} H = (\frac{1}{3})(20 \text{ ft}) = 6.67$ ft. The factor of safety (F) for sliding is equal to

$$F = \frac{N \tan \delta + P_p}{P_H} = \frac{(W + P_{AE} \sin \phi_w) \tan \delta + P_p}{P_{AE} \cos \phi_w} \tag{14.9}$$

Substituting values into Eq. 14.9:

$$F = \frac{(3000 + 9000 + 10{,}400 \sin 30°) \tan 30° + 750}{10{,}400 \cos 30°} = 1.19$$

The factor of safety for overturning is as follows:

$$F = \frac{Wa}{\frac{1}{3}\,HP_{AE} \cos \phi_w - P_{AE} \sin \phi_w e} \tag{14.10}$$

Substituting values into Eq. 14.10:

$$F = \frac{55{,}500}{\frac{1}{3}\,(20)(10{,}400)(\cos 30°) - (10{,}400)(\sin 30°)(7)} = 2.35$$

d) Summary of values:

The values from the static and earthquake analyses using $k_h = a_{max}/g = 0.2$ are summarized as follows:

Type of condition		P_E or P_{AE} (lb/ft)	Location of P_E or P_{AE} above base of wall (ft)	Factor of safety for sliding	Factor of safety for overturning
Static (see Sec. 11.3.2)		$P_E = 0$	—	1.69*	3.73
Earthquake ($k_h = 0.2$)	Eq. 14.2	$P_E = 2540$	$^2/_3\,H = 13.3$	1.17	1.14
	Seed and Whitman	$P_E = 3300$	0.6 H = 12	1.07	1.02
	Mononobe-Okabe	$P_{AE} = 10{,}400$	$^1/_3\,H = 6.7$	1.19	2.35

*Factor of safety for sliding using Eq. 11.8.

For the analysis of sliding and overturning of the retaining wall, it is common to accept a lower factor of safety (1.1 to 1.2) under the combined static and earthquake loads. Thus the retaining wall would be considered marginally stable for the earthquake sliding and overturning conditions.

Note in the above table that the factor of safety for overturning is equal to 2.35 based on the Mononobe-Okabe method. This factor of safety is much larger than the other two methods. This is because the force P_{AE} is assumed to be located at a distance of $^1/_3\,H$ above the base of the wall. Kramer (1996) suggests that it is more appropriate to assume that P_E is located at a distance of 0.6 H above the base of the wall (i.e., $P_E = P_{AE} - P_A$, see Eq. 14.4).

Although the calculations are not shown, it can be demonstrated that the resultant location of N for the earthquake condition is outside the middle third of the footing. Depending on the type of material beneath the footing, this condition could cause a bearing capacity failure or excess settlement at the toe of the footing during the earthquake.

14.4.3 Retaining Wall Analyses for Liquefied Soil

Introduction. The types of retaining walls most susceptible to damage and collapse due to earthquakes are port and wharf facilities, which are often located in areas susceptible to liquefaction. The ports and wharves often contain major retaining structures, such as seawalls, anchored bulkheads, gravity and cantilever walls, and sheet-pile cofferdams, that allow large ships to moor adjacent the retaining walls and then load or unload their cargo. There are often three different types of liquefaction effects that can damage the retaining wall, as follows:

1. The first is liquefaction of soil in front of the retaining wall. In this case, the passive pressure in front of the retaining wall is reduced.

2. In the second case, the soil behind the retaining wall liquefies and the pressure exerted on the wall is greatly increased. Cases 1 and 2 can act individually or together and they can initiate an overturning failure of the retaining wall or cause the wall to slide outward or tilt towards the water. Another possibility is that the increased pressure exerted on the wall could exceed the strength of the wall, resulting in a structural failure of the wall.

   Liquefaction of the soil behind the retaining wall can also affect tieback anchors. For example, the increased pressure due to liquefaction of the soil behind the wall could break the tieback anchors or reduce their passive resistance.

3. The third case is liquefaction below the bottom of the wall. In this case, the bearing capacity or slide resistance of the wall is reduced, resulting in a bearing capacity failure or promoting rotational movement of the wall.

Some spectacular examples of damage to waterfront structures due to liquefaction occurred during the Kobe earthquake on January 17, 1995. Particular details concerning the Kobe earthquake are as follows (EQE Summary Report, 1995; EERC, 1995):

• The Kobe earthquake, also known as the Hyogo-ken Nanbu Earthquake, had a moment magnitude M_w of 6.9.
• The earthquake occurred in a region with a complex system of previously mapped active faults.
• The focus of the earthquake was at a depth of approximately 15 to 20 km (9 to 12 mi). The focal mechanism of the earthquake indicated right-lateral strike-slip faulting on a nearly vertical fault that runs from Awaji Island through the city of Kobe.
• Ground rupture due to the right-lateral strike-slip faulting was observed on Awaji Island, which is located to the southwest of the epicenter. In addition, the Akashi Kaikyo Bridge, which was under construction at the time of the earthquake, suffered vertical and lateral displacement between the north and south towers. This is the first time that a structure of this size was offset by a fault rupture.
• Peak ground accelerations as large as $0.8g$ were recorded in the near-fault region on alluvial sites in Kobe.
• In terms of regional tectonics, Kobe is located on the southeastern margin of the Eurasian plate, where the Philippine Sea plate is being subducted beneath the Eurasian plate (see Fig. 13.1).
• More than 5000 people perished, more than 26,000 people were injured, and about $200 billion in damage were attributed to this earthquake.

Damage was especially severe at the relatively new Port of Kobe. In terms of damage to the port, the EERC (1995) stated:

The main port facilities in Kobe harbor are located primarily on reclaimed land along the coast and on two man-made islands, Port Island and Rokko Island, which are joined by bridges to the mainland. The liquefaction and lateral spread-induced damage to harbor structures on the islands disrupted nearly all of the container loading piers, and effectively shut down the Port of Kobe to international shipping. All but 6 of about 187 berths were severely damaged.

Concerning the damage caused by liquefaction, the EERC (1995) concluded:

Extensive liquefaction of natural and artificial fill deposits occurred along much of the shoreline on the north side of the Osaka Bay. Probably the most notable were the liquefaction failures of relatively modern fills on the Rokko and Port islands. On the Kobe mainland, evidence of liquefaction extended along the entire length of the waterfront, east and west of Kobe, for a distance of about 20 km [12 mi]. Overall, liquefaction was a principal factor in the extensive damage experienced by the port facilities in the affected region.
Most of the liquefied fills were constructed of poorly compacted decomposed granite soil. This material was transported to the fill sites and loosely dumped in water. Compaction was generally only applied to materials placed above water level. As a result, liquefaction occurred within the underwater segments of these poorly compacted fills.
Typically, liquefaction led to pervasive eruption of sand boils and, on the islands, to ground settlements on the order of as much as 0.5 m. The ground settlement caused surprisingly little damage to high- and low-rise buildings, bridges, tanks, and other structures supported on deep foundations. These foundations, including piles and shafts, performed very well in supporting superstructures where ground settlement was the principal effect of liquefaction. Where liquefaction generated lateral ground displacements, such as near island edges and in other waterfront areas, foundation performance was typically poor. Lateral displacements fractured piles and displaced pile caps, causing structural distress to several bridges. In a few instances, such as the Port Island Ferry Terminal, strong foundations withstood the lateral ground displacement with little damage to the foundation or the superstructure.

There were several factors that apparently contributed to the damage at the Port of Kobe, as follows (EQE Summary Report, 1995; EERC, 1995):

1. *Design criteria.*   The area had been previously considered to have a relatively low seismic risk, hence the earthquake design criteria were less stringent than in other areas of Japan.

2. *Earthquake shaking.*   There was rupture of the strike-slip fault directly in downtown Kobe. Hence the release of energy along the earthquake fault was close to the port. In addition, the port is located on the shores of a large embayment, which has a substantial thickness of soft and liquefiable sediments. This thick deposit of soft soil caused an amplification of the peak ground acceleration and an increase in the duration of shaking.

3. *Construction of the port.*   The area of the port was built almost entirely on fill and reclaimed land. As previously mentioned, the fill and reclaimed land material often consisted of decomposed granite soils that were loosely dumped into the water. The principal factor in the damage at the Port of Kobe was attributed to liquefaction, which caused lateral deformation of the retaining walls.

4. *Man-made islands.*   On Rokko and Port Islands, retaining walls were constructed by using caissons, which consisted of concrete box structures, up to 15 m wide and 20 m deep, with two or more interior cells. The first step was to prepare the seabed by installing a sand layer. Then the caissons were towed to the site, submerged in position to form the retaining wall, and then the interior cells were backfilled with sand. Once in place, the area behind the caisson retaining walls was filled in with soil in order to create the man-made islands.

    During the Kobe earthquake, a large number of these caisson retaining walls rotated and slid outward (lateral spreading). This outward movement of the retaining walls by as much as 3 m (10 ft) caused lateral displacement and failure of the loading dock cranes.

5. *Buildings on deep foundations.*   In some cases, the buildings adjacent the retaining walls had deep foundations consisting of piles or piers. Large differential movement occurred between the relatively stable buildings having piles or piers and the port retaining walls, which settled and deformed outward.

6. *Lateral spreading.*   Similar to the lateral spreading of slopes, there was also damage due to the lateral spreading of retaining walls.

Design Pressures.   The first step in the analysis is to determine the factor of safety against liquefaction for the soil behind the retaining wall, in front of the retaining wall, and below the bottom the wall. The analysis presented in Sec. 13.4 can be used to determine the factor of safety against liquefaction. The retaining wall may exert significant shear stress into the underlying soil, which can decrease the factor of safety against liquefaction for loose soils. Likewise, there could be sloping ground in front of the wall or behind the wall, in which case the factor of safety against liquefaction may need to be adjusted (see Sec. 13.5.2).

    After the factor of safety against liquefaction has been calculated, the next step is to determine the design pressures that act on the retaining wall, as follows:

1. *Passive pressure.*   For those soils that will be subjected to liquefaction in the passive zone, one approach is to assume that the liquefied soil has zero shear strength. In essence, the liquefied zones no longer provide sliding or overturning resistance.

2. *Active pressure.*   For those soils that will be subjected to liquefaction in the active zone, the pressure exerted on the face of the wall will increase. One approach is to assume zero shear strength of the liquefied soil (i.e., $\phi' = 0$). There are two possible conditions, as follows:

    a. *Water level located only behind the retaining wall.*   In this case, the wall and the ground beneath the bottom of the wall are relatively impermeable. In addition, there is a groundwater table behind the wall with dry conditions in front of the wall. The thrust on the wall due to liquefaction of the backfill can be calculated by using Eq. 11.1 with $k_A = 1$ (i.e., for $\phi' = 0$, $k_A = 1$, see Eq. 11.2) and $\gamma_t = \gamma_{sat}$ (i.e., γ_{sat} = saturated unit weight of the soil).

    b. *Water levels are approximately the same on both sides of the retaining wall.*   The more common situation is where the elevation of the groundwater table behind the wall is approximately at the same elevation as the water level in front of the wall. The thrust on the wall due to liquefaction of the soil can be calculated by using Eq. 11.1 with $k_A = 1$ (i.e., for $\phi' = 0$, $k_A = 1$, see Eq. 11.2) and using γ_b (buoyant unit weight) in place of γ_t.

The only difference between the two cases is that the first case includes the unit weight of water (i.e., $\gamma_{sat} = \gamma_b + \gamma_w$), while the second case does not include γ_w because it is located on both sides of the wall and hence its effect is canceled out.

In addition to the increased pressure acting on the retaining wall due to liquefaction, there may also be a reduction in support and/or resistance of the tieback anchors.

3. *Bearing soil.*   The analysis presented in Sec. 6.5 is used for the liquefaction of the bearing soil.

Sheet Pile Walls.   As previously mentioned, higher pressures will be exerted on the back face of a sheet pile wall if the soil behind the wall should liquefy. Likewise, there will be less passive resistance if the soil in front of the sheet pile wall will liquefy during the design earthquake. The prior discussion of design pressures can be used as a guide in the selection of the pressures exerted on the sheet pile wall during the earthquake. Once these earthquake-induced pressures behind and in front of the wall are known, then the factor of safety for toe failure and the anchor pull force can be calculated in the same manner as outlined in Sec. 11.7.

Example Problem 14.4   Using the sheet pile wall diagram shown in Fig. 11.14, assume that the soil behind and in front of the sheet wall is uniform sand with a friction angle $\phi' = 33°$, buoyant unit weight $\gamma_b = 64$ pcf, and above the groundwater table, the total unit weight $\gamma_t = 120$ pcf. Also assume that the sheet pile wall has $H = 30$ ft and $D = 20$ ft, the water level in front of the wall is at the same elevation as the groundwater table which is located 5 ft below the ground surface, and the tieback anchor is located 4 ft below the ground surface. In the analysis, neglect wall friction. Calculate the factor of safety for toe kick-out and the tieback anchor force for static conditions.

Solution

$$\text{Equation 11.2: } k_A = \tan^2 (45° - \tfrac{1}{2} \phi) = \tan^2 [45° - (\tfrac{1}{2})(33°)] = 0.295$$

$$\text{Equation 11.5: } k_p = \tan^2 (45° + \tfrac{1}{2} \phi) = \tan^2 [45° + (\tfrac{1}{2})(33°)] = 3.39$$

$$\text{From 0 to 5 ft: } P_{1A} = \tfrac{1}{2}k_A\gamma_t (5)^2 = \tfrac{1}{2}(0.295)(120)(5)^2 = 400 \text{ lb/ft}$$

$$\text{From 5 to 50 ft: } P_{2A} = k_A\gamma_t (5)(45) + \tfrac{1}{2} k_A\gamma_b (45)^2 = (0.295)(120)(5)(45)$$

$$+ \tfrac{1}{2}(0.295)(64)(45)^2 = 8000 + 19{,}100 = 27{,}100$$

$$P_A = P_{1A} + P_{2A} = 400 + 27{,}100 = 27{,}500 \text{ lb/ft}$$

$$\text{Equation 11.4 with } \gamma_b: P_p = \tfrac{1}{2} k_p\gamma_bD^2 = \tfrac{1}{2} (3.39)(64)(20)^2 = 43{,}400 \text{ lb/ft}$$

Moment due to passive force = $(43{,}400)[26 + \tfrac{2}{3}(20)] = 1.71 \times 10^6$

Neglecting P_{1A}, moment due to active force (at the tieback anchor) =

$$(8000)[1 + (45/2)] + (19{,}100)[1 + \tfrac{2}{3}(45)] = 7.8 \times 10^5$$

F = resisting moment/destabilizing moment = $(1.71 \times 10^6)/(7.8 \times 10^5) = 2.19$

$$A_p = P_A - P_p/F = 27{,}500 - 43{,}400/2.19 = 7680 \text{ lb/ft}$$

For a 10-ft spacing, therefore $A_p = (10)(7680) = 76{,}800$ lb = 76.8 kips

Example Problem 14.5   Using the data from the prior example problem, perform an earthquake analysis and assume that the sand behind, beneath, and in front of the wall has a factor of safety against liquefaction that is greater than 2.0. The design earthquake condition is $a_{max} = 0.20g$. Using the pseudostatic approach (i.e., Sec. 14.4.2), calculate the factor of safety for toe kick-out and the tieback anchor force.

Solution   With a factor of safety against liquefaction greater than 2.0, the soil will not weaken during the earthquake and hence the pseudostatic method can be utilized. Since the effect of the water pressure tends to cancel out on both sides of the wall, Eq. 14.2 can be used with P_E based on the buoyant unit weight ($\gamma_b = 64$ pcf), or:

$$P_E = \frac{1}{2}(k_A)^{1/2}\,(a_{max}/g)\,H^2\,\gamma_b = \frac{1}{2}\,(0.295)^{1/2}\,(0.20)(50)^2\,(64) = 8690\ \text{lb/ft}$$

P_E acts at a distance of $\frac{2}{3}(H + D)$ above the bottom of the sheet pile wall

Moment due to $P_E = (8690)[(\frac{1}{3})(50) - 4] = 1.10 \times 10^5$

Total destabilizing moment $= 7.80 \times 10^5 + 1.10 \times 10^5 = 8.90 \times 10^5$

Moment due to passive force $= 1.71 \times 10^6$

F = resisting moment/destabilizing moment $= (1.71 \times 10^6)/(8.90 \times 10^5) = 1.92$

$$A_p = P_A + P_E - P_p/F = 27{,}500 + 8690 - 43{,}400/1.92 = 13{,}600\ \text{lb/ft}$$

For a 10 ft spacing, therefore $A_p = (10)(13{,}600) = 136{,}000$ lb $= 136$ kips

---

Example Problem 14.6   For this earthquake analysis, assume that the sand located behind the retaining wall has a factor of safety against liquefaction greater than 2.0. Also assume that the upper 10 ft of sand located in front of the retaining wall will liquefy during the design earthquake, while the sand located below a depth of 10 ft has a factor of safety greater than 2.0. Calculate the factor of safety for toe kick-out and the tieback anchor force.

Solution   For the passive wedge:

From 0 to 10 ft: Passive resistance = 0

At 10-ft depth: Passive resistance $= (k_p)(\gamma_b)(d) = (3.39)(64)(10) = 2170$ psf

At 20-ft depth: Passive resistance $= (k_p)(\gamma_b)(d) = (3.39)(64)(20) = 4340$ psf

Passive force $= [(2170 + 4340)/2](10) = 32{,}600$ lb/ft

Moment due to passive force

$$= (2170)(10)(45 - 4) + [(4340 - 2170)/2](10)[40 + (\tfrac{2}{3})(10) - 4]$$
$$= 890{,}000 + 463{,}000 = 1.35 \times 10^6$$

Including a pseudostatic force in the analysis:

$$P_E = \frac{1}{2}(k_A)^{1/2}\,(a_{max}/g)\,H^2\,\gamma_b = \frac{1}{2}(0.295)^{1/2}\,(0.20)(50)^2\,(64) = 8690\ \text{lb/ft}$$

P_E acts at a distance of $\frac{2}{3}(H + D)$ above the bottom of the sheet pile wall

Moment due to $P_E = (8690)\,[(\frac{1}{3})(50) - 4] = 1.10 \times 10^5$

(*Continued*)

Total destabilizing moment = $7.80 \times 10^5 + 1.10 \times 10^5 = 8.90 \times 10^5$

Moment due to passive force = 1.35×10^6

F = resisting moment/destabilizing moment = $(1.35 \times 10^6)/(8.90 \times 10^5) = 1.52$

$$A_p = P_A + P_E - P_p/F = 27{,}500 + 8690 - 32{,}600 \,/\, 1.52 = 14{,}700 \text{ lb/ft}$$

For a 10-ft spacing, therefore $A_p = (10)(14{,}700) = 147{,}000 \text{ lb} = 147 \text{ kips}$

Example Problem 14.7   For this earthquake analysis, assume that the sand located in front of the retaining wall has a factor of safety against liquefaction greater than 2.0. However, assume that the submerged sand located behind the retaining will liquefy during the earthquake. Further assume that the tieback anchor will be unaffected by the liquefaction. Calculate the factor of safety for toe kick-out.

Solution   When the water levels are approximately the same on both sides of the retaining wall, use Eq. 11.1 with $k_A = 1$ (i.e., for $\phi' = 0$, $k_A = 1$, see Eq. 11.2) and using γ_b (buoyant unit weight) in place of γ_t. As an approximation, assume that the entire 50 ft of soil behind the sheet pile wall will liquefy during the earthquake.

Using Eq. 11.1, with $k_A = 1$ and $\gamma_b = 64$ pcf

$$P_L = \tfrac{1}{2}k_A\gamma_b(H + D)^2 = \tfrac{1}{2}(1.0)(64)(50)^2 = 80{,}000 \text{ lb/ft}$$

Moment due to liquefied soil = $80{,}000[\tfrac{2}{3}(50) - 4] = 2.35 \times 10^6$

Moment due to passive force = 1.71×10^6

F = resisting moment/destabilizing moment = $(1.71 \times 10^6)/(2.35 \times 10^6) = 0.73$

Summary of values:

Example problem		Factor of safety for toe kick-out	A_p (kips)
Static analysis		2.19	76.8
Earthquake	Pseudostatic method (Eq. 14.2)	1.92	136
	Partial passive wedge liquefaction*	1.52	147
	Liquefaction of soil behind wall	0.73	—

*Pseudostatic force included for the active wedge.

As indicated by the values in this summary table, the sheet pile wall would not fail for partial liquefaction of the passive wedge. However, liquefaction of the soil behind the retaining wall would cause failure of the wall.

Summary.   As discussed in the previous sections, the liquefaction of soil can affect the retaining wall in many different ways. It is also possible that even with a factor of safety against liquefaction greater than 1.0, there could still be significant weakening of the soil leading to a retaining wall failure. In summary, the type of analysis should be based on the factor of safety against liquefaction (FS_L), as follows:

1. *$FS_L \leq 1.0$.* In this case, the soil is expected to liquefy during the design earthquake and thus the design pressures acting on the retaining wall must be adjusted as previously discussed.

2. *$FS_L > 2.0$.* If the factor of safety against liquefaction is greater than about 2.0, the pore water pressures generated by the earthquake-induced contraction of the soil are usually small enough so that they can be neglected. In this case, it could be assumed that the earthquake does not weaken the soil and the pseudostatic analyses outlined in Sec. 14.4.2 could be performed.

3. *$1.0 < FS_L \leq 2.0$.* For this case, the soil is not anticipated to liquefy during the earthquake. However, as the loose granular soil contracts during the earthquake, there could still be a substantial increase in pore water pressure and hence weakening of the soil. Figure 4.25 can be used to estimate the pore water pressure ratio r_u for various values of the factor of safety against liquefaction (FS_L). The analysis would vary depending on the location of the increase in pore water pressure, as follows:

   a. Passive wedge. If the soil in the passive wedge has a factor of safety against liquefaction greater than 1.0 but less than 2.0, then the increase in pore water pressure would decrease the effective shear strength and the passive resisting force would be reduced [i.e., passive resistance = $(P_p)(1 - r_u)$].

   b. Bearing soil. For an increase in the pore water pressure in the bearing soil, use the analysis in Sec. 6.5.

   c. Active wedge. In addition to the pseudostatic force P_E and the active earth pressure resultant force P_A, include an additional force that is equivalent to the anticipated earthquake-induced pore water pressure.

14.4.4   Retaining Wall Analysis for Weakened Soil

Besides the liquefaction of soil, there are many other types of soil that can be weakened during the earthquake. In general, there are three cases, as follows:

1. *Weakening of backfill soil.* In this case, only the backfill soil will be weakened during the earthquake. An example would be backfill soil that is susceptible to strain softening during the earthquake. As the backfill soil weakens during the earthquake, the force exerted on the back face of the wall will increase. One design approach would be to estimate the shear strength corresponding to the weakened condition of the backfill soil and then use this strength to calculate the force exerted on the wall. The bearing pressure, factor of safety for sliding, factor of safety for overturning, and the location of the resultant vertical force could then be calculated for this weakened backfill soil condition.

2. *Reduction in the soil resistance.* In this case, the soil beneath the bottom of the wall or the soil in the passive wedge will be weakened during the earthquake. For example, the bearing soil could be susceptible to strain softening during the earthquake. As the bearing soil weakens during the earthquake, the wall foundation could experience additional settlement, a bearing capacity failure, sliding failure, or overturning failure. In addition, the weakening of the ground beneath or in front of the wall could result in a shear failure beneath the retaining wall. One design approach would be to reduce the shear strength of the bearing soil or passive wedge soil to account for its weakened state during the earthquake. The settlement, bearing capacity, factor of safety for sliding, factor of safety for overturning, and the factor of safety for a shear failure beneath the bottom of the wall would then be calculated for this weakened soil condition.

3. *Weakening of the backfill soil and reduction in the soil resistance.* This is the most complicated case and would require combined analyses of both items 1 and 2 as outlined previously.

14.4.5   Restrained Retaining Walls

As mentioned in Sec. 11.2, in order for the active wedge to be developed, there must be sufficient movement of the retaining wall. There are many cases where movement of the retaining wall is restricted. Examples include massive bridge abutments, rigid basement walls, and retaining walls that are anchored in non-yielding rock. These cases are often described as *restrained retaining walls.*

For earthquake conditions, restrained retaining walls will usually be subjected to larger forces as compared to those retaining walls that have the ability to develop the active wedge. Provided the soil is not weakened during the earthquake, one approach is to use the pseudostatic method in order to calculate the earthquake force, with an increase to compensate for the unyielding wall conditions, or:

$$P_{ER} = P_E(k_o/k_A) \tag{14.11}$$

where P_{ER} = pseudostatic force acting upon a restrained retaining wall (lb per linear ft or kN per linear m of wall length)

P_E = pseudostatic force assuming the wall that has the ability to develop the active wedge, i.e., use Eqs. 14.2, 14.3, or 14.4 (lb per linear ft or kN per linear m of wall length)

k_o = the coefficient of earth pressure at rest (dimensionless)

k_A = active earth pressure coefficient, calculated from Eq. 11.2 or using the k_A equation in Fig. 11.4 (dimensionless)

Example Problem 14.8   Use example problem from Sec. 11.5 (i.e., Fig. 11.5) and assume that it is an unyielding bridge abutment. Determine the static and earthquake resultant forces acting on the restrained retaining wall. Neglect friction between the wall and backfill (i.e., $\delta = \phi_w = 0$).

Solution

A) Static analysis:

Using a value of $k_o = 0.5$ and substituting k_o for k_A in Eq. 11.1, the static earth pressure resultant force exerted on the restrained retaining wall has been calculated in Sec. 11.5, or: $P = 11,000$ lb per linear foot of wall. The location of this static force is at a distance of $1/3 H = 6.7$ ft above the base of the wall.

B) Earthquake analysis:

Using the psuedostatic method and since $k_A = 0.333$, the value of $P_E = 2540$ lb per linear ft of wall length (from Eq. 14.2). Using Eq. 14.11:

$P_{ER} = P_E (k_o/k_A) = (2540)(0.5/0.333) = 3800$ lb per linear ft of wall

The location of this pseudostatic force is assumed to act at a distance of $2/3 H = 13.3$ ft above the base of the wall.

In summary, the resultant earth pressure forces acting on the restrained retaining wall are static $P = 11,000$ lb/ft acting at a distance of 6.7 ft above the base of the wall and earthquake $P_{ER} = 3800$ lb/ft acting at a distance of 13.3 ft above the base of the wall.

14.5 FOUNDATION ALTERNATIVES TO MITIGATE EARTHQUAKE EFFECTS

14.5.1 Introduction

The usual approach for settlement analyses is to first estimate the amount of earthquake-induced total settlement (ρ_{max}) of the foundation. Because of variable soil conditions and structural loads, the earthquake-induced settlement is rarely uniform. A common assumption is that the maximum differential settlement Δ of the foundation will be equal to 50 to 75 percent of ρ_{max} (i.e., $0.5\,\rho_{max} \leq \Delta \leq 0.75\,\rho_{max}$). If the anticipated total settlement ρ_{max} and/or the maximum differential settlement Δ are deemed to be unacceptable, then remedial measures are needed. One alternative is soil improvement, which will be discussed in Chap. 15.

Instead of soil improvement, the foundation can be designed to resist the anticipated soil movement caused by the earthquake. For example, mat foundations or posttensioned slabs may enable the building to remain intact, even with substantial movements. Another option is a deep foundation system that transfers the structural loads to adequate bearing material in order to bypass a compressible or liquefiable soil layer. A third option is to construct a floating foundation, which is a special type of deep foundation where the weight of the structure is balanced by the removal of soil and construction of an underground basement. A floating foundation could help reduce the amount of rocking settlement caused by the earthquake.

14.5.2 Shallow Foundations

In southern California, the most common types of shallow foundations used for single-family houses and other lightly loaded structures are a raised wood floor foundation and a concrete slab-on-grade, as follows:

Raised Wood Floor Foundation.   The typical raised wood floor foundation consists of continuous concrete perimeter footings and interior (isolated) concrete pads. The floor beams span between the continuous perimeter footings and the isolated interior pads. The continuous concrete perimeter footings are typically constructed so that they protrude about 1 to 2 ft (0.3 to 0.6 m) above adjacent pad grade. The interior concrete pad footings are not as high as the perimeter footings, and short wood posts are used to support the floor beams. The perimeter footings and interior posts elevate the wood floor and provide for a crawl space below the floor. In general, damage caused by southern California earthquakes has been more severe to houses having a raised wood floor foundation. There may be several different reasons for this behavior:

1. *Lack of shear resistance of wood posts.*   As previously mentioned, in the interior, the raised wood floor beams are supported by short wood posts bearing on interior concrete pads. During the earthquake, these short posts are vulnerable to collapse or tilting during the earthquake.

2. *No bolts or inadequate bolted condition.*   Because in many cases the house is not adequately bolted to the foundation, it can slide or even fall off of the foundation during the earthquake. In other cases the bolts are spaced too far apart and the wood sill plate splits, allowing the house to slide off the foundation.

3. *Age of residence.*   The houses having this type of raised wood floor foundation are often older. The wood is more brittle and in some cases weakened due to rot or termite damage. In some cases, the concrete perimeter footings are nonreinforced or have been weakened due to prior soil movement, making them more susceptible to cracking during the earthquake.

4. *Crawl space vents.*   In order to provide ventilation to the crawl space, long vents are often constructed just above the concrete foundation. Such vents provide areas of weakness just above the foundation.

All of these factors can contribute to the detachment of the house from the foundation. Besides determining the type of foundation to resist earthquake related effects, the foundation engineer could also be involved with the retrofitting of existing structures. For example, to prevent the house from sliding off of the foundation, bolts or tie-down anchors could be installed to securely attach the wood framing to the concrete foundation. Wood bracing or plywood could also be added to the open areas between posts to give the foundation more shear resistance.

Slab-on-Grade Foundation.   In southern California, the concrete slab-on-grade is the most common type of foundation that is presently used for houses and other lightly loaded structures. Typical slab-on-grade foundations consist of either the conventional or the posttensioned foundation.

The conventional slab-on-grade foundation consists of perimeter and interior continuous footings, interconnected by the slab-on-grade. Construction of the conventional slab-on-grade begins with the excavation of the interior and perimeter continuous footings. Steel reinforcing bars are commonly centered in the footing excavations and wire mesh or steel bars are used as reinforcement for the slab. The concrete for both the footings and the slab are usually placed at the same time, in order to create a monolithic foundation. Unlike the raised wood floor foundation, the slab-on-grade does not have a crawl space.

In general, for those houses with a slab-on-grade, the wood sill plate is securely bolted to the concrete foundation. In many cases, an earthquake can cause the development of an exterior crack in the stucco at the location where the sill plate meets the concrete foundation. In some cases, the crack can be found on all four sides of the house. The crack develops when the house framing bends back and forth during the seismic shaking.

For raised wood floor foundations and the slab-on-grade foundations subjected to similar earthquake intensity and duration, those houses having a slab-on-grade generally have the best performance. This is because the slab-on-grade is typically stronger due to steel reinforcement and monolithic construction, the houses are newer (less wood rot and concrete deterioration), there is greater frame resistance because of the construction of shear walls, and the wood sill plate is in continuous contact with the concrete foundation.

It should be mentioned that although the slab-on-grade generally has the best performance, these houses could be severely damaged. In many cases, these houses do not have adequate shear walls, there are numerous wall openings, or there is poor construction. The construction of a slab-on-grade by itself is not enough to protect a structure from collapse if the structural frame above the slab does not have adequate shear resistance.

14.5.3   Deep Foundations

Deep foundations are one of the most effective means of mitigating foundation movement during an earthquake. For example, the Niigata Earthquake resulted in dramatic damage due to liquefaction of the sand deposits in the low-lying areas of Niigata City. At the time of the Niigata Earthquake, there were approximately 1500 reinforced concrete buildings in Niigata City and about 310 of these buildings were damaged, of which approximately 200 settled or tilted rigidly without appreciable damage to the superstructure. The damaged concrete buildings were built on very shallow foundations or friction piles in loose soil. Similar concrete buildings founded on piles bearing on firm strata at a depth of 66 ft (20 m) did not suffer damage.

There are several important earthquake design considerations when using deep foundations, such as piles or piers, as follows:

1. *Connection between pile and cap.*   It is important to have an adequate connection between the top of the pile and the pile cap. This can be accomplished by using steel reinforcement to connect the pile to the pile cap. Without this reinforced connection, the pile will be susceptible to separation at the pile cap during the earthquake.

2. *Downdrag loads due to soil liquefaction.*   The pile-supported structure may remain relatively stationary, but the ground around the piles may settle as the pore water pressures dissipate in the liquefied soil. The settlement of the ground relative to the pile will induce downdrag loads onto the pile. The piles should have an adequate capacity to resist the downdrag loads.

   The relative movement between the relatively stationary structure and the settling soil can also damage utilities. To mitigate damage to utilities, flexible connections can be provided at the location where the utilities enter the building.

3. *Passive resistance for liquefiable soil.*   A common assumption is that the liquefied soil will be unable to provide any lateral resistance. If a level-ground site contains an upper layer of nonliquefiable soil that is of sufficient thickness to prevent ground fissuring and sand boils, then this layer may provide passive resistance for the piles, caps, and grade beams.

4. *Liquefaction of sloping ground.*   For liquefaction of sloping ground, there will often be lateral spreading of the ground, which could shear-off the piles. One mitigation measure consists of the installation of compaction piles, in order to create a zone of nonliquefiable soil around and beneath the foundation.

NOTATION

The following notation is used in this chapter:

a = horizontal distance from W to the toe of the footing

a_{max}, a_p = maximum horizontal acceleration at the ground surface (also known as the peak ground acceleration)

A_p = anchor pull force (sheet pile wall)

CSR = cyclic stress ratio

D = portion of the sheet pile wall anchored in soil (Fig. 11.14)

D_r = relative density

e = lateral distance from P_v to the toe of the retaining wall

E_m = hammer efficiency

F = factor of safety

F = lateral force reacting to the earthquake-induced base shear (Fig. 14.12)

FS_L = factor of safety against liquefaction

g = acceleration of gravity

H = initial thickness of the soil layer (Sec. 14.3.4)

H = height of the retaining wall (Sec. 14.4.2)

H = unsupported face of the sheet pile wall (Fig. 11.14)

H_1 = thickness of the surface layer that does not liquefy

H_2 = thickness of the soil layer that will liquefy during the earthquake

ΔH = change in height of the soil layer

k_A = active earth pressure coefficient

k_{AE} = combined active plus earthquake coefficient of pressure (Mononobe-Okabe equation)

k_h = seismic coefficient, also known as the pseudostatic coefficient

k_o = coefficient of earth pressure at rest

k_p = passive earth pressure coefficient

L = length of the active wedge at the top of the retaining wall

M_w = moment magnitude of the earthquake

N = sum of the wall weights W plus, if applicable, P_v

N = uncorrected SPT blow count (blows per foot)

N_{corr} = value added to $(N_1)_{60}$ to account for fines in the soil

N_1 = Japanese standard penetration test value for Fig. 14.7

$(N_1)_{60}$ = N value corrected for field testing procedures and overburden pressure

OCR = overconsolidation ratio = $\sigma'_{vm} / \sigma'_{vo}$

P_A = active earth pressure resultant force

P_{AE} = sum of the active and earthquake resultant forces

P_E = pseudostatic horizontal force acting on the retaining wall

P_{ER} = pseudostatic horizontal force acting on a restrained retaining wall

P_L = lateral force due to liquefied soil

P_p = passive resultant force

P_v = vertical component of the active earth pressure resultant force

P_{1A}, P_{2A} = forces acting on the sheet pile wall

q_{c1} = cone resistance corrected for overburden pressure

r_u = pore water pressure ratio

u_e = excess pore water pressure

V = base shear induced by the earthquake (Fig. 14.12)

W = total weight of the active wedge (Sec. 14.4.2)

W = resultant of the vertical retaining wall loads (Sec. 14.4.2)

β = slope inclination behind the retaining wall

δ, ϕ_w = friction angle between the vear side of the wall and the soil backfill

δ = friction angle between the bottom of the wall and the underlying soil

Δ = earthquake-induced maximum differential settlement of the foundation

ε_v = volumetric strain

ϕ = friction angle based on a total stress analysis

ϕ' = friction angle based on an effective stress analysis

γ_{max} = maximum shear strain

γ_t = total unit weight of the soil

γ_b = buoyant unit weight of the soil

γ_{sat} = saturated unit weight of the soil

γ_w = unit weight of water

θ = inclination of the vear of the retaining wall

ρ_{max} = earthquake-induced total settlement of the foundation

σ'_{vm} = maximum past pressure, also known as the preconsolidation pressure

σ'_{vo} = vertical effective stress

$\Delta\sigma_v$ = increase in foundation pressure due to the earthquake

ψ = equal to $\tan^{-1}(a_{max}/g)$

PROBLEMS

Solutions to the problems are presented in App. C of this book. The problems have been divided into basic categories as indicated below:

Liquefaction-Induced Settlement

14.1   Assume a site has clean sand and a groundwater table near ground surface. The following data was determined for the site:

Layer depth	Cyclic stress ratio (CSR)	$(N_1)_{60}$
2–3 m	0.18	10
3–5 m	0.20	5
5–7 m	0.22	7

Using Figs. 14.7 and 14.8, calculate the total liquefaction-induced settlement of these layers caused by a magnitude 7.5 earthquake.

ANSWER: Based on Fig. 14.7 there will be 22 cm of settlement and using Fig. 14.8 there will be 17 cm of settlement.

14.2   Use the data from Problem 13.1 and the subsoil profile shown in Fig. 13.24. Ignore any possible settlement of the soil above the groundwater table (i.e., ignore settlement from ground surface to a depth of 1.5 m). Also ignore any possible settlement of the soil located below a depth of 21 m. Using Figs. 14.7 and 14.8, calculate the earthquake-induced settlement of the sand located below the groundwater table.

ANSWER: Based on Fig. 14.7 there will be 61 cm of settlement and using Fig. 14.8 there will be 53 cm of settlement.

14.3   Use the data from Problem 13.4 and the subsoil profile shown in Fig. 13.25. Ignore any possible settlement of the surface soil (i.e., ignore settlement from ground surface to a depth of 1.2 m). Also ignore any possible settlement of soil located below a depth of 20 m. Using Figs. 14.7 and 14.8, calculate the earthquake-induced settlement of the sand located below the groundwater table.

ANSWER: Based on Fig. 14.7 there will be 22 cm of settlement and using Fig. 14.8 there will be 17 cm of settlement.

14.4   Use the data from Problem 13.1 and Fig. 13.24. Assume that there has been soil improvement from ground surface to a depth 15 m. Also assume that for this zone of soil, the factor of safety against liquefaction is greater than 2.0. A mat foundation for a heavy building will be constructed such that the bottom of the mat is at a depth of 1.0 m. The mat foundation is 20 m long and 10 m wide and according to the structural engineer, the foundation will impose a net stress of 50 kPa onto the soil (the 50 kPa includes earthquake-related seismic load). Calculate the earthquake-induced settlement of the heavy building using Figs. 14.7 and 14.8.

ANSWER: Based on Fig. 14.7 there will be 17 cm of settlement and using Fig. 14.8 there will be 19 cm of settlement.

14.5   Use the data from Problem 13.4 and Fig. 13.25. A sewage disposal tank will be installed at a depth of 2 to 4 m below ground surface. Assuming the tank is empty at the time of the design earthquake, calculate the liquefaction-induced settlement of the tank.

ANSWER: Since the tank is in the middle of a liquefied soil layer, it is expected that the empty tank will not settle, but rather float to the ground surface.

Liquefaction-Induced Ground Damage

14.6   A soil deposit has a 6 m thick surface layer of unliquefiable soil underlain by a 4 m thick layer that is expected to liquefy during the design earthquake. The design earthquake condition is a peak ground acceleration a_{max} equal to 0.40g. Will there be liquefaction induced ground damage for this site?

ANSWER: Based on Fig. 14.10, liquefaction induced ground damage would be expected for this site.

14.7   Use the data from Problem 13.1 and Fig. 13.24. Assume that the groundwater table is unlikely to rise above its present level. Using a peak ground acceleration a_{max} equal to 0.20g and the standard penetration test data, will there be liquefaction induced ground damage for this site?

ANSWER: Based on Fig. 14.10, liquefaction induced ground damage would be expected for this site.

14.8   Assume an oil tank will be constructed at a level-ground site and the subsurface soil conditions are shown in Fig. 14.13. The groundwater table is located at a depth of 1 m below ground surface. The standard penetration test values shown in Fig. 14.13 are uncorrected N values. Assume a

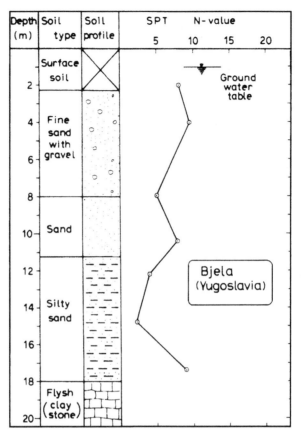

FIGURE 14.13 Subsoil profile, Bjela, Yugoslavia. (*Reproduced from Ishihara, 1985.*)

hammer efficiency E_m of 60 percent, a boring diameter of 100 mm, and the length of the drill rods is equal to the depth of the SPT below ground surface. The design earthquake conditions are a peak ground acceleration a_{max} of 0.20g and a magnitude of 7.5. For the materials shown in Fig. 14.13, assume the following:

a. The surface soil layer (0 to 2.3 m) is clay having an undrained shear strength s_u of 50 kPa. The total unit weight of the soil above the groundwater table γ_t is 19.2 kN/m³ and the buoyant unit weight γ_b is equal to 9.4 kN/m³.

b. The fine sand with gravel layer (2.3 to 8 m) has low gravel content and can be considered to be essentially a clean sand (γ_b = 9.7 kN/m³).

c. The sand layer (8 to 11.2 m) has less than 5 percent fines (γ_b = 9.6 kN/m³).

d. The silty sand layer (11.2 to 18 m) meets the requirements for a potentially liquefiable soil and has 35 percent fines (γ_b = 9.6 kN/m³).

e. The Flysh claystone (> 18 m) is essentially solid rock and it is not susceptible to earthquake-induced liquefaction or settlement.

Assume the oil tank will be constructed at ground surface and will have a diameter of 20 m, an internal storage capacity equal to a 3 m depth of oil (unit weight of oil = 9.4 kN/m³), and the actual weight of the tank can be ignored in the analysis. Determine the factor of safety against liquefaction and the amount of fill that must be placed at the site to prevent liquefaction-induced ground surface fissuring and sand boils. With the fill layer in place, determine the liquefaction-induced settlement of the tank and calculate the factor of safety against a bearing capacity failure of the tank. Assume that the fill will be obtained from a borrow site that contains clay and when compacted, the clay will have an undrained shear strength s_u of 50 kPa.

ANSWER: Zone of liquefaction extends from 2.3 to 18 m, thickness of required fill layer at site = 0.7 m, liquefaction-induced settlement of the oil tank = 54 to 66 mm based on Figs. 14.7 and 14.8, and factor of safety against a bearing capacity failure = 1.06.

Retaining Walls

14.9   Using the retaining wall shown in Fig. 11.2, assume $H = 4$ m, the thickness of the reinforced concrete wall stem = 0.4 m, the reinforced concrete wall footing is 3 m wide by 0.5 m thick, the ground surface in front of the wall is level with the top of the wall footing, and the unit weight of concrete = 23.5 kN/m³. The wall backfill will consist of sand having $\phi = 32°$ and $\gamma_t = 20$ kN/m³. Also assume that there is sand in front of the footing with these same soil properties. The friction angle between the bottom of the footing and the bearing soil (δ) = 38°. For the condition of a level backfill and neglecting the wall friction on the back side of the wall and the front side of the footing, determine the factor of safety for sliding and factor of safety for overturning for earthquake conditions using the pseudostatic method (Eq. 14.2) and $a_{max} = 0.20g$.

ANSWER: F for sliding = 0.86 and F for overturning = 1.29.

14.10   Solve Problem 14.9 using Eq. 14.3.

ANSWER: F for sliding = 0.78 and F for overturning = 1.18.

14.11   Solve Problem 14.9 but include wall friction in the analysis. Assume the friction angle between the back side of the retaining wall and the backfill is equal to ¾ of ϕ (i.e., $\phi_w = ¾\,\phi = 24°$). Use Eq. 14.4 for the earthquake analysis.

ANSWER: F for sliding = 1.26 and F for overturning = ∞.

14.12   Using the retaining wall shown at the top of Fig. 11.6B (i.e., a cantilevered retaining wall), assume $H = 4$ m, the thickness of the reinforced concrete wall stem = 0.4 m and the wall stem is located at the centerline of the footing, the reinforced concrete wall footing is 2 m wide by 0.5 m thick, the ground surface in front of the wall is level with the top of the wall footing, and the unit weight of concrete = 23.5 kN/m³. The wall backfill will consist of sand having $\phi = 32°$ and $\gamma_t = 20$ kN/m³. Also assume that there is sand in front of the footing with these same soil properties. The friction angle between the bottom of the footing and the bearing soil (δ) = 24°. For the condition of a level backfill and assuming total mobilization of the shear strength along the vertical plane at the heel of the wall, calculate the factor of safety for sliding and factor of safety for overturning for earthquake conditions using the pseudostatic method (Eq. 14.4) and $a_{max} = 0.20g$.

ANSWER: F for sliding = 1.17 and F for overturning = 29.

14.13   Using the mechanically stabilized earth retaining wall shown in Fig. 11.12, let $H = 20$ ft, the width of the mechanically stabilized earth retaining wall = 14 ft, the depth of embedment at the front of the mechanically stabilized zone = 3 ft, and there is a level backfill with no surcharge pressures (i.e., $P_2 = 0$). Assume that the soil behind and in front of the mechanically stabilized zone is a clean sand having a friction angle $\phi = 30°$, a total unit weight $\gamma_t = 110$ pcf, and there will be no shear stress (i.e., $\delta = 0°$) along the vertical back and front sides of the mechanically stabilized zone. For the mechanically stabilized zone, assume the soil will have a total unit weight $\gamma_t = 120$ pcf and $\delta = 23°$ along the bottom of the mechanically stabilized zone. Also use a reduction factor = 2 for the passive resultant force. Calculate the factor of safety for sliding and the factor of safety for overturning for the earthquake conditions using the pseudostatic method (Eq. 14.2) and $a_{max} = 0.20g$.

ANSWER: F for sliding $= 1.52$ and F for overturning $= 2.84$.

14.14   A braced excavation will be used to support the vertical sides of a 20 ft deep excavation (i.e., $H = 20$ ft in Fig. 11.18). If the site consists of a sand with a friction angle $\phi = 32°$ and a total unit weight $\gamma_t = 120$ pcf, calculate the resultant earth pressure force acting on the braced excavation for the static condition and the earthquake condition (using Eq. 14.1) for $a_{max} = 0.20g$. Assume the groundwater table is well below the bottom of the excavation.

ANSWER: Static condition: resultant force $= 9600$ lb per linear ft of wall length. Earthquake condition: $P_E = 2700$ lb per linear ft of wall length.

FOUNDATION CONSTRUCTION

CHAPTER 15
GRADING AND OTHER SOIL IMPROVEMENT METHODS

15.1 GRADING

15.1.1 Introduction

Chapters 5 through 14 (Part 2 of the book) have dealt with the geotechnical aspects of foundation engineering design. Part 3 of the book (Chaps. 15 to 17) deals with foundation construction and includes such topics as grading, soil improvement, foundation excavations, foundation underpinning, field load-testing of foundations, geosynthetics, and instrumentation.

Since most building sites start out as raw land, the first step in site construction work usually involves the grading of the site. Grading basically consists of the cutting or filling of the ground in order to create a level building pad upon which the foundation and structure can be built. The three types of level building pads that are created by the grading operations are cut lots, cut-fill transition lots, and fill lots as illustrated in Fig. 15.1. Appendix A (Glossary 4) presents a list of common grading terms and their definitions.

15.1.2 Grading Operation

The typical steps in a grading operation are as follows:

1. *Easements.* The first step in the grading operation is to determine the location of any on-site utilities and easements. The on-site utilities and easements often need protection so that they are not damaged during the grading operation.
2. *Clearing, brushing, and grubbing.* Clearing, brushing, and grubbing are defined as the removal of vegetation (grass, brush, trees, and similar plant types) by mechanical means. This debris is often stockpiled at the site and it is important that this debris be removed from the site and not accidentally placed within the structural fill mass. Figure 15.2 shows one method of dealing with vegetation, where a large mechanical grinder has been brought to the site and the trees and brush are being ground-up into wood chips. The wood chips will be removed from the site and then recycled.
3. *Cleanouts.* Once the site has been cleared of undesirable material, the next step is the removal of unsuitable bearing material at the site, such as loose or porous alluvium, colluvium, and uncompacted fill.
4. *Benching (hillside areas).* Benching is defined as the excavation of relatively level steps into earth material on which fill is to be placed. The benches provide favorable (i.e., not out-of-slope) frictional contact between the structural fill mass and the horizontal portion of the bench.
5. *Canyon subdrain.* A *subdrain* is defined as a pipe and gravel or similar drainage system placed in the alignment of canyons or former drainage channels. The purpose of a canyon subdrain is to intercept groundwater and to not allow it to build up within the fill mass.

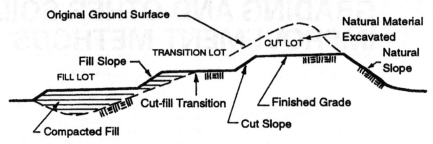

FIGURE 15.1   Three types of building pads created during the grading operation.

6. *Scarifying and recompaction.*   In flat areas that have not been benched, scarifying and recompaction of the ground surface is performed by compaction equipment in order to get a good bond between the in-place material and compacted fill.

7. *Cut and fill rough grading operations.*   Rough grading operations involve the cutting or excavation of earth materials and the compaction of this material as fill in conformance with the grading plans. The location of the excavated earth material is often referred to as the *borrow area*. During the rough grading operation, fill is placed in horizontal lifts and then each lift of fill is compacted to create a uniformly compacted material such as shown in Fig. 15.3.

   Other activities that could be performed during rough grading operations are as follows:

   a. *Ripping or blasting of rock.*   Large rock fragments can be removed from the site or disposed of in windrows. Ripping has been previously covered in Sec. 2.7, where it was discussed that the seismic wave velocity can be used to determine if rock is rippable or nonrippable. Figure 15.4 shows a Caterpillar D10 tractor/ripper that can be used to excavate rock.

   b. *Removal of rock fragments.*   Large rock size fragments interfere with the compaction process and are usually an undesirable material in structural fill. The large rock size fragments

FIGURE 15.2   A large mechanical grinder has been brought to the site and the trees and brush are being ground-up into wood chips. The wood-chips will be removed from the site and then recycled.

FIGURE 15.3 A lift of fill has been placed and compacted into a dense state.

may become nested, creating open voids within the fill mass. Figures 15.5 and 15.6 show one method used to remove large rock size fragments. A screen is set up as shown in Fig. 15.5 and then a loader is used to dump the material on top of the screen. As shown in Fig. 15.6, the large rock size fragments roll off of the screen while the material that passes through the screen is used as structural fill.

c. Cut-fill transition. Figure 15.1 illustrates a cut-fill transition. It is the location in a building pad where on one side the ground surface has been cut down exposing natural or rock material, while on the other side, fill has been placed.

FIGURE 15.4 A Caterpillar D10 tractor/ripper that can be used to excavate rock.

FIGURE 15.5   A screen has been set up in order to remove large-size rock fragments from the soil.

d. *Slope stabilization.*   Examples of slope stabilization using earth materials include stabilization fill, buttress fill, drainage buttress, and shear keys. Such devices should be equipped with back drain systems.
e. *Fill slopes.*   During the grading process, fill slopes can be created out of earth materials. Figure 15.1 shows the construction of fill and cut slopes.
f. *Revision of grading operations.*   Every grading job is different and there could be a change in grading operations based on field conditions.

FIGURE 15.6   A loader is in the process of depositing material on top of the screen in order to separate the large size rock fragments.

8. *Fine grading.*   Fine grading is also known as *precise grading.* At the completion of the rough grading operations, fine grading is performed in order to obtain the finish elevations that are in accordance with the precise grading plan.

9. *Slope protection.*   Upon completion of the fine grading, slope protection and permanent erosion control devices are installed.

10. *Trench excavations.*   Utility trenches are excavated in the proposed road alignments and building pads for the installation of the on-site utilities. The excavation and compaction of utility trenches is often part of the grading process. Once the utility lines are installed, scarifying and recompaction of the road subgrade is performed and base material is placed and compacted. Figures 15.7 and 15.8 show trench excavations for the installation of storm drainage systems. The trenches should be either sloped or shored in order to prevent a cave-in.

11. *Foundation construction.*   Although usually not a part of the grading operation, the footing and foundation elements can be excavated at the completion of grading in accordance with the foundation plans.

Most projects involve grading and it is an essential part of geotechnical engineering. For many projects, it is usually necessary to prepare a set of grading specifications. These specifications are often

FIGURE 15.7   Utility trench excavation for a storm-drain line.

FIGURE 15.8   Another example of a utility trench excavation for a storm-drain line.

used to develop the grading plans, which are basically a series of maps that indicate the type and extent of grading work to be performed at the site. Often the grading specifications will be included as an appendix in the preliminary or feasibility report prepared by the geotechnical engineer.

15.1.3   Grading Equipment

Common types of equipment used during the grading operations are as follows:

1. *Bulldozer (Fig. 15.9).*   The bulldozer is used to clear the land of debris and vegetation (clearing, brushing, and grubbing), excavate soil from the borrow area, cut haul roads, spread out dumped fill, rip rock, and compact the soil.

FIGURE 15.9   The bulldozer is in the process of spreading out a layer of fill for compaction.

FIGURE 15.10 The scraper is used to excavate material from the borrow area, transport it to the site, dump it at the site, and then compact the soil.

2. *Scraper (Fig. 15.10).* The scraper is used to excavate (scrape up) soil from the borrow area, transport it to the site, dump it at the site, and the rubber-tires of the scraper can be used to compact the soil. Push-pull scrapers can be used in tandem in order to provide additional energy to excavate hard soil or soft rock.

3. *Loader (Fig. 15.11).* Similar to the scraper, the loader can be used to excavate soil from the borrow area, transport the soil, and then compact it as structural fill.

4. *Excavator (Fig. 15.12).* This type of equipment is ideally suited to excavating narrow trenches for the construction of utilities such as storm drain lines and sewer lines. This equipment is also used to excavate footings and other foundation elements.

FIGURE 15.11 The loader can be used to move soil about the job site and compact the soil as structural fill.

FIGURE 15.12   The excavator is ideally suited to excavating narrow trenches for the construction of utilities such as storm drain lines and sewer lines.

5. *Dump Trucks and Water Trucks* (*Fig. 15.13*).   If the borrow area is quite a distance from the site, then dump trucks may be required to transport the borrow soil to the site. Dump trucks are also needed to transport soil on public roads or to import select material.

Especially in the southwestern United States, the near surface soil can be in a dry and powdery state and water must be added to the soil in order to approach the optimum moisture content. A water truck, such as shown in Fig. 15.13, is often used to add water to the fill during the grading operation.

The *Caterpillar Performance Handbook* (1997), which is available at Caterpillar dealerships, is a valuable reference because it not only lists rippability versus types of equipment, but also indicates

FIGURE 15.13   The water truck is adding water to fill that is in the process of being compacted.

types and models of compaction equipment, equipment sizes and dimensions, and performance specifications. Compaction equipment can generally be grouped into five main categories, as follows:

1. *Static weight or pressure.*   This type of compaction equipment applies a static or relatively uniform pressure to the soil. Examples include the compaction by the rubber-tires of a scraper, from the tracks of a bulldozer, and by using smooth drum rollers.

2. *Kneading action or manipulation.*   The sheepsfoot roller, which has round or rectangular shaped protrusions or feet, is ideally suited to applying a kneading action to the soil. This has proven to be effective in compacting silts and clays.

3. *Impact or a sharp blow.*   There are compaction devices, such as the high-speed tamping foot and the Caterpillar tamping foot, that compact the soil by imparting impacts or sharp blows to the soil.

4. *Vibration or shaking.*   Nonplastic sands and gravels can be effectively compacted by vibrations or shaking. An example is the smooth drum vibratory soil compactor.

5. *Chopper wheels.*   This type of compaction equipment has been specially developed for the compaction of waste products at municipal landfills.

Table 15.1 presents a summary of different types of compaction equipment best suited to compact different types of soil.

A common objective of the grading operations is to balance the volume of cut and fill. This means that just enough earth material is cut from the high areas to fill in the low areas. A balanced cut and fill operation means that no soil needs to be imported or exported from the site, leading to a reduced cost of the grading operation.

When developing a site so that the cut and fill is balanced, consideration must be given to the bulking or shrinkage factor associated with the compaction operation. *Bulking* is defined as an increase in volume of soil or rock caused by its excavation. For example, very dense soil will increase in volume upon excavation and when compacted, the compacted soil may have a dry unit weight that is less than existed at the borrow area. Conversely, when loose material is excavated from a borrow area and worked into a compacted state, the compacted soil usually has a dry unit weight that is greater than existed at the borrow area. The *shrinkage factor* is often defined as the ratio of the volume of compacted material to the volume of borrow material (based on dry unit weight).

Fill placement should proceed in thin lifts, i.e., 6 to 8 in. (15 to 20 cm) loose thickness. Each lift should be moisture conditioned and thoroughly compacted. The desired moisture condition should be maintained or reestablished, where necessary, during the period between successive lifts. Selected lifts should be tested to ascertain that the desired compaction is being achieved, which will be discussed in the next section.

There are many excellent publications on field compaction equipment. For example, *Moving the Earth* (Nichols and Day, 1999) presents an in-depth discussion of the practical aspects of earth moving equipment and earthwork operations.

Example Problem 15.1   Soil at the borrow area has a total unit weight γ_t of 120 pcf and a water content w of 15 percent. The soil from the borrow area will be used as structural fill and compacted to an average dry unit weight γ_d of 110 pcf. Determine the shrinkage factor.

Solution   At the borrow area, the dry unit weight is determined from Eq. 3.3:

$$\gamma_d = \gamma_t/(1 + w) = 120/(1 + 0.15) = 104 \text{ pcf}$$

The shrinkage factor is the ratio of the volume of compacted material to the volume of borrow material (based on dry unit weight), or:

Shrinkage factor = 104/110 = 0.945

In terms of the percent shrinkage:

Percent shrinkage = (110 − 104)/110 = 0.055 = 5.5 percent

TABLE 15.1 Characteristics of Compacted Subgrade for Roads and Airfields

Major Divisions	Sub-divisions	USCS symbol	Name	Value as subgrade (no frost action)	Potential frost action	Compressibility	Drainage properties	Compaction equipment	Typical dry densities		CBR	Sub. mod*, pci
									pcf	Mg/m³		
Coarse-grained soils	Gravel and gravelly soils	GW	Well-graded gravels or gravel-sand mixtures, little or no fines	Excellent	None to very slight	Almost none	Excellent	Crawler-type tractor, rubber-tired roller, steel-wheeled roller	125–140	2.00–2.24	40–80	300–500
		GP	Poorly graded gravels or gravelly sands, little or no fines	Good to excellent	None to very slight	Almost none	Excellent	Crawler-type tractor, rubber-tired roller, steel-wheeled roller	110–140	1.76–2.24	30–60	300–500
		GM	Silty gravels, gravel-sand-silt mixtures	Good to excellent	Slight to medium	Very slight to slight	Fair to very poor	Rubber-tired roller, sheepsfoot roller	115–145	1.84–2.32	20–60	200–500
		GC	Clayey gravels, gravel-sand-clay mixtures	Good	Slight to medium	Slight	Poor to very poor	Rubber-tired roller, sheepsfoot roller	130–145	2.08–2.32	20–40	200–500
	Sand and sandy soils	SW	Well-graded sands or gravelly sands, little or no fines	Good	None to very slight	Almost none	Excellent	Crawler-type tractor, rubber-tired roller	110–130	1.76–2.08	20–40	200–400
		SP	Poorly graded sands or gravelly sands, little or no fines	Fair to good	None to very slight	Almost none	Excellent	Crawler-type tractor, rubber-tired roller	105–135	1.68–2.16	10–40	150–400
		SM	Silty sands, sand-silt mixtures	Fair to good	Slight to high	Very slight to medium	Fair to poor	Rubber-tired roller, sheepsfoot roller	100–135	1.60–2.16	10–40	100–400
		SC	Clayey sands, sand-clay mixtures	Poor to fair	Slight to high	Slight to medium	Poor to very poor	Rubber-tired roller, sheepsfoot roller	100–135	1.60–2.16	5–20	100–300

		Description	Value as subgrade	Potential frost action	Compressibility and expansion	Drainage characteristics	Compaction equipment	Dry weight (lb/ft³)	(g/cm³)	CBR	Subgrade modulus*
Fine-grained soils	Silts and clays with liquid limit less than 50	ML	Inorganic silts, rock flour, silts of low plasticity	Poor to fair	Medium to very high	Slight to medium	Fair to poor	Rubber-tired roller, sheepsfoot roller	90–130	1.44–2.08	15 or less	100–200
		CL	Inorganic clays of low plasticity, gravelly clays, sandy clays, etc.	Poor to fair	Medium to high	Medium	Practically impervious	Rubber-tired roller, sheepsfoot roller	90–130	1.44–2.08	15 or less	50–150
		OL	Organic silts and organic clays of low plasticity	Poor	Medium to high	Medium to high	Poor	Rubber-tired roller, sheepsfoot roller	90–105	1.44–1.68	5 or less	50–100
	Silts and clays with liquid limit 50 or greater	MH	Inorganic silts, micaceous silts, silts of high plasticity	Poor	Medium to very high	High	Fair to poor	Sheepsfoot roller, rubber-tired roller	80–105	1.28–1.68	10 or less	50–100
		CH	Inorganic clays of high plasticity, fat clays, silty clays, etc.	Poor to fair	Medium	High	Practically impervious	Sheepsfoot roller, rubber-tired roller	90–115	1.44–1.84	15 or less	50–150
		OH	Organic silts and organic clays of high plasticity	Poor to very poor	Medium	High	Practically impervious	Sheepsfoot roller, rubber-tired roller	80–110	1.28–1.76	5 or less	25–100
Peat	Highly organic	PT	Peat and other highly organic soils	Not suitable	Slight	Very high	Fair to poor	Compaction not practical	—	—	—	—

Source: U.S. Army Waterways Experiment Station, 1960.

*Subgrade Modulus.

15.13

15.2 COMPACTION

15.2.1 Introduction

Grading work usually involves fill compaction, which is defined as the densification of soil by mechanical means. This physical process of getting the soil into a dense state can increase the shear strength, decrease the compressibility, and decrease the permeability of the soil.

There are many different types of fill, such as hydraulic fill, structural fill, dumped or uncompacted fill, debris fill, and municipal landfill. Table 15.2 presents a discussion of the typical characteristics, uses, and possible engineering problems for these types of fill.

It is always desirable to construct foundations on structural fill. For example, structural fill is used for all types of earthwork projects, such as during grading operations in order to create level building pads, slope buttresses, and shear keys. Structural fill is used for highway embankments, earth dams, and for backfill material of retaining walls and utility trenches. Structural fill is also used to create mechanically stabilized earth retaining walls and road subgrade. If it is not possible to support a foundation on structural fill, then a deep foundation system that penetrates the compressible soil and bears on solid material would be required.

TABLE 15.2 Different Types of Fill

Main category	Typical characteristics and uses	Possible engineering problems
Hydraulic fill	This refers to a fill placed by transporting soils through a pipe using large quantities of water. These fills are generally loose because they have little or no mechanical compaction during construction.	Subject to significant compression and hydraulically placed sands are susceptible to liquefaction.
Structural fill	Used for all types of earthwork projects, such as during grading operations in order to create level building pads, slope buttresses, and shear keys. Also used for highway embankments, earth dams, and for backfill material of retaining walls and utility trenches. Structural fill is also used to create mechanically stabilized earth retaining walls and road subgrade. Often the individual fill lifts can be identified.	Upper surface of structural fill may have become loose or weathered.
Dumped or uncompacted fill	This refers to fill that was not documented with compaction testing as it was placed or fill that may have been compacted but there is no documentation of testing or the amount of effort that was used to perform the compaction. Dumped or uncompacted fill often consists of random soil deposits and chunks of different types and/or sizes of rock fragments.	Susceptible to compression and collapse. Dumped or uncontrolled fill should not be used to support structures.
Debris fill	This refers to fill that contains pieces of debris, such as concrete, brick, and wood fragments. When consisting of dumped material, the debris fill is usually loose and compressible. A debris fill can be turned into a structural fill by removing the compressible and degradable material, and then recompacting the material with no nesting of oversize particles.	Susceptible to compression and collapse. A dumped debris fill should not be used to support structures.
Municipal landfill	Contains debris and waste products such as household garbage or yard trimmings. Soil is typically used to encase layers of garbage and cap the municipal landfill.	Significant compression and gas from organic decomposition

As previously discussed, heavy compaction equipment is commonly used to create structural fill for mass grading operations. But many projects require backfill compaction where the construction space is too small to allow for such heavy equipment. Hand-operated vibratory plate compactors, such as shown in Fig. 15.14, are ideally suited for compacting fill in small or tight spaces. The hand-operated compactors can be used for all types of restricted access areas, such as the compaction of fill in utility trenches, behind basement walls, or around storm drains. For many projects having small or tight spaces, the backfill is simply dumped in place or compacted with minimal compaction effort. These factors of limited access, poor compaction process, and lack of compaction testing frequently lead to backfill settlement.

There are four basic factors that affect compaction of structural fill as follows:

1. *Soil type.*   Nonplastic (i.e., cohesionless) soil, such as sands and gravels, can be effectively compacted by using a vibrating or shaking type of compaction operation. Plastic (i.e., cohesive) soil, such as silts and clays, are more difficult to compact and require a kneading or manipulation type of compaction operation. If the soil contains oversize particles, such as coarse gravel and cobbles, then they tend to interfere with the compaction process and reduce the effectiveness of compaction for the finer soil particles. Typical values of dry density for different types of compacted soil are listed in Table 15.1.

2. *Material gradation.*   Those soils that have a well-graded grain size distribution can generally be compacted into a denser state than a poorly graded soil that is composed of soil particles of about the same size. For example, a well-graded decomposed granite (DG) can have a maximum dry density of 137 pcf (2.2 Mg/m^3), while a poorly graded sand can have a maximum dry density of only 100 pcf (1.6 Mg/m^3, modified Proctor).

3. *Water content.*   The water content is an important parameter in the compaction of soil. Water tends to lubricate the soil particles thus helping them slide into dense arrangements. However, too much water and the soil becomes saturated and often difficult to compact. There is an optimum water content at which the soil can be compacted into its densest state for a given compaction energy. Typical optimum moisture contents (modified Proctor) for different soil types are as follows:

   a. Clay of high plasticity (CH).   Optimum moisture content $\geq$ 18 percent

   b. Clay of low plasticity (CL).   Optimum moisture content = 12 to 18 percent

FIGURE 15.14   Hand-operated vibratory plate compactor used for compacting fill in tight spaces.

  c. Well-graded sand (SW). Optimum moisture content = 10 percent

  d. Well-graded gravel (GW). Optimum moisture content = 7 percent

   Some soils may be relatively insensitive to compaction water content. For example, open-graded gravels and clean coarse sands are so permeable that water simply drains out of the soil or is forced out of the soil during the compaction process. These types of soil can often be placed in a dry state and then vibrated into dense particle arrangements.

 4. *Compaction effort (or energy).* The compaction effort is a measure of the mechanical energy applied to the soil. Usually the greater the amount of compaction energy applied to a soil, the denser the soil will become. There are exceptions, such as pumping soils (discussed in Sec. 15.2.5) that cannot be densified by an increased compaction effort. Compactors are designed to use one or a combination of the following types of compaction effort:

  a. Static weight or pressure

  b. Kneading action or manipulation

  c. Impact or a sharp blow

  d. Vibration or shaking

15.2.2 Relative Compaction

  The most common method of assessing the quality of the field compaction is to calculate the *relative compaction* (RC) of the fill, defined as

$$RC = \frac{100 \, \rho_d}{\rho_{d\text{max}}} \tag{15.1}$$

where $\rho_{d\text{max}}$ is the laboratory maximum dry density (pcf or Mg/m^3) and ρ_d is the field dry density (pcf or Mg/m^3).

  In California, the typical mass grading specification for structural fill is a minimum relative compaction of 90 percent using the modified Proctor laboratory compaction test. In some cases, such as for the compaction of roadway base or for the lower portions of deep fill, a higher compaction of a minimum relative compaction of 95 percent is often specified.

  As discussed in Sec. 3.6, the maximum dry density $\rho_{d\text{max}}$ is obtained from the laboratory compaction tests, such as by using the modified Proctor test procedures (ASTM D 1557-02, 2004) or the standard Proctor test procedures (ASTM D 698-00, 2004). The objective of the laboratory compaction test is to obtain the compaction curve, with the peak point of the compaction curve corresponds to the laboratory maximum dry density (e.g., see Fig. 3.52).

  In addition to the maximum dry density, the field dry density of the compacted soil must also be determined in order to calculate the relative compaction. Field dry density tests are discussed in the next section.

15.2.3 Field Density Tests

  In order to determine ρ_d for Eq. 15.1, a field density test must be performed. Field density tests can be classified as either destructive or nondestructive tests (Holtz and Kovacs, 1981).

  Probably the most common destructive method for determining the field dry density is through the use of the sand cone apparatus (ASTM D 1556-00, 2004). The test procedure consists of excavating a hole in the ground, filling the hole with sand using the sand cone apparatus, and then determining the volume of the hole based on the amount of sand required to fill the hole. By knowing the wet mass of soil removed from the hole divided by the volume of the hole, the wet density of the fill can be calculated. The water content w of the soil extracted from the hole can be determined and thus the dry density ρ_d can then be calculated.

  Another type of destructive test for determining the field dry density is the drive cylinder (ASTM D 2937-00, 2004). This method involves the driving of a steel cylinder of known volume into the soil. Based on the mass of soil within the cylinder, the wet density can be calculated. Once the water content w of the soil is obtained, the dry density ρ_d of the fill can be calculated.

Probably the most common type of nondestructive field test is the nuclear method described in ASTM D 2922-01, 2004, "Standard Test Methods for Density of Soil and Soil-Aggregate in Place by Nuclear Methods (Shallow Depth)." In this method, the wet density is determined by the attenuation of gamma radiation. The water content is often determined by the thermalization or slowing of fast neutrons, described in ASTM D 3017-01, 2004, "Standard Test Method for Water Content of Soil and Rock in Place by Nuclear Methods (Shallow Depth)." The most common approach is to use the *backscatter method*, where the source and detector remain on the ground surface. In general, the nuclear method is much quicker than any of the destructive tests. However, disadvantages of this test are that special equipment is required and the equipment is much more expensive than the equipment required for the other types of field density tests. In addition, the equipment uses radioactive materials that could be hazardous to the health of the user. There are special governmental regulations concerning the storage, transportation, and use of this equipment and these safety requirements are beyond the scope of this book. A final disadvantage is that the equipment is subject to long-term aging of radioactive materials that may change the relationship between count rate and soil density. Hence, the equipment will need to be periodically calibrated (usually a block of material of known density is used as a calibration device). Because of these limitations, sand cone tests or drive cylinder tests should be used as a check on the results from the nuclear method.

NAVFAC DM-7.2 (1982) presents guidelines on the number of field density tests for different types of grading projects, as follows:

- One test for every 500 yd^3 (380 m^3) of material placed for embankment construction
- One test for every 500 to 1000 yd^3 (380 to 760 m^3) of material for canal or reservoir linings or other relatively thin fill sections
- One test for every 100 to 200 yd^3 (75 to 150 m^3) of backfill in trenches or around structures, depending upon total quantity of material involved
- At least one test for every full shift of compaction operations on mass earthwork
- One test whenever there is a definite suspicion of a change in the quality of moisture control or effectiveness of compaction

There are many other guidelines concerning the number of field density tests for specific grading activities.

It is rare for the licensed geotechnical engineer to perform field density testing on a daily basis because of the repetitive and time-consuming nature of such work. For large mass grading operations, it is common to have technicians performing the field density testing. The technician will have to be able to perform the field density tests, classify different soil types (based on visual and tactile methods), and insist on remedial measures when compaction falls below the specifications.

As indicated earlier, the number of field density tests per volume of compacted fill is often very low (e.g., one field density test per 500 yd^3 of fill). It is important that the field technician perform the density tests on areas where compaction is suspect. For example, the technician should not perform field compaction tests in the haul road area, because this path receives continuous traffic and will usually be in a dense compacted state. Likewise, testing in the wheel paths of the compaction equipment will yield high values. Often the field technician uses a metal rod to probe for possible poorly compacted fill zones. Field density tests would then be performed in these areas of possible poor compaction.

15.2.4 Types of Structural Fill

There are four general types of structural fill, as follows:

1. *Select import.* Select import refers to a processed material. The material may be derived from several different sources, then screened and mixed to provide a material of specified gradation. Table 15.3 presents different methods that can be used to produce a select import material. A common type of select import is granular base material, which may have to meet specifications

TABLE 15.3   Methods Used to Produce a Select Import Material

Method	Description	Effect
Screening	Material processed over vibrating screens (can be combined with spray washing on screens).	Divide by particle size.
Crushing	Material run through a crusher.	Produces angular shape.
Log washers	Material is run through an inclined unit with dual rotating shafts mounted with paddles. Continuous flow of water carries fine material out of low end of the unit while cleaned aggregate is discharged at the upper end.	Removes deleterious material (e.g., clay) present in the aggregate or removes coating on aggregates.
Sand classifying unit	Continuous flow of water containing sand is fed into horizontal unit. Coarse sand settles first; finer sands later; finer contaminants are carried out of far end by the flow of the water.	Divides sand into fractions based on particle size.
Screw classifier	Water and sand are fed into the low end of an inclined unit having a rotating screw auger. Sand is moved up the unit and out of the water by the screws. Waste water at the low end carries off fines and lightweight contaminants.	Removes lightweight material and fine contaminants.
Rotary scrubber	Water and aggregate are fed into a revolving, inclined drum equipped with lifting angles. The aggregate tumbles upon itself as it proceeds through the scrubber.	Capable of removing large quantities of soluble contaminants.
Jig benefaction	Mechanical or air pulses agitate water, allowing material to sink to the bottom of the unit and form layers of different density.	Separates aggregate on the basis of specific gravity.
Heavy separator	Aggregate fed into medium of given specific gravity. Denser particles sink; lighter particles float or are suspended in medium.	Precise separation on basis of specific gravity of medium.

Source: Rollings and Rollings (1996).

for gradation, wear resistance, and shear strength (Standard Specifications for Public Works Construction, 2003). Other uses for select import include backfill for retaining walls and utilities, and even for mass-graded fill.

The main characteristics of select import are a well-graded granular soil, which has a high laboratory maximum dry density, typically in the range of 125 to 135 pcf (2.0 to 2.2 Mg/m^3). As a processed material, the particle size gradation for each batch of fill should be similar. Usually an import material will have all laboratory maximum dry density values within 3 pcf (0.05 Mg/m^3) and a standard deviation of 1 pcf (0.02 Mg/m^3) or less.

2. *Uniform borrow.*   Uniform borrow typically refers to a natural material that will consistently have the same soil classification and similar grain size distribution. An example of a possible uniform borrow could be a natural deposit of beach sand. Other uniform borrow could be formational rock, such as deposits of sandstone or siltstone. As the name implies, the main characteristic of the material is its uniformity. Usually a uniform borrow material has consistently the same soil classification, with all laboratory maximum dry density values within 8 pcf (0.13 Mg/m^3) and a standard deviation of 3 pcf (0.05 Mg/m^3) or less.

Figure 15.15 presents laboratory test results on a uniform borrow material. The fill was derived from a formational rock, classified as a weakly cemented shale. When used as fill, the material is classified as silty clay, having a liquid limit between 41 and 50. The laboratory maximum dry density varies from 115 to 123 pcf (1.84 to 1.97 Mg/m^3), with an average value of 120 pcf (1.92 Mg/m^3) and a standard deviation of 2.2 pcf (0.035 Mg/m^3).

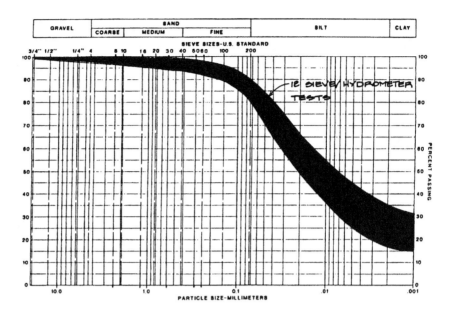

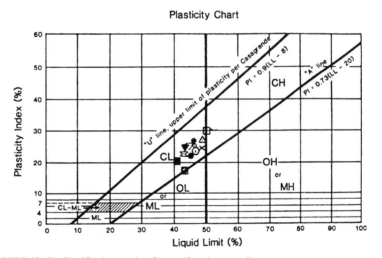

FIGURE 15.15   Classification test data for a uniform borrow soil.

3. *Mixed borrow.*   Mixed borrow contains material of different classifications. For example, mixed borrow could be a deposit of alluvium, that contains alternating layers of sand, silt, and clay. Mixed borrow could also be formational rock that contains thin alternating layers of sandstone and claystone. The main characteristics of mixed borrow are that each load of fill could have soils with significantly different grain size distributions and soil classifications. The fill commonly contains many different soil types, all jumbled up and mixed together.

One method to deal with mixed borrow material is to thoroughly mix each load of import and then perform a laboratory maximum dry density test on that batch of import soil. Another option is to thoroughly mix each batch of mixed borrow material and then perform a *one-point Proctor test* (see Sec. 3.6.3) in order to estimate the laboratory maximum dry density.

4. *Borrow with oversize particles.* The last basic type of fill is borrow having oversize particles, which are typically defined as those particles retained on the $3/4$ in. (19 mm) U.S. standard sieve, i.e., coarse gravel and cobble size particles. The *soil matrix* is defined as those soil particles that pass the $3/4$ in. (19 mm) U.S. standard sieve. When a field density test (such as a sand cone test) is performed, the soil excavated for the test can be sieved on the $3/4$ in. (19 mm) sieve in order to determine the mass of oversize particles. The elimination method (Day, 1989) can then be used to mathematically eliminate the volume of oversize particles in order to calculate the dry density of the matrix material. The relative compaction is calculated by dividing the dry density of the matrix material by the laboratory maximum dry density, where the laboratory compaction test is performed on the matrix material. By using the elimination method to calculate the relative compaction of the matrix material, the compaction state of the matrix soil is controlled. This is desirable because it is the matrix soil (not the oversize particles) that usually govern the compressibility, shear strength, and permeability of the soil mass.

If the matrix soil can be considered to be a uniform borrow material, then the procedure for selecting the laboratory maximum dry density in the field is the same as previously discussed for a uniform borrow material. If the matrix material is a mixed borrow material, then the procedure for selecting the laboratory maximum dry density in the field is the same as previously discussed for a mixed borrow material. Other methods have been developed to deal with fill containing oversize particles (Saxena et al., 1984; Houston and Walsh, 1993).

15.2.5 Pumping of Saturated Clay

Pumping is a form of bearing capacity failure that occurs during compaction of fill. A commonly used definition of pumping is the softening and squeezing of clay from underneath the compaction equipment. Continual passes of the compaction equipment can cause a decrease in the undrained shear strength of the wet clay and the pumping may progressively worsen. Figure 15.16 shows the pumping of a wet clay subgrade.

FIGURE 15.16 Pumping of clay subgrade during construction.

Pumping is dependent on the penetration resistance of the compacted clay. Figure 15.17 (from Turnbull and Foster, 1956) presents data on the California bearing ratio (CBR) of compacted clay and shows that the penetration resistance approaches zero (i.e., the clay can exhibit pumping) when the clay has a water content that is wet of optimum. Also note in the lower part of Fig. 15.17 that the laboratory maximum dry density increases and the optimum moisture content decreases as the compaction energy increases (more blows per layer).

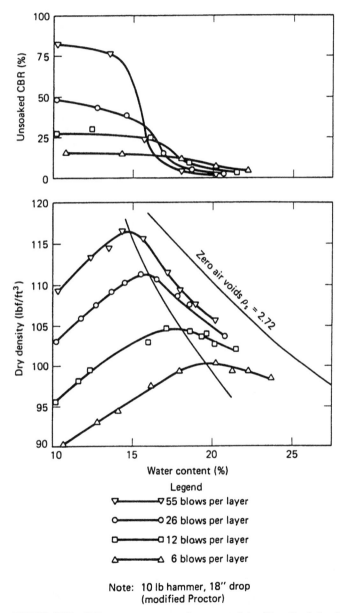

FIGURE 15.17   CBR versus water content for compacted clay. (*From Turnbull and Foster, 1956; reprinted with permission from the American Society of Civil Engineers.*)

There are many different methods to stabilize pumping soil. The most commonly used method is to simply allow the plastic soil to dry out. Other methods include adding a chemical agent (such as lime) to the clay or placing a geotextile on top of the pumping clay to stabilize its surface (Winterkorn and Fang, 1975).

Another common procedure to stabilize pumping clay is to add gravel to the clay. The typical procedure is to dump angular gravel at ground surface and then work it in from the surface. The angular gravel produces a granular skeleton that increases the undrained shear strength and penetration resistance of the mixture (Day, 1996b).

15.3 SOIL IMPROVEMENT METHODS

If the expected settlement for a proposed structure is too large, then different foundation support or soil stabilization options must be evaluated. As discussed in Chap. 5, one alternative is a deep foundation system that can transfer structural loads to adequate bearing material in order to bypass a compressible soil layer. Another option is to construct a floating foundation, which is a special type of deep foundation where the weight of the structure is balanced by the removal of soil and construction of an underground basement. Other alternatives include site improvement methods that are summarized in Table 15.4 and discussed below.

15.3.1 Soil Replacement

There are basically two types of soil replacement methods: (1) removal and replacement, and (2) displacement. The first method is the most common approach and it consists of the removal of the compressible soil layer and replacement with structural fill during the grading operations. Usually the removal and replacement grading option is only economical if the compressible soil layer is near the ground surface and the groundwater table is below the compressible soil layer or the groundwater table can be economically lowered.

15.3.2 Water Removal

Table 15.4 lists several different types of water removal site improvement techniques. If the site contains an underlying compressible cohesive soil layer, the site can be surcharged with a fill layer placed at ground surface. Vertical drains (such as wick drains or sand drains) can be installed in the compressible soil layer to reduce the drainage path and speed up the consolidation process. Once the compressible cohesive soil layer has sufficient consolidation, the fill surcharge layer is removed and the building is constructed.

15.3.3 Site Strengthening

There are many different methods that can be used to strengthen the on-site soil (see Table 15.4). Examples are as follows:

- *Dynamic compaction methods.*   Heavy tamping consists of using a crane that repeatedly lifts and drops a large weight onto the ground surface in order to vibrate the ground and increase the density of near surface granular soils. Although this method can increase the density of soil to a depth of 60 ft (18 m), it is usually only effective to depths of approximately 20 to 30 ft (6 to 9 m). In addition, this method requires the filling of impact craters and releveling of the ground surface.
- *Compaction piles.*   Large displacement piles, such as precast concrete piles or hollow steel piles with a closed end, can be driven into the ground in order to increase the density of the soil. The soil is densified by both the actual displacement of the soil and the vibration of the ground that occurs during the driving process. The piles are typically left in place, which makes this method

TABLE 15.4 Site Improvement Methods

Method	Technique	Principles	Suitable Soils	Remarks
Soil replacement methods	Remove and replace	Excavate weak or undesirable material and replace with better soils	Any	Limited depth and area where cost-effective; generally ≤ 30 ft
	Displacement	Overload weak soils so that they shear and are displaced by stronger fill	Very soft	Problems with mud-waves and trapped compressible soil under the embankment; highly dependent on specific site
Water removal methods	Trenching	Allows water drainage	Soft, fine-grained soils and hydraulic fills	Effective depth up to 10 ft; speed dependent on soil and trench spacing; resulting desiccated crust can improve site mobility
	Precompression	Loads applied prior to construction to allow soil consolidation	Normally consolidated fine-grained soil, organic soil, fills	Generally economical; long time may be needed to obtain consolidation; effective depth limited only by ability to achieve needed stresses
	Precompression with vertical drains	Shortens drainage path to speed consolidation	Same as above	More costly; effective depth usually limited to ≤ 100 ft
	Electroosmosis	Electric current causes water to flow to cathode	Normally consolidated silts and silty clays	Expensive; relatively fast; usable in confined area; not usable in conductive soils; best for small area
Site strengthening methods	Dynamic compaction	Large impact loads applied by repeated dropping of a 5- to 35-ton weight; larger weights have been used	Cohesionless best; possible use for soils with fines; cohesive soils below groundwater table give poorest results	Simple and rapid; usable above and below the groundwater table; effective depths up to 60 ft; moderate cost; potential vibration damage to adjacent structures
	Vibrocompaction	Vibrating equipment densifies soils	Cohesionless soils with < 20 percent fines	Can be effective up to 100-ft depth; can achieve good density and uniformity; grid spacing of holes critical, relatively expensive

(Continued)

TABLE 15.4 Site Improvement Methods (*Continued*)

Method	Technique	Principles	Suitable Soils	Remarks
Site strengthening method, (*Continued*)	Vibroreplacement	Jetting and vibration used to penetrate and remove soil; compacted granular fill then placed in hole to form support columns surrounded by undisturbed soil	Soft cohesive soils (s_u = 15 to 50 kPa)	Relatively expensive
	Vibrodisplacement	Similar to vibroreplacement except soil is displaced laterally rather than removed from the hole	Stiffer cohesive soils (s_u = 30 to 60 kPa)	Relatively expensive
Grouting	Injection of grout	Fill soil voids with cementing agents to strengthen and reduce permeability	Wide spectrum of coarse- and fine-grained soils	Expensive; more expensive grouts needed for fine-grained soils, may use pressure injection, soil fracturing, or compaction techniques
	Deep mixing	Jetting or augers used to physically mix stabilizer and soil	Wide spectrum of coarse- and fine-grained soils	Jetting poor for highly cohesive clays and some gravelly soils; deep mixing best for soft soils up to 165 ft deep
Thermal	Heat	Heat used to achieve irreversible strength gain and reduced water susceptibility	Cohesive soils	High energy requirements; cost limits practicality
	Freezing	Moisture in soil frozen to hold particles together and increase shear strength and reduce permeability	All soils below the groundwater table; cohesive soils above the groundwater table	Expensive; highly effective for excavations and tunneling; high groundwater flows troublesome; slow process
Geosynthetics	Geogrids, geotextiles, geonets, and geomembranes	Use geosynthetic materials for filters, erosion control, water barriers, drains, or soil reinforcing (see Chap. 17)	Effective filters for all soils; reinforcement often used for soft soils	Widely used to accomplish a variety of tasks; commonly used in conjunction with other methods (e.g., strip drain with surcharge or to build a construction platform for site access)

Source: Rollings and Rollings (1996).

more expensive than the other methods. In addition, there must be relatively close spacing of the piles in order to provide meaningful densification of soil between the piles.

- *Blasting.*   Deep densification of the soil can be accomplished by blasting. This method has a higher risk of injury and damage to adjacent structures. There may be local restrictions on the use of such a method.

- *Compaction with vibratory probes.*   As indicated in Table 15.4, there are many different types of vibratory methods, such as vibrocompaction, vibrodisplacement, and vibrodisplacement. The equipment used for these deep vibratory techniques is illustrated in Fig. 15.18. Vibrodisplacement is considered to be one of the most reliable and comprehensive methods for the mitigation of liquefaction hazard when liquefiable soils occur at depth (Seed, 1991).

- *Vertical gravel drains.*   Vibrofloation or other methods are used to make a cylindrical vertical hole, which is filled with compacted gravel or crushed rock. These columns of gravel or crushed rock have a very high permeability and can quickly dissipate earthquake-induced pore water pressures in the surrounding soil. This method can be effective in reducing the loss of shear strength during earthquakes, but it will not prevent overall site settlements. In addition, the method can be effective in relatively free-draining soils, but the vertical columns must be closely spaced to provide meaningful pore pressure dissipation. If the drain capacity is exceeded by the rate of pore water pressure increase, there will be no partial mitigation of liquefiable soils (Seed, 1991).

15.3.4 Grouting

There are many grouting methods that can be used to strengthen the on-site soil (see Table 15.4). For example, in order to stabilize the ground, fluid grout can be injected into the ground to fill in joints, fractures, or underground voids (Graf, 1969; Mitchell, 1970). For the releveling of existing structures, one option is mudjacking, which has been defined as a process whereby a water and soil-cement or soil-lime cement grout is pumped beneath the slab, under pressure, to produce a lifting force that literally floats the slab to the desired position (Brown, 1992). Other site improvement grouting methods are as follows:

- *Compaction grouting.*   A commonly used site improvement technique is compaction grouting, which consists of intruding a mass of very thick consistency grout into the soil, which both displaces and compacts the loose soil (Brown and Warner, 1973; Warner, 1978, 1982). Compaction

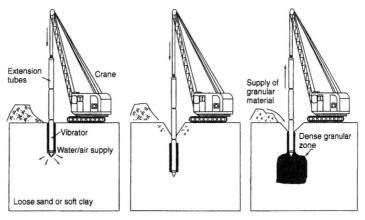

FIGURE 15.18 Equipment used for deep vibratory techniques. (*From Rollings and Rollings 1996; reprinted with permission of McGraw-Hill, Inc.*)

grouting has proved successful in increasing the density of poorly compacted fill, alluvium, and compressible or collapsible soil. The advantages of compaction grouting are less expense and disturbance to the structure than foundation underpinning, and it can be used to relevel the structure. The disadvantages of compaction grouting are that it is difficult to analyze the results, it is usually ineffective near slopes or for near-surface soils because of the lack of confining pressure, and there is the danger of filling underground pipes with grout (Brown and Warner, 1973).

- *Jet grouting (columnar).* This process is used to create columns of grouted soil. The grouted columns are often brittle and may provide little or no resistance to lateral movements and may be broken by lateral ground movements (Seed, 1991).

- *Deep mixing.* Jetting or augers are used to physically mix the stabilizer and soil. There can be overlapping of treated columns in order to create a more resistant treated zone.

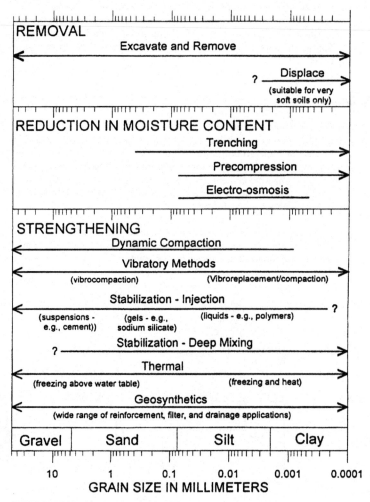

FIGURE 15.19 Site improvement methods as a function of soil grain size. (*From Rollings and Rollings 1996; reprinted with permission of McGraw-Hill, Inc.*)

15.3.5 Thermal

As indicated in Table 15.4, the thermal site improvement method consists of either heating or freezing the soil in order to improve its shear strength and reduce its permeability. These types of soil improvement methods are usually very expensive and thus have limited uses.

15.3.6 Summary

Figure 15.19 presents a summary of site improvement methods as a function of soil grain size. For an in-depth discussion of soil improvement methods, see Lawton (1996).

Whatever method of soil improvement is selected, the final step should be to check the results in the field using such methods as the cone penetration test (CPT) or standard penetration test (SPT). If the soil improvement is unsatisfactory, then it should be repeated until the desired properties are attained.

NOTATION

The following notation is used in this chapter:

RC = relative compaction

w = water content (also known as the moisture content)

γ_d = dry unit weight of the soil

γ_t = total unit weight of the soil

ρ_d = dry density of the soil

ρ_{dmax} = laboratory maximum dry density

PROBLEMS

Solutions to the problems are presented in App. C of this book.

15.1   During grading, a sand cone test was performed on fill. The following data were obtained:

Volume of hole: 2000 cm³

Mass of soil removed from hole: 4.0 kg

Water content of soil: 8.3 percent

Determine the field dry density ρ_d.

ANSWER: $\rho_d = 1.85$ Mg/m³.

15.2   Project specifications require a relative compaction of 95 percent (modified Proctor). Construction of a highway embankment requires 10,000 yd³ of fill. Assume the borrow soil has an in situ dry density of 94 pcf. Also assume that this borrow soil has a laboratory maximum dry density of 122.5 pcf. Determine the total volume of soil that must be excavated from the borrow area.

ANSWER: 12,380 yd³.

15.3   Project specifications require a relative compaction of 90 percent (modified Proctor). Construction of a building pad requires 5000 yd³ of fill. Assume the borrow soil has an in situ wet density of 128 pcf and an in situ water content of 6.5 percent. Also assume that this borrow soil has a laboratory maximum dry density of 122.5 pcf. Determine the total volume of soil that must be excavated from the borrow area.

ANSWER: 4590 yd³.

15.4   Use the data from Problem 15.2. Assume the water content of the in situ borrow soil is 8.0 percent and that the embankment fill must be compacted at optimum moisture content ($w_{opt} = 11.0$ percent). How much water (gallons) must be added to the soil during compaction?

ANSWER: 113,000 gal.

CHAPTER 16
FOUNDATION EXCAVATION, UNDERPINNING, AND FIELD LOAD TESTS

16.1 INTRODUCTION

Chapter 15 has dealt with site grading work, which includes the testing of fill to determine if the compaction meets the project specifications. Site grading work is a major construction activity that involves geotechnical engineers and engineering geologists. At the end of grading, there is often additional fieldwork, such as tests performed on the compacted road subgrade (i.e., CBR, R value tests, and so forth) or the testing of the rough-graded building pads to determine the presence of expansive soils.

There are many types of services performed by the geotechnical engineer during the actual construction of the project. Examples of these types of services are as follows:

1. *Excavation of the foundation.* During the construction of the project, the geotechnical engineer will often be asked or required to review the foundation excavation. This type of service involves measuring the dimension of geotechnical elements (such as the depth and width of footings) to make sure that they conform to the requirements of the construction plans. This service is often performed at the same time as the field observation of the foundation bearing conditions, such as confirming the presence of dense soil or intact rock.

   In many cases, field observations to confirm bearing conditions and check foundation dimensions will be required by the local building department. In addition, a letter indicating the outcome of the observations must be prepared by the engineer to satisfy the local building department. Building departments often refer to these types of reports as *foundation inspection reports.* The local building department often considers these reports to be so important that they may not issue a certificate of occupancy until the reports have been submitted and accepted.

   Foundation excavations will be discussed in Sec. 16.2.

2. *Field load or performance tests.* There are numerous types of field load or performance tests. For example, load tests are common for pile foundations and are used to determine their load carrying capacity. Field load tests will be discussed in Sec. 16.3.

   Besides load tests, there can be all types of performance tests during construction that will need to be observed by the geotechnical engineer. An example is the field-testing of tieback anchors as discussed in Sec. 11.8.3 and Table 11.3.

3. *Underpinning.* There are many different situations where a structure may need to be underpinned. Common reasons for underpinning include supporting a structure that is sinking or tilting due to ground subsidence or instability of the structure. Foundation underpinning will be discussed in Sec. 16.4

4. *Observational method.* There are numerous types of subsurface conditions that can lead to delays and additional expenses during construction. If it is anticipated that there will be earth failure,

FIGURE 16.1   Slope failure during the grading of a site. The backpack located in the middle of the photograph provides a scale for the size of the ground cracks.

ground subsidence, or groundwater seepage, then it is important that the geotechnical engineer plan for such site conditions. Earth movement and failure can affect all types of construction projects. For example, Fig. 16.1 shows a slope failure during the grading of a project. The process of grading can undermine the toe of a slope or surcharge the head of a slope, leading to a failure such as shown in Fig. 16.1. The observational method is one approach that can be used to anticipate such conditions and modify the design if needed during construction. The observational method will be discussed in Sec. 16.5.

16.2   FOUNDATION EXCAVATION AND CONSTRUCTION

16.2.1   Introduction

There are many different types of excavations performed during the construction of a project. For example, soil may be excavated from the cut or borrow area and then used as fill (Chap. 15). Another example is the excavation of a shear key or buttress that will be used to stabilize a slope or landslide. Other examples are excavations for the construction of foundations, which will be discussed in this section. An important aspect of the excavation may be groundwater control, which will be discussed in Sec. 16.2.4.

16.2.2   Shallow Footing Excavations

A shallow foundation is often selected when the structural load will not cause excessive settlement or lateral movement of the underlying soil layers. Shallow foundations are also used when there are adequate bearing strata at shallow depth. In general, shallow foundations are more economical to construct than deep foundations. Common types of shallow foundations have been described in Sec. 5.3.
   Important considerations in shallow footing excavations are as follows:

1. *Dimensions of footings.*   The geotechnical engineer will often be required to confirm the dimensions of the footings per the building plans. The depth of the footing should always be referenced from the final grade, which may be different from the grade at the time of the footing observation. Usually it is acceptable if the footings have a width and/or depth that is greater than as indicated on the foundation plans. Footings often end up wider than planned because mechanical equipment is used to excavate the footings.

2. *Bearing conditions.* The bearing conditions exposed in the footings should be checked with the conditions anticipated during the design of the project. If the bearing soil or rock is substantially weaker than that assumed during the design phase, the footings might need to be deepened. The footings may also have to be deepened or the footing design revised if unanticipated conditions are encountered, such as uncompacted fill, loose soil, expansive material, or unstable soils. A metal probe can be used to locate loose soil at the bottom of the footing excavation.

During the excavation, the bottom of the footings often become disturbed, creating a loose soil zone. This disturbance occurs during the actual excavation, especially when mechanical equipment is used such as shown in Fig. 16.2. The bottom of the excavation can also be disturbed when workers descend into the footings in order to install the steel reinforcement, such as shown in Fig. 16.3. Also, debris such as loose soil or rock fragments may be inadvertently knocked into the footing trench after completion of the excavation. Even a thin zone of disturbed and loose soil can lead to settlement that is greatly in excess of calculated values.

It is important that undistured soil (i.e., natural ground) or adequately compacted soil be present at the bottom of the footing excavations and that the footings be cleaned of all loose debris prior to placement of concrete. Any loose rocks at the bottom of the footing excavation should also be removed and the holes filled with concrete (during placement of concrete for the footings).

3. *Groundwater conditions.* The presence of groundwater can impact bearing conditions. For example, groundwater in a footing excavation may cause the side of the hole to cave or loose slough to accumulate at the bottom of the footing. The groundwater table may need to be lowered in order to cleanout any loose debris at the bottom of the footing.

If the footings should become flooded, such as from a heavy rainstorm, then loose debris can be washed into the excavations. This loose or soft soil will have to be removed prior to placing the concrete for the footings. Water can also soften the soil located at the bottom of the excavations. For example, clayey soil may absorb water and swell, producing a layer of very soft and compressible material. An example of such a condition is shown in Fig. 16.4, where the footing excavation has become flooded during a heavy rainstorm. Note also in Fig. 16.4 that the steel

FIGURE 16.2 Excavation of footings.

FIGURE 16.3 Installation of steel reinforcement in footings. The arrow points to the workers in the footing excavation in the process of installing the steel reinforcement.

reinforcement has been prefabricated at ground surface and once the footings are dry and clean, the prefabricated steel reinforcement sections will be lifted and lowered into the footing trench.

4. *Steel reinforcement inspection.*   Sometimes the geotechnical engineer may be required to inspect the type and location of steel reinforcement in the footings.

5. *Local building department requirements.*   In many cases, field observations to confirm bearing conditions and check foundation dimensions will be required by the local building department. In addition, a memo or letter indicating the outcome of the observations must be prepared by the engineer to satisfy the local building department. Building departments often refer to these types

FIGURE 16.4   Footing excavations. Note the flooded condition of the footing trenches.

of reports as final inspection reports. An example of such a report for the construction of a foundation is as follows:

Footing Inspection: The footings at the site have been inspected and are generally in conformance with the approved building plans. Additionally, the footings have been approved for installing steel reinforcement and the soil conditions are substantially in conformance with those observed during the subsurface exploration. Furthermore, the footing excavations extend to the proper depth and bearing strata. Care should be taken to keep all loose soil and debris out of the footing excavations prior to placement of concrete.

16.2.3 Open and Braced Excavations

An *open excavation* is defined as an excavation that has stable and unsupported side slopes. Table 16.1 presents a discussion of the general factors that control the excavation stability and Table 16.2 lists factors that control the stability of excavation slopes in some problem soils.

A *braced excavation* is defined as an excavation where the sides are supported by retaining structures. Figure 11.17 shows common types of retaining systems for braced excavations. Table 16.3 lists the design considerations for braced excavations and Table 16.4 indicates factors that are involved in the choice of a support system for a deep excavation.

16.2.4 Groundwater Control

Groundwater and seepage, including the lowering of the groundwater table by pumping from wells, has been discussed in Sec. 4.7. Groundwater can cause or contribute to foundation failure because of excess saturation, seepage pressures, or uplift forces. It has been stated that uncontrolled saturation and seepage causes many billions of dollars a year in damage (Cedergren, 1989). Common types of geotechnical and foundation problems due to groundwater are as follows (Harr, 1962; Collins and Johnson, 1988; Cedergren, 1989):

- Piping failures of dams, levees, and reservoirs
- Seepage pressures that cause or contribute to slope failures and landslides
- Deterioration and failure of roads due to the presence of groundwater in the base or subgrade

TABLE 16.1   General Factors that Control the Stability of the Excavation Slopes

Construction activity	Objectives	Comments
Dewatering (also see Sec. 16.2.4)	In order to prevent boiling, softening, or heave of the excavation bottom, reduce lateral pressures on sheeting, reduce seepage pressures on face of open cut, and eliminate piping of fines through sheeting.	Investigate soil compressibility and effect of dewatering on settlement of nearby structures; consider recharging or slurry wall cutoff. Examine for presence of lower aquifer and need to dewater. Install piezometers if needed. Consider effects of dewatering in cavity-laden limestone. Dewater in advance of excavation.
Excavation and grading (also see Chap. 15)	Utility trenches, basement excavations, and site grading.	Analyze safe slopes (see Chap. 10) or bracing requirements, and effects of stress reduction on overconsolidated, soft, or swelling soils and shales. Consider horizontal and vertical movements in adjacent areas due to excavation and effect on nearby structures. Keep equipment and stockpiles a safe distance from the top of the excavation.
Excavation wall construction	To support vertical excavation walls, and to stabilize trenching in limited space.	See Chap. 11 for retaining wall design. Reduce earth movements and bracing stresses, where necessary, by installing lagging on front flange of soldier pile. Consider effect of vibrations due to driving sheet piles or soldier piles. Consider dewatering requirements as well as wall stability in calculating sheeting depth. Movement monitoring may be warranted.
Blasting	To remove or to facilitate the removal of rock in the excavation.	Consider the effect of vibrations on settlement or damage to adjacent areas. Design and monitor or require the contractor to design and monitor blasting in critical areas, and require a pre-construction survey of nearby structures.
Anchor or strut installation	To obtain support system stiffness and interaction.	Major excavations require careful installation and monitoring, e.g., case anchor holes in collapsible soil, measure stress in ties and struts and the like.

Sources: NAVFAC DM-7.2, 1982; Clough and Davidson, 1977; and Departments of the Army and the Air Force, 1979.

- Highway and other fill foundation failures caused by perched groundwater
- Earth embankment and foundation failures caused by excess pore water pressures
- Retaining wall failures caused by hydrostatic water pressures
- Canal linings, dry-docks, and basement or spillway slabs uplifted by groundwater pressures
- Soil liquefaction, caused by earthquake shocks, because of the presence of loose granular soil that is below the groundwater table
- Transportation of contaminants by the groundwater

Proper drainage design and construction of drainage facilities can mitigate many of these groundwater problems. In addition to the groundwater problems described earlier, drainage design and construction facilities are usually required for deep foundation excavations that are below the groundwater table.

The groundwater table (also known as the *phreatic surface*) is the top surface of underground water, the location of which is often determined from piezometers. A perched groundwater table refers to groundwater occurring in an upper zone separated from the main body of groundwater by underlying unsaturated rock or soil. An artesian condition refers to groundwater that is under pressure and is confined by impervious material. If trapped pressurized water is released, such as by digging an excavation, the water will rise above the groundwater table and may even rise above the ground surface. Figure 16.5 shows an example of an artesian condition, where a test pit has been excavated into a pavement and the released water has flowed out of the test pit.

TABLE 16.2 Factors that Control the Stability of Excavation Slopes in Some Problem Soils

Topic	Discussion
General discussion	The depth and slope of an excavation, and groundwater conditions control the overall stability and movements of open excavations. Factors that control the stability of the excavation for different material types are as follows:
	1. Rock: For rock, stability is controlled by depths and slopes of excavation, particular joint patterns, in situ stresses, and groundwater conditions.
	2. Granular soils: For granular soils, instability usually does not extend significantly below the bottom of the excavation provided that seepage forces are controlled.
	3. Cohesive Soils: For cohesive soils, stability typically involves side slopes but may also include the materials well below the bottom of the excavation. Instability of the bottom of the excavation, often referred to as *bottom heave*, is affected by soil type and strength, depth of cut, side slope and/or berm geometry, groundwater conditions, and construction procedures.
Stiff-fissured clays and shales	Field shear resistance may be less than suggested by laboratory testing. Slope failures may occur progressively and shear strengths are reduced to the residual value compatible with relatively large deformations. Some case histories suggest that the long-term performance is controlled by the drained residual friction angle. The most reliable design would involve the use of local experience and recorded observations.
Loess and other collapsible soil	Such soils have a strong potential for collapse and erosion of relatively dry materials upon wetting. Slopes in loess are frequently more stable when cut vertical to prevent water infiltration. Benches at intervals can be used to reduce effective slope angles. Evaluate potential for collapse as described in Sec. 7.2.
Residual soil	Depending on the weathering profile from the parent rock, residual soil can have a significant local variation in properties. Guidance based on recorded observations provides a prudent basis for design.
Sensitive clay	Very sensitive and quick clays have a considerable loss of strength upon remolding, which could be generated by natural or man-made disturbance. Minimize disturbance and use total stress analysis based on undrained shear strength from unconfined compression tests or field vane tests.
Talus	Talus is characterized by loose aggregation of rock that accumulates at the foot of rock cliffs. Stable slopes are commonly between 1.25:1 to 1.75:1 (horizontal : vertical). Instability is often associated with abundance of water, mostly when snow is melting.
Loose sands	Loose sands may settle under blasting vibrations, or liquefy, settle, and lose shear strength if saturated. Such soils are also prone to erosion and piping.
Engineering evaluation	Methods described in Chap. 10 (slope stability analyses) may be used to evaluate the stability of open excavations in soils where the behavior of such soils can be reasonably determined by field investigations, laboratory testing, and engineering analysis. As described earlier, in certain geologic formations stability is controlled by construction procedures, side effects during and after excavation, and inherent geologic planes of weaknesses.

Sources: NAVFAC DM-7.2, 1982 and Clough and Davidson, 1977.

TABLE 16.3 Design Considerations for Braced Excavations

Design factor	Comments
Water loads	Often greater than earth loads on an impervious wall. Recommend piezometers during construction to monitor water levels. Should also consider possible lower water pressures as a result of seepage of water through or under the wall. Dewatering can be used to reduce the water loads. Seepage under the wall reduces the passive resistance.
Stability	Consider the possible instability in any berm or exposed slope. The sliding potential beneath the wall or behind the tiebacks should also be evaluated. For weak soils, deep-seated bearing failure due to the weight of the supported soil should be checked. Also include in stability analysis the weight of surcharge or weight of other facilities in close proximity to the excavation.
Piping	Piping due to a high groundwater table causes a loss of ground, especially for silty and fine sands. Difficulties occur due to flow of water beneath the wall, through bad joints in the wall, or through unsealed sheet pile handling holes. Dewatering may be required.
Movements	Movements can be minimized through the use of a stiff wall supported by preloaded tiebacks or a braced system.
Dewatering and recharge	Dewatering reduces the loads on the wall system and minimizes the possible loss of ground due to piping. Dewatering may cause settlements and in order to minimize settlements, there may be the need to recharge outside of the wall system.
Surcharge	Construction materials are usually stored near the wall systems. Allowances should always be made for surcharge loads on the wall system.
Prestressing of tieback anchors	In order to minimize soil and wall movements, it is useful to remove slack by prestressing tieback anchors.
Construction sequence	The amount of wall movement is dependent on the depth of the excavation. The amount of load on the tiebacks is dependent on the amount of wall movement that occurs before they are installed. Movements of the wall should be checked at every major construction stage. Upper struts should be installed as early as possible.
Temperature	Struts may be subjected to load fluctuations due to temperature differences. This may be important for long struts.
Frost penetration	In cold climates, frost penetration can cause significant loading on the wall system. Design of the upper portion of the wall system should be conservative. Anchors may have to be heated. Freezing temperatures also can cause blockage of flow of water and thus unexpected buildup of water pressure.
Earthquakes	Seismic loads may be induced during an earthquake.
Factors of safety	The following are suggested minimum factors of safety (F) for overall stability. Note that these values are suggested guidelines only. Design factors of safety depend on project requirements. Earth berms: permanent, $F = 2.0$; temporary, $F = 1.5$ Cut slopes: permanent, $F = 1.5$; temporary, $F = 1.3$ General stability: permanent, $F = 1.5$; temporary, $F = 1.3$ Bottom heave: permanent, $F = 2.0$; temporary, $F = 1.5$

Source: NAVFAC DM-7.2, 1982.

TABLE 16.4 Factors that are Involved in the Choice of a Support System for a Deep Excavation

Requirements	Type of support system	Comments
Open excavation area	Tiebacks or rakers. For shallow excavation, use cantilever walls.	Consider design items listed in Table 16.3.
Low initial cost	Soldier pile or sheet pile walls. Consider combined soil slope and wall.	Consider design items listed in Table 16.3.
Use as part of permanent structure	Diaphragm or pier walls.	Diaphragm wall is the most common type of permanent wall.
Subsurface conditions of deep, soft clay	Struts or rakers that support a diaphragm or pier wall.	Tieback capacity not adequate in soft clays.
Subsurface conditions of dense, gravelly sands or clay	Soldier pile, diaphragm wall, or pier wall.	Sheet piles may lose interlock on hard driving.
Subsurface conditions of overconsolidated clays	Struts, long tiebacks, or combination of tiebacks and struts.	High in situ lateral stresses are relieved in overconsolidated soil. Lateral movements may be large and extend deep into the soil.
Avoid dewatering	Use diaphragm walls or possibly sheet pile walls in soft subsoils.	Soldier pile wall is too pervious for this application.
Minimize lateral movements of wall	Use high preloads on stiff strutted or tieback walls.	Analyze the stability of the bottom of the excavation.
Wide excavation (greater than 65 ft wide)	Use tiebacks or rackers.	Tiebacks are preferable except in very soft clay soils.
Narrow excavation (less than 65 ft wide)	Use cross-excavation struts.	Struts are more economical, but tiebacks still may be preferred in order to keep the excavation open.

Note: Deep excavation is defined as an excavation that is more than 20 ft (6 m) below ground surface.
Source: NAVFAC DM-7.2, 1982.

FIGURE 16.5 Groundwater exiting a test pit excavated into a pavement.

TABLE 16.5 Methods of Groundwater Control

Method	Soils suitable for treatment	Uses	Comments
Sump pumping	Clean gravels and coarse sands.	Open shallow excavations.	Simplest pumping equipment. Fines easily removed from the ground. Encourages instability of formation. See Fig. 16.6.
Wellpoint system with suction pump	Sandy gravels down to fine sands (with proper control can also be used in silty sands).	Open excavations including utility trench excavations.	Quick and easy to install in suitable soils. Suction lift is limited to about 18 ft (5.5 m). If greater lift is needed, multistage installation is necessary. See Figs. 16.7 and 16.8.
Deep wells with electric submersible pumps	Gravels to silty fine sands, and water bearing rocks.	Deep excavation in, through, or above water bearing formations.	No limitation on depth of drawdown. Wells can be designed to draw water from several layers throughout its depth. Wells can be sited clear of working area.
Jetting system	Sands, silty sand, and sandy silts.	Deep excavations in confined space where multistage wellpoints cannot be used.	Jetting system uses high-pressure water to create vacuum as well as to lift the water. No limitation on depth of drawdown.
Sheet piling cutoff wall	All types of soil (except boulder beds).	Practically unrestricted use.	Tongue and groove wood sheeting utilized for shallow excavations in soft and medium soils. Steel sheet piling for other cases. Well-understood method and can be rapidly installed. Steel sheet piling can be incorporated into permanent works or recovered. Interlock leakage can be reduced by filling interlock with bentonite, cement, grout, or similar materials.
Slurry trench cutoff wall	Silts, sands, gravels, and cobbles.	Practically unrestricted use. Extensive curtain walls around open excavations.	Rapidly installed. Can be keyed into impermeable strata such as clays or soft shales. May be impractical to key into hard or irregular bedrock surfaces, or into open gravels.
Freezing: ammonium and brine refrigerant	All types of saturated soils and rock.	Formation of ice in void spaces stops groundwater flow.	Treatment is effective from a working surface outwards. Better for large applications of long duration. Treatment takes longer time to develop.
Freezing: liquid nitrogen refrigerant	All types of saturated soils and rock.	Formation of ice in voids spaces stops groundwater flow.	Better for small applications of short duration where quick freezing is required. Liquid nitrogen is expensive and requires strict site control. Some ground heave could occur.

(Continued)

TABLE 16.5 Methods of Groundwater Control (*Continued*)

Method	Soils suitable for treatment	Uses	Comments
Diaphragm structural walls: structural concrete	All soil types including those containing boulders.	Deep basements, underground construction, and shafts.	Can be designed to form a part of the permanent foundation. Particularly efficient for circular excavations. Can be keyed into rock. Minimum vibration and noise. Can be used in restricted space. Also can be installed very close to the existing foundation.
Diaphragm structural walls: bored piles or mixed in place piles	All soil types, but penetration through boulders may be difficult and costly.	Deep basements, underground construction, and shafts.	A type of diaphragm wall that is rapidly installed. It can be keyed into impermeable strata such as clays or soft shales.

Sources: NAVFAC DM-7.2 (1982), based on the work by Cashman and Harris (1970).

As indicated in Table 16.5, there are many different methods of groundwater control. For example, sump pumping, such as illustrated in Fig. 16.6, can be used to lower the groundwater table in the excavation. Another commonly used method of groundwater control for excavations is the wellpoint system with suction pumps. The purpose of this method is to lower the groundwater table by installing a system of perimeter wells. As illustrated in Fig. 16.7, the system consists of closely spaced wellpoints installed around the excavation. A wellpoint is a small-diameter pipe having perforations at the bottom end. A pump is used to extract water from the pipe that lowers the groundwater table as illustrated in Fig. 16.7. Additional details on the two-stage wellpoint system and the combined wellpoint and deepwell system are presented in Fig. 16.8.

It is important to consider the possible damage to adjacent structures caused by the lowering of the groundwater table at the site. For example, a lowering of the groundwater table could lead to consolidation of soft clay layers or rotting of wood piling.

16.2.5 Pier and Grade Beam Support

The previous sections have described shallow foundation excavation and construction, foundation construction in open and braced excavations, and groundwater control. The next two sections will discuss the excavation and construction of deep foundations.

A common type of deep foundation support is through the use of piers and grade beams. The typical steps in the construction of a foundation consisting of piers and grade beams are as follows:

1. *Excavation of piers.* Figures 16.9 to 16.11 show the excavation of the piers using a truck-mounted auger drill rig. This type of equipment can quickly and economically excavate the piers to the desired depth. In Figs. 16.9 to 16.11, a 30 in. (0.76 m) diameter auger is being used to excavate the pier holes.

2. *Cleaning of the bottom of the excavation.* Piers are often designed as end-bearing members. For example, there may be a loose or compressible upper soil zone with the piers excavated through this material and into competent material. The ideal situation is to have the groundwater table below the bottom of the piers. This will then allow for a visual inspection of the bottom of the

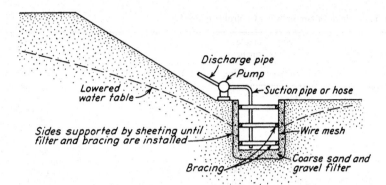

FIGURE 16.6   Groundwater control: example of a sump being used to lower the ground-water table. (*From Peck, Hanson, and Thornburn 1974; reproduced with permission of John Wiley & Sons.*)

pier excavation. Often an experienced driller will be able to clean out most of the bottom of the pier by quickly spinning the auger.

A light can then be lowered into the pier hole to observe the embedment conditions (i.e., see Fig. 16.12). A worker should not descend into the hole to clean out the bottom, but rather any loose material at the bottom of the pier should be pushed to one side and then scraped into a bucket lowered into the pier hole. If it is simply not possible to clean out the bottom of the pier, then the

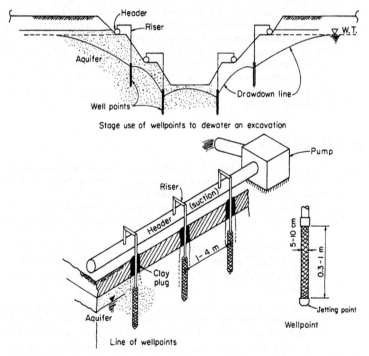

FIGURE 16.7   Groundwater control: wellpoint system with suction pump. (*From Bowles 1982; reprinted with permission of McGraw-Hill, Inc.*)

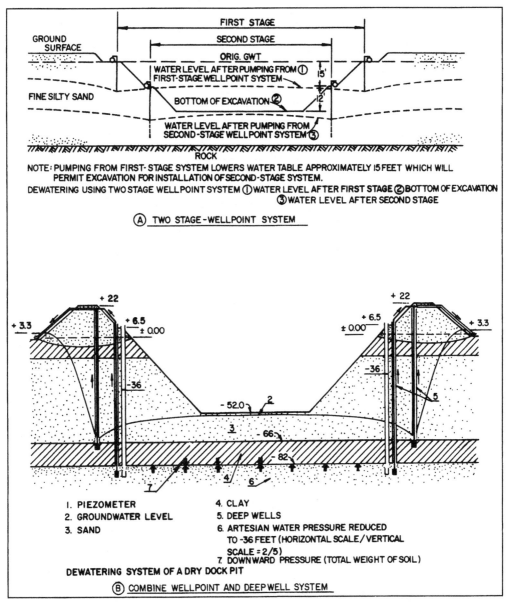

FIGURE 16.8 Groundwater control: two stage wellpoint system and a combined wellpoint and deepwell system. (*From NAVFAC DM-7.2, 1982.*)

pier resistance could be based solely on skin friction in the bearing strata with the end-bearing resistance assumed to be equal to zero.

3. *Steel cage and concrete.* Once the bottom of the pier hole has been cleaned, a steel reinforcement cage is lowered into the pier hole. Small concrete blocks can be used to position the steel cage within the hole. Care should be used when inserting the steel cage so that soil is not knocked off

FIGURE 16.9   Truck-mounted auger drill rig used to excavate piers.

of the sides of the hole. Once the steel cage is in place, the hole is filled with concrete. Figure 16.13 shows the completion of the pier with the steel reinforcement extending out of the top of the pier.

4. *Grade beam construction.*   The next step is to construct the grade beams that span between the piers. Figure 16.14 shows the excavation of a grade beam between two piers. Figure 16.15 shows the installation of steel for the grade beam. Similar to the piers, small concrete blocks are used to position the steel reinforcement within the grade beam. A visqueen moisture barrier is visible on the left side of Fig. 16.15. Figure 16.16 shows a pier located at the corner of the building. The steel reinforcement from the grade beams is attached to the steel reinforcement from the piers. Once the steel reinforcement is in place, the final step is to place the concrete for the grade beams.

FIGURE 16.10   Close-up of auger being pushed into the soil.

FIGURE 16.11 Close-up of auger being extracted from the ground with soil lodged within its groves.

Figure 16.17 shows the finished grade beams. The steel reinforcement protruding out of the grade beams will be attached to the steel reinforcement in the floor slab.

5. *Floor slab.* Before placement of the floor slab, a visqueen moisture barrier and a gravel capillary break should be installed. Then the steel reinforcement for the floor slab is laid out, such as shown in Fig. 16.18. Although not shown in Fig. 16.18, small concrete blocks will be used to elevate the steel reinforcement off the subgrade and the steel will be attached to the steel from the grade beams. The final step is to place the concrete for the floor slab. Figure 16.19 shows the completed floor slab.

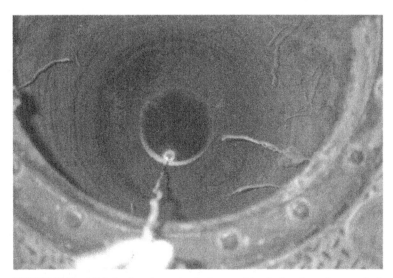

FIGURE 16.12 A light has been lowered to the bottom of the pier in order to observe embedment conditions.

FIGURE 16.13   The pier hole has been filled with concrete. The steel reinforcement from the pier will be attached to the steel reinforcement in the grade beam.

FIGURE 16.14   Excavation for the grade beam that will span between the two piers.

FIGURE 16.15 Steel reinforcement is being installed within the grade beam excavation.

FIGURE 16.16 Corner of the building where the steel reinforcement from the two grade beams has been attached to the steel reinforcement from the pier.

FIGURE 16.17   The concrete for the grade beams has been placed. The steel reinforcement from the grade beams will be attached to the steel reinforcement in the floor slab.

6. *Columns.*   When designing the building, the steel columns that support the superstructure can be positioned directly over the center of the piers. For example, Fig. 16.20 shows the location where the bottom of a steel column is aligned with the top of a pier. A steel column having an attached base plate will be bolted to the concrete. Then the steel reinforcement from the pier (see Fig. 16.20) will be positioned around the bottom of the steel column. Once filled with concrete, the final product will be essentially a fixed-end column condition having a high lateral resistance.

FIGURE 16.18   Positioning of the steel reinforcement for the floor slab.

FIGURE 16.19   Concrete for the floor slab has been placed.

The main advantage of this type of foundation is that there are no open joints or planes of weakness that can be exploited by soil movement or seismic shaking. The strength of the foundation is due to its monolithic construction, with the floor slab attached and supported by the grade beams, which are in turn anchored to the piers. In addition, the steel columns of the superstructure can be constructed so that they bear directly on top of the piers and have fixed end connections. This monolithic foundation and the solid connection between the steel columns and piers will enable the structure to resist soil movement or seismic shaking.

Usually the structural engineer will design this foundation system. The geotechnical engineer provides various design parameters, such as the estimated depth of the bearing strata, the allowable

FIGURE 16.20   Location where a steel column will be attached to the top of a pier.

end-bearing resistance, allowable skin friction in the bearing material, allowable passive resistance of the bearing material, and any anticipated downdrag loads that could be induced on the piers if the upper loose or compressible soil should settle under its own weight or during the anticipated earthquake. The geotechnical engineer will also need to inspect the foundation during construction in order to confirm the embedment conditions of the piers.

This type of foundation can also be used to resist the effects of expansive soil. However, when dealing with expansive soils, it is often preferred to use a raised concrete slab or raised wood floor to elevate the first floor off of the expansive subgrade. Loads from the raised floor are transferred to the grade beams and then to the piers. To mitigate uplift loads on the grade beams, a layer of easily deformable material, such as Styrofoam can be placed beneath the grade beams.

16.2.6 Prestressed Concrete Piles

Common types of prestressed concrete piles are shown in Fig. 5.5. Prestressed piles are typically produced at a manufacturing plant and the first step is to set-up the form, which contains the prestressed strands that are surrounded by wire spirals. The concrete is then placed within the form and allowed to cure. Once the concrete has reached an adequate strength, the tensioning force is released, which induces a compressive stress into the pile. The prestressed piles are then loaded onto trucks, transported to the site, and stockpiled such as shown in Fig. 16.21.

Solid square concrete piles, such as shown in Fig. 16.21, are the most commonly used type of prestressed piles. As shown in Fig. 16.21, the end of the pile that will be driven into the ground is flush, while at the opposite end; the strands protrude from the concrete. The main advantage of prestressed concrete piles is that they can be manufactured to meet site conditions. For example, the prestressed concrete piles shown in Fig. 16.21 were manufactured to meet the following specifications:

12 in. (0.3 m) square piles

Design load = 70 tons (620 kN) per pile

Required prestress = 700 psi (5 MPa)

28 day compressive stress = 6000 psi (40 MPa)

Maximum water-cement ratio = 0.38

Portland cement type V (i.e., high sulfate content in the soil)

FIGURE 16.21   Prestressed concrete piles stockpiled at the job site.

Large pile-driving equipment, such as shown in Fig. 16.22, is required in order to drive the piles into place. If the piles are to be used as end-bearing piles and the depth to the bearing strata is variable, then the first step is to drive indicator piles. An indicator pile is essentially a prestressed pile that is manufactured so that it is longer than deemed necessary. For example, if the depth to adequate bearing material is believed to be at a depth of 30 ft (9 m), then an indicator pile could be manufactured so that it is 35 ft (11 m) long. Usually about 10 to 20 percent of the piles will be indicator piles. The indicator piles are used to confirm embedment conditions and thus some indicator piles may be driven near the locations of prior borings, while other indicator piles are driven in areas where there is uncertainty as to the depth of the bearing strata. Once the indicator piles have been driven, the remainder of the prestressed piles are manufactured with the lengths of the piles based on the depths to bearing strata as determined from the indicator piles.

It is always desirable for the geotechnical engineer to observe the driving conditions for the prestressed piles. Prior to driving the piles, basic pile driving information should be recorded (see Table 16.6). In addition, during the actual driving of the piles, the number of blows per foot of penetration should be recorded. The pile-driving contractor typically marks the pile in one-foot increments so that the number of blows per foot can be easily counted.

Table 16.7 presents actual data during the driving of a prestressed pile. At this site, soft and liquefiable soil was encountered at a depth of about 15 to 30 ft (4.6 to 9.2 m) below ground surface. Although the blows per foot at this depth were reduced to about one per foot, the driving contractor actually allowed the hammer to free fall and thus the energy supplied to the top of the pile was significantly less than at the other depths. For the data in Table 16.7, the very high blow counts recorded at a depth of 31 ft (9.5 m) are due to the presence of hard bedrock that underlies the soft and loose

FIGURE 16.22 Pile driving equipment. A prestressed concrete pile is in the process of being hoisted into position.

TABLE 16.6 Example of Pile-Driving Information that should be Recorded for the Project

Pile Driving Record
- Date: March 7, 2001
- Project name and number: Grossmont Healthcare, F.N. 22132.06
- Name of contractor: Foundation Pile Inc.
- Type of pile and date of casting: Precast concrete, cast 2-6-01
- Pile location: See pile driving records (Table 16.7)
- Sequence of driving in pile group: Not applicable
- Pile dimensions: 12 in. by 12 in. cross section, lengths vary
- Ground elevation: varies
- Elevation of tip after driving: See total depth on the driving record
- Final tip and cutoff elevation of pile after driving pile group: Not applicable
- Records of redriving: No redriving
- Elevation of splices: No splices
- Type, make, model, and rated energy of hammer: D30 DELMAG
- Weight and stroke of hammer: Piston weight = 6615 lb. Double action hammer, maximum stroke = 9 ft
- Type of pile-driving cap used: Wood blocks
- Cushion material and thickness: Wood blocks approximately 1 ft thick
- Actual stroke and blow rate of hammer: Varies, but stoke did not exceed 9 ft
- Pile-driving start and finish time; and total driving time: See driving record (Table 16.7)
- Time, pile-tip elevation, and reason for interruptions: No interruptions
- Record of number of blows per foot: See driving record (Table 16.7)
- Pile deviations from location and plumb: No deviations
- Record preboring, jetting, or special procedures used: No preboring, jetting, or special procedures
- Record of unusual occurrences during pile driving: None

soil. Figure 16.23 shows the completed installation of the prestressed concrete pile. The wood block shown on the top of the concrete pile in Fig. 16.23 was used as a cushion in order to protect the pile top from being crushed during the driving operation.

A major disadvantage of prestressed concrete piles is that they can break during the driving process. The most common reason for the breakage of a prestressed concrete pile is because it strikes an underground obstruction, such as a boulder or large piece of debris that causes the pile to deflect laterally and break. For example, Fig. 16.24 shows the lateral deflection of a prestressed concrete pile as it was driven into the ground. In some cases, the fact that the pile has broken will be obvious. In Fig. 16.25, the prestressed concrete pile hit an underground obstruction, displaced laterally and

TABLE 16.7 Actual Blow Count Record Obtained during Driving of a Prestressed Concrete Pile

Blow count record		
Location: M–14.5		
Start time:	8:45 a.m.	
End time:	8:58 a.m.	
Blows per foot:	0 to 5 ft	= 1, 2, 3, 5, 9
	5 to 10 ft	= 9, 9, 11, 10, 9
	10 to 15 ft	= 7, 5, 4, 3, 2
	15 to 20 ft	= 2, 2, 1, 1, 1
	20 to 25 ft	= 1, 1, 1, 1, 1
	25 to 30 ft	= 1, 1 for 2 ft, 1, 2
	>30 ft	= 8, 50 for 10 in.
Total depth		= 31.8 ft

FIGURE 16.23 A prestressed concrete pile has been successfully driven to the bearing strata. The wood block shown on the top of the concrete pile was used as a cushion in order to protect the pile top from being crushed during the driving operation.

then broke near ground surface. In other cases where the pile breaks well below ground surface, the telltale signs will be a continued lateral drifting of the pile and low blow counts at the bearing strata. If a pile should break during installation, the standard procedure is to install another pile adjacent to the broken pile. Often the new pile will be offset a distance of 5 ft (1.5 m) from the broken pile. Grade beams are often used to tie together the piles and thus the location of the new pile should be in-line with the proposed grade beam location. The structural engineer will need to redesign the grade beam for its longer span.

FIGURE 16.24 Lateral displacement of a prestressed concrete pile during the driving operations.

FIGURE 16.25 This prestressed concrete pile struck an underground obstruction, displaced laterally, and broke near ground surface. The arrow points to the location of the breakage.

FIGURE 16.26 Prestressed concrete piles have been installed and the excavations for the pile caps and grade beams are complete. The strands at the top of the pile will be connected to the steel reinforcement in the pile cap and grade beam.

After the piles have been successfully installed, the next step is to construct the remainder of the foundation, as follows:

1. *Cut-off top of piles.*   Especially for the indicator piles, the portion of the pile extending above ground surface may be much longer than needed. In this case, the pile can be cut-off or the concrete chipped-off by using a jackhammer, such as shown in Fig. 16.26.

2. *Grade beam excavation.*   The next step is to excavate the ground for the grade beams that span between the piles. Figure 16.27 shows the excavation of a grade beam between two piles. For the foundation shown in Fig. 16.27, there is only one pile per cap, thus the pile caps are relatively small as compared to the size of the grade beams.

   Those prestressed piles that broke during installation should also be incorporated into the foundation. For example, in Fig. 16.27, the pile located at the bottom of the picture is the same broken pile shown in Fig. 16.25. The replacement pile, which was successfully installed to the bearing strata, is located at a distance of 5 ft (1.5 m) from the broken pile (i.e., the pile near the center of Fig. 16.27). As previously mentioned, replacement piles should be installed in-line with the grade beam. As shown in Fig. 16.27, both the broken pile and the replacement pile will be attached to the grade beam; however, the broken pile will be assumed to have no support capacity.

FIGURE 16.27   The prestressed concrete pile at the bottom of the picture is the same pile shown in Fig. 16.25. The pile near the center of the photograph is the replacement pile. The broken pile and the replacement pile will be attached to the grade beam.

Once the grade beams have been excavated, the next step is to trim the top of the prestressed piles such that they are relatively flush, such as shown in Figs. 16.28 and 16.29. The strands at the top of the pile are not cut off because they will be tied to the steel reinforcement in the grade beam in order to make a solid connection at the top of the pile.

3. *Installation of steel in grade beams.*   After the pile caps and grade beams have been excavated, the next step is to install the steel reinforcement. Figure 16.30 shows a close-up view of the top of a prestressed concrete pile with the steel reinforcement from the grade beam positioned on top of the pile. Note in Fig. 16.30 that the strands from the prestressed pile are attached to the reinforcement steel in the grade beams. This will provide for a solid connection between the pile and the grade beam. Figure 16.31 presents an overview of the grade beam with the steel reinforcement in place and the grade beam ready for the placement of concrete.

4. *Floor slab.*   Before placement of the floor slab, the visqueen moisture barrier and a gravel capillary break should be installed. Then the steel reinforcement for the floor slab is laid-out and the final step is to place the concrete for the floor slab.

5. *Columns.*   When designing the building, the steel columns that support the superstructure can be positioned directly over the center of the pile caps.

Similar to the pier and grade beam foundation, the main advantage of the prestressed pile foundation is that there are no open joints or planes of weakness that can be exploited by soils movement

FIGURE 16.28   The excavation for the grade beams is complete and the top of the prestressed piles are trimmed so that they are relatively flush.

FIGURE 16.29 Close-up view of one of the presstressed piles showing a trimmed top surface with the strands extending out of the top of the pile.

or seismic shaking. The strength of the foundation is due to its monolithic construction, with the floor slab attached and supported by the grade beams, which are in turn anchored by the pile caps and the prestressed piles. In addition, the steel columns of the superstructure can be constructed so that they bear directly on top of the pile caps and have fixed-end connections. This monolithic foundation and the solid connection between the steel columns and piles will enable the structure to resist soils movement and seismic shaking.

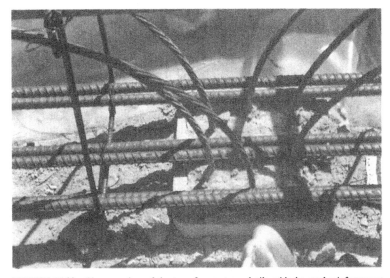

FIGURE 16.30 Close-up view of the top of a prestressed pile with the steel reinforcement from the grade beam positioned on top of the pile. The strands from the pile are attached to the steel reinforcement in the grade beam.

FIGURE 16.31   Overview of the steel reinforcement positioned within the grade beam excavation.

Usually the structural engineer will design this foundation system. The geotechnical engineer provides various design parameters, such as the estimated depth to the bearing strata, the allowable end-bearing resistance, allowable skin friction in the bearing material, allowable passive resistance of the bearing material, and any anticipated downdrag loads that could be induced on the piles if the upper loose or compressible soil should settle under its own weight or during an anticipated earthquake. The geotechnical engineer should also perform pile load tests and inspect the foundation during construction in order to confirm the design recommendations.

16.3   FIELD LOAD TESTS

16.3.1   Introduction

There are numerous types of field load or performance tests. An example is the field testing of tieback anchors as discussed in Table 11.3.

Load tests are also common for pile foundations because they are considered to be the most accurate method for determining their vertical and/or lateral load carrying capacity. ASTM standards for the field-testing of piles have been developed. Typical pile load tests and the applicable ASTM standard are listed below:

- Static Axial Compressive Load Test (ASTM D 1143-94, "Standard Test Method for Piles Under Static Axial Compression Load," 2004)
- Static Axial Compressive Load Test for Piles in Permafrost (ASTM D 5780-02, "Standard Test Method for Individual Piles in Permafrost Under Static Axial Compressive Load," 2004)
- Static Axial Tension Load Test (ASTM D 3689-95, "Standard Test Method for Individual Piles Under Static Axial Tensile Load," 2004)
- Lateral Load Test (ASTM D 3966-95, "Standard Test Method for Piles Under Lateral Loads," 2004)
- Dynamic Testing of Piles (ASTM D 4945-00, "Standard Test Method for High-Strain Dynamic Testing of Piles," 2004)

16.3.2   Pile Load Tests

Pile load tests have been previously discussed in Secs. 6.3.1 and 6.4.2. As shown in Fig. 6.19, the pile load test can take a considerable amount of time and effort to properly set-up. Thus only one or two load tests are often recommended for a particular site.

The location of the pile load tests should be at the most critical area of the site, such as where the bearing stratum is deepest or weakest. The first step involves driving or installing the pile to the desired depth. In Fig. 16.32, the small arrows point to the prestressed concrete piles that have been installed and are founded on the bearing strata. The next step is to install the anchor piles, which are used to hold the reaction frame in place and provide resistance to the load applied to the test piles. The most common type of pile load test to determine its vertical load capacity is the simple compression load test (i.e., see "Standard Test Method for Piles Under Static Axial Compressive Load," ASTM D 1143-94, 2004). A schematic set-up for this test is shown in Fig. 16.33 and includes the test pile, anchor piles, test beam, hydraulic jack, load cell, and dial gauges. Figure 16.34 shows an actual load test where the reaction frame has been installed on top of the anchor piles and the hydraulic loading

FIGURE 16.32   Pile load test. The small arrows point to the prestressed concrete piles that will be subjected to a load test. The large arrow points to one of the six anchor piles.

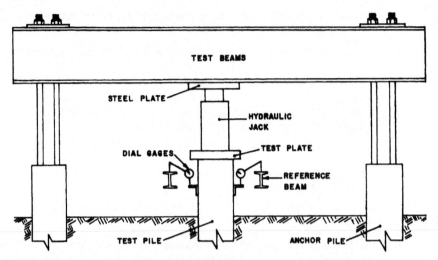

FIGURE 16.33   Schematic setup for applying vertical load to the test pile using a hydraulic jack acting against an anchored reaction frame. (*Reproduced from ASTM D 1143-94, 2004, with permission from the American Society for Testing and Materials.*)

jack is in place. A load cell is used to measure the force applied to the top of the pile. Dial gauges, such as shown in Fig. 16.35, are used to record the vertical displacement of the piles during testing.

As the load is applied to the pile, the deformation behavior of the pile is measured. The pile is often subjected to a vertical load that is at least two times the design value. In most cases, the objective is not to break the pile or load the pile until a bearing capacity failure occurs, but rather to confirm that the design end-bearing parameters used for the design of the piles are adequate. The advantage of this type of approach is that the piles that are load-tested can be left in-place and used as part of the foundation. Figure 16.36 presents the actual load test data for the pile load test shown

FIGURE 16.34   Pile load tests. The reaction frame has been set up and the hydraulic jack and load cell are in place.

FIGURE 16.35   Pile load tests. This photograph shows one of the dial gauges that are used to record the vertical displacement of the top of the pile during testing.

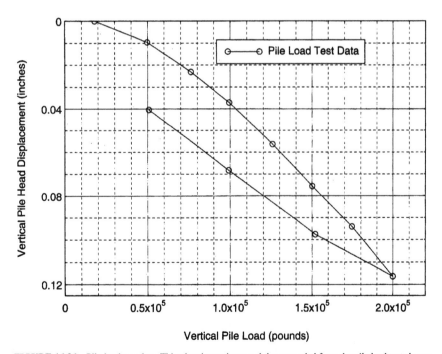

FIGURE 16.36   Pile load test data. This plot shows the actual data recorded from the pile load test shown in Figs. 16.34 and 16.35. The vertical deformation is the average displacement recorded by the dial gauges. The axial load is determined from a load cell.

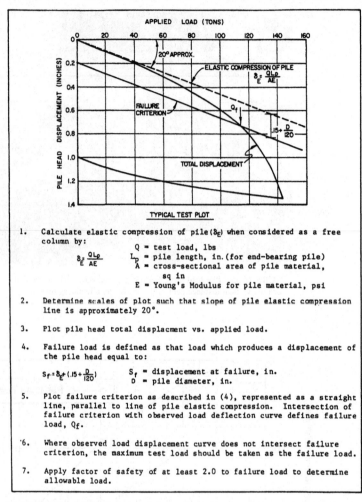

1. Calculate elastic compression of pile(δ_E) when considered as a free column by:

   $$\delta_E = \frac{QL_p}{AE}$$

   Q = test load, lbs
   L_p = pile length, in. (for end-bearing pile)
   A = cross-sectional area of pile material, sq in
   E = Young's Modulus for pile material, psi

2. Determine scales of plot such that slope of pile elastic compression line is approximately 20°.

3. Plot pile head total displacment vs. applied load.

4. Failure load is defined as that load which produces a displacement of the pile head equal to:

   $$S_f = \delta_E + (.15 + \frac{D}{120})$$

   S_f = displacement at failure, in.
   D = pile diameter, in.

5. Plot failure criterion as described in (4), represented as a straight line, parallel to line of pile elastic compression. Intersection of failure criterion with observed load deflection curve defines failure load, Q_f.

6. Where observed load displacement curve does not intersect failure criterion, the maximum test load should be taken as the failure load.

7. Apply factor of safety of at least 2.0 to failure load to determine allowable load.

FIGURE 16.37 Method of analysis for static axially compressive or tension load testing of piles. (*Reproduced from NAVFAC DM-7.2, 1982.*)

in Figs. 16.34 and 16.35. For this project, the prestressed concrete piles were founded on solid bedrock and thus the data in Fig. 16.36 show very little compression of the pile. In fact, the recorded displacement of the pile was almost entirely due to elastic compression of the pile itself, instead of deformation of the bearing strata.

Figure 16.37 shows a commonly used method of analysis for static axially compressive or tension load testing of piles.

16.4 FOUNDATION UNDERPINNING

16.4.1 Introduction

There are many different situations where a structure may need to be underpinned. Tomlinson (1986) lists several underpinning possibilities:

- To support a structure that is sinking or tilting due to ground subsidence or instability of the superstructure
- As a safeguard against possible settlement of a structure when excavating close to and below its foundation level
- To support a structure while making alterations to its foundation or main supporting members
- To enable the foundations to be deepened for structural reasons, for example to construct a basement beneath a building
- To increase the width of a foundation to permit heavier loads to be carried, for example, when increasing the story height of a building
- To enable a building to be moved bodily to a new site

Tomlinson (1986) also indicates that each underpinning project is unique and requires highly skilled personnel, and therefore it should only be attempted by experienced firms. Because each job is different, individual consideration of the most economical and safest scheme is required for each project. Common methods of underpinning include the construction of continuous strip foundations, piers, and piles. To facilitate the underpinning process, the ground can be temporarily stabilized by freezing the ground or by injecting grout or chemicals into the soil (Tomlinson, 1986). A further discussion is presented in *Underpinning* by Prentis and White (1950) and *Underpinning* by Thorburn and Hutchison (1985).

16.4.2 Underpinning with a New Foundation

Underpinning is often required to support a structure that is sinking or tilting due to ground subsidence or instability of the superstructure. The most expensive and rigorous method of underpinning would be to entirely remove the existing foundation and install a new foundation. This method of repair is usually only reserved for projects where there is a large magnitude of soil movement or when the foundation is so badly damaged or deteriorated that it cannot be saved.

Figure 16.38 shows the manometer survey (i.e., a floor level survey) of a building containing two condominium units at a project called Timberlane in Scripps Ranch, California. The building shown in Fig. 16.38 was constructed in 1977 and was underlain by poorly compacted fill that increased in depth toward the front of the building. In 1987, the amount of fill settlement was estimated to be 4 in. (100 mm) at the rear of the building and 8 in. (200 mm) at the front of the building. As shown in Fig. 16.38, the fill settlement caused 3.2 in. (80 mm) of differential settlement for the conventional slab-on-grade and 3.9 in. (99 mm) for the second floor. The reason the second floor had more differential settlement was because it extended out over the garage. Note in Fig. 16.38 that the foundation tilts downward from the rear to the front of the building, or in the direction of deepening fill. Typical damage consisted of cracks in the slab-on-grade, exterior stucco cracks, interior wallboard damage, ceiling cracks, and racked door frames. Using Table 7.3, the damage was classified as severe. Due to on-going fill settlement, the future (additional) differential settlement of the foundation was estimated to be 4 in. (100 mm).

In order to reduce the potential for future damage due to the anticipated fill settlement, it was decided to underpin the building by installing a new foundation. The type of new foundation for the building was a reinforced mat, 15 in. (380 mm) thick, and reinforced with number 7 bars, 12 in. (305 mm) on center, each way, top and bottom. In order to install the reinforced mat, the connections between the building and the existing slab-on-grade were severed and the entire building was raised about 8 ft (2.4 m). Figure 16.39 shows the building in its raised condition. Steel beams, passing through the entire building, were used to lift the building during the jacking process. After the building was raised, the existing slab-on-grade foundation was demolished. The formwork for the construction of the reinforced mat is shown in Fig. 16.39. The mat was designed and constructed so that it sloped 2 in. (50 mm) upward from the back to the front of the building. It was anticipated that with future settlement, the front of the building would settle 4 in. (100 mm) such that the mat would eventually slope 2 in. (50 mm) downward from the back to the front of the building.

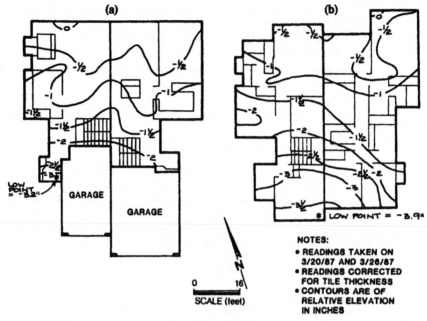

FIGURE 16.38   Manometer survey: (a) first floor; (b) second floor.

After placement and hardening of the new concrete for the mat, the building was lowered onto its new foundation. The building was then attached to the mat and the interior and exterior damages were repaired. Flexible utility connections were used to accommodate the difference in movement between the building and settling fill.

Another underpinning option is to remove the existing foundation and install a mat supported by piers. The mat transfers building loads to the piers that are embedded in a firm bearing material.

FIGURE 16.39   Raised building.

For a condition of soil settlement, the piers will usually be subjected to downdrag loads from the settling soil.

The piers are usually at least 2 ft (0.6 m) in diameter to enable downhole logging to confirm end-bearing conditions. The piers can either be built within the building or the piers can be constructed outside the building with grade beams used to transfer loads to the piers. Given the height of a drill rig, it is usually difficult to drill within the building (unless it is raised). The advantages of constructing the piers outside the building are that the height restriction is no longer a concern and a powerful drill rig can be used to quickly and economically drill the holes for the piers.

Figure 16.40 shows a photograph of the conditions at an adjacent building at Timberlane. Given the very large magnitude of the estimated future differential settlement for this building, it was decided to remove the existing foundation and then construct a mat supported by 2.5 ft (0.76 m) diameter piers. The arrow in Fig. 16.41 points to one of the piers.

In order to construct the mat supported by piers, the building was raised and then the slab-on-grade was demolished. With the building in a raised condition, a drill rig was used to excavate the piers. The piers were drilled through the poorly compacted fill and into the underlying bedrock. The piers were belled at the bottom in order to develop additional end-bearing resistance. After drilling and installation of the steel reinforcement consisting of eight No. 6 bars with No. 4 ties at 1 ft (0.3 m) spacing, the piers were filled with concrete to near ground surface. Figure 16.42 shows a close-up of the pier indicated in Fig. 16.41. To transfer loads from the mat to the piers, the steel reinforcement (No. 6 bars) at the top of the pier is connected to the steel reinforcement in the mat.

16.4.3 Underpinning of the Existing Foundation

The most common type of underpinning involves using the existing foundation and then underpinning the foundation to provide deeper support. This is usually a much less expensive and rigorous method of underpinning than removing the entire foundation. For example, a common type of underpinning is by using continuous strip footings that are connected to the existing foundation. This method of underpinning is often used when the problem soil is near ground surface and the existing foundation is not badly damaged or deteriorated.

Underpinning by using continuous strip footings is a common type of repair for damage caused by expansive soil (Chen, 1988). For example, Fig. 16.43 shows a cross section of a typical design for underpinning using a continuous strip footing. The construction of the footing starts with the

FIGURE 16.40 Construction of the mat foundation.

FIGURE 16.41   Construction of a mat supported by piers. The arrow points to one of the piers.

excavation of slots in order to install the hydraulic jacks. The hydraulic jacks are used to temporarily support the existing foundation until the new underpinning portion is installed. Steel reinforcement is usually tied to the existing foundation by using dowels. The final step is to fill the excavation with concrete and the jacks are left in place during the placement of the concrete. Figure 16.44 shows the installation of a continuous strip footing that is being used to underpin the existing foundation.

An existing foundation can also be underpinned with piles or piers (Brown, 1992). As discussed in Sec. 5.4.3, helical anchors can also be used to underpin existing foundations. These types of deep foundation underpinning are often used when the problem soil is located well below the bottom of

FIGURE 16.42   Close-up view of Fig. 16.41.

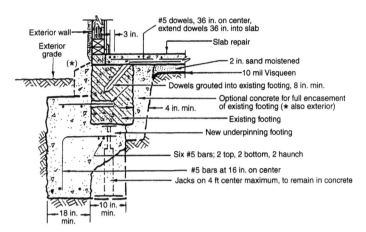

FIGURE 16.43   Underpinning of an existing foundation by using continuous strip footings.

the foundation. Greenfield and Shen (1992) present a list of the advantages and disadvantages of pile and pier underpinning installations.

In addition to underpinning the foundation, repair work may be needed to fix the damaged foundation or strengthen the foundation so that the damage does not reoccur. Figure 16.45 shows the strip replacement method, which is one type of repair for concrete slab-on-grade cracks. The construction of the strip replacement starts by saw-cutting out the area containing the concrete crack. As indicated in Fig. 16.45, the concrete should be saw-cut at a distance of about 1 ft (0.3 m) on both sides of the concrete crack. This is to provide enough working space to install reinforcement and the dowels. After the new reinforcement (No. 3 bars) and dowels are installed, the area is filled with a new portion of concrete.

FIGURE 16.44   Underpinning of a structure using continuous strip footings.

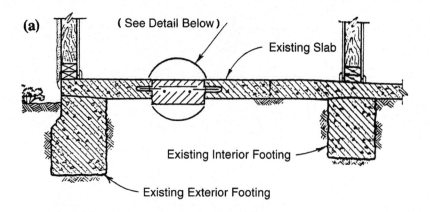

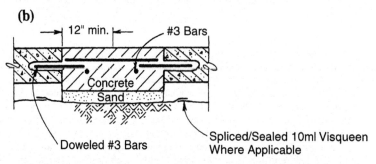

Saw-Cut 12" Each Side of Crack, Dowel #3 Bars 6"(min.) Into
Existing Slab. Provide 5" Concrete Section with #3 Bars, 12"
O.C. Both Ways. Underlay with 2" Moist Sand. Where Visqueen
Exists, Splice/Seal in a Replacement Section

FIGURE 16.45   Concrete crack repair: (a) strip replacement of floor cracks; (b) strip replacement
detail.

Another option is to patch the existing concrete cracks. The objective is to return the concrete slab
to a satisfactory appearance and provide structural strength at the cracked areas. It has been stated
(Transportation Research Board, 1977) that a patching material must meet the following requirements:

1. Be at least as durable as the surrounding concrete.
2. Require a minimum of site preparation.
3. Be tolerant of a wide range of temperature and moisture conditions.
4. Not harm the concrete through chemical incompatibility.
5. Preferably be similar in color and surface texture to the surrounding concrete.

Figure 16.46 shows a typical detail for concrete crack repair. If there is differential movement at
the crack, then the concrete may require grinding or chipping to provide a smooth transition across
the crack. The material commonly used to fill the concrete crack is epoxy. Epoxy compounds consist of a resin, a curing agent or hardener, and modifiers that make them suitable for specific uses.
The typical range (500 to 5000 psi, 3400 to 35,000 kPa) in tensile strength of epoxy is similar to its
range in compressive strength (Schutz, 1984). Performance specifications for epoxy have been

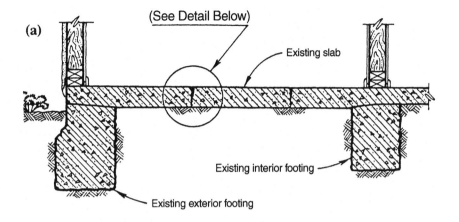

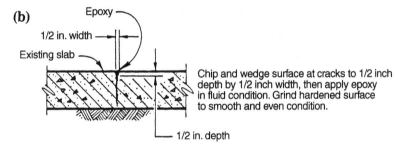

FIGURE 16.46 Concrete crack repair: (a) epoxy repair of floor cracks; (b) detail of crack repair with epoxy.

developed (e.g., ASTM C 881, "Standard Specification for Epoxy-Resin-Base Bonding Systems for Concrete," 2004). In order for the epoxy to be effective, it is important that the crack faces be free of contaminants (such as dirt) that could prevent bonding. In many cases, the epoxy is injected under pressure so that it can penetrate the full depth of the concrete crack. Figure 16.47 shows the installation of pressure-injected epoxy into concrete slab cracks.

16.4.4 Underpinning Alternatives

The previous discussion has dealt with the strengthening and underpinning of the foundation in order to resist future soil movement, bypass the problem soil, or relevel the foundation. There are many other types of underpinning and soil treatment alternatives (Brown, 1990, 1992; Greenfield and Shen, 1992; Lawton, 1996). In some cases, the magnitude of soil movement may be so large that the only alternative is to demolish the structure. For example, movement of the Portuguese Bend Landslide in Palos Verdes, California has destroyed about 160 homes. But a few homeowners refuse to abandon their homes as they slowly slide downslope. Some owners have underpinned their house foundations with steel beams that are supported by hydraulic jacks that are periodically used to relevel the house. Other owners have tried bizarre underpinning methods, such as supporting the house on huge steel drums.

An alternate method to underpinning is to treat the problem soil. Section 15.3 presents soil improvement methods that may be more effective and economical than underpinning the foundation.

For expansive soil, underpinning alternatives include horizontal or vertical moisture barriers to reduce the cyclic wetting and drying around the perimeter of the structure (Nadjer and Werno, 1973;

FIGURE 16.47   Pressure injection of epoxy into concrete slab cracks.

Snethen, 1979; Williams, 1965). Drainage improvements and the repair of leaky water lines are also performed in conjunction with the construction of the moisture barriers. Other expansive soil stabilization options include chemical injection (such as a lime slurry) into the soil below the structure. The goal of such mitigation measures is to induce a chemical mineralogical change of the clay particles that will reduce the soils tendency to swell.

16.5   OBSERVATIONAL METHOD

The observational method is an important tool that can be used for the design and construction of new or underpinned foundations. Concerning the observational method, Terzaghi and Peck (1967) state:

> Design on the basis of the most unfavorable assumptions is inevitably uneconomical, but no other procedure provides the designer in advance of construction with the assurance that the soil-supported structure will not develop unanticipated defects. However, if the project permits modifications of the design during construction, important savings can be made by designing on the basis of the most probable rather than the most unfavorable possibilities. The gaps in the available information are filled by observations during construction, and the design is modified in accordance with the findings. This basis of design may be called the observational procedure.
>
> . . . In order to use the observational procedure in earthwork engineering, two requirements must be satisfied. First of all, the presence and general characteristics of the weak zones must be disclosed by the results of the subsoil exploration in advance of construction. Secondly, special provisions must be made to secure quantitative information concerning the undesirable characteristics of these zones during construction before it is too late to modify the design in accordance with the findings.

As mentioned earlier, the observational method is used during the construction of the foundation. It is a valuable technique because it allows the geotechnical engineer, based on observations and testing during construction, to revise the design and provide a more economical foundation or earth structure. The method is often used during the installation of deep foundations, where field performance testing or observations are essential in confirming that the foundation is bearing on the appropriate strata.

In some cases, the observational method can be misunderstood or misused. For example, at one project the geotechnical engineer discovered the presence of a shallow groundwater table and indicated in the feasibility report that the best approach would be to use the observational method, where the available information on the groundwater table would be supplemented by observations during construction. When the excavation for the underground garage was made, extensive groundwater control was required and an expensive dewatering system was installed. But the client had not anticipated the cost of the expensive dewatering system. The client was very upset at the high cost of the dewatering system because a simple design change of using the first floor of the building as the garage (i.e., above-grade garage) and adding an extra floor to the building would have been much less expensive than dealing with the groundwater. As the client stated: "if I had known that the below-grade garage was going to cost this much, I would never have attempted to construct it."

As this case illustrates, the observational method must not be used in place of a plan of action, but rather to make the plan more economical based on observed subsurface conditions during construction. The client must understand that the savings associated with the observational method may be offset by construction delays due to the redesign of the foundation.

CHAPTER 17
GEOSYNTHETICS AND INSTRUMENTATION

17.1 INTRODUCTION

The purpose of this chapter is to provide an introduction to the use of geosynthetics and instrumentation for foundation engineering.

A *geosynthetic* is defined as a planar product manufactured from polymeric material and typically placed in soil to form an integral part of a drainage, reinforcement, or stabilization system. Common types of geosynthetics used during construction that are discussed in this chapter are geogrids, geotextiles, geomembranes, geonets, geocomposites, and geosynthetic clay liners (Rollings and Rollings, 1996; Fluet, 1988; Richardson and Koerner, 1990; Koerner, 1998). Geosynthetics will be discussed in Sec.17.2.

A common type of construction service performed by geotechnical engineers is the installation of monitoring devices. There are many types of monitoring devices used by geotechnical engineers and the usual purpose of the monitoring devices is to measure the performance of the foundation and structure as it is being built. Monitoring devices are also installed to monitor existing adjacent structures, groundwater conditions, or slopes that may be impacted by the new construction. Instrumentation will be discussed in Sec. 17.3.

17.2 GEOSYNTHETICS

17.2.1 Geogrids

Figure 17.1 shows a photograph of a *geogrid*, which contains relatively high-strength polymer grids consisting of longitudinal and transverse ribs connected at their intersections. Geogrids have a large and open structure and the openings (i.e., apertures) are usually 0.5 to 4 in. (1.3 to 10 cm) in length and/or width. Geogrids can be either biaxial or uniaxial depending on the size of the apertures and shape of the interconnecting ribs. Geogrids are principally used as follows:

1. *Soil reinforcement.* Used for subgrade stabilization, slope reinforcement, erosion control (reinforcement), and mechanically stabilized earth-retaining walls. Also used to strengthen the junction between the top of soft clays and overlying embankments.

2. *Asphalt overlays.* Used in asphalt overlays to reduce reflective cracking.

The most common usage of geogrids is as soil reinforcement. Compacted soil tends to be strong in compression but weak in tension. The geogrid is just the opposite, strong in tension but weak in compression. Thus, layers of compacted soil and geogrid tend to compliment each other and produce a soil mass having both high compressive and tensile strength. The open structure of the geogrid (see Fig. 17.1) allows the compacted soil to bond in the open geogrid spaces. Geogrids provide soil reinforcement by transferring local tensile stresses in the soil to the geogrid. Because geogrids are

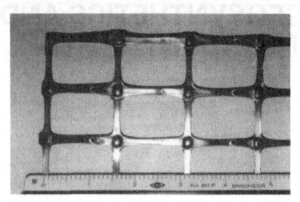

FIGURE 17.1 Photograph of a geogrid. (*From Rollings and Rollings 1996; reprinted with permission of McGraw-Hill, Inc.*)

continuous, they also tend to transfer and redistribute stresses away from areas of high stress concentrations (such as beneath a wheel load). Figure 11.10 shows geogrids being used as soil reinforcement for a mechanically stabilized earth retaining wall.

Similar to other geosynthetics, geogrids are transported to the site in 3 ft (0.9 m) to 12 ft (3.7 m) wide rolls. It is generally not feasible to connect the ends of the geogrid, and it is typically overlapped at joints. Typical design methods for using geogrids are summarized by Koerner (1998).

Some of the limitations of geogrids are as follows:

1. *Ultraviolet light.* Even geogrids produced of carbon black (i.e., ultraviolet stabilized geogrids) can degrade when exposed to long-term ultraviolet light. It is important to protect the geogrid from sunlight and cover the geogrid with fill as soon as possible.

2. *Nonuniform tensile strength.* Geogrids often have different tensile strengths in different directions as a result of the manufacturing process. For example, a Tensar SS-2 (BX1200) biaxial geogrid has an ultimate tensile strength of 2100 lb/ft in the main direction and only 1170 lb/ft in the minor (perpendicular) direction. It is essential that the engineer always check the manufacturer's specification and determine the tensile strength in the main and minor directions.

3. *Creep.* Polymer material can be susceptible to creep. Thus, it is important to use an allowable tensile strength that does allow for creep of the geosynthetic. Tensile strengths are often determined by using ASTM test procedures, such as ASTM D 6637-01 ("Standard Test Method for Determining Tensile Properties of Geogrids by the Single or Multi-Rib Tensile Method," 2004) and ASTM D 5262-04 ("Standard Test Method for Evaluating the Unconfined Tension Creep Behavior of Geosynthetics," 2004).

Many manufacturers will provide their recommended long-term design tensile strength for a specific type of geogrid. This recommended long-term design tensile strength from the manufacturer is usually much less than the ultimate strength of the geogrid. For example, for a Tensar SS-2 (BX1200) biaxial geogrid, the manufacturer's recommended long-term design tensile strength is about 300 lb/ft, which is only one-seventh the ultimate tensile strength (2100 lb/ft). The engineer should never apply an arbitrary factor of safety to the ultimate tensile strength, but rather obtain the recommended long-term design tensile strength from the manufacturer.

17.2.2 Geotextiles

Geotextiles are the most widely used type of geosynthetic and they are often referred to as *fabric*. For example, common construction terminology for geotextiles includes geofabric, filter fabric, construction fabric, synthetic fabric, and road-reinforcing fabric. As shown in Figs. 17.2 and 17.3, geotextiles

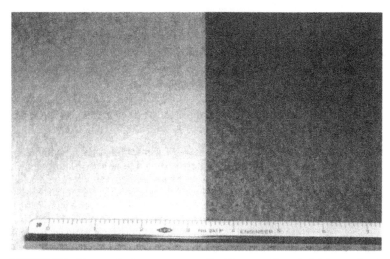

FIGURE 17.2   Photograph of nonwoven geotextiles. The geotextile on the left has no ultraviolet protection, while the geotextile on the right has ultraviolet protection. (*From Rollings and Rollings 1996; reprinted with permission of McGraw-Hill, Inc.*)

are usually categorized as being either woven or nonwoven depending on the type of manufacturing process. Geotextiles are principally used as follows:

1. *Soil reinforcement.*   Used for subgrade stabilization, slope reinforcement, and mechanically stabilized earth-retaining walls. Also used to strengthen the junction between the top of soft clays and overlying embankments.

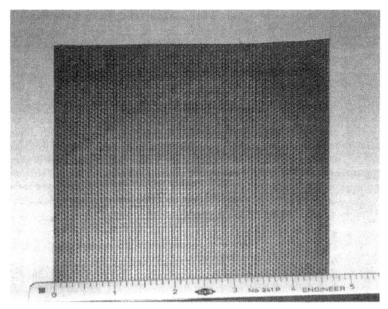

FIGURE 17.3   Photograph of a woven geotextile. (*From Rollings and Rollings 1996; reprinted with permission of McGraw-Hill, Inc.*)

2. *Sediment control.* Used as silt fences to trap sediment on-site.

3. *Erosion control.* Installed along channels, under riprap, and used for shore and beach protection.

4. *Asphalt overlays.* Used in asphalt overlays to reduce reflective cracking.

5. *Separation.* Used between two dissimilar materials, such as an open-graded base and a clay subgrade, in order to prevent contamination.

6. *Filtration and drainage.* Used in place of a graded filter where the flow of water occurs across (perpendicular to) the plane of the geotextile. For drainage applications, the water flows within the geotextile.

Probably the most common usage of geotextiles is for filtration (i.e., flow of water through the geotextile). For filtration, the geotextile should be at least 10 times more permeable than the soil. In addition, the geotextile must always be placed between a less permeable (i.e., the soil) and a more permeable (i.e., the open-graded gravel) material. An inappropriate use of a geotextile would be to place it around the drainage pipe, because then it would have more permeable material on both sides of the geotextile and it would tend to restrict flow.

Geotextiles to be used as filtration devices must have adequate hydraulic properties that allow the water to flow through them and they must also retain the soil particles. Important hydraulic properties are as follows:

1. *Percent open area.* Although geotextiles have been developed that limit the open area of filtration to 5 percent or less, it is best to have a larger open area to develop an adequate flow capacity.

2. *Permittivity or flow rate.* Manufactures typically provide the flow capacity of a geotextile in terms of its permittivity or flow rate. These hydraulic properties are often determined by using ASTM test procedures, such as ASTM D 4491-99, "Standard Test Methods for Water Permeability of Geotextiles by Permittivity Soil Retention Capability," 2004.

3. *Apparent opening size.* The apparent opening size (AOS), also known as the *effective opening size* (EOS), determines the soil retention capability. The AOS is often expressed in terms of opening size (mm) or equivalent sieve size (e.g., AOS = 40–70 indicates openings equivalent to the No. 40 to No. 70 sieves). The test procedures in ASTM D 4751-99, "Standard Test Method for Determining Apparent Opening Size of a Geotextile," can be used to determine the AOS. Obviously, if the geotextile openings are larger than the largest soil particle diameter, then all of the soil particles will migrate through the geotextile and clog the drainage system. A common recommendation is that the required AOS be less than or equal to D_{85} (grain size corresponding to 85 percent passing).

Some of the limitations of geotextiles are as follows:

1. *Ultraviolet light.* Geotextiles that have no ultraviolet light protection can rapidly deteriorate. For example, certain polypropylene geotextiles lost 100 percent of their strength after only 8 weeks of exposure (Raumann, 1982; Koerner, 1998). Manufacturers will often list the ultraviolet light resistance after 500 h of exposure in terms of the percentage of remaining tensile resistance based on the test procedures in ASTM D 4355-02 "Standard Test Method for Deterioration of Geotextiles by Exposure to Light, Moisture and Heat in a Xenon Arc Type Apparatus," 2004.

2. *Sealing of the geotextile.* When used for filtration, an impermeable soil layer can develop adjacent to the geotextile if it has too low an open area or too small an AOS.

3. *Construction problems.* Some of the more common problems related to construction with geotextiles are as follows (Richardson and Wyant, 1987):
   a. Fill placement or compaction techniques damage the geotextile.
   b. Installation loads are greater than design loads, leading to failure during construction.
   c. Construction environment leads to a significant reduction in assumed fabric properties, causing failure of the completed project.

d. Field seaming or overlap of the geotextile fails to fully develop desired fabric mechanical properties.

e. Instabilities during various construction phases may render a design inadequate even though the final product would have been stable.

17.2.3 Geomembranes

Common construction terminology for *geomembranes* includes liners, membranes, visqueen, plastic sheets, and impermeable sheets. Geomembranes are most often used as barriers to reduce water or vapor migration through soil (see Fig. 17.4). For example, Fig. 12.20 shows design specifications for a below foundation moisture barrier that includes a 6-mil visqueen vapor barrier (i.e., a geomembrane). In the United States, 1 mil is one-thousandth of an inch. Another common usage for geomembranes is for the lining and capping systems in municipal landfills. For liners in municipal landfills, the thickness of the geomembrane is usually at least 80 mil. The surface of the geomembrane can be textured in order to provide more frictional resistance between the soil and geomembrane surface.

Some of the limitations of geomembranes are as follows:

1. *Puncture resistance.* The geomembrane must be thick enough so that it is not punctured during installation and subsequent usage. The puncture strength of a geomembrane can be determined by using the test procedures outlined in ASTM D 4833-00 "Standard Test Method for Index Puncture Resistance of Geotextiles, Geomembranes, and Related Products," 2004.

2. *Slide resistance.* Slope failures have developed in municipal liners because of the smooth and low frictional resistance between the geomembrane and overlying or underlying soil. Textured geomembranes (such as shown in Fig. 17.4) have been developed to increase the frictional resistance of the geomembrane surface.

3. *Sealing of seams.* A common cause of leakage through geomembranes is due to inadequate sealing of seams. The following are different methods commonly used to seal geomembrane seams (Rollings and Rollings, 1996):

a. Extrusion welding: Suitable for all polyethylenes. A ribbon of molten polymer is extruded over the edge (filet weld) or between the geomembrane sheets (flat weld). This melts the adjacent surfaces that are then fused together upon cooling.

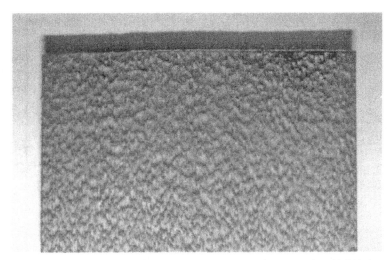

FIGURE 17.4 Photograph of a geomembrane that has a surface texture for added friction. (*From Rollings and Rollings 1996; reprinted with permission of McGraw-Hill, Inc.*)

b. Thermal fusion: Suitable for thermoplastics. Adjacent surfaces are melted and then pressed together. Commercial equipment is available that uses a heated wedge (most common) or hot air to melt the materials. Also, ultrasonic energy can be used for melting rather than heat.

c. Solvent-based systems: Suitable for materials that are compatible with the solvent. A solvent is used with pressure to join adjacent surfaces. Heating may be used to accelerate the curing. The solvent may contain some of the geomembrane polymer already dissolved in the solvent liquid (bodied solvent) or an adhesive to improve the seam quality.

d. Contact adhesive: Primarily suitable for thermosets. Solution is brushed onto surfaces to be joined, and pressure is applied to ensure good contact. Upon curing, the adhesive bonds the surfaces together.

17.2.4 Geonets and Geocomposites

Geonets are three-dimensional netlike polymeric materials used for drainage (i.e., flow of water within the geosynthetic). Figure 17.5 shows a photograph of a *geonet*. Geonets are usually used in conjunction with a geotextile and/or geomembrane, hence geonets are technically a *geocomposite*.

Depending on the particular project requirements, different types of geosynthetics can be combined together to form a geocomposite. For example, a geocomposite consisting of a geotextile and a geomembrane provides for a barrier that has increased tensile strength and resistance to punching and tearing. Figure 17.6 shows a photograph of a geocomposite consisting of a textured geomembrane, geonet, and geotextile (filter fabric).

17.2.5 Geosynthetic Clay Liners

Geosynthetic clay liners are frequently used as liners for municipal landfills. The geosynthetic clay liner typically consists of dry bentonite sandwiched between two geosynthetics. When moisture infiltrates the geosynthetic clay liner, the bentonite swells and creates a soil layer having a very low hydraulic conductivity, transforming it into an effective barrier to moisture migration.

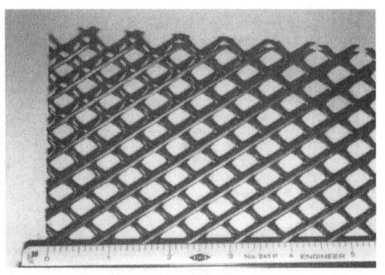

FIGURE 17.5 Photograph of a geonet. (*From Rollings and Rollings 1996; reprinted with permission of McGraw-Hill, Inc.*)

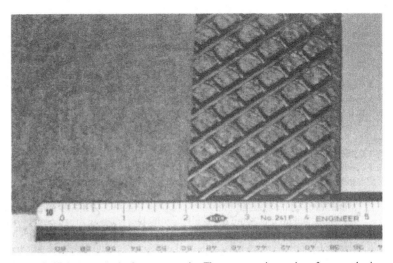

FIGURE 17.6 Photograph of a geocomposite. The geocomposite consists of a geonet having a textured geomembrane on top, and a filter fabric (geotextile) on the bottom. (*From Rollings and Rollings 1996; reprinted with permission of McGraw-Hill, Inc.*)

17.3 *INSTRUMENTATION*

17.3.1 Introduction

Another broad category of construction services performed by geotechnical engineers is the installation of monitoring devices. There are many types of monitoring devices used by geotechnical engineers. The usual purpose of the installation of monitoring devices is to measure the performance of the structure as it is being built. Monitoring devices could also be installed to monitor existing adjacent structures, groundwater conditions, or slopes that may be impacted by the new construction.

Monitoring devices are especially important in urban areas where there are often adjacent structures that could be damaged by the construction activities. Damage to an adjacent structure can result in a lawsuit. Frequent causes of damage to an adjacent structure include the lowering of the groundwater table or lateral movement of temporary underground shoring systems. Monitoring devices are essential for adjacent historic structures, which tend to be brittle and easily damaged, and can be very expensive to repair. For example, Feld and Carper (1997) describe the construction of the John Hancock tower that damaged the adjacent historic Trinity Church, in Boston, Massachusetts. According to court records, the retaining walls (used for the construction of the John Hancock basement) moved 33 in. (84 cm) as the foundation was under construction in 1969. This movement of the retaining walls caused the adjacent street to sink 18 in. (46 cm) and caused the foundation of the adjacent Trinity Church to shift, which resulted in structural damage and 5 in. (13 cm) of tilting of the central tower. The resulting lawsuit was settled in 1984 for about $12 million. An interesting feature of this lawsuit was that the Trinity Church was irreparably damaged, and the damage award was based on the cost of completely demolishing and reconstructing the historic masonry building (ASCE, 1987).

17.3.2 Commonly Used Monitoring Devices

Some of the more common monitoring devices are as follows:

Inclinometers.   The horizontal movement preceding or during the movement of slopes can be investigated by successive surveys of the shape and position of flexible vertical casings installed in the ground (Terzaghi and Peck, 1967). The surveys are performed by lowering an inclinometer probe into the flexible vertical casing. The inclinometer probe is capable of measuring its deviation from the vertical. An initial survey (base reading) is performed and then successive surveys are compared to the base reading to obtain the horizontal movement of the slope.

Figure 17.7 shows a sketch of the inclinometer probe in the casing and the calculations used to obtain the lateral deformation. Inclinometers are often installed to monitor the performance of earth dams and during the excavation and grading of slopes where lateral movement might affect off-site structures. Inclinometers are also routinely installed to monitor the lateral ground movement due to the excavation of building basements and underground tunnels.

Piezometers.   Piezometers are installed in order to monitor pore water pressures in the ground. Several different types are commercially available, including borehole, embankment, or push-in piezometers. Figure 17.8 shows an example of a borehole piezometer.

In their simplest form, piezometers can consist of a standpipe that can be used to monitor groundwater levels and obtain groundwater samples. Figure 17.9 shows an example of a standpipe piezometer. It is standard procedure to install piezometers when an urban project requires dewatering in order to make excavations below the groundwater table. Piezometers are also used to monitor the performance of earth dams and dissipation of excess pore water pressure associated with the consolidation of soft clay deposits.

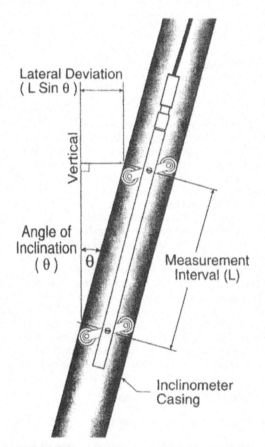

FIGURE 17.7   Inclinometer probe in a casing. (*Reprinted with permission from the Slope Indicator Company.*)

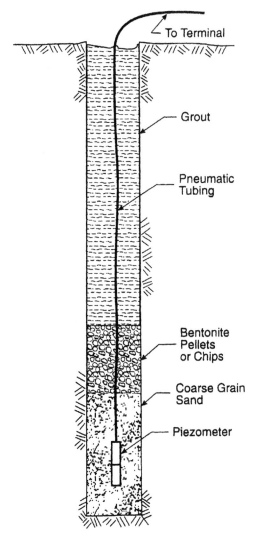

FIGURE 17.8 Pneumatic piezometer installed in a bore-hole. (*Reprinted with permission from the Slope Indicator Company.*)

Settlement Monuments or Cells.   Settlement monuments or settlement cells can be used to monitor settlement or heave. Figure 17.10 shows a diagram of the installation of a pneumatic settlement cell and plate. More advanced equipment include settlement systems installed in borings that can not only measure total settlement, but also the incremental settlement at different depths.

Settlement monuments or cells are often installed to measure the deformation of the foundation or embankments during construction or to monitor the movement of existing structures that are located adjacent to the area of construction.

Pressure and Load Cells.   A total pressure cell measures the sum of the effective stress and pore water pressure. The total pressure cell can be manufactured from two circular plates of stainless

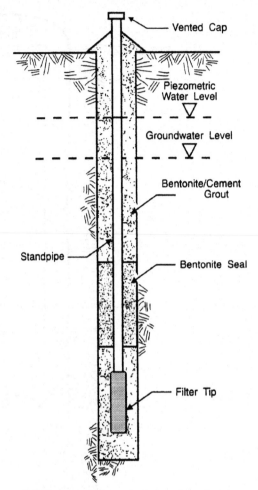

FIGURE 17.9 Standpipe (Casagrande) piezometer. *(Reprinted with permission from the Slope Indicator Company.)*

steel. The edges of the plates are welded together to form a sealed cavity that is filled with fluid. Then a pressure transducer is attached to the cell. The total pressure acting on the sensitive surface is transmitted to the fluid inside the cell and measured by the pressure transducer (Slope Indicator Company, 1998).

Total pressure cells are often used to monitor total pressure exerted on a structure to verify design assumptions and to monitor the magnitude, distribution, and orientation of stresses. For example, load cells are commonly installed during the construction of earth dams to monitor the stresses within the dam core. During construction of the earth dam, the total pressure cells are often installed in arrays with each cell being placed in a different orientation and then covered with compacted fill. For the monitoring of earth pressure on retaining walls, the total pressure cell is typically placed into a recess so that the sensitive side is flush with the retaining wall surface.

Load cells are similar in principle to total pressure cells. They can be used for many different types of geotechnical engineering projects. For example, Fig. 17.11 shows a center-hole load cell that

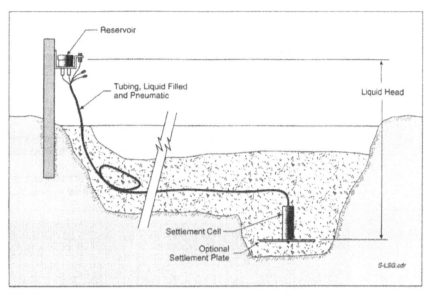

FIGURE 17.10   Pneumatic settlement cell installation. (*Reprinted with permission from the Slope Indicator Company.*)

is designed to measure loads in tiebacks. This center-hole load cell can also be used to measure loads in rock bolts and cables. As shown in Fig. 17.11, for best results, the load cell is centered on the tieback bar and bearing plates are placed above and below the cell. The bearing plates must be able to distribute the load without bending or yielding.

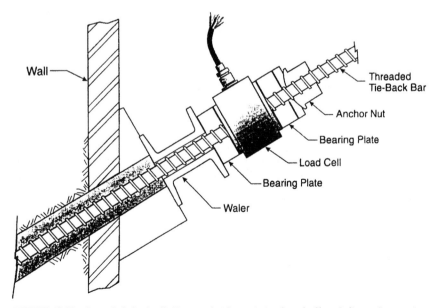

FIGURE 17.11   Center-hole load cell. (*Reprinted with permission from the Slope Indicator Company.*)

Crack Monitoring Devices. For construction in congested urban areas, it is essential to monitor the performance of adjacent buildings, especially if they already have existing cracking. This can often be the case in historic districts of cities, where old buildings may be in a weakened or cracked state. Monitoring of existing cracks in adjacent buildings should be performed where there is pile driving or blasting at the construction site. The blasting of rock could be for the construction of an underground basement or for the construction of road cuts. People are often upset by the noise and vibrations from pile driving and blasting and will claim damage due to the vibrations from these construction activities. By monitoring the width of existing cracks, the geotechnical engineer will be able to evaluate these claims of damage.

A simple method to measure the widening of cracks in concrete or brickwork is to install crack pins on both sides of the crack. By periodically measuring the distance between the pins, the amount of opening or closing of the crack can be determined.

Other crack monitoring devices are commercially available. For example, Fig. 17.12 shows two types of crack monitoring devices. For the Avongard crack monitoring device, there are two installation

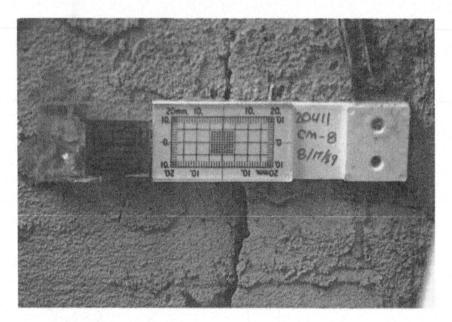

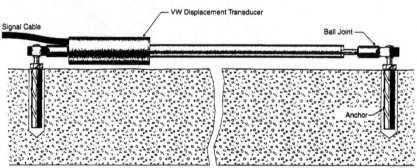

FIGURE 17.12 Crack monitoring devices. The upper photograph shows the Avongard crack monitoring device. The lower diagram shows the VW Crackmeter (*Reprinted with permission from the Slope Indicator Company*).

procedures: (1) the ends of the device are anchored by the use of bolts or screws, or (2) the ends of the device are anchored with epoxy adhesive. The center of the Avongard crack monitoring device is held together with clear tape that is cut once the ends of the monitoring device have been securely fastened with bolts, screws, or epoxy adhesive.

Other Monitoring Devices.   There are many other types of monitoring devices that can be used by the geotechnical engineer. Some commercially available devices include borehole and tape extensometers, soil strainmeters, beam sensors and tiltmeters, and strain gauges. See *Geotechnical Instrumentation for Monitoring Field Performance* (Dunnicliff, 1993) for a further discussion of monitoring devices.

17.3.3 Development of an Instrumentation Program

Prior to the construction of the project, an instrumentation program may need to be developed. Figure 17.13 shows an example of an instrumentation program for a proposed 75 ft (23 m) deep excavation adjacent to a building that has a shallow foundation system. The subsurface conditions at the site consist of there layers of soil (sand/gravel layer, soft/medium clay layer, and a sand/gravel layer) overlying rock. The groundwater table is located in the upper sand/gravel layer. Prior to starting the excavation, concrete diaphragm walls will be installed from ground surface and anchored in the underlying rock. As shown in Fig. 17.13, the instrumentation program consists of the following:

- *Settlement monuments.* Seven settlement monuments will be installed along the exterior of the building. The settlement monuments will be used to monitor the settlement of the building.

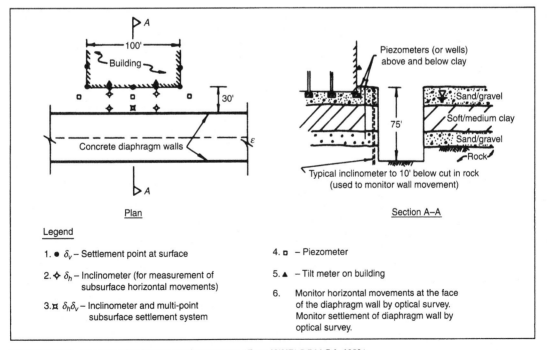

FIGURE 17.13 Example of an instrumentation program. (*From NAVFAC DM-7.1, 1982.*)

- *Inclinometers.*   Four inclinometers are proposed and the inclinometers will be installed from ground surface to below the bottom of the proposed excavation. The inclinometers will be used to monitor lateral movement of the concrete diaphragm wall.

- *Inclinometer and multi-point subsurface settlement system.*   One special inclinometer is proposed that can both record lateral movement as well as determine settlement at various depths below ground surface.

- *Piezometers.*   Three sets of piezometers are proposed, with each set of piezometers having one shallow and one deep piezometer to monitor the pore water pressures in the soil layers located above and below the clay layer. If the pore water pressures should drop, then there could be consolidation of the clay layer and settlement of the building.

- *Tiltmeters.*   Two tiltmeters are proposed on the face of the building closest to the excavation to monitor possible tilting of the building facade.

- *Optical survey.*   The purpose of the optical survey will be to monitor horizontal and vertical movements of the diaphragm wall.

As this example illustrates, the monitoring program could be quite extensive and consist of many different types of devices. An additional part of the monitoring program often consists of documenting the condition of the building prior to the start of the excavation. This is usually accomplished by inspecting the building and taking notes and photographs of observed cracks or other types of damage. If the owner of the building claims damage caused by the excavation, then the pictures can be used to compare the preexisting condition versus the damaged condition of the building.

2009 INTERNATIONAL BUILDING CODE

CHAPTER 18
INTERNATIONAL BUILDING CODE REGULATIONS FOR SOILS

18.1 INTRODUCTION

Building codes are developed to safeguard the public health, safety, and general welfare. In this regard, building codes typically present minimum building regulations. For example, the preface of the *International Building Code* (2009) states: "The comprehensive building code establishes minimum regulations for building systems using prescriptive and performance-related provisions." Although building codes provide minimum building regulations, it should be recognized that the design engineer, for a variety of reasons, might provide more stringent design recommendations.

This part of the book (i.e. Chaps. 18 and 19) will deal with the geotechnical aspects of foundation engineering as specified by the *International Building Code* (2009). A companion volume is the *International Residential Code* (2009), which presents regulations for the construction of detached one- and two-family dwellings and multiple single-family dwellings (townhouses) not more than three stories above grade plane in height. The *International Existing Building Code* (2009) is applicable for existing buildings undergoing repair, alterations, or additions and change of occupancy.

In order to simplify the presentation, this book will deal only with the *International Building Code*. Important chapters in the *International Building Code* that specifically cover the geotechnical aspects of foundation engineering are as follows:

- *Structural tests and inspections (Chapters 1 and 17).* Section 110.3.1 presents footing and foundation inspections to be performed by the building official. Sections 1704.7 to 1704.10 present a discussion of special inspections related to site soil conditions, driven deep foundations, cast-in-place deep foundations, and helical pile foundations. The project geotechnical engineer, engineering geologist, and field technician often perform these inspections. The inspection regulations, as related to geotechnical engineering, are summarized in Table 18.1 of this book.

- *Structural design (Chapter 16).* Although this chapter is predominately concerned with structural engineering, there are some geotechnical issues. For example, the geotechnical engineer will need to determine the site class based on the results of the soils investigation. This chapter also provides soil lateral loads for foundation walls and retaining walls. The site class and soil lateral loads will be discussed in Chap. 19 of this book.

- *Soils and foundations (Chapter 18).* This is the most important part of the *Code* in terms of the geotechnical aspects of foundation engineering. Both Chaps. 18 and 19 in this part of the book will make reference to sections within Chapter 18 of the *International Building Code*.

- *Concrete (Chapter 19).* Since most foundations are made of concrete, the geotechnical engineer will often need to provide concrete material recommendations. For example, the soils engineer often determines the soluble sulfate of the soil that will be in contact with the concrete foundation and then provides recommendations concerning the concrete material properties required to resist sulfate attack (see Table 18.2).

TABLE 18.1   Inspection Regulations

Activity	Regulations
Site Soil Conditions	"Section 1704.7 Soils. Special inspections for existing site soil conditions, fill placement and load-bearing requirements shall be as required by this section and Table 1704.7. The approved geotechnical report and the construction documents prepared by the registered design professionals shall be used to determine compliance. During fill placement, the special inspector shall determine that proper materials and procedures are used in accordance with the provisions of the approved geotechnical report. Exception: Where Section 1803 does not require reporting of materials and procedures for fill placement, the special inspector shall verify that the in-place dry density of the compacted fill is not less than 90 percent of the maximum dry density at optimum moisture content determined in accordance with ASTM D 1557."
	Table 1704.7 titled Required Verification and Inspection of Soil lists continuous tasks and tasks performed periodically. The continuous task is to: "Verify use of proper materials, densities and lift thicknesses during placement and compaction of compaction of compacted fill." Periodically performed tasks are as follows: "(1) Verify that materials below shallow foundations are adequate to achieve the design bearing capacity. (2) Verify that excavations are extended to proper depth and have reached proper material. (3) Perform classification and testing of compaction fill materials. (4) Prior to placement of compacted fill, observe subgrade and verify that site has been prepared properly."
Driven Deep Foundations	"Section 1704.8 Driven deep foundations. Special inspections shall be performed during installation and testing of driven deep foundation elements as required by Table 1704.8. The approved geotechnical report, and the construction documents prepared by the registered design professionals, shall be used to determine compliance."
	Table 1704.8 titled *Required Verification and Inspection of Driven Deep Foundation Elements* lists the following continuous tasks that must be performed: "(1) Verify that element materials, sizes and lengths comply with the requirements. (2) Determine capacities of test elements and conduct additional load tests, as required. (3) Observe driving operations and maintain complete and accurate records for each element. (4) Verify placement locations and plumbness, confirm type and size of hammer, record number of blows per foot of penetration, determine required penetrations to achieve design capacity, record tip and butt elevations and document any damage to foundation element."
Cast-in-place Deep Foundations	"Section 1704.9 Cast-in-place deep foundations. Special inspections shall be performed during installation and testing of cast-in-place deep foundation elements as required by Table 1704.9. The approved geotechnical report, and the construction documents prepared by the registered design professionals, shall be used to determine compliance."
	Table 1704.9 titled *Required Verification and Inspection of Cast-In-Place Deep Foundation Elements* lists the following continuous tasks that must be performed: "(1) Observe drilling operations and maintain complete and accurate records for each element. (2) Verify placement locations and plumbness, confirm element diameters, bell diameters (if applicable), lengths, embedment into bedrock (if applicable) and adequate end-bearing strata capacity. Record concrete or grout volumes."
Helical Pile Foundations	"Section 1704.10 Helical pile foundations. Special inspections shall be performed continuously during installation of helical pile foundations. The information recorded shall include installation equipment used, pile dimensions, tip elevations, final depth, final installation torque and other pertinent installation data as required by the registered design professional in responsible charge. The approved geotechnical report and the construction documents prepared by the registered design professional shall be used to determine compliance."

Source: Sections 1704.7 to 1704.10 of the *International Building Code.*

TABLE 18.2 Requirements for Concrete Exposed to Sulfate-Containing Solutions

Sulfate exposure	Water soluble sulfate (SO_4) in soil, percent by weight	Sulfate (SO_4) in water (ppm)	Cement type			Maximum water-cementitious materials ratio, by weight, normal-weight aggregate concrete[*]	Minimum f_c' normal-weight and lightweight aggregate concrete (psi)[*]
			ASTM C 150	ASTM C 595	ASTM C 1157		
Negligible	0.00–0.10	0–150	—	—	—	—	—
Moderate[†]	0.10–0.20	150–1,500	II	II, IP (MS), IS (MS), P (MS), I (PM)(MS), I (SM)(MS)	MS	0.50	4,000
Severe	0.20–2.00	1,500–10,000	V	—	HS	0.45	4,500
Very severe	Over 2.00	Over 10,000	V plus pozzolan[‡]	—	HS plus pozzolan[§]	0.45	4,500

Notes: For SI: 1.0 psi = 6.89 kPa

[*]A lower water-cementitious materials ratio or higher strength may be required for low permeability or for protection against corrosion of embedded items or freezing and thawing (see Tables 1904.2.2 of the *International Building Code*).

[†]Seawater.

[‡]Pozzolan that has been determined by test or service record to improve sulfate resistance when used in concrete containing Type V cement.

[§]Pozzolan that has been determined by test or service record to improve sulfate resistance when used in concrete containing Type HS blended cement.

Source: Table 1904.3 of the *International Building Code* 2003.

• *Safeguards during construction (Chapter 33).* This chapter of the *Code* includes Section 3304 (Site Work) and discusses excavation and fill, slope limits, and surcharge loads. Section 3304 from the *International Building Code* is reproduced in Table 18.3 of this book.

• *Grading (Appendix J).* This appendix of the *Code* specifically deals with site grading. However, provisions contained in a code appendix are not mandatory unless specifically referenced in the adopting ordinance.

The remaining portion of this chapter of the book will deal with commonly used soil regulations as outlined in Chapter 18 of the *International Building Code*. Chapter 19 will deal with the *International Building Code* regulations for foundations.

It is important to point out that Chaps. 18 and 19 of this book will not discuss every detail of the *International Building Code* that is applicable to foundation engineering, but will rather present the more commonly used *Code* regulations. For the sake of brevity, some *Code* sections will be summarized or condensed, exceptions eliminated, or references to related material omitted. To obtain a complete understanding of the *Code* regulations, the reader is encouraged to study the applicable sections of the *International Building Code*.

18.2 SOILS INVESTIGATION

Section 1803 of the *International Building Code* deals with many different soils investigation issues. The seismic issues raised in Section 1803 will be discussed in Sec. 19.5 of this book. General categories of information as presented in Section 1803 are as follows:

TABLE 18.3   Site Work

Topics	Regulations
Excavation and fill	"Section 3304.1 Excavation and fill. Excavation and fill for buildings and structures shall be constructed or protected so as not to endanger life or property. Stumps and roots shall be removed from the soil to a depth of at least 12 inches (305 mm) below the surface of the ground in the area to be occupied by the building. Wood forms which have been used in placing concrete, if within the ground or between foundation sills and the ground, shall be removed before a building is occupied or used for any purpose. Before completion, loose or casual wood shall be removed from direct contact with the ground under the building."
Slope inclinations	"Section 3304.1.1 Slope limits. Slopes for permanent fill shall not be steeper than one unit vertical in two units horizontal (50-percent slope). Cut slopes for permanent excavations shall not be steeper than one unit vertical in two units horizontal (50-percent slope). Deviation from the foregoing limitations for cut slopes shall be permitted only upon the presentation of a soil investigation report acceptable to the building official."
Surcharge and excavations	"Section 3304.1.2 Surcharge. No fill or other surcharge loads shall be placed adjacent to any building or structure unless such building or structure is capable of withstanding the additional loads caused by the fill or surcharge. Existing footings or foundations which can be affected by any excavation shall be underpinned adequately or otherwise protected against settlement and shall be protected against later movement."

Source: Section 3304 of the *International Building Code*.

18.2.1   Groundwater Table

As discussed in this book, one of the main purposes of subsurface exploration is to determine the depth of the groundwater table. Concerning the groundwater table, the *International Building Code* states:

Section 1803.5.4 Groundwater table. A subsurface soil investigation shall be performed to determine whether the existing groundwater table is above or within 5 feet (1524 mm) below the elevation of the lowest floor level where such floor is located below the finished ground level adjacent to the foundation.

In a related topic, waterproofing for walls and floors is presented in Section 1805 of the *International Building Code*.

18.2.2   Rock Strata

The depth to rock can be very important especially for projects, such as bridge foundations with high loads that must be supported by rock. Concerning rock strata, the *International Building Code* states:

Section 1803.5.6 Rock strata. Where subsurface explorations at the project site indicate variations or doubtful characteristics in the structure of the rock upon which foundations are to be constructed, a sufficient number of borings shall be made to a depth of not less than 10 feet (3048 mm) below the level of the foundations to provide assurance of the soundness of the foundation bed and its load-bearing capacity.

18.2.3   Soil Classification

The *International Building Code* indicates in Section 1803.5.1 that the classification of soil materials shall be in accordance with ASTM D 2487, "Standard Practice for Classification of Soils for Engineering Purposes (Unified Soil Classification System)." Chapter 4 of this book has presented an in-depth discussion of the Unified Soil Classification System.

18.2.4 Site Investigation

The site investigation includes document review, subsurface exploration, and laboratory testing. The site investigation has been covered in Chap. 2 (Subsurface Exploration) and Chap. 3 (Laboratory Testing). The *International Building Code* states:

> Section 1803.3 Basis of investigation. Soil classification shall be based on observation and any necessary tests of the materials disclosed by borings, test pits or other subsurface exploration made in appropriate locations. Additional studies shall be made as necessary to evaluate slope stability, soil strength, position and adequacy of load-bearing soils, the effect of moisture variation on soil-bearing capacity, compressibility, liquefaction and expansiveness.
>
> Section 1803.3.1 Scope of investigation. The scope of the geotechnical investigation including the number and types of borings or soundings, the equipment used to drill or sample, the in-situ testing equipment, and the laboratory testing program shall be determined by a registered design professional.
>
> Section 1803.4 Qualified representative. The investigation procedure and apparatus shall be in accordance with generally accepted engineering practice. The registered design professional shall have a fully qualified representative on site during all boring or sampling operations.

18.2.5 Reports

Concerning report preparation, the *International Building Code* states:

> Section 1803.6 Reporting. Where geotechnical investigations are required, a written report of the investigation shall be submitted to the building official by the owner or authorized agent at the time of permit application. The geotechnical report shall include, but need not be limited to, the following information:
>
> 1. A plot showing the location of the soil investigations.
> 2. A complete record of the soil boring and penetration test logs and soil samples.
> 3. A record of the soil profile.
> 4. Elevation of the water table, if encountered.
> 5. Recommendations for foundation type and design criteria, including but not limited to: bearing capacity of natural or compacted soil; provisions to mitigate the effects of expansive soils; mitigation of the effects of liquefaction, differential settlement and varying soil strength; and the effects of adjacent loads.
> 6. Expected total and differential settlement.
> 7. Deep foundation information in accordance with Section 1803.5.5.
> 8. Special design and construction provisions for foundations of structures founded on expansive soils, as necessary.
> 9. Compacted fill material properties and testing in accordance with Section 1803.5.8.

The local building department or governing agency may require that additional items be included in the foundation engineering report. The geotechnical engineer should always inquire about local building department or governing agency requirements concerning report preparation.

18.3 EXCAVATION, GRADING, AND FILL

In this book, grading and fill are covered in Chap. 15 and a discussion of excavations is presented in Chap. 16. The primary section dealing with excavation, grading, and fill placement in the *International Building Code* is Section 1804. Other sections of the *International Building Code* that deal with excavation, grading, and fill compaction are as follows:

• *Special inspections (Section 1704.7).* Special inspections for grading and fill compaction (site soil inspections) as presented in the *International Building Code* are summarized in Table 18.1 of this book.

- *Site work (Section 3304.1).* Excavation and fills are also discussed in Chapter 33 (Safeguards During Construction). The information in Section 3304.1 of the *International Building Code* has been summarized in Table 18.3 of this book.

- *Grading (Appendix J).* Grading regulations are presented in Appendix J of the *International Building Code* and will be briefly discussed in Sec. 18.3.3 of this book.

18.3.1   Excavation and Grading

For excavation and grading, the *International Building Code* states:

> Section 1804.1 Excavation near foundations. Excavation for any purpose shall not remove lateral support from any foundation without first underpinning or protecting the foundation against settlement or lateral translation.
>
> Section 1804.2 Placement of backfill. The excavation outside the foundation shall be backfilled with soil that is free of organic material, construction debris, cobbles and boulders or with a controlled low-strength material (CLSM). The backfill shall be placed in lifts and compacted in a manner that does not damage the foundation or the waterproofing or dampproofing material. Exception: CLSM need not be compacted.
>
> Section 1804.3 Site grading. The ground immediately adjacent to the foundation shall be sloped away from the building at a slope of not less than one unit vertical in 20 units horizontal (5-percent slope) for a minimum distance of 10 feet (3048 mm) measured perpendicular to the face of the wall. If physical obstructions or lot lines prohibit 10 feet (3048 mm) of horizontal distance, a 5-percent slope shall be provided to an approved alternative method of diverting water away from the foundation. Swales used for this purpose shall be sloped a minimum of 2 percent where located within 10 feet (3048 mm) of the building foundation. Impervious surfaces within 10 feet (3048 mm) of the building foundation shall be sloped a minimum of 2 percent away from the building. Exception: Where climatic or soil conditions warrant, the slope of the ground away from the building foundation shall be permitted to be reduced to not less than one unit vertical in 48 units horizontal (2-percent slope).
>
> The procedure used to establish the final ground level adjacent to the foundation shall account for additional settlement of the backfill.

This section of the *Code* also provides a regulation that restricts grading which will result in an increase in flood levels during the occurrence of the design flood (Section 1804.4). The earlier quote refers to a controlled low-strength material (CLSM), which is also known as *soil-cement slurry, soil-cement grout, flowable fill, controlled density fill, unshrinkable fill, K-Krete,* and other similar names (see ASTM D 4832). CLSM is typically used as backfill material around structures.

The geotechnical engineer will often provide much more extensive grading specifications then those listed earlier.

18.3.2   Compacted Fill

For compacted fill material, the *International Building Code* states:

> Section 1804.5 Compacted fill material. Where shallow foundations will bear on compacted fill material, the compacted fill shall comply with the provisions of an approved geotechnical report, as set forth in Section 1803. Exception: Compacted fill material 12 inches (305 mm) in depth or less need not comply with an approved report, provided the in-place dry density is not less than 90 percent of the maximum dry density at optimum moisture content in accordance with ASTM D 1557. The compaction shall be verified by special inspection in accordance with Section 1704.7.

Compaction regulations are also provided for foundations on a controlled low-strength material (Section 1804.6). Compacted fill material is also discussed in Section 1803.5.8 of the *International Building Code*, as follows:

Section 1803.5.8 Compacted fill material. Where shallow foundations will bear on compacted fill material more than 12 inches (305 mm) in depth, a geotechnical investigation shall be conducted and shall include all of the following:

1. Specifications for the preparation of the site prior to placement of compacted fill material.
2. Specifications for material to be used as compacted fill.
3. Test methods to be used to determine the maximum dry density and optimum moisture content of the material to be used as compacted fill.
4. Maximum allowable thickness of each lift of compacted fill material.
5. Field test method for determining the in-place dry density of the compacted fill.
6. Minimum acceptable in-place dry density expressed as a percentage of the maximum dry density determined in accordance with Item 3.
7. Number and frequency of field tests required to determine compliance with Item 6.

18.3.3 Appendix J: Grading

As previously mentioned, the *International Building Code* specifically states that the provisions contained in an appendix are not mandatory unless specifically referenced in the adopting ordinance. Appendix J of the *International Building Code* does provide additional grading recommendations, which may be required if they are adopted by local ordinance.

In terms of fill compaction, Appendix J of the *Code* states that all fill material shall be compacted to a relative compaction of 90 percent based on the Modified Proctor. This is the usual requirement in California, where for structural fill, the minimum relative compaction is 90 percent, based on the laboratory maximum dry density being determined by using the Modified Proctor specifications (i.e., ASTM D 1557-02, 2004). For some types of construction projects, such as the compaction of road subgrade or for the lower portions of deep canyon fill, a minimum relative compaction of 95 percent based on the Modified Proctor is often recommended by the geotechnical engineer.

Appendix J also provides regulations concerning the construction of fill slopes. For example, the construction of a fill slope over an existing slope that is steeper than an inclination of 5:1 (horizontal:vertical) requires the construction of a key and benching, such as shown in Fig. 18.1.

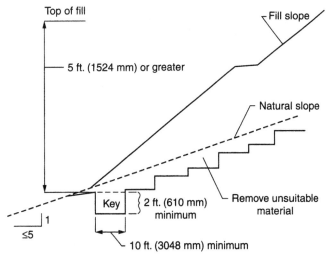

FIGURE 18.1 Benching details (*From International Building Code*)

Appendix J also deals with permits, inspections, excavations, setbacks, drainage, terracing, and erosion control regulations. If adopted by local ordinance, the reader should review Appendix J of the *Code* concerning these items.

18.4   PRESUMPTIVE LOAD-BEARING VALUES OF SOILS

The presumptive load-bearing values of soils are covered in Section 1806 of the *International Building Code*. These soil parameters are very important and are often used for foundation design and retaining wall design. The foundation report as prepared by the geotechnical engineer will usually specify these load-bearing values. For the load-bearing values, the *International Building Code* states:

> Section 1806.2 Presumptive load-bearing values. The load-bearing values used in design for supporting soils near the surface shall not exceed the values specified in Table 1806.2 unless data to substantiate the use of higher values are submitted and approved. Where the building official has reason to doubt the classification, strength or compressibility of the soil, the requirements of Section 1803.5.2 [i.e. a required geotechnical investigation] shall be satisfied.
>
> Presumptive load-bearing values shall apply to materials with similar physical characteristics and dispositions. Mud, organic silt, organic clays, peat or unprepared fill shall not be assumed to have a presumptive load-bearing capacity unless data to substantiate the use of such a value are submitted. Exception: A presumptive load-bearing capacity shall be permitted to be used where the building official deems the load-bearing capacity of mud, organic silt or unprepared fill is adequate for the support of lightweight or temporary structures.

Table 1806.2 from the *International Building Code* has been reproduced as Table 18.4 in this book. The presumptive load-bearing values listed in Table 18.4 are based on the type of rock (crystalline bedrock or sedimentary and foliated rock) and classification of soil using the Unified Soil Classification System. According to Section 1806.1 of the *International Building Code*, when using the alternative basic load combinations that include wind and earthquake loads, the vertical foundation pressures and lateral bearing pressures in Table 18.4 can be increased by one-third.

18.4.1   Vertical Foundation Pressure

As indicated in Table 18.4, the vertical foundation pressure can vary from 12,000 psf for crystalline bedrock to 1,500 psf for cohesive soil. Some of the limitations of the vertical foundation pressures listed in Table 18.4 are as follows:

- *Settlement considerations.*   Different structures may have different requirements in terms of the maximum amount of acceptable settlement. But the vertical foundation pressures in Table 18.4 do not consider the amount of settlement that would occur under the applied foundation loads.

- *Material density.*   The vertical foundation pressures do not consider the weathered condition of the rock or the density of the soil. For example, the values listed in Table 18.4 may be too high if the rock is highly fractured or weathered. Likewise, the values may be too high for soil that is in a loose or soft state.

- *Size and depth of footing.*   The vertical foundation pressures in Table 18.4 do not consider the size or depth of the footing. For example, as indicated in Chap. 6, for granular soils, the bearing capacity increases as the footing size increases and the depth of the footing increases (see Eq. 6.1). Thus for large size footings or where the bottom of the footing is well below ground surface, the values in Table 18.4 may be too low.

- *Seismic loads.*   An increase of one-third in the vertical foundation pressures in Table 18.4 is permitted for seismic loads. But as explained in Sec. 6.5.1 of this book, for some types of soils and rock, there may actually be a reduction in shear strength of the material during an earthquake. Such materials that lose strength during an earthquake should not be permitted to have a one-third increase in foundation bearing pressure and, in fact, a reduction may be appropriate.

TABLE 18.4   Presumptive Load-Bearing Values

Class of materials	Vertical foundation pressure (psf)	Lateral bearing pressure (psf/f below natural grade)	Lateral sliding resistance	
			Coefficient of friction[a]	Cohesion (psf)[b]
1. Crystalline bedrock	12,000	1,200	0.70	—
2. Sedimentary and foliated rock	4,000	400	0.35	—
3. Sandy gravel and/or gravel (GW and GP)	3,000	200	0.35	—
4. Sand, silty sand, clayey sand, silty gravel and clayey gravel (SW, SP, SM, SC, GM and GC)	2,000	150	0.25	—
5. Clay, sand clay, silty clay, clayey silt, silt and sandy silt (CL, ML, MH and CH)	1,500	100	—	130

For SI: 1 pound per square foot (psf) = 0.0479 kPa. 1 pound per square foot per foot of depth (psf/f) = 0.157 kPa/m.
[a]Coefficient to be multiplied by the dead load.
[b]Cohesion value to be multiplied by the contact area, as limited by Section 1806.3.2.
Source: Table 1806.2 of the *International Building Code*.

- *Footing weight.*   Based on Table 18.4, it is not clear if these are net or gross vertical foundation pressure values. In other words, it is not clear how the weight of the below-grade part of the foundation should be considered in the foundation bearing calculations. One approach is to recommend in the foundation engineering report that the weight of the below-grade concrete foundation elements can be neglected when using the vertical foundation pressure values in Table 18.4. The reason is that the unit weight of excavated soil or rock is typically assumed to be approximately equal to the unit weight of concrete.

18.4.2   Presumptive Lateral Bearing Pressure and Lateral Sliding Resistance

Table 18.4 also presents lateral bearing pressures and lateral sliding resistance values. Concerning these values, the *International Building Code* states:

> Section 1806.3.1. Combined resistance. The total resistance to lateral loads shall be permitted to be determined by combining the values derived from the lateral bearing pressure and the lateral sliding resistance specified in Table 1806.2 [see Table 18.4].
> Section 1806.3.2 Lateral sliding resistance limit. For clay, sandy clay, silty clay, clayey silt, silt and sandy silt, in no case shall the lateral sliding resistance exceed one-half the dead load.
> Section 1806.3.3 Increase for depth. The lateral bearing pressures specified in Table 1806.2 [see Table 18.4] shall be permitted to be increased by the tabular value for each additional foot (305 mm) of depth to a maximum of 15 times the tabular value.

Thus the lateral bearing pressure can be increased in a linear fashion as the footing depth is increased. Some of the limitations of the values in Table 18.4 are as follows:

- *Lateral deformation.*   In order to develop passive pressure in soils, the footing must deform laterally into the soil. Especially for clayey soils, the amount of lateral movement needed to develop the lateral bearing pressure values in Table 18.4 can be quite large.

- *Material density.*   Similar to the discussion on vertical foundation pressures, the lateral bearing pressures in Table 18.4 do not consider the weathered condition of the rock or the density of the soil. For example, the lateral bearing pressures in Table 18.4 may be too high if the rock is highly fractured or weathered. Likewise, the lateral bearing pressures may be too high for soil that is in a loose or soft state.

18.5   EXPANSIVE SOIL

The final section in this chapter deals with expansive soil, which has been discussed in Chap. 9 of this book. The *International Building Code* has regulations concerning the identification, treatment, and foundation construction on expansive soil.

18.5.1   Identification of Expansive Soil

In terms of identifying expansive soil, the *International Building Code* states (Section 1803.5.3): "In areas likely to have expansive soil, the building official shall require soil tests to determine where such soils do exist." The type of tests are identified in Section 1803.5.3, as follows:

> Soils meeting all four of the following provisions shall be considered expansive, except that tests to show compliance with Items 1, 2 and 3 shall not be required if the test prescribed in Item 4 is conducted:
>
> **1.** Plasticity index (PI) of 15 or greater, determined in accordance with ASTM D 4318.
> **2.** More than 10 percent of the soil particles pass a No. 200 sieve (75 μm), determined in accordance with ASTM D 422.
> **3.** More than 10 percent of the soil particles are less than 5 micrometers in size, determined in accordance with ASTM D 422.
> **4.** Expansion index greater than 20, determined in accordance with ASTM D 4829.

As the above regulation indicates, the Expansion Index Test (ASTM D 4829) can be used to determine the presence of expansive soils, which are defined as those soils having an expansion index greater than 20. The Expansion Index Test has been discussed in Sec. 9.2.2 of this book. It should also be mentioned that for Item Nos. 2 and 3, checking item No. 2 (i.e. > 10 percent passing No. 200 sieve) is unnecessary because item No. 3 (i.e. > 10 percent finer than 5 μm) will always govern.

According to the above quote, there are two options for determining the presence of expansive soils at a site:

First Option.   The first option would be to simply perform an Expansion Index Test (ASTM D 4829) and if the expansion index is greater than 20, the site contains expansive soil.

Second Option.   This option entails performing several different laboratory tests: Atterberg Limits tests (ASTM D 4318) in order to determine the plasticity index of the soil, sieve and hydrometer tests (ASTM D 422) in order to obtain the gradation curve, and finally an Expansion Index Test. In order for the soil to be classified as expansive, it would have to meet all four of the criteria listed earlier. Because a plasticity index of 15 often corresponds to an expansion index of around 40 to 50 (see Figure 9.11), this more rigorous option could actually be less restrictive.

18.5.2   Treatment of Expansive Soil

After having identified that the site contains expansive soils, the next step would be to take action to treat the soils or to build adequate foundations to resist the expansive soil forces. In terms of the treatment of expansive soils, the *International Building Code* provides two alternatives, as follows:

Alternative 1: Removal of Expansive Soil.   For this option, the *Code* provision (Section 1808.6.3) states: "soil shall be removed to a depth sufficient to ensure a constant moisture content in the remaining soil. Fill material shall not contain expansive soils." As discussed in Sec. 9.3.1, the depth to constant moisture content is known as the *depth of seasonal moisture change*, also referred to as the *depth of the active zone*. This depth can be quite large depending on the nature of the expansive soil and the climate conditions. Once the expansive soil is removed to the depth of seasonal moisture change, the soil must be replaced with fill material that does not contain expansive material. Section 1808.6.3 of the *Code* mentions an exception, as follows:

> Exception: Expansive soil need not be removed to the depth of constant moisture, provided the confining pressure in the expansive soil created by the fill and supported structure exceeds the swell pressure.

This exception requires the determination of the swell pressure from laboratory tests performed on the expansive soil (see Fig. 9.28 of this book). Then the depth H of expansive soil removal could be calculated as follows:

$$\sigma_s = \sigma_z + \sigma_v = \sigma_z + [(H - D)\gamma_t] \qquad (18.1)$$

where σ_s = swell pressure obtained from laboratory tests (psf or kPa)
  σ_z = dead load of the structure, converted to an average pressure exerted by the foundation, and then projected to the required depth by using a stress distribution method such as the 2:1 Approximation (psf or kPa)
  σ_v = vertical pressure of the column of soil located from the bottom of the foundation to the depth H (psf or kPa)
  D = depth from ground surface to the bottom of the foundation (ft or m)
  H = total depth of removal of expansive soil (ft or m)
  γ_t = total unit weight of imported and compacted nonexpansive fill (pcf or kN/m^3)

Equation 18.1 assumes that the groundwater table is located below a depth H. The known values in Eq. 18.1 are the swelling pressure σ_s, dead load of the structure, depth of the proposed foundation below ground surface (D), and total unit weight γ_t of the imported nonexpansive fill to be placed and compacted below the proposed foundation. The unknown in Eq. 18.1 is the total depth of removal (H). However, since a stress distribution method (such as the 2:1 Approximation) must be utilized, trial and error will be needed to solve for H. The trial and error process will begin by assuming a value of H, then calculating σ_z, and ultimately calculating the sum of $\sigma_z + \sigma_v$ and comparing it to σ_s. See Example Problem 18.1 on the following page.

Alternative 2: Expansive Soil Stabilization.   The second alternative for the treatment of expansive soil is to use stabilization techniques. Various stabilization methods have been presented in Table 9.4. The *International Building Code* states that the soil in the active zone (i.e., depth of seasonal moisture change) shall be stabilized by chemical, dewatering, presaturation, or equivalent techniques.

18.5.3   Foundations on Expansive Soil

Instead of the treatment of expansive soil, special foundation systems can be constructed to resist the expansive soil forces. *The International Building Code* states:

> Section 1808.6.1 Foundations. Foundations placed on or within the active zone of expansive soils shall be designed to resist differential volume changes and to prevent structural damage to the supported structure. Deflection and racking of the supported structure shall be limited to that which will not interfere with the usability and serviceability of the structure.
>
> Foundations placed below where volume change occurs or below expansive soil shall comply with the following provisions:
>
> **1.** Foundations extending into or penetrating expansive soils shall be designed to prevent uplift of the supported structure.
> **2.** Foundations penetrating expansive soils shall be designed to resist forces exerted on the foundation due to soil volume changes or shall be isolated from the expansive soil.

Example Problem 18.1   Assume the following:

1. Expansive soil having a uniform swell pressure versus depth of 2100 psf (100 kPa)
2. Mat foundation, where the bottom of the mat will be located 2 ft (0.6 m) below ground surface (i.e., $D = 2$ ft). The size of the mat foundation is 100 ft by 100 ft (30 m by 30 m).
3. Dead load of building, including weight of foundation, divided by the area of the foundation = 1000 psf (48 kPa). In essence, the bottom of the mat foundation will exert a pressure of 1000 psf (48 kPa) onto the soil.
4. Total unit weight of imported and compacted nonexpansive soil = 130 pcf (20 kN/m³)

Determine the depth of expansive soil removal (H).

Solution

Assume a depth of removal (H) = 12 ft (3.7 m). Thus the distance from the bottom of the mat to the top of the expansive soil = 12 – 2 = 10 ft (3.0 m). Using the 2:1 Approximation equation:

$$\sigma_z = P/[(B + z)(L + z)]$$

where P = total dead load of the structure = (1000 psf)(100 ft)(100 ft) = 10,000,000 lb
    B = width of the mat = 100 ft
    L = length of the mat = 100 ft
    z = distance from bottom of mat to top of expansive soil = 10 ft
    $\sigma_z = P/[(B + z)(L + z)] = (10,000,000)/[(100 + 10)(100 + 10)] = 825$ psf

Using Eq. 18.1:

$$\sigma_z + \sigma_v = \sigma_z + [(H - D)\gamma_t] = 825 + [(12 - 2)(130)] = 2125 \text{ psf } (102 \text{ kPa})$$

Since this value slightly exceeds the swell pressure, a depth of removal of 12 ft (3.7 m) would be required. It should be noted that a swell pressure of 2100 psf (100 kPa) is rather low and highly expansive soils often have much higher swell pressures. Based on Eq. 18.1 the depth of expansive soil removals could become quite excessive.

Section 1808.6.2 Slab-on-ground foundations. Moments, shears and deflections for use in designing slab-on-ground, mat or raft foundations on expansive soils shall be determined in accordance with *WRI/CRSI Design of Slab-on-Ground Foundations* or *PTI Standard Requirements for Analysis of Shallow Concrete Foundations on Expansive Soils*. Using the moments, shears and deflections determined above, nonprestressed slabs-on-ground, mat or raft foundations on expansive soils shall be designed in accordance with *WRI/CRSI Design of Slab-on-Ground Foundations* and post-tensioned slab-on-ground, mat or raft foundations on expansive soil shall be designed in accordance with *PTI Standard Requirements for Design of Shallow Post-Tensioned Concrete Foundations on Expansive Soils*. It shall be permitted to analyze and design such slabs by other methods that account for soil-structure interaction, the deformed shape of the soil support, the plate or stiffened plate action of the slab as well as both center lift and edge lift conditions. Such alternative methods shall be rational and the basis for all aspects and parameters of the method shall be available for peer review.

Chapter 9 of this book has presented a discussion of post-tensioned slabs on expansive soil.

CHAPTER 19
INTERNATIONAL BUILDING CODE REGULATIONS FOR FOUNDATIONS

19.1 INTRODUCTION

Chapter 18 of this book has presented a discussion of some regulations that are directly applicable to the geotechnical aspects of foundations. For example, Table 18.1 presents special inspection regulations for deep foundations. Likewise, for the sake of continuity, Sec. 18.5 has discussed regulations for foundations supported by expansive soil.

This chapter will mainly deal with the other foundation regulations contained in Chapter 18 (Soils and Foundations) of the *International Building Code*. Topics will include general regulations for footings and foundations, foundations adjacent slopes, retaining walls, and geotechnical earthquake engineering.

19.2 GENERAL REGULATIONS FOR FOOTINGS AND FOUNDATIONS

The main regulations for foundations in the *International Building Code* are located in Section 1808 (Foundations), Section 1809 (Shallow Foundations), and Section 1810 (Deep Foundations). Specific general regulations for foundations in the *International Building Code* are as follows:

- *Design for capacity and settlement (Section 1808.2).* This *Code* section states: "Foundations shall be so designed that the allowable bearing capacity of the soil is not exceeded, and that differential settlement is minimized."

- *Design loads (Section 1808.3).* This *Code* section requires that foundations be designed for the most unfavorable effects due to the combinations of loads.

- *Vibratory loads (Section 1808.4).* This *Code* section states: "Where machinery operations or other vibrations are transmitted through the foundation, consideration shall be given in the foundation design to prevent detrimental disturbances of the soil."

- *Shifting or moving soils (Section 1808.5).* This *Code* section states: "Where it is known that the shallow subsoils are of a shifting or moving character, foundations shall be carried to a sufficient depth to ensure stability."

19.2.1 Shallow Foundations

The *International Building Code* requires that shallow foundations be built on undisturbed soil, compacted fill, or CLSM material (Section 1809.2). During construction, it is common for loose soil or

19.1

debris to be knocked into the footing excavations during the construction process or during the installation of the steel reinforcement. It is important that prior to placing concrete, the footings be cleaned of loose debris so that the foundation will bear on undisturbed soil or compacted fill.

In terms of the top and bottom surfaces of shallow foundations, the *International Building Code* states (Section 1809.3):

> The top surface of footings shall be level. The bottom surface of footings shall be permitted to have a slope not exceeding one unit vertical in 10 units horizontal (10-percent slope). Footings shall be stepped where it is necessary to change the elevation of the top surface of the footing or where the surface of the ground slopes more than one unit vertical in 10 units horizontal (10-percent slope).

The *International Building Code* also states that the minimum depth of shallow footings below the undisturbed ground surface shall be 12 in. (305 mm) and that the foundations be protected from frost (Sections 1809.4 and 1809.5). Regulations for footings on granular soils are as follows:

> 1809.6 Location of footings. Footings on granular soil shall be so located that the line drawn between the lower edges of adjoining footings shall not have a slope steeper that 30 degrees (0.52 rad) with the horizontal, unless that material supporting the higher footing is braced or retained or otherwise laterally supported in an approved manner or a greater slope has been properly established by engineering analysis.

19.2.2 Deep Foundations

Deep foundations have been covered in Secs. 5.4, 6.3, 6.4, 16.2, and 16.3 of this book. In terms of the general requirements for deep foundations, the *International Building Code* states:

> Section 1803.5.5 Deep foundations. Where deep foundations will be used, a geotechnical investigation shall be conducted and shall include all of the following, unless sufficient data upon which to base the design and installation is otherwise available:
>
> **1.** Recommended deep foundation types and installed capacities.
> **2.** Recommended center-to-center spacing of deep foundation elements.
> **3.** Driving criteria.
> **4.** Installation procedures.
> **5.** Field inspection and reporting procedures (to include procedures for verification of the installed bearing capacity where required).
> **6.** Load test requirements.
> **7.** Suitability of deep foundation materials for the intended environment.
> **8.** Designation of bearing stratum or strata.
> **9.** Reductions for group action, where necessary.

In terms of determining the allowable load, the *International Building Code* states in Section 1810.3.3: "The allowable axial and lateral loads on a deep foundation element shall be determined by an approved formula, load tests or method of analysis." For additional details on the design and construction of deep foundations, see Section 1810 of the *International Building Code*.

19.3 FOUNDATIONS ADJACENT SLOPES

There are very few regulations in the *International Building Code* for slope stability. As discussed in Chap. 10 of this book, slope stability is an important part of geotechnical and foundation engineering. In order to assess the safety of a slope, the geotechnical engineer will need to perform a slope stability analysis. The slope stability analysis should include likely changes that will develop during and after the proposed construction, such as a rise in the groundwater table that would decrease the factor of safety of the slope. As discussed in Chap. 10, the minimum acceptable factor of safety for permanent slopes is 1.5. A lower factor of safety may be acceptable for temporary slopes.

In general, the main regulation in the *International Building Code* (Section 3304.1.1) states that permanent fill slopes and permanent cut slopes shall not be steeper than one unit vertical in two units horizontal (50-percent slope). Another regulation applies to minimum foundation setback, which will be discussed later in this section. Simply having a maximum slope inclination of 50 percent with minimum foundation setbacks will not insure the safety of a site. The geotechnical engineer should perform slope stability analyses, such as those presented in Chap. 10 of this book, to determine the factor of safety of the slope and evaluate the potential for lateral movement. When subjected to seismic shaking, the stability of slopes is reduced; and geotechnical earthquake engineering analyses for slopes should be performed, as indicated in Sec. 13.5 of this book.

Section 1808.7 of the *International Building Code* deals with foundation setbacks for slopes. This section of the *International Building Code* makes reference to Figure 1808.7.1, which has been reproduced in this book as Figure 19.1. Section 1808.7 of the *International Building Code* is reproduced below:

Section 1808.7 Foundations on or adjacent to slopes. The placement of buildings and structures on or adjacent to slopes steeper than one unit vertical in three units horizontal (33.3-percent slope) shall comply to Sections 1808.7.1 through 1808.7.5.

Section 1808.7.1 Building clearance from ascending slopes. In general, buildings below slopes shall be set a sufficient distance from the slope to provide protection from slope drainage, erosion and shallow failures. Except as provided for in Section 1808.7.5 and Figure 1808.7.1 [see Figure 19.1], the following criteria will be assumed to provide this protection. Where the existing slope is steeper than one unit vertical in one unit horizontal (100-percent slope), the toe of the slope shall be assumed to be at the intersection of a horizontal plane drawn from the top of the foundation and a plane drawn tangent to the slope at an angle of 45 degrees (0.79 rad) to the horizontal. Where a retaining wall is constructed at the toe of the slope, the height of the slope shall be measured from the top of the wall to the top of the slope.

Section 1808.7.2 Foundation setback from descending slope surface. Foundations on or adjacent to slope surfaces shall be founded in firm material with an embedment and set back from the slope surface sufficient to provide vertical and lateral support for the foundation without detrimental settlement. Except as provided for in Section 1808.7.5 and Figure 1808.7.1 [see Figure 19.1], the following setback is deemed adequate to meet the criteria. Where the slope is steeper than 1 unit vertical in 1 unit horizontal (100-percent slope), the required setback shall be measured from an imaginary plane 45 degrees (0.79 rad) to the horizontal, projected upward from the toe of the slope.

Section 1808.7.5 Alternate setback and clearance. Alternate setbacks and clearances are permitted, subject to the approval of the building official. The building official shall be permitted to require a geotechnical investigation as set forth is Section 1803.5.10.

Section 1803.5.10 Alternate setback and clearance. Where setbacks or clearances other than those required in Section 1808.7 are desired, the building official shall be permitted to require a geotechnical investigation by a registered design professional to demonstrate that the intent of Section 1808.7 would be satisfied. Such an investigation shall include consideration of material, height of slope, slope gradient, load intensity and erosion characteristics of slope material.

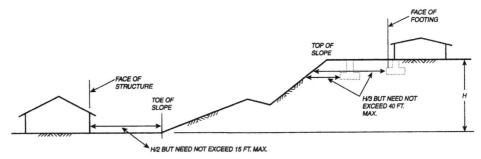

FIGURE 19.1 Foundation clearances from slopes. (From the *International Building Code.*)

As previously mentioned, the *International Building Code* requires that both fill and cut slopes have maximum inclinations of one unit vertical in two units horizontal (50 percent slope). Thus for most slopes, the discussion in Section 1808.7.1 and 1808.7.2 dealing with slopes steeper than 1 unit vertical in 1 unit horizontal (100-percent slope) will not be applicable. For the usual situation of a 2:1 (50-percent slope) or flatter slope, the setback requirements will be as shown in Figure 1808.7.1 (i.e. Figure 19.1 in this book).

As shown in Figure 19.1, the setback for structures at the toe of the slope is easy to determine and simply consists of a horizontal distance "at least the smaller of H/2 and 15 ft [4.6 m]," where H = height of the slope. At the top of the slope, the required setback is more complicated. The horizontal setback is measured from the face of the footing to the face of the slope and must be "at least the smaller of H/3 and 40 ft [12 m]." If, because of property size constraints, the building must be close to the top of slope, then the perimeter footing can be simply deepened in order to meet the requirements of Figure 19.1.

Note that as mentioned in the earlier discussion, there is no building code regulation for a minimum factor of safety for slope stability. Nevertheless, the geotechnical engineer should evaluate the stability of slopes that will potentially impact the proposed development.

19.4   RETAINING WALLS

Retaining walls have been covered in Chap. 11 and Sec. 14.4 of this book. The main regulations in the *International Building Code* for retaining walls are Section 1610 (Soil Lateral Loads) and Section 1807 (Foundation Walls, Retaining Walls and Embedded Posts and Poles).

Table 19.1 presents lateral soils loads for the design of foundation walls and retaining walls per the *International Building Code*. Concerning these lateral soil loads, the *International Building Code* states:

> Section 1610.1 General. Foundation walls and retaining walls shall be designed to resist lateral soil loads. Soil loads specified in Table 1610.2 [see Table 19.1] shall be used as the minimum design lateral soil loads unless determined otherwise by a geotechnical investigation in accordance with Section 1803. Foundation walls and other walls in which horizontal movement is restricted at the top shall be designed for at-rest pressures. Retaining walls free to move and rotate at the top shall be permitted to be designed for active pressure. Design lateral pressure from surcharge loads shall be added to the lateral earth pressure load. Design lateral pressure shall be increased if soils at the site are expansive. Foundation walls shall be designed to support the weight of the full hydrostatic pressure of undrained backfill unless a drainage system is installed in accordance with Sections 1805.4.2 and 1805.4.3. Exception: Foundation walls extending not more than 8 feet (2438 mm) below grade and laterally supported at the top by flexible diaphragms shall be permitted to be designed for active pressure.

Although the *Code* (see Table 19.1) allows the use of clayey soils (i.e. GC, SM-SC, SC, ML, ML-CL, and CL) as backfill materials, clayey soils should generally not be used as retaining wall backfill material because of the following reasons:

- *Predictable behavior.* Import granular backfill generally has a more predictable behavior in terms of the earth pressure exerted on the wall.

- *Expansive soil forces.* Expansive soil related forces would not be generated by clean granular soil. However, if clay backfill is used, the seepage of water into the backfill could cause swelling pressures well in excess of the at-rest values listed in Table 19.1.

- *Excessive rotation of the top of wall.* As indicated in Table 11.1 of this book, the rotation Y/H (where Y = wall displacement and H = height of wall) to reach the active state for dense cohesionless soil is 0.0005, while the value of Y/H is 0.02 for soft cohesive soil. Hence, given a wall of the same height, the top of the wall will need to move horizontally about 40 times more for soft cohesive backfill as compared to dense granular backfill in order to reach the active state.

TABLE 19.1 Lateral Soil Load

Description of backfill material[c]	Unified soil classification	Design lateral soil load[a] (pound per square foot per foot of depth)	
		Active pressure	At-rest pressure
Well-graded, clean gravels; gravel-sand mixes	GW	30	60
Poorly graded clean gravels; gravel-sand mixes	GP	30	60
Silty gravels, poorly graded gravel-sand mixes	GM	40	60
Clayey gravels, poorly graded gravel-and-clay mixes	GC	45	60
Well-graded, clean sands; gravelly sand mixes	SW	30	60
Poorly graded clean sands; sand-gravel mixes	SP	30	60
Silty sands, poorly graded sand-silt mixes	SM	45	60
Sand-silt clay mix with plastic fines	SM-SC	45	100
Clayey sands, poorly graded sand-clay mixes	SC	60	100
Inorganic silts and clayey silts	ML	45	100
Mixture of inorganic silt and clay	ML-CL	60	100
Inorganic clays of low to medium plasticity	CL	60	100
Organic silts and silt clays, low plasticity	OL	Note b	Note b
Inorganic clayey silts, elastic silts	MH	Note b	Note b
Inorganic clays of high plasticity	CH	Note b	Note b
Organic clays and silty clays	OH	Note b	Note b

For SI: 1 pound per square foot per foot of depth = 0.157 kPa/m. 1 foot = 304.8 mm.

[a]Design lateral soil loads are given for moist conditions for the specific soils at their optimum densities. Actual field conditions shall govern. Submerged or saturated soil pressures shall include the weight of the buoyant soil plus the hydrostatic loads.

[b]Unsuitable as backfill material.

[c]The definition and classification of soil materials shall be in accordance with ASTM D 2487 (Unified Soil Classification System).

Source: Table 1610.1 of the *International Building Code.*

- *Drainage system.* Retaining walls usually are constructed with drainage systems to prevent the buildup of hydrostatic water pressure on the retaining wall. The drainage system will be more effective if highly permeable soil, such as clean granular soil, is used instead of clayey backfill.

- *Frost action.* If freezing temperatures prevail, the backfill soil can be susceptible to frost action, where ice lenses will form parallel to the wall. Backfill soil consisting of clean granular soil and the installation of a drainage system at the heel of the wall will be much more effective in preventing frost action then using clayey backfill.

Additional regulations concerning retaining walls are presented in Section 1807 of the *International Building Code.* Concerning the design of retaining walls, the *International Building Code* states:

> Section 1807.2.1 General. Retaining walls shall be designed to ensure stability against overturning, sliding, excessive foundation pressure and water uplift. Where a keyway is extended below the wall base with the intent to engage passive pressure and enhance sliding stability, lateral soil pressures on both sides of the keyway shall be considered in the sliding analysis.
>
> Section 1807.2.2 Design lateral soil loads. Retaining walls shall be designed for the lateral soil loads set forth in Section 1610 [see Table 19.1].
>
> Section 1807.2.3 Safety factor. Retaining walls shall be designed to resist the lateral action of soil to produce sliding and overturning with a minimum safety factor of 1.5 in each case. The load combinations of Section 1605 shall not apply to this requirement. Instead, design shall be based on 0.7 times nominal earthquake loads, 1.0 times other nominal loads, and investigation with one or more variable loads set to zero. The safety factor against lateral sliding shall be taken as the available soil resistance at the base of

the retaining wall foundation divided by the net lateral force applied to the retaining wall. Exception: Where earthquake loads are included, the minimum safety factor for retaining wall sliding and overturning shall be 1.1.

19.5   GEOTECHNICAL EARTHQUAKE ENGINEERING

19.5.1   Introduction

Geotechnical earthquake engineering has been covered in Chaps. 13 and 14 of this book. This last section presents a discussion of the role of building codes in geotechnical earthquake engineering. The geotechnical engineer should always review local building codes and other regulatory specifications that may govern the seismic design of the project. These local requirements may be more stringent than the regulations contained in the *International Building Code*.

Types of information that could be included in the building code or other regulatory documents are as follows:

1. *Earthquake Potential.*   Local building requirements may specify the earthquake potential for a given site. The seismic potential often changes as new earthquake data is evaluated. For example, as discussed in Sec. 14.4.3, one of the main factors that contributed to the damage at the Port of Kobe during the Kobe Earthquake was that the area had been previously considered to have a relatively low seismic risk; hence the earthquake design criteria was less stringent than in other areas of Japan.

2. *General Requirements.*   The building code could also specify general requirements that must be fulfilled by the geotechnical engineer. For example, the *International Building Code* states the geotechnical investigation shall include the following items (Section 1803.5.12):

   a. The determination of lateral pressures on foundation walls and retaining walls due to earthquake motions.

   b. The potential for liquefaction and soil strength loss evaluated for site peak ground accelerations, magnitudes and source characteristics consistent with the design earthquake ground motions. Peak ground acceleration shall be permitted to be determined based on a site-specific study taking into account soil amplification effects.

   c. An assessment of potential consequences of liquefaction and soil strength loss, including estimation of differential settlement, lateral movement, lateral loads on foundations, reduction in foundation soil-bearing capacity, increases in lateral pressures on retaining walls and flotation of buried structures.

   d. Discussion of mitigation measures such as, but not limited to, ground stabilization, selection of appropriate foundation type and depths, selection of appropriate structural systems to accommodate anticipated displacements and forces, or any combination of these measures and how they shall be considered in the design of the structure.

3. *Detailed Analyses.*   The building code could also provide detailed seismic analyses. For example, Table 19.2 presents data that can be used to determine the site class definition per the *International Building Code*.

The site class is based on the average condition of the material that exists at the site from ground surface to a depth of 100 ft (30 m). The best site class is class A, consisting of hard rock, and the worst site class is class F, where there are soil profiles that may liquefy during the earthquake or there are soft soils that can increase the peak ground acceleration (see Sec. 13.3.4 of this book).

If the ground surface will be raised or lowered by grading operations, then the site class analysis should be based on the final as-built conditions. As indicated in Table 19.2, the selection of the site class is based on the material type and engineering properties, such as the shear wave velocity, standard penetration test values, and the undrained shear strength.

Profiles containing distinctly different soil and/or rock layers should be subdivided into layers with the average conditions in the upper 100 feet (30 m) of the profile based on the thickness of the individual layers. Equations (16-40 to 16-43) in the *International Building Code* can be used to calculate average values when there are distinctly different soil and/or rock layers at the site. The procedure for determining the site class is as follows:

1. *Site class F*:   Start with the four categories listed under site class F (see Table 19.2). If the site meets any one of these four categories, then the site is designated as site class F.

2. *Site class E*:   If a site is not a site class F, then check to see if the site meets the criteria for the definition of site class E in Table 19.2, i.e. a soft clay layer of more that 10 feet in thickness meeting the plasticity index, moisture content, and undrained shear strength criteria.

3. *Site classes C, D, and E*:   If a site does not conform to the two previous items, then determine the average shear wave velocity, standard penetration resistance, and/or the undrained shear strength. As indicated in Table 19.2, site class C has the best soil properties (i.e. very dense soil), while site class E has poor soil properties (i.e. soft soil profile). Engineering properties of the soil are used to evaluate the site class as follows:

Shear Wave Velocity: The shear wave velocity can be measured in situ by using several different geophysical techniques, such as the uphole, down-hole, or cross-hole methods (see Sec. 2.7 of this book). When using the shear wave velocity, it is best to use V_{1s}, (ft/s), which is corrected for the overburden pressure (see Eq. 6.9, Day (2002), *Geotechnical Earthquake Engineering Handbook*).

Standard Penetration Test (SPT): The *International Building Code* states that the standard penetration resistance (ASTM D 1586), as directly measured in the field without corrections, should be used for Table 19.2. However, it is best to use the SPT values that are corrected for both sam-

TABLE 19.2   Site Class Definitions

Site Class	Soil Profile Name	Shear wave velocity, V_{1s}, (ft/s)	Average Properties in Top 100 feet*	
			Standard penetration resistance, $(N_1)_{60}$	Soil undrained shear strength, s_u, (psf)
A	Hard rock	$V_{1s} > 5,000$	N/A	N/A
B	Rock	$2,500 < V_{1s} \leq 5,000$	N/A	N/A
C	Very dense soil and soft rock	$1,200 < V_{1s} \leq 2,500$	$(N_1)_{60} > 50$	$s_u > 2,000$
D	Stiff soil profile	$600 \leq V_{1s} \leq 1,200$	$15 \leq (N_1)_{60} \leq 50$	$1,000 \leq s_u \leq 2,000$
E	Soft soil profile	$V_{1s} < 600$	$(N_1)_{60} < 15$	$s_u < 1,000$
E	Any profile with more than 10 feet of soil having the following characteristics: plasticity index PI > 20, moisture content w ≥ 40%, and undrained shear strength s_u < 500 psf			
F	Any profile containing soils having one or more of the following characteristics:			

Any profile containing soils having one or more of the following characteristics:

1. Soils vulnerable to potential failure or collapse under seismic loading, such as liquefiable soils, quick and highly sensitive clays, collapsible weakly cemented soils.
2. Peats and/or highly organic clays (H > 10 feet of peat and/or highly organic clay, where H = thickness of soil).
3. Very high plasticity clays (H > 25 feet with plasticity index PI > 75).
4. Very thick soft/medium stiff clays (H > 120 feet).

Note: *See Section 1613.5.5 for further details. For SI: 1 foot = 304.8 mm, 1 square foot = 0.0929 m², 1 pound per square foot = 0.0479 kPa. N/A = Not applicable

Source: Table 1613.5.2 of the *International Building Code*.

pling procedures and overburden pressure, i.e. $(N_1)_{60}$ values (see Eq. 2.5 in this book), because the $(N_1)_{60}$ value is a more reliable indicator of the density of granular soil than uncorrected SPT values.

Undrained Shear Strength: The undrained shear strength has been discussed in Sec. 4.6.2 of this book. The *International Building Code* states that the undrained shear strength (s_u) is to be determined in accordance with ASTM D 2166 (unconfined compression test) or ASTM D 2850 (unconsolidated-undrained triaxial compression test).

4. *Site classes A and B.*   The rock categories A and B should not be used if there are more than 10 feet (3 m) of soil between the rock surface and the bottom of the foundation.

The *International Building Code* provides figures (e.g. Figure 1613.5) that delineate 0.2-second spectral acceleration values. These spectral accelerations are considered to be applicable for firm rock sites (i.e. site class B material). In the discussion printed on Figure 1613.5 of the *International Building Code*, there is information on how to obtain coefficients that allow the user to adjust the spectral response acceleration for different site classes. The structural engineer will use the spectral response acceleration in the seismic design of the building.

19.5.2   Code Development

One of the most important methods of code development is to observe the performance of structures during earthquakes. There must be a desire to improve conditions and not simply accept the death and destruction from earthquakes as inevitable. Two examples of the impact of earthquakes on codes and regulations are as follows:

1. *March 10, 1933 Long Beach earthquake in California.*   This earthquake brought an end to the practice of laying brick masonry without reinforcing steel. Prior to this earthquake, the exterior walls of building were often of brick, or in some cases hollow clay tile. Wood was used to construct the roofs and floors that were supported by the brick walls. This type of construction was used for schools and the destruction to these schools was some of the most spectacular damage during the 1933 Long Beach Earthquake. Fortunately, the earthquake occurred after school hours and a catastrophic loss of life was averted. However, the destruction was so extensive and had such dire consequences that the California legislature passed the Field Law on April 10, 1933. This law required that all new public schools be constructed so that they are highly resistant to earthquakes. The Field Law also required that there be field supervision during the construction of schools.

2. *February 9, 1971 San Fernando Earthquake in California.*   Because of the damage caused by this earthquake, building codes were strengthened and the California legislature passed the Alquist Priolo Special Studies Zone Act in 1972. The purpose of this act is to prohibit the construction of structures for human occupancy across the traces of active faults. The goal of this legislation is to mitigate the hazards caused by fault rupture.

There has also been a considerable amount of federal legislation in response to earthquake damage. For example, the Federal Emergency Management Agency (1994) states:

> At the federal level, there are two important pieces of legislation relating to local seismic hazard assessment. These are Public Law 93-288, amended in 1988 as the Stafford Act, which establishes basic rules for federal disaster assistance and relief, and the Earthquake Hazards Reduction Act of 1977, amended in 1990, which establishes the National Earthquake Hazards Reduction Program (NEHRP).
>   The Stafford Act briefly mentions "construction and land use" as possible mitigation measures to be used after a disaster to forestall repetition of damage and destruction in subsequent events. However, the final rules promulgated by the Federal Emergency Management Agency (FEMA) to implement the Stafford Act (44 CFR Part 206, Subparts M and N) require post-disaster state-local hazard mitigation plans to be prepared as a prerequisite for local governments to receive disaster assistance funds to repair and restore damaged or destroyed public facilities. Under the regulations implementing Section 409 of the Stafford

Act, a city or county must adopt a hazard mitigation plan acceptable to FEMA if it is to receive facilities restoration assistance authorized under Section 406.

The overall purpose of the National Earthquake Hazards Reduction Act is to reduce risks to life and property from earthquakes. This is to be carried out through activities such as: hazard identification and vulnerability studies; development and dissemination of seismic design and construction standards; development of an earthquake prediction capability; preparation of national, state and local plans for mitigation, preparedness and response; conduct basic and applied research into causes and implications of earthquake hazards; and, education of the public about earthquakes. While this bears less directly on earthquake preparation for a particular local government, much of the growing body of earthquake-related scientific and engineering knowledge has been developed through NEHRP funded research, including this study.

19.5.3 Limitations of Building Codes

Common limitations of building codes are that they may not be up to date or may underestimate the potential for earthquake shaking at a particular area. In addition, the building codes may not be technically sound or they may contain loopholes that can be exploited by developers. For example, in terms of the collapse of structures caused by the Chi-chi Earthquake in Taiwan on September 21, 1999, Hands (1999) states:

> Why then were so many of these collapses occurring in 12-story buildings? Was it, as the local media suggested, a result of seismic waves hitting just the right resonant frequency to take them out? Professor Chern dismisses this as bordering on superstition. "Basically Taiwan has a lot of 12-story buildings, especially central Taiwan. You hardly see any 20-story high-rises in those areas hit by the quake. The reason for this is that buildings under 50 meters in height don't have to go to a special engineering committee to be approved, so 12 stories is just right." Approval of a structure by qualified structural engineers, and correct enforcement of the building codes, is the crux of the problem, Chern believes.

Another example is the Kobe Earthquake in Japan on January 17, 1995. It was observed that a large number of 20-year and older high-rise buildings collapsed at the fifth floor. The cause of these building collapses was apparently an older version of the building code that allowed a weaker superstructure beginning at the fifth floor.

Even with a technically sound building code without loopholes, there could be many other factors that are needed to produce earthquake-resistance structures, as follows:

1. *Qualified engineers.* There must be qualified structural and geotechnical engineers that can prepare seismic designs and building plans. However, the availability of a professional engineering group will not insure adequate designs. For example, concerning the collapse of structures caused by the Chi-chi Earthquake in Taiwan on September 21, 1999, Hands (1999) states:

> Professor Chern is particularly damning of some of his fellow engineers, and the professional associations to which they belong. "In 1997 we had 6,300 registered civil engineers. Three hundred of them are working in their own consultancies, and 2,800 are employed by building contractors.
>
> That means that the other 3,300, or more than half, are possibly renting their licenses." Asked to explain further, Chern said that it was common practice for an engineer to rent his engineer's license to a building contractor, so that the contractor could then claim the architectural drawings had been approved by a qualified engineer, without the engineer even having seen the blueprints. Chern sees the problem as stemming from the way the engineers' professional associations are run. "When they elect a president of the association, the candidate who favors license-renting will get all the votes from those people and win the election, and then he won't be willing to do anything about the problem."

2. *Permit process.* After the engineers have prepared the structural plans and specifications, they must be reviewed and approved by the governing agency. The local jurisdiction should have qualified engineers that review the designs to ensure that proper actions are taken to mitigate the impact of seismic hazards, to evaluate structural and nonstructural seismic design and construction practices so that they minimize earthquake damage in critical facilities, and to prevent the total collapse of any structure designed for human occupancy. An important aspect of the permit process is that the governing agency has the power to deny construction of the project if it is deemed to be below the standard of practice.

3. *Inspection during construction.*   Similar to the permit process, there must be adequate inspection during the construction of the project to ensure that the approved building plans and specifications are being followed. Any proposed changes to the approved building plans and specifications would have to be reviewed by the governing agency. The project engineers should issue final reports in order to certify that the structure was built in conformance with the approved building plans.

4. *Construction industry.*   An experienced workforce that will follow the approved plans and specifications is needed during construction. In addition, there must be available materials that meet project requirements in terms of quality, strength, and the like. An example of lax construction is as follows (Hands, 1999):

> Professor Chern said the construction industry is riddled with problems from top to bottom. Even the concrete has problems. "In Taiwan we have quite narrow columns with a lot of rebar in them. This makes it difficult to pour the concrete and get it through and into all the spaces between the bars. Just imagine it—you usually have a small contractor doing the pouring, maybe five men with one pumping car, with two doing the vibrating. They pour 400 cubic meters in one day, and only make NT$5,000 for one morning's work."
>
> It's also a manpower quality problem, he said. "You have low quality workers on low pay, so everything is done quickly. Very good concrete is viscous, so they add water to ready-mixed concrete to make it flow better. But then you get segregation of the cement and aggregate, and the bonding of the concrete and rebar is poor. We've seen that in a lot of the collapsed buildings. Adding water is the usual practice," Chern said. "They even bring along a water tank for the purpose." And although structural engineers are wont to criticize architects for designing pretty buildings that fall down in quakes, perhaps the opposite extreme should also be avoided. "If I had my way all buildings would be squat concrete cubes with no windows," joked Vincent Borov, an engineer with the EQE team.

APPENDICES

APPENDIX A
GLOSSARY

The following is a list of commonly used geotechnical engineering and engineering geology terms and definitions. The glossary has been divided into five main categories:

1. Subsurface exploration terminology
2. Laboratory testing terminology
3. Terminology for engineering analysis and computations
4. Compaction, grading, and construction terminology
5. Geotechnical earthquake engineering terminology

Basic Terms

Civil engineer   A professional engineer who is registered to practice in the field of civil works.

Civil engineering   The application of the knowledge of the forces of nature, principles of mechanics, and the properties of materials for the evaluation, design, and construction of civil works for the beneficial uses of mankind.

Earthquake engineering   Deals with the design of structures to resist the forces exerted on the structure by the seismic energy of the earthquake.

Engineering geologist   A geologist who is experienced and knowledgeable in the field of engineering geology.

Engineering geology   The application of geologic knowledge and principles in the investigation and evaluation of naturally occurring rock and soil for use in the design of civil works.

Foundation engineering   In general, foundation engineering applies the knowledge of geology, soil mechanics, rock mechanics, and structural engineering to the design and construction of foundations for buildings and other structures. The most basic aspect of foundation engineering deals with the selection of the type of foundation, such as using a shallow or deep foundation system. Another important aspect of foundation engineering involves the development of design parameters, such as the bearing capacity or estimated settlement of the foundation. Foundation engineering could also include the actual foundation design, such as determining the type and spacing of steel reinforcement in concrete footings.

Geologist   An individual educated and trained in the field of geology.

Geotechnical engineer   A licensed individual who performs an engineering evaluation of earth materials including soil, rock, groundwater, and man-made materials and their interaction with earth retention systems, structural foundations, and other civil engineering works.

Geotechnical engineering   A subdiscipline of civil engineering. Geotechnical engineering requires the knowledge of engineering laws, formulas, construction techniques, and the performance of civil engineering works influenced by earth materials. Geotechnical engineering encompasses many of the engineering aspects of soil mechanics, rock mechanics, foundation engineering, geology, geophysics, hydrology, and related sciences.

Rock mechanics The application of the knowledge of the mechanical behavior of rock to engineering problems dealing with rock. Rock mechanics overlaps with engineering geology.

Soil mechanics The application of the laws and principles of mechanics and hydraulics to engineering problems dealing with soil as an engineering material.

Soils engineer Synonymous with geotechnical engineer (see Geotechnical engineer).

Soils engineering Synonymous with geotechnical engineering (see Geotechnical engineering).

A.1 SUBSURFACE EXPLORATION TERMINOLOGY

Abrasion The mechanical weathering, grinding, scraping, or rubbing away of rock surfaces by friction and/or impact.

Adobe Sun-dried bricks composed of mud and straw. Adobe is commonly used for construction in the southwestern United States and in Mexico.

Aeolian (or eolian) Particles of soil that have been deposited by the wind. Aeolian deposits include dune sands and loess.

Alluvium Detrital deposits resulting from the flow of water, including sediments deposited in river beds, canyons, flood plains, lakes, fans at the foot of slopes, and estuaries.

Aquiclude A relatively impervious rock or soil strata that will not transmit groundwater fast enough to furnish an appreciable supply of water to a well or spring.

Aquifer A relatively pervious rock or soil strata that will transmit groundwater fast enough to furnish an appreciable supply of water to a well or spring.

Artesian Groundwater that is under pressure and is confined by impervious material. If the trapped pressurized water is released, such as by drilling a well, the water will rise above the groundwater table and may even rise above the ground surface.

Ash Fine fragments of rock, between 4 and 0.25 mm in size, that originated as air borne debris from explosive volcanic eruptions. Nonwelded tuff has an engineering behavior similar to volcanic ash. These materials have been used as mineral filler in highways and other earth-rock construction. Some types of volcanic ash have been used as pozzolanic cement and as admixtures in concrete to retard undesired reactions between cement alkalies and aggregates. Also see Tuff.

Badlands An area, large or small, characterized by extremely intricate and sharp erosional sculpture. Badlands occur chiefly in arid or semiarid climates where the rainfall is concentrated in sudden heavy showers. They may, however, occur in humid regions where vegetation has been destroyed, or where soil and coarse detritus are lacking.

Bedding The arrangement of rock in layers, strata, or beds.

Bedrock A more or less solid, relatively undisturbed rock in place either at the surface or beneath deposits of soil.

Bentonite A soil or formational material that has a high concentration of the clay mineral montmorillonite. It is derived from the alteration of volcanic tuff or ash. Because bentonite consists almost exclusively of montmorillonite, it will swell, shrink, and cause more expansive soil-related damage than any other type of soil.

The term bentonite also refers to manufactured products that have a high concentration of montmorillonite. Examples include bentonite pellets and products that are used as impermeable barriers, such as geosynthetic clay liners (GCL), which are bentonite/geosynthetic composites.

Bit A device that is attached to the end of the drill stem and is used as a cutting tool to bore into soil and rock.

Bog A peat covered area with a high groundwater table. The surface is often covered with moss and it tends to be nutrient poor and acidic.

Boring A method of investigating subsurface conditions by drilling a hole into the earth materials. Usually soil and rock samples are extracted from the boring. Field tests, such as the standard penetration test (SPT) and the vane shear test (VST), can also be performed in the boring. A boring is also referred to as a borehole.

Boring log A written record of the materials penetrated during the subsurface exploration.

Caliche This type of material is common in arid or semiarid parts of the southwestern United States and consists of soil that is normally cemented together by calcium carbonate. When water evaporates near or at ground surface, the calcium carbonate is deposited in the void spaces between soil particles. Caliche is generally strong and stable in an undisturbed state, but it can become unstable if the cementing agents are leached away by water from leaky pipes or sewers or from the infiltration of irrigation water.

California bearing ratio (CBR) The CBR can be determined for soil in the field or soil compacted in the laboratory. The CBR is frequently used for the design of roads and airfields.

Casing A steel pipe that is temporarily inserted into a boring or drilled shaft in order to prevent the adjacent soil from caving.

Cohesionless soil A soil, such as clean gravel or sand, that when unconfined, will fall apart in either a wet or dry state.

Cohesive soil A soil, such as a silt or clay, that when unconfined, has considerable shear strength when dried, and will not fall apart in a saturated state. Cohesive soil is also known as a plastic soil, or a soil that has a plasticity index.

Colluvium Generally loose deposits usually found near the base of slopes and brought there chiefly by gravity through slow continuous downhill creep.

Cone penetration test (CPT) A field test used to identify and determine the in situ properties of soil deposits and soft rock.

> **Electric cone** A cone penetrometer that uses electric-force transducers built into the apparatus for measuring cone resistance and friction resistance.

> **Mechanical cone** A cone penetrometer that uses a set of inner rods to operate a telescoping penetrometer tip and to transmit the resistance force to the surface for measurement.

> **Mechanical-friction cone** A cone penetrometer with the additional capability of measuring the local side friction component of penetration resistance.

> **Piezocone** A cone penetrometer with the additional capability of measuring pore water pressure generated during the penetration of the cone.

Core drilling Also known as diamond drilling, the process of cutting out cylindrical rock samples in the field.

Core recovery (RQD) The RQD is computed by summing the lengths of all pieces of the rock core (NX size) equal to or longer than 4 in. (10 cm) and then dividing by the total length of the core run. The RQD is usually multiplied by 100 and then expressed as a percentage.

Deposition The geologic process of laying down or accumulating natural material into beds, veins, or irregular masses. Deposition includes mechanical settling (such as sedimentation in lakes), precipitation (such as the evaporation of surface water to form halite), and the accumulation of dead plants (such as in a peat bog).

Detritus Any material worn or broken down from rocks by mechanical means.

Diatomaceous earth   Diatomaceous earth usually consists of fine, white, siliceous powder, composed mainly of diatoms and their remains. Diatoms secrete outer shells of silica, called frustules, in a great variety of forms, which can accumulate in sediments in enormous amounts. Deposits of diatoms have low dry density and high moisture content because of this structure. Industrial uses of diatomaceous earth are as filters to remove impurities, as abrasives to polish soft metals, and when mixed with nitroglycerin, as an absorbent in the production of dynamite.

Drilling fluid   A fluid that can be used to remove drill bit cuttings, to clean and cool the drill bit, and to prevent caving in of the borehole.

Erosion   The wearing away of the ground surface, caused by the movement of wind, water, or ice.

Fold   Bending or flexure of a layer or layers of rock. Examples of folded rock include anticlines and synclines. Usually folds are created by the massive compression of rock layers.

Fracture   Visible break in a rock mass. Examples include joints, faults, and fissures.

Geophysical techniques   Various methods of determining subsurface soil and rock conditions without performing subsurface exploration. A common geophysical technique is to induce a shock wave into the earth and then measure the seismic velocity of the wave's travel through the earth material. The seismic velocity has been correlated with the rippability of the earth material.

Groundwater table (also known as phreatic surface)   The top surface of underground water, the location of which is often determined from piezometers, such as an open standpipe. A perched groundwater table refers to groundwater occurring in an upper zone separated from the main body of groundwater by underlying unsaturated rock or soil.

Hardpan   A hard and impervious layer, often consisting of clay that is cemented together. Hardpan does not become plastic when mixed with water and thus it restricts the flow of water or roots through it.

Horizon   One of the layers of a soil profile that can be distinguished by its texture, color, and structure.

> **"A" Horizon**   The uppermost layer of a soil profile, which often contains remnants of organic life. Inorganic colloids and soluble materials are often leached from this horizon.

> **"B" Horizon**   The layer of a soil profile in which material leached from the overlying "A" horizon is accumulated.

> **"C" Horizon**   Undisturbed parent material from which the overlying soil profile has been developed.

Humus   The portion of the soil that contains organic matter. Humus is black or brown and is formed by the decomposition of vegetation or animal matter.

Inclinometer   The horizontal movement preceding or during the movement of slopes can be investigated by successive surveys of the shape and position of flexible vertical casings installed in the ground. Lowering an inclinometer probe into the flexible vertical casing performs the surveys.

In situ   Used in reference to the original in-place (or in situ) condition of the soil or rock.

Interstitial   A term that refers to the space between soil particles or the pores in rock.

Iowa borehole shear test (BST)   A field test where the device is lowered into an uncased borehole and then expanded against the sidewalls. The force required to pull the device toward ground surface is measured and much like a direct shear test, the shear strength properties of the in situ soil can then be determined.

Joint   A break or fracture in rock occurring singly, but more frequently as a set or system of joints. A joint does not have movement parallel to its surface.

Karst topography   A type of landform developed in a region of easily soluble limestone. It is characterized by vast numbers of depressions of all sizes, sometimes by great outcrops of limestone ledges, sinks, and other solution passages, an almost total lack of surface streams, and large springs in the deeper valleys.

Kelly   A heavy tube or pipe, usually square or rectangular in cross section, which is used to provide a downward load when excavating an auger borehole.

Landslide   Mass movement of soil or rock that involves shear displacement along one or several rupture surfaces, which are either visible or may be reasonably inferred.

Landslide debris   Material, generally porous and of low density, produced from instability of natural or man-made slopes.

Leaching   The removal of soluble materials in soil or rock caused by percolating or moving groundwater.

Loess   A wind deposited silt often having a high porosity and low density, which can be susceptible to collapse of its soil structure upon wetting. Loess is widespread in the central portion of the United States. It consists of uniform cohesive wind-blown silt, commonly light brown, yellow, or gray in color, with most of the particle sizes between 0.01 and 0.05 mm. The cohesion is commonly due to calcareous cement, which binds the particles together. An unusual feature of loess is the presence of vertical root holes and fractures that make it much more permeable in the vertical direction than the horizontal direction. Another unusual feature of loess is that it can form near vertical slopes, but when saturated, the cohesion is lost and the slope will fail or the ground surface will settle.

Marl   A calcareous clay that usually contains between 35 and 65 percent calcium carbonate.

Marsh   A wetland that is characterized by a grassy surface that is interspersed with open water. A marsh can also have a closed canopy of grasses, sedges, or other plants.

Mineral   An inorganic substance that has a definite chemical composition and distinctive physical properties. Most minerals are crystalline solids.

Muskeg   A level and practically treeless area characterized by dense growth consisting primarily of grasses. The surface is covered with a layer of partially decayed grass and is usually wet and soft when not frozen.

Observation well   Usually a small diameter well used to measure changes in the groundwater level.

Overburden   The soil that overlies bedrock. In other cases, it refers to all material overlying a point of interest in the ground, such as the overburden pressure exerted on a clay layer.

Peat   A naturally occurring highly organic deposit derived primarily from plant materials, where the remains of leaves, stems, twigs, and roots can be identified. The places where peat accumulates are known as peat bogs or peat moors. Its color ranges from light brown to black. Peat is unusual because it has a very high water content, which makes it extremely compressible. This almost always makes it unsuitable for supporting foundations.

Penetration resistance   See Standard penetration test.

Percussion drilling   A drilling process in which a borehole is advanced by using a series of impacts to the drill rods and attached bit.

Permafrost   Often defined as perennially frozen soil. Also defined as ground that remains below freezing temperatures for two or more years. The bottom of permafrost lies at depths ranging from a few feet to over a thousand feet. The *active layer* is defined as the upper few inches to several feet of ground that is frozen in winter but thawed in summer.

Piezometer   A device installed for measuring the pore water pressure (or pressure head) at a specific point within the soil mass.

Pit (or test pit)   An excavation made for the purpose of observing subsurface conditions, performing field tests, and obtaining soil samples. A pit also refers to an excavation in the surface of the earth from which ore is extracted, such as an open pit mine.

Pressuremeter test (PMT)   A field test that involves the expansion of a cylindrical probe within an uncased borehole.

Quick clay   A type of clay where the water content is often greater than the liquid limit (i.e., liquidity index greater than 1). Quick clays have unstable bonds between particles. As long as these unstable bonds are not broken, the clay can support a heavy load. But once remolded, the bonding is destroyed and the shear strength is substantially reduced. For example, sensitive Leda clay, from Ottawa, Ontario, has high shear strength in the undisturbed state, but once remolded, the clay is essentially a fluid (no shear strength). There are reports of entire hillsides of quick clays becoming unstable and then simply flowing away.

Refusal   During subsurface exploration, refusal means an inability to excavate any deeper with the boring equipment. Refusal could be due to many different factors, such as hard rock, boulders, or a layer of cobbles.

Residual soil   A type of soil derived by in-place weathering of the underlying material.

Rock   A relatively solid mass that has permanent and strong bonds between the minerals. Rock can be classified as sedimentary, igneous, or metamorphic.

Rock flour (or Bull's liver)   This soil consists predominately of silt size particles, but has little or no plasticity. Nonplastic rock flour contains particles of quartz, ground to a very fine state by the abrasive action of glaciers. Because of its fine particle size, this soil is often mistaken as clay. In describing this soil, the term *bull's liver* apparently comes from its in situ appearance. It has been observed that in a saturated state, it quakes like jelly from shock or vibration and can even flow like a liquid.

Rotary drilling   A drilling process in which a borehole is advanced by rotation of a drill bit under constant pressure without impact.

Rubble   Rough stones of irregular shape and sizes that are naturally or artificially broken from larger masses of rock. Rubble is often created during quarrying, stone cutting, and blasting.

Screw plate compressometer (SPC)   A field test that involves a plate that is screwed down to the desired depth, and then as pressure is applied, the settlement of the plate is measured.

Seep   A small area where water oozes from the soil or rock.

Sensitive clay   See Quick clay.

Slaking   The crumbling and disintegration of earth materials when exposed to air or moisture. Slaking can also refer to the breaking up of dried clay when submerged in water, due either to compression of entrapped air by inwardly migrating water or to the progressive swelling and sloughing off of the outer layers.

Slickensides   Surfaces within a soil mass, which have been smoothed and striated by shear movements on these surfaces.

Slope wash   Soil and/or rock material that has been transported down a slope by mass wasting assisted by runoff water not confined by channels (also see Colluvium).

Soil   Sediments or other accumulations of mineral particles produced by the physical and chemical disintegration of rocks. Inorganic soil does not contain organic matter, while organic soil contains organic matter.

Soil sampler   A device used to obtain soil samples during subsurface exploration. Based on the inside clearance ratio and the area ratio, soil samples can be either disturbed or undisturbed.

Standard penetration test (SPT)   A field test that consists of driving a thick-walled sampler (I.D. = 1.5 in., O.D. = 2 in.) into the soil by using a 140 lb. hammer falling 30 in. The number of blows to drive the sampler 18 in. is recorded. The N value (penetration resistance) is defined as the number of blows that drive the sampler from a depth interval of 6 to 18 in.

Strike and dip Strip and dip refer to a planar structure, such as a shear surface, fault, or bed. The strike is the compass direction of a level line drawn on the planar structure. The dip angle is measured between the planar structure and a horizontal surface.

Subgrade modulus This value is often obtained from field plate load tests and is used in the design of pavements and airfields. The subgrade modulus is also known as the modulus of subgrade reaction.

Subsoil profile Developed from subsurface exploration, a cross section of the ground that shows the soil and rock layers. A summary of field and laboratory tests could also be added to the subsoil profile.

Swamp A forested or shrub covered wetland where standing or gently flowing water persists for long periods of time.

Talus Rock fragments, often mixed with soil, which separates from a natural slope and then accumulates at the foot of the slope.

Till Material created directly by glaciers, without transportation or sorting by water. Till often consists of a wide range in particle sizes, including boulders, gravel, sand, and clay.

Topsoil The fertile upper zone of soil, which contains organic matter and is usually darker in color and loose.

Transported soil A type of soil that has been transported from its place of origin by the action of wind, water, or ice.

Trench Usually a long, narrow, and near vertical sided cut in rock or soil used for subsurface exploration or for the placement of utility lines, pipes, and culverts.

Tuff A pyroclastic rock, originating as air borne debris from explosive volcanic eruptions. An important aspect of tuff is the degree of welding, which can be described as nonwelded, partially welded to varying degrees, or densely welded. Welding is generally caused by fragments that are hot when deposited, and because of this heat, the sticky glassy fragments may actually fuse together. There are distinct changes in the original shards and pumice fragments, such as the union and elongation of the glassy shards and flattening of the pumice fragments, which is characteristic of completely welded tuff. The degree of welding depends on many factors, such as type of fragments, plasticity of the fragments (which depends on the emplacement temperature and chemical composition), thickness of the resulting deposit, and rate of cooling.

Vane shear test (VST) An in situ field test that consists of inserting a four-bladed vane into the borehole and then pushing the vane into the clay deposit located at the bottom of the borehole. Once inserted into the clay, the maximum torque required to rotate the vane and shear the clay is measured. Based on the dimensions of the vane and the maximum torque, the undrained shear strength s_u of the clay can be calculated.

Varved clay A lake deposit with alternating thin layers of sand and clay. Each varve represents the deposition during a year, where the lower sandy part is deposited during the summer, and the upper clayey part is then deposited during the winter when the surface of the lake is frozen and the water is tranquil. This causes an unusual variation in shear strength in the soil, where the horizontal shear strength along the clay portion of the varve is much less than the vertical shear strength. This can cause the stability of structures founded on varved clay to be overestimated, resulting in a bearing capacity failure.

Varved silt Similar to varved clay, but consisting of thin alternating layers of sand and silt.

Weathering The chemical and/or physical processes by which materials (such as rock) at or near the earth's surface are broken apart and disintegrated. The material can have a change in color, texture, composition, density, and form due to the processes of weathering.

Wetland Land that has a groundwater table at or near the ground surface, or land that is periodically under water, and supports various types of vegetation that are adapted to a wet environment.

A.2   *LABORATORY TESTING TERMINOLOGY*

Absorption   Defined as the mass of water in the aggregate divided by the dry mass of the aggregate. Absorption is used in soil mechanics for the study of oversize particles or in concrete mix design.

Activity of clay   The ratio of plasticity index to percent dry mass of the minus No. 40 sieve material that is smaller than 0.002 mm in grain size. This property is related to the types of clay minerals in the soil.

Angle of internal friction   See Friction angle

Atterberg limits   Water contents corresponding to different behavior conditions of plastic soil.

Liquid limit   The water content corresponding to the behavior change between the liquid and plastic state of a soil. The liquid limit is arbitrarily defined as the water content that a pat of soil, cut by a groove of standard dimensions, will flow together for a distance of 0.5 in. (12.7 mm) under the impact of 25 blows in a standard liquid limit device.

Plastic limit   The water content corresponding to the behavior change between the plastic and semisolid state of a soil. The plastic limit is arbitrarily defined as the water content where the soil will just begin to crumble when rolled into a thread approximately $\frac{1}{8}$ in. (3.2 mm) in diameter.

Shrinkage limit   The water content corresponding to the behavior change between the semisolid to solid state of a soil. The shrinkage limit is also defined as the water content where any further reduction in water content will not result in a decrease in volume of the soil mass.

Average degree of consolidation   The ratio, expressed as a percentage, of the settlement at any given time to the primary consolidation.

Binder (soil binder)   Typically clay size particles that can bind together or provide cohesion between soil particles. Organic matter and precipitation of cementing minerals can also bind together soil particles.

Boulder   A large detached rock fragment with an average dimension greater than 12 in. (30 cm).

Capillarity   Also known as capillary action and capillary rise, the rise of water through a soil due to the fluid property known as surface tension. Due to capillarity, the pore water pressures are less than atmospheric because of the surface tension of pore water acting on the meniscus formed in void spaces between the soil particles. The height of capillary rise is inversely proportional to the pore size of the soil.

Cation exchange capacity   The capacity of clay size particles to exchange cations with the double layer. Also see Double layer.

Clay minerals   The three most common clay minerals are listed below, with their respective activity *A* values:

Kaolinite   (A = 0.3 to 0.5). The kaolin minerals are a group of clay minerals consisting of hydrous aluminum silicates. A common kaolin mineral is kaolinite, having the general formula $Al_2Si_2O_5(OH)_4$. Kaolinite is usually formed by alteration of feldspars and other aluminum-bearing minerals. Kaolinite is usually a large clay mineral of low activity and often plots below the A-line in the plasticity chart. Kaolinite is a relatively inactive clay mineral and even though it is technically clay, it behaves more like a silt material. Kaolinite has many industrial uses including the production of china, medicines and cosmetics.

Montmorillonite   (Na-montmorillonite, A = 4 to 7 and Ca-montmorillonite, A = 1.5). A group of clay minerals that are characterized by weakly bonded layers. Each layer consists of two silica sheets with an aluminum (gibbsite) sheet in the middle. Water and exchangeable cations (e.g., Na, Ca) can enter and separate the layers, creating a very small crystal that has a strong attraction for water. Montmorillonite has the highest activity and it can have the highest water content, greatest compressibility, and lowest shear strength of all the clay minerals. Montmorillonite plots just below the U-line in the plasticity chart. Montmorillonite often forms as the result of the weathering of ferromagnesian minerals, calcic feldspars, and volcanic materials. For example, sodium montmorillonite is often formed from the weathering of volcanic ash. Other environments that are likely to form montmorillonite are alkaline conditions with a supply of magnesium ions and a lack of leaching. Such conditions are often present in semiarid regions.

Illite   ($A = 0.5$ to 1.3). This clay mineral has a structure similar to montmorillonite, but the layers are more strongly bonded together. In terms of cation exchange capacity, in ability to absorb and retain water, and in physical characteristics such as plasticity index, illite is intermediate in activity between clays of the kaolin and montmorillonite groups. Illite often plots just above the A-line in the plasticity chart.

Clay size particles   Clay size particles are finer than 0.002 mm. Most clay particles are flat or plate-like in shape, and as such they have a large surface area. The most common clay minerals belong to the kaolin, montmorillonite, and illite groups.

Coarse-grained soil   According to the Unified Soil Classification System, coarse-grained soils have more than 50 percent soil particles (by dry mass) retained on the No. 200 U.S. standard sieve.

Cobble   A rock fragment, usually rounded or semirounded, with an average dimension between 3 and 12 in. (75 and 300 mm).

Coefficient of compressibility   It is defined as the change in void ratio divided by the corresponding change in vertical effective stress.

Coefficient of consolidation   A coefficient used in the theory of consolidation. It is obtained from laboratory consolidation tests and is used to predict the time-settlement behavior of field loading of fine-grained soil.

Coefficient of curvature and coefficient of uniformity   These two parameters are used for the classification of coarse-grained soils in the Unified Soil Classification System. These two parameters are used to distinguish a well-graded soil from a uniformly graded soil.

Coefficient of permeability   See Hydraulic conductivity.

Cohesion   Two types of cohesion are: (1) cohesion in terms of total stress and (2) cohesion in terms of effective stress. For total cohesion c, the soil particles are predominately held together by capillary tension. For effective stress cohesion c', there must be actual bonding or attraction forces between the soil particles.

Cohesionless soil   See Nonplastic soil.

Cohesive soil   See Plastic soil.

Colloidal soil particles   Generally refers to clay size particles (finer than 0.002 mm) where the surface activity of the particle has an appreciable influence on the properties of the soil.

Compaction (laboratory)

Compaction curve   For a given compaction energy, a curve showing the relationship between the dry density and the water content of a soil.

Compaction test   A laboratory compaction procedure whereby a soil at known water content is compacted into a mold of specific dimensions. The procedure is repeated for various water contents to establish the compaction curve. The most common testing procedures (compaction energy, number of soil layers in the mold, and the like) are the Modified Proctor (ASTM D 1557) or Standard Proctor (ASTM D 698). The objective of the laboratory compaction test is to obtain the laboratory maximum dry density and the optimum moisture content for the tested soil.

Relative compaction   The degree of compaction (expressed as a percentage) defined as the field dry density divided by the laboratory maximum dry density.

Compression index For a consolidation test, the slope of the linear portion of the vertical pressure versus void ratio curve on a semilog plot. The compression index is calculated for the virgin consolidation curve.

Compressive strength See Unconfined compressive strength.

Consistency of clay Generally refers to the firmness of a cohesive soil. For example, a cohesive soil can have a consistency that varies from *very soft* up to *hard.*

Consolidated drained triaxial compression test See Triaxial test.

Consolidated undrained triaxial compression test See Triaxial test.

Consolidation test A laboratory test used to measure the consolidation properties of saturated cohesive soil. The specimen is laterally confined in a ring and is compressed between porous plates (oedometer apparatus). Also see Consolidation in Sec. A. 3.

Contraction (during shear) During the shearing of soil, the tendency of loose soil to decrease in volume (or contract).

Controlled strain test A laboratory test where the load is applied so as to control the rate of strain. For triaxial compression test on a soil specimen, shearing is performed at a specific rate of axial strain.

Controlled stress test A laboratory test where the load is applied in increments. The consolidation test is often performed by subjecting the soil specimen to an incremental increase in load, with the soil specimen subjected to each load for a period of 24 h.

Creep For laboratory tests, drained creep occurs when a plastic soil experiences continued deformation under constant effective stress. For example, secondary compression is often referred to as drained creep.

Deflocculating agent Used during the hydrometer test, a compound such as sodium hexametaphosphate that prevents clay size particles from coalescing into flocs.

Density It is defined as mass per unit volume. In the International System of Units (SI), typical units for the density of soil are Mg/m^3.

Deviator stress Difference between the major and minor principal stress in a triaxial test.

Dilation (during shear) The tendency of dense soil to increase in volume (or dilate) during shear.

Direct shear test A laboratory test used to obtain the effective shear strength properties (c' and ϕ') of the soil. The test consists of applying a vertical pressure to the laterally confined soil specimen, submerging the soil specimen in distilled water, allowing the soil to consolidate, and then shearing the soil specimen by moving the top of the shear box relative to the fixed bottom. The soil specimen must be sheared at a slow enough rate so that excess pore water pressures do not develop.

Dispersing agent See Deflocculating agent.

Double layer A grossly simplified interpretation of the positively charged water layer, together with the negatively charged surface of the particle itself. Two reasons for the attraction of water to the clay particle are: (1) dipolar structure of water molecule which causes it to be electrostatically attracted to the surface of the clay particle, and (2) the clay particles attract cations which contribute to the attraction of water by the hydration process. The *adsorbed water layer* consists of water molecules that are tightly held to the clay particle face, such as by the process of hydrogen bonding.

Exchange capacity See Cation exchange capacity.

Fabric (of soil) Definitions vary, but in general the fabric of soil often refers only to the geometric arrangement of the soil particles. In contrast, the soil structure refers to both the geometric arrangement of soil particles and the interparticle forces, which may act between them.

Fine grained soil   Per the Unified Soil Classification System, a fine grained soil contains 50 percent or more (by dry mass) of particles finer than the No. 200 sieve.

Fines   Refers to the silt and clay size particles in the soil, i.e., soil particles that are finer than the No. 200 U.S. standard sieve.

Flocculation   When suspended in water, the process of fines attracting each other to form a larger particle or floc. In the hydrometer test, a dispersing agent is added to prevent flocculation of fines.

Friction angle   In terms of effective shear stress, the soil friction is usually considered to be due to the interlocking of the soil or rock grains and the resistance to sliding between the grains. A relative measure of the soil's frictional shear strength is the friction angle. Friction angle is also known as the *angle of internal friction* and *angle of shear resistance*.

Grain size distribution   See Particle size distribution.

Gravel size fragments   Rock fragments and soil particles that will pass the 3 in. (76 mm) sieve and be retained on a No. 4 (4.75 mm) U.S. standard sieve.

Hydraulic conductivity (or coefficient of permeability)   For laminar flow of water in soil, both terms are synonymous and indicate a measure of the soil's ability to allow water to flow through its pores. The hydraulic conductivity is often measured in a constant head or falling head permeameter.

Illite   See Clay minerals.

Index test   Index tests are the most basic types of laboratory tests performed on soil samples. Index tests include the water content test, wet density determinations, specific gravity tests, particle size distributions, Atterberg limits, and tests specifically labeled as index tests, such as the Expansion Index Test.

Kaolinite   See Clay minerals.

Laboratory maximum dry density   The peak point of the compaction curve (see Compaction).

Liquid imit   See Atterberg limits.

Liquidity index   The liquidity index can be used to distinguish quick clays (liquidity index usually greater than 1.0) from highly desiccated clays (negative liquidity index).

Log of time method   Using data from the laboratory consolidation test, a plot of the vertical deformation versus time on a semilog graph. The log of time method is used to determine the coefficient of consolidation. Also see Square root of time method.

Moisture content (or water content)   Moisture content and water content are synonymous. The definition of moisture content is the ratio of the mass of water in the soil divided by the dry mass of the soil, usually expressed as a percentage.

Montmorillonite   See Clay minerals.

Nonplastic soil   A granular soil that cannot be rolled or molded at any water content. A nonplastic soil has a plasticity index equal to zero, or the plastic limit is greater than the liquid limit. A nonplastic soil is known as a cohesionless soil.

Optimum moisture content   The moisture content, determined from a laboratory compaction test, at which the maximum dry density of a soil is obtained using specific compaction energy. Also see Compaction.

Organic soil   Soils that partly or predominately consist of organic matter.

Overconsolidation ratio (OCR)   The ratio of the preconsolidation vertical effective stress to the current vertical effective stress.

Oversize particles   For fill compaction, the oversize particles are the gravel and cobble size particles retained on the $3/4$ in. or No. 4 (4.75 mm) U.S. standard sieve. Also see Soil matrix.

Particle size distribution   Also known as grain size distribution or gradation, the distribution of particle sizes in the soil based on dry mass.

Peak shear strength   The maximum shear strength along a shear failure surface.

Permeability   The ability of water (or other fluid) to flow through a soil by traveling through the void spaces. A high permeability indicates flow occurs rapidly, and vice versa. A measure of the soil's permeability is the hydraulic conductivity, also known as the coefficient of permeability.

Plastic limit   See Atterberg limits.

Plastic soil   A soil that exhibits plasticity, i.e., the ability to be rolled and molded without breaking apart. A measure of a soil's plasticity is the plasticity index. A plastic soil is also known as a cohesive soil.

Plasticity   Term applied to silt and clay, to indicate the soil's ability to be rolled and molded without breaking apart. A measure of the soil's plasticity is the plasticity index.

Plasticity index   The plasticity index is defined as the liquid limit minus the plastic limit, often expressed as a whole number (also see Atterberg limits).

Pore water pressure   See Pore water pressure in Sec. A.3.

Principal planes and principal stresses   See Sec. A.3.

Sample disturbance   Through soil sampling or other actions, the soil may become remolded and thus experience sample disturbance. Sample disturbance causes a reduction in effective stress, a reduction in the interparticle bonds, and a rearrangement of the soil particles. Some of the factors that can cause soil disturbance are pieces of hard gravel or shell fragments in the soil, which can cause voids to develop along the sides of the sampling tube during the sampling process; soil adjustment caused by stress relief when making a borehole; disruption of the soil structure due to hammering or pushing the sampling tube into the soil stratum; expansion of gas during retrieval of the sampling tube; jarring or banging the sampling tube during transportation to the laboratory; roughly removing the soil from the sampling tube; and crudely cutting the soil specimen to a specific size for a laboratory test.

Sand equivalent (SE)   A measure of the amount of silt or clay contamination in fine aggregate as determined by ASTM D 2419 test procedures.

Sand size particles   Soil particles that will pass the No. 4 (4.75 mm) sieve and be retained on the No. 200 (0.075 mm) U.S. standard sieve.

Secant modulus   On a stress-strain plot, the slope of the line from the origin to a given point on the curve. The data for the stress-strain plot are often obtained from a laboratory triaxial compression test.

Shear strength   The maximum shear stress that a soil or rock can sustain. Shear strength of soil is based on total stresses (i.e., undrained shear strength) or effective stresses (i.e., effective shear strength).

Shear strength in terms of total stress   Shear strength of soil based on total stresses. The undrained shear strength of soil could be expressed in terms of the undrained shear strength s_u, or by using the failure envelope that is defined by total cohesion c and total friction angle ϕ.

Effective shear strength   Shear strength of soil based on effective stresses. The effective shear strength of soil could be expressed in terms of the failure envelope that is defined by effective cohesion c' and effective friction angle ϕ'.

Shear strength tests (laboratory)   There are many types of shear strength tests that can be performed in the laboratory. The objective is to obtain the shear strength of the soil. Laboratory tests can generally be divided into two categories:

Shear strength tests based on total stress   The purpose of these laboratory tests is to obtain the undrained shear strength of the soil or the failure envelope in terms of total stresses. An example is the unconfined compression test, which is also known as an unconsolidated-undrained test.

Shear strength tests based on effective stress   The purpose of these laboratory tests is to obtain the effective shear strength of the soil based on the failure envelope in terms of effective stress. An example is a direct shear test where the saturated, submerged, and consolidated soil specimen is sheared slow enough that excess pore water pressures do not develop (this test is known as a consolidated-drained test).

Shrinkage limit   See Atterberg limits.

Sieve   Laboratory equipment consisting of a pan with a screen at the bottom. U.S. standard sieves are used to separate particles of a soil sample into their various sizes.

Silt size particles   That portion of a soil that is finer than the No. 200 sieve (0.075 mm) and coarser than 0.002 mm. Silt and clay size particles are considered to be *fines*.

Soil matrix   For fill compaction, the matrix is that portion of the soil that is finer than the $3/_4$ in. or No. 4 (4.75 mm) U.S. standard sieve. Also see Oversize particles.

Soil structure   Definitions vary, but in general, the soil structure refers to both the geometric arrangement of the soil particles and the interparticle forces, which may act between them. Common soil structures are as follows:

Cluster structure   Soil grains that consist of densely packed silt or clay size particles.

Dispersed structure   The clay size particles are oriented parallel to each other.

Flocculated (or cardhouse) structure   The clay size particles are oriented in edge-to-face arrangements.

Honeycomb structure   Loosely arranged bundles of soil particles, having a structure that resembles a honeycomb.

Single-grained structure   An arrangement composed of individual soil particles. This is a common structure of sands.

Skeleton structure   An arrangement where coarser soil grains form a skeleton with the void spaces partly filled by a relatively loose arrangement of soil fines.

Specific gravity   The specific gravity of soil or oversize particles can be determined in the laboratory. Specific gravity is generally defined as the ratio of the density of the soil particles divided by the density of water.

Square root of time method   Using data from the laboratory consolidation test, a plot of the vertical deformation versus square root of time. The square root of time method is used to determine the coefficient of consolidation. Also see Log of time method.

Tangent modulus   On a stress-strain plot, the slope of the line tangent to the stress-strain curve at a given stress value. The stress value used to obtain the tangent modulus is often the stress value that is equal to one-half of the compressive strength. The data for the stress-strain plot can be obtained from a laboratory triaxial compression test.

Tensile test   A laboratory test in which a geosynthetic is stretched in one direction to determine the force-elongation characteristics, breaking force, and the breaking elongation.

Texture (of soil)   The term texture refers to the degree of fineness of the soil, such as smooth, gritty, or sharp, when the soil is rubbed between the fingers.

Thixotropy   The property of remolded clay that enables it to stiffen (gain shear strength) in a relatively short time.

Torsional ring shear test   A laboratory test where a relatively thin soil specimen of circular or annular cross-section is consolidated and then sheared at a slow rate in order to obtain the drained residual friction angle.

Triaxial test   A laboratory test in which a cylindrical specimen of soil or rock encased in an impervious membrane is subjected to a confining pressure and then loaded axially to failure. Different types of commonly used triaxial tests are as follows:

> **Consolidated drained triaxial compression test**   A triaxial test in which the cylindrical soil specimen is first saturated and consolidated by the effective confining pressure. Then the soil specimen is sheared by increasing the axial load. During shearing, drainage is provided to the soil specimen and it is sheared slow enough so that the shear induced pore water pressures can dissipate.

> **Consolidated undrained triaxial compression test**   A triaxial test in which the cylindrical soil specimen is first saturated and consolidated by the effective confining pressure. Then the soil specimen is sheared by increasing the axial load. During shearing, drainage is not provided to the soil specimen and hence it is an undrained test. The shear induced pore water pressures can be measured during the shearing process.

> **Unconsolidated undrained triaxial compression test**   A triaxial test in which the cylindrical soil specimen retains its initial water content throughout the test (i.e., the water content remains unchanged during both the application of the confining pressure and during shearing). Since drainage is not provided during both the application of the confining pressure and during shearing, the soil specimen is unconsolidated and undrained during shearing.

Unconfined compressive strength   The vertical stress that causes the shear failure of a cylindrical specimen of a plastic soil or rock in a simple compression test. For the simple compression test, the undrained shear strength s_u of the plastic soil is defined as one-half the unconfined compressive strength.

Unconsolidated undrained triaxial compression test   See Triaxial test.

Unit weight   Unit weight is defined as weight per unit volume. In the International System of Units (SI), unit weight has units of kN/m^3. In the United States Customary System, unit weight has units of pcf (pounds-force per cubic foot).

Water content   See Moisture content.

Zero air voids curve   On the laboratory compaction curve, the zero air voids curve is often included. It is the relationship between water content and dry density for a condition of saturation ($S = 100$ percent) for a specified specific gravity.

A.3   TERMINOLOGY FOR ENGINEERING ANALYSIS AND COMPUTATIONS

Adhesion   Shearing resistance between two different materials. For example, for piles driven into clay deposits, there is adhesion between the surface of the pile and the surrounding clay.

Allowable bearing pressure   Allowable bearing pressure is the maximum pressure that can be imposed by a foundation onto soil or rock supporting the foundation. It is derived from experience and general usage, and provides an adequate factor of safety against shear failure and excessive settlement.

Anisotropic soil   A soil mass having different properties in different directions at any given point referring primarily to stress-strain or permeability characteristics.

Arching   The transfer of stress from an unconfined area to a less-yielding or restrained structure. Arching is important in the design of pile or pier walls that have open gaps between the members.

Bearing capacity:

Allowable bearing capacity   The maximum allowable bearing pressure for the design of foundations.

Ultimate bearing capacity   The bearing pressure that causes failure of the soil or rock supporting the foundation.

Bearing capacity failure   A foundation failure that occurs when the shear stresses in the adjacent soil exceed the shear strength.

Collapsible formations   Examples of collapsible formations include limestone formations and deep mining of coal beds. Limestone can form underground caves and caverns, which can gradually enlarge resulting in a collapse of the ground surface and the formation of a sinkhole. Sites that are underlain by coal or salt mines could also experience ground surface settlement when the underground mine collapses.

Collapsible soil   Collapsible soil can be broadly classified as soil that is susceptible to a large and sudden reduction in volume upon wetting. Collapsible soil usually has a low dry density and low moisture content. Such soil can withstand a large applied vertical stress with a small compression, but then experience much larger settlements after wetting, with no increase in vertical pressure. Collapsible soil can include fill compacted dry of optimum and natural collapsible soil, such as alluvium, colluvium, or loess.

Compressibility   A decrease in volume that occurs in the soil mass when it is subjected to an increase in loading. Some highly compressible soils are loose sands, organic clays, sensitive clays, highly plastic and soft clays, uncompacted fills, municipal landfills, and permafrost soils.

Consolidation   The consolidation of a saturated clay deposit is generally divided into three separate categories:

Initial or immediate settlement   The initial settlement of the structure caused by undrained shear deformations, or in some cases contained plastic flow, due to two- or three-dimensional loading.

Primary consolidation   The compression of clays under load that occurs as excess pore water pressures slowly dissipate with time.

Secondary compression   The final component of settlement, which is that part of the settlement that occurs after essentially all of the excess pore water pressures have dissipated.

Creep   An imperceptibly slow and more or less continuous movement of slope-forming soil or rock debris.

Critical height   Critical height refers to the maximum height at which a vertical excavation or slope will stand unsupported.

Critical slope   Critical slope refers to the maximum angle at which a sloped bank of soil or rock of given height will stand unsupported.

Crown   Generally, the highest point. For tunnels, the crown is the arched roof. For landslides, the crown is the area above the main scarp of the landslide.

Dead load   Structural loads due to the weight of beams, columns, floors, roofs, and other fixed members. Does not include nonstructural items such as furniture, snow, occupants, or inventory.

Debris flow   An initial shear failure of a soil mass that transforms itself into a fluid mass that can move rapidly over the ground surface.

Depth of seasonal moisture change   A layer of expansive soil subjected to shrinkage during the dry season and swelling during the wet season. This zone extends from ground surface to the depth of significant moisture fluctuation. Also known as the active zone.

Desiccation   The process of shrinkage of clays. The process involves a reduction in volume of the grain skeleton and subsequent cracking of the clay caused by the development of capillary stresses in the pore water as the soil dries.

Design load All forces and moments that are used to proportion a foundation. The design load includes the dead weight of a structure, and in most cases, also includes live loads. Considerable judgment and experience are required to determine the design load that is to be used to proportion a foundation.

Downdrag force A force induced on deep foundations resulting from the downward movement of adjacent soil relative to the foundation element. Also referred to as negative skin friction.

Earth pressure Usually used in reference to the lateral pressure imposed by a soil mass against an earth-supporting structure such as a retaining wall or basement wall.

Active earth pressure (k_A) The horizontal pressure for a condition where the retaining wall has yielded sufficiently to allow the backfill to mobilize its shear strength.

At-rest earth pressure (k_o) The horizontal pressure for a condition where the retaining wall has not yielded or compressed into the soil. This would also be applicable to a soil mass in its natural state.

Passive earth pressure (k_p) The horizontal pressure due to a retaining wall footing that has moved into and compressed the soil sufficiently to develop its maximum lateral resistance.

Effective stress The effective stress is defined as the total stress minus the pore water pressure.

Equipotential line A line connecting points of equal total head.

Equivalent fluid pressure Horizontal pressures of soil, or soil and water in combination, which increases linearly with depth and are equivalent to those that would be produced by a soil of a given density. Equivalent fluid pressure is often used in the design of retaining walls.

Excess pore water pressure See Pore water pressure.

Exit gradient The hydraulic gradient near the toe of a dam or the bottom of an excavation through which groundwater seepage is exiting the ground surface.

Finite element A soil and structure profile subdivided into regular geometrical shapes for the purpose of numerical stress analysis.

Flow line The path of travel traced by moving groundwater as it flows through a soil mass.

Flow net A graphical representation used to study the flow of groundwater through a soil. A flow net is composed of flow lines and equipotential lines.

Head From Bernoulli's energy equation, the total head is defined as the sum of the velocity head, pressure head, and elevation head. Head has units of length. For seepage problems in soil, the velocity head is usually small enough to be neglected and thus for laminar flow in soil, the total head h is equal to the sum of the pressure head h_p and elevation head h_e.

Heave The upward movement of foundations or other structures caused by frost heave or expansive soil and rock. Frost heave refers to the development of ice layers or lenses within the soil that causes the ground surface to heave upward. Heave due to expansive soil and rock is caused by an increase in water content of clays or rocks, such as shale or slate.

Homogeneous soil A soil that exhibits essentially the same physical properties at every point throughout its mass.

Hydraulic gradient Difference in total head at two points divided by the distance between them. Hydraulic gradient is used in seepage analyses.

Hydrostatic pore water pressure See Pore water pressure.

Isotropic soil A soil mass having essentially the same properties in all directions at any given point, referring primarily to stress-strain or permeability characteristics.

Laminar flow Groundwater seepage in which the total head loss is proportional to the velocity.

Live load   Structural loads due to nonstructural members, such as furniture, occupants, inventory, and snow.

Maximum past pressure   See Preconsolidation pressure.

Modulus of elasticity   The ratio of stress to strain for a material under given loading conditions. The modulus of elasticity is numerically equal to the slope of the tangent or secant of the stress-strain curve.

Mohr circle   A graphical representation of the stresses acting on the various planes at a given point in the soil.

Negative pore water pressure   See Pore water pressure.

Negative skin friction   See Downdrag

Normally consolidated   The condition that exists if a soil deposit has never been subjected to an effective stress greater than the existing overburden pressure and if the deposit is completely consolidated under the existing overburden pressure.

Overconsolidated   The condition that exists if a soil deposit has been subjected to an effective stress greater than the existing overburden pressure. A soil can become overconsolidated by a reduction in total stress (e.g., removal of overburden), a decrease in pore water pressure (e.g., desiccation due to drying), or a change in soil structure.

Phase relationships   Phase relationships are the basic soil relationships used in geotechnical engineering. They mathematically relate the three basic parts of soil (i.e., solid, liquid, and gas).

Piping   The movement of soil particles as a result of unbalanced seepage forces produced by percolating water, leading to the development of ground surface boils or underground erosion voids and channels.

Plastic equilibrium   The state of stress of a soil mass that has been loaded and deformed to such an extent that its ultimate shearing resistance is mobilized at one or more points.

Poisson's ratio   A ratio between linear strain changes perpendicular to and in the direction of a uniaxial stress change. A Poisson's ratio of 0.5 is often assumed for loading of saturated clay.

Pore water pressure   The water pressure that exists in the soil void spaces.

   Excess pore water pressure   The increment of pore water pressures greater than hydrostatic values, produced by consolidation stress in compressible materials or by shear strain.

   Hydrostatic pore water pressure   Pore water pressure or groundwater pressures exerted under conditions of no flow where the magnitude of pore pressures increase linearly with depth below the groundwater table.

   Negative pore water pressure   Pore water pressure that is less than atmospheric. An example is capillary rise, which can induce a negative pore water pressure in the soil. Another example is the undrained shearing of dense or highly overconsolidated soils, where the soil wants to dilate during shear, resulting in negative pore water pressures.

Porosity   The ratio, usually expressed as a percentage, of the volume of voids divided by the total volume of the soil or rock.

Preconsolidation pressure   The greatest vertical effective stress to which a soil, such as a clay layer, has been subjected. Also known as the maximum past pressure.

Pressure (or stress)   The load divided by the area over which it acts.

Principal planes   Each of three mutually perpendicular planes through a point in the soil mass on which the shearing stress is zero. For soil mechanics, compressive stresses are positive.

   Major principal plane   The plane normal to the direction of the major principal stress (highest stress in the soil).

Intermediate principal plane   The plane normal to the direction of the intermediate principal stress.

Minor principal plane   The plane normal to the direction of the minor principal stress (lowest stress in the soil).

Principal stresses   The stresses that occur on the principal planes. Also see Mohr circle.

Progressive failure   Formation and development of localized stresses which lead to fracturing of the soil, which spreads and eventually forms a continuous rupture surface and a failure condition. Stiff fissured clay slopes are especially susceptible to progressive failure.

Quick condition (or Quicksand)   A condition in which groundwater is flowing upward with a sufficient hydraulic gradient to produce a zero effective stress condition in the sand deposit.

Relative density   Term applied to a sand deposit to indicate its relative density state, defined as the ratio of (1) the difference between the void ratio in the loosest state and the in situ void ratio, to (2) the difference between the void ratios in the loosest and in the densest states.

Saturation (degree)   The degree of saturation is calculated as the volume of water in the void space divided by the total volume of voids. It is usually expressed as a percentage. A completely dry soil has a degree of saturation of 0 percent and a saturated soil has a degree of saturation of 100 percent.

Seepage   The infiltration or percolation of water through soil and rock.

Seepage analysis   An analysis to determine the quantity of groundwater flowing through a soil deposit. For example, by using a flow net, the quantity of groundwater flowing through or underneath an earth dam can be determined.

Seepage force   The frictional drag of water flowing through the soil voids.

Seepage velocity   The velocity of flow of water in the soil, while the superficial velocity is the velocity of flow into or out of the soil.

Sensitivity   The ratio of the undrained shear strength of the undisturbed plastic soil to the remolded shear strength of the same plastic soil.

Settlement   The permanent downward vertical movement experienced by structures as the underlying soil consolidates, compresses, or collapses due to the structural load or secondary influences.

Differential settlement   The difference in settlement between two foundation elements or between two points on a single foundation.

Total settlement   The absolute vertical movement of the foundation.

Shear failure   A failure in a soil or rock mass caused by shearing strain along one or more slip (rupture) surfaces.

General shear failure   A failure in which the shear strength of the soil or rock is mobilized along the entire slip surface.

Local shear failure   A failure in which the shear strength of the soil or rock is mobilized only locally along the slip surface.

Progressive shear failure   See Progressive failure.

Punching shear failure   Shear failure where the foundation pushes (or punches) into the soil due to the compression of soil directly below the footing as well as vertical shearing around the footing perimeter.

Shear plane (or slip surface)   A plane along which failure of soil or rock occurs by shearing.

Shear stress   A stress that acts parallel to the surface element.

Slope stability analyses   Determination of the factor of safety for slope stability. Common types of stability analyses are as follows:

Gross slope stability   The stability of slope material below a plane approximately 3 to 4 ft (0.9 to 1.2 m) deep measured from and perpendicular to the slope face.

Surficial slope stability   The stability of the outer 3 to 4 ft (0.9 to 1.2 m) of slope material measured from and perpendicular to the slope face.

Strain   The change in shape of soil when it is acted upon by stress.

Normal stain   A measure of compressive or tensile deformations, and is defined as the change in length divided by the initial length. In geotechnical engineering, strain is positive when it results in compression of the soil.

Shear strain   A measure of the shear deformation of soil.

Stress distribution   Term commonly applied to the use of charts or equations for the purpose of determining the increase in pressure at depth due to a surface loading. Stress distribution methods can vary from approximate methods (such as the 2:1 approximation) to charts and equations based on the theory of elasticity.

Subsidence   Settlement of the ground surface over a very large area, such as caused by the extraction of oil from the ground or the pumping of groundwater from wells.

Swell   Increase in soil volume, typically referring to volumetric expansion of clay due to an increase in water content.

Time factor (*T*)   A dimensionless factor, used in the Terzaghi theory of consolidation or swelling of cohesive soil.

Total stress   The total stress is defined as the effective stress plus the pore water pressure. The vertical total stress for uniform soil and a level ground surface can be calculated by multiplying the total unit weight of the soil times the depth below ground surface.

Underconsolidation   The condition that exists if a soil deposit is not fully consolidated under the existing overburden pressure and excess pore water pressures exist within the soil. Underconsolidation occurs in areas where a cohesive soil is being deposited very rapidly and not enough time has elapsed for the soil to consolidate under its own weight.

Void ratio   The void ratio is defined as the volume of voids divided by the volume of soil solids.

A.4   COMPACTION, GRADING, AND CONSTRUCTION TERMINOLGY

Aggregate   A granular material used for a pavement base, wall backfill, and the like.

Coarse aggregate   Gravel or crushed rock that is retained on the No. 4 sieve (4.75 mm).

Fine aggregate   Often refers to sand (passes the No. 4 sieve and is retained on the No. 200 U.S. standard sieve).

Open-graded aggregate   Generally refers to gravel that does not contain any soil particles finer than the No. 4 sieve.

Angle of repose   The maximum inclination that a slope can assume through natural processes. The term is typically only used for cohesionless soil. The angle of repose for dry sand will be equal to its friction angle.

Apparent opening size   The approximate largest particle size that would effectively pass through a geotextile.

Approval   A written engineering or geologic opinion by the responsible engineer, geologist of record, or responsible principal of the engineering company concerning the process and completion of the work unless it specifically refers to the building official.

Approved plans The current grading plans that bear the stamp of approval of the building official.

Approved testing agency A facility the testing operations of which are controlled and monitored by a the testing operation of which registered civil engineer and which is equipped to perform and certify the tests as required by the local building code or building official.

Armor The man-made facing of riverbanks, shorelines, or embankments with cobbles or boulders in order to resist erosion or scour. Also see Riprap.

As-graded (or As-built) The surface conditions at the completion of grading.

Asphalt A dark brown to black cementitious material where the main ingredient is bitumen (high molecular hydrocarbons) that occurs in nature or is obtained from petroleum processing.

Asphalt concrete (AC) A mixture of asphalt and aggregate that is compacted into a dense pavement surface. Asphalt concrete is often prepared in a batch plant.

Backdrain Generally a pipe and gravel or similar drainage system placed behind earth retaining structures such as buttresses, stabilization fills, and retaining walls.

Backfill Soil material placed behind or on top of an area that has been excavated. For example, backfill is placed behind retaining walls and in utility trench excavations.

Base A layer of specified or selected material of planned thickness constructed on the subgrade or subbase for the purpose of providing support to the overlying concrete or asphalt concrete surface of roads and airfields. Also known as the base course.

Bell The enlarged portion of the bottom of a drilled shaft foundation. A bell is used to increase the end bearing resistance. Not all drilled shafts have bells.

Bench A relatively level step excavated into earth material on which fill is to be placed.

Berm A raised bank or path of soil. For example, a berm is often constructed at the top of slopes to prevent water from flowing over the top of the slope.

Borrow Earth material acquired from an off-site location for use in grading on a site.

Brooming Crushing or separation of wood fibers at the butt (top of the pile) of a timber pile while it is being driven.

Building official The city engineer, director of the local building department or a duly delegated representative.

Bulking An increase in volume of soil or rock caused by its excavation. For example, rock or dense soil will increase in volume upon excavation or by being dumped into a truck for transportation.

Buttress fill A fill mass, the configuration of which is designed by engineering calculations to stabilize a slope exhibiting adverse geologic features. A buttress is generally specified by minimum key width and depth and by maximum backcut angle. A buttress normally contains a backdrainage system.

Caisson Sometimes large diameter piers are referred to as caissons. Another definition is a large structural chamber utilized to keep soil and water from entering into a deep excavation or construction area. Caissons may be installed by being sunk in place or by excavating the bottom of the unit as it slowly sinks to the desired depth.

Cat A term that is slang for Caterpillar grading or construction equipment.

Clearing, brushing, and grubbing The removal of vegetation (grass, brush, trees, and similar plant types) by mechanical means.

Clogging For a geotextile, a decrease in permeability due to soil particles that have either lodged in the geotextile openings or have built up a restrictive layer on the surface of the geotextile.

Compaction The densification of a fill by mechanical means. Also see Compaction in Sec. A. 2.

Compaction equipment   Compaction equipment can be grouped generally into five different types or classifications: sheepsfoot, vibratory, pneumatic, high-speed tamping foot, and chopper wheels (for municipal landfill). Combinations of these types are also available.

Compaction production   Compaction production is expressed in compacted cubic meters (m^3) or compacted cubic yards (yd^3) per hour.

Concrete   A mixture of aggregates (sand and gravel) and paste (cementitious materials and water). The paste binds the aggregates together into a rocklike mass as the paste hardens because of the chemical reactions between the cement and the water.

Contractor   A person or company under contract or otherwise retained by the client to perform demolition, grading, and other site improvements.

Cut   See Excavation.

Cut-fill transition   The location in a building pad where on one side the pad has been cut down exposing natural or rock material, while on the other side, fill has been placed.

Dam   A structure built to impound water or other fluid products such as tailing waste, wastewater effluent, and the like.

    Homogeneous earth dam   An earth dam, the embankment of which is formed of one soil type without a systematic zoning of fill materials.

    Zoned earth dam   An earth dam embankment zoned by the systematic distribution of soil types according to their strength and permeability characteristics, usually with a central impervious core and shells of coarser materials.

Debris   All products of clearing, grubbing, demolition, or contaminated soil material that is unsuitable for reuse as compacted fill and/or any other material so designated by the geotechnical engineer or building official.

Dewatering   The process used to remove water from a construction site, such as pumping from wells in order to lower the groundwater table during a foundation excavation.

Down drain   A device for collecting water from a swale or ditch located on or above a slope and safely delivering it to an approved drainage facility.

Dozer   A term that is slang for bulldozer construction equipment.

Drainage   The removal of surface water from the site.

Drawdown   The lowering of the groundwater table that occurs in the vicinity of a well that is in the process of being pumped.

Earth material   Any rock, natural soil, or fill, or any combination thereof.

Electro-osmosis   A method of dewatering, applicable for silts and clays, in which an electric field is established in the soil mass to cause the movement by electro-osmotic forces of pore water to well-point cathodes.

Erosion control devices (temporary)   Devices that are removable and can rarely be salvaged for subsequent reuse. In most cases they will last no longer than one rainy season. They include sandbags, gravel bags, plastic sheeting (visqueen), silt fencing, straw bales, and similar items.

Erosion control system   A combination of desilting facilities, and erosion protection, including effective planting to protect adjacent private property, watercourses, public facilities, and receiving waters from any abnormal deposition of sediment or dust.

Essential facility   Essential facilities can be defined as those structures or buildings that must be safe and usable for emergency purposes after an earthquake or other natural disaster in order to preserve the health and safety of the general public. Typical examples of essential facilities are hospitals and other medical facilities having surgery or emergency treatment areas, fire and police stations, and municipal government disaster operations and communication centers deemed to be vital in emergencies. According to the *Standard Specifications for Highway Bridges* (AASHTO, 1996), other facilities that could be classified as essential are military bases, supply depots, and National Guard installations; facilities such as schools and arenas which could provide shelter or be converted to aid stations; major airports; defense industries and those that could easily or logically be converted to such; refineries, fuel storage, and distribution centers; major railroad terminals, railheads, docks, and truck terminals; major power plants including nuclear power facilities and hydroelectric centers at major dams, and other facilities that the state considers important from a national defense viewpoint or during emergencies resulting from natural disasters or other unforeseen circumstances. Essential bridges are defined as those that must continue to function after an earthquake. Transportation routes to critical facilities such as hospitals, police, fire stations, and communication centers must continue to function and bridges required for this purpose should be classified as essential. In addition, a bridge that has the potential to impede traffic if it collapses onto an essential route should also be classified as essential.

Excavation   The mechanical removal of earth material, also referred to as cut material.

Fill   A deposit of earth material placed by artificial means. An engineered (or structural) fill refers to a fill in which the geotechnical engineer has, during grading, made sufficient tests to enable the conclusion that the fill has been placed in substantial compliance with the recommendations of the geotechnical engineer and the governing agency requirements.

Footing   A structural member typically installed at a shallow depth that is used to transmit structural loads to the soil or rock strata. Common types of footings include combined footings, spread (or pad) footings, and strip (or wall) footings.

Forms   The purpose of a form is to confine and support the fluid concrete until it hardens. For excavated footings in soil or rock material, the sides and bottom of the excavation serves as the form, provided the soil can remain stable during construction. In other cases, forms are usually constructed out of wood.

Foundation   That part of the structure that supports the weight of the structure and transmits the load to underlying soil or rock.

   Deep foundation   A foundation that derives its support by transferring loads to soil or rock at some depth below the structure.

   Shallow foundation   A foundation that derives its support by transferring load directly to soil or rock at a shallow depth.

Freeze   Also known as setup, an increase in the load capacity of a pile after it has been driven. Freeze is caused primarily by the dissipation of excess pore water pressures.

Geosynthetic   A planar product manufactured from polymeric material and typically placed in soil to form an integral part of a drainage, reinforcement, or stabilization system. Types include geotextiles, geogrids, geonets, and geomembranes.

Geotextile   A permeable geosynthetic composed solely of textiles.

Grade   The vertical location of the ground surface.

   Existing grade   The ground surface prior to grading.

   Finished grade   The final grade of the site, which conforms to the approved plan.

   Lowest adjacent grade   The lowest point in elevation of the finished surface of the ground, paving, or sidewalk that is adjacent to the structure.

   Natural grade   The ground surface unaltered by artificial means.

   Rough grade   The stage at which the grade approximately conforms to the approved plan.

Grading   Any operation consisting of excavation, filling, or a combination thereof.

Grading contractor A contractor licensed and regulated who specializes in grading work or is otherwise licensed to do grading work.

Grading permit An official document or certificate issued by the building official authorizing grading activity as specified by approved plans and specifications.

Grouting The process of injecting grout into soil or rock formations to change their physical characteristics. Common examples include grouting to decrease the permeability of a soil or rock strata, or compaction grouting to densify loose soil or fill.

Hillside site A site that entails cut and/or fill grading of a slope which may be adversely affected by drainage and/or stability conditions within or outside the site, or which may cause an adverse affect on adjacent property.

Hydraulic fill A fill placed by transporting soils through a pipe using large quantities of water. These fills are generally loose because they have little or no mechanical compaction during construction.

Inspection See Special inspection.

Jetting The use of a water jet to facilitate the installation of a pile. It can also refer to the fluid placement of soil, such as jetting in the soil for a utility trench.

Key A designed compacted fill placed in a trench excavated in earth material beneath the toe of a proposed fill slope.

Keyway An excavated trench into competent earth material beneath the toe of a proposed fill slope.

Lift During compaction operations, a lift is a layer of soil that is dumped by the construction equipment and then subsequently compacted as structural fill.

Mixed-in-place pile A soil-cement pile that is created by forcing grout through a hollow shaft in the ground. As the grout is forced into the soil, an auger-like head (that is attached to the hollow shaft) mixes the soil to create the soil cement.

Necking A reduction in cross-sectional area of a drilled shaft as a result of the inward movement of the adjacent soils.

Open cut An excavation in rock or soil that is made through a hill or other topographic feature in order to construct a highway, railroad, or waterway. The open cut can consist of a single cut slope, multiple cut slopes, and/or benches.

Owner Any person, agency, firm, or corporation having a legal or equitable interest in a given real property.

Permanent erosion control devices Improvements that remain throughout the life of the development. They include terrace drains, down-drains, slope landscaping, channels, storm drains, and the like.

Permit An official document or certificate issued by the building official authorizing performance of a specified activity.

Pier A deep foundation system, similar to a cast-in-place pile, that consists of column-like reinforced concrete members. Piers are often of large enough diameter to enable down-hole inspection. Piers are also commonly referred to as drilled shafts, bored piles, or drilled caissons.

Pile A deep foundation system, consisting of relatively long, slender, column-like members that are often driven into the ground.

> **Batter pile** A pile driven in at an angle inclined to the vertical to provide higher resistance to lateral loads.

> **Combination end-bearing and friction pile** A pile that derives its capacity from combined end-bearing resistance developed at the pile tip and frictional and/or adhesion resistance on the pile perimeter.

End-bearing pile   A pile, the support of which capacity is derived principally from the resistance of the foundation material on which the pile tip rests.

Friction pile   A pile, the support capacity of which is derived principally from the resistance of the soil friction and/or adhesion mobilized along the side of the embedded pile.

Pozzolan   For concrete mix design, a siliceous or siliceous and aluminous material that will chemically react with calcium hydroxide within the cement paste to form compounds having cementitious properties.

Precise grading permit   A permit that is issued on the basis of approved plans which show the precise structure location, finish elevations, and all on-site improvements.

Relative compaction   The degree of compaction (expressed as a percentage) defined as the field dry density divided by the laboratory maximum dry density.

Ripping or rippability   The characteristic of rock or dense and rocky soils that can be excavated without blasting. Ripping is accomplished by using equipment such as a Caterpillar ripper, ripper-scarifiers, tractor-ripper, or impact ripper. Ripper performance has been correlated with the seismic wave velocity of the soil or rock.

Riprap   Rocks that are generally less than 2 tons (1800 kg) in mass that are placed on the ground surface, on slopes or at the toe of slopes, or on top of structures to prevent erosion by wave action or strong currents.

Running soil or running ground   In tunneling or trench excavations, a granular material that tends to flow or *run* into the excavation.

Sand boil   Also known as sand blows, sand volcanoes, or silt volcanoes. The ejection of sand at ground surface, usually forming a cone shape, caused by underground piping. Sand boils can also form at ground surface when there has been liquefaction of underlying soil during an earthquake.

Shaft   Usually a vertical or near vertical excavation that extends from ground surface and is constructed in order to access tunnels, chambers, or other underground works.

Shear key   Similar to a buttress, however, it is generally constructed by excavating a slot within a natural slope in order to stabilize the upper portion of the slope without grading encroachment into the lower portion of the slope. A shear key is also often used to increase the factor of safety of an ancient landslide.

Shotcrete   Mortar or concrete pumped through a hose and projected at high velocity onto a surface. Shotcrete can be applied by a *wet* or *dry* mix method.

Shrinkage factor   When the loose material is worked into a compacted state, the shrinkage factor (SF) is the ratio of the volume of compacted material to the volume of borrow material.

Site   The particular lot or parcel of land where grading or other development is performed.

Slope   An inclined ground surface. For graded slopes, the steepness is generally specified as a ratio of horizontal:vertical (e.g., 2:1 slope) or as a percentage (e.g., 50 percent slope). Common types of slopes include natural (unaltered) slopes, cut slopes, false slopes (temporary slopes generated during fill compaction operations), and fill slopes.

Slough   Loose, noncompacted fill material generated during grading operations. Slough can also refer to a shallow slope failure, such as sloughing of the slope face.

Slump   In the placement of concrete, the slump is a measure of consistency of freshly mixed concrete as measured by the slump test. In geotechnical engineering, a slump could also refer to a slope failure.

Slurry seal   In the construction of asphalt pavements, a slurry seal is a fluid mixture of bituminous emulsion, fine aggregate, mineral filler, and water. A slurry seal is applied to the top surface of an asphalt pavement in order to seal its surface and prolong its wearing life.

Soil stabilization   The treatment of soil to improve its properties. There are many methods of soil stabilization such as adding gravel, cement, or lime to the soil. The soil could also be stabilized by using geotextiles, by drainage, or through the use of compaction.

Special inspection   A type of inspection that requires special expertise to ensure compliance with the approved construction documents and plans.

Specification   A precise statement in the form of specific requirements. The requirements could be applicable to a material, product, system, or engineering service.

Stabilization fill   Similar to a buttress fill, the configuration of which is typically related to slope height and is specified by the standards of practice for enhancing the stability of locally adverse conditions. A stabilization fill is normally specified by minimum key width and depth and by maximum backcut angle. A stabilization fill usually has a backdrainage system.

Staking   During grading, staking is the process where a land surveyor places wood stakes that indicate the elevation of existing ground surface and the final proposed elevation per the grading plans.

Structure   A structure is defined as that which is built or constructed, an edifice or building of any kind, or any piece of work artificially built up or composed of parts joined together in some definite manner.

Subdrain (for canyons)   A pipe and gravel or similar drainage system placed in the alignment of canyons or former drainage channels. After placement of the subdrain, structural fill is placed on top of the subdrain.

Subgrade   For roads and airfields, the subgrade is defined as the underlying soil or rock that supports the pavement section (subbase, base, and wearing surface). The subgrade is also referred to as the basement soil or foundation soil.

Substructure   The foundation of a building or other structure.

Sulfate (SO_4)   A chemical compound occurring in some soils which, at above certain levels of concentration, can have a corrosive effect on concrete and some metals.

Sump   A small pit excavated in the ground or through the basement floor to serve as a collection basin for surface runoff or groundwater. A sump pump is used to periodically drain the pit when it fills with water.

Superstructure   The portion of the structure located above the foundation and includes beams, columns, floors, and other structural and architectural members.

Tack coat   In the construction of asphalt pavements, the tack coat is a bituminous material that is applied to an existing surface to provide a bond between different layers of the asphalt concrete.

Tailings   In terms of grading, tailings are nonengineered fill that accumulate on or adjacent to equipment haul-roads. Tailings could also be the waste products generated during a mining operation.

Terrace   A relatively level step constructed in the face of a graded slope surface for drainage control and maintenance purposes.

Tremie   Material placed under water through a tremie pipe in such a manner that it rests on the bottom without mixing with the water. An example is concrete placed at the bottom of a slurry filled trench.

Underpinning   Piles or other types of foundations built to provide new support for an existing foundation. Underpinning is often used as a remedial measure.

Vibrodensification   The densification or compaction of cohesionless soils by imparting vibrations into the soil mass so as to rearrange soil particles resulting in fewer voids in the overall mass.

Walls:

    Bearing wall   In a general sense, a bearing wall carries and supports an overlying load, such as the weight of the roof. The removal of a bearing wall could cause partial or complete collapse of the structure. In the *International Building Code*, a bearing wall is defined as any metal or wood stud wall that supports more than 100 lb per linear ft of vertical load in addition to its own weight and any masonry or concrete wall that supports more than 200 lb per linear ft of vertical load in addition to its own weight.

Cutoff wall The construction of tight sheeting or a barrier of impervious material extending downward to an essentially impervious lower boundary to intercept and block the path of ground-water seepage. Cutoff walls are often used in dam construction.

Nonbearing wall A wall, such as a partition wall, that does not support the overlying floors. The removal of a nonbearing wall should have no significant effect on the strength of the building.

Retaining wall A wall designed to resist the lateral displacement of soil or other materials.

Shear wall A wall designed to resist lateral forces parallel to the plane of the wall. Shear walls are used to resist the lateral force induced by an earthquake (Also see Shear wall in Sec.A. 5).

Water-cement ratio For concrete mix design, the ratio of the mass of water (exclusive of that part absorbed by the aggregates) to the mass of cement.

Water-cementitious materials ratio Similar to the water-cement ratio, the ratio of the mass of water (exclusive of that part absorbed by the aggregates) to the mass of cementitious materials in the concrete mix. Commonly used cementitious materials for the concrete mix include Portland cement, fly ash, pozzolan, slag, and silica fume.

Well point During the pumping of groundwater, the well point is the perforated end section of a well pipe where the groundwater is drawn into the pipe.

Windrow A string of large rocks buried within engineered fill in accordance with guidelines set forth by the geotechnical engineer or governing agency requirements.

Workability of concrete The ability to manipulate a freshly mixed quantity of concrete with a minimum loss of homogeneity.

A.5 GEOTECHNICAL EARTHQUAKE ENGINEERING TERMINOLGY

Active fault See Fault.

Aftershock An earthquake that follows a larger earthquake or main shock and originates in or near the rupture zone of the larger earthquake. Generally, major earthquakes are followed by a large number of aftershocks, usually decreasing in frequency with time.

Amplitude The maximum height of a wave crest or depth of a trough.

Anticline Layers of rock that have been folded in a generally convex upward direction. The core of an anticline contains the older rocks.

Array An arrangement of seismometers or geophones that feed data into a central receiver.

Arrival The appearance of seismic energy on a seismic record.

Arrival time The time at which a particular wave phase arrives at a detector.

Aseismic A term that indicates the event is not due to an earthquake. An example is an aseismic zone, which indicates an area that has no record of earthquake activity.

Asthenosphere The layer or shell of the earth below the lithosphere. Magma can be generated within the asthenosphere.

Attenuation relationship A relationship that is used to estimate the peak horizontal ground acceleration at a specified distance from the earthquake. Numerous attenuation relationships have been developed. Many attenuation relationships relate the peak horizontal ground acceleration to the earthquake magnitude and closest distance between the site and the focus of the earthquake. Attenuation relationships have also been developed assuming *soft soil* or *hard rock* sites.

Base shear   The earthquake induced total design lateral force or shear assumed to act on the base of the structure.

Body wave magnitude scales (m_b and M_B)   The body wave magnitude scales are based on the amplitude of the first few P waves to arrive at the seismograph.

Body waves   A seismic wave that travels through the interior of the earth. P waves and S waves are body waves.

Continental drift   The theory, first advanced by Alfred Wegener, that the earth's continents were originally one land mass. Pieces of the land mass split off and migrated to form the continents.

Core (of the earth)   The innermost layers of the earth. The inner core is solid and has a radius of about 1300 km. The outer core is fluid and is about 2300 km thick. S-waves cannot travel through the outer core.

Crust   The thin outer layer of the earth's surface, averaging about 10 km thick under the oceans and up to about 50 km thick on the continents.

Cyclic mobility   The concept of cyclic mobility is used to describe large-scale lateral spreading of slopes. In this case, the static driving forces do not exceed the shear strength of the soil along the slip surface, and thus the ground is not subjected to a flow slide. Instead, the driving forces only exceed the resisting forces during those portions of the earthquake that impart net inertial forces in the downslope direction. Each cycle of net inertial forces in the downslope direction cause the driving forces to exceed the resisting forces along the slip surface, resulting in progressive and incremental lateral movement. Often the lateral movement and ground surface cracks first develop at the unconfined toe and then the slope movement and ground cracks progressively move upslope.

Design response spectrum   For the design of structures, the response spectrum is an elastic response spectrum that includes viscous damping and is used to represent the dynamic effects of the earthquake. The response spectrum could be a site-specific spectrum based on a study of the geologic, tectonic, seismological, and soil characteristics of the site.

Dip   See Strike and dip

Earthquake   Shaking of the earth caused by the sudden rupture along a fault or weak zone in the earth's crust or mantle. Earthquakes can also be caused by other events, such as a volcanic eruption.

Earthquake swarm   A series of minor earthquakes, none of which may be identified as the main shock, occurring in a limited area and time.

En échelon   A geologic feature that has a staggered or overlapping arrangement. An example would be surface fault rupture, where the rupture is in a linear form, but there are individual features that are oblique to the main trace.

Epicenter   The location on the ground surface that is directly above the point where the initial earthquake motion originated.

Fault   A fracture or weak zone in the earth's crust or upper mantle along which movement has occurred. Earthquakes cause faults, and earthquakes are likely to recur on preexisting faults. Although definitions vary, a fault is often considered to be active if movement has occurred within the last 11,000 years (Holocene geologic time). Typical terms used to describe different types of faults are as follows:

   Strike-slip fault   A strike-slip fault is defined as a fault on which the movement is parallel to the strike of the fault.

   Transform fault   A strike-slip fault of plate-boundary dimensions that transforms into another plate-boundary structure at its terminus.

Normal fault A normal fault is defined as a fault where the hanging wall block has moved downward with respect to the footwall block. The hanging wall is defined as the overlying side of a nonvertical fault.

Reverse fault A reverse fault would be defined as a fault where the hanging wall block has moved upward with respect to the footwall block.

Thrust fault A thrust fault is defined as a reverse fault where the dip is less than or equal to 45°.

Blind fault A blind fault is defined as a fault that has never extended upward to the ground surface. Blind faults often terminate in the upward region of an anticline.

Blind thrust fault A blind reverse fault where the dip is less than or equal to 45°.

Longitudinal step fault A series of parallel faults. These parallel faults develop when the main fault branches upward into several subsidiary faults.

Dip-slip fault A fault which experiences slip only in the direction of its dip, or in other words, the movement is perpendicular to the strike. Thus a fault could be described as a *dip-slip normal fault*, which would indicate that it is a normal fault with the slip only in the direction of its dip.

Oblique-slip fault A fault that experiences components of slip in both its strike and dip directions. A fault could be described as a *oblique-slip normal fault*, which would indicate that it is a normal fault with components of slip in both the strike and dip directions.

Fault scarp This generally only refers to a portion of the fault that has been exposed at ground surface due to ground surface fault rupture. The exposed portion of the fault often consists of a thin layer of *fault gouge*, which is a clayey seam that has formed during the slipping or shearing of the fault and often contains numerous slickensides.

First arrival The first recorded data attributed to seismic waves generated by the fault rupture.

Flow slide If liquefaction occurs in or under a sloping soil mass, the entire mass could flow or translate laterally to the unsupported side in a phenomenon termed a flow slide. Such slides tend to develop in loose, saturated, cohesionless materials that liquefy during the earthquake.

Focal depth The distance between the focus and epicenter of the earthquake.

Focus Also known as the *hypocenter* of an earthquake, the location within the earth that coincides with the initial slip of the fault. In essence, the focus is the location where the earthquake was initiated.

Foreshock A small tremor that commonly precedes a larger earthquake or main shock by seconds to weeks and that originates in or near the rupture zone of the larger earthquake.

Gouge The exposed portion of the fault often consists of a thin layer of *fault gouge*, which is a clayey seam that has formed during the slipping or shearing of the fault and often contains numerous slickensides.

Graben The dropping of a crustal block along faults. The crustal block usually has a length that is much greater than its width, resulting in the formation of a long narrow valley. A graben can also be used to describe the down-dropping of the ground surface, such as a graben area associated with a landslide.

Hazard A risk. An object or situation that has the possibility of injury or damage.

Hypocenter See Focus.

Inactive fault Definitions vary, but in general, an inactive fault has had no displacement over a sufficiently long period of time in the geologic past so that displacements in the foreseeable future are considered unlikely.

Intensity (of an earthquake) The intensity of an earthquake is based on the observations of damaged structures and the presence of secondary effects, such as earthquake induced landslides, liquefaction, and ground cracking. The intensity of an earthquake is also based on the degree to which individuals felt the earthquake, which is determined through interviews. The most commonly used scale for the determination of the intensity of an earthquake is the modified Mercalli intensity scale.

Isolator unit A horizontally flexible and vertically stiff structural element that allows for large lateral deformation under the seismic load.

Isoseismal line A line connecting points on the earth's surface at which earthquake intensity is the same. It is usually a closed curve around the epicenter.

Leaking mode A surface seismic wave that is imperfectly trapped so that its energy leaks or escapes across a layer boundary, causing some attenuation or loss of energy.

Liquefaction The sudden and large decrease of shear strength of a submerged cohesionless soil caused by contraction of the soil structure, produced by shock or earthquake induced shear strains, associated with a sudden but temporary increase of pore water pressures. Liquefaction occurs when the increase in pore water pressures causes the effective stress to become equal to zero and the soil behaves as a liquid.

Lithosphere The outermost layer of the earth. It commonly includes the crust and the more rigid part of the upper mantle.

Love wave Surface waves that are analogous to S waves in that they are transverse shear waves that travel close to the ground surface. It is named after A. E. H. Love, the English mathematician, who discovered it.

Low-velocity zone Any layer in the earth in which seismic wave velocities are lower than in the layers above and below.

Magnitude (of the earthquake) The magnitude of an earthquake is a measure of the size of the earthquake at its source. There are many different methods used to determine the magnitude of an earthquake, such as the local magnitude scale, surface wave magnitude scale, the body wave magnitude scales, and the moment magnitude scale.

Major earthquake An earthquake having a magnitude of 7.0 or larger on the Richter scale.

Mantle The layer of material that lies between the crust and the outer core of the earth. It is approximately 2900 km thick and is the largest of the earth's major layers.

Maximum credible earthquake The maximum credible earthquake (MCE) is often considered to be the largest earthquake that can reasonably be expected to occur based on known geologic and seismologic data. In essence, the maximum credible earthquake is the maximum earthquake that an active fault can produce, considering the geologic evidence of past movement and recorded seismic history of the area. Other terms that have been used to describe similar worst-case levels of shaking include safe shutdown earthquake (used in the design of nuclear power plants), maximum capable earthquake, maximum design earthquake, contingency level earthquake, safe level earthquake, credible design earthquake, and contingency design earthquake. In general, these terms are used to describe the upper-most level of earthquake forces in the design of essential facilities. The maximum credible earthquake is determined for particular earthquakes or levels of ground shaking. As such, the analysis used to determine the maximum credible earthquake is typically referred to as a *deterministic method*.

Maximum probable earthquake There are many different definitions of the maximum probable earthquake. The maximum probable earthquake is based on a study of nearby active faults. By using attenuation relationships, the maximum probable earthquake magnitude and maximum probable peak ground acceleration can be determined. A commonly used definition of maximum probable earthquake is the largest earthquake a fault is predicted capable of generating within a specified time period of concern such as 50 or 100 years. Maximum probable earthquakes are most likely to occur within the time span of most developments, and therefore, are commonly used in assessing seismic risk. Another commonly used definition of a maximum probable earthquake is an earthquake that will produce a peak ground acceleration a_{max} with a 50 percent probability of exceedence in 50 years. Other terms that have been used to describe earthquakes of similar size are operating basis earthquake, operating level earthquake, probable design earthquake, and strength level earthquake.

Microearthquake   An earthquake having a magnitude of 2 or less on the Richter scale.

Modified mercalli intensity scale   See Intensity (of an earthquake).

Mohorovicic discontinuity (or moho discontinuity)   The boundary surface or sharp seismic-velocity discontinuity that separates the earth's crust from the underlying mantle. Named for Andrija Mohorovicic, the Croatian seismologist who first suggested its existence.

Normal fault   (see Fault)

P wave   A body wave that is also known as the primary wave, compressional wave, or longitudinal wave. It is a seismic wave that causes a series of compressions and dilations of the materials through which it travels. The P wave is the fastest wave and is the first to arrive at a site. Being a compression-dilation type wave, P waves can travel through both solids and liquids. Because soil and rock are relatively resistant to compression-dilation effects, the P wave usually has the least impact on ground surface movements.

Paleomagnetism   The natural magnetic traces that reveal the intensity and direction of the earth's magnetic field in the geologic past. Also defined as the study of these magnetic traces.

Paleoseismology   The study of ancient (i.e., prehistoric) earthquakes.

Peak ground acceleration (PGA)   The peak ground acceleration is also known as the maximum horizontal ground acceleration. The peak ground acceleration can be based on an analysis of historical earthquakes or based on probability. An attenuation relationship is used to relate the peak ground acceleration to the earthquake magnitude and closest distance between the site and the focus of the earthquake.

Period   The time interval between successive crests in a wave train. The period is the inverse of the frequency.

Plate boundary   The location where two or more plates in the earth's crust meet.

Plate tectonics   According to the plate tectonic theory, the earth's surface contains tectonic plates, also known as lithosphere plates, with each plate consisting of the crust and the more rigid part of the upper mantle. Depending on the direction of movement of the plates, there are three types of plate boundaries: divergent boundary, convergent boundary, and transform boundary.

Pseudostatic analysis   A method that ignores the cyclic nature of the earthquake and treats it as if it applies an additional static force upon the slope or retaining wall.

Rayleigh wave   Surface waves that have been described as being similar to the surface ripples produced by a rock thrown into a pond. These seismic waves produce both vertical and horizontal displacement of the ground as the surface waves propagate outward. They are usually felt as a rolling or rocking motion and in the case of major earthquakes, can be seen as they approach. They are named after Lord Rayleigh, the English physicist who predicted their existence.

Recurrence interval   The approximate length of time between earthquakes in a specific seismically active area.

Resonance   A condition where the frequency of the structure is equal to the natural frequency of the vibrating ground. At resonance, the structure will experience the maximum horizontal displacement.

Response spectrum   See Design response spectrum.

Richter magnitude scale   Also known as the local magnitude scale, a system used to measure the strength of an earthquake. Professor Charles Richter developed this earthquake magnitude scale in 1935 as a means of categorizing local earthquakes.

Rift valley   A divergent boundary between tectonic plates can create a rift valley, which is defined as a long and linear valley formed by tectonic depression accompanied by extension. Earthquakes at a rift valley are often due to movement on normal faults. Examples of rift valleys are the East African Rift and the Rhine Graben.

Risk   See Seismic risk.

Rupture zone   The area of the earth through which faulting occurred during an earthquake. For great earthquakes, the rupture zone may extend several hundred kilometers in length and tens of kilometers in width.

S wave   A body wave that is also known as the secondary wave, shear wave, or transverse wave. The S wave causes shearing deformations of the materials through which it travels. Because liquids have no shear resistance, S waves can only travel through solids. The shear resistance of soil and rock is usually less than the compression-dilation resistance, and thus a S wave travels slower through the ground than a P wave. Soil is weak in terms of its shear resistance and S waves typically have the most impact on ground surface movements.

Sand boil   Also known as sand blows, sand volcanoes, or silt volcanoes. The ejection of sand at ground surface, usually forming a cone shape, caused by liquefaction of underlying soil during an earthquake. Piping can also cause sand boils (see Sec.A. 4).

Seiche   Identical to tsunami, except that it occurs in an inland body of water, such as a lake. It can be caused by lake-bottom earthquake movements or by volcanic eruptions and landslides within the lake. A seiche has been described as being similar to the sloshing of water in a bathtub.

Seismic or seismicity   Dealing with earthquake activity.

Seismic belt   An elongated earthquake zone. Examples include the circum-Pacific, Mediterranean, and Rocky Mountain seismic belts.

Seismic risk   The probability of human and property loss due to an earthquake.

Seismogram   See Seismograph.

Seismograph   A seismograph is defined as an instrument that records, as a function of time, the motion of the earth's surface due to the seismic waves generated by the earthquake. The actual record of ground shaking from the seismograph, known as a seismogram, can provide information about the nature of the earthquake. The simplest seismographs can consist of a pendulum or a mass attached to a spring, and they are used to record the horizontal movement of the ground surface. For the pendulum type seismograph, a pen is attached to the bottom of the pendulum, and the pen is in contact with a chart that is firmly anchored to the ground. When the ground shakes during an earthquake, the chart moves, but the pendulum and its attached pen tend to remain more or less stationary because of the effects of inertia. The pen then traces the horizontal movement between the relatively stationary pendulum and the moving chart. After the ground shaking has ceased, the pendulum will want to return to a stable position, and thus could indicate false ground movement. Therefore a pendulum damping system is required so that the ground displacements recorded on the chart will produce a record that is closer to the actual ground movement. Most modern seismographs (accelerographs) use an electronic transducer that produces an output voltage that is proportional to the acceleration. This output voltage is recorded and then converted to acceleration and plotted versus time.

Seismology   The study of earthquakes.

Shear wall   Sometimes referred to as a vertical diaphragm or structural wall, a shear wall is designed to resist lateral forces parallel to the plane of the wall. Shear walls are used to resist the lateral forces induced by the earthquake.

Spreading center   An elongated region where two plates are being pulled away from each other. New crust is formed as molten rock is forced upward into the gap. An example is sea floor spreading, which has created the Mid-Atlantic ridge. Another example is a rift valley, such as the East African Rift.

Strike and dip (of a fault plane)   Strike is the azimuth of a horizontal line drawn on the fault plane. The dip is measured in a direction perpendicular to the strike and is the angle between the inclined fault plane and a horizontal plane. The strike and dip provide a description of the orientation of the fault plane in space.

Strike-slip fault   See Fault.

Subduction zone   An elongated region along which a plate descends relative to another plate. An example is the descent of the Nazca plate beneath the South American plate along the Peru-Chile Trench.

Syncline   Layers of rock that have been folded in a generally concave upward direction. The core of an syncline contains the younger rocks.

Travel time   The time required for a seismic wave train to travel from its source to a point of observation.

Tsunami   Tsunami is a Japanese word that when translated into English means *harbor wave*. Tsunami is an ocean wave that is created by a disturbance that vertically displaces a column of seawater. Tsunami can be generated by an oceanic meteorite impact, submarine landslide, or earthquake if the sea floor abruptly deforms and vertically displaces the overlying water. Earthquakes generated at sea-floor subduction zones are particularly effective in generating tsunamis. Waves are formed as the displaced water mass, which acts under the influence of gravity, attempts to regain its equilibrium. A tsunami is different from a normal ocean wave in that it has a long period and wavelength. For example, tsunami can have a wavelength in excess of 60 mi (100 km) and a period on the order of 1 h. In the Pacific Ocean, where the typical water depth is about 13,000 ft (4000 m), tsunami travels at about 650 ft/sec (200 m/sec). Because the rate at which a wave loses its energy is inversely related to its wavelength, tsunamis not only propagate at high speed, they can also travel long transoceanic distances with limited energy losses. The tsunami is transformed as it leaves the deep water of the ocean and travels into the shallower water near the coast. The tsunami's speed diminishes as it travels into the shallower coastal water and its height grows. While the tsunami may be imperceptible at sea, the shoaling effect near the coast causes the tsunami to grow to be several meters or more in height. When it finally reaches the coast, the tsunami may develop into a rapidly rising or falling tide, a series of breaking waves, or a tidal bore.

REFERENCES

AASHTO (2002). *Standard Specifications for Highway Bridges*. 17th ed., Prepared by the American Association of State Highway and Transportation Officials (AASHTO), Washington, DC.

ASTM (2004). *Annual Book of ASTM Standards*, Vol. 04.08, *Soil and Rock (I)*. Standard No. D 653-03, "Standard Terminology Relating to Soil, Rock, and Contained Fluids." Terms prepared jointly by the American Society of Civil Engineers and ASTM, West Conshohocken, PA, pp. 46–78.

ASTM (2000). *Annual Book of ASTM Standards*, Vol. 04.09, *Soil and Rock (II), Geosynthetics*. Standard No. D 4439-99, "Standard Terminology for Geosynthetics," West Conshohocken, PA, pp. 852–855.

Asphalt Institute (1984). *Thickness Design-Asphalt Pavements for Highways and Streets*. Published by The Asphalt Institute, College Park, MD, 80 pp.

Caterpillar Performance Handbook (1997). 28th ed., Prepared by Caterpillar, Inc., Peoria, IL, 1006 pp.

Coduto, D. P. (1994). *Foundation Design, Principles and Practices*. PrenticeHall, Englewood Cliffs, NJ, 796 pp.

Holtz, R. D. and Kovacs, W. D. (1981). *An Introduction to Geotechnical Engineering*, PrenticeHall, Englewood Cliffs, NJ, 733 pp.

International Building Code (2006). International Code Council, Country Club Hills, IL.

Kramer, S.L. (1996). *Geotechnical Earthquake Engineering*. Prentice Hall, Englewood Cliffs, NJ.

Krinitzsky, E. L., Gould, J. P., and Edinger, P. H. (1993). *Fundamentals of Earthquake-Resistant Construction.* Wiley, New York, NY, 299 pp.

Lambe, T. W. and Whitman, R. V. (1969). *Soil Mechanics.* Wiley, New York, NY, 553 pp.

McCarthy, D. F. (1977*). Essentials of Soil Mechanics and Foundations.* Reston Publishing Company, Reston Virginia, VA, 505 pp.

NAVFAC DM-7.1 (1982). *Soil Mechanics, Design Manual 7.1,* Department of the Navy, Naval Facilities Engineering Command, Alexandria, VA, 364 pp.

NAVFAC DM-7.2 (1982). *Foundations and Earth Structures, Design Manual 7.2,* Department of the Navy, Naval Facilities Engineering Command, Alexandria, VA, 253 pp.

NAVFAC DM-7.3 (1983). *Soil Dynamics, Deep Stabilization, and Special Geotechnical Construction, Design Manual 7.3,* Department of the Navy, Naval Facilities Engineering Command, Alexandria, VA, 106 pp.

Orange County Grading Manual (1993). *Orange County Grading Manual,* part of the *Orange County Grading and Excavation Code,* prepared by Orange County, CA.

Stokes, W. L., and Varnes, D. J. (1955). Glossary of Selected Geologic Terms with Special Reference to Their Use in Engineering. *Colorado Scientific Society Proceedings,* Vol. 16, Denver, CO, 165 pp.

Terzaghi, K., and Peck, R. B. (1967). *Soil Mechanics in Engineering Practice,* 2nd ed., Wiley & New York, NY, 729 pp.

United States Geological Survey (2000). *Glossary of Some Common Terms in Seismology.* USGS Earthquake Hazards Program, National Earthquake Information Center, World Data Center for Seismology, Denver, CO, Glossary Obtained from the Internet, 8 pp.

Yeats, R. S., Sieh, K, and Allen, C. R. (1997). *The Geology of Earthquakes.* Oxford University Press, New York, NY, 568 pp.

APPENDIX B
EXAMPLE OF A FOUNDATION ENGINEERING REPORT

B.1 INTRODUCTION

This report presents the results of our geotechnical investigation for the proposed self-supporting telecom tower and adjacent equipment shelter building. The purpose of this investigation was to obtain engineering information to evaluate the site subsurface conditions and to provide foundation recommendations for the proposed development. The site is located near the top of a mountain in the southwestern corner of Imperial County. The site is known as *Hendrix Peak*, which is located near the town of Jacumba, California. Figure B.1 shows the approximate location of the site.

Based upon our review of documents and discussion with the structural engineer, it is our understanding that the proposed development will consist of the following:

1. *Telecom tower.* The telecom tower will be self-supporting. The tower will be about 160 ft (49 m) in height and will be constructed as a 3-legged structure.

2. *Equipment shelter room.* Adjacent to the telecom tower, an equipment shelter building will be constructed. This building will house the equipment needed for the operation of the telecomm tower.

Figure B.2 shows a site plan that indicates the location of the proposed telecom tower and equipment shelter room. Figure B.3 represents a cross section of the area and shows the proposed telecom tower as well as existing towers at the site.

The scope of work performed during our investigation included the following:

- Review of available literature and maps pertaining to geotechnical conditions at the site and surrounding area.
- Excavation of two backhoe pits located at the building site. Figure B.2 shows the locations of the backhoe pits.
- Laboratory testing of samples collected from the test excavations to estimate engineering properties of site materials.
- Earthquake analyses using the EQSEARCH and EQFAULT computer programs.
- Engineering analysis to develop geotechnical recommendations and design parameters.
- Preparation of this report including conclusions and recommendations for development.

B.2 SITE OBSERVATIONS

The site is located near the top of *Hendrix Peak* in the southwestern corner of Imperial County. The site is just south of Interstate 8 and can be accessed by a dirt road that connects the site to Interstate 8. The area near Interstate 8 is relatively flat and is at an elevation of about 3200 ft (975 m) above

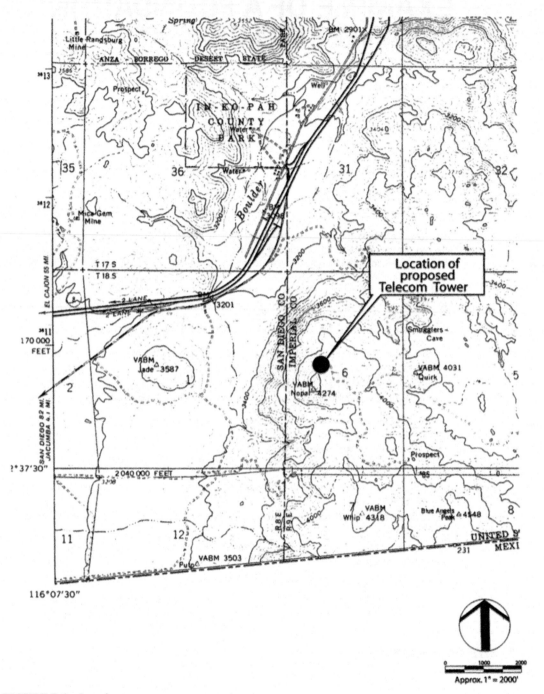

FIGURE B.1 Location map.

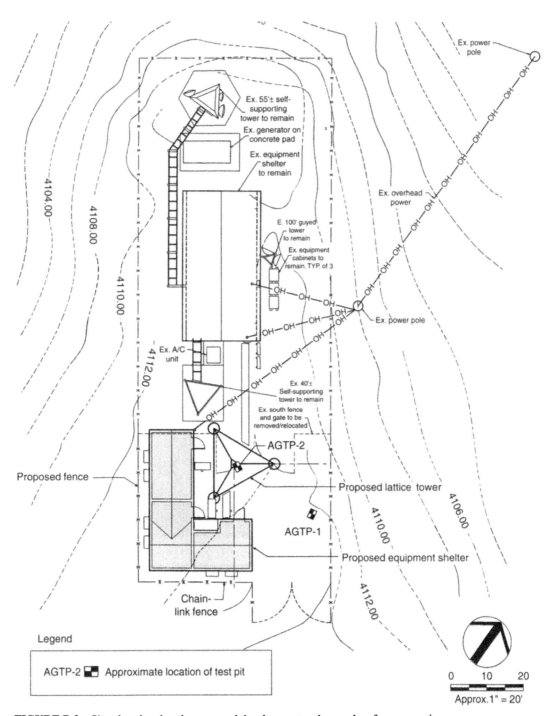

FIGURE B.2   Site plan showing the proposed development and our subsurface excavations.

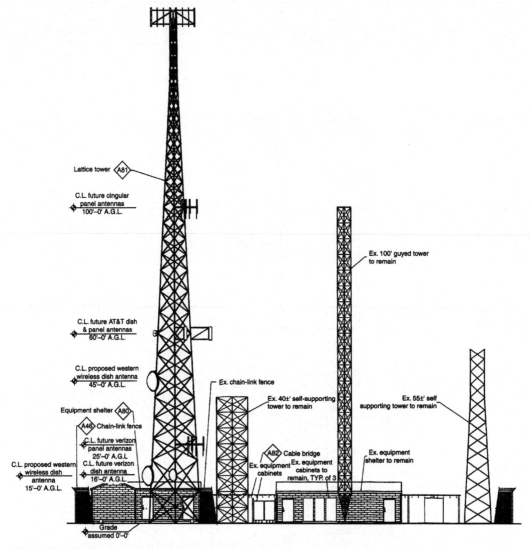

FIGURE B.3   Cross section through the site showing the proposed telecom tower.

sea level. The mountain rises rather steeply from an elevation of about 3200 ft (975 m) up to a maximum height of about 4270 ft (1300 m). The site is located slightly to the north of the peak elevation of the mountain with the site at an elevation of 4112 ft (1254 m). Although the site is located near the top of a mountain, the area of development is relatively flat and slope stability is not considered to be an issue at the site.

B.3   GEOLOGY

The site is located within the Peninsular Range geomorphic province. The Peninsular Range Province is one of the largest geologic units in the western United States. It is bounded by the Santa Monica Mountains to the north and the Colorado Desert to the east. The Peninsular Ranges consist of a series of northwest to southeast trending blocks of Cretaceous age igneous and older metamorphic rocks separated by paralleling, and predominately strike-slip faults. The igneous and metamorphic rocks forming the mountain ranges give way to post-Cretaceous marine and non-marine sedimentary, and volcanic rocks deposited along the coastline.

The subject property is located near the mountaintop that is informally named *Hendrix Peak*. This peak occurs within the Jacumba Mountain range located just east of the San Diego—Imperial County line. The site has a thin layer of coarse sandy alluvium that is underlain by Mesozoic granite rocks. Hard crystalline bedrock was encountered between 1.5 to 2 ft (0.46 to 0.6 m) below ground surface in our test excavations.

No groundwater was observed during our subsurface exploration and groundwater is not expected to be a factor in the proposed development. However, it should be noted that surface or shallow perched groundwater conditions can and may develop in areas where no such condition existed prior to development. This can occur due to changes made to the natural drainage patterns during development, increased irrigation, heavy rainfall, and/or other reasons. Because the introduction of water is usually the triggering mechanism for most common types of soil problems, it is important to provide adequate surface drainage and drainage for proposed improvements such as foundation and slab areas, and other improvements that could be adversely affected by water. Recommendations for surface drainage are provided in Sec. B.7.

Geologic hazards that occur within the region include seismic shaking, liquefaction, rock falls, landslides, and flooding events. Liquefaction and flooding are unlikely to affect the site due to the topographic elevation, the coarse-grained granular nature of the existing soil, and the shallow depth to bedrock at the site. Landsliding is not considered a credible hazard because the site is underlain essentially with crystalline igneous rock not subject to landslide failure. Rock falls/rockslides undoubtedly occur along the flanks of the mountain peak. However, because the site is located near the top of the peak, any rock slides or falls that occur are unlikely to impact the site.

The most likely geologic hazard to affect the site would be ground shaking in the event of a large earthquake. The nearest active fault to the site is the Coyote Mountain segment of the Elsinore Fault zone, which lies approximately 11.7 mi to the northeast of the project site. The maximum magnitude earthquake M_w estimated to occur along this fault segment is a 6.8. Table B.1 lists the peak horizontal ground accelerations estimated for the site from various nearby faults using attenuation relationships. The peak horizontal ground accelerations generated from the computer modeling do not take into account topographic amplification effects that are likely to increase the ground accelerations that could occur at the top of the mountain peak. Therefore we estimate a peak horizontal ground acceleration of 0.30g for the site.

Even if the structural engineer provides designs in accordance with applicable codes for seismic design, the possibility of damage occurring cannot be ruled out if moderate-to-strong ground shaking occurs as a result of a large earthquake. This is the case for essentially all structures in southern California. The structures should be designed in accordance with the latest *International Building Code* criteria for seismic design. For design purposes, the site should be considered to be Site Class A.

TABLE B.1   Selected Faults and Estimated Maximum Earthquake Magnitudes/Ground Accelerations

Fault name	Approximate distance (mi)	Estimated maximum earthquake event	
		Maximum earthquake magnitude M_w	Peak site acceleration (g)
Elsinore-Coyote Mountain	11.7	6.8	0.25
Laguna Salada	14.5	7.0	0.23
Elsinore-Julian	27.5	7.1	0.13
Superstition Mountain (San Jacinto)	27.0	6.6	0.09
San Andreas—Southern	54.6	7.4	0.08
San Andreas—Coachella	54.6	7.1	0.06

B.4   SITE INVESTIGATION

The site investigation included a subsurface investigation of the site soil conditions and laboratory testing. The subsurface exploration consisted of two backhoe excavations. The approximate locations of the excavations are depicted on Fig. B.2. The purpose of the excavations was to delineate the subsurface conditions in the area of the proposed improvements. A registered geologist logged the soil and rock conditions encountered in the test excavations. During the excavation, soil samples were collected and transported to our laboratory for testing. Each excavation was then backfilled upon completion.

In general, the excavations revealed shallow decomposed granite (DG) soil about 1.5 to 2 ft (0.46 to 0.6 m) thick overlying intact granite. The underlying granite bedrock was found to be in a very hard state. This granite can best be described as massive crystalline bedrock. Using a backhoe, we encountered refusal on the granite at a depth of 1.5 ft (0.46 m) at test pit number 1 and at a depth of 2 ft (0.6 m) at test pit number 2. Groundwater was not encountered in any of the excavations.

The test pits were logged by visual and tactile methods, selectively sampled, and backfilled upon completion. Soil samples recovered from the test pits were placed in moisture resistant containers and transported to the laboratory for testing. The laboratory-testing program consisted of field density and moisture content tests. In addition, we have sent the DG soil to an independent laboratory to be tested for resistivity, sulfate, and chloride content. We will send an addendum letter with the laboratory test results when they are completed.

B.5   CONCLUSIONS

Based upon the results of our geotechnical investigation of the subject site, no geotechnical conditions were encountered which would preclude the proposed development, provided the following conclusions and recommendations are incorporated into the project plans and specifications. Foundation recommendations have been provided to help reduce the potential for problems associated with the soil conditions encountered. The actual recommendations are discussed in detail in the following sections.

The results of our subsurface investigation indicate that there is intact granite at a shallow depth of about 1.5 to 2 ft (0.46 to 0.6 m) below ground surface. As such, it is recommended that the foundations for the telecom tower and the equipment shelter building be entirely founded in intact granite. Specific foundation recommendations are presented in Sec. B.6.

B.6   FOUNDATION RECOMMENDATIONS

As previously mentioned, our subsurface exploration has revealed the presence of DG soil underlain by granite. It is recommended that the foundation for the telecom tower and the equipment shelter building be entirely founded in intact granite. Specific foundation recommendations are as follows:

B.6.1   Telecom Tower

It is our understanding that two different types of foundations are being considered for the self-supporting telecom tower: (1) a mat foundation where the weight of the mat is used to resist overturning wind and earthquake loads, or (2) isolated spread footings under each leg of the tower with each spread footing equipped with tie-down anchors to resist overturning wind and earthquake loads. Design parameters are as follows:

1. *Allowable bearing pressure.*   The recommended allowable bearing capacity is 4000 psf (190 kPa) for the intact granite (based on a 1 ft wide footing embedded 1 ft in granite). This allowable bearing capacity can be increased by 20 percent for each additional 1 ft increase in width and/or 1 ft increase in depth, up to a maximum value of 12,000 psf (570 kPa). For example, if 5-ft wide square footings are used under each leg of the tower and the bottom of the footings are at a depth of 2 ft below lowest adjacent grade, then the allowable bearing capacity is 8000 psf (380 kPa). If necessary, the width of the footings and/or depth of the footings can be increased in order to increase their load capacity. For short term loading, such as wind loads or seismic loads, the allowable bearing capacity can be increased by a factor of $\frac{1}{3}$.

2. *Eccentric loads.* It is likely that the foundation will be subjected to eccentric loads. There are several different methods that can be used to determine the bearing capacity for eccentrically loaded foundations. For example, one commonly used method is the *reduced area method*, where the foundation width that is available for bearing resistance is defined as B' (calculated as $B' = B - 2e$, where B = actual width of foundation and e = eccentricity of the load).

3. *Passive and sliding resistance.* Lateral resistance of the foundation can be obtained from friction and passive resistance of the granite. The recommended allowable coefficient of sliding friction for concrete cast on intact granite = 0.70 and the coefficient of friction should be applied to dead-load forces only. The recommended allowable passive pressure for that part of the foundation embedded in the granite is 1200 pcf (190 kN/m^3) (equivalent fluid pressure). For short-term loading, such as wind loads or seismic loads, the allowable passive resistance can be increased by a factor of $\frac{1}{3}$.

4. *Minimum embedment.* The foundation should be embedded at least one foot (0.3 m) into intact granite. If the granite is too hard to excavate, then less embedment may be used provided we review the foundation excavations and confirm the hard nature of the granite. If the mat foundation option is selected and the granite is too hard to excavate, then a portion of the mat could extend above ground surface.

5. *Tie-down anchors.* For the foundation option consisting of tie-down anchors, the allowable frictional resistance between the grout in the tie-down hole and intact granite is 10 psi (70 kPa). It is also recommended that all of the tie-down anchors be tested by subjecting them to a pullout force that is equal to 200 percent of the uplift design load. We would also recommend that the tie-down anchors be locked off at 120 percent of the uplift design load. We should be on-site to observe the pullout testing of the tie-down anchors.

6. *Concrete type.* It is recommended that the foundation concrete be at least Type II with a maximum water/cement ratio of 0.50 by weight. Minimum unconfined compressive strength and reinforcement conditions should be in accordance with structural engineering requirements. In addition, ACI recommendations should be followed in terms of minimum reinforcement conditions for shrinkage and temperature stresses.

7. *Foundation excavations.* All foundation excavations will need to be reviewed by the geotechnical engineer to confirm embedment within intact granite.

B.6.2 Equipment Shelter Building

The following are the recommendations for the equipment shelter building foundation:

1. *Foundation type.* It is recommended that the equipment shelter building have either a mat foundation or a slab-on-grade foundation. The entire foundation should be bearing on granite. Based on our subsurface exploration, it is anticipated that the depth to granite should be rather shallow, on the order of 1.5 to 2 ft (0.46 to 0.6 m) below existing grade.

2. *Minimum dimensions and reinforcement.* For the proposed equipment shelter building, the mat or floor slab should be at least 4 in. (10 cm) thick. The structural engineer should determine the steel reinforcement required for the mat or floor slab based on structural loadings, shrinkage, and temperature stresses. As a minimum, the floor slab should have No. 3 bars placed 16 in. (0.4 m) apart (on-center), in both directions, with the No. 3 bars positioned at the center of the concrete section. Wire mesh is not recommended. Pulling up the steel reinforcement after placement of concrete is also not recommended. Instead, concrete chairs should be used to ensure that the steel

reinforcement is properly placed within the concrete slab. It is also recommended that the slab concrete be at least Type II with a maximum water/cement ratio of 0.50 by weight.

3. *Moisture barrier.*   Depending on the type of equipment in the shelter building, it may be important to reduce moisture intrusion through the foundation. Since it is desirable to have the entire foundation bearing on granite, the best approach would be to install a surface sealer to the top of the concrete foundation to act as the moisture barrier. We can provide recommendations for a surface sealer if desired.

Because the telecom tower and the equipment shelter building will have their foundations entirely bearing on granite, the maximum differential settlement is expected to be less than $1/4$ in. Settlement should consist of deformation of the granite due to dead loads and should occur during construction.

B.7   OTHER CONSIDERATIONS

Other design considerations are as follows:

1. *Surface drainage.*   For the telecom tower and equipment shelter building, it is recommended that positive drainage be provided so that water flows away from the foundations. Two percent drainage away from the foundation is recommended. Surface water should be directed to a suitable disposal area.

2. *Flatwork.*   Concrete flatwork (such as for sidewalks) should have adequate joints for crack control (see ACI specifications), a minimum of 4 in. in thickness, and reinforced with a minimum of No. 3 bars at 16 in. on-center, each way (place steel rebar at mid-height of the concrete section).

3. *Inspections.*   Supplemental consulting will include review of final plans and geotechnical services during construction. This office should be contacted for review of the final plans for improvements. In addition, the soils engineer should be involved during construction to monitor the geotechnical aspects of the development (i.e., foundation excavations). During construction, it is recommended that this office verify site geotechnical conditions and conformance with the intentions of the recommendations for construction. For example, all foundation excavations will need to be reviewed by the soils engineer in order to confirm embedment within intact granite. Although not all possible geotechnical observation and testing services are required by the governing agencies, the more site reviews performed, the lower the risk of future problems.

4. *Site safety.*   The contractor is the party responsible for providing a safe site. The geotechnical engineer will not direct the contractor's operations and cannot be responsible for the safety of personnel other than his own representatives on-site. The contractor should notify the owner if unsafe conditions are anticipated. At the time of construction, if the geotechnical consultant considers conditions unsafe, the contractor's, as well as the owner's representatives, will be notified.

B.8   CLOSURE

The geotechnical investigation was performed using the degree of care and skill ordinarily exercised, under similar circumstances, by geotechnical engineers practicing in this or similar localities. No warranty, expressed or implied, is made as to the conclusions and professional advice included in this report.

The samples taken and used for testing and the observations are believed to be representative of the entire area. However, soil and geologic conditions can vary significantly between test pits. As in many developments, conditions revealed by excavations may be at variance with preliminary findings. If this occurs, the changed conditions must be evaluated by the geotechnical engineer and designs adjusted or alternate designs recommended.

APPENDIX C
SOLUTIONS TO PROBLEMS

Chapter 2

2.1 $D_s = D_t - 2t$ (where t = wall thickness)

$D_s = 3.00 - 2(0.065) = 2.87$ in.

$D_w = D_t = 3.0$ in.

Inside clearance ratio = $100(D_s - D_e)/D_e = 100(2.87 - 2.84)/2.84 = 1.06\%$

Area ratio = $100(D_w^2 - D_e^2)/D_e^2 = 100(3^2 - 2.84^2)/2.84^2 = 11.6\%$

Clearance ratio = 1.06%, area ratio = 11.6%, and it is close to meeting the criteria for undisturbed soil sampling.

2.2 N value = $8 + 9 = 17$

For 100 mm borehole diameter, $C_b = 1.0$

For drill rod length = 5 m, $C_r = 0.85$

$N_{60} = C_b C_r N(E_m/60) = (1.0)(0.85)(17)(60/60) = 14.5$

$(N_1)_{60} = C_N N_{60} = (100/\sigma'_{vo})^{0.5} N_{60} = (100/50)^{0.5}(14.5) = 20.4$

Per Table 2.6, for $(N_1)_{60} = 20.4$, the sand is in a dense condition.

2.3 $H = 4$ in. = 0.333 ft $D = 2$ in. = 0.167 ft

$s_u = T_{max}/[\pi(0.5\ D^2 H + 0.167\ D^3)]$

$= 8.5/[\pi(0.5(0.167)^2\ 0.333 + 0.167(0.167)^3)] = 500$ psf

2.4 Given: $T_1 = T_2 = 0.04$, $V_1 = 800$ ft/sec, $d' = 50$ ft, and the intersection of the clay and rock portions of the graph occur at a distance from the shot = 120 ft.

Time corresponding to the intersection of the clay and topsoil curves = d'/V_1

= 50/800 = 0.0625 sec

$V_2 = (50\ \text{ft})/(0.0625 - 0.04\ \text{sec}) = 2220$ ft/sec

$\sin \alpha = V_1/V_2 = 800/2220 = 0.36$ or $\alpha = 21.1°$

$H_1 = [(T_1 V_1)/(2 \cos \alpha)] = [(0.04)(800)]/(2 \cos 21.1°) = 17.1$ ft

Time corresponding to the intersection of the rock and clay curves = $0.04 + (120/V_2) =$

$0.04 + (120/2220) = 0.094$ sec

$V_3 = (120\ \text{ft})/(0.094 - 0.08\ \text{sec}) = 8570$ ft/sec

$\sin \beta = V_2/V_3 = 2220/8570 = 0.26$ or $\beta = 15.0°$

$H_2 = [(T_2 V_2)/(2 \cos \beta)] = [(0.04)(2220)]/(2 \cos 15.0°) = 46.0$ ft

Answers: $H_1 = 17.1$ ft and $H_2 = 46.0$ ft

2.5 D11R tractor/ripper, therefore use Fig. 2.35. For granite with a seismic velocity = 12,000 to 15,000 ft/sec, it is nonrippable.

Chapter 3

3.1 Mass of water = 530.8 − 483.7 = 47.1 g

Mass of dry soil = 483.7 − 105.6 = 378.1 g

Using Eq. 3.1

$w(\%) = 100(M_w/M_s) = 100(47.1/378.1) = 12.5\%$

3.2 Using Eq. 3.2, total density $(\rho_t) = M/V$

$= (530.8 − 105.6)/225 = 1.89$ g/cm^3 = 1.89 Mg/m^3

Total unit weight $(\gamma_t) = (\rho_t)(g) = (1.89$ Mg/m$^3)(9.81$ m/sec$^2) = 18.5$ kN/m^3

Using Eq. 3.3, dry unit weight $(\gamma_d) = \gamma_t/(1 + w) = 18.5/(1 + 0.125) = 16.5$ kN/m^3

3.3 Using Eq. 3.4, $\gamma_b = \gamma_{sat} − \gamma_w = 19.5 − 9.8 = 9.7$ kN/m^3

3.4 Using Eq. 3.5:

$$G_s = \frac{\rho_s}{\rho_w} = \frac{M_s/V_s}{\rho_w} = \frac{102.2 \text{ g}/38.9 \text{ cm}^3}{1.0 \text{ g/cm}^3} = 2.63$$

3.5 (A) Dry mass of the soil specimen:

Water content: 8.3%   Wet mass: 1386.9 g   Initial dry mass (M_s): 1280.6 g

(B) Dry mass of the soil specimen after washing on the No. 200 sieve:

Mass of empty evaporating dish: 234.8 g   Mass of dish plus dry soil: 1350.8 g

Mass of dry soil retained on the No. 200 sieve (M_R): 1116.0 g

(C) Sieve analysis:

Sieve No.	Mass retained for each sieve (g)	Cumulative mass retained R_{DS} (g)	Percent finer (Eq. 3.6)
2-in.	0	0	100
1½ -in.	0	0	100
1-in.	93.3	93.3	92.7
3/4 -in.	71.9	165.2	87.1
½ -in.	114.3	279.5	78.2
3/8 -in.	135.7	415.2	67.6
No. 4	182.2	597.4	53.3
No. 10	150.1	747.5	41.6
No. 20	142.2	889.7	30.5
No. 40	112.8	1002.5	21.7
No. 60	47.8	1050.3	18.0
No. 100	29.6	1079.9	15.7
No. 200	35.9	1115.8	12.9
Pan	0.1	1115.9	—

Check: Cumulative retained on Pan: 1115.9 versus M_R: 1116.0

3.6

A) Water contents for liquid limit:

Trial number	1	2	3	4	5
Container number	#1	#2	#4		
Container mass (M_c)	10.92	10.84	11.33		
Container + wet soil (M_{wc})	20.89	22.90	24.07		
Container + dry soil (M_{dc})	16.36	17.63	18.80		
Mass of water (M_w)	4.53	5.27	5.27		
Mass of solids (M_s)	5.44	6.79	7.47		
Water content = M_w/M_s	83.3	77.6	70.5		
Number of blows	15	19	30		

B) Water contents for plastic limit:

Trial number	1	2	3	4	5
Container number	#3	#5			
Container mass (M_c)	11.25	10.98			
Container + wet soil (M_{wc})	13.15	13.21			
Container + dry soil (M_{dc})	12.81	12.80			
Mass of water (M_w)	0.34	0.41			
Mass of solids (M_s)	1.56	1.82			
Water content = M_w/M_s	21.8	22.5			

C) Summary:

Liquid limit (LL) = 73	Plastic limit (PL) = 22	Plasticity index (PI) = 51

Note: All mass values are in grams.

3.7   In Fig. 3.12, using $N_{60} = 5$ and $\sigma_v' = 43$ kPa, $\phi' = 30°$

3.8   In Fig. 3.13, using $q_c = 40$ kg/cm^2 and $\sigma_v' = 43$ kPa, $\phi' = 40°$

3.9   Nonplastic silty sand (SM)

Per Fig. 3.19 for B-1 at depth of 6 ft, dry unit weight = 93.4 pcf

Entering Fig. 3.14 with a dry density of 93.4 pcf and intersecting SM line, $\phi' = 30°$

3.10   Nonplastic silty sand (SM)

Per Fig. 3.19 for B-2 at depth of 10 ft, dry unit weight = 104 pcf

Entering Fig. 3.14 with a dry density of 104 pcf and intersecting SM line, $\phi' = 36°$

3.11   Cohesionless soil, therefore $c' = 0$

Specimen diameter = 6.35 cm, therefore area $(A) = 0.00317$ m^2

First test: $\sigma_n' = N/A = 150/0.00317 = 47{,}300$ Pa $= 47.3$ kPa

$\tau_f = T/A = 94/0.00317 = 29{,}700$ Pa $= 29.7$ kPa

Using Eq. 3.10:

$\tau_f = \sigma_n' \tan \phi'$

$\tan \phi' = (29.7/47.3) = 0.628$

Solving for $\phi' = 32°$

Second test: $\sigma_n' = N/A = 300/0.00317 = 94{,}600$ Pa $= 94.6$ kPa

$\tau_f = T/A = 188/0.00317 = 59{,}300$ Pa $= 59.3$ kPa

Using Eq. 3.10:

$\tau_f = \sigma_n' \tan \phi'$

$\tan \phi' = (59.3/94.6) = 0.627$

Solving for $\phi' = 32°$

3.12 $P_f = 24.8$ lb

$H_o = 6.0$ in.

$\Delta H = 0.8$ in.

$D_o = 2.5$ in. and therefore $A_o = 4.91$ in^2

$\varepsilon_f = \Delta H/H_o = 0.8/6.0 = 0.133$

Using Eq. 3.24:

$A_f = A_o/(1 - \varepsilon_f) = 4.91/(1 - 0.133) = 5.66$ in^2

Using Eq. 3.25:

$\sigma_1 = q_u = P_f/A_f = 24.8/5.66 = 4.38$ psi $= 630$ psf

Using Eq. 3.26:

$s_u = q_u/2 = 630/2 = 315$ psf

3.13   Per Fig. 3.30, failure (i.e., maximum value of $\sigma_1 - \sigma_3$) occurs at 16.01 psi which corresponds to $\varepsilon = 12.28\%$ axial strain. At $\varepsilon = 12.28\%$, $p = 100.9$ psi and $q = 8.00$ psi. Note that p must be adjusted for the back pressure, or adjusted $p = 100.9 - 85.9 = 15.0$ psi. Since $c = 0$, then $a = 0$. Using $p = 15.0$ psi and $q = 8.0$ psi, then:

$\tan \alpha = q/p = 8.0/15.0 = 0.533$

Solving for $\alpha = 28°$

Using Eq. 3.17:

$\sin \phi = \tan \alpha = \tan 28°$

Solving for $\phi = 32°$

3.14 Per Fig. 3.30, failure (i.e., maximum value of $\sigma_1 - \sigma_3$) occurs at 16.01 psi which corresponds to $\varepsilon = 12.28\%$ axial strain. At $\varepsilon = 12.28\%$, $p = 100.9$ psi and $q = 8.00$ psi. Note that p must be adjusted for the back pressure, or adjusted $p = 100.9 - 85.9 = 15.0$ psi.

$q = a + p \tan \alpha$

Using $p = 15.0$ psi and $q = 8.0$ psi, then:

$8.0 = a + 15.0 \tan \alpha$

Using Eq. 3.17:

$a = c \cos \phi = 2.0 \cos \phi$ and $\tan \alpha = \sin \phi$

Substituting these two equations in the above equation:

$8.0 = 2.0 \cos \phi + 15.0 \sin \phi$

and solving for $\phi = 24°$

3.15 At failure (Point E): $\sigma_1 = \sigma_v = 50 + 80 = 130$ kPa

$\sigma_3 = \sigma_h = 50$ kPa

$p = 0.5(\sigma_1 + \sigma_3) = 0.5(130 + 50) = 90$ kPa

$q = 0.5(\sigma_1 - \sigma_3) = 0.5(130 - 50) = 40$ kPa

$q = a + p \tan \alpha$

Using Eq. 3.17:

$a = c \cos \phi$ and $\tan \alpha = \sin \phi$

Substituting these two equations in the above equation:

$q = c \cos \phi + p \sin \phi$

$40 = 5 \cos \phi + 90 \sin \phi$

And solving for $\phi = 23°$

3.16 Using Eq. 3.20:

$B = \Delta u / \Delta \sigma_c = 99.8/100 = 0.998$

Since B is approximately $= 1.0$, use Eq. 3.21:

$A_f = \Delta u / \Delta \sigma_1 = 6.7/80 = 0.08$

3.17 At Point B, $E_u = \Delta \sigma_v / \varepsilon = 20/(0.13/10.67) = 1600$ kPa

3.18 At failure (Point E):

$\sigma_1' = \sigma_v' = 130 - 6.7 = 123.3$ kPa

$\sigma_3' = \sigma_h' = 50 - 6.7 = 43.3$ kPa

Using Eq. 3.16:

$p' = 0.5(\sigma_1' + \sigma_3') = 0.5(123.3 + 43.3) = 83.3$ kPa

$q = 0.5(\sigma_1' - \sigma_3') = 0.5(123.3 - 43.3) = 40$ kPa

$q = a' + p' \tan \alpha'$

$40 = 2 + 83.3 \tan \alpha'$

Solving for $\alpha' = 24.5°$

3.19   Using Eq. 3.18:

$\tan \alpha' = \sin \phi'$

$\tan 24.5° = \sin \phi'$

Solving for $\phi' = 27°$

$a' = c' \cos \phi'$

$2.0 = c' \cos 27°$

Solving for $c' = 2.2$ kPa

3.20   Using Eq. 3.24:

$A_f = A_o/(1 - \varepsilon_f) = 9.68/[1 - (1.48/11.7)] = 11.1$ cm^2

3.21   Using Eq. 3.8:

$\sigma_1' = \sigma_v' = \sigma_1 - u = [100 + (0.0484/0.00111)] - 45.6 = 98$ kPa

$\sigma_3' = \sigma_h' = \sigma_3 - u = 100 - 45.6 = 54.4$ kPa

3.22   Using Eq. 3.16:

$p' = \frac{1}{2}(\sigma_1' + \sigma_3') = (0.5)(98 + 54.4) = 76.2$ kPa

$q = \frac{1}{2}(\sigma_1' - \sigma_3') = (0.5)(98 - 54.4) = 21.8$ kPa

3.23   For $c' = 0$, $a' = 0$

$q = p' \tan \alpha'$

$\tan \alpha' = q/p' = 21.8/76.2 = 0.286$

Solving for $\alpha' = 16°$

Using Eq. 3.18:

$\sin \phi' = \tan \alpha' = \tan 16°$

Solving for $\phi' = 17°$

3.24   Assuming $B = 1.0$ and using Eq. 3.21:

$A_f = \Delta u/\Delta \sigma_1 = 45.6/(0.0484/0.00111) = 1.05$

3.25   Based on the low effective friction angle and the high A value at failure, the most likely type of inorganic soil would be normally consolidated clay of high plasticity (CH).

3.26   Using Eq. 3.16:

$p' = \frac{1}{2}(\sigma_1' + \sigma_3') = (0.5)(333 + 126) = 230$ kPa

$q = \frac{1}{2}(\sigma_1' - \sigma_3') = (0.5)(333 - 126) = 104$ kPa

Since the clay is normally consolidated, $c' = 0$ and therefore $a' = 0$

$q = p' \tan \alpha'$

$\tan \alpha' = q/p' = 104/230 = 0.452$

Solving for $\alpha' = 24°$

Using Eq. 3.18:

$\sin \phi' = \tan \alpha' = \tan 24°$

Solving for $\phi' = 27°$

3.27 Using Eq. 3.16:

$p' = \frac{1}{2}(\sigma'_1 + \sigma'_3) = (0.5)(238 + 83) = 161$ kPa

$q = \frac{1}{2}(\sigma'_1 - \sigma'_3) = (0.5)(238 - 83) = 78$ kPa

Assume the clay has an effective friction angle $(\phi') = 27°$

Using Eq. 3.18:

$\sin \phi' = \tan \alpha'$

$\sin 27° = \tan \alpha'$

Solving for $\alpha' = 24°$

$q = a' + p' \tan \alpha'$

$78 = a' + 161 \tan 24°$

Solving for $a' = 6.3$ kPa

$a' = c' \cos \phi'$

$6.3 = c' \cos 27°$

Solving for $c' = 7$ kPa

3.28 See Fig. 3.62 for the effective stress path, where $a' = 0.5$ psi and $\alpha' = 27°$.

3.29 Using Eq. 3.18:

$\sin \phi' = \tan \alpha'$

$\sin \phi' = \tan 27°$

Solving for $\phi' = 31°$

$a' = c' \cos \phi'$

$0.5 = c' \cos 31°$

Solving for $c' = 0.6$ psi

3.30 A value at failure $= 0.018$ (i.e., maximum value of $S1/S3 = \sigma'_1/\sigma'_3$) and therefore the soil type is a heavily overconsolidated clay.

3.31 Area $= \pi D_o^2/4 = 4.91$ in^2 = 0.0341 ft^2

Test No. 1: Vertical effective stress $= 1.7/0.0341 = 50$ psf

Maximum shear stress $= 5.8/0.0341 = 170$ psf

Test No. 2: Vertical effective stress $= 8.5/0.0341 = 250$ psf

Maximum shear stress $= 12/0.0341 = 350$ psf

Test No. 3: Vertical effective stress $= 15/0.0341 = 440$ psf

Maximum shear stress $= 15/0.0341 = 440$ psf

Test No. 4: Vertical effective stress $= 32/0.0341 = 940$ psf

Maximum shear stress $= 27/0.0341 = 790$ psf

Test No. 5: Vertical effective stress $= 70/0.0341 = 2050$ psf

Maximum shear stress $= 41/0.0341 = 1200$ psf

Test No. 6: Vertical effective stress $= 133/0.0341 = 3900$ psf

Maximum shear stress $= 64/0.0341 = 1880$ psf

See Fig. 3.63 for the effective shear strength envelope. For $\sigma'_n > 700$ psf, $\phi' = 21°$ and $c' = 450$ psf. For $\sigma'_n < 700$ psf, the effective shear strength envelope is nonlinear.

3.32 Using Eq. 3.9 with $c' = 0$

$\tan \phi'_r = \tau_f / \sigma'_n = 2400/14{,}600 = 0.164$

Solving for $\phi'_r = 9.3°$

3.33

Test No.	Dry Soil	Dry density
1	4.14/(1 + 0.11) = 3.73 lb	3.73/(1/30) = 111.9 pcf
2	4.26/(1 + 0.125) = 3.79 lb	3.79/(1/30) = 113.7 pcf
3	4.37/(1 + 0.140) = 3.83 lb	3.83/(1/30) = 115.0 pcf
4	4.33/(1 + 0.155) = 3.75 lb	3.75/(1/30) = 112.5 pcf

By plotting the dry density versus water content, $\rho_{dmax} = 115$ pcf (1.84 Mg/m³) and $w_{opt} = 14.0\%$

3.34 Higher compaction energy results in a higher laboratory maximum dry density and lower optimum moisture content.

3.35 Using Eq. 3.29:

$$\rho_z = \frac{G_s \rho_w}{1 + G_s w}$$

Substituting $\rho_z = 115$ pcf, $G_s = 2.65$, and $\rho_w = 62.4$ pcf into Eq. 3.29:

$(115)[1 + (2.65)(w)] = (2.65)(62.4)$

Solving for the water content $(w) = 16.5\%$.

Therefore, for $\rho_{dmax} = 115$ pcf, the water content corresponding to the zero air voids curve = 16.5%. The difference between the optimum water content ($w_{opt} = 14.0\%$) and the water content corresponding to the zero air voids curve ($w = 16.5\%$) is equal to 2.5%.

3.36 In Fig. 3.52, the line of optimums can be drawn through the peak point of the compaction curve and parallel to the zero air voids curve. Then a second line is drawn through the point defined by dry density = 117 pcf and water content = 8.0% and parallel to the left side of the compaction curve. The intersection of this line and the line of optimums is at a dry density of 120 pcf. Therefore: $\rho_{dmax} = 120$ pcf (1.92 Mg/m³).

3.37 $Q = 782$ mL = 782 cm³

$t = 31$ sec

$L = 2.54$ cm

$\Delta h = 2.0$ m = 200 cm

$D = 6.35$ cm and therefore $A = 31.67$ cm²

Using Eq. 3.31:

$k = QL/(\Delta h A t) = [(782)(2.54)]/[(200)(31.67)(31)] = 0.01$ cm/sec

3.38 For the standpipe, diameter = 0.635 cm, therefore $a = 0.317$ cm²

For the specimen, diameter = 6.35 cm, therefore $A = 31.7$ cm²

$h_o = 1.58$ m

$h_f = 1.35$ m

$t = 11$ h $= 39,600$ sec

$L = 2.54$ cm

Using Eq. 3.32:

$k = 2.3 [(a L)/(At)] \log(h_o/h_f)$

$= 2.3 [(0.317)(2.54)/(31.7)(39,600)] \log (1.58/1.35) = 1.0 \times 10^{-7}$ cm/sec

Chapter 4

4.1 35% passes the No. 40 sieve, and therefore the clay size fraction $= 20/35 = 57\%$

Using Eq. 4.6:

Activity $=$ PI/%clay fraction $= (93 - 18)/57 = 1.3$

4.2 Plasticity index $=$ liquid limit $-$ plastic limit $= 60 - 20 = 40$

Entering Fig. 4.2 with liquid limit $= 60$ and plasticity index $= 40$, the predominant clay mineral in the soil is montmorillonite.

4.3 Using Eq. 4.1:

$C_u = D_{60}/D_{10} = 15/0.075 = 200$

Using Eq. 4.2:

$C_c = D_{30}^2/(D_{10}D_{60}) = (2.5)^2/[(0.075)(15)] = 5.6$

4.4 Using the data from Problem 4.3 and recognizing that 0.075 mm is the opening of the No. 200 sieve, therefore percent passing No. 200 sieve $= 10\%$. Since $D_{50} = 12$ mm, which is a larger size than the No. 4 sieve, the majority of the soil particles are gravel. Since C_c does not meet the requirements for a well-graded gravel and the limits plot below the A-line, per the USCS, there is a dual classification (because of the 10% fines) of GP-GM.

4.5 Based on the values of C_c and C_u, the sand is well-graded. Because of 4% nonplastic fines, the classification is SW.

4.6 Since all the soil particles pass the No. 200 sieve, the soil is fine-grained. The limits plot below the A-line and the LL is less than 50, therefore per the USCS system, the classification is ML.

4.7 The plasticity index $= 40$. Based on a LL $= 60$ and a PI $= 40$, the limits plot above the A-line and the clay is classified as a CH for the USCS system.

4.8 The LL (oven dry) divided by the LL (not dried) $= 40/65 = 0.61$ and therefore the soil is an organic soil. Since the LL is greater than 50, the classification per the USCS is OH.

4.9 Per the USCS, more than 50% of the soil particles are retained on the No. 200 sieve and therefore it is a coarse grained soil. The majority of the soil particles are of sand size and since there are greater than 12% fines with the limits (LL $= 85$, PI $= 67$) plotting above the A-line, the classification is SC per the USCS.

4.10 Soil is from small boring number 14 at a depth of 8 to 12 ft (SB-14 @ 8-12 ft) and the data are summarized in Fig. 4.34.

Unified soil classification system (USCS):

Since 28.4% of the soil particles pass the No. 200 sieve (i.e., 28.4% fines), the soil is coarse-grained. Of the soil particles retained on the No. 200 sieve, 50.6% are gravel-size particles (i.e., 36.2/0.716 = 50.6%) and 49.4% are sand-size particles (i.e., 35.4/0.716 = 49.4%). Since the larger fraction consists of gravel size particles, the primary soil classification is gravel (Table 4.1). Since there are more than 12% fines and the soil is nonplastic (PI = 0), the soil classification is GM (silty gravel).

AASHTO soil classification system:

Since there are less than 35% passing the No. 200 sieve, the soil is classified as a granular material. Using the particle size distribution (Fig. 4.34), the percent passing is as follows:

Percent passing No. 10 sieve = 55%

Percent passing No. 40 sieve = 42%

Percent passing No. 200 sieve = 28%

For these percent passing values, and because the soil is nonplastic (PI = 0), all of the criteria are met for group A-2-4 (see Table 4.2).

Summary:

USCS: Coarse-grained soil, silty gravel (GM)

AASHTO: Granular material, silty gravel and sand (A-2-4)

4.11 Soil is from small boring number 20 at a depth of 0 to 4 ft (SB-20 @ 0-4 ft) and the data are summarized in Fig. 4.34.

Unified soil classification system (USCS):

Since 27.8% of the soil particles pass the No. 200 sieve (i.e., 27.8% fines), the soil is coarse-grained. Of the soil particles retained on the No. 200 sieve, 10.8% are gravel-size particles (i.e., 7.8 /0.722 = 10.8%) and 89.2% are sand-size particles (i.e., 64.4/0.722 = 89.2%). Since the larger fraction consists of sand size particles, the primary soil classification is sand (Table 4.1). Since there are more than 12% fines and the soil is nonplastic (PI = 0), the soil classification is SM (silty sand).

AASHTO soil classification system:

Since there are less than 35% passing the No. 200 sieve, the soil is classified as a granular material. Using the particle size distribution (Fig. 4.34), the percent passing is as follows:

Percent passing No. 10 sieve = 82%

Percent passing No. 40 sieve = 53%

Percent passing No. 200 sieve = 28%

For these percent passing values, and because the soil is nonplastic (PI = 0), all of the criteria are met for group A-2-4 (see Table 4.2).

Summary:

USCS: Coarse-grained soil, silty sand (SM)

AASHTO: Granular material, silty sand (A-2-4)

4.12 Soil is from small boring number 25 at a depth of 4 to 8 ft (SB-25 @ 4-8 ft) and the data are summarized in Fig. 4.34.

Unified soil classification system (USCS):

Since 40.2% of the soil particles pass the No. 200 sieve (i.e., 40.2% fines), the soil is coarse-grained. Of the soil particles retained on the No. 200 sieve, 25.3% are gravel-size particles (i.e., 15.1/0.598 = 25.3%) and 74.7% are sand-size particles (i.e., 44.7/0.598 = 74.7%). Since the larger fraction consists of sand size particles, the primary soil classification is sand (Table 4.1). Since there are more than 12% fines and the soil is nonplastic (PI = 0), the soil classification is SM (silty sand).

AASHTO soil classification system:

Since there are more than 35% passing the No. 200 sieve, the soil is classified as a silt-clay material. Because the soil in nonplastic (PI = 0, LL = 0), all of the criteria are met for group A-4. By inserting values into the group index equation (see Table 4.2), a negative value is obtained and thus the group index = 0.

Summary:

USCS: Coarse-grained soil, silty sand (SM)

AASHTO: Silt-clay material, silty soil A-4 (0)

4.13 Soil is from small boring number 29 at a depth of 4 to 8 ft (SB-29 @ 4-8 ft) and the data are summarized in Fig. 4.34.

Unified soil classification system (USCS):

Since 14.8% of the soil particles pass the No. 200 sieve (i.e., 14.8% fines), the soil is coarse-grained. Of the soil particles retained on the No. 200 sieve, 18.0% are gravel-size particles (i.e., 15.3/0.852 = 18.0%) and 82.0% are sand-size particles (i.e., 69.9/0.852 = 82.0%). Since the larger fraction consists of sand size particles, the primary soil classification is sand (Table 4.1). Since there are more than 12% fines and the soil is nonplastic (PI = 0), the soil classification is SM (silty sand).

AASHTO soil classification system:

Since there are less than 35% passing the No. 200 sieve, the soil is classified as a granular material. Using the particle size distribution (Fig. 4.34), the percent passing is as follows:

Percent passing No. 10 sieve = 77%

Percent passing No. 40 sieve = 39%

Percent passing No. 200 sieve = 15%

For these percent passing values, and because the soil is nonplastic (PI = 0), all of the criteria are met for group A-1-b (see Table 4.2).

Summary:

USCS: Coarse-grained soil, silty sand (SM)

AASHTO: Granular material, gravel and sand mixture (A-1-b)

4.14 Soil is from small boring number 38 at a depth of 4 to 8 ft (SB-38 @ 4-8 ft) and the data are summarized in Fig. 4.34.

Unified soil classification system (USCS):

Since 52.8% of the soil particles pass the No. 200 sieve (i.e., 52.8% fines), the soil is fine-grained. Since the Atterberg limits (i.e., liquid limit = 50 and plasticity index = 35) plot above the A-line in Fig. 4.1, the soil is classified as sandy clay of high plasticity (CH).

AASHTO soil classification system:

Since there are more than 35% passing the No. 200 sieve, the soil is classified as a silt-clay material. With a plasticity index = 35, a liquid limit = 50, and since the LL − 30 is less than the plasticity index, the soil meets all the classification requirements for group A-7-6 (clayey soils). Using the group index equation listed in Table 4.2, with F = 52.8, LL = 50, and PI = 35, the group index = 13.9. The soil classification is therefore an A-7-6 (14), clayey soil.

USDA textural classification system:

The percent sand, percent silt, and percent clay size particles are as follows:

Sand (2 to 0.05 mm) = (99 − 50.5)/0.99 = 49%

Silt (0.05 to 0.002 mm) = (50.5 − 31.5)/0.99 = 19%

Clay (finer than 0.002 mm) = 31.5/0.99 = 32%

Using Fig. 4.5 with the percent sand = 49%, percent silt = 19%, and percent clay = 32%, the soil classification is sandy clay loam.

Summary:

USCS: Fine-grained soil, sandy clay of high plasticity (CH)

AASHTO: Silt-clay material, clayey soil, A-7-6 (14)

USDA: Sandy clay loam

4.15   Soil is from small boring number 39 at a depth of 4 to 8 ft (SB-39 @ 4-8 ft) and the data are summarized in Fig. 4.35.

Unified soil classification system (USCS):

Since 39.0% of the soil particles pass the No. 200 sieve (i.e., 39.0% fines), the soil is coarse-grained. Of the soil particles retained on the No. 200 sieve, 1.5% are gravel-size particles (i.e., 0.9/0.610 = 1.5%) and 98.5% are sand-size particles (i.e., 60.1/0.610 = 98.5%). Since the larger fraction consists of sand size particles, the primary soil classification is sand (Table 4.1). The soil has greater than 12% fines and since Atterberg limits (i.e., liquid limit = 58 and plasticity index = 41) plot above the A-line in Fig. 4.1, the soil is classified as clayey sand (SC) per Table 4.1.

AASHTO soil classification system:

Since there are more than 35% passing the No. 200 sieve, the soil is classified as a silt-clay material. With a plasticity index = 41, a liquid limit = 58, and LL − 30 = 58 − 30 = 28 which is less than the plasticity index, the soil meets all the classification requirements for group A-7-6 (clayey soils). Using the group index equation listed in Table 4.2, with F = 39.0, LL = 58, and PI = 41, the group index = 8.6. The soil classification is therefore an A-7-6 (9), clayey soil.

USDA textural classification system:

The percent sand, percent silt, and percent clay size particles are as follows:

Sand (2 to 0.05 mm) = (98 − 38)/0.98 = 61%

Silt (0.05 to 0.002 mm) = (38 − 26)/0.98 = 12%

Clay (finer than 0.002 mm) = 26/0.98 = 27%

Using Fig. 4.5 with the percent sand = 61%, percent silt = 12%, and percent clay = 27%, the soil classification is sandy clay loam.

Summary:

USCS: Coarse-grained soil, clayey sand (SC)

AASHTO: Silt-clay material, clayey soil, A-7-6 (9)

USDA: Sandy clay loam

4.16   Soil is from small boring number 42 at a depth of 4 to 8 ft (SB-42 @ 4-8 ft) and the data are summarized in Fig. 4.35.

Unified soil classification system (USCS):

Since 71.5% of the soil particles pass the No. 200 sieve (i.e., 71.5% fines), the soil is fine-grained. Since the Atterberg limits (i.e., liquid limit = 33 and plasticity index = 13) plot above the A-line in Fig. 4.1, the soil is classified as silty clay of low plasticity (CL).

AASHTO soil classification system:

Since there are more than 35% passing the No. 200 sieve, the soil is classified as a silt-clay material. With a plasticity index = 13, and a liquid limit = 33, the soil meets all the classification requirements for group A-6 (clayey soils). Using the group index equation listed in Table 4.2, with $F = 71.5$, LL = 33, and PI = 13, the group index = 7.7. The soil classification is therefore an A-6 (8), clayey soil.

USDA textural classification system:

The percent sand, percent silt, and percent clay size particles are as follows:

Sand (2 to 0.05 mm) = (96 – 64)/0.96 = 33%

Silt (0.05 to 0.002 mm) = (64 – 19)/0.96 = 47%

Clay (finer than 0.002 mm) = 19/0.96 = 20%

Using Fig. 4.5 with the percent sand = 33%, percent silt = 47%, and percent clay = 20%, the soil classification is loam.

Summary:

USCS: Fine-grained soil, silty clay of low plasticity (CL)

AASHTO: Silt-clay material, clayey soil, A-6 (8)

USDA: Loam

4.17 Soil is from small boring number 44 at a depth of 4 to 8 ft (SB-44 @ 4-8 ft) and the data are summarized in Fig. 4.35.

Unified soil classification system (USCS):

Since 15.3% of the soil particles pass the No. 200 sieve (i.e., 15.3% fines), the soil is coarse-grained. Of the soil particles retained on the No. 200 sieve, 17.7% are gravel-size particles (i.e., 15.0/0.847 = 17.7%) and 82.3% are sand-size particles (i.e., 69.7/0.847 = 82.3%). Since the larger fraction consists of sand size particles, the primary soil classification is sand (Table 4.1). Since there are more than 12% fines and the soil is nonplastic (PI = 0), the soil classification is SM (silty sand).

AASHTO soil classification system:

Since there are less than 35% passing the No. 200 sieve, the soil is classified as a granular material. Using the particle size distribution (Fig. 4.35), the percent passing is as follows:

Percent passing No. 10 sieve = 75%

Percent passing No. 40 sieve = 37%

Percent passing No. 200 sieve = 15%

For these percent passing values, and because the soil is nonplastic (PI = 0), all of the criteria are met for group A-1-b.

Summary:

USCS: Coarse-grained soil, silty sand (SM)

AASHTO: Granular material, gravel and sand mixture (A-1-b)

4.18 Soil is from small boring number 45 at a depth of 4 to 8 ft (SB-45 @ 4-8 ft) and the data are summarized in Fig. 4.35.

Unified soil classification system (USCS):

Since 33.4% of the soil particles pass the No. 200 sieve (i.e., 33.4% fines), the soil is coarse-grained. Of the soil particles retained on the No. 200 sieve, 26.3% are gravel-size particles (i.e., 17.5/0.666 = 26.3%) and 73.7% are sand-size particles (i.e., 49.1/0.666 = 73.7%). Since the larger fraction consists of sand size particles, the primary soil classification is sand (Table 4.1). The soil has greater than 12% fines and since the Atterberg limits (i.e., liquid limit = 30 and plasticity index = 15) plot above the A-line in Fig. 4.1, the soil is classified as clayey sand (SC) per Table 4.1.

AASHTO soil classification system:

Since there are less than 35% passing the No. 200 sieve, the soil is classified as a granular material. With a plasticity index = 15 and a liquid limit = 30, the soil meets all the classification requirements for group A-2-6 (clayey gravel and sand). Using only the PI portion of the group index equation listed in Table 4.2, with F = 33.4 and PI = 15, the group index = 0.92. The soil classification is therefore an A-2-6 (1), clayey soil.

Summary:

USCS: Coarse-grained soil, clayey sand (SC)

AASHTO: Granular material, clayey sand and gravel, A-2-6 (1)

4.19   Soil is from test pit number 1 at a depth of 1 to 2.5 ft (TP-1 @ 1-2.5 ft) and the data are summarized in Fig. 4.36.

Unified soil classification system (USCS):

Since 34.2% of the soil particles pass the No. 200 sieve (i.e., 34.2% fines), the soil is coarse-grained. Of the soil particles retained on the No. 200 sieve, 34.5% are gravel-size particles (i.e., 22.7/0.658 = 34.5%) and 65.5% are sand-size particles (i.e., 43.1/0.658 = 65.5%). Since the larger fraction consists of sand size particles, the primary soil classification is sand (Table 4.1). The soil has greater than 12% fines and since the Atterberg limits (i.e., liquid limit = 34 and plasticity index = 17) plot above the A-line in Fig. 4.1, the soil is classified as clayey sand (SC) per Table 4.1.

AASHTO soil classification system:

Since there are less than 35% passing the No. 200 sieve, the soil is classified as a granular material. With a plasticity index = 17 and a liquid limit = 34, the soil meets all the classification requirements for group A-2-6 (clayey gravel and sand). Using only the PI portion of the group index equation listed in Table 4.2, with F = 34.2 and PI = 17, the group index = 1.3. The soil classification is therefore an A-2-6 (1), clayey soil.

Summary:

USCS: Coarse-grained soil, clayey sand (SC)

AASHTO: Granular material, clayey sand and gravel, A-2-6 (1)

4.20   Soil is from test pit number 15 at a depth of 1 to 2 ft (TP-15 @ 1-2 ft) and the data are summarized in Fig. 4.36.

Unified soil classification system (USCS):

Since 52.8% of the soil particles pass the No. 200 sieve (i.e., 52.8% fines), the soil is fine-grained. Since the Atterberg limits (i.e., liquid limit = 35 and plasticity index = 22) plot above the A-line in Fig. 4.1, the soil is classified as sandy clay of low plasticity (CL).

AASHTO soil classification system:

Since there are more than 35% passing the No. 200 sieve, the soil is classified as a silt-clay material. With a plasticity index = 22, and a liquid limit = 35, the soil meets all the classification requirements for group A-6 (clayey soils). Using the group index equation listed in Table 4.2, with F = 52.8, LL = 35, and PI = 22, the group index = 7.7. The soil classification is therefore an A-6 (8), clayey soil.

Summary:

USCS: Fine-grained soil, sandy clay of low plasticity (CL)

AASHTO: Silt-clay material, clayey soil, A-6 (8)

4.21   Soil is from test pit number 2 at a depth of 1 to 2 ft (TP-2 @ 1-2 ft) and the data are summarized in Fig. 4.36.

Unified soil classification system (USCS):

Since 82.1% of the soil particles pass the No. 200 sieve (i.e., 82.1% fines), the soil is fine-grained. Since the Atterberg limits (i.e., liquid limit = 76 and plasticity index = 61) plot above the *A*-line in Fig. 4.1, the soil is classified as clay of high plasticity (CH).

AASHTO soil classification system:

Since there are more than 35% passing the No. 200 sieve, the soil is classified as a silt-clay material. With a plasticity index = 61, a liquid limit = 76, and since the LL − 30 is less than the plasticity index, the soil meets all the classification requirements for group A-7-6 (clayey soils). Using the group index equation listed in Table 4.2, with $F = 82.1$, LL = 76, and PI = 61, the group index = 52. The soil classification is therefore an A-7-6 (52), clayey soil.

USDA textural classification system:

The percent sand, percent silt, and percent clay size particles are as follows:

Sand (2 to 0.05 mm) = (95 − 79)/0.95 = 17%

Silt (0.05 to 0.002 mm) = (79 − 74)/0.95 = 5%

Clay (finer than 0.002 mm) = 74/0.95 = 78%

Using Fig. 4.5 with the percent sand = 17%, percent silt = 5%, and percent clay = 78%, the soil classification is clay.

Summary:

USCS: Fine-grained soil, clay of high plasticity (CH)

AASHTO: Silt-clay material, clayey soil, A-7-6 (52)

USDA: Clay

4.22   The two soils from small boring number 2 at a depth of 5 ft (SB-2 @ 5 ft) and small boring number 2 at a depth of 31 ft (SB-2 @ 31 ft) are nearly identical. The data are summarized in Fig. 4.37.

Unified soil classification system (USCS):

Since more than 50% of the soil particles pass the No. 200 sieve, the soil is fine-grained. Since the Atterberg limits (i.e., LL = 48, PI = 28 and LL = 46 and PI = 27) plot above the *A*-line in Fig. 4.1 and because the LL is less than 50, the soil is classified as silty clay of low plasticity (CL).

AASHTO soil classification system:

Since there are more than 35% passing the No. 200 sieve, the soil is classified as a silt-clay material. With a plasticity index = 28, a liquid limit = 48, and since the LL − 30 is less than the plasticity index, the soil meets all the classification requirements for group A-7-6 (clayey soils). Using the group index equation listed in Table 4.2, with $F = 78.4$, LL = 48, and PI = 28, the group index = 21.8 for SB-2 @ 5 ft. Using the group index equation listed in Table 4.2, with $F = 81.4$, LL = 46, and PI = 27, the group index = 22.0 for SB-2 @ 31 ft. Thus for both soils, the soil classification is A-7-6 (22), clayey soil.

USDA textural classification system:

The percent sand, percent silt, and percent clay size particles are as follows:

For SB-2 @ 5 ft:

Sand (2 to 0.05 mm) = 100 − 69 = 31%

Silt (0.05 to 0.002 mm) = 69 − 28 = 41%

Clay (finer than 0.002 mm) = 28%

For SB-2 @ 31 ft:

Sand (2 to 0.05 mm) = 100 − 69 = 31%

Silt (0.05 to 0.002 mm) = 69 − 33 = 36%

Clay (finer than 0.002 mm) = 33%

Using Fig. 4.5 with the above percentages of sand, silt, and clay, the soil classification for both soils is clay loam.

Summary (for both soils):

USCS: Fine-grained soil, silty clay of low plasticity (CL)

AASHTO: Silt-clay material, clayey soil, A-7-6 (22)

USDA: Clay loam

4.23   Soil is from core number 2 at a depth of 0.6 to 0.8 ft (AGC-2 @ 0.6 − 0.8 ft) and the data are summarized in Fig. 4.38.

Unified soil classification system (USCS):

Since 15.7% of the soil particles pass the No. 200 sieve (i.e., 15.7% fines), the soil is coarse-grained. All of the soil particles retained on the No. 200 sieve are of sand size and thus the primary soil classification is sand (Table 4.1). Since there are more than 12% fines and the soil is nonplastic (PI = 0), the soil classification is SM (silty sand).

AASHTO soil classification system:

Since there are less than 35% passing the No. 200 sieve, the soil is classified as a granular material. Using the particle size distribution (Fig. 4.38), the percent passing is as follows:

Percent passing No. 10 sieve = 97%

Percent passing No. 40 sieve = 61%

Percent passing No. 200 sieve = 15.7%

For these percent passing values, and because the soil is nonplastic (PI = 0), all of the criteria are met for group A-2-4 (silty sand).

USDA textural classification system:

The percent sand, percent silt, and percent clay size particles are as follows:

Sand (2 to 0.05 mm) = (97 − 12)/0.97 = 88%

Silt (0.05 to 0.002 mm) = (12 − 4)/0.97 = 8%

Clay (finer than 0.002 mm) = 4/0.97 = 4%

Using Fig. 4.5 with the percent sand = 88%, percent silt = 8%, and percent clay = 4%, the soil classification is sand.

Summary:

USCS: Coarse-grained soil, silty sand (SM)

AASHTO: Granular material, silty sand (A-2-4)

USDA: Sand

4.24   Soil is from core number 2 at a depth of 0.8 to 1.4 ft (AGC-2 @ 0.8 − 1.4 ft) and the data are summarized in Fig. 4.38.

Unified soil classification system (USCS):

Since 28.3% of the soil particles pass the No. 200 sieve (i.e., 28.3% fines), the soil is coarse-grained. Of the soil particles retained on the No. 200 sieve, 10.6% are gravel-size particles (i.e., 7.6 /0.717 = 10.6%) and 89.4% are sand-size particles (i.e., 64.1/0.717 = 89.4%). Since the larger fraction consists of sand size particles, the primary soil classification is sand (Table 4.1). The soil has greater than 12% fines and since Atterberg limits (i.e., liquid limit = 42 and plasticity index = 20) plot above the A-line in Fig. 4.1, the soil is classified as clayey sand (SC) per Table 4.1.

AASHTO soil classification system:

Since there are less than 35% passing the No. 200 sieve, the soil is classified as a granular material. With a plasticity index = 20 and a liquid limit = 42, the soil meets all the classification requirements for group A-2-7 (clayey sand). Using only the PI portion of the group index equation listed in Table 4.2, with $F = 28.3$, and PI = 20, the group index = 1.3. The soil classification is therefore an A-2-7 (1), clayey soil.

Summary:

USCS: Coarse-grained soil, clayey sand (SC)

AASHTO: Granular material, clayey sand, A-2-7 (1)

4.25 Soil type number 1 is described as crushed limestone from Tennessee.

Unified soil classification system (USCS):

Since 11% of the soil particles pass the No. 200 sieve (i.e., 11% fines), the soil is coarse-grained. Of the soil particles retained on the No. 200 sieve, 62% are gravel-size particles (i.e., 55/0.89 = 62%) and 38% are sand-size particles (i.e., 34/0.89 = 38%). Since the larger fraction consists of gravel size particles, the primary soil classification is gravel. Extending the grain size curve, assume D_{10} is approximately equal to 0.05 mm. Using $D_{60} = 8.6$ mm and $D_{30} = 2.0$ mm, then $C_u = 170$ and $C_c = 9.3$ (Eqs. 4.1 and 4.2). Per Table 4.1, the soil is poorly graded and since there are 11% fines, dual symbols are required. Since the limits plot above the A-line and the PI is greater than 4, the soil classification is GP-GC (poorly graded gravel and clayey gravel).

AASHTO soil classification system:

Since there are less than 35% passing the No. 200 sieve, the soil is classified as a granular material. With a LL = 18 and a PI = 7, the soil is classified per Table 4.2 as a group A-2-4 (silty or clayey gravel or sand).

Summary:

USCS: Coarse-grained soil, poorly graded gravel and clayey gravel (GP-GC)

AASHTO: Granular material, silty or clayey gravel, and sand (A-2-4)

4.26 Soil type number 2 is described as silty gravel derived from the weathering of Gabbro (Igneous Rock), from Oman.

Unified soil classification system (USCS):

Since 10% of the soil particles pass the No. 200 sieve (i.e., 10% fines), the soil is coarse-grained. Of the soil particles retained on the No. 200 sieve, 58% are gravel-size particles (i.e., 52/0.90 = 58%) and 42% are sand-size particles (i.e., 38/0.90 = 42%). Since the larger fraction consists of gravel size particles, the primary soil classification is gravel. Using $D_{60} = 9$ mm, $D_{30} = 1.4$ mm, and $D_{10} = 0.08$ mm, then $C_u = 112$ and $C_c = 2.7$ (Eqs. 4.1 and 4.2). Per Table 4.1, the soil is well-graded and since there are 10% fines, dual symbols are required. Since the Atterberg Limits plot above the A-line, the soil classification is GW-GC (well-graded gravel and clayey gravel).

Note that ASTM D 2487-00 (2004) "Standard Practice for Classification of Soils for Engineering Purposes (Unified Soil Classification System)" requires that the fines classification for coarse-grained soil be based on Fig. 4.1. Using Fig. 4.1 and a LL = 23 and PI = 3, the data plot above the A-line but in the ML category. Thus per ASTM D 2487-00 the fines would be classified as ML, and hence the soil classification would be GW-GM (well-graded gravel and silty gravel).

AASHTO soil classification system:

Since there are less than 35% passing the No. 200 sieve, the soil is classified as a granular material. With 34% passing the No.10 sieve, 19% passing the No. 40 sieve, 10% passing the No. 200 sieve, and the PI = 3, the soil meets all the classification requirements per Table 4.2 as a group A-1-a (stone or gravel fragments).

Summary:

USCS: Coarse-grained soil, well-graded gravel, and clayey gravel (GW-GC)

Per ASTM D 2487-00, well-graded gravel, and silty gravel (GW-GM)

AASHTO: Granular material, stone or gravel fragments (A-1-a)

4.27 Soil type number 3 is described as alluvial gravelly sand from Mississippi. This soil is non-plastic (i.e., PI = 0).

Unified soil classification system (USCS):

Since 0% of the soil particles pass the No. 200 sieve (i.e., 0% fines), the soil is coarse-grained. Of the soil particles retained on the No. 200 sieve, 44% are gravel-size particles and 56% are sand-size particles. Since the larger fraction consists of sand-size particles, the primary soil classification is sand. Using $D_{60} = 6.4$ mm, $D_{30} = 0.53$ mm, and $D_{10} = 0.36$ mm, then $C_u = 18$ and $C_c = 0.1$ (Eqs. 4.1 and 4.2). Per Table 4.1, the soil is poorly-graded and since there are 0% fines and the PI = 0, the soil classification is SP (poorly-graded gravelly sand).

AASHTO soil classification system:

Since there are less than 35% passing the No. 200 sieve, the soil is classified as a granular material. With 50.1% passing the No.10 sieve, 19% passing the No. 40 sieve, 0% passing the No. 200 sieve, and a PI = 0, the soil meets all the classification requirements for group A-1-b (gravel and sand mixtures).

Summary:

USCS: Coarse-grained soil, poorly-graded gravelly sand (SP)

AASHTO: Granular material, gravel and sand mixtures (A-1-b)

4.28 Soil type number 4 is described as eolian sand from Oman. This soil is nonplastic (i.e., PI = 0).

Unified soil classification system (USCS):

Since 6% of the soil particles pass the No. 200 sieve (i.e., 6% fines), the soil is coarse-grained. All of the soil particles retained on the No. 200 sieve are sand-size particles and thus the primary soil classification is sand. Using $D_{60} = 0.40$ mm, $D_{30} = 0.27$ mm, and $D_{10} = 0.10$ mm, then $C_u = 4$ and $C_c = 1.8$ (Eqs. 4.1 and 4.2). Per Table 4.1, the soil is poorly-graded and since there are 6% fines, dual symbols are required. Since PI = 0, the soil classification is SP-SM (poorly-graded sand and silty sand).

AASHTO soil classification system:

Since there are less than 35% passing the No. 200 sieve, the soil is classified as a granular material. With 100% passing the No.10 sieve, 76% passing the No. 40 sieve, 6% passing the No. 200 sieve, and a PI = 0, the soil meets all the classification requirements for Group A-3 (fine sand that is nonplastic).

USDA textural classification system:

Extending the grain size curve, the percent particles finer than 0.05 mm is approximately = 4%. The remainder of the soil particles are of sand size. Therefore the percent sand = 96%, percent silt = 4%, and percent clay = 0%. Using Fig. 4.5, the soil classification is sand.

Summary:

USCS: Coarse-grained soil, poorly-graded sand and silty sand (SP-SM)

AASHTO: Granular material, fine sand that is nonplastic (A-3)

USDA: Sand

4.29 Soil type number 5 is described as glacial till from Illinois. This soil has a liquid limit = 25 and a plasticity index = 10.

Unified soil classification system (USCS):

Since 62% of the soil particles pass the No. 200 sieve (i.e., 62% fines), the soil is fine-grained. Since the liquid limit is less than 50, the soil has a low plasticity. The PI > 7 and the data plot above the A-line, and thus the soil is classified as silty clay of low plasticity (CL).

AASHTO soil classification system:

Since there are more than 35% passing the No. 200 sieve, the soil is classified as a silt-clay material. With a plasticity index = 10 and a liquid limit = 25, the soil meets all the classification requirements for group A-4 (silty soils). Using the group index equation listed in Table 4.2, with $F = 62$, LL = 25, and PI = 10, the group index = 3.4. The soil classification is therefore an A-4 (3), silty soil.

USDA textural classification system:

The percent sand, percent silt, and percent clay size particles are as follows:

Sand (2 to 0.05 mm) = (90 − 59)/0.9 = 34%

Silt (0.05 to 0.002 mm) = (59 − 15)/0.9 = 49%

Clay (finer than 0.002 mm) = 15/0.9 = 17%

Using Fig. 4.5 with the percent sand = 34%, percent silt = 49%, and percent clay = 17%, the soil classification is loam.

Summary:

USCS: Fine-grained soil, silty clay of low plasticity (CL)

AASHTO: Silt-clay material, silty soil, A-4 (3)

USDA: Loam

4.30 Soil type number 6 is described as Wewahitchka sandy clay from Florida. This soil has a liquid limit = 65 and a plasticity index = 41.

Unified soil classification system (USCS):

Since 70% of the soil particles pass the No. 200 sieve (i.e., 70% fines), the soil is fine-grained. Since the liquid limit is greater than 50, the soil has a high plasticity. The PI > 7 and the data plot above the A-line, and thus the soil is classified as a sandy clay of high plasticity (CH).

AASHTO soil classification system:

Since there are more than 35% passing the No. 200 sieve, the soil is classified as a silt-clay material. With a plasticity index = 41, a liquid limit = 65, and since the LL − 30 is less than the plasticity index, the soil meets all the classification requirements for group A-7-6 (clayey soils). Using the group index equation listed in Table 4.2, with $F = 70$, LL = 65, and PI = 41, the group index = 28.4. The soil classification is therefore an A-7-6 (28), clayey soil.

USDA textural classification system:

The percent sand, percent silt, and percent clay size particles are as follows:

Sand (2 to 0.05 mm) = 100 − 65 = 35%

Silt (0.05 to 0.002 mm) = 65 − 44 = 21%

Clay (finer than 0.002 mm) = 44%

Using Fig. 4.5 with the percent sand = 35%, percent silt = 21%, and percent clay = 44%, the soil classification is clay.

Summary:

USCS: Fine-grained soil, sandy clay of high plasticity (CH)

AASHTO: Silt-clay material, clayey soil, A-7-6 (28)

USDA: Clay

4.31 Soil type number 7 is described as loess from Mississippi. This soil has a liquid limit = 29 and a plasticity index = 5.

Unified soil classification system (USCS):

Since 96% of the soil particles pass the No. 200 sieve (i.e., 96% fines), the soil is fine-grained. Since the liquid limit is less than 50, the soil has a low plasticity. The data plot below the A-line in Fig. 4.1, and thus the soil is classified as a silt of low plasticity (ML).

AASHTO soil classification system:

Since there are more than 35% passing the No. 200 sieve, the soil is classified as a silt-clay material. With a plasticity index = 5 and a liquid limit = 29, the soil meets all the classification requirements for group A-4 (silty soil). Using the group index equation listed in Table 4.2, with $F = 96$, LL = 29, and PI = 5, the group index = 4.8. The soil classification is therefore an A-4 (5), silty soil.

USDA textural classification system:

The percent sand, percent silt, and percent clay size particles are as follows:

Sand (2 to 0.05 mm) = 100 − 90 = 10%

Silt (0.05 to 0.002 mm) = 90 − 5 = 85%

Clay (finer than 0.002 mm) = 5%

Using Fig. 4.5 with the percent sand = 10%, percent silt = 85%, and percent clay = 5%, the soil classification is silt.

Summary:

USCS: Fine-grained soil, silt of low plasticity (ML)

AASHTO: Silt-clay material, silty soil, A-4 (5)

USDA: Silt

4.32 Soil type number 8 is described as backswamp deposit from the Mississippi River. This soil has a liquid limit = 59 and a plasticity index = 41.

Unified soil classification system (USCS):

Since 95% of the soil particles pass the No. 200 sieve (i.e., 95% fines), the soil is fine-grained. Since the liquid limit is greater than 50, the soil has a high plasticity. The data plot above the A-line in Fig. 4.1, and thus the soil is classified as silty clay of high plasticity (CH).

AASHTO soil classification system:

Since there are more than 35% passing the No. 200 sieve, the soil is classified as a silt-clay material. With a plasticity index = 41, a liquid limit = 59, and since the LL – 30 is less than the plasticity index, the soil meets all the classification requirements for group A-7-6 (clayey soil). Using the group index equation listed in Table 4.2, with $F = 95$, $LL = 59$, and $PI = 41$, the group index = 42.5. The soil classification is therefore an A-7-6 (43), clayey soil.

USDA textural classification system:

The percent sand, percent silt, and percent clay size particles are as follows:

Sand (2 to 0.05 mm) = 100 – 90 = 10%

Silt (0.05 to 0.002 mm) = 90 – 42 = 48%

Clay (finer than 0.002 mm) = 42%

Using Fig. 4.5 with the percent sand = 10%, percent silt = 48%, and percent clay = 42%, the soil classification is silty clay.

Summary:

USCS: Fine-grained soil, silty clay of high plasticity (CH)

AASHTO: Silt-clay material, clayey soil, A-7-6 (43)

USDA: Silty clay

4.33 Soil 5: $PI = 10$ and clay fraction = 15%. The percent passing the No. 40 sieve = 80%. Therefore the activity = $10/(15/0.8) = 0.53$

Soil 6: $PI = 41$ and clay fraction = 44%. Almost all the soil particles pass the No. 40 sieve. Therefore the activity = $41/44 = 0.93$

Soil 7: $PI = 5$ and clay fraction = 5%. All of the soil particles pass the No. 40 sieve. Therefore the activity = $5/5 = 1.0$

Soil 8: $PI = 41$ and clay fraction = 42%. All of the soil particles pass the No. 40 sieve. Therefore the activity = $41/42 = 0.98$

4.34 Using Table 4.4 and since there is visible ice that is less than 1-in. (25-mm) thick, the main group symbol is V. Because it was observed that there are ice coatings on the individual soil particles, the group symbol is V_c.

4.35 For United States customary system, $V = 1$ ft^3

$\gamma_d = \gamma_t/(1 + w) = 121.4/(1 + 0.295) = 93.7$ pcf

$\gamma_d = W_s/V$ or: $W_s = \gamma_d V = (93.7 \text{ pcf})(1 \text{ ft}^3) = 93.7$ lb

$M_s = W_s = 93.7$ lb

$M_w = M_s w = (93.7 \text{ lb})(0.295) = 27.7$ lb

$V_s = W_s/[(G_s)(\gamma_w)] = 93.7/[(2.70)(62.4)] = 0.56$ ft^3

$V_w = M_w/\rho_w = 27.7 \text{ lb}/62.4 \text{ pcf} = 0.44$ ft^3

$V_a = V - V_s - V_w = 1 - 0.56 - 0.44 = 0$ ft^3

For SI, $V = 1$ m^3

$\gamma_d = \gamma_t/(1 + w) = 19.1/(1 + 0.295) = 14.7$ kN/m^3

$\gamma_d = W_s/V$ or: $W_s = \gamma_d V = (14.7 \text{ kN/m}^3)(1 \text{ m}^3) = 14.7$ kN

$W_s = M_s a$ or:

$M_s = W_s/a = 14.7 \text{ kN}/9.81 \text{ m/sec}^2 = 1.50$ Mg

$M_w = M_s w = (1.50 \text{ Mg})(0.295) = 0.44 \text{ Mg}$

$V_s = W_s / [(G_s)(\gamma_w)] = 14.7/[(2.70)(9.81)] = 0.56 \text{ m}^3$

$V_w = M_w/\rho_w = 0.44 \text{ Mg}/1.0 \text{ Mg/m}^3 = 0.44 \text{ m}^3$

$V_a = V - V_s - V_w = 1 - 0.56 - 0.44 = 0 \text{ m}^3$

Void ratio, porosity, and degree of saturation:

$e = V_v/V_s = (V_w + V_a)/V_s = (0.44 + 0)/0.56 = 0.79$

$n = V_v/V = (0.44 + 0)/1.0 = 0.44 \text{ or } 44\%$

$S = [(G_s)(w)]/e = [(2.70)(0.295)]/0.79 = 1.0 \text{ or } 100\%$

4.36 $D_r = 100 (e_{max} - e)/(e_{max} - e_{min}) = 100 (0.85 - 0.79)/(0.85 - 0.30) = 11\%$

4.37   Per Part (a) of Fig. 3.30:

Water content = 23.2%

Total unit weight = 116.7 pcf

Dry unit weight = 94.7 pcf

Assuming $V = 1 \text{ ft}^3$

$V_s = W_s / [(G_s)(\gamma_w)] = 94.7/[(2.70)(62.4)] = 0.56 \text{ ft}^3$

$V_w = M_w/\rho_w = (116.7 - 94.7)/62.4 \text{ pcf} = 0.35 \text{ ft}^3$

$V_a = V - V_s - V_w = 1 - 0.56 - 0.35 = 0.09 \text{ ft}^3$

$e = V_v/V_s = (V_w + V_a)/V_s = (0.35 + 0.09)/0.56 = 0.79$

$n = V_v/V = (0.35 + 0.09)/1.0 = 0.44 \text{ or } 44\%$

$S = [(G_s)(w)]/e = [(2.70)(0.232)]/0.79 = 0.79 \text{ or } 79\%$

4.38   Given the following values:

Void ratio $(e) = 1.16$

Specific gravity $(G_s) = 2.72$

Water content $(w) = 42.7\% = 0.427$

Using the total unit weight (γ_t) equation from Table 4.8:

$$\gamma_t = \frac{G_s\gamma_w(1+w)}{1+e} = \frac{(2.72)(62.4)(1+0.427)}{1+1.16} = 112 \text{ pcf}$$

Using the dry unit weight (γ_d) equation from Table 4.8:

$$\gamma_d = \frac{\gamma_t}{1+w} = \frac{112}{1+0.427} = 78.6 \text{ pcf}$$

Using the saturated unit weight (γ_{sat}) equation from Table 4.8:

$$\gamma_{sat} = \frac{(G_s + e)\gamma_w}{1+e} = \frac{(2.72+1.16)(62.4)}{1+1.16} = 112 \text{ pcf}$$

Note: For this problem, the total unit weight (γ_t) is equal to the saturated unit weight (γ_{sat}) because all the void spaces are filled with water (i.e., S = 100%).

Using the buoyant unit weight (γ_b) equation from Table 4.8:

$\gamma_b = \gamma_{sat} - \gamma_w = 112 - 62.4 = 49.6 \text{ pcf}$

Check by using the following equation from Table 4.8:

$$\gamma_b = \frac{\gamma_w(G_s - 1)}{1 + e} = \frac{62.4\,(2.72 - 1)}{1 + 1.16} = 49.6 \text{ pcf}$$

4.39 Using Eq. 4.15:

$\sigma_v = \gamma_t z = (19.5)(6) = 117$ kPa or 2480 psf

Using Eq. 4.16:

$u = \gamma_w z = (9.81)(6) = 59$ kPa or 1250 psf

Using Eq. 4.14:

$\sigma_v' = \sigma_v - u = 117 - 59 = 58$ kPa or 1230 psf

4.40 Using Eq. 4.15:

$\sigma_v = \gamma_t z = (19.5)(6) = 117$ kPa or 2480 psf

Using Eq. 4.16:

$u = \gamma_w z = (9.81)(4.5) = 44$ kPa or 940 psf

Using Eq. 4.14:

$\sigma_v' = \sigma_v - u = 117 - 44 = 73$ kPa or 1540 psf

4.41 Using Eq. 4.15 and letting z_1 = depth of the lake and z_2 = depth below lake bottom:

$\sigma_v = \gamma_w z_1 + \gamma_t z_2 = (9.81)(3) + (19.5)(6) = 146$ kPa or 3100 psf

Using Eq. 4.16:

$u = \gamma_w(z_1 + z_2) = (9.81)(3 + 6) = 88$ kPa or 1870 psf

Using Eq. 4.14:

$\sigma_v' = \sigma_v - u = 146 - 88 = 58$ kPa or 1230 psf

4.42 Using Eq. 4.15 and letting z_1 = depth below ground surface and z_2 = distance above ground-water table:

$\sigma_v = \gamma_t z_1 = (19.5)(1.5) = 29$ kPa or 620 psf

Using Eq. 4.17:

$u = -\gamma_w z_2 = (9.81)(3 - 1.5) = -15$ kPa or –310 psf

Using Eq. 4.14:

$\sigma_v' = \sigma_v - u = 29 - (-15) = 44$ kPa or 930 psf

4.43 Using Eq. 4.14:

$\sigma_v' = \sigma_v - u = 117 - (3)(9.81) = 88$ kPa or 1860 psf

4.44 In Fig. 2.39, the total unit weight values are shown on the left side of the figure. Using $\gamma_t = 122$ pcf for the sand-gravel layer, with z = thickness of each soil layer, then from Eq. 4.15:

$\sigma_v = \gamma_t z = (125)(21 - 12.5) + (101)(12.5 + 6) + (122)(4) + (119)(30) +$

$(117)(25) + (123)(25) = 13{,}000$ psf or 6.3 kg/cm^2

Using Eq. 4.16:

$u = \gamma_w d = (62.4)(12 + 90) = 6400$ psf or 3.1 kg/cm^2

Using Eq. 4.14:

$\sigma_v' = \sigma_v - u = 13{,}000 - 6400 = 6600$ psf or 3.2 kg/cm^2

4.45   For one-dimensional loading, using Eq. 4.21 with h = thickness of the fill layer:

$\Delta\sigma_v = h\gamma_t = (3)(18.7) = 56$ kPa or 1190 psf

4.46   Using Eq. 4.25:

$P = BL\sigma_o = BL\gamma_t H = (6)(10)(18.7)(3) = 3370$ kN

For the 2:1 approximation, use Eq. 4.26. For the loaded area, B = 6 m and L = 10 m. At a depth z = 12 m:

$\Delta\sigma_v = P/[(B + z)(L + z)] = 3370/[(6 + 12)(10 + 12)] = 8.5$ kPa or 180 psf

4.47   Using Eq. 4.29 with the following values:

$P = Q = 3370$ kN

$r = 0$

$z = 12$ m

$\Delta\sigma_v = [3Qz^3]/[2\pi(r^2 + z^2)^{5/2}] = [(3)(3370)(12)^3]/[2\pi(12^2)^{5/2}] = 11$ kPa or 230 psf

4.48 $q_o = \gamma_t H = 56$ kPa

Dividing the loaded area into four squares and determining the increase in stress beneath the corner:

$m = x/z = 3/12 = 0.25$

$n = y/z = 5/12 = 0.42$

From Fig. 4.8 with m = 0.25 and n = 0.42, I = 0.044

Using the equation in Fig. 4.8 and multiplying by 4 to account for the four areas:

$\Delta\sigma_v = \sigma_z = 4q_o I = (4)(56)(0.044) = 9.9$ kPa or 210 psf

4.49 $q = \gamma_t H = 56$ kPa

Dividing the loaded area into four squares and determining the increase in stress beneath the corner:

$m = x/z = 3/12 = 0.25$

$n = y/z = 5/12 = 0.42$

From Fig. 4.15 with m = 0.25 and n = 0.42, I = 0.027

Using the equation in Fig. 4.15 and multiplying by 4 to account for the four areas:

$\Delta\sigma_v = \sigma_z = 4qI = (4)(56)(0.027) = 6.1$ kPa or 130 psf

4.50   Using Fig. 4.13 with the distance AB = 12 m, draw a rectangle with width = 6/12 = 0.5 AB and length = 10/12 = 0.83 AB, with the center of the rectangle at the center of Fig. 4.13. Counting the number of blocks within the rectangle = 34, therefore using Eq. 4.31:

$\Delta\sigma_v = \sigma_z = q_o IN = (56)(0.005)(34) = 9.5$ kPa or 200 psf

Summary of Values		
Problem number	Method of analysis	$\Delta\sigma_v$
4.46	2:1 approximation	8.5 kPa (180 psf)
4.47	Concentrated load	11 kPa (230 psf)
4.48	Newmark Chart (Fig. 4.8)	9.9 kPa (210 psf)
4.49	Westergaard Chart	6.1 kPa (130 psf)
4.50	Newmark Chart (Fig. 4.13)	9.5 kPa (200 psf)

4.51   Using the 2:1 approximation with $B = 6$ m, $L = 10$ m:

$\Delta\sigma_v = 0.1 \, q_o = 5.6$ kPa

Using Eq. 4.26:

$\Delta\sigma_v = P/[(B + z)(L + z)]$

$5.6 = 3370/[(6 + z)(10 + z)]$

Solving for $z = 17$ m

4.52   Using the following values:

$d = D = 5.0$ cm

$q = 1000/33 = 30.3$ cm^3/sec

$H_c = 4$ m $= 400$ cm

For case C (Fig. 4.27) and a constant head condition:

$k = q/[(2.75)(D)(H_c)] = 30.3/[(2.75)(5.0)(400)] = 0.006$ cm/sec

4.53   Using the following values:

$d = D = 0.5$ cm

$H_1 = 4.0$ m

$H_2 = 3.1$ m

$t_2 - t_1 = 60$ sec

For case C (Fig. 4.27) and a variable head condition:

$k = \pi D/[11(t_2 - t_1)] \ln(H_1/H_2) = \pi(5.0)/[(11)(60)] \ln(4/3.1) = 0.006$ cm/sec

4.54   A doubling of h_w (i.e., Δh), doubles Q per Eq. 4.42.

4.55   14.5 drops, therefore Δh lost $= (14.5/18)(10) = 8.06$ m

Pore water pressure head $= 10 - 8.06 + 12 = 13.9$ m

Converting to a pore water pressure $= (13.9)(9.81) = 137$ kPa

4.56   Using Eq. 4.14:

$\sigma_v' = \sigma_v - u = (12)(19.8) - 137 = 101$ kPa

4.57   The length (L) of square labeled 18 is approximately $= 4$ m, therefore:

$i_e = \Delta h/L = h'/L = (10/18)/4 = 0.14$

$i_c = \gamma_b/\gamma_w = (19.8 - 9.81)/9.81 = 1.02$

$F = i_c/i_e = 1.02/0.14 = 7$

4.58   Since the strata is sand, assume $G_s = 2.65$

Using the following equation from Table 4.8:

$\gamma_b = \gamma_w(G_s - 1)/(1 + e)$

$(19.8 - 9.81) = [(9.81)(2.65 - 1)]/(1 + e)$

Solving for $e = 0.62$

Using Eq. 4.10:

$n = e/(1 + e) = 0.62/(1 + 0.62) = 0.38$

Using Eq. 4.40:

$v_s = ki/n = (0.1)(0.14)/0.38 = 0.037$ m/day

The seepage in an approximately upward direction.

4.59   Since the equipotential drops are all equal, the highest seepage velocity occurs where the length of the flow net square is the smallest, or at the sheet pile tip.

4.60   Using Eq. 4.42 and the following values:

$k = 1 \times 10^{-8}$ m/sec

$n_f = 10$

$n_d = 14$

$\Delta h = 20$ m

$t = 1$ day $= 86,400$ sec

$Q = k\Delta ht(n_f/n_d)$

$Q = (1 \times 10^{-8}$ m/sec$)(20$ m$)(86,400$ sec$)(10/14) = 0.012$ m^3 per unit length

Times a length of 200 m $= 2.5$ m^3 of water per day

4.61   Number of equipotential drops $= 14$ (Note: for the uppermost flow channel, one of the equipotential drops occurs in the drainage filter).

$h' = h_w/n_d = 20/14 = 1.43$ m

4.62   Since the equipotential drops are all equal, the highest seepage velocity occurs where the length of the flow net squares is the smallest, or in the soil that is located in front of the longitudinal drainage filter.

4.63   The soil located in front of the longitudinal drainage filter has the highest seepage velocity (v_s) and this is the most likely location for piping of soil into the drainage filter.

4.64   At a point located at the centerline of the dam and 20 m below the top of the dam, the number of equipotential drops $= 6.2$.

For 6.2 drops, therefore Δh lost $= (6.2/14)(20) = 8.86$ m

Pore water pressure head $= 20 - 8.86 = 11.1$ m

Converting to pore water pressure $= (11.1)(9.81) = 110$ kPa

4.65   Using Eq. 4.14:

$\sigma'_v = \sigma_v - u = (20)(20) - 110 = 290$ kPa

4.66   Using Eq. 3.9:

$\tau_f = c' + \sigma'_n \tan \phi'$

In this case with a horizontal slip surface, $\sigma'_n = \sigma'_v = 290$ kPa

$\tau_f = 2 + (290) \tan 28° = 156$ kPa

4.67 $F =$ shear strength divided by shear stress

$F = \tau_f/\tau = 156/83 = 1.88$

Chapter 5

5.1   A deep foundation system consisting of piles or piers embedded in the sandstone.

5.2   Assuming the weight of the soil excavated for the basement is approximately equal to the weight of the two-story structure, a floating foundation would be desirable.

5.3   Because the upper 10 ft of the site consists of overconsolidated clay, it would be desirable to use a shallow foundation system based on the assumption of light building loads.

5.4   No, because of the very high sensitivity of the clay, high displacement piles will remold the clay and result in a loss of shear strength. The preferred option is to install low displacement piles or use predrilled, cast-in-place concrete.

Chapter 6

6.1 $\gamma_b = \gamma_{sat} - \gamma_w = 125 - 62.4 = 62.6$ pcf (9.89 kN/m³)

Using Eq. 6.7:

$\gamma_a = \gamma_b + [(h' - D_f)/B](\gamma_t - \gamma_b)$

$= 62.6 + [(2 - 2)/4](125 - 62.6) = 62.6$ pcf

From Fig. 6.5, for $\phi = 30°$, $N_\gamma = 15$ and $N_q = 19$. Using Eq. 6.1 with $c' = 0$:

$q_{ult} = \frac{1}{2}\, \gamma_a B N_\gamma + \gamma_t D_f N_q = \frac{1}{2}(62.6)(4)(15) + (125)(2)(19) = 6600$ psf

Using Eq. 6.2 with factor of safety = 3:

$q_{all} = q_{ult}/3 = 6600/3 = 2200$ psf

$Q_{all} = (q_{all})(B) = (2200$ psf$)(4$ ft$) = 8800$ lb/ft $= 8.8$ kips per linear foot

6.2   Using Eq. 6.7:

$\gamma_b = \gamma_{sat} - \gamma_w = 125 - 62.4 = 62.6$ pcf

$\gamma_a = \gamma_b + [(h' - D_f)/B](\gamma_t - \gamma_b) = 62.6 + [(4 - 2)/4](125 - 62.6) = 93.9$ pcf

From Fig. 6.5, for $\phi = 30°$, $N_\gamma = 15$ and $N_q = 19$. Using Eq. 6.3 with $c' = 0$:

$q_{ult} = 0.4\gamma_a B N_\gamma + \gamma_t D_f N_q = 0.4(93.9)(4)(15) + (125)(2)(19) = 7000$ psf

Using Eq. 6.2 with factor of safety = 3:

$q_{all} = q_{ult}/3 = 7000/3 = 2300$ psf

$Q_{all} = (q_{all})(B)^2 = (2300$ psf$)(4$ ft$)^2 = 37{,}000$ lb $= 37$ kips

6.3   From Eq. 6.1:

$q_{ult} = (Q_{ult})/[(B)(L)]$

From Eq. 6.2:

$Q_{all} = Q_{ult}/F$

$Q_{ult} = Q_{all}F$

Substituting into the first equation:

$q_{ult} = [(Q_{all})(F)/[(B)(L)]$

Using the following values:

$Q_{all} = 150$ kN

$F = 3$

Assuming $L = 1$ m, therefore:

$q_{ult} = [(Q_{all})(F)]/[(B)(L)] = [(150)(3)]/[(B)(1)] = 450/B$

Using Eq. 6.1 with $c' = 0$ and total unit weight = 19.7 kN/m³:

$q_{ult} = \frac{1}{2}\gamma_t BN_\gamma + \gamma_t D_f N_q$

$450/B = \frac{1}{2}(19.7)(B)(15) + (19.7)(0.6)(19)$

$0 = \frac{1}{2}(19.7)(B^2)(15) + (19.7)(0.6)(19)(B) - 450$

$0 = B^2 + 1.52\ B - 3.05$

Solving for $B = 1.14$ m

6.4 For a relative density (D_r) of 65%, the sand is at the boundary between the medium and dense states (see Table 2.6). At the top of Fig. 6.5, the boundary between the medium and dense states corresponds to a friction angle of 36°. Using $\phi' = 36°$, then $N_\gamma = 45$ and $N_q = 38$ from Fig. 6.5. Using Eq. 6.3 with $c' = 0$:

$q_{ult} = 0.4\gamma_t BN_\gamma + \gamma_t D_f N_q = 0.4(120)(10)(45) + (120)(5)(38) = 44,400$ psf

Using Eq. 6.2 with factor of safety = 3:

$q_{all} = q_{ult}/3 = 44,400/3 = 14,800$ psf

$Q_{all} = (q_{all})(B)^2 = (14,800\text{ psf})(10\text{ ft})^2 = 1,480,000$ lb = 1480 kips

6.5 Using Eq. 6.8:

$q_{ult} = 5.5s_u + \gamma_t D_f$

$q_{ult} = (5.5)(20) + (19.7)(0.6) = 122$ kPa

Using Eq. 6.2 with a factor of safety = 3:

$q_{all} = q_{ult}/F = 122/3 = 41$ kPa

For the 1.2-m wide strip footing:

$Q_{all} = (q_{all})(B) = (41)(1.2) = 49$ kN per linear meter of footing length

6.6 From Eq. 6.1:

$q_{ult} = (Q_{ult})/[(B)(L)]$

From Eq. 6.2:

$Q_{all} = Q_{ult}/F$

$Q_{ult} = Q_{all}F$

Substituting into the first equation:

$q_{ult} = [(Q_{all})(F)]/[(B)(L)]$

Using the following values:

$Q_{all} = 50$ kN

$F = 3$

Assuming $L = 1$ m, therefore:

$q_{ult} = [(Q_{all})(F)]/[(B)(L)] = [(50)(3)]/[(B)(1)] = 150/B$

Using Eq. 6.8 and total unit weight = 19.7 kN/m³:

$q_{ult} = 5.5c + \gamma_t D_f$

$150/B = (5.5)(14.5) + (19.7)(0.6)$

$150 = (80)(B) + (12)(B)$

Solving for $B = 1.64$ m

6.7 Total stress analysis:

Using a total unit weight $= 19.7$ kN/m^3 and Eq. 6.8:

$q_{ult} = 5.5s_u + \gamma_t D_f = (5.5)(200) + (19.7)(0.6) = 1100$ kPa

Effective stress analysis:

From Fig. 6.6, for $\phi' = 28°$, $N_c = 30$, $N_\gamma = 15$, and $N_q = 18$

Using $c' = 5$ kPa, $\gamma_a = \gamma_b$, and Eq. 6.1:

$q_{ult} = c'N_c + \frac{1}{2}\gamma_a BN_\gamma + \gamma_t D_f N_q$

$= (5)(30) + \frac{1}{2}(19.7 - 9.81)(1.2)(15) + (19.7)(0.6)(18) = 450$ kPa

Therefore the effective stress analysis governs:

$Q_{all} = (q_{ult})(B)/F = (450)(1.2)/3 = 180$ kN per linear meter of footing length

6.8 The middle one-third of the footing is $1.2/3 = 0.4$ m, or a distance of 0.2 m from the center-line of the footing. Since $e = 0.15$ is less than 0.2 m, the eccentricity is within the middle one-third of the footing and Eqs. 6.12 and 6.13 can be used:

$q' = Q(B + 6\,e)/B^2$

$q'' = Q(B - 6\,e)/B^2$

$q' = (100)[1.2 + (6)(0.15)]/(1.2)^2 = 146$ kPa

$q'' = (100)[1.2 - (6)(0.15)]/(1.2)^2 = 21$ kPa

and since $q' = 146$ kPa and $q_{all}, = 120$ kPa, then $q' > q_{all}$ and q' is unacceptable

6.9 From Fig. 6.8:

$B' = B - 2e = 1.2 - (2)(0.15) = 0.9$ m

$q = Q/[(B')(L)]$

Using $L = 1$ m

$q = 100/[(0.9)(1)] = 111$ kPa

And since $q = 111$ kPa and $q_{all} = 120$ kPa, then $q < q_{all}$ and q is acceptable by this method.

6.10 Q per linear meter of footing $= Q/B = 100$ kN/1.2 m $= 83.3$ kN/m

The middle one-third of the footing is $1.2/3 = 0.4$ m, or a distance of 0.2 m from the center-line of the footing. Since $e = 0.15$ is less than 0.2 m, the eccentricity is within the middle one-third of the footing and Eqs. 6.12 and 6.13 can be used:

$q' = Q(B + 6\,e)/B^2$

$q'' = Q(B - 6\,e)/B^2$

$q' = (83.3)[1.2 + (6)(0.15)]/(1.2)^2 = 122$ kPa

$q'' = (83.3)[1.2 - (6)(0.15)]/(1.2)^2 = 17$ kPa

and since $q' = 122$ kPa and $q_{all} = 120$ kPa, then $q' > q_{all}$ and q' is unacceptable.

6.11 From Fig. 6.8:

$B' = B - 2e = 1.2 - (2)(0.15) = 0.9$ m

$q = Q/[(B')(L)]$

Since $L = 1.2$ m

$q = 100/[(0.9)(1.2)] = 93$ kPa

and since $q = 93$ kPa and $q_{all} = 120$ kPa, then $q < q_{all}$ and q is acceptable by this method.

6.12 Using Eq. 6.16:

$q_{ult} = Q_p/(\pi r^2) = \sigma'_v N_q$

Rearranging the terms:

$N_q = Q_p/(\sigma'_v \pi r^2) = 250/[(87)(\pi)(0.3/2)^2] = 40.7$

For the 0.4 m diameter pile:

$Q_p = (N_q)(\pi r^2 \sigma'_v) = (40.7)(\pi)(0.4/2)^2(87) = 444$ kN

Using a factor of safety of 3:

$Q_{all} = Q_p/F = 444/3 = 148$ kN

Check: Based on areas (i.e., $0.4^2/0.3^2 = 1.78$), $Q_p = 1.78 (250) = 444$ kN

6.13 Assuming the term: $k \tan \phi'_w$ is the same for the larger and smaller diameter pile, the frictional resistance is proportional to the surface area (i.e., $0.4/0.3 = 1.33$):

$Q_s = 1.33(250) = 333$ kN

Using a factor of safety of 3:

$Q_{all} = Q_s/F = 333/3 = 111$ kN

6.14 End bearing $= (0.6)(250) = 150$ kN

The end-bearing resistance is proportional to pile tip area (i.e., $0.4^2/0.3^2 = 1.78$), therefore:

$Q_p = (1.78)(150) = 267$ kN

Side friction $= (0.4)(250) = 100$ kN

The side friction resistance is proportional to a side area (i.e., $0.4/0.3 = 1.33$), therefore:

$Q_s = (1.33)(100) = 133$ kN

$Q_p + Q_s = 267 + 133 = 400$ kN

Using a factor of safety of 3:

$Q_{all} = (Q_p + Q_s)/F = 400/3 = 133$ kN

6.15 Using Fig. 6.6, for $\phi' = 40°$, $N_c = 84$, $N_\gamma = 100$, and $N_q = 72$

Using Eq. 6.4 with $c' = 50$ kPa, $B = 1$ m, and assuming sandstone at elevation -19 m:

$q_{ult} = 1.3\, c'N_c + 0.4\, \gamma_b BN_\gamma + \gamma_b D_f N_q$

$= (1.3)(50)(84) + (0.4)(11.7)(1)(100) + [(9.2)(9) + (11.7)(3)](72) = 14{,}400$ kPa

Using Eq. 6.16:

$Q_p = (q_{ult})(\pi r^2) = (14{,}400)(\pi)(1.0/2)^2 = 11{,}300$ kN

Using a factor of safety of 3:

$Q_{all} = Q_p/F = 11{,}300/3 = 3770$ kN

6.16 Using Eq. 6.21:

$Q_D = 2\pi R L_1 \sigma'_v k \tan \phi_w$

For the sand layer, average $\sigma'_v = (0.25)(9.2) = 2.3$ kPa

$Q_D = 2\pi(1.0/2)(0.5)(2.3)(0.5) \tan 20° = 0.7$ kN

For the 3-m thick silt-peat layer, average $\sigma'_v = (0.5)(9.2) + (1.5)(9.2) = 18.4$ kPa

$Q_D = 2\pi(1.0/2)(3)(18.4)(0.4) \tan 15° = 18.6$ kN

Total $Q_D = 0.7 + 18.6 = 19.3$ kN

6.17 Assume the pile cap is from elevation +20 to +21 and ignore its weight since it is approximately compensated by the weight of removed soil. From Fig. 2.39, use an average $s_u = c = 0.6$ kg/cm² (1200 psf) from elevation –10 to –50 ft and an average $s_u = c = 0.4$ kg/cm² (800 psf) at elevation –50 ft. Using Fig. 6.13, for $c = 1200$ psf, $c_A/c = 0.62$ (average curve, all piles) and therefore $c_A = 740$ psf. Using Eq. 6.19:

Q_{ult} = end bearing + side adhesion = $9\pi cR^2 + 2\pi c_A Rz$

$= 9\pi(800)(1.5/2)^2 + 2\pi(740)(1.5/2)(40)$

$= 13,000 + 140,000 = 153,000$ lb = 153 kips

For the pile group, use Fig. 6.14. The length $(L) = 70/1.5 = 47$ pile diameters, therefore use $L = 48$ curve. For spacing in pile diameters = 3, $G_e = 0.705$.

Using the equation in Fig. 6.14:

Ultimate load of group = $G_e n Q_{ult} = (0.705)(81)(153) = 8700$ kips

Using factor of safety = 3:

$Q_{all} = 8700/3 = 2900$ kips

6.18 From Fig. 2.40, for pile adhesion, use an average undrained shear strength $(s_u) = 350$ psf. For end bearing, use an average undrained shear strength $(s_u) = 400$ psf. Based on Fig. 6.13, use $c_A = c = s_u = 350$ psf. Using Eq. 6.19:

Q_{ult} = end bearing + side adhesion = $9\pi cR^2 + 2\pi c_A Rz$

$= 9\pi(400)(1.5/2)^2 + 2\pi(350)(1.5/2)(30)$

$= 6400 + 49,500 = 56,000$ lb = 56 kips

$Q_{all} = Q_{ult}/F = 56/3 = 19$ kips

6.19 For the pile group, use Fig. 6.14. The length $(L) = 35/1.5 = 23$ pile diameters, therefore use $L = 24$ curve. For spacing in pile diameters = 1.5, $G_e = 0.41$. Using the equation in Fig. 6.14:

Ultimate load of group = $G_e n Q_{ult} = (0.41)(81)(56) = 1900$ kips

Using factor of safety = 3:

$Q_{all} = 1900/3 = 630$ kips

6.20 From Fig. 2.41, for pile adhesion and end bearing, use an average undrained shear strength $(s_u) = 0.6$ kg/cm² (1200 psf). Based on Fig. 6.13, use $c_A/c = 0.67$ (average curve for concrete piles) and therefore $c_A = (0.67)(1200) = 800$ psf. Using Eq. 6.19:

Q_{ult} = end bearing + side adhesion = $9\pi cR^2 + 2\pi c_A Rz$

$= 9\pi(1200)(1.0/2)^2 + 2\pi(800)(1.0/2)(40)$

$= 8000 + 100,000 = 108,000$ lb = 108 kips

$Q_{all} = Q_{ult}/F = 108/3 = 36$ kips

6.21 At a pile head deflection of 25 mm (1-in.), the lateral load = 38 kN

Using ½ of this lateral load, or allowable lateral load = (½)(38) = 19 kN

6.22 Using the following values:

$F = 5.0$

$T = 2.5$ m

$\tau_f = 50$ kPa $= 50$ kN/m^2

$P = 500$ kN

$B = L$

Inserting the above values into Eq. 6.31:

$F = R/P = [2(B + L)(T)(\tau_f)]/P$

$5.0 = [2(2\ B)(2.5\ \text{m})(50\ \text{kN/m}^2)]/500\ \text{kN}$

Solving for B:

$B = L = 5$ m

6.23 Using the following values:

$F = 5.0$

$T = 2.5$ m

$\tau_f = 50$ kPa $= 50$ kN/m^2

$B = L = 2$ m

Inserting the above values into Eq. 6.31:

$F = [2(B + L)(T)(\tau_f)]/P$

$5.0 = [2(2 + 2)(2.5\ \text{m})(50\ \text{kN/m}^2)]/P$

Solving for P:

$P = 200$ kN

6.24 Using the following values:

$F = 5.0$

$T = 2.5$ m

$\tau_f = 10$ kPa $= 10$ kN/m^2

Inserting the above values into Eq. 6.30:

$F = (2T\tau_f)/P$

$5.0 = [2(2.5\ \text{m})(10\ \text{kN/m}^2)]/P$

Solving for P:

$P = 10$ kN/m

6.25 Using the following values:

$F = 5.0$

$T = 2.5$ m

$\tau_f = 10$ kPa $= 10$ kN/m^2

$B = L = 2$ m

Inserting the above values into Eq. 6.31:

$F = [2(B + L)\ T\tau_f]/P$

$5.0 = [2(2 + 2)(2.5\ \text{m})(10\ \text{kN/m}^2)]/P$

Solving for P:

$P = 40$ kN

6.26 Using the following values:

$F = 5.0$

$B = L = 1$ m

$T = 2.7$ m (i.e., total thickness of the unliquefiable soil layer minus footing embedment depth $= 3$ m $- 0.3$ m $= 2.7$ m)

Using Eq. 4.14:

$\sigma'_v = \sigma_v - u$

Since the soil is above the groundwater table, assume $u = 0$. Using a total unit weight $= 18.3$ kN/m^3 and an average depth of 1.65 m [i.e., $(0.3 + 3.0)/2 = 1.65$ m]:

$\sigma'_v = (18.3$ kN/m$^3)(1.65$ m$) = 30$ kN/m$^2 = 30$ kPa

Using Eq. 6.34:

$\tau_f = k_o \sigma'_v \tan \phi' = (0.5)(30$ kPa$)(\tan 33°) = 9.8$ kPa $= 9.8$ kN/m^2

Using Eqs. 6.30 and 6.31 and the above values:

For the strip footings:

$P = q_{all} B = 2T\tau_f/F$

$q_{all} = (2T\tau_f)/[(F)(B)] = [(2)(2.7$ m$)(9.8$ kPa$)]/[(5)(1$ m$)] = 10$ kPa

For the spread footings:

$P = q_{all} B^2 = 2(B + L)T\tau_f/F$

$q_{all} = 2(B + L)T\tau_f/(FB^2) = [2(1 + 1)(2.7$ m$)(9.8$ kPa$)]/[(5)(1$ m$)^2] = 21$ kPa

6.27 Using the following values:

$F = 5.0$

$B = L = 1$ m

$T = 2.7$ m (i.e., total thickness of the unliquefiable soil layer minus footing embedment depth $= 3$ m $- 0.3$ m $= 2.7$ m)

$\tau_f = s_u = 20$ kPa

Using Eqs. 6.30 and 6.31 and the above values:

For the strip footings:

$P = q_{all} B = 2T\tau_f/F$

$q_{all} = (2T\tau_f)/[(F)(B)] = [(2)(2.7$ m$)(20$ kPa$)]/[(5)(1$ m$)] = 21.6$ kPa

Use $q_{all} = 20$ kPa

For the spread footings:

$P = q_{all} B^2 = 2(B + L)T\tau_f/F$

$q_{all} = 2(B + L)T\tau_f/(FB^2) = [2(1 + 1)(2.7$ m$)(20$ kPa$)]/[(5)(1$ m$)^2] = 43.2$ kPa

Use $q_{all} = 40$ kPa

6.28 For the sand, $c = 0$ and neglecting the third term in Eq. 6.1, therefore:

$q_{ult} = \frac{1}{2} \gamma_t B N_\gamma$

Using Fig. 6.5, for $\phi = 33°$, $N_\gamma = 26$

$T = 2.7$ m (i.e., total thickness of the unliquefiable soil layer minus footing embedment depth = 3 m − 0.3 m = 2.7 m)

Since $T/B = 2.7/1.0 = 2.7$, a reduction in N_γ would tend to be small for such a high ratio of T/B.

For the strip footings:

$q_{ult} = \frac{1}{2}\gamma_t B N_\gamma = \frac{1}{2}(18.3 \text{ kN/m}^3)(1 \text{ m})(26) = 238$ kPa

$q_{all} = q_{ult}/F = 238/5 = 48$ kPa

For the spread footings (using Eq. 6.3):

$q_{ult} = 0.4\gamma_t B N_\gamma = 0.4 (18.3 \text{ kN/m}^3)(1 \text{ m})(26) = 190$ kPa

$q_{all} = q_{ult}/F = 190/5 = 38$ kPa

Summary:

$q_{all} = 48$ kPa for the 1-m wide strip footings and $q_{all} = 38$ kPa for the 1 m by 1 m spread footings. For the design of the footings, use the lower values calculated in Problem 6.26.

6.29 $T = 2.7$ m (i.e., total thickness of the unliquefiable soil layer minus footing embedment depth = 3 m − 0.3 m = 2.7 m)

Since $T/B = 2.7/1.0 = 2.7$, $N_c = 5.5$ per Fig. 6.7.

For the strip footings (using Eq. 6.8 and neglecting the embedment term):

$q_{ult} = s_u N_c = (20 \text{ kPa})(5.5) = 110$ kPa

$q_{all} = q_{ult}/F = 110/5 = 22$ kPa

For the spread footings (using Eq. 6.9 and neglecting the embedment term):

$q_{ult} = s_u N_c [1 + 0.3(B/L)] = 1.3(20 \text{ kPa})(5.5) = 143$ kPa

$q_{all} = q_{ult}/F = 143/5 = 29$ kPa

Summary:

$q_{all} = 22$ kPa for the 1-m wide strip footings and $q_{all} = 29$ kPa for the 1 m by 1 m spread footings. For the design of the strip footings, use the value from Problem 6.27 (i.e., $q_{all} = 20$ kPa). For the design of the spread footings, use the lower value calculated in this problem (i.e., $q_{all} = 29$ kPa).

6.30 Using Eq. 6.31:

In order to calculate the allowable bearing pressure for the spread footings, the following values are used:

$F = 5.0$

$T = 2.7$ m (i.e., total thickness of the unliquefiable soil layer minus footing embedment depth = 3 m − 0.3 m = 2.7 m)

$\tau_f = s_u = 20$ kPa

$B = L = 3$ m

Inserting the above values into Eq. 6.31:

$P = q_{all}B^2 = 2(B + L)T\tau_f/F$

$q_{all} = 2(B + L)T\tau_f/(FB^2) = [2(3 + 3)(2.7 \text{ m})(20 \text{ kPa})]/[(5)(3 \text{ m})^2] = 14$ kPa

Using Fig. 6.7:

In order to calculate the allowable bearing pressure for the spread footings, the following values are used:

$T = 2.7$ m (i.e., total thickness of the unliquefiable soil layer minus footing embedment depth = 3 m − 0.3 m = 2.7 m)

$c_1 = s_{u1} = 20$ kPa

Since $T/B = 2.7/3.0 = 0.9$ and $c_2/c_1 = c_2/s_u = 0/20 = 0$

Then $N_c = 2.3$ per Fig. 6.7

Using Eq. 6.11 and neglecting the second term:

$q_{ult} = N_c c_1 [1 + 0.3 (B/L)] = (2.3)(20 \text{ kPa})(1.3) = 60$ kPa

$q_{all} = q_{ult}/F = 60/5 = 12$ kPa

Summary:

From Eq. 6.31, $q_{all} = 14$ kPa. Using Fig. 6.7, $q_{all} = 12$ kPa. Use the lower value of 12 kPa for the design of the 3 m by 3 m spread footings.

6.31 Using Eq. 6.31:

$F = 2(B + L)T\tau_f / P$

Where $B = L = 20$ m

$T = 2$ m (distance from the pile tips to the top of the liquefied soil layer)

$P = 50$ MN $= 5 \times 10^4$ kN

Using Eq. 6.34:

$\tau_f = k_o \sigma'_{vo} \tan \phi' = (0.6)(168 \text{ kN/m}^2)(\tan 34°) = 68$ kN/m^2

$F = [(2)(20 \text{ m} + 20\text{m})(2 \text{ m})(68 \text{ kN/m}^2)]/(5 \times 10^4 \text{ kN}) = 0.22$

Therefore, the pile foundation will punch down into the liquefied soil layer located at a depth of 17 to 20 m below ground surface.

6.32 Based on Eq. 6.31:

$F = 2(B' + L')T\tau_f/P$

To determine B' and L', use the following (per the 2:1 approximation):

$z = 1/3\ L = 1/3\ (15 \text{ m}) = 5$ m

$L' = L + z = 20 + 5 = 25$ m

$B' = B + z = 20 + 5 = 25$ m

$T = 2$ m (distance from the pile tips to the top of the liquefied soil layer)

$P = 50$ MN $= 5 \times 10^4$ kN

Using Eq. 6.34:

$\tau_f = k_o \sigma'_{vo} \tan \phi' = (0.6)(168 \text{ kN/m}^2)(\tan 34°) = 68$ kN/m^2

$F = [(2)(25 \text{ m} + 25 \text{ m})(2 \text{ m})(68 \text{ kN/m}^2)]/(5 \times 10^4 \text{ kN}) = 0.27$

Therefore, the pile foundation will punch down into the liquefied soil layer located at a depth of 17 to 20 m below ground surface.

6.33 Strip footing (using Eq. 6.12):

$e = 0.10$ m (for middle one-third of footing, e cannot exceed 0.17 m, and therefore e is within the middle one-third of the footing)

$T = 2.5$ m (i.e., total thickness of the unliquefiable soil layer minus footing embedment depth = 3 m − 0.5 m = 2.5 m)

$c_1 = s_{u1} = 50$ kPa $= 50$ kN/m² (upper cohesive soil layer)

$c_2 = 0$ kPa $= 0$ kN/m² (liquefied soil layer)

$B = 1$ m

Using Fig. 6.7 with $T/B = 2.5/1.0 = 2.5$ and $c_2/c_1 = 0$, therefore:

$N_c = 5.5$

Using the Terzaghi bearing capacity equation to calculate q_{ult}, or from Eq. 6.10 (neglecting the second term):

$q_{ult} = c_1 N_c = s_{u1} N_c = (50$ kN/m²$)(5.5) = 275$ kN/m²

$F = q_{ult}/q'$, therefore:

$q' = 275$ kN/m²$/5 = 55$ kN/m²

Using Eq. 6.12:

$q' = Q(B + 6e)/B^2$

55 kN/m² $= Q [1 + (6)(0.1)]/(1)^2$

Solving for $Q = Q_{all} = 34$ kN/m

$e = M_{all}/Q_{all}$

$M_{all} = eQ_{all} = (0.1)(34) = 3.4$ kN-m/m

Strip footing (using Fig. 6.8):

$B' = B - 2e = 1 - 2 (0.10) = 0.8$ m

$T = 2.5$ m

$c_1 = s_{u1} = 50$ kPa $= 50$ kN/m² (upper cohesive soil layer)

$c_2 = 0$ kPa $= 0$ kN/m² (liquefied soil layer)

Using Fig. 6.7 with $T/B = 2.5/1.0 = 2.5$ and $c_2/c_1 = 0$, therefore:

$N_c = 5.5$

Using the Terzaghi bearing capacity equation to calculate q_{ult}, or from Eq. 6.10 (neglecting the second term):

$q_{ult} = c_1 N_c = s_{u1} N_c = (50$ kN/m²$)(5.5) = 275$ kN/m²

$Q_{ult} = q_{ult} B' = (275$ kN/m²$)(0.8$ m$) = 220$ kN/m

$F = Q_{ult}/Q_{all}$, therefore:

$Q_{all} = Q_{ult}/F = 220$ kN/m$/5 = 44$ kN/m

$e = M_{all}/Q_{all}$

$M_{all} = eQ_{all} = (0.1)(44) = 4.4$ kN-m/m

Use the lower values of $Q_{all} = 34$ kN/m and $M_{all} = 3.4$ kN-m/m calculated by using Eq. 6.12.

Spread footing (using Eq. 6.12):

$e = M/Q = 0.30$ m (for middle one-third of footing, e cannot exceed 0.33 m, and therefore e is within the middle one-third of the footing)

$T = 2.5$ m

$c_1 = s_{u1} = 50$ kPa $= 50$ kN/m² (upper cohesive soil layer)

$c_2 = 0$ kPa $= 0$ kN/m² (liquefied soil layer)

$B = 2$ m

Using Fig. 6.7 with $T/B = 2.5/2 = 1.25$ and $c_2/c_1 = 0$, therefore:

$N_c = 3.2$

Using the Terzaghi bearing capacity equation to calculate q_{ult}, or from Eq. 6.11 (neglecting the second term):

$q_{ult} = N_c s_{u1} [1 + 0.3(B/L)] = (N_c)(s_{u1})(1.3) = (3.2)(50 \text{ kN/m}^2)(1.3) = 208 \text{ kN/m}^2$

$F = q_{ult}/q'$, therefore:

$q' = 208 \text{ kN/m}^2/5 = 41.6 \text{ kN/m}^2$

Using Eq. 6.12:

$q' = Q(B + 6\ e)/B^2$

$41.6 \text{ kN/m}^2 = Q[2 + (6)(0.3)]/(2)^2$

Solving for $Q = 43.8$ kN/m

Converting Q to a load per the entire length of footing, or:

$Q_{all} = (43.8 \text{ kN})(2 \text{ m}) = 88 \text{ kN}$

$e = M_{all}/Q_{all}$

$M_{all} = eQ_{all} = (0.3)(88) = 26 \text{ kN-m}$

Spread footing (using Fig. 6.8):

$B' = B - 2e = 2 - 2\ (0.30) = 1.4$ m

$L' = L = 2$ m (moment only in B direction of the footing)

$T = 2.5$ m

$c_1 = s_{u1} = 50 \text{ kPa} = 50 \text{ kN/m}^2$ (upper cohesive soil layer)

$c_2 = 0 \text{ kPa} = 0 \text{ kN/m}^2$ (liquefied soil layer)

Using Fig. 6.7 with $T/B = 2.5/2 = 1.25$ and $c_2/c_1 = 0$, therefore:

$N_c = 3.2$

Using the Terzaghi bearing capacity equation to calculate q_{ult}, or from Eq. 6.11 (neglecting the second term):

$q_{ult} = N_c s_{u1}[1 + 0.3\ (B'/L')] = (N_c)(s_{u1})(1.2) = (3.2)(50 \text{ kN/m}^2)(1.2) = 190 \text{ kN/m}^2$

$Q_{ult} = q_{ult} B'L' = (190 \text{ kN/m}^2)(1.4 \text{ m})(2 \text{ m}) = 530 \text{ kN}$

$F = Q_{ult}/Q_{all}$, therefore:

$Q_{all} = 530 \text{ kN}/5 = 106 \text{ kN}$

$e = M_{all}/Q_{all}$

$M_{all} = eQ_{all} = (0.3)(106) = 32 \text{ kN-m}$

Use the lower values of $Q_{all} = 88$ kN and $M_{all} = 26$ kN-m calculated by using Eq. 6.12.

6.34 $B' = B - 2e = 2 - 2\ (0.30) = 1.4$ m

$L' = L - 2e = 2 - 2\ (0.30) = 1.4$ m

$T = 2.5$ m

$c_1 = s_{u1} = 50 \text{ kPa} = 50 \text{ kN/m}^2$ (upper cohesive soil layer)

$c_2 = 0 \text{ kPa} = 0 \text{ kN/m}^2$ (liquefied soil layer)

Using Fig. 6.7 with $T/B = 2.5/2 = 1.25$ and $c_2/c_1 = 0$, therefore:

$N_c = 3.2$

Using the Terzaghi bearing capacity equation to calculate q_{ult}, or from Eq. 6.11 (neglecting the second term):

$$q_{ult} = N_c s_{u1}[1 + 0.3\ (B'/L')] = (N_c)(s_{u1})(1.3) = (3.2)(50\ kN/m^2)(1.3) = 208\ kN/m^2$$
$$Q_{ult} = q_{ult}\ B'L' = (208\ kN/m^2)(1.4\ m)(1.4\ m) = 408\ kN$$
$$F = Q_{ult}/Q = 408\ kN/500\ kN = 0.82$$

Chapter 7

7.1   Perimeter footing: $(60)(30)(2) + (60)(42)(2) = 8640\ kN$

Interior columns: 6-m spacing, therefore 24 interior columns, or:

Interior columns $= (24)(900) = 21{,}600\ kN$

Floor slab: 6 kPa

Include live load from stored items: 30 kPa

$\sigma_o = (8640 + 21{,}600)/[(30)(42)] + 6 + 30 = 60\ kPa$

7.2   Using Eq. 7.1:

$\%C = 100\Delta h/h_o$

$\Delta h = (h_o)(\%C/100)$

At the centerline of the canyon:

$\Delta h = (h_o)(\%C/100) = (5\ ft)(5/100) = 0.25\ ft$

At the opposite side of the building:

$\Delta h = (h_o)(\%C/100) = (1\ ft)(5/100) = 0.05\ ft$

The difference is the differential settlement (Δ):

$\Delta = 0.25 - 0.05 = 0.2\ ft = 2.4\ in.$

7.3   Using Eq. 7.1:

$\%C = 100\Delta h/h_o$

$\Delta h = \rho_{max} = (h_o)(\%C/100)$

Since all layers are 4 ft thick ($h_o = 4\ ft$):

$\rho_{max} = (4\ ft)[(0.1/100) + (0.25/100) + (0.63/100) + (1.14/100)]$

$\rho_{max} = 0.08\ ft = 1.0\ in.$

7.4   Using Eq. 7.1:

$\%C = I_e = 100\Delta h/h_o = (100)(2.8)/25.4 = 11\%$

Per Table 7.1, the degree of specimen collapse = severe

7.5   Settlement analysis:

Using the Terzaghi and Peck method, the calculated settlement is 1.0 in. The net pressure that causes this settlement is 6000 psf.

Bearing capacity analysis:

A bearing capacity analysis has already been performed (see Problem 6.4). Using a factor of safety of 3, the allowable bearing pressure = 14,800 psf. This value of 14,800 psf is much greater than the pressure that results in 1.0 in. of settlement and hence settlement (not bearing capacity) governs the design.

7.6   Per Table 18.4, the allowable foundation pressure is 2000 psf for sand.

7.7 Using Eq. 7.9:

$S = qBI(1 - \mu^2)/E_s$

As indicated in Sec. 7.3.3, for uniform coarse sand:

$E_s/N_{60} = 10$

Since $N_{60} = 20$

$E_s = (N_{60})(10) = (20)(10) = 200$ tsf $= 400,000$ psf $= 20,000$ kPa

For a square and flexible loaded area on an elastic half-space of infinite depth, Fig. 7.10 indicates that $I = 1.12$ (center) and $I = 0.56$ (corner). Using the above equation with $E_s = 20,000$ kPa and $\mu = 0.3$, therefore:

$S = (30$ kPa$)(20$ m$)(1.12)(1 - 0.3^2)/20,000$ kPa $= 0.031$ m $= 3.1$ cm (center)

$S = (30$ kPa$)(20$ m$)(0.56)(1 - 0.3^2)/20,000$ kPa $= 0.015$ m $= 1.5$ cm (corner)

ρ_{max} occurs at the center $= 3.1$ cm

$\Delta = 3.1 - 1.5 = 1.6$ cm

7.8 $q = P/B^2 = (230)(1000)/(6)^2 = 6400$ psf $= 3.2$ tsf

Using Eq. 7.9:

$S = qBI(1 - \mu^2)/E_s$

As indicated in Sec. 7.3.3, for clean fine to medium sand:

$E_s/N_{60} = 7$

Since $N_{60} = 30$

$E_s = (N_{60})(10) = (30)(7) = 210$ tsf

For a square and rigid loaded area on an elastic half-space of infinite depth, Fig. 7.10 indicates that $I = 0.82$. Using the above equation with $E_s = 210$ tsf and $\mu = 0.3$, therefore:

$S = \rho_{max} = (3.2$ tsf$)(6$ ft$)(0.82)(1 - 0.3^2)/(210$ tsf$) = 0.068$ ft $= 0.82$ in.

7.9 From Fig. 7.8, for $B = 6$ ft and $N_{60} = 30$, $q = 3.4$ tsf for 1 in. settlement. Since actual $q = 3.2$ tsf, ρ_{max} will be slightly less than 1 in. (Note: for this problem, the theory of elasticity and Fig. 7.8 provide similar answers).

7.10 Solve by trial and error:

Assuming $B = 8.5$ ft, from Fig. 7.8 for $N_{60} = 30$, $q = 3.2/2 = 1.6$ tsf

Maximum load $= (8.5)(8.5)(1.6) = 116$ tons $= 230$ kips

Thus for an 8.5 ft by 8.5 ft footing subjected to a load of 230 kips, there will be 1 in. of settlement.

7.11 $\sigma_0 = 200 - 4 (19.3) = 123$ kPa

Divide the sand below the foundation into two layers, as follows:

Layer no.	Layer thickness	Depth to center of layer	σ'_{vo} at the center of the layer	$\Delta\sigma_v$ at the center of the layer	Settlement
1	8 m	8 m	130 kPa	98.5 kPa	11 cm
2	8 m	16 m	220 kPa	67.5 kPa	1 cm

For layer no. 1:

From Fig. 7.28 (8 m specimen):

At $\sigma'_{vo} = 130$ kPa, $e_i = 0.482$

At $\sigma'_{vo} + \Delta\sigma_v = 130 + 98.5 = 229$ kPa, $e_f = 0.462$

$\Delta e = e_i - e_f = 0.482 - 0.462 = 0.020$

Using Eq. 7.13:

$S = \Delta e\, H_o/(1 + e_o) = (0.020)(8)/(1 + 0.5) = 0.11$ m $= 11$ cm

For layer no. 2:

From Fig. 7.28 (16 m specimen):

At $\sigma'_{vo} = 220$ kPa, $e_i = 0.395$

At $\sigma'_{vo} + \Delta\sigma_v = 220 + 67.5 = 288$ kPa, $e_f = 0.393$

$\Delta e = e_i - e_f = 0.395 - 0.393 = 0.002$

Using Eq. 7.13:

$S = \Delta e H_o/(1 + e_o) = (0.002)(8)/(1 + 0.4) = 0.01$ m $= 1$ cm

Total settlement $(\rho_{max}) = 11 + 1 = 12$ cm

7.12   Per Table 7.2, for a continuous steel frame:

$\Delta = 0.002\, L = (0.002)(6\text{ m}) = 0.012$ m $= 1.2$ cm

7.13   For sensitive machinery, per Fig. 7.25, $\delta/L = 1/750$

Assuming $\delta = \Delta$ therefore $\Delta/L = 1/750$ and for $L = 6$ m:

$\Delta = (1/750)(6\text{ m}) = 0.008$ m $= 0.8$ cm

7.14   As discussed in Sec. 7.6, a house having a conventional slab-on-grade foundation will experience gypsum wallboard cracking when $\Delta = 1.25$ in., therefore:

Allowable $\Delta = 1.25/2.5 = 0.5$ in.

7.15   The maximum differential settlement (Δ) will occur between the center and edge of the tank. Thus $L = 15$ m and assuming $\delta = \Delta$, therefore:

$\Delta/L = 1/200$

$\Delta = (1/200)(15\text{ m}) = 0.075$ m $= 7.5$ cm

Chapter 8

8.1   Using Eq. 8.9:

$s_i = qBI(1 - \mu^2)/E_u$

For a square and flexible loaded area on an elastic half-space of infinite depth, Fig. 7.10 indicates that $I = 1.12$ (center) and $I = 0.56$ (corner). Using the above equation with $E_u = 20,000$ kPa and $\mu = 0.5$ (saturated cohesive soil), therefore:

$s_i = (30\text{ kPa})(20\text{ m})(1.12)(1 - 0.5^2)/20,000\text{ kPa} = 0.025$ m $= 2.5$ cm (center)

$s_i = (30\text{ kPa})(20\text{ m})(0.56)(1 - 0.5^2)/20,000\text{ kPa} = 0.013$ m $= 1.3$ cm (corner)

8.2   Using Eq. 8.9:

$s_i = qBI(1 - \mu^2)/E_u$

For a circular and flexible loaded area on an elastic half-space of infinite depth, Fig. 7.10 indicates that $I = 1.0$ (center). Using the above equation with $E_u = 40{,}000$ kPa and $\mu = 0.4$, therefore:

$s_i = (50 \text{ kPa})(10 \text{ m})(1.0)(1 - 0.4^2)/40{,}000 \text{ kPa} = 0.010 \text{ m} = 1.0 \text{ cm (center)}$

8.3 Immediate settlement s_i due to undrained creep of the soft saturated clay was not included in the original settlement analysis by the design engineer.

8.4 Using the Casagrande construction technique (see Fig. 8.2), the maximum past pressure (σ'_{vm}) = 20 kPa (solid line) and = 30 kPa (dashed line).

8.5 For the consolidation test with a solid line:

From Fig. 8.22:

$e = 3.3$ at $\sigma'_{vc1} = 20$ kPa

$e = 1.6$ at $\sigma'_{vc2} = 80$ kPa

Using Eq. 8.6:

$C_c = \Delta e/\log(\sigma'_{vc2}/\sigma'_{vc1})$

$C_c = (3.3 - 1.6)/\log(80/20) = 2.8$

For the consolidation test with a dashed line:

From Fig. 8.22:

$e = 2.7$ at $\sigma'_{vc1} = 20$ kPa

$e = 1.6$ at $\sigma'_{vc2} = 80$ kPa

Using Eq. 8.6:

$C_c = \Delta e/\log(\sigma'_{vc2}/\sigma'_{vc1})$

$C_c = (2.7 - 1.6)/\log(80/20) = 1.8$

8.6 For the consolidation test on the undisturbed soil specimen:

From Fig. 8.3:

$\varepsilon = 0.02$ at $\sigma'_{vc1} = 2$ kg/cm^2

$\varepsilon = 0.27$ at $\sigma'_{vc2} = 10$ kg/cm^2

$C_{c\varepsilon} = \Delta\varepsilon/\log(\sigma'_{vc2}/\sigma'_{vc1})$

$C_{c\varepsilon} = (0.27 - 0.02)/\log(10/2) = 0.36$

For the consolidation test on the disturbed soil specimen:

From Fig. 8.3:

$\varepsilon = 0.09$ at $\sigma'_{vc1} = 2$ kg/cm^2

$\varepsilon = 0.26$ at $\sigma'_{vc2} = 10$ kg/cm^2

$C_{c\varepsilon} = (0.26 - 0.09)/\log(10/2) = 0.24$

8.7 For the recompression curve:

From Fig. 8.7, over one log cycle, $\Delta\varepsilon = 0.04$

$C_{r\varepsilon} = \Delta\varepsilon/\log(\sigma'_{vc2}/\sigma'_{vc1})$

$C_{r\varepsilon} = (0.04)/\log(100/10) = 0.04$

For the virgin consolidation curve:

From Fig. 8.7:

$\varepsilon = 0.03$ at $\sigma'_{vc1} = 70$ kPa

$\varepsilon = 0.1$ at $\sigma'_{vc2} = 100$ kPa

$C_{c\varepsilon} = \Delta\varepsilon/\log(\sigma'_{vc2}/\sigma'_{vc1})$

$C_{c\varepsilon} = (0.1 - 0.03)/\log(100/70) = 0.45$

8.8 From the example problem, $\Delta\sigma'_v = (\frac{1}{2})(50 \text{ kPa}) = 25$ kPa

As indicated in the example problem, use the following values:

$C_c = 0.83$

$H_o = 2$ m

$e_o = 1.10$

$\sigma'_{vo} = 150$ kPa

For underconsolidated soil (OCR < 1), use Eq. 8.11:

$s_c = C_c[H_o/(1 + e_o)][\log (\sigma'_{vo} + \Delta\sigma)/\sigma'_{vo} + \log \sigma'_{vo}/(\sigma'_{vo} - \Delta\sigma'_v)]$

$s_c = (0.83)[(2 \text{ m})/(1 + 1.1)][\log (150 + 100)/150 + \log 150/(150 - 25)]$

$s_c = 0.24$ m (9.4 in.)

8.9 As indicated in the example problem, use the following values:

$C_c = 0.83$

$H_o = 2$ m

$e_o = 1.10$

$\sigma'_{vo} = 150$ kPa

For normally consolidated soil (OCR = 1), use Eq. 8.12:

$s_c = C_c[H_o/(1 + e_o)] \log [(\sigma'_{vo} + \Delta\sigma_v)/\sigma'_{vo})$

$s_c = (0.83)[(2 \text{ m})/(1 + 1.1)] \log [(150 \text{ kPa} + 100 \text{ kPa})/150 \text{ kPa}]$

$s_c = 0.18$ m (6.9 in.)

8.10 As indicated in the example problem, use the following values:

$C_r = 0.05$

$H_o = 2$ m

$e_o = 1.10$

$\sigma'_{vo} = 150$ kPa

Since $\Delta\sigma_v = 100$ kPa and $\Delta\sigma_v + \sigma'_{vo} = \sigma'_{vm}$, use Eq. 8.13:

$s_c = C_r[H_o/(1 + e_o)] \log [(\sigma'_{vo} + \Delta\sigma_v)/\sigma'_{vo})]$

$s_c = (0.05)[(2 \text{ m})/(1 + 1.1)] \log [(150 \text{ kPa} + 100 \text{ kPa})/150 \text{ kPa}]$

$s_c = 0.011$ m (0.42 in.)

8.11 The original 2-m thick clay layer is reduced in thickness by 10 cm due to primary consolidation from the 50 kPa surcharge load. Ignore both the decrease in thickness of the clay layer and the increase in buoyant unit weight due to consolidation. Using Eq. 4.14, and for the condition after placement of the surcharge load and complete consolidation of the clay layer:

$\sigma'_v = \sigma_v - u$

$\sigma_v' = 50$ kPa $+ (5$ m$)(18.7$ kN/m$^3) + (5$ m$)(19.7 - 9.81$ kN/m$^3) + (1$ m$)(7.9$ kN/m$^3)$

$\sigma_v' = 200$ kPa

Because of complete consolidation of the clay layer, $\sigma_v' = \sigma_{vm}' = 200$ kPa

Determining σ_v' after the permanent rise of the groundwater table and the complete equilibrium of the clay layer:

$\sigma_v' = 50$ kPa $+ (2$ m$)(18.7$ kN/m$^3) + (8$ m$)(19.7 - 9.81$ kN/m$^3) + (1$ m$)(7.9$ kN/m$^3)$

$\sigma_v' = 174$ kPa

Using Eq. 8.10:

OCR $= \sigma_{vm}'/\sigma_{vo}' = 200/174 = 1.15$

8.12 The original 2-m thick clay layer is reduced in thickness by 10 cm due to primary consolidation from the 50 kPa surcharge load. Ignore both the decrease in thickness of the clay layer and the increase in buoyant unit weight due to consolidation. Using Eq. 4.14, and for the condition after placement of the surcharge load and complete consolidation of the clay layer:

$\sigma_v' = \sigma_v - u$

$\sigma_v' = 50$ kPa $+ (5$ m$)(18.7$ kN/m$^3) + (5$ m$)(19.7 - 9.81$ kN/m$^3) + (1$ m$)(7.9$ kN/m$^3)$

$\sigma_v' = 200$ kPa

Because of complete consolidation of the clay layer, $\sigma_v' = \sigma_{vm}' = 200$ kPa

Determining σ_v' after the removal of the surcharge and the complete equilibrium of the clay layer:

$\sigma_v' = (5$ m$)(18.7$ kN/m$^3) + (5$ m$)(19.7 - 9.81$ kN/m$^3) + (1$ m$)(7.9$ kN/m$^3)$

$\sigma_v' = 150$ kPa

Using Eq. 8.10:

OCR $= \sigma_{vm}'/\sigma_{vo}' = 200/150 = 1.33$

8.13 $\Delta\sigma_v = 4\gamma_w - 4(\Delta\gamma_t) = (4)(9.81) - (4)(19.7 - 18.9) = 36$ kPa (Note: $\Delta\sigma_v$ can also be calculated as the change in vertical effective stress from the initial to the final condition, i.e., after permanent lowering of the groundwater table).

$\sigma_{vo}' = 150 + 24 = 174$ kPa

Using Eq. 8.12 (OCR = 1):

$s_c = [C_c H_o/(1 + e_o)] \log [(\sigma_{vo}' + \Delta\sigma_v)/\sigma_{vo}']$

$s_c = [(0.83)(2$ m$)/(1 + 1.1)] \log [(174$ kPa $+ 36$ kPa$)/174$ kPa$]$

$s_c = 0.064$ m $= 6.4$ cm

8.14 The original 2-m thick clay layer is reduced in thickness by 5.1 cm due to primary consolidation from the building load. Ignore both the decrease in thickness of the clay layer and the increase in buoyant unit weight due to consolidation. Divide the 2-m clay layer into two 1-m thick layers as follows:

Layer no.	Layer thickness	Depth to center of layer	σ_{vo}' at the center of the layer	$\Delta\sigma_v$ at the center of the layer	Primary consolidation settlement (s_c)
1	1 m	10.5 m	171 kPa	10 kPa	0.98 cm
2	1 m	11.5 m	178 kPa	30 kPa	2.67 cm

For Layer 1 (depth of 10 to 11 m below ground surface):

Using the 2:1 approximation (Eq. 4.26) with $z = 10.5$ m:

$\sigma_z = \Delta\sigma_v = P/[(B + z)(L + z)]$

$= [(50 \text{ kPa})(20)(30)]/[(20 + 10.5)(30 + 10.5)] = 24.3$ kPa

Determine σ'_{vo} at the end of primary consolidation due to the building load:

$\sigma'_{vo} = 24.3$ kPa $+ (5 \text{ m})(18.7 \text{ kN/m}^3) + (5 \text{ m})(19.7 - 9.81 \text{ kN/m}^3)$

$+ (0.5 \text{ m})(7.9 \text{ kN/m}^3)$

$\sigma'_{vo} = 171$ kPa

For a linear distribution of $\Delta\sigma_v = 0$ at the top of the clay layer and 40 kPa at the bottom of the clay layer, at the center of layer no. 1, $\Delta\sigma_v = 10$ kPa. Using Eq. 8.12 (OCR = 1):

$s_c = [C_c H_o/(1 + e_o)] \log [(\sigma'_{vo} + \Delta\sigma_v)/\sigma'_{vo}]$

$= [(0.83)(1.0)/(1 + 1.1)] \log [(171 + 10)/171] = 0.0098$ m $= 0.98$ cm

For Layer 2 (depth of 11 to 12 m below ground surface):

Using the 2:1 approximation (Eq. 4.26) with $z = 11.5$ m:

$\sigma_z = \Delta\sigma_v = P/[(B + z)(L + z)]$

$= [(50 \text{ kPa})(20)(30)]/[(20 + 11.5)(30 + 11.5)] = 22.9$ kPa

Determine σ'_{vo} at the end of primary consolidation due to the building load:

$\sigma'_{vo} = 22.9$ kPa $+ (5 \text{ m})(18.7 \text{ kN/m}^3) + (5 \text{ m})(19.7 - 9.81 \text{ kN/m}^3)$

$+ (1.5 \text{ m})(7.9 \text{ kN/m}^3)$

$\sigma'_{vo} = 178$ kPa

For a linear distribution of $\Delta\sigma_v = 0$ at the top of the clay layer and 40 kPa at the bottom of the clay layer, at the center of Layer no. 2, $\Delta\sigma_v = 30$ kPa. Using Eq. 8.12 (OCR = 1):

$s_c = [C_c H_o/(1 + e_o)] \log [(\sigma'_{vo} + \Delta\sigma_v)/\sigma'_{vo}]$

$= [(0.83)(1.0)/(1 + 1.1)] \log [(178 + 30)/178] = 0.0267$ m $= 2.67$ cm

Total primary consolidation settlement $(s_c) = 0.98 + 2.67 = 3.7$ cm

8.15 Divide the clay into four layers, as follows:

Layer no.	Layer thickness	Depth to center of layer	σ'_{vo} at the center of the layer	$\Delta\sigma_v$ at the center of the layer	Primary consolidation settlement (s_c)
1	10 ft	76 ft	4600 psf	610 psf	0.10 ft
2	10 ft	86 ft	5100 psf	575 psf	0.085 ft
3	10 ft	96 ft	5700 psf	545 psf	0.073 ft
4	10 ft	106 ft	6300 psf	520 psf	0.063 ft

Note that the vertical effective stress σ'_{vo} is obtained from Fig. 2.39. In addition, the oedometer test data in Fig. 2.39 indicate that the clay from elevation –50 to –90 ft is essentially normally consolidated (OCR = 1.0).

Layer no. 1:

Using the 2:1 approximation (Eq. 4.26) with $z = 76$ ft:

$\sigma_z = \Delta\sigma_v = P/[(B + z)(L + z)]$

$= [(1000 \text{ psf})(400 \text{ ft})(200 \text{ ft})]/[(400 + 76)(200 + 76)] = 610$ psf

Using Eq. 8.12 (OCR = 1):

$s_c = [C_c H_0/(1 + e_0)] \log [(\sigma'_{vo} + \Delta\sigma_v)/\sigma'_{vo}]$

$= [(0.35)(10)/(1 + 0.9)] \log [(4600 + 610)/4600] = 0.10$ ft

Layer no. 2:

Using the 2:1 approximation (Eq. 4.26) with $z = 86$ ft:

$\sigma_z = \Delta\sigma_v = P/[(B + z)(L + z)]$

$= [(1000 \text{ psf})(400 \text{ ft})(200 \text{ ft})]/[(400 + 86)(200 + 86)] = 575$ psf

Using Eq. 8.12 (OCR = 1):

$s_c = [C_c H_0/(1 + e_0)] \log [(\sigma'_{vo} + \Delta\sigma_v)/\sigma'_{vo}]$

$= [(0.35)(10)/(1 + 0.9)] \log [(5100 + 575)/5100] = 0.085$ ft

Layer no. 3:

Using the 2:1 approximation (Eq. 4.26) with $z = 96$ ft:

$\sigma_z = \Delta\sigma_v = P/[(B + z)(L + z)]$

$= [(1000 \text{ psf})(400 \text{ ft})(200 \text{ ft})]/[(400 + 96)(200 + 96)] = 545$ psf

Using Eq. 8.12 (OCR = 1):

$s_c = [C_c H_0/(1 + e_0)] \log [(\sigma'_{vo} + \Delta\sigma_v)/\sigma'_{vo}]$

$= [(0.35)(10)/(1 + 0.9)] \log [(5700 + 545)/5700] = 0.073$ ft

Layer no. 4:

Using the 2:1 approximation (Eq. 4.26) with $z = 106$ ft:

$\sigma_z = \Delta\sigma_v = P/[(B + z)(L + z)]$

$= [(1000 \text{ psf})(400 \text{ ft})(200 \text{ ft})]/[(400 + 106)(200 + 106)] = 520$ psf

Using Eq. 8.12 (OCR = 1):

$s_c = [C_c H_0/(1 + e_0)] \log [(\sigma'_{vo} + \Delta\sigma_v)/\sigma'_{vo}]$

$= [(0.35)(10)/(1 + 0.9)] \log [(6300 + 520)/6300] = 0.063$ ft

Total primary consolidation settlement:

$s_c = 0.10 + 0.085 + 0.073 + 0.063 = 0.32$ ft $= 3.9$ in.

8.16 Divide the clay into four layers, as follows:

Layer no.	Layer thickness	z	σ'_{vo} at the center of the layer	$\Delta\sigma_v$ at the center of the layer	Primary consolidation settlement (s_c)
1	10 ft	18 ft	4600 psf	940 psf	0.149 ft
2	10 ft	28 ft	5100 psf	675 psf	0.099 ft
3	10 ft	38 ft	5700 psf	510 psf	0.069 ft
4	10 ft	48 ft	6300 psf	400 psf	0.049 ft

Note that the vertical effective stress σ'_{vo} is obtained from Fig. 2.39. In addition, the oedometer test data in Fig. 2.39 indicate that the clay from elevation –50 to –90 ft is essentially normally consolidated (OCR = 1.0). For the pile group, $B = L = 37.5$ ft

Layer no. 1:

z = distance from elevation –37 ft to the center of the clay layer, or:

$z = 55 - 37 = 18$ ft

Using the 2:1 approximation (Eq. 4.26) with $z = 18$ ft:

$\sigma_z = \Delta\sigma_v = P/[(B + z)(L + z)]$

$= [(2900 \text{ kips})(1000)]/[(37.5 + 18)(37.5 + 18)] = 940$ psf

Using Eq. 8.12 (OCR = 1):

$s_c = [C_c H_o/(1 + e_o)] \log [(\sigma'_{vo} + \Delta\sigma_v)/\sigma'_{vo}]$

$= [(0.35)(10)/(1 + 0.9)] \log [(4600 + 940)/4600] = 0.149$ ft

Layer no. 2:

$z =$ distance from elevation -37 ft to the center of the clay layer, or:

$z = 65 - 37 = 28$ ft

Using the 2:1 approximation (Eq. 4.26) with $z = 28$ ft:

$\sigma_z = \Delta\sigma_v = P/[(B + z)(L + z)]$

$= [(2900 \text{ kips})(1000)]/[(37.5 + 28)(37.5 + 28)] = 675$ psf

Using Eq. 8.12 (OCR = 1):

$s_c = [C_c H_o/(1 + e_o)] \log [(\sigma'_{vo} + \Delta\sigma_v)/\sigma'_{vo}]$

$= [(0.35)(10)/(1 + 0.9)] \log [(5100 + 675)/5100] = 0.099$ ft

Layer no. 3:

$z =$ distance from elevation -37 ft to the center of the clay layer, or:

$z = 75 - 37 = 38$ ft

Using the 2:1 approximation (Eq. 4.26) with $z = 38$ ft:

$\sigma_z = \Delta\sigma_v = P/[(B + z)(L + z)]$

$= [(2900 \text{ kips})(1000)] [(37.5 + 38)(37.5 + 38)] = 510$ psf

Using Eq. 8.12 (OCR = 1):

$s_c = [C_c H_o/(1 + e_o)] \log [(\sigma'_{vo} + \Delta\sigma_v)/\sigma'_{vo}]$

$= [(0.35)(10)/(1 + 0.9)] \log [(5700 + 510)/5700] = 0.069$ ft

Layer no. 4:

$z =$ distance from elevation -37 ft to the center of the clay layer, or:

$z = 85 - 37 = 48$ ft

Using the 2:1 approximation (Eq. 4.26) with $z = 48$ ft:

$\sigma_z = \Delta\sigma_v = P/[(B + z)(L + z)]$

$= [(2900 \text{ kips})(1000)]/[(37.5 + 48)(37.5 + 48)] = 400$ psf

Using Eq. 8.12 (OCR = 1):

$s_c = [C_c H_o/(1 + e_o)] \log [(\sigma'_{vo} + \Delta\sigma_v)/\sigma'_{vo}]$

$= [(0.35)(10)/(1 + 0.9)] \log [(6300 + 400)/6300] = 0.049$ ft

Total primary consolidation settlement:

$s_c = 0.149 + 0.099 + 0.069 + 0.049 = 0.37$ ft $= 4.4$ in.

8.17   From Fig. 8.23:

$d_o = 0.1$ mm

$d_{100} = 2.0$ mm

$d_{50} = (d_o + d_{100})/2 = (0.1 + 2.0)/2 = 1.05$ mm

At $d_{50} = 1.05$ mm, $t_{50} = 150$ min $= 9000$ sec

$H_{dr} = (10.05 - 1.05)/2 = 4.5$ mm (double drainage)

Using Eq. 8.20:

$c_v = (T)(H_{dr})^2/t$

From Table 8.2, for $U_{avg} = 50\%$, $T = 0.197$

$c_v = (0.197)(0.45 \text{ cm})^2/(9000 \text{ sec}) = 4.4 \times 10^{-6}$ cm²/sec

8.18 Per Example Problem 8.7, for a normally consolidated clay layer subjected to a fill surcharge of 50 kPa, the primary consolidation settlement s_c is equal to 10 cm. Using Eq. 8.21:

$U_{avg} = 100 s_t/s_c = (100)(5/10) = 50\%$

From Table 8.2, for $U_{avg} = 50\%$, $T = 0.197$

$t = 180$ days $= 0.5$ year

Double drainage, hence $H_{dr} = 1$ m

Using Eq. 8.20:

$c_v = (T)(H_{dr})^2/t_{50}$

$c_v = (0.197)(1 \text{ m})^2/(0.5 \text{ year}) = 0.4$ m²/year

8.19 $u_o = \Delta\sigma_v = 50$ kPa

$u_e = (9.92 - 6)(9.81) = 38.5$ kPa

Using Eq. 8.19:

$U_z = 1 - (u_e/u_o) = 1 - (38.5/50) = 0.23$

Using Eq. 8.18:

$Z = z/H_{dr} = 0.5 H/0.5 H = 1.0$

Entering Fig. 8.11 with $U_z = 0.23$ and $Z = 1.0$, therefore $T = 0.20$

$t = 180$ days $= 0.5$ year

$H_{dr} = 1$ m (double drainage)

Using Eq. 8.20:

$c_v = (T)(H_{dr})^2/t$

$c_v = (0.20)(1 \text{ m})^2/(0.5 \text{ year}) = 0.4$ m²/year

8.20 Since the initial specimen height is about equal to 10 mm, then the vertical axis in Fig. 8.23 can be converted to vertical strain ε_v by dividing the scale by 10.

From Fig. 8.23:

At 3000 min, $\varepsilon_v = 0.206$

At 30,000 min, $\varepsilon_v = 0.213$

$C_\alpha = \Delta\varepsilon_v/\Delta\log t = (0.213 - 0.206)/[\log (30,000) - \log (3000)] = 0.007$

8.21 Using Eq. 8.20:

$t = (T)(H_{dr})^2/c_v$

Based on $T = 1.0$, the time for the end of primary consolidation:

$t = (1.0)(1\ \text{m})^2/0.32\ \text{m}^2/\text{year} = 3.1\ \text{year}$

Using Eq. 8.24:

$s_s = C_\alpha H_o \Delta \log t$

$s_s = (0.01)(2)\ [\log (50) - \log (3.1)] = 0.024\ \text{m} = 2.4\ \text{cm}$

Using Eq. 8.1:

$\rho_{\text{max}} = s_i + s_c + s_s = 0 + 5.1 + 2.4 = 7.5\ \text{cm}$

Chapter 9

9.1   Expansion potential for the laboratory data shown in Fig. 3.6:

Expansion potential based on percent clay size: Per Fig. 3.6, percent clay size particles = 25.7%, and therefore high expansion potential from Table 9.1.

Expansion potential based on plasticity index: From Fig. 3.6, the plasticity index = 51 and the percent passing the No. 40 sieve = 72%. Thus the plasticity index of the whole sample = (51)(0.72) = 37. Using a PI of the whole sample = 37, the soil has a very high expansion potential based on Table 7.1.

Expansion potential from Fig. 9.1: The activity of the clay fraction = 51/(25.7/0.72) = 1.43. Using the percent clay size = 25.7% and the activity = 1.43, the data plot just barely as a high expansion potential in Fig. 9.1.

Expansion potential from Fig. 9.2: Using Fig. 9.2 with percent of clay in whole sample = 25.7%, plasticity index of whole sample = 37, and activity = 1.43, the data plot as a high expansion potential.

9.2   Expansion potential for the laboratory data shown in Fig. 4.34 (i.e., SB-38 at 4-8 ft):

Expansion potential based on percent clay size: Per Fig. 4.34, percent clay size particles = 31.5%, and therefore high expansion potential from Table 9.1.

Expansion potential based on plasticity index: Per Fig. 4.34, the plasticity index = 35 and the percent passing the No. 40 sieve = 79%. Thus the plasticity index of the whole sample = (35)(0.79) = 28. Using a PI of the whole sample = 28, the soil has a high expansion potential based on Table 9.1.

Expansion potential from Fig. 9.1: The activity of the clay fraction = 35/(31.5/0.79) = 0.88. Using the percent clay size = 31.5% and the activity = 0.88, the data plot as a medium expansion potential in Fig. 9.1.

Expansion potential from Fig. 9.2: Using Fig. 9.2 with percent of clay in whole sample = 31.5%, plasticity index of whole sample = 28, and activity = 0.88, the data plot as a high expansion potential.

9.3   Expansion potential for the laboratory data shown in Fig. 4.35 (i.e., SB-39 at 4-8 ft):

Expansion potential based on percent clay size: Per Fig. 4.35 percent clay size particles = 25.7%, and therefore a high expansion potential from Table 9.1.

Expansion potential based on plasticity index: Per Fig. 4.35, the plasticity index = 41 and the percent passing the No. 40 sieve = 73%. Thus the plasticity index of the whole sample = (41)(0.73) = 30. Using a PI of the whole sample = 30, the soil has a high expansion potential from Table 9.1.

Expansion potential from Fig. 9.1: The activity of the clay fraction = 41/(25.7/0.73) = 1.16. Using the percent clay size = 25.7% and the activity = 1.16, the data plot as a medium expansion potential in Fig. 9.1.

Expansion potential from Fig. 9.2: Using Fig. 9.2 with percent of clay in whole sample = 25.7%, plasticity index of whole sample = 30, and activity = 1.16, the data plot as a high expansion potential.

9.4 Expansion potential for the laboratory data shown in Fig. 4.35 (i.e., SB-42 at 4-8 ft):

Expansion potential based on percent clay size: Per Fig. 4.35, percent clay size particles = 19.0%, and therefore a medium expansion potential from Table 9.1.

Expansion potential based on plasticity index: Per Fig. 4.35, the plasticity index = 13 and the percent passing the No. 40 sieve = 86%. Thus the plasticity index of the whole sample = (13)(0.86) = 11.2. Using a PI of the whole sample = 11.2, the soil has a low expansion potential from Table 9.1.

Expansion potential from Fig. 9.1: The activity of the clay fraction = 13/(19.0/0.86) = 0.59. Using the percent clay size = 19.0% and the activity = 0.59, the data plot as a low expansion potential in Fig. 9.1.

Expansion potential from Fig. 9.2: Using Fig. 9.2 with percent of clay in whole sample = 19.0%, plasticity index of whole sample = 11.2, and activity = 0.59, the data plot as a low expansion potential.

9.5 Expansion potential for the laboratory data shown in Fig. 4.35 (i.e., SB-45 at 4-8 ft):

Expansion potential based on percent clay size: Per Fig. 4.35, percent clay size particles = 14.1%, and therefore a low expansion potential from Table 9.1.

Expansion potential based on plasticity index: Per Fig. 4.35, the plasticity index = 15 and the percent passing the No. 40 sieve = 47%. Thus the plasticity index of the whole sample = (15)(0.47) = 7. Using a PI of the whole sample = 7, the soil has a very low expansion potential based on Table 9.1.

Expansion potential from Fig. 9.1: The activity of the clay fraction = 15/(14.1/0.47) = 0.50. Using the percent clay size = 14.1% and the activity = 0.50, the data plot as a low expansion potential in Fig. 9.1.

Expansion potential from Fig. 9.2: Using Fig. 9.2 with percent of clay in whole sample = 14.1%, plasticity index of whole sample = 7, and activity = 0.50, the data plot as a low expansion potential.

9.6 Expansion potential for the laboratory data shown in Fig. 4.36 (i.e., TP-1 at 1-2.5 ft):

Expansion potential based on percent clay size: Per Fig. 4.36, percent clay size particles = 6.2%, and therefore a very low expansion potential from Table 9.1.

Expansion potential based on plasticity index: Per Fig. 4.36, the plasticity index = 17 and the percent passing the No. 40 sieve = 54%. Thus the plasticity index of the whole sample = (17)(0.54) = 9.2. Using a PI of the whole sample = 9.2, the soil has a very low expansion potential based on Table 9.1.

Expansion potential from Fig. 9.1: The activity of the clay fraction = 17/(6.2/0.54) = 1.48. Using the percent clay size = 6.2% and the activity = 1.48, the data plot as a low expansion potential in Fig. 9.1

Expansion potential from Fig. 9.2: Using Fig. 9.2 with percent of clay in whole sample = 6.2%, plasticity index of whole sample = 9.2, and activity = 1.48, the data plot as a low expansion potential.

9.7 Expansion potential for the laboratory data shown in Fig. 4.36 (i.e., TP-15 at 1-2 ft):

Expansion potential based on percent clay size: Per Fig. 4.36, percent clay size particles = 22.1%, and therefore a medium expansion potential from Table 9.1.

Expansion potential based on plasticity index: Per Fig. 4.36, the plasticity index = 22 and the percent passing the No. 40 sieve = 80%. Thus the plasticity index of the whole sample = (22)(0.80) = 17.6. Using a PI of the whole sample = 17.6, the soil has a medium expansion potential based on Table 9.1.

Expansion potential from Fig. 9.1: The activity of the clay fraction = 22/(22.1/0.80) = 0.80. Using the percent clay size = 22.1% and the activity = 0.80, the data plot as a low expansion potential in Fig. 9.1.

Expansion potential from Fig. 9.2: Using Fig. 9.2 with percent of clay in whole sample = 22.1%, plasticity index of whole sample = 17.6, and activity = 0.80, the data plot as a medium expansion potential.

9.8 Expansion potential for the laboratory data shown in Fig. 4.36 (i.e., TP-2 at 1-2 ft):

Expansion potential based on percent clay size: Per Fig. 4.36, percent clay size particles = 73.7%, and therefore a very high expansion potential from Table 9.1.

Expansion potential based on plasticity index: Per Fig. 4.36, the plasticity index = 61 and the percent passing the No. 40 sieve = 90%. Thus the plasticity index of the whole sample = (61)(0.90) = 55. Using a PI of the whole sample = 55, the soil has a very high expansion potential based on Table 9.1.

Expansion potential from Fig. 9.1: The activity of the clay fraction = 61/(73.7/0.90) = 0.74. Using the percent clay size = 73.7% and the activity = 0.74, the data plot as a very high expansion potential in Fig. 9.1.

Expansion potential from Fig. 9.2: Using Fig. 9.2 with percent of clay in whole sample = 73.7%, plasticity index of whole sample = 55, and activity = 0.74, the data plot as a very high expansion potential.

9.9 Expansion potential for the laboratory data shown in Fig. 4.37 (i.e., SB-2 at 5 ft):

Expansion potential based on percent clay size: Per Fig. 4.37, percent clay size particles = 28.1%, and therefore a high expansion potential from Table 9.1.

Expansion potential based on plasticity index: Per Fig. 4.37, the plasticity index of the whole sample = 28 (i.e., nearly 100% of the soil passes the No. 40 sieve). Using a PI of the whole sample = 28, the soil has a high expansion potential based on Table 9.1.

Expansion potential from Fig. 9.1: The activity of the clay fraction = 28/28.1 = 1.0. Using the percent clay size = 28.1% and the activity = 1.0, the data plot as a medium expansion potential in Fig. 9.1.

Expansion potential from Fig. 9.2: Using Fig. 9.2 with percent of clay in whole sample = 28.1%, plasticity index of whole sample = 28, and activity = 1.0, the data plot as a high expansion potential.

9.10 Expansion potential for the laboratory data shown in Fig. 4.38 (i.e., AGC-2 at 0.8–1.4 ft):

Expansion potential based on percent clay size: Based on problem statement, percent clay size particles = 12%, and therefore a low expansion potential from Table 9.1.

Expansion potential based on plasticity index: Per Fig. 4.38, the plasticity index = 20 and the percent passing the No. 40 sieve = 45%. Thus the plasticity index of the whole sample = (20)(0.45) = 9. Using a PI of the whole sample = 9, the soil has a very low expansion potential based on Table 9.1.

Expansion potential from Fig. 9.1: The activity of the clay fraction = 20/(12/0.45) = 0.75. Using the percent clay size = 12% and the activity = 0.75, the data plot as a low expansion potential in Fig. 9.1.

Expansion potential from Fig. 9.2: Using Fig. 9.2 with percent of clay in whole sample = 12%, plasticity index of whole sample = 9, and activity = 0.75, the data plot as a low expansion potential.

9.11 Expansion potential for the laboratory data shown in Fig. 4.39 (i.e., soil no. 5, glacial till from Illinois):

Expansion potential based on percent clay size: Per Fig. 4.39, percent clay size particles is slightly less than 15%, and therefore a low expansion potential from Table 9.1.

Expansion potential based on plasticity index: Per Fig. 4.39, the plasticity index = 10 and the percent passing the No. 40 sieve = 80%. Thus the plasticity index of the whole sample = (10)(0.80) = 8. Using a PI of the whole sample = 8, the soil has a very low expansion potential based on Table 9.1.

Expansion potential from Fig. 9.1: The activity of the clay fraction = 10/(15/0.8) = 0.53. Using the percent clay size = 15% and the activity = 0.53, the data plot as a low expansion potential in Fig. 9.1

Expansion potential from Fig. 9.2: Using Fig. 9.2 with percent of clay in whole sample = 15%, plasticity index of whole sample = 8, and activity = 0.53, the data plot as a low expansion potential.

9.12 Expansion potential for the laboratory data shown in Fig. 4.39 (i.e., soil no. 6, Wewahitchka sandy clay from Florida):

Expansion potential based on percent clay size: Per Fig. 4.39, percent clay size particles of the whole sample = 44%, and therefore a very high expansion potential from Table 9.1.

Expansion potential based on plasticity index: Per Fig. 4.39, the plasticity index = 41 and nearly all of the soil passes the No. 40 sieve. Thus using a PI of the whole sample = 41, the soil has a very high expansion potential based on Table 9.1.

Expansion potential from Fig. 9.1: The activity of the clay fraction = 41/44 = 0.93. Using the percent clay size = 44% and the activity = 0.93, the data plot as a high expansion potential in Fig. 9.1.

Expansion potential from Fig. 9.2: Using Fig. 9.2 with percent of clay in whole sample = 44%, plasticity index of whole sample = 41, and activity = 0.93, the data plot as a very high expansion potential.

9.13 Expansion potential for the laboratory data shown in Fig. 4.39 (i.e., soil no. 7, loess from Mississippi):

Expansion potential based on percent clay size: Per Fig. 4.39, percent clay size particles of the whole sample = 5%, and therefore a very low expansion potential from Table 9.1.

Expansion potential based on plasticity index: Per Fig. 4.39, the plasticity index = 5 and all of the soil passes the No. 40 sieve. Thus using a PI of the whole sample = 5, the soil has a very low expansion potential based on Table 9.1.

Expansion potential from Fig. 9.1: The activity of the clay fraction = 5/5 = 1.0. Using the percent clay size = 5% and the activity = 1.0, the data plot as a low expansion potential in Fig. 9.1.

Expansion potential from Fig. 9.2: Using Fig. 9.2 with percent of clay in whole sample = 5%, plasticity index of whole sample = 5, and activity = 1.0, the data plot as a low expansion potential.

9.14 Expansion potential for the laboratory data shown in Fig. 4.39 (i.e., soil no. 8, backswamp deposit from the Mississippi river):

Expansion potential based on percent clay size: Per Fig. 4.39, percent clay size particles of the whole sample = 42%, and therefore a very high expansion potential from Table 9.1.

Expansion potential based on plasticity index: Per Fig. 4.39, the plasticity index = 41 and all of the soil passes the No. 40 sieve. Thus using a PI of the whole sample = 41, the soil has a very high expansion potential based on Table 9.1.

Expansion potential from Fig. 9.1: The activity of the clay fraction = 41/42 = 0.98. Using the percent clay size = 42% and the activity = 0.98, the data plot as a high expansion potential in Fig. 9.1.

Expansion potential from Fig. 9.2: Using Fig. 9.2 with percent of clay in whole sample = 42%, plasticity index of whole sample = 41, and activity = 0.98, the data plot as a very high expansion potential.

9.15 Using Fig. 9.34, the intersection of the two straight-line segments corresponds to a percent swell of approximately 14%. Using Eq. 9.1, the end of primary expansion index = 140 and per Table 9.1, the soil has a very high expansion potential.

9.16 For the expansion index test, there is infiltration from both ends of the specimen, hence H_{dr} = one-half the specimen height or 1.31 cm. In Fig. 9.8, the end of primary swell occurs at 6.2% swell and hence 50% of the primary swell corresponds to 3.1% swell. Entering Fig. 9.8 at 3.1% swell, the time for 50% swell (t_{50}) = 6.5 min. From Table 8.2, the time factor (T) = 0.197 for 50% swell or consolidation. Using Eq. 9.4:

$$c_s = TH_{dr}^2/t = (0.197)(1.31 \text{ cm})^2/(6.5 \text{ min}) = 0.052 \text{ cm}^2/\text{min} = 2.7 \text{ m}^2/\text{year}$$

The field clay layer is 2-m thick with water infiltration from only the ground surface and hence the value of H_{dr} = 2 m. From Table 8.2, the time factor (T) = 0.848 for 90% swell or consolidation. Using Eq. 9.4:

$$t = TH_{dr}^2/c_s = (0.848)(2 \text{ m})^2/2.7 \text{ m}^2/\text{year} = 1.2 \text{ year}$$

9.17 For double boundary water infiltration, H_{dr} = 1m

Using Eq. 9.4:

$$t = TH_{dr}^2/c_s = (0.848)(1 \text{ m})^2/2.7 \text{ m}^2/\text{year} = 0.3 \text{ year}$$

9.18 The equation used to calculate the secondary compression ratio (Sec. 8.6) can be used to calculate the secondary swell ratio $(C_{\alpha s})$, or:

$$C_{\alpha s} = \Delta\varepsilon_v/\Delta\log t$$

From Fig. 9.8:

At t = 100 min, ε_v = 0.063

At t = 1000 min, ε_v = 0.066

$$C_{\alpha s} = \Delta\varepsilon_v/\Delta\log t = (0.066 - 0.063)/(\log 1000 - \log 100) = 0.003$$

9.19 Allowable foundation heave = 0.9 in. = 0.075 ft

Entering the right side graph in Fig. 9.24 at total swell = 0.075 ft and intersecting the total swell curve, the depth of undercut = 2 ft.

9.20 Plotting swell versus depth, the total swell is the area under the percent swell curve (from a depth of 1 to 10 ft), or:

Total swell = $\frac{1}{2}(10 - 1)(4.5/100)$ = 0.203 ft = 2.4 in.

9.21 P_o = 100 kPa

From Fig. 9.28:

$\Delta h = [(h_o C_s)/(1 + e_o)] \log (P_f/P_o)$

Layer no. 1:

$\Delta h = [(500)(0.1)/(1 + 1.0)] \log (4.5/100) = 33.7$ mm

Layer no. 2:

$\Delta h = [(500)(0.1)/(1 + 1.0)] \log (13.5/100) = 21.7$ mm

Layer no. 3:

$\Delta h = [(1000)(0.1)/(1 + 1.0)] \log (27/100) = 28.4$ mm

Total heave of foundation $= 33.7 + 21.7 + 28.4 = 84$ mm

9.22 $C_s = 0.20$

From Fig. 9.28:

$\Delta h = [(h_o C_s)/(1 + e_o)] \log (P_f/P_o)$

Layer no. 1:

$\Delta h = [(500)(0.2)/(1 + 1.0)] \log (4.5/200) = 82.4$ mm

Layer no. 2:

$\Delta h = [(500)(0.2)/(1 + 1.0)] \log (13.5/200) = 58.6$ mm

Layer no. 3:

$\Delta h = [(1000)(0.2)/(1 + 1.0)] \log (27/200) = 87.0$ mm

Total heave of foundation $= 82.4 + 58.6 + 87.0 = 228$ mm

9.23 The initial condition is a uniform P_s' condition $= 200$ kPa and the final condition is a groundwater table at a depth of 0.5 m.

From Fig. 9.28:

$\Delta h = [(h_o C_s)/(1 + e_o)] \log (P_f/P_o)$

Layer no. 1:

$P_f = \sigma_v' = (0.25)(18) = 4.5$ kPa

$\Delta h = [(500)(0.1)/(1 + 1.0)] \log (4.5/200) = 41.2$ mm

Layer no. 2:

$P_f = \sigma_v' = (0.5)(18) + (0.25)(18 - 9.81) = 11.0$ kPa

$\Delta h = [(500)(0.1)/(1 + 1.0)] \log (11.0/200) = 31.5$ mm

Layer no. 3:

$P_f = \sigma_v' = (0.5)(18) + (1.0)(18 - 9.81) = 17.2$ kPa

$\Delta h = [(1000)(0.1)/(1 + 1.0)] \log (17.2/200) = 53.3$ mm

Total heave of foundation $= 41.2 + 31.5 + 53.3 = 126$ mm

9.24 The initial condition is a uniform P_s' condition $= 200$ kPa and the final condition is a groundwater table at a depth of 0.5 m.

From Fig. 9.28:

$\Delta h = [(h_o C_s)/(1 + e_o)] \log (P_f/P_o)$

Layer no. 2:

$P_f = \sigma'_v = 25 + (0.25)(18 - 9.81) = 27.0$ kPa

$\Delta h = [(500)(0.1)/(1 + 1.0)] \log (27.0/200) = 21.7$ mm

Layer no. 3:

$P_f = \sigma'_v = 25 + (1.0)(18 - 9.81) = 33.2$ kPa

$\Delta h = [(1000)(0.1)/(1 + 1.0)] \log (33.2/200) = 39.0$ mm

Total heave of center of mat foundation = 21.7 + 39.0 = 61 mm

9.25 The initial condition is a uniform P'_s condition = 200 kPa and the final condition is a groundwater table at a depth of 0.5 m. The square footing does not exert one-dimensional pressures, so the net pressure (net σ_o) must be used in the analysis, or:

net $\sigma_o = 50 - (18)(0.5) = 41$ kPa

From Fig. 9.28:

$\Delta h = [(h_o C_s)/(1 + e_o)] \log (P_f/P_o)$

Layer no. 2:

$P_f = \sigma'_v = [(41)(1.2)^2/(1.2 + 0.25)^2] + (18)(0.5) + (0.25)(18 - 9.81)$

$= 39.1$ kPa

$\Delta h = [(500)(0.1)/(1 + 1.0)] \log (39.1/200) = 17.7$ mm

Layer no. 3:

$P_f = \sigma'_v = [(41)(1.2)^2/(1.2 + 1.0)^2] + (18)(0.5) + (1.0)(18 - 9.81)$

$= 29.4$ kPa

$\Delta h = [(1000)(0.1)/(1 + 1.0)] \log (29.4/200) = 41.6$ mm

Total heave of square footing = 17.7 + 41.6 = 59 mm

9.26 Layer no. 3:

$P_f = \sigma'_v = [(41)(1.2)^2/(1.2 + 1.0)^2] + (18)(0.5) + (1.0)(18 - 9.81)$

$= 29.4$ kPa

From Fig. 9.28:

$\Delta h = [(h_o C_s)/(1 + e_o)] \log (P_f/P_o)$

$\Delta h = [(1000)(0.1)/(1 + 1.0)] \log (29.4/100) = 26.6$ mm

Total heave of square footing = 17.7 + 26.6 = 44 mm

9.27 Depth to bottom of piers below ground surface:

From Fig. 6.13, for $s_u = c = 50$ kPa (1040 psf), the value of $c_A/c = 0.75$ (average curve for concrete piles), or:

$c_A = (0.75)(50) = 37.5$ kPa

Using Eq. 9.12:

$T_u = c_A 2\pi R Z_a$

Piers start at bottom of grade beam (0.4 m):

$T_u = (37.5)(2)(\pi)(0.3/2)(2 - 0.4) = 57$ kN

Using Eq. 9.13:

$T_r = P + c_A 2\pi R Z_{na}$

Equating T_u and T_r, therefore:

$57 = 20 + (80)(2)(\pi)(0.3/2)(Z_{na})$

Solving for $Z_{na} = 0.48$ m

Use a factor of safety = 2, therefore:

Depth to bottom of piers below ground surface

$= 2$ m $+ (2)(0.48$ m$) = 3$ m

Air gap below grade beams:

From Fig. 9.28, total heave of layer nos. 2 and 3 = 73 mm. Using a factor of safety = 1.5, air gap = (1.5)(73) = 110 mm.

Chapter 10

10.1 From Eq. 10.2:

$\sigma_n' = \gamma_b d \cos^2\alpha = (10)(1.2)(\cos^2 33.7°) = 8.3$ kPa

Using the linear extrapolated shear strength envelope and in terms of effective stresses (from Eq. 3.9):

$\tau_f = c' + \sigma_n' \tan\phi' = 25 + (8.3) \tan 16° = 27.4$ kPa

For the actual nonlinear portion of the shear strength envelop at low effective stress (Fig. 10.11):

$\sigma_n' = 8.3$ kPa and the shear strength $(\tau_f) = 10.7$ kPa

10.2 For the linear extrapolated shear strength envelope ($c' = 25$ kPa, $\phi' = 16°$) and using Eq. 10.2:

$F = (c' + \gamma_b d \cos^2 \alpha \tan\phi')/(\gamma_t d \cos \alpha \sin \alpha)$

$F = [25 + (10.0)(1.2)(\cos^2 26.6°)(\tan 16°)]/[(19.8)(1.2)(\cos 26.6°)(\sin 26.6°)]$

$F = 2.9$

For the actual nonlinear shear strength envelope (Fig. 10.11b):

From Eq. 10.2:

$\sigma_n' = \gamma_b d \cos^2\alpha = (10)(1.2)(\cos^2 26.6°) = 9.6$ kPa

From Fig. 10.11b, for $\sigma_n' = 9.6$ kPa, $\tau_f = 13.0$ kPa

$F = (13)/[(19.8)(1.2)(\cos 26.6°)(\sin 26.6°)] = 1.37$

10.3 Substituting $(W + Q)$ for W in Eqs. 10.3 and 10.4:

Total stress analysis:

$$F = \frac{cL + (W + Q) \cos\alpha \tan\phi}{(W + Q)\sin\alpha}$$

Effective stress analysis:

$$F = \frac{c'L + [(W + Q) \cos\alpha - uL] \tan\phi'}{(W + Q)\sin\alpha}$$

10.4 Component of the earthquake force normal to the slip surface $= -bW \sin\alpha$

Component of the earthquake force parallel to the slip surface $= bW \cos\alpha$

Adjust Eqs. 10.3 and 10.4 to include these two components.

Total stress analysis:

$$F = \frac{cL + (W\cos\alpha - bW\sin\alpha)\tan\phi}{W\sin\alpha + bW\cos\alpha}$$

Effective stress analysis:

$$F = \frac{c'L + (W\cos\alpha - bW\sin\alpha - uL)\tan\phi'}{W\sin\alpha + bW\cos\alpha}$$

10.5   Total stress example problem:

Width of wedge at the top of slope = (3)(30) − (2)(30) = 30 ft

Q = (200 psf)(30 ft) = 6000 lb per linear ft of slope length

$W + Q$ = 56,700 + 6000 = 62,700 lb/ft

Using the total stress equation from Problem 10.3:

F = [(70)(95) + (62,700)(cos 18.4°)(tan 29°]/[(62,700) sin 18.4°]

F = 2.00

Effective stress example problem:

$W + Q$ = 56,700 + 6000 = 62,700 lb/ft

uL = (50 psf)(95 ft) = 4750 lb/ft

Using the effective stress equation from Problem 10.3:

F = [(70)(95) + (62,700 cos 18.4° − 4750) tan 29°]/(62,700 sin 18.4°)

F = 1.87

10.6   Total stress example problem:

Horizontal earthquake force = bW

= (0.1)(56,700) = 5670 lb per linear ft of slope length

Using the total stress equation from Problem 10.4:

F = [(70)(95) + (56,700 cos 18.4° − 5670 sin 18.4°) tan 29°]/

(56,700 sin 18.4° + 5670 cos 18.4°)

F = 1.52

Effective stress example problem:

uL = (50 psf)(95 ft) = 4750 lb/ft

Using the effective stress equation from Problem 10.4:

F = [(70)(95) + (56,700 cos 18.4° − 5670 sin 18.4° − 4750) tan 29°]/

(56,700 sin 18.4° + 5670 cos 18.4°)

F = 1.41

10.7   Using Fig. 10.19, length (L) of slip surface (for $\Delta x = 1$) = 1/cos 26°

Shear stress:

$\tau = W\sin\alpha/L$ = (100)(sin 26°)/(1/cos 26°) = 39.4 kPa

$\sigma'_n = W\cos\alpha/L$ = (100)cos² 26° = 80.8 kPa

Using Eq. 3.9:

$\tau_f = c' + \sigma'_n \tan \phi'$

$\tau_f = c' + \sigma'_n \tan \phi' = 2 + 80.8 \tan 25° = 39.7$ kPa

10.8 $F = \tau_f / \tau = 39.7/39.4 = 1.01$. Yes, progressive failure is likely for this slope.

10.9 The slope is deforming laterally on a slip surface located about 7 m below ground surface.

10.10 Depth of seasonal moisture changes = 4.5 m, therefore because the slope inclination is 2:1 (horizontal:vertical), the setback = (2)(4.5) = 9 m.

Chapter 11

11.1 Using Eq. 11.2:

$k_A = \tan^2 (45° - \frac{1}{2}\phi) = \tan^2 [45° - (\frac{1}{2})(32°)] = 0.307$

Using Eq. 11.5:

$k_p = \tan^2 (45° + \frac{1}{2}\phi) = \tan^2 [45° + (\frac{1}{2})(32°)] = 3.25$

Using Eq. 11.1:

$P_A = \frac{1}{2} k_A \gamma_t H^2 = \frac{1}{2}(0.307)(20 \text{ kN/m}^3)(4 \text{ m})^2 = 49.2$ kN/m

Using Eq. 11.4:

$P_p = \frac{1}{2} k_p \gamma_t D^2 = \frac{1}{2}(3.25)(20 \text{ kN/m}^3)(0.5 \text{ m})^2 = 8.14$ kN/m

With reduction factor = 2, allowable P_p = 4.07 kN/m

11.2 From Table 11.1, $Y/H = 0.0005$ for dense sand, therefore:

$Y = (0.0005)(4 \text{ m}) = 0.002 \text{ m} = 0.2$ cm

11.3 Per Fig. 11.3, the active wedge is inclined at:

$45° + \phi/2 = 45° + (32°/2) = 61°$

Width of active wedge = $H/\tan 61° = 4$ m/tan $61° = 2.2$ m

11.4 Footing weight = (3 m)(0.5 m)(23.5 kN/m³) = 35.3 kN/m

Stem weight = (0.4 m)(3.5 m)(23.5 kN/m³) = 32.9 kN/m

N = weight of concrete wall = 35.3 + 32.9 = 68.2 kN/m

Take moments about the toe of the wall to determine $\bar{x}$, or:

$Nx = - P_A(4/3) + (W)$(moment arms)

$(68.2)\bar{x} = - (49.2)(4/3) + (35.3)(3/2) + (32.9)(2.8)$

Solving for $\bar{x} = 79.5/68.2 = 1.165$ m

11.5 Using Eqs. 6.12 and 6.13:

$q' = Q(B + 6 e)/B^2$

$q'' = Q(B - 6 e)/B^2$

e = eccentricity = 1.5 - 1.165 = 0.335 m

$q' = (68.2)[3 + (6)(0.335)]/(3)^2 = 37.9$ kPa

$q'' = (68.2)[3 - (6)(0.335)]/(3)^2 = 7.5$ kPa

11.6 For $\phi' = 32°$, from Fig. 6.5, $N_\gamma = 21$ and $N_q = 23$

Using Eq. 6.1 with $c = 0$:

$q_{ult} = \frac{1}{2} \gamma_t BN_\gamma + \gamma_t D_f N_q$

$q_{ult} = \frac{1}{2} (20 \text{ kN/m}^3)(3 \text{ m})(21) + (20 \text{ kN/m}^3)(0.5 \text{ m})(23) = 860 \text{ kPa}$

$q_{all} = 860/3 = 290 \text{ kPa}$

Since $q' = 37.9 \text{ kPa}$ and $q_{all} = 290 \text{ kPa}$, then $q_{all} > q'$ and q' is acceptable

11.7 Using Eq. 11.6:

$F = (N \tan \delta + P_p)/P_A$

$F = (68.2 \tan 24° + 4.07)/49.2 = 0.70$

11.8 Using Eq. 11.7:

$F = (W)(a)/[(1/3)(P_A)(H)]$

$F = [(35.3)(3/2) + (32.9)(2.8)]/[(1/3)(49.2)(4)] = 2.2$

11.9 Using Eq. 11.6:

$F = (N \tan \delta + P_p)/P_A$

$1.5 = (68.2 \tan 24° + P_p)/49.2$

Solving for $P_p = 43.5 \text{ kN/m}$

Double P_p to account for reduction factor and use Eq. 11.4:

$P_p = \frac{1}{2} k_p \gamma_t D^2$

$(43.5 \text{ kN/m})(2) = \frac{1}{2}(3.25)(20 \text{ kN/m}^3)(D)^2$

Solving for $D = 1.64 \text{ m}$

11.10 Using the following values:

$\delta = \phi_w = 24°$

$\phi = 32°$

$\theta = 0°$

$\beta = 0°$

Inserting the above values into Coulomb's equation (static condition) from Fig. 11.4:

$k_A = 0.275$

Using Eq. 11.1:

$P_A = \frac{1}{2} k_A \gamma_t H^2$

$P_A = \frac{1}{2} (0.275)(20 \text{ kN/m}^3)(4 \text{ m})^2 = 43.9 \text{ kN/m}$

$P_H = P_A \cos 24° = (43.9 \text{ kN/m})(\cos 24°) = 40.1 \text{ kN/m}$

$P_v = P_A \sin 24° = (43.9 \text{ kN/m})(\sin 24°) = 17.9 \text{ kN/m}$

11.11 Footing weight = $(3 \text{ m})(0.5 \text{ m})(23.5 \text{ kN/m}^3) = 35.3 \text{ kN/m}$

Stem weight = $(0.4 \text{ m})(3.5 \text{ m})(23.5 \text{ kN/m}^3) = 32.9 \text{ kN/m}$

N = weight of concrete wall + P_v = 35.3 + 32.9 + 17.9 = 86.1 kN/m

Take moments about the toe of the wall to determine $\bar{x}$:

$N x = - P_H(4/3) + (W)(\text{moment arms}) + (P_v)(3)$

$(86.1)\ \bar{x} = - (40.1)(4/3) + (35.3)(3/2) + (32.9)(2.8) + (17.9)(3)$

Solving for $\bar{x} = 145/86.1 = 1.69$ m

11.12 Using Eqs. 6.12 and 6.13:

$q' = Q(B + 6e)/B^2$

$q'' = Q(B - 6e)/B^2$

$e = \text{eccentricity} = 1.5 - 1.69 = -0.19$ m

$q' = (86.1)[3 + (6)(0.19)]/(3)^2 = 39.6$ kPa

$q'' = (86.1)[3 - (6)(0.19)]/(3)^2 = 17.8$ kPa

11.13 The allowable bearing pressure is the same as Problem 11.6 ($q_{all} = 290$ kPa) and since $q_{all} > q'$, the design is acceptable in terms of bearing pressures.

11.14 Using Eq. 11.8:

$F = (N \tan \delta + P_p)/P_H$

$N = W + P_v = 86.1$ kN/m

$F = (86.1 \tan 24° + 4.07)/40.1 = 1.06$

11.15 Overturning moment $= (P_H)(1/3)(H) - (P_v)(3)$

$= (40.1\ \text{kN/m})(1/3)(4\ \text{m}) - (17.9\ \text{kN/m})(3\ \text{m}) = -0.23$ kN-m/m

Therefore: $F = \infty$

11.16 Using Eq. 11.8:

$F = (N \tan \delta + P_p)/P_H$

$N = W + P_v = 86.1$ kN/m

$F = (86.1 \tan 24° + P_p)/40.1$

$1.5 = (86.1 \tan 24° + P_p)/40.1$

Solving for $P_p = 21.8$ kN/m

Double P_p to account for reduction factor and use Eq. 11.4:

$P_p = \tfrac{1}{2} k_p \gamma_t D^2$

$(21.8\ \text{kN/m})(2) = \tfrac{1}{2}(3.25)(20\ \text{kN/m}^3)(D)^2$

Solving for $D = 1.16$ m

11.17 Using the following values:

$\delta = \phi_w = 32°$

$\phi = 32°$

$\theta = 0°$

$\beta = 0°$

Inserting the above values into Coulomb's equation (static condition) from Fig. 11.4:

$k_A = 0.277$

Using Eq. 11.5:

$k_p = \tan^2(45° + \frac{1}{2}\phi) = \tan^2[45° + (\frac{1}{2})(32°)] = 3.25$

Using Eq. 11.1:

$P_A = \frac{1}{2}k_A\gamma_t H^2$

$P_A = \frac{1}{2}(0.277)(20\text{ kN/m}^3)(4\text{ m})^2 = 44.3\text{ kN/m}$

$P_v = P_A \sin 32° = (44.3)(\sin 32°) = 23.5\text{ kN/m}$

$P_H = P_A \cos 32° = (44.3)(\cos 32°) = 37.6\text{ kN/m}$

Using Eq. 11.4:

$P_p = \frac{1}{2}k_p\gamma_t D^2 = \frac{1}{2}(3.25)(20\text{ kN/m}^3)(0.5\text{ m})^2 = 8.14\text{ kN/m}$

With reduction factor = 2, allowable $P_p = 4.07\text{ kN/m}$

11.18   Footing weight = $(2\text{ m})(0.5\text{ m})(23.5\text{ kN/m}^3) = 23.5\text{ kN/m}$

Stem weight = $(0.4\text{ m})(3.5\text{ m})(23.5\text{ kN/m}^3) = 32.9\text{ kN/m}$

Soil weight on top of footing = $(0.8\text{ m})(3.5\text{ m})(20\text{ kN/m}^3) = 56.0\text{ kN/m}$

N = weights + $P_v = 23.5 + 32.9 + 56.0 + 23.5 = 135.9\text{ kN/m}$

Take moments about the toe of the wall to determine $\bar{x}$, or:

$N\bar{x} = -P_H(4/3) + (W)(\text{moment arms}) + (P_v)(2)$

$(135.9)\bar{x} = -(37.6)(4/3) + (56.4)(1) + (56)(1.6) + (23.5)(2)$

Solving for $\bar{x} = 143/135.9 = 1.05\text{ m}$

11.19   Using Eqs. 6.12 and 6.13:

$q' = Q(B + 6e)/B^2$

$q'' = Q(B - 6e)/B^2$

e = eccentricity = $1.0 - 1.05 = -0.05\text{ m}$

$q' = (135.9)[2 + (6)(0.05)]/(2)^2 = 78.1\text{ kPa}$

$q'' = (135.9)[2 - (6)(0.05)]/(2)^2 = 57.8\text{ kPa}$

11.20   For $\phi' = 32°$, from Fig. 6.5, $N_\gamma = 21$ and $N_q = 23$

Using Eq. 6.1 with $c = 0$:

$q_{ult} = \frac{1}{2}\gamma_t BN_\gamma + \gamma_t D_f N_q$

$q_{ult} = \frac{1}{2}(20\text{ kN/m}^3)(2\text{ m})(21) + (20\text{ kN/m}^3)(0.5\text{ m})(23) = 650\text{ kPa}$

$q_{all} = 650/3 = 220\text{ kPa}$

Since $q' = 78.1\text{ kPa}$ and $q_{all} = 220\text{ kPa}$, then $q_{all} > q'$ and q' is acceptable

11.21   Using Eq. 11.8:

$F = (N\tan\delta + P_p)/P_H$

$N = W + P_v = 135.9\text{ kN/m}$

$F = (135.9\tan 24° + 4.07)/37.6 = 1.72$

11.22   Taking moments about the toe of the wall:

Overturning moment = $(P_H)(1/3)(H) - (P_v)(2)$

$= (37.6\text{ kN/m})(1/3)(4\text{ m}) - (23.5\text{ kN/m})(2\text{ m}) = 3.1\text{ kN-m/m}$

Moment of weights = (56.4 kN/m)(1 m) + (56 kN/m)(1.6 m) = 146 kN-m/m

$F = 146/3.1 = 47$

11.23 Using Eq. 11.8:

$F = (N \tan \delta + P_p)/P_H$

$N = W + P_v = 135.9$ kN/m

$F = (135.9 \tan 24° + P_p)/37.6$

$2.0 = (135.9 \tan 24° + P_p)/37.6$

Solving for $P_p = 14.7$ kN/m

Double P_p to account for reduction factor and use Eq. 11.4:

$P_p = \frac{1}{2} k_p \gamma_t D^2$

$(14.7 \text{ kN/m})(2) = \frac{1}{2} (3.25)(20 \text{ kN/m}^3)(D)^2$

Solving for $D = 0.95$ m

11.24 Using Eq. 11.3:

$P_Q = QHk_A = (200 \text{ psf})(20 \text{ ft})(0.297) = 1190$ lb/ft

P_Q is located at $H/2 = 10$ ft

$P_{QH} = 1190 \cos \phi_w = 1190 \cos 30° = 1030$ lb/ft

$P_{Qv} = 1190 \sin \phi_w = 1190 \sin 30° = 595$ lb/ft

$N = 15{,}270 + 595 = 15{,}870$ lb/ft

$P_H = 5660 + 1030 = 6690$ lb/ft

Factor of safety for sliding:

Using Eq. 11.8:

$F = (N \tan \delta + P_p)/P_H$

$\delta = \phi_{cv} = 30°$

$F = (15{,}870 \tan 30° + 750)/6690 = 1.48$

Factor of safety for overturning:

Overturning moment = 14,900 + (1030)(10) − (595)(7) = 21,030 ft-lb/ft

Moment of weight = 55,500 ft-lb/ft

Therefore: $F = 55{,}500/21{,}030 = 2.64$

Location of N:

$\bar{x} = (55{,}500 - 21{,}030)/15{,}870 = 2.17$ ft

Middle one-third of the foundation: $\bar{x} = 2.33$ to 4.67 ft

Therefore N is not within the middle one-third of the foundation.

11.25 Using the following values:

$\delta = \phi_w = 30°$

$\phi = 30°$

$\theta = 0°$

$\beta = 18.4°$

Inserting the above values into Coulomb's equation (static condition) from Fig. 11.4:

$k_A = 0.4065$

Using Eq. 11.1:

$P_A = \frac{1}{2} k_A \gamma_t H^2$

$P_A = \frac{1}{2}(0.4065)(110 \text{ pcf})(20 \text{ ft})^2 = 8940 \text{ lb/ft}$

$P_v = P_A \sin 30° = (8940 \text{ lb/ft})(\sin 30°) = 4470 \text{ lb/ft}$

$P_H = P_A \cos 30° = (8940 \text{ lb/ft})(\cos 30°) = 7740 \text{ lb/ft}$

$N = 12,000 + 4470 = 16,470 \text{ lb/ft}$

Factor of safety for sliding:

Using Eq. 11.8:

$F = (N \tan \delta + P_p)/P_H$

$\delta = \phi_{cv} = 30°$

$F = (16,470 \tan 30° + 750)/7740 = 1.32$

Factor of safety for overturning:

Overturning moment $= (7740)(20/3) - (4470)(7) = 20,310 \text{ ft-lb/ft}$

Moment of weight $= 55,500 \text{ ft-lb/ft}$

Therefore: $F = 55,500/20,310 = 2.73$

Location of N:

$\bar{x} = (55,500 - 20,310)/16,470 = 2.14 \text{ ft}$

For the middle 1/3 of the retaining wall foundation, $\bar{x} = 2.33$ to 4.67 ft.

Therefore, N is not within the middle one-third of the retaining wall foundation.

11.26 The depth of seasonal moisture change $= 4.5$ m. Assume that the depth of slope creep is along a plane that is at a depth of 4.5 m at the top of slope and this plane passes through the toe of the slope. Therefore the angle of inclination of this plane $= \tan^{-1} [(20 \text{ m} - 4.5 \text{ m})/50 \text{ m}] = 17.2°$

Neglecting c' and using Eq. 11.10 with $\beta = -17.2°$ and $\phi' = 28°$, therefore:

$k_p = 1.61$

Using Eq. 11.4:

$P_p = \frac{1}{2} k_p \gamma_t D^2 = \frac{1}{2}(1.61)(19 \text{ kN/m}^3)(D)^2$

With a reduction factor $= 2$

$(10 \text{ kN/m})(2) = \frac{1}{2}(1.61)(19 \text{ kN/m}^3)(D)^2$

Solving for $D = 1.1$ m

For a retaining wall to be constructed at the top of slope, the total depth to the bottom of the foundation $= 4.5$ m $+ 1.1$ m $= 5.6$ m

11.27 Initial active earth pressure resultant force:

Using Eq. 11.2:

$k_A = \tan^2 (45° - \frac{1}{2} \phi) = \tan^2 [45° - (\frac{1}{2})(30°)] = 0.333$

Using Eq. 11.1:

$P_A = \frac{1}{2} k_A \gamma_t H^2 = \frac{1}{2}(0.333)(20 \text{ kN/m}^3)(3 \text{ m})^2 = 30 \text{ kN/m}$

Total force acting on wall due to rise in groundwater level:

Water pressure $= \frac{1}{2}\gamma_w H^2 = \frac{1}{2}(9.81 \text{ kN/m}^3)(1.5 \text{ m})^2 = 11.0 \text{ kN/m}$

Above the groundwater table:

$P_{1A} = \frac{1}{2}(0.333)(20 \text{ kN/m}^3)(1.5 \text{ m})^2 = 7.5 \text{ kN/m}$

Below the groundwater table:

$P_{2A} = (0.333)(20 \text{ kN/m}^3)(1.5 \text{ m})^2 + \frac{1}{2}k_A\gamma_b (1.5)^2$

$= 15.0 \text{ kN/m} + \frac{1}{2}(0.333)(20 - 9.81)(1.5)^2 = 18.8 \text{ kN/m}$

Total Force $= 11.0 + 7.5 + 18.8 = 37.3 \text{ kN/m}$

Percent increase in force:

% increase $= 100 (37.3 - 30)/30 = 24\%$

11.28 Using Eq. 4.19:

$k_o = 1 - \sin \phi' = 1 - \sin 32° = 0.47$

Based on Eq. 11.1 (with k_o substituted for k_A):

$P = \frac{1}{2}k_o\gamma_t H^2 = \frac{1}{2}(0.47)(20 \text{ kN/m}^3)(4 \text{ m})^2 = 75.2 \text{ kN/m}$

% increase $= 100 (75.2 - 49.2)/49.2 = 53\%$

11.29 Using Eq. 11.2:

$k_A = \tan^2 (45° - \frac{1}{2}\phi) = \tan^2 [45° - (\frac{1}{2})(30°)] = 0.333$

Using Eq. 11.5:

$k_p = \tan^2 (45° + \frac{1}{2}\phi) = \tan^2 [45° + (\frac{1}{2})(30°)] = 3.0$

Using Eq. 11.1:

$P_A = \frac{1}{2}k_A\gamma_t H^2 = \frac{1}{2}(0.333)(110 \text{ pcf})(20 \text{ ft})^2 = 7330 \text{ lb/ft}$

Using Eq. 11.4:

$P_p = \frac{1}{2}k_p\gamma_t D^2 = \frac{1}{2}(3.0)(110 \text{ pcf})(3 \text{ ft})^2 = 1490 \text{ lb/ft}$

With reduction factor $= 2$, allowable $P_p = 740 \text{ lb/ft}$

11.30 Per Fig. 11.3, the active wedge is inclined at $45° + \phi/2 = 45° + (30°/2) = 60°$

Width of active wedge $= H/\tan 60° = (20 \text{ ft})/\tan 60° = 11.5 \text{ ft}$

As measured from the upper left corner: $11.5 + 14 = 25.5 \text{ ft}$

11.31 $N =$ weight of reinforced soil mass zone $= (H)(L)(\gamma_t)$

$= (20 \text{ ft})(14 \text{ ft})(120 \text{ pcf}) = 33,600 \text{ lb/ft}$

Taking moments about the toe of the wall (Point O) to determine $\bar{x}$:

$N\bar{x} = - P_A(20/3) + (W)(L/2)$

$(33,600) \bar{x} = - (7330)(20/3) + (33,600)(7)$

Solving for $\bar{x} = 186,000/33,600 = 5.55 \text{ ft}$

11.32 Using Eqs. 6.12 and 6.13:

$q' = Q(B + 6 e)/B^2$

$q'' = Q(B - 6 e)/B^2$

e = eccentricity = $7.0 - 5.55 = 1.45$ ft

$q' = (33,600)[14 + (6)(1.45)]/(14)^2 = 3890$ psf

$q'' = (33,600)[14 - (6)(1.45)]/(14)^2 = 910$ psf

11.33 From Fig. 6.5, for $\phi = 30°$, $N_\gamma = 15$ and $N_q = 19$. Using Eq. 6.1 with $c' = 0$:

$q_{ult} = \frac{1}{2}\gamma_t BN_\gamma + \gamma_t D_f N_q$

$q_{ult} = \frac{1}{2}(120 \text{ pcf})(14 \text{ ft})(15) + (110 \text{ pcf})(3 \text{ ft})(19) = 18,900$ psf

$q_{all} = 18,900/3 = 6300$ psf

Since $q' = 3890$ psf and $q_{all} = 6300$ psf, then $q_{all} > q'$ and the design is acceptable in terms of bearing pressures.

11.34 Using Eq. 11.6:

$F = (N \tan \delta + P_p)/P_A$

$F = (33,600 \tan 23° + 740)/7330 = 2.05$

11.35 Taking moments about the toe of the wall:

Overturning moment = $(P_A)(H/3) = (7330 \text{ lb/ft})(20 \text{ ft}/3) = 48,900$ ft-lb/ft

Moment of weight = $(33,600 \text{ lb/ft})(14 \text{ ft}/2) = 235,000$ ft-lb/ft

Therefore: $F = 235,000/48,900 = 4.81$

11.36 Using Eq. 11.3:

$P_Q = P_2 = QHk_A = (200 \text{ psf})(20 \text{ ft})(0.333) = 1330$ lb/ft

P_2 is located at $H/2 = 10$ ft

$P_H = P_A + P_2 = 7330 + 1330 = 8660$ lb/ft

From Problem 11.31 : $N = 33,600$ lb/ft

Factor of safety for sliding:

Using Eq. 11.6:

$F = (N \tan \delta + P_p)/P_H$

$F = (33,600 \tan 23° + 740)/8660 = 1.73$

Factor of safety for overturning:

Overturning moment = $(7330)(20/3) + (1330)(10) = 62,200$ ft-lb/ft

Moment of weight (from Problem 11.35) = $235,000$ ft-lb/ft

Therefore: $F = 235,000/62,200 = 3.78$

Maximum pressure exerted by the bottom of the wall:

$\bar{x} = (235,000 - 62,200)/33,600 = 5.14$ ft

Using Eq. 6.12:

$q' = Q(B + 6 e)/B^2$

e = eccentricity = $7 - 5.14 = 1.86$ ft

$q' = (33,600) [14 + (6)(1.86)]/(14)^2 = 4300$ psf

11.37 Using the following values:

$\delta = 0°$

$\phi = 30°$

$\theta = 0°$

$\beta = 18.4°$

Inserting the above values into Coulomb's equation (static condition) from Fig. 11.4:

$k_A = 0.427$

Using Eq. 11.1:

$P_A = \frac{1}{2} k_A \gamma_t H^2 = \frac{1}{2}(0.427)(110 \text{ pcf})(20 \text{ ft})^2 = 9390 \text{ lb/ft}$

Factor of safety for sliding:

Using Eq. 11.6:

$F = (N \tan \delta + P_p)/P_A$

$F = (33,600 \tan 23° + 740)/9390 = 1.60$

Factor of safety for overturning:

Overturning moment = $(9390 \text{ lb/ft})(20/3) = 62,600 \text{ ft-lb/ft}$

Moment of weight = $235,200 \text{ ft-lb/ft}$

Therefore: $F = 235,200/62,600 = 3.76$

Maximum pressure exerted by the wall foundation:

$\bar{x} = (235,200 - 62,600)/33,600 = 5.14 \text{ ft}$

Using Eq. 6.12:

$q' = Q(B + 6 e)/B^2$

$e = \text{eccentricity} = 7 - 5.14 = 1.86 \text{ ft}$

$q' = (33,600)[14 + (6)(1.86)]/(14)^2 = 4310 \text{ psf}$

11.38 For the failure wedge:

$W = \frac{1}{2} (20 \text{ ft})(11.1 \text{ ft})(120 \text{ pcf}) = 13,300 \text{ lb/ft}$

$R = 12,000 \text{ lb/ft}$

Based on Equation 10.4 with $c' = 0$ and $u = 0$ and including a horizontal resistance force = R:

$F = [(W \cos \alpha + R \sin \alpha) \tan \phi']/(W \sin \alpha - R \cos \alpha)$

$F = [(13,300 \cos 61° + 12,000 \sin 61°) \tan 32°]/(13,300 \sin 61° - 12,000 \cos 61°)$

$F = 1.82$

11.39 Using Eq. 11.2:

$k_A = \tan^2 (45° - \frac{1}{2} \phi) = \tan^2 [45° - (\frac{1}{2})(33°)] = 0.295$

Using Eq. 11.5:

$k_p = \tan^2 (45° + \frac{1}{2} \phi) = \tan^2 [45° + (\frac{1}{2})(33°)] = 3.39$

From 0 to 5 ft:

$P_{1A} = \frac{1}{2} k_A \gamma_t (5)^2 = \frac{1}{2}(0.295)(120 \text{ pcf})(5 \text{ ft})^2 = 400 \text{ lb/ft}$

From 5 to 50 ft:

$P_{2A} = k_A \gamma_t (5)(45) + \frac{1}{2} k_A \gamma_b (45)^2$

$= (0.295)(120 \text{ pcf})(5 \text{ ft})(45 \text{ ft}) + \frac{1}{2}(0.295)(64 \text{ pcf})(45 \text{ ft})^2$

$= 8000 + 19,100 = 27,100 \text{ lb/ft}$

$P_A = P_{1A} + P_{2A} = 400 + 27{,}100 = 27{,}500$ lb/ft

Using Eq. 11.4 with γ_b substituted for γ_t:

$P_p = \frac{1}{2} k_p \gamma_b D^2 = \frac{1}{2}(3.39)(64 \text{ pcf})(20 \text{ ft})^2 = 43{,}400$ lb/ft

11.40 Obtaining moments at the tieback anchor:

Moment due to passive force

$= (43{,}400) [26 + (2/3)(20)] = 1.71 \times 10^6$ ft-lb/ft

Neglecting P_{1A}, moment due to active force

$= (8000) [1 + (45/2)] + (19{,}100) [1 + (2/3)(45)]$

$= 7.8 \times 10^5$ ft-lb/ft

$F = $ resisting moment/destabilizing moment

$F = (1.71 \times 10^6)/(7.8 \times 10^5) = 2.19$

11.41 Using Eq. 11.14:

$A_p = P_A - P_p/F = 27{,}500 - 43{,}400/2.19 = 7680$ lb/ft

For 10 ft spacing, therefore:

$A_p = (10 \text{ ft})(7680 \text{ lb/ft}) = 76{,}800$ lb $= 76.8$ kips

11.42 Using Eq. 11.4 with γ_b substituted for γ_t:

$P_p = \frac{1}{2} k_p \gamma_b D^2 = \frac{1}{2}(3.39)(64 \text{ pcf})(D)^2 = 108 D^2$

Moment arm $= 26 + 2/3\ D$

$P_{2A} = k_A \gamma_t (5)(25 + D) + \frac{1}{2} k_A \gamma_b (25 + D)^2 = (177)(25 + D) + (9.44)(25 + D)^2$

Moment arm for first term: $1 + \frac{1}{2}(25 + D)$

Moment arm for second term: $1 + (2/3)(25 + D)$

Taking moments at the tie back anchor:

$F = 1.5 = \{(108\ D^2) [26 + (2/3)(D)]\}/$

$\{(177)(25 + D)[1 + \frac{1}{2}(25 + D)] + (9.44)(25 + D)^2[1 + (2/3)(25 + D)]\}$

Solving for $D = 14.6$ ft

11.43 Using Eq. 11.3:

$P_Q = Q H k_A = (200 \text{ psf})(50 \text{ ft})(0.295) = 2950$ lb/ft

Located at $\frac{1}{2}(D + H) = \frac{1}{2}(20 \text{ ft} + 30 \text{ ft}) = 25$ ft

Moment due to active forces

$= 7.8 \times 10^5 + (2950)(25 - 4) = 8.4 \times 10^5$ ft-lb/ft

$F = $ resisting moment/destabilizing moment

$F = (1.71 \times 10^6)/(8.4 \times 10^5) = 2.03$

11.44 Using the following values:

$\delta = 0°$

$\phi = 33°$

$\theta = 0°$

$\beta = 26.6°$

Inserting the above values into Coulomb's equation (static condition) from Fig. 11.4:

$k_A = 0.443$

Destabilizing moment = $(0.443/0.295)(7.8 \times 10^5) = 1.17 \times 10^6$ ft-lb/ft

F = resisting moment/destabilizing moment

$F = (1.71 \times 10^6)/(1.17 \times 10^6) = 1.46$

11.45 Using Eq. 11.10 with $\beta = -18.4°$ and $\phi' = 33°$, therefore:

$k_p = 1.83$

Resisting moment = $(1.83/3.39)(1.71 \times 10^6) = 9.23 \times 10^5$ ft-lb/ft

F = resisting moment/destabilizing moment

$F = (9.23 \times 10^5)/(7.8 \times 10^5) = 1.18$

11.46 Using Eq. 11.2:

Upper sand layer: $k_A = \tan^2(45° - \frac{1}{2}\phi) = \tan^2[45° - (\frac{1}{2})(33°)] = 0.295$

Lower sand layer: $k_A = \tan^2(45° - \frac{1}{2}\phi) = \tan^2[45° - (\frac{1}{2})(30°)] = 0.333$

Using Eq. 11.5:

$k_p = \tan^2(45° + \frac{1}{2}\phi) = \tan^2[45° + (\frac{1}{2})(30°)] = 3.0$

From 0 to 5 ft:

$P_{1A} = \frac{1}{2}k_A\gamma_t(5)^2 = \frac{1}{2}(0.295)(120 \text{ pcf})(5 \text{ ft})^2 = 440$ lb/ft

From 5 to 30 ft:

$P_{2A} = k_A\gamma_t(5)(25) + \frac{1}{2}k_A\gamma_b(25)^2$

$= (0.295)(120 \text{ pcf})(5 \text{ ft})(25 \text{ ft}) + \frac{1}{2}(0.295)(64 \text{ pcf})(25 \text{ ft})^2$

$= 4400 + 5900 = 10,300$ lb/ft

From 30 to 50 ft:

$P_{3A} = k_A[\gamma_t(5) + \gamma_b(25)](20) + \frac{1}{2}k_A\gamma_b(20)^2$

$= (0.333)[(120 \text{ pcf})(5 \text{ ft}) + (64 \text{ pcf})(25 \text{ ft})](20 \text{ ft}) + \frac{1}{2}(0.333)(60 \text{ pcf})(20 \text{ ft})^2$

$= 14,700 + 4000 = 18,700$ lb/ft

$P_A = P_{1A} + P_{2A} + P_{3A} = 440 + 10,300 + 18,700 = 29,400$ lb/ft

Using Eq. 11.4 with γ_b substituted for γ_t:

$P_p = \frac{1}{2}k_p\gamma_b D^2 = \frac{1}{2}(3.0)(60 \text{ pcf})(20 \text{ ft})^2 = 36,000$ lb/ft

11.47 Moment due to passive force = $(36,000)[26 + (2/3)(20)] = 1.42 \times 10^6$ ft-lb/ft

Neglecting P_{1A}, moment due to active force

$= (4400)[1 + \frac{1}{2}(25)] + (5900)[1 + (2/3)(25)] + (14,700)[26 + \frac{1}{2}(20)]$

$+ (4000)[26 + (2/3)(20)]$

$= 8.50 \times 10^5$ ft-lb/ft

F = resisting moment/destabilizing moment

$F = (1.42 \times 10^6)/(8.50 \times 10^5) = 1.67$

11.48 Using Eq. 11.14:

$A_p = P_A - P_p/F = 29,400 - 36,000/1.67 = 7840$ lb/ft

For 10 ft spacing, therefore:

$A_p = (10 \text{ ft})(7840 \text{ lb/ft}) = 78,400 \text{ lb} = 78.4 \text{ kips}$

11.49 Using Eq. 11.3:

For the upper sand layer:

$P_Q = QHk_A = (200 \text{ psf})(30 \text{ ft})(0.295) = 1770 \text{ lb/ft}$

Located at $D + \frac{1}{2}H = 20 \text{ ft} + \frac{1}{2}(30 \text{ ft}) = 35 \text{ ft}$ above wall bottom

For the lower sand layer:

$P_Q = QHk_A = (200 \text{ psf})(20 \text{ ft})(0.333) = 1330 \text{ lb/ft}$

Located at $\frac{1}{2}D = \frac{1}{2}(20 \text{ ft}) = 10 \text{ ft}$ above wall bottom

Moment due to active forces

$= 8.5 \times 10^5 \text{ ft-lb/ft} + (1770 \text{ lb/ft})(11 \text{ ft}) + (1330 \text{ lb/ft})(36 \text{ ft})$

$= 9.17 \times 10^5 \text{ ft-lb/ft}$

F = resisting moment/destabilizing moment

$F = (1.42 \times 10^6)/(9.17 \times 10^5) = 1.55$

11.50 Neglecting the effective cohesion (c') in the analysis:

Moment due to passive force:

For the clay layer, using Eq. 11.5:

$k_p = \tan^2(45° + \frac{1}{2}\phi) = \tan^2[45° + (\frac{1}{2})(25°)] = 2.46$

Using Eq. 11.4 with γ_b substituted for γ_t:

$P_p = \frac{1}{2}k_p\gamma_bD^2 = \frac{1}{2}(2.46)(60 \text{ pcf})(20 \text{ ft})^2 = 29,500 \text{ lb/ft}$

Taking moments about the tie back anchor

$= (29,500 \text{ lb/ft})[26 \text{ ft} + (2/3)(20 \text{ ft})]$

$= 1.16 \times 10^6 \text{ ft-lb/ft}$

Moment due to active force:

For the clay layer, using Eq. 11.2:

$k_A = \tan^2(45° - \frac{1}{2}\phi) = \tan^2[45° - (\frac{1}{2})(25°)] = 0.406$

Active force, from 30 to 50 ft:

$P_{3A} = k_A[\gamma_t(5 \text{ ft}) + \gamma_b(25 \text{ ft})](20 \text{ ft}) + \frac{1}{2}k_A\gamma_b(20 \text{ ft})^2$

$= (0.406)[(120 \text{ pcf})(5 \text{ ft}) + (64 \text{ pcf})(25 \text{ ft})](20 \text{ ft}) + \frac{1}{2}(0.406)(60 \text{ pcf})(20 \text{ ft})^2$

$= 17,860 + 4870 = 22,700 \text{ lb/ft}$

Neglecting P_{1A} and taking moments about the tieback anchor

$= (4400)[1 + \frac{1}{2}(25)] + (5900)[1 + (2/3)(25)] + (17,860)[26 + \frac{1}{2}(20)]$

$+ (4870)[26 + (2/3)(20)]$

$= 9.98 \times 10^5 \text{ ft-lb/ft}$

F = resisting moment/destabilizing moment

$F = (1.16 \times 10^6)/(9.98 \times 10^5) = 1.16$

11.51 Using the following values:

$k_A = 0.333$

$k_p = 3.0$

$k_p/k_A = (3.0)/(0.333) = 9$

$\alpha = 1.0$

$H = 13.8$ ft

$\gamma_b = 57$ pcf

From Fig. 11.16, the depth ratio = 1.5 and the moment ratio = 1.1. Therefore, using $H = 13.8$ ft, the value of D for a factor of safety of 1 is equal to: $(1.5)(13.8) = 20.7$ ft. Using a 30% increase in embedment depth, the required embedment depth $(D) = (20.7 \text{ ft})(1.3) = 27$ ft.

For a moment ratio = 1.1

$M_{max}/(\gamma_b k_A H^3) = 1.1$

$M_{max} = (1.1)(57 \text{ pcf})(0.333)(13.8 \text{ ft})^3 = 55{,}000$ ft-lb/ft

11.52 Using the following values:

$k_A = 0.20$

$k_p = 8.65$

$k_p/k_A = (8.65)/(0.20) = 43$

$\alpha = 0$

$H = 13.8$ ft

$\gamma_b = 57$ pcf

From Fig. 11.16, the depth ratio = 0.43 and the moment ratio = 0.20. Therefore, using $H = 13.8$ ft, the value of D for a factor of safety of 1 is equal to: $(0.43)(13.8) = 6$ ft. Using a 30% increase in embedment depth, the required embedment depth $(D) = (6 \text{ ft})(1.3) = 8$ ft.

For a moment ratio of 0.20

$M_{max}/(\gamma_b k_A H^3) = 0.20$

$M_{max} = (0.20)(57 \text{ pcf})(0.2)(13.8 \text{ ft})^3 = 6000$ ft-lb/ft

11.53 Using Eq. 11.2:

$k_A = \tan^2 (45° - 1/2\, \phi) = \tan^2 [45° - (1/2)(32°)] = 0.307$

From Fig. 11.18 for sand:

$\sigma_h = 0.65 k_A \gamma_t H = (0.65)(0.307)(120 \text{ pcf})(20 \text{ ft}) = 480$ psf

Resultant force $= \sigma_h H = (480 \text{ psf})(20 \text{ ft}) = 9600$ lb/ft

11.54 From Fig. 11.18 for clay:

$N_o = \gamma_t H/c = (120 \text{ pcf})(20 \text{ ft})/300 \text{ psf} = 8$

Therefore use case (b) in Fig. 11.18

$k_A = 1 - m(4\, c)/(\gamma_t H)$ use $m = 1$, therefore:

$k_A = 1 - [(4)(300 \text{ psf})]/[(120 \text{ pcf})(20 \text{ ft})] = 0.5$

From Fig. 11.18:

$\sigma_h = k_A \gamma_t H = (0.5)(120 \text{ pcf})(20 \text{ ft}) = 1200$ psf

Using the earth pressure distribution shown in Fig. 11.18, i.e., case (b):

Resultant force $= 1/2(1200 \text{ psf})(0.25)(20 \text{ ft}) + (1200 \text{ psf})(0.75)(20 \text{ ft})$

$= 21{,}000$ lb/ft

11.55 From Fig. 11.18 for clay:

$N_o = \gamma_t H / c = (120 \text{ pcf})(20 \text{ ft})/1200 = 2$

Therefore use case (c) in Fig. 11.18:

$\sigma_{h2} = 0.4\,\gamma_t H = (0.4)(120 \text{ pcf})(20 \text{ ft}) = 960 \text{ psf}$

Using the earth pressure distribution shown in Fig. 11.18, i.e., case (c):

Resultant force $= \frac{1}{2}(960 \text{ psf})(0.5)(20 \text{ ft}) + (960 \text{ psf})(0.5)(20 \text{ ft})$

$= 14,400 \text{ lb/ft}$

11.56 Using Eq. 11.2:

$k_A = \tan^2 (45° - \frac{1}{2}\phi) = \tan^2 [45° - (\frac{1}{2})(32°)] = 0.307$

Using Eq. 11.1:

$P_A = \frac{1}{2} k_A \gamma_t H^2 = \frac{1}{2}(0.307)(120 \text{ pcf})(20 \text{ ft})^2 = 7370 \text{ lb/ft}$

For 7 ft spacing, the resultant force

$= (7370 \text{ lb/ft})(7 \text{ ft})$

$= 51,600 \text{ lb on each steel I-beam}$

11.57 Using Eq. 11.2:

$k_A = \tan^2 (45° - \frac{1}{2}\phi) = \tan^2 [45° - (\frac{1}{2})(32°)] = 0.307$

From Fig. 11.18 for sand:

$\sigma_h = 0.65\,k_A \gamma_t H = (0.65)(0.307)(120 \text{ pcf})(20 \text{ ft}) = 480 \text{ psf}$

Resultant force $= \sigma_h H = (480 \text{ psf})(20 \text{ ft}) = 9600 \text{ lb/ft}$

For 7-ft spacing $= (9600 \text{ lb/ft})(7 \text{ ft}) = 67,200 \text{ lb}$

11.58 Using Eq. 11.15, with $\tau_f / F = 10 \text{ psi}$

$L_b = PF/\pi d\tau_f = [(20 \text{ kips})(1000)]/[(\pi)(6 \text{ in.})(10 \text{ psi})]$

$L_b = 106 \text{ in.} = 8.8 \text{ ft}$

Chapter 13

13.1

Depth (m)	Cyclic stress ratio (CSR)					N value corrections						FS = CRR/CSR
	σ_v (kPa)	σ_v' (kPa)	σ_v/σ_v'	r_d	CSR	N value	C_r	N_{60}	C_N	$(N_1)_{60}$	CRR	
1.5	27.5	27.5	1.00	0.98	0.10	8	0.75	6.0	1.91	11	0.12	1.18
2.5	47.0	37.2	1.26	0.97	0.13	5	0.75	3.8	1.64	6.2	0.07	0.55
3.5	66.5	46.9	1.42	0.96	0.14	4	0.75	3.0	1.46	4.4	0.05	0.35
4.5	86.0	56.6	1.52	0.95	0.15	5	0.85	4.3	1.33	5.7	0.06	0.40
5.5	105	66.3	1.58	0.93	0.15	9	0.85	7.7	1.23	9.5	0.11	0.72
6.5	125	76.0	1.64	0.92	0.16	10	0.95	9.5	1.15	11	0.12	0.76
7.5	144	85.7	1.68	0.91	0.16	12	0.95	11	1.08	12	0.13	0.82
8.5	164	95.4	1.72	0.90	0.16	12	0.95	11	1.02	11	0.12	0.75
9.5	183	105	1.74	0.89	0.16	15	0.95	14	0.98	14	0.16	1.00
10.5	203	115	1.77	0.87	0.16	11	1.00	11	0.93	10	0.11	0.69
11.5	222	124	1.79	0.86	0.16	23	1.00	23	0.90	21	0.23	1.44

(Continued)

Depth (m)	Cyclic stress ratio (CSR)					N value corrections					CRR	FS = CRR/CSR
	σ_v (kPa)	σ'_v (kPa)	σ_v/σ'_v	r_d	CSR	N value	C_r	N_{60}	C_N	$(N_1)_{60}$		
12.5	242	134	1.81	0.85	0.16	11	1.00	11	0.86	9.5	0.11	0.69
13.5	261	144	1.81	0.84	0.16	10	1.00	10	0.83	8.3	0.09	0.57
14.5	281	154	1.82	0.83	0.16	10	1.00	10	0.81	8.1	0.09	0.57
15.5	300	163	1.84	0.81	0.16	25	1.00	25	0.78	20	0.23	1.48
16.5	320	173	1.85	0.80	0.15	27	1.00	27	0.76	21	0.24	1.56
17.5	339	183	1.85	0.79	0.15	4	1.00	4	0.74	3.0	0.03	0.20
18.5	359	192	1.87	0.78	0.15	5	1.00	5	0.72	3.6	0.04	0.26
19.5	378	202	1.87	0.77	0.15	3	1.00	3	0.70	2.1	0.02	0.13
20.5	398	212	1.88	0.75	0.15	38	1.00	38	0.69	26	0.30	2.05

Notes: Cyclic stress ratio: $a_{max} = 0.16g$, r_d from Eq. 13.10
N value corrections: $E_m = 60\%$, $C_b = 1.0$, C_N from Eq. 2.5
CRR from Fig. 13.17

13.2

Depth below ground surface (m)	Cyclic resistance ratio (CRR) from laboratory tests	Cyclic stress ratio (CSR) from Problem 13.1	FS = CRR/CSR
2.3	0.18	0.13	1.39
5.5	0.16	0.15	1.07
7.0	0.15	0.16	0.94
8.5	0.17	0.16	1.06
9.5	0.17	0.16	1.06
11.5	0.16	0.16	1.00
13.0	0.16	0.16	1.00
14.5	0.24	0.16	1.50

13.3 The standard penetration test data indicate that there are three zones of liquefaction from about 2 to 11 m, 12 to 15 m, and 17 to 20 m below ground surface. The laboratory cyclic strength tests indicate that there are two zones of liquefaction from about 6 to 8 m and 10 to 14 m below ground surface.

13.4

Depth (m)	Cyclic stress ratio (CSR)					N value corrections					CRR	FS = CRR/CSR
	σ_v (kPa)	σ'_v (kPa)	σ_v/σ'_v	r_d	CSR	N value	C_r	N_{60}	C_N	$(N_1)_{60}$		
1.2	22.9	15.1	1.52	0.99	0.16	4	0.75	3.0	2.57	7.7	0.09	0.56
2.2	42.4	24.8	1.71	0.97	0.17	6	0.75	4.5	2.01	9.0	0.10	0.59
3.2	61.9	34.5	1.79	0.96	0.18	5	0.75	3.8	1.70	6.5	0.07	0.39
4.2	81.5	44.2	1.84	0.95	0.18	8	0.85	6.8	1.50	10	0.11	0.61
5.2	101	53.9	1.87	0.94	0.18	7	0.85	6.0	1.36	8.2	0.09	0.50
6.2	120	63.6	1.89	0.93	0.18	13	0.95	12	1.25	15	0.16	0.89
7.2	140	73.3	1.91	0.91	0.18	36	0.95	34	1.17	40	>0.5	>2.8
8.2	159	83.0	1.92	0.90	0.18	24	0.95	23	1.09	25	0.29	1.61
9.2	179	92.7	1.93	0.89	0.18	35	0.95	33	1.04	34	>0.5	>2.8
10.2	199	102	1.95	0.88	0.18	30	1.00	30	0.99	30	0.50	2.78
11.2	218	112	1.95	0.87	0.18	28	1.00	28	0.94	26	0.30	1.67
12.2	238	122	1.95	0.85	0.17	32	1.00	32	0.91	29	0.45	2.65
13.2	257	131	1.96	0.84	0.17	16	1.00	16	0.87	14	0.16	0.94

(Continued)

Depth (m)	Cyclic stress ratio (CSR)					N value corrections					CRR	FS = CRR/CSR
	σ_v (kPa)	σ'_v (kPa)	σ_v/σ'_v	r_d	CSR	N value	C_r	N_{60}	C_N	$(N_1)_{60}$		
14.2	277	141	1.96	0.83	0.17	28	1.00	28	0.84	24	0.28	1.65
15.2	296	151	1.96	0.82	0.17	27	1.00	27	0.81	22	0.25	1.47
16.2	316	161	1.96	0.81	0.17	23	1.00	23	0.79	18	0.20	1.18
17.2	335	170	1.97	0.79	0.16	38	1.00	38	0.77	29	0.45	2.81
18.2	355	180	1.97	0.78	0.16	32	1.00	32	0.75	24	0.28	1.75
19.2	374	190	1.97	0.77	0.16	47	1.00	47	0.73	34	>0.5	>3.1

Notes: Cyclic stress ratio: $a_{max} = 0.16g$, r_d from Eq. 13.10
N value corrections: $E_m = 60\%$, $C_b = 1.0$, C_N from Eq. 2.5
CRR from Fig. 13.17

13.5

Depth below ground surface (m)	Cyclic resistance ratio (CRR) from laboratory tests	Cyclic stress ratio (CSR) from Problem 13.4	FS = CRR/CSR
2.0	0.20	0.17	1.18
3.5	0.20	0.18	1.11
5.0	0.21	0.18	1.17
8.0	0.28	0.18	1.56
11.0	0.29	0.18	1.61

13.6 The standard penetration test data indicate that there are two zones of liquefaction from about 1.2 to 6.7 m and 12.7 to 13.7 m below ground surface. The laboratory cyclic strength tests indicate that the soil has a factor of safety against liquefaction in excess of 1.0.

13.7

	Before improvement											
Depth (m)	Cyclic stress ratio (CSR)					N value corrections					CRR (see Notes)	FS = CRR/CSR
	σ_v (kPa)	σ'_v (kPa)	σ_v/σ'_v	r_d	CSR	N value	C_r	N_{60}	C_N	$(N_1)_{60}$		
0.5	9.45	9.45	1.00	0.99	0.26	16	0.75	12	3.24	39	>0.5	>2
1.5	29.1	19.3	1.51	0.98	0.38	11	0.75	8.3	2.28	19	0.25	0.66
2.5	48.7	29.1	1.67	0.97	0.42	9	0.75	6.8	1.85	13	0.18	0.43
3.5	68.3	38.9	1.76	0.96	0.44	12	0.75	9.0	1.60	14	0.19	0.42
4.5	87.9	48.7	1.80	0.95	0.44	19	0.85	16	1.43	23	0.33	0.76
5.5	108	58.5	1.85	0.93	0.45	11	0.85	9.4	1.31	12	0.16	0.36
6.5	127	68.3	1.86	0.92	0.44	9	0.95	8.6	1.21	10	0.14	0.32
7.5	147	78.1	1.88	0.91	0.44	19	0.95	18	1.13	20	0.26	0.59
8.5	166	87.9	1.89	0.90	0.44	10	0.95	9.5	1.07	10	0.14	0.32
9.5	186	97.7	1.90	0.89	0.44	10	0.95	9.5	1.01	10	0.14	0.32
10.5	206	107	1.93	0.88	0.44	11	1.00	11	0.97	11	0.15	0.34
11.5	225	117	1.93	0.86	0.43	4	1.00	4.0	0.92	3.7	0.07	0.16
12.5	245	127	1.93	0.85	0.43	10	1.00	10	0.89	8.9	0.13	0.30
13.5	264	137	1.93	0.84	0.42	11	1.00	11	0.85	9.4	0.13	0.31
14.5	284	147	1.93	0.83	0.42	12	1.00	12	0.82	10	0.14	0.33
15.5	304	156	1.95	0.81	0.41	13	1.00	13	0.80	10	0.14	0.34

Notes: Cyclic stress ratio: $a_{max} = 0.40g$, r_d from Eq. 13.10
N value corrections: $E_m = 60\%$, $C_b = 1.0$, C_N from Eq. 2.5
CRR from Fig. 13.17 (silty sand with 15% fines). The values of CRR from Fig. 13.17 were multiplied by a magnitude scaling factor = 0.89 (see Table 13.5)

Depth (m)	σ_v (kPa)	σ'_v (kPa)	σ_v/σ'_v	r_d	CSR	N value	C_r	N_{60}	C_N	$(N_1)_{60}$	CRR (see Notes)	FS = CRR/CSR
0.5	9.45	9.45	1.00	0.99	0.26	19	0.75	14	3.24	46	>0.5	>2
1.5	29.1	19.3	1.51	0.98	0.38	21	0.75	16	2.28	36	>0.5	>1.3
2.5	48.7	29.1	1.67	0.97	0.42	19	0.75	14	1.85	26	>0.5	>1.2
3.5	68.3	38.9	1.76	0.96	0.44	21	0.75	16	1.60	25	>0.5	>1.1
4.5	87.9	48.7	1.80	0.95	0.44	25	0.85	21	1.43	30	>0.5	>1.1
5.5	108	58.5	1.85	0.93	0.45	31	0.85	26	1.31	35	>0.5	>1.1
6.5	127	68.3	1.86	0.92	0.44	33	0.95	31	1.21	38	>0.5	>1.1
7.5	147	78.1	1.88	0.91	0.44	31	0.95	29	1.13	33	>0.5	>1.1
8.5	166	87.9	1.89	0.90	0.44	37	0.95	35	1.07	38	>0.5	>1.1
9.5	186	97.7	1.90	0.89	0.44	41	0.95	39	1.01	39	>0.5	>1.1
10.5	206	107	1.93	0.88	0.44	35	1.00	35	0.97	34	>0.5	>1.1
11.5	225	117	1.93	0.86	0.43	36	1.00	36	0.92	33	>0.5	>1.2
12.5	245	127	1.93	0.85	0.43	39	1.00	39	0.89	35	>0.5	>1.2
13.5	264	137	1.93	0.84	0.42	45	1.00	45	0.85	38	>0.5	>1.2
14.5	284	147	1.93	0.83	0.42	40	1.00	40	0.82	33	>0.5	>1.2

Notes: Cyclic stress ratio: $a_{max} = 0.40g$, r_d from Eq. 13.10
N value corrections: $E_m = 60\%$, $C_b = 1.0$, C_N from Eq. 2.5
CRR from Fig. 13.17 (silty sand with 15% fines). The values of CRR from Fig. 13.17 were multiplied by a magnitude scaling factor = 0.89 (see Table 13.5)

13.8 The area of the wedge is first determined from simple geometry and is equal to:

$\frac{1}{2}(10\text{ m})(30\text{ m}) - \frac{1}{2}(10\text{ m})(20\text{ m}) = 50\text{ m}^2$

For a unit length of the slope, the total weight (W) of the wedge equals the area times total unit weight, or:

$(50\text{ m}^2)(18\text{ kN/m}^3) = 900\text{ kN}$ per meter of slope length.

Static case:

Using Eq. 10.3 and the following values:

$c = 15\text{ kPa} = 15\text{ kN/m}^2$

$\phi = 0$

Length of slip surface $(L) = 31.6\text{ m}$

Slope inclination $(\alpha) = 18.4°$

$$F = \frac{cL + (W\cos\alpha)\tan\phi}{W\sin\alpha} = \frac{(15\text{ kN/m}^2)(31.6\text{ m})}{(900\text{ kN/m})\sin 18.4°} = 1.67$$

Earthquake case:

Using the equation derived in Problem 10.4 where $F_h = bW$:

$$F = \frac{cL + (W\cos\alpha - F_h\sin\alpha)\tan\phi}{W\sin\alpha + F_h\cos\alpha}$$

Since $\phi = 0$, therefore:

$$F = \frac{(15\text{ kN/m}^2)(31.6\text{ m})}{(900)\sin 18.4° + (0.3)(900)\cos 18.4°} = 0.88$$

13.9   From Problem 13.8, the total weight (W) of the wedge = 900 kN per meter of slope length.

Static case:

Using Eq. 10.4 and the following values: $c' = 4$ kPa = 4 kN/m^2

$\phi' = 29°$

Length of slip surface (L) = 31.6 m

Slope inclination (α) = 18.4°

Average pore water pressure acting on the slip surface (u) = 5 kPa = 5 kN/m^2

$$F = \frac{c'L + (W \cos\alpha - uL)\tan\phi'}{W \sin\alpha}$$

$$F = \frac{(4\,\text{kN/m}^2)(31.6\,\text{m}) + [(900\cos 18.4°) - (5)(31.6\,\text{m})]\tan 29°}{(900\,\text{kN/m})\sin 18.4°} = 1.80$$

Earthquake case:

Using the equation derived in Problem 10.4 where $F_h = bW$:

$$F = \frac{c'L + (W \cos\alpha - F_h \sin\alpha - uL)\tan\phi'}{W \sin\alpha + F_h \cos\alpha}$$

$$F = \frac{(4\,\text{kN/m}^2)(31.6\,\text{m}) + [(900\cos 18.4° - (0.2)(900)\sin 18.4° - (5)(31.6)]\tan 29°}{(900)\sin 18.4° + (0.2)(900)\cos 18.4°}$$

$$F = 1.06$$

13.10 $W = (20\text{ ft})(10\text{ ft})(140\text{ pcf}) = 28{,}000$ lb/ft

$F_h = (0.50)(28{,}000) = 14{,}000$ lb/ft

Using the equation derived in Problem 10.4 where $F_h = bW$:

$$F = \frac{c'L + (W \cos\alpha - F_h \sin\alpha - uL)\tan\phi'}{W \sin\alpha + F_h \cos\alpha}$$

Since $c' = 0$, $u = 0$, and $\alpha = 0$, the above equation reduces to:

$$F = \frac{W \tan\phi'}{F_h} = \frac{28{,}000\tan 40°}{14{,}000} = 1.68$$

13.11   Using the equation derived in Problem 10.4 where $F_h = bW$:

$$F = \frac{cL + (W \cos\alpha - F_h \sin\alpha)\tan\phi}{W \sin\alpha + F_h \cos\alpha}$$

Since $\phi = 0$ and solving for k_h when $F = 1.0$, therefore:

$$F = 1.0 = \frac{(15\,\text{kN/m}^2)(31.6\,\text{m})}{(900)\sin 18.4° + k_h(900)\cos 18.4°}$$

$k_h = a_y/g = 0.22$

From Problem 13.8, $a_{max} = 0.30g$

Therefore: $a_y/a_{max} = 0.22g/0.30g = 0.733$

Using Eq. 13.13 with $a_y/a_{max} = 0.733$

$\log_{10} d = 0.90 + \log_{10} [(1 - a_y/a_{max})^{2.53} (a_y/a_{max})^{-1.09}]$

$\log_{10} d = 0.90 + \log_{10} [(1 - 0.733)^{2.53} (0.733)^{-1.09}]$

$\log_{10} d = 0.90 - 1.30 = -0.40$

Solving for $d = 0.4$ cm

13.12   Since pseudostatic $F > 1.0$, $d = 0$.

13.13   The slope height = 9 m + 1.5 m = 10.5 m. Consider possible liquefaction of the soil from ground surface to a depth of 10.5 m. Per Problem 13.1, the soil will liquefy from a depth of about 2 to 11 m. Thus most of the 3:1 sloping river bank is expected to liquefy during the design earthquake. As indicated in Sec. 13.5.2, if most of the sloping mass is susceptible to failure, a mass liquefaction flow slide can be expected during the earthquake.

Chapter 14

14.1   Solution using Fig. 14.7:

First determine the factor of safety against liquefaction (FS) using Fig. 13.17 to determine the cyclic resistance ratio (clean sand, therefore use the less than 5% fines curve), or:

Layer depth	CSR	CRR (Fig. 13.17)	FS = CRR/CSR
2–3 m	0.18	0.11	0.61
3–5 m	0.20	0.06	0.30
5–7 m	0.22	0.08	0.36

Assume that the Japanese N_1 value is approximately equal to the $(N_1)_{60}$ value. Using Fig. 14.7 with the above factors of safety against liquefaction and the given $(N_1)_{60}$ values:

For the 2 to 3 m layer: $\varepsilon_v = 3.5\%$

Settlement = (0.035)(1.0 m) = 0.035 m

For the 3 to 5 m layer: $\varepsilon_v = 4.8\%$

Settlement = (0.048)(2.0 m) = 0.096 m

For the 5 to 7 m layer: $\varepsilon_v = 4.3\%$

Settlement = (0.043)(2.0 m) = 0.086 m

Total settlement = 0.035 + 0.096 + 0.086 = 0.22 m = 22 cm

Solution using Fig. 14.8:

Enter the curve with the given $(N_1)_{60}$ and CSR values:

For the 2 to 3 m layer: $\varepsilon_v = 2.6\%$

Settlement = (0.026)(1.0 m) = 0.026 m

For the 3 to 5 m layer: $\varepsilon_v = 4.2\%$

Settlement = (0.042)(2.0 m) = 0.084 m

For the 5 to 7 m layer: $\varepsilon_v = 3.2\%$

Settlement = (0.032)(2.0 m) = 0.064 m

Total settlement = 0.026 + 0.084 + 0.064 = 0.174 m = 17 cm

14.2

	Figure 14.7					Figure 14.8				
Depth (m)	$(N_1)_{60}$	FS	ε_v (%)	H (m)	Settlement (cm)	$(N_1)_{60}$	CSR	ε_v (%)	H (m)	Settlement (cm)
1.5	11	1.18	0.6	0.5	0.3	11	0.10	0.1	0.5	0.05
2.5	6.2	0.55	4.5	1.0	4.5	6.2	0.13	3.7	1.0	3.7
3.5	4.4	0.35	5.1	1.0	5.1	4.4	0.14	4.5	1.0	4.5
4.5	5.7	0.40	4.6	1.0	4.6	5.7	0.15	4.0	1.0	4.0
5.5	9.5	0.72	3.6	1.0	3.6	9.5	0.15	2.7	1.0	2.7
6.5	11	0.76	3.3	1.0	3.3	11	0.16	2.4	1.0	2.4
7.5	12	0.82	3.1	1.0	3.1	12	0.16	2.2	1.0	2.2
8.5	11	0.75	3.3	1.0	3.3	11	0.16	2.4	1.0	2.4
9.5	14	1.00	1.1	1.0	1.1	14	0.16	1.2	1.0	1.2
10.5	10	0.69	3.5	1.0	3.5	10	0.16	2.6	1.0	2.6
11.5	21	1.44	0.2	1.0	0.2	21	0.16	0	1.0	0
12.5	9.5	0.69	3.6	1.0	3.6	9.5	0.16	2.7	1.0	2.7
13.5	8.3	0.57	3.9	1.0	3.9	8.3	0.16	2.9	1.0	2.9
14.5	8.1	0.57	4.0	1.0	4.0	8.1	0.16	2.9	1.0	2.9
15.5	20	1.48	0.2	1.0	0.2	20	0.16	0	1.0	0
16.5	21	1.56	0.2	1.0	0.2	21	0.15	0	1.0	0
17.5	3.0	0.20	5.5	1.0	5.5	3.0	0.15	6.0	1.0	6.0
18.5	3.6	0.26	5.3	1.0	5.3	3.6	0.15	5.0	1.0	5.0
19.5	2.1	0.13	6.0	1.0	6.0	2.1	0.15	8.0	1.0	8.0
20.5	26	2.05	0	1.0	0	26	0.15	0	1.0	0
					Total = 61 cm					Total = 53 cm

Notes: $(N_1)_{60}$, FS, and CSR obtained from Problem 13.1
For Fig. 14.7, assume Japanese $N_1 = (N_1)_{60}$
H = thickness of soil layer

14.3

	Figure 14.7					Figure 14.8				
Depth (m)	$(N_1)_{60}$	FS	ε_v (%)	H (m)	Settlement (cm)	$(N_1)_{60}$	CSR	ε_v (%)	H (m)	Settlement (cm)
1.2	7.7	0.56	4.1	0.5	2.0	7.7	0.16	3.2	0.5	1.6
2.2	9.0	0.59	3.7	1.0	3.7	9.0	0.17	2.8	1.0	2.8
3.2	6.5	0.39	4.4	1.0	4.4	6.5	0.18	3.6	1.0	3.6
4.2	10	0.61	3.5	1.0	3.5	10	0.18	2.6	1.0	2.6
5.2	8.2	0.50	4.0	1.0	4.0	8.2	0.18	3.0	1.0	3.0
6.2	15	0.89	1.8	1.0	1.8	15	0.18	1.7	1.0	1.7
7.2	40	>2.8	0	1.0	0	40	0.18	0	1.0	0
8.2	25	1.61	0.2	1.0	0.2	25	0.18	0	1.0	0
9.2	34	>2.8	0	1.0	0	34	0.18	0	1.0	0
10.2	30	2.78	0	1.0	0	30	0.18	0	1.0	0
11.2	26	1.67	0.1	1.0	0.1	26	0.18	0	1.0	0
12.2	29	2.65	0	1.0	0	29	0.17	0	1.0	0
13.2	14	0.94	1.6	1.0	1.6	14	0.17	1.6	1.0	1.6
14.2	24	1.65	0.1	1.0	0.1	24	0.17	0	1.0	0
15.2	22	1.47	0.2	1.0	0.2	22	0.17	0	1.0	0
16.2	18	1.18	0.5	1.0	0.5	18	0.17	0.1	1.0	0.1
17.2	29	2.81	0	1.0	0	29	0.16	0	1.0	0
18.2	24	1.75	0.1	1.0	0.1	24	0.16	0	1.0	0
19.2	34	>3.1	0	1.3	0	34	0.16	0	1.3	0
					Total = 22 cm					Total = 17 cm

Notes: $(N_1)_{60}$, FS, and CSR obtained from Problem 13.4
For Fig. 14.7, assume Japanese $N_1 = (N_1)_{60}$
H = thickness of soil layer

14.4 Assume with soil improvement, there will be no settlement in the upper 15 m of the soil deposit. The only remaining liquefiable soil layer is at a depth of 17 to 20 m. The stress increase due to the building load of 50 kPa at a depth of 17 m can be estimated from the 2:1 approximation, or:

$$\Delta\sigma_v = qBL/[(B + z)(L + z)]$$

where z = depth from the bottom of the footing to the top of the liquefied soil layer

$z = 17$ m $- 1$ m $= 16$ m

$L = 20$ m

$B = 10$ m

$q = 50$ kPa

$\Delta\sigma_v = (50 \text{ kPa})(20 \text{ m})(10 \text{ m})/[(10 \text{ m} + 16 \text{ m}) (20 \text{ m} + 16 \text{ m})] = 10.7$ kPa

Or in terms of a percent increase in σ'_{vo}:

Percent increase in $\sigma'_{vo} = 10.7/178 = 6\%$

This is a very low percent increase in vertical stress due to the foundation load. Thus the shear stress caused by the building load should not induce any significant additional settlement of the liquefied soil. Using the data from Problem 14.2 at a depth of 15 to 20 m:

	Figure 14.7					Figure 14.8				
Depth (m)	$(N_1)_{60}$	FS	ε_v (%)	H (m)	Settlement (cm)	$(N_1)_{60}$	CSR	ε_v (%)	H (m)	Settlement (cm)
15.5	20	1.48	0.2	1.0	0.2	20	0.16	0	1.0	0
16.5	21	1.56	0.2	1.0	0.2	21	0.15	0	1.0	0
17.5	3.0	0.20	5.5	1.0	5.5	3.0	0.15	2.9	1.0	6.0
18.5	3.6	0.26	5.3	1.0	5.3	3.6	0.15	5.0	1.0	5.0
19.5	2.1	0.13	6.0	1.0	6.0	2.1	0.15	8.0	1.0	8.0
20.5	26	2.05	0	1.0	0	26	0.15	0	1.0	0
					Total = 17 cm					Total = 19 cm

Notes: All data obtained from Problem 14.2

14.5 Since the tank is in the middle of a liquefied soil layer, it is expected that the empty tank will not settle, but rather float to the ground surface.

14.6 The thickness of the liquefiable sand layer (H_2) is equal to 4 m. Entering Fig. 14.10 with $H_2 = 4$ m and intersecting the $a_{max} = 0.4g$ curve, the minimum thickness of the surface layer (H_1) needed to prevent surface damage is 8.3 m. Since the surface layer of unliquefiable soil is only 6 m thick, then there will be liquefaction induced ground damage.

14.7 Problem 13.1 was solved based on a peak ground acceleration $(a_{max}) = 0.16g$. For a peak ground acceleration $a_{max} = 0.2g$, the factor of safety against liquefaction at a depth of 1.5 m is as follows:

CSR = (0.20/0.16)(0.10) = 0.13, therefore:

FS = CRR/CSR = 0.12/0.13 = 0.92

For a peak ground acceleration = $0.2g$, the zone of liquefaction will extend from a depth of 1.5 to 11 m. The thickness of the liquefiable sand layer (H_2) is equal to 9.5 m. Entering Fig. 14.10 with $H_2 = 9.5$ m and extending the $a_{max} = 0.2g$ curve, the minimum thickness of the surface layer (H_1) needed to prevent surface damage is 3 m. Since the surface layer of unliquefiable soil is only 1.5 m thick, there will be liquefaction induced ground damage.

14.8 Zone of liquefaction:

The following table shows the liquefaction calculations. The data indicate that the zone of liquefaction extends from a depth of 2.3 to 18 m.

Fill layer:

Since the zone of liquefaction extends from a depth of 2.3 to 18 m, the thickness of the liquefiable sand layer (H_2) is equal to 15.7 m. Entering Fig. 14.10 with $H_2 = 15.7$ m and extending the $a_{max} = 0.2g$ curve, the minimum thickness of the surface layer (H_1) needed to prevent surface damage is 3 m. Since the surface layer of unliquefiable soil is 2.3 m thick, there will be liquefaction induced ground damage. The required fill layer to be added at ground surface is equal to 0.7 m (i.e., 3 m – 2.3 m = 0.7 m).

Settlement:

The following table shows the calculations for the liquefaction-induced settlement of the ground surface. The calculated settlement is 54 to 66 mm using Figs. 14.7 and 14.8. The settlement calculations should include the 0.7 m fill layer, but its effect is negligible. The settlement calculations should also include the weight of the oil in the tank, which could cause the oil tank to punch through or deform downward the upper clay layer, resulting in substantial additional settlement. As indicated in the next section, the factor of safety for a bearing capacity failure is only 1.06, and thus the expected liquefaction-induced settlement will be significantly greater than 66 mm.

Bearing capacity:

$$P = \pi(D^2/4)(H)(\gamma_{oil})$$

Where D = diameter of the tank, H = height of oil in the tank, and γ_{oil} = unit weight of oil. Therefore:

$$P = \pi[(20 \text{ m})^2/4](3 \text{ m})(9.4 \text{ kN/m}^3) = 8860 \text{ kN}$$

$$\tau_f = 50 \text{ kPa} = 50 \text{ kN/m}^2$$

For the circular tank:

$$F = R/P = \pi DT\tau_f/P = [\pi(20 \text{ m})(3 \text{ m})(50 \text{ kN/m}^2)]/8860 \text{ kN} = 1.06$$

Liquefaction analysis

Depth (m)	σ_v (kPa)	σ'_v (kPa)	σ_v/σ'_v	r_d	CSR	N value	C_r	N_{60}	C_N	$(N_1)_{60}$	CRR	FS = CRR/CSR
4	77.3	47.4	1.631	0.952	0.202	9	0.85	7.7	1.45	11	0.12	0.59
8	155	86.7	1.788	0.904	0.210	5	0.95	4.8	1.07	5.2	0.06	0.29
10.4	202	110	1.836	0.875	0.209	8	1.00	8	0.95	7.6	0.08	0.38
12.1	235	126	1.865	0.855	0.207	4	1.00	4	0.89	3.6	0.10	0.48
14.8	287	152	1.888	0.822	0.202	2	1.00	2	0.81	1.6	0.07	0.35
17.4	338	177	1.910	0.791	0.196	9	1.00	9	0.75	6.8	0.15	0.77

Liquefaction-induced settlement analysis

Depth (m)	$(N_1)_{60}$	FS	ε_v (%)	H (m)	Settlement (cm)	$(N_1)_{60}$	CSR	ε_v (%)	H (m)	Settlement (cm)
	Figure 14.7					Figure 14.8				
4	11	0.59	3.3	3.7	12.2	11	0.202	2.5	3.7	9.3
8	5.2	0.29	4.8	3.2	15.4	5.2	0.210	4.2	3.2	13.4
10.4	7.6	0.38	4.1	2.0	8.2	7.6	0.209	3.2	2.0	6.4
12.1	6.4	0.48	4.4	2.3	10.1	6.4	0.207	3.6	2.3	8.3

(Continued)

			Liquefaction-induced settlement analysis						
		Figure 14.7					Figure 14.8		
Depth (m)	$(N_1)_{60}$	FS	ε_v (%)	H (m)	Settlement (cm)	$(N_1)_{60}$	CSR	ε_v (%)	H (m)	Settlement (cm)
14.8	4.4	0.35	5.1	2.6	13.3	4.4	0.202	4.5	2.6	11.7
17.4	9.6	0.77	3.6	1.9	6.8	9.6	0.196	2.6	1.9	4.9
					Total = 66 cm					Total = 54 cm

Notes: Cyclic stress ratio: $a_{max} = 0.20g$, r_d from Eq. 13.10
N value corrections: $E_m = 60\%$, $C_b = 1.0$, C_N from Eq. 2.5
CRR from Fig. 13.17
Settlement analysis: Assume Japanese $N_1 = (N_1)_{60}$ with $N_{corr} = 2.8$ for the silty sand layer. H = thickness of soil layer in meters.

14.9 The retaining wall in this problem is identical to Problem 11.1, except that the friction between the bottom of the footing and underlying soil is larger. Using Eq. 14.2 with $a_{max}/g = 0.20$:

$$P_E = \frac{1}{2}(k_A)^{1/2}\ (a_{max}/g)H^2\gamma_t$$

$$P_E = \frac{1}{2}(0.307)^{1/2}\ (0.20)(4\text{ m})^2\ (20\text{ kN/m}^3) = 17.7\text{ kN/m}$$

P_E is located at a distance of 2/3 H above base of wall, or: 2/3 H = 2/3 (4 m) = 2.67 m

Factor of safety for sliding:

Using Eq. 14.6 to determine the factor of safety for sliding:

$$F = (N\tan\delta + P_p)/(P_H + P_E)$$

Obtaining $P_H = P_A$ from Problem 11.1 and N from Problem 11.4:

$$F = (68.2\tan 38° + 4.07\text{ kN/m})/(49.2\text{ kN/m} + 17.7\text{ kN/m})$$

$$F = 0.86$$

Factor of safety for overturning:

Using Eq. 14.7 with $P_v = 0$:

$$F = (Wa)/(1/3\ P_H H + 2/3\ HP_E)$$

$$F = [(35.3)(3/2) + (32.9)(2.8)]/[(1/3)(49.2)(4) + (2/3)(4)(17.7)]$$

$$F = 1.29$$

14.10 Using Eq 14.3 with $a_{max}/g = 0.20$:

$$P_E = 3/8(a_{max}/g)H^2\gamma_t$$

$$P_E = 3/8(0.20)(4\text{ m})^2\ (20\text{ kN/m}^3) = 24\text{ kN/m}$$

P_E is located at a distance of 0.6 H above base of wall, or: 0.6 H = 0.6 (4) = 2.4 m

Factor of safety for sliding:

Using Eq. 14.6 to determine the factor of safety for sliding:

$$F = (N\tan\delta + P_p)/(P_H + P_E)$$

$$F = (68.2\tan 38° + 4.07\text{ kN/m})/(49.2\text{ kN/m} + 24\text{ kN/m}) = 0.78$$

Factor of safety for overturning:

Using Eq. 14.8 with $P_v = 0$:

$$F = (Wa)/(1/3\ P_H H + 0.6\ HP_E)$$

$$F = [(35.3)(3/2) + (32.9)(2.8)]/[(1/3)(49.2)(4) + (0.6)(4)(24)] = 1.18$$

14.11 Using Eq. 14.5 and $a_{max}/g = 0.20$:

$$\psi = \tan^{-1}k_h = \tan^{-1}(a_{max}/g) = \tan^{-1}(0.20) = 11.3°$$

Using the k_{AE} equation in Fig. 11.4, with the following values:

$\delta = \phi_w = 24°$

$\phi = 32°$

$\theta = 0°$

$\beta = 0°$

Therefore: $k_{AE} = 0.428$

Using Eq. 14.4:

$P_{AE} = P_A + P_E = \frac{1}{2}k_{AE}H^2\gamma_t$

$P_{AE} = \frac{1}{2}(0.428)(4 \text{ m})^2 (20 \text{ kN/m}^3) = 68.5 \text{ kN/m}$

Factor of safety for sliding:

Using Eq. 14.9 to determine the factor of safety for sliding:

$F = (N \tan \delta + P_p)/(P_{AE} \cos \delta)$

Where: $N = W + P_{AE} \sin \delta = 96.1 \text{ kN/m}$

$F = (96.1 \tan 38° + 4.07)/(68.5 \cos 24°) = 1.26$

Factor of safety for overturning:

Overturning moment $= (P_{AE} \cos \delta)(1/3)(H) - (P_{AE} \sin \delta)(3 \text{ m})$

$= (68.5 \text{ kN/m} \cos 24°)(1/3)(4 \text{ m}) - (68.5 \text{ kN/m} \sin 24°)(3 \text{ m})$

$= -0.15 \text{ kN-m/m}$

Therefore: $F = \infty$

Summary:

Problem no.		Static analysis					Earthquake analysis			
	N (kN/m)	Location of N from toe (m)	q' (kPa)	q'' (kPa)	F sliding	F over-turning	P_E or P_{AE} (kN/m)	Location of N from toe (m)	F sliding	F over-turning
14.9	68.2	1.16	37.9	7.5	1.17	2.2	17.7	0.48	0.86	1.29
14.10	68.2	1.16	37.9	7.5	1.17	2.2	24.0	0.33	0.78	1.18
14.11	86.1	1.69	39.6	17.8	1.78	∞	68.5	1.51	1.26	∞

14.12   Using Eq. 14.5 and $a_{max}/g = 0.20$:

$\psi = \tan^{-1}k_h = \tan^{-1}(a_{max}/g) = \tan^{-1}(0.20) = 11.3°$

Using the k_{AE} equation in Fig. 11.4, with the following values:

$\delta = \phi_w = 32°$

$\phi = 32°$

$\theta = 0°$

$\beta = 0°$

Therefore: $k_{AE} = 0.445$

Using Eq. 14.4:

$P_{AE} = P_A + P_E = \frac{1}{2}k_{AE}H^2\gamma_t$

$P_{AE} = \frac{1}{2}(0.445)(4 \text{ m})^2 (20 \text{ kN/m}^3) = 71.2 \text{ kN/m}$

Factor of safety for sliding:

Using Eq. 14.9 to determine the factor of safety for sliding:

$F = (N \tan \delta + P_p)/(P_{AE} \cos \delta)$

Where: $N = W + P_{AE} \sin \delta$

Using Problem 11.18 to obtain $W = 23.5 + 32.9 + 56.0 = 112.4$ kN/m

$N = W + P_{AE} \sin \delta = 112.4$ kN/m $+ (71.2$ kN/m$)(\sin 32°) = 150.1$ kN/m

$F = (150.1 \tan 24° + 4.07)/(71.2 \cos 32°) = 1.17$

Factor of safety for overturning:

Overturning moment $= (P_{AE} \cos \delta)(1/3)(H) - (P_{AE} \sin \delta)(2$ m$)$

$= (71.2 \cos 32°)(1/3)(4$ m$) - (71.2 \sin 32°)(2$ m$) = 5.1$ kN-m/m

Moment of weights $= (56.4)(1$ m$) + (56)(1.6$ m$) = 146$ kN-m/m

$F = 146/5.1 = 29$

14.13 $k_A = \tan^2 (45° - \frac{1}{2}\phi) = \tan^2 [45° - (\frac{1}{2})(30°)] = 0.333$

$k_p = \tan^2 (45° + \frac{1}{2}\phi) = \tan^2 [45° + (\frac{1}{2})(30°)] = 3.0$

$P_A = P_H = \frac{1}{2}k_A\gamma_t H^2 = \frac{1}{2}(0.333)(110$ pcf$)(20$ ft$)^2 = 7330$ lb/ft

$P_p = \frac{1}{2}k_p\gamma_t D^2 = \frac{1}{2}(3.0)(110$ pcf$)(3$ ft$)^2 = 1490$ lb/ft

With reduction factor $= 2$, allowable $P_p = 740$ lb/ft

$W = N = HL\gamma_t = (20$ ft$)(14$ ft$)(120$ pcf$)$

$= 33,600$ lb per linear foot of wall length

Using Eq. 14.2:

$P_E = \frac{1}{2}(k_A)^{1/2} (a_{max}/g)H^2\gamma_t$

$= \frac{1}{2}(0.333)^{1/2} (0.20)(20$ ft$)^2(110$ pcf$) = 2540$ lb/ft

For sliding analysis, use Eq. 14.6:

$$F = \frac{N \tan \delta + P_p}{P_H + P_E} = \frac{33,600 \tan 23° + 740}{7330 + 2540} = 1.52$$

For overturning analysis, use Eq 14.7 with $P_v = 0$:

$$F = \frac{Wa}{1/3\, P_H H + 2/3\, HP_E} = \frac{(33,600)(7)}{1/3(7330)(20) + 2/3\,(20)(2540)} = 2.84$$

14.14 For the static condition, see the solution to Problem 11.53:

Resultant force $= \sigma_h H = (480$ psf$)(20$ ft$) = 9600$ lb/ft

For the earthquake condition, use Eq. 14.1 with $a_{max}/g = 0.20$:

$P_E = \frac{1}{2}(k_A)^{1/2} (a_{max}/g)H^2\gamma_t$

$P_E = \frac{1}{2}(0.307)^{1/2} (0.20)(20$ ft$)^2 (120$ pcf$) = 2700$ lb/ft

Chapter 15

15.1 Wet density of soil $= M/V$

$= 4$ kg/2000 cm$^3 = 0.002$ kg/cm$^3 = 2.0$ Mg/m^3

$\rho_d = 2.0/(1 + 0.083) = 1.85$ Mg/m^3

15.2   Total volume = $(10,000 \text{ yd}^3)(0.95)(122.5/94) = 12,380 \text{ yd}^3$

15.3 $\rho_d = 128/(1 + 0.065) = 120.2 \text{ pcf}$
Total volume. = $(5000 \text{ yd}^3)(0.90)(122.5/120.4) = 4590 \text{ yd}^3$

15.4   Total dry mass = $(12,380 \text{ yd}^3)(94 \text{ pcf})(27 \text{ ft}^3/\text{yd}^3) = 3.14 \times 10^7 \text{ lb}$
Water to be added = $(3.14 \times 10^7)(0.11 - 0.08) = 943,000 \text{ lb}$
Number of gallons = $943,000 \text{ lb}/8.34 \text{ lb/gal.} = 113,000 \text{ gal.}$

APPENDIX D

CONVERSION FACTORS

From	Multiply by*	Converts to†
Area, acres	4046.9	square meters
Area, square yards	0.8361	square meters
Area, square feet	0.0929	square meters
Area, square inches	0.0006451	square meters
Bending moment, pounds-foot	1.3558	newton-meter
Density, pounds/cubic yard	0.5932	kilograms/cubic meter
Density, pounds/cubic foot	16.0185	kilograms/cubic meter
Force, kips	4.4482	kilonewtons
Force, pounds	4.4482	newtons
Length, miles	1609.344	meter
Length, yards	0.9144	meter
Length, feet	0.3048	meter
Length, inches	0.0254	meter
Force/length, pounds/foot	14.5939	newtons/meter
Force/length, pounds/inch	175.127	newtons/meter
Mass, tons	907.184	kilogram
Mass, pounds	0.4536	kilogram
Mass, ounces	28.35	gram
Pressure or stress, pounds/square foot	47.8803	pascal
Pressure or stress, pounds/square inch	6.8947	kilopascal
Temperature, °F	$(t_F^\circ - 32)/1.8 = t_C^\circ$	°C
Volume, cubic yards	0.7646	cubic meters
Volume, cubic feet	0.02831	cubic meters
Volume, cubic inches	1.6387×10^{-5}	cubic meters

*The precision of a measurement converted to other units can never be greater than that of the original. To go from SI units to U.S. customary system units, divide by the given constant. ASTM E 380 provides guidance on use of SI.

†The common SI prefixes are

mega	M	1,000,000
kilo	k	1000
centi	c	0.01
milli	m	0.001
micro	μ	0.000001

APPENDIX E
BIBLIOGRAPHY

AASHTO (1996). *Standard Specifications for Highway Bridges*. 16th ed. American Association of State Highway and Transportation Officials (AASHTO), Washington, DC.

ACI (1980). *Control of Cracking in Concrete Structures*. ACI 224-80, Report by ACI Committee 224, American Concrete Institute, Detroit, MI.

ACI (1982). *Guide to Durable Concrete*. ACI Committee 201, American Concrete Institute, Detroit, MI.

ACI (1990). *ACI Manual of Concrete Practice, Part 1, Materials and General Properties of Concrete*. American Concrete Institute, Detroit, MI.

Arnold, C. and Reitherman, R. (1982). *Building Configuration and Seismic Design*. Wiley, New York, NY.

ASCE (1964). *Design of Foundations for Control of Settlement*. American Society of Civil Engineers, New York, NY, 592 pp.

ASCE (1972). "Subsurface Investigation for Design and Construction of Foundations of Buildings." Task Committee for Foundation Design Manual. Part I, *Journal of the Soil Mechanics and Foundations Division*, ASCE, Vol. 98, No. SM5, pp. 481–490; Part II, No. SM6, pp. 557–578; Parts III and IV, No. SM7, pp. 749–764.

ASCE (1976). *Subsurface Investigation for Design and Construction of Foundations of Buildings*. Manual No. 56. American Society of Civil Engineers, New York, NY, 61 pp.

ASCE (1978). *Site Characterization & Exploration*, Proceedings of the Specialty Workshop at Northwestern University, C. H. Dowding, ed., New York, NY, 395 pp.

ASCE (1987). *Civil Engineering Magazine*, American Society of Civil Engineers, New York, NY, April.

ASTM (1970). "Special Procedures for Testing Soil and Rock for Engineering Purposes." *ASTM Special Technical Publication 479*, Philadelphia, PA, 630 pp.

ASTM (1971). "Sampling of Soil and Rock." *ASTM Special Technical Publication 483*, Philadelphia, PA, 193 pp.

ASTM (2004). *Annual Book of ASTM Standards: Concrete and Aggregates, Vol. 04.02*, Standard No. C 127-93, "Standard Test Method for Specific Gravity and Absorption of Coarse Aggregate," West Conshohocken, PA.

ASTM (2004). *Annual Book of ASTM Standards: Concrete and Aggregates, Vol. 04.02*, Standard No. C 881-90, "Standard Specification for Epoxy-Resin-Base Bonding Systems for Concrete," West Conshohocken, PA.

ASTM (2004). *Annual Book of ASTM Standards, Vol. 04.08, Soil and Rock (1)*. Standard No. D 420-03, "Standard Guide to Site Characterization for Engineering, Design, and Construction Purposes," West Conshohocken, PA, pp. 1–7.

ASTM (2004). *Annual Book of ASTM Standards, Vol. 04.08, Soil and Rock (1)*. Standard No. D 422-02, "Standard Test Method for Particle-Size Analysis of Soils," West Conshohocken, PA, pp. 10–17.

ASTM (2004). *Annual Book of ASTM Standards, Vol. 04.08, Soil and Rock (1)*. Standard No. D 427-98, "Standard Test Method for Shrinkage Factors of Soils by the Mercury Method," West Conshohocken, PA, pp. 22–25.

ASTM (2004). *Annual Book of ASTM Standards, Vol. 04.08, Soil and Rock (1)*. Standard No. D 653-03, "Standard Terminology Relating to Soil, Rock, and Contained Fluids."American Society of Civil Engineers and ASTM, West Conshohocken, PA, pp. 46–78.

ASTM (2004). *Annual Book of ASTM Standards, Vol. 04.08, Soil and Rock (1)*. Standard No. D 698-00, "Standard Test Methods for Laboratory Compaction Characteristics of Soil Using Standard Effort," West Conshohocken, PA, pp. 81–91.

ASTM (2004). *Annual Book of ASTM Standards, Vol. 04.08, Soil and Rock (1)*. Standard No. D 854-02, "Standard Test Methods for Specific Gravity of Soil Solids by Water Pycnometer," West Conshohocken, PA, pp. 96–102.

ASTM (2004). *Annual Book of ASTM Standards, Vol. 04.08, Soil and Rock (1)*. Standard No. D 1143-94, "Standard Test Method for Piles Under Static Axial Compressive Load," West Conshohocken, PA, pp. 107–117.

ASTM (2003). *Annual Book of ASTM Standards, Vol. 04.08, Soil and Rock (I)*. Standard No. D 1194-94, "Standard Test Method for Bearing Capacity of Soil for Static Load and Spread Footings." West Conshohocken, PA, pp. 115–117. Note: this test method was withdrawn from ASTM in December 2002.

ASTM (2004). *Annual Book of ASTM Standards, Vol. 04.08, Soil and Rock (I)*. Standard No. D 1556-00, "Standard Test Method for Density and Unit Weight of Soil in Place by the Sand-Cone Method," West Conshohocken, PA, pp. 127–133.

ASTM (2004). *Annual Book of ASTM Standards, Vol. 04.08, Soil and Rock (I)*. Standard No. D 1557-02, "Standard Test Methods for Laboratory Compaction Characteristics of Soil Using Modified Effort," West Conshohocken, PA, pp. 134–143.

ASTM (2004). *Annual Book of ASTM Standards, Vol. 04.08, Soil and Rock (I)*. Standard No. D 1586-99, "Standard Test Method for Penetration Test and Split-Barrel Sampling of Soils," West Conshohocken, PA, pp. 147–151.

ASTM (2004). *Annual Book of ASTM Standards, Vol. 04.08, Soil and Rock (I)*. Standard No. D 1587-00, "Standard Practice for Thin-Walled Tube Sampling of Soils for Geotechnical Purposes," West Conshohocken, PA, pp. 152–155.

ASTM (2004). *Annual Book of ASTM Standards, Vol. 04.08, Soil and Rock (I)*. Standard No. D 2113-99, "Standard Practice for Rock Core Drilling and Sampling of Rock for Site Investigation," West Conshohocken, PA, pp. 182–201.

ASTM (2004). *Annual Book of ASTM Standards, Vol. 04.08, Soil and Rock (I)*. Standard No. D 2166-00, "Standard Test Method for Unconfined Compressive Strength of Cohesive Soil," West Conshohocken, PA, pp. 202–207.

ASTM (2004). *Annual Book of ASTM Standards, Vol. 04.08, Soil and Rock (I)*. Standard No. D 2216-98, "Standard Test Method for Laboratory Determination of Water (Moisture) Content of Soil and Rock by Mass," West Conshohocken, PA, pp. 220–224.

ASTM (2004). *Annual Book of ASTM Standards, Vol. 04.08, Soil and Rock (I)*. Standard No. D 2434-00, "Standard Test Method for Permeability of Granular Soils (Constant Head)," West Conshohocken, PA, pp. 234–238.

ASTM (2004). *Annual Book of ASTM Standards, Vol. 04.08, Soil and Rock (I)*. Standard No. D 2435-03, "Standard Test Methods for One-Dimensional Consolidation Properties of Soils Using Incremental Loading," West Conshohocken, PA, pp. 239–248.

ASTM (2004). *Annual Book of ASTM Standards, Vol. 04.08, Soil and Rock (I)*. Standard No. D 2487-00, "Standard Practice for Classification of Soils for Engineering Purposes (Unified Soil Classification System)," West Conshohocken, PA, pp. 249–260.

ASTM (2004). *Annual Book of ASTM Standards, Vol. 04.08, Soil and Rock (I)*. Standard No. D 2488-00, "Standard Practice for Description and Identification of Soils (Visual-Manual Procedure)," West Conshohocken, PA, pp. 261–271.

ASTM (2004). *Annual Book of ASTM Standards, Vol. 04.08, Soil and Rock (I)*. Standard No. D 2573-01, "Standard Test Method for Field Vane Shear Test in Cohesive Soil," West Conshohocken, PA, pp. 272–280.

ASTM (2004). *Annual Book of ASTM Standards, Vol. 04.08, Soil and Rock (I)*. Standard No. D 2844-01, "Standard Test Method for Resistance R-Value and Expansion Pressure of Compacted Soils," West Conshohocken, PA, pp. 285–292.

ASTM (2004). *Annual Book of ASTM Standards, Vol. 04.08, Soil and Rock (I)*. Standard No. D 2850-03, "Standard Test Method for Unconsolidated-Undrained Triaxial Compression Test on Cohesive Soils," West Conshohocken, PA, pp. 300–305.

ASTM (2004). *Annual Book of ASTM Standards, Vol. 04.08, Soil and Rock (I)*. Standard No. D 2922-01, "Standard Test Methods for Density of Soil and Soil-Aggregate in Place by Nuclear Methods (Shallow Depth)," West Conshohocken, PA, pp. 310–315.

ASTM (2004). *Annual Book of ASTM Standards, Vol. 04.08, Soil and Rock (I)*. Standard No. D 2937-00, "Standard Test Method for Density of Soil in Place by the Drive-Cylinder Method," West Conshohocken, PA, pp. 319–323.

ASTM (2004). *Annual Book of ASTM Standards, Vol. 04.08, Soil and Rock (I)*. Standard No. D 2974-00, "Standard Test Methods for Moisture, Ash, and Organic Matter of Peat and Other Organic Soils," West Conshohocken, PA, pp. 331–334.

ASTM (2004). *Annual Book of ASTM Standards, Vol. 04.08, Soil and Rock (I)*. Standard No. D 3017-01, "Standard Test Method for Water Content of Soil and Rock in Place by Nuclear Methods (Shallow Depth)." West Conshohocken, PA, pp. 344–348.

ASTM (2004). *Annual Book of ASTM Standards, Vol. 04.08, Soil and Rock (I)*. Standard No. D 3080-03, "Standard Test Method for Direct Shear Test of Soils Under Consolidated Drained Conditions," West Conshohocken, PA, pp. 349–355.

ASTM (2004). *Annual Book of ASTM Standards, Vol. 04.08, Soil and Rock (I)*. Standard No. D 3148-02, "Standard Test Method for Elastic Moduli of Intact Rock Core Specimens in Uniaxial Compression," West Conshohocken, PA, pp. 356–361.

ASTM (2004). *Annual Book of ASTM Standards, Vol. 04.08, Soil and Rock (I)*. Standard No. D 3282-97, "Standard Practice for Classification of Soils and Soil-Aggregate Mixtures for Highway Construction Purposes," West Conshohocken, PA, pp. 377–382.

ASTM (2004). *Annual Book of ASTM Standards, Vol. 04.08, Soil and Rock (I)*. Standard No. D 3441-98, "Standard Test Method for Mechanical Cone Penetration Tests of Soil," West Conshohocken, PA, pp. 401–405.

ASTM (2004). *Annual Book of ASTM Standards, Vol. 04.08, Soil and Rock (I)*. Standard No. D 3550-01, "Standard Practice for Thick Wall, Ring-Lined, Split Barrel, Drive Sampling of Soils," West Conshohocken, PA, pp. 406–410.

ASTM (2004). *Annual Book of ASTM Standards, Vol. 04.08, Soil and Rock (I)*. Standard No. D 3689-95, "Standard Test Method for Individual Piles Under Static Axial Tensile Load," West Conshohocken, PA, pp. 413–423.

ASTM (2004). *Annual Book of ASTM Standards, Vol. 04.08, Soil and Rock (I)*. Standard No. D 3966-95, "Standard Test Method for Piles Under Lateral Loads," West Conshohocken, PA, pp. 437–451.

ASTM (2004). *Annual Book of ASTM Standards, Vol. 04.08, Soil and Rock (I)*. Standard No. D 4083-01, "Standard Practice for Description of Frozen Soils (Visual-Manual Procedure)," West Conshohocken, PA, pp. 508–513.

ASTM (2004). *Annual Book of ASTM Standards, Vol. 04.08, Soil and Rock (I)*. Standard No. D 4220-00, "Standard Practices for Preserving and Transporting Soil Samples," West Conshohocken, PA, pp. 537–547.

ASTM (2004). *Annual Book of ASTM Standards, Vol. 04.08, Soil and Rock (I)*. Standard No. D 4253-00, "Standard Test Methods for Maximum Index Density and Unit Weight of Soils Using a Vibratory Table," West Conshohocken, PA, pp. 556–570.

ASTM (2004). *Annual Book of ASTM Standards, Vol. 04.08, Soil and Rock (I)*. Standard No. D 4254-00, "Standard Test Methods for Minimum Index Density and Unit Weight of Soils and Calculation of Relative Density," West Conshohocken, PA, pp. 571–579.

ASTM (2004). *Annual Book of ASTM Standards, Vol. 04.08, Soil and Rock (I)*. Standard No. D 4318-00, "Standard Test Methods for Liquid Limit, Plastic Limit, and Plasticity Index of Soils," West Conshohocken, PA, pp. 580–593.

ASTM (2004). *Annual Book of ASTM Standards, Vol. 04.08, Soil and Rock (I)*. Standard No. D 4394-98, "Standard Test Method for Determining the In Situ Modulus of Deformation of Rock Mass Using the Rigid Plate Loading Method," West Conshohocken, PA, pp. 619–627.

ASTM (2004). *Annual Book of ASTM Standards, Vol. 04.08, Soil and Rock (I)*. Standard No. D 4395-98, "Standard Test Method for Determining the In Situ Modulus of Deformation of Rock Mass Using the Flexible Plate Loading Method," West Conshohocken, PA, pp. 628–636.

ASTM (2004). *Annual Book of ASTM Standards, Vol. 04.08, Soil and Rock (I)*. Standard No. D 4452-02, "Standard Test Methods for X-Ray Radiography of Soil Samples," West Conshohocken, PA, pp. 685–698.

ASTM (2004). *Annual Book of ASTM Standards, Vol. 04.08, Soil and Rock (I)*. Standard No. D 4546-03, "Standard Test Methods for One-Dimensional Swell or Settlement Potential of Cohesive Soils," West Conshohocken, PA, pp. 736–742.

ASTM (2004). *Annual Book of ASTM Standards, Vol. 04.08, Soil and Rock (I)*. Standard No. D 4555-01, "Standard Test Method for Determining Deformability and Strength of Weak Rock by an In Situ Uniaxial Compressive Test," West Conshohocken, PA, pp. 756–759.

ASTM (2004). *Annual Book of ASTM Standards, Vol. 04.08, Soil and Rock (I)*. Standard No. D 4643-00, "Standard Test Method for Determination of Water (Moisture) Content of Soil by the Microwave Oven Heating," West Conshohocken, PA, pp. 811–815.

ASTM (2004). *Annual Book of ASTM Standards, Vol. 04.08, Soil and Rock (I)*. Standard No. D 4648-00, "Standard Test Method for Laboratory Miniature Vane Shear Test for Saturated Fine-Grained Clayey Soil," West Conshohocken, PA, pp. 838–844.

ASTM (2004). *Annual Book of ASTM Standards, Vol. 04.08, Soil and Rock (I)*. Standard No. D 4719-00, "Standard Test Method for Prebored Pressuremeter Testing in Soils," West Conshohocken, PA, pp. 897–905.

ASTM (2004). *Annual Book of ASTM Standards, Vol. 04.08, Soil and Rock (I)*. Standard No. D 4767-02, "Standard Test Method for Consolidated Undrained Triaxial Compression Test for Cohesive Soils," West Conshohocken, PA, pp. 923–935.

ASTM (2004). *Annual Book of ASTM Standards, Vol. 04.08, Soil and Rock (I)*. Standard No. D 4829-03, "Standard Test Method for Expansion Index of Soils," West Conshohocken, PA, pp. 936–939.

ASTM (2004). *Annual Book of ASTM Standards, Vol. 04.08, Soil and Rock (I)*. Standard No. D 4832-02, "Standard Test Method for Preparation and Testing of Controlled Low Strength Material (CLSM) Test Cylinders," West Conshohocken, PA, pp. 940–944.

ASTM (2004). *Annual Book of ASTM Standards, Vol. 04.08, Soil and Rock (I)*. Standard No. D 4943-02, "Standard Test Method for Shrinkage Factors of Soils by the Wax Method," West Conshohocken, PA, pp. 971–975.

ASTM (2004). *Annual Book of ASTM Standards, Vol. 04.08, Soil and Rock (I)*. Standard No. D 4945-00, "Standard Test Method for High-Strain Dynamic Testing of Piles," West Conshohocken, PA, pp. 981–987.

ASTM (2004). *Annual Book of ASTM Standards, Vol. 04.08, Soil and Rock (I)*. Standard No. D 5079-02, "Standard Practices for Preserving and Transporting Rock Core Samples," West Conshohocken, PA, pp. 1009–1015.

ASTM (2004). *Annual Book of ASTM Standards, Vol. 04.08, Soil and Rock (I)*. Standard No. D 5084-00, "Standard Test Methods for Measurement of Hydraulic Conductivity of Saturated Porous Materials Using a Flexible Wall Permeameter," West Conshohocken, PA, pp. 1024–1046.

ASTM (2004). *Annual Book of ASTM Standards, Vol. 04.08, Soil and Rock (I)*. Standard No. D 5298-03, "Standard Test Method for Measurement of Soil Potential (Suction) Using Filter Paper," West Conshohocken, PA, pp. 1139–1144.

ASTM (2004). *Annual Book of ASTM Standards, Vol. 04.08, Soil and Rock (I)*. Standard No. D 5333-03, "Standard Test Method for Measurement of Collapse Potential of Soils," West Conshohocken, PA, pp. 1213–1216.

ASTM (2004). *Annual Book of ASTM Standards, Vol. 04.08, Soil and Rock (I)*. Standard No. D 5434-03, "Standard Guide for Field Logging of Subsurface Explorations of Soil and Rock," West Conshohocken, PA, pp. 1268–1271.

ASTM (2004). *Annual Book of ASTM Standards, Vol. 04.08, Soil and Rock (I)*. Standard No. D 5550-00, "Standard Test Method for Specific Gravity of Soil Solids by Gas Pycnometer," West Conshohocken, PA, pp. 1340–1343.

ASTM (2004). *Annual Book of ASTM Standards, Vol. 04.09, Soil and Rock (II)*. Standard No. D 5778-00, "Standard Test Method for Performing Electronic Friction Cone and Piezocone Penetration Testing of Soils," West Conshohocken, PA.

ASTM (2004). *Annual Book of ASTM Standards, Vol. 04.09, Soil and Rock (II)*. Standard No. D 5780-02, "Standard Test Method for Individual Piles in Permafrost Under Static Axial Compressive Load," West Conshohocken, PA.

ASTM (2004). *Annual Book of ASTM Standards, Vol. 04.09, Soil and Rock (II)*. Standard No. D 6066-96, "Standard Practice for Determining the Normalized Penetration Resistance of Sands for Evaluation of Liquefaction Potential," West Conshohocken, PA.

ASTM (2004). *Annual Book of ASTM Standards, Vol. 04.09, Soil and Rock (II)*. Standard No. D 6467-99, "Standard Test Method for Torsional Ring Shear Test to Determine Drained Residual Shear Strength of Cohesive Soils," West Conshohocken, PA.

ASTM (2004). *Annual Book of ASTM Standards, Vol. 04.13, Geosynthetics*. Standard No. D 4355-02, "Standard Test Method for Deterioration of Geotextiles by Exposure to Light, Moisture and Heat in a Xenon Arc Type Apparatus," West Conshohocken, PA.

ASTM (2004). *Annual Book of ASTM Standards, Vol. 04.13, Geosynthetics*. Standard No. D 4439-99, "Standard Terminology for Geosynthetics," West Conshohocken, PA.

ASTM (2004). *Annual Book of ASTM Standards, Vol. 04.13, Geosynthetics*. Standard No. D 4491-99, "Standard Test Methods for Water Permeability of Geotextiles by Permittivity Soil Retention Capability," West Conshohocken, PA.

ASTM (2004). *Annual Book of ASTM Standards, Vol. 04.13, Geosynthetics*. Standard No. D 4751-99, "Standard Test Method for Determining Apparent Opening Size of a Geotextile," West Conshohocken, PA.

ASTM (2004). *Annual Book of ASTM Standards, Vol. 04.13, Geosynthetics*. Standard No. D 4833-00, "Standard Test Method for Index Puncture Resistance of Geotextiles, Geomembranes, and Related Products," West Conshohocken, PA.

ASTM (2004). *Annual Book of ASTM Standards, Vol. 04.13, Geosynthetics*. Standard No. D 5262-04, "Standard Test Method for Evaluating the Unconfined Tension Creep Behavior of Geosynthetics," West Conshohocken, PA.

ASTM (2004). *Annual Book of ASTM Standards, Vol. 04.13, Geosynthetics*. Standard No. D 6637-01, "Standard Test Method for Determining Tensile Properties of Geogrids by the Single or Multi-Rib Tensile Method," West Conshohocken, PA.

Abbott, P. L. (1996). *Natural Disasters*. Brown Publishing, Dubuque, IA, p. 438.

A. B. Chance Co. (1989). *Chance Anchor Design & Practice, A Technical Manual for Design and Application of Residential and Commercial Underpinning and Tieback Anchors*, A. B. Chance, Centralia, MO.

Aberg, B. (1996). "Grain-Size Distribution for Smallest Possible Void Ratio." *Journal of Geotechnical Engineering*, ASCE, Vol. 122, No. 1, pp. 74–77.

Abramson, L. W., Lee, T. S., Sharma, S., and Boyce, G. M. (1996). *Slope Stability and Stabilization Methods*. Wiley, New York, NY, 629 pp.

Al-Homoud, A. S., Basma, A. A., Husein Malkawi, A. I., and Al Bashabsheh, M. A. (1995). "Cyclic Swelling Behavior of Clays." *Journal of Geotechnical Engineering*, ASCE, Vol. 121, No. 7, pp. 562–565.

Al-Homoud, A. S., Basma, A. A., Husein Malkawi, A. I., and Al Bashabsheh, M. A. (1997). Closure of "Cyclic Swelling Behavior of Clays." *Journal of Geotechnical and Geoenvironmental Engineering*, ASCE, Vol. 123, No. 8, pp. 783–786.

Al-Khafaji, A. W. N. and Andersland, O. B. (1981). "Ignition Test for Soil Organic Content Measurement." *Journal of Geotechnical Engineering Division*, ASCE, Vol. 107, No. 4, pp. 465–479.

Allen, L. R., Yen, B. C., and McNeill, R. L. (1978). "Stereoscopic X-ray Assessment of Offshore Soil Samples." *Offshore Technology Conference*, Vol. 3, pp. 1391–1399.

Alpan, I. (1967). "The Empirical Evaluation of the Coefficients K_o and K_{oR}." *Soils and Foundations*, Vol. 7, No. 1, pp. 31–40.

Ambraseys, N. N. (1960). "On the Seismic Behavior of Earth Dams." *Proceedings of the Second World Conference on Earthquake Engineering*, Vol. 1, Tokyo and Kyoto, Japan, pp. 331–358.

Ambraseys, N. N. (1988). "Engineering Seismology." *Earthquake Engineering and Structural Dynamics*, Vol. 17, pp. 1–105.

Ambraseys, N. N. and Menu, J. M. (1988). "Earthquake-Induced Ground Displacements." *Earthquake Engineering and Structural Dynamics*, Vol. 16, pp. 985–1006.

American Geological Institute (1982). *AGI Data Sheets for Geology in the Field, Laboratory, and Office*. American Geological Institute, Falls Church, VA, 61 data sheets.

Anderson, S. A. and Sitar, N. (1995). "Analysis of Rainfall-Induced Debris Flows." *Journal of Geotechnical Engineering*, ASCE, Vol. 121, No. 7, pp. 544–552.

Anderson, S. A. and Sitar, N. (1996). Closure of "Analysis of Rainfall-Induced Debris Flows." *Journal of Geotechnical Engineering*, ASCE, Vol. 122, No. 12, pp. 1025–1027.

Andrus, R. D. and Stokoe, K. H. (1997). "Liquefaction Resistance Based on Shear Wave Velocity." *Proceedings, NCEER Workshop on Evaluation of Liquefaction Resistance of Soils, Technical Report NCEER-97-0022*, T. L. Youd and I. M. Idriss, eds., National Center for Earthquake Engineering Research, Buffalo, NY, pp. 89–128.

Andrus, R. D. and Stokoe, K. H. (2000). "Liquefaction Resistance of Soils from Shear-Wave Velocity." *Journal of Geotechnical and Geoenvironmental Engineering*, Vol. 126, No. 11, pp. 1015–1025.

Arango, I. (1996). "Magnitude Scaling Factors for Soil Liquefaction Evaluations." *Journal of Geotechnical Engineering*, ASCE, Vol. 122, No. 11, pp. 929–936.

Arnold, C. and Reitherman, R. (1982). *Building Configuration and Seismic Design*. Wiley, New York, NY.

Asphalt Institute (1984). *Thickness Design-Asphalt Pavements for Highways and Streets*. Asphalt Institute, College Park, MD, 80 pp.

Asphalt Institute (1989). *The Asphalt Handbook*. Asphalt Institute, Lexington, KY, 607 pp.

Atterberg, A. (1911). "The Behavior of Clays with Water, Their Limits of Plasticity and Their Degrees of Plasticity." *Kungliga Lantbruksakademiens Handlingar och Tidskrift*, Vol. 50, No. 2, pp. 132–158.

Aubeny, C. P. and Lytton, R. L. (2004). "Shallow Slides in Compacted High Plasticity Clay Slopes." *Journal of Geotechnical and Geoenvironmental Engineering*, ASCE, Vol. 130, No. 7, pp. 717–727.

Baldwin, J. E., Donley, H. F., and Howard, T. R. (1987). "On Debris Flow/Avalanche Mitigation and Control, San Francisco Bay Area, California." *Debris Flow/Avalanches: Process, Recognition, and Mitigation*, The Geological Society of America, Boulder, CO, pp. 223–236.

Bartlett, S. F. and Youd, T. L. (1995). "Empirical Prediction of Liquefaction-Induced Lateral Spread." *Journal of Geotechnical Engineering*, ASCE, Vol. 121, No. 4, pp. 316–329.

Bell, F. G. (1983). *Fundamentals of Engineering Geology*. Butterworths, London, England, 648 pp.

Bellport, B. P. (1968). "Combating Sulphate Attack on Concrete on Bureau of Reclamation Projects." *Performance of Concrete, Resistance of Concrete to Sulphate and Other Environmental Conditions*, University of Toronto Press, Toronto, Canada, pp. 77–92.

Biddle, P. G. (1979). "Tree Root Damage to Buildings—An Arboriculturist's Experience." *Arboricultural Journal*, Vol. 3, No. 6, pp. 397–412.

Biddle, P. G. (1983). "Patterns of Soil Drying and Moisture Deficit in the Vicinity of Trees on Clay Soils." *Geotechnique*, London, England, Vol. 33, No. 2, pp. 107–126.

Bishop, A. W. (1955). "The Use of the Slip Circle in the Stability Analysis of Slopes." *Geotechnique*, London, England, Vol. 5, No. 1, pp. 7–17.

Bishop, A. W. and Henkel, D. J. (1962). *The Measurement of Soil Properties in the Triaxial Test.* 2nd ed., Edward Arnold, London, 228 pp.

Bjerrum, L. (1963). "Allowable Settlement of Structures." *Proceedings of the Third European Conference on Soil Mechanics and Foundation Engineering*, Vol. 2, Wiesbaden, Germany, pp. 135–137.

Bjerrum, L. (1967a). "The Third Terzaghi Lecture: Progressive Failure in Slopes of Overconsolidated Plastic Clay and Clay Shales." *Journal of the Soil Mechanics and Foundations Division*, ASCE, Vol. 93, No. SM5, Part 1, pp. 1–49.

Bjerrum, L (1967b). "Engineering Geology of Norwegian Normally Consolidated Marine Clays as Related to Settlements of Buildings." Seventh Rankine Lecture, *Geotechnique*, Vol. 17, No. 2, London, England, pp. 81–118.

Bjerrum, L. (1972). "Embankments on Soft Ground." *Proceeding of the ASCE Specialty Conference on Performance of Earth and Earth-Supported Structures.* Purdue University, West Lafayette, IN, Vol. 2, pp. 1–54.

Bjerrum, L (1973). "Problems of Soil Mechanics and Construction on Soft Clays and Structurally Unstable Soils." Session 4, *Proceedings of the Eighth International Conference on Soil Mechanics and Foundation Engineering*, Moscow, Vol. 3, pp. 111–159.

Blake, T. F. (2000a). EQFAULT Computer Program, Version 3.00, Computer Program for Deterministic Estimation of Peak Acceleration from Digitized Faults. Thousand Oaks, CA.

Blake, T. F. (2000b). EQSEARCH Computer Program, Version 3.00, Computer Program for Estimation of Peak Acceleration from California Earthquake Catalogs. Thousand Oaks, CA.

Blight, G. E. (1965). "The Time-Rate of Heave of Structures on Expansive Clays." *Moisture Equilibria and Moisture Changes in Soils Beneath Covered Areas*, G. D. Aitchison, ed., Butterworth, Sydney, Australia, pp. 78–88.

Boardman, B. T. and Daniel, D. E. (1996). "Hydraulic Conductivity of Desiccated Geosynethic Clay Liners." *Journal of Geotechnical Engineering*, ASCE, Vol. 122, No. 3, pp. 204–208.

Bolt, B. A. (1988). *Earthquakes.* W. H. Freeman, New York, NY, 282 pp.

Bonilla, M. G. (1970). "Surface Faulting and Related Effects." Chapter 3 of *Earthquake Engineering*, Robert L. Wiegel, Coordinating Editor. PrenticeHall, Englewood Cliffs, NJ, pp. 47–74.

Boone, S. T. (1996). "Ground-Movement-Related Building Damage." *Journal of Geotechnical Engineering*, ASCE, Vol. 122, No. 11, pp. 886–896.

Boone, S. T. (1998). Closure to "Ground-Movement-Related Building Damage." *Journal of Geotechnical and Geoenvironmental Engineering*, ASCE, Vol. 124, No. 5, pp. 463–465.

Boscardin, M. D. and Cording, E. J. (1989). "Building Response to Excavation-Induced Settlement." *Journal of Geotechnical Engineering*, ASCE, Vol. 115, No. 1, pp. 1–21.

Boussinesq, J. (1885). *Application des Potentiels à L' Étude de L' Équilibre et du Mouvement des Solides Élastiques*, Gauthier-Villars, Paris, France.

Bowles, J. E. (1974). *Analytical and Computer Methods In Foundation Engineering*, McGraw-Hill, New York, NY, 519 pp.

Bowles, J. E. (1982). *Foundation Analysis and Design.* 3rd ed., McGraw-Hill, New York, NY, 816 pp.

Bray, J. D. et al. (2004). "Subsurface Characterization at Ground Failure Sites in Adapazari, Turkey." *Journal of Geotechnical and Geoenvironmental Engineering*, ASCE, Vol. 130, No. 7, pp. 673–685.

Brewer, H. W. (1965). "Moisture Migration—Concrete Slab-on-Ground Construction." *Journal of the PCA Research and Development Laboratories*, May, pp. 2–17.

Briaud, J. L., Smith, T., and Mayer, B. (1984). Laterally Loaded Piles and the Pressuremeter: Comparison of Existing Methods," *Laterally Loaded Deep Foundations*, ASTM, Reston, VA, STP 835, pp. 97–111.

Bromhead, E. N. (1984). *Ground Movements and their Effects on Structures*, Chapter 3, "Slopes and Embankments." Attewell, P. B. and Taylor, R. K., eds., Surrey University Press, London, England, p. 63.

Broms, B. B. (1965). "Design of Laterally Loaded Piles." *Journal of the Soil Mechanics and Foundations Division*, ASCE, Vol. 91, No. SM3, pp. 79–99.

Broms, B. B. (1972). "Stability of Flexible Structures (Piles and Pile Groups)." *Proceedings of the Fifth European Conference on Soil Mechanics and Foundation Engineering*, Madrid, Vol. 2, pp. 239–269.

Brooker, E. W. and Ireland, H. O. (1965). "Earth Pressures at Rest Related to Stress History." *Canadian Geotechnical Journal*, Vol. 2, No. 1, pp. 1–15.

Brown, D. A., Morrison, C., and Reese, L. C. (1988). "Lateral Load Behavior of Pile Group in Sand." *Journal of Geotechnical Engineering*, ASCE, Vol. 114, No. 11, pp. 1261–1276.

Brown, D. A., Reese, L. C., and O'Neill, M. W. (1987). "Cyclic Lateral Loading of a Large-Scale Pile Group." *Journal of Geotechnical Engineering*, ASCE, Vol. 113, No. 11, pp. 1326–1343.

Brown, D. R. and Warner, J. (1973). "Compaction Grouting." *Journal of the Soil Mechanics and Foundations Division*, ASCE, Vol. 99, SM8, pp. 589–601.

Brown, R. W. (1990). *Design and Repair of Residential and Light Commercial Foundations*. McGraw-Hill, New York, NY, 241 pp.

Brown, R. W. (1992). *Foundation Behavior and Repair, Residential and Light Commercial*. McGraw-Hill, New York, NY, 271 pp.

Bruce, D. A. and Jewell, R. A. (1987). "Soil Nailing: The Second Decade." *International Conference on Foundations and Tunnels*, London, England, pp. 68–83.

Bruneau, M. (1999). "Structural Damage: Kocaeli, Turkey Earthquake, August 17, 1999." MCEER Deputy Director and Professor, Department of Civil, Structural and Environmental Engineering, University at Buffalo. Posted on the Internet.

Bureau of Reclamation (1990). *Earth Manual*, Part 2, Department of Interior, Government Printing Office, Washington, DC.

Burland, J. B., Broms, B. B., and de Mello, V. F. B. (1977). "Behavior of Foundations and Structures." State-of-the-Art Report, *Proceedings of the Ninth International Conference on Soil Mechanics and Foundation Engineering*, Japanese Geotechnical Society, Tokyo, Japan, Vol. 1, pp. 495–546.

Burland, J. B. and Burbidge, M. C. (1985). "Settlement of Foundations on Sand and Gravel." *Proceedings, Institution of Civil Engineers*, Part 1, Vol. 78, pp. 1325–1381.

Butt, T. K. (1992). "Avoiding and Repairing Moisture Problems in Slabs on Grade." *The Construction Specifier*, December, pp. 17–27.

California Department of Water Resources (1967). "Earthquake Damage to Hydraulic Structures in California." *California Department of Water Resources, Bulletin 116-3*, Sacramento, CA.

Carlson, R. W. (1938). "Drying Shrinkage of Concrete as Affected by Many Factors." *Proceedings of the Forty-First Annual Meeting of the American Society for Testing and Materials*, Vol. 38, Part II, Technical Papers, American Society for Testing and Materials, Philadelphia, PA, pp. 419–440.

Casagrande, A. (1932a). "Research on the Atterberg Limits of Soils." *Public Roads*, Vol. 13, pp. 121–136.

Casagrande, A. (1932b). Discussion of "A New Theory of Frost Heaving," edited by A. C. Benkelman and F. R. Ohlmstead. *Proceedings of the Highway Research Board*, Vol. 11, pp. 168–172.

Casagrande, A. (1936). "The Determination of the Pre-Consolidation Load and Its Practical Significance." Discussion D-34, *Proceedings of the First International Conference on Soil Mechanics and Foundation Engineering*, Cambridge, MA, Vol. 3, pp. 60–64.

Casagrande, A. (1940). "Seepage Through Dams." *Contributions to Soil Mechanics, 1925–1940*. Boston Society of Civil Engineers, Boston, MA, pp. 295–336. Originally published in the *Journal of the New England Water Works Association*, June, 1937.

Casagrande, A. (1948). "Classification and Identification of Soils." *Transactions*, ASCE, Vol. 113, pp. 901–930.

Casagrande, A (1975). "Liquefaction and Cyclic Deformation of Sands, A Critical Review." *Proceedings of the Fifth Panamerican Conference on Soil Mechanics and Foundation Engineering*, Buenos Aires, Argentina, Vol. 5, pp. 79–133.

Cashman, P. M. and Harris, E. T. (1970). *Control of Groundwater by Water Lowering*. Conference on Ground Engineering, Institute of Civil Engineers, London.

Castro, G., Seed, R. B., Keller, T. O., and Seed, H. B. (1992). "Steady-State Strength Analysis of Lower San Fernando Dam Slide." *Journal of Geotechnical Engineering*, ASCE, Vol. 118, No. 3, pp. 406–427.

Caterpillar Performance Handbook (1997). 28th ed., Caterpillar, Peoria, IL, 1,006 pp.

Cedergren, H. R. (1989). *Seepage, Drainage, and Flow Nets*. 3d ed., Wiley, New York, NY, 465 pp.

Cernica, J. N. (1995a). *Geotechnical Engineering: Soil Mechanics*. Wiley, New York, NY, 453 pp.

Cernica, J. N. (1995b). *Geotechnical Engineering: Foundation Design*. Wiley, New York, NY, 486 pp.

Cheeks, J. R. (1996). "Settlement of Shallow Foundations on Uncontrolled Mine Spoil Fill." *Journal of Performance of Constructed Facilities*, ASCE, Vol. 10, No. 4, pp. 143–151.

Chen, F. H. (1988). *Foundations on Expansive Soils*, 2d ed., Elsevier, New York, NY, 463 pp.

Chen, F. H. (2000). *Soil Engineering: Testing, Design, and Remediation*, CRC Press, Boca Raton, FL, 288 pp.

Chen, W. F. (1995). *The Civil Engineering Handbook*. Chapter 17, "Groundwater and Seepage," M. E. Harr, ed., CRC, Boca Raton, FL.

Cheney, J. E. and Burford, D. (1975). "Damaging Uplift to a Three-Story Office Block Constructed on a Clay Soil Following Removal of Trees." *Proceedings, Conference on Settlement of Structures*, Pentech Press, London, England, pp. 337–343.

Christensen, D. H. (2000). "The Great Prince William Sound Earthquake of March 28, 1964." Geophysical Institute, University of Alaska, Fairbanks, Report Obtained from the Internet.

Christensen, D. H. and Ruff, L. J. (1988). "Seismic Coupling and Outer Rise Earthquakes." *Journal of Geophysical Research*, Vol. 93, No. 13, pp. 421–444.

Cleveland, G. B. (1960). "Geology of the Otay Clay Deposit, San Diego County, California." *California Division of Mines Special Report 64*, Sacramento, CA, 16 pp.

Clough, G. W. and Davidson, R. R. (1977). "Effects of Construction on Geotechnical Performance." Specialty Session No. 3, Relationship Between Design and Construction in Soil Engineering, *Proceedings of the Ninth International Conference on Soil Mechanics and Foundation Engineering*, Tokyo, Vol. 3, pp. 479–485.

Cluff, L. S. (1971). "Peru Earthquake of May 31, 1970; Engineering Geology Observations." *Bulletin of the Seismological Society of America*, Vol. 61, pp. 511–533.

Coduto, D. P. (1994). *Foundation Design, Principles, and Practices*. Prentice Hall, Englewood Cliffs, NJ, 796 pp.

Collins, A. G. and Johnson, A. I. (1988). *Ground-Water Contamination, Field Methods*, Symposium papers published by American Society for Testing and Materials, Philadelphia, PA, 491 pp.

Converse Consultants Southwest, Inc. (1990). *Soil and Foundation Investigation, Proposed 80-Acre Clayton-Alexander Parcel, North Las Vegas, Nevada*. CCSW Project No. 90-33264-01, Las Vegas, NV, 46 pp.

County of San Diego (1983). Below Slab Moisture Barrier Specifications (DPL No. 65B). Department of Planning and Land Use, County of San Diego, CA.

Cousins, B. F. (1978). "Stability Charts for Simple Earth Slopes." *Journal of the Geotechnical Engineering Division*, ASCE, Vol. 104, No. GT2, pp. 267–279.

Cox, J. B. (1968). "A Review of the Engineering Characteristics of the Recent Marine Clays in South East Asia." Asian Institute of Technology, Research Report No. 6, Bangkok.

Cox, W. R., Dixon, D. A., and Murphy, B. S. (1984). "Lateral Load Tests of 5.4 mm Diameter Piles in Very Soft Clay in Side-by-Side and In-Line Groups." *Laterally Loaded Deep Foundations: Analysis and Performance*, ASTM, West Conshohocken, PA.

Cutler, D. F. and Richardson, I. B. K. (1989). *Tree Roots and Buildings*. 2nd ed., Longman, England, pp. 1–67.

Dakoulas, P. and Gazetas, G. (1986). "Seismic Shear Strains and Seismic Coefficients in Dams and Embankments." *Soil Dynamics and Earthquake Engineering*, Vol. 5, No. 2, pp. 75–83.

David, D. and Komornik, A. (1980). "Stable Embedment Depth of Piles in Swelling Clays." *Fourth International Conference on Expansive Soils*, ASCE, Vol. 2, Denver, CO, pp. 798–814.

Davis, E. H. and Poulos, H. G. (1972). "Rate of Settlement Under Two- and Three-Dimensional Conditions." *Geotechnique*, Vol. 22, No. 1, pp. 95–114.

Day, R. W. (1980). *Engineering Properties of the Orinoco Clay*. Thesis Submitted for the Master of Science in Civil Engineering (MCE) and Civil Engineer (CE) degrees. Massachusetts Institute of Technology, Cambridge, MA, 140 pp.

Day, R. W. (1989). "Relative Compaction of Fill Having Oversize Particles." *Journal of Geotechnical Engineering*, ASCE, Vol. 115, No. 10, pp. 1487–1491.

Day, R. W. (1990a). "Differential Movement of Slab-on-Grade Structures." *Journal of Performance of Constructed Facilities*, ASCE, Vol. 4, No. 4, pp. 236–241.

Day, R. W. (1990b). "Index Test for Erosion Potential." *Bulletin of the Association of Engineering Geologists*, Vol. 27, No. 1, pp. 116–117.

Day, R. W. (1991a). Discussion of "Collapse of Compacted Clayey Sand." *Journal of Geotechnical Engineering*, ASCE, Vol. 117, No. 11, pp. 1818–1821.

Day, R. W. (1991b). "Expansion of Compacted Gravelly Clay." *Journal of Geotechnical Engineering*, ASCE, Vol. 117, No. 6, pp. 968–972.

Day, R. W. (1992a). "Swell Versus Saturation for Compacted Clay." *Journal of Geotechnical Engineering*, ASCE, Vol. 118, No. 8, pp. 1272–1278.

Day, R. W. (1992b). "Walking of Flatwork on Expansive Soils." *Journal of Performance of Constructed Facilities*, ASCE, Vol. 6, No. 1, pp. 52–57.

Day, R. W. (1992c). "Moisture Migration through Concrete Floor Slabs." *Journal of Performance of Constructed Facilities*, ASCE, Vol. 6, No. 1, pp. 46–51.

Day, R. W. (1993a). "Expansion Potential According to the Uniform Building Code." *Journal of Geotechnical Engineering*, ASCE, Vol. 119, No. 6, pp. 1067–1071.

Day, R. W. (1993b). "Surficial Slope Failure: A Case Study." *Journal of Performance of Constructed Facilities*, ASCE, Vol. 7, No. 4, pp. 264–269.

Day, R. W. (1994a). Discussion of "Evaluation and Control of Collapsible Soils." *Journal of Geotechnical Engineering*, ASCE, Vol. 120, No. 5, pp. 924–925.

Day, R. W. (1994b). "Swell-Shrink Behavior of Compacted Clay." *Journal of Geotechnical Engineering*, ASCE, Vol. 120, No. 3, pp. 618–623.

Day, R. W. (1994c). "Performance of Slab-on-Grade Foundations on Expansive Soil." *Journal of Performance of Constructed Facilities*, ASCE, Vol. 8, No. 2, pp. 129–138.

Day, R. W. (1994d). "Weathering of Expansive Sedimentary Rock due to Cycles of Wetting and Drying." *Bulletin of the Association of Engineering Geologists*, Vol. 31, No. 3, pp. 387–390.

Day, R. W. (1994e). "Surficial Stability of Compacted Clay: Case Study." *Journal of Geotechnical Engineering*, ASCE, Vol. 120, No. 11, pp. 1980–1990.

Day, R. W. (1994f). "Moisture Migration through Basement Walls." *Journal of Performance of Constructed Facilities*, ASCE, Vol. 8, No. 1, pp. 82–86.

Day, R. W. (1995a). "Effect of Maximum Past Pressure on Two-Dimensional Immediate Settlement." *Journal of Environmental and Engineering Geoscience*, Vol. 1, No. 4, pp. 514–517.

Day, R. W. (1995b). "Reactivation of an Ancient Landslide." *Journal of Performance of Constructed Facilities*, ASCE, Vol. 9, No. 1, pp. 49–56.

Day, R. W. (1996a). "Study of Capillary Rise and Thermal Osmosis." *Journal of Environmental and Engineering Geoscience*, Joint Publication, AEG and GSA, Vol. 2, No. 2, pp. 249–254.

Day, R. W. (1996b). "Effect of Gravel on Pumping Behavior of Compacted Soil." *Journal of Geotechnical Engineering*, ASCE, Vol. 122, No. 10, pp. 863–866.

Day, R. W. (1997a). Discussion of "Grain-Size Distribution for Smallest Possible Void Ratio." *Journal of Geotechnical and Geoenvironmental Engineering*, Vol. 123, No. 1, p. 78.

Day, R. W. (1997b). "Hydraulic Conductivity of a Desiccated Clay Upon Wetting." *Journal of Environmental and Engineering Geoscience*, Joint Publication, AEG and GSA, Vol. 3, No. 2, pp. 308–311.

Day, R. W. (1998). Discussion of "Ground-Movement-Related Building Damage." *Journal of Geotechnical and Geoenvironmental Engineering*, ASCE, Vol. 124, No. 5, pp. 462–463.

Day, R. W. (1999a). *Geotechnical and Foundation Engineering: Design and Construction*, McGraw-Hill, New York, NY, 808 pp.

Day, R. W. (1999b). *Forensic Geotechnical and Foundation Engineering*. McGraw-Hill, New York, NY, 461 pp.

Day, R. W. (2000a). *Geotechnical Engineer's Portable Handbook*, McGraw-Hill, New York, NY, 750 pp.

Day, R. W. (2000b). "Structural Foundations and Retaining Walls," Chapter 16 of *Forensic Structural Engineering Handbook*, Robert T. Ratay, ed., McGraw-Hill, New York, NY, pp. 16.1–16.78.

Day, R. W. (2001a). *Soil Testing Manual: Procedures, Classification Data, and Sampling Practices*. Published in Hardcopy and in the Digital Engineering Library (DEL), McGraw-Hill, New York, NY, 619 pp.

Day, R. W. (2001b). "Soil Mechanics and Foundations." Section 6 in *Building Design and Construction Handbook*, 6th ed., Frederick S. Merritt and Jonathan T. Ricketts, eds., McGraw-Hill, New York, NY, pp. 6.1–6.121.

Day, R. W. (2002). *Geotechnical Earthquake Engineering Handbook*. Sponsored by the International Conference of Building Officials (ICBO), published in Hardcopy and in the Digital Engineering Library (DEL), McGraw-Hill, New York, NY, 585 pp.

Day, R. W. (2005). "Foundations and Retaining Walls," Chapter 14 in *Structural Condition Assessment*, Robert T. Ratay, ed., Wiley, New York, NY, pp. 461–494.

Day, R. W. and Axten, G. W. (1989). "Surficial Stability of Compacted Clay Slopes." *Journal of Geotechnical Engineering*, ASCE, Vol. 115, No. 4, pp. 577–580.

Day, R. W. and Axten, G. W. (1990). "Softening of Fill Slopes Due to Moisture Infiltration." *Journal of Geotechnical Engineering*, ASCE, Vol. 116, No. 9, pp. 1424–1427.

Day, R. W. and Marsh, E. T. (1995). "Triaxial A-Value versus Swell or Collapse for Compacted Soil." *Journal of Geotechnical Engineering*, ASCE, Vol. 121, No. 7, pp. 566–570.

Day, R. W. and Poland, D. M. (1996). "Damage Due to Northridge Earthquake Induced Movement of Landslide Debris." *Journal of Performance of Constructed Facilities*, ASCE, Vol. 10, No. 3, pp. 96–108.

Day, R. W. and Thoeny, S. (1998). "Reactivation of a Portion of an Ancient Landslide." *Journal of the Environmental and Engineering Geoscience*, Joint Publication, AEG and GSA, Vol. 4, No. 2, pp. 261–269.

DeGaetano, A. T., Wilks, D. S., and McKay, M. (1997). "Extreme-Value Statistics for Frost Penetration Depths in Northeastern United States." *Journal of Geotechnical and Geoenvironmental Engineering*, ASCE, Vol. 123, No. 9, pp. 828–835.

de Mello, V. F. B. (1971). "The Standard Penetration Test." State-of-the-Art Report, *Fourth Panamerican Conference on Soil Mechanics and Foundation Engineering*, San Juan, Puerto Rico, Vol. 1, pp. 1–86.

Department of the Army (1970). *Engineering and Design, Laboratory Soils Testing (Engineer Manual EM 1110-2-1906)*. U.S. Army Engineer Waterways Experiment Station, Department of the Army, Washington, DC, 282 pp.

Departments of the Army and the Air Force (1979). *Soils and Geology, Procedures for Foundation Design of Buildings and Other Structures (Except Hydraulic Structures)*. TM 5/818-1/AFM 88-3, Chapter 7, Washington, DC.

Detwiler, R. J., Taylor, P. C., Powers, L. J., Corley, W. G., Delles, J. B., and Johnson, B. R. (2000). "Assessment of Concrete in Sulfate Soils." *Journal of Performance of Constructed Facilities*, ASCE, Vol. 14, No. 3, pp. 89–96.

Diaz, C. F., Hadipriono, F. C., and Pasternack, S. (1994). "Failure of Residential Building Basements in Ohio." *Journal of Performance of Constructed Facilities*, ASCE, Vol. 8, No. 1, pp. 65–80.

Division of Mines and Geology (1997). *Guidelines for Evaluating and Mitigating Seismic Hazards in California, Special Publication 117*. Department of Conservation, Division of Mines and Geology, California, 74 pp.

Driscol, R. (1983). "The Influence of Vegetation on the Swelling and Shrinking of Clay Soils in Britain." *Geotechnique*, London, England, Vol. 33, No. 2, pp. 1–67.

Dudley, J. H. (1970). "Review of Collapsing Soils." *Journal of the Soil Mechanics and Foundations Division*, ASCE, Vol. 96, No. SM3, pp. 925–947.

Duke, C. M. (1960). "Foundations and Earth Structures in Earthquakes." *Proceedings of the Second World Conference on Earthquake Engineering*, Vol. 1, Tokyo and Kyoto, Japan, pp. 435–455.

Duncan, J. M. (1993). "Limitations of Conventional Analysis of Consolidation Settlement." *Journal of Geotechnical Engineering*, ASCE, Vol. 119, No. 9, pp.1331–1359.

Duncan, J. M. (1996). "State-of-the-Art: Limit Equilibrium and Finite-Element Analysis of Slopes." *Journal of Geotechnical Engineering*, ASCE, Vol. 122, No. 7, pp. 577–596.

Duncan, J. M. and Buchignani, A. L. (1976). "An Engineering Manual for Settlement Studies." *Geotechnical Engineering Report*, University of California at Berkeley, CA, 94 pp.

Duncan, J. M., Williams, G. W., Sehn, A. L., and Seed, R. B. (1991). "Estimation Earth Pressures due to Compaction." *Journal of Geotechnical Engineering*, ASCE, Vol. 117, No. 12, pp. 1833–1847.

Dunnicliff, J. (1993). *Geotechnical Instrumentation for Monitoring Field Performance*. Wiley, New York, NY, 608 pp.

Dyni, R. C. and Burnett, M. (1993). "Speedy Backfilling for Old Mines." *Civil Engineering Magazine*, ASCE, Vol. 63, No. 9, pp. 56–58.

EERC (1995). "Geotechnical Reconnaissance of the Effects of the January 17, 1995, Hyogoken-Nanbu Earthquake, Japan." *Report No. UCB/EERC-95/01*, Earthquake Engineering Research Center, College of Engineering, University of California, Berkeley.

EERC (2000). Photographs and Descriptions from the Steinbrugge Collection, EERC, University of California, Berkeley. Photographs and description on the Internet.

Ehlig, P. L. (1986). "The Portuguese Bend Landslide: Its Mechanics and a Plan for its Stabilization." *Landslides and Landslide Mitigation in Southern California*. 82nd Annual Meeting of the Cordilleran Section of the Geological Society of America, Los Angeles, CA, pp. 181–190.

Ehlig, P. L. (1992). "Evolution, Mechanics, and Migration of the Portuguese Bend Landslide, Palos Verdes Peninsula, California." *Engineering Geology Practice in Southern California*. B. W. Pipkin and R. J. Proctor, eds., Star Publishing Company, Association of Engineering Geologists, Southern California Section, Special Publication No. 4, pp. 531–553.

El-Garhy, B. M. and Wray, W. K. (2004). "Method for Calculating the Edge Moisture Variation Distance." *Journal of Geotechnical and Geoenvironmental Engineering*, ASCE, Vol. 130, No. 9, pp. 945–955.

Ellen, S. D., and Fleming, R. W. (1987). "Mobilization of Debris Flows from Soil Slips, San Francisco Bay Region, California." *Debris Flows/Avalanches: Process, Recognition, and Mitigation*, The Geological Society of America, Boulder, CO, pp. 31–40.

EQE Summary Report (1995). "The January 17, 1995 Kobe Earthquake." Report Posted on the EQE Internet Site.

Evans, D. A. (1972). *Slope Stability Report*, Slope Stability Committee, Department of Building and Safety, Los Angeles, CA.

Federal Emergency Management Agency (1994). "Phoenix Community Earthquake Hazard Evaluation, Maricopa County, Arizona." State of Arizona, Department of Emergency and Military Affairs, Division of Emergency Management, FEMA/NEHRP, Federal Emergency Management Agency Cooperative Agreement No. AZ102EPSA.

Feld, J. (1965). "Tolerance of Structures to Settlement." *Journal of the Soil Mechanics and Foundations Division*, ASCE, Vol. 91, No. SM3, pp. 63–77.

Feld, J. and Carper, K. L. (1997). *Construction Failure*. 2nd. ed., Wiley, New York, NY, 512 pp.

Fellenius, W. (1936). "Calculation of the Stability of Earth Dams." *Proceedings of the Second Congress on Large Dams*, Vol. 4, Washington, DC, pp. 445–463.

Fields of Expertise (undated). Joint Publication Prepared by the Civil Engineers and Engineering Geology Committee, Appointed by the Professional Engineers and Geologist and Geophysicist Boards of Registration, CA, 5 pp.

Finn, W. D. L., Bransby, P. L., and Pickering, D. J. (1970). "Effect of Strain History on Liquefaction of Sands." *Journal of the Soil Mechanics and Foundations Division*, ASCE, Vol. 96, No. SM6, pp. 1917–1934.

Fityus, S. G., Smith, D. W., and Allman, M. A. (2004). "Expansive Soil Test Site Near Newcastle." *Journal of Geotechnical and Geoenvironmental Engineering*, ASCE, Vol. 130, No. 7, pp. 686–695.

Floor Seal Technology (1998). "Concrete Vapor Emission and Alkalinity Control for Flooring Materials and Coated Surfaces." Floor Seal Technology, San Jose, CA, 40 pp.

Florensov, N. A. and Solonenko, V. P. (eds.) (1963). "Gobi-Altayskoye Zemletryasenie." *Iz. Akad. Nauk SSSR.*; also 1965, *The Gobi-Altai Earthquake*, U.S. Department of Commerce (English translation), Washington, DC.

Fluet, J. E. (1988). "Geosynthetics for Soil Improvement: A General Report and Keynote Address." *Geosynthetics for Soil Improvement*, R. D. Holtz, ed., Geotechnical Special Publication No. 18, American Society of Civil Engineers, New York, NY.

Foott, R. and Ladd, C. C. (1981). "Undrained Settlement of Plastic and Organic Clays." *Journal of the Geotechnical Engineering Division*, ASCE, Vol. 107, GT8, pp. 1079–1094.

Foshee, J. and Bixler, B. (1994). "Cover-Subsidence Sinkhole Evaluation of State Road 434, Longwood, Florida." *Journal of Geotechnical Engineering*, ASCE, Vol. 120, No. 11, pp. 2026–2040.

Foster, C. R. and Ahlvin, R. G. (1954). "Stresses and Deflections Induced by a Uniform Circular Load." *Proceeding of the Highway Research Board*, Vol. 33, Washington, DC, pp. 467–470.

Fourie, A. B. (1989). "Laboratory Evaluation of Lateral Swelling Pressure." *Journal of Geotechnical Engineering*, ASCE, Vol. 115, No. 10, pp. 1481–1486.

Fowler, C. M. R. (1990). *The Solid Earth: An Introduction to Global Geophysics*. Cambridge University Press, Cambridge, England, 472 pp.

Franklin, A. G., Orozco, L. F., and Semrau, R. (1973). "Compaction and Strength of Slightly Organic Soils." *Journal of Soil Mechanics and Foundations Division*, ASCE, Vol. 99, No. 7, pp. 541–557.

Fredlund, D. G. (1983). "Prediction of Ground Movements in Swelling Clays." Presented at the *Thirty-first Annual Soil Mechanics and Foundation Engineering Conference*, ASCE, Invited Lecture, University of Minnesota, MN.

Fredlund, D. G. and Rahardjo, H. (1993). *Soil Mechanics for Unsaturated Soil*. Wiley, New York, NY, 517 pp.

Geologist and Geophysicist Act (1986). Board of Registration for Geologists and Geophysicists, Department of Consumer Affairs, Sacramento, CA, 55 pp.

Geo-Slope (1991). *User's Guide, SLOPE/W for Slope Stability Analysis*. Version 2, Published by Geo-Slope International, Calgary, Canada, 444 pp.

Geo-Slope (1992). *User's Guide, SEEP/W for Finite Element Seepage Analysis*. Version 2, Published by Geo-Slope International, Calgary, Canada, 349 pp.

Geo-Slope (1993). *User's Guide, SIGMA/W for Finite Element Stress/Deformation Analysis*. Version 2, Published by Geo-Slope International, Calgary, Canada, 407 pp.

Gere, J. M. and Shah, H. C. (1984). *Terra Non Firma*. W. H. Freeman , New York, NY, 203 pp.

Gill, L. D. (1967). "Landslides and Attendant Problems." *Mayor's Ad Hoc Landslide Committee Report*, Los Angeles, CA.

gINT (1991). *Geotechnical Integrator Software Manual*. Computer Software and Manual Developed by Geotechnical Computer Applications (GCA), Santa Rosa, CA, 800 pp.

Goh, A. T. C. (1993). "Behavior of Cantilever Retaining Walls." *Journal of Geotechnical Engineering*, ASCE, Vol. 119, No.11, pp. 1751–1770.

Graf, E. D. (1969). "Compaction Grouting Techniques," *Journal of the Soil Mechanics and Foundations Division*, ASCE, Vol. 95, SM5, pp. 1151–1158.

Grant, R., Christian, J. T., and Vanmarcke, E. H. (1974). "Differential Settlement of Buildings." *Journal of the Geotechnical Engineering Division*, ASCE, Vol. 100, No. GT9, pp. 973–991.

Grantz, A., Plafker, G., and Kachadoorian, R. (1964). *Alaska's Good Friday Earthquake*, March 27, 1964. Department of the Interior, Geological Survey Circular 491, Washington, DC.

Gray, R. E. (1988). "Coal Mine Subsidence and Structures." *Mine Induced Subsidence: Effects on Engineered Structures, Geotechnical Special Publication 19*, ASCE, New York, NY, pp. 69–86.

Greenfield, S. J. and Shen, C. K. (1992). *Foundations in Problem Soils.* Prentice Hall, Englewood Cliffs, NJ, 240 pp.

Gromko, G. J. (1974). "Review of Expansive Soils." *Journal of the Geotechnical Engineering Division*, ASCE, Vol. 100, No. GT6, pp. 667–687.

"Guajome Ranch House, Vista, California." (1986). *National Historic Landmark Condition Assessment Report*, Preservation Assistance Division, National Park Service, Washington, DC.

Hammer, M. J. and Thompson, O. B. (1966). "Foundation Clay Shrinkage Caused by Large Trees." *Journal of the Soil Mechanics and Foundations Division*, ASCE, Vol. 92, No. SM6, pp. 1–17.

Hands, S. (1999). "Soft Stories Teach Hard Lesson to Taiwan's Construction Engineers." *Taipei Times*, Saturday, October 9, 1999.

Hanks, T. C. and Kanamori, H. (1979). "A Moment Magnitude Scale." *Journal of Geophysical Research*, Vol. 84, pp. 2348–2350.

Hansbo, S. (1975). *Jordmateriallära*, Almqvist & Wiksell Förlag AB, Stockholm, Sweden, 218 pp.

Hansen, B. (1961). "The Bearing Capacity of Sand, Tested by Loading Circular Plates." *Proceeding of the Fifth International Conference on Soil Mechanics and Foundation Engineering*, Paris, France, Vol. 1.

Hansen, M. J. (1984). "Strategies for Classification of Landslides," in *Slope Instability*, Wiley, New York, NY, pp. 1–25.

Hansen, T. C. and Mattock, A. H. (1966). "Influence of Size and Shape of Member on the Shrinkage and Creep of Concrete." *Development Department Bulletin DX103*, Portland Cement Association, Skokie, IL.

Hansen, W. R. (1965). *Effects of the Earthquake of March 27, 1964 at Anchorage, Alaska.* Geological Survey Professional Paper 542-A, U.S. Department of the Interior, Washington, DC.

Hanson, J. A. (1968). "Effects of Curing and Drying Environments on Splitting Tensile Strength of Concrete." *Development Department Bulletin DX141*, Portland Cement Association, Skokie, IL.

Harr, M. E. (1962). *Groundwater and Seepage.* McGraw-Hill, New York, NY, 315 pp.

Harrison, B. A. and Blight, G. E. (2000). "A Comparison of In Situ Soil Suction Measurements." *Unsaturated Soils for Asia, Singapore*, H. Rahardo, D. Toll, and E. Leong, eds., Balkema, Rotterdam, The Netherlands, pp. 281–285.

Hawkins, A. B. and Pinches, G. (1987). "Expansion due to Gypsum Crystal Growth." *Proceedings of the Sixth International Conference on Expansive Soils*, Vol. 1, New Delhi, India, pp. 183–188.

Hawkins, A. B. and Privett, K. D. (1985). "Measurement and Use of Residual Shear Strength of Cohesive Soils." *Ground Engineering*, Vol. 18, No. 8, pp. 22–29.

Heaton, T. H., Tajima, F., and Mori, A. W. (1982). *Estimating Ground Motions Using Recorded Accelerograms.* Report by Dames & Moore to Exxon Production Research Company, Houston, TX.

Hollingsworth, R. A. and Grover, D. J. (1992). "Causes and Evaluation of Residential Damage in Southern California." *Engineering Geology Practice in Southern California*, Pipkin, B. W. and Proctor, R. J., eds., Star Publishing Company, Belmont, CA, pp. 427–441.

Hollingsworth, R. A. and Kovacs, G. S. (1981). "Soil Slumps and Debris Flows: Prediction and Protection." *Bulletin of the Association of Engineering Geologists*, Vol. 18, No. 1, pp. 17–28.

Holtz, R. D. and Kovacs, W. D. (1981). *An Introduction to Geotechnical Engineering*, PrenticeHall, Englewood Cliffs, NJ, 733 pp.

Holtz, W. G. (1959). "Expansive Clays–Properties and Problems." *Quarterly of the Colorado School of Mines*, Vol. 54, No. 4, pp. 89–125.

Holtz, W. G. (1984). "The Influence of Vegetation on the Swelling and Shrinkage of Clays in the United States of America." *The Influence of Vegetation on Clays*, Thomas Telford LTD, London, England, pp. 69–73.

Holtz, W. G. and Gibbs, H. J. (1956a). "Shear Strength of Pervious Gravelly Soils." *Proceedings*, ASCE, Paper No. 867.

Holtz, W. G. and Gibbs, H. J. (1956b). "Engineering Properties of Expansive Clays." *Transactions*, ASCE, Vol. 121, pp. 641–677.

Horn, H. M. and Deere, D. U. (1962). "Frictional Characteristics of Minerals." *Geotechnique*, Vol. 12, London, England, pp. 319–335.

Hough, B. K. (1969). *Basic Soils Engineering*. Ronald Press, New York, NY, 634 pp.

Housner, G., ed. (1963). "An Engineering Report on the Chilean Earthquakes of May 1960." *Bulletin of the Seismological Society of America*, Vol. 53, No. 2.

Housner, G. W. (1970). "Strong Ground Motion." Chapter 4 of *Earthquake Engineering*, Robert L. Wiegel, Coordinating Editor. PrenticeHall, Englewood Cliffs, NJ, pp. 75–92.

Houston, S. L. and Walsh, K. D. (1993). Comparison of Rock Correction Methods for Compaction of Clayey Soils." *Journal of Geotechnical Engineering*, ASCE, Vol. 119, No. 4, pp. 763–778.

Howard, A. K. (1977). *Laboratory Classification of Soils–Unified Soil Classification System*. Earth Sciences Training Manual No. 4, U.S. Bureau of Reclamation, Denver, CO, 56 pp.

Hurst, W. D. (1968). "Experience in the Winnipeg Area with Sulphate-Resisting Cement Concrete." *Performance of Concrete, Resistance of Concrete to Sulphate and Other Environmental Conditions*, University of Toronto Press, Toronto, Canada, pp. 125–134.

Hvorslev, M. J. (1949). *Subsurface Exploration and Sampling of Soils for Civil Engineering Purposes: Report on a Research Project of the Committee on Sampling and Testing Soil Mechanics and Foundations Division, American Society of Civil Engineers*, Sponsored by Engineering Foundation, Graduate School of Engineering at Harvard University, and Waterways Experiment Station, Corps of Engineers, U.S. Army. Edited and Published by Waterways Experiment Station, Vicksburg, MS, 521 pp.

Hvorslev, M. J. (1951). "Time Lag and Soil Permeability in Ground Water Measurements," Corps of Engineers Waterways Experiment Station, Vicksburg, MI, *Bulletin 36*, 50 pp.

Hynes-Griffin, M. E., and Franklin, A. G. (1984). "Rationalizing the Seismic Coefficient Method." *Miscellaneous Paper GL-84-13*, U.S. Army Corps of Engineers Waterways Experiment Station, Vicksburg, MS, 21 pp.

Idriss, I. M. (1985). "Evaluating Seismic Risk in Engineering Practice." *Proceedings of the Eleventh International Conference on Soil Mechanics and Foundation Engineering*, San Francisco, Vol. 1, pp. 255–320.

Idriss, I. M. (1999). "Presentation Notes: An Update of the Seed-Idriss Simplified Procedure for Evaluating Liquefaction Potential." *Proceedings, TRB Workshop on New Approaches to Liquefaction Analysis, Publication No. FHWA-RD-99-165*, Federal Highway Administration,Washington, DC.

International Building Code (2009). International Code Council, Country Club Hills, IL.

International Existing Code (2009). International Code Council, Country Club Hills, IL.

International Residential Code (2009). International Code Council, Country Club Hills, IL.

Ishihara, K. (1985). "Stability of Natural Deposits During Earthquakes." *Proceedings of the Eleventh International Conference on Soil Mechanics and Foundation Engineering*, Vol. 1, San Francisco, pp. 321–376.

Ishihara, K. (1993). "Liquefaction and Flow Failure During Earthquakes." *Geotechnique*, Vol. 43, No. 3, London, England, pp. 351–415.

Ishihara, K, Sodekawa, M, and Tanaka, Y. (1978). "Effects of Overconsolidation on Liquefaction Characteristics of Sands Containing Fines." *Dynamic Geotechnical Testing, ASTM Special Technical Publication 654*, ASTM, Philadelphia, PA, pp. 246–264.

Ishihara, K. and Yoshimine, M. (1992). "Evaluation of Settlements in Sand Deposits Following Liquefaction During Earthquakes." *Soils and Foundations*, Vol. 32, No. 1, pp. 173–188.

Jaky, J. (1944). "The Coefficient of Earth Pressure at Rest." (in Hungarian), *Journal of the Society of Hungarian Architects and Engineers*, Vol. 78, No. 22, pp. 355–358.

Jaky, J. (1948). "Earth Pressure in Silos." *Proceedings of the Second International Conference on Soil Mechanics and Foundation Engineering*, Rotterdam, Netherlands, Vol. 1, pp. 103–107.

Jamiolkowski, M, Ladd, C. C., Germaine, J. T., and Lancellotta, R. (1985). "New Developments in Field and Laboratory Testing of Soils." *Proceedings of the Eleventh International Conference on Soil Mechanics and Foundation Engineering*, San Francisco, Vol. 1, pp. 57–153.

Janbu, N. (1957). "Earth Pressure and Bearing Capacity Calculation by Generalized Procedure of Slices." *Proceedings of the Fourth International Conference on Soil Mechanics and Foundation Engineering*, London, England, Vol. 2, pp. 207–212.

Janbu, N. (1968). "Slope Stability Computations." *Soil Mechanics and Foundation Engineering Report*, The Technical University of Norway, Trondheim, Norway.

Jennings, J. E. (1953). "The Heaving of Buildings on Desiccated Clay." *Proceedings of the Third International Conference on Soil Mechanics and Foundation Engineering*, Vol. 1, Zurich, Switzerland, pp. 390–396.

Jennings, J. E. and Knight, K. (1957). "The Additional Settlement of Foundations Due to a Collapse Structure of Sandy Subsoils on Wetting." *Proceedings of the Fourth International Conference on Soil Mechanics and Foundation Engineering*, Vol. 1, London, England, pp. 316–319.

Jimenez Salas, J. A. (1948). "Soil Pressure Computations: A Modification of Newmark's Method." *Proceedings of the Second International Conference on Soil Mechanics and Foundation Engineering*, Rotterdam.

Johnpeer, G. D. (1986). "Land Subsidence Caused by Collapsible Soils in Northern New Mexico." *Ground Failure*, National Research Council, Committee on Ground Failure Hazards, Vol. 3, Washington, DC, 24 pp.

Johnson, A. M. and Hampton, M. A. (1969). "Subaerial and Subaqueous Flow of Slurries." *Final Report, U.S. Geological Survey (USGS) Contract No. 14-08-0001-10884*, USGS, Boulder, CO.

Johnson, A. M. and Rodine, J. R. (1984). "Debris Flow." Chapter 8 of *Slope Instability*, Wiley, New York, NY, pp. 257–361.

Johnson, L. D. (1980). "Field Test Sections on Expansive Soil." *Fourth International Conference on Expansive Soils*, ASCE, Denver, CO, pp. 262–283.

Jones, D. E. and Jones, K. A. (1987). "Treating Expansive Soils." *Civil Engineering Magazine*, ASCE, Vol. 57, No. 8, August Issue.

Jones, D. E. and Holtz, W. G. (1973). "Expansive Soils—The Hidden Disaster." *Civil Engineering Magazine*, ASCE, Vol. 43, No. 8, pp. 49–51.

Jubenville, D. M. and Hepworth, R. C. (1981). "Drilled Pier Foundations in Shale, Denver Colorado Area." *Proceedings of the Session on Drilled Piers and Caissons*, ASCE, New York, pp. 66–81.

Kanamori, H. (1977). "The Energy Release in Great Earthquakes," *Journal of Geophysical Research*, Vol. 82, pp. 2981–2987.

Kaplar, C. W. (1970). "Phenomenon and Mechanism of Frost Heaving." *Highway Research Record 304*, pp. 1–13.

Kassiff, G. and Baker, R. (1971). "Aging Effects on Swell Potential of Compacted Clay." *Journal of the Soil Mechanics and Foundations Division*, ASCE, Vol. 97, No. SM3, pp. 529–540.

Kayan, R. E., Mitchell, J. K., Seed, R. B., Lodge, A., Nishio, S., and Coutinho, R. (1992). "Evaluation of SPT-, CPT-, and Shear Wave-Based Methods for Liquefaction Potential Assessments Using Loma Prieta Data." *Proceedings, Fourth Japan-U.S. Workshop on Earthquake Resistant Design of Lifeline Facilities and Countermeasures for Soil Liquefaction; NCEER-92-0019*, National Center for Earthquake Engineering, Buffalo, NY, pp. 177–192.

Keefer, D. K. (1984). "Landslides Caused by Earthquakes." *Geological Society of America Bulletin*, Vol. 95, No. 2, pp. 406–421.

Kennedy, M. P. (1975). "Geology of the Western San Diego Metropolitan Area, California." *Bulletin 200*, California Division of Mines and Geology, Sacramento, CA, 39 pp.

Kennedy, M. P. and Tan, S. S. (1977). "Geology of National City, Imperial Beach and Otay Mesa Quadrangles, Southern San Diego Metropolitan Area, California." *Map Sheet 29*, California Division of Mines and Geology, Sacramento, CA.

Kenney, T. C. (1964). "Sea-Level Movements and the Geologic Histories of the Post-Glacial Marine Soils at Boston, Nicolet, Ottawa, and Oslo." *Geotechnique*, Vol. 14, No. 3, London, England, pp. 203–230.

Kerwin, S. T. and Stone, J. J. (1997). "Liquefaction Failure and Remediation: King Harbor Redondo Beach, California." *Journal of Geotechnical and Geoenvironmental Engineering*, ASCE, Vol. 123, No. 8, pp. 760–769.

Koerner, R. (1998). *Designing with Geosynthetics*, 4th ed., Prentice Hall, Englewood Cliffs, NJ, 761 pp.

Kononova, M. M. (1966). *Soil Organic Matter*, 2nd ed., Pergamon Press, Oxford, England.

Krahn, J. and Fredlund, D. G. (1972). "On Total Matric and Osmotic Suction." *Journal of Soil Science*, Vol. 114, No. 5, pp. 339–348.

Kramer, S. L. (1996). *Geotechnical Earthquake Engineering*. PrenticeHall, Englewood Cliffs, NJ, 653 pp.

Kratzsch, H. (1983). *Mining Subsidence Engineering*. Springer-Verlag, Berlin, Germany, 543 pp.

Krinitzsky, E. L., Gould, J. P., and Edinger, P. H. (1993). *Fundamentals of Earthquake-Resistant Construction*. Wiley, New York, NY, 299 pp.

Ladd, C. C. (1971). "Strength Parameters and Stress-Strain Behavior of Saturated Clays." Massachusetts Institute of Technology, Cambridge, MA, MIT Research Report R71-23.

Ladd, C. C. (1973). "Settlement Analysis for Cohesive Soil." *MIT Research Report R71-2, No. 272*, Department of Civil Engineering, Massachusetts Institute of Technology, Cambridge, MA, 115 pp.

Ladd, C. C., Azzouz, A. S., Martin, R. T., Day, R. W., and Malek, A. M. (1980). *Evaluation of Compositional and Engineering Properties of Offshore Venezuelan Soils*. Vol. 1, MIT Research Report R80-14, No. 665, Cambridge, MA, 286 pp.

Ladd, C. C., Foote, R., Ishihara, K., Schlosser, F., and Poulos, H. G. (1977). "Stress-Deformation and Strength Characteristics." *State-of-the-Art Report, Proceedings, Ninth International Conference on Soil Mechanics and Foundation Engineering*, Japanese Society of Soil Mechanics and Foundation Engineering, Tokyo, Japan, Vol. 2, pp. 421–494.

Ladd, C. C. and Lambe, T. W. (1961). "The Identification and Behavior of Expansive Clays." *Proceedings, Fifth International Conference on Soil Mechanics and Foundation Engineering*, Paris, France, Vol. 1.

Ladd, C. C. and Lambe T. W. (1963). "The Strength of 'Undisturbed' Clay Determined from Undrained Tests," *STP 361*, ASTM, Philadelphia, PA, pp. 342–371.

Lambe, T. W. (1951). *Soil Testing for Engineers*. Wiley, New York, NY, 165 pp.

Lambe, T. W. (1958a). "The Structure of Compacted Clay." *Journal of the Soil Mechanics and Foundations Division*, ASCE, Vol. 84, No. SM2, pp. 1654-1 to 1654-34.

Lambe, T. W. (1958b). "The Engineering Behavior of Compacted Clay." *Journal of the Soil Mechanics and Foundations Division*, ASCE, Vol. 84, No. SM2, pp. 1655-1 to 1655-35.

Lambe, T. W. (1967). "Stress Path Method." *Journal of the Soil Mechanics and Foundations Division*, ASCE, Vol. 93, No. SM6, pp. 309–331.

Lambe, T. W., and Whitman, R. V. (1969). *Soil Mechanics*. Wiley, New York, NY, 553 pp.

Lane et al. (1947). "Modified Wentworth Scale." *Transactions of the American Geophysical Union*, Vol. 28, pp. 936–938.

Lawson, A. C. et al. (1908). *The California Earthquake of April 18, 1906–Report of the State Earthquake Investigation Commission*, Vol. 1, Part 1, pp. 1–254; Part 2, pp. 255–451. Carnegie Institution of Washington, Publication 87.

Lawton, E. C. (1996). "Nongrouting Techniques." *Practical Foundation Engineering Handbook*. Robert W. Brown. ed., McGraw-Hill, New York, NY, Section 5, pp. 5.3–5.276.

Lawton, E. C., Fragaszy, R. J., and Hardcastle, J. H. (1989). "Collapse of Compacted Clayey Sand." *Journal of Geotechnical Engineering*, ASCE, Vol. 115, No. 9, pp. 1252–1267.

Lawton, E. C., Fragaszy, R. J., and Hardcastle, J. H. (1991). "Stress Ratio Effects on Collapse of Compacted Clayey Sand." *Journal of Geotechnical Engineering*, ASCE, Vol. 117, No. 5, pp. 714–730.

Lawton, E. C., Fragaszy, R. J., and Hetherington, M. D. (1992). "Review of Wetting-Induced Collapse in Compacted Soil." *Journal of Geotechnical Engineering*, ASCE, Vol. 118, No. 9, pp. 1376–1394.

Lea, F. M. (1971). *The Chemistry of Cement and Concrete*, First American Edition, Chemical Publishing Company, New York, NY.

Leonards, G. A. (1962). *Foundation Engineering*. McGraw-Hill, New York, NY, 1136 pp.

Leonards, G. A. (1976). "Estimating Consolidation Settlements of Shallow Foundations on Overconsolidated Clay." *Estimation of Consolidation Settlement*, TRB Special Report 163, National Research Council, pp. 13–16.

Leonards, G. A. and Altschaeffl, A. G. (1964). "Compressibility of Clay." *Journal of the Soil Mechanics and Foundations Division*, ASCE, Vol. 90, No. SM5, pp. 133–156.

Leonards, G. A. and Ramiah, B. K. (1959). "Time Effects in the Consolidation of Clay." *Special Technical Publication No. 254*, ASTM, Philadelphia, PA, pp. 116–130.

Lin, G., Bennett, R. M., Drumm, E. C., and Triplett, T. L. (1995). "Response of Residential Test Foundations to Large Ground Movements." *Journal of Performance of Constructed Facilities*, ASCE, Vol. 9, No. 4, pp. 319–329.

Lowe, J. and Zaccheo, P. F. (1975). "Subsurface Explorations and Sampling." Chapter 1 of *Foundation Engineering Handbook*, Hans F. Winterkorn and Hsai-Yang Fang, eds., Van Nostrand Reinhold Company, New York, NY, pp. 1–66.

Lytton, R. L. (1977). "Foundations in Expansive Soils." *Numerical Methods in Geotechnical Engineering*, McGraw-Hill, New York, NY, pp. 427–457.

Lytton, R. L. and Dyke, L. D. (1980). "Creep Damage to Structures on Expansive Clays Slopes." *Fourth International Conference on Expansive Soils*, ASCE, Vol. 1, New York, NY, pp. 284–301.

Mabsout, M. E., Reese, L. C. and Tassoulas, J. L. (1995). "Study of Pile Driving by Finite-Element Method." *Journal of Geotechnical Engineering*, ASCE, Vol. 121, No. 7, pp. 535–543.

Maksimovic, M. (1989a). "On the Residual Shearing Strength of Clays." *Geotechnique*, London, England, Vol. 39, No. 2, pp. 347–351.

Maksimovic, M. (1989b). "Nonlinear Failure Envelope for Soils." *Journal of Geotechnical Engineering*, ASCE, Vol. 115, No. 4, pp. 581–586.

Marcuson, W. F. (1981). "Moderator's Report for Session on 'Earth Dams and Stability of Slopes Under Dynamic Loads'." *Proceedings, International Conference on Recent Advances in Geotechnical Earthquake Engineering and Soil Dynamics*, St. Louis, MO, Vol. 3, 1175 pp.

Marcuson, W. F. and Hynes, M. E. (1990). "Stability of Slopes and Embankments during Earthquakes." *Proceedings*, Geotechnical Seminar Sponsored by ASCE and the Pennsylvania Department of Transportation, Hershey, PA.

Marino, G. G., Mahar, J. W., and Murphy, E. W. (1988). "Advanced Reconstruction for Subsidence-Damaged Homes." *Mine Induced Subsidence: Effects on Engineered Structures*. H. J. Siriwardane, ed., ASCE, New York, NY, pp. 87–106.

Marsh, E. T. and Thoeny, S. A. (1999). "Damage and Distortion Criteria for Residential Slab-on-Grade Structures." *Journal of Performance of Constructed Facilities*, ASCE, Vol. 13, No. 3, pp. 121–127.

Marsh, E. T. and Walsh, R. K. (1996). "Common Causes of Retaining-Wall Distress: Case Study." *Journal of Performance of Constructed Facilities*, ASCE, Vol. 10, No. 1, pp. 35–38.

Masia, M. J., Totoev, Y. Z., and Kleeman, P. W. (2004). "Modeling Expansive Soil Movements Beneath Structures." *Journal of Geotechnical and Geoenvironmental Engineering*, ASCE, Vol. 130, No. 6, pp. 572–579.

Massarsch, K. R. (1979). "Lateral Earth Pressure in Normally Consolidated Clay." *Proceedings of the Seventh European Conference on Soil Mechanics and Foundation Engineering*, Brighton, England, Vol. 2, pp. 245–250.

Massarsch, K. R., Holtz, R. D., Holm, B. G., and Fredricksson, A. (1975). "Measurement of Horizontal In Situ Stresses." *Proceedings of the ASCE Specialty Conference on In Situ Measurement of Soil Properties*, Raleigh, North Carolina, Vol. 1, pp. 266–286.

Mather, B. (1968). "Field and Laboratory Studies of the Sulphate Resistance of Concrete." *Performance of Concrete, Resistance of Concrete to Sulphate and Other Environmental Conditions*, University of Toronto Press, Toronto, Canada, pp. 66–76.

Matlock, H. (1970). "Correlations for Design of Laterally-Loaded Piles in Soft Clay." *Proceeding of the Second Offshore Technology Conference*, Vol. 1, OTC, Houston, Texas, pp. 577–594.

McCarthy, D. F. (1977). *Essentials of Soil Mechanics and Foundations*. Reston Publishing Company, Reston Virginia, 505 pp.

Meehan, R. L. and Karp, L. B. (1994). "California Housing Damage Related to Expansive Soils." *Journal of Performance of Constructed Facilities*, ASCE, Vol. 8, No. 2, pp. 139–157.

Mehta, P. K. (1976). Discussion of "Combating Sulfate Attack in Corps of Engineers Concrete Construction." T. J. Reading, ed., *ACI Journal Proceedings*, Vol. 73, No. 4, pp. 237–238.

Merfield, P. M. (1992). "Surficial Slope Failures: the Role of Vegetation and Other Lessons from Rainstorms." *Engineering Geology Practice in Southern California*. B. W. Pipkin and R. J. Proctor, eds., Star Publishing Company, Association of Engineering Geologists, Southern California Section, Special Publication No. 4, pp. 613–627.

Meyerhof, G. G. (1951). "The Ultimate Bearing Capacity of Foundations." *Geotechnique*, Vol. 2, No. 4, pp. 301–332.

Meyerhof, G. G. (1953). "Bearing Capacity of Foundations under Eccentric and Inclined Loads." *Proceedings of the Third International Conference on Soil Mechanics and Foundation Engineering*, Vol. 1, Zurich, pp. 440–445.

Meyerhof, G. G. (1961). Discussion of "Foundations Other than Piled Foundations." *Proceedings of the Fifth International Conference on Soil Mechanics and Foundation Engineering*, Vol. 3, Paris, p. 193.

Meyerhof, G. G. (1963). "Some Recent Research on the Bearing Capacity of Foundations." *Canadian Geotechnical Journal*, Vol. 1, No. 1.

Meyerhof, G. G. (1965). "Shallow Foundations." *Journal of the Soil Mechanics and Foundations Division*, ASCE, Vol. 91, No. SM2, pp. 21–31.

Mitchell, J. K. (1970). "In-Place Treatment of Foundation Soils." *Journal of the Soil Mechanics and Foundations Division*, ASCE, Vol. 96, No. SM1, pp. 73–110.

Mitchell, J. K. (1976). *Fundamentals of Soil Behavior*. Wiley, New York, NY, 422 pp.

Mitchell, J. K. (1978). "In Situ Techniques for Site Characterization." *Site Characterization & Exploration. Proceedings of Specialty Workshop*, Northwestern University, C. H. Dowding, ed., ASCE, New York, NY, pp. 107–130.

Mononobe, N. and Matsuo, H. (1929). "On the Determination of Earth Pressures during Earthquakes." *Proceedings, World Engineering Congress*, 9 pp.

Morgenstern, N. R. and Price, V. E. (1965). "The Analysis of the Stability of General Slip Surfaces." *Geotechnique*, Vol. 15, No. 1, London, England, pp. 79–93.

Mottana, A., Crespi, R., and Liborio, G. (1978). *Rocks and Minerals*. Simon & Schuster, New York, NY, 607 pp.

Munsell Soil Color Charts (1975). Munsell Color, A Division of Kollmorgen Corporation, Baltimore, MD.

Myslivec, A. and Kysela, Z. (1978). *The Bearing Capacity of Building Foundations*, Developments in Geotechnical Engineering 21, Elsevier, New York, NY, 237 pp.

Nadim, F. (1982). "A Numerical Model for Evaluation of Seismic Behavior of Gravity Retaining Walls." *Research Report R82-33*, Department of Civil Engineering, Massachusetts Institute of Technology, Cambridge, MA.

Nadim, F. and Whitman, R. V. (1984). "Coupled Sliding and Tilting of Gravity Retaining Walls During Earthquakes." *Proceedings, Eighth World Conference on Earthquake Engineering*, San Francisco, Vol. 3, pp. 477–484.

Nadjer, J. and Werno, M. (1973). "Protection of Buildings on Expansive Clays." *Proceeding of the Third International Conference on Expansive Soils*, Haifa, Israel, Vol. 1, pp. 325–334.

National Coal Board (1975). *Subsidence Engineers Handbook*. National Coal Board Mining Department, National Coal Board, Hobart House, Grosvenor Square, London, England, 111 pp.

National Information Service for Earthquake Engineering (2000). "Building and Its Structure Should Have a Uniform and Continuous Distribution of Mass, Stiffness, Strength and Ductility." University of California, Berkeley. Report obtained from the Internet.

National Research Council (1985a). *Reducing Losses from Landsliding in the United States*. Committee on Ground Failure Hazards, Commission on Engineering and Technical Systems, National Academy Press, Washington, DC, 41 pp.

National Research Council (1985b). *Liquefaction of Soils During Earthquakes*, National Academy Press, Washington, DC, 240 pp.

National Science Foundation (NSF). (1992). "Quantitative Nondestructive Evaluation for Constructed Facilities." *Announcement Fiscal Year 1992*. Directorate for Engineering, Division of Mechanical and Structural Systems, Washington, DC.

NAVFAC DM-7 (1971). *Soil Mechanics, Foundations, and Earth Structures*, Design Manual DM-7, Department of the Navy, Naval Facilities Engineering Command, Alexandria, Va., 280 pp.

NAVFAC DM-7.1 (1982). *Soil Mechanics, Design Manual 7.1*, Department of the Navy, Naval Facilities Engineering Command, Alexandria, VA, 364 pp.

NAVFAC DM-7.2 (1982). *Foundations and Earth Structures, Design Manual 7.2*, Department of the Navy, Naval Facilities Engineering Command, Alexandria, VA, 253 pp.

NAVFAC DM-7.3 (1983). *Soil Dynamics, Deep Stabilization, and Special Geotechnical Construction, Design Manual 7.3*, Department of the Navy, Naval Facilities Engineering Command, Alexandria, VA, 106 pp.

Neary, D. G. and Swift, L. W. (1987). "Rainfall Thresholds for Triggering a Debris Avalanching Event in the Southern Appalachian Mountains." *Debris Flows/Avalanches: Process, Recognition, and Mitigation*, The Geological Society of America, Boulder, CO, pp. 81–92.

Nelson, J. D. and Miller, D. J. (1992). *Expansive Soils, Problems and Practice in Foundation and Pavement Engineering*. Wiley, New York, NY, 259 pp.

Nelson, N. A. (2000). "Slope Stability, Triggering Events, Mass Wasting Hazards." Tulane University, Natural Disasters Internet Site.

Newmark, N. M. (1935). "Simplified Computation of Vertical Pressures in Elastic Foundations." *University of Illinois Engineering Experiment Station Circular 24*, Urbana, IL, 19 pp.

Newmark, N. M. (1942). "Influence Charts for Computation of Stresses in Elastic Foundations." University of Illinois Engineering Experiment Station Bulletin, Series No. 338, Vol. 61, No. 92, Urbana, IL, reprinted 1964, 268 pp.

Newmark, N. M. (1965). "Effects of Earthquakes on Dams and Embankments." *Geotechnique*, London, Vol. 15, No. 2, pp. 139–160.

Nichols, H. L. and Day, D. A. (1999). *Moving the Earth: The Workbook of Excavation*. 4th ed., McGraw-Hill, New York, NY, 1400 pp.

Noorany, I. (1984). "Phase Relations in Marine Soils." *Journal of Geotechnical Engineering*, ASCE, Vol. 110, No. 4, pp. 539–543.

Ohsaki, Y (1969). "The Effects of Local Soil Conditions upon Earthquake Damage." Paper Presented at Session No. 2, Soil Dynamics, *Proceedings of the Seventh International Conference on Soil Mechanics and Foundation Engineering*, Mexico, Vol. 3, pp. 421–422.

Okabe, S. (1926). "General Theory of Earth Pressures." *Journal of the Japan Society of Civil Engineering*, Vol. 12, No. 1.

Oldham, R. D. (1899). "Report on the Great Earthquake of 12th June, 1897." *India Geologic Survey Memorial Publication 29*, 379 pp.

Oliver, A. C. (1988). *Dampness in Buildings*. Internal and Surface Waterproofers, Nichols Publishing, New York, NY, 221 pp.

Olson, R. E. and Langfelder, L. J. (1965). "Pore-Water Pressures in Unsaturated Soils." *Journal of the Soil Mechanics and Foundations Division*, ASCE, Vol. 91, No. SM4, pp. 127–160.

Olson, S. M., Stark, T. D., Walton, W. H., and Castro, G. (2000). "1907 Static Liquefaction Flow Failure of the North Dike of Wachusett Dam." *Journal of Geotechnical and Geoenvironmental Engineering*, ASCE, Vol. 126, No. 12, pp. 1184–1193.

O'Neill, M. W. and Reese, L. C. (1970). *Behavior of Axially Loaded Drilled Shafts in Beaumont Clay*, Research Report 89-8, Part 1, State-of-the-Art Report, Center for Highway Research, University of Texas at Austin.

Orange County Grading Manual (1993). *Orange County Grading Manual*, part of the *Orange County Grading and Excavation Code*, Orange County, CA.

Ortigao, J. A. R., Loures, T. R. R., Nogueiro, C., and Alves, L. S. (1997). "Slope Failures in Tertiary Expansive OC Clays." *Journal of Geotechnical and Geoenvironmental Engineering*, ASCE, Vol. 123, No. 9, pp. 812–817.

Osterberg, J. O. (1957). "Influence Values for Vertical Stresses in a Semi-infinite Mass Due to an Embankment Loading." *Proceedings of the Fourth International Conference on Soil Mechanics and Foundation Engineering*, Vol. 1, London, England, pp. 393–394.

Pacheco, J. F. and Sykes, L. R. (1992). "Seismic Moment Catalog of Large Shallow Earthquakes, 1900 to 1989." *Bulletin of the Seismological Society of America*, Vol. 82, pp. 1306–1349.

PCA (1994). *Design and Control of Concrete Mixtures*. Fourth Printing of the Thirteenth Edition, Portland Cement Association, Stokie, IL, 205 pp.

Peck, R. B., Hanson, W. E., and Thornburn, T. H. (1974). *Foundation Engineering*, Wiley, New York, NY, 515 pp.

Peckover, F. L. (1975). "Treatment of Rock Falls on Railway Lines." *American Railway Engineering Association, Bulletin 653*, Chicago, IL, pp. 471–503.

Peng, S. S. (1986). *Coal Mine Ground Control*. 2nd ed., Wiley, New York, NY, 491 pp.

Peng, S. S. (1992). *Surface Subsidence Engineering*. Society for Mining, Metallurgy and Exploration, Littleton, CO.

Perloff, W. H. and Baron, W. (1976). *Soil Mechanics, Principles and Applications*. Wiley, New York, NY, 745 pp.

Piteau, D. R., and Peckover, F. L. (1978). "Engineering of Rock Slopes." *Landslides, Analysis and Control, Special Report 176*, Transportation Research Board, National Academy of Sciences, Chapter 9, pp. 192–228.

Plafker, G. (1972). "Alaskan Earthquake of 1964 and Chilean Earthquake of 1960: Implications for Arc Tectonics." *Journal of Geophysical Research*, Vol. 77, pp. 901–925.

Plafker, G., Ericksen, G. E., and Fernandez, C. J. (1971). "Geological Aspects of the May 31, 1970, Peru Earthquake." *Bulletin of the Seismological Society of America*, Vol. 61, pp. 543–578.

Post-Tensioning Institute (1996). "Recommendations for Prestressed Rock and Soil Anchors." Post Tensioning Manual, 4th ed., Phoenix, AZ, 41 pp.

Post-Tensioning Institute (1996). "Design and Construction of Post-tensioned Slabs-on-Ground," report, 2nd ed., Phoenix, AZ, 101 pp.

Poulos, H. G. and Davis, E. H. (1974). *Elastic Solutions for Soil and Rock Mechanics*. Wiley, New York, NY, 411 pp.

Poulos, S. J., Castro, G., and France, J. W. (1985). "Liquefaction Evaluation Procedure." *Journal of Geotechnical Engineering*, ASCE, Vol. 111, No. 6, pp. 772–792.

Pradel, D. and Raad, G. (1993). "Effect of Permeability on Surficial Stability of Homogeneous Slopes." *Journal of Geotechnical Engineering*, ASCE, Vol. 119, No. 2, pp. 315–332.

Prentis, E. A. and White, L. (1950). *Underpinning*. 2d ed., Columbia University Press, New York, NY.

Price, N. J. (1966). *Fault and Joint Development in Brittle and Semi-Brittle Rock*. Pergamon Press, Oxford.

Proctor, R. R. (1933). "Fundamental Principles of Soil Compaction." *Engineering News-Record*, Vol. 111, Nos. 9, 10, 12, and 13.

Purkey, B. W., Duebendorfer, E. M., Smith, E. I., Price, J. G., Castor, S. B. (1994). *Geologic Tours in the Las Vegas Area*, Nevada Bureau of Mines and Geology, Special Publication 16, Las Vegas, NV, 156 pp.

Pyke, R., Seed, H. B., and Chan, C. K. (1975). "Settlement of Sands under Multidirectional Shaking." *Journal of the Geotechnical Engineering Division*, ASCE, Vol. 101, No. GT 4, pp. 379–398.

Raschke, S. A. and Hryciw, R. D. (1997). "Vision Cone Penetrometer for Direct Subsurface Soil Observation." *Journal of Geotechnical and Geoenvironmental Engineering*, ASCE, Vol. 123, No. 11, pp. 1074–1076.

Rathje, W. L. and Psihoyos, L. (1991). "Once and Future Landfills." *National Geographic*, Vol. 179, No. 5, pp. 116–134.

Raumann, G. (1982). "Outdoor Exposure Tests on Geotextiles." *Proceedings of the Second International Conference on Geotextiles*, Industrial Fabrics Association International, St. Paul, MN.

Ravina, I. (1984). "The Influence of Vegetation on Moisture and Volume Changes." *The Influence of Vegetation on Clays*, Thomas Telford, London, England, pp. 62–68.

Reading, T. J. (1975). "Combating Sulfate Attack in Corps of Engineering Concrete Construction." *Durability of Concrete, SP47*, American Concrete Institute, Detroit, MI, pp. 343–366.

Reed, M. A., Lovell, C. W., Altschaeffl, A. G., and Wood, L. E. (1979). "Frost Heaving Rate Predicted from Pore Size Distribution," *Canadian Geotechnical Journal*, Vol. 16, No. 3, pp. 463–472.

Reese, L. C., Cox, W. R., and Koop, F. D. (1974). "Analysis of Laterally Loaded Piles in Sand." *Proceedings of the Fifth Offshore Technology Conference*, OTC, Houston, TX.

Reese, L. C., Cox, W. R., and Koop, F. D. (1975). "Field Testing and Analysis of Laterally Loaded Piles in Stiff Clay." *Proceedings of the Seventh Offshore Technology Conference*, OTC, Houston, TX.

Richardson, G. N., and Koerner, R. M. (1990). *A Design Primer: Geotextiles and Related Materials*, Industrial Fabrics Association, St. Paul, MN.

Richardson, G. N. and Wyant, D. C. (1987). "Geotextiles Construction Criteria." *Geotextile Testing and the Design Engineer*, J. E. Fluet, ed., ASTM Special Technical Publication 952, American Society for Testing and Materials, Philadelphia, PA, pp. 125–138.

Richter, C. F. (1935). "An Instrumental Earthquake Magnitude Scale." *Bulletin of the Seismological Society of America*, Vol. 25, pp. 1–32.

Richter, C. F. (1958). *Elementary Seismology*. W. H. Freeman, San Francisco, 768 pp.

Ritchie, A. M. (1963). "Evaluation of Rockfall and Its Control." *Highway Research Record 17*, Highway Research Board, Washington, DC, pp. 13–28.

Robertson, P. K. and Campanella, R. G. (1983). "Interpretation of Cone Penetration Tests: Parts 1 and 2." *Canadian Geotechnical Journal*, Vol. 20, pp. 718–745.

Rogers, J. D. (1992). "Recent Developments in Landslide Mitigation Techniques." Chapter 10 of *Landslides/Landslide Mitigation*. J. E. Slosson, G. G. Keene, and J. A. Johnson, eds., The Geological Society of America, Boulder, CO, pp. 95–118.

Rollings, M. P. and Rollings, R. S. (1996). *Geotechnical Materials in Construction*. McGraw-Hill, New York, NY, 525 pp.

Rollins, K. M., Peterson, K. T., and Weaver, T. J. (1998). "Lateral Load Behavior of Full-Scale Pile Group in Clay." *Journal of Geotechnical and Geoenvironmental Engineering*, ASCE, Vol. 124, No. 6, pp. 468–478.

Rollins, K. M., Rollins, R. L., Smith, T. D., and Beckwith, G. H. (1994). "Identification and Characterization of Collapsible Gravels." *Journal of Geotechnical Engineering*, ASCE, Vol. 120, No. 3, pp. 528–542.

Rutledge, P. C. (1944). "Relation of Undisturbed Sampling to Laboratory Testing," *Transactions*, ASCE, Vol. 109, pp. 1162–1163.

Sanglerat, G. (1972). *The Penetrometer and Soil Exploration*. Elsevier, New York, NY, 464 pp.

Savage, J. C. and Hastie, L. M. (1966). "Surface Deformation Associated with Dip-Slip Faulting." *Journal of Geophysical Research*, Vol. 71, No. 20, pp. 4897–4904.

Saxena, S. K., Lourie, D. E., and Rao, J. S. (1984). "Compaction Criteria for Eastern Coal Waste Embankments." *Journal of Geotechnical Engineering*, ASCE, Vol. 110, No. 2, pp. 262–284.

Schmertmann, J. H. (1970). "Static Cone to Compute Static Settlement Over Sand." *Journal of the Soil Mechanics and Foundations Division*, ASCE, Vol. 96, SM3, pp. 1011–1043.

Schmertmann, J. H. (1975). "Measurement of In Situ Shear Strength." State-of-the-Art Report, *Proceedings of the ASCE Specialty Conference on In Situ Measurement of Soil Properties*, Raleigh, North Carolina, Vol. 2, pp. 57–138.

Schmertmann, J. H. (1977). *Guidelines for Cone Penetration Test, Performance and Design*. Published by the U.S. Department of Transportation, Federal Highway Administration, Washington, DC, 145 pp.

Schmertmann, J. H., Hartman, J. P., and Brown, P. R. (1978). "Improved Strain Influence Factor Diagrams." *Journal of the Geotechnical Engineering Division*, Vol. 104, No. GT8, pp. 1131–1135.

Schnitzer, M. and Khan, S. U. (1972). *Humic Substances in the Environment*. Marcel Dekker, New York, NY.

Schutz, R. J. (1984). "Properties and Specifications for Epoxies Used in Concrete Repair." *Concrete Construction Magazine*, Published by Concrete Construction Publications, Inc., Addison, IL, pp. 873–878.

Seed, H. B. (1970). "Soil Problems and Soil Behavior." Chapter 10 of *Earthquake Engineering*, Robert L. Wiegel, Coordinating Editor. PrenticeHall, Englewood Cliffs, NJ, pp. 227–252.

Seed, H. B. (1979a). "Soil Liquefaction and Cyclic Mobility Evaluation for Level Ground During Earthquakes." *Journal of the Geotechnical Engineering Division*, ASCE, Vol. 105, No. GT2, pp. 201–255.

Seed, H. B. (1979b). "Considerations in the Earthquake-Resistant Design of Earth and Rockfill Dams." *Geotechnique*, Vol. 29, No. 3, pp. 215–263.

Seed, H. B. and Idriss, I. M. (1971). "Simplified Procedure for Evaluating Soil Liquefaction Potential." *Journal of the Soil Mechanics and Foundations Division*, ASCE, Vol. 97, No. SM9, pp. 1249–1273.

Seed, H. B. and Idriss, I. M. (1982). *Ground Motions and Soil Liquefaction during Earthquakes*, Earthquake Engineering Research Institute, University of California, Berkeley, 134 pp.

Seed, H. B., Idriss, I. M., and Arango, I. (1983). "Evaluation of Liquefaction Potential Using Field Performance Data." *Journal of Geotechnical Engineering*, ASCE, Vol. 109, No. 3, pp. 458–482.

Seed, H. B. and Martin, G. R. (1966). "The Seismic Coefficient in Earth Dam Design." *Journal of the Soil Mechanics and Foundations Division*, ASCE, Vol. 92, No. SM3, pp. 25–58.

Seed, H. B., Mori, K., and Chan, C. K. (1975). "Influence of Seismic History on the Liquefaction Characteristics of Sands." *Report EERC 75-25*, Earthquake Engineering Research Center, University of California, Berkeley, 21 pp.

Seed, H. B. and Peacock, W. H. (1971). "Test Procedures for Measuring Soil Liquefaction Characteristics." *Journal of the Soil Mechanics and Foundations Division*, ASCE, Vol. 97, No. SM8, pp. 1099–1119.

Seed, H. B. and Silver, M. L. (1972). "Settlement of Dry Sands during Earthquakes." *Journal of the Soil Mechanics and Foundations Division*, ASCE, Vol. 98, No. SM4, pp. 381–397.

Seed, H. B., Tokimatsu, K., Harder, L. F., and Chung, R. (1985). "Influence of SPT Procedures in Soil Liquefaction Resistance Evaluations." *Journal of Geotechnical Engineering*, ASCE, Vol. 111, No. 12, pp. 1425–1445.

Seed, H. B. and Whitman, R. V. (1970). "Design of Earth Retaining Structures for Dynamic Loads." *Proceedings, ASCE Specialty Conference on Lateral Stresses in the Ground and Design of Earth Retaining Structures*, ASCE, pp. 103–147.

Seed, H. B., Woodward, R. J., and Lundgren, R. (1962). "Prediction of Swelling Potential for Compacted Clays." *Journal of the Soil Mechanics and Foundations Division*, ASCE, Vol. 88, No. SM3, pp. 53–87.

Seed, R. B. (1991). *Liquefaction Manual*, Course Notes for CE 275: Geotechnical Earthquake Engineering, College of Engineering, University of California, Berkeley, CA, 82 pp.

Seed, R. B. et al. (1990). "Preliminary Report on the Principal Geotechnical Aspects of the October 17, 1989 Loma Prieta Earthquake." *Report UCB/EERC-90/05*, Earthquake Engineering Research Center, University of California, Berkeley, CA, 137 pp.

Seed, R. B. and Harder, L. F. (1990). "SPT-Based Analysis of Cyclic Pore Pressure Generation and Undrained Residual Strength." J. M. Duncan, ed., *Proceedings, H. Bolton Seed Memorial Symposium*, University of California, Berkeley, CA, Vol. 2, pp. 351–376.

Seismic Safety Study (1995). City of San Diego, Development Services Department, San Diego, CA.

Shannon and Wilson, Inc. (1964). *Report on Anchorage Area Soil Studies, Alaska, to U.S. Army Engineer District, Anchorage, Alaska*. Seattle, Washington.

Sherard, J. L., Woodward, R. J., Gizienski, S. F., and Clevenger, W. A. (1963). *Earth and Earth-Rock Dams*. Wiley, New York, NY, 725 pp.

Sherif, M. A. and Fang, Y. S. (1984a). "Dynamic Earth Pressures on Wall Rotating About the Top." *Soils and Foundations*, Vol. 24, No. 4, pp. 109–117.

Sherif, M. A. and Fang, Y. S. (1984b). "Dynamic Earth Pressures on Wall Rotating About the Base." *Proceedings, Eighth World Conference on Earthquake Engineering*, San Francisco, Vol. 6, pp. 993–1000.

Sherif, M. A., Ishibashi, I., and Lee, C. D. (1982). "Earth Pressures Against Rigid Retaining Walls." *Journal of the Geotechnical Engineering Division*, ASCE, Vol. 108, No. GT5, pp. 679–695.

Siddharthan, R., Ara, S., and Norris, G. M. (1992). "Simple Rigid Plastic Model for Seismic Tilting of Rigid Walls." *Journal of Structural Engineering*, ASCE, Vol. 118, No. 2, pp. 469–487.

Silver, M. L. and Seed, H. B. (1971). "Volume Changes in Sands during Cyclic Loading." *Journal of the Soil Mechanics and Foundations Division*, ASCE, Vol. 97, No. SM9, pp. 1171–1182.

Singh, J. (1994). "The Built Environment and the Development of Fungi." *Building Mycology: Management of Decay and Health in Buildings*. J. Singh, ed., E&FN Spon, London, U.K., pp. 34–54.

Singh, J. and White, N. (1997). "Timber Decay in Buildings: Pathology and Control." *Journal of Performance of Constructed Facilities*, ASCE, Vol. 11, No. 1, pp. 3–12.

Skempton, A. W. (1953). "The Colloidal Activity of Clays." *Proceeding of the Third International Conference on Soil Mechanics and Foundation Engineering*, Zurich, Switzerland, Vol. 1, pp. 57–61.

Skempton, A. W. (1954). "The Pore-Pressure Coefficients A and B." *Geotechnique*, Vol. 4, No. 4, pp. 143–147.

Skempton, A. W. (1961). "Effective Stress in Soils, Concrete and Rock." *Proceedings of Conference on Pore Pressure and Suction in Soils*, Butterworths, London, pp. 4–16.

Skempton, A. W. (1964). "Long-term Stability of Clay Slopes." *Geotechnique*, London, England, Vol. 14, No. 2, pp. 75–101.

Skempton, A. W. (1985). "Residual Strength of Clays in Landslides, Folded Strata and the Laboratory." *Geotechnique*, London, England, Vol. 35, No.1, pp. 3–18.

Skempton, A. W. (1986). "Standard Penetration Test Procedures and the Effects in Sands of Overburden Pressure, Relative Density, Particle Size, Aging and Overconsolidation." *Geotechnique*, Vol. 36, No. 3, pp. 425–447.

Skempton, A. W. and Bjerrum, L. (1957). "A Contribution to the Settlement Analysis of Foundations on Clay." *Geotechnique*, Vol. 7, No. 4, pp. 168–178.

Skempton, A. W. and Henkel, D. J. (1953). "The Post-Glacial Clays of the Thames Estuary at Tilbury and Shellhaven." *Proceeding of the Third International Conference on Soil Mechanics and Foundation Engineering*, Switzerland, Vol. 1, pp. 302–308.

Skempton, A. W. and Hutchinson, J. (1969). "State-of-the-art Report: Stability of Natural Slopes and Embankment Foundations." *Proceedings of the Seventh International Conference on Soil Mechanics and Foundation Engineering*, Mexico, pp. 291–340.

Skempton, A. W. and MacDonald, D. H. (1956). "The Allowable Settlements of Buildings." *Proceedings of the Institution of Civil Engineers*, part III. The Institution of Civil Engineers, London, England, No. 5, pp. 727–768.

Slope Indicator Company (1998). *Geotechnical and Structural Instrumentation*, Slope Indicator Company, Bothell, WA, 98 pp.

Smith, E. A. (1962). "Pile Driving Analysis by the Wave Equation." *Transactions*, ASCE, Vol. 127, Part 1, pp. 1145–1193.

Snethen, D. R. (1979). *Technical Guidelines for Expansive Soils in Highway Subgrades*. U.S. Army Engineering Waterway Experiment Station, Vicksburg, MS, Report No. FHWA-RD-79-51.

Sokolowski, T. J. (2000). "The Great Alaskan Earthquake & Tsunamis of 1964." West Coast & Alaska Tsunami Warning Center, Palmer, Alaska, Report Obtained from the Internet.

Sorensen, C. P., and Tasker, H. E. (1976). "Cracking in Brick and Block Masonry." *Technical Study No. 43*, 2d ed., Department of Housing and Construction Experimental Building Station.

Southern Nevada Building Code Amendments (1997). Amendments adopted by Clark County, Boulder City, North Las Vegas, City of Las Vegas, City of Mesquite, and City of Henderson. Available at the Clark County Department of Building, Permit Application Center, Las Vegas, NV.

Sowers, G. B. and Sowers, G. F. (1970). *Introductory Soil Mechanics and Foundations*. 3d ed., Macmillan, New York, NY, 556 pp.

Sowers, G. F. (1962). "Shallow Foundations," Chapter Six from *Foundation Engineering*, G. A. Leonards, ed., McGraw-Hill, New York, NY.

Sowers, G. F. (1979). *Soil Mechanics and Foundations: Geotechnical Engineering*, 4th ed., Macmillan, New York, NY.

Sowers, G. F. (1997). *Building on Sinkholes: Design and Construction of Foundations in Karst Terrain*, ASCE Press, New York, NY.

Sowers, G. F. and Royster, D. L. (1978). "Field Investigation." Chapter 4 of *Landslides, Analysis and Control, Special Report 176*, Transportation Research Board, National Academy of Sciences, R. L. Schuster and R. J. Krizek, eds.,Washington, DC, pp. 81–111.

Spencer, E. (1967). "A Method of Analysis of the Stability of Embankments Assuming Parallel Interslice Forces." *Geotechnique*, Vol. 17, No. 1, London, England, pp. 11–26.

Spencer, E. (1968). "Effect of Tension on Stability of Embankments." *Journal of the Soil Mechanics and Foundations Division*, ASCE, Vol. 94, SM5, pp. 1159–1173.

Standard Specifications For Public Works Construction (2003). 12th ed., Published by BNi Building News, Anaheim, CA, Commonly known as the "Greenbook."

Stark, T. D. and Eid, H. T. (1994). "Drained Residual Strength of Cohesive Soils." *Journal of Geotechnical Engineering*, ASCE, Vol. 120, No. 5, pp. 856–871.

Stark, T. D. and Mesri, G. (1992). "Undrained Shear Strength of Liquefied Sands for Stability Analysis." *Journal of Geotechnical Engineering*, ASCE, Vol. 118, No. 11, pp. 1727–1747.

Stark, T. D. and Olson, S. M. (1995). "Liquefaction Resistance Using CPT and Field Case Histories." *Journal of Geotechnical Engineering*, ASCE, Vol. 121, No. 12, pp. 856–869.

Steedman, R. S. and Zeng, X. (1990). "The Seismic Response of Waterfront Retaining Walls." *Proceedings, ASCE Specialty Conference on Design and Performance of Earth Retaining Structures, Special Technical Publication 25*, Cornell University, Ithaca, New York, NY, pp. 872–886.

Steinbrugge, K. V. (1970). "Earthquake Damage and Structural Performance in the United States." Chapter 9 of *Earthquake Engineering*, Robert L. Wiegel, Coordinating Editor. PrenticeHall, Englewood Cliffs, NJ, pp. 167–226.

Stokes, W. L. and Varnes, D. J. (1955). *Glossary of Selected Geologic Terms*. Colorado Scientific Society Proceedings, Vol. 16, Denver, CO, 165 pp.

Stone, W. C., Yokel, F. Y., Celebi, M., Hanks, T., and Leyendecker, E. V. (1987). "Engineering Aspects of the September 19, 1985 Mexico Earthquake." *NBS Building Science Series 165*, National Bureau of Standards, Washington, DC, 207 pp.

Tadepalli, R. and Fredlund, D. G. (1991). "The Collapse Behavior of a Compacted Soil during Inundation." *Canadian Geotechnical Journal*, Vol. 28, No. 4, pp. 477–488.

Taniguchi, E. and Sasaki, Y. (1986). "Back Analysis of Landslide due to Naganoken Seibu Earthquake of September 14, 1984." *Proceedings, Eleventh International Soil Mechanics and Foundation Engineering Conference, Session 7B, San Francisco, California*. University of Missouri, Rolla, MO.

Taylor, D. W. (1937). "Stability of Earth Slopes." *Journal of the Boston Society of Civil Engineers*, Vol. 24, No. 3, pp. 197–246.

Taylor, D. W. (1948). *Fundamentals of Soil Mechanics*. Wiley, New York, NY, 700 pp.

Technical Guidelines for Soil and Geology Reports (1993). Part of the *Orange County Grading Manual* and *Orange County Grading and Excavation Code*, Orange County, CA.

Terzaghi, K. (1925). *Erdbaumechanik*. Franz Deuticke, Vienna, 399 pp.

Terzaghi, K. (1938). "Settlement of Structures in Europe and Methods of Observation," *Transactions*, ASCE, Vol. 103, p. 1432.

Terzaghi, K. (1943). *Theoretical Soil Mechanics*. Wiley, New York, NY, 510 pp.

Terzaghi, K. (1950). "Mechanisms of Landslides." *Engineering Geology Volume*, Geological Society of America.

Terzaghi, K. and Peck, R. B. (1967). *Soil Mechanics in Engineering Practice*, 2nd ed., Wiley, New York, NY, 729 pp.

Thorburn, S. and Hutchison, J. F., eds. (1985). *Underpinning*. Surrey University Press, London, England, 296 pp.

Thornwaite, C. W. (1948). "An Approach Toward a Rational Classification of Climate." *Geographical Review*, Vol. 38, No. 1, pp. 55–94.

Tokimatsu, K. and Seed, H. B. (1984). *Simplified Procedures for the Evaluation of Settlements in Sands due to Earthquake Shaking*. Report No. UCB/EERC-84/16, Report sponsored by the National Science Foundation, Earthquake Engineering Research Center, College of Engineering, University of California, Berkeley, CA, 41 pp.

Tokimatsu, K. and Seed, H. B. (1987). "Evaluation of Settlements in Sands due to Earthquake Shaking." *Journal of Geotechnical Engineering*, ASCE, Vol. 113, No. 8, pp. 861–878.

Tomlinson, M. J. (1986). *Foundation Design and Construction*. 5th ed., Longman, Essex, England, 842 pp.

Transportation Research Board (1977). *Rapid-Setting Materials for Patching Concrete*. National Cooperative Highway Research Program Synthesis of Highway Practice 45, Published by National Academy of Sciences, Washington, DC.

Tucker, R. L. and Poor, A. R., (1978). "Field Study of Moisture Effects on Slab Movements." *Journal of the Geotechnical Engineering Division*, ASCE, Vol. 104, No. GT4, pp. 403–414.

Turnbull, W. J. and Foster, C. R. (1956). "Stabilization of Materials by Compaction." *Journal of the Soil Mechanics and Foundations Division*, ASCE, Vol. 82, No. SM2, pp. 934.1–934.23.

Tuthill, L. H. (1966). "Resistance to Chemical Attack-Hardened Concrete." *Significance of Tests and Properties of Concrete and Concrete-Making Materials*, STP-169A, ASTM, Philadelphia, PA, pp. 275–289.

United States Army Engineer Waterways Experiment Station (1960). *The Unified Soil Classification System*. Technical Memorandum No. 3-357. U.S. Army Engineer Waterways Experiment Station, Vicksburg, MS.

United States Department of Agriculture (1975). *Agriculture Handbook No. 436*. Published by the U.S. Department of Agriculture, Washington, DC.

United States Department of Housing and Urban Development (1971). "Expansive Soils-Identification and Classification." Region IX, HUD and FHA Regulations.

United States Department of the Interior (1947). "Long-Time Study of Cement Performance in Concrete—Tests of 28 Cements Used in the Parapet Wall of Green Mountain Dam." *Materials Laboratories Report No. C-345*, U.S. Department of the Interior, Bureau of Reclamation, Denver, CO.

United States Department of the Interior (1987). *Engineering Geology Field Manual*, Published by the U.S. Department of the Interior, Bureau of Reclamation, Washington, DC, 598 pp.

United States Geological Survey (1975). *Encinitas Quadrangle, San Diego, California*. Topographic Map was Mapped, Edited, and Published by the U.S. Geological Survey, Denver, CO.

United States Geological Survey (1996). *Peak Ground Acceleration Maps*. Maps prepared by the United States Geological Survey, National Seismic Hazards Mitigation Project, Maps obtained from the Internet.

United States Geological Survey (2000a). *Glossary of Some Common Terms in Seismology*. USGS Earthquake Hazards Program, National Earthquake Information Center, World Data Center for Seismology, Denver, CO, Glossary Obtained from the Internet, 8 pp.

United States Geological Survey (2000b). Earthquake Internet Site, United States Geological Survey, National Earthquake Information Center, Golden, CO.

United States Geological Survey (2002) "Interpolated Probabilistic Ground Motion for the Conterminous 48 States by Latitude and Longitude, 2002 Data." Internet Site.

USS Steel Sheet Piling Design Manual (1984). U.S. Department of Transportation, Washington, DC, 132 pp.

Van der Merwe, C. P. and Ahronovitz, M. (1973). "The Behavior of Flexible Pavements on Expansive Soils." *Third International Conference on Expansive Soil*, Haifa, Israel.

Van der Merwe, D. H. (1964). "The Prediction of Heave From the Plasticity Index and the Percentage Clay Fraction of Soils." *Civil Engineering*, South Africa, 6:103-107.

Varnes, D. J. (1978). "Slope Movement Types and Processes." *Landslides, Analysis and Control*, Transportation Research Board, National Academy of Sciences, Washington, DC, Special Report 176, Chapter 2, pp. 11–33.

Vesic, A. S. (1963). "Bearing Capacity of Deep Foundations in Sand." *Highway Research Record, 39*, National Academy of Sciences, National Research Council, Washington, DC, pp. 112–153.

Vesic, A. S. (1967). "Ultimate Loads and Settlements of Deep Foundations in Sand." *Proceedings of the Symposium on Bearing Capacity and Settlement of Foundations*, Duke University, Durham, NC, 53 pp.

Vesic, A. S. (1975). "Bearing Capacity of Shallow Foundations." Chapter 3 of *Foundation Engineering Handbook*, Hans F. Winterkorn and Hsai-Yang Fang, eds., Van Nostrand Reinhold Company, New York, NY, pp. 121–147.

Wahls, H. E. (1994). "Tolerable Deformations." *Vertical and Horizontal Deformations of Foundations and Embankments, Geotechnical Special Publication No. 40*, ASCE, New York, NY, pp. 1611–1628.

Waldron, L. J. (1977). "The Shear Resistance of Root-Permeated Homogeneous and Stratified Soil." *Soil Science Society of America*, Vol. 41, No. 5, pp. 843–849.

Warner, J. (1978). "Compaction Grouting—A Significant Case History." *Journal of the Geotechnical Engineering Division*, ASCE, Vol. 104, No. GT7, pp. 837–847.

Warner, J. (1982). "Compaction Grouting—The First Thirty Years." *Proceedings of the Conference on Grouting in Geotechnical Engineering*, W. H. Baker, ed., ASCE, New York, NY, pp. 694–707.

Watry, S. M. and Ehlig, P. L. (1995). "Effect of Test Method and Procedure on Measurements of Residual Shear Strength of Bentonite from the Portuguese Bend Landslide." *Clay and Shale Slope Instability*, W. C. Haneberg and S. A. Anderson, eds., Geological Society of America, Reviews in Engineering Geology, Volume 10, Boulder, CO, pp. 13–38.

Wellington, A. M. (1888). "Formulae for Safe Loads of Bearing Piles." *Engineering News*, No. 20, pp. 509–512.

Westergaard, H. M. (1938). "A Problem of Elasticity Suggested by a Problem in Soil Mechanics: A Soft Material Reinforced by Numerous Strong Horizontal Sheets." In *Contributions to the Mechanics of Solids, Stephen Timoshenko 60th Anniversary Volume*, Macmillan, New York, NY, pp. 268–277.

WFCA (1984). "Moisture Guidelines for the Floor Covering Industry." WFCA Management Guidelines, Western Floor Covering Association, Los Angeles, CA.

Whitaker, T. (1957). "Experiments with Model Piles in Groups." *Geotechnique*, London, England.

Whitlock, A. R. and Moosa, S. S. (1996). "Foundation Design Considerations for Construction on Marshlands." *Journal of Performance of Constructed Facilities*, ASCE, Vol. 10, No. 1, pp. 15–22.

Whitman, R. V. (1990). "Seismic Design Behavior of Gravity Retaining Walls." Proceedings, *ASCE Specialty Conference on Design and Performance of Earth Retaining Structures, Geotechnical Special Publication 25*, ASCE, New York, NY, pp. 817–842.

Whitman, R. V. and Bailey, W. A. (1967). "Use of Computers for Slope Stability Analysis." *Journal of the Soil Mechanics and Foundations Division*, ASCE, Vol. 93, No. SM4, pp. 475–498.

Williams, A. A. B. (1965). "The Deformation of Roads Resulting From Moisture Changes in Expansive Soils in South Africa." *Moisture Equilibria and Moisture Changes in Soils Beneath Covered Areas*, G. D. Aitchison, ed., Symposium Proceedings, Butterworths, Australia, pp. 143–155.

Wilson, R. C. and Keefer, D. K. (1985). "Predicting Areal Limits of Earthquake-Induced Lansliding," in *Evaluating Earthquake Hazards in the Los Angeles Region*, J. I. Ziony, ed., U.S. Geological Survey, Reston, VA, Professional Paper 1360, pp. 317–345.

Winterkorn, H. F. and Fang, H. (1975). *Foundation Engineering Handbook*. Van Nostrand Reinhold, New York, NY, 751 pp.

Woodward, R. J., Gardner, W. S., and Greer, D. M. (1972). "Design Considerations," *Drilled Pier Foundations*, D. M. Greer, ed., McGraw-Hill, New York, NY, pp 50–52.

Wu, T. H., Randolph, B. W., and Huang, C. (1993). "Stability of Shale Embankments." *Journal of Geotechnical Engineering*, ASCE, Vol. 119, No. 1, pp. 127–146.

Wray, W. K. (1989). "Migration of Damage to Structures Supported on Expansive Soils." *Final Report*, National Science Foundation, Grant No. ECE-8320493, Vol. 1, 325 pp.

Wray, W. K. (1997). "Using Soil Suction to Estimate Differential Soil Shrink or Heave." *Unsaturated Soil Engineering Practice*, Geotechnical Special Publication No. 68, S. L. Houston and D. G. Fredlund, eds., ASCE, New York, NY, pp. 66–87.

Yeats, R. S., Sieh, K, and Allen, C. R. (1997). *The Geology of Earthquakes*. Oxford University Press, New York, 568 pp.

Yong, R. N., and Warkentin, B. P. (1975). *Soil Properties and Behavior*, Elsevier, New York, NY, 449 pp.

Yoshimi, Y., Tokimatsu, K., and Hasaka, Y. (1989). "Evaluation of Liquefaction Resistance of Clean Sands Based on High-Quality Undisturbed Samples." *Soils and Foundations*, Vol. 29, No. 1, pp. 93–104.

Youd, T. L. (1978). "Major Cause of Earthquake Damage is Ground Failure." *Civil Engineering Magazine*, ASCE, Vol. 48, No. 4, pp. 47–51.

Youd, T. L. (1984). "Geologic Effects—Liquefaction and Associated Ground Failures." *Proceedings, Geologic, and Hydrologic Hazards Training Program*, Open File Report 84-760, U.S. Geological Survey, Menlo Park, CA, pp. 210–232.

Youd, T. L. and Gilstrap, S. D. (1999). "Liquefaction and Deformation of Silty and Fine-grained Soils." *Earthquake Geotechnical Engineering*, 2nd ed., Balkema, Rotterdam, pp. 1013–1020.

Youd, T. L., Hansen, C. M., and Bartlett, S. F. (2002). "Revised Multilinear Regression Equations for Prediction of Lateral Spread Displacement." *Journal of Geotechnical and Geoenvironmental Engineering*, ASCE, Vol. 128, No. 12, pp. 1007–1017.

Youd, T. L. and Hoose, S. N. (1978). "Historic Ground Failures in Northern California Triggered by Earthquakes." *Professional Paper 933*, U.S. Geological Survey, Washington, DC.

Youd, T. L. and Idriss, I. M., eds. (1997). *Proceedings, NCEER Workshop Evaluation of Liquefaction Resistance of Soils*, National Center for Earthquake Engineering Research, State University of New York at Buffalo.

Youd, T. L. and Idriss, I. M. (2001). "Liquefaction Resistance of Soils: Summary Report From the 1996 NCEER and 1998 NCEER/NSF Workshops on Evaluation of Liquefaction Resistance of Soils." *Journal of Geotechnical and Geoenvironmental Engineering*, ASCE, Vol. 127, No. 4, pp. 297–313.

Youd, T. L. and Noble, S. K. (1997). "Liquefaction Criteria Based on Statistical and Probabilistic Analyses." *Proceedings, NCEER Workshop on Evaluation of Liquefaction Resistance of Soils, Technical Report NCEER-97-0022*, T. L. Youd and I. M. Idriss, eds., National Center for Earthquake Engineering Research, Buffalo, NY, pp. 201–215.

Zaruba, Q. and Mencl, V. (1969). *Landslides and Their Control*. Elsevier, New York, NY, 205 pp.

Zornberg, J. G., Sitar, N., and Mitchell, J. K. (1998). "Limit Equilibrium as Basis for Design of Geosynthetic Reinforced Slopes." *Journal of Geotechnical and Geoenvironmental Engineering*, ASCE, Vol. 124, No. 8, pp. 684–698.

INDEX

CPSIA information can be obtained
at www.ICGtesting.com
Printed in the USA
BVOW07*1513260717

489957BV00017B/60/P